Springer Proceedings in Physics

Volume 238

Indexed by Scopus

The series Springer Proceedings in Physics, founded in 1984, is devoted to timely reports of state-of-the-art developments in physics and related sciences. Typically based on material presented at conferences, workshops and similar scientific meetings, volumes published in this series will constitute a comprehensive up-to-date source of reference on a field or subfield of relevance in contemporary physics. Proposals must include the following:

- name, place and date of the scientific meeting
- a link to the committees (local organization, international advisors etc.)
- scientific description of the meeting
- list of invited/plenary speakers
- an estimate of the planned proceedings book parameters (number of pages/articles, requested number of bulk copies, submission deadline).

More information about this series at http://www.springer.com/series/361

N. A. Orr · M. Ploszajczak ·
F. M. Marqués · J. Carbonell
Editors

Recent Progress in Few-Body Physics

Proceedings of the 22nd International Conference on Few-Body Problems in Physics

Springer

Editors
N. A. Orr
LPC-Caen, ENSICAEN, IN2P3/CNRS
Université de Caen Normandie
Caen, France

M. Ploszajczak
GANIL
CEA/DRF-IN2P3/CNRS
Caen, France

F. M. Marqués
LPC-Caen, ENSICAEN, IN2P3/CNRS
Université de Caen Normandie
Caen, France

J. Carbonell
IPN-Orsay, IN2P3/CNRS
Université Paris-Sud, Université
Paris-Saclay
Orsay, France

ISSN 0930-8989 ISSN 1867-4941 (electronic)
Springer Proceedings in Physics
ISBN 978-3-030-32356-1 ISBN 978-3-030-32357-8 (eBook)
https://doi.org/10.1007/978-3-030-32357-8

This Springer imprint is published by the registered company Springer Nature Switzerland AG
The registered company address is: Gewerbestrasse 11, 6330 Cham, Switzerland

Preface

The 22nd International Conference on Few-Body Problems in Physics (FB22) was held in Caen, Normandy, France, from the 9 to 13 July 2018. This meeting was the most recent in a long series which had its beginnings in London in 1959. While the early conferences concentrated on nucleon scattering, the very lightest nuclei and hadrons, more recent editions have grown to encompass a wide and interdisciplinary range of topics spanning the very broadest possible energy scales from molecular and atomic physics to high-energy particle physics. FB22 continued this tradition and brought together some 270 participants from 33 different countries to share and discuss ideas in few-body physics from both the experimental and theoretical perspectives.

The conference comprised 10 plenary sessions together with 32 parallel sessions organised thematically:

- *Few-body aspects of atomic and molecular physics.*
- *Hadron and related high-energy physics.*
- *Strange and exotic matter, including hypernuclear physics.*
- *Few-nucleon systems, including QCD inspired approaches.*
- *Few-body aspects of nuclear physics and nuclear astrophysics.*
- *Interdisciplinary aspects of few-body physics and techniques.*

Of particular note were two parallel invited sessions focussing on the topical subjects of lattice QCD and neutron clusters. A dedicated poster session was also held with contributions from over 50 participants.

The programme was also marked by improved participation with respect to earlier meetings from the female members of the few-body community, encouraged in part by the scientific programme committee's choice of plenary and invited speakers. Similarly, the younger members of the community (graduate students and postdoctoral researchers) provided a significant level of contributions to the programme. The present volume is composed of papers covering the majority of the presentations (oral and poster).

FB22 was the venue, during a special plenary session, for the presentation of the inaugural Faddeev medal. The award, which honours the renowned theorist Ludvig Faddeev, was established by the European (ERCFBG) and North American (APS-GRB) few-body physics communities, with sponsorship from Springer and the APS-GFB, to recognize achievements in few-body physics. The award was shared between two laureates: Vitaly Efimov "For the theoretical discovery of a series of weakly-bound three-body quantum states known as Efimov states" and Rudolf Grimm "In recognition of his ground-breaking experiments confirming the Efimov effect". Many of the contributions to the present volume reflect the importance and ground breaking character of the work of Vitaly and Rudolf.

The conference was held at the "Centre de Congrès de Caen", located near the historical centre of Caen. Financial support was provided by the IUPAP, LPC-Caen, GANIL, IPN-Orsay, the IN2P3/CNRS, the CEA, l'université de Caen Normandie and the local government authority "Caen la mer".

N. A. Orr (LPC-Caen)
Chair Organising Committee
Caen, France

M. Ploszajczak (GANIL)
Chair Scientific Programme Committee
Caen, France

J. Carbonell (IPN-Orsay)
Chair International Advisory Committee
Orsay, France

F. M. Marqués (LPC-Caen)
Scientific Secretary
Caen, France

Contents

Part VI Strange and Exotic Matter

Contributors

Yukihiro Abe Research Center for Nuclear Physics (RCNP), Osaka University, Ibaraki, Osaka, Japan

N. L. Achouri LPC Caen, Normandie Université, ENSICAEN, Université de Caen, CNRS/IN2P3, Caen, France

N. Added Instituto de Física, Universidade de São Paulo, São Paulo, SP, Brazil

P. Adsley Department of Subatomic Physics, iThemba LABS, Somerset West, South Africa;
Department of Physics, University of Stellenbosch, Matieland, Stellenbosch, South Africa

A. N. Agnihotri CIMAP, CEA—CNRS—ENSICAEN—UNICAEN, Normandie Université, Caen Cedex 04, France

Mohammad W. Ahmed North Carolina Central University, Durham, NC, USA;
Triangle Universities Nuclear Laboratory, Durham, NC, USA

D. S. Ahn RIKEN Nishina Center, Wako, Saitama, Japan

T. Akieda Department of Physics, Tohoku University, Sendai, Miyagi, Japan;
Cyclotron and Radioisotope Center (CYRIC), Tohoku University, Sendai, Miyagi, Japan;
High Energy Accelerator Research Organization (KEK), Tsukuba, Ibaraki, Japan;
National Institute of Radiological Science, Chiba, Japan

J. Alcantara-Nunez Instituto de Física, Universidade de São Paulo, São Paulo, SP, Brazil

A. Amirkhani Politecnico di Milano, Dipartimento di Elettronica, Informazione e Bioingegneria and INFN Sezione di Milano, Milan, Italy

Miron Ya Amusia Racah Institute of Physics, Hebrew University, Jerusalem, Israel;
A. F. Ioffe Physical-Technical Institute, St. Petersburg, Russian Federation

Lorenzo Ugo Ancarani Université de Lorraine, CNRS, LPCT, Metz, France

Shung-Ichi Ando Sun Moon University, Asan, Chungman, South Korea

Paolo Andreatta Department of Physics, University of Trento, Trento, Italy; INFN-TIFPA Trento Institute of Fundamental Physics and Applications, Povo-Trento, Italy

K. Aoki IPNS, KEK, Tsukuba, Japan

S. Appannababu Instituto de Física, Universidade de São Paulo, São Paulo, SP, Brazil

T. Arai Department of Physics, Hokkaido University, Sapporo, Japan

Adam Arslanaliev V. N. Karazin Kharkiv National University, Kharkov, Ukraine

D. A. Artemenkov Joint Institute for Nuclear Research (JINR), Dubna, Russia

M. Assunção Departamento de Física, Universidade Federal de São Paulo, Diadema, SP, Brazil

D. Atta Government General Degree College, Kharagpur II, Ambigeria, West Bengal, India

T. Aumann Institut für Kernphysik, Technische Universität Darmstadt, Darmstadt, Germany;
ExtreMe Matter Institute EMMI and Research Division, GSI Helmholtzzentrum für Schwerionenforschung GmbH, Darmstadt, Germany

G. Authelet Département de Physique Nucléaire, IRFU, CEA, Université Paris-Saclay, Saint-Aubin, France

Alexandr Vladimirovich Averyanov Joint Institute for Nuclear Research, Dubna, Moscow Region, Russia

H. Baba RIKEN Nishina Center, Wako, Saitama, Japan

James F. Babb ITAMP, Center for Astrophysics | Harvard & Smithsonian, Cambridge, MA, USA

Sonia Bacca Institut für Kernphysik and PRISMA Cluster of Excellence, Johannes-Gutenberg-Universität Mainz, Mainz, Germany;
TRIUMF, Vancouver, BC, Canada;
Department of Physics and Astronomy, University of Manitoba, Winnipeg, MB, Canada

J. Balibrea-Correa Università degli Studi di Napoli Federico II, Naples, Italy; INFN, Sezione di Napoli, Naples, Italy

L. P. L. Baloyi School of Physics, University of the Witwatersrand, Johannesburg, South Africa

A. Baniahmad Politecnico di Milano, Dipartimento di Elettronica, Informazione e Bioingegneria and INFN Sezione di Milano, Milan, Italy

Nir Barnea Racah Institute of Physics, The Hebrew University, Jerusalem, Israel

D. N. Basu Variable Energy Cyclotron Centre, HBNI, Kolkata, India

E. F. Batista Departamento de Ciências Exatas e Naturais—DCEN, Universidade Estadual do Sudoeste da Bahia—UESB, Itapetinga, BA, Brazil

M. T. Bayat KVI-CART, University of Groningen, Groningen, The Netherlands

D. Baye Physique Quantique, and Physique Nucléaire Théorique et Physique Mathématique, C.P. 229, Université libre de Bruxelles (ULB), Brussels, Belgium

Betzalel Bazak The Racah Institute of Physics, The Hebrew University, Jerusalem, Israel

M. Bazzi INFN, Laboratori Nazionali di Frascati, Frascati, Rome, Italy

D. Beaumel Institut de Physique Nucléaire Orsay, IN2P3-CNRS, Orsay Cedex, France

Marco A. Bedolla INFN, Sezione di Genova, Genova, Italy;
Instituto de Física y Matemáticas, Universidad Michoacana de San Nicolás de Hidalgo, Morelia, Michoacán, Mexico

G. Bellotti Politecnico di Milano, Dipartimento di Elettronica, Informazione e Bioingegneria and INFN Sezione di Milano, Milan, Italy

Pavel A. Belov St. Petersburg State University, St. Petersburg, Russian Federation

S. Bera Department of Physics, Presidency University, Kolkata, India

C. A. Bertulani Department of Physics and Astronomy, Texas A&M University-Commerce, Commerce, TX, USA

C. Berucci INFN, Laboratori Nazionali di Frascati, Frascati, Rome, Italy;
Stefan-Meyer-Institut für Subatomare Physik, Vienna, Austria

A. Best Università degli Studi di Napoli Federico II, Naples, Italy;
INFN, Sezione di Napoli, Naples, Italy

S. Bhattacharyya Department of Physics, Scottish Church College 1 & 3 Urquhart Square, Kolkata, India

Elmar P. Biernat CFTP, Instituto Superior Técnico, Universidade de Lisboa, Lisbon, Portugal

M. Böhmer Technische Universität München, Garching, Germany

J. Bonnard IPNO, Université Paris-Sud, IN2P3-CNRS, Orsay, France

K. Boretzky RIKEN Nishina Center, Wako, Saitama, Japan;
GSI Helmholtzzentrum für Schwerionenforschung, Darmstadt, Germany

D. Bosnar Department of Physics, Faculty of Science, University of Zagreb, Zagreb, Croatia

Alain Boudard IRFU, CEA, Université Paris-Saclay, Gif-sur-Yvette, France

Eric Braaten Department of Physics, The Ohio State University, Columbus, OH, USA

Alexandru Bragadireanu Horia Hulubei National Institute of Physics and Nuclear Engineering, Magurele, Romania

M. Bragadireanu Horia Hulubei National Institute of Physics and Nuclear Engineering (IFIN-HH), Magurele, Romania

N. A. Burkova Al-Farabi Kazakh National University, Almaty, Kazakhstan

M. Burrows Department of Physics and Astronomy, Ohio University, Athens, OH, USA

Arslan Butt Politecnico di Milano, Dipartimento di Elettronica, Informazione e Bioingegneria, Milan, Italy;
INFN Sezione di Milano, Milan, Italy

M. Caamaño Departamento de Física de Partículas and IGFAE, Universidade de Santiago de Compostela, Santiago de Compostela, Spain

C. Caesar Institut für Kernphysik, Technische Universität Darmstadt, Darmstadt, Germany

D. Calvet Département de Physique Nucléaire, IRFU, CEA, Université Paris-Saclay, Saint-Aubin, France

E. N. Cardozo Departamento de Física, Universidade Federal Fluminense, Niterói, RJ, Brazil

M. Cargnelli Stefan-Meyer-Institut für Subatomare Physik, Vienna, Austria

B. V. Carlson Departamento de Física, Instituto Tecnológico de Aeronaútica, DCTA, São José dos Campos, SP, Brazil

J. Carter School of Physics, University of the Witwatersrand, Johannesburg, South Africa

J. Casal European Centre for Theoretical Studies in Nuclear Physics and Related Areas (ECT*), Trento, Italy;
Dipartimento di Fisica e Astronomia "G. Galilei" and INFN—Sezione di Padova, Padua, Italy

A. Cassimi CIMAP, CEA—CNRS—ENSICAEN—UNICAEN, Normandie Université, Caen Cedex 04, France

Francesca Cavanna Istituto Nazionale di Fisica Nucleare Sezionde di Genova, Genova, Italy

B. Chakrabarti Department of Physics, Presidency University, Kolkata, India

Lei Chang School of Physics, Nankai University, Tianjin, China

S. Chebotaryov Department of Physics, Kyungpook National University, Daegu, Korea

Hua-Xing Chen School of Physics and Beijing Key Laboratory of Advanced Nuclear Materials and Physics, Beihang University, Beijing, China

Muyang Chen Nankai University, Tianjin, China

S. Chen Department of Physics, The University of Hong Kong, Hong Kong, China

Wei Chen School of Physics, Sun Yat-Sen University, Guangzhou, China

Evgenyi Vasilievich Chernykh Joint Institute for Nuclear Research, Dubna, Moscow region, Russia

Larissa V. Chernysheva A. F. Ioffe Physical-Technical Institute, St. Petersburg, Russian Federation

N. Chiga RIKEN Nishina Center, Wako, Saitama, Japan

G. F. Ciani Gran Sasso Science Institute, L'Aquila, Italy;
INFN, Laboratori Nazionali del Gran Sasso, Assergi, L'Aquila, Italy

I. Ciepał Institute of Nuclear Physics PAN, Kraków, Poland;
Henryk Niewidniczański Institute of Nuclear Physics PAS, Kraków, Poland

A. Corsi Département de Physique Nucléaire, IRFU, CEA, Université Paris-Saclay, Saint-Aubin, France

M. L. Cortés RIKEN Nishina Center, Wako, Saitama, Japan

D. Cortina Departamento de Física de Partículas and IGFAE, Universidade de Santiago de Compostela, Santiago de Compostela, Spain

L. Csedreki Gran Sasso Science Institute, L'Aquila, Italy;
INFN, Laboratori Nazionali del Gran Sasso, Assergi, L'Aquila, Italy

Joseph Cugnon AGO department, University of Liège, Liège, Belgium

Er-Liang Cui School of Physics and Beijing Key Laboratory of Advanced Nuclear Materials and Physics, Beihang University, Beijing, China

C. Curceanu INFN, Laboratori Nazionali di Frascati, Frascati (Roma), Italy

P. Danielewicz NSCL/FRIB Laboratory, Department of Physics and Astronomy, Michigan State University, East Lansing, MI, USA

Jean-Christophe David IRFU, CEA, Université Paris-Saclay, Gif-sur-Yvette, France

Zohreh Davoudi Maryland Center for Fundamental Physics and Department of Physics, University of Maryland, College Park, MD, USA;
RIKEN Center for Accelerator-based Sciences, Wako, Japan

A. Dawood Butt Politecnico di Milano, Dipartimento di Elettronica, Informazione e Bioingegneria and INFN Sezione di Milano, Milan, Italy

A. L. de Lara Instituto de Física, Universidade de São Paulo, São Paulo, SP, Brazil

Francois de Oliveira Santos GANIL, CEA/DSM-CNRS/IN2P3, Caen Cedex 5, France

L. De Paolis INFN, Laboratori Nazionali di Frascati, Frascati (Roma), Italy; Department of Physics, Faculty of Science MM.FF.NN., University of Rome 2 (Tor Vergata), Rome, Italy

R. Del Grande INFN, Laboratori Nazionali di Frascati, Frascati, Rome, Italy; Universit degli Studi di Roma Tor Vergata, Rome, Italy

P. Delahaye GANIL, CEA/DSM-CNRS/IN2P3, Caen, France

F. Delaunay LPC Caen, Normandie Université, ENSICAEN, Université de Caen, CNRS/IN2P3, Caen, France

A. Delbart Département de Physique Nucléaire, IRFU, CEA, Université Paris-Saclay, Saint-Aubin, France

Pierre Descouvemont Physique Nucléaire Théorique et Physique Mathématique, C.P. 229, Université Libre de Bruxelles (ULB), Brussels, Belgium

Francesca De Mori Università degli Studi di Torino and INFN Sezione di Torino, Torino, Italy

Lucio Di Paolo INFN, Laboratori Nazionali del Gran Sasso, L'Aquila, Italy

Minghui Ding European Centre for Theoretical Studies in Nuclear Physics and Related Areas (ECT*) and Fondazione Bruno Kessler, Villazzano (TN), Italy

Jérémy Dohet-Eraly Physique nucléaire et Physique quantique, Université libre de Bruxelles (ULB), Brussels, Belgium

L. M. Donaldson Department of Subatomic Physics, iThemba LABS, Somerset West, South Africa

Chen Dong College of Information and Communication, National University of Defense Technology, Xi'an, China;
State Key Laboratory of Cryptology, Beijing, China

P. Doornenbal RIKEN Nishina Center, Wako, Saitama, Japan

J. Douady CIMAP, CEA—CNRS—ENSICAEN—UNICAEN, Normandie Université, Caen Cedex 04, France

C. A. Douma KVI-CART, University of Groningen, Groningen, The Netherlands

M. Dozono Center for Nuclear Study, University of Tokyo, Tokyo, Japan

Jerry P. Draayer Department of Physics and Astronomy, Louisiana State University, Baton Rouge, LA, USA

Menglin Du Helmholtz-Institut für Strahlen- und Kernphysik and Bethe Center for Theoretical Physics, Universität Bonn, Bonn, Germany

S. B. Dubovichenko Fesenkov Astrophysical Institute "NCSRT", Almaty, Kazakhstan

F. Dufter Technische Universität München, Garching, Germany

Tomas Dytrych Department of Physics and Astronomy, Louisiana State University, Baton Rouge, LA, USA;
Nuclear Physics Institute, Academy of Sciences of the Czech Republic, Řež, Czech Republic

A. V. Dzhazairov-Kakhramanov Fesenkov Astrophysical Institute "NCSRT", Almaty, Kazakhstan

Gernot Eichmann CFTP, Instituto Superior Técnico, Universidade de Lisboa, Lisboa, Portugal

Andreas Ekström Department of Physics, Chalmers University of Technology, Gothenburg, Sweden

Nils Elander Department of Physics, Stockholm University, Stockholm, Sweden

Z. Elekes RIKEN Nishina Center, Wako, Saitama, Japan;
MTA ATOMKI, Debrecen, Hungary

Serdar Elhatisari Karamanoglu Mehmetbey University, Karaman, Turkey;
Helmholtz-Institut für Strahlen- und Kernphysik (Theorie) and Bethe Center for Theoretical Physics, Universität Bonn, Bonn, Germany

Ch. Elster Institute of Nuclear and Particle Physics, and Department of Physics and Astronomy, Ohio University, Athens, OH, USA

Dan Enache Advanced Research Institute for Electrical Engineering, Bucharest, Romania

David R. Entem Grupo de Física Nuclear and Instituto de Física Fundamental y Matemáticas (IUFFyM) Universidad de Salamanca, Salamanca, Spain

Evgeny Epelbaum Fakultät für Physik und Astronomie, Institut für Theoretische Physik II, Ruhr-Universität Bochum, Bochum, Germany

Jutta E. Escher Lawrence Livermore National Laboratory, Livermore, CA, USA

M. Eslami-Kalantari Department of Physics, School of Science, Yazd University, Yazd, Iran

D. Eto Department of Physics, Tohoku University, Sendai, Miyagi, Japan

D. Etoh Department of Physics, Tohoku University, Sendai, Miyagi, Japan;
Cyclotron and Radioisotope Center (CYRIC), Tohoku University, Sendai, Miyagi, Japan;
High Energy Accelerator Research Organization (KEK), Tsukuba, Ibaraki, Japan;
National Institute of Radiological Science, Chiba, Japan

L. Fabbietti Excellence Cluster "Origin and Structure of the Universe", Garching, Germany;
Physik Department E12, Technische Universität München, Garching, Germany

Albert Feijoo Nuclear Physics Institute, Řež, Czech Republic

J. Feng Department of Physics, Peking University, Beijing, China;
RIKEN Nishina Center, Wako, Saitama, Japan;
School of Physics and State Key Laboratory of Nuclear Physics and Technology, Peking University, Beijing, China

Francisco Fernández Grupo de Física Nuclear and Instituto de Física Fundamental y Matemáticas (IUFFyM) Universidad de Salamanca, Salamanca, Spain

B. Fernández-Domínguez Departamento de Física de Partículas and IGFAE, Universidade de Santiago de Compostela, Santiago de Compostela, Spain

Fabrizio Ferrari Ruffino Department of Physics, University of Trento, Trento, Italy

C. Fiorini Stefan-Meyer-Institut für Subatomare Physik, Vienna, Austria;
Politecnico di Milano, Dip. di Elettronica, Informazione eBioingegneria, Milano, Italy;
INFN Sezione di Milano, Milano, Italy

F. Flavigny IPNO, Université Paris-Sud, IN2P3-CNRS, Orsay, France

X. Fléchard Normandie University, ENSICAEN, UNICAEN, CNRS/IN2P3, LPC Caen, Caen, France

Alba Formicola INFN, Laboratori Nazionali del Gran Sasso, L'Aquila, Italy

U. Forsberg Department of Physics, University of York, York, UK;
Department of Physics, Lund University, Lund, Sweden

K. Fossez NSCL/FRIB Laboratory, Michigan State University, East Lansing, MI, USA

T. Frederico Departamento de Física, Instituto Tecnológico de Aeronaútica, DCTA, São José dos Campos, SP, Brazil

Jason Fry Institute of Nuclear and Particle Physics, University of Virginia, Charlottesville, VA, USA

H. Fujimura Research Center for Electron Photon Science (ELPH), Tohoku University, Sendai, Japan

T. Fujino Department of Physics, Rikkyo University, Toshima, Tokyo, Japan

H. Fujioka Department of Physics, Tokyo Institute of Technology, Meguro, Japan

H. Fukasawa Research Center for Electron Photon Science (ELPH), Tohoku University, Sendai, Japan

N. Fukuda RIKEN Nishina Center, Wako, Saitama, Japan

Kenji Fukukawa Suma Gakuen, Kobe, Hyogo, Japan

H. O. U. Fynbo Department of Physics and Astronomy, Aarhus University, Aarhus, Denmark

I. Gašparić RIKEN Nishina Center, Wako, Saitama, Japan;
Ruđer Bošković Institut (RBI), Zagreb, Croatia

M. Gattobigio CNRS, Institut de Physique de Nice, Université Côte d'Azur, Nice, France

Z. Ge RIKEN Nishina Center, Wako, Saitama, Japan

Lisheng Geng School of Physics, Beihang University, Beijing, China

R. Gernhäuser Technische Universität München, Garching, Germany

B. Gervais CIMAP, CEA—CNRS—ENSICAEN—UNICAEN, Normandie Université, Caen Cedex 04, France

J. M. Gheller Département de Physique Nucléaire, IRFU, CEA, Université Paris-Saclay, Saint-Aubin, France;
CEA, Centre de Saclay, IRFU, Gif-sur-Yvette, France

F. Ghio INFN Sezione di Roma I, Rome, Italy;
Istituto Superiore di Sanit, Rome, Italy

Alessandro Giachino Istituto Nazionale di Fisica Nucleare (INFN), Sezione di Genova, Genova, Italy

J. Gibelin LPC Caen, Normandie Université, ENSICAEN, Université de Caen, CNRS/IN2P3, Caen, France

A. Giganon Département de Physique Nucléaire, IRFU, CEA, Université Paris-Saclay, Saint-Aubin, France

A. Gillibert Département de Physique Nucléaire, IRFU, CEA, Université Paris-Saclay, Saint-Aubin, France;
CEA, Centre de Saclay, IRFU, Gif-sur-Yvette, France

L. Girlanda Dipartimento di Matematica e Fisica, University of Salento, Lecce, Italy;
INFN, Sezione di Lecce, Lecce, Italy

B. Głowacz August Chełkowski Institute of Physics, University of Silesia, Chorzow, Poland

Alex Gnech Gran Sasso Science Institute, L'Aquila, Italy;
Istituto Nazionale di Fisica Nucleare, Pisa, Italy

B. M. Godoy LPC Caen, ENSICAEN, Université de Caen, CNRS/IN2P3, Caen Cedex, France

J. Golak M. Smoluchowski Institute of Physics, Jagiellonian University, Kraków, Poland

J. Gómez-Camacho Departamento de Física Atómica, Molecular y Nuclear, Facultad de Física, Universidad de Sevilla, Seville, Spain;
Centro Nacional de Aceleradores (CNA), U. Sevilla, J. Andalucía, CSIC, Seville, Spain

S. Goto Department of Physics, Kyushu University, Fukuoka, Japan

V. A. Gradusov St. Petersburg State University, St. Petersburg, Russian Federation

G. Grams Departamento de Física, CFM, Universidade Federal de Santa Catarina, Florianópolis, Brazil

Harald W. Grießhammer Department of Physics, The George Washington University, Washington, DC, USA

L. V. Grigorenko Flerov Laboratory of Nuclear Reactions, JINR, Dubna, Russia;
National Research Nuclear University "MEPhI", Moscow, Russia;
National Research Centre "Kurchatov Institute", Moscow, Russia

C. Guaraldo INFN, Laboratori Nazionali di Frascati, Frascati, Rome, Italy

S. Guillous CIMAP, CEA—CNRS—ENSICAEN—UNICAEN, Normandie Université, Caen Cedex 04, France

V. Guimarães Instituto de Física, Universidade de São Paulo, São Paulo, SP, Brazil

Yuriy Vitalievich Gurchin Joint Institute for Nuclear Research, Dubna, Moscow Region, Russia

Kyriakos Hadjiyiannakou The Cyprus Institute, Nicosia, Cyprus

G. Hagen Physics Division, Oak Ridge National Laboratory, Oak Ridge, TN, USA;
Department of Physics and Astronomy, University of Tennessee, Knoxville, TN, USA

Mohammad Haidar Laboratoire Kastler Brossel, Sorbonne Université, CNRS, ENS—Université PSL, Collége de France, Paris, France

Johann Haidenbauer Institute for Advanced Simulation, Institut für Kernphysik, and Jülich Center for Hadron Physics, Jülich, Germany

M. Haiduc Institute of Space Science, Magurele, Romania

Z. Halász MTA ATOMKI, Debrecen, Hungary

Koji Harada Faculty of Arts and Science, Kyushu University, Fukuoka, Japan

T. Harada RIKEN Nishina Center, Wako, Saitama, Japan;
Toho University, Tokyo, Japan

M. N. Harakeh KVI-CART, University of Groningen, Groningen, The Netherlands

K. Hasegawa Department of Physics, Tohoku University, Sendai, Japan

R. Hashimoto Research Center for Electron Photon Science (ELPH), Tohoku University, Sendai, Japan

K. Hatanaka Research Center for Nuclear Physics, Osaka University, Osaka, Ibaraki, Japan

Ryugo Hayano The University of Tokyo, Tokyo, Japan

Q. He Research Center for Electron Photon Science (ELPH), Tohoku University, Sendai, Japan

Oscar J. Hernandez Institut für Kernphysik and PRISMA Cluster of Excellence, Johannes-Gutenberg-Universität Mainz, Mainz, Germany;
Department of Physics and Astronomy, University of British Columbia, Vancouver, BC, Canada;
TRIUMF, Vancouver, BC, Canada

Renato Higa Instituto de FíSica, Universidade de São Paulo, São Paulo, SP, Brazil

Laurent Hilico Laboratoire Kastler Brossel, Sorbonne Université, CNRS, ENS—Université PSL, Collége de France, Paris, France;
Université d'Evry-Val d'Essonne, Université Paris Saclay, Evry, France

Yasuhisa Hiratsuka Preparation School of HLF Ltd., Kiryu, Gunma, Japan

A. Hirayama RIKEN Nishina Center, Wako, Saitama, Japan;
Department of Physics, Tokyo Institute of Technology, Meguro, Tokyo, Japan

Jason Hirtz IRFU, CEA, Université Paris-Saclay, Gif-sur-Yvette, France;
Space Research and Planetary Sciences, Physics Institute, University of Bern, Bern, Switzerland

L. Hlophe National Superconducting Cyclotron Laboratory and Department of Physics and Astronomy, Michigan State University, East Lansing, MI, USA

Y. Honda Research Center for Electron Photon Science (ELPH), Tohoku University, Sendai, Japan

Ki-Hoon Hong Department of Physics, Inha University, Incheon, Republic of Korea

W. Horiuchi Department of Physics, Hokkaido University, Sapporo, Japan

Atsushi Hosaka Research Center for Nuclear Physics (RCNP), Osaka University, Ibaraki, Osaka, Japan;
Advanced Science Research Center, Japan Atomic Energy Agency, Ibaraki, Japan

T. Hotta RCNP, Osaka University, Ibaraki, Japan

Calvin R. Howell Duke University, Durham, NC, USA;
Triangle Universities Nuclear Laboratory, Durham, NC, USA

Ming-Qiu Huang College of Liberal Arts and Sciences, National University of Defense Technology, Changsha, China

S. W. Huang RIKEN Nishina Center, Wako, Saitama, Japan;
School of Physics and State Key Laboratory of Nuclear Physics and Technology, Peking University, Beijing, China

N. Hubbard Department of Physics and Astronomy, Aarhus University, Aarhus, Denmark;
Department of Physics, University of York, York, UK

M. S. Hussein Instituto de Estudos Avançados, Universidade de São Paulo, São Paulo, SP, Brazil;
Instituto de Física, Universidade de São Paulo, São Paulo, SP, Brazil;
Departamento de Física, Instituto Tecnológico de Aeronaútica, DCTA, São José dos Campos, SP, Brazil

J. W. Hwang Department of Physics and Astronomy, Seoul National University, Gwanak-gu, Seoul, Republic of Korea

Tetsuo Hyodo Yukawa Institute for Theoretical Physics, Kyoto University, Kyoto, Japan

M. Iliescu INFN, Laboratori Nazionali di Frascati, Frascati, Rome, Italy

N. Inabe RIKEN Nishina Center, Wako, Saitama, Japan

T. Ino High Energy Accelerator Research Organization (KEK), Tsukuba, Ibaraki, Japan

A. Inoue Research Center for Nuclear Physics, Osaka University, Osaka, Ibaraki, Japan

Y. Inoue Department of Physics, Tohoku University, Sendai, Miyagi, Japan;
Cyclotron and Radioisotope Center (CYRIC), Tohoku University, Sendai, Miyagi, Japan;

High Energy Accelerator Research Organization (KEK), Tsukuba, Ibaraki, Japan; National Institute of Radiological Science, Chiba, Japan

T. Ishikawa Research Center for Electron Photon Science (ELPH), Tohoku University, Sendai, Japan

W. Iskandar Chemical Science Division, Lawrence Berkeley National Laboratory, Berkeley, USA

T. Isobe RIKEN Nishina Center, Wako, Saitama, Japan

Alexandr Yurievich Isupov Joint Institute for Nuclear Research, Dubna, Moscow region, Russia

Naoyuki Itagaki Yukawa Institute for Theoretical Physics, Kyoto University, Kyoto, Japan

K. Itahashi RIKEN Nishina Center, Wako, Japan

M. Ito Cyclotron and Radioisotope Center (CYRIC), Tohoku University, Sendai, Miyagi, Japan

M. Itoh Cyclotron and Radioisotope Center (CYRIC), Tohoku University, Sendai, Miyagi, Japan

M. Iwasaki RIKEN, The Institute of Physics and Chemical Research, Saitama, Japan

T. Iwata Department of Physics, Yamagata University, Yamagata, Japan

E. Jacquet CIMAP, CEA—CNRS—ENSICAEN—UNICAEN, Normandie Université, Caen Cedex 04, France

Marian Janek Zilina University, Žilina, Slovakia

G. Jansen Physics Division, Oak Ridge National Laboratory, National Center for Computational Sciences, Oak Ridge, TN, USA

A. S. Jensen Department of Physics and Astronomy, Aarhus University, Aarhus C, Denmark

Chen Ji Key Laboratory of Quark and Lepton Physics (MOE) and Institute of Particle Physics, Central China Normal University, Wuhan, China

B. Juliá-Díaz Departament de Física Quàntica i Astrofísica, Facultat de Física, Universitat de Barcelona, Barcelona, Spain

Ju-Hyun Jung Institute of Physics, University of Graz, Graz, Austria

Matthias Junker INFN, Laboratori Nazionali del Gran Sasso, L'Aquila, Italy

R. S. Kabatayeva IETP, Al-Farabi Kazakh National University, Almaty, Kazakhstan;
International Information Technology University, Almaty, Kazakhstan

J. Kahlbow RIKEN Nishina Center, Wako, Saitama, Japan;
Institut für Kernphysik, Technische Universität Darmstadt, Darmstadt, Germany

S. Kaida Research Center for Electron Photon Science (ELPH), Tohoku University, Sendai, Japan

N. Kalantar-Nayestanaki KVI-CART, University of Groningen, Groningen, The Netherlands

Hiroyuki Kamada Department of Physics, Faculty of Engineering, Kyushu Institute of Technology, Kitakyushu, Japan

D. Kameda RIKEN Nishina Center, Wako, Saitama, Japan

Yoshiko Kanada-En'yo Department of Physics, Kyoto University, Kyoto, Japan

Y. Kanaya Department of Applied Physics, University of Miyazaki, Miyazaki, Japan

H. Kanda RCNP, Osaka University, Ibaraki, Japan;
Department of Physics, Tohoku University, Sendai, Japan

Daekyoung Kang Key Laboratory of Nuclear Physics and Ion-beam Application (MOE) and Institute of Modern Physics, Fudan University, Shanghai, China

D. Kanno Department of Physics, Tokyo Institute of Technology, Meguro, Tokyo, Japan

R. Kanungo Astronomy and Physics Department, Saint Mary's University, Halifax, NS, Canada;
TRIUMF, Vancouver, BC, Canada

Julia-Tatiana Karachuk Advanced Research Institute for Electrical Engineering, Bucharest, Romania

V. A. Karmanov Lebedev Physical Institute, Moscow, Russia

Jean-Philippe Karr Laboratoire Kastler Brossel, Sorbonne Université, CNRS, ENS—Université PSL, Collége de France, Paris, France;
Université d'Evry-Val d'Essonne, Université Paris Saclay, Evry, France

Kiyoshi Katō Nuclear Reaction Data Centre, Hokkaido University, Sapporo, Japan

K. Kawahara Department of Physics, Tohoku University, Sendai, Miyagi, Japan;
Cyclotron and Radioisotope Center (CYRIC), Tohoku University, Sendai, Miyagi, Japan;
High Energy Accelerator Research Organization (KEK), Tsukuba, Ibaraki, Japan;
National Institute of Radiological Science, Chiba, Japan

H. Kawai Department of Physics, Chiba University, Chiba, Japan

S. Kawakami Department of Applied Physics, University of Miyazaki, Miyazaki, Japan;
Faculty of Engineering, University of Miyazaki, Miyazaki, Japan

N. Kawamura Department of Physics, Hokkaido University, Sapporo, Japan

A. Kawano Department of Information Science, Tohoku Gakuin University, Sendai, Japan

S. Kawase Center for Nuclear Study, University of Tokyo, Tokyo, Japan

R. Ya. Kezerashvili New York City College of Technology, Graduate School and University Center, City University of New York, New York City, NY, USA

Anatolyi Nikolaevich Khrenov Joint Institute for Nuclear Research, Dubna, Moscow region, Russia

A. Kievsky Istituto Nazionale di Fisica Nucleare, Pisa, Italy

Yuma Kikuchi General Education, Tokuyama College, National Institute of Technology, Shunan, Japan

June-Young Kim Department of Physics, Inha University, Incheon, Republic of Korea

Hyun-Chul Kim Department of Physics, Inha University, Incheon, Republic of Korea;
School of Physics, Korea Institute for Advanced Study (KIAS), Seoul, Republic of Korea;
Advanced Science Research Center, Japan Atomic Energy Agency, Tokai, Ibaraki, Japan

D. Kim Department of Physics, Ehwa Womans University, Seoul, South Korea

S. Kim Department of Physics and Astronomy, Seoul National University, Seoul, Republic of Korea;
Department of Physics, Ehwa Womans University, Seoul, Korea

Sang-Ho Kim Center for Extreme Nuclear Matters (CENuM), Korea University, Seoul, Republic of Korea;
Department of Physics, Pukyong National University (PKNU), Busan, Republic of Korea

St. Kistryn Institute of Physics, Jagiellonian University, Kraków, Poland

Y. Kiyokawa Center for Nuclear Study, University of Tokyo, Tokyo, Japan

William H. Klink Department of Physics and Astronomy, University of Iowa, Iowa City, USA

B. Kłos Institute of Physics, University of Silesia, Chorzów, Poland

M. A. Knösel Institut für Kernphysik, Technische Universität Darmstadt, Darmstadt, Germany

M. Kobayashi Center for Nuclear Study, University of Tokyo, Tokyo, Japan

N. Kobayashi Department of Physics, Tokyo Institute of Technology, Meguro, Tokyo, Japan;
Research Center for Nuclear Physics, Osaka University, Osaka, Ibaraki, Japan

T. Kobayashi Department of Physics, Tohoku University, Sendai, Miyagi, Japan

E. A. Kolganova BLTP JINR, Dubna, Moscow Region, Russia;
Dubna State University, Moscow region, Russia

H. Kon Department of Physics, Cyclotron and Radioisotope Center (CYRIC), Tohoku University, Sendai, Miyagi, Japan;
High Energy Accelerator Research Organization (KEK), Tsukuba, Ibaraki, Japan;
National Institute of Radiological Science, Chiba, Japan

Y. Kondo Department of Physics, Tokyo Institute of Technology, Meguro, Tokyo, Japan

Sebastian König Institut für Kernphysik, Technische Universität Darmstadt, Darmstadt, Germany;
ExtreMe Matter Institute EMMI, GSI Helmholtzzentrum für Schwerionenforschung GmbH, Darmstadt, Germany

A. S. Kopenbayeva IETP, Al-Farabi Kazakh National University, Almaty, Kazakhstan

Z. Korkulu RIKEN Nishina Center, Saitama, Japan

N. K. Kornegrutsa Joint Institute for Nuclear Research (JINR), Dubna, Russia

A. A. Korobitsin BLTP JINR, Dubna, Moscow Region, Russia

Vladimir I. Korobov Bogoliubov Laboratory of Theoretical Physics, Joint Institute for Nuclear Research, Dubna, Russia

P. Koseoglou Institut für Kernphysik, Technische Universität Darmstadt, Darmstadt, Germany;
GSI Helmholtzzentrum für Schwerionenforschung, Darmstadt, Germany

Yan Kostylenko V. N. Karazin National University, Kharkov, Ukraine

D. A. Kostyleva II. Physikalisches Institut, Justus-Liebig-Universität, Giessen, Germany;
GSI Helmholtzzentrum für Schwerionenforschung GmbH, Darmstadt, Germany

S. Koyama Department of Physics, University of Tokyo, Tokyo, Japan

A. Kozela Henryk Niewidniczański Institute of Nuclear Physics PAS, Kraków, Poland

Adam Kozela Institute of Nuclear Physics Polish Academy of Sciences, Kraków, Poland

Hermann Krebs Fakultät für Physik und Astronomie, Institut für Theoretische Physik II, Ruhr-Universität Bochum, Bochum, Germany

G. Krein Instituto de Física Teórica, Universidade Estadual Paulista, São Paulo, SP, Brazil

Dmitryi Olegovich Krivenkov Joint Institute for Nuclear Research, Dubna, Moscow region, Russia

T. Kubo RIKEN Nishina Center, Wako, Saitama, Japan

Joanna Kuboś Institute of Nuclear Physics PAS, Kraków, Poland

Y. Kubota Center for Nuclear Study, University of Tokyo, Tokyo, Japan; RIKEN Nishina Center, Wako, Saitama, Japan

V. I. Kukulin Skobeltsyn Institute of Nuclear Physics, Moscow State University, Moscow, Russia

P. Kulessa Henryk Niewidniczański Institute of Nuclear Physics PAS, Kraków, Poland

V. Kumar CIMAP, CEA—CNRS—ENSICAEN—UNICAEN, Normandie Université, Caen Cedex 04, France

Pavel Konstantievich Kurilkin Joint Institute for Nuclear Research, Dubna, Moscow region, Russia

V. O. Kurmangaliyeva Al-Farabi Kazakh National University, Almaty, Kazakhstan

I. Kuti MTA ATOMKI, Debrecen, Hungary

S. Kuwasaki Research Center for Electron Photon Science (ELPH), Tohoku University, Sendai, Japan

Vladimir Petrovich Ladygin Joint Institute for Nuclear Research, Dubna, Moscow region, Russia

Nadezhda Ladygina Joint Institute for Nuclear Research, Dubna, Russia

Nadezhda Borisovna Ladygina Joint Institute for Nuclear Research, Dubna, Moscow region, Russia

Ranjan Laha PRISMA Cluster of Excellence and Mainz Institute for Theoretical Physics, Johannes Gutenberg-Universität Mainz, Mainz, Germany

V. Lapoux Département de Physique Nucléaire, IRFU, CEA, Université Paris-Saclay, Saint-Aubin, France

Åsa Larson Department of Physics, Stockholm University, Stockholm, Sweden

K. D. Launey Department of Physics and Astronomy, Louisiana State University, Baton Rouge, LA, USA

Rimantas Lazauskas Université de Strasbourg, CNRS, IPHC UMR 7178, Strasbourg, France

S. Leblond LPC Caen, Normandie Université, ENSICAEN, Université de Caen, CNRS/IN2P3, Caen, France

J. Lee RIKEN Nishina Center, Wako, Saitama, Japan

J. H. Lee Dong-Eui University, Busan, Korea

Su Youn Lee Dong-Eui University, Busan, Korea

Young Jun Lee Dong-Eui University, Busan, Korea

C. Lehr RIKEN Nishina Center, Wako, Saitama, Japan;
Institut fuïr Kernphysik, Technische Universität Darmstadt, Darmstadt, Germany

Jin Lei Institute of Nuclear and Particle Physics, and Department of Physics and Astronomy, Ohio University, Athens, OH, USA

W. Leidemann Department of Physics, University of Trento, Trento, Italy;
INFN-TIFPA Trento Institute of Fundamental Physics and Applications, Trento, Italy

Sofia Leitão CFTP, Instituto Superior Técnico, Universidade de Lisboa, Lisbon, Portugal

M. L. Lekala Department of Physics, University of South Africa, Pretoria, South Africa

Sylvie Leray IRFU, CEA, Université Paris-Saclay, Gif-sur-Yvette, France

P. Levi Sandri INFN, Laboratori Nazionali di Frascati, Frascati, Rome, Italy

Ingo Leya Space Research and Planetary Sciences, Physics Institute, University of Bern, Bern, Switzerland

P. J. Li Department of Physics, The University of Hong Kong, Hong Kong, China

R. Lichtenthaler Filho Instituto de Física, Universidade de São Paulo, São Paulo, SP, Brazil

E. Liénard Normandie Univ, ENSICAEN, UNICAEN, CNRS/IN2P3, LPC Caen, Caen, France

R. Linares Departamento de Física, Universidade Federal Fluminense, Niterói, RJ, Brazil

Xiang Liu School of Physical Science and Technology, Lanzhou University, Lanzhou, China;
Research Center for Hadron and CSR Physics, Lanzhou University and Institute of Modern Physics of CAS, Lanzhou, China

Y. Liu School of Physics and State Key Laboratory of Nuclear Physics and Technology, Peking University, Beijing, China;
RIKEN Nishina Center, Wako, Saitama, Japan

Yong-Lu Liu College of Liberal Arts and Sciences, National University of Defense Technology, Changsha, China

Alexei Nikolaevich Livanov Joint Institute for Nuclear Research, Dubna, Moscow region, Russia

Cédric Lorcé Centre de Physique Théorique, École polytechnique, CNRS, Université Paris-Saclay, Palaiseau, France

A. Lovato INFN-TIFPA Trento Institute of Fundamental Physics and Applications, Trento, Italy;
Physics Division, Argonne National Laboratory, Argonne, IL, USA

Junxu Lu School of Physics, Beihang University, Beijing, China;
Groupe de Physique Théorique, IPN (UMR8608), Université Paris-Sud 11, Orsay, France

J. Lubian Departamento de Física, Universidade Federal Fluminense, Niterói, RJ, Brazil

R. Machleidt Department of Physics, University of Idaho, Moscow, ID, USA

K. Maeda Department of Physics, Tohoku University, Sendai, Japan

Y. Maeda Department of Applied Physics, Faculty of Engineering, University of Miyazaki, Miyazaki, Japan

A. Magiera Institute of Physics, Jagiellonian University, Kraków, Poland

M. Mahjour-Shafiei Department of Physics, University of Tehran, Tehran, Iran

Davide Mancusi Den-Service d'étude des réacteurs et de mathématiques appliquées (SERMA), CEA, Université Paris-Saclay, Gif-sur-Yvette, France

Carlo A. Manzata Department of Physics, University of Trento, Trento, Italy

L. E. Marcucci Dipartimento di Fisica "E. Fermi", Pisa University, Pisa, Italy;
Istituto Nazionale di Fisica Nucleare, Sezione di Pisa, Pisa, Italy

P. Maris Department of Physics and Astronomy, Iowa State University, Ames, IA, USA

F. M. Marqués LPC Caen, Normandie Université, ENSICAEN, Université de Caen, CNRS/IN2P3, Caen, France

J. Marton INFN, Laboratori Nazionali di Frascati, Frascati, Rome, Italy;
Stefan-Meyer-Institut für Subatomare Physik, Vienna, Austria

S. Masumoto Department of Physics, University of Tokyo, Tokyo, Japan

S. Masuoka Center for Nuclear Study, University of Tokyo, Wako, Saitama, Japan

J. Matsumoto Department of Chemistry, Tokyo Metropolitan University, Tokyo, Japan

M. Matsumoto RIKEN Nishina Center, Wako, Saitama, Japan;
Department of Physics, Tokyo Institute of Technology, Meguro, Tokyo, Japan

Y. Matsumura RCNP, Osaka University, Ibaraki, Japan

Hideaki Matsuno Department of Physics, Kyoto University, Kyoto, Japan

J. Mayer Institut für Kernphysik, Universität zu Köln, Cologne, Germany

Judith A. McGovern School of Physics and Astronomy, The University of Manchester, Manchester, UK

Mehmandoost-Khajeh-dad Department of Physics, University of Sistan and Baluchestan, Zahedan, Iran

Vladimir S. Melezhik Bogoliubov Laboratory of Theoretical Physics, Joint Institute for Nuclear Research, Dubna, Moscow Region, Russian Federation;
Peoples Friendship University of Russia (RUDN University), Moscow, Russian Federation

D. P. Menezes Depto de Física, CFM, Universidade Federal de Santa Catarina, Florianópolis, Brazil

A. Mercenne Department of Physics & Astronomy, Louisiana State University, Tower Dr. Baton Rouge, LA, USA

A. Méry CIMAP, CEA—CNRS—ENSICAEN—UNICAEN, Normandie Université, Caen Cedex 04, France

J. G. Messchendorp KVI-CART, University of Groningen, Groningen, The Netherlands

Olena Mezhenska Pavol Jozef Safarik University, Košice, Slovakia

Cédric Mezrag Istituto Nazionale di Fisica Nucleare, Sezione di Roma, Rome, Italy;
IRFU, CEA, Université Paris-Saclay, Gif-sur-Yvette, France

N. Michel GANIL, CEA/DSM—CNRS/IN2P3, Caen Cedex, France;
Grand Accélérateur National d'Ions Lourds (GANIL), Caen Cedex, France

Miha Mihovilovič Jožef Stefan Institute, Ljubljana, Slovenia;
Faculty of Mathematics and Physics, University of Ljubljana, Ljubljana, Slovenia;
Institut für Kernphysik, Johannes Gutenberg-Universität Mainz, Mainz, Germany

H. Miki Department of Physics, Tokyo Institute of Technology, Meguro, Tokyo, Japan

K. Miki Department of Physics, Tohoku University, Sendai, Miyagi, Japan

M. Miliucci INFN, Laboratori Nazionali di Frascati, Frascati (Roma), Italy;
Department of Physics, Faculty of Science MM.FF.NN., University of Rome 2 (Tor Vergata), Rome, Italy

E. Milman Department of Physics, Kyungpook National University, Daegu, Korea

R. Minakata Department of Physics, Tokyo Institute of Technology, Meguro, Tokyo, Japan

E. Mitsova Joint Institute for Nuclear Research (JINR), Dubna, Russia;
Southwestern University, Blagoevgrad, Bulgaria

S. Mitsumoto Department of Physics, Kyushu University, Fukuoka, Japan

M. Miwa Toho University, Tokyo, Japan;
RIKEN Nishina Center, Wako, Saitama, Japan

M. Miyabe Research Center for Electron Photon Science (ELPH), Tohoku University, Sendai, Japan

F. Miyahara Research Center for Electron Photon Science (ELPH), Tohoku University, Sendai, Japan

Kenta Miyahara Department of Physics, Graduate School of Science, Kyoto University, Kyoto, Japan

S. Miyata Department of Physics, University of Tokyo, Bunkyo, Japan

T. Miyazaki Department of Physics, University of Tokyo, Tokyo, Japan

K. Mochizuki Research Center for Electron Photon Science (ELPH), Tohoku University, Sendai, Japan

M. K. Mohamed Department of Physics, University of the Western Cape, Bellville, South Africa

M. Mohammadi-Dadkan KVI-CART, University of Groningen, Groningen, The Netherlands;
Department of Physics, University of Sistan and Baluchestan, Zahedan, Iran

Abhishek Mohapatra Department of Physics, Duke University, Durham, NC, USA

Glòria Montaña Departament de Física Quàntica i Astrofísica, Institut de Ciències del Cosmos (ICCUB), Universitat de Barcelona, Barcelona, Spain

B. Monteagudo LPC Caen, ENSICAEN, Université de Caen, CNRS/IN2P3, Caen, France

D. S. Monteiro Universidade da Integração Latino-Americana, UNILA, Foz do Iguaçu, PR, Brazil;
Department of Physics, University of Notre Dame, Notre Dame, IN, USA

V. Morcelle Universidade Federal Rural do Rio de Janeiro, Seropédica, RJ, Brazil

P. Moskal Institute of Physics, Jagiellonian University, Kracow, Poland

T. Motobayashi RIKEN Nishina Center, Wako, Saitama, Japan

T. Mukai Department of Physics, Tohoku University, Sendai, Miyagi, Japan;
Cyclotron and Radioisotope Center (CYRIC), Tohoku University, Sendai, Miyagi, Japan;
High Energy Accelerator Research Organization (KEK), Tsukuba, Ibaraki, Japan;
National Institute of Radiological Science, Chiba, Japan

S. Mukhopadhyay Variable Energy Cyclotron Centre, HBNI, Kolkata, India

D. Murai Department of Physics, Rikkyo University, Toshima, Tokyo, Japan

T. Murakami Department of Physics, Kyoto University, Kyoto, Japan

N. Muramatsu Research Center for Electron Photon Science (ELPH), Tohoku University, Sendai, Japan

I. Murray RIKEN Nishina Center, Wako, Saitama, Japan

K. Muto Department of Physics, Tohoku University, Sendai, Miyagi, Japan

Takayuki Myo General Education, Faculty of Engineering, Osaka Institute of Technology, Osaka, Japan

T. Nagahisa Department of Physics, Hokkaido University, Sapporo, Japan

Yasuyuki Nagashima Department of Physics, Tokyo University of Science, Shinjuku, Tokyo, Japan

S. Nakai Department of Physics, Tohoku University, Sendai, Miyagi, Japan;
Cyclotron and Radioisotope Center (CYRIC), Tohoku University, Sendai, Miyagi, Japan;
High Energy Accelerator Research Organization (KEK), Tsukuba, Ibaraki, Japan;
National Institute of Radiological Science, Chiba, Japan

A. Nakamura Research Center for Electron Photon Science (ELPH), Tohoku University, Sendai, Japan

S. Nakamura Research Center for Nuclear Physics, Osaka University, Osaka, Ibaraki, Japan

T. Nakamura Department of Physics, Tokyo Institute of Technology, Meguro, Tokyo, Japan

T. Nakashima Department of Physics, Tokyo Institute of Technology, Meguro, Tokyo, Japan

N. Nakatsuka Department of Physics, Kyoto University, Kyoto, Japan

E. S. Nathanson Georgia Gwinnett College, Lawerenceville, GA, USA

A. Navin GANIL, CEA/DRF-CNRS/IN2P3, Caen Cedex 5, France

K. Nawa Research Center for Electron Photon Science (ELPH), Tohoku University, Sendai, Japan

W. Nazarewicz Department of Physics and Astronomy and NSCL/FRIB Laboratory, Michigan State University, East Lansing, MI, USA

R. Neveling Department of Subatomic Physics, iThemba LABS, Somerset West, South Africa

Nir Nevo-Dinur TRIUMF, Vancouver, BC, Canada

S. Niedźwiecki Jagiellonian University, Kraków, Poland

S. Nishi Department of Physics, Tokyo Institute of Technology, Meguro, Tokyo, Japan

Y. Nishio Department of Physics, Kyushu University, Fukuoka, Japan

A. Nogga IAS-4, IKP-3, JHCP, and JARA-HPC, Forschungszentrum Jülich, Jülich, FRG, Germany

J. H. Alvarenga Nogueira Instituto Tecnológico de Aeronáutica, DCTA, São José dos Campos, Brazil;
Dipartimento di Fisica, Università di Roma "La Sapienza" & INFN, Roma, Italy

Kenneth M. Nollett Department of Physics, San Diego State University, San Diego, CA, USA

F. M. Nunes National Superconducting Cyclotron Laboratory and Department of Physics and Astronomy, Michigan State University, East Lansing, MI, USA

J. C. Nzobadila-Ondze Department of Subatomic Physics, iThemba LABS, Somerset West, South Africa;
Department of Physics, University of the Western Cape, Bellville, South Africa

G. G. O'Neill Department of Subatomic Physics, iThemba LABS, Somerset West, South Africa;
Department of Physics, University of the Western Cape, Bellville, South Africa

A. Obertelli Département de Physique Nucléaire, IRFU, CEA, Université Paris-Saclay, Saint-Aubin, France;
Institut für Kernphysik, Technische Universität Darmstadt, Darmstadt, Germany

Myagmarjav Odsuren School of Engineering and Applied Sciences and Nuclear Research Center, National University of Mongolia, Ulaanbaatar, Mongolia

S. Ogoshi Department of Physics, Tokyo Institute of Technology, Meguro, Tokyo, Japan

S. Ogushi Research Center for Electron Photon Science (ELPH), Tohoku University, Sendai, Japan

A. Ohkura Department of Physics, Kyushu University, Fukuoka, Japan

H. Ohnishi ELPH, Tohoku University, Sendai, Japan

H. Ohshiro Department of Physics, Kyushu University, Fukuoka, Japan

S. Okada RIKEN, The Institute of Physics and Chemical Research, Saitama, Japan

Y. Okada Research Center for Electron Photon Science (ELPH), Tohoku University, Sendai, Japan

H. Okamura RCNP, Osaka University, Osaka, Ibaraki, Japan

K. Okamura Research Center for Electron Photon Science (ELPH), Tohoku University, Sendai, Japan

H. J. Ong Research Center for Nuclear Physics, Osaka University, Osaka, Ibaraki, Japan

Y. Onodera Research Center for Electron Photon Science (ELPH), Tohoku University, Sendai, Japan

G. Orlandini Department of Physics, University of Trento, Trento, Italy; INFN-TIFPA Trento Institute of Fundamental Physics and Applications, Trento, Italy

N. A. Orr LPC Caen, Normandie Université, ENSICAEN, Université de Caen, CNRS/IN2P3, Caen, France

Pablo G. Ortega Grupo de Física Nuclear and Instituto de Física Fundamental y Matemáticas (IUFFyM) Universidad de Salamanca, Salamanca, Spain

Shinsho Oryu Tokyo University of Science, Noda, Chiba, Japan

S. Ota Center for Nuclear Study, University of Tokyo, Tokyo, Japan

H. Otsu RIKEN Nishina Center, Wako, Saitama, Japan

T. Ozaki Department of Physics, Tokyo Institute of Technology, Tokyo, Japan

K. Ozawa Institute of Particle and Nuclear Studies, High Energy Accelerator Research Organization (KEK), Tsukuba, Japan

Emanuele Pace Università di Roma “Tor Vergata” and INFN, Sezione di Roma Tor Vergata, Rome, Italy

T. Pałasz August Chełkowski Institute of Physics, University of Silesia, Chorzow, Poland

P. K. Panda Department of Physics, Utkal University, Bhubaneswar, India

V. Panin RIKEN Nishina Center, Wako, Saitama, Japan

T. Papenbrock Physics Division, Oak Ridge National Laboratory, Oak Ridge, TN, USA;
Department of Physics and Astronomy, University of Tennessee, Knoxville, TN, USA

P. Papka Department of Subatomic Physics, iThemba LABS, Somerset West, South Africa;
Department of Physics, University of Stellenbosch, Matieland, Stellenbosch, South Africa

S. Park Department of Physics, Ehwa Womans University, Seoul, Korea

M. Parlog LPC Caen, ENSICAEN, Université de Caen, CNRS/IN2P3, Caen Cedex, France

W. Parol Institute of Nuclear Physics PAN, Kraków, Poland;
Henryk Niewidniczański Institute of Nuclear Physics PAS, Kraków, Poland

S. Paschalis Institut für Kernphysik, Technische Universität Darmstadt, Darmstadt, Germany;
Department of Physics, University of York, York, UK

Barbara Pasquini Dipartimento di Fisica, Università degli Studi di Pavia, Pavia, Italy;
Istituto Nazionale di Fisica Nucleare, Sezione di Pavia, Pavia, Italy

L. Pellegri Department of Subatomic Physics, iThemba LABS, Somerset West, South Africa;
School of Physics, University of the Witwatersrand, Johannesburg, South Africa

M. T. Peña CFTP, Instituto Superior Técnico, Universidade de Lisboa, Lisbon, Portugal

N. G. Peresadko Lebedev Physical Institute, Russian Academy of Sciences, Moscow, Russia

V. Pesudo Department of Subatomic Physics, iThemba LABS, Somerset West, South Africa;
Department of Physics, University of the Western Cape, Bellville, South Africa

Daniel R. Phillips Department of Physics and Astronomy, Institute of Nuclear and Particle Physics, Ohio University, Athens, OH, USA;
Institut für Kernphysik, TU Darmstadt, Darmstadt, Germany;
ExtreMe Matter Institute EMMI, GSI Helmholtzzentrum für Schwerionenforschung GmbH, Darmstadt, Germany

Maria Piarulli Washington University, St. Louis, MO, USA

D. Pietreanu INFN, Laboratori Nazionali di Frascati, Frascati, Rome, Italy; Stefan-Meyer-Institut für Subatomare Physik, Vienna, Austria

Dorel Pietreanu Horia Hulubei National Institute of Physics and Nuclear Engineering, Magurele, Romania

K. C. C. Pires Instituto de Física, Universidade de São Paulo, São Paulo, SP, Brazil

K. Piscicchia INFN, Laboratori Nazionali di Frascati, Frascati, Rome, Italy; CENTRO FERMI—Museo Storico della Fisica e Centro Studi e Ricerche "Enrico Fermi", Rome, Italy

Semen Mikhailovich Piyadin Joint Institute for Nuclear Research, Dubna, Moscow region, Russia

Lucas Platter Department of Physics, University of Tennessee, Knoxville, TN, USA;
Physics Division, Oak Ridge National Laboratory, Oak Ridge, TN, USA

Willibald Plessas University of Graz, Institute of Physics, Universitätsplatz 5, Graz, Austria

M. Płoszajczak GANIL, CEA/DSM—CNRS/IN2P3, Caen Cedex, France;
Grand Accélérateur National d'Ions Lourds (GANIL), Caen Cedex, France

E. C. Pollacco Département de Physique Nucléaire, IRFU, CEA, Université Paris-Saclay, Saint-Aubin, France

A. Polls Departament de Física Quàntica i Astrofísica, Facultat de Física, Universitat de Barcelona, Barcelona, Spain

W. N. Polyzou The University of Iowa, Iowa City, IA, USA

B. Pons Université de Bordeaux—CNRS—CEA, CELIA, Talence, France

G. Popa Department of Physics and Astronomy, Ohio University, Athens, OH, USA

M. Potlog Institute of Space Sciences, Magurele, Romania

C. Providência CFisUC, Department of Physics, University of Coimbra, Coimbra, Portugal

G. Quéméner Normandie Univ, ENSICAEN, UNICAEN, CNRS/IN2P3, LPC Caen, Caen, France

Udit Raha Department of Physics, Indian Institute of Technology Guwahati, Assam, India

G. Ramalho LFTC, Universidade Cruzeiro do Sul, São Paulo, Brazil

A. Ramazani Moghaddam Arani Faculty of Physics, University of Kashan, Kashan, Iran

R. Ramazani-Sharifabadi KVI-CART, University of Groningen, Groningen, The Netherlands;
Department of Physics, University of Tehran, Tehran, Iran

Àngels Ramos Departament de Física Quàntica i Astrofísica, Institut de Ciències del Cosmos (ICCUB), Universitat de Barcelona, Barcelona, Spain

G. J. Rampho Department of Physics, University of South Africa, Pretoria, South Africa

Juan Martín Randazzo División Física Atómica, Molecular y Óptica, Centro Atómico Bariloche, S. C. de Bariloche, Argentina;
CONICET, Buenos Aires, Argentina

J. Rangama CIMAP, CEA—CNRS—ENSICAEN—UNICAEN, Normandie Université, Caen Cedex 04, France

Christophe Rappold GSI Helmholtz Center for Heavy Ion Research, Darmstadt, Germany

B. Rebeiro Department of Physics, University of the Western Cape, Bellville, South Africa

Nikita Reichelt Institute of Physics, University of Graz, Graz, Austria

S. Reichert Department of Physics, Technische Universitt Munchen, München, Germany;
Technische Universität München, Garching, Germany

Patrick Reinert Fakultät für Physik und Astronomie, Institut für Theoretische Physik II, Ruhr-Universität Bochum, Bochum, Germany

Xiulei Ren Institut für Theoretische Physik II, Ruhr-Universität Bochum, Bochum, Germany

János Révai MTA Wigner RCP, Budapest, Hungary

A. Revel Grand Accélérateur National d'Ions Lourds (GANIL), CEA/DRF, CNRS/IN2P3, Caen, France

Sergei Grigorievich Reznikov Joint Institute for Nuclear Research, Dubna, Moscow region, Russia

Jean-Marc Richard Institut de Physique Nucléaire de Lyon, IN2P3-CNRS–UCBL, Université de Lyon, Villeurbanne, France

Craig D. Roberts School of Physics, Nanjing University, Nanjing, Jiangsu, China;
Institute for Nonperturbative Physics, Nanjing University, Nanjing, Jiangsu, China

N. Rocco Physics Division, Argonne National Laboratory, Argonne, IL, USA; Theoretical Physics Department, Fermi National Accelerator Laboratory, Batavia, IL, USA

Jose Luis Rodríguez-Sánchez IRFU, CEA, Université Paris-Saclay, Gif-sur-Yvette, France;
Universidad de Santiago de Compostela, Santiago de Compostela, Spain;
GSI-Helmholtzzentrum für Schwerionenforschung GmbH, Darmstadt, Germany

F. Ropars CIMAP, CEA—CNRS—ENSICAEN—UNICAEN, Normandie Université, Caen Cedex 04, France

D. S. Rosa Instituto de Física Teórica, Universidade Estadual Paulista, São Paulo, SP, Brazil

D. Rossi Institut für Kernphysik, Technische Universität Darmstadt, Darmstadt, Germany

J. Rotureau NSCL/FRIB Laboratory, Michigan State University, East Lansing, MI, USA

V. A. Roudnev St. Petersburg State University, St. Petersburg, Russian Federation

Vladimir Roudnev St. Petersburg State University, St. Petersburg, Russian Federation

J. Y. Rousse Département de Physique Nucléaire, IRFU, CEA, Université Paris-Saclay, Saint-Aubin, France

O. A. Rubtsova Skobeltsyn Institute of Nuclear Physics, Moscow State University, Moscow, Russia

E. Ruiz Arriola Departamento de Física Atómica, Molecular Y Nuclear, Facultad de Ciencias, Universidad de Granada—UGR, Granada, Andalucía, Spain

V. V. Rusakova Joint Institute for Nuclear Research (JINR), Dubna, Russia

C. P. Safvan Inter University Accelerator Center, New Delhi, India

D. A. Sailaubek Faculty of Physics and Technical Sciences, L. N. Gumilyov Eurasian National University, Nur-Sultan, Kazakhstan

Takehiko R. Saito GSI Helmholtz Center for Heavy Ion Research, Darmstadt, Germany;
Helmholtz Institute Mainz, Mainz, Germany;
High Energy Nuclear Physics Laboratory RIKEN, Saitama, Japan;
Institute of Modern Physics, Lanzhou, China

A. Saito Department of Physics, Tokyo Institute of Technology, Meguro, Tokyo, Japan

A. T. Saito Department of Physics, Tokyo Institute of Technology, Tokyo, Japan

S. Sakaguchi Department of Physics, Kyushu University, Fukuoka, Japan

D. Sakai Department of Physics, Tohoku University, Sendai, Miyagi, Japan; Cyclotron and Radioisotope Center (CYRIC), Tohoku University, Sendai, Miyagi, Japan;
High Energy Accelerator Research Organization (KEK), Tsukuba, Ibaraki, Japan;
National Institute of Radiological Science, Chiba, Japan

H. Sakai RIKEN Nishina Center, Wako, Saitama, Japan

N. Sakamoto RIKEN Nishina Center, Wako, Saitama, Japan

Y. Sakamoto Department of Information Science, Tohoku Gakuin University, Sendai, Japan

M. Sako RIKEN Nishina Center, Saitama, Japan

Giovanni Salmè INFN, Sezione di Roma, Rome, Italy

F. Sammarruca Department of Physics, University of Idaho, Moscow, ID, USA

Paolo Levi Sandri INFN, Laboratori Nazionali di Frascati, Frascati, Rome, Italy

C. Santamaria RIKEN Nishina Center, Saitama, Japan

Elena Santopinto Istituto Nazionale di Fisica Nucleare (INFN), Sezione di Genova, Genova, Italy

A. M. Santos Depto de Física, CFM, Universidade Federal de Santa Catarina, Florianópolis, Brazil

O. C. B. Santos Instituto de Física, Universidade de São Paulo, São Paulo, SP, Brazil

Satoru Sasabe Department of Physics, Kyushu University, Fukuoka, Japan

M. Sasano RIKEN Nishina Center, Wako, Saitama, Japan

H. Sato RIKEN Nishina Center, Wako, Saitama, Japan

M. Sato Research Center for Electron Photon Science (ELPH), Tohoku University, Sendai, Japan

Y. Satou Department of Physics and Astronomy, Seoul National University, Gwanak-gu, Seoul, Republic of Korea;
Rare Isotope Science Project, Institute for Basic Science, Yuseong-gu, Daejeon, Republic of Korea

H. Scheit Institut für Kernphysik, Technische Universität Darmstadt, Darmstadt, Germany

F. Schindler Technische Universität München, Garching, Germany

Christiane H. Schmickler Institut Für Kernphysik, Technische Universität Darmstadt, Darmstadt, Germany;
RIKEN Nishina Center, RIKEN, Saitama, Japan

Regina A. Schmidt University of Graz, Institute of Physics, Universitätsplatz 5, Graz, Austria

Georg Schnabel IRFU, CEA, Université Paris-Saclay, Gif-sur-Yvette, France

Karin Schönning Department. of Physics and Astronomy, Uppsala University, Uppsala, Sweden

Wolfgang Schweiger Institute of Physics, University of Graz, Graz, Austria

Sergio Scopetta Università di Perugia and INFN, Sezione di Perugia, Perugia, Italy

A. Scordo INFN, Laboratori Nazionali di Frascati, Frascati, Rome, Italy

Jorge Segovia Departamento de Sistemas Físicos, Químicos y Naturales, Universidad Pablo de Olavide, Seville, Spain

K. Sekiguchi Department of Physics, Cyclotron and Radioisotope Center (CYRIC), Tohoku University, Sendai, Miyagi, Japan;
High Energy Accelerator Research Organization (KEK), Tsukuba, Ibaraki, Japan;
National Institute of Radiological Science, Chiba, Japan

Claude Semay Service de Physique Nucléaire et Subnucléaire, Université de Mons, UMONS Research Institute for Complex Systems, Mons, Belgium

M. A. Shalchi Instituto de Física Teórica (IFT), Universidade Estadual Paulista (UNESP), São Paulo, SP, Brazil

Aleksandr Shebeko National Science Center "Kharkov Institute of Physics and Technology", Kharkiv, Ukraine

Oleksandr Shebeko National Science Center "Kharkov Institute of Physics and Technology", Kharkov, Ukraine

Nina Shevchenko Nuclear Physics Institute, Řež, Czech Republic

H. Shi INFN, Laboratori Nazionali di Frascati, Frascati, Rome, Italy;
Stefan-Meyer-Institut für Subatomare Physik, Vienna, Austria

S. Shibuya Department of Physics, Tohoku University, Sendai, Miyagi, Japan;
Cyclotron and Radioisotope Center (CYRIC), Tohoku University, Sendai, Miyagi, Japan;
High Energy Accelerator Research Organization (KEK), Tsukuba, Ibaraki, Japan;
National Institute of Radiological Science, Chiba, Japan

M. Shikata Department of Physics, Tokyo Institute of Technology, Tokyo, Japan

T. Shimada Department of Physics, Tokyo Institute of Technology, Meguro, Tokyo, Japan

H. Shimizu Research Center for Electron Photon Science (ELPH), Tohoku University, Sendai, Japan

Y. Shimizu RIKEN Nishina Center, Wako, Saitama, Japan

S. Shimoura Center for Nuclear Study, University of Tokyo, Wako, Saitama, Japan

Y. Shindo Department of Physics, Kyushu University, Fukuoka, Japan

Y. Shiokawa Department of Physics, Tohoku University, Sendai, Miyagi, Japan;
Cyclotron and Radioisotope Center (CYRIC), Tohoku University, Sendai, Miyagi, Japan;
High Energy Accelerator Research Organization (KEK), Tsukuba, Ibaraki, Japan;
National Institute of Radiological Science, Chiba, Japan

H. Shiromaru Department of Chemistry, Tokyo Metropolitan University, Tokyo, Japan

Guillaume Sicorello Service de Physique Nucléaire et Subnucléaire, Université de Mons, UMONS Research Institute for Complex Systems, Mons, Belgium

M. Silarski Institute of Physics, Jagiellonian University, Kraków, Poland

Diana Laura Sirghi INFN, Laboratori Nazionali di Frascati, Frascati, Rome, Italy;
Horia Hulubei National Institute of Physics and Nuclear Engineering, Magurele, Romania

D. Sirghi INFN, Laboratori Nazionali di Frascati, Frascati, Rome, Italy;
Horia Hulubei National Institute of Physics and Nuclear Engineering (IFIN-HH), Magurele, Romania

F. Sirghi INFN, Laboratori Nazionali di Frascati, Frascati, Rome, Italy;
Horia Hulubei National Institute of Physics and Nuclear Engineering (IFIN-HH), Magurele, Romania

Yaroslav Tarasovich Skhomenko Joint Institute for Nuclear Research, Dubna, Moscow Region, Russia;
Belgorod State University, Belgorod, Russia

R. Skibiński M. Smoluchowski Institute of Physics, Jagiellonian University, Kraków, Poland

M. Skurzok Institute of Physics, Jagiellonian University, Kraków, Poland;
INFN, Laboratori Nazionali di Frascati, Frascati, Rome, Italy

I. Skwira-Chalot Faculty of Physics, University of Warsaw, Warsaw, Poland

F. D. Smit Department of Subatomic Physics, iThemba LABS, Somerset West, South Africa

A. Spallone INFN, Laboratori Nazionali di Frascati, Frascati, Rome, Italy

Alfred Stadler Departamento de Física, Universidade de Évora, Évora, Portugal; CFTP, Instituto Superior Técnico, Universidade de Lisboa, Lisbon, Portugal

R. Stanoeva Southwestern University, Blagoevgrad, Bulgaria;
Institute for Nuclear Research and Nuclear Energy, Sofia, Bulgaria

I. Stefan Institut de Physique Nucléaire Orsay, IN2P3-CNRS, Orsay Cedex, France

Margarita Stepanova St. Petersburg State University, St. Petersburg, Russia

E. Stephan Institute of Physics, University of Silesia, Chorzow, Poland;
Marian Smoluchowski Institute of Physics, Jagiellonian University, Kraków, Poland

G. F. Steyn Department of Subatomic Physics, iThemba LABS, Somerset West, South Africa

S. Storck Institut für Kernphysik, Technische Universität Darmstadt, Darmstadt, Germany

L. Stuhl RIKEN Nishina Center, Saitama, Japan;
Center for Nuclear Study, University of Tokyo, Wako, Saitama, Japan

H. Sugai Research Center for Electron Photon Science (ELPH), Tohoku University, Sendai, Japan

Jung-Min Suh Department of Physics, Inha University, Incheon, Republic of Korea

O. M. Sukhareva Omsk State Technical University, Omsk, Russia

T. Sumikama RIKEN Nishina Center, Saitama, Japan

H. Suzuki RIKEN Nishina Center, Wako, Saitama, Japan

K. Suzuki Research Center for Electron Photon Science (ELPH), Tohoku University, Sendai, Japan

J. A. Swartz Department of Physics and Astronomy, Aarhus University, Aarhus, Denmark

D. Symochko Institut für Kernphysik, Technische Universität Darmstadt, Darmstadt, Germany

S. Szpigel Centro de Rádio-Astronomia e Astrofísica Mackenzie—CRAAM, Universidade Presbiteriana Mackenzie, São Paulo, SP, Brazil

M. Tabata Department of Physics, Kyushu University, Fukuoka, Japan;
Department of Physics, Chiba University, Chiba, Japan

T. Taguchi Department of Physics, Cyclotron and Radioisotope Center (CYRIC), Tohoku University, Sendai, Miyagi, Japan;
High Energy Accelerator Research Organization (KEK), Tsukuba, Ibaraki, Japan;
National Institute of Radiological Science, Chiba, Japan

Y. Tajima Department of Physics, Yamagata University, Yamagata, Japan

K. Takahashi Department of Physics, Tohoku University, Sendai, Miyagi, Japan

S. Takahashi Research Center for Electron Photon Science (ELPH), Tohoku University, Sendai, Japan

H. Takeda RIKEN Nishina Center, Wako, Saitama, Japan

Masayuki Takeda Tokyo University of Science, Noda, Chiba, Japan

S. Takeuchi Research Center for Nuclear Physics (RCNP), Osaka University, Ibaraki, Osaka, Japan;
Theoretical Research Division, Nishina Center, RIKEN, Saitama, Japan;
Japan College of Social Work, Kiyose, Tokyo, Japan

N. Takibayev Al-Farabi Kazakh National University, Almaty, Kazakhstan

Makoto Takizawa Theoretical Research Division, Nishina Center, RIKEN, Saitama, Japan;
Showa Pharmaceutical University, Machida, Tokyo, Japan;
J-PARC Branch, KEK Theory Center, Institute for Particle and Nuclear Studies, KEK, Ibaraki, Japan

A. Tamii Research Center for Nuclear Physics, Osaka University, Osaka, Ibaraki, Japan

R. Tanaka Department of Physics, Tokyo Institute of Technology, Meguro, Tokyo, Japan

J. Tanaka Institut für Kernphysik, Technische Universität Darmstadt, Darmstadt, Germany;
GSI Helmholtzzentrum für Schwerionenforschung, Darmstadt, Germany

T. L. Tang RCNP, Osaka University, Osaka, Ibaraki, Japan

Y. Taniguchi Research Center for Electron Photon Science (ELPH), Tohoku University, Sendai, Japan

H. Tatsuno Lund Univeristy, Lund, Sweden

H. Tavakoli-Zaniani KVI-CART, University of Groningen, Groningen, The Netherlands;
Department of Physics, School of Science, Yazd University, Yazd, Iran

Arkadiy Arkadievich Terekhin Joint Institute for Nuclear Research, Dubna, Moscow region, Russia

N. Timofeyuk Department of Physics, University of Surrey, Guildford, Surrey, UK

Vladimir Timoshenko St. Petersburg State University, St. Petersburg, Russian Federation

V. S. Timóteo Grupo de Óptica e Modelagem Numérica—GOMNI, Faculdade de Tecnologia—FT, Universidade Estadual de Campinas—UNICAMP, Limeira, SP, Brazil

Aleksei Viktorovich Tishevsky Joint Institute for Nuclear Research, Dubna, Moscow region, Russia

A. S. Tkachenko Fesenkov Astrophysical Institute "NCSRT", Almaty, Kazakhstan;
Al-Farabi Kazakh National University, Almaty, Kazakhstan

Y. Togano ExtreMe Matter Institute EMMI and Research Division, GSI Helmholtzzentrum für Schwerionenforschung GmbH, Darmstadt, Germany;
Department of Physics, Tokyo Institute of Technology, Tokyo, Japan;
Department of Physics, Rikkyo University, Toshima, Tokyo, Japan;
RIKEN Nishina Center, Wako, Saitama, Japan

Yoshio Togawa Tokyo University of Science, Noda, Chiba, Japan

A. O. Tokiyasu ELPH, Tohoku University, Sendai, Japan

T. Tomai RIKEN Nishina Center, Wako, Saitama, Japan;
Department of Physics, Tokyo Institute of Technology, Meguro, Tokyo, Japan

Lauro Tomio Instituto de Física Teórica (IFT), Universidade Estadual Paulista (UNESP), São Paulo, SP, Brazil;
Instituto Tecnológico de Aeronáutica, Departamento de Ciência e Tecnologia Aeroespacial (DCTA), São José dos Campos, Brazil

K. Topolnicki M. Smoluchowski Institute of Physics, Jagiellonian University, Kraków, Poland

Werner Tornow Duke University, Durham, NC, USA;
Triangle Universities Nuclear Laboratory, Durham, NC, USA

H. T. Törnqvsit RIKEN Nishina Center, Wako, Saitama, Japan;
Institut für Kernphysik, Technische Universität Darmstadt, Darmstadt, Germany

S. Triambak Department of Physics, University of the Western Cape, Bellville, South Africa

J. Tscheuschner Institut für Kernphysik, Technische Universität Darmstadt, Darmstadt, Germany

J. Tsubota Department of Physics, Tokyo Institute of Technology, Tokyo, Japan

Y. Tsuchikawa Department of Physics, Nagoya University, Nagoya, Japan; Research Center for Electron Photon Science (ELPH), Tohoku University, Sendai, Japan

A. G. Tuff Department of Physics, University of York, Heslington, York, UK

S. A. Turakulov Institute of Nuclear Physics, Uzbekistan Academy of Sciences, Ulugbek, Tashkent, Uzbekistan

Nurbek Tursunbayev Joint Institute for Nuclear Researches, Dubna, Russia; Dubna State University, Dubna, Russia; Institute of Nuclear Physics, Almaty, Kazakhstan

E. M. Tursunov Institute of Nuclear Physics, Uzbekistan Academy of Sciences, Ulugbek, Tashkent, Uzbekistan

T. Ueda ELPH, Tohoku University, Sendai, Japan

T. Uesaka RIKEN Nishina Center, Wako, Saitama, Japan

Tomohiro Uesaka Nishina Center for Accelerator-Based Science, RIKEN, Wako, Japan

U. Umbelino Instituto de Física, Universidade de São Paulo, São Paulo, SP, Brazil

Jozef Urban Pavol Jozef Safarik University, Košice, Slovakia

Yuriy Uzikov Joint Institute for Nuclear Researches, Dubna, Russia; Dubna State University, Dubna, Russia; Department of Physics, M.V. Lomonosov Moscow State University, Moscow, Russia

Alfredo Valcarce Departamento de Física Fundamental and IUFFyM, Universidad de Salamanca, Salamanca, Spain

M. Vandebrouck Institut de Physique Nucléaire, Université Paris-Sud, IN2P3/CNRS, Orsay, France

V. S. Vasilevsky Bogolyubov Institute for Theoretical Physics, Kiev, Ukraine

O. Vazquez Doce INFN, Laboratori Nazionali di Frascati, Frascati, Rome, Italy; Excellence Cluster Universe, Technische Universiät München, Garching, Germany

Javier Vijande Departamento de Física Atómica, Molecular y Nuclear, IFIC (UV-CSIC) and IRIMED Joint Research Unit (IIS Hospital La Fe-UV), Universidad de Valencia (UV), Valencia, Spain

M. Viviani Istituto Nazionale di Fisica Nucleare, Sezione di Pisa, Pisa, Italy

Y. Wada Department of Physics, Cyclotron and Radioisotope Center (CYRIC), Tohoku University, Sendai, Miyagi, Japan;
High Energy Accelerator Research Organization (KEK), Tsukuba, Ibaraki, Japan;
National Institute of Radiological Science, Chiba, Japan

V. Wagner Institut für Kernphysik, Technische Universität Darmstadt, Darmstadt, Germany

T. Wakasa Department of Physics, Kyushu University, Fukuoka, Japan

T. Wakui National Institute of Radiological Sciences, Chiba, Chiba, Japan

Takashi Watanabe Tokyo University of Science, Noda, Chiba, Japan

A. Watanabe Department of Physics, Tohoku University, Sendai, Miyagi, Japan

M. Watanabe Department of Physics, Cyclotron and Radioisotope Center (CYRIC), Tohoku University, Sendai, Miyagi, Japan;
High Energy Accelerator Research Organization (KEK), Tsukuba, Ibaraki, Japan;
National Institute of Radiological Science, Chiba, Japan

Wolfram Weise Physik-Department, Technische Universität München, Garching, Germany

S. P. Weppner Natural Sciences, Eckerd College, St. Petersburg, FL, USA

E. Widmann Stefan-Meyer-Institut für Subatomare Physik, Vienna, Austria

A. Wilczek Institute of Physics, University of Silesia, Chorzow, Poland

K. Wimmer Center for Nuclear Study, University of Tokyo, Wako, Saitama, Japan

H. Witała M. Smoluchowski Institute of Physics, Jagiellonian University, Kraków, Poland

Jin-Yun Wu College of Information and Communication, National University of Defense Technology, Xi'an, China

Sławomir Wycech National Centre for Nuclear Research, Warsaw, Poland

B. Włoch Henryk Niewidniczański Institute of Nuclear Physics PAS, Kraków, Poland

Masanobu Yahiro Department of Physics, Kyushu University, Fukuoka, Japan

Ulugbek Yakhshiev Department of Physics, Inha University, Incheon, Republic of Korea

K. Yako Center for Nuclear Study, University of Tokyo, Tokyo, Japan

S. L. Yakovlev St. Petersburg State University, St. Petersburg, Russian Federation

Sergey Yakovlev St. Petersburg State University, St. Petersburg, Russia

H. Yamada Department of Physics, Tokyo Institute of Technology, Meguro, Tokyo, Japan

Yasuhiro Yamaguchi Theoretical Research Division, Nishina Center, RIKEN, Saitama, Japan;
RIKEN Nishina Center, Hirosawa, Wako, Saitama, Japan

T. Yamamoto Faculty of Engineering, University of Miyazaki, Miyazaki, Japan

Nodoka Yamanaka IPNO, Université Paris-Sud, CNRS/IN2P3, Orsay, France

M. T. Yamashita Instituto de Física Teórica, Universidade Estadual Paulista, São Paulo, SP, Brazil

Takeshi Yamazaki Faculty of Pure and Applied Sciences, University of Tsukuba, Tsukuba, Ibaraki, Japan;
Center for Computational Sciences, University of Tsukuba, Tsukuba, Ibaraki, Japan

H. Yamazaki Research Center for Electron Photon Science (ELPH), Tohoku University, Sendai, Japan

R. Yamazaki Research Center for Electron Photon Science (ELPH), Tohoku University, Sendai, Japan

Y. Yanagisawa RIKEN Nishina Center, Wako, Saitama, Japan

B. Yang RIKEN Nishina Center, Wako, Saitama, Japan;
School of Physics and State Key Laboratory of Nuclear Physics and Technology, Peking University, Beijing, China

Z. H. Yang Research Center for Nuclear Physics, Osaka University, Ibaraki, Osaka, Japan;
RIKEN Nishina Center, Wako, Saitama, Japan

L. Yang Center for Nuclear Study, University of Tokyo, Wako, Saitama, Japan

E. A. Yarevsky St. Petersburg State University, St. Petersburg, Russian Federation

J. Yasuda Department of Physics, Kyushu University, Fukuoka, Japan

Y. Yasuda RIKEN Nishina Center, Wako, Saitama, Japan;
Department of Physics, Tokyo Institute of Technology, Meguro, Tokyo, Japan

E. Ydrefors Instituto Tecnológico de Aeronáutica, DCTA, São José dos Campos, Brazil

K. Yoneda RIKEN Nishina Center, Wako, Saitama, Japan

C. Yoshida ELPH, Tohoku University, Sendai, Japan

H. Y. Yoshida Department of Physics, Yamagata University, Yamagata, Japan

Chang-Zheng Yuan Institute of High Energy Physics, Chinese Academy of Sciences, Beijing, China;
University of Chinese Academy of Sciences, Beijing, China

A. A. Zaitsev Joint Institute for Nuclear Research (JINR), Dubna, Russia;
Lebedev Physical Institute, Russian Academy of Sciences, Moscow, Russia

L. Zanetti RIKEN Nishina Center, Wako, Saitama, Japan;
Institut für Kernphysik, Technische Universität Darmstadt, Darmstadt, Germany

D. Zanuttini CIMAP, CEA—CNRS—ENSICAEN—UNICAEN, Normandie Université, Caen Cedex 04, France

P. I. Zarubin Joint Institute for Nuclear Research (JINR), Dubna, Russia;
Lebedev Physical Institute, Russian Academy of Sciences, Moscow, Russia

I. G. Zarubina Joint Institute for Nuclear Research (JINR), Dubna, Russia

J. Zejma August Chełkowski Institute of Physics, University of Silesia, Chorzow, Poland

J. Zenihiro RIKEN Nishina Center, Wako, Saitama, Japan

E. O. N. Zevallos Instituto de Física, Universidade de São Paulo, São Paulo, SP, Brazil

K. A. Zhaksybekova IETP, Al-Farabi Kazakh National University, Almaty, Kazakhstan

Jian-Rong Zhang College of Liberal Arts and Sciences, National University of Defense Technology, Changsha, China

Xilin Zhang Department of Physics, University of Washington, Seattle, WA, USA;
Department of Physics, The Ohio State University, Columbus, OH, USA

Shi-Lin Zhu School of Physics and State Key Laboratory of Nuclear Physics and Technology, Center of High Energy Physics, Peking University, Beijing, China;
Collaborative Innovation Center of Quantum Matter, Beijing, China

M. V. Zhukov Department of Physics, Chalmers University of Technology, Göteborg, Sweden

M. A. Zhusupov IETP, Al-Farabi Kazakh National University, Almaty, Kazakhstan

Johann Zmeskal Stefan-Meyer-Institut für Subatomare Physik, Vienna, Austria

J. Zmeskal INFN, Laboratori Nazionali di Frascati, Frascati, Rome, Italy;
Stefan-Meyer-Institut für Subatomare Physik, Vienna, Austria

Part I
Atomic and Molecular Physics

Chapter 1
Positronium Negative Ions: The Simplest Three Body State Composed of a Positron and Two Electrons

Yasuyuki Nagashima

Abstract The positronium negative ion is a three-body system of a positron and two electrons bound via Coulomb interactions. Recently new experiments have been accomplished for this ion, including observations of its photodetachment and shape resonance, based on its efficient formation using alkali-metal coated surfaces.

1.1 Introduction

Several kinds of exotic systems composed of three particles with masses of the same order of magnitude and bound via Coulomb interaction, $p^+\mu^-p^+$, $d^+\mu^-d^+$, $p^+\mu^-d^+$ and the positronium negative ion ($e^-\ e^+\ e^-$, Ps^-), have been studied theoretically for many years [1–3]. The stability of Ps^- was first discussed by Wheeler in 1946 [1]. He found out that this system has ground-state energy of -6.96 eV and a mean lifetime of the order of 0.1 ns. After this work, a number of theoretical studies on Ps^- were performed. For example, the energy and the annihilation rate in vacuum were calculated precisely [1–8]. The resonance states and the photodetachment cross sections of Ps^- were also studied [9–13].

The first observation of Ps^- was made by Mills in 1981 [14]. Following this observation, a few groups determined its decay rate [15–17]. In 2008, an efficient production method for Ps^- was developed and new experimental investigations have been accomplished for these ten years [18, 19].

This paper reviews the developments of these experimental investigations.

Y. Nagashima (✉)
Department of Physics, Tokyo University of Science, 1-3 Kagurazaka, Shinjuku, Tokyo 162-8601, Japan
e-mail: ynaga@rs.kagu.tus.ac.jp

N. A. Orr et al. (eds.), *Recent Progress in Few-Body Physics*,
Springer Proceedings in Physics 238,
https://doi.org/10.1007/978-3-030-32357-8_1

1.2 Properties of Ps^-

Table 1.1 summarises the properties of Ps^-. Those of positronium ($e^- e^+$, Ps) are also shown. The Ps^- properties have been studied only theoretically except for its decay rate. The decay rate has also been obtained experimentally [15–17] and compared theoretical values [8].

1.3 First Observation of Ps^- and Measurements of Its Decay Rate

The first observation of Ps^- was made by Mills [14] in 1981. Slow positrons with energy of 470 eV, produced using a ^{58}Co source and a Cu(111)+S moderator, were transported along a magnetic field to a 3.7 nm thick carbon film. A grid, positively biased at 0.5–4.5 kV, was placed 2.5 mm from the film so that any Ps^- ions emitted from the downstream side of the film were accelerated. The Ps^- produced self-annihilated, with its short lifetime, within a few mm from the grid. A Ge detector monitored the annihilation and blue-shifted γ-rays from the on-coming Ps^-. Small bumps on the tails of the γ-ray peaks due to pair annihilation of positrons indicated the emission of Ps^- from the carbon film. The Ps^- formation efficiency was as low as 0.028%. In his paper, the production of high-velocity Ps by the photodetachment of accelerated Ps^- was discussed.

Mills also measured the decay rate of Ps^- by changing the acceleration voltages and the distance between the carbon film and the acceleration grid [15]. The obtained

Table 1.1 Properties of Ps^- and Ps

Properties	Ps^-	Ps [20]
Binding energy (eV)	0.011 981 051 (1) au [6] = 0.326 021 13 (3) eV (e^-–Ps)	0.25 au = 6.8 eV (ground state)
Excited states	No excited states [21] There are many resonance states	There are many excited states
Lifetime (ns)	0.478 933 6 (3) (th) [8] 0.479 0 (11) (exp) [17]	0.125 (para-Ps, ground state) 142 (ortho-Ps, ground state)
Principal decay mode	2γ	2γ (para-Ps) 3γ (ortho-Ps)
Mean distance between e^+ and e^-	$5.49a_0$ [7]	$3a_0$
Mean distance between two electrons	$8.55a_0$ [7]	–

decay rate was 2.09(9) ns^{-1}. Its precision has been improved using a stripping-based detection technique, which was originally developed by Mills et al. [22], to 2.086(6) ns^{-1} [16]. This technique was later combined with an intense slow positron beam from a research reactor in Munich, to give a more precise decay rate value of 2.0875(50) ns^{-1} [17]. All these values agree with the theoretical values [7, 8].

1.4 Efficient Production of Ps^-

When low-energy positrons impinge on metal surfaces, they are thermalized in bulk and a significant fraction of them diffuse back to the surface. The positrons may be spontaneously emitted if the positron work function ϕ_+ is negative. They may also be emitted as Ps. The energy required to emit Ps is written as $\phi_{Ps} = \phi_+ + \phi_- - 6.8$ eV, where ϕ_- is the electron work function. The values of ϕ_{Ps} for most metals are negative; hence, Ps atoms are spontaneously emitted. The energy required to emit Ps^- can be written as $\phi_{Ps^-} = \phi_+ + 2\phi_- - 7.13$ eV. For tungsten, this value is negative and spontaneous emission of Ps^- is expected [23].

Figure 1.1a shows an experimental setup for the observation of Ps^- emitted from a tungsten surface. Positrons with energy of 0.1 keV were guided along a magnetic field and impinged on the negatively biased target through a grounded grid. Emitted Ps^-

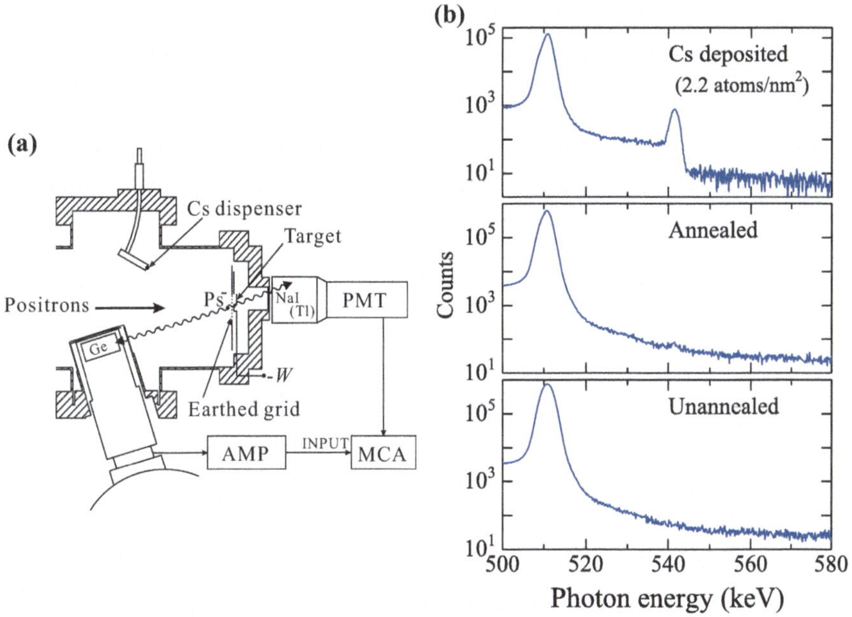

Fig. 1.1 **a** Experimental setup for the observation of Ps^- emission from a tungsten surface. **b** Obtained γ-ray energy spectra. Adapted from [18]

ions were accelerated and annihilated. The emitted γ-rays were monitored using a Ge detector. Figure 1.1b shows the obtained γ-ray energy spectra [18]. After annealing the tungsten, a small bump which indicates Ps^- emission from the surface appeared. The Ps^- ions emission efficiency for this experiment was 0.005%, which was lower than that obtained using a carbon film. However, after Cs deposition, the efficiency increased up to 1.25%.

After the Cs coating, the electron and positron work functions can be written as $\phi_- - D$ and $\phi_+ + D$, respectively, where D is the effect of Cs coating. The value of D is 3 eV for the Cs coating of 0.2–0.3 monolayer on tungsten. Accordingly the value of ϕ_{Ps^-} changes to $\phi_{Ps^-} - D$ due to Cs deposition. This means that Ps^- emission from Cs coated surfaces is more favourable than from clean surfaces.

The Ps^- emission efficiency decreased in only half a day because Cs is chemically reactive. We used Na, which is less chemically reactive than Cs, to obtain durability. While the durability of Ps^- emission has been improved to a few days, the emission efficiency was still as high as that of Cs coated surfaces [24].

1.5 Observation of Ps^- Photodetachment

The efficient formation of Ps^- has enabled the observation of Ps^- photodetachment [25]. Figure 1.3a shows an experimental setup conducted to observe this phenomenon. In the setup, pulsed slow positrons produced using linac in KEK were transported and impinged on Na-coated tungsten through grounded grids. The produced Ps^- ions were accelerated and blue-shifted γ-rays were monitored by two Ge detectors. The Ps^- ions were irradiated with pulsed photons from Nd YAG laser synchronised with the linac.

The relative amount of para-Ps to ortho-Ps formed from the Ps^- photodetachment was 1:3. Therefore, a fraction of the Ps^- ions were converted to ortho-Ps and did not contribute to the Ps^- Doppler-shifted peak when the photodetachment occurred. Figure 1.3b shows the obtained γ-ray energy spectra. It is shown that the peak obtained from the annihilation of accelerated Ps^- decreased by the laser irradiation. This indicates that the Ps^- photodetachment occurred (Fig. 1.2).

1.6 Observation of Ps^- Shape Resonance

The process of Ps^- photodetachment can also be observed by detecting produced Ps atoms, which have almost the same velocity as photodetached Ps^-, with a microchannel plate. This method provides clean data with low background [26].

Using this technique, we observed a peak due to the shape resonance of Ps^- in a spectrum of photodetachment (Fig. 1.3) [27]. The resonant energy was 5.437(1) eV, which is consistent with the theoretical predictions [10, 11, 13]. Further experiments, e.g. observation of Ps^- Feshbach resonance, may feasible using this method.

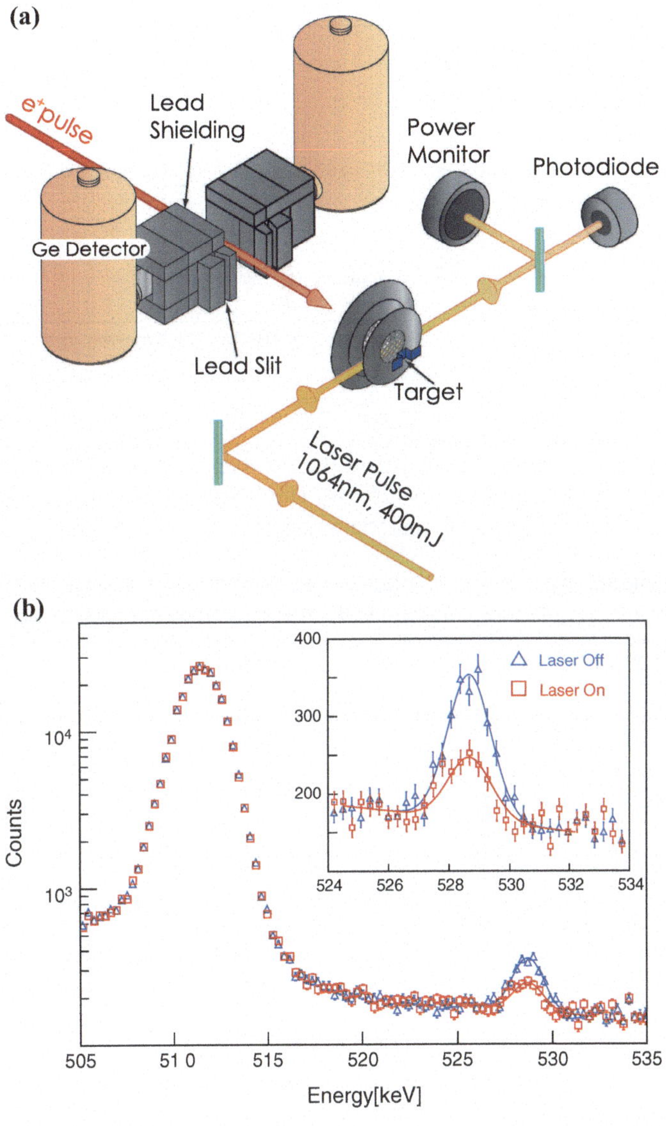

Fig. 1.2 **a** Experimental setup for the observation of Ps^- photodetachment. **b** Obtained spectra using (**a**). Plots of the square (red) and triangle (blue) are the experimental data for laser on and off, respectively. Reprinted with permission from [25]. Copyright 2018 by the American Physical Society

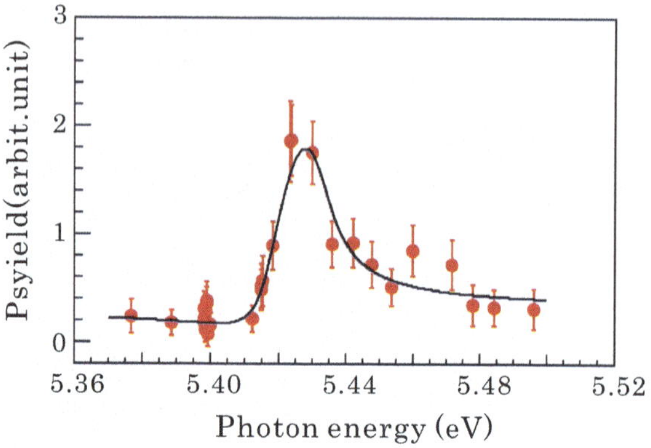

Fig. 1.3 The yield of Ps atoms produced in the Ps^- photodetachment plotted against the photon energy. Adapted from [27]

Moreover, the Ps atoms produced in the photodetachment of accelerated Ps^- provided an energy tunable Ps beam [26, 28]. This beam can provide new information on the basic science of Ps and in material science.

Acknowledgements The author would like to thank K. Michishio, L. Chiari, Y. Nagata, T. Tachibana, S. Kuma, T. Azuma, I. Mochizuki, K. Wada, T. Hyodo and many graduate students who participated in the Ps^- experiments at Tokyo University of Science. This work was supported by JSPS KAKENHI Grant Numbers JP24221006 and JP17H01074.

References

1. Wheeler, J.A.: Polyelectrons. Ann. New York Acad. Sci. **48**, 219 (1946)
2. Kolos, W., Roothaan, C.C.J., Sack, R.A.: Ground state of systems of three particles with Coulomb interaction. Rev. Mod. Phys. **32**, 178 (1960)
3. Frost, A.A., Inokuti, M., Lowe J.P.: Approximate series solutions of nonseparable Shrödinger equations. II. General three-particle system with Coulomb interaction. J. Chem. Phys. **41**, 482 (1964)
4. Bhatia, A.K., Drachman, R.J.: New calculation of the properties of the positronium ion. Phys. Rev. A **28**, 2523 (1983)
5. Korobov, V.I.: Coulomb three-body bound-state problem: variational calculations of nonrelativistic energies. Phys. Rev. A **61**, 064503 (2000)
6. Drake, G.W.F., Grigorescu, M.: Binding energy of the positronium negative ion: relativistic and QED energy shifts. J. Phys. B At. Mol. Opt. Phys. **38**, 3377 (2005)
7. Frolov, A.M.: Positron annihilation in the positronium negative ion Ps^-. Phys. Lett. A **342**, 430 (2005)
8. Puchalski, M., Czarnecki, A., Karshenboim, S.G.: Positronium-ion decay. Phys. Rev. Lett. **99**, 203401 (2007)
9. Ho, Y.K.: Autoionization states of the positronium negative ion. Phys. Rev. A **19**, 2347 (1979)
10. Bhatia, A.K., Drachman, R.J.: Photodetachment of the positronium negative ion. Phys. Rev. A **32**, 3745 (1985)
11. Botero, J., Greene, C.H.: Resonant photodetachment of the positronium negative ion. Phys. Rev. Lett. **56**, 1366 (1986)

12. Ward, S.J., Humberston, J.W., McDowell, M.R.C.: Elastic scattering of electrons (or positrons) from positronium and the photodetachment of the positronium negative ion. J. Phys. B At. Mol. Phys. **20**, 127 (1987)
13. Igarashi, A., Shimamura, I., Toshima, N.: Photodetachment cross sections of the positronium negative ion. New J. Phys. **2**, 17 (2000)
14. Mills Jr., A.P.: Observation of the positronium negative ion. Phys. Rev. Lett. **46**, 717 (1981)
15. Mills Jr., A.P.: Measurement of the decay rate of the positronium negative ion. Phys. Rev. Lett. **50**, 671 (1983)
16. Fleischer, F., et al.: Measurement of the decay rate of the negative ion of positronium (Ps^-). Phys. Rev. Lett. **96**, 063401 (2006)
17. Ceeh, H., et al.: Precision measurement of the decay rate of the negative positronium ion Ps^-. Phys. Rev. A **84**, 062508 (2011)
18. Nagashima, Y., Hakodate, T., Miyamoto, A., Michishio, K.: Efficient emission of positronium negative ions from Cs deposited W(100) surfaces. New J. Phys. **10**, 123029 (2008)
19. Nagashima, Y.: Experiments on positronium negative ions. Phys. Rep. **545**, 95 (2014)
20. Charlton, M., Humberston, J.W.: Positron Physics. Cambridge University Press, Cambridge (2001)
21. Mills Jr., A.P.: Probable nonexistence of a $^3P^e$ metastable excited state of the positronium negative ion. Phys. Rev. A **24**, 3242 (1981)
22. Mills, A.P, Jr., Friedman, P.G., Zuckerman, D.M.: Decay rate and other properties of the positronium negative ion. In: Drachman R.J. (ed.) Proceedings of a Workshop held at NASA Goddard Space Flight Center, July 19–21, pp. 213–221. NASA Conference Publication 3058 (1989)
23. Nagashima, Y., Sakai, T.: First observation of positronium negative ions emitted from tungsten surfaces. New J. Phys. **8**, 319 (2006)
24. Terabe, H., Michishio, K., Tachibana, T., Nagashima, Y.: Durable emission of positronium negative ions from Na- and K-coated W(100) surfaces. New J. Phys. **14**, 015003 (2012)
25. Michishio, K., et al.: Photodetachment of positronium negative ions. Phys. Rev. Lett. **106**, 153401 (2011)
26. Michishio, K., et al.: An energy-tunable positronium beam produced using the photodetachment of the positronium negative ion. Appl. Phys. Lett. **100**, 254102 (2012)
27. Michishio, K., et al.: Observation of a shape resonance of the positronium negative ion. Nat. Commun. **10**, 1038 (2016)
28. Michishio, K., et al.: A high-quality and energy-tunable positronium beam system employing a trap-based positron beam. Rev. Sci. Instrum. **90**, 023305 (2019)

Chapter 2
Time Evolution of a Three-Body Wave Function in Two Dimensions

D. S. Rosa, M. T. Yamashita, G. Krein and A. S. Jensen

Abstract In this article we study a three-body system in two dimensions formed by a light particle and two identical heavy dipoles in the Born-Oppenheimer approximation. We present the time evolution of the dipoles wave function using imaginary and real time propagation. In the first case we study the three-body ground state. In the second case the three-body system is a superposition of two different states. The time evolution shows that the tunneling of the wave function may be used to simulate a qubit.

2.1 Introduction

Recently, significant progress has been made in experimental techniques to confine atoms in ultracold traps [1]. The ability to produce such ultracold atomic clouds, tuning the interatomic potentials through Feshbach's resonance [2] and squeezing the cloud through an asymmetric change of lasers and magnetic fields, putting the system in different dimensions [3], has been made ultracold atoms good candidates to develop and test different theories in several areas in science.

As shown in [4], a three-body system consisting of two heavy dipoles and a light atom may present, depending on their interactions, some interesting phenomena, such as, avoided crossing in its observables and tunnel effect in its three-body wave function. Particularly, the possibility to tunnel the three-body wave function may provide a playground to simulate a qubit.

D. S. Rosa (✉) · M. T. Yamashita · G. Krein
Instituto de Física Teórica, Universidade Estadual Paulista,
São Paulo, SP 01140-070, Brazil
e-mail: derick.s.rosa@gmail.com

A. S. Jensen
Department of Physics and Astronomy, Aarhus University,
8000 Aarhus C, Denmark

N. A. Orr et al. (eds.), *Recent Progress in Few-Body Physics*,
Springer Proceedings in Physics 238,
https://doi.org/10.1007/978-3-030-32357-8_2

2.2 Formalism

In this section we show the formalism used to solve the three-body system formed by two heavy dipoles with masses m_A and an atom with mass m_B. The Hamiltonian is given by

$$H = -\frac{\hbar^2}{2\mu_{AA}}\nabla_R^2 - \frac{\hbar^2}{2\mu_{B,AA}}\nabla_r^2 + V_B(\mathbf{R}) + V_A(\mathbf{r} - \frac{\mu_{AA}}{m_A}\mathbf{R}) + V_A(\mathbf{r} + \frac{\mu_{AA}}{m_A}\mathbf{R}), \tag{2.1}$$

where $\mu_{AA} = m_A/2$ and $\mu_{B,AA} = 2m_A m_B/\,(2m_A + m_B)$. Here we are using an odd-man-out notation that is V_A and V_B denote, respectively, the AB (atom/dipole) and AA (dipole/dipole) two-body interactions. The distance $\mathbf{R}$ is the separation vector of the dipoles and $\mathbf{r}$ is the distance of the light atom with respect to the center of mass of the two heavy dipoles. We are considering a severe mass asymmetry, $m_B \ll m_A$, in such a way we can apply the Born-Oppenheimer approximation. This enable us to write two independent wave function for the fast light atom and for the two heavy dipoles.

Heavy-particle equation After making the change of variable $R = (a_0/2)\,x$, and considering the two heavy atoms as two identical dipoles in the xy-plane with dipole moments $\mathbf{D}$ forming an angle θ with the z-axis and an azimuthal angle ϕ we can write

$$\left[\frac{\partial^2}{\partial x^2} - \frac{(l^2 - 1/4)}{x^2}\right]\chi(x) + \left[\frac{a_0}{2\mathcal{Q}_{\text{eff}}^2}\epsilon(x) + \frac{\lambda x_0^3}{x^3 + x_0^3}\right]\chi(x) = -\frac{a_0}{2\mathcal{Q}_{\text{eff}}^2}E_3\chi(x), \tag{2.2}$$

$$\lambda \equiv e^{-\gamma}CD^2\sqrt{\frac{32\mu_{AA}^4|E_2|}{\mu_{B,AA}\hbar^6}}(1 - 3\sin^2\theta\cos^2\phi), \tag{2.3}$$

where $\chi(x) = \sqrt{R}\phi(R)$, $a_0 = \hbar^2/\mu_{AA}\mathcal{Q}_{\text{eff}}^2$ and $\mathcal{Q}_{\text{eff}}^2 = \sqrt{2e^{-2\gamma}\hbar^2|E_2|/\mu_{,AA}}$. The cubic divergence at $x = 0$ was regularized by modifying the dipole interaction for distances smaller than a constant x_0, which can be thought as the dipole length. For more details see [4].

2.2.1 *Time Dependent Schroedinger Equation*

We want to solve the time dependent Schroedinger equation given by

$$H\phi(x,t) = i\frac{\partial}{\partial t}\phi(x,t) \quad \longrightarrow \quad \phi(x,t) = \phi(x,t_0)e^{-iH(t-t_0)}. \tag{2.4}$$

Let us start from the following two differential equations

$$U\phi_1(x,t) = i\frac{\partial}{\partial t}\phi_1(x,t) \quad \longrightarrow \quad \phi_1(x,t) = \phi_1(x,t_0)e^{-iU(t-t_0)},$$
$$K\phi_2(x,t) = i\frac{\partial}{\partial t}\phi_2(x,t) \quad \longrightarrow \quad \phi_2(x,t) = \phi_2(x,t_0)e^{-iK(t-t_0)}, \tag{2.5}$$

where $K + U = H$. Choosing the initial condition for ϕ_2 like $\phi_2(x,t_0) = \phi_1(x,t)$ we have

$$\phi_2(x,t) = \phi_1(x,t_0)e^{-iH(t-t_0)}, \tag{2.6}$$

which differs from (2.4) only by an initial condition for ϕ_1. For $(t - t_0) \approx 0$ the error in this approximation is going to be proportional to $[K, U] \triangle t^2$. The time evolution can then be performed studying the solution for $\phi_2(x,t)$, which is found using Split-Step Crank-Nicolson Method.

Imaginary time propagation The imaginary time propagation method has long been used as a way to find ground state wavefunctions. It involves a replacement of the real time by an imaginary time in the Schroedinger equation. This leads to the

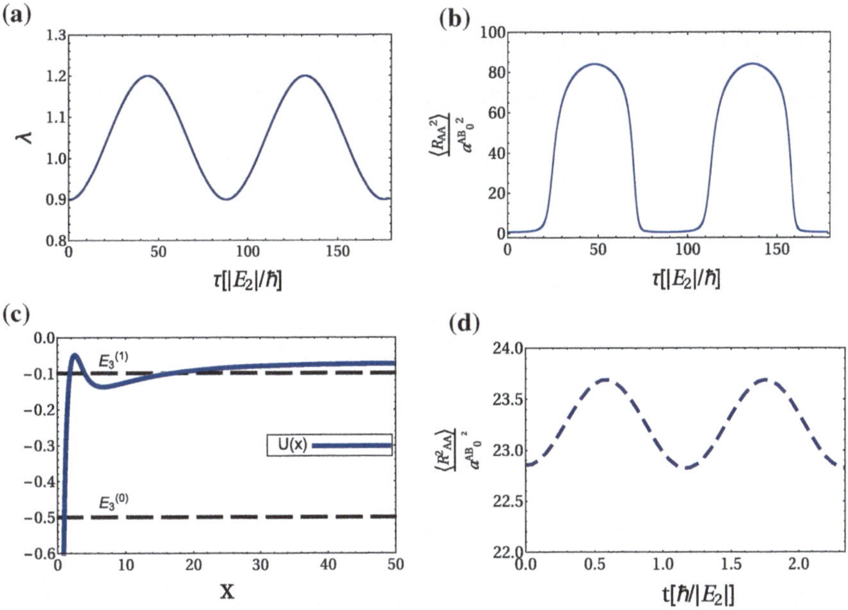

Fig. 2.1 **a**: Imaginary time evolution of the strength of the potential. **b**: Imaginary time evolution of the MSR for the "*Oscillatory behaviour*" of the strength. **c**: Total potential (blue solid). The horizontal lines are the ground states and first excited states $\overline{E}^n = (a_0/2Q^2_{\text{eff}})E^n$ for $\lambda = 0.6$ and $x_0 = 3.16$. **d**: MSR as a function of $t[\hbar/|E_2|]$. Initial wave function choosed to be a combination of the ground and first excited states fixing $\lambda = 0.6$ and $x_0 = 3.16$

derivation of an imaginary time operator which results in a exponential decay of all states excepting the ground state.

Real time propagation The real time propagation is used in this work to study the behavior of the system when we choose the initial wave function to be a combination of two eigenstates of the system. Writing the initial wave function like a combination of two states of the hamiltonian and computing $\langle R^2 \rangle$, which is estimate like

$$\langle \psi(x,t)|R^2|\psi(x,t)\rangle = \frac{1}{2}\langle \psi_0|R^2|\psi_0\rangle + \frac{1}{2}\langle \psi_1|R^2|\psi_1\rangle + \langle \psi_0|R^2|\psi_1\rangle \cos(E_0 - E_1)t. \quad (2.7)$$

We see that this quantity is going to oscillate with a defined frequency and amplitude, around the value of MSR of these two states (Fig. 2.1).

Conclusions The possibility to localize the dipoles wave function in the two minima by tuning the interactions may provide an interesting possibility to simulate a qubit.

References

1. Bloch, I., et al.: Nat. Phys. **8** (2012)
2. Pethick, C.J., Smith, H.: Bose-Einstein condensation in dilute gases. Cambridge University Press, New York (2008)
3. Rychtarik, D., et al.: Phys. Rev. Lett. **92**, 173003 (2004)
4. Rosa, D.S., et al.: Phys. Rev. A **94**, 062707 (2016)

Chapter 3
Discrete Scaling and Scattering Properties from Atom-Dimer Collision

Lauro Tomio and M. A. Shalchi

Abstract A recent proposal on a new perspective to detect the Efimov-like discrete scaling in ultra-cold binary mixtures with strong mass-imbalanced atomic species is communicated. The discrete scaling can be identified by the energy dependence of an atom-molecule elastic cross-section, when colliding a heavy atom with mass m_α into a weakly-bound heavy-light (m_α, m_β) dimer. In the extreme mass-imbalanced system it was verified that the s-wave elastic cross section presents zeros at several specific energies of the projectile, in which emerges a discrete behavior from the ratio between these energies. This discrete behavior is identified with the same one predicted for the trimer excited bound-states, when considering the same mass-imbalanced system in the unitary limit.

3.1 Introduction

In this presentation, we are reporting some results obtained in a recent work in which it was verified that the usual Efimov scaling factor [1], obtained in the spectrum of a mass-imbalanced three-body system near the unitary limit, can also be well identified in scattering observables with a mixture of two-species strong mass-imbalanced atomic system. In order to show that, the scattering solutions for a system with two heavy and one light particles ($\alpha\alpha\beta$) are obtained by considering the heaviest one (α) colliding with the dimer formed the other two particles [2]. Here, we restrict the presentation to some indicative results related to discrete scaling, which are obtained by using zero-range interactions.

L. Tomio (✉) · M. A. Shalchi
Instituto de Física Teórica (IFT), Universidade Estadual Paulista (UNESP), São Paulo, SP 01140-070, Brazil
e-mail: lauro.tomio@unesp.br

L. Tomio
Instituto Tecnológico de Aeronáutica, Departamento de Ciência e Tecnologia Aeroespacial (DCTA), São José dos Campos 12228-900, Brazil

N. A. Orr et al. (eds.), *Recent Progress in Few-Body Physics*,
Springer Proceedings in Physics 238,
https://doi.org/10.1007/978-3-030-32357-8_3

First, we should remind that the usual Efimov scaling factor is predicted to occur in the unitary limit, when the three-boson bound-state levels are related by an exponential scaling factor $\exp(2\pi/s_0)$, where s_0 is a mass-ratio dependent constant. For $m_\alpha = m_\beta$, the energy-ratio is predicted to be ~ 515, such that it will be quite difficult for an experimental verification. However, more favorable conditions for possible observations are expected with strong mass-imbalanced cold-atom mixtures [3, 4]. In this regard, we should notice that the ratio between consecutive levels of the bound-state energy spectrum is reduced to $\exp(2\pi/s_0) \approx 4.7$ for the case that $m_\alpha = 100 m_\beta$ [5]. In such extreme mass-imbalanced case, one can also rely on an analytical approach based on the Born-Oppenheimer (BO) approximation, as in [6]. This BO approach was extended to the scattering region, in [2], considering the heavy α particle colliding with the $\alpha\beta$ bound system, from where it was possible to identify the same discrete Efimov scaling factor when examining the colliding energy positions corresponding to the zeros of the s-wave cross-sections. Therefore, in view of recent experiments in cold-atom laboratories [3, 7], considering systems such as Lithium-Caesium (LiCs) or Lithium-Yterbium (LiYb) where the mass-ratios m_β/m_α are, respectively, 0.045 for LiCs and 0.034 for LiYb, we understand that studies with low-energy collisions of heavy atoms in weakly-bound molecules can be relevant to establish the predicted discrete scaling from scattering observables.

3.2 Model Description and Results

The scattering observables we consider are the s-wave phase shifts δ_0 and the corresponding cross-sections σ, which are given as functions of the colliding energies E, in the center-of-mass of the atom-dimer system. In order to simplify the units, all energies are given in terms of the ground-state energy $\mathcal{B}_3$ of the corresponding three-body system. The phase-shifts are expressed in terms of the function $k \cot \delta_0$, where k is the relative momentum with respect to the heavy-light dimer, given in terms of E by $k \equiv \sqrt{2\mu_{\alpha,\alpha\beta}E}$ (our units are such that $\hbar = 1$), with $\mu_{\alpha,\alpha\beta} \equiv m_\alpha(1+A)/(2+A)$ being the heavy-(light-heavy) reduced mass, where $A = m_\beta/m_\alpha$. Another relevant observable in our treatment is the two-body dimer energy $B_{\alpha\beta} \equiv 1/(2\mu_{\alpha\beta}a_{\alpha\beta}^2)$, where $\mu_{\alpha\beta}$ is the light-heavy reduced mass and $a_{\alpha\beta}$ the corresponding scattering length, assumed to be bound and close to the unitary limit ($a_{\alpha\beta} \to \infty$).

Given $\mathcal{B}_3$ as a scale for the three-body system and $B_{\alpha\beta}$ the dimer energy, with E being the colliding energy of the heavy particle α into the heavy-light $\alpha\beta$ bound subsystem, the s-wave phase shift δ_0, the cross-section σ, as well as the absorption parameter $\eta \le 1$ are obtained by using the on-shell scattering amplitude given by $h_\alpha = (S_\alpha - 1)/(2\mathrm{i}k)$, where $S_\alpha = \eta \exp(2\mathrm{i}\delta_0)$ is the scattering matrix for the s-wave channel. With δ_0 defined from S_α for the elastic case ($\eta = 1$) we obtain $k \cot \delta_0$ and the corresponding cross-section as

$$h_\alpha = 1/(k \cot \delta_0 - \mathrm{i}k), \quad \sigma = 4\pi |h_\alpha|^2. \tag{3.1}$$

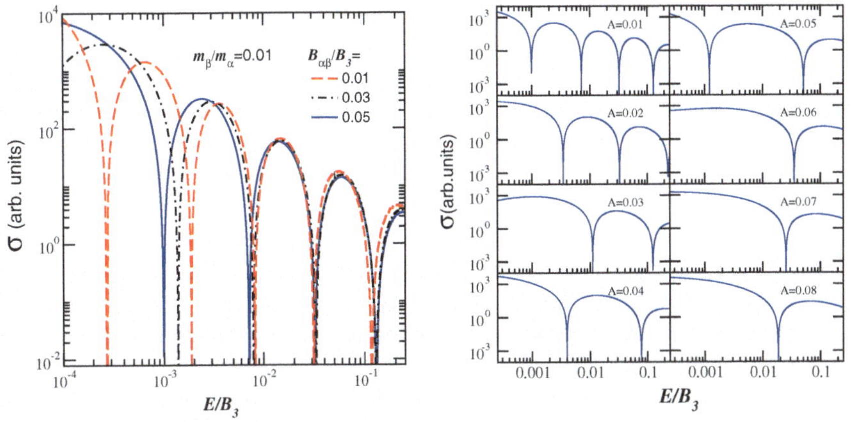

Fig. 3.1 Results for the s-wave cross-section σ (in arbitrary units), in term of the colliding energies E (given in units of $\mathcal{B}_3$), obtained by using zero-range two-body interactions. In the left-side frame, the results are for fixed mass-ratio $A = m_\beta/m_\alpha = 0.01$, considering $B_{\alpha\beta}/\mathcal{B}_3 = 0.05$ (solid-blue), 0.03 (dot-dashed-black) and 0.01 (dashed-red). In the right side we have a set with eight panels, with the mass-ratio A varying from 0.01 till 0.08 (as indicated inside the frames), for fixed $B_{\alpha\beta}/\mathcal{B}_3 = 0.05$

In terms of the observables and parameters, one can also express the cross-section as a scaling function, which will be obtained numerically by solving the three-body atom-dimer scattering equation and depends on the three- and two-body bound-state energies, mass ratio A, and colliding energy E, such that

$$\sigma(E; B_{\alpha\beta}, \mathcal{B}_3, A) = \mathcal{S}\left(E/\mathcal{B}_3\,,\, B_{\alpha\beta}/\mathcal{B}_3\,,\, A\right)\mathcal{B}_3^{-1}. \tag{3.2}$$

Results—In Fig. 3.1 we present s-wave cross-section results obtained for the scattering of particle α by the dimer $\alpha\beta$, considering zero-range two-body interactions. In the left-side frame, we fix the mass-ratio at $A = 0.01$, considering three values for $B_{\alpha\beta}/\mathcal{B}_3$ (=0.05, 0.03 and 0.01), such that we observe that the zeros obtained for σ are for energies E such that the ratios between consecutive ones can be identified with the corresponding ratio (~ 4.7), which are verified for the case of excited three-body bound state in the unitary limit ($B_{\alpha\beta} \to 0$), when considering the same mass-imbalanced system. In the right-side frame of Fig. 3.1, we have a set of eight panels for mass-ratios A varying from 0.01 till 0.08, for a fixed dimer energy, $B_{\alpha\beta}/B_3 = 0.05$. In this way, we can appreciate how the sequence of zeros for the cross-section σ are emerging as the mass difference between the α and β particles are increasing from 0.08 (bottom-right) to 0.01 (top-left). In this case with $B_{\alpha\beta} = 0.05\mathcal{B}_3$, one should noticed that, for the less-pronounced mass-imbalanced cases ($A \sim 0.06$–0.08), only one zero for σ is observed within the given energy range. A second minimum can be verified in these cases by decreasing the two-body binding $B_{\alpha\beta}$, as pointed out in [2]. But relevant to notice it that the number of observed zeros in σ is fast decreasing as A increases.

Within a similar study for a mass-imbalanced system with two light and one heavy particles ($A > 1$), the relevance of a pole in $k \cot \delta_0$ (corresponding to a zero in σ) was investigated in [10], in the nuclear physics context for the neutron$-^{19}$C scattering. Related to this case and considering the results obtained in [2], partially reproduced here, it is interesting to notice that only for extreme mass-imbalanced systems with two heavy and one light particles, and close to the unitary limit of the two-body system, one can verify the occurrence of a sequence of zeros in the cross-section matching with the corresponding sequence of Efimov three-body excited bound-state levels, in the relevant physical interval with $E < \mathcal{B}_3$. This behavior can be explained from the fact that when the masses of the particles α and β approach each other (being comparable) the Efimov discrete scaling factor, identified in the unitary limit ($B_{\alpha\beta} \approx 0$) of the three-body spectrum, implies in much larger energy differences for consecutive levels, as detailed for the mass-imbalanced systems in [5].

3.3 Summary

The results reported here are obtained by using the zero-range approach within a Faddeev three-body formalism for atom-dimer collision, in order to show another possible way to detect discrete scaling in three-body systems. As pointed out, the observation of a sequence of zeros in the cross section can only be verified for quite large mass ratios, such that the main focus for recent ultra-cold atomic experiments are binary condensed systems combining atomic species such as Li, Yb or Rb. By considering the mass ratio between Li and Yb, $A = 0.034$, the cross-section for the Yb + LiYb collision can in principle present a couple of zeros. More details on the result reported here, where it was also considered range effects, can be obtained in [2]. Another extension of this investigation on mass-imbalanced atom-dimer scattering was also reported recently in [11], by considering the scattering of ^{4}He on ^{4}He$-^{6,7}$Li and ^{4}He$-^{23}$Na molecules.

Acknowledgements We thank the Brazilian agencies FAPESP (2017/05660-0), CNPq (Procs. 306191/2014-8) and CAPES for partial support.

References

1. Efimov, V.: Energy levels arising from resonant two-body forces in a three-body systems. Phys. Lett. B **33**, 563–564 (1970)
2. Shalchi, M.A., Yamashita, M.T., Hadizadeh, M.R., Garrido, E., Tomio, L., Frederico, T.: Probing Efimov discrete scaling in an atom-molecule collision. Phys. Rev. A **97**, 012701 (2018)
3. Ulmanis, J., Häfner, S., Kuhnle, E.D., Weidemüller, M.: Heteronuclear Efimov resonances in ultra-cold quantum gases. Natl. Sci. Rev. **3**, 174–188 (2016)

4. Blume, D., Yan, Y.: Generalized Efimov scenario for heavy-light mixtures. Phys. Rev. Lett. **113**, 213201 (2014)
5. Braaten, E., Hammer, H.-W.: Universality in few-body systems with large scattering length. Phys. Rep. **428**, 259–390 (2006)
6. Fonseca, A., Redish, E., Shanley, P.E.: Efimov effect in an analytically solvable model. Nucl. Phys. A **320**, 273–288 (1979)
7. Hara, H., Takasu, Y., Yamaoka, Y., Doyle, J.M., Takahashi, Y.: Quantum degenerate mixtures of alkali and alkaline-earth-like atoms. Phys. Rev. Lett. **106**, 205304 (2011)
8. Hansen, A.H., Khramov, A., Dowd, W.H., Jamison, A.O., Ivanov, V.V., Gupta, S.: Quantum degenerate mixture of ytterbium and lithium atoms. Phys. Rev. A **84**, 011606(R) (2011)
9. Makrides, C., Hazra, J., Pradhan, G.B., Petrov, A., Kendrick, B.K., González-Lezana, T., Balakrishnan, N., Kotochigova, S.: ultra-cold chemistry with alkali-metal-rare-earth molecules. Phys. Rev. A **91**, 012708 (2015)
10. Shalchi, M.A., Yamashita, M.T., Hadizadeh, M.R., Frederico, T., Tomio, L.: Neutron$-^{19}$C scattering: Emergence of universal properties in a finite-range potential. Phys. Lett. B **764**, 196–202 (2017)
11. Shalchi, M.A., Delfino, A., Frederico, T., Tomio, L.: Scattering of cold ^{4}He on ^{4}He$-^{6,7}$Li and ^{4}He$-^{23}$Na. Phys. Rev. A **98**, 032705 (2018)

Chapter 4
Interaction Potential for a System of N Charged Particles Near a Spherical Interface

Juan Martín Randazzo and Lorenzo Ugo Ancarani

Abstract By integration of the electrostatic energy density of the system, we provide analytical expressions of the potential for a system of N charged particles in the presence of a surface which separates two continuous media, conductor, dielectric or vacuum. Such situations correspond, for example, to a number of electrons inside a spherical cavity of a conductor or interacting with a dielectric sphere. Bound states are analyzed for different values of the sphere radius.

4.1 The Potential for N Charged Particles Near a Spherical Interface

When under the influence of an environment, the study of N charges (*e.g.*, an atom) may differ considerably from the isolated case (in a vacuum). The energy spectrum and other properties depend on the nature of the interaction between the charges and the surrounding like, for example, in a plasma environment, confined systems or close to metal surfaces [1, 2].

We consider here N charged particles q_i, placed at $\mathbf{r}_i$ $(i = 1, \ldots, N)$ near a spherical surface of radius R which divides two continuous media. The $N = 2$ case is schematized in Fig. 4.1.

The potential energy can be approximated by the integration of the electrostatic energy density, which depends on the electric and displacement fields, for a fixed configuration of the system (this procedure is described in [3] (pp. 41–42) for two Coulomb particles). The resulting interaction potential can be subdivided in two contributions:

J. M. Randazzo (✉)
Divisi6n Física Atómica, Molecular y Óptica, Centro Atómico Bariloche,
8400 S. C. de Bariloche, Argentina

CONICET, Buenos Aires, Argentina
e-mail: randazzo@cab.cnea.gov.ar

L. U. Ancarani
Université de Lorraine, CNRS, LPCT, F-57000 Metz, France

N. A. Orr et al. (eds.), *Recent Progress in Few-Body Physics*,
Springer Proceedings in Physics 238,
https://doi.org/10.1007/978-3-030-32357-8_4

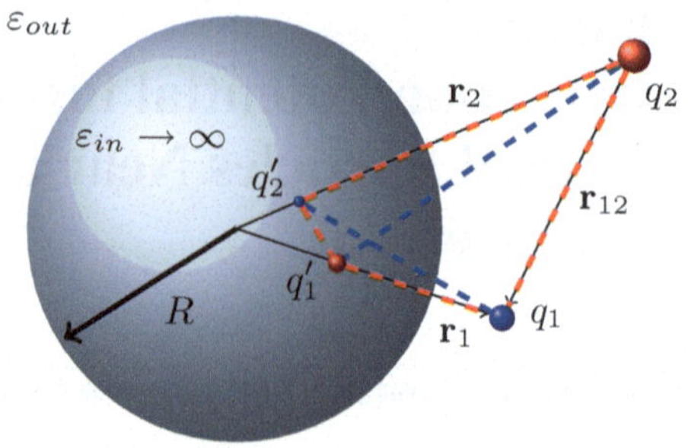

Fig. 4.1 Schematic representation of two charged particles q_1 and q_2 in front of a metallic ball; $q'_1 = -q_1 R/r_1$ and $q'_2 = -q_2 R/r_2$ are their image charges

$$V(\mathbf{r}_1, \mathbf{r}_2, ..., \mathbf{r}_N) = \sum_{i=1}^{N} V_i(r_i) + \sum_{i=1, j>i}^{N} V_{ij}(r_i, r_j, \theta_{ij})$$

where θ_{ij} is the angle between $\mathbf{r}_i$ and $\mathbf{r}_j$.

It turns out that we can express the central potentials V_i in terms of the functional form: $U_{\pm}(x) = \frac{1}{1-x^2} \pm \frac{\text{atanh}(x)}{x}$. For the spherical surface dividing two dielectric media with constant ε_{in} and ε_{out} we have:

$$V_i(r_i) = \frac{q_i^2}{2R}(\varepsilon_{out} - \varepsilon_{in}) \begin{cases} \frac{R^2}{r_i^2 \varepsilon_{out}^2} U_-\left(\frac{R}{r_i}\right) & \text{if } r_i > R \\ \frac{1}{\varepsilon_{in}^2} U_+\left(\frac{r_i}{R}\right) & \text{if } r_i < R \end{cases},$$

while for the conducting sphere or the cavity in a conductor:

$$V_i(r_i) = -\frac{q_i^2}{2R} \begin{cases} \frac{R^2}{r_i^2} \frac{1}{\varepsilon_{out}} U_+\left(\frac{R}{r_i}\right) & \text{if } r_i > R \text{ (metallic sphere)} \\ \frac{1}{\varepsilon_{in}} U_-\left(\frac{r_i}{R}\right) & \text{if } r_i < R \quad \text{(cavity in conductor)} \end{cases},$$

where ε is the dielectric constant of the media inside the cavity or outside the sphere. The correlation terms for the dielectric-dielectric case have the following partial wave expansion:

$$V_{ij}(r_i, r_j, \theta_{ij}) = q_i q_j \sum_{l=0}^{\infty} P_l(\cos\theta_{12}) F_l(r_<, r_>, R, \varepsilon_{in}, \varepsilon_{out}), \tag{4.1}$$

where $r_< = \text{Min}[\text{r}_\text{i}, \text{r}_\text{j}]$ and $r_> = \text{Max}[\text{r}_\text{i}, \text{r}_\text{j}]$, and

$$F_l(r_<, r_>, R, \varepsilon_{in}, \varepsilon_{out}) = \begin{cases} \frac{1}{\varepsilon_{out}} \left(\frac{r_<^l}{r_>^{l+1}} + \frac{l(\varepsilon_{out}-\varepsilon_{in})}{(\varepsilon_{in} l + \varepsilon_0 (l+1))} \frac{R^{2l+1}}{r_<^{l+1} r_>^{l+1}} \right) & \text{if } r_< > R \\ \frac{(2l+1)}{(l(\varepsilon_{in}+\varepsilon_{out})+\varepsilon_{out})} \frac{r_<^l}{r_>^{l+1}} & \text{if } r_> \geq R \geq r_< \\ \frac{1}{\varepsilon_{in}} \left(\frac{r_<^l}{r_>^{l+1}} - \frac{(l+1)(\varepsilon_{out}-\varepsilon_{in})}{(l(\varepsilon_{in}+\varepsilon_{out})+\varepsilon_{out})} \frac{r_>^l r_<^l}{R^{2l+1}} \right) & \text{if } r_> < R. \end{cases}$$

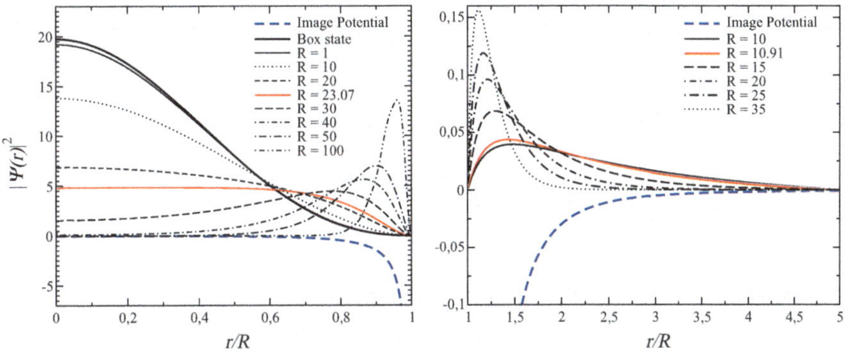

Fig. 4.2 Bound states distribution for a particle interacting with a spherical conducting surface: cavity (left) and sphere (right). The (negative) image potential is also represented

If $\varepsilon_{in} = \varepsilon_{out}$ we obtain, as expected, the pure Coulomb interaction $1/r_{ij}$ (its partial wave expansion). The case of a metallic sphere (respectively, cavity in a conductor) is obtained through the limit $\varepsilon_{in} \to \infty$ (respectively, $\varepsilon_{out} \to \infty$) in (4.1). Note that for the isolated conducting sphere with a charge Q, one must further add the electrostatic term $U(r_i, r_j)$ which does not depend on θ_{ij}.

4.2 Bound States Dependence on the Radius R

Figure 4.2 shows, for different values of the radius R, radial eigenfunctions of the Schrödinger equation for a single electron placed at a distance r from the center of a cavity in a conductor (left) and from the center of a metallic sphere (right). For the single particle case, only the central potential is present (correlation terms appear for $N \geq 2$ cases). Homogeneous boundary conditions are imposed at $r = R$ in both cases and at $r = 5R$ for the sphere. For the cavity we observe the passage from the boxed (where the kinetic energy overcomes the potential energy) to the conductor regimes, and a critical value $R \simeq 23.07$ a.u. where $E < 0$; for the sphere the critical value is $R \simeq 10.91$ a.u. In both cases, the states energies decrease as R increases, and the distributions become similar to that of a plane conductor when $R \to \infty$.

4.3 Summary

We derived analytically the potential energy for a system of N charged particles interacting with a spherical surface separating two continuous media. The result involves a central potential for the interaction of each particle with the sphere, and a correlation between the charged particles; it can be interpreted as the Coulomb interaction distorted by the polarization of the sphere. Details of the potentials derivation together with an analysis of the solutions for the more challenging $N = 2$ case will be presented in an overcoming publication, where we will compare the present expressions with the work needed to completely separate the particles from the sphere.

Acknowledgements We acknowledge FONCYT (PICT-2014-1400) and CONICET for funding.

References

1. Jaskólski, W.: Confined many-electron systems. Phys. Rep. **271**, 1 (1996)
2. Echenique, P.M., et al.: Image-potential-induced states at metal surfaces. J. Electron. Spectrosc. Relat. Phenom. **126**, 163–175 (2002)
3. Jackson, J.D.: Classical electrodynamics, 3rd edn. Wiley, New York (1999)

Chapter 5
Potential Splitting Approach for Faddeev-Merkuriev Equations: Application to $e^- - \bar{\mathrm{H}}$ (e^+–H) and e^+–He^+ Multichannel Scattering

V. A. Gradusov, V. A. Roudnev, E. A. Yarevsky and S. L. Yakovlev

Abstract We solve the three-particle multichannel Coulomb scattering problem with rearrangement channels without partial wave decomposition on the base of Faddeev-Merkuriev equations. Highly efficient numerical approach allowed us to perform the detailed calculations of all possible S-wave cross-sections in systems $e^+e^-\bar{p}$ and $e^+e^-\mathrm{He}^{++}$ for low-energy binary collisions.

5.1 Introduction

Highly accurate calculations of scattering processes and resonances in few-body systems of charged particles require an approach which is sound both theoretically and computationally. The Faddeev-Merkuriev formalism [1] which fulfills this requirement is used to calculate all possible S-wave cross sections in systems $e^+e^-\bar{p}$ and $e^+e^-\mathrm{He}^{++}$ in the low-energy region of binary collisions. A special emphasis is made on antihydrogen formation by antiproton impact of positronium which is currently used in experiments on antimatter at CERN (see [2] and references therein). As in [2] we have found Gailitis-Damburg oscillations [3] of the antihydrogen formation cross sections for energies just above the first antihydrogen excitation threshold. Our high-quality results allowed us to verify proper spacing of maxima of those oscillations quantitatively.

5.2 Theory

The Faddeev-Merkuriev equations for three charged particles of masses m_α and charges Z_α, $\alpha = 1, 2, 3$ in reduced Jacobi coordinates read:

V. A. Gradusov (✉) · V. A. Roudnev · E. A. Yarevsky · S. L. Yakovlev
St. Petersburg State University, St. Petersburg 198504, Russian Federation
e-mail: s.yakovlev@spbu.ru

N. A. Orr et al. (eds.), *Recent Progress in Few-Body Physics*,
Springer Proceedings in Physics 238,
https://doi.org/10.1007/978-3-030-32357-8_5

$$\{T_\alpha + V_\alpha(x_\alpha) + \sum_{\beta\neq\alpha} V_\beta^{(l)}(x_\beta) - E\}\psi_\alpha = -V_\alpha^{(s)}(x_\alpha)\sum_{\beta\neq\alpha}\psi_\beta. \tag{5.1}$$

Here the kinetic energy operators are of the form $T_\alpha = -\Delta_{x_\alpha} - \Delta_{y_\alpha}$ and the components of the wave function ψ_α sum up to the solution of the Schrödinger equation. The pairwise Coulomb potentials $V_\alpha(x_\alpha) = \sqrt{2m_\beta m_\gamma/(m_\beta + m_\gamma)} Z_\beta Z_\gamma / x_\alpha$ $(\beta, \gamma \neq \alpha)$ are split into the interior $V_\alpha^{(s)}$ and long-range tail parts $V_\alpha^{(l)}$ as proposed in [4]:

$$V_\alpha(x_\alpha) = V_\alpha^{(s)}(x_\alpha) + V_\alpha^{(l)}(x_\alpha). \tag{5.2}$$

The equations are supplied with boundary conditions of the form given in [5].

5.3 Results

We have calculated cross sections of all possible scattering processes for the total energy from −0.49973 a.u. to −0.05553 a.u. for $e^+e^-\bar{p}$ system (6 open channels including $\bar{H}$ (n = 1, 2) and Ps(n = 1, 2) binary channels) and for the total energy from −1.9997 a.u. to −0.12496 a.u. for $e^+e^-\text{He}^{++}$ system (7 open channels including He^+(1, 2, 3) and Ps(1) binary channels). Hereafter $\bar{H}$(n) and $\bar{H}$(n, l) (and similar for He ion) are shortcuts for the atom states with principal quantum number n and angular momentum ℓ. Cross sections of all the processes involving positronium and antiproton (or helium ion) are presented in Figs. 5.1 and 5.2. The energy step of calculation is 0.0007 a.u. for cross sections shown in Figs. 5.1 and 5.2 and more fine

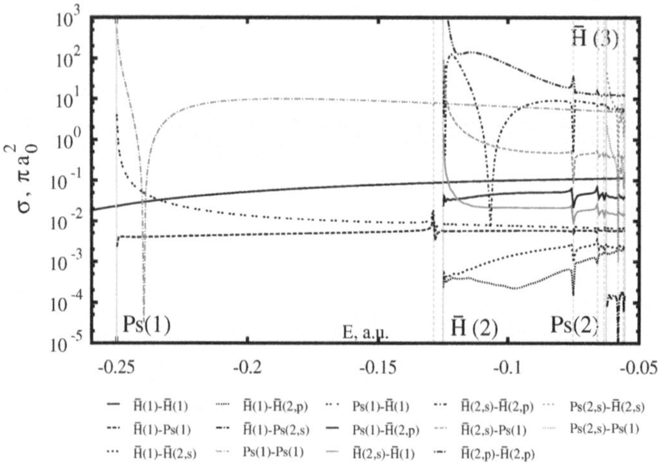

Fig. 5.1 Cross sections for $e^-e^+\bar{p}$ system. Vertical solid lines denote binary thresholds, vertical dashed lines mark resonance positions

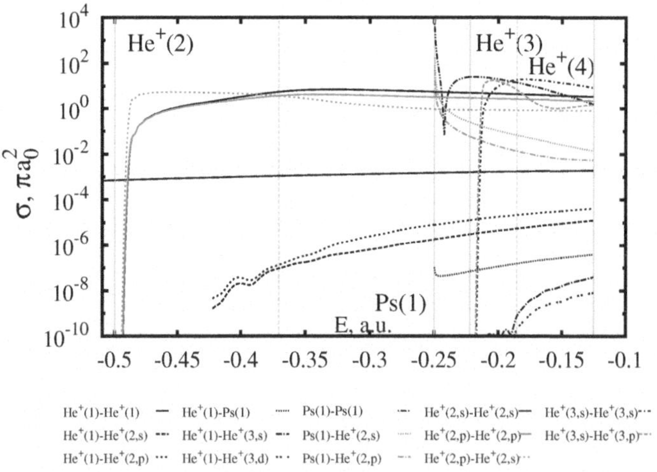

Fig. 5.2 Cross sections for $e^-e^+\mathrm{He}^{++}$ system

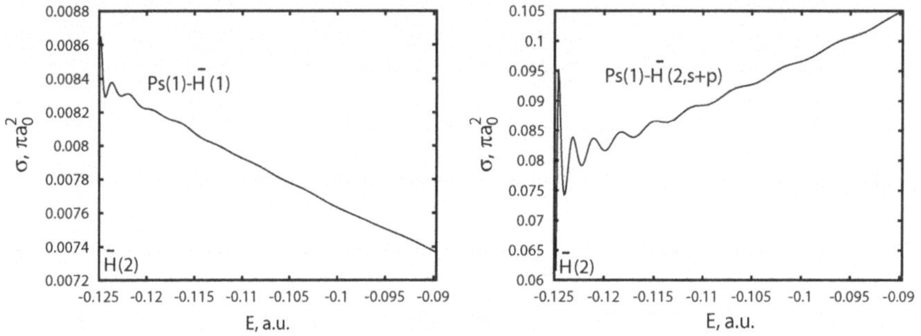

Fig. 5.3 Cross sections of antihydrogen formation above the H̄(2) threshold

step of 0.0001 a.u. is used for cross sections of Fig. 5.3. The calculated cross sections of antihydrogen formation in Fig. 5.3 exhibit prominent oscillations just above the H̄(2) threshold, which are identified as Gailitis-Damburg oscillations [5].

In all figures the known resonances [6, 7] (and references therein) manifest themselves as singularities in cross sections except for the two broad resonances of $e^-e^+\mathrm{He}^{++}$ system at -0.371 a.u. and -0.188 a.u. [7]. We have found these broad resonances, however, by using complex rotation method applied to the Schrödinger equation [8]. The positions and widths (E_r, Γ) are $(-0.3704, 0.1297)$ and $(-0.1857, 0.0395)$ (in a.u.), where $E_r = \Re\mathrm{e}E$ and $\Gamma = 2\Im\mathrm{m}E$.

Acknowledgements This work is supported by RFBR grant No. 18-02-00492. The calculations were performed on resources of the Computational Center of St Petersburg State University.

References

1. Faddeev, L.D., Merkuriev, S.P.: Quantum scattering theory for several particle systems. Kluwer, Dordrech (1993)
2. Valdes, M., Dufour, M., Lazauskas, R., Hervieux, P.A.: Phys. Rev. A **97**, 012,709 (2018)
3. Gailitis, M., Damburg, R.: Proc. Phys. Soc. **82**, 192–200 (1963)
4. Gradusov, V.A., Roudnev, V.A., Yakovlev, S.L.: Atoms **4**, 9 (2016)
5. Gradusov, V.A., Roudnev, V., Yarevsky, E.A., Yakovlev, S.L.: ArXiV: 1810.11123 (2018)
6. Umair, M., Jonsell, S.: J. Phys. B **47**, 225,001 (2014)
7. Igarashi, A., Shimamura, I.: Phys. Rev. A **70**, 012,706 (2004)
8. Elander, N., Yarevsky, E.: Phys. Rev. A **57**, 3119 (1998)

Chapter 6
Energy Levels of Excitons in Square Quantum Wells

Pavel A. Belov and Sergey L. Yakovlev

Abstract Numerical modeling of energy levels and corresponding wave functions of exciton states in heterostructures with GaAs/$Al_{0.30}Ga_{0.70}As$ quantum wells of different widths is performed. The modeling is based on an accurate finite-difference solution of the time-independent three-dimensional Schrödinger equation. A dependence of energy levels on a quantum well width as a parameter is studied. The calculated exciton states are classified according to their quantum-confinement and Coulomb nature. The radiative decay rates of the modeled states are also obtained.

6.1 Introduction

The electron-hole bound states, excitons, in semiconductor quantum wells (QW) are remarkable examples of Coulomb systems, in which the external potential of the heterostructure plays a role of "the third particle", thus making such a system to be of a three-body nature. The exciton states and the exciton-light coupling in heterostructures with QWs have been experimentally and theoretically studied for several decades [1–8]. A quality of heterostructures is permanently growing and experimental samples with excellent properties have become available now [9–13]. Measurements of the reflectance spectra for the high-quality heterostructures show that the accurate data on the exciton energies and radiative as well as nonradiative broadenings can be easily obtained [10, 11]. In this context, the high quality of samples requires the improved precision of theoretical modeling of the exciton states and resonances. Therefore, such problems gradually draw attention of the few-body community [14–18].

In this report, we present the results of an accurate modeling of the energy levels and corresponding wave functions of the exciton states. The energies of the ground

P. A. Belov · S. L. Yakovlev (✉)
St. Petersburg State University, 198504 St. Petersburg, Russian Federation
e-mail: s.yakovlev@spbu.ru

P. A. Belov
e-mail: pavelbelov@gmail.com

N. A. Orr et al. (eds.), *Recent Progress in Few-Body Physics*,
Springer Proceedings in Physics 238,
https://doi.org/10.1007/978-3-030-32357-8_6

and excited states of excitons in heterostructures with $GaAs/Al_{0.30}Ga_{0.70}As$ finite square QWs of various widths (up to 80 nm) are calculated. This type of QWs is widely experimentally and theoretically studied now as a model heterostructure due to a well-known band structure allowing to employ the envelope function approximation and perturbative methods [5, 19]. Determination of the exciton states is achieved by solving the time-independent three-dimensional Schrödinger equation. The corresponding eigenvalue problem is solved numerically using the finite-difference discretization scheme [11] and properly taking into account the discontinuities of the material parameters at the heterointerfaces [20]. Our numerical method is asymptotically exact and allows us to obtain accurate energies of the exciton states for a wide range of QW widths and potential profiles [21]. The calculated bound states of electron-hole pairs in $GaAs/Al_{0.30}Ga_{0.70}As$ QWs are classified according to their quantum-confinement and Coulomb nature [22, 23]. The accurate radiative decay rates for the calculated s-like exciton states are also obtained [5, 7]. Calculated data are in agreement with the experimental reflectance spectra measured for high-quality $GaAs/Al_{0.30}Ga_{0.70}As$ heterostructures with QWs [11].

6.2 Theory

In the envelope-function approximation [5], the conduction band dispersion is parabolic and the valence band is described by the Luttinger Hamiltonian [19, 24]. We simplify the consideration by studying the uncoupled heavy-hole exciton in QW. Then, the time-independent configuration space Schrödinger equation with the effective Hamiltonian for s-like heavy-hole exciton states in QW reads

$$\left(K - \frac{e^2}{\epsilon\sqrt{\rho^2 + (z_e - z_h)^2}} + V_e(z_e) + V_h(z_h)\right) \chi(z_e, z_h, \rho) = E_X\, \chi(z_e, z_h, \rho) \tag{6.1}$$

where K is the kinetic energy term:

$$K = -\frac{\hbar^2}{2m_e}\frac{\partial^2}{\partial z_e^2} - \frac{\hbar^2}{2m_{hz}}\frac{\partial^2}{\partial z_h^2} - \frac{\hbar^2}{2\mu}\left(\frac{\partial^2}{\partial \rho^2} - \frac{1}{\rho}\frac{\partial}{\partial \rho} + \frac{1}{\rho^2}\right).$$

The energy E_X denotes the exciton energy with respect to a value E_g of the energy gap. The function χ is related to the exciton wave function as $\chi(z_e, z_h, \rho) = \psi(z_e, z_h, \rho)/\rho$, where $\rho = \sqrt{(x_e - x_h)^2 + (y_e - y_h)^2}$ is the radius in the QW plane. The masses m_e and m_{hz} are the isotropic effective mass of the electron and the effective mass of the hole along the z axis, respectively. The mass μ is the reduced effective mass of the electron and the hole in the QW plane. We omit here discontinuities of effective masses and the dielectric constant ϵ for simplicity [20]. Potentials $V_{e,h}(z_{e,h})$ are the external finite square QW potentials

$$V_{e,h}(z_{e,h}) = \begin{cases} 0 & \text{if } |z_{e,h}| < L/2 \\ V_{e,h} & \text{if } |z_{e,h}| \geq L/2 \end{cases} \tag{6.2}$$

of the heterostructure and L is the QW width.

A numerical modeling is based on a solution of the eigenvalue problem for (6.1). Using the finite-difference method and the implicitly restarted Arnoldi algorithm we discretize the problem and obtain a part of the matrix spectrum [11, 20, 22, 23, 25]. An additional analysis of the convergence of numerical results allows us to guarantee a good accuracy of the obtained energy levels.

6.3 Results

The energies and wave functions of s-like quantum states of the heavy-hole exciton in QW were calculated for various QW widths, L, from 4 nm up to 80 nm.

For narrow QWs ($L < 15$ nm) the external potentials (6.2) of the heterostructure dominate and the exciton is effectively two-dimesional one. The exciton wave function can be approximated by a product of three one-dimensional functions: the quantum-confinement wave functions of the uncoupled electron and hole in QW and the radial function of the 2D Coulomb potential $-1/\rho$. We classified the exciton states by three indices (ei, hj, Ns), where i and j correspond to these quantum-confinement wave functions and N is the principal number of the Coulomb function. This classification is convenient, though it is only approximate for wider QWs since the 2D Coulomb potential gradually changes to 3D one with increase of QW width. The 3D Coulomb potential, in turn, couples the factorized states.

The calculated energies as a function of L as a parameter are shown in Fig. 6.1 with respect to the lower boundary of the continuous spectrum. This boundary is defined by a sum of the lowest quantum-confinement energies $E_{e1} + E_{h1}$. Since the exciton binding energy is defined as $E_b = E_{e1} + E_{h1} - E_X$, the shown zero binding energy corresponds to the exciton energy $E_{e1} + E_{h1}$. Anticrossings of energy levels of the same symmetry of the quantum-confinement wave functions are denoted by circles. The radiative decay rates [5] of the calculated states are additionally shown in μeV by numbers in *italic*. These values characterize the magnitude of the exciton-light coupling, thus allow us to distinguish the optically active exciton states [7].

Our numerical results can be compared with the reflectance spectroscopy data for heterostructures with GaAs/$Al_{0.30}Ga_{0.70}$As QWs of $L = 14$ nm and $L = 20$ nm presented in [11] (Fig. 4). The experimental data for such QWs readily show only the exciton ground state. This is in accordance with our calculations since for $L < 20$ nm there is only one optically active electron-hole bound state, namely the ground one.

Acknowledgements Financial support from RFBR (grants No. 18-32-00568 and No. 18-02-00492) is acknowledged. The calculations were performed on resources of the Computational Center of St. Petersburg State University.

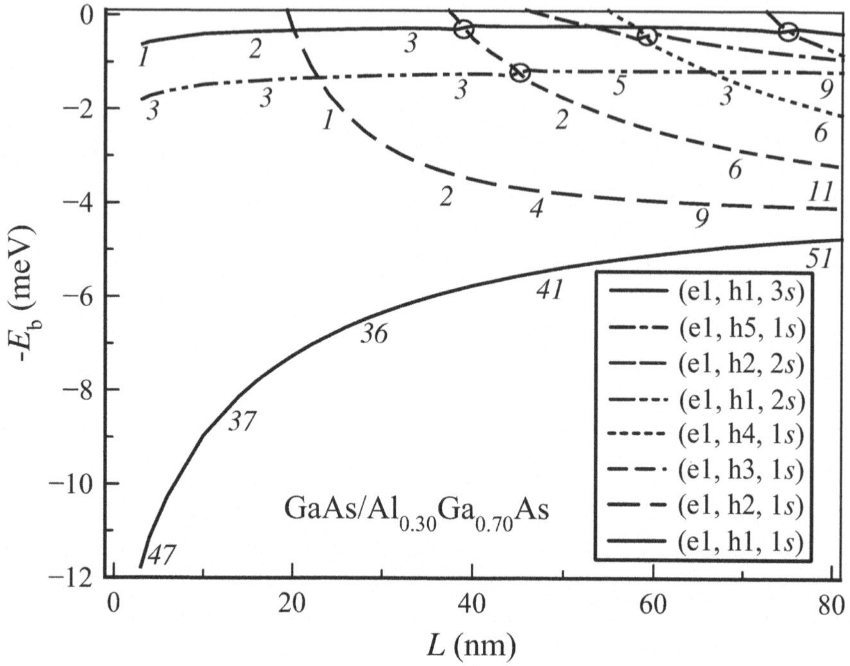

Fig. 6.1 Energy levels of the heavy-hole exciton in $GaAs/Al_{0.30}Ga_{0.70}As$ QW as a function of the QW width L. The electron-hole bound states are shown with respect to the lower boundary of the continuous spectrum which is denoted by the zero binding energy. Discontinuities of the material parameters are taken into account in calculations. Anticrossings of energy levels are denoted by circles. The radiative decay rates in μeV are given by numbers in *italic*

References

1. Bastard, G., et al.: Phys. Rev. B **26**, 1974 (1982)
2. Greene, R.L., et al.: Phys. Rev. B **29**, 1807 (1984)
3. Iotti, R.C., Andreani, L.C.: Phys. Rev. B **56**, 3922 (1997)
4. Alferov, ZhI: Semiconductors **32**, 1 (1998)
5. Ivchenko, E.L.: Optical spectroscopy of semiconductor nanostructures. Alpha Science International, Harrow (2005)
6. Hassanabadi, H., et al.: Few-Body Syst. **45**, 71 (2009)
7. Kavokin, A.V., et al.: Microcavities. Oxford University Press, Oxford (2017)
8. Tsintzos, S.I., et al.: Phys. Rev. Lett. **121**, 037401 (2018)
9. Poltavtsev, S.V., et al.: Solid State Commun. **199**, 47 (2014)
10. Trifonov, A.V., et al.: Phys. Rev. B **91**, 115307 (2015)
11. Khramtsov, E.S., et al.: J. Appl. Phys. **119**, 184301 (2016)
12. Shamirzaev, T.S., et al.: Phys. Rev. B **94**, 045411 (2016)
13. Abramkin, D.S., et al.: Semiconductors **52**, 1484 (2018)
14. Kezerashvili R.Y., Tsiklauri, S.M.: Few-Body Syst. **58**, 18 (2017)
15. Berman, O.L., et al.: Phys. Rev. B **96**, 014505 (2017)
16. Combescot, R.: Phys. Rev. X **7**, 041035 (2017)
17. Nader, D.J., et al.: Few-Body Syst. **58**, 116 (2017)

18. Brunetti, M.N., et al.: Phys. Rev. B **98**, 125406 (2018)
19. Luttinger, J.M.: Phys. Rev. **102**, 1030 (1956)
20. Belov, P.A., Khramtsov, E.S.: J. Phys. Conf. Ser. **816**, 012018 (2017)
21. Grigoryev, P.S., et al.: Superlatt. Microstr. **97**, 452 (2016)
22. Belov, P.A.: Semiconductors **52**, 551 (2018)
23. Belov, P.A.: Semiconductors **52**, 1791 (2018)
24. Glinskii, G.F., Mironova, M.S.: Semiconductors **48**, 1324 (2014)
25. Belov, P.A., et al.: J. Phys.: Conf. Ser. **929**, 012035 (2017)

Chapter 7
Ultracold Neon Trimer via Faddeev Differential Equations

A. A. Korobitsin and E. A. Kolganova

Abstract The Faddeev differential equation in the total angular momentum representation are used to calculate the spectrum of the neon trimer. The Ne-Ne interatomic interaction is described by modern realistic potentials. A comparison with previously published results is done.

7.1 Introduction

Three-atomic clusters comprise a large class of molecules whose interaction is described by means of van der Waals potentials. Studies of these clusters have been attracting attention of theoretical and experimental physicists for a long time and are still important today, especially due to such a unique quantum phenomenon as the Efimov effect [1]. This effect was theoretically predicted in 1970, and it perfectly shows a variety of opportunities of transition from a two-body problem to a three-particle problem. When there are at least two subsystems of zero binding energy, the three-body system has an infinite number of weakly bound states—this is the essence of the Efimov effect. Calculations of ultracold three-body clusters require accurate methods suitable for solving three-body bound states and scattering problems [2]. The aim of this paper is to develop a numerical algorithm for solving three-dimensional Faddeev differential equations in the total angular momentum representation [3, 4] and apply it to calculation of the binding energies of the $^{20}Ne_3$ triatomic system.

A. A. Korobitsin (✉) · E. A. Kolganova
BLTP JINR, 141980 Dubna, Moscow Region, Russia
e-mail: koroaa@theor.jinr.ru

E. A. Kolganova
Dubna State University, 141980 Dubna, Moscow region, Russia

N. A. Orr et al. (eds.), *Recent Progress in Few-Body Physics*,
Springer Proceedings in Physics 238,
https://doi.org/10.1007/978-3-030-32357-8_7

7.2 Three-Atomic Clusters

One of the effective methods of studying three-particle systems is based on the differential Faddeev equations in the total angular momentum representation [3–5]. The three-body system is described by the Hamiltonian:

$$H = H_0 + \sum_{\alpha} V_\alpha(x_\alpha),$$

where H_0 stands for the kinetic energy of three particles, $V_\alpha(x_\alpha)$ is the interaction potential acting in the pair α. We consider the states with zero total angular momentum. The angular degrees of freedom corresponding to collective rotation of the three-body system can be separated and the kinetic energy operator reduces to

$$\begin{aligned} H_0 = -\rho^{-5}\partial_\rho \rho^5 \partial_\rho - \tfrac{4}{\rho^2}\sin^{-2}\chi_\alpha \times \\ \times\, (\partial_{\chi_\alpha}\sin^2\chi_\alpha + \sin^{-1}\theta_\alpha \partial_{\theta_\alpha}\sin\theta_\alpha \partial_{\theta_\alpha}), \end{aligned} \tag{7.1}$$

where ρ, χ_α and θ_α are the hyperspherical coordinates expressed through standard Jacobi variables $\mathbf{x}_\alpha$, $\mathbf{y}_\alpha$, $\alpha = 1, 2, 3$:

$$\begin{aligned} \rho = \sqrt{x_\alpha^2 + y_\alpha^2}, \quad \tan\chi_\alpha/2 = y_\alpha/x_\alpha, \quad \cos\theta_\alpha = \tfrac{(\mathbf{x}_\alpha, \mathbf{y}_\alpha)}{x_\alpha y_\alpha}, \\ x_\alpha = |\mathbf{x}_\alpha|, \quad y_\alpha = |\mathbf{y}_\alpha|, \quad \rho \in [0, \infty), \qquad \{\chi_\alpha, \theta_\alpha\} \in [0, \pi] \otimes [0, \pi]. \end{aligned} \tag{7.2}$$

The total wave function Ψ of a three-body system can be written as the sum of the Faddeev components Ψ_α

$$\Psi(\rho, \chi_\alpha, \theta_\alpha) = \sum_\alpha \Psi_\alpha(\rho, \chi_\alpha, \theta_\alpha), \tag{7.3}$$

which satisfy the Faddeev equations

$$(H_0 + V_\alpha - E)\,\Psi_\alpha(\rho, \chi_\alpha, \theta_\alpha) = -V_\alpha \sum_{\beta \neq \alpha} \Psi_\beta(\rho, \chi_\beta, \theta_\beta), \tag{7.4}$$

where E is the total energy of the system. For numerical solution it is suitable to use scaled components Φ_α:

$$\Phi_\alpha = \rho^{5/2}\sin\chi_\alpha \sin\theta_\alpha \Psi_\alpha, \tag{7.5}$$

which satisfy the boundary conditions:

$$\Phi_\alpha(\rho = 0) = \Phi_\alpha(\chi_\alpha = 0, \pi) = \Phi_\alpha(\theta_\alpha = 0, \pi) = 0. \tag{7.6}$$

Equation (7.4) can be written in terms of the scaled components Φ_α as follows:

$$\left(\tilde{H}_0 + V_\alpha(x) - E\right)\Phi_\alpha(\rho,\chi,\theta) = -V_\alpha(x)\sum_{\beta\neq\alpha}\Phi_\beta\left(\rho,\chi_\beta,\theta_\beta\right), \tag{7.7}$$

where $\chi \equiv \chi_\alpha, \theta \equiv \theta_\alpha, x = \rho\cos(\chi_\alpha/2)$. The differential operator $\tilde{H}_0$ is given by

$$\tilde{H}_0 = -\partial_\rho^2 - \frac{4}{\rho^2}\left[\partial_\chi^2 + \sin^{-2}\chi\left(\partial_\theta^2 - \cot\theta\partial_\theta + \sin^{-2}\theta\right) + \frac{1}{16}\right].$$

For calculations of the spectrum of the neon trimer we use the finite-difference approximation for solving the differential equations (7.7) with the Dirichlet boundary conditions. To increase the speed of calculation, we used the Eigen library [6] for linear algebra, OpenMP and CUDA technology. The developed algorithm has no physical approximations and was applied in [7] for calculation of the ground and first excited state energies of the neon trimer. Here we applied this numerical algorithm to calculate the spectrum of the ^{20}Ne three-atomic system. Atomic masses for the neon isotope were taken from [8]. To describe the interatomic interaction, the realistic potential models TT [9] and HFD-B [10] were used. Table 7.1 contains the spectrum for the neon trimer, which is in good agreement with the results obtained by other authors using different methods.

Investigation of the ground and first excited state energy convergence with respect to the number of grid points demonstrates that $N_\rho = 250$ is sufficient for obtaining precision up to four significant digits (see Fig. 7.1).

Table 7.1 The bound state energies (in K) for $^{20}Ne_3$ calculated with TT [9] and HFD-B [10] potentials

E(K)	HFD-B	TT	[11][a]	[12][a]
E_0	74.13	74.07	74.10	74.11
E_1	52.44	52.37	52.41	52.43
E_2	49.25	49.19	49.23	49.24
E_3	45.53	45.49	45.51	45.52
E_4	40.37	40.31	40.34	40.35
E_5	34.67	34.62	34.65	34.66
E_6	32.33	32.27	32.30	32.31
E_7	31.54	31.48	31.51	31.52
E_8	27.66	27.61	27.64	27.65
E_9	26.20	26.16	26.17	26.18
E_{10}	24.95	24.92	24.93	25.03

[a]The HFD-B [10] potential model used

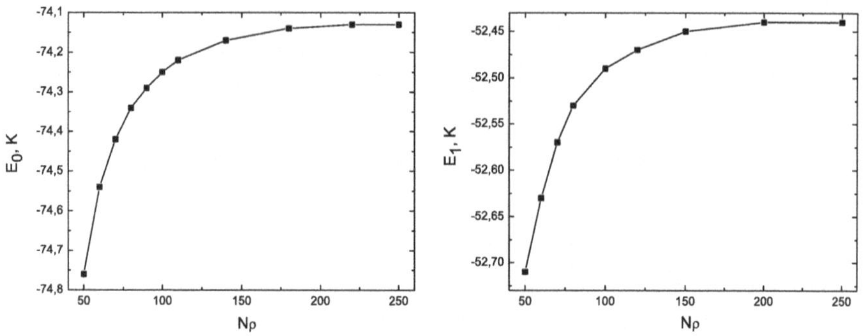

Fig. 7.1 Convergence of the neon trimer ground and first excited state energies on the grid points N_ρ for fixed value of $N_\theta = 100$ and $N_\chi = 10$

7.3 Conclusion

In the course of this work we have calculated the spectrum for the neon trimer using the realistic potentials TT [9] and HFD-B [10]. In order to perform calculations, we developed a numerical algorithm for solving the differential Faddeev equations in the total angular momentum representation. This algorithm has been realized in the programming language C++. Our results are in good agreement with the results obtained by other methods. The developed numerically effective computational scheme, especially in combination with the option of using multiple processors, makes it possible to calculate a wide range of three-body problems.

References

1. Efimov, V.N.: Energy levels arising from resonant two-body forces in a three-body system. Phys. Lett. B **33**, 563–564 (1970). https://doi.org/10.1016/0370-2693(70)90349-7
2. Kolganova, E.A., Motovilov, A.K., Sandhas, W.: The ^{4}He trimer as an Efimov system. Few-Body Syst. **51**, 249–257 (2011). https://doi.org/10.1007/s00601-011-0233-x
3. Kostrykin, V.V., Kvitsinsky, A.A., Merkuriev, S.P.: Faddeev approach to the three-body problem in total-angular-momentum representation. Few-Body Syst. **6**, 97–113 (1989). https://doi.org/10.1007/BF01080553
4. Kvitsinsky, A.A., Hu, C.-Y.: Solution of three-dimensional Faddeev equations for three-body Coulomb bound states. Few-Body Syst. **12**, 7–19 (1992). https://doi.org/10.1007/BF01080193
5. Roudnev, V.A., Yakovlev, S.L., Sofianos, S.A.: Bound-state calculations for three atoms without explicit partial wave decomposition. Few-Body Syst. **37**, 179–196 (2005). https://doi.org/10.1007/s00601-005-0114-2
6. http://eigen.tuxfamily.org
7. Korobitsin, A.A., Kolganova, E.A.: A theoretical study of van der Waals neon trimer using Faddeev equations. Phys. Part. Nuclei Lett. **14**, 971–974 (2017). https://doi.org/10.1134/S1547477117070081
8. Mills, I., Cvita, T., Homann, K., Kallay, N., Kuchitsu, K.: The grid: quantities. Units and symbols in physical chemistry. Blackwell Scientific Publications, Oxford (1993)

9. Tang, K.T., Toennies, J.P.: The van der Waals potentials between all the rare gas atoms from He to Rn. J. Chem. Phys. **118**, 4976–4983 (2003). https://doi.org/10.1063/1.1543944
10. Aziz, R.A., Slaman, M.J.: The Ne-Ne interatomic potential revisited. J. Chem. Phys. **130**, 187–194 (1989). https://doi.org/10.1016/0301-0104(89)87048-X
11. Salci, M., Levin, S.B., Elander, N., Yarevsky, E.: A theoretical study of the rovibrational levels of the bosonic van der Waals neon trimer. J. Chem. Phys. **129**(134304), 1–7 (2008). https://doi.org/10.1063/1.2955736
12. Suno, H.: A theoretical study of Ne3 using hyperspherical coordinates and a slow variable discretization approach. J. Chem. Phys. **135**(134312), 1–10 (2011). https://doi.org/10.1063/1.3645183 ibid., Geometrical structure of helium triatomic systems: comparison with the neon trimer. J. Phys. B 49, 014003 (2016). https://doi.org/10.1088/0953-4075/49/1/014003

Chapter 8
QED in the Clothed-Particle Representation

Adam Arslanaliev, Yan Kostylenko and Aleksandr Shebeko

Abstract We have extended [1] our previous applications of the method of unitary clothing transformations (UCTs) in mesodynamics to quantum electrodynamics (QED). An analytical expression for the QED Hamiltonian in the clothed-particle representation (CPR) has been derived. Its distinctive feature is the appearance of a new family of the Hermitian and energy independent interaction operators built up in the e^2-order for the clothed electrons and positrons instead the primary canonical interaction between electromagnetic and electron-positron fields. The problem of describing the bound states in QED in case of the positronium system has been considered. Finally, the first correction to the energy of the ground state of the para-positronium has been calculated by using the new interaction in e^-e^+-system.

8.1 UCT Method in Action

When realizing the clothing procedure elaborated in [2] we are starting with the QED interaction Hamiltonian given by [3]:

$$V_{qed} = \int d\mathbf{x} j_k(\mathbf{x}) a^k(\mathbf{x}) + V_{Coul} = V^{(1)} + V_{Coul},$$

with the electron-positron current density operator

A. Arslanaliev (✉) · Y. Kostylenko
V.N. Karazin National University, Kharkov, Ukraine
e-mail: arslanaliev.kh@gmail.com

Y. Kostylenko
e-mail: nlokost@gmail.com

A. Shebeko
National Science Center "Kharkov Institute of Physics & Technology", Kharkiv, Ukraine
e-mail: shebeko@kipt.kharkov.ua

N. A. Orr et al. (eds.), *Recent Progress in Few-Body Physics*,
Springer Proceedings in Physics 238,
https://doi.org/10.1007/978-3-030-32357-8_8

$$j_\mu(\mathbf{x}) = e\bar{\psi}(\mathbf{x})\gamma_\mu\psi(\mathbf{x})$$

and the Coulomb part

$$V_{Coul} =: \frac{1}{2}\int d\mathbf{x}\int d\mathbf{y}\frac{j^0(\mathbf{x})j^0(\mathbf{y})}{4\pi|\mathbf{x}-\mathbf{y}|}e^{-\lambda|\mathbf{x}-\mathbf{y}|} :,$$

where the exponential factor is introduced to deal with infrared divergences. Admittedly, the parameter $\lambda > 0$ is set to zero at the end of all calculations. Note also that here we use the Coulomb gauge (CG), where the photon field a_μ being transverse, has two independent polarizations.

Following [1, 2] after removing the so-called bad terms (in our case $V^{(1)}$ operator) via the corresponding UCT we get the QED Hamiltonian in CPR

$$H \equiv K(\alpha_c) = K_F(\alpha_c) + K_I(\alpha_c),$$

where α_c denotes the set of all creation and destruction operators for the clothed particles included. It is proved, that in the e^2-order the interaction part $K_I(\alpha_c)$ is approximated by

$$K_I^{(2)}(\alpha_c) = K(e^-e^- \to e^-e^-) + K(e^+e^+ \to e^+e^+) + K(e^+e^- \to e^+e^-) +$$
$$+K(e^+e^- \to 2\gamma) + K(2\gamma \to e^+e^-) + K(e^-\gamma \to e^-\gamma) + K(e^+\gamma \to e^+\gamma),$$

where the separate contributions in the r.h.s. correspond to the different interactions between clothed particles.

8.2 Interaction Operator for the $e^-e^+ \to e^-e^+$ Scattering

Among them we will consider the interaction operator between clothed electrons and positrons

$$K(e^-e^+ \to e^-e^+) = \int d1'd2'd1d2V_{e^-e^+}(1',2';1,2)b_c^\dagger(1')d_c^\dagger(2')b_c(1)d_c(2),$$

with

$$V_{e^-e^+}(1',2';1,2) = \frac{e^2m^2}{(2\pi)^3}\delta(\mathbf{p}_2' + \mathbf{p}_1' - \mathbf{p}_2 - \mathbf{p}_1)\Big[\upsilon_S(1',2';1,2) + \upsilon_A(1',2';1,2)\Big],$$

where m the physical electron (positron) mass, $b_c(d_c)$ is the destruction operator for the clothed electron (positron), the argument j of $b_c(j)$ includes the particle momentum and its polarization, etc. In addition, we introduce the scattering (annihilation) term υ_S (υ_A). The latter have the structure

$$v_{S/A} = v_{S/A}(\text{Feynman-like}) + v_{S/A}(\text{off-energy-shell}).$$

The quasipotential is

$$\langle b_c^\dagger(1')d_c^\dagger(2')\Omega|K(e^-e^+ \to e^-e^+)|b_c^\dagger(1)d_c^\dagger(2)\Omega\rangle,$$

where Ω the physical vacuum state. Finally, we would like to show explicit expressions

$$v_S(\text{Feynman-like}) = -\bar{u}(1')\gamma^\mu u(1)\frac{1}{2}\left\{\frac{1}{(p_1'-p_1)^2} + \frac{1}{(p_2'-p_2)^2}\right\}\bar{v}(2)\gamma_\mu v(2'),$$

$$v_S(\text{off-energy-shell}) = \frac{(p_1'+p_2'-p_1-p_2)}{(\mathbf{p}_1'-\mathbf{p}_1)^2+\lambda^2}\bar{u}(1')\gamma^0 u(1)$$
$$\times\frac{1}{2}\left\{\frac{(p_1'-p_1)}{(p_1'-p_1)^2} + \frac{(p_2'-p_2)}{(p_2'-p_2)^2}\right\}\bar{v}(2)\gamma^0 v(2'),$$

$$v_A(\text{Feynman-like}) = \bar{u}(1')\gamma^\mu v(2')\frac{1}{2}\left\{\frac{1}{(p_1+p_2)^2} + \frac{1}{(p_1'+p_2')^2}\right\}\bar{v}(2)\gamma_\mu u(1),$$

$$v_A(\text{off-energy-shell}) = \frac{(p_1'+p_2'-p_1-p_2)}{(\mathbf{p}_1'+\mathbf{p}_2')^2+\lambda^2}\bar{u}(1')\gamma^0 v(2')$$
$$\times\frac{1}{2}\left\{\frac{(p_1'+p_2')}{(p_1'+p_1)^2} - \frac{(p_1+p_2)}{(p_1+p_2)^2}\right\}\bar{v}(2)\gamma^0 u(1).$$

Such a decomposition implies that only the Feynman-like part survives on the energy shell, i.e. on the condition $p_1'^0 + p_2'^0 = p_1^0 + p_2^0$. Of course, all momenta included are defined on the mass-shell: $p^2 = p_0^2 - \mathbf{p}^2 = m^2$.

8.3 Corrections to the Positronium Ground State Energy

Positronium consisting of an electron and a positron is the simplest bound system in QED. Its ground state (g.s.) has two possible configurations with total spin values S = 0, 1. The singlet (triplet) lowest-energy state with S = 0 (S = 1) is known as the para-positronium (ortho-positronium). It is our first attempt to describe the positronium properties within the CPR. For this exposition, we will restrict ourselves to the consideration of the para-positronium (p-Ps) system. The corresponding g.s., being the H eigenvector, viz.,

$$H|\mathbf{P};\text{p-Ps}\rangle = E|\mathbf{P};\text{p-Ps}\rangle,$$

can be represented as

$$|\mathbf{P};\text{p-Ps}\rangle = \int d1d2\,\Psi_{00}(\mathbf{P};1,2)b_c^\dagger(1)d_c^\dagger(2)|\Omega\rangle,$$

where $\Psi_{00}(\mathbf{p}) \equiv \Psi_{00}(\mathbf{P} = 0;\ 1, 2)$ (since we work in c.m.s.) satisfies

$$(2p_0 - m_{\text{p-Ps}})\Psi_{00}(\mathbf{p}) + \int d\mathbf{p}' V(p', p)\Psi_{00}(\mathbf{p}') = 0.$$

Naturally, the positronium mass is connected with its bind energy $\varepsilon_{\text{p-Ps}}$ by

$$m_{\text{p-Ps}} = m_{e^-} + m_{e^+} + \varepsilon_{\text{p-Ps}} \approx 2m + \varepsilon_{\text{p-Ps}}.$$

We have

$$V(p', p) = \frac{e^2}{(2\pi)^3 p'_0 p_0}\left(V_{\text{Feynman-like}} + V_{\text{off-shell}}\right),$$

$$V_{\text{Feynman-like}} = \frac{2p_0 p'_0 - m^2}{(p' - p)^2 - \lambda^2}, \quad V_{\text{off-shell}} = -\frac{1}{2}\frac{(p'_0 - p_0)^2}{(\mathbf{p}' - \mathbf{p})^2 + \lambda^2}\frac{(p'_- + p)^2}{(p' - p)^2 - \lambda^2}.$$

Doing so we note $V_{\text{off-shell}}$ disappears when $p'_0 = p_0$ i.e. on the energy shell. Moreover, in the non-relativistic limit ($p_0 = p'_0 = m$) the eigenvalue equation reduces to the ordinary Schrödinger equation for the Coulomb potential in momentum space

$$\frac{\mathbf{p}^2}{m}\Psi_{00}(\mathbf{p}) + \int d\mathbf{p}' V_C(p', p)\Psi_{00}(\mathbf{p}') = \varepsilon_{\text{p-Ps}}\Psi_{00}(\mathbf{p}).$$

Therefore, we come to the well-known Coulomb problem with the energy spectrum determined by

$$\varepsilon_{\text{p-Ps}} = -\frac{m\alpha^2}{4n^2} \qquad (n = 1, 2, \cdots),$$

where $\alpha = e^2/4\pi$ the fine-structure constant and the ground state ($n = 1$) belongs to the eigenvalue $\varepsilon_{\text{g.s.}} \approx -6.8\,\text{eV}$. In this context let us rewrite our eigenvalue equation in the form

$$\frac{\mathbf{p}^2}{m}\Psi_{00}(\mathbf{p}) + \int d\mathbf{p}'\big(V_C(p', p) + U(p', p) + W(p', p)\big)\Psi_{00}(\mathbf{p}') = \varepsilon_{\text{p-Ps}}\Psi_{00}(\mathbf{p}),$$

where

$$U(p', p) = V(p', p) - V_C(p', p),$$

$$W(p', p) = (2p_0 - 2m - \frac{\mathbf{p}^2}{m})\delta(\mathbf{p} - \mathbf{p}').$$

By considering the quantity $U(p', p) + W(p', p)$ as a perturbation and using the non-perturbative wave function of the ground state $\Psi_{00}(\mathbf{p})$ from Appendix C in [4] we have computed the energy shift $\Delta\varepsilon$. It leads to the following corrections:

$$\langle V_{\text{Feynman-like}} - V_C \rangle \approx -7.19711 \cdot 10^{-4}\,\text{eV},$$

$$\langle V_{\text{off-shell}} \rangle \approx 3.5878 \cdot 10^{-4}\,\text{eV}, \quad \langle W \rangle \approx -1.12314 \cdot 10^{-4}\,\text{eV}.$$

So we get the total shift

$$\Delta\varepsilon = -4.7325 \cdot 10^{-4}\,\text{eV}.$$

This result is comparable with those estimations given in [4] (see formula (1.1) therein) where, for example,

$$\Delta\varepsilon = -\frac{21}{64} m\alpha^4 \approx -4.7547 \cdot 10^{-4}\,\text{eV}.$$

But they are due to the completely different physical ingredients such as the fine and hyperfine structure of the positronium energy levels. Against the background such a relativistic correction as $\langle V_{\text{off-shell}} \rangle$ contributes considerably.

To conclude we remark that further applications of our approach in the positronium spectroscopy and lifetime are under way.

References

1. Arslanaliev, A.: The method of unitary clothing transformations: application in quantum electrodynamics, magistr diploma. V. N. Karazin National University, Kharkov, Ukraine (2017)
2. Shebeko, A.V., Shirokov, M.I., Phys. Part. Nucl. **32** 31, (2001). arXiv: nucl-th/0102037
3. Weinberg, S.: The quantum theory of fields, vol. 1. University Press, Cambridge (1995)
4. Smith, C.: Bound state description in quantum electrodynamics and chromodynamics. Louvain-la-Neuve (2002)

Chapter 9
Low-Dimensional Few-Body Processes in Confined Geometry of Atomic and Hybrid Atom-Ion Traps

Vladimir S. Melezhik

Abstract We have developed an efficient approach for treating low-dimensional few-body processes in confined geometry of atomic and hybrid atom-ion traps. It based on the split-operator method in 2D discrete-variable representation (DVR) suggested by V. Melezhik for integration of the few-dimensional time-dependent Schrödinger equation. We give a brief review of the application to resonant ultracold atomic processes and discuss our latest results on hybrid atomic-ion systems. Prospects for the application of the method in other hot problems of the physics of low-dimensional few-particle systems are also discussed.

9.1 Introduction

Impressive progress of the physics of ultracold quantum gases has stimulated the necessity of detailed and comprehensive investigations of collisional processes in the confined geometry of atomic and ionic traps. The traditional free-space scattering theory is no longer valid here and the development of the low-dimensional few-body theory including the influence of the confinement is needed. In our works we have developed quantitative models [1–4] for pair collisions in tight atomic waveguides and have found several novel effects in its application: the confinement-induced resonances (CIRs) in multimode regimes including effects of transverse excitations and deexcitations [2], the so-called dual CIR yielding a complete suppression of quantum scattering [1], and resonant molecule formation with a transferred energy to center-of-mass excitation while forming molecules [5]. Last effect was confirmed experimentally in [6]. Our calculations have also been used for planning and

V. S. Melezhik (✉)
Bogoliubov Laboratory of Theoretical Physics, Joint Institute for Nuclear Research, Dubna, Moscow Region 141980, Russian Federation

Peoples Friendship University of Russia (RUDN University), Miklukho-Maklaya str. 6, Moscow 117199, Russian Federation
e-mail: melezhik@theor.jinr.ru
URL: http://theor.jinr.ru/~melezhik/

N. A. Orr et al. (eds.), *Recent Progress in Few-Body Physics*, Springer Proceedings in Physics 238,
https://doi.org/10.1007/978-3-030-32357-8_9

interpretation of the Innsbruck experiment where CIRs in ultracold Cs gas were observed [7]. Mention also the calculation of the Feshbach resonance shifts and widths induced by atomic waveguides [8]. In the frame of our approach we have predicted dipolar CIRs [9] which may pave the way for the experimental realization of, e.g., Tonks-Girardeau-like or super-Tonks-Girardeau-like phases in effective one-dimensional dipolar gases.

Our latest results on hybrid atomic-ion systems and prospects are discussed in this report.

9.2 Atom-Ion Collisions in Hybrid Atom-Ion Traps

Recently, we have predicted the atom-ion CIRs [10] which are important for a hot problem of control of the confined hybrid atom-ion systems having many promising applications [11]. The condition of appearance of CIR in a atom-ion collision confined in a harmonic waveguide-like trap was found in [10] in "static" ion approximation. This approach, when one neglects by the ion motion, is well defined for for the Li-Yb^+ collision considered in [10]. However, in real experiments an actual problem is controlling of the unremovable effect of ion micromotion in the ion Paul traps [11].

To evaluate the effect of the ion motion on the CIR we performed full quantum calculation for the ^{6}Li-atom scattering by $^{171}Yb^+$ for a special case of harmonic transversal traps with $\omega_A \ll \omega_I$ frequencies for atom (A) and ion (I)

$$U(\rho_A, \rho_I) = \frac{1}{2}(m_A\omega_A^2\rho_A^2 + m_I\omega_I^2\rho_I^2)$$

(where $\rho_i = r_i sin\theta_i$). For that, we have integrated the 4D time-dependent Schrödinger equation with the Hamiltonian

$$H(\rho_R, \mathbf{r}) = H_{CM}(\rho_R) + H_{rel}(\mathbf{r}) + W(\rho_R, \mathbf{r}). \quad (9.1)$$

by using computational scheme developed earlier for confined distinguishable atom collisions [1, 3, 5]. Here ($\hbar = 1$)

$$H_{CM} = -\frac{1}{2M}(\frac{\partial^2}{\partial\rho_R^2} + \frac{1}{\rho_R^2}\frac{\partial^2}{\partial\phi^2} + \frac{1}{4\rho_R^2}) + \frac{1}{2}(m_A\omega_A^2 + m_I\omega_I^2)\rho_R^2 \quad (9.2)$$

and

$$H_{rel} = -\frac{1}{2\mu}\frac{\partial^2}{\partial r^2} + \frac{L^2(\theta,\phi)}{2\mu r^2} + \frac{\mu^2}{2}(\frac{\omega_A^2}{m_A} + \frac{\omega_I^2}{m_I})\rho^2 + V_{AI}(r) \quad (9.3)$$

describe the CM and relative (rel) atom-ion motions. The potential $V_{AI}(r)$ describes the atom-ion interaction, ρ_R and $\mathbf{r} = \mathbf{r_A} - \mathbf{r_I} \mapsto (\mathbf{r}, \theta, \phi) \mapsto (\rho, \phi, \mathbf{z})$ are the polar

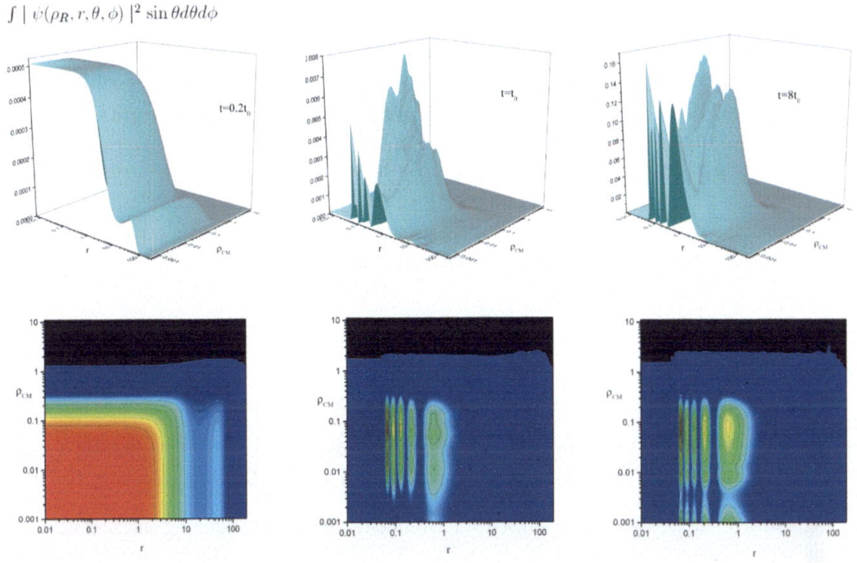

Fig. 9.1 Evolution in time of the atom-ion probability density distribution $W(r, \rho_R, t) = \int |\psi(\mathbf{r}, \rho_R, t)|^2 \sin\theta d\theta d\phi$ in the process of the confined collision of the ^{6}Li atoms with ^{171}Yb$^+$ ions. Top row of graphs demonstrates the probability density $W(r, \rho_R, t)$ calculated at three time points during the collision and the bottom row is the corresponding contour plots of these probabilities $W(r, \rho_R, t)$. The time unit is $t_0 = 2\pi/\omega_A$, $\rho_R = \rho_{CM}$

radial CM and the relative coordinates and $M = m_A + m_I$, $\mu = m_A m_I / M$. The term $\frac{L^2(\theta,\phi)}{2\mu r^2}$ represents the angular part of the kinetic energy operator of the relative atom-ion motion. The term

$$W(\rho_R, \mathbf{r}) = \mu(\omega_A^2 - \omega_I^2) r \rho_R \sin\theta \cos\phi \tag{9.4}$$

leads to a coupling of the CM and relative motion, i.e. to the nonseparability of the quantum two-body problem in confined geometry of the harmonic trap.

In Fig. 9.1 we present the calculated time-evolution of the probability density distribution of ^{6}Li and ^{171}Yb$^+$ near the CIR in the harmonic waveguides. This quantum calculation confirms the surviving of the CIR in the case of ion-motion and demonstrates the molecule ion LiYb$^+$ formation during this collision.

9.3 Conclusion

The efficiency of the splitting-up method based on the 2D DVR for the time-dependent Schrödinger equation makes the method promising in application to actual problems of low-dimensional few-body physics in atomic and atom-ion traps. One can mention the problem of ultracold atomic collisions in anharmonic and asymmetric waveguides and in quasi-2D confining traps. Of great interest in connection with possible important applications is the two-center (and N-center) problem in a confining trap [12, 13]. Note also a collisional three-body problem in tight traps, and non-linear time-dependent Schrödinger equation with a few spatial variables arising in physics of Bose-Einstein condensates.

Acknowledgements This work was supported by the Russian Foundation for Basic Research, Grant No. 18-02-00673 and the "RUDN University Program 5-100".

References

1. Melezhik, V.S., Kim, J.I., Schmelcher, P.: Wave-packet dynamical analysis of ultracold scattering in cylindrical waveguides. Phys. Rev. A 76, 053611-1-15 (2007). https://doi.org/10.1103/PhysRevA.76.053611
2. Saeidian, S., Melezhik, V.S., Schmelcher, P.: Multichannel atomic scattering and confinement-induced resonances in waveguides. Phys. Rev. A 77, 042721-1-15 (2008). https://doi.org/10.1103/PhysRevA.77.042721
3. Melezhik, V.S.: Mathematical modeling of ultracold few-body processes in atomic traps. EPJ Web of Conf. 108, 01008-1-9 (2016). https://doi.org/10.1051/epjconf/201610801008; Mathematical modeling of resonant processes in confined geometry of atomic and atom-ion traps. EPJ Web of Conf. 173, 01008-1-8 (2018). https://doi.org/10.1051/epjconf/201817301008
4. Saeidian, S., Melezhik, V.S.: Multichannel scattering problem with a nonseparable angular part as a boundary-value problem. Phys. Rev. E 96, 053302-1-8 (2017). https://doi.org/10.1103/PhysRevE.96.053302
5. Melezhik V.S., Schmelcher, P.: Quantum dynamics of resonant molecule formation in waveguides. New J. Phys. 11, 073031-1-11 (2009). https://doi.org/10.1088/1367-2630/11/7/073031
6. Sala, S., Zürn, G., Lompe, T., Wenz, A.N., Murmann, S., Serwane, F., Jochim, S., Saenz, A.: Coherent molecule formation in anharmonic potentials near confinement-induced resonances. Phys. Rev. Lett. 110, 203202-1-5 (2013). https://doi.org/10.1103/PhysRevLett.110.203202
7. Haller, E., Mark, M.J., Hart, R., Danzl, J.G., Reichsöllner, L., Melezhik, V., Schmelcher, P., Nägerl, H.-C.: Confinement-induced resonances in low-dimensional quantum systems. Phys. Rev. Lett. 104, 153203-1-4 (2010). https://doi.org/10.1103/PhysRevLett.104.153203
8. Saeidian, S., Melezhik, V.S., Schmelcher, P.: Shifts and widths of Feshbach resonances in atomic waveguides. Phys. Rev. A 86, 062713-1-9 (2012). https://doi.org/10.1103/PhysRevA.86.062713; Shifts and widths of p-wave confinement induced resonances in atomic waveguides. J. Phys B 48, 155301-1-9 (2015). https://doi.org/10.1088/0953-4075/48/15/155301
9. Giannakeas, P., Melezhik, V.S., Schmelcher, P.: Dipolar confinement-induced resonances of ultracold gases in waveguides. Phys. Rev. Lett. 111, 183201-1-5 (2013). https://doi.org/10.1103/PhysRevLett.111.183201
10. Melezhik, V.S., Negretti, A.: Confinement-induced resonances in ultracold atom-ion systems. Phys. Rev. A 94, 022704-1-8 (2016). https://doi.org/10.1103/PhysRevA.94.022704
11. Tomza, M., Jachymski, K., Gerritsma, R., Negretti, A., Calarco, T., Idziaszek, Z., Julienne, P.S.: Cold hybrid atom-ion systems. arXiv:1708.07832. https://arxiv.org/abs/1708.07832
12. Sroczyska, M., Wasak, T., Jachymski, K., Calarco, T., Idziaszek, Z.: Trap-induced shape resonances in an ultracold few-body system of an atom and static impurities. Phys. Rev. A 98, 012708-1-10 (2018). https://doi.org/10.1103/PhysRevA.98.012708
13. Shadmehri, S., Melezhik, V.S.: Confinement-induced resonances in two-center problem via pseudopotential approach. Phys. Rev. A 99, 032705-1-11 (2019). https://doi.org/10.1103/PhysRevA.99.032705

Chapter 10
Mesoscopic Bose-Einstein Condensate in Anharmonic Trap: Concept of Transition Exponent

M. L. Lekala, S. Bera, G. J. Rampho, B. Chakrabarti and S. Bhattacharyya

Abstract We utilize a two-body correlated basis function and van der Waals interaction to describe interacting bosons in the anharmonic trap. We analyze the behaviour of specific heat capacity near the transition temperature in mesoscopic region. We calculate the transition exponent to define the quasi phase transition for different anharmonicity. Comparison with pure harmonic trap is also addressed.

A true phase transition is signalled by a singularity in thermodynamic potential [1]. A finite discontinuity in the first derivative of the thermodynamic potentials is usually termed as first-order phase transition. Whereas if the first derivatives are continuous but second derivatives are discontinuous, the transition is categorized as the higher-order. One can also define critical exponent, which characterize the nature of the phase transition and universal scaling [1, 2]. However for the experimentally achieved Bose-Einstein condensation (BEC), the inter-atomic interaction makes a profound effect on the condensate properties [3]. In particular, the BEC statistics with mesoscopic number of atoms where $N \sim 10^2 - 10^7$ is challenging [4–6]. The usually employed grand-canonical-ensemble can not take care of critical fluctuations [7, 8]. It is necessary to solve the system in more complicated way in canonical ensemble. However in vast attempt, both experimentally and theoretically, it is established that the transition from condensate phase to Bose gas is smooth, the system does not exhibit true phase transition.

The motivation of our present article is to revisit the problem of mesoscopic BEC in an anharmonic trap modelled as $V(r) = \frac{1}{2}m\omega^2 r^2 + \lambda_0 r^4$, $\lambda_0 > 0$. For the tight quartic confinement, the trap becomes very tight, the transition temperature increases as expected. However our focus is to study the transition near the transition

M. L. Lekala (✉) · G. J. Rampho
Department of Physics, University of South Africa, P.O. Box 392, Pretoria 0003, South Africa
e-mail: lekalml@unisa.ac.za

S. Bera · B. Chakrabarti
Department of Physics, Presidency University, 86/1 College Street, Kolkata 700073, India

S. Bhattacharyya
Department of Physics, Scottish Church College 1 & 3 Urquhart Square, Kolkata 700006, India

N. A. Orr et al. (eds.), *Recent Progress in Few-Body Physics*,
Springer Proceedings in Physics 238,
https://doi.org/10.1007/978-3-030-32357-8_10

temperature for different anharmonicity. Although Bose-Einstein condensation is more favourable in tight trap, the most important issue of phase transition has not been addressed before. The main finding of the article is to discover that no dramatic change in the lambda-structure in specific heat is possible in the tight trap. We further define the concept of transition exponent to define this quasi phase transition and also calculate the value of the exponent for different anharmonic parameter.

In this present communication, we adopt an *ab initio*, many-body technique called, potential harmonic expansion method (PHEM) [9, 10]. PHEM takes care of all possible two-body correlations between the atoms and thus goes beyond the Gross-Pitaevskii theory. In our earlier calculations of thermodynamic properties and condensate fluctuations [11, 12], PHEM was able to provide the correct effective potential of weakly interacting BEC. As for the dilute condensate, three-body collisions are ignored, the Faddeev component ψ_{ij} of the (ij)-th interacting pair is expanded as a function of $(\mathbf{r}_{ij})$ and the global hyperradius r as [9, 10]

$$\phi_{ij}(\mathbf{r}_{ij}, r) = r^{-(\frac{3N-1}{2})} \sum_{K} \mathcal{P}^{lm}_{2K+l}(\Omega^{ij}_{N})\eta(r_{ij})u^{l}_{K}(r). \tag{10.1}$$

where $\mathcal{P}^{lm}_{2K+l}(\Omega^{ij}_{N})$ is the potential harmonics basis (PH-basis) in terms of which we expand the Faddeev component. Note we add an additional correlation function $\eta(r_{ij})$ to reach fast convergence [13]. Where the global hyperradius is defined through the Jacobi coordinates as $r = \left[\sum_{i=1}^{A} \zeta_i^2\right]^{\frac{1}{2}}$. Substituting of above equation into the many-body Schrödinger equation and projection on to a particular PH basis we reach in the set of coupled differential equation (CDE):

$$\Big[- \frac{\hbar^2}{m}\frac{d^2}{dr^2} + \frac{\hbar^2}{mr^2}\{\mathcal{L}(\mathcal{L}+1) + 4K(K+\alpha+\beta+1)\} + V_{trap}(r) - E_R\Big]U_{Kl}(r) + \sum_{K'} f_{Kl}V_{KK'}(r)f_{K'l}U_{K'l}(r) = 0, \tag{10.2}$$

where V_{trap} is the external trap potential. $V_{KK'}$ is the potential matrix element [9]. The set of CDEs are further solved by hyperspherical adiabatic approximation (HAA) [10]. In HAA, potential matrix together with hypercentrifugal repulsion is diagonalized for fixed value of r. The lowest eigenvalue gives the lowest eigenpotential $\omega_0(r)$ as a parametric function of r. The energy and wavefunctions are further calculated by solving the adiabatically separated hyperradial equation for the collective motion in the effective potential. Our present work considers ^{87}Rb atoms in JILA trap and interatomic interaction is characterized by the scattering length $a = 100$ Bohr. We choose the realistic van der Waals potential with a hard core of radius r_c and attractive $\frac{1}{r^6}$ tail. To solve the complete differential equation and to obtain the energy eigen spectrum in the many-body effective potential we follow the numerical steps documented in our earlier calculations [9–13]. However in this

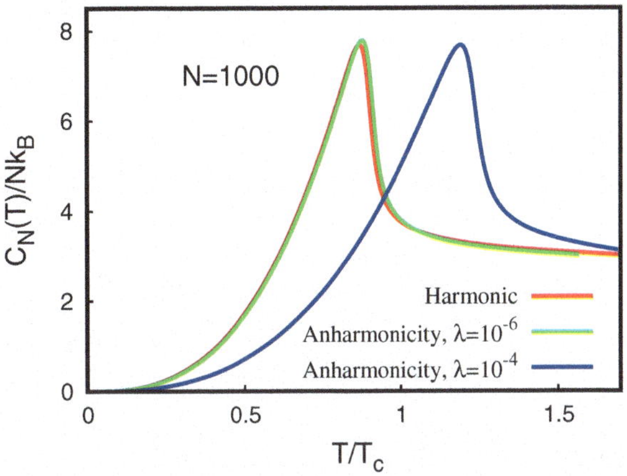

Fig. 10.1 Plot of specific heat of interacting bosons both in anharmonic and harmonic trap with $N = 1000$

present calculation the gas is confined in the anharmonic trap with $\lambda_0 = 10^{-6}$ and 10^{-4}. All quantities are in oscillator units.

Next utilizing the Bose-distribution function calculated from the energy spectrum of the many-body effective potential and following our earlier calculations we calculate the specific heat $C_N(T)$ for fixed N. In Fig. 10.1, we plot $\frac{C_N(T)}{Nk_B}$ for $N = 1000$ and for two different choices of anharmonic parameters. For comparison we also present the result for harmonic trap with $\lambda_0 = 0$. We observe that transition temperature increases in anharmonic trap. Thus achievement of BEC is more favourable, however the lambda-structure in the specific heat does not change due to this anharmonic affect. The broad maxima obtained in this mesoscopic region truly rulled out the possibility of any true phase transition as in the case of harmonic trap. However we define it as a quasi-phase transition and instead of critical exponent which is valid for true phase transition, we define a transition exponent. We define the reduced temperature as $t = \frac{T-T_C}{T_C}$ and the critical exponent for any thermodynamic function is usually defined as

$$\lambda_c = \lim_{t\to 0} \frac{\ln|F(t)|}{\ln|t|}, \tag{10.3}$$

or it can be written as $F(t) \sim t^{\lambda_c}$, it basically represent the asymptotic behaviour of the thermodynamic function at $t \to 0$. Thus at the transition temperature $F(t)$ either vanishes or be singular. However for mesoscopic BEC, as $C_N(T)$ becomes sharper smoothly with increase in particle number N, neither vanishes nor become singular. Instead we follow the following prescription [1],

$$F(t) = F(0) + b|t|^{\lambda_c} + ..., \tag{10.4}$$

where λ_c is the leading exponent.

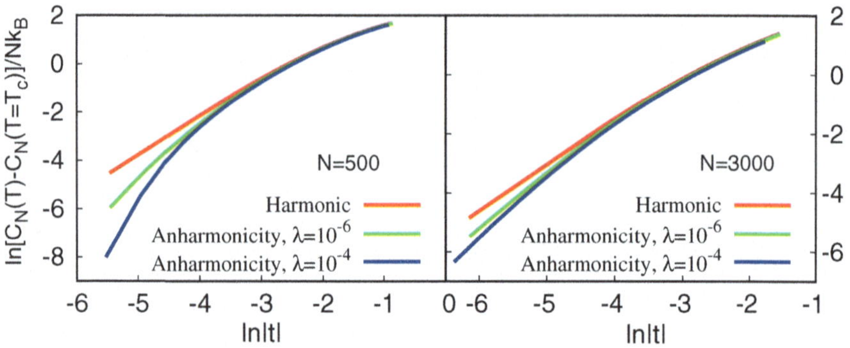

Fig. 10.2 Exponent curves for the ^{87}Rb atoms with $N = 500$ in [panel(a)] and $N = 3000$ in [panel(b)] both in harmonic and anharmonic trap

We define λ_c as the transition exponent

$$\lambda_c = \lim_{t \to 0} \frac{\ln |F(t) - F(0)|}{\ln |t|}. \tag{10.5}$$

In Fig. 10.2(a) and (b) we plot $\ln [\frac{C_N(T)-C_N(T=T_C)}{Nk_B}]$ as a function of $\ln |t|$ for $N =$ 500 and 3000. Each graph reports both cases of anharmonicity together with harmonic trap. For truly finite sized systems, when the number of bosons is order of few hundreds, away from the transition temperature (close to the zero value of $\ln |t|$), the plots shows some curvature. However for large negative values of $\ln |t|$, we fit a straight line for the results for harmonic trap. We observe a single exponent $\lambda_c =$ 1.775 fits well. For anharmonic trap, we still find a curvature near the large negative values of $\ln |t|$ and we are unable to fit any straight line. However for higher particle number (Fig. 10.2b), we fit a straight line and obtain the exponents. These are 1.798 for $\lambda_0 = 10^{-6}$ and 1.826 for $\lambda_0 = 10^{-4}$.

In the present work, we study the quasi phase transition of mesoscopic condensate having few hundred to few thousands particles in the anharmonic trap. We utilize two-body correlated basis function which clearly takes care of finite sized effect and the methodology goes beyond the mean-field theory. Utilizing Bose-Einstein distribution function we further calculate the specific heat capacity for finite sized BEC. We observe, like harmonic trap, the width of specific heat curve, only decreases with increase in particle number. We do not observe any sharp transition near the transition temperature. We call it quasi phase transition and define the transition exponent to characterize the smooth transition from condensate to Bose gas.

Acknowledgements Sangita Bera wants to acknowledge DST for giving financial support through INSPIRE fellowship to complete this research work.

References

1. Yeomans, J.M.: Statistical mechanics of phase transitions. Clarendon Press, Oxford (1992)
2. Huang, K.: Statistical mechanics, 2nd edn. Wiley, New York (1987)
3. Dalfovo, F., et al.: Theory of Bose-Einstein condensation in trapped gases. Rev. Mod. Phys. **71**, 463 (1999)
4. Kocharovsky, V.V., Kocharovsky, V.V.: Analytical theory of mesoscopic Bose-Einstein condensation in an ideal gas. Phys. Rev. A **81**, 033615 (2010)
5. Tarasov, S.V., et al.: Universal scaling in the statistics and thermodynamics of a Bose-Einstein condensation of an ideal gas in an arbitrary trap. Phys. Rev. A **90**, 033605 (2014)
6. Berrada, T., et al.: Integrated Mach-Zehner interferometer for Bose-Einstein condensates. Nat. Commun. **4**, 2077 (2013)
7. Landsberg, P.T.: Thermodynamics—with quantum statistical illustrations. Interscience, New York (1961)
8. Mullin, W.J., Fernandez, J.P.: Bose-Einstein condensation, fluctuations, and recurrence relations in statistical mechanics. Am. J. Phys. **71**, 661 (2003)
9. Das, T.K., Chakrabarti, B.: Potential harmonics expansion method for trapped interacting bosons: inclusion of two-body correlation. Phys. Rev. A **70**, 063601 (2004)
10. Das, T.K., et al.: Behavior of a Bose-Einstein condensate containing a large number of atoms interacting through a finite-range interatomic interaction. Phys. Rev. A **78**, 042705 (2007)
11. Bhattacharyya, S., et al.: Effects of interaction on thermodynamics of a repulsive Bose-Einstein condensate. Phys. Rev. A **88**, 053614 (2013)
12. Bhattacharyya, S., et al.: Energy fluctuation of a finite number of interacting bosons: a correlated many-body approach. Phys. Rev. A **93**, 033624 (2016). Biswas, A.: Effect of realistic interatomic interactions and two-body correlation on the heat capacity of a trapped BEC. J. Phys. B **42**, 215302 (2009); Goswami, S., Das, T.K., Biswas, A.: Thermodynamic properties of ultracold Bose gas: transition exponents and universality. J. Low Temp. Phys. **172**, 184 (2013)
13. Das, T.K., et al.: ^{85}Rb Bose-Einstein condensate with tunable interaction: a quantum many body approach. Phys. Lett. A **373**, 258 (2009)

Chapter 11
Study of Helium and Lithium Atomic Systems with the Discrete Variable Representations

Vladimir Timoshenko and Evgeny Yarevsky

Abstract Significant computational resources are necessary to calculate bound and resonance states of weakly bound systems with high accuracy. The discrete-variable representation method allows to reduce the computational complexity of determining the matrix elements of operators in the frame of variational methods. In this work, the binding energies of the Li-He_2 and He_3 systems are calculated and compared with other results.

11.1 Introduction

The study of quantum-mechanical systems consisting of few particles can be a complicated problem. This is true for weakly bound systems, e.g. systems consisting of helium and lithium atoms. In order to perform calculations faster, the discrete-variable representations (DVR) [1, 2] of different kinds are used in this paper. The binding energies of the Li-He_2 and He_3 systems are calculated and compared with theoretical results of other authors.

11.2 Formulation of the Problem and the Discrete-Variable Representation

For states with zero angular momentum, the Hamiltonian of a three-particle system in the Jacobi coordinates is written as [3]

$$\left(-\frac{1}{\mu_{1,23}y}\frac{\partial^2}{\partial y^2}y + \frac{1}{\mu_{23}x}\frac{\partial^2}{\partial x^2}x\right) + \left(\frac{1}{\mu_{1,23}y^2}+\frac{1}{\mu_{23}x^2}\right)\left(\frac{\partial^2}{\partial \theta^2} + \mathrm{ctg}\theta\frac{\partial}{\partial \theta}\right) + V(x, y, \theta).$$

V. Timoshenko · E. Yarevsky (✉)
St. Petersburg State University, St. Petersburg 198504, Russian Federation
e-mail: e.yarevsky@spbu.ru

N. A. Orr et al. (eds.), *Recent Progress in Few-Body Physics*,
Springer Proceedings in Physics 238,
https://doi.org/10.1007/978-3-030-32357-8_11

Here, the potential $V = V(x, y, \theta)$ is a sum of two-particle potentials that depend on interparticle distances only. Let us rewrite the kinetic-energy operator in terms of $z = \cos\theta$ and apply the DVR method [2, 4] to the variable z. The DVR functions $\varphi_i(z)$ and their derivatives $\varphi_i'(z)$ are constructed with orthogonal polynomials $P_n(z)$ and are related to the Gauss-type quadrature formulas

$$\varphi_i(z) = \frac{P_n(z)}{P_n'(z_i)(z - z_i)}, \quad \varphi_i'(z_k) = \frac{P_n'(z_k)}{P_n'(z_i)(z_k - z_i)}, \; i \neq k, \quad \varphi_i'(z_i) = -\frac{P_n''(z_i)}{2P_n'(z_i)}.$$

Here $z_1, \dots, z_n$ are the roots of the polynomial $P_n(z)$. For the Jacobi polynomials $P_n^{(\alpha,\beta)}$, the values of the derivatives are written as

$$\varphi_i'(z_i) = \frac{\beta - \alpha - (\alpha + \beta + 2)z_i}{2(1 - z_i^2)}. \tag{11.1}$$

The wave function is represented as a linear combination of the DVR functions, $f(z) = \sum_i c_i \varphi_i(z)$. Using expression for the derivatives, we obtain the matrix elements of the kinetic energy matrix T

$$T_{ij} = \langle \varphi_i | T | \varphi_j \rangle = \sum_k \frac{w_k}{\rho(z_k)} \frac{\varphi_i'(z_k)}{\sqrt{w_i}} \frac{\varphi_j'(z_k)}{\sqrt{w_j}} (1 - z_k^2).$$

Here $\rho(z) = (1 - z)^\alpha (1 + z)^\beta$ is the weight function for the Jacobi polynomials, and w_i are the weights of the Gauss quadrature formula. As $\varphi_i(z_k) = \delta_{ik}$, the matrix elements of the potential energy are diagonal:

$$V_{ij} = \langle \varphi_i | V | \varphi_j \rangle = \sum_k \frac{w_k}{\rho(z_k)} V(x, y, z_k) \frac{\varphi_i(z_k)}{\sqrt{w_i}} \frac{\varphi_j(z_k)}{\sqrt{w_j}} = \frac{V(x, y, z_i)}{\rho(z_i)} \delta_{ij}.$$

11.3 Results and Discussion

The approach combining the finite-element method [3] for the coordinates x and y, and the DVR method for the coordinate z has been developed for calculating the binding energies of three particle quantum systems. The energy levels of weakly bound systems ^{6}Li–He$_2$ and ^{7}Li–He$_2$ were calculated with the DVR method based on the Legendre polynomials, $\alpha = \beta = 0$. Due to the use of the DVR, the computation time has been significantly reduced without loss of accuracy. The results for binding energies, computation times and relative inaccuracies $\delta E = |(E_{\text{exact}} - E)/E_{\text{exact}}|$ for the ^{7}Li–He$_2$ system are presented in Table 11.1. One can see that the results of calculations without the DVR demonstrate the variational behaviour while the DVR results approach the exact value from below. Both results converge to the same exact value when accuracy increases.

Table 11.1 The binding energies E_1, computation times t, and relative errors δE_1 for the different number n of functions in the expansion. Results for the ^{7}Li–He$_2$ system are shown

n	5	10	15	20	25
Legendre polynomial expansion without the DVR					
E_1, cm^{-1}	$-3.44{\cdot}10^{-7}$	$-3.00{\cdot}10^{-2}$	$-3.94{\cdot}10^{-2}$	$-4.07{\cdot}10^{-2}$	$-4.12{\cdot}10^{-2}$
δE_1	1.00	$2.75{\cdot}10^{-1}$	$4.67{\cdot}10^{-2}$	$1.49{\cdot}10^{-2}$	$2.26{\cdot}10^{-3}$
t, sec	6.0	36.6	119.6	287.9	553.5
Legendre polynomial expansion with the DVR					
E_1, cm^{-1}	$-4.81{\cdot}10^{-2}$	$-4.67{\cdot}10^{-2}$	$-4.19{\cdot}10^{-2}$	$-4.15{\cdot}10^{-2}$	$-4.14{\cdot}10^{-2}$
δE_1	$1.63{\cdot}10^{-1}$	$1.31{\cdot}10^{-1}$	$1.41{\cdot}10^{-2}$	$5.10{\cdot}10^{-3}$	$1.96{\cdot}10^{-3}$
t, sec	2.2	8.2	19.1	35.2	65.8

Table 11.2 Binding energies of the Li–He$_2$ system (cm^{-1}) for different interparticle potentials: TTY [8], LM2M2 [9], Cvetko [10]

	He-He potential	Li-He potential	^{6}Li–He$_2$	^{7}Li–He$_2$
Paper [5]	TTY	KTTY	$-2.18{\cdot}10^{-2}$	$-3.18{\cdot}10^{-2}$
This work	TTY	KTTY	$-3.71{\cdot}10^{-2}$	$-5.41{\cdot}10^{-2}$
Paper [6]	LM2M2	Cvetko	$-3.61{\cdot}10^{-2}$	$-5.10{\cdot}10^{-2}$
This work	LM2M2	Cvetko	$-2.62{\cdot}10^{-2}$	$-4.07{\cdot}10^{-2}$
Paper [7]	LM2M2	KTTY	$-2.46{\cdot}10^{-2}$	$-3.54{\cdot}10^{-2}$
This work	LM2M2	KTTY	$-3.71{\cdot}10^{-2}$	$-5.41{\cdot}10^{-2}$

Calculated binding energies for the ^{6}Li–He$_2$ and ^{7}Li–He$_2$ systems are shown in Table 11.2 together with the comparison to the theoretical results of other authors [5–7]. For the TTY+KTTY and LM2M2+KTTY potentials, our binding energies are considerably deeper. As our results are almost variational, we believe that they are closer to the exact values. The situation with the potential [10] should be further investigated.

The DVR approach has also been implemented with the DVR functions constructed with the Jacobi polynomials $P_n^{(\alpha,\beta)}(z)$. These polynomials make it possible to choose parameters α and β such that the weight $\rho(z)$ countervails the interaction potential. This approach has been used for calculating the binding energy of the helium trimer. The results are presented in Table 11.3. The most accurate energy values are calculated with the Chebyshev polynomials of the first kind, $\alpha = \beta = -0.5$.

Table 11.3 Binding energies of He$_3$ and relative inaccuracies δE for different parameters of the Jacobi polynomials $P_n^{(\alpha,\beta)}$ in the DVR method

$\alpha = \beta$	-0.75	-0.50	-0.25	0.0	0.25	0.50	0.75
E, cm$^{-1}\cdot 10^{-2}$	-8.648	-8.372	-7.164	-9.716	-8.850	-8.627	-8.544
δE	0.049	0.015	0.131	0.178	0.073	0.046	0.035

Acknowledgements This work is supported by RFBR grant No. 18-02-00492. The calculations were performed on resources of the Computational Center of St. Petersburg State University.

References

1. Baye, D.: Phys. Rep. **565**, 1107 (2015)
2. Shizgal, B.: Spectral methods in chemistry and physics. Springer, Netherlands (2015)
3. Salci, M., Levin, S.B., Elander, N., Yarevsky, E.A.: J. Chem. Phys. **129**, 134304 (2008)
4. Light, J.C., Tucker Jr., C.: Adv. Chem. Phys. **114**, 263 (2000)
5. Yuan, J., Lin, C.D.: J. Phys. B At. Mol. Opt. Phys. **31**, 647 (1998)
6. Baccarelli, I., et al.: Phys. Chem. Chem. Phys. **2**, 4067 (2000)
7. Kolganova, E.A.: Few-Body Syst. **58**, 57 (2017)
8. Tang, K., Toennies, J., Yiu, C.: Phys. Rev. Lett. **74**, 1546 (1995)
9. Aziz, R., Slaman, M.: J. Chem. Phys. **84**, 8047 (1991)
10. Cvetko, D., Lausi, A., Morgante, A., Tommasini, F.: J. Chem. Phys. **100**, 2052 (1994)

Chapter 12
Potential Splitting Approach for Atomic and Molecular Systems

Evgeny Yarevsky, Sergey L. Yakovlev, Nils Elander and Åsa Larson

Abstract In order to describe few-body scattering in the case of the Coulomb interaction, an approach based on splitting the reaction potential into a finite range part and a long range tail part is presented. The resulting driven Schrödinger equation with asymptotic outgoing waves is solved with the exterior complex scaling. The approach is illustrated with calculations of the electron scattering on the hydrogen atom and the positive helium ion. The scattering processes in the $H^+ - H_2^+$ system with one-state electronic energy surface have also been studied.

12.1 Introduction

Systems with Coulomb interactions are often considered in nuclear, atomic, and molecular physics. Scattering problems for them are of great interest for many physical processes. The complicated boundary conditions at large distances are a major difficulty for these problems [1]. Several methods have been developed for constructing solutions to the three-body scattering problem [2]. Some of them avoid using the explicit asymptotic form of the wave function.

In several recent studies, we have reported a method which is capable to correctly treat the Coulomb scattering problem using exterior complex scaling (ECS) [3–6]. The key point of this method is splitting the long-range Coulomb potential into the core and tail parts. The tail part is used to construct the distorted incident wave, which is responsible for the asymptotic Coulomb dynamics. The core part of the potential generates an inhomogeneous term in the Schrödinger equation making possible the application of ECS for solving the equation. Here we outline the potential splitting approach and present its application to study atomic systems and molecular systems with *ab initio* potentials.

E. Yarevsky (✉) · S. L. Yakovlev
St. Petersburg State University, St. Petersburg 198504, Russian Federation
e-mail: e.yarevsky@spbu.ru

N. Elander · Å. Larson
Department of Physics, Stockholm University, 106 91 Stockholm, Sweden

N. A. Orr et al. (eds.), *Recent Progress in Few-Body Physics*,
Springer Proceedings in Physics 238,
https://doi.org/10.1007/978-3-030-32357-8_12

12.2 Theoretical Approach

The three-body quantum system is described with the Schrödinger equation in Jacobi coordinates $\mathbf{x}_\alpha, \mathbf{y}_\alpha$ with the Hamiltonian

$$H = T_{\mathbf{x}_\alpha} + T_{\mathbf{y}_\alpha} + V(\mathbf{x}_\alpha, \mathbf{y}_\alpha),$$

where $T_{\mathbf{x}_\alpha}$, $T_{\mathbf{y}_\alpha}$ are the kinetic energy operators, and $V(\mathbf{x}_\alpha, \mathbf{y}_\alpha)$ is the full interaction in the system. The total wave function Ψ of the system is written as the sum $\Psi = \Phi + \Psi^R$, and the function Φ satisfies the driven equation

$$(H - E)\,\Phi = -V_R \Psi^R. \tag{12.1}$$

Here the distorted incident wave $\Psi^R(\mathbf{x}_\alpha, \mathbf{y}_\alpha)$ is introduced as a solution to the scattering problem with the sum of the potential in the target pair $V_\alpha(\mathbf{x}_\alpha)$ and the tail potential $V^R(\mathbf{x}_\alpha, \mathbf{y}_\alpha)$. The full potential is split into the sum of $V_\alpha(\mathbf{x}_\alpha)$, the core $V_R(\mathbf{x}_\alpha, \mathbf{y}_\alpha)$ and the tail V^R parts: $V(\mathbf{x}_\alpha, \mathbf{y}_\alpha) = V_\alpha(\mathbf{x}_\alpha) + V_R + V^R$. Here $V_R = V - V_\alpha$ for $y_\alpha \le R$, and $V_R = 0$ for $y_\alpha > R$. Details of the splitting procedure is described in the papers [5, 6]. The right hand side of equation (12.1) is of finite range with respect to the variable y_α. Thus this equation can be solved numerically with the ECS transformation [7].

In order to find scattering amplitudes and cross sections, the asymptotic form of the scattered wave function at large distances is used [1]. Projecting the asymptotic representation on the two body wave functions, the local representation for the partial amplitude can be derived [6].

12.3 Applications of the Potential Splitting Method

12.3.1 Atomic Systems

First, our approach is applied to the scattering problem in the electron-H and electron-He^+ systems. These systems are of fundamental importance in atomic physics and have been studied by various methods and approximations, see [8] and references therein. The numerical solution of the driven equation (12.1) is performed with the finite element method, which is described in details in [9]. The grid used in our calculations is given in [6].

Our results for the singlet 1s→ns cross sections for the e-H and e-He^+ scattering are presented in Fig. 12.1. In both cases the calculated cross sections display complicated resonance behaviour. The resonance structure of the e-He^+ cross section occur at lower energies due to the presence of the asymptotic Coulomb interaction. The comparison of our data with other theoretical results [8, 10] shows the relative difference less than 10^{-3}.

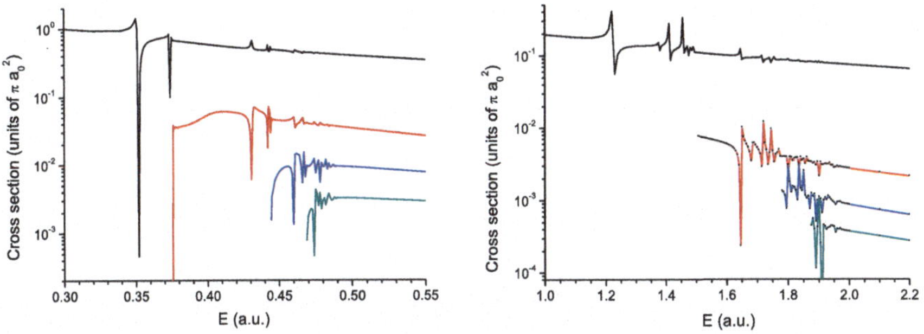

Fig. 12.1 The singlet (spin weight included) 1s→1s, 2s, 3s, 4s cross sections (from above) for the e-H (left) and e-He^+ (right) scattering as functions of the electron energy, Adapted from [6]

12.3.2 Molecular Systems

Molecular systems cannot be exactly studied with few-body analytical methods as the total number of particles is too large. To apply such methods, additional approximations are necessary. The most obvious one is the Born-Oppenheimer approximation where the electron degrees of freedom do not participate in the dynamical equations but are averaged to the potential energy surfaces. For accurate calculation of processes, numerical *ab-initio* potentials calculated with quantum chemistry approaches must be used. These potentials depend on all internuclear distances, and are given numerically. The ECS approach can be used to calculate scattering processes with this type of potentials provided that the scaling radius is larger than the internuclear distance where the potential is numerically calculated. For systems with asymptotic Coulomb interactions, the potential splitting approach has been used.

In this work, we have considered the $H^+ - H_2^+$ scattering. The potential energy surface $U(r_1, r_2, \theta)$ of the electronic ground state of H_3^{2+} depends on the two bond-lengths r_1, r_2, and the angle θ between them. The potential is computed using the aug-cc-pVQZ basis set of Dunning [11]. The *ab initio* calculations are carried out using internal coordinates where the bond-lengths are varied in the range $0.8a_0 \leq r_i \leq 20.0a_0$, and the angle θ is varied in $[0, \pi]$. These calculations have been performed with the MOLPRO program [12]. For the regions where two nuclei come close together and the asymptotic regions at large internuclear distances, the *ab initio* potential energy surface is extrapolating and interpolated by first introducing the function

$$U(r_1, r_2, \theta) = V(r_1, r_2, \theta) - \sum_{i=1}^{3} E_{H_2^+}(r_i). \tag{12.2}$$

where r_3 is calculated as $r_3 = \sqrt{r_1^2 + r_2^2 - 2r_1 r_2 \cos\theta}$. The energy $E_{H_2^+}(r)$ is the energy of the Coulomb two-centre problem with the electron and two charges +1

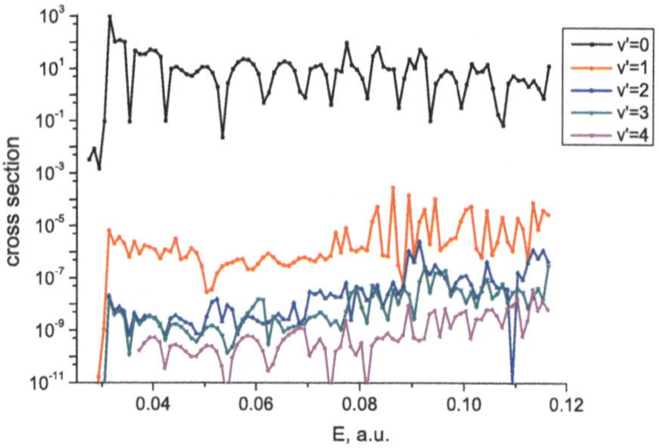

Fig. 12.2 The elastic and excitation cross sections for the $H^+ - H_2^+$ scattering

each placed at the distance r. The function $U(r_1, r_2, \theta)$ has no singularities at $r_i = 0$ and hence is much easier to interpolate and extrapolate.

Using the united atom approximations, we can find the values of U when the distances between two or three protons approach zero. For bond-lengths in the range $r_i \leq 20.0a_0$, the potential energy surface is calculated at an arbitrary point with the 3D spline interpolation.

As we use the ECS, the potential energy must be represented by an analytical function at large distances. In order to make this representation, we use a combination of the asymptotic Coulomb potential and the polarization potential $1/r_i^2$. The coefficients are chosen to make the total potential energy continuous on the boundary of the numerical grid.

We have calculated the elastic and excitation scattering cross sections $H_2^+(v = 0, J = 0) + H^+ \to H_2^+(v', J' = 0) + H^+$ with the constructed potential energy surface. In the calculations, we use a rectangular product grid. For the reaction coordinate y_α, five finite elements have been used at short distance [0–4] a.u., 44 elements for intermediate region, and ten elements of total length 40 a.u. for the discretization beyond the splitting point $R = 31$ a.u. For the coordinate x_α, 19, 9, and 4 elements respectively have been used for the regions mentioned above. One element has been used for the angular variable θ_α.

Our results for the elastic and excitation $H_2^+(v = 0, J = 0) \to H_2^+(v', J' = 0)$ cross sections for the $H^+ - H_2^+$ scattering are presented in Fig. 12.2, where the energy E is the incident energy of H^+. The structure in the cross section appears because of large number of states in the H_2^+ molecule.

12.4 Conclusions

We have proposed the mathematically sound approach for calculations of scattering processes. The potential splitting approach allows for the solution of the scattering problem with the Coulomb interaction. Besides systems with analytically given interactions, molecular systems with numerically defined *ab initio* potentials can be studied with the combined ECS and splitting potential approaches.

Acknowledgements This work is supported by RFBR grant No. 18-02-00492. ÅL acknowledges support from the Swedish Research Council under project number 2014–4164.

References

1. Faddeev, L.D., Merkuriev, S.P.: Quantum Scattering Theory for Several Particle Systems. Kluwer, Dordrecht (1993)
2. Bray, I., et al.: Phys. Rep. **520**, 135 (2012)
3. Volkov, M.V., et al.: Phys. Rev. A **83**, 032722 (2011)
4. Volkov, M.V., et al.: EPL **110**, 30006 (2015)
5. Yarevsky, E., et al.: J. Phys. B: At. Mol. Opt. Phys. **48**, 115002 (2015)
6. Yarevsky, E., et al.: J. Phys. B: At. Mol. Opt. Phys. **50**, 055001 (2017)
7. Rescigno, T.N., et al.: Phys. Rev. A **55**, 4253 (1997)
8. Bartlett, P.L.: J. Phys. B: At. Mol. Opt. Phys. **39**, R379 (2006)
9. Elander, N., et al.: Phys. Rev. A **67**, 062508 (2003)
10. Wang, Y.D., Callaway, J.: Phys. Rev. A **48**, 2058 (1993)
11. Dunning Jr., T.H.: J. Chem. Phys. **90**, 1007 (1989)
12. Werner, H.-J. et al.: MOLPRO, version 2010.1, a package of ab initio programs (2010). http://www.molpro.net

Chapter 13
Fragmentation Dynamics of Atomic and Molecular Clusters

A. Méry, A. N. Agnihotri, J. Douady, X. Fléchard, B. Gervais, S. Guillous, W. Iskandar, E. Jacquet, V. Kumar, J. Matsumoto, J. Rangama, F. Ropars, C. P. Safvan, H. Shiromaru, D. Zanuttini and A. Cassimi

Abstract We report our results from collisions between multiply charged Ar^{9+} and Xe^{20+} ions and atomic or molecular dimers. In such systems, the presence of a surrounding environment may give rise to specific energy relaxation mechanisms. For atomic dimers, we found that the low electron mobility along the dimer results in an asymmetry in the charge repartition among the two ionic fragments. Specific relaxation process such as radiative charge transfer has also been identified. For molecular nitrogen dimers, the role of the environment on molecular fragmentation has been investigated by comparing the fragmentation of multiply ionized monomers N_2^{2+} and dimers $(N_2)_2^{(3)+}$.

13.1 Introduction

Atomic or molecular clusters constitute an intermediate step from gas phase to condensed matter. The properties of such systems under ion irradiation are of growing interest in the fields of atmospheric science, hadron-therapy or astrophysics. Here we focus on dimers consisting of two identical argon atoms or two identical nitrogen

A. Méry (✉) · A. N. Agnihotri · J. Douady · B. Gervais · S. Guillous · E. Jacquet · V. Kumar · J. Rangama · F. Ropars · D. Zanuttini · A. Cassimi
CIMAP, CEA -CNRS -ENSICAEN -UNICAEN, Normandie Université, BP5133, F -14050 Caen Cedex 04, France
e-mail: mery@ganil.fr

X. Fléchard
Normandie University, ENSICAEN, UNICAEN, CNRS/IN2P3, LPC Caen, 14000 Caen, France

W. Iskandar
Chemical Science Division, Lawrence Berkeley National Laboratory, Berkeley 94720, USA

J. Matsumoto · H. Shiromaru
Department of Chemistry, Tokyo Metropolitan University, 1-1 Minami-Osawa, Hachioji, 192-0397 Tokyo, Japan

C. P. Safvan
Inter University Accelerator Center, Aruna Asaf Ali Marg, 110067 New Delhi, India

N. A. Orr et al. (eds.), *Recent Progress in Few-Body Physics*, Springer Proceedings in Physics 238,
https://doi.org/10.1007/978-3-030-32357-8_13

molecules. Low energy highly charged ions are used as ionizing particles. These projectile ions have energies of about 3 keV/nucleon corresponding to velocities of 0.4 atomic units. In this velocity regime, electron capture is by far the dominant process and allows producing multiply ionized targets with low amount of excitation [1]. Moreover, the electron capture radii can be much larger than the dimer bond length so that electrons may be captured from the two atomic or molecular sites of the dimer. The resulting ionized target further dissociates either by direct coulomb explosion if dissociative states have been populated or through intermediate relaxation processes such as radiative charge transfer (RCT). In the case of molecular dimers, molecular fragmentation can also occur if more than one electron have been removed from one of the constitutive molecules resulting in both van der Waals and covalent bond cleavages. Such dissociation pathways allow to investigate the role of the environment on molecular fragmentation.

13.2 Experiment

The weakly bound van der Waals dimers are produced in the supersonic expansion of a gas jet and irradiated by low energy keV ions delivered by the ARIBE-GANIL beam line. The supersonic jet target is composed mainly by monomers and contains only about 1% of dimers. A COLTRIMS (COLd Target Recoil Ion Momentum Spectroscopy) setup is used to measure in coincidence the three-dimension momentum vectors of the charged fragments produced in the collision [2]. Coincidence measurements and specially momentum conservation laws allow to make a very clean data selection and to take out random coincidences originating from monomer targets ionization. For each fragmentation channel, we then derive several relevant observables such as the Kinetic Energy Release (KER) and the relative orientation of fragment momentum.

13.3 Ar_2 Rare Gas Dimers

Argon dimers are weakly bound van der Waals atomic systems resulting from the induced dipole interaction between the two atoms. The van der Waals bond length $R_{Ar_2} = 3.8$Å [3] is significantly larger than the atomic radius and there is very small overlap between the two atomic electron clouds. As a consequence, low electron mobility is expected between the two centers of the dimer in contrast to covalently bound molecules. This property has been used to shed new light on the electron capture process. However, some cluster specific relaxation pathways such as RCT can lead to charge transfer between the two sites of the dimer [4].

13.3.1 Electron Capture Process

Following the collision, several fragmentation channels could be identified depending on the charge state of each detected fragment. The following notation $(i, j)_F$ is used to refer to the fragmentation channel $Ar_2^{(i+j)+} \rightarrow Ar^{i+} + Ar^{j+}$. As an example, $(3, 1)_F$ refers to asymmetric fragmentation $Ar_2^{4+} \rightarrow Ar^{3+} + Ar^{+}$ following quadruple electron capture. After data selection, the branching ratios can be directly deduced by integrating the number of events in each channel [4] and are presented in Table 13.1. Low charge state channels clearly dominate because multiple electron capture requires smaller impact parameters and has thus a lower probability to occur.

In the case of quadruple electron capture, two different fragmentation channels can be observed: a symmetric $(2, 2)_F$ and an asymmetric $(3, 1)_F$ one. It appears that asymmetric fragmentation dominates. This is an unusual behavior in comparison to the fragmentation of diatomic covalent molecule such as N_2 for which symmetric dissociation is known to be by far the dominant process [5]. In covalent diatomic molecules, fast intramolecular electron rearrangement usually takes place prior dissociation leading to equal charge sharing. Contrarily, in van der Waals molecules, charge rearrangement is expected to be small. If at least one electron is removed from each atom, the potential energy curves (PECs) are purely dissociative and no charge exchange is expected during the dissociation process. However, the very asymmetric capture channel ($Ar^{4+} - Ar^{0}$) has curve crossings with excited states $Ar^{3+*} - Ar^{1+}$ and also contributes to the $(3, 1)_F$ channel [4]. Nevertheless, when slow highly charged ions collide with rare gas dimers, the low electron mobility allows to keep the memory of the initial capture process. Thus, for quadruple electron capture, symmetric (resp. asymmetric) electron capture will give rise to symmetric (resp. asymmetric) fragmentation. As a consequence, the predominance of asymmetric fragmentation reflects the predominance of asymmetric electron capture. This last point is possible because the internuclear distance in Ar dimers $R_{Ar_2} = 3.8$Å is of the same order of magnitude as the projectile capture radii. These values have been calculated using an extended Classical Over the Barrier Model (COBM) and are listed in Table 13.2 [6]. The experimental branching ratios are fairly reproduced in the model by considering the dimer as two independent atoms and by integrating over all possible orientation and impact parameters using a Monte-Carlo approach (MC-COBM).

The MC-COBM also gives access to the impact parameter distribution associated to each capture channel. For asymmetric channels, it appears that the projectile pref-

Table 13.1 Relative fragmentation branching ratios for the $Ar^{9+} + Ar_2$ and $Xe^{20+} + Ar_2$ collision systems

Channel	$(1, 1)_F$	$(2, 1)_F$	$(3, 1)_F$	$(2, 2)_F$
Ar^{9+}	0.42	0.30	0.19	0.09
Xe^{20+}	0.41	0.39	0.11	0.09

Table 13.2 Capture radii estimated from the Classical Over the Barrier Model for the $Ar^{9+} + Ar_2$ and $Xe^{20+} + Ar_2$ collision systems (see details in [6]). R_i^{in} values are given in Å

Projectile	$R_1^{in}[3p^{-1}]$	$R_2^{in}[3p^{-2}]$	$R_3^{in}[3p^{-3}]$
Ar^{9+}	6.4	5.5	4.7
Xe^{20+}	9.1	7.6	6.5

erentially captures electrons from the atom closest to its trajectory while symmetric channels are essentially produced when the projectile ion trajectory is in the median plan of the dimer [7]. Even if the projectile impact parameter is not directly accessible from the experimental data, this information can be deduced from the direction of the scattered projectile with respect to the emission of the mostly charged fragment. This observable can also be derived from the MC-COBM and shows a very nice quantitative agreement with experimental measurement proving that the electron capture process can be successfully described assuming that the dimer is the juxtaposition of two independent atoms [7].

13.3.2 Relaxation Pathways

The experimental KER distributions are used to get information about the fragmentation dynamics of each channel. For the $(2, 1)_F$, $(2, 2)_F$, and $(3, 1)_F$ channels associated to the $Ar^{9+} + Ar_2$ collision system, the KER spectra show only one peak corresponding to the direct Coulomb Explosion (CE) at the equilibrium interatomic distance of the neutral dimer. This is consistent with the fact that the potential energy curves of these dissociative states are purely Coulombic. As already mentioned, the $(3, 1)_F$ fragmentation channel is also fed by the $Ar^{4+} - Ar^0$ capture channel but, as curve crossings take place in the Franck-Condon region, these events also ends up with the same Coulombic KER value.

In contrast, the KER spectrum of the $(1, 1)_F$ channel shows two different contributions (Fig. 13.1). The low energy peak corresponds to Coulomb Explosion (CE) following two-site single electron capture for which the ground state $Ar^+ - Ar^+$ potential energy curve is also purely Coulombic. The high energy peak corresponds to shorter interatomic distance when dissociation occurs. These events have been identified as the result of one-site double electron capture for which the $Ar^{2+} - Ar^0$ PEC is attractive in the Frank-Condon region and has a minimum close to 2.9 Å. This state can decay through RCT to the dissociative $Ar^+ - Ar^+$ state within a nanosecond timescale thus giving rise to the high energy peak in the KER spectrum. Thus, even if leading to the same $(1, 1)_F$ dissociation channel, two-site (CE peak) and one-site (RCT peak) electron capture can be distinguished using the KER spectra. The relative intensity of the two processes depends on the projectile charge state and two-site capture is more likely to occur for projectiles with higher charge state such as Xe^{20+} ions (see Fig. 13.1). This can also be qualitatively explained by geometrical

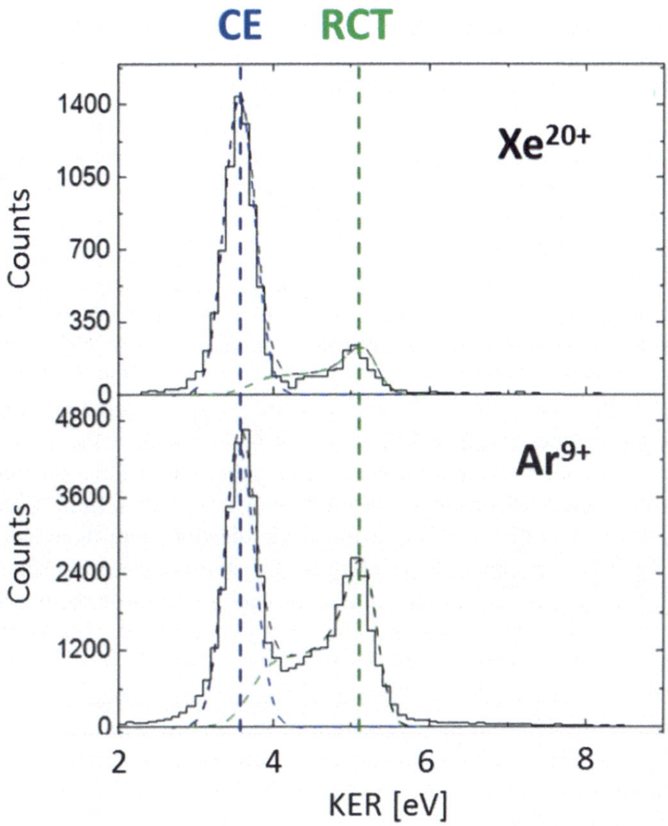

Fig. 13.1 KER spectra corresponding to $(1, 1)_F$ fragmentation channel from the $Xe^{20+} + Ar_2$ (upper panel) and $Ar^{9+} + Ar_2$ (lower panel) collision systems. Vertical dashed lines indicate the central position of the CE and RCT peaks. Dashed curves correspond to theoretical calculations of the KER distribution associated to the CE and RCT processes as described in [4]

considerations taking into account that Xe^{20+} projectiles offer larger capture radii than Ar^{9+} (see Table 13.2).

13.4 $(N_2)_2$ Nitrogen Dimers

Van der Waals nitrogen $(N_2)_2$ dimers are also bound by the induced dipole interaction between the two non-polar N_2 molecules. The intermolecular distance is predicted to be about 4 Å but the three dimensional conformation of the two molecules is still not well known [8]. Quantum chemistry calculations of the potential energy surface of the neutral N_2 dimer have shown that it might exist several isomeric conformations with very close interaction energies [9, 10].

We measured the KER distribution of the dimer fragmentation $(N_2)_2^{2+} \rightarrow N_2^+ + N_2^+$ following two-site single electron capture. The spectrum exhibits one main peak located at 3.35 eV associated to an equilibrium intermolecular distance of $R_e = 4.3$ Å. This value would more likely correspond to either a S or T shape structure according to theoretical predictions [9, 10].

In comparison with atomic dimers, when several electron are removed from one site, the resulting molecular ion also likely undergoes fragmentation. Such a molecular dimer thus constitutes a clean experimental system for the investigation of the role of the environment on molecular fragmentation. Here we focus on the three-body fragmentation channel $(N_2)_2^{3+} \rightarrow N_2^+ + N^+ + N^+$ resulting from a triple electron capture event where one electron is removed from the first molecule and two electrons from the second molecule. As most of the N_2^{2+} dication states are dissociative [5, 11], the multi ionized target will mainly undergo a concerted fragmentation in which both the van der Waals and the covalent bonds break at the same time. In order to get information on the influence of a neighbor ion on the fragmentation of the N_2^{2+} dication, the KER spectra of the three-body dissociation from dimer targets has been compared to the KER spectrum of the dication from monomer targets: $N_2^{2+} \rightarrow N^+ + N^+$. A complete description of this study is given in [12]. It shows that the three-body spectrum exhibits the same peak structure as the one from the monomers but is shifted towards higher energies by about 6.7 eV. This shift is interpreted using a pure Coulomb model where all three ions are considered as point charges located initially at their equilibrium distance from each other. The additional potential energy results from the interaction of the N_2^+ ion with the two N^+ ions. For this particular fragmentation channel, the expected extra potential energy calculated within this Coulomb model is: $2 \times 3.35 = 6.7\,\text{eV}$, in perfect agreement with the experimental measurement [12].

13.5 Conclusion

Low energy multiply charged ions are a powerful tool to produce multiply ionized targets. Here, collisions with rare gas Ar_2 dimers have allowed to shed light on both the electron capture process and on cluster specific relaxation mechanisms. For such a collision system, projectiles mainly capture electrons from the nearest atom. Moreover, the low electron mobility along the dimer allows to keep the memory of the initial capture process and lead to the predominance of asymmetric fragmentation as predicted by MC-COBM calculations. Diatomic molecular dimers offer a rather simple system to study the effect of a surrounding environment on the molecular fragmentation dynamics. For $(N_2)_2$ molecular dimers, it has been demonstrated that the presence of a neighbor N_2^+ ion does not affect much the molecular fragmentation of the dication and that the initial capture process populates the same electronic states of the dication as for monomer N_2 targets. The present study will be extended to other diatomic molecular dimers such as $(CO)_2$ in which concerted and sequential fragmentation has recently been observed [13].

References

1. Walch, B., et al.: Phys. Rev. Lett. **72**, 1439 (1994)
2. Matsumoto, J., et al.: Phys. Scr. **144**, 014016 (2011)
3. Tang, K.T., Toennies, J.P.: J. Chem. Phys. **118**, 4976 (2003)
4. Matsumoto, J., et al.: Phys. Rev. Lett. **105**, 263202 (2010)
5. Remscheid, A., et al.: J. Phys. B: At. Mol. Opt. Phys **29**, 515 (1996)
6. Iskandar, W., et al.: Phys. Rev. A **98**, 012701 (2018)
7. Iskandar, W., et al.: Phys. Rev. Lett. **110**, 143201 (2014)
8. McKellar, A.R.W.: J. Chem. Phys. **88**, 4190 (1988)
9. Aquilanti, V., et al.: J. Chem. Phys. **117**, 615 (2002)
10. Gomez, L., et al.: Chem. Phys. Lett. **445**, 99 (2007)
11. Baldit, E., et al.: Phys. Rev. A **71**, 021403 (2005)
12. Méry, A., et al.: Phys. Rev. Lett. **118**, 233402 (2017)
13. Ding, X., et al.: Phys. Rev. Lett. **118**, 153001 (2017)

Chapter 14
Precision Calculations for Three-Body Molecular Bound States

Jean-Philippe Karr, Mohammad Haidar, Laurent Hilico and Vladimir I. Korobov

Abstract Although they do not lend themselves to analytical resolution, three-body atomic or molecular systems are still simple enough to allow for very precise theoretical predictions of their energy levels, which makes them attractive candidates for fundamental tests and determination of fundamental physical constants. Focusing on the hydrogen molecular ions (H_2^+, HD^+, D_2^+), we outline the methods which have been used to improve the theoretical accuracy by several orders of magnitude over the last two decades. The three-body Schrödinger equation can be solved with extreme precision by variational methods with trial functions involving exponentials of interparticle distances. Quantum electrodynamics (QED) corrections are evaluated in the framework of nonrelativistic QED (NRQED). The current status of theory and possibilities of further improvement are briefly sketched.

14.1 Introduction

Precision spectroscopy of two-body (hydrogenlike) atoms, combined with QED calculations of their energy levels, has been a very successful way to test fundamental physics at a low-energy scale and has led to precise determinations of the Rydberg constant and proton charge radius [1]. In the last decade, the still unresolved discrepancy between results from H spectroscopy [2, 3], muonic hydrogen spectroscopy [4],

J.-P. Karr · M. Haidar · L. Hilico
Laboratoire Kastler Brossel, Sorbonne Université, CNRS, ENS -Université PSL, Collége de France, 4 place Jussieu, 75005 Paris, France

J.-P. Karr (✉) · L. Hilico
Université d'Evry-Val d'Essonne, Université Paris Saclay,
Boulevard François Mitterrand, 91025 Evry, France
e-mail: karr@lkb.upmc.fr
URL: http://www.lkb.upmc.fr/iontrap/

V. I. Korobov
Bogoliubov Laboratory of Theoretical Physics, Joint Institute for Nuclear Research,
141980 Dubna, Russia

N. A. Orr et al. (eds.), *Recent Progress in Few-Body Physics*,
Springer Proceedings in Physics 238,
https://doi.org/10.1007/978-3-030-32357-8_14

and electron-proton scattering experiments [5], known as the "proton-radius puzzle", has been a subject of intense activity. Three-body atoms or molecules have rich potential for further investigations in this field. Although theoretically more complex, they can have experimentally favorable features like the existence of narrow transitions, and can be sensitive to different physics or fundamental constants. For example, spectroscopy of antiprotonic helium has been used to determine the antiproton-to-electron mass ratio and test the *CPT* symmetry [6], and experiments in pionic helium [7] are underway to determine the charged pion mass. Measurements in He have recently allowed extracting the $^3He{-}^4$He nuclear charge radius difference [8, 9], to be compared with the value deduced from muonic helium spectroscopy [10]; future theoretical progress would allow extraction of individual radii [11]. Spectroscopy of ro-vibrational transitions in the hydrogen molecular ions (HMI), H_2^+ and its isotopes HD^+ and D_2^+, can be used for determination of the proton-electron and deuteron-electron mass ratios [12–14], and, if an appropriate set of transitions is measured, of the proton and deuteron charge radii and Rydberg constant [15]. Measurements in the antihydrogen molecular ion $\bar{H}_2^-$ compared with its normal matter counterpart have also been proposed for improved *CPT* symmetry tests [16].

The main purpose of this paper is to introduce the reader to the theoretical methods used to calculate the energy levels of three-body Coulomb bound states with very high accuracy, with emphasis on the case of HMI. The first part is devoted to the Schrödinger equation and its resolution by variational methods, and the second part deals with the calculation of QED corrections. An outline of the NRQED approach is given, after which the current status of theoretical predictions in HMI and perspectives of further improvement are briefly discussed.

14.2 Variational Solutions of the Schrödinger Equation

HMI play the role of benchmark systems in quantum chemistry, and the calculation of their nonrelativistic energy levels have been studied in hundreds of theoretical papers. Calculations were initially performed in the framework of the Born-Oppenheimer approximation; nonadiabatic calculations appeared in the 1970s [17], motivated by the first precise spectroscopic measurements in HD^+ [18]. Since then, refinements in theoretical methods and constant increase of available computing power have allowed improving the precision by many orders of magnitude, as illustrated in Table 14.1 for the ground-state energy of H_2^+. A majority of the most accurate results have been obtained using the variational approach and different variants of "exponential" basis sets [19–23], which we will now present.

Let us consider a rovibrational state of a HMI supported by the ground $1s\sigma_g$ electronic curve. Its total spatial parity is $\Pi = (-1)^L$ where L is the rotational quantum number. The wavefunction ψ^{Π}_{LM} of such a state may be written using the following separation of angular and radial degrees of freedom [24]:

Table 14.1 Selected theoretical results for the nonrelativistic energy of the H_2^+ ground state (i.e. the ($v = 0, L = 0$) rovibrational state supported by the $1s\sigma_g$ electronic curve). The CODATA 1986 value of the mass ratio, $m_p/m_e = 1836.152\,701$, was used in all works excepted for [17] where $m_e/m_p = 5.446\,17\,10^{-4}$. N is the number of terms in the expansion of the wave function. Stars (*) signal cases where the eigenvalue problem in (14.6) involves sparse-band matrices, which greatly reduces its complexity

Author (year)	References	Method	N	Energy (a.u.)
Bishop (1977)	[17]	Var. elliptic	515	−0.5971390625
Moss (1993)	[25]	Transformed H		−0.59713906312(5)
Grémaud (1998)	[26]	Var. perimetric	31746*	−0.597139063123(1)
Moss (1999)	[27]	Var. elliptic		−0.5971390631234(1)
Hilico (2000)	[28]	Var. perimetric	66046*	−0.59713906312340(1)
Korobov (2000)	[19]	Var. exponential	2200	−0.597139063123405074
Bailey (2002)	[20]	Var. exponential	3500	−0.59713906312340507483
Cassar (2004)	[21]	Var. exponential	1052	−0.5971390631234050748334338(3)
Li (2007)	[22]	Var. exponential	8381	−0.597139063123405074834134096026(5)
Hijikata (2009)	[29]	Free complement	19286	−0.5971390631234050748341340960260
Ning (2014)	[23]	Var. exponential	3806	−0.5971390631234050748341340960261899(1)

$$\psi_{LM}^{\Pi}(\mathbf{R},\mathbf{r_1}) = \sum_{l_1+l_2=L} \mathcal{Y}_{LM}^{l_1 l_2}(\mathbf{R},\mathbf{r_1})\, G_{l_1 l_2}^{L\Pi}(R,r_1,r_2), \tag{14.1}$$

$$\mathcal{Y}_{LM}^{l_1 l_2}(\mathbf{R},\mathbf{r_1}) = r_1^{l_1} r_2^{l_2} \left\{Y_{l_1} \otimes Y_{l_2}\right\}_{LM}. \tag{14.2}$$

Here, $\mathbf{r_i}$ ($i = 1, 2$) is the position of the electron with respect to nucleus i, and $\mathbf{R} = \mathbf{r_1} - \mathbf{r_2}$ is the internuclear vector. The radial functions $G_{l_1 l_2}^{L\Pi}(R, r_1, r_2)$ are then expanded in a basis set involving exponentials of inter-particle distances. Two different types of expansion have been used with particular success. The first one uses pure exponential functions [19, 20]:

$$G_{l_1 l_2}^{L\Pi}(R,r_1,r_2) = \sum_{n=1}^{N_{l_2}} \left[C_n \mathrm{Re}(e^{-\alpha_n R-\beta_n r_1-\gamma_n r_2}) + D_n \mathrm{Im}(e^{-\alpha_n R-\beta_n r_1-\gamma_n r_2})\right] \tag{14.3}$$

where the exponents α_n, β_n, γ_n are complex numbers. In practice, it is essential to use complex α_n in order to reproduce the oscillatory behavior of the vibrational part of the wavefunction, but β_n and γ_n can be kept real. The real and imaginary parts of exponents are generated pseudo-randomly in several intervals, the bounds of which play the role of variational parameters and need to be optimized. The second type of expansion is [21, 23, 30]

$$G_{l_1 l_2}^{L\Pi}(R,r_1,r_2) = \sum_{p=1}^{2} \sum_{i,j=0}^{\Omega} \sum_{k=\Omega_{\mathrm{low}}}^{\Omega_{\mathrm{high}}} C_{ijk}^{(p)}\, r_1^i r_2^j R^k\, e^{-\alpha_p R-\beta_p r_1-\gamma_p r_2} \tag{14.4}$$

where all exponents are real, and their values are fully optimized through calculation of the first derivatives of the energy with respect to α_p, β_p, γ_p.

According to the basic variational theorem, the quantity

$$E = \frac{\langle \psi | H | \psi \rangle}{\langle \psi | \psi \rangle} \tag{14.5}$$

provides an upper bound for the exact ground-state energy E_0. Finding the extrema of E with respect to the linear parameters (C_n, D_n in (14.3) or $C_{ijk}^{(p)}$ in (14.4)) is equivalent to solving the generalized eigenvalue problem

$$\mathbf{A}c = \lambda \mathbf{B}c \tag{14.6}$$

where c is a column vector of coefficients ($\psi = \sum_{i=1}^{N} c_i \psi_i$), $\mathbf{A}$ the Hamiltonian matrix ($A_{ij} = \langle \psi_i | H | \psi_j \rangle$), and $\mathbf{B}$ the overlap matrix ($B_{ij} = \langle \psi_i | \psi_j \rangle$). The lowest eigenvalue λ_0 is an upper bound to E_0; a less widely known property called the Hylleraas-Undheim-MacDonald theorem [31, 32], crucial for the applicability of variational methods to excited states, is that the other eigenvalues $\lambda_1, \lambda_2 \ldots$ are also upper bounds to the exact energies $E_1, E_2 \ldots$.

An essential property of the above "exponential" expansions allowing to get high-precision results is that the matrix elements of the Hamiltonian, as well as those appearing in calculation of relativistic and QED corrections, can be evaluated analytically [30, 33, 34].

The efficiency of variational methods is not limited to the first few excited states, contrary to a common misconception. For example, in the recent work [35], the energies of all bound and quasibound states of H_2^+ supported by the $1s\sigma_g$ electronic curve were calculated with an uncertainty of $10^{-7}\,\text{cm}^{-1}$, using the complex coordinate rotation method for quasibound states. An important technical point is that the quasi-adiabaticity of HMI greatly helps reducing the complexity of the calculation for high rotational states as it allows restricting the sum in (14.1) to low values of l_2 (which is related to the electronic orbital momentum).

One may wonder what kind of accuracy is actually required for fundamental metrology applications. Typically, an uncertainty of 10^{-20} a.u. on the nonrelativistic energy level allows calculation of leading-order (α^2) corrections with 10 significant digits, and with an absolute uncertainty $< 10^{-14}$ a.u. The resulting relative uncertainty on rovibrational transition frequencies ($\nu \sim 0.01$ a.u.) is $< 10^{-12}$, which is still smaller than the uncertainty from unevaluated high-order QED corrections (see next Section). Since such (or even better) accuracies can be obtained with modest computing resources, the nonrelativistic three-body bound-state problem may be regarded as solved from a practical viewpoint.

14.3 Calculation of QED Corrections

We now give a basic introduction to nonrelativistic quantum electrodynamics (NRQED) and its application to HMI. The NRQED approach, originally proposed by Caswell and Lepage [36] and further developed by other authors (see e.g. [37–39]) is a powerful tool to study QED corrections in weakly bound few-body systems. In brief, it consists in constructing from QED a nonrelativistic Lagrangian describing the interaction of an electron or a nucleus with the electromagnetic field, and then using it to calculate the QED corrections by applying the nonrelativistic perturbation theory. One way of constructing the NRQED Lagrangian [36, 37] is to write all possible interactions satisfying the required symmetries, such as gauge invariance, parity invariance, time reversal, Galilean invariance, hermiticity, and locality. In principle this is an infinite expansion, but it may be truncated according to the order of the correction one wants to calculate. Its first few terms are

$$\mathcal{L}=\psi^{\dagger}\left\{iD_t+\frac{\mathbf{D}^2}{2m}+\frac{\mathbf{D}^4}{8m^3}+c_F q\frac{\boldsymbol{\sigma}\cdot\mathbf{B}}{2m}+c_D q\frac{\nabla\cdot\mathbf{E}}{8m^2}+ic_S q\frac{\boldsymbol{\sigma}\cdot(\mathbf{D}\times\mathbf{E}-\mathbf{E}\times\mathbf{D})}{8m^2}\right\}\psi \tag{14.7}$$

where q, m are the particle's mass and charge, $D_t = \partial_t + iqA_0$, and $\mathbf{D} = \nabla - iq\mathbf{A}$.

The coefficients of this Lagrangian are regularized by introducing a cutoff on the photon momentum of the order of the electron's rest energy. Physics at relativistic energy scales is incorporated in the theory in the form of contact terms. Finally, the coefficients of the Lagrangian are fixed by imposing that the NRQED and QED scattering amplitudes coincide up to the desired order—this is the so-called "matching" procedure. An alternative approach [38, 39] is to obtain the NRQED Hamiltonian directly from the Dirac Hamiltonian through a Foldy-Wouthuysen transformation.

It is important to note that the matching is the only stage of the method which involves calculation of QED diagrams. It is done using only the scattering of free particles, and does not involve any bound states. This separation of the matching from the bound-state calculations is a key simplification allowed by the NRQED approach.

After the Lagrangian has been constructed, the next step is to apply the nonrelativistic perturbation theory. This can be formalized in terms of NRQED Feynman rules similarly to QED (see Fig. 3 in [37]). The number of interaction vertices is much higher than in QED since there are many terms in the NRQED Lagrangian, but each diagram is much simpler to calculate. To illustrate this procedure, all diagrams contributing to the leading-order (Breit-Pauli) hyperfine Hamiltonian of HMI are shown in Fig. 14.1. The interaction potential corresponding to each diagram is directly obtained in impulse space through application of NRQED Feynman rules, and Fourier transform gives the potentials in coordinate space (see equation (2) of [40]).

Application of the NRQED approach to HMI has allowed calculation of leading-order relativistic and radiative corrections at the $m\alpha^4$ and $m\alpha^5$ orders [42]. Among these, the most difficult contribution to evaluate is the Bethe logarithm, which may

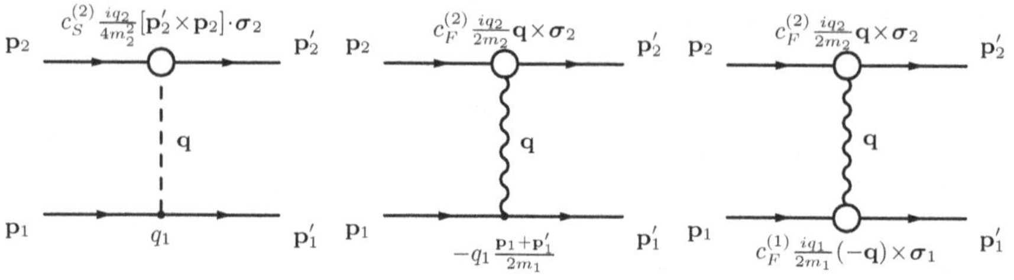

Fig. 14.1 Leading-order NRQED diagrams contributing to the hyperfine structure of HMI. Particles 1 and 2 may be either an electron or one of the nuclei. Left: Coulomb photon exchange (dashed line) where 1 and 2 respectively interact via Coulomb and spin-orbit vertices. Center: transverse photon exchange (wiggly line) where 1 and 2 interact via dipole and Fermi vertices. Right: transverse photon exchange where both particles interact via a Fermi vertex. The photon impulse is $\mathbf{q} = \mathbf{p}_2' - \mathbf{p}_2$. The matching with QED yields $c_F^{(e)} = 1 + a_e$, $c_S^{(e)} = 1 + 2a_e$ for an electron [37], and $c_F^{(N)} = Z + a_N$, $c_S^{(N)} = Z + 2a_N$ for a spin $1/2$ nucleon of charge Z [41], where a_e (resp. a_N) is the anomalous magnetic moment of the electron (resp. nucleus)

be expressed as a sum over intermediate states converging very slowly as the maximal energy of the states included in the sum is increased. Its evaluation therefore requires an accurate representation of scattering states lying high in the continuum. An efficient numerical scheme for calculating the Bethe logarithm is presented in [43].

Corrections of order $m\alpha^6$, $m\alpha^7$, and (partially) $m\alpha^8$ have been calculated in the framework of the adiabatic approximation [44–46], including only corrections to the bound electron in a two-center potential. Spin-averaged rovibrational transition frequencies are now predicted with a relative uncertainty of 7.6 10^{-12} [46], which already allows for an improved determination of the proton-electron mass ratio if a measurement of similar accuracy is performed in ongoing Doppler-free spectroscopy experiments. Further progress in accuracy by a factor 2–3 would allow an independent cross-check of the values of the Rydberg constant and nuclear radii [15].

The highest precision reached in experiments so far is 3.8 10^{-10} on the fundamental rotational transition in HD^+ [13]. In this case, a single hyperfine component was measured, and comparison with theory is limited by hyperfine structure calculations [13, 47]. Calculation of $m\alpha^6(m/M)$ order corrections to the spin-orbit and spin-spin tensor interactions in the three-body framework is currently in progress to improve this, following previous work on the spin-spin contact Fermi interaction [48]. Spin-averaged interaction potentials at the same order have been recently derived [49].

Regarding spin-averaged transition frequencies, the largest source of theoretical uncertainty is the one-loop self-energy contribution at the $m\alpha^8$ order [35] which has not been calculated yet even in hydrogenlike systems. One possible way forward would be to calculate the one-loop self-energy of a bound electron in a two-center potential, without performing the expansion in powers of the binding potential, as previously done in hydrogenlike atoms [50, 51].

References

1. Mohr, P.J., et al.: Rev. Mod. Phys. **88**, 035009 (2016)
2. Beyer, A., et al.: Science **358**, 79 (2017)
3. Fleurbaey, H., et al.: Phys. Rev. Lett. **120**, 183001 (2018)
4. Antognini, A., et al.: Science **339**, 417 (2013)
5. Arrington, J., Sick, I.: J. Phys. Chem. Ref. Data **44**, 031204 (2015)
6. Hori, M., et al.: Science **354**, 610 (2016)
7. Hori, M., et al.: Phys. Rev. A **89**, 042515 (2014)
8. Zheng, X., et al.: Phys. Rev. Lett. **119**, 263002 (2017)
9. Rengelink, R.J., et al.: Nat. Phys. **14**, 1132 (2018)
10. Antognini, A., et al.: EPJ Web Conf. **113**, 01006 (2016)
11. Pachucki, K., et al.: Phys. Rev. A **95**, 062510 (2017)
12. Biesheuvel, J., et al.: Nat. Commun. **7**, 10385 (2016)
13. Alighanbari, S., et al.: Nat. Phys. **14**, 555 (2018)
14. Karr, J-Ph, et al.: J. Phys. Conf. Ser. **723**, 012048 (2016)
15. Karr, J-Ph, et al.: Phys. Rev. A **94**, 050501(R) (2016)
16. Myers, E.G.: Phys. Rev. A **98**, 010101 (2018)
17. Bishop, D.M., Cheung, L.M.: Phys. Rev. A **16**, 640 (1977)
18. Wing, W.H., et al.: Phys. Rev. Lett. **36**, 1488 (1976)
19. Korobov, V.I.: Phys. Rev. A **61**, 064503 (2000)
20. Bailey, D.H., Frolov, A.M.: J. Phys. B At. Mol. Opt. Phys. **35**, 4287 (2002)
21. Cassar, M.M., Drake, G.W.F.: J. Phys. B At. Mol. Opt. Phys. **37**, 2485 (2004)
22. Li, H., et al.: Phys. Rev. A **75**, 012504 (2007)
23. Ning, Y., Yan, Z.-C.: Phys. Rev. A **90**, 032516 (2014)
24. Schwartz, C.: Phys. Rev. **123**, 1700 (1961)
25. Moss, R.E.: Mol. Phys. **80**, 1541 (1993)
26. Grémaud, B., et al.: J. Phys. B At. Mol. Opt. Phys. **31**, 383 (1998)
27. Moss, R.E.: J. Phys. B At. Mol. Opt. Phys. **32**, L89 (1999)
28. Hilico, L., et al.: Eur. Phys. J. D **12**, 449 (2000)
29. Hijikata, Y., et al.: J. Chem. Phys. **130**, 024102 (2009)
30. Drake, G.W.F. (ed.): Springer handbook of atomic, molecular, and optical physics, 2nd edn. Springer, Berlin (2006)
31. Hylleraas, E.A., Undheim, B.: Z. Phys. **65**, 759 (1930)
32. MacDonald, J.K.L.: Phys. Rev. **43**, 830 (1933)
33. Korobov, V.I.: J. Phys. B At. Mol. Opt. Phys. **35**, 1959 (2002)
34. Harris, F.E., et al.: J. Chem. Phys. **121**, 6323 (2004)
35. Korobov, V.I.: Mol. Phys. **116**, 93 (2017)
36. Caswell, W., Lepage, G.: Phys. Lett. B **167**, 437 (1986)
37. Kinoshita, T., Nio, M.: Phys. Rev. D **53**, 4909 (1996)
38. Pachucki, K.: Phys. Rev. A **71**, 012503 (2005)
39. Jentschura, U.D., et al.: Phys. Rev. A **72**, 062102 (2005)
40. Korobov, V.I., et al.: Phys. Rev. A **74**, 040502(R) (2006)
41. Hill, R.J., et al.: Phys. Rev. D **87**, 053017 (2013)
42. Korobov, V.I.: Phys. Rev. A **74**, 052506 (2006)
43. Korobov, V.I.: Phys. Rev. A **85**, 042514 (2012)
44. Korobov, V.I.: Phys. Rev. A **77**, 022509 (2008)
45. Korobov, V.I., et al.: Phys. Rev. A **89**, 032511 (2014)
46. Korobov, V.I., et al.: Phys. Rev. Lett. **118**, 233001 (2017)
47. Bakalov, D., et al.: Phys. Rev. Lett. **97**, 243001 (2006)
48. Korobov, V.I., et al.: Phys. Rev. A **79**, 012501 (2009)
49. Zhong, Z.-X., et al.: Phys. Rev. A **98**, 032502 (2018)
50. Jentschura, U.D., et al.: Phys. Rev. Lett. **82**, 53 (1999)
51. Jentschura, U.D., et al.: Phys. Rev. A **63**, 042512 (2001)

Part II
Nuclear Physics and Astrophysics

Chapter 15
Tensor Correlations in α Clustering Studied with Antisymmetrized Quasi Cluster Model

Yoshiko Kanada-En'yo, Hideaki Matsuno and Naoyuki Itagaki

Abstract We investigate the tensor and short-range correlations in ^{4}He and ^{8}Be with the AQCM-T framework using a realistic nuclear interaction of the G3RS force and an effective interaction. The tensor suppression is found to contribute to the α-α repulsion at short distances.

The tensor correlation in many-nucleon systems is one of the long-standing problems in nuclear physics. Thanks to recent development of *ab initio* approaches, remarkable progress has been made in study of the tensor correlation. In physics of nuclear clustering, the tensor correlation is considered to play important roles in cluster structures. Various approaches have been developed to explicitly treat the tensor correlation in cluster states [1–6]. Recently, a new framework using imaginary centroid Gaussians has been proposed to efficiently deal with the tensor correlation in ^{4}He firstly by Itagaki et al. [5], and later by Myo et al. [6]. The framework is a version of antisymmetrized quasi cluster model (AQCM), which has been developed by one of the authors (N. I.) and extended for the tensor correlation (see [5] and references therein). Very recently, we newly proposed an improved version of the AQCM so as to explicitly treat the tensor correlation in the two-nucleon pairs [7] and call it "AQCM-T". It is also regarded as a specific version of the antisymmetrized molecular dynamics (AMD) and called high-momentum AMD by Myo et al. [6]. In this paper, we apply the AQCM-T method to ^{4}He and ^{8}Be using a realistic *NN* interaction of the G3RS [8] containing the tensor part and a repulsive core for the central part, and compare results of a new effective *NN* interaction with the tensor interaction proposed in [7] to discuss the *NN* tensor and short-range correlations in α clusters.

Y. Kanada-En'yo (✉) · H. Matsuno
Department of Physics, Kyoto University, Kitashirakawa Oiwake -Cho, Kyoto 606-8502, Japan
e-mail: yenyo@ruby.scphys.kyoto-u.ac.jp

N. Itagaki
Yukawa Institute for Theoretical Physics, Kyoto University, Kitashirakawa Oiwake-Cho, Kyoto 606-8502, Japan

N. A. Orr et al. (eds.), *Recent Progress in Few-Body Physics*, Springer Proceedings in Physics 238,
https://doi.org/10.1007/978-3-030-32357-8_15

The procedures of the calculation based on AQCM-T is, in principle, the same as those presented in [7]. In order to take into account the *NN* correlations, a basis wave function for ^{4}He is given with the parameter k in the momentum vector $\boldsymbol{k} = (0, 0, k)$ and the channel $\beta = \{^1S, ^3S, \text{and } ^3D\}$ as

$$\Phi^{\text{AQCM-T}}_{^4\text{He},0^+}(k, \beta) \ = \widehat{P}^{0+}\mathcal{A}\{\phi_{\frac{ik}{2\nu}}\phi_{-\frac{ik}{2\nu}}\phi_0\phi_0 \otimes \mathcal{X}_\beta\}, \tag{15.1}$$

where the single-particle wave function $\phi_{\boldsymbol{Z}}$ is the imaginary-centroid Gaussian wave packet (see [7] for the detailed definition). The total wave function of ^{4}He is expressed by linear combination of various k values and β channels.

For ^{8}Be we place an α cluster at $\boldsymbol{R} = \frac{\boldsymbol{d}}{2}$ and the other α at $\boldsymbol{R}' = -\frac{\boldsymbol{d}}{2}$ with $\boldsymbol{d} = (d\sin\theta_\alpha, 0, d\cos\theta_\alpha)$. The total wave function of 2α is given with a correlated α (α_k) wave function and an uncorrelated α (α_0) one as

$$\Phi^{\text{AQCM-T}}_{2\alpha,0^+}(k, \beta, \boldsymbol{d}) = \widehat{P}^{0+}\mathcal{A}\left\{\Phi_{\alpha_k}(k, \beta, \boldsymbol{R})\Phi_{\alpha_0}(\boldsymbol{R}')\right\}, \tag{15.2}$$

where the α_k is expressed using the AQCM-T wave function for ^{4}He as

$$\Phi_{\alpha_k}(k, \beta, \boldsymbol{R}) = \frac{1 + \widehat{P}_k}{2}\mathcal{A}\{\phi_{\boldsymbol{R}+\frac{ik}{2\nu}}\phi_{\boldsymbol{R}-\frac{ik}{2\nu}}\phi_{\boldsymbol{R}}\phi_{\boldsymbol{R}} \otimes \mathcal{X}_\beta\}. \tag{15.3}$$

Here the operator $\widehat{P}_k$ is the transformation operator of the intrinsic parity of the α_k cluster. It should be noted that the present AQCM-T wave function for ^{8}Be is equivalent to the symmetric state $\alpha_0(\boldsymbol{R}) \otimes \alpha_k(\boldsymbol{R}') + \alpha_k(\boldsymbol{R}') \otimes \alpha_0(\boldsymbol{R})$ of two α clusters because of the total parity projection. In order to obtain the total wave function of ^{8}Be, we perform calculation of the generator coordinate method (GCM) by superposing all the AQCM-T wave functions with various k, θ_α, β, and d values. We also perform the fixed-d calculation by superposing various k, θ_α, β for a given value of d (the relative distance) .

We use the G3RS interaction of "case 1" (G3RS1) and "case 2" (G3RS2) parameters. The G3RS1 (G2RS2) is designed so as to reproduce the *NN*-scattering phase shift up to 600 MeV (150 MeV) and its central-even parts have repulsive cores of 2000 MeV (500–630 MeV) height. We also use an effective interaction of V2m-3R introduced in [7] containing the tensor part but no repulsive core for the central part. For the tensor term of the G3RS1, the radial function is approximated by the three-range Gaussian function [7]. We use the same function of the tensor part for cases of the G3RS2 and V2m interactions.

Table 15.1 shows the results of ^{4}He. For ^{4}He, the G3RS2 and V2m interactions give similar results for contribution of the tensor interaction and D-state probability. Figure 15.1 shows the pair wave functions $\phi^{\beta}_{NN}(r)$ for the $\beta = {}^3S$ and ^3D channels. In the G3RS1 and G3RS2 results, $\phi^{^3D}_{NN}(r)$ shows a peak in the middle-range region, $r \sim 1$ fm. $\phi^{^3S}_{NN}(r)$ is also peaked in the middle-range ($r \sim 1$ fm) region corresponding to the potential pocket of the central-even interaction but strongly suppressed in the $r < 1$ fm region because of the short-range repulsion. The short-range

Table 15.1 Energies (total energy, kinetic, central, and tensor terms) and $0s$ and D-state probabilities ($\mathcal{P}_{0s,{}^3D}$) of ^{4}He obtained by AQCM-T. The result of the $(0s)^4$ configuration with the Volkov No. 2 ($m = 0.60$) interaction (an effective interaction without the tensor interaction) is also shown for comparison (labeled by V2:0s).

	V2m	G3RS2	G3RS1	V2:0s	Exp.
E (MeV)	−30.3	−26.5	−16.2	−27.9	−28.296
T (MeV)	64.6	72.3	70.9	46.7	
V_c (MeV)	−56.7	−58.4	−54.6	−75.3	
V_t (MeV)	−39.9	−41.7	−33.9		
$\mathcal{P}_{0s}$	0.901	0.881	0.897		
$\mathcal{P}_{^3D}$	0.077	0.082	0.063		

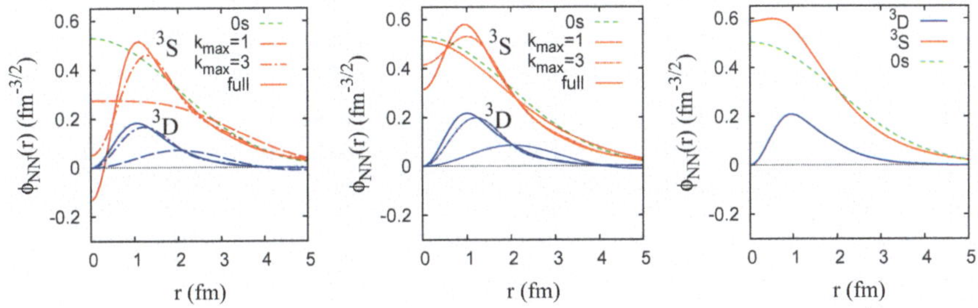

(a) G3RS1 (b) G3RS2 (c)V2m

Fig. 15.1 Pair wave functions $\phi_{NN}(r)$ of ^{4}He(g.s.) for the ^3D, ^3S, and $0s$ component obtained with the full configurations. The result obtained with the truncated model space of the momentum parameter $k \le k_{\max}$ for the cases of $k_{\max} = 1$ fm^{-1} and 3 fm^{-1} are also shown in the G3RS1 and G3RS2 results.

suppression and middle-range peak in the ^3S channel is a unique feature of the realistic interactions, G3RS1 and G3RS2, but cannot be seen in the V2m-3R result for the case of effective interactions with no repulsive core of the central part. We found that medium momentum ($k \sim 2$ fm^{-1}) components contribute to the tensor correlation in the ^3D channel, whereas high-momentum ($k = 3 \sim 4$ fm^{-1}) components are essential for the short-range correlation in the $^{1,3}S$ channel as seen in the results of the momentum truncation $k \le k_{\max}$ shown in Fig. 15.1a and b.

Table 15.2 shows the results of the fixed-d and GCM calculations of ^{8}Be obtained with the G3RS2 and V2m interactions. At short distances, the tensor suppression occurs as can be seen in the reduction of the tensor energy and ^3D probability. This tensor suppression gives a repulsive effect to the two-α system in the region of the distance $d \lesssim 2$ fm. The G3RS2-3R result shows additional repulsive effect in the same d region, which comes from the reduction of the central-even interaction. Such the repulsive effect from the central-even interaction cannot be seen in V2m-3R without the repulsive core of the central part.

The total energy of ^{8}Be obtained with the GCM calculation is 5–8 MeV higher than the 2α threshold energy. This overestimation of the total energy mainly originates in

Table 15.2 Energies (MeV), and 0s and ^3D probabilities of ^{8}Be obtained with the fixed-d and GCM calculations. For the values at $d = 8$ fm, the total and kinetic energies after the subtraction of the relative kinetic ($\hbar\omega/4 = 5.2$ MeV) energy and Coulomb energy (0.72 MeV) are also shown in parentheses

Interaction	d (fm)	E	T	V_c	V_t	$\mathcal{P}_{0s}$	$\mathcal{P}_{^3D}$
G3RS2	2	−2.7	157.8	−121.0	−43.8	0.89	0.09
	4	−39.0	144.5	−117.6	−69.9	0.84	0.12
	8	−40.7 (−46.6)	139.2 (134.0)	−113.4	−69.7	0.84	0.12
	GCM	−44.2	136.2	−114.1	−69.8	0.84	0.12
	2α thres.	−52.0	147.2	−118.3	−83.6	0.78	
V2m	2	−18.6	149.1	−132.3	−40.0	0.90	0.09
	4	−47.4	132.3	−119.7	−64.4	0.86	0.12
	8	−47.3 (−53.2)	126.7 (121.5)	−113.6	−64.1	0.86	0.12
	GCM	−51.3	124.2	−115.4	−64.2	0.86	0.12
	2α thres.	−56.4	129.7	−114.9	−74.5	0.82	
V2:0s	2	−45.7	132.1	−181.4	0	1	0
	4	−53.0	104.1	−160.2	0	1	0
	8	−49.9 (−55.8)	98.5 (93.3)	−150.7	0	1	0
	GCM	−55.5	100.7	−159.2	0	1	0
	2α thres.	−55.8	93.3	−150.7	0	1	0

the overestimation of the 2α energy in the asymptotic region in the present framework. The present model for ^{8}Be omits the second-order correlation, where both of two α clusters contain the D-state component, even though it significantly contains correlations beyond $\alpha_0(\boldsymbol{R}) \otimes \alpha_k(\boldsymbol{R}')$, where one of the α clusters stays merely as α_0. Because of the missing part, it is not be able to completely describe the exact solution of the asymptotic 2α state with the present 2α wave function, and the asymptotic 2α energy is artificially raised by several MeV from the value evaluated by twice of the energy of the ^{4}He ground state.

In summary, we studied the tensor and short-range correlations in ^{4}He and ^{8}Be with AQCM-T and showed important contributions of the NN correlations to the α-α repulsion at short distances.

Acknowledgements This work was supported by JSPS KAKENHI Grant Numbers 26400270 (Y. K-E.), 17K05440 (N. I.), 18J13400 (H. M.), and 18K03617 (Y. K-E.).

References

1. Neff, T., Feldmeier, H.: Nucl. Phys. A **713**, 311 (2003)
2. Dote, A., Kanada-En'yo, Y., Horiuchi, H., Akaishi, Y., Ikeda, K.: Prog. Theor. Phys. **115**, 1069 (2006)
3. Itagaki, N., Masui, H., Ito, M., Aoyama, S., Ikeda, K.: Phys. Rev. C **73**, 034310 (2006)

4. Myo, T., Toki, H., Ikeda, K., Horiuchi, H., Suhara, T.: PTEP **2015**(7), 073D02 (2015)
5. Itagaki, N., Tohsaki, A.: Phys. Rev. C **97**(1), 014304 (2018)
6. Myo, T., Toki, H., Ikeda, K., Horiuchi, H., Suhara, T., Lyu, M., Isaka, M., Yamada, T.: PTEP **2017**, 111 (2017)
7. Matsuno, H., Kanada-En'yo, Y., Itagaki, N.: Phys. Rev. C **98**(5), 054306 (2018)
8. Tamagaki, R.: Prog. Theor. Phys. **39**, 91 (1968)

Chapter 16
The First Unbound States in the $A = 9$ Mirror Nuclei ^{9}B and ^{9}Be

Myagmarjav Odsuren, Yuma Kikuchi, Takayuki Myo and Kiyoshi Katō

Abstract The structures of the first $1/2^+$ unbound states of $A = 9$ mirror nuclei are studied in the complex scaled $\alpha + \alpha + N$ three-body model. The results indicate that the $1/2^+$ state of ^{9}Be is a virtual state of the s-wave neutron around the ^{8}Be $(\alpha - \alpha)$ cluster. But the corresponding analog ^{9}B$(1/2^+)$ state is predicted as a resonant state due to the Coulomb barrier.

16.1 Introduction

Studies of mirror nuclei are important in understanding of the nuclear structure. However, in spite of considerable experimental [1–5] and theoretical [6] efforts, the location of the first excited state of the ^{9}B nuclei has not yet determined and the structures of the first excited $1/2^+$ states in mirror nuclei ^{9}Be and ^{9}B are still an open problem. The nuclei ^{9}Be and ^{9}B are considered to have an $\alpha + \alpha + N$ ($N =$ neutron or proton) structure according to the lowest three-body thresholds, and to be the Borromean systems. Except for the ground state $3/2^-$ of ^{9}Be, all states including the ground state of ^{9}B are unbound. Therefore, in order to study the first excited states of ^{9}Be and ^{9}B, it is necessary to calculate the $\alpha + \alpha + N$ three-cluster systems under a correct boundary condition for unbound states.

M. Odsuren (✉)
School of Engineering and Applied Sciences and Nuclear Research Center, National University of Mongolia, Ulaanbaatar 210646, Mongolia
e-mail: odsuren@seas.num.edu.mn

Y. Kikuchi
General Education, Tokuyama College, National Institute of Technology, Shunan 745-8585, Japan

T. Myo
General Education, Faculty of Engineering, Osaka Institute of Technology, Osaka 535-8585, Japan

K. Katō
Nuclear Reaction Data Centre, Hokkaido University, Sapporo 060-0810, Japan

N. A. Orr et al. (eds.), *Recent Progress in Few-Body Physics*, Springer Proceedings in Physics 238,
https://doi.org/10.1007/978-3-030-32357-8_16

Recently, we investigated the structure of the first unbound state of ^{9}Be located just above the $\alpha + \alpha + n$ threshold energy applying the $\alpha + \alpha + n$ three-body model [6] and the complex scaling method (CSM) [7, 8]. In the results, we found that the ^{9}Be$(1/2^+)$ state has a virtual-state character of an s-wave neutron around the ^{8}Be(0^+) cluster.

It is particularly interesting to study the ^{9}B$(1/2^+)$ state using the same model that reproduces the observed photo-disintegration cross section of ^{9}Be$(1/2^+)$ [9] on the basis of the picture of the virtual state for ^{9}Be$(1/2^+)$.

In this study, we discuss the structure of the first excited $1/2^+$ state of ^{9}B together with the mirror state in ^{9}Be. Considering the virtual and resonance characters of the states for $A = 9$ systems, we investigate the nature of first excited $1/2^+$ states of ^{9}Be and ^{9}B mirror nuclei.

16.2 Three-Body Model and Complex Scaling Method

The Hamiltonian for the relative motion of the $\alpha + \alpha + N$ three-body system is given as

$$\hat{H} = \sum_{i=1}^{3} t_i - T_{\text{c.m.}} + \sum_{i=1}^{2} V_{\alpha N}(\boldsymbol{\xi}_i) + V_{\alpha\alpha} + V_{\text{PF}} + V_3, \tag{16.1}$$

where t_i and $T_{\text{c.m.}}$ are kinetic energy operators for each particle and the center-of-mass of the system, respectively. The interactions between the valence nucleon and the i-th α particle is given as $V_{\alpha N}(\boldsymbol{\xi}_i)$ [10], where $\boldsymbol{\xi}_i$ is the relative coordinate between them. The inter-α potential is expressed by $V_{\alpha\alpha}$ [11], and we introduce the three-cluster potential V_3 to reproduce the energies of ^{9}Be states measured from the $\alpha + \alpha + n$ threshold energy. The Pauli-forbidden states for $\alpha - \alpha$ and $\alpha - N$ are projected out by using the pseudo-potential V_{PF}. The details of this model are explained in [6, 9].

In the present calculation, we use the $\alpha + \alpha + N$ three-body potential V_3 given as

$$V_3 = v_3 \exp(-\mu\rho^2), \tag{16.2}$$

where ρ is the hyperradius of the $\alpha + \alpha + N$ system, and defined as

$$\rho^2 = 2r^2 + \frac{8}{9}R^2, \tag{16.3}$$

where r is the distance between two α's and R is that between the nucleon and the center-of-mass of the $\alpha + \alpha$ subsystem.

We investigate the properties of the unbound states of $A = 9$ mirror nuclei using the CSM [7, 8]. The complex-scaled Schrödingier equation is expressed by using the complex-scaled Hamiltonian H^θ as

$$\hat{H}^\theta \Psi_J^\nu(\theta) = E_\nu^\theta \Psi_J^\nu(\theta). \tag{16.4}$$

The relative coordinates of the three-body system $\alpha + \alpha + N$ are transformed as $\mathbf{r} \to \mathbf{r}\,e^{i\theta}$ and $\mathbf{R} \to \mathbf{R}\,e^{i\theta}$ with a real parameter θ. By solving the complex-scaled Schrödingier equation with appropriate L^2-basis functions, we obtain the energy eigenvalues and eigenstates [8].

16.3 Results

The $3/2^-$ ground state of ^{9}Be is calculated at -1.57 MeV from the $\alpha + \alpha + n$ threshold with the repulsive three-body potential, whose parameters are given as $v_3 = 1.10$ MeV and $\mu = 0.02$ fm^2 in (16.2) [6]. The Hamiltonian in (16.1) reproduces all the threshold energies of three-body and two-body subsystems. Using this Hamiltonian, we obtain the ground state energy 0.277 MeV with the width 1.73×10^{-3} MeV of ^{9}B, which well correspond to the observed 0.277 MeV and 0.54×10^{-3} MeV [2], respectively. On the other hand, the $1/2^+$ resonance of ^{9}Be is not found in our previous work [6]. However, the photo-disintegration cross section from the ground $3/2^-$ to the unbound $1/2^+$ states of ^{9}Be is calculated as the sharp peak just above the ^{8}Be + n threshold, and well reproduces the observed distribution [12, 13]. The origin of the low-lying peak of the cross section above the ^{8}Be + n threshold is investigated to be a consequence of a ^{8}Be + n virtual state but not resonant one.

In this calculation we try to obtain the $1/2^+$ state in ^{9}B based on our well-defined methodology considering its analog state in ^{9}Be [6] as well. In Fig. 16.1, we show the distribution of the energy eigenvalues for the $1/2^+$ state calculated with $v_3 = -1.02$ MeV, which reproduces the photo-disintegration of ^{9}Be, in the $\alpha + \alpha + p$ model. We find a resonance solution at the energy $E^{res} = 2.42$ MeV with a decay width $\Gamma = 1.61$ MeV, which is presented with a triangle in Fig. 16.1. In addition to the resonance solution, the $\alpha + \alpha + p$ three-body, ^{8}Be$(0^+) + p$, ^{8}Be$(2^+) + p$ and ^{5}Li$(3/2^-) + p$ two-body continuum solutions are obtained separately.

We investigate the consistency between the results of the ^{9}Be$(1/2^+)$ and the experiments of ^{9}B. First we calculate the ^{9}B$(1/2^+)$ state without three-body potential

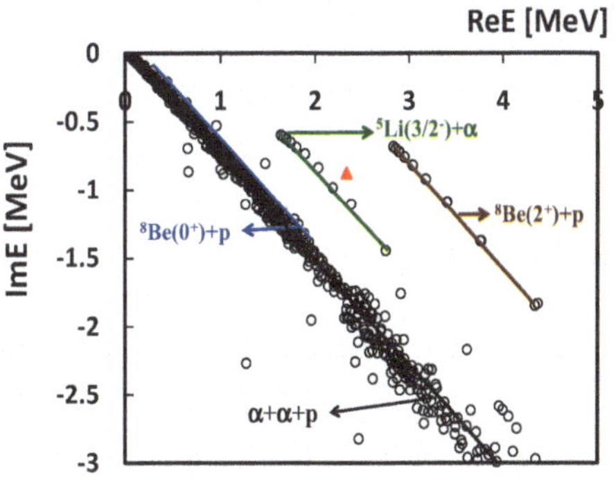

Fig. 16.1 Distribution of energy eigenvalues of the ^{9}B$(1/2^+)$ state. Red triangle displays the resonance. The continuum states of the three- and two-body subsystems are given

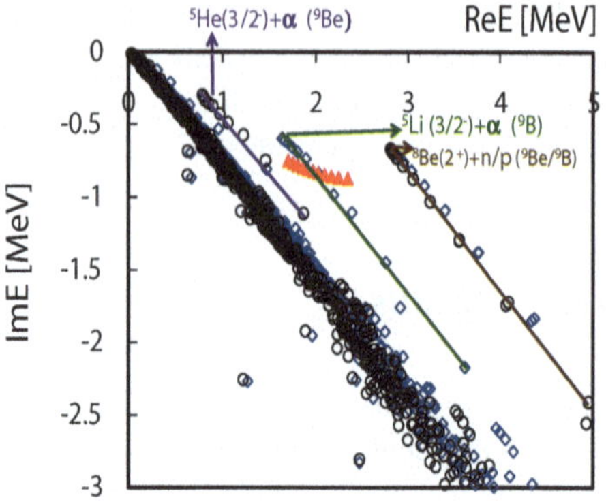

Fig. 16.2 The open circles and open diamonds are represents distribution of energy eigenvalues of the $^9B(1/2^+)$ and $^9Be(1/2^+)$ states, respectively

($v_3 = 0$), and obtain $E^{res} = 2.59$ MeV with a decay width $\Gamma = 1.51$ MeV. This implies that the $^9B(1/2^+)$ solution has a weak dependence on the three-body potential. The reason is considered to come from the additional Coulomb interaction due to the valence proton in the $\alpha + \alpha + p$ system.

We perform the calculation of ^{9}B by reducing the charge of the valence proton gradually. In Fig. 16.2, the calculated energies by reducing the charge of the valence proton is shown. The open triangles and open circles display the calculated eigenvalues of the $1/2^+$ states of ^{9}Be and ^{9}B, respectively. The filled red triangles present the resonance solutions, are obtained decreasing the point charge of the valence proton up to 50%. At present, the analytical continuation to zero charge was not succeeded to evaluate the solutions of the 1/2$^+$ state.

16.4 Summary

The structures of the first excited states of mirror ^{9}Be and ^{9}B nuclei are studied by using the $\alpha + \alpha + N$ three-body model and the complex scaling method. The first excited states of ^{9}Be nuclei observed in photo-disintegration cross sections showing a sharp peak just above the $^8Be(0^+) + n$ threshold [6]. Its analog $^9B(1/2^+)$ state is obtained as the resonance with the resonance energy $E^{res} = 2.42$ MeV with the decay width $\Gamma = 1.61$ MeV.

References

1. Akimune, H., Fujimura, M., Fujiwara, M., Hara, K., Ihikawa, T., Utsunomiya, H., Yamagata, T., Yamasaki, K., Yosoi, M.: Phys. Rev. C **64**, 041305 (2001)
2. Tiede, M.A., Kemper, K.W., Fletcher, N.R., Robson, D., Caussyn, D.D., Bennett, S.J., Brown, J.D., Carford, W.N., Jones, C.D., Watson, D.L., Rae, W.D.M.: Phys. Rev. C **52**, 1315 (1995)
3. Catford, W.N., Fifield, L.K., Reber, E.L., Kemper, K.W., Brown, J.D.: Phys. A **550**, 517 (1992)

4. Fortune, H.T., Sherr, R.: Nucl. Phys. A **898**, 78 (2013)
5. Baldwin, T.D., Catford, W.N., Mahboub, D., Timis, C.N., Ashwood, N.I., Clarke, N.M., Curtis, N., Ziman, V., Brown, T.A.D., Fox, S.P., Fulton, B.R., Groombridge, D., Watson, D.L., Pucknell, V.F.E., Weisser, D.C.: Phys. Rev. C **86**, 034330 (2012)
6. Odsuren, M., Kikuchi, K., Myo, T., Aikawa, M., Katō, K.: Phys. Rev. C **92**, 014322 (2015)
7. Ho, K.: Phys. Rep. **99**, 1 (1983)
8. Myo, T., Kikuchi, Y., Masui, H., Katō, K.: Prog. Part. Nucl. Phys. **79**, 1 (2014)
9. Kikuchi, K., Odsuren, M., Myo, T., Katō, K.: Phys. Rev. C **93**, 054605 (2016)
10. Kanada, H., Kaneko, T., Nagata, S., Nomoto, M.: Prog. Theor. Phys. **61**, 1327 (1979)
11. Kurokawa, C., Kato, K.: Nucl. Phys. A **792**, 87 (2007)
12. Arnold, C.W., Clegg, T.B., Iliadis, C., Karwowski, H.J., Rich, G.C., Tompkins, J.R., Howell, C.R.: Phys. Rev. C **85**, 044605 (2012)
13. Utsunomiya, H., Katayama, S., Gheorghe, I., Imai, S., Yamaguchi, H., Kahl, D., Sakaguchi, Y., Shima, T., Takahisa, K., Miyamoto, S.: Phys. Rev. C **92**, 064323 (2015)

Chapter 17
Gamow Shell Model Description of ^{4}He(d,d)

N. Michel, A. Mercenne and M. Płoszajczak

Abstract The Gamow Shell Model (GSM) has been recently developed to describe the structure of drip-line nuclei within a configuration interaction framework. The Resonating Group Method (RGM), implemented in GSM and applied to the description of one-nucleon reactions, is extended to deuteron scattering in the context of GSM-RGM. Applications consist in the study of the ^{4}He(d,d) scattering cross sections.

17.1 Introduction

GSM [1–4] is a configuration interaction approach based on the use of the Berggren basis, which comprises one-body bound, resonant and scattering states [5]. For the description of nuclear reactions, GSM-RGM has firstly been formulated with one-nucleon projectiles [6–10], and is extended here to reactions involving deuteron projectiles. As a first application of GSM-RGM with deuteron projectiles, the elastic scattering of deuteron on α-particle at low center of mass (CM) energies will be considered.

N. Michel (✉) · M. Płoszajczak
GANIL, CEA/DSM - CNRS/IN2P3, BP 55027 14076, Caen Cedex, France
e-mail: nicolas.lj.michel@gmail.com

A. Mercenne
Department of Physics & Astronomy, Louisiana State University,
264 Nicholson Hall, Tower Dr. Baton Rouge, LA 70803-4001, USA

N. A. Orr et al. (eds.), *Recent Progress in Few-Body Physics*,
Springer Proceedings in Physics 238,
https://doi.org/10.1007/978-3-030-32357-8_17

17.2 Cluster States Definition in Relative and Laboratory Frames

Projectile states read:

$$|\Psi_p^{J_p}\rangle^{HO} = \left[|N_{CM}, L_{CM}\rangle^{HO} \otimes |K_{int}, J_{int}\rangle^{HO}\right]^{J_p}_{M_p} \tag{17.1}$$

where the HO (harmonic oscillator) indicates that a basis of HO states is used therein, N_{CM}^{HO} and L_{CM} are respectively the linear momentum and angular momenta of the HO CM state, K_{int} is the intrinsic linear momentum of the projectile, and J_{int} represent the intrinsic angular momenta, so that we have $\mathbf{J_p} = \mathbf{J_{int}} + \mathbf{L_{CM}}$. The use of a basis of HO states at this level is justified by the finite range of the nuclear interaction, as (17.1) will be used only to generate the GSM-RGM nuclear Hamiltonian matrix. $|K_{int}, J_{int}\rangle^{HO}$ projectile intrinsic states are then firstly calculated from the Berggren basis diagonalization of the relative deuteron Hamiltonian $\hat{H}_{int}$, to provide with $|K_{int}, J_{int}\rangle$ eigenstates, and projected afterwards on a basis of relative HO states. Its expansion in terms of Slater determinants is then obtained using the Brody-Moshinsky transformation.

In order to properly deal with the asymptotic behavior of the scattering states solutions of the coupled-channel equations of the Hamiltonian, we will compute projectile Berggren basis states of the form $|K_{CM}, L_{CM}\rangle\,|K_{int}, J_{int}\rangle$, where $|K_{CM}, L_{CM}\rangle$ must be generated by a finite-range potential. For this, one uses the basis-generating projectile Hamiltonian $\hat{H}_p = \hat{H}_{int} + \hat{H}_{CM}$, with $\hat{H}_{CM} = \dfrac{\mathbf{P_{CM}}^2}{2M_p} + \hat{U}_{CM}$, where P_{CM} is the CM linear momentum of the projectile, M_p is its mass and $\hat{U}_{CM}$ is a spherical potential acting on $|K_{CM}, L_{CM}\rangle$, issued from a mean-field approximation of the projectile Hamiltonian at cluster approximation level.

17.3 Hamiltonian Coupled-Channel Equations

We consider an A-body state decomposed in reaction channels:

$$|(c, R)\rangle = \hat{\mathcal{A}}\,|\{|\Psi_T^{J_T}\rangle \otimes |R\;L\;J_{int}\;J_p\rangle\}^{J_A}_{M_A}\rangle \tag{17.2}$$

$$|\Psi^{J_A}_{M_A}\rangle = \sum_c \int_0^{+\infty} |(c, R)^{J_A}_{M_A}\rangle\, \frac{u_c(R)}{R} R^2\, dR\,, \tag{17.3}$$

where the channel index c stands for the $\{A-a, J_T; a, L, J_{int}, J_p\}$ quantum numbers, is an antimmetrized tensor product of the $|\Psi_T^{J_T}\rangle$ target state, and projectile channel state $|R\;L\;J_{int}\;J_p\rangle$, where $\mathbf{J_p} = \mathbf{J_{int}} + \mathbf{L}$ and $\mathbf{J_A} = \mathbf{J_p} + \mathbf{L}$, the CM subscript is dropped for convenience and $u_c(R)$ is the radial amplitude to be determined.

The coupled-channel equations can be formally derived from the Schrödinger equation $H\,|\Psi^{J_A}_{M_A}\rangle = E\,|\Psi^{J_A}_{M_A}\rangle$, and read:

$$\sum_c \int_0^\infty R^2 \left(H_{cc'}(R,R') - E N_{cc'}(R,R')\right) \frac{u_c(R)}{R} = 0 \qquad (17.4)$$

where $H_{cc'}(R,R')$ and $N_{cc'}(R,R')$ are calculated by integrating out all degrees of freedom except R, so that:

$$\begin{aligned} H^{J_A M_A}_{cc'}(R,R') &= \frac{\hbar^2}{2M_p}\left(-\frac{d^2}{dR^2} + \frac{L(L+1)}{R^2}\right)\frac{\delta(R-R')}{RR'}\delta_{cc'} \\ &+ (E_T + E_{int})\frac{\delta(R-R')}{RR'}\delta_{cc'} + U^L_{CM}(R,R')\,\delta_{cc'} + \tilde{V}^{J_A M_A}_{cc'}(R,R') \\ N^{J_A M_A}_{cc'}(R,R') &= \frac{\delta(R-R')}{RR'}\delta_{cc'} + \Delta N^{J_A M_A}_{cc'}(R,R') \end{aligned} \qquad (17.5)$$

where $\tilde{V}^{J_A M_A}_{cc'}$ and $\Delta N^{J_A M_A}_{cc'}(R,R')$ include the remaining short-range terms of the Hamiltonian and norm kernels. They are handled using shell model formulas as both target and projectiles are expanded in Slater determinants. The treatment of the non-orthogonality of channels is the same as in the one-nucleon projectile case [6, 7].

Equation (17.4) is conveniently solved by representing H with the Berggren basis of cluster CM states $|K_{CM}, L_{CM}\rangle$ (see Sect. 17.2). Indeed, as H becomes a matrix therein, bound, resonant and scattering states are obtained with matrix diagonalization and linear system solutions, respectively, so that the integro-differential equations of (17.4) are quickly and precisely calculated.

17.4 Differential Cross Sections

The internal structure of deuteron is taken into account using the N^{3}LO interaction [11], building intrinsic Berggren bases bearing $(J^\pi)_{int} = 1^+, 3^+$, which are the most important channels. CM parts of deuteron projectiles bear $L_{CM} \leq 4$ for $(J^\pi)_{int} = 1^+$, $L_{CM} = 0$ for $(J^\pi)_{int} = 3^+$, and $J_p \leq 3$ [12].

Our GSM model space consists of a ^{4}He core with two valence nucleons, using partial waves up to $\ell = 4$. The ^{4}He core is mimicked by a Wood-Saxon potential and the interaction used is that of FHT type [13–15], where two sets of FHT parameters are used. In the first set of parameters, denoted as FHT(E), one fits only the energies of the lowest states of ^{6}Li. However, by doing so, the asymptotic normalization coefficient (ANC) of its ground state, sensitive to its S and D contents, are poorly reproduced, probably because of the lack of three-body forces in our model. Consequently, in the second set of parameters, denoted as FHT(ANC), the S and D ANCs are fitted instead, so that the ground state of ^{6}Li is overbound by about 1 MeV.

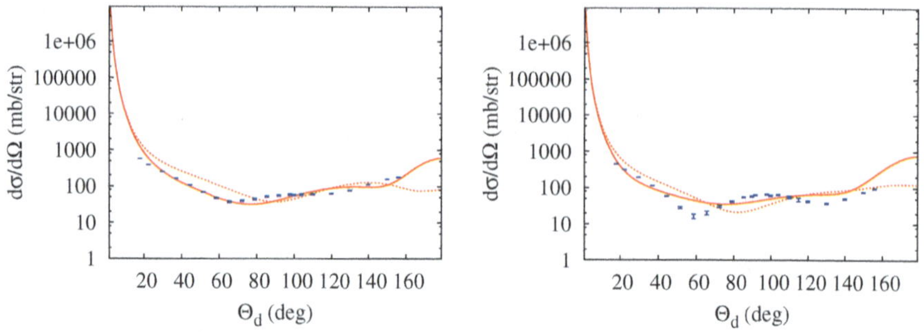

Fig. 17.1 Center of mass angular distributions of the ^{4}He(^{2}H,^{2}H) elastic scattering reaction calculated at two different energies (laboratory frame) using both the FHT(E) (dashed line) and FHT(ANC) (solid line) interactions are compared with the experimental data (symbols) [16, 17]

The center of mass differential cross sections of the ^{4}He(^{2}H,^{2}H) elastic scattering process have been calculated using both FHT(E) and FHT(ANC) interactions at two different deuteron kinetic energies in laboratory frame: 6.476 and 8.971 MeV (see Fig.17.1). The ability of the FHT(ANC) interaction to describe experimental data better than the FHT(E) interaction is very clear. The only discrepancy therein is that minima and maxima are slightly shifted to larger angles, and not deep enough for the first minimum at 8.971 MeV.

Results are of comparable quality with those obtained in [12]. However, an additional fit of the ANCs of ^{6}Li was necessary in our approach, pointing out the absence of three-body force.

17.5 Conclusion

GSM incorporates both continuum and nucleon inter-correlations degrees of freedom by using the Berggren basis in a configuration interaction picture. While initially formuled for structure [4], it has been successfully extended to the study of reaction cross sections with one-nucleon projectiles using the GSM-RGM approach [6, 7, 9]. The inclusion of many-nucleon clusters was nevertheless more difficult to devise, that due to their internal structure, which demands a precise treatment of their relative and CM parts. Applications have been considered with the example of the ^{4}He(d,d) elastic scattering reaction. Experimental data are well reproduced with our fitting methods, but they suggest that three-body forces should be more efficiently included of in our model.

We wish to thank P. Descouvemont and G. Hupin for useful discussions.

References

1. Michel, N., et al.: Phys. Rev. Lett. **89**, 042502 (2002)
2. Betan, R.M. Id, et al.: Phys. Rev. Lett. **89**, 042501 (2002)
3. Michel, N., et al.: Phys. Rev. C **67**, 054311 (2003)
4. Michel, N., et al.: J. Phys. G: Nucl. Part. Phys. **36**, 013101 (2009)
5. Berggren, T.: Nucl. Phys. A **109**, 265 (1968)
6. Jaganathen, Y., et al.: J. Phys.: Conf. Ser. **403**, 012022 (2012)
7. Jaganathen, Y., Michel, N., Płoszajczak, M.: Phys. Rev. C **89**, 034624 (2014)
8. de Grancey, F., et al.: Phys. Lett. B **758**, 26 (2016)
9. Fossez, K., et al.: Phys. Rev. C **91**, 034609 (2015)
10. Dong, G.X., et al.: J. Phys. G: Nucl. Part. Phys. **44**, 045201 (2017)
11. Entem, D.R., Machleidt, R.: Phys. Rev. C **68**, 041001 (2003)
12. Hupin, G., Quaglioni, S., Navrátil, P.: Phys. Rev. Lett. **114**, 212502 (2015)
13. Furutani, H., Horiuchi, H., Tamagaki, R.: Prog. Theor. Phys. **62**, 981 (1979)
14. Furutani, H., et al.: Prog. Theor. Phys. Suppl. **68**, 193 (1980)
15. Jaganathen, Y., et al.: Phys. Rev. C **96**, 054316 (2017)
16. Jett, J.H., Detch, J.L., Jarmie, N.: Phys. Rev. C **3**, 1769 (1971)
17. Senhouse, L.S., Tombrello, T.A.: Nucl. Phys. **57**, 624 (1964)

Chapter 18
Boron Isotopes at the Drip-Line: The ^{19}B Case

J. Gibelin, S. Leblond, F. M. Marqués, N. A. Orr, Y. Kondo, T. Nakamura, J. Bonnard, N. Michel, N. L. Achouri, T. Aumann, H. Baba, F. Delaunay, P. Doornenbal, N. Fukuda, J. W. Hwang, N. Inabe, T. Isobe, D. Kameda, D. Kanno, S. Kim, N. Kobayashi, T. Kobayashi, T. Kubo, J. Lee, R. Minakata, T. Motobayashi, D. Murai, T. Murakami, K. Muto, T. Nakashima, N. Nakatsuka, A. Navin, S. Nishi, S. Ogoshi, H. Otsu, H. Sato, Y. Satou, Y. Shimizu, H. Suzuki, K. Takahashi, H. Takeda, S. Takeuchi, R. Tanaka, Y. Togano, A. G. Tuff, M. Vandebrouck and K. Yoneda

Abstract We report here results for SAMURAI Dayone campaign which, by using knock-out reaction and invariant mass method, allowed to obtain the first measurement of exited states of the most neutron rich boron isotope ^{19}B.

J. Gibelin (✉) · S. Leblond · F. M. Marqués · N. A. Orr · N. L. Achouri · F. Delaunay
LPC Caen, ENSICAEN, Université de Caen, CNRS/IN2P3, Caen, France
e-mail: gibelin@lpccaen.in2p3.fr

Y. Kondo · T. Nakamura · D. Kanno · N. Kobayashi · R. Minakata · T. Nakashima · S. Nishi · S. Ogoshi · R. Tanaka · Y. Togano
Department of Physics, Tokyo Institute of Technology, Tokyo, Japan

J. Bonnard · M. Vandebrouck
IPNO, Université Paris-Sud, IN2P3-CNRS, Orsay, France

N. Michel
NSCL, Michigan State University, East Lansing, USA

School of Physics, Peking University, Beijing, China

T. Aumann
Institut für Kernphysik, Technische Universität Darmstadt, Darmstadt, Germany

T. Aumann · Y. Togano
ExtreMe Matter Institute EMMI and Research Division, GSI, Darmstadt, Germany

H. Baba · P. Doornenbal · N. Fukuda · N. Inabe · T. Isobe · D. Kameda · T. Kubo · J. Lee · T. Motobayashi · H. Otsu · H. Sato · Y. Shimizu · H. Suzuki · H. Takeda · S. Takeuchi · K. Yoneda
RIKEN Nishina Center, Wako, Japan

N. A. Orr et al. (eds.), *Recent Progress in Few-Body Physics*,
Springer Proceedings in Physics 238,
https://doi.org/10.1007/978-3-030-32357-8_18

18.1 Introduction

Neutron rich light nuclei, especially those with halos, are one of the main themes of the physics of nuclear structure in the past decades. These studies have, however, been largely limited to helium, lithium and beryllium isotopes due to the difficulty in producing heavier nuclei around the drip-line. We report here the first measurement of excited states of ^{19}B for which little is known.

18.2 Experimental Stup

The experiment, performed at the Radioactive Isotope Beam Factory (RIBF) of the RIKEN Nishina Center, was part of the SAMURAI [1–3] DayOne campaign aiming to investigate the structure of light neutron-rich nuclei at and beyond the neutron drip-line. Details on the experimental setup were presented in several articles [4–6] and Ph.D. theses [7–10] and the reader is referred to these publications for details. Here the ^{19}B of interest was produced via proton knock out of a secondary ^{20}C beam at 280*A* MeV, which impinged on a 1.8 g/cm^2 carbon reaction target.

18.3 Analysis and Interpretation

General analysis procedures for SAMURAI are given in great details in [4–10]. Here, in order to obtain information on the unbound excited states of ^{19}B we rebuilt the relative energy (E_r) between the core of ^{17}B and two neutrons, from their respective four-momenta. In this particular case, special attention is paid in rejecting the cross-talk that arises when one neutron diffuses from one detector to another. We present in Fig. 18.1 the experimental E_r distribution. We clearly see a sharp structure at low energy and a wider one around 3 MeV.

J. W. Hwang · S. Kim · Y. Satou
Department of Physics and Astronomy, Seoul National University, Seoul, Republic of Korea

T. Kobayashi · K. Muto · K. Takahashi
Department of Physics, Tohoku University, Sendai, Japan

D. Murai
Department of Physics, Rikkyo University, Tokyo, Japan

T. Murakami · N. Nakatsuka
Department of Physics, Kyoto University, Kyoto, Japan

A. Navin
GANIL, CEA/DSM-CNRS/IN2P3, Caen, France

A. G. Tuff
Department of Physics, University of York, York, UK

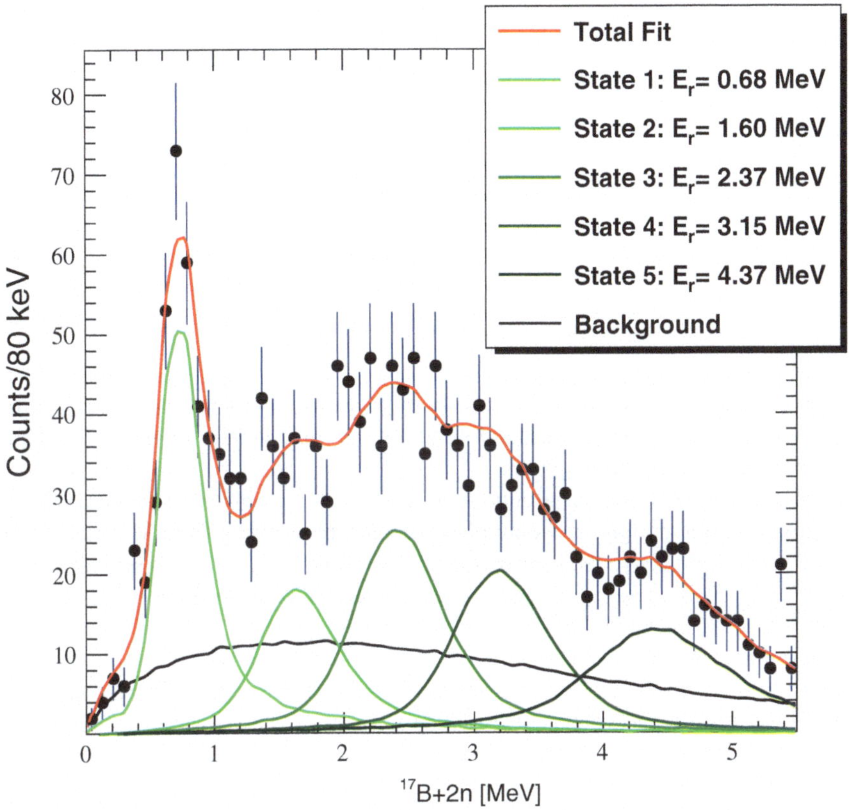

Fig. 18.1 ^{17}B + 2n relative energy spectrum adjusted using five Breit-Wigner distributions and a uncorrelated distribution. All contributions have been folded with the experimental response. Error bars are statistical

Analysis of the coincidence between relative energy and gamma emission shows that the first excited state decays to the ground state of ^{17}B and a clear structure around $E_r = 3.5$ MeV decays to the excited state of ^{17}B (measured here at 1110 ± 25 keV). The systematic study of the Dalitz distributions for all energies between 0 and 6 MeV (similarly to what is described in [11]) suggests also that another state is located around 1.5 MeV.

Since Gamow shell model (GSM) [12–14] calculations performed here predict widths qualitatively smaller than 0.5 MeV, we adjusted the spectrum with five Breit-Wigner distributions (green) and an uncorrelated—simulated—background (black). All Breit-Wigner distributions include the energy dependence of the observed width, as well as the level-shift term and these contributions are convoluted with the experimental resolutions. Results of the fit are listed in the first column of Table 18.1.

In order to interpret our results we compare them with the configuration-mixing shell model (SM) calculations [15] with the two-body interactions YSOX [16]. The

Table 18.1 Summary of the relative energies measured, with their assigned excitation energies and spin-parity

E_r (MeV)	E_x (MeV)	J^π
0.68 ± 0.01	0.82 ± 0.40	$1/2^-$
1.60 ± 0.01	1.74 ± 0.40	$3/2^-$
2.37 ± 0.01	2.51 ± 0.40	$3/2^-$
3.12 ± 0.02 } 4.38 ± 0.01 }	4.39 ± 0.42	$1/2^-, 3/2^-$

considered active space is composed of the $(0p_{3/2}, 0p_{1/2}, 0d_{5/2}, 1s_{1/2}, 0d_{3/2})$ orbits, and the center-of-mass spurious contributions are removed *via* the Lawson prescription. Configurations corresponding up to 5 particle-hole excitations $(5\hbar\omega)$ are included in the many-body space.

Calculations show that the ground state is $J^\pi = 3/2^-$, which directly comes from the proton configuration, as the number of neutrons is even. The SM gives a first excited state right above the neutron threshold, with $J^\pi = 1/2^-$; and several states. Due to the small energy differences and our experimental resolution, we can not conclude on the nature of the ^{19}B states solely on energy basis. Spectroscopic factors suggest that coming from ^{20}C mainly $J^\pi = 1/2^-$ and $J^\pi = 3/2^-$ are populated. We then tentatively assign a $J^\pi = 3/2^-$ to the second and third states observed here. As for the two remaining structures, since they are measured at an energy difference of $\approx$1.1 MeV compatible with the energy of the first excited state of ^{17}B and that the lowest one does decay to this excited state, we assumed they belong to the same state around $E_x = 4.4$ MeV. This is also dictated by the energy (>7 MeV) of the next $3/2^-$. Results are summarized in Table 18.1.

18.4 Conclusion

We presented for the first time experimental results for the structure of neutron-rich ^{19}B isotope above its two neutron emission threshold. Our results show a first excited state close to this threshold and three other states above. Comparison with state-of-the-art shell model calculations allow us to tentatively determine their nature.

Acknowledgements This work was supported in part by JSPS KAKENHI Grant No. 24740154, MEXT KAKENHI Grant No. 24105005, the WCU (R32-2008-000-10155-0), the GPF (NRF-2011-0006492) programs of NRF Korea, the HIC for FAIR, and the US Department of Energy, Office of Science, Office of Nuclear Physics under Awards No. DE-SC0013365 (MSU) and No. DE-SC0018083 (NUCLEI SciDAC-4 Collaboration). We acknowledge partial support from the Franco-Japanese LIA-International Associated Laboratory for Nuclear Structure Problems as well as JSPS Invitation fellowship program for long term research in Japan.

References

1. Otsu, H., Koyama, S., Chiga, N., et al.: Nucl. Instrum. Meth. B **376**, 175 (2016)
2. Kobayashi, T., Chiga, N., Isobe, T., et al.: Nucl. Instrum. Meth. B **317**, 294 (2013)
3. Sato, H., Kubo, T., Yano, Y., et al.: IEEE T. Appl Supercon. **23**, 4500308 (2013)
4. Kondo, Y., Nakamura, T., Tanaka, R., et al.: Phys. Rev. Lett. **116**, 102503 (2016)
5. Hwang, J., Kim, S., Satou, Y., et al.: Phys. Lett. B **769**, 503 (2017)
6. Togano, Y., Nakamura, T., Kondo, Y., et al.: Phys. Lett. B **761**, 412 (2016)
7. Hwang, J.: Study of ^{19}C by one-neutron knockout reaction with a carbon target. Ph.D. thesis, Department of Physics and Astronomy College of Natural Science, Seoul National University (2015). http://dcollection.snu.ac.kr/jsp/common/DcLoOrgPer.jsp?sItemId=000000067486
8. Kim, S.: Spectroscopy of ^{17}C via one-neutron knockout reaction. Ph.D. thesis, Department of Physics and Astronomy College of Natural Science, Seoul National University (2015). http://dcollection.snu.ac.kr/jsp/common/DcLoOrgPer.jsp?sItemId=000000067319
9. Leblond, S.: Structure of neutron-rich Isotopes of Boron and Carbon at the limits of stability. Ph.D. thesis, Normandie Université, France (2015). https://tel.archives-ouvertes.fr/tel-01289381
10. Deshayes, Q.: Nitrogen isotopes beyond the neutron drip line: ^{23}N, ^{24}N and ^{25}N. Ph.D. thesis, Normandie Université, Université Caen Normandie (2017). https://tel.archives-ouvertes.fr/tel-01696611
11. Revel, A., Marqués, F.M., Sorlin, O., et al.: Phys. Rev. Lett. **120**, 152504 (2018)
12. Jaganathen, Y., Betan, R.M.I., Michel, N., et al.: Phys. Rev. C **96**, 054316 (2017)
13. Wang, S.M., Michel, N., Nazarewicz, W., et al.: Phys. Rev. C **96**, 044307 (2017)
14. Fossez, K., Rotureau, J., Michel, N., et al.: Phys. Rev. C **94**, 054302 (2016)
15. Caurier, E., Martínez-Pinedo, G., Nowacki, F., et al.: Rev. Mod. Phys. **77**, 427 (2005)
16. Yuan, C., Suzuki, T., Otsuka, T., et al.: Phys. Rev. C **85**, 064324 (2012)

Chapter 19
Elastic α-^{12}C Scattering at Low Energies with the Bound States of ^{16}O in EFT

Shung-Ichi Ando

Abstract We study the elastic α-^{12}C scattering at low energies, including the energies of excited bound states of ^{16}O, in effective field theory. After fitting the parameters to the phase shift data, we calculate the asymptotic normalization constants of the 0_2^+, 1_1^-, 2_1^+, 3_1^- states of ^{16}O and compare them with those in the previous studies.

19.1 Introduction

The elastic α-^{12}C scattering at low energies is an important reaction to fix some model parameters for the study of the radiative α capture on carbon-12, $^{12}C(\alpha,\gamma)^{16}$O, which determines the C/O ratio synthesized in the stars [1]. The reaction rate of the radiative capture process at the Gamow peak energy, $T_G = 0.3$ MeV, cannot be determined in experiment due to the Coulomb barrier. A theoretical model is necessary to employ in order to extrapolate the reaction rate down to T_G by fitting model parameters to experimental data. Accurate measurements of the elastic scattering have been reported in [2, 3]. Elastic scattering data at low energies in general can be used for deducing an asymptotic normalization constant (ANC), which determines an overall strength of a nuclear reaction involving bound states [4–6]. ANCs for the excited bound states of ^{16}O are important inputs for an estimate of the reaction rate of the radiative capture process.

Effective field theories (EFTs) provide us a model independent and systematic method for theoretical calculations, in which one introduces a scale to separate relevant degrees of freedom at low energies from irrelevant degrees of freedom at high energies for a reaction in question. For review, one may refer to [7, 8]. In this talk, we briefly discuss a construction of EFT for the radiative capture process and apply it to the study of the elastic α-^{12}C scattering process below resonance energies [9]. We then calculate the ANCs for the bound states of ^{16}O and compare them with those reported in the previous studies [10, 11].

S.-I. Ando (✉)
Sun Moon University, Asan, Chungman 31460, South Korea
e-mail: sando@sunmoon.ac.kr

N. A. Orr et al. (eds.), *Recent Progress in Few-Body Physics*,
Springer Proceedings in Physics 238,
https://doi.org/10.1007/978-3-030-32357-8_19

19.2 Formalism

In the study of the radiative capture process, $^{12}C(\alpha,\gamma)^{16}O$, at $T_G = 0.3$ MeV employing an EFT, at such a low energy, the ground states of α and ^{12}C can be regarded as point-like particles whereas the first excited states of α and ^{12}C are chosen as irrelevant degrees of freedom, by which a large scale of the theory is determined. The effective Lagrangian for the process is constructed in terms of two spinless scalar fields for α and ^{12}C, and the terms of the Lagrangian are expanded in terms of the number of derivatives. An expression of the effective Lagrangian has been obtained in Eq. 1 in [9]. The expansion parameter of the theory is $Q/\Lambda_H \sim 1/3$ where Q denotes a typical momentum scale $Q \sim k_G$: k_G is the Gamow peak momentum, $k_G = \sqrt{2\mu T_G} \simeq 41$ MeV, where μ is the reduced mass of α and ^{12}C. Λ_H denotes a large momentum scale $\Lambda_H \sim 150$ MeV determined from the first excited energy of α or ^{12}C.

In the study of the elastic α-^{12}C scattering in EFT, the amplitudes of the elastic scattering are calculated from diagrams depicted in Figs. 19.1 and 19.2. The scattering amplitudes for l-th partial wave states are obtained as [9, 12, 13]

$$A_l = \frac{2\pi}{\mu} \frac{(2l+1)P_l(\cos\theta)e^{2i\sigma_l}W_l(\eta)C_\eta^2}{K_l(k) - 2\kappa H_l(k)}, \tag{19.1}$$

where k is the magnitude of relative momentum between α and ^{12}C and θ is the scattering angle in the C.M. frame. In addition, η is the Sommerfeld parameter, $\eta = \kappa/k$, where κ is the inverse of the Bohr radius, $\kappa = Z_2 Z_6 \mu\alpha$, and $C_\eta^2 = \frac{2\pi\eta}{e^{2\pi\eta}-1}$, $W_l(\eta) = \frac{\kappa^{2l}}{(l!)^2}\prod_{n=0}^{l}\left(1+\frac{n^2}{\eta^2}\right)$, $H_l(k) = W_l(\eta)H(\eta)$, with $H(\eta) = \psi(i\eta) + \frac{1}{2i\eta} - \ln(i\eta)$. $\psi(z)$

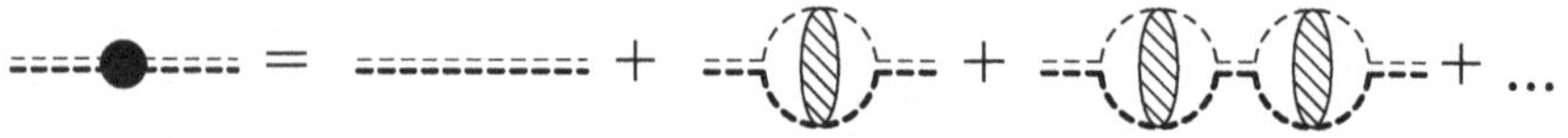

Fig. 19.1 Diagrams for dressed ^{16}O propagator. A thick (thin) dashed line represents a propagator of ^{12}C (α), and a thick and thin double dashed line with and without a filled blob represent a dressed and bare ^{16}O propagator, respectively. A shaded blob represents a set of diagrams for non-perturbative Coulomb interaction

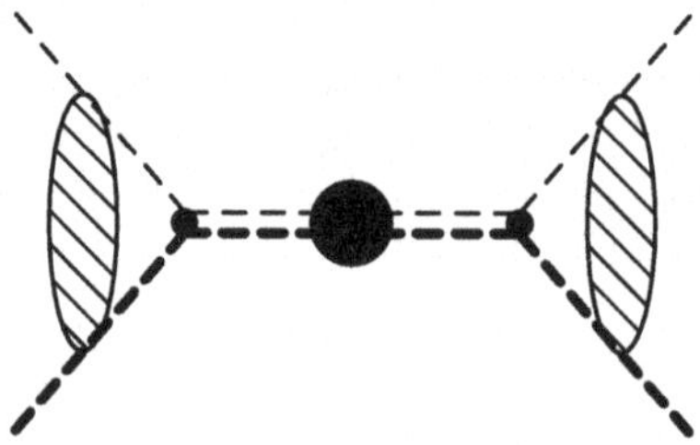

Fig. 19.2 Diagram of the scattering amplitude. See the caption of Fig. 19.1 as well

is the digamma function, $P_l(x)$ are the Legendre polynomials, and σ_l are the Coulomb phase shifts. The function $K_l(k)$ represents the interaction due to the short range nuclear force, which is obtained in terms of the effective range parameters as

$$K_l(k) = -\frac{1}{a_l} + \frac{1}{2}r_l k^2 - \frac{1}{4}P_l k^4 + Q_l k^6 - R_l k^8 + \cdots \tag{19.2}$$

where a_l are fixed by using the binding energies of ^{16}O, and the other effective range parameters are fitted to the phase shift δ_l by using a relation

$$W_l(\eta)C_\eta^2 k \cot\delta_l = K_l(k) - 2\kappa\, Re H_l(k)\,. \tag{19.3}$$

Because of the modification of the counting rules discussed in [10], we include three effective range parameters, r_l, P_l, and Q_l, for $l = 0, 1, 2$, and four effective range parameters, r_3, P_3, Q_3, and R_3, for $l = 3$.

19.3 Results and Discussion

The effective range parameters are fitted to the phase shift data [3] where we employ three data sets, which have different energy ranges below the resonance energies, denoted as $L0$, $L1$, and $L2$ where $L = S, P, D, F$ for $l = 0, 1, 2, 3$, respectively [10]. To estimate the ANC, $|C_b|$, for the 0_2^+, 1_1^-, 2_1^+, 3_1^- states of ^{16}O, we employ the definition of $|C_b|$ from (14) in [14]:

$$|C_b| = \gamma_l^l \frac{\Gamma(l+1+|\eta_b|)}{l!}\left(\left|-\frac{d[K_l(k) - 2\kappa H_l(k)]}{dk^2}\right|_{k^2=-\gamma_l^2}\right)^{-\frac{1}{2}} \quad (\mathrm{fm}^{-1/2}), \tag{19.4}$$

where $\eta_b = \kappa/k_b$. We show our results for the ANCs in Table 19.1.

For the ANC for the 1_1^- state, one finds that the result of the present study, $|C_b| = (1.6 - 1.9) \times 10^{14}$ (fm$^{-1/2}$), is in good agreement with experimental values, $(2.10 \pm 0.14) \times 10^{14}$, $(2.00 \pm 0.35) \times 10^{14}$, $(2.08 \pm 0.20) \times 10^{14}$, obtained by Avila et al. [15], Oulebsir et al. [16], Brune et al. [17], respectively. One also finds good agreement with theoretical estimates, $(2.22 - 2.24) \times 10^{14}$, obtained from a potential model calculation by Katsuma [18], and $2.14(6) \times 10^{14}$ and 2.073×10^{14}

Table 19.1 ANC for the 0_2^+, 1_1^-, 2_1^+, and 3_1^- states of ^{16}O in the unit of fm$^{-1/2}$ estimated from the fitted parameters by using data of the three data sets $L0$, $L1$, and $L2$

	$\lvert C_b(0_2^+)\rvert \times 10^2$	$\lvert C_b(1_1^-)\rvert \times 10^{14}$	$\lvert C_b(2_1^+)\rvert \times 10^4$	$\lvert C_b(3_1^-)\rvert \times 10^2$
$L0$	6.8(16)	1.9(4)	2.4(3)	1.2(1)
$L1$	7.4(15)	1.8(1)	2.3(2)	1.3(1)
$L2$	6.4(7)	1.6(1)	2.1(1)	1.5(1)

from the new method of the parameterization by Ramirez Suarez and Sparenberg [19] and by Orlov et al. [20], respectively. Thus, estimates of the ANC for the 1_1^- state appear converging in both theory and experiment.

For the ANC for the 2_1^+ state, the result of the present study, $|C_b| = (2.1 - 2.4) \times 10^4$ (fm$^{-1/2}$), underestimates the experimental values, $(12.2 \pm 0.7) \times 10^4$ [15], $(14.4 \pm 2.8) \times 10^4$ [16], $(11 \pm 1) \times 10^4$ [17]. On the other hand, the result of the present study is in good agreement with theoretical estimates, $(2.41 \pm 0.38) \times 10^4$ and 2.106×10^4, from the effective range analysis up to the r_2 term by Konig et al. [21] and up to the P_2 term by Orlov et al. [22], respectively, and underestimates the other theoretical results, $(14.45 \pm 0.85) \times 10^4$ from the supersymmetric potential model by Sparenberg [23] and $(12.6 \pm 0.5) \times 10^4$ from the R matrix analysis with a microscopic cluster model by Dufour and Descouvemont [24] and 5.050×10^4 from the new method of the parameterization by Orlov et al. [20]. Thus, the effective range expansion may not be reliable for estimates of the ANC for the 2_1^+ state.

For the ANCs for the 0_2^+ and 3_1^- states, the result of the present study, $|C_b| = (6.4 - 7.4) \times 10^2$ (fm$^{-1/2}$) for the 0_2^+ state underestimates an experimental value, $(15.6 \pm 1.0) \times 10^2$ [15] and overestimates a theoretical value, 4.057×10^2 [20]. Meanwhile, the result of the present study, $|C_b| = (1.2 - 1.5) \times 10^2$ (fm$^{-1/2}$) for the 3_1^- state is in good agreement to the experimental value, $(1.39 \pm 0.09) \times 10^2$, recently reported by Avila et al. [15]. Further studies would be necessary for the ANCs for the 0_2^+ and 3_1^- states.

This work was supported by the Basic Science Research Program through the National Research Foundation of Korea funded by the Ministry of Education of Korea Grant No. NRF-2016R1D1A1B03930122.

References

1. Fowler, W.A.: Rev. Mod. Phys. **56**, 149 (1984)
2. Plaga, R., et al.: Nucl. Phys. A **465**, 291 (1987)
3. Tischhauser, P., et al.: Phys. Rev. C **79**, 055803 (2009)
4. Mukhamedzhanov, A.M., Tribble, R.E.: Phys. Rev. C **59**, 3418 (1999)
5. Blokhintsev, L.D., Yeremenko, Y.O.: Phys. Atom. Nucl. **71**, 1219 (2008)
6. Tribble, R.E., et al.: Rep. Prog. Phys. **77**, 106901 (2014)
7. Bedaque, P.F., van Kolck, U.: Ann. Rev. Nucl. Part. Sci. **52**, 339 (2002)
8. Hammer, H.-W., Ji, C., Phillips, D.R.: J. Phys. G **44**, 103002 (2017)
9. Ando, S.-I.: Eur. Phys. J. A **52**, 130 (2016)
10. Ando, S.-I.: Phys. Rev. C **97**, 014604 (2018)
11. Ando, S.-I.: J. Korean Phys. Soc. **73**, 1452–1457 (2018)
12. Ando, S., Shin, J.W., Hyun, C.H., Hong, S.W.: Phys. Rev. C **76**, 064001 (2007)
13. Ando, S.-I.: Eur. Phys. J. A **33**, 185 (2007)
14. Sparenberg, J.-M., Capel, P., Baye, D.: Phys. Rev. C **81**, 011601 (2010)
15. Avila, M.L., et al.: Phys. Rev. Lett. **114**, 071101 (2015)
16. Oulebsir, N., et al.: Phys. Rev. C **85**, 035804 (2012)
17. Brune, C.R., et al.: Phys. Rev. Lett. **83**, 4025 (1999)
18. Katsuma, M.: Phys. Rev. C **78**, 034606 (2008)
19. Ramirez Suarez, O.L., Sparenberg, J.-M.: Phys. Rev. C **96**, 034601 (2017)

20. Orlov, Y.V., Irgaziev, B.F., Nabi, J.-U.: Phys. Rev. C **96**, 025809 (2017)
21. Konig, S., Lee, D., Hammer, H.-W.: J. Phys. G: Nucl. Part. Phys. **40**, 045106 (2013)
22. Orlov, Y.V., Irgaziev, B.F., Nikitina, L.I.: Phys. Rev. C **93**, 014612 (2016)
23. Sparenberg, J.-M.: Phys. Rev. C **69**, 034601 (2004)
24. Dufour, M., Descouvemont, P.: Phys. Rev. C **78**, 015808 (2008)

Chapter 20
The S_{E1} Factor of Radiative α Capture on ^{12}C in Effective Field Theory

Shung-Ichi Ando

Abstract The S_{E1} factor of radiative α capture on ^{12}C is studied in effective field theory. We briefly discuss the strategy for the calculation of the reaction and report a first result of S_{E1} at the Gamow-peak energy, $E_G = 0.3$ MeV.

20.1 Introduction

The radiative α capture on carbon-12, $^{12}C(\alpha, \gamma)^{16}$O, is a fundamental reaction in nuclear-astrophysics, which determines the C/O ratio created in the stars [1]. The reaction rate, equivalently the astrophysical S-factor, of the process at the Gamow peak energy, $E_G = 0.3$ MeV, however, cannot be determined in experiment due to the Coulomb barrier. A theoretical model is necessary to employ in order to extrapolate the reaction rate down to E_G by fitting model parameters to experimental data typically measured at a few MeV. In constructing a model for the study, one needs to take account of excited states of ^{16}O [2], particularly, two excited bound states for $l^{\pi}_{i\text{-}th} = 1^-_1$ and 2^+_1 just below the α-^{12}C breakup threshold at $E = -0.045$ and -0.24 MeV,[1] respectively, as well as two resonant (second excited) 1^-_2 and 2^+_2 states at $E = 2.42$ and 2.68 MeV, respectively. The capture reaction to the ground state of ^{16}O at E_G is expected to be $E1$ and $E2$ transition dominant due to the subthreshold 1^-_1 and 2^+_1 states. See [2, 3] for review.

Theoretical frameworks employed for the previous studies are mainly categorized into two [3]: the cluster models using generalized coordinate method [4] or potential model [5] and the phenomenological models using the parameterization of Breit-Wigner, R-matrix [6], or K-matrix [7]. A recent trend of the study is to rely on intensive numerical analysis, in which a large amount of the experimental data relevant to the study are accumulated, and a significant number of parameters of the

[1] The energy E denotes that of the α-^{12}C system in center of mass frame.

S.-I. Ando (✉)
Sunmoon University, Asan, Chungman 31460, South Korea
e-mail: sando@sunmoon.ac.kr

N. A. Orr et al. (eds.), *Recent Progress in Few-Body Physics*,
Springer Proceedings in Physics 238,
https://doi.org/10.1007/978-3-030-32357-8_20

models are fitted to the data by using computational power [3, 8, 9]. In the present work, we discuss an alternative approach to estimate the S-factor at E_G; we employ a new method for the study and briefly discuss a calculation of the S_{E1} factor at E_G based on an effective field theory [10, 11].

20.2 Diagrams

In the study of the radiative capture process, $^{12}C(\alpha,\gamma)^{16}O$, at $E_G = 0.3$ MeV employing an EFT, one may regard the ground states of α and ^{12}C as point-like particles whereas the first excited states of α and ^{12}C are chosen as irrelevant degrees of freedom, from which a large scale of the theory is determined [12]. Thus the expansion parameter of the theory is $Q/\Lambda_H \sim 1/3$ where Q denotes a typical momentum scale $Q \sim k_G$; k_G is the Gamow peak momentum, $k_G = \sqrt{2\mu E_G} \simeq 41$ MeV, where μ is the reduced mass of α and ^{12}C. Λ_H denotes a large momentum scale $\Lambda_H \simeq 150$ MeV obtained from the first excited energy of α or ^{12}C. An effective Lagrangian for the study is obtained in (1) in [13].

The capture amplitudes are calculated from the Feynman diagrams depicted in Figs. 20.1 and 20.2. One can find an expression of the amplitudes in (6), (7), (8), and (9) in [13]. We note that the loop diagrams (a) and (b) in Fig. 20.1 are finite whereas those (d), (e), and (f) diverge. The divergence terms are renormalized by a counter term $h^{(1)}$ in a contact vertex in the diagram (c). Six parameters remain in the amplitudes. Four of them are effective range parameters of elastic α-^{12}C scattering

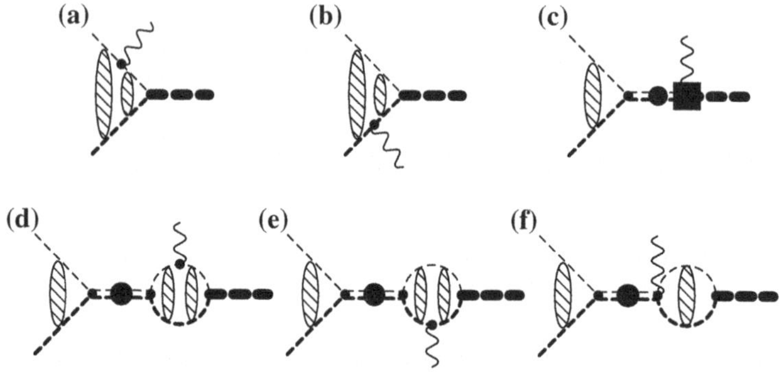

Fig. 20.1 Diagrams for the radiative capture process from the initial p-wave α-^{12}C state. A wavy line denotes the outgoing photon, and the same part of the diagram displayed in Fig. 20.2 is the dressed composite propagator for $l = 1$ state. A thick dashed line in the final state denotes the ground (0_1^+) state of ^{16}O. See the caption of Fig. 20.2 as well

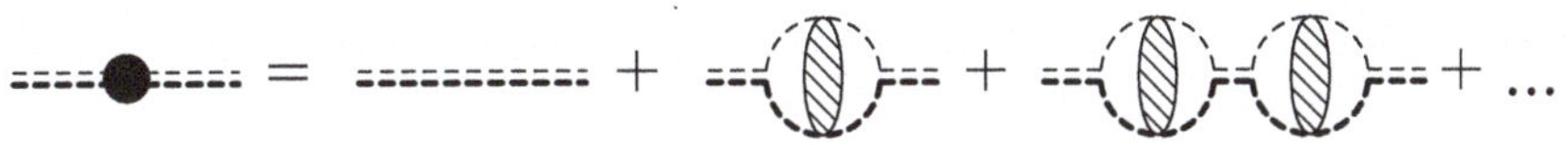

Fig. 20.2 Diagrams for dressed ^{16}O propagator. A thick (thin) dashed line represents a propagator of ^{12}C (α), and a thick and thin double dashed line with and without a filled blob represent a dressed and bare ^{16}O propagator, respectively. A shaded blob represents a set of diagrams for non-perturbative Coulomb interaction

for $l = 1$ [14]. One of them is fixed by using the binding energy of the subthreshold $l = 1$ state of ^{16}O, and the others are fitted to the phase shift data of the elastic scattering [15].

20.3 Result

In the left panel of Fig. 20.3, we show the data and the fitted curve of the phase shift and find that the fitted curve well reproduces the data. The remaining two parameters, $h^{(1)R}$ and $y^{(0)}$, in the amplitudes are fitted to the S_{E1} data [3], and we obtain $h^{(1)R} = -6.95(11) \times 10^2$ MeV^3 and $y^{(0)} = 0.495(18)$ MeV$^{-1/2}$, where the number of the data is $N = 151$ and $\chi^2/N \simeq 1.72$. In the right panel of Fig. 20.3, we show the data and the fitted curve for S_{E1}. At the Gamow peak energy, $E_G = 0.3$ MeV, thus, we obtain $S_{E1} \simeq 58$ keV b. An error estimate of S_{E1} is now under investigation.

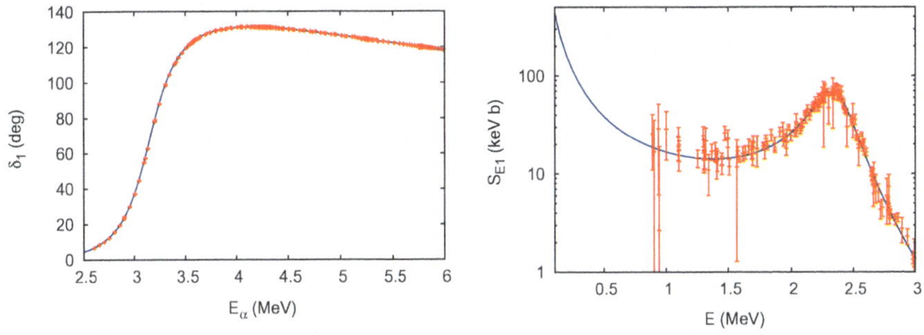

Fig. 20.3 (Left panel) Phase shift, δ_1, plotted by using the fitted effective range parameters as a function of E_α. (Right panel) S_{E1} factor plotted by using the fitted parameters as a function of E. The experimental data are also displayed in the figures

Acknowledgements This work was supported by the Basic Science Research Program through the National Research Foundation of Korea funded by the Ministry of Education of Korea Grant No. NRF-2016R1D1A1B03930122.

References

1. Fowler, W.A.: Rev. Mod. Phys. **56**, 149 (1984)
2. Buchmann, L.R., Barnes, C.A.: Nucl. Phys. A **777**, 254 (2006)
3. deBoer, R.J., et al.: Rev. Mod. Phys. **89**, 035007 (2017)
4. Descouvemont, P., Baye, D., Heenen, P.-H.: Nucl. Phys. A **430**, 426 (1984)
5. Langanke, K., Koonin, S.E.: Nucl. Phys. A **439**, 384 (1985)
6. Lane, A.M., Thomas, R.G.: Rev. Mod. Phys. **30**, 257 (1958)
7. Humblet, J., Dyer, P., Zimmerman, B.A.: Nucl. Phys. A **271**, 210 (1976)
8. Xu, Y., et al.: Nucl. Phys. A **918**, 61 (2013)
9. An, Z.-D., et al.: Phys. Rev. C **92**, 045802 (2015)
10. Bedaque, B.F., van Kolck, U.: Ann. Rev. Nucl. Part. Sci. **52**, 339 (2002)
11. Braaten, E., Hammer, H.-W.: Phys. Rept. **428**, 259 (2006)
12. Ando, S.-I.: Eur. Phys. J. A **52**, 130 (2016)
13. Ando, S.-I.: arXiv:1806.09073
14. Ando, S.-I.: Phys. Rev. C **97**, 014604 (2018)
15. Tischhauser, P., et al.: Phys. Rev. C **79**, 055803 (2009)

Chapter 21
Astrophysical S-Factor of the Direct $\alpha(d, \gamma)^6$Li Capture Reaction in a Three-Body Model

E. M. Tursunov, D. Baye and S. A. Turakulov

Abstract At the long-wavelength approximation, electric dipole transitions are forbidden between isospin-zero states. In an $\alpha + n + p$ model with $T = 1$ contributions, the $\alpha(d, \gamma)^6$Li astrophysical S-factor is in agreement with the experimental data of the LUNA collaboration, without adjustable parameter. The exact-masses prescription used to avoid the disappearance of $E1$ transitions in potential models is not founded at the microscopic level.

21.1 Introduction

A radiative-capture reaction is an electromagnetic transition between an initial scattering state and a final bound state. Astrophysical collision energies can be very low with respect to the Coulomb barrier and cross sections are then tiny. The dominant multipolarity is $E1$ in general. In the special case of reactions between $N = Z$ nuclei however, $E1$ transitions are forbidden by an isospin selection rule at the long-wavelength approximation (LWA) and $E2$ transitions become crucial. Nevertheless, $E1$ transitions are not exactly forbidden since isospin is an approximate symmetry. The analysis of the recent LUNA data [1, 2] for the $\alpha(d, \gamma)^6$Li reaction indicates that $E1$ cross sections dominate the $E2$ cross sections below about 0.1 MeV. Since $E1$ transitions vanish in many models, recent calculations use the exact-masses prescription to avoid their disappearance [3]. Here we present results of the simplest model allowing $E1$ transitions thanks to small $T = 1$ components, i.e. the $\alpha + n + p$ three-body model.

E. M. Tursunov (✉) · S. A. Turakulov
Institute of Nuclear Physics, Uzbekistan Academy of Sciences,
100214 Ulugbek, Tashkent, Uzbekistan
e-mail: tursune@inp.uz

D. Baye
Physique Quantique, and Physique Nucléaire Théorique et Physique Mathématique,
C.P. 229, Université libre de Bruxelles (ULB), 1050 Brussels, Belgium

N. A. Orr et al. (eds.), *Recent Progress in Few-Body Physics*,
Springer Proceedings in Physics 238,
https://doi.org/10.1007/978-3-030-32357-8_21

21.2 Isospin-Forbidden $E1$ Transitions

The electric multipole operators read at the LWA,

$$\mathcal{M}_\mu^{E\lambda} = e\sum_{j=1}^{A}\left(\frac{1}{2} - t_{j3}\right) r_j'^{\lambda} Y_{\lambda\mu}(\Omega_j'), \tag{21.1}$$

where A is the number of nucleons, $\boldsymbol{r}'_j = (r'_j, \Omega'_j)$ is the coordinate of nucleon j with respect to the centre of mass of the system, and t_{j3} is the third component of its isospin operator $\boldsymbol{t}_j$. The isoscalar (IS) part of the $E1$ operator vanishes at the LWA since $\sum_{j=1}^{A} r'_j Y_{1\mu}(\Omega'_j) = 0$ and this operator becomes an isovector (IV),

$$\mathcal{M}_\mu^{E1} = -e\sum_{j=1}^{A} t_{j3} r'_j Y_{1\mu}(\Omega'_j) = \mathcal{M}_\mu^{E1,\mathrm{IV}}. \tag{21.2}$$

At the LWA, $E1$ matrix elements thus vanish between isospin-zero states. This leads to the total isospin T selection rule in $N = Z$ nuclei and reactions: $T_i = 0 \rightarrow T_f = 0$ is forbidden. But $E1$ transitions are not exactly forbidden in these $N = Z$ systems because isospin is not an exact quantum number. Small $T = 1$ admixtures appear in the wave functions. The main isovector $E1$ contributions are due to $T_f = 1$ admixtures in the final state or to $T_i = 1$ admixtures in the initial state. Moreover, the isoscalar $E1$ operator reads beyond the LWA,

$$\mathcal{M}_\mu^{E1,\mathrm{IS}} \approx -\frac{1}{60} e k_\gamma^2 \sum_{j=1}^{A} r_j'^3 Y_{1\mu}(\Omega'_j), \tag{21.3}$$

up to terms that should give only a small contribution [4], contrary to other expressions often used in the literature. The isoscalar $E1$ contribution to the capture involves the $T = 0$ parts of the wave functions.

21.3 Three-Body Model of $\alpha(d, \gamma)^6$Li Reaction

The present wave functions [5] are adapted from the $\alpha + n + p$ model of [3]. The $J_f = 1^+$ final bound state is described in hyperspherical coordinates and the initial scattering states are described in Jacobi coordinates. Three-body effective $E1$ and $E2$ operators are constructed which assume that the α particle or cluster is in its 0^+ ground state. For example, the isovector part of the effective three-body $E1$ operator reads at the LWA,

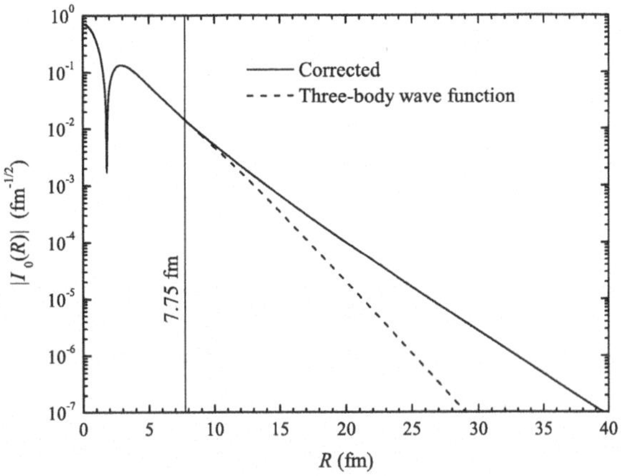

Fig. 21.1 $E2$ overlap integral $|I_0(R)|$ with and without correction beyond 7.75 fm

$$\widetilde{\mathcal{M}}_{\mu}^{E1,\text{IV}} = \frac{1}{2} e r Y_{1\mu}(\Omega_r), \tag{21.4}$$

where $\boldsymbol{r}$ is the Jacobi coordinate between n and p. The expressions of the isoscalar part of the $E1$ operator beyond the LWA and of the $E2$ operator can be found in [5].

The three-body states contain $S = 0$ and 1 components. Because of the isospin zero of the α particle and the antisymmetry of the $n + p$ subsystem with orbital momentum l, the components with $l + S$ odd correspond to $T = 0$ and those with $l + S$ even to $T = 1$. The initial scattering state is described by the product of a frozen deuteron wave function ($l_i = 0$, $S_i = 1$) and $\alpha + d$ L partial scattering waves. Hence, it is purely $T_i = 0$. The $J_f = 1^+$ final bound state contains a small $T_f = 1$ component (about 0.5 %). The $E1$ transitions start from $L_i = 1$ and the $E2$ transitions from $L_i = 0$ and 2.

This model requires an asymptotic correction to the $E2$ matrix elements. Indeed the overlap integrals $I_L(R)$ of the deuteron and $\alpha + n + p$ final wave functions decrease too fast beyond 10 fm as shown for $L = 0$ by Fig. 21.1. This is corrected by matching at 7.75 fm the overlap integrals with the exact Whittaker asymptotic function multiplied by realistic asymptotic normalization coefficients.

Total $E1 + E2$ astrophysical S factors calculated in the three-body model with the $E2$ correction are compared in Fig. 21.2 with experimental data. The isoscalar $E1$ capture contribution is small and can be neglected in first approximation. The isovector $E1$ contribution dominates below about 0.1 MeV. Model A and B correspond to different $\alpha + N$ potentials (see [5] for details).

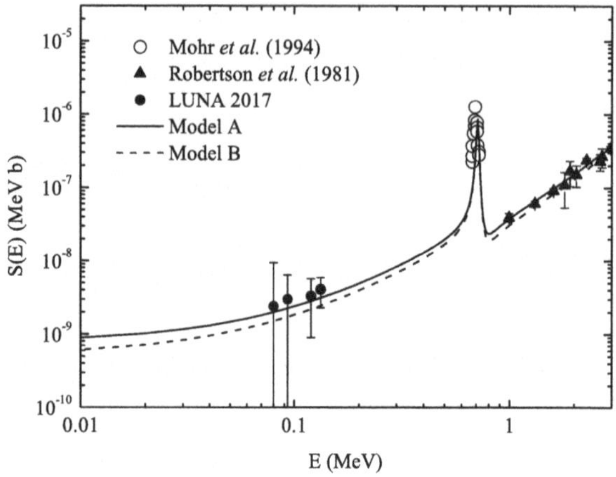

Fig. 21.2 Total $E1 + E2$ astrophysical S factor (full and dashed lines). Experimental data from [6] (triangles), [7] (open circles), and [2] (full circles). Adapted from [5]

21.4 Comment on the Exact-Masses Prescription

To obtain non-vanishing $E1$ transitions in the two-body or potential model, experimental masses are used in the effective charge of $N = Z$ nuclei,

$$Z_{\mathrm{eff}}^{(\mathrm{E1})} \propto \left(\frac{Z_1}{A_1} - \frac{Z_2}{A_2}\right) \to Z_{\mathrm{eff}}^{(\mathrm{E1})} \propto m_N \left(\frac{Z_1}{M_1} - \frac{Z_2}{M_2}\right) \tag{21.5}$$

where m_N is the nucleon mass and $Z_{1,2}$, $A_{1,2}$ and $M_{1,2}$ are the charges, mass numbers and experimental masses of the colliding nuclei, respectively. This exact-masses prescription is unfounded.

(i) $E1$ transitions would remain exactly forbidden in the $d(d,\gamma)^4$He reaction, in contradiction with ab initio calculations.

(ii) Using the mass expression $M = Am_N + (N - Z)\frac{1}{2}(m_n - m_p) - B(A, Z)/c^2$, effective charges would depend on the binding energies $B(A_{1,2}, Z_{1,2})$,

$$m_N \left(\frac{Z_1}{M_1} - \frac{Z_2}{M_2}\right) \approx \frac{1}{2m_N c^2} \left(\frac{B(A_1, Z_1)}{A_1} - \frac{B(A_2, Z_2)}{A_2}\right). \tag{21.6}$$

Binding energies per nucleon $B(A, Z)/A$ mostly depend on the main $T = 0$ components of the wave functions and not on the small $T = 1$ components physically responsible for the non vanishing of "forbidden" $E1$ transitions.

(iii) $E1$ matrix elements would be unphysically sensitive to the long $T_f = 0$ $\alpha + d$ tail of the ^{6}Li wave function.

21.5 Conclusion

Isovector $E1$ transitions with $T = 1$ admixtures in the final state and $E2$ transitions explain the order of magnitude of the LUNA data [1, 2], without any adjustable parameter. Isoscalar $E1$ transitions beyond the LWA are negligible for the $\alpha(d,\gamma)^6$Li reaction. The exact-masses prescription is not founded and should not be trusted for reactions between $N = Z$ nuclei, such as $\alpha(d,\gamma)^6$Li and ^{12}C$(\alpha,\gamma)^{16}$O. A three-body model with $T = 1$ admixtures in both initial and final states should be developed. Microscopic six-body and ab initio calculations are difficult but possible and necessary for a deeper understanding of this reaction.

References

1. Anders, M., et al.: Phys. Rev. Lett. **113**, 042501 (2014)
2. Trezzi, D., et al.: Astropart. Phys. **89**, 57 (2017)
3. Tursunov, E.M., et al.: Phys. Rev. C **94**, 015801 (2016)
4. Baye, D.: Phys. Rev. C **86**, 034306 (2012)
5. Baye, D., Tursunov, E.M.: J. Phys. G **45**, 085102 (2018)
6. Robertson, R.G.H., et al.: Phys. Rev. Lett. **47**, 1867 (1981)
7. Mohr, P., et al.: Phys. Rev. C **50**, 1543 (1994)

Chapter 22
Observation of New Neutron Resonances in ^{17,19}C

Y. Satou, S. Kim, J. W. Hwang, N. A. Orr, T. Nakamura, Y. Kondo, J. Gibelin, N. L. Achouri, T. Aumann, H. Baba, F. Delaunay, P. Doornenbal, N. Fukuda, N. Inabe, T. Isobe, D. Kameda, D. Kanno, N. Kobayashi, T. Kobayashi, T. Kubo, S. Leblond, J. Lee, F. M. Marqués, R. Minakata, T. Motobayashi, D. Murai, T. Murakami, K. Muto, T. Nakashima, N. Nakatsuka, A. Navin, S. Nishi, S. Ogoshi, H. Otsu, H. Sato, Y. Shimizu, H. Suzuki, K. Takahashi, H. Takeda, S. Takeuchi, R. Tanaka, Y. Togano, A. G. Tuff, M. Vandebrouck and K. Yoneda

Abstract High-lying isolated neutron resonances have been observed in ^{17,19}C. Analyses of the decay properties utilizing shell-model calculations suggest that they could be interpreted as possible variants of proton core excited states. The spins of these states are likely to be originating from anti-parallel coupling of spins of the $d_{5/2}$ neutron and the proton 2^+ state in the assumed respective core nuclei, ^{16,18}C.

Y. Satou (✉) · S. Kim · J. W. Hwang
Department of Physics and Astronomy, Seoul National University,
1 Gwanak-ro, Gwanak-gu, Seoul 08826, Republic of Korea
e-mail: ysatouyster@gmail.com

N. A. Orr · J. Gibelin · N. L. Achouri · F. Delaunay · S. Leblond · F. M. Marqués
LPC Caen, Normandie Université, ENSICAEN, Université de Caen, CNRS/IN2P3,
14050 Caen, France

T. Nakamura · Y. Kondo · D. Kanno · N. Kobayashi · R. Minakata · T. Nakashima · S. Nishi · S. Ogoshi · R. Tanaka
Department of Physics, Tokyo Institute of Technology, 2-12-1 O-Okayama,
Meguro, Tokyo 152-8551, Japan

T. Aumann
Institut für Kernphysik, Technische Universität Darmstadt,
64289 Darmstadt, Germany

T. Aumann · Y. Togano
ExtreMe Matter Institute EMMI and Research Division, GSI Helmholtzzentrum für Schwerionenforschung GmbH, 64291 Darmstadt, Germany

H. Baba · P. Doornenbal · N. Fukuda · N. Inabe · T. Isobe · D. Kameda · T. Kubo · J. Lee · T. Motobayashi · H. Otsu · H. Sato · Y. Shimizu · H. Suzuki · H. Takeda · S. Takeuchi · K. Yoneda
RIKEN Nishina Center, Hirosawa 2-1, Wako, Saitama 351-0198, Japan

T. Kobayashi · K. Muto · K. Takahashi
Department of Physics, Tohoku University, Aoba, Sendai, Miyagi 980-8578, Japan

N. A. Orr et al. (eds.), *Recent Progress in Few-Body Physics*,
Springer Proceedings in Physics 238,
https://doi.org/10.1007/978-3-030-32357-8_22

22.1 Introduction

There is currently a wide interest in investigating the structure of nuclei under extreme conditions, i.e., in terms of high isospin T_z and excitation energy E_x. Firstly, the spectroscopy of nuclei with $A < 20$ meets a frontier of ab initio structure calculations including continuum and three-body forces [1]. Secondly, nucleon-nucleon correlations, which manifest in the reduction of spectroscopic factors, C^2S, can best be examined in such nuclei [2, 3]. Thirdly, new insights into the nuclear many-body dynamics can be obtained from decay properties of deep hole states of both major and minor nucleon species [4, 5]. This study focuses on reporting new neutron resonances in ^{17,19}C, observed, respectively, by neutron-knockout from ^{18}C (245 MeV/nucleon) and proton-knockout from ^{20}N (257 MeV/nucleon) channels on carbon. A high-lying state populated by the (p, p') inelastic scattering at 70 MeV/nucleon [6] is revisited.

22.2 Experiment and Results

The SAMURAI spectrometer [7] at RIKEN-RIBF [8] was utilized for the measurement, which was based on the invariant mass method involving detection of a neutron and a charged fragment emitted from the unbound knockout residue. The experimental setup shared that described in [9–11].

The new states were observed above the one-neutron $(1n)$ threshold of $S_n = 0.735$ MeV in ^{17}C: $E_x \sim 1.55$ MeV [12], and the three-neutron threshold of $S_{3n} = 5.5$ MeV in ^{19}C: $E_x \sim 6.6$ MeV [13], in the $1n$ decay channels. The $E_x \sim 1.55$-MeV state did not appear as an isolated resonance; its presence was inferred from a comparison of the spectra obtained with and without requiring coincidence detection of the ^{16}C(2^+) de-excitation γ rays, with the consequence for this state to appear only in the former

D. Murai
Department of Physics, Rikkyo University, 3 Nishi-Ikebukuro, Toshima, Tokyo 171-8501, Japan

T. Murakami · N. Nakatsuka
Department of Physics, Kyoto University, Kyoto 606-8502, Japan

A. Navin
GANIL, CEA/DRF-CNRS/IN2P3, 14076 Caen Cedex 5, France

A. G. Tuff
Department of Physics, University of York, Heslington, York YO105DD, UK

M. Vandebrouck
Institut de Physique Nucléaire, Université Paris-Sud, IN2P3/CNRS, 91406 Orsay, France

Y. Satou
Rare Isotope Science Project, Institute for Basic Science, 102, Somunsanseong-gil, Yuseong-gu, Daejeon 34000, Republic of Korea

inclusive relative energy spectrum [12]. The $E_x \sim 6.6$-MeV state in ^{19}C was observed to reside on a structure-less continuum background, and seemingly to feed $^{18}C_{g.s.}$ in its decay. Such a high-lying isolated state reminds us of a state at $E_x = 6.13(9)$ MeV ($\Gamma_r = 0.26^{+0.4}_{-0.26}$ MeV) in ^{17}C, observed in a (p, p') study, to be feeding $^{16}C(2^+)$ after neutron emission [6].

22.3 Discussion

Analyses on the above three states based on shell model and eikonal/DWBA reaction model, taking account of the resonance energy, population strength, and decay properties were attempted to characterize their microscopic structures. The results are summarized in Table 22.1. Branching ratio B.R.th (values for the most and second most dominant branches are given together with the decay energy E^{th}_{rel} in Table 22.1) and width Γ^{th} for $1n$ emission to energetically allowed states in the daughter nucleus were calculated using the analytic formulas assuming a square well potential with centrifugal barrier [16]:

$$\Gamma_{s.p.} \approx 2\hbar^2/(\mu R^2) \cdot kR \cdot v_l(kR) \cdot (2l-1)/(2l+1) \quad (l > 0, \quad kR < l^{1/2}), \tag{22.1}$$

$$\Gamma_{s.p.} \approx 2\hbar^2/(\mu R^2) \cdot kR \cdot v_0(kR) \quad (l = 0), \tag{22.2}$$

$$\Gamma_i = \Sigma_j C^2 S_i(j) \cdot \Gamma_{s.p.}, \quad \Gamma^{th} = \Sigma_i \Gamma_i, \quad \text{B.R.}^{th} = \Gamma_i/\Gamma^{th}. \tag{22.3}$$

The radius R was that of the daughter nucleus. The wave number was calculated by $k = (2\mu E^{th}_{rel})^{1/2}/\hbar c$, with μ the reduced mass of the decay products. v_l deals in the transmission of a neutron through the centrifugal barrier. $C^2 S_i(j)$ refers to the spectroscopic overlap between the initial state and the i-th daughter state, with j the

Table 22.1 Experimental resonance parameters compared with those from shell model utilizing the YSOX interaction [14], as calculated using NUSHELLX@MSU [15]. Theoretical excitation energy E^{th}_x refers to the energy interval from the predicted lowest energy state. The experimental values for the $E^{exp}_{rel} = 3.63(9)$-MeV state in ^{17}C are from [6]. σ^{th} refers to the theoretical population cross section for the relevant reaction

Nucleus	E^{exp}_{rel} (MeV)	E^{exp}_x (MeV)	σ^{exp} (mb)	J^{π}_{SM}	E^{th}_x (MeV)	B.R.th(%)	E^{th}_{rel} (MeV)	Γ^{th} (MeV)	σ^{th} (mb)
^{17}C	~0.81	~1.55	~0.57	$5/2^+_2$	1.69	100 (0^+_1)	0.96	3.2 × 10^{-4}	0.43
	3.63(9)	6.13(9)	0.8(1)	$1/2^+_4$	6.85	18 (0^+_1) 76 (2^+_1)	6.11 4.35	0.33	0.14
^{19}C	~6.0	~6.6	~1.5	$1/2^+_3$	7.51	54 (0^+_1) 24 (2^+_1)	6.93 5.34	1.53	2.11

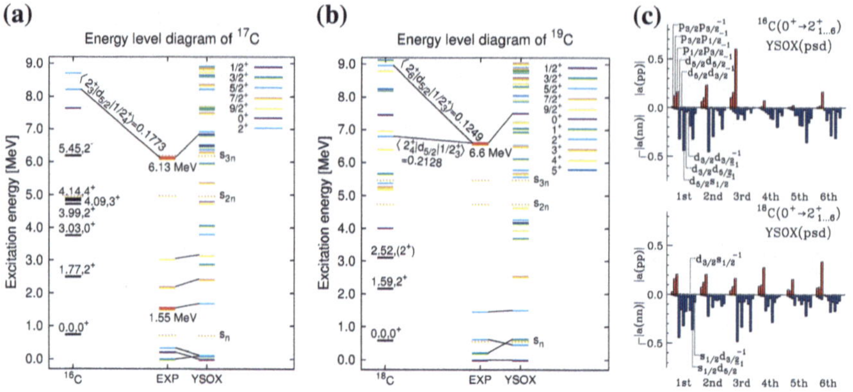

Fig. 22.1 Energy level diagrams of **a** ^{17}C and **b** ^{19}C compared with YSOX shell-model spectroscopy. States discussed are marked by red. **c** One-body transition densities in the proton-neutron representation |a(pp/nn)| for quadrupole transitions in ^{16}C (top) and ^{18}C (bottom)

angular momentum involved in the decay. The dependence on j of $\Gamma_{\text{s.p.}}$ (through l) is implicit in the first relationship of (22.3). σ^{th} for the $E_x = 6.13$-MeV state was calculated as in [6] including the quadrupole correction [17], while those for the other states as in [11].

Figure 22.1 shows the comparison in excitation energy between experiment and shell model for the states in (a) ^{17}C and (b) ^{19}C. Energy levels of the $1n$ decay products ^{16,18}C are also depicted (shell-model excitation energies, colored, substitute experimental ones in the E_x region where the latter are unavailable). A reported negative parity 2^- state at 5.45 MeV in ^{16}C [18] has been incorporated in Fig. 22.1a. Shell-model one-body transition density coefficients for the $0_1^+ \to 2^+$ quadrupole transitions in ^{18,20}C are depicted in panel (c), where the decomposed strengths of proton and neutron amplitudes facilitate identifying the dominant excitation mode for each 2^+ state.

The $E_x \sim 1.55$-MeV state in ^{17}C was well explained in terms of the excitation energy, population cross section, and decay scheme by assuming $5/2_2^+$. Decay pattern for the $E_x = 6.13$-MeV state turned out to fit well with that of the $1/2_4^+$ state at $E_x^{\text{th}} = 6.85$ MeV (YSOX [14]). The finite width of this state could be due to low spin of $J = 1/2$. For the state with such a low spin, the coupling to high-lying proton core excited 2^+ states in ^{16}C (especially 2_3^+), which are well above the two neutron threshold of ^{16}C and could have appreciable widths, via s wave is hampered due to the angular momentum selection rule, leading to less chances for its destruction. A large calculated d-wave spectroscopic overlap (C^2S) between the 2_3^+ state in ^{16}C and the $1/2_4^+$ state in ^{17}C: $\langle 2_3^+ | d_{5/2} | 1/2_4^+ \rangle = 0.1773$ is noted. The predicted (p, p') cross section ($\sigma^{\text{th}}_{(p,p')} = 0.14$ mb) seems to be not fully consistent with the observation. It is noted that the corresponding cross section leading to $1/2_3^+$ predicted by WBT [19] within the psd model space (in WBT adopted in [6], the third $1/2^+$ state exhibited the observed decay pattern) is calculated to be $\sigma^{\rm th}_{(p,p')} = 0.17$ mb, slightly closer to the experimental value of 0.8(1) mb.

The knockout of a $p_{3/2}$ proton from ^{20}N, whose $J^{\pi}_{\rm g.s.} = 2^-$ [20–22], creates a variety of states in the excitation energy range of $E_x = 6 - 10$ MeV in ^{19}C: $^{20}{\rm N}(2^-) \otimes \pi p^{-1}_{3/2} \rightarrow 1/2^+, 3/2^+, 5/2^+, 7/2^+$. Among them, again the low-spin $J = 1/2$ state is preferred as the candidate for the narrow width state at $E_x \sim 6.6$ MeV in ^{19}C, since it couples to the high-lying proton core exited 2^+ states in ^{18}C only via d wave. The $7/2^+$ state is calculated to populate via $1n$ decay largely higher spin states predicted above S_n in ^{18}C, and thus is unlikely to be observed in the $1n$ emission channel. It turned out that in the YSOX spectroscopy the third $1/2^+$ well matches the $E_x \sim$ 6.6-MeV state in the light of the present comparisons between theory and experiment. Large d-wave spectroscopic overlaps (C^2S) between the $2^+_{4,6}$ states in ^{18}C, which exhibit enhanced proton amplitudes (as seen, especially, in the $\pi p_{1/2} p^{-1}_{3/2}$ amplitude), and the $1/2^+_3$ state in ^{19}C have been calculated as $\langle 2^+_4 | d_{5/2} | 1/2^+_3 \rangle = 0.2128$ and $\langle 2^+_6 | d_{5/2} | 1/2^+_3 \rangle = 0.1249$.

22.4 Summary

An attempt to characterize high-lying isolated states, which appeared above $E_x = 6$ MeV in ^{17,19}C and were observed to decay by $1n$ emission, was made referring to shell-model excitation energy, population strength, and spectroscopic overlap (with both high- and low-lying states in daughter nuclei). Tentative $1/2^+$ assignments would be preferred for them. The $E_x = 6.13$-MeV state in ^{17}C was formerly suggested to be $5/2^+_4$ in WBT [6], thus a new candidate for its J^{π} was added in this study. The first neutron unbound state in ^{17}C was also noted. The decay pattern of the presently observed high-lying $(1/2^+)$ states resembles that of the Auger effect, where the disintegration of an atomic inner hole state is associated with emission of an Auger electron. Detailed and systematic investigation of this category of (proton core exited) state will provide new insights into the aspect of nuclei as a highly correlated open quantum system, and into the evolution of deeply bound proton orbits in neutron-rich isotopic chains.

Acknowledgements The present work was supported by the WCU (R32-2008-000-10155-0), NRF grants (NRF-2011-0006492, NRF-2018R1A5A1025563), and RISP (2013 M7A1A 1075764) of IBS in Korea, and JSPS KAKENHI Grant Number 16H02179 and MEXT KAKENHI Grant Number 24105005 in Japan. NLA, FD, JG, FMM, and NAO acknowledge partial support from the French-Japanese LIA-International Associated Laboratory for Nuclear Structure Problems. AN and JG would like to acknowledge the JSPS Invitation fellowship programme for long term research in Japan at the Tokyo Institute of Technology and RIKEN respectively. SL gratefully acknowledges the support provided by the RIKEN International Associate Programme and the hospitality of the Nishina Center staff during his sojourn.

References

1. Smalley, D., et al.: Phys. Rev. C **92**, 064314 (2015)
2. Tostevin, J.A., Gade, A.: Phys. Rev. C **90**, 057602 (2014)
3. Atar, L., et al.: Phys. Rev. Lett. **120**, 052501 (2018)
4. Gottardo, A., et al.: Phys. Lett. B **772**, 359 (2017)
5. Macchiavelli, A.O., et al.: Phys. Rev. C **90**, 067305 (2014)
6. Satou, Y., et al.: Phys. Lett. B **660**, 320 (2008)
7. Kobayashi, T., et al.: Nucl. Instrum. Methods Phys. Res. Sect. B **317**, 294 (2013)
8. Yano, Y.: Nucl. Instrum. Methods Phys. Res. Sect. B **261**, 1009 (2007)
9. Kondo, Y., et al.: Phys. Rev. Lett. **116**, 102503 (2016)
10. Togano, Y., et al.: Phys. Lett. B **761**, 412 (2016)
11. Hwang, J.W., et al.: Phys. Lett. B **769**, 503 (2017), ibid. **774**, 723 (2017)
12. Kim, S., et al.: AIP Conf. Proc. **1947**, 020006 (2018)
13. Hwang, J.W., et al.: In preparation
14. Yuan, C., et al.: Phys. Rev. C **85**, 064324 (2012)
15. Brown, B.A., Rae, W.D.M.: Nucl. Data Sheets **120**, 115 (2014)
16. Bohr, A., Mottelson, B.R.: Nuclear Structure I. World Scientific, Singapore (1998)
17. Sagawa, H., et al.: Phys. Rev. C **70**, 054316 (2004)
18. Satou, Y., et al.: Phys. Lett. B **728**, 462 (2014)
19. Warburton, E.K., Brown, B.A.: Phys. Rev. C **46**, 923 (1992)
20. Sauvan, E., et al.: Phys. Lett. B **491**, 1 (2000)
21. Sohler, D., et al.: Phys. Rev. C **77**, 044303 (2008)
22. Rodriguez-Tajes, C., et al.: Phys. Rev. C **83**, 064313 (2011)

Chapter 23
Two-Nucleon Emitters Within a Pseudostate Approach

J. Casal and J. Gómez-Camacho

Abstract A method to identify and characterize three-body resonances in a discrete basis is discussed in the context of two-nucleon emitters. For this purpose, a resonance operator is introduced and diagonalized in a basis of energy pseudostates within the hyperspherical formalism. Then, the energy and width of the resonance are obtained from its time dependence. The approach is illustrated for ^{16}Be (^{14}Be $+ n + n$).

23.1 Introduction

The study of two-nucleon correlations beyond the driplines has gained renewed attention since the recent experimental observation of direct two-proton [1, 2] and two-neutron [3, 4] decays. These are typically discussed in terms of different possible paths: The so-called sequential, direct and democratic decays [5]. From the theoretical point of view, a proper description of three-body resonances can help in understanding these correlations [6, 7]. The description of few-body resonant states, however, is not an easy task. In this work, we briefly describe a robust approach to identify and characterize three-body resonances in a discrete basis within the hyperspherical formalism, and we apply the method to ^{16}Be (^{14}Be $+ n + n$). For further details, see [8].

J. Casal (✉)
European Centre for Theoretical Studies in Nuclear Physics and Related Areas (ECT*), Strada delle Tabarelle 286, 38123 Trento, Italy
e-mail: casal@pd.infn.it

Dipartimento di Fisica e Astronomia "G. Galilei" and INFN - Sezione di Padova, Via Marzolo 8, 35131 Padua, Italy

J. Gómez-Camacho
Departamento de Física Atómica, Molecular y Nuclear, Facultad de Física, Universidad de Sevilla, Apartado 1065, 41080 Seville, Spain

Centro Nacional de Aceleradores (CNA), U. Sevilla, J. Andalucía, CSIC, Tomas A. Edison 7, 41092 Seville, Spain

N. A. Orr et al. (eds.), *Recent Progress in Few-Body Physics*, Springer Proceedings in Physics 238,
https://doi.org/10.1007/978-3-030-32357-8_23

23.2 Resonance Operator

In general, resonances correspond to a range of continuum energy eigenstates whose probabilities are concentrated within the potential well. Since these continuum structures will be very sensitive to changes in the potential, we introduce the resonance operator

$$\widehat{M} = \widehat{H}^{-1/2}\widehat{V}\widehat{H}^{-1/2}, \quad \widehat{M}|\psi\rangle = m|\psi\rangle \tag{23.1}$$

whose eigenstates $|\psi\rangle$ can be expanded in Hamiltonian pseudostates $|n\rangle$,

$$|\psi\rangle = \sum_n C_n|n\rangle, \quad \widehat{H}|n\rangle = \varepsilon_n|n\rangle. \tag{23.2}$$

Note that, if the system has no bound states, all energies ε are positive. Therefore, since $m \sim V/\varepsilon$, resonances can be identified from the eigenstates of $\widehat{M}$ corresponding to large negative eigenvalues.

We apply the method to study ^{16}Be(0^+) states in a three-body (^{14}Be $+ n + n$) model using the potentials in [6, 7]. Calculations are performed within the hyperspherical formalism [9], where the relevant parameters are the maximum hypermomentum K_{max} (which determines the number of angular components in the wave-function expansion), the number of basis functions N, and the scale parameter γ controlling the radial extension of the basis [10]. Here, smaller γ values correspond to more extended basis functions and a larger concentration of energy pseudostates just above the breakup threshold. In Fig. 23.1, we present the spectra of $\widehat{M}$ for three different bases, where the lowest eigenstate is stable and clearly separated from the

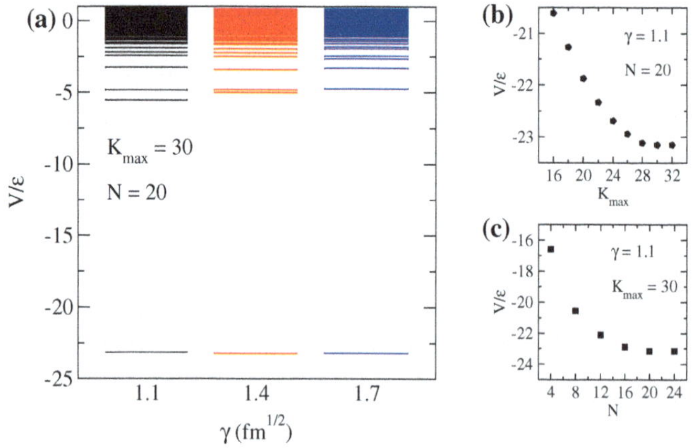

Fig. 23.1 **a** Eigenvalues of $\widehat{M}$ for ^{16}Be(0^+) as a function of the basis parameter γ. The right panel shows the convergence of the lowest eigenstate as a function of **b** the maximum hypermomentum K_{max} and **c** the number of basis functions N. In each case, the other two parameters are fixed

rest. This state represents the ground-state resonance of ^{16}Be, which shows a fast convergence with respect to the size of the model space, as shown in the right panel. As discussed in [7, 8], the corresponding wave function presents a dominant dineutron component, which favors the picture of a correlated two-neutron emission from the ground state of ^{16}Be.

23.3 Time Dependence and Resonance Parameters

As time evolves, the state given by (23.2) loses its character, and we can define a time-dependent amplitude

$$|\psi(t)\rangle = \sum_n C_n e^{-i\varepsilon_n t}|n\rangle, \quad a(t) = \langle\psi(0)|\psi(t)\rangle = \sum_n |C_n|^2 e^{-i\varepsilon_n t}. \tag{23.3}$$

For an ideal Breit-Wigner resonance, we would expect

$$a_r(t) = e^{-i\varepsilon_r t - \frac{\Gamma}{2}t}, \tag{23.4}$$

given by the resonance energy ε_r and its width Γ. These parameters can be chosen so that (23.4) is as close as possible to the amplitude in (23.3). We define a resonance quality parameter with the meaning of a quadratic deviation

$$\delta^2(\varepsilon_r, \Gamma) = \frac{\int_0^\infty e^{-xt}\,|a(t) - a_r(t)|^2\,dt}{\int_0^\infty e^{-xt}\,|a(t)|^2\,dt}, \tag{23.5}$$

where $1/x$ corresponds to a relevant time scale for the resonance formation. Thus, small values of x will be related to long times associated to the decay. The resonance parameters, as a function of x, can be obtained by minimizing (23.5),

$$\frac{\partial}{\partial\varepsilon_r}\delta^2(\varepsilon_r, \Gamma) = 0, \quad \frac{\partial}{\partial\Gamma}\delta^2(\varepsilon_r, \Gamma) = 0. \tag{23.6}$$

Details on the derivation of the relevant equations can be found in [8].

For the 0^+ ground state of ^{16}Be, we obtain the results presented in Fig. 23.2. Here we show the convergence of the resonance parameters $\varepsilon_r(x)$ and $\Gamma(x)$ with respect to K_{max}, N and γ as in the preceding section. With $K_{max} = 30$ and $N = 20$, we show that the resonance energy and width are fully converged. Both functions follow approximately a linear trend, with a small slope, for small values of x. Then, a sudden drop of the width is observed close to zero. As discussed in [8], this occurs when a pseudostate energy ε_n matches the resonance energy and is a consequence of the discrete nature of the basis. By increasing the level density near the threshold (i.e., choosing smaller values of γ), the linear trend explores smaller values of x. Therefore, it is reasonable to fix the resonance parameters ε_r and Γ by extrapolating this linear

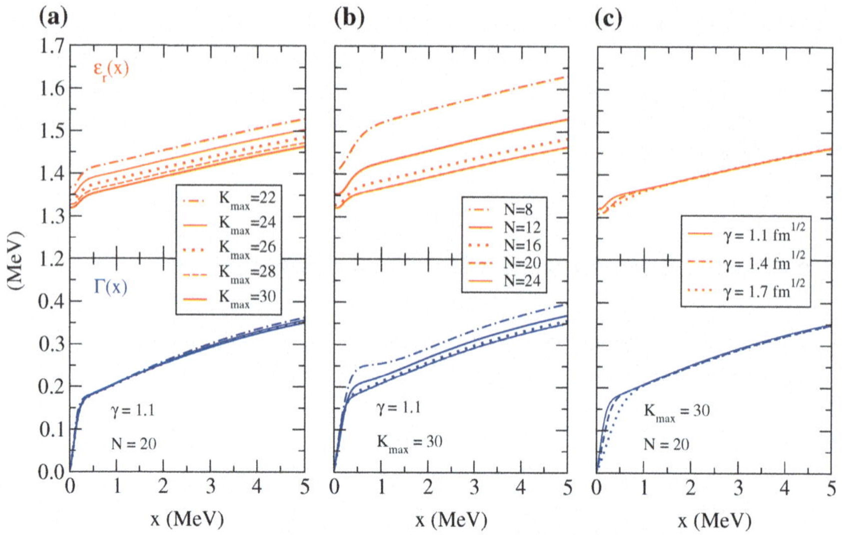

Fig. 23.2 Convergence of the resonance parameters ε_r and Γ with respect to **a** K_{max}, **b** N and **c** γ. In each case, the other two parameters are fixed

Table 23.1 Convergence of ε_r and Γ as a function of K_{max} and N, with $\gamma = 1.1$ fm$^{1/2}$

K_{max}	ε_r (MeV)	Γ (MeV)	N	ε_r (MeV)	Γ (MeV)
22	1.403	0.163	8	1.489	0.228
24	1.379	0.160	12	1.399	0.188
26	1.363	0.159	16	1.359	0.168
28	1.347	0.159	20	1.342	0.160
30	1.341	0.160	24	1.341	0.160

behavior to $x = 0$. Following this prescription, for an energy of $\varepsilon_r(0^+) = 1.341$ MeV we obtain $\Gamma(0^+) = 0.160$ MeV. The convergence of these values with respect to K_{max} and N is shown in Table 23.1. The computed width is consistent with the results in [6] from the three-body eigenphases within the hyperspherical R-matrix method to solve the actual three-body scattering problem. This is an indication that the method here presented provides a reasonable description of three-body resonances in a discrete basis.

Acknowledgements This work has been supported by the Spanish Ministerio de Ciencia, Innovación y Universidades and FEDER funds under Projects No. FIS2017-88410-P, FPA2016-77689-C2-1-R and FIS2014-51941-P, and by the European Union's Horizon 2020 research and innovation program under grant agreement No. 654002.

References

1. Giovinazzo, J., et al.: Phys. Rev. Lett. **89**, 102501 (2002)
2. Grigorenko, L.V., et al.: Phys. Lett. B **677**, 30 (2009)
3. Spyrou, A., et al.: Phys. Rev. Lett. **108**, 102501 (2012)
4. Kohley, Z., et al.: Phys. Rev. Lett. **110**, 152501 (2013)
5. Pfützner, M., et al.: Rev. Mod. Phys. **84**, 567 (2012)
6. Lovell, A.E., Nunes, F.M., Thompson, I.J.: Phys. Rev. C **95**, 034605 (2017)
7. Casal, J.: Phys. Rev. C **97**, 034613 (2018)
8. Casal, J., Gómez-Camacho, J.: Phys. Rev. C **99**, 014604 (2019)
9. Zhukov, M.V., et al.: Phys. Rep. **231**, 151 (1993)
10. Casal, J., Rodríguez-Gallardo, M., Arias, J.M.: Phys. Rev. C **88**, 014327 (2013)

Chapter 24
The Hoyle State in Relativistic ^{12}C Dissociation

D. A. Artemenkov, M. Haiduc, N. K. Kornegrutsa, E. Mitsova, N. G. Peresadko, V. V. Rusakova, R. Stanoeva, A. A. Zaitsev, P. I. Zarubin and I. G. Zarubina

Abstract Production of α-particle triples in the Hoyle state (HS) in dissociation of ^{12}C nuclei at 3.65 and 0.42 A GeV in nuclear track emulsion is revealed by the invariant mass approach. Contribution of the HS to the dissociation $^{12}\text{C} \rightarrow 3\alpha$ is $(11 \pm 3)\%$. Reanalysis of data on coherent dissociation $^{16}\text{O} \rightarrow 4\alpha$ at 3.65 A GeV is revealed the HS contribution of $(22 \pm 2)\%$.

Events of dissociation of relativistic light nuclei observable in detail in the nuclear track emulsion (NTE) contain holistic information on ensembles of lightest nuclei which is of interest to the nuclear cluster physics [1]. The best spatial resolution provided by the NTE technique turns out to be a decisive factor for recognition relativistic ^{8}Be and ^{9}B decays among the projectile fragments [2]. The decays are identified by the invariant mass M^* defined by the sum of all products of 4-momenta P_i of relativistic fragments He and H. Subtracting the sum of the residual masses M is a matter of convenience $Q = M^* - M$. The components P_i are determined by the fragment emission angles under the assumption of conservation a projectile momentum per nucleon. In such an approach the contribution of the decays $^9\text{B} \rightarrow {}^8\text{Be}p \rightarrow 2\alpha p$ in relativistic dissociation of the isotopes ^{10}B and ^{10,11}C is revealed and, then, an indication to the resonance around 4 MeV in the channel $^{10}\text{C} \rightarrow {}^9\text{B}p$ found [2].

D. A. Artemenkov · N. K. Kornegrutsa · E. Mitsova · V. V. Rusakova · A. A. Zaitsev · P. I. Zarubin (✉) · I. G. Zarubina
Joint Institute for Nuclear Research (JINR), Dubna, Russia
e-mail: zarubin@lhe.jinr.ru

M. Haiduc
Institute of Space Science, Magurele, Romania

E. Mitsova · R. Stanoeva
Southwestern University, Blagoevgrad, Bulgaria

N. G. Peresadko · A. A. Zaitsev · P. I. Zarubin
Lebedev Physical Institute, Russian Academy of Sciences, Moscow, Russia

R. Stanoeva
Institute for Nuclear Research and Nuclear Energy, Sofia, Bulgaria

N. A. Orr et al. (eds.), *Recent Progress in Few-Body Physics*,
Springer Proceedings in Physics 238,
https://doi.org/10.1007/978-3-030-32357-8_24

This experience gave confidence to search in the relativistic dissociation $^{12}C \to 3\alpha$ for the Hoyle state (HS) or the second ^{12}C exited state exceeding just 378 keV the 3α-threshold [3–5]. The HS studies are reviewed in [6]. This search is inspired by the concept of the α-particle Bose-Einstein condensate whose status is discussed most recently in [6]. In the 90s, the ^8Be contribution was determined by smallest α-pair opening angles in "white" stars (coherent dissociation not accompanied by any target fragment or produced meson) $^{12}C \to 3\alpha$ [7] and $^{16}O \to 4\alpha$ [8] at energy of 3.65 *A* GeV. Nowadays, the persisted NTE plates and the relevant data files [7, 8] are reused to obtain distributions over the α-triple invariant mass $Q_{3\alpha}$. Data on the 72 (G.M. Chernov's group, Tashkent) and 114 "white" stars $^{12}C \to 3\alpha$ (A.Sh. Gaitinov's group, Alma-Ata) underwent reanalysis. Besides, additional data on 238 3α stars including 130 "white" ones are obtained recently in this exposure. Then, the NTE layers exposed to 420 *A* MeV ^{12}C nuclei allow one to verify the invariant mass approach [5]. In the latter case the α-particle emission angles are measured in the 86 found 3α events including the 36 "white" stars.

The distribution $Q_{3\alpha}$ for all 510 stars is shown in Fig. 24.1. There is a peak in the region $Q_{3\alpha} < 1$ MeV for the 51 stars where HS signal is expected. For events at 3.65 *A* GeV the mean value for the events contributed in the peak $\langle Q_{3\alpha} \rangle$ (RMS) is 397 ± 26 (166) keV, and at 420 *A* MeV, respectively, 346 ± 28 (85) keV. According to the condition $Q_{3\alpha} < 0.7$ MeV 42 (of 424) events at 3.65 *A* GeV can be attributed to HS and 420 *A* MeV–9 (of 86), including 5 "white" stars (of 36). In sum, the contribution of the HS decays to the dissociation of $^{12}C \to 3\alpha$ is (11 ± 3)%.

HS could emerge in the dissociation $^{16}O \to {}^{12}C^* (\to 3\alpha) + \alpha$. Figure 24.2 shows the distribution $Q_{3\alpha}$ of all 3α combinations in the 641 "white" stars [8]. While its main part limited $Q_{3\alpha} < 10$ MeV is described by the Rayleigh distribution there is the peak at $Q_{3\alpha} < 700$ keV. The condition $Q_{2\alpha} < 200$ keV meaning at least one ^8Be per 4α-event does not affect the statistics in this region. The main part contribution in the peak estimated at 8% is excluded. The remaining 139 events have an average value $\langle Q_{3\alpha} \rangle = (349 \pm 14)$ keV corresponding to HS and RMS 174 keV defined by the

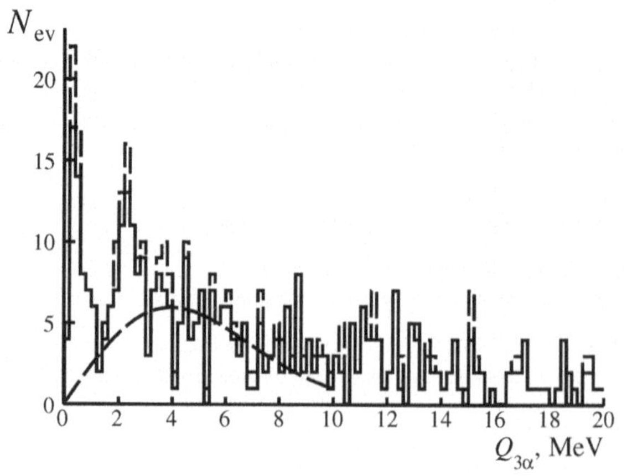

Fig. 24.1 Distribution over invariant mass $Q_{3\alpha}$ of all α-triples in dissociation of $^{12}C \to 3\alpha$ at 3.65 *A* GeV (shaded) and 420 *A* MeV (added by dashed line); line—approximation by the Rayleigh distribution with the parameter $\sigma_{Q_{3\alpha}}$ = (3.9 ± 0.4) MeV

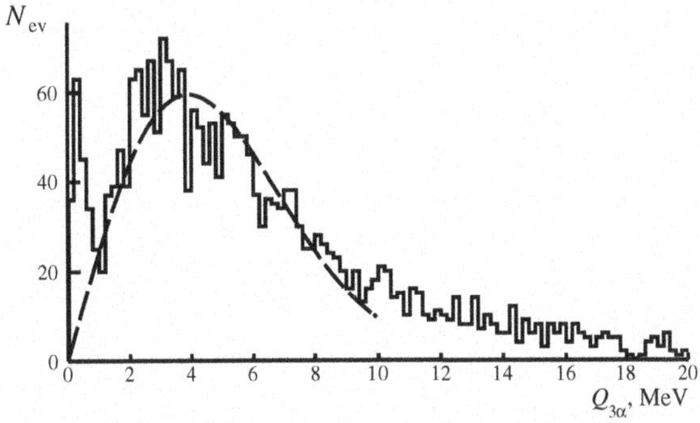

Fig. 24.2 Distribution over invariant mass $Q_{3\alpha}$ of all 3α combinations in 641 events of "white" stars $^{16}O \to 4\alpha$; line—approximation by the Rayleigh distribution with the parameter $\sigma_{Q_{3\alpha}} = (3.8 \pm 0.2)$ MeV

method resolution. In 9 of them more than one 3α-combination meets the condition $Q_{3\alpha} < 700$ keV. In sum, the contribution of HS decays to coherent dissociation of $^{16}O \to 4\alpha$ is $(22 \pm 2)\%$. Besides, HS can arise the α-decay product of the ^{16}O 0_6^+ state [9] (in an analogy with the HS decay $^8Be + \alpha$). The condition $Q_{4\alpha} < 1$ MeV allocates 9 events satisfying with a mean value $\langle Q_{4\alpha} \rangle = (624 \pm 84)$ keV at RMS 252 keV. Then, the estimate of the 0_6^+ contribution is $(7 \pm 2)\%$.

Thus, HS identified in the relativistic ^{12}C dissociation is manifested in the ^{16}O case. These observations indicate that it is not reduced to the unusual ^{12}C excitation and, like ^{8}Be, is a more universal object of nuclear molecular nature. The closest confirmation of this assumption would be the HS observation in relativistic fragmentation $^{14}N \to 3\alpha$. The analysis of the NTE layers exposed to relativistic ^{14}N nuclei [10] is resumed in the HS context.

References

1. Zarubin, P.I.: Lect. Notes Phys. Clust. Nuclei **875**(3), 51–93 (2014). Springer International Publishing. arXiv: 1309.4881
2. Artemenkov, D.A., Zaitsev, A.A., Zarubin, P.I.: Phys. Part. Nucl. **48**, 147–157 (2017). arXiv: 1607.08020
3. Artemenkov, D.A., Mamatkulov, K.Z., Kharlamov, S.P., Zaitsev, A.A., Zarubin, P.I.: Few-Body Syst. **58**, 89 (2017)
4. Belaga, V.V., Gerasimov, S.G., Dronov, V.A., Peresadko, N.G., Pisetskaya, A.V., Rusakova, V.V., Fetisov, V.N., Kharlamov, S.P., Shesterkina, L.N.: Phys. At. Nucl. **80**, 650–656 (2017)
5. Artemenkov, D.A., et al.: Rad. Meas. **119**, 119–203 (2018). arXiv:1812.09096
6. Freer, M., Fynbo, H.O.U.: Prog. Part. Nucl. Phys. **78**, 1–23 (2014)
7. Belaga, V.V., Benjaza, A.A., Rusakova, V.V., Salomov, D.A., Chernov, G.M.: Phys. At. Nucl. **58**, 1905 (1995). arXiv:1109.0817
8. Andreeva, N.P., et al.: Phys. At. Nucl. **59**, 102–108 (1996). arXiv:1109.3007
9. Schuck, P.: arXiv:1811.11580
10. Shchedrina, T.V., et al.: Phys. At. Nucl. **70**, 1230–1234 (2007). arXiv: nucl-ex/0605022

Chapter 25
Differential Cross Sections and Analyzing Powers of the Three-Body Break-Up Channel in Deuteron-Deuteron Scattering at 65 MeV/nucleon

Reza Ramazani-Sharifabadi, M. T. Bayat, N. Kalantar-Nayestanaki, St. Kistryn, A. Kozela, M. Mahjour-Shafiei, J. G. Messchendorp, M. Mohammadi-Dadkan, A. Ramazani Moghaddam Arani, E. Stephan and H. Tavakoli Zaniani

Abstract In this contribution, the results for the differential cross sections and a rich set of vector and tensor analyzing powers are presented for various configurations of the three-body break-up channel of ^{2}H(**d**, dp)n. The data were obtained at KVI using a polarised deuteron beam with a beam energy of 65 MeV/nucleon and a liquid deuterium target in combination with a 4π detection system carrying the name **B**ig-**I**nstrument for **N**uclear-Polarization **A**nalysis (BINA). This work extends the results from an earlier study reported in [1]. In particular, additional kinematical configurations and spin observables were analyzed and obtained.

R. Ramazani-Sharifabadi (✉) · M. T. Bayat · N. Kalantar-Nayestanaki · J. G. Messchendorp · M. Mohammadi-Dadkan · H. Tavakoli Zaniani
KVI-CART, University of Groningen, Groningen, The Netherlands
e-mail: r.ramazani.sharifabadi@rug.nl

R. Ramazani-Sharifabadi · M. Mahjour-Shafiei
Department of Physics, University of Tehran, Tehran, Iran

St. Kistryn
Institute of Physics, Jagiellonian University, Cracow, Poland

A. Kozela
Institute of Nuclear Physics, Polish Academy of Sciences, Cracow, Poland

M. Mohammadi-Dadkan
Department of Physics, University of Sistan and Baluchestan, Zahedan, Iran

A. Ramazani Moghaddam Arani
Faculty of Physics, University of Kashan, Kashan, Iran

E. Stephan
Institute of Physics, University of Silesia, Chorzow, Poland

H. Tavakoli Zaniani
Department of Physics, School of Science, Yazd University, Yazd, Iran

N. A. Orr et al. (eds.), *Recent Progress in Few-Body Physics*,
Springer Proceedings in Physics 238,
https://doi.org/10.1007/978-3-030-32357-8_25

Compared to the three nucleon (3N) systems, there is a limited experimental database for the four nucleon (4N) at intermediate energies. To enrich the experimental database, different scattering experiments including a study of the deuteron-deuteron elastic and inelastic scattering processes [2–8] were carried out at various laboratories. The new data allow to study various aspects of three-nucleon force (3NF) effects, especially for its spin-dependent part. In this paper, we present the results of an investigation of the ^{2}H($\mathbf{d}$, dp)n break-up scattering process at a deuteron-beam energy of 65 MeV/nucleon.

The prerequisites of the analysis of the three-body break-up channel are the particle identification (PID) in which the time-of-flight information of the particles was used, and the energy calibration of the E-scintillator bars in which two energy calibration procedures have been exploited to extract the observables. To measure the number of counts for each configuration, the obtained S-Curve (energy correlation between outgoing deuteron and proton) [9] after PID is divided into S-bins with a width of 10 MeV. The number of break-up events for each S-bin is extracted by subtracting the background to obtain the signal after the projection of events on the axis perpendicular to the S-Curve. The extracted number of counts is corrected for efficiencies of the system such as the computer live-time of 40%, MWPC efficiencies for the deuteron of 99% and the proton of 97%, hadronic interactions of 84%, and down-scale factor that was used in the hardware trigger to measure the cross section [1].

To extract the analyzing powers, the spectrum of the ratio of the cross section of the reaction using a pure vector (tensor) polarized deuteron beam to the cross section of an unpolarized beam as a function of ϕ was used to fit a $\cos\phi$-dependent ($\cos 2\phi$-dependent with a possible offset from one) of the azimuthal asymmetry. The amplitude of the $\cos\phi$ ($\cos 2\phi$) is proportional to Re(iT_{11}) (Re(T_{22})). Also, Re(T_{20}) is proportional to the offset from one for the $\cos 2\phi$ azimuthal asymmetry.

25.1 Experimental Results

The preliminary results of the differential cross sections and analyzing powers of the three-body breakup process in the ^{2}H($\mathbf{d}$, dp)n reaction as a function of S for a number of kinematic configurations ($\theta_d = 28°$, $\theta_p = 28°$, $\phi_{dp} = 180°$, 160°, and 140°) have been extracted; see Fig. 25.1. Two energy calibration methods have been used to extract the observables. The preliminary results of the [9] had a normalization mistake of a factor of 2000 which is corrected in the present analysis. The results of the two calibrations have been averaged to yield the final results and the difference between the two calibrations (2σ) is indicated by a gray band for each configuration. The scattering angles of each configuration are shown at the top of the figures.

The results of the coplanar configuration in the Fig. 25.1 are compared with the recent theoretical calculations based on SSA by using CD-Bonn+Δ potential. In SSA, the break-up amplitude is calculated using the first term in the Neumann

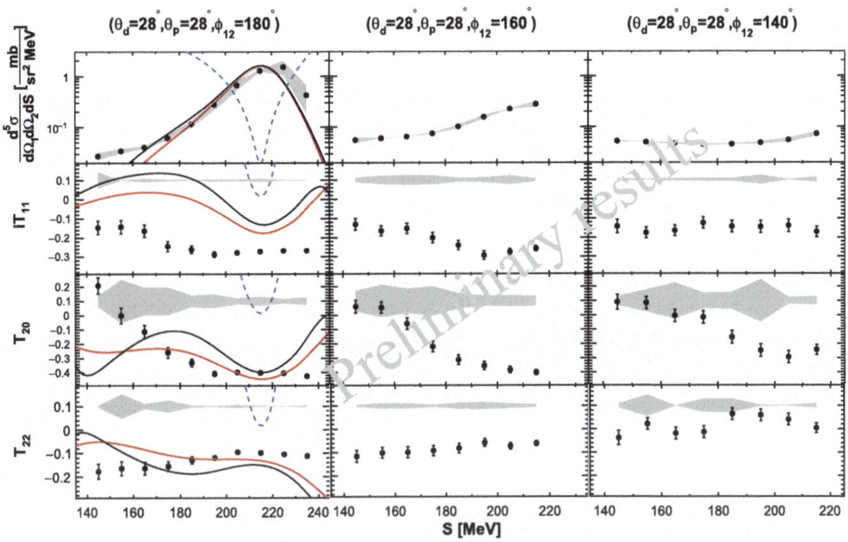

Fig. 25.1 The results of the cross section and analyzing powers for the kinematic configurations ($\theta_d = 28°, \theta_p = 28°, \phi_{dp} = 180°, 160°$, and $140°$) of the ${}^2\mathrm{H}(\mathbf{d}, dp)n$ reaction. The red (black) solid line shows the SSA1 (SSA4) approximation. The blue dashed-line indicates the energy of the neutron in units of MeV using the same scale on the left axis. The gray bands show the systematic uncertainty due to the calibration processes

series expansion of the corresponding exact four-nucleon scattering. We expect this single-scattering approximation to provide a rough estimation of three-body break-up observables in quasi-free configurations. The single-scattering break-up amplitude consists of four terms. Two of them are related to the break-up of the incident deuteron. The remaining two terms are related to the break-up of the target deuteron. In SSA1, only the term that corresponds to the breaking of the deuteron target with a spectator neutron is included, while in the SSA4 approximation, all terms are included [10–12]. The red (black) lines correspond to SSA1 (SSA4). The blue dashed lines indicate the energy of the outgoing neutron in units of MeV using the same scale on the left axis. The quasi-free scattering (QFS) region corresponds to the cases at which the neutron energy is at its minimum (less than 0.3 MeV). Clearly, the cross section peaks at this point and the SSA1 calculations coincide with the predictions of SSA4. Moreover, the measured cross section matches very well at the corresponding value of S. Also, the two measured tensor analyzing powers are well described by the calculation at the QFS regime. Strikingly, the vector analyzing power data deviate strongly from the predictions. In general, we conclude that the SSA gives a reasonable description of the QFS regime.

References

1. Ramazani-Moghaddam-Arani, A., et al.: Phys. Rev. C **83**, 024002 (2011)
2. Ermisch, K., et al.: Phys. Rev. Lett. **86**, 5862 (2001)
3. Ermisch, K., et al.: Phys. Rev. C **68**, 051001(R) (2003)
4. Ermisch, K., et al.: Phys. Rev. C **71**, 064004 (2005)
5. Sakai, H., et al.: Phys. Rev. Lett. **84**, 5288 (2000)
6. Kalantar-Nayestanki, N., et al.: Rep. Prog. Phys. **75**, 016301 (2012)
7. Kistryn, S., et al.: J. Phys. G **6**, 40 (2013)
8. Sekiguchi, K., et al.: Phys. Rev. C **65**, 034003 (2002)
9. Ramazani-Moghaddam-Arani, A.: Ph.D. thesis, University of Groningen (2009)
10. Deltuva, A., et al.: Phys. Rev. C **68**, 024005 (2003)
11. Deltuva, A., et al.: Phys. Rev. C **92**, 024001 (2015)
12. Deltuva, A., et al.: Phys. Rev. C **93**, 044001 (2016)

Chapter 26
Analyzing Powers of the Proton-Deuteron Break-Up Reaction at Large Proton Scattering Angles at 135 MeV

M. T. Bayat, M. Eslami-Kalantari, N. Kalantar-Nayestanaki, St. Kistryn, A. Kozela, J. G. Messchendorp, M. Mohammadi-Dadkan, R. Ramazani-Sharifabadi, E. Stephan and H. Tavakoli-Zaniani

Abstract A measurement of the analyzing powers for the ^{2}H($\mathbf{p}$, pp)n break-up reaction was carried out at KVI exploiting a polarized-proton beam at 135 MeV. The results are compared with theoretical calculations. The theoretical calculations describe the main features of the measured distributions, but none of them can reproduce the details. This indicates the need for further development of the three-nucleon force (3NF) models.

The underlying dynamics of three-nucleon forces (3NF) is mainly investigated by the measurements of differential cross sections and polarization observables (vector and tensor analyzing powers, spin-correlation coefficients and polarization transfer coefficients) in elastic Nd scattering and break-up of the deuteron in its collision with a nucleon. In the past three decades, many measurements have been carried out at KVI and at other laboratories to obtain high-precision and rich data sets. An overview

M. T. Bayat (✉) · N. Kalantar-Nayestanaki · J. G. Messchendorp · M. Mohammadi-Dadkan · R. Ramazani-Sharifabadi · H. Tavakoli-Zaniani
KVI-CART, University of Groningen, Groningen, The Netherlands
e-mail: m.t.bayat@rug.nl

M. Eslami-Kalantari · H. Tavakoli-Zaniani
Department of Physics, School of Science, Yazd University, Yazd, Iran

St. Kistryn
Institute of Physics, Jagiellonian University, 30059 Cracow, Poland

A. Kozela
Institute of Nuclear Physics, PAN, 31342 Cracow, Poland

M. Mohammadi-Dadkan
Department of Physics, University of Sistan and Baluchestan, Zahedan, Iran

R. Ramazani-Sharifabadi
Department of Physics, University of Tehran, Tehran, Iran

E. Stephan
Institute of Physics, University of Silesia, 40007 Chorzow, Poland

N. A. Orr et al. (eds.), *Recent Progress in Few-Body Physics*,
Springer Proceedings in Physics 238,
https://doi.org/10.1007/978-3-030-32357-8_26

of the results can be found in [1, 2]. The discrepancies between the measured data and theory predictions with currently available 3NF models demonstrate that spin-dependent parts of the 3NFs are not completely understood [3].

Based on these observations, and considering the rich phase space in the break-up reaction, it was decided to extend the analysis of the data taken in 2006 at KVI. These measurements were performed with the use of the Big Instrument for Nuclear-polarization Analysis (BINA). In this work, we extended the earlier measurements [1] that were done for kinematical configurations in which protons scatter to small forward angles up to 35° angles by analyzing configurations at which one of the final-state protons scatters to the backward part of BINA.

BINA is composed of two parts: The Wall (forward) and the Ball (backward). The Wall consists of a Multi-Wire Proportional Chamber (MWPC) and a segmented hodoscope of thin (2 mm) and thick (12 cm) scintillators for determination of the energy of the charged reaction products. The Wall covers the polar angles, θ, between 10° and 35°, and the full range of the azimuthal angle ϕ. The Ball has 149 phoswich scintillator elements which are glued together and cover polar angles between 40° and 160°. These detectors also act as the scattering chamber holding the target in the middle. A beam of polarized protons produced in an atomic-beam type polarized ion source from the AGOR accelerator impinged on liquid-deuterium target with a thickness of 3.85 ± 0.2 mm which was mounted in the center of the Ball of BINA; see the Fig. 26.1.

In this work, break-up events were selected in which one of the final-state protons was detected in the Wall and the second one in the Ball. For each configuration θ_1, θ_2, $\phi_{12} = \phi_1 - \phi_2$ of the two outgoing protons, the kinematical spectra E_1 versus E_2 were built. The angular bins for events were chosen to be $\Delta\theta_1 = 20°$, $\Delta\theta_2 = 4°$ and $\Delta\phi_{12} = 20°$. These bin sizes reflect the angular resolution of the detector. The bin size along the kinematical S-curve is set to be ~10 MeV. The background of accidental coincidences and the hadronic contribution, which originates from hadronic interactions inside the detector or in the material between the target and the detector are considered as a background and are subtracted in the procedure. The hadronic interactions can, subsequently, be corrected for using simulations.

The experimental results for two analyzing powers, A_x and A_y for some selected configurations are presented in Fig. 26.2. In this figure, the error bars are statistical and they arise mainly from the number of counts for the break-up coincidences. The systematic error stems primarily from the uncertainty in the measurement of the beam polarization via the proton-deuteron elastic-scattering and the corresponding values were ~3 and ~6% for up and down polarizations, respectively. The maximum systematic uncertainty was less than 7%. The theoretical predictions obtained based on the realistic charge-dependent (CD) Bonn potential only (red dash-dotted lines) and within the coupled-channel framework with the CD Bonn+Δ without (green dashed lines) and with (black solid lines) Coulomb force. These curves take the effects of the finite size of solid angles into account.

The experimental results for A_x show in general good agreement with the theoretical calculations where the effects of the 3NF are also shown to be small. For A_y,

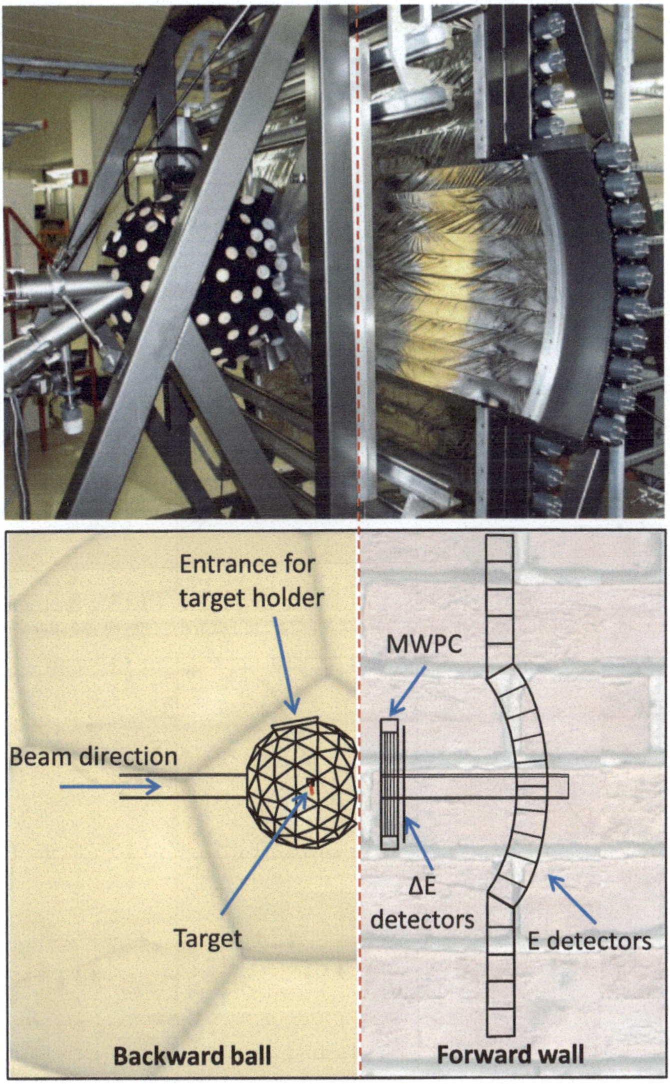

Fig. 26.1 A side view of BINA. The top panel shows a photograph of BINA side-view and the bottom one presents schematic drawing of the forward wall and the backward ball

however, there are sizable discrepancies for small ϕ_{12} and the inclusion of the 3NF makes the discrepancies even larger.

In conclusion, it can be claimed that the spin-structure of the 3NF is not well understood. Therefore, the origin of these discrepancies must lie in the treatment of three-nucleon potentials (3NPs).

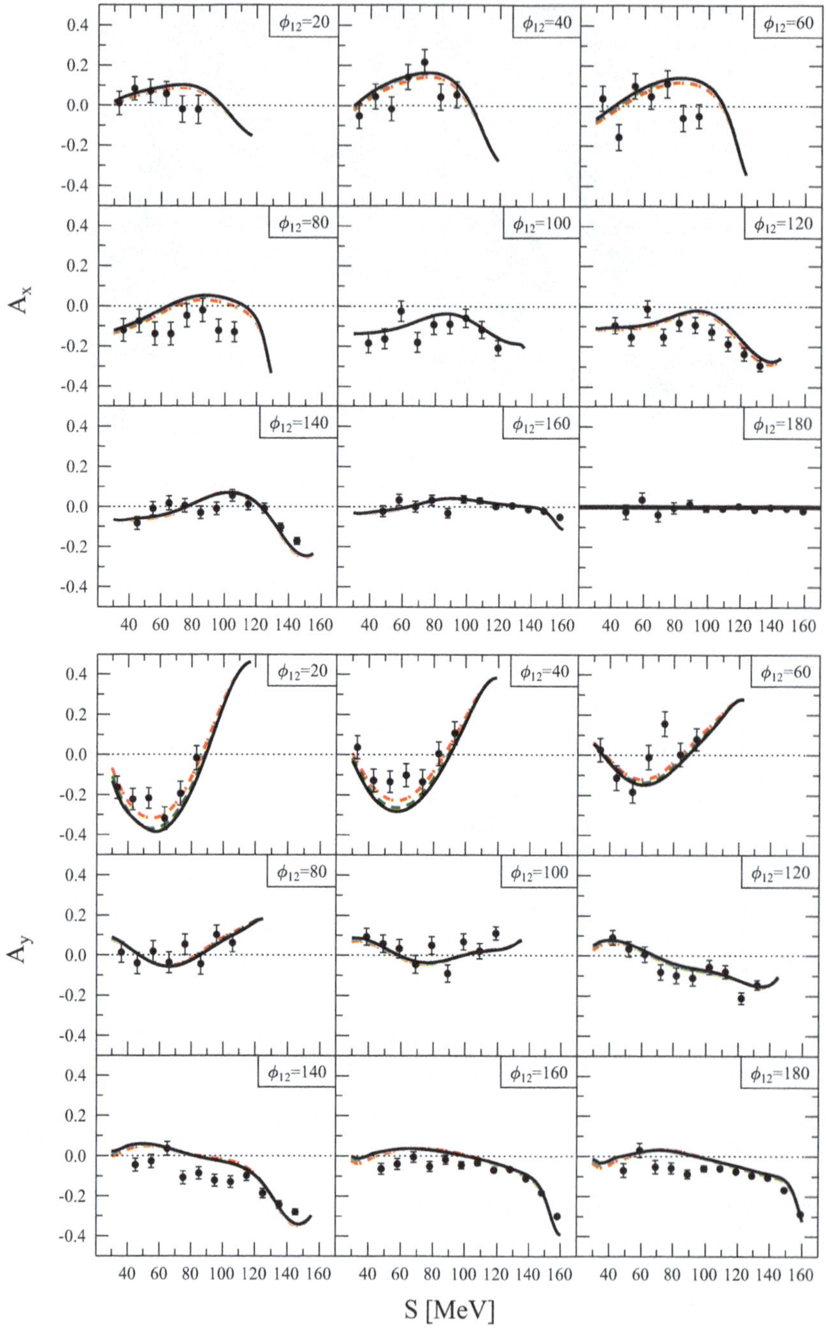

Fig. 26.2 The analyzing powers at $(\theta_1, \theta_2) = (50°, 28°)$ as a function of S for different azimuthal opening angles. Error bars reflect only statistical uncertainties. See the text for details

References

1. Kalantar-Nayestanaki, N., Messchendorp, J.G., Epelbaum, E., Nogga, A.: Rep. Prog. Phys. **75**, 016301 (2012)
2. Stephan, E., Kistryn, S.: J. Phys. G Nucl. Part. Phys. **40**, 063101 (2013)
3. Ermisch, K., Amir-Ahmadi, H.R., van den Berg, A.M., Castelijns, R., Davids, B., Deltuva, A., Epelbaum, E., Glöckle, W., Golak, J., Harakeh, M.N., Hunyadi, M., de Huu, M.A., Kalantar-Nayestanaki, N., Kamada, H., Kiš, M., Mahjour-Shafiei, M., Nogga, A., Sauer, P.U., Skibiński, R., Witała, H., Wörtche, H.J.: Phy. Rev. C **71**, 064004 (2005)

Chapter 27
Ab initio Folding Potentials for Proton-Nucleus Scattering with NCSM Nonlocal One-Body Densities

Ch. Elster, M. Burrows, S. P. Weppner, K. D. Launey, P. Maris and G. Popa

Abstract Based on the spectator expansion of the multiple scattering series we employ a nonlocal translationally invariant nuclear density derived from a chiral next-to-next-to-leading order (NNLO) and the very same interaction for consistent full-folding calculations of the effective (optical) potential for nucleon-nucleus scattering for light nuclei.

The theoretical approach to the elastic scattering of a nucleon from a nucleus, pioneered by Watson [1], made familiar by Kerman, McManus, and Thaler (KMT) [2] and further developed as the spectator expansion [3, 4] is now being applied together with ab initio structure calculations to obtain effective folding nucleon-nucleus (NA) potentials. The spectator expansion is predicated upon the idea that at projectile energies higher than about 100 MeV the two-body interactions between the projectile and the nucleons in the target dominate elastic scattering, for which a transition operator can be defined as

$$T_{el} \equiv PTP = PUP + PUPG_0(E)T_{el}. \tag{27.1}$$

The projector $P = \frac{|\Phi_A\rangle\langle\Phi_A|}{\langle\Phi_A|\Phi_A\rangle}$ is conventionally taken to project on the elastic channel so that $[G_0, P] = 0$. Here $|\Phi_A\rangle$ stands for the ground state of the target so that $H_A|\Phi_A\rangle = E_A|\Phi_A\rangle$, and $G_0(E) = (E - H_0 + i\varepsilon)^{-1}$, where $H_0 = h_0 + H_A$, being

Ch. Elster (✉) · M. Burrows · G. Popa
Department of Physics and Astronomy, Ohio University, Athens, OH 45701, USA
e-mail: elster@ohio.edu

S. P. Weppner
Natural Sciences, Eckerd College, St. Petersburg, FL 33711, USA

K. D. Launey
Department of Physics and Astronomy, Louisiana State University, Baton Rouge, LA 70803, USA

P. Maris
Department of Physics and Astronomy, Iowa State University, Ames, IA 50011, USA

N. A. Orr et al. (eds.), *Recent Progress in Few-Body Physics*, Springer Proceedings in Physics 238,
https://doi.org/10.1007/978-3-030-32357-8_27

the free propagator for the projectile+target system. The effective (optical) potential is given by

$$U = V + VG_0(E)QU, \tag{27.2}$$

where the operator Q is defined via the relation $P + Q = 1$. The first-order term involves two-body interactions between the projectile and one of the target nucleons, i.e. $U = \sum_{i=1}^{A} \tau_i$, where the operator τ_i is derived to be

$$\begin{aligned} \tau_i &= v_{0i} + v_{0i} G_0(E) Q \tau_i \\ &= v_{0i} + v_{0i} G_0(E) \tau_i - v_{0i} G_0(E) P \tau_i \\ &= \hat{\tau}_i - \hat{\tau}_i G_0(E) P \tau_i . \end{aligned} \tag{27.3}$$

Here $\hat{\tau}_i$ is the NN t-matrix and is defined as the solution of $\hat{\tau}_i = v_{0i} + v_{0i} G_0(E)\hat{\tau}_i$. In lowest order $\hat{\tau}_i \approx t_{0i}$, which corresponds to the conventional impulse approximation. Here the operator t_{0i} stands for the standard solution of a Lippmann-Schwinger equation with the NN interaction as driving term.

For elastic scattering only $P\tau_i P$ needs to be considered,

$$\langle \Phi_A | \tau_i | \Phi_A \rangle = \langle \Phi_A | \hat{\tau}_i | \Phi_A \rangle - \langle \Phi_A | \hat{\tau}_i | \Phi_A \rangle \frac{1}{(E - E_A) - h_0 + i\varepsilon} \langle \Phi_A | \tau_i | \Phi_A \rangle, \tag{27.4}$$

and this matrix element determines the full-folding effective (optical) potential when summing over all target nuclei,

$$\langle \mathbf{k}' | U | \mathbf{k} \rangle = \langle \mathbf{k}' \Phi_A | \sum_i \tau_i | \mathbf{k} \Phi_A \rangle. \tag{27.5}$$

Since $\langle \mathbf{k}' | U | \mathbf{k} \rangle$ is the solution of the sum of one-body integral equations represented by (27.4), it is sufficient to consider the driving term

$$\langle \mathbf{k}' | \hat{U} | \mathbf{k} \rangle = \langle \mathbf{k}' \Phi_A | \sum_i \hat{\tau}_i | \mathbf{k} \Phi_A \rangle, \tag{27.6}$$

where $\hat{\tau}_i \approx t_{0i}$ when considering the first order single scattering term. Inserting a complete set of momenta for the struck target nucleon before and after the collision, representing the sum over target protons and neutrons by α, evaluating the momentum conserving delta function and changing variables to $\mathbf{q} = \mathbf{k}' - \mathbf{k}$, $\mathbf{K} = \frac{1}{2}(\mathbf{k} + \mathbf{k}')$ and $\mathbf{P} = \frac{1}{2}(\mathbf{p}' + \mathbf{p}) + \frac{\mathbf{K}}{A}$, the final expression for the full-folding effective potential is given by

$$\begin{aligned} \hat{U}(\mathbf{q}, \mathbf{K}) = \sum_{\alpha = p,n} \int d^3\mathbf{P}\, \eta(\mathbf{P}, \mathbf{q}, \mathbf{K})\, \hat{\tau}_\alpha \left(\mathbf{q}, \frac{1}{2}\left(\frac{A+1}{A}\mathbf{K} - \mathbf{P} \right); \epsilon \right) \\ \times \rho_\alpha \left(\mathbf{P} - \frac{A-1}{A}\frac{\mathbf{q}}{2}, \mathbf{P} + \frac{A-1}{A}\frac{\mathbf{q}}{2} \right). \end{aligned} \tag{27.7}$$

Here $\eta(\mathbf{P}, \mathbf{q}, \mathbf{K})$ is the Møller factor for the frame transformation [5] relating the NN zero-momentum frame to the NA zero-momentum frame. Further details can be found in [6]. The quantity ρ_α, with $\alpha = p(n)$, represents a nonlocal one-body density matrix (OBD) for the proton (neutron) distribution, and must be given in a translationally invariant fashion [7]. An important product of this work is that the NN t-matrix and OBD now use the same underlying NN interaction. For this we choose the optimized chiral NN interaction at next-to-next-to-leading order NNLO$_{\text{opt}}$ from [8]. In the $A = 3, 4$ nucleon systems the contributions of the 3NFs are smaller than in most other parameterizations of chiral interactions. From this point of view, the NNLO$_{\text{opt}}$ interaction is very well suited for our calculations, since the first-order folding potential does not contain any explicit 3NF contributions. We calculated the full-folding integral for the first-order effective (optical) potential for NA scattering ab initio. i.e. they are based consistently on one single NN interaction, in our case the chiral NNLO$_{\text{opt}}$ interaction from [8], which is fitted to NN data up to 125 MeV laboratory kinetic energy with $\chi^2 \approx 1$ per degree of freedom. Based on this interaction the one-body nonlocal, translationally invariant nuclear densities are calculated as laid out in [7]. Further details of the calculations of the effective potential and the NA scattering are described in [6].

First, we want to illustrate the difference in employing U or $\hat{U}$ as effective NA potential in Fig. 27.1 for proton scattering from ^{4}He and ^{16}O at 200 MeV projectile laboratory kinetic energy. The figure shows that taking into account the effect of the operator Q by solving (27.4) to obtain U is clearly visible for the light nucleus ^{4}He, while very small for a heavier nucleus like ^{16}O.

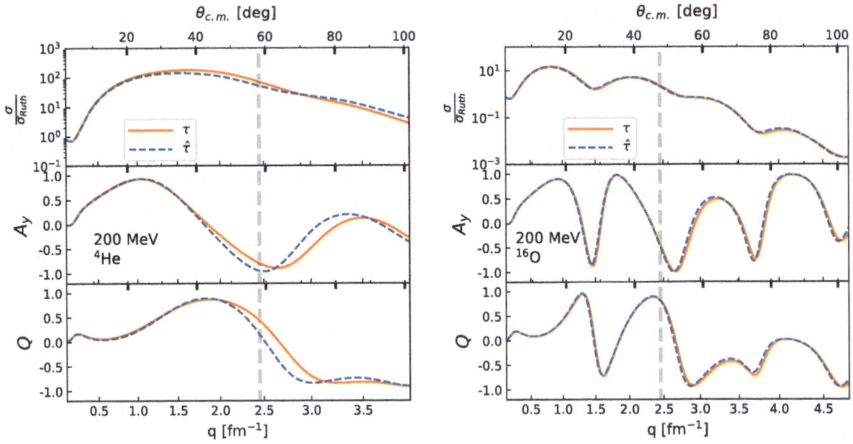

Fig. 27.1 Angular distribution of the differential cross section divided by the Rutherford cross section, analyzing power A_y and spin rotation function Q for elastic proton scattering from ^{4}He (left) and ^{16}O (right) as function of the momentum transfer and the c.m. angle calculated with the NNLO$_{\text{opt}}$ chiral interaction [8]. The solid line represents the calculation based on τ_i (27.5), the dashed line the one based on $\hat{\tau}_i$ (27.6)

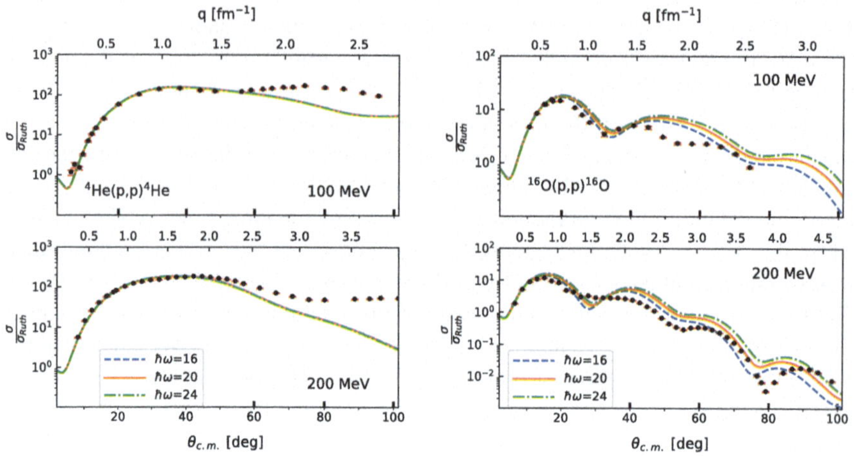

Fig. 27.2 Angular distribution of the differential cross section divided by the Rutherford cross section for elastic proton scattering from ^{4}He (left) and ^{16}O (right) as function of the c.m. angle calculated with the NNLO$_{\text{opt}}$ chiral interaction [8]. The calculations for ^{4}He are carried out with $N_{\text{max}} = 18$, while the ones for ^{16}O with $N_{\text{max}} = 10$. The values of $\hbar\omega$ are indicated in the lower panels. The ^{4}He data for 100 MeV are taken from [9] and for 200 MeV from [10]. The ^{16}O data for 100 MeV are taken from [11] and for 200 MeV from [12]

As examples for elastic proton scattering based on an ab initio effective potential we show in Fig. 27.2 the angular distributions (divided by the Rutherford amplitude) for ^{4}He and ^{16}O for energies between 100 and 200 MeV laboratory kinetic energy. We find them in very good agreement with the data in the angle and momentum transfer regime where the first order term of the full-folding effective potential should be valid. We also want to point out that the first order term in the multiple scattering expansion does not explicitly contain any 3NF contributions, thus the choice of the NNLO$_{\text{opt}}$ interaction works well with the theoretical content of the effective potential. Further in the future with different interactions will have to shed more light on the effect of including 3NFs in the one-body density for the first-order effective potential.

Acknowledgements Partial support for this work is given by the U.S. DoE under DE-FG02-93ER40756, DE-SC0018223, DE-AC02-05CH11231, and the U.S. NSF under OIA-1738287, ACI-1713690, OCI-0725070, and ACI-1238993.

References

1. Watson, K.M.: Phys. Rev. **89**, 575 (1953)
2. Kerman, A.K., McManus, H., Thaler, R.M.: Ann. Phys. **8**, 551 (1959)
3. Siciliano, E.R., Thaler, R.M.: Phys. Rev. **C16**, 1322 (1977)
4. Chinn, C.R., Elster, Ch., Thaler, R.M., Weppner, S.P.: Phys. Rev. C **52**, 1992 (1995)
5. Joachain, C.J.: Quantum Collision Theory. Elsevier Science Ltd. (1984)

6. Burrows, M., Elster, Ch., Weppner, S.P., Launey, K.D., Maris, P., Nogga, A., Popa, G.: arXiv:1810.06442 (2018)
7. Burrows, M., Elster, Ch., Popa, G., Launey, K.D., Nogga, A., Maris, P.: Phys. Rev. C **97**(2), 024325 (2018)
8. Ekström, A., et al.: Phys. Rev. Lett. **110**, 192502 (2013)
9. Goldstein, N.P., Held, A., Stairs, D.G.: Can. J. Phys. **48**(21), 2629 (1970)
10. Moss, G.A., et al.: Phys. Rev. C **21**, 1932 (1980)
11. Seifert, H.: Energy dependence of the effective interaction for nucleon-nucleus scattering. Ph.D. thesis, University of Maryland (1990)
12. Glover, C.W., et al.: Phys. Rev. C **31**, 1 (1985)

Chapter 28
Few-Body Reactions and Processes in Neutron Star Envelopes

N. Takibayev, V. O. Kurmangaliyeva, Kiyoshi Katō and V. S. Vasilevsky

Abstract We consider the processes and reactions in the crusts of neutron star envelopes and analyze the density oscillations in the layers of this volume-centered crystalline structure, and the dynamics of stimulated reactions, including the transformations of nuclei. The action of acoustic waves that spread in the crystalline structure and can start up the different threshold reactions are considered. It is stated that the threshold reactions with capture of Fermi electrons in own cells by even-even nuclei stop after emission of two neutrinos. Other types of nuclei have also the own threshold cells, where the direct and inverse reactions can generate neutrino and anti-neutrino for a very long time. We believe this effect intensively contributes to the process of the neutron star rapid cooling.

28.1 Introduction

Let us first consider the processes and reactions which can be stimulated by huge pressure in the neutron star envelopes. In the outer crust of envelopes free neutrons decay $n \rightarrow p + e^- + \overline{\nu}_e$, but in the inner crust the reactions $p + e^- \rightarrow n + \nu_e$ create free neutrons because the pressure is enough high. The ordered structure allows us to consider the behavior and properties of the crystalline cells as the function of the layer depth [1]. We investigate the nuclear reactions initiated by the capture of Fermi electrons by nucleus in own cells with emission of neutrinos by the new nucleus. Each cell contains a nucleus A(Z, N) and Z electrons with zero total charge of the

N. Takibayev (✉) · V. O. Kurmangaliyeva
Al-Farabi Kazakh National University, Almaty, Kazakhstan
e-mail: takibayev@gmail.com

K. Katō
Hokkaido University, Sapporo, Japan
e-mail: kato-iku@gd6.so-net.ne.jp

V. S. Vasilevsky
Bogolyubov Institute for Theoretical Physics, Kiev, Ukraine
e-mail: vsvasilevsky@gmail.com

N. A. Orr et al. (eds.), *Recent Progress in Few-Body Physics*,
Springer Proceedings in Physics 238,
https://doi.org/10.1007/978-3-030-32357-8_28

cell. Every type of nuclei has its own threshold energy for the reaction in which a Fermi electron can be captured by a nucleus in own cell. It means that the energy of the Fermi electron must be equal or exceed this threshold energy, i.e. the pressure waves in the layer must be strong enough.

The properties of matter in the inner crusts are very important: creation of neutron rich nuclei, the appearance of free neutrons inside of the crystalline layers in inner crusts, and the neutron resonances of the few-body type. It is very important to take into account the long-term generation of neutrino and anti-neutrino near the threshold layers of the certain nuclei.

28.2 Two Types of Electron-Nucleus Threshold Reactions

In the case of the parent even-even nucleus like ${}^{56}_{26}Fe$ the two-step chain of successive reactions with capture electrons and emission of neutrinos happened in own cell that leads to irreversible transformation of the nuclei: ${}^{56}_{26}Fe \rightarrow {}^{56}_{25}Mn \rightarrow {}^{56}_{24}Cr$ (see also Table 28.1).

Table 28.1 The chains of some reactions generated by Fermi electrons with even-even nuclei. Here and below the energies are given in MeV. Intermediate states have to be considered like the virtual ones. All pairs of successive reactions are stopped

Main reactions	$E_{e;th}$		Daughter reactions	$E_{e;th}$
${}^{28}_{14}Si^{0+} + e^- \rightarrow {}^{28}_{13}Al^{3+} + \nu_e$	4.643	$\rightarrow$	${}^{28}_{13}Al^{3+} + e^- \rightarrow {}^{28}_{12}Mg^{0+} + \nu_e$	2.131
${}^{52}_{24}Cr^{0+} + e^- \rightarrow {}^{52}_{23}V^{3+} + \nu_e$	3.97	$\rightarrow$	${}^{52}_{23}V^{3+} + e^- \rightarrow {}^{52}_{22}Ti^{0+} + \nu_e$	1.977
${}^{56}_{26}Fe^{0+} + e^- \rightarrow {}^{56}_{25}Mn^{3+} + \nu_e$	3.695	$\rightarrow$	${}^{56}_{25}Mn^{3+} + e^- \rightarrow {}^{56}_{24}Cr^{0+} + \nu_e$	1.629
${}^{64}_{30}Zn^{0+} + e^- \rightarrow {}^{64}_{29}Cu^{1+} + \nu_e$	0.579	$\rightarrow$	${}^{64}_{29}Cu^{1+} + e^- \rightarrow {}^{64}_{28}Ni^{0+} + \nu_e$	−1.675

Table 28.2 The one-step reactions generated by Fermi electrons with odd-odd, odd-even and even-odd nuclei. The symbol $\hookleftarrow$ means that the second reactions are impossible, but the way back becomes open, i.e. when the pressure is down, Fermi-electron and an anti-neutrino can be emitted by daughter nuclei, for example: ${}^{55}_{24}Cr^{3/2-} \rightarrow {}^{55}_{25}Mn^{5/2-} + e^- + \overline{\nu}_e$

Main reactions	$E_{e;th}$		Daughter reactions	$E_{e;th}$
${}^{45}_{21}Sc^{7/2-} + e^- \rightarrow {}^{45}_{20}Ca^{7/2-} + \nu_e$	0.256	$\hookleftarrow$	${}^{45}_{20}Ca^{7/2-} + e^- \rightarrow {}^{45}_{19}K^{3/2+} + \nu_e$	4.203
${}^{55}_{25}Mn^{5/2-} + e^- \rightarrow {}^{55}_{24}Cr^{3/2-} + \nu_e$	2.603	$\hookleftarrow$	${}^{55}_{24}Cr^{3/2-} + e^- \rightarrow {}^{55}_{23}V^{7/2-} + \nu_e$	5.957
${}^{63}_{29}Cu^{3/2-} + e^- \rightarrow {}^{63}_{28}Ni^{1/2-} + \nu_e$	0.066	$\hookleftarrow$	${}^{63}_{28}Ni^{1/2-} + e^- \rightarrow {}^{63}_{27}Co^{7/2-} + \nu_e$	3.673
${}^{69}_{31}Ga^{3/2-} + e^- \rightarrow {}^{69}_{30}Zn^{1/2-} + \nu_e$	0.909	$\hookleftarrow$	${}^{69}_{30}Zn^{1/2-} + e^- \rightarrow {}^{69}_{29}Cu^{3/2-} + \nu_e$	2.682

Meanwhile, the reactions with the odd-even, odd-odd and even-odd nuclei demonstrate some specific features in the particular layers of the crystalline envelopes: these are the one-step reactions (Table 28.2). In the phase of cell compression, a nucleus captures a Fermi-electron and emits a neutrino. Then in the expansion phase of the cell, the new nucleus emits a Fermi electron and an anti-neutrino transforming itself back into the previous nucleus. It is quite remarkable that this process can repeat many times.

The pressure oscillation $P = P_0(N) + \Delta P$ is the current pressure, where $P_0(N)$ is the average pressure in the N-layer of the crust, and ΔP corresponds to the value of the oscillations in this layer. Acoustic waves create pressure oscillations that lead to the oscillations of the Fermi electron energy inside the cell.

It is very important that the behavior of the above mentioned nuclei shows the exclusive properties when the layers are situated in the area of the critical states for these nuclei. Then in the first phase, the nucleus starts to capture the Fermi electron and emits the neutrino when the pressure increases. After that, the behavior of even-even nuclei becomes different in the second phase as compared to other types of nuclei.

Numerous collisions of the Fermi electron with the nucleus inside the cell lead to virtual rearrangement in the nucleus preserving the total sum of all quantum numbers in the cell. When the energy E_F of the Fermi electron increases nearly approaching to the inelastic reaction threshold E_{th}, the exponential tail: $exp(-[k_{th} - k]R)$ called the "excitation of threshold swelling" [2] is formed in the elastic reactions: $e_F^- + A(Z, N) \rightarrow e_F^- + A(Z, N)$, where $k \leq k_{th}$. Here k and k_{th} are the wave number of electron and the threshold one, $R = R(P)$ is the current radius of the cell, respectively.

Real changes occur only when the energy of the Fermi electron is over the threshold energy of the inelastic reaction. In this case, the nucleus, by absorbing the Fermi electron, emits a neutrino that has an energy equal to the difference between the Fermi energy of the electron and the threshold energy of electron capture by the nucleus.

Then, we can consider the two-step reactions of the even-even parent nuclei, when each of them sequentially captures consecutively two Fermi electrons:

(1) If the energy threshold is open, the daughter reaction proceeds also. Here the threshold of the daughter reaction is significantly lower than the energy threshold of the parent reaction. That is why two neutrinos are sequentially emitted. In this case, the energy of the second neutrino is high enough and it can be registered in some cases. After that the chain of reactions is ruptured.

(2) The reaction ceases because the third threshold is already much higher, and it could go on only in the deeper layers of the crystalline envelopes where the Fermi energy of the electrons can be sufficient. However, such reactions are impossible within the given layer of envelopes.

In the case of the parent even-even nuclei, the two-steps chain of successive reactions with capture of electrons and emission of neutrinos leads to irreversible transformation of the nuclei $A(Z, N)$:

$$e_F^- + A(Z, N) \to A'(Z-1, N+1) + \nu_e\,, \quad E_F \geq E_{th}; \tag{28.1}$$

$$e_F^- + A'(Z-1, N+1) \to A''(Z-2, N+2) + \nu_e\,, \quad E_F \geq E'_{th}. \tag{28.2}$$

Each even-even nuclide $A(Z, N)$ has its own threshold energy to capture one Fermi electron and emits the neutrino of small energy. A new nuclide $A'(Z-1, N+1)$ turns immediately into another nuclide, which has the smaller threshold energy, and then easily captures the second Fermi electron and also emits the second neutrino of higher energy. After that, the reactions cannot continue in this cell, except for the extra case when additional energy comes from outside of the cell.

The capture reactions of Fermi electrons by even-even nuclei demonstrate the exclusive property connected with the two-step successive reactions, where the intermediate nuclei can be considered as virtual swelled states with indeterminate quantum numbers. These reactions cannot go back because the even-even nucleus turns to another nucleus and takes part in the next electron capture reaction in this cell. These reactions happen at the first stage of the neutron star cooling.

Table 28.1 presents the reaction chains with parental even-even nuclei and the Table 28.2 presents the chains of the other types of reactions (the tables include only a small part of the reactions).

Now, let us consider the other types of nuclei: even-odd, odd-even and odd-odd. The nuclei of these groups capture the Fermi electrons in own cells and emit neutrinos like in the equation (28.1). But in the second step, when the pressure is down, every daughter nucleus emits an electron and an anti-neutrino and becomes the same initial nucleus.

$$A'(Z-1, N+1) \to A(Z, N) + e_F^- + \overline{\nu}_e\,, \quad E'_{th} > E_F. \tag{28.3}$$

The difference in the behavior of the first and the second phase is connected with properties of nuclei and the action of acoustic waves that spread in the crystalline structure. The acoustic waves in the layers of the crystalline crusts create the oscillations of density in the layers of the neutron star envelopes. The changing of pressure stimulates the direct and reverse reactions in the threshold cells: at squeeze of the cell the nucleus captures the Fermi electron and emits the neutrino (28.1), then on the expansion phase of the cell the daughter nucleus emits the Fermi electron and an anti-neutrino and returns back to the previous state (28.3).

The reactions with such nuclei have their own unique properties—they support the cyclic processes. Then, only the one-step and reverse reactions are realized in these cells: the neutron captures the Fermi electron and emits a neutrino, and then in the next phase the daughter nucleus emits Fermi electron and an anti-neutrino to return to the previous state, for example: ${}^{57}_{25}Mn \rightarrow {}^{57}_{26}Fe + e^- + \overline{\nu}_e$. These processes can continue for a long time. These types of the nuclei take part in direct and reverse reactions in the correspondent critical cells and emit successively neutrinos and anti-neutrinos.

28.3 Conclusion

Density oscillations in the critical layers lead to direct and reverse reactions that arise in cells with the nucleus whose energy threshold is crossed by the Fermi electron energy. So, the increase in pressure squeezes the cell, and the energy of the Fermi electron rises in the cell. In the opposite phase at the drop in cell pressure, the Fermi-electron energy decreases. Each type of the nuclei with the odd number of protons or neutrons takes part in such direct and reverse reactions in corresponding critical layers and generates neutrinos and anti-neutrinos. In this case, the direct and reverse reactions can work as a permanent generator of neutrinos and anti-neutrinos for a very long time. Such emissions, therefore, contribute to fast cooling of neutron stars.

Neutron resonances of a few-body type, for example, in three-body systems: $n + A_1 + A_2$, lead to own oscillations in neighboring cells of the inner crust which propagate in the envelopes [3].

Acknowledgements The researches are supported by the program IRN BR05236494 and the grant AP05130476 of MES RK.

References

1. Takibayev, N.: Few-body effects in neutron star matter. Few-Body Syst. **57–58**, 279–282 (2017)
2. Baz, A., Zeldovich, Ya., Perelomov, A.: Scattering, Reactions, and Decay in the Nonrelativistic Quantum Mechanics. Israel Programme for Scientific Translations, Jerusalem (1969)
3. Takibayev, N.: Neutron resonances in systems of few nuclei and their possible role in radiation of overdence star. Few-Body Syst. **50**, 311–314 (2011)

Chapter 29
Five-Nucleon Systems with an Hyperspherical Harmonic Method

Jérémy Dohet-Eraly, Michele Viviani, Alejandro Kievsky, Laura Elisa Marcucci and Alex Gnech

Abstract A bound-state variational method based on fully-antisymmetric hyperspherical harmonics and an hyperradial Lagrange mesh is presented and applied to the five-nucleon system with an effective nucleon-nucleon interaction, which artificially bounds this system.

29.1 Introduction

The hyperspherical harmonic (HH) method has been proved to be an efficient tool for studying bound states, via the Rayleigh-Ritz variational principle, and low-energy scattering, via the Kohn variational principle, of three- and four-nucleon systems (see [1] for a review). Although the method can be applied to systems made of an arbitrary number of nucleons (A), the fast increase with A of computational needs, time and memory, makes the application of the HH method to five- and more-nucleon systems particularly challenging. This requires not only a well-balanced parallelization of the computational tasks but also the improvement of the existing algorithms or the development of new ones, valid for any number of nucleons. We report here some progress made in the building of orthonormal bases of fully-antisymmetrized HH, which have enabled us to apply the HH approach to five-body systems. In particular, we present here the study of the five-nucleon ground state obtained from a central

J. Dohet-Eraly (✉)
Physique nucléaire et Physique quantique, Université libre de Bruxelles (ULB), Brussels, Belgium
e-mail: jdoheter@ulb.ac.be

M. Viviani · A. Kievsky · L. E. Marcucci
INFN, Sezione di Pisa, Pisa, Italy

L. E. Marcucci
Department of Physics, University of Pisa, Pisa, Italy

A. Gnech
Gran Sasso Science Institute, L'Aquila, Italy

N. A. Orr et al. (eds.), *Recent Progress in Few-Body Physics*,
Springer Proceedings in Physics 238,
https://doi.org/10.1007/978-3-030-32357-8_29

nucleon-nucleon potential [2], which unrealistically bounds this system. This work is a preliminary step before the application of the fully-antisymmetric HH method to the $\alpha + n$ scattering.

29.2 Hyperspherical Harmonics

For a five-nucleon system, the internal Jacobi coordinates can be defined as

$$\boldsymbol{x}_{5-j} = \sqrt{\frac{2j}{j+1}} \left(\boldsymbol{r}_{j+1} - \frac{1}{j} \sum_{i=1}^{j} \boldsymbol{r}_i \right), \tag{29.1}$$

where $\boldsymbol{r}_i$ is the coordinate of nucleon i and $j = 1, 2, 3, 4$. From them, the hyperradius ρ reads

$$\rho = \sqrt{x_1^2 + x_2^2 + x_3^2 + x_4^2}. \tag{29.2}$$

The hyperangular coordinates (Ω) are constituted by the angular parts of the Jacobi vectors $(\hat{x}_1, \hat{x}_2, \hat{x}_3, \hat{x}_4)$ and by the hyperangles (ϕ_2, ϕ_3, ϕ_4) defined by

$$\cos \phi_i = \frac{x_i}{\sqrt{\sum_{j=1}^{i} x_j^2}}, \tag{29.3}$$

where x_i is the module of $\boldsymbol{x}_i$ and $i = 2, 3, 4$. In hyperspherical coordinates, the internal kinetic-energy operator of five-nucleon systems reads [1]

$$T = -\frac{\hbar^2}{m} \sum_{i=1}^{4} \Delta_{\boldsymbol{x}_i} = -\frac{\hbar^2}{m} \left(\frac{\partial^2}{\partial \rho^2} + \frac{11}{\rho} \frac{\partial}{\partial \rho} + \frac{\Lambda^2(\Omega)}{\rho^2} \right), \tag{29.4}$$

where m is the nucleon mass, $\Delta_{\boldsymbol{x}_i}$ is the Laplacian associated with $\boldsymbol{x}_i$ and $\Lambda^2(\Omega)$ is the grand angular momentum operator. The hyperspherical harmonics $\mathcal{Y}_{[K]}(\Omega)$ are eigenvectors of the grand angular momentum operator,

$$\Lambda^2(\Omega) \mathcal{Y}_{[K]}(\Omega) = -K(K+10) \mathcal{Y}_{[K]}(\Omega), \tag{29.5}$$

where K is the grand angular momentum. They are explicitly given by Kievsky et al. [1]

$$\mathcal{Y}_{[K]}^{KLM}(\Omega) = [[[Y_{l_1}(\hat{x}_1) Y_{l_2}(\hat{x}_2)]_{L_2} Y_{l_3}(\hat{x}_3)]_{L_3} Y_{l_4}(\hat{x}_4)]_{LM} \prod_{j=2}^{4} {}^{(j)}\mathcal{P}_{n_j}^{K_{j-1}, l_j}(\phi_j), \tag{29.6}$$

where L is the total orbital angular momentum, M is its projection on the z axis, the functions Y_l are the spherical harmonics, $K_j = l_1 + \sum_{i=2}^{j}(2n_i + l_i)$, the functions ${}^{(j)}\mathcal{P}_{n_j}^{K_{j-1},l_j}$ are related to some Jacobi polynomials (see [1] for their explicit definitions), the notation $[K]$ stands for $l_1, l_2, l_3, l_4, L_2, L_3, n_2, n_3, n_4$, and $K = n_2 + n_3 + n_4 + l_{\rm sum}$ with $l_{\rm sum} = l_1 + l_2 + l_3 + l_4$. The HH have a well-defined parity given by $\pi = (-1)^{l_{\rm sum}}$.

The spin functions $\chi_{S_2S_3S_4}^{SM_S}$ with total spin S and total spin projection M_S can be built by coupling the individual spin functions $\chi_{1/2,\pm 1/2}$ of each nucleon,

$$\chi_{S_2S_3S_4}^{SM_S} = [[[[\chi_{1/2}\chi_{1/2}]_{S_2}\chi_{1/2}]_{S_3}\chi_{1/2}]_{S_4}\chi_{1/2}]_{SM_S}. \tag{29.7}$$

The isospin functions $\xi_{T_2T_3T_4}^{TM_T}$ with total isospin T and total isospin projection M_T can be built in a similar way. Taking the spin and isospin into account, the hyperspherical harmonics functions are defined by

$$\mathbb{Y}_{[KST]}^{KLSTMM_SM_T}(\Omega) = \mathcal{Y}_{[K]}^{KLM}(\Omega)\chi_{[S]}^{SM_S}\xi_{[T]}^{TM_T}, \tag{29.8}$$

where $[S]$ stands for S_2, S_3, S_4, $[T]$ for T_2, T_3, T_4, and $[KST]$ for $[K][S][T]$. These functions are orthonormal and constitute a complete basis. However, they do not respect the Pauli principle. To build antisymmetric states, one can apply to the HH the antisymmetrization operator $\mathcal{A} = \sum_P \mathrm{sign}(P)P$, where the sum is over all permutations of the nucleons. This requires to compute $P\chi_{[S]}^{SM_S}$ and $P\xi_{[T]}^{TM_T}$, which is easily done by uncoupling the spin and isospin functions, and $P\mathcal{Y}_{[K]}^{KLM}$. The grand angular momentum K being invariant with respect to any permutation, one has

$$P\mathcal{Y}_{[K]}^{KLM}(\Omega) = \sum_{[K']} a_{[K];[K']}^{KL}\mathcal{Y}_{[K']}^{KLM}(\Omega). \tag{29.9}$$

The transformation coefficients $a_{[K];[K']}^{KL}$ are obtained by recurrence relations [3], which generalize the approach of [4] to an arbitrary number of particles. The antisymmetric HH defined by $\mathbb{Y}_{[KST]}^{KLSTMM_SM_T,\mathcal{A}} = \mathcal{A}\mathbb{Y}_{[KST]}^{KLSTMM_SM_T}$ are not orthogonal and are linearly dependent. The set of antisymmetric HH $\mathbb{Y}_{[KST]}^{KLSTMM_SM_T,\mathcal{A}}$ is highly redundant as shown in Table 29.1. Considering only the antisymmetric HH allows a reduction of the basis size by a factor 60, approximately. To get an orthonormal basis of antisymmetric states, a new strategy has been developed [3]. First, a complete set of linearly-independent antisymmetric HH is extracted by a method based on their evaluation on random coordinates. At this step, the transformation coefficients are not needed. Then, the selected antisymmetric HH are built after having computed the associated transformation coefficients. Finally, they are orthonormalized by means of the Gram-Schmidt algorithm. The resulting HH are denoted by $\mathbb{Y}_i^{KLSTMM_SM_T,\mathcal{A},\perp}$ where i varies from 1 to the dimension of the antisymmetric space for quantum

Table 29.1 Number of HH functions $\mathbb{Y}_{[KST]}^{KLSTMM_SM_T}$ antisymmetric with respect to the exchange of nucleons 1 and 2, i.e. with $l_4 + S_2 + T_2$ odd, compared to the number of linearly-independent fully-antisymmetric HH functions $\mathbb{Y}_{[KST]}^{KLSTMM_SM_T,\mathcal{A}}$ for $(L, S, T)^\pi = (1, 1/2, 1/2)^-$ and given (M, M_S, M_T), as a function of K

K	1	3	5	7	9	11	13
HH ($l_4 + S_2 + T_2$ odd)	49	447	2241	8182	24,264	62040	141,800
HH ($\mathcal{A}$)	1	7	36	134	399	1025	2349
Ratio	49	64	62	61	61	61	60

numbers $(K, L, S, T, M, M_S, M_T)$. In practice, a truncated antisymmetric space can also be used by, for instance, considering only the space spanned by the functions $\mathcal{A}\mathbb{Y}_{[KST]}^{KLSTMM_SM_T}$ with l_{sum} smaller than some value l_{max}.

29.3 Variational Approach

The five-nucleon system is described by a microscopic Hamiltonian

$$H = T + \sum_{i>j=1}^{5} v(|\boldsymbol{r}_i - \boldsymbol{r}_j|), \tag{29.10}$$

where v is the Volkov potential [2]. With this potential, the five-nucleon system has one bound state characterized by $(L, S, T)^\pi = (1, 1/2, 1/2)^-$ [5]. The Schrödinger equation is solved variationally by expanding the wave function ψ as

$$\psi = \sum_{Kij} c_{Kij} \mathbb{Y}_i^{KLSTMM_SM_T,\mathcal{A},\perp}(\Omega) \frac{f_j(\rho)}{\rho^{11/2}}, \tag{29.11}$$

where the functions $f_j(\rho)$ constitue a Lagrange-Laguerre mesh regularized by $\sqrt{\rho}$ [6–9]. Using the Gauss-Laguerre quadrature associated with this mesh to compute the potential matrix elements leads to a sparse Hamiltonian matrix. The ground-state energies are given in Table 29.2 (see the caption for computational details) and compared with the results obtained with a nonsymmetrized HH basis [5]. The small difference between both approaches is due to the truncation over the antisymmetric HH space, restricted to $l_{\text{sum}} = 1$.

Table 29.2 Ground-state energy in MeV of the five-nucleon system computed with the Volkov potential [2] and $\hbar^2/m = 41.47$ MeV fm. The HH expansion (29.11) is restricted to $K \leq K_{\max}$. Twenty-five Lagrange-Laguerre functions have been used. The antisymmetric HH spaces are truncated at $l_{\max} = 1$

$K_{\max}$	$l_{\max} = 1$	[5]
1	−39.635	−39.635
3	−40.001	−40.001
5	−41.022	−41.022
7	−41.785	−41.785
9	−42.383	−42.384
11	−42.681	−42.682
13	−42.867	−42.868
15	−42.951	−42.952
17	−42.990	−42.996

References

1. Kievsky, A., Rosati, S., Viviani, M., Marcucci, L.E., Girlanda, L.: J. Phys. G **35**, 063101 (2008)
2. Volkov, A.: Nucl. Phys. **74**, 33 (1965)
3. Dohet-Eraly, J., Viviani, M.: In preparation
4. Viviani, M.: Few-Body Syst. **25**, 177 (1998)
5. Gattobigio, M., Kievsky, A., Viviani, M.: Phys. Rev. C **83**, 024001 (2011)
6. Baye, D., Heenen, P.-H.: J. Phys. A **19**, 2041 (1986)
7. Vincke, M., Malegat, L., Baye, D.: J. Phys. B **26**, 811 (1993)
8. Baye, D.: Phys. Rep. **565**, 1 (2015)
9. Dohet-Eraly, J.: Eur. Phys. J. Plus **132**, 362 (2017)

Chapter 30
S-factor and Scattering Parameters from ^{3}He + ^{4}He → ^{7}Be + γ Data

Xilin Zhang, Kenneth M. Nollett and Daniel R. Phillips

Abstract We use the next-to-leading-order (NLO) amplitude in an effective field theory (EFT) for $^3\text{He} + ^4\text{He} \to ^7\text{Be} + \gamma$ to perform the extrapolation of higher-energy data to solar energies. At this order the EFT describes the capture process using an s-wave scattering length and effective range, the asymptotic behavior of ^{7}Be and its excited state, and short-distance contributions to the E1 capture amplitude. We use a Bayesian analysis to infer the multi-dimensional posterior of these parameters from capture data below 2 MeV. The total S factor $S(0) = 0.578^{+0.015}_{-0.016}$ keV b at 68% degree of belief. We also find significant constraints on ^{3}He–^{4}He scattering parameters.

The solar-fusion reaction $^3\text{He} + ^4\text{He} \to ^7\text{Be} + \gamma$ has not been measured directly at solar energies, due to the exponential suppression of the cross section there. Solar models use cross sections for it based on extrapolations that are derived using potential models or R-matrix, and constrained by S-factor and ^{3}He–^{4}He scattering data, as well as ^{7}Be bound-state properties. Adelberger et al. [1] reviews the most prominent efforts before 2011; additional evaluations have emerged since [2, 3].

X. Zhang
Department of Physics, University of Washington, Seattle, WA 98195, USA
e-mail: zhang.10038@osu.edu

Department of Physics, The Ohio State University, Columbus, OH 43210, USA

K. M. Nollett
Department of Physics, San Diego State University, 5500 Campanile Drive, San Diego, CA 92182-1233, USA
e-mail: kenneth.nollett@sdsu.edu

D. R. Phillips (✉)
Institute of Nuclear and Particle Physics and Department of Physics and Astronomy, Ohio University, Athens, OH 45701, USA
e-mail: phillid1@ohio.edu

Institut für Kernphysik, TU Darmstadt, 64289 Darmstadt, Germany

ExtreMe Matter Institute EMMI, GSI Helmholtzzentrum für Schwerionenforschung GmbH, 64291 Darmstadt, Germany

N. A. Orr et al. (eds.), *Recent Progress in Few-Body Physics*, Springer Proceedings in Physics 238,
https://doi.org/10.1007/978-3-030-32357-8_30

30.1 Formalism for $E1$ Capture

We use Halo Effective Field Theory (EFT) [4], treating ^{3}He ($\frac{1}{2}^+$) and ^{4}He (0^+) as fundamental degrees of freedom and ^{7}Be (ground state, GS, $\frac{3}{2}^-$) and ^{7}Be* (excited state, ES, $\frac{1}{2}^-$) as shallow p-wave bound states of the two. From the breakup energies of ^{3}He and ^{4}He we infer an EFT breakdown scale Λ of about 200 MeV. The energy range $E \lesssim 2$ MeV implies a low-momentum scale Q of 70–80 MeV, thus we have $Q/\Lambda \approx 0.4$. We systematically expand both scattering and reaction amplitudes in this small parameter. The NLO S-factor for E1 capture to the ^{7}Be GS can then be written [5]

$$S_{P_{3/2}}(E) = \frac{e^{2\pi\eta}}{e^{2\pi\eta}-1}\frac{8\pi}{9}\left(e\,Z_{eff}\right)^2 k_C \omega^3 C^2_{(P_{3/2})}\left(|\,\mathcal{S}\,|^2 + 2\,|\,\mathcal{D}\,|^2\right), \tag{30.1}$$

with the same result, *mutatis mutandis* for capture to the $P_{1/2}$ ES. This is analogous to our results for $^7\text{Be} + p \to {}^8\text{B} + \gamma$ [6, 7]. Here, $k_C \equiv \alpha_{\text{em}} Z^2 M_{\text{R}}$ with M_{R} the reduced mass of the ^{3}He–^{4}He system; $\eta \equiv k_C/p$ is the well-known Sommerfeld parameter; ω is the energy of the photon produced in the reaction; and the "effective charge" $Z_{eff} \equiv (Z/M_4 - Z/M_3)\,M_{\text{R}}$. The factors $C^2_{(P_{3/2})}$ ($C^2_{(P_{1/2})}$) are the squared p-wave asymptotic normalization coefficients (ANCs) of the GS (ES) [7]. The two reduced matrix elements, $\mathcal{S}$ and $\mathcal{D}$, are for the E1 transition from initial s- and d-wave states. At NLO, $\mathcal{S}$ consists of the well-known external capture contributions plus a short-distance piece similar to R-matrix internal capture. We parameterize the latter contribution to capture to the GS (ES) by a single number, $\overline{L}$ ($\overline{L}_*$). The d-wave reduced matrix element $\mathcal{D}$ is given by the standard asymptotic expression for external capture, but integrated all the way to zero radius. Explicit formulae for $\mathcal{S}$ and $\mathcal{D}$ can be found in [5, 7].

Capture reactions to the ground and excited state share the same initial state for s-waves ($\frac{1}{2}^+$), so $\mathcal{S}$ depends on the scattering length, a_0 and effective range, r_0. Up to NLO there are then 6 EFT parameters, henceforth denoted as the vector $\mathbf{g}$: $C^2_{(P_{3/2})}$ (fm^{-1}), $C^2_{(P_{1/2})}$ (fm^{-1}), a_0 (fm), r_0 (fm), $\overline{L}$ (fm), and $\overline{L}_*$ (fm).

30.2 Data, Bayesian Analysis, and Results

There are six total S-factor data sets, here labeled Seattle, Weizmann, Luna, Erna, Notre Dame (ND), and Atomki. There are four branching-ratio data sets: Seattle, Luna, Erna, and Notre Dame. In order to ensure that the data used are within the domain of validity of the EFT we only employ data with $E \leq 2$ MeV. This, together with other data-selection criteria, yields 59 S-factor and 32 Br data, see Fig. 30.1. (Details, including original references and a full listing of these data, will appear in [5].) To account for the common-mode errors we introduce normalization corrections,

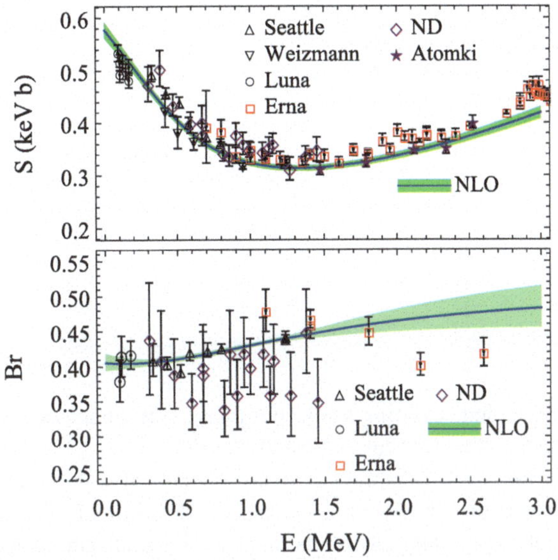

Fig. 30.1 Total S-factor and branching-ratio results. The data is denoted in the legend, and summarized in [2]. The green band shows the 68% interval for $S(E)$ and $Br(E)$ in our NLO Halo EFT analysis. The mean is denoted by the blue line

$\{\xi_J\colon J = 1 \ldots N_{\text{exp}}\}$, for the S-factor data. Such errors mostly cancel for Br data, so this correction is not used for them.

We take the EFT expressions such as (30.1) and employ Bayesian analysis—implemented via Markov-Chain-Monte-Carlo (MCMC) sampling—to infer probability distribution functions (PDFs) for the EFT parameters **g** from these data. Taking box priors with ranges considerably larger than those suggested by naive dimensional analysis for **g**, and gaussian priors for the ξ_J's, we can write the desired PDF as $\text{pr}\left(\mathbf{g}, \{\xi_J\}|D; T; I\right) \equiv c \exp\left(-\chi^2/2\right)$. The χ^2 is non-standard because it includes not only contributions from S-factor and branching-ratio measurements, but also the effect of the normalization corrections [5, 7].

The MCMC sampling produces the full twelve-dimensional pdf for **g** and $\{\xi_J\}$. These samples can then be used to compute a histogram for $S(E)$ and $Br(E)$ at any energy, E. Figure 30.1 shows the resulting 68% intervals: the mean is denoted by the blue line. The data is shown without re-scaling by the factors $1/(1 - \xi_J)$, so Fig. 30.1 under-represents the quality of our final result. Adopting values for the ξ_J's that maximize their posterior pdf produces a distribution of χ^2's in our MCMC sample that peaks at 1.04 per degree of freedom.

At NLO we have $S(0) = 0.578^{+0.015}_{-0.016}$ keV b. The recommended value from [1] is $0.56 \pm 0.02(\text{exp}) \pm 0.02(\text{theory})$—consistent with our result, but with an uncertainty that is almost a factor of two larger. Other recent analyses are broadly consistent, but also have somewhat bigger errors [2, 3]. We also find $Br(0) = 0.406^{+0.013}_{-0.011}$. There are two essential differences between this paper and another, recent, EFT evaluation of the same reaction [8]. First, we employ Bayesian methods. Second, we do

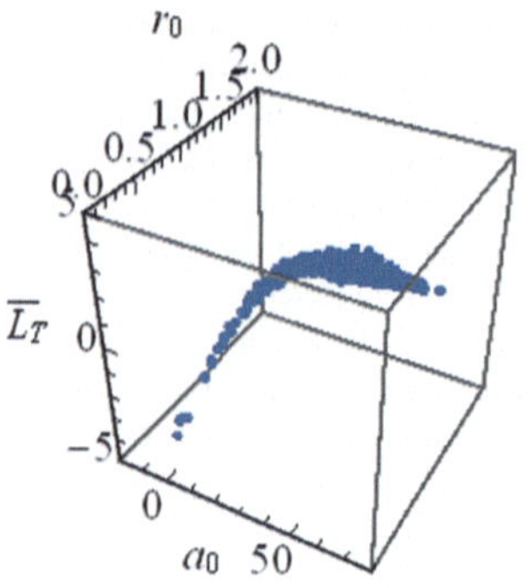

Fig. 30.2 a_0–r_0–$\overline{L}_T$ (all in fm) 3D scatter plot based on MCMC samples

not include existing scattering phase shift analyses in our constraints because their systematic errors are poorly quantified.

Figure 30.2 displays a three-dimensional scatter plot of the NLO MCMC samples, projected to the a_0–r_0–$\overline{L}_T$ subspace. Projecting further onto the a_0-r_0 subspace shows that significant constraints on ^{3}He–^{4}He scattering parameters can be obtained from the extant radiative capture data—in contrast to cases such as ^{7}Be(p, γ) [6]. The corresponding effective-range function can be tested against future high-quality ^{3}He–^{4}He scattering data at low energy.

We conclude that data on $^3\text{He} + {}^4\text{He} \to {}^7\text{Be} + \gamma$ already tightly constrain important aspects of the dynamics needed for extrapolation of this reaction's S-factor: we find quite small uncertainties on the s-wave elastic scattering parameters and the ANCs of the final states. Better measurements of scattering cross sections will test the EFT approach to the reaction presented here.

Acknowledgements This work is supported by the US DoE under grants DE-FG02-93ER-40756 (DP), DE-FG02-97ER-41014 (XZ), DE-SC0019257 (KMN), through MSU subcontract RC107839-OSU for the NUCLEI SciDAC collaboration (XZ), by the US NSF via grants PHY-1614460 (XZ) and PHY-1630782, N3AS FRHTP (DP), and by the ExtreMe Matter Institute EMMI at the GSI Helmholtzzentrum für Schwerionenphysik (DP).

References

1. Adelberger, E.G., et al.: Solar fusion cross sections II: the pp chain and CNO cycles. Rev. Mod. Phys. **83**, 195 (2011). https://doi.org/10.1103/RevModPhys.83.195
2. deBoer, R.J., et al.: Monte Carlo uncertainty of the He3(α, γ)Be7 reaction rate. Phys. Rev. C **90**, 035804 (2014). https://doi.org/10.1103/PhysRevC.90.035804
3. Iliadis, C., et al.: Bayesian estimation of thermonuclear reaction rates. Astrophys J. **831**, 107 (2016). https://doi.org/10.3847/0004-637X/831/1/107
4. Hammer, H.-W., Ji, C., Phillips, D.R.: Effective field theory description of halo nuclei. J. Phys. G **44**, 103002 (2017). https://doi.org/10.1088/1361-6471/aa83db
5. Zhang, X., Nollett, K.M., Phillips, D.R.: S-factor and scattering-parameter extractions from $^3\text{He} + {}^4\text{He} \to {}^7\text{Be} + \gamma$. arXiv:1909.07287 [nucl-th]

6. Zhang, X., Nollett, K.M., Phillips, D.R.: Halo effective field theory constrains the solar ^{7}Be + p $\rightarrow$ ^{8}B + γ rate. Phys. Lett. B **751**, 535 (2015). https://doi.org/10.1016/j.physletb.2015.11.005
7. Zhang, X., Nollett, K.M., Phillips, D.R.: Models, measurements, and effective field theory: proton capture on ^{7}Be at next-to-leading order. Phys. Rev. C **98**, 034616 (2018). https://doi.org/10.1103/PhysRevC.98.034616
8. Higa, R., Ruak, G., Vaghani, A.: Radiative^{3}He(α, γ)^{7}Be reaction in halo effective field theory. Eur. Phys. J. A **54**, 89 (2018). https://doi.org/10.1140/epja/i2018-12486-5

Chapter 31
Analysis of $^{74-82}$Se Within the Framework of the IBM and the Few-Body Aspects

Su Youn Lee, Young Jun Lee and J. H. Lee

Abstract An analysis of $^{74-82}$Se within the framework of the interacting boson model (IBM) and the few-body aspects are presented. Selenium isotopes with mass number (A) ranging from 74 to 82 are typical vibrational nuclei that can be explained by the U(5) limit of the IBM. However, they are not fully symmetrical nuclei, and are represented by adding a perturbation term to account for the symmetry breaking. These nuclei tend to exhibit properties of a harmonic oscillator and an asymmetric deformed rotor. The low-energy level and B(E2) ratios are calculated using the IBM via direct diagonalization of the perturbed Hamiltonian and the E(5) symmetry breaking. Comparisons between the theoretical and experimental results of the even-even selenium isotopes with A ranging from 74 to 82 are presented.

31.1 Introduction

Selenium compound, recognized by the World Health Organization and the U.N. Food and Agriculture Organization in 1978, is composed of antioxidants and is an important part of the immune system, especially in cancer prevention and anti-cancer treatment. Selenium isotopes are created during the process of nuclear fission or beta decay of uranium and plutonium, both of which are used as nuclear fuels in the reactors.

The interacting boson model (IBM) proposed by Arima and Ichcello [1, 2] is algebraically described in terms of the unitary group in six-dimensional space, with U(6) as the highest group, and is divided into the three limits of dynamic symmetry through group reduction with U(6) as the group, which are the U(5) [2], SU(3) [3] and O(6) limits [4]. They correspond to the geometric models of the harmonic oscillator nuclei, deformed rotator nuclei, and asymmetric deformed rotor nuclei, respectively. U(5) and O(6) limits have many similar properties in their eigenstate classification

S. Y. Lee (✉) · Y. J. Lee · J. H. Lee
Dong-Eui University, Busan 614-714, Korea
e-mail: syyi@deu.ac.kr
URL: http://physics.deu.ac.kr

N. A. Orr et al. (eds.), *Recent Progress in Few-Body Physics*,
Springer Proceedings in Physics 238,
https://doi.org/10.1007/978-3-030-32357-8_31

and energy spectra [5], and in some cases, two limits are mixed rather than separate symmetries in a given nucleus. As the most noticeable difference between the two limits, U(5) symmetry shows almost triplets of similar energy in 4_1^+, 2_2^+ and 0_2^+ states, while 0_2^+ has a high energy of nearly 3_1^+ state in O(6) symmetry. The energy ratio between 2_1^+ and 4_1^+ is 2 in U(5) limit and 2.5 in O(6) limit [5].

$^{74-82}$Se tend to show both limits [2, 6, 7]. The Hamiltonian in this work is composed by adding the pairing term with Casimir operator of the O(6) limit to the U(5) Hamiltonian [8] to explain the structure of the even-even selenium isotopes. We measured the degree of symmetry breaking of $^{74-82}$Se nuclei and compared it to the results of the experiment.

31.2 Theoretical Approach

In the IBM, collective properties about the low-lying energy level of an even-even nuclei is explained by using the two types of boson creation and annihilation operators, which combines s-boson with zero angular momentum and d-boson with angular momentum of two. The Hamiltonian of U(5) limit is [5]

$$H_{U(5)} = E_0 + \epsilon'(d^\dagger \cdot \tilde{d}) + \sum_{L=0,2,4} \frac{1}{2}(2L+1)^{\frac{1}{2}} c_L \left[[d^\dagger \times d^\dagger]^{(L)} \times [\tilde{d} \times \tilde{d}]^{(L)} \right]_0^{(0)} \tag{31.1}$$

where $d^\dagger$ and $\tilde{d}$ represent creation and annihilation of d-boson respectively. The basis state expressed by the quantum number corresponding to the subgroup of group chain in the U(5) limit of IBM is

$$|[N]\, n_d\; v\; n_\Delta\; L\rangle\,. \tag{31.2}$$

where N is total boson number, n_d is d-boson number, while v is the number of coupled d-bosons whose angular momentum is not zero; n_Δ is triplet number of d-boson whose angular momentum is zero and L is angular momentum.

The energy eigenvalue of U(5) limit is [5]

$$E = E_0 + \epsilon n_d + \alpha n_d(n_d + 4) + 2\beta v(v+3) + 2\gamma L(L+1.) \tag{31.3}$$

where, E_0 is binding energy, while the relation between the parameters in (31.1) and (31.3) is defined by [5]

$$\begin{aligned}
\epsilon &= \tfrac{1}{190}(190\epsilon' + 589c_0 - 790c_2 + 676c_4),\\
\alpha &= \tfrac{1}{190}(133c_0 - 170c_2 + 132c_4),\\
\beta &= \tfrac{1}{380}(19c_0 + 30c_2 - 11c_4),\\
\gamma &= -\tfrac{1}{38}(c_2 - c_4).
\end{aligned} \tag{31.4}$$

In this work, the Hamiltonian that represents the nucleus within the framework of the IBM, where the properties of a harmonic oscillator and an asymmetric deformed rotor occur simultaneously, is given as

$$H = H_{U(5)} + \lambda P_6, \tag{31.5}$$

to which the perturbation term Casimir operator P_6 of O(6) limit is added

$$P_6 = \frac{1}{4}\sum_m \left[(-)^m d_m^\dagger d_m^\dagger - \frac{1}{2}s^\dagger s^\dagger\right] \times \sum_{m'} \left[(-)^{m'} d_{m'} d_{-m'} - \frac{1}{2}ss\right], \tag{31.6}$$

where, $s^\dagger(s)$ is s-boson creation (annihilation) operator.

The eigenvalue is produced by SU(1,1) [2, 9], which is a boson quasi-spin group, while the Casimir operator P_6 of O(6) limit is the pairing operator of O(6) group.

The perturbed wave function is [8]

$$\begin{aligned} |[N]n_d v n_\Delta L\rangle_{\text{per}} &= [N]n_d v n_\Delta L\rangle \\ &+ \sum_{n'_d v' n'_\Delta L'} \frac{\langle [N]n'_d v' n'_\Delta L' |\lambda P_6| [N]n_d v n_\Delta L\rangle}{E-E'} \times |[N]n'_d v' n'_\Delta L'\rangle, \end{aligned} \tag{31.7}$$

this corresponds only to first-order. Electric quadrupole operator is needed for calculation of $E2$ transition rates

$$T_\mu^{(E2)} = \alpha_2[(d^\dagger \times \tilde{s}) + (s^\dagger \times \tilde{d})]_\mu^{(2)} + \beta_2[d^\dagger \times \tilde{d}]_\mu^{(2)}, \tag{31.8}$$

and the general expression of $E2$ transition rates between $|i\rangle$ and $|f\rangle$ is

$$B(E2; i, L_i \to f, L_f) = \frac{\left|\langle f, L_f || T^{(E2)} || i, L_i\rangle\right|^2}{2L_i + 1}. \tag{31.9}$$

The calculation results of the intraband and interband transition rates by the perturbed effect in this work are represented in [8, 10].

31.3 Nuclear Structures of $^{74-82}$Se Isotopes

The calculations of this study were applied to $^{74-82}$Se corresponding to the typical nuclei, in which the two properties of U(5) and O(6) symmetries in IBM appear simultaneously, and were compared with the experimental values. ^{74}Se isotopes consist of three proton bosons and five neutron bosons, i.e., eight bosons in total, in outside excluding core. $^{76-82}$Se can also be used to calculate the total boson number. The energy level corresponding to each spin is expressed with the parameters of (31.4). The parameters can be determined from the experimental data of $^{74-82}$Se isotopes and the parameters for $^{74-82}$Se are given in Table 31.1. The energy levels calculated

Table 31.1 Values of parameters for the energy spectra of $^{74-82}$Se (keV)

	N	ϵ	α	β	γ	λ
^{74}Se	8	596.554	−10.199	5	6.093	−82.487
^{76}Se	7	570.101	−0.803	−4	8.887	87.126
^{78}Se	6	654.657	15.871	−23	14.846	−1758.1
^{80}Se	5	1025.043	−47.265	0.135	7.794	544.57
^{82}Se	4	1005.311	−47.775	13.243	3.198	100.34

Table 31.2 Energy spectra of $^{74-82}$Se of the present model and experimental data [13–17] (keV)

	^{74}Se		^{76}Se		^{78}Se		^{80}Se		^{82}Se	
Spin	Theor.	Exp.	Theor.	Exp.	Theor.	Exp.	Theor.	Exp.	Theor.	Exp.
2_1^+	659	635	641	559	728	614	883	666	911	655
4_1^+	1414	1363	1406	1332	1634	1503	1797	1701	1830	1735
6_1^+	2267	2231	2296	2262	2716	2547	2742	2896	2758	3145
8_1^+	3217	3198	3310	3270	3976	4121	3718	3636	3695	3519
2_2^+	1244	1269	1157	1216	1218	1309	1579	1731	1741	1732
4_2^+	1999	2108	1905	2026	2063	2191	2399	2550	–	–
6_2^+	2852	2987	2777	2976	3086	3140	–	–	–	–
3_1^+	1902	1884	1763	1689	1826	1854	2274	2121	–	–
5_1^+	2705	2662	2564	2489	2729	2735	–	–	–	–
7_1^+	3606	3525	3490	3432	3810	3704	–	–	–	–
0_2^+	1071	854	1131	1122	1500	1499	1483	1479	1437	1440
2_3^+	1689	1838	–	–	–	–	–	–	–	–

using (31.3), by substituting the values of the determined parameters from Table 31.1, are displayed in Table 31.2 and Fig. 31.1 in comparison to the experimental results.

The Hamiltonian proposed in this study was successfully reproduced in the energy spectrum of the selenium nucleus. The energy spectrum of the nucleus corresponding to the typical U(5) limit is similar in $E(4_1^+)$, $E(2_2^+)$ and $E(0_2^+)$, but ^{78}Se shows that $E(0_2^+)$ has a higher value near $E(3_1^+)$. This means that ^{78}Se tends to have the U(5) limit energy spectrum. Additionally, the trend of ratio between each energy level, $R_{4/2}(E(4_1^+)/E(2_1^+))$ of ^{74}Se is 2.15, and the experimental value is 2.15. In the U(5) limit, $R_{4/2}$ is 2 and in the O(6) limit, $R_{4/2}$ is 2.5. Then, $R_{0/2}(E(0_2^+)/E(2_1^+))$ is 2.06 for ^{78}Se and 2 and 4.5 for U(5) and O(6) limit, respectively. ^{78}Se has $R_{4/2}$ of 2.24 and an experimental value of 2.45. ^{78}Se values are between the U(5) and O(6) limit.

The $E2$ transition rates are calculated from [10] and arranged in Table 31.3. The parameters α_2 and β_2 of (31.8) can be determined from the experimental values of the $E2$ transition rate and the electric quadrupole moment. However, in this study, the determination of parameters α_2 and β_2 was not necessary due to the transition rate between the states and the ratio between $B(E2; 2_1^+ \rightarrow 0_1^+)$. The parameter λ

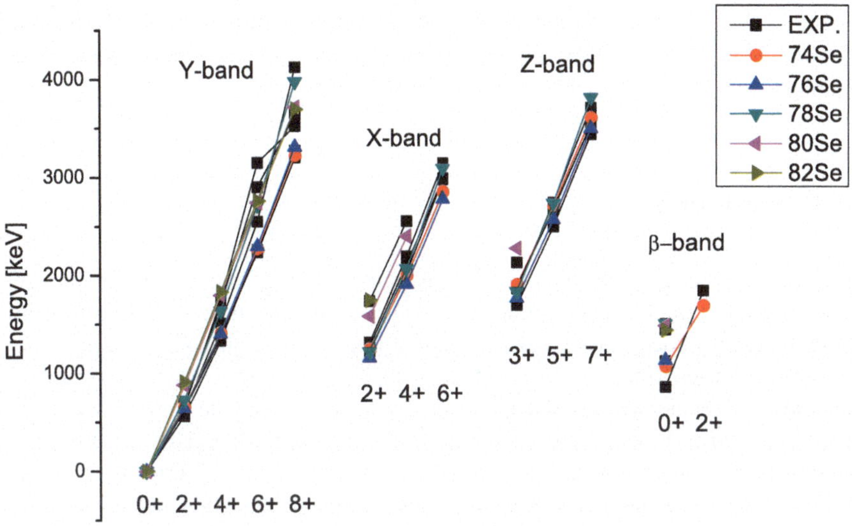

Fig. 31.1 Energy spectra of the U(5) → O(6) limit and the corresponding experimental data for $^{74-82}$Se [13–17]

Table 31.3 Transition rates $\frac{B(E2;L_i\rightarrow L_f)}{B(E2;2_1^+\rightarrow 0_1^+)}$ in $^{74-82}$Se, compared to the calculations of this work and the experimental data [18–20]

	$L_i = 4_1^+$	$L_f = 2_1^+$	$L_i = 2_2^+$	$L_f = 2_1^+$	$L_i = 0_2^+$	$L_f = 0_1^+$
	Theor.	Exp.	Theor.	Exp.	Theor.	Exp.
^{74}Se	2.01	1.92	2.01	1.14	0.83	1.94
^{76}Se	1.62	1.62	1.62	1.43	0.55	0.11
^{78}Se	1.18	1.45	1.18	0.99	0.69	0.86
^{80}Se	1.44	1.42	1.62	1.43	0.67	0.28
^{82}Se	1.43	1.1	–	–	0.37	0.21

that is included in the $E2$ transition rate was determined by comparing with the experimental data in $\frac{B(E2;L_i\rightarrow L_f)}{B(E2;2_1^+\rightarrow 0_1^+)}$ for $^{74-82}$Se. λ is a parameter for the perturbation effect as shown in Table 31.1. This means that ^{78}Se has a greater symmetry breaking than ^{74,76}Se. The calculation of ^{74,76}Se can be done through a process similar to that for ^{78}Se. The results are presented in Fig. 31.1 and Table 31.3. $R_{4/2}$ of ^{80,82}Se is 2.03 and 2.01 respectively, and $R_{0/2}$ is 1.68 and 1.58, respectively. Particularly noticeable from the theoretical values of the energy of $^{74-82}$Se is that the greater the spin, the greater the $B(E2)$ value, which is one of the characteristics [12] shown in the critical point symmetric nucleus [11] between the U(5) and O(6) limit. This trend is a characteristic that cannot be explained by a single limit such as the U(5) and O(6) limit, but it has been reproduced by adding perturbation term in this study. ^{80,82}Se are nuclei that exhibit typical U(5) symmetry and break symmetry in the O(6) direction. From the λ values calculated in this study, the values of λ for ^{74,76}Se are 82.487 keV and 87.126 keV, respectively. A comparison shows that ^{78}Se is more symmetry breaking in the U(5)-O(6) direction than ^{74,76}Se.

31.4 Summary

To account for the energy levels and $E2$ transition rates of $^{74-82}$Se, a perturbation term P_6 was added to the U(5) Hamiltonian. The U(5) symmetry proposed here does not affect the energy spectrum by considering the first-order perturbation, but it allows for some $E2$ transitions that are forbidden in U(5) symmetry through symmetry breaking. $^{74-82}$Se are identified from the energy spectrum and the parameter λ of the perturbation term, in which the properties in the direction of the U(5) to O(6) limit are shown. There was not sufficient experimental data on the $E2$ transition of ^{82}Se to discuss the U(5) symmetry breaking. Calculations using IBM-2(proton-neutron interacting boson model) to explain the nuclear structure that separates proton bosons from neutron bosons will be an interesting subject for future research.

Acknowledgements This work was supported by Basic Science Research Program through the National Research Foundation of Korea (NRF) funded by the Ministry of Education (NRF-2017R1D1A3A03080968).

References

1. Arima, A., Iachello, F.: Collective nuclear states as representations of a SU(6) group. Phys. Rev. Lett. **35**, 1069 (1975)
2. Arima, A., Iachello, F.: Interacting boson model of collective states I. The vibrational limit. Ann. Phys. **99**, 253 (1976)
3. Arima, A., Iachello, F.: Interacting boson model of collective nuclear states II. The rotational limit. Ann. Phys. **111**, 201 (1978)
4. Arima, A., Iachello, F.: Interacting boson model of collective nuclear states IV. The O(6) limit. Ann. Phys. **123**, 468 (1979)
5. Iachello, F., Arima, A.: The Interacting Boson Model. Cambridge University Press, Cambridge (1987)
6. Casten, R.F., Von Brentano, P.: An extensive region of O(6)-like nuclei near A = 130. Phys. Lett. **152**, 22 (1985)
7. Rikovska, J., Stone, N.J., Walters, W.B.: Dynamical symmetries in even-even Te nuclides. Phys. Rev. C **36**, 2162 (1987)
8. Woo, Z.-H.: Master's thesis, Busan National University (2002)
9. Ui, H.: SU(1,1) quasi-spin formalism of the many-boson system in a spherical field. Ann. Phys. **49**, 69 (1968)
10. Lee, J.H., Lee, S.Y., Woo, Z.-H.: Nuclear structures of tellurium and xenon nuclei for A = 128, 130 within the Framework of the interacting boson model. SAEMULLI **65**, 36 (2015)
11. Iachello, F.: Dynamic symmetries at the critical point. Phys. Rev. Lett. **85**, 3580 (2000)
12. Lee, S.Y., Lee, J.H.: E(5) Critical point symmetry in ^{128,130}Xe. SAEMULLI **63**, 766 (2013)
13. Singh, Balraj, Farhan, Ameenah R.: Nuclear data sheets for A = 74. Nucl. Data Sheets **107**, 1923 (2006)

14. Singh, Balraj: Nuclear data sheets update for A = 76. Nucl. Data Sheets **74**, 63 (1995)
15. Farhan, A.R., Singh, B.: Nuclear data sheets for A = 78. Nucl. Data Sheets **110**, 1917 (2009)
16. Singh, Balraj: Nuclear data sheets for A = 80. Nucl. Data Sheets **105**, 223 (2005)
17. Tuli, J.K.: Nuclear data sheets for A = 82. Nucl. Data Sheets **98**, 209 (2003)
18. Radhi, F.S., Stewart, N.M.: An IBM description of 76Se and neighbouring Se-isotopes. Z. Phys. A **356**, 145 (1996)
19. Türkan, N., Uluer, Y.: IBM-2 calculations of some even-even selenium nuclei. CEJP **4**, 124 (2006)
20. Böyükata, M., Uluer, I.: Investigation of some even-even selenium isotopes within the interacting boson model-2. Cent. Eur. J. Phys. **6**, 518 (2008)

Chapter 32
Microscopic Optical Potentials from First Principles

J. Rotureau, P. Danielewicz, G. Hagen, G. Jansen, F. Nunes and T. Papenbrock

Abstract We construct nucleonic microscopic optical potentials by combining the Green's function approach with the coupled-cluster method for ^{40}Ca and ^{48}Ca. We work with the chiral nucleon-nucleon and three-nucleon interaction NNLO$_{\text{sat}}$ which reproduces the charge radii of ^{40}Ca and ^{48}Ca. The overall form of the neutron scattering cross section is reproduced for both nuclei, but the imaginary part of the calculated potential, which reflects the loss of flux in the elastic channel, is negligible. The latter points to many-body correlations that would appear beyond the coupled-cluster truncation level considered in this work. We show that, by artificially increasing the parameter η in the Green's function, practical results can be further improved.

32.1 Introduction

Nuclear reactions are the primary experimental tool to study nuclei. With the recent progress in the development of rare-isotopes beams (RIBs), regions of the nuclear chart, previously out of reach, are now becoming accessible. Even more progress is expected, with future projects at RIB facilities, to explore systems far from stability [1–3]. In parallel to the progress on the experimental side, efforts should be pursued

J. Rotureau (✉) · P. Danielewicz · F. Nunes
NSCL/FRIB Laboratory, Michigan State University, East Lansing, MI 48824-1321, USA
e-mail: rotureau@nscl.msu.edu

P. Danielewicz · F. Nunes
Department of Physics and Astronomy, Michigan State University, East Lansing, MI 48824-1321, USA

G. Hagen · G. Jansen · T. Papenbrock
Physics Division, Oak Ridge National Laboratory, Oak Ridge, TN 37831, USA

G. Hagen · T. Papenbrock
Department of Physics and Astronomy, University of Tennessee, Knoxville, TN 37996, USA

G. Jansen
National Center for Computational Sciences, Oak Ridge National Laboratory, Oak Ridge, TN 37831, USA

N. A. Orr et al. (eds.), *Recent Progress in Few-Body Physics*,
Springer Proceedings in Physics 238,
https://doi.org/10.1007/978-3-030-32357-8_32

on the theoretical front to develop or extend reaction models to nuclei far from stability. In reaction theory, one usually reduces the many-body problem to a few-body one where only the most relevant degrees of freedom are retained [4]. In that case, the Hamiltonian is given in terms of effective interactions, the so-called optical potentials, among the different clusters. Traditionally, the optical potentials have been constrained by data, especially data on stable isotopes. As a consequence, using these potentials to study exotic nuclei is unreliable and has uncontrolled uncertainties. It is then critical, in order to advance the field of reactions, to connect the effective interaction to an underlying microscopic theory, so that extrapolations to exotic regions are better under control, together with rigorous assessment of uncertainties.

In this paper, we present ab-initio calculations of nucleon-nucleus optical potentials by combining the Green's function (GF) approach with the coupled-cluster (CC) method. We will show results for the doubly magic nuclei ^{40}Ca and ^{48}Ca with the chiral nucleon-nucleon (NN) and three-nucleon (3NFs) interaction NNLO$_{\text{sat}}$. The optical potential enters the Dyson equation together with the one-body Green's function. Assuming some approximations for the self-energy, the standard way of obtaining the optical potential is to iterate the non-linear Dyson equation until a self-consistent solution is obtained. This is known as the self-consistent Green's function approach [5–8]. Our approach here differs in that the optical potential is obtained directly by inverting the Dyson equation [9, 10].

This paper is organized as follows. In Sect. 32.2, we briefly revisit the formalism of the Green's function and the coupled-cluster method. In Sect. 32.3 we present results for the bound states in ^{41}Ca and ^{49}Ca and cross sections results for the neutron elastic scattering on ^{40}Ca and ^{48}Ca. Finally, we conclude and discuss future possible applications in Sect. 32.4.

32.2 Formalism

Let us consider a nucleus with A nucleons. The corresponding single-particle Green's function has matrix elements

$$
\begin{aligned}
G(\alpha, \beta, E) = & \left\langle \Psi_0 | a_\alpha \frac{1}{E - (H - E_{gs}^A) + i\eta} a_\beta^\dagger | \Psi_0 \right\rangle \\
& + \left\langle \Psi_0 | a_\beta^\dagger \frac{1}{E - (E_{gs}^A - H) - i\eta} a_\alpha | \Psi_0 \right\rangle ,
\end{aligned}
\tag{32.1}
$$

where α and β represent single-particle states, and $|\Psi_0\rangle$ is the ground state of the nucleus with energy E_{gs}^A solution of the Hamiltonian H. We work here with the single-particle states of the Hartree–Fock (HF) potential generated by H and we recall that the HF basis is a good starting point for coupled-cluster calculations and that the HF Green's function denoted as $G^{(0)}$ is a first order approximation to the

Green's function (32.1). By definition, the parameter η is such that in the physical limit $\eta \to 0^+$.

The Green's function fulfills the Dyson equation

$$G(\alpha,\beta,E) = G^{(0)}(\alpha,\beta,E) + \sum_{\gamma,\delta} G^{(0)}(\alpha,\gamma,E)\Sigma^*(\gamma,\delta,E)G(\delta,\beta,E). \quad (32.2)$$

Here, $\Sigma^*(\gamma,\delta,E)$ is the self energy, which can be obtained from the inversion of (32.2). Finally, one obtains the optical potential as $\Sigma' \equiv \Sigma^* + U$ where U is the HF potential. For $E \leq E^A_{gs}$, Σ' in (32.2) has a discrete number of solutions which correspond to the bound states in the A + 1 nucleus. For $E \geq E^A_{gs}$, Σ' is the optical potential for the elastic scattering from the ground state of the target A [11, 12]. The optical potential is non-local, energy-dependent and complex [12]; for $E \geq E^A_{gs}$, its imaginary component describes, by construction, the loss of flux due to absorption into channels other than the elastic channel.

In this paper, the optical potential is obtained by inverting the Dyson equation (32.2) after a direct computation of the Green's function (32.1) with the coupled-cluster method [13]. In coupled-cluster theory, the ground state is represented as

$$|\Psi_0\rangle = e^T|\Phi_0\rangle, \quad (32.3)$$

where T denotes the cluster operator which is expanded in the number of particle-hole excitations. In practice, T is truncated. In this paper all results are obtained from coupled-cluster with singles and doubles (CCSD) for which all operators beyond $2p-2h$ are neglected. For a more detailed account on CC and the construction of the GF with the CC we refer the reader to [9, 10, 13, 14].

Our goal is to compute the optical potential for elastic scattering at arbitrary energies. However, as $\eta \to 0^+$, the Greens' function (see (32.1)) has poles at energies $E = (E^{A+1}_i - E^A_{gs})$ (with E^{A+1}_i the eigenvalues of the A+1 system), which make the numerical calculation unstable. In order to avoid this issue, we consider [9], an analytic continuation of the Green's function in the complex-energy plane by working in the complex Berggren basis (generated from the HF potential), which includes bound-, resonant, and discretized non-resonant continuum states [15–19]. In that case, the Green's function matrix elements for $E \geq 0$ smoothly will converge to a finite value as $\eta \to 0^+$.

32.3 Results

Before presenting results for elastic scattering, we focus on the bound states in ^{41}Ca and ^{49}Ca. As we wrote previously, we work with the NNLO$_{\text{sat}}$ chiral interaction [20] which reproduces the binding energy and charge radius in both systems [21, 22]. Table 32.1 shows the calculated energy for bound states in ^{41}Ca and ^{49}Ca as

Table 32.1 Energies (in MeV) for bound states in ^{41}Ca and ^{49}Ca calculated with the particle-attached equation of motion (PA-EOM) [13, 14] CCSD using the chiral NNLO$_{sat}$ interaction as a function of N_{max} (see text)

	N_{max}	$E(7/2^-)$	$E(3/2^-)$	$E(1/2^-)$
^{41}Ca				
	12	−7.35	−3.47	−1.31
	14	−7.62	−3.87	−1.80
	14 ($N_3 = 16$)	−7.84	−4.07	−2.15
Exp		−8.36	−6.42	−4.74
^{49}Ca		$E(3/2^-)$	$E(1/2^-)$	$E(5/2^-)$
	12	−3.88	−2.02	−0.37
	14	−4.35	−2.40	−1.00
	14 ($N_3 = 16$)	−4.56	−2.45	−1.42
Exp		−5.14	−3.12	−1.56

a function of N_{max}. First, we perform HF calculations in a single-particle basis that employs a mixed representation of harmonic oscillator (we include harmonic oscillator shells with $2n + l \leq N_{max}$) and discretized Berggren states [10]. We denote N_2 and N_3, the cutoffs of the interaction terms defined respectively as the maximum number of quanta allowed in the relative motion of two nucleons and three nucleons. In all calculations here, we always take $N_2 = N_{max}$ and N_3 is taken equal to N_{max}, except for the most extensive calculations considered here where $N_{max} = 14$ and $N_3 = 16$. Moreover, the three-nucleon terms in NNLO$_{sat}$ are truncated at the normal-ordered two-body level in the HF basis [23]. We work at $\hbar\omega = 16$ MeV. For both nuclei, there are only three bound states supported by the NNLO$_{sat}$ Hamiltonian. As expected (see Table 32.1), convergence is slower for the higher-energy states: in the case of ^{41}Ca, the difference between the energies for $(N_{max}, N_3) = (14, 14)$ and $(14, 16)$ is $\sim$220 keV for the ground-state, whereas it is $\sim$350 keV for the $J^\pi = 1/2^-$ second excited state. For ^{49}Ca, the difference is $\sim$210 keV for the ground state and $\sim$420 keV for the $J^\pi = 5/2^-$ excited state. Even though the absolute binding energy is underestimated in the CCSD approximation, when compared to experiment the neutron separation energies are consistently within 600 keV of the experimental values.

Let us now discuss predictions for the neutron elastic cross section for ^{40}Ca and ^{48}Ca at respectively 5.17 and 7.81 MeV and compare with experimental data. In both cases, one expects the scattering phase shifts obtained from the calculated optical potential to have a finite imaginary part which reflects the loss of flux in the elastic channel. For instance, in the case of neutron scattering on ^{40}Ca at E = 5.17 MeV, there is a potential absorption due to excitation of ^{40}Ca to either its first excited state E(0^+) = 3.35 MeV or second excited state E(3^-) = 3.74 MeV. However, our calculations yield a negligible value for the absorption in all partial waves [10]. The position of the 0^+ excited state (which is known to have significant $4p - 4h$ components) is not properly captured in the EOM-CCSD approximation: its calculated energy is at 15.98 MeV above the ground state. This implies that absorption due to the excitation to the 0^+ excited state cannot be captured at that level of truncation in the CC. On the other hand, the energy of the second excited state is well reproduced at the EOM-CCSD level with $E_{EOM-CCSD}(3^-) = 3.94$ MeV. Consequently, potential excitation to the second excited state could in principle be accounted for. The fact that the calculated absorption is nevertheless negligible implies that the CCSD and PA-EOM wave functions are not sufficiently correlated. In other words, correlations beyond the singles and doubles truncation level are needed to account for the absorption due to the target excitation. A similar situation occurs for the neutron scattering off ^{48}Ca at 7.81 MeV: in that case, the position of the first excited state $E(2^+) = 3.83$ MeV is fairly well reproduced at the EOM-CCSD level with $E_{EOM-CCSD}(2^+) = 4.65$ MeV, but the calculated absorption is still negligible pointing out again to a lack of correlations in the CCSD and PA-EOM wavefunctions.

In order to make up for the lack of absorption at the CCSD level, one could artificially increase it by considering finite values of η instead of taking the limit $\eta \to 0^+$ (see (32.1)). In the following, we explore the impact of using finite η values for the neutron elastic scattering on ^{40}Ca and ^{48}Ca.

The calculated differential elastic cross sections for neutron scattering on ^{40}Ca at E= 5.17 MeV and on ^{48}Ca at 7.81 MeV are shown in Fig. 32.1. All calculations here correspond to the largest model space discussed previously that is, $N_{max} = 14$ and $N_3 = 16$. For comparison, we also show the angular distributions calculated with the phenomenological Koning–Delaroche (KD) potential [24] and also the measured cross sections (errors on the data are smaller than the symbols). As expected, when η increases, the elastic scattering cross section decreases with a more pronounced (relative) reduction at larger angles and the agreement with data improves. The level of disagreement between the experimental data and the result obtained with KD is an illustration of the level of accuracy that can be expected from a phenomenological interaction. We want to point out that the value of η we use should not be interpreted as the effective width of the states, but rather as a means to compensate for the truncations inherent to our approach.

32.4 Conclusions

We construct microscopic nucleon-nucleus optical potentials by combining the Green's function approach with the coupled-cluster method at the singles and doubles truncation level. By working in the Berggren basis, we obtain an analytical continuation of the Green's function in the complex-energy plane which allows to efficiently solve the Dyson equation and compute the optical potential.

We showed applications for ^{40}Ca and ^{48}Ca with the chiral NN and 3NF interaction $NNLO_{sat}$. The choice of this interaction was motivated by the fact that it allows for a good description of masses and radii in a wide mass-range and, furthermore, it

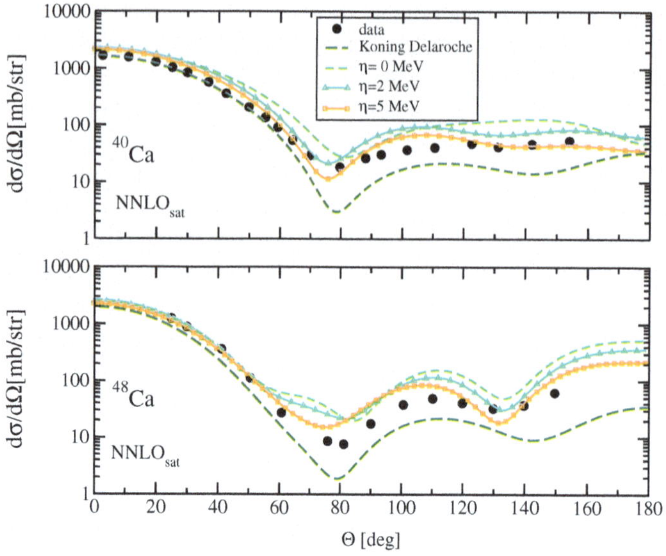

Fig. 32.1 Differential elastic cross section for ^{40}Ca(n, n)^{40}Ca at 5.17 MeV (top) and ^{48}Ca(n, n)^{48}Ca at 7.81 MeV (bottom) calculated with the NNLO$_{\text{sat}}$ interaction. Calculations are shown for $\eta = 0, 2, 5$ MeV. and results obtained using the KD potential are shown for comparison. Data points are taken from [24]

reproduces the charge radii of ^{40}Ca and ^{48}Ca. We first showed results for the bound states in ^{41}Ca and ^{49}Ca and then presented applications to the neutron scattering. We have seen that the overall form of the scattering cross section is reproduced for both nuclei. However, at that level of truncation, the absorption is practically negligible which points to a lack of many-body correlations in the wave functions of the coupled-cluster method at the singles and doubles approximation level. We showed that, by increasing the parameter η in the Green's function, results can be somewhat improved.

This work can be extended further by considering higher-order correlations in the coupled-cluster Green's function calculations as was recently done for the dipole response of ^{48}Ca [25], and excited states in ^{101}Sn [26]. It would also be interesting to investigate optical potentials constructed by starting with the singles and doubles coupled-cluster Green's function potential and add an *ad hoc* polarization terms which would effectively account for the missing physics such as collective excitations and formation of compound nucleus.

Acknowledgements We thank K. Hebeler for providing us with matrix elements in Jacobi coordinates for the NNN interaction at next-to-next-to-leading order. This work was supported by the National Science Foundation under Grant PHY-1403906, the Department of Energy under Contract No. DE-FG52- 08NA28552, by the Office of Science, U.S. Department of Energy under Award Number DE-SC0013365 and by the Office of Nuclear Physics, U.S. Department of Energy, under Grants DE-SC0008499 (SciDAC-3 NUCLEI), DE-SC0018223 (SciDAC-4 NUCLEI), the Field

Work Proposal ERKBP57 and ERKBP72 at Oak Ridge National Laboratory (ORNL). An award of computer time was provided by the Institute for Cyber-Enabled Research at Michigan State University and part of this research used resources of the Oak Ridge Leadership Computing Facility located at ORNL, which is supported by the Office of Science of the Department of Energy under Contract No. DE- AC05-00OR22725.

References

1. http://www.nupecc.org/index.php?display=lrp2016/main
2. Bollen, G.: Frib—facility for rare isotope beams. AIP Conf. Proc. **1224**, 432–441 (2010)
3. Kester, O., Stöcker, H.: FAIR Project at GSI. World Scientific Publishing Co. (2016)
4. Thompson, I.J., Nunes, F.M.: Nuclear Reactions for Astrophysics. Cambridge University Press, Cambridge (2009)
5. Dickhoff, W.H., Barbieri, C.: Self-consistent green's function method for nuclei and nuclear matter. Prog. Part. Nucl. Phys. **52**(2), 377–496 (2004)
6. Barbieri, C., Jennings, B.K.: Nucleon-nucleus optical potential in the particle-hole approach. Phys. Rev. C **72**, 014613 (2005)
7. Mahzoon, M.H., Charity, R.J., Dickhoff, W.H., Dussan, H., Waldecker, S.J.: Phys. Rev. Lett. **112**, 162503 (2014)
8. Barbieri, C., Carbone, A.: Self-Consistent Green's Function Approaches, pp. 571–644. Springer International Publishing, Cham (2017)
9. Rotureau, J., Danielewicz, P., Hagen, G., Nunes, F., Papenbrock, T.: Optical potential from first principles. Phys. Rev. C **95**(2), 024315 (2017)
10. Rotureau, J., Danielewicz, P., Hagen, G., Jansen, G.R., Nunes, F.M.: Microscopic optical potentials for calcium isotopes. Phys. Rev. C **98**, 044625 (2018)
11. Capuzzi, F., Mahaux, C.: Projection operator approach to the self-energy. Ann. Phys. **245**(1), 147–208 (1996)
12. Dickhoff, W.H., Van Neck, D.: Many-Body Theory Exposed!. World Scientific, Singapore (2007)
13. Hagen, G., Papenbrock, T., Hjorth-Jensen, M., Dean, D.J.: Coupled-cluster computations of atomic nuclei. Rep. Prog. Phys. **77**(9), 096302 (2014)
14. Bartlett, R.J., Musiał, M.: Coupled-cluster theory in quantum chemistry. Rev. Mod. Phys. **79**, 291–352 (2007)
15. Berggren, T.: On the use of resonant states in eigenfunction expansions of scattering and reaction amplitudes. Nucl. Phys. A **109**(2), 265–287 (1968)
16. Michel, N., Nazarewicz, W., Płoszajczak, M., Bennaceur, K.: Gamow shell model description of neutron-rich nuclei. Phys. Rev. Lett. **89**, 042502 (2002)
17. Id Betan, R., Liotta, R.J., Sandulescu, N., Vertse, T.: Two-particle resonant states in a many-body mean field. Phys. Rev. Lett. **89**, 042501 (2002)
18. Hagen, G., Vaagen, J.S., Hjorth-Jensen, M.: The contour deformation method in momentum space, applied to subatomic physics. J. Phys. A: Math. Gen. **37**(38), 8991 (2004)
19. Hagen, G., Vaagen, J.S.: Study of resonant structures in a deformed mean field by the contour deformation method in momentum space. Phys. Rev. C **73**, 034321 (2006)
20. Ekström, A., Jansen, G.R., Wendt, K.A., Hagen, G., Papenbrock, T., Carlsson, B.D., Forssén, C., Hjorth-Jensen, M., Navrátil, P., Nazarewicz, W.: Accurate nuclear radii and binding energies from a chiral interaction. Phys. Rev. C **91**, 051301 (2015)
21. Hagen, G., Ekström, A., Forssén, C., Jansen, G.R., Nazarewicz, W., Papenbrock, T., Wendt, K.A., Bacca, S., Barnea, N., Carlsson, B., Drischler, C., Hebeler, K., Hjorth-Jensen, M., Miorelli, M., Orlandini, G., Schwenk, A., Simonis, J.: Neutron and weak-charge distributions of the ^{48}Ca nucleus. Nat. Phys. **12**, 186 (2016)

22. Garcia Ruiz, R.F., Bissell, M.L., Blaum, K., Ekström, A., Frömmgen, N., Hagen, G., Hammen, M., Hebeler, K., Holt, J.D., Jansen, G.R., Kowalska, M., Kreim, K., Nazarewicz, W., Neugart, R., Neyens, G., Nörtershä, W., Papenbrock, T., Papuga, J., Schwenk, A., Simonis, J., Wendt, K.A., Yordanov, D.T.: Unexpectedly large charge radii of neutron-rich calcium isotopes. Nat. Phys. **12**, 594–598 (2016)
23. Hagen, G., Hjorth-Jensen, M., Jansen, G.R., Papenbrock, T.: Emergent properties of nuclei from ab initio coupled-cluster calculations. Phys. Scr. **91**(6), 063006 (2016)
24. Koning, A.J., Delaroche, J.P.: Local and global nucleon optical models from 1 kev to 200 mev. Nucl. Phys. A **713**(3), 231–310 (2003)
25. Miorelli, M., Bacca, S., Hagen, G., Papenbrock, T.: Computing the dipole polarizability of ^{48}Ca with increased precision. Phys. Rev. C **98**, 014324 (2018)
26. Morris, T.D., Simonis, J., Stroberg, S.R., Stumpf, C., Hagen, G., Holt, J.D., Jansen, G.R., Papenbrock, T., Roth, R., Schwenk, A.: Structure of the lightest tin isotopes. Phys. Rev. Lett. **120**, 152503 (2018)

Chapter 33
Beryllium-9 in Cluster Effective Field Theory

Paolo Andreatta, Carlo A. Manzata, Chen Ji, Winfried Leidemann and Giuseppina Orlandini

Abstract It is aimed to describe the nucleus of ^{9}Be in an α-α-neutron cluster model, where the α-α and α-neutron interactions are determined via an effective field theory (EFT) for halo nuclei (halo-EFT). Here, only the α-α case is discussed more detailed and hence results are given for α-α phase shifts and for the ground-state energy of ^{12}C as a three α cluster. The calculation of the ^{12}C wave function is made via an expansion on nonsymmetrized hyperspherical harmonics (NSHH) in momentum space.

33.1 Introduction

The nucleus of ^{9}Be is a typical example of a multi-nucleon system that can be described by a cluster ansatz, namely by two α-particles plus an additional neutron. Excitations of ^{9}Be can still be treated in such an approach if the excitation energy is sufficiently lower than the first excitation energy of the α-particle of about 20 MeV. Our final aim is to make use of such a cluster ansatz not only for ^{9}Be but also for its low-energy excitations into the n-α-α continuum with a future plan to calculate low-energy ^{9}Be photodisintegration.

In order to determine α-α and α-n interactions we use an effective field theory, named cluster (or halo) EFT [1, 2]. The theoretical ansatz to describe these potentials is given in Sect. 33.2, where also the momentum space NSHH calculation is

P. Andreatta (✉) · C. A. Manzata · W. Leidemann · G. Orlandini
Department of Physics, University of Trento, Trento, Italy
e-mail: paolo.Andreatta-1@unitn.it

P. Andreatta · W. Leidemann · G. Orlandini
INFN-TIFPA Trento Institute of Fundamental Physics and Applications,
Via Sommarive 14, 38123 Povo-Trento, Italy

C. Ji
Key Laboratory of Quark and Lepton Physics (MOE) and Institute of Particle Physics,
Central China Normal University, Wuhan 430079, China

N. A. Orr et al. (eds.), *Recent Progress in Few-Body Physics*,
Springer Proceedings in Physics 238,
https://doi.org/10.1007/978-3-030-32357-8_33

briefly illustrated. Section 33.3 contains results for α-α phase shifts in comparison to experimental data and for the binding energy of ^{12}C as a three-α cluster. In both cases the regulation cutoff dependence of the results is discussed.

33.2 Halo-EFT Approach

In the n-α-α system the interaction between the n-α is dominated by the $3/2^-$ p-wave interaction; while α-α interaction is driven by a combination of s-wave short-range interaction and a long-range Coulomb repulsion. The strengths and ranges of the inter-cluster interactions determine the two-body scattering phase shifts, which can be extracted from scattering experiments. In general, one can introduce partial-wave decomposition to the momentum-space potential to emphasize on the dominant partial-wave channel with an angular momentum l,

$$V_l(p,p') = p^l p'^l g(p) g(p') \sum_{i,j=0}^{1} (2l+1) p^{2i} \Omega_{ij} p'^{2j} P_l(\hat{p}\cdot\hat{p}') , \tag{33.1}$$

where the Ω matrix is defined as

$$\Omega = \begin{pmatrix} \lambda_0 & \lambda_1 \\ \lambda_1 & 0 \end{pmatrix} . \tag{33.2}$$

The potential $V_l(p,p')$ is in a separable form, which represents only the short-range parts of the interactions. There one includes the regulator $g(p)$, which modifies the large-momentum (short-distance) behavior of $V_l(p,p')$, such that $g(p=0)=1$ and $g(p\to\infty)=0$. In principle, the indices i and j appearing in (33.1) could be larger than 1. Since we are only interested in the potential at very low energies they are limited in order to get a phase shift expansion up to the effective range order setting $l=0$ and $l=1$ for α-α and α-n interactions, respectively. Besides V_l, a long-range Coulomb interaction exists in the α-α system.

The parameters λ_i were determined by solving the corresponding scattering states with the Lippmann-Schwinger equation. For the α-α interaction, in order to include the Coulomb final-state interactions, we proceed along the lines of [3] separating the T-matrix in two parts

$$T(\mathbf{p},\mathbf{p}') = T_{\mathrm{C}}(\mathbf{p},\mathbf{p}') + T_{\mathrm{SC}}(\mathbf{p},\mathbf{p}'), \tag{33.3}$$

where $T_{\mathrm{C}}(\mathbf{p},\mathbf{p}')$ is a genuine Coulomb T-matrix, whereas $T_{\mathrm{SC}}(\mathbf{p},\mathbf{p}')$ originates from a Coulomb-modified short-range amplitude. To facilitate the calculation we take here the step function $g(p)=\Theta(\Lambda-p)$ as regulator. Finally, λ_0 and λ_1 are fitted to describe the experimental data for scattering length and effective range of α-α scattering. More details on the calculation are given in [4].

The NSHH expansion in momentum space is similar to that in coordinate space [5, 6], but instead of a hyperradius ρ one has a hypermomentum Q. Since our EFT potentials do not depend only on the relative momentum $\mathbf{p}$, but also on a relative momentum $\mathbf{p}'$ one needs to introduce a hypermomentum Q' and additional primed hyperspherical variables.

33.3 Results

In Fig. 33.1 we show results for the α-α phase shifts obtained with our EFT potential. One finds a very nice agreement with experimental data [7] for the regulation cutoffs with $\Lambda = 120$ and 160 MeV, whereas the phase shifts for $\Lambda = 90$ MeV start to deviate from the experimental data with the increase of scattering energy due to the remaining regularization effects. Note that there appears an upper limit for Λ due to the Wigner bound condition. In Fig. 33.1 one also sees the effect of the ^{8}Be resonance at an energy of about 0.2 MeV. In fact at this energy the phase shift jumps from 0 to $\pi/2$ and a narrow resonance peak appears in the α-α s-wave scattering cross section (see inset of figure). In Fig. 33.2 we show the α-α-α ground-state energy, there is a rather flat minimum at about $\Lambda = 160$ MeV with a binding energy of almost 1 MeV and one observes that below $\Lambda = 110$ MeV and above about $\Lambda = 195$ MeV the three-α system becomes unbound.

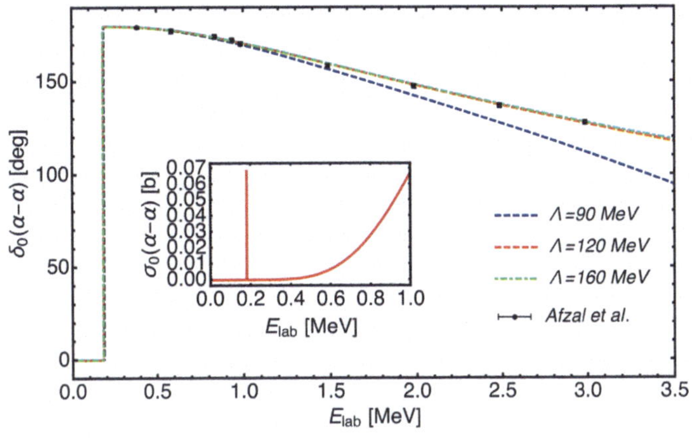

Fig. 33.1 α-α phase shift δ_0 obtained with the α-α potential in comparison to experimental data [7]; in the inset α-α s-wave cross section for $\Lambda = 200$ MeV

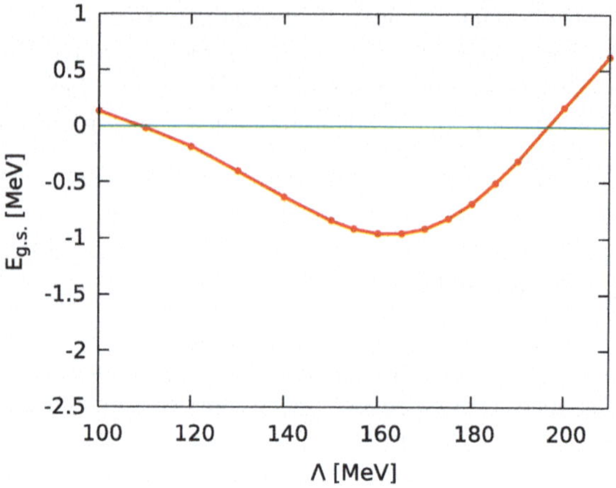

Fig. 33.2 Ground-state energy $E_{g.s.}$ of the α-α-α system as function of Λ

References

1. Higa, R., Hammer, H.-W., van Kolck, U.: Nucl. Phys. A **809**, 171 (2008)
2. Bertulani, C.A., Hammer, H.-W., van Kolck, U.: Nucl. Phys. **A712**, 37 (2002); Bedaque, P.F., Hammer, H.-W., van Kolck, U.: Phys. Lett. **B569**, 159 (2003)
3. Kong, X., Ravndal, F.: Nucl. Phys. A **665**, 137 (2000)
4. Manzata C.A.: Master thesis, University of Trento (2016)
5. Deflorian, S., et al.: Few-Body Syst. **54**, 1879 (2013)
6. Ferrari-Ruffino, F., et al.: Few-Body Syst. **58**, 113 (2017)
7. Afzal, S.A., Ahmad, A.A.Z., Ali, S.: Rev. Mod. Phys. **41**, 247 (1969)

Chapter 34
Cluster Configuration Effects in the Elastic Scattering of Boron Isotopes ^{8}B, ^{10}B, ^{11}B and ^{12}B on ^{58}Ni

V. Guimarães, E. O. N. Zevallos, E. N. Cardozo, J. Lubian, O. C. B. Santos, R. Linares, M. Assunção, J. Alcantara-Nunez, A. L. de Lara, R. Lichtenthaler Filho, K. C. C. Pires, U. Umbelino, S. Appannababu, N. Added, D. S. Monteiro and V. Morcelle

Abstract In this contribution, we report on effects due to the cluster configuration of boron isotopes, ^{8}B, ^{10}B, ^{11}B and ^{12}B, in the elastic scattering on ^{58}Ni target at energies close to the Coulomb barrier. The results of coupled channel calculations, where the different configurations of the boron isotopes, $^{8}B = ^{7}Be+p$, $^{10}B = ^{6}Li+\alpha$, $^{11}B = ^{7}Li+\alpha$ and $^{12}B = ^{11}B+n$, have been attributed to different effects in the angular distributions for the elastic scattering process, are present.

34.1 Introduction

The investigation of nuclear structure and reaction mechanisms induced by radioactive nuclei (proton-rich and neutron-rich) has attracted much interest in the recent years [1, 2]. In particular, elastic scattering and breakup measurements are good tools to investigate the unusual static features, as extended halos and neutron skin, present in some light nuclei, and as dynamic effects related to coupling between

V. Guimarães (✉) · E. O. N. Zevallos · O. C. B. Santos · J. Alcantara-Nunez · A. L. de Lara · R. Lichtenthaler Filho · K. C. C. Pires · U. Umbelino · S. Appannababu · N. Added
Instituto de Física, Universidade de São Paulo, São Paulo, SP, Brazil
e-mail: valdirg@if.usp.br

E. N. Cardozo · J. Lubian · R. Linares
Departamento de Física, Universidade Federal Fluminense, Niterói, RJ, Brazil

M. Assunção
Departamento de Física, Universidade Federal de São Paulo, Diadema, SP, Brazil

D. S. Monteiro
Universidade da Integração Latino-Americana, UNILA, Foz do Iguaçu, PR, Brazil

Department of Physics, University of Notre Dame, Notre Dame, IN, USA

V. Morcelle
Universidade Federal Rural do Rio de Janeiro, Seropédica, RJ, Brazil

N. A. Orr et al. (eds.), *Recent Progress in Few-Body Physics*,
Springer Proceedings in Physics 238,
https://doi.org/10.1007/978-3-030-32357-8_34

Table 34.1 The configuration of each nucleus of the boron isotope chain and its relation with the important effect and/or channel in the reaction

Projectile	Spin	Cluster configuration	Binding energy (MeV)	Quadrupole moment (fm^2)	Channel
^{8}B	2^+	^{7}Be+p	0.138	6.83	Breakup
^{10}B	3^+	^{6}Li+α	4.462	8.47	Spin-orbit
^{11}B	$3/2^+$	^{7}Li+α	8.866	4.07	Inelastic + reo
^{12}B	1^+	^{11}B+n	3.337	1.32	Transfer

different channels [1, 3, 4]. These statics and dynamics effects are manifested in the angular distribution for the elastic scattering process at energies close to the Coulomb barrier. By comparing the experimental angular distributions with coupled channel calculations it is possible to infer about the importance of these effects. For instance, for weakly-bound projectile, with halo structure, correlations of the valence particles can significantly distort the shell structure and/or the collective properties of the nucleus with drastic effect in the angular distribution. Cluster configurations can produce strong couplings to the continuum introducing characteristic dynamic polarization (attractive or repulsive) in the optical potential, which is not present for elastic induced by strongly-bound projectiles. For nuclei with halo structure and low binding energy, the breakup and transfer reactions may strongly compete with the elastic scattering process. The effect of the breakup on the elastic scattering would be damping the Fresnel peak in the elastic angular distribution at energies close to the Coulomb barrier [5]. Tightly-bound nuclei can also have cluster configurations, but in this case, the effect in the angular distribution may be more related to the deformation. Elastic scattering induced by deformed nuclei may have strong effect of couplings with inelastic channels. Reorientation effects can also play a role in the description of the elastic process. In this contribution we report on the investigation of these static and dynamic features in the elastic scattering of the boron isotopes, ^{8}B, ^{10}B, ^{11}B and ^{12}B on ^{58}Ni target. The boron isotopes chain consists of nuclei with quite diverse cluster configuration as shown in Table 34.1. For instance, the proton-rich nucleus ^{8}B has a very low proton binding energy, $S_p = 0.138$ MeV, for the ^{7}Be+p decay, while ^{11}B is quite tightly-bound with binding energy of $S_\alpha = 8.884$ MeV. The ^{10}B is a well deformed nucleus, with the largest spin (3^+) among boron isotope chain. The ^{12}B nucleus is radioactive with a low deformation, low spin and with binding energy of $S_n = 3.370$ MeV for the ^{11}B+n decay. Each of the configuration of the boron isotopes can lead to a different prevalent effect in elastic scattering. In this contribution, we present a discussion about possible correlation between configuration of the projectile and dynamic effects in the angular distribution for the elastic process. In addition, new preliminary experimental data on the elastic scattering of ^{12}B+^{58}Ni, are present.

34.2 Systematic for Boron Isotopes Elastic Scattering

The data for the elastic scattering of ^{8}B+^{58}Ni is presented in [4]. The ^{8}B has already been established to have halo configuration. Due to its low binding energy ($S_p = 0.138$ MeV), this nucleus can easily breaks up (^{7}Be+p) in the presence of an electromagnetic interaction of the target. The effect of the breakup channel in the angular distributions of the elastic scattering on ^{58}Ni target was investigated through CDCC (Continuum Discretization Coupled Channel) calculations in the [7, 8]. This effect showed to be very important, indicating that the most striking feature in the elastic scattering distribution for the ^{8}B+^{58}Ni system is the low binding energy of the ^{8}B projectile. The ^{10}B nucleus, by its turn, is described as ^{6}Li+α cluster configuration, with a binding energy of $S_\alpha = 4.461$ MeV. It also has the largest spin among the Boron isotopes, $J^\pi = 3^+$. The results of the coupled channel calculation analysis for the elastic scattering of ^{10}B+^{58}Ni system, at energies close to the Coulomb barrier, are reported in [9]. Despite of the large deformation, $Q(^{10}\mathrm{B}) = 8.47\,\mathrm{fm}^2$, of this nucleus, the most important effect for the description of the elastic scattering was the spin-orbit interaction [9]. The other stable boron isotope, ^{11}B, is a tightly-bound nucleus with predominant ^{7}Li+α cluster configuration ($S_\alpha = 8.664$ MeV). This nucleus can be considered one of the most strongly bound nucleus and this high binding energy prevents the breakup channel to be important. Data for elastic scattering of ^{11}B+^{58}Ni were obtained at energies close to the barrier and reported in [10]. The analysis of the elastic scattering angular distributions with coupled channel calculation showed that inelastic channels and reorientation of the ^{11}B, due to its quadrupole deformation, $Q(^{11}\mathrm{B}) = 4.07\,\mathrm{fm}^2$, were important to describe the elastic scattering data. To complete the systematic investigation on the elastic scattering with boron isotopes on nickel target, we recently measured angular distributions for the ^{12}B+^{58}Ni system at energies close to the barrier (30.0 and 33.0 MeV). The neutron-rich ^{12}B isotope can be described as ^{11}B+n cluster configuration ($S_n = 3.370$ MeV). This nucleus is radioactive and with lifetime of about 20 ms. Its configuration resembles more the proton-rich isotope ^{8}B than the other two stable isotopes, ^{10}B and ^{11}B. The neutron binding energy put this nucleus in the border between weakly- and tightly-bound nuclei. In this contribution we present some partial and preliminary results for the elastic scattering of ^{12}B+^{58}Ni. The full analysis is being published in [11]. The ^{12}B radioactive beam was produced by the RIBRAS system installed in the Physics Institute of the University of São Paulo, Brazil [12]. The angular distributions were analyzed in the terms of coupled channel calculations. By considering the couplings to inelastic channels, reorientation and spin-orbit terms, we were able to describe the angular distribution measured at 30.0 MeV. However, these channels were not sufficient to describe the experimental angular distribution measured at the energy of 33.0 MeV. A better agreement between the calculation and data at 33.0 MeV was obtained when transfer reactions were considered with CRC (Coupled Reaction Coupled) calculations. The main results for these calculations are shown in Fig. 34.1. As can be observed, the 1-n transfer, (^{12}B,^{11}B) $Q = +5.629$ MeV, has been found to be of particular importance due to the ^{11}B+n configuration for the ^{12}B.

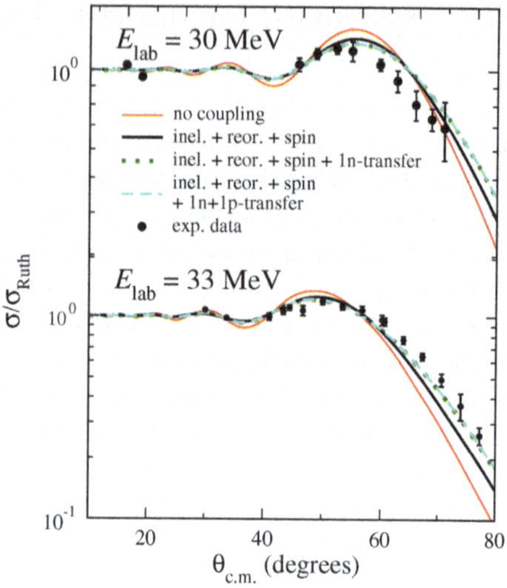

Fig. 34.1 Angular distributions for the $^{12}B+^{58}Ni$ system at 30.0 and 33.0 MeV. The curves correspond to coupled channel calculations with the channels and effect indicated

34.3 Summary

We have investigated the cluster configuration effects in the elastic scattering of boron isotopes, ^{8}B, ^{10}B, ^{11}B and ^{12}B, nuclei as projectiles on ^{58}Ni target, with coupled-channel calculations. The proton and neutron rich boron isotope, ^{8}B and ^{12}B are radioactive with short lifetime, of 770 and 20 ms, respectively. The possibility of producing radioactive beams out of these nuclei has offered new and unique opportunities for research in nuclear physics. The description of the cross sections for elastic scattering has shown to be sensitive to the cluster configurations of the projectiles. The different configurations of the boron isotopes have shown to produce different effects in the angular distributions for the elastic scattering.

Acknowledgements The authors would like to thank São Paulo Research Foundation (FAPESP) (Grants 2016/02864-4, 2016/17612-7, 2016/21434-7 and 2013/22100-7) and the Conselho Nacional de Desenvolvimento Cientifico (CNPq) (Grant 304961/2017-5) for the financial support. V. M. thanks FAPERJ. The authors also thank INCT-FNA (research project 464898/2014-5).

References

1. Kolata, J.J., Guimarães, V., Aguilera, E.F.: Eur. Phys. J. A **52**, 123 (2016)
2. Canto, L.F., Gomes, P.R.S., et al.: Phys. Rep. **596**, 1 (2015)
3. Guimarães, V., Kolata, J., et al.: Phys. Rev. Lett. **84**, 1862 (2000)
4. Aguilera, E.F., et al.: Phys. Rev. Lett. **107**, 092701 (2011)
5. Keeley, N., Kemper, K.W., Rusek, K.: Eur. Phys. J. A **50**, 145 (2014)
6. Aguilera, E.F., et al.: Phys. Rev. C **79**, 021601 (2009)
7. Aguilera, E.F., et al.: Phys. Rev. C **83**, 021601(R) (2011)

8. Lubian, J., et al.: Phys. Rev. C **79**, 064605 (2009)
9. Scarduelli, V., Crema, E., Guimarães, V., et al.: Phys. Rev. C **96**, 054610 (2017)
10. Deshmukh, N.N., Guimarães, V., Crema, E., et al.: Phys. Rev. C **92**, 054615 (2015)
11. Zevallos, E.O.N., Guimarães, V., et al.: Phys. Rev. C (2019). Submitted
12. Lepine, A., Lichtenthaler, R., Guimarães, V.: Eur. Phys. J. A **50**, 128 (2014)

Chapter 35
Inclusive Breakup Reaction of a Two-Cluster Projectile on a Two-Fragment Target: A Genuine Four-Body Problem

M. S. Hussein, C. A. Bertulani, B. V. Carlson and T. Frederico

Abstract We develop a four-body model for the inclusive breakup of two-fragment projectiles colliding with two-fragment targets. In the case of a short lived projectiles, such as halo nuclei, on a deuteron target, the model allows the extraction of the neutron capture cross section of such projectiles. We supply examples.

35.1 Introduction

The ongoing research on the reaction of radioactive nuclei has supplied us with invaluable information about the structure of nuclei near the drip line. Further, they produced important capture reactions and other direct reactions needed to fill the gaps in the chain of reactions in the r and s processes in astrophysics. The neutron capture reactions referred to above involve capture by stable nuclei. Neutron capture reactions on radioactive nuclei, especially near the drip nuclei are not available. A possible way to obtain these cross sections is through indirect hybrid reactions. One such method is the Surrogate Method [1]. So far this method was mostly used to obtain neutron capture cross section of fast neutrons by actinide nuclei for use in research in fast breeder reactors. Recently, the Surrogate Method was proposed to obtain the neutron capture cross section of radioactive nuclei [2]. A recent review gives an account of the

M. S. Hussein (✉)
Instituto de Estudos Avançados, Universidade de São Paulo, Caixa Postal 72012,
São Paulo, SP 05508-970, Brazil
e-mail: hussein@if.usp.br

Instituto de Física, Universidade de São Paulo, Caixa Postal 66318,
São Paulo, SP 05314-970, Brazil

M. S. Hussein · B. V. Carlson · T. Frederico
Departamento de Física, Instituto Tecnológico de Aeronaútica, DCTA,
São José dos Campos, SP 12228-900, Brazil

C. A. Bertulani
Department of Physics and Astronomy, Texas A&M University-Commerce,
Commerce, TX 75429-3011, USA

N. A. Orr et al. (eds.), *Recent Progress in Few-Body Physics*,
Springer Proceedings in Physics 238,
https://doi.org/10.1007/978-3-030-32357-8_35

(d, p) inclusive breakup reaction which is the basis of the Surrogate Method [3]. The theory employed for this is the Inclusive Nonelastic Breakup (INEB) Reaction theory [4–8]. In this contribution we report on recent work that extends the application of the INEB to the case of capture by a radioactive target or projectile. In our approach we consider first the three-body case of a non-cluster projectile interacting with a two-cluster target, such as the deuteron. In this case the reaction is a neutron pickup. Through the measurement of the inclusive proton spectrum one is able to extract the neutron capture cross section. This cross section is not the free capture cross section as several factors come into play owing to the fact that the neutron is bound in the deuteron. The second case we consider is the four-body one involving three-cluster projectile and no-cluster target [9]. In this contribution we propose an extension of the theory of [9] to the case of a two-fragment projectile on a two-fragment target. One such reaction involves the one proton halo nucleus ^{8}B, $^8\mathrm{B} + \mathrm{d} \rightarrow \mathrm{p} + {}^9\mathrm{B}$ or $\mathrm{p} + ({}^7\mathrm{Be} + \mathrm{d})$. So the inclusive proton spectrum will exhibit two groups a low proton energy one associated with the incomplete fusion $^7\mathrm{Be} + \mathrm{d}$ and a higher proton energy group connected with the capture reaction. We also consider the one-neutron halo projectiles on the deuteron target, $^{11}\mathrm{Be} + \mathrm{d}$ and the $^{19}\mathrm{C} + \mathrm{d}$. Our work reported here should be useful to assess the applicability of the INEB theory to isotopes such as ^{135}Xe whose lifetime is 9.8 h, which is a notorious nuclear reactor poison as its thermal neutron capture cross section is huge, 2.5×10^6 barns. Several other nuclei exhibit very large thermal neutron capture cross sections [10], whose explanation was attempted in [11]. Our aim is to use ^{135}Xe as a benchmark to test the inclusive proton spectrum in a reaction of the type $\mathrm{d} + {}^{135}\mathrm{Xe} \rightarrow \mathrm{p} + {}^{136}\mathrm{Xe}$.

35.2 Two-Fragment Projectile on a One-Fragment Target Nucleus

We will consider the scattering of a radioactive projectile with a two-cluster target (deuteron). Let us take ^{135}Xe as an example. Its life time is 9.6 h (very long). The reaction we want to describe is $^{135}\mathrm{Xe} + \mathrm{d} \rightarrow \mathrm{p} + {}^{136}\mathrm{Xe}$. A pickup reaction. The spectrum of the protons is measured, and the theory for this inclusive reaction is available. The quantity which is extracted from the measurement and the analysis is the total reaction cross section $\mathrm{n} + {}^{135}\mathrm{Xe} \rightarrow {}^{136}\mathrm{Xe}$, the neutron capture reaction. The cross section is given by the Austern or Hussein–McVoy (HM) expression

$$\frac{d^2\sigma_p}{dE_p d\Omega_p} = \rho_p(E_p)\hat{\sigma}_R^n, \tag{35.1}$$

where $\hat{\sigma}_R^n$ is the medium-modified total reaction cross section of the process $\mathrm{n} + \mathrm{A}$,

$$\hat{\sigma}_R^n = \hat{\sigma}_R(n + A). \tag{35.2}$$

More explicitly, the Inclusive Nonelastic Breakup theory gives for the reaction a + A → b + (x + A)

$$\frac{d^2\sigma_b^{\text{INEB}}}{dE_b d\Omega_b} = \hat{\sigma}_R^x \, \rho_b(E_b), \tag{35.3}$$

where $\hat{\sigma}_R^x$ is the total reaction cross section of the interacting fragment, x, and

$$\rho_b(E_b) \equiv \frac{d\mathbf{k}_b}{(2\pi)^3} \frac{1}{dE_b d\Omega_b} = \frac{\mu_b k_b}{(2\pi)^3 \hbar^2} \tag{35.4}$$

is the density of state of the observed, spectator fragment, b. The reaction cross section $\hat{\sigma}_R^x$ is given by [12]

$$\hat{\sigma}_R^x = -\frac{k_x}{E_x} \langle \hat{\rho}_x(\mathbf{r}_x) \, | W_x(\mathbf{r}_x) | \, \hat{\rho}_x(\mathbf{r}_x) \rangle, \tag{35.5}$$

where W_x is the imaginary part of the complex optical potential, U_x, of the interacting fragment, x, in the field of the target, A. The source function $\hat{\rho}_x(\mathbf{r}_x)$ is the overlap of the distorted wave of the interacting fragment, x, and the total wave function of the incident channel. In the DWBA limit of the this latter wave function and using the post form of the interaction, V_{xb}, the source function in the HM approach [7], is just

$$\hat{\rho}_x(\mathbf{r}_x) = (\chi_b^{(-)} | \chi_a^{(+)} \Phi_a > (\mathbf{r}_x). \tag{35.6}$$

In the IAV theory [4], based on the post form of the interaction, V_{xb}, the source function contains a Green's function referring to the propagation of x,

$$\hat{\rho}_x(\mathbf{r}_x) = \frac{1}{E_x - U_x + i\varepsilon} (\chi_b^{(-)} | V_{xb} | \chi_a^{(+)} \Phi_a \rangle. \tag{35.7}$$

The cross section in (35.5), according to the Hussein–McVoy model [7], can be decomposed into partial waves giving

$$\frac{E_x}{k_x} \hat{\sigma}_R^x = \int d\mathbf{r}_x |\hat{S}_b(\mathbf{r}_x)|^2 W(\mathbf{r}_x) |\chi_x^{(+)}(\mathbf{r}_x)|^2, \tag{35.8}$$

where

$$\hat{S}_b(\mathbf{r}_x) \equiv \int d\mathbf{r}_b \langle \chi_b^{(-)} | \chi_b^{(+)} \rangle (\mathbf{r}_b) \Phi_a(\mathbf{r}_b, \mathbf{r}_x), \tag{35.9}$$

and $\Phi_a(\mathbf{r}_b, \mathbf{r}_x)$ is the internal wave function of the projectile which carries the observed spectator fragment, b. The above formalism has recently been employed to calculate the (d, p) inclusive proton spectrum in (d, p) reactions [13–18].

In applying the above formalism to the reaction involving the deuteron as a projectile and ^{135}Xe as the target, or vice versa, one is reminded once again of the lifetime of the latter, 9.8 h. So there is the practical question which of these two reactions is feasible. In any case the final result in either case is the medium-modified total reaction cross section of the system n + ^{135}Xe. The capture cross section is the difference between this cross section and the contributions of other direct reactions, such as inelastic excitation of ^{135}Xe. In passing we remind the reader once again that in free space the thermal neutron capture cross sections of several nuclei is abnormally large, [10, 11].

35.3 Inclusive Non-elastic Breakup Reactions of Three Fragment Projectiles

Recently we have developed the theory of INEB involving a three-fragment projectiles, $a = b + x_1 + x_2$, such as ^{9}Be = ^{4}He + ^{4}He + n and Borromean nuclei such as ^{11}Li = ^{9}Li + n + n, The cross section for this four-body process, $b + x_1 + x_2 + A$, where b is the observed spectator fragment and x_1 and x_2 are the interacting participants fragments, is

$$\frac{d^2\sigma_b^{INEB}}{dE_b d\Omega_b} = \rho_b(E_b)\sigma_R^{4B}, \tag{35.10}$$

$$\sigma_R^{4B} = \frac{k_a}{E_a}\left[\frac{E_{x_1}}{k_{x_1}}\sigma_R^{x_1} + \frac{E_{x_2}}{k_{x_2}}\sigma_R^{x_2} + \frac{E_{CM}(x_1, x_2)}{(k_{x_1} + k_{x_2})}\sigma_R^{3B}\right], \tag{35.11}$$

where, the form of the reaction or fusion cross section as derived in [12] is used,

$$\sigma_R^{x_1} = \frac{k_{x_1}}{E_{x_1}}\langle\hat{\rho}_{x_1,x_2}|W_{x_1}|\hat{\rho}_{x_1,x_2}\rangle, \tag{35.12}$$

$$\sigma_R^{x_2} = \frac{k_{x_2}}{E_{x_2}}\langle\hat{\rho}_{x_1,x_2}|W_{x_2}|\hat{\rho}_{x_1,x_2}\rangle, \tag{35.13}$$

and,

$$\sigma_R^{3B} = \frac{(k_{x_1} + k_{x_2})}{E_{CM}(x_1, x_2)}\langle\hat{\rho}_{x_1,x_2}|W_{3B}|\hat{\rho}_{x_1,x_2}\rangle, \tag{35.14}$$

is a three-body, $x_1 + x_2 + A$, reaction cross section. The energies of the different fragments are defined through the beam energy, since the projectiles we are considering are weakly bound and thus the binding energy is marginally important in deciding the energies of the three fragments. Thus, e.g., $E_{x_1,Lab} = E_{a,Lab}(M_{x_1}/M_a)$, where by M_a and M_{x_1} we mean the mass numbers of the projectile and fragment x_1, respectively. The three-body source function, $\hat{\rho}_{x_1,x_2}$, is a generalisation of the two-body source function in (35.6), (35.7),

$$\hat{\rho}_{x_1,x_2}(\mathbf{r}_{x_1}, \mathbf{r}_{x_2}) = (\chi_b^{(-)}(\mathbf{r}_b)|\chi_a^{(+)}(\mathbf{r}_b, \mathbf{r}_{x_1}, \mathbf{r}_{x_2})\Phi_a(\mathbf{r}_b, \mathbf{r}_{x_1}, \mathbf{r}_{x_2})\rangle. \tag{35.15}$$

The cross sections $\sigma_R^{x_1}$ and $\sigma_R^{x_2}$ are the reaction cross sections of x_1 + A and x_2 + A individually, while the other fragments, x_2 and x_1 respectively, are scattered and not observed.

$$\frac{E_{x_1}}{k_{x_1}}\sigma_R^{x_1} = \int d\mathbf{r}_{x_1} d\mathbf{r}_{x_2} |\hat{S}_b(\mathbf{r}_{x_1}, \mathbf{r}_{x_2})|^2 |\chi_{x_2}^{(+)}(\mathbf{r}_{x_2})|^2 W(\mathbf{r}_{x_1}) |\chi_{x_1}^{(+)}(\mathbf{r}_{x_1})|^2, \qquad (35.16)$$

$$\frac{E_{x_2}}{k_{x_2}}\sigma_R^{x_2} = \int d\mathbf{r}_{x_1} d\mathbf{r}_{x_2} |\hat{S}_b(\mathbf{r}_{x_1}, \mathbf{r}_{x_2})|^2 |\chi_{x_1}^{(+)}(\mathbf{r}_{x_1})|^2 W(\mathbf{r}_{x_2}) |\chi_{x_2}^{(+)}(\mathbf{r}_{x_2})|^2. \qquad (35.17)$$

35.4 The Case of a Two-Fragment Projectile on a Two-Fragment Target

In the following we treat another four-body breakup problem: the case of two-fragment projectile and two-fragment target. Both projectile and target can break into their two fragments. This is a genuine four-body scattering problem. In principle the formalism of [9] can be applied after several modifications. Thus the target is a = b + x_2, and the projectile is A = x_1 + B. Thus the inclusive spectrum of b will contain breakup of the projectile with x_1 interacting with the target a, x_1 + d, and the breakup of the target with x_2 interacting with the projectile, x_2 + A. In principle this process is a complicated four-body reaction. Here, however we take a simpler approach and treat the process as a two three-body problems. As such we have the breakup of the projectile without affecting the target and the breakup of the target without affecting the projectile. In the calculation of the inclusive non-elastic breakup, one would obtain two distinct groups of detected spectator fragments, one related to the target and the other to the projectile. This method would be valuable in the case of a projectile being an exotic, neutron or proton-rich nucleus.

In the following we consider the reaction ^{8}B + d, which leads to p + (n + ^{8}B) → p + ^{9}B, and p + (^{7}Be + d). We remind the reader that ^{8}B is a one proton halo with a halo separation energy of 0.137 MeV. The first reaction results in the neutron capture by a one-proton halo nucleus, while the second reaction results in the incomplete fusion of the core of this halo nucleus with the deuteron target. The inclusive non-elastic proton spectrum can be written as (denoting the proton originating from the radioactive projectile by p_1 and that from the deuteron target breakup by p_2)

$$\frac{d^2\sigma_p}{dE_p d\Omega_p} = \rho(E_{p_2})\hat{\sigma}_R(n + {}^8\text{B}) + \rho(E_{p_1})\hat{\sigma}_R(d + {}^7\text{Be}) + \cdots \qquad (35.18)$$

The first term on the RHS of the above equation contains the neutron capture cross section of the halo nucleus and would be concentrated at higher proton energy (the proton separation energy of the deuteron is 2.22 MeV) in its spectrum, while the second term corresponds to the incomplete fusion, ^{7}Be + d, which involves the emission of the halo proton in ^{8}B and the collision of its core ^{7}Be with the deuteron. This process should dominate the low energy part of the inclusive proton spectrum.

In the case of a one-neutron halo projectile such as ^{11}Be or ^{19}C, with halo neutron separation energies, $E_s = 0.501$ MeV and $E_s = 0.530$ MeV, respectively, the same type of reaction will results in an inclusive proton spectrum which should exhibit a now low energy peak related to the target deuteron breakup at 2.22 MeV, and a higher energy and weaker peak connected with removing a proton from the tightly bound cores, ^{10}Be, ^{18}C.

$$\frac{d^2\sigma_p}{dE_p d\Omega_p} = \rho(E_{p_2})\hat{\sigma}_R(n + {}^{11}\text{Be}) + \rho(E_{p_1})\hat{\sigma}_R(d + {}^{10}\text{Be}) + \cdots \tag{35.19}$$

$$\frac{d^2\sigma_p}{dE_p d\Omega_p} = \rho(E_{p_2})\hat{\sigma}_R(n + {}^{19}\text{C}) + \rho(E_{p_1})\hat{\sigma}_R(d + {}^{18}\text{B}) + \cdots \tag{35.20}$$

The cross sections, $\hat{\sigma}_R(n+{}^{11}\text{Be})$, $\hat{\sigma}_R(n+{}^{19}\text{C})$, $\hat{\sigma}_R(d+{}^{10}\text{Be})$, $\hat{\sigma}_R(d+{}^{18}\text{B})$, are given by expressions similar to (35.8). One needs the S-matrix elements, $\hat{S}_{p_1}(\mathbf{r}_{p_1})$ and $\hat{S}_{p_2}(\mathbf{r}_{p_2})$ in order to evaluate the above cross sections. These matrix elements can be evaluated once appropriate optical potentials for protons on deuteron and on the different halo projectiles are given. Further, optical potentials for the projectile target systems are needed, as well as those for the generation of the participant fragment distorted waves. These are $n+{}^{11}$Be, $n+{}^{19}$C, $d+{}^{10}$Be, $d+{}^{18}$B.

For the proton halo nucleus ^{8}B, we need similar ingredients: $\hat{S}_{p_1}(\mathbf{r}_{p_1})$ for p + d elastic scattering and $\hat{S}_{p_2}(\mathbf{r}_{p_2})$ for p + ^{8}B. Similarly one needs the d + ^{8}B optical potential and the n + ^{8}B and d + ^{7}Be optical potentials. These potentials in principle are known from elastic scattering data.

Once the incomplete fusion cross sections are calculated from fusion theory [19], the neutron capture cross sections can be obtained from the general form of the breakup cross sections, (35.18)–(35.20). Thus the Inclusive Non-Elastic Breakup is a potentially powerful method to extract the neutron capture cross section of short-lived radioactive nuclei [20–25].

The coefficients A and B are related to the density of states of the observed proton,

$$\rho_p(E_p) = \frac{m_p k_p}{(2\pi)^3 \hbar^2} \tag{35.21}$$

In ^{8}B + d, due to the low value of the halo proton separation energy, of E_s, in the inclusive nonelastic breakup reaction, we expect a low energy peak in the inclusive proton spectrum connected with the incomplete fusion d + ^{7}Be, and a higher energy peak connected with the neutron capture n + ^{8}B reaction.

In ^{11}Be + d, with $E_s = 0.5$ MeV, we expect a lower energy peak associated with the neutron capture n + ^{11}Be and a much higher energy peak connected with the incomplete fusion d + ^{10}Li. The higher energy peak is connected to the proton emitted from the core, ^{10}Be with a separation energy of $E_s = 5$ MeV Similarly, for ^{19}C + d: a low energy peak n + ^{19}C with a higher energy peak d + ^{18}B.

35.5 Outline of the Derivation of (35.18)–(35.20)

Here we present an outline of the derivation of (35.18)–(35.20). We take the projectile A to be a bound system of two fragments, x_1 and B, and the target a as similarly composed of a bound system of two fragments, x_2 and b. In this derivation we follow the works of [4, 8, 9].

We invoke the spectator model in the sense that the observed fragment is only optically scattered from the projectile or target. Thus we take the Hamiltonian to be

$$H = K_{x_1} + K_{x_2} + K_b + K_a + V_{x_1x_2} + V_{x_2A} + V_{x_2b} + V_{x_1b} + U_{x_1A} + U_{bA}. \quad (35.22)$$

The steps to be followed to obtain (35.18)–(35.20) are lengthy but rest on a generalization of the case of three-fragment projectile breakup formalison of [9], and will be reported elsewhere [26].

Acknowledgements This work was partly supported by the US-NSF and by the Brazilian agencies, Fundação de Amparo à Pesquisa do Estado de São Paulo (FAPESP), the Conselho Nacional de Desenvolvimento Científico e Tecnológico (CNPq) and INCT-FNA project 464898/2014-5. CAB acknowledges a Visiting Professor support from FAPESP and MSH acknowledges a Senior Visiting Professorship granted by the Coordenação de Aperfeiçoamento de Pessoal de Nível Superior (CAPES), through the CAPES/ITA-PVS program. CAB also acknowledges support by the U.S. NSF Grant No. 1415656 and the U.S. DOE Grant No. DE-FG02-08ER41533.

References

1. Escher, J.E., et al.: Rev. Mod. Phys. **84**, 353 (2012)
2. Escher, J.E., et al.: Phys. Rev. Lett. **121**, 052501 (2018)
3. Potel, G., et al.: Eur. Phys. J. A **53**, 178 (2017)
4. Ichimura, M., et al.: Phys. Rev. C **32**, 431 (1985)
5. Kasano, A., Ichimura, M.: Phys. Lett. B **115**, 81 (1982)
6. Udagawa, T., Tamura, T.: Phys. Rev. C **24**, 1348 (1981)
7. Hussein, M.S., McVoy, K.W.: Nucl. Phys. A **445**, 124 (1985)
8. Austern, N., et al.: Phys. Rep. **154**, 125 (1987)
9. Carlson, B.V., Frederico, T., Hussein, M.S.: Phys. Lett. B **767**, 53 (2017)
10. Mughabghab S. F.: Thermal neutron capture cross sections resonance integrals and g-factors, Int. At. Energy Agency, INDC(NDS)-440 (2003)
11. Carlson, B.V., Hussein, M.S., Kerman, A.K.: Acta Phys. Pol. B **47**, 491 (2016)
12. Hussein, M.S.: Phys. Rev. C **30**, 1962 (1984)
13. Potel, G., Nunes, F.M., Thompson, I.J.: Phys. Rev. C **92**, 034611 (2015)
14. Ducasse, Q., et al.: Phys. Rev. C **94**, 024614 (2016)
15. Lei, J., Moro, A.M.: Phys. Rev. C **92**, 061602(R) (2015)

16. Lei, J., Moro, A.M.: Phys. Rev. C **92**, 044616 (2015)
17. Carlson, B.V., Capote, R., Sin, M.: Few-Body Syst. **57**, 307 (2016)
18. Lei, J., Moro, A.M.: Phys. Rev. C **95**, 044605 (2017)
19. Canto, L.F., Gomes, P.R.S., Donangelo, R., Hussein, M.S.: Phys. Rep. **424**, 1 (2006)
20. Canto, L.F., Hussein, M.S.: Scattering Theory of Molecules, Atoms and Nuclei. World Scientific, Singapore (2013)
21. Bertulani, C.A., Hussein, M.S., Typel, S.: Phys. Lett. B **776**, 217 (2018)
22. Spitaleri, C., et al.: Phys. At. Nucl. **74**, 1725 (2011)
23. Tumino, A., et al.: Few Body Syst. **54**, 745 (2012)
24. Pizzone, R.G., et al.: Eur. Phys. J. **86**, 00034 (2015)
25. Hussein, M.S.: Two-step nuclear reactions: the surrogate method, the trojan horse method and their common foundations. EPJA **53**, 110 (2017)
26. Bertulani C. A., Carlson B. V., Frederico T., Hussein M. S.: In preparation

Chapter 36
Glauber Model Analysis for the ^{22}C Nuclear Radius

T. Nagahisa and W. Horiuchi

Abstract In this contribution, we report our recent analysis of the nuclear radius of ^{22}C with total reaction cross sections on ^{12}C and ^{1}H targets. The total reaction cross sections are calculated consistently for both the targets within a reliable microscopic framework, the Glauber theory. The multiple-scattering processes within the Glauber theory is fully taken into account using a Monte Carlo technique. We show that the simultaneous reproduction of the two recent cross section data are not feasible within the error bar and unlikely to obtain the huge matter radius $\sim$5 fm.

36.1 Introduction

A ^{22}C nucleus is the neutron dripline of carbon isotopes and is the heaviest two-neutron halo nucleus which has been found so far. This nucleus is the so-called Borromean in which neither of the ^{20}C-n and n-n subsystems are bound and has attracted much attention not only to nuclear physics but also atomic physics in connection to the Efimov physics. However, the nuclear radius, which is one of the most basic properties of atomic nuclei, has been under discussion: The two recent interaction cross section measurements show the quite different ^{22}C radii, 5.4 ± 0.9 fm [1] and 3.44 ± 0.08 fm [2]. Since the nuclear radius has often used as one of the inputs to some theoretical models, this demands appropriate evaluation of the nuclear radius.

In this contribution, we present our recent analysis of the nuclear radius of ^{22}C with total reaction cross sections to resolve the radius problem [3]. The nuclear radius of unstable nuclei has often been determined by the total reaction or interaction cross section measurement incident at several tens MeV to 1 GeV on a stable target nucleus such as ^{12}C and ^{1}H. We prepare the ^{22}C wave function based on a simple ^{20}C+$n + n$ picture and calculate the total reaction cross sections as a function of its nuclear radius. The total reaction cross sections are obtained in a reliable microscopic

T. Nagahisa · W. Horiuchi (✉)
Department of Physics, Hokkaido University, Sapporo 060-0810, Japan
e-mail: whoriuchi@nucl.sci.hokudai.ac.jp

N. A. Orr et al. (eds.), *Recent Progress in Few-Body Physics*,
Springer Proceedings in Physics 238,
https://doi.org/10.1007/978-3-030-32357-8_36

theoretical framework, the Glauber theory, consistently for ^{12}C and ^{1}H target nuclei. We carefully evaluate the recent interaction cross sections measured in [1, 2] and discuss possible uncertainties in the radius determination of ^{22}C.

36.2 Models

Here we consider a high-energy collision of the projectile (P) and target (T) nuclei with mass numbers A_P and A_T, respectively. The Glauber theory [5] is a microscopic multiple-scattering and widely accepted theory to describe high-energy nucleus-nucleus collisions. In this theory, the final state wave function of a projectile and target system is greatly simplified by the adiabatic and eikonal approximations, and the total reaction cross section is evaluated by

$$\sigma_R = \int d\boldsymbol{b}\, \left(1 - |e^{i\chi(\boldsymbol{b})}|^2\right), \tag{36.1}$$

where $\boldsymbol{b}$ is the impact parameter vector perpendicular to the beam direction z. Here we only need to evaluate the optical phase-shift function or the Glauber amplitude, $e^{i\chi(\boldsymbol{b})}$, which includes all information of the elastic processes in the high-energy nuclear collision

$$e^{i\chi(\boldsymbol{b})} = \left\langle \Phi_0^P \Phi_0^T \right| \prod_{j=1}^{A_P} \prod_{k=1}^{A_T} \left[1 - \Gamma_{NN}(\boldsymbol{b} + \hat{\boldsymbol{s}}_j^P - \hat{\boldsymbol{s}}_k^T)\right] \left| \Phi_0^P \Phi_0^T \right\rangle, \tag{36.2}$$

where $\hat{\boldsymbol{s}}_j^P$ ($\hat{\boldsymbol{s}}_k^T$) denotes the two-dimensional single-particle coordinate operator projected onto the xy-plane of the jth (kth) nucleon from the center-of-mass of the projectile (target), and Γ_{NN} is the so-called profile function which is determined based on the nucleon-nucleon scattering cross sections. In this contribution, we employ incident-energy-dependent parameter sets listed in [6]. The other inputs to the theory are the ground-state wave function of the projectile, Φ_0^P, and target and Φ_0^T nuclei. Once these inputs are set, the theory has no adjustable parameter. We do not consider the Coulomb breakup contributions since the effects are negligible in systems involving small Z nuclei.

In the present work, we consider the three nuclei, ^{12}C, ^{20}C, and ^{22}C. A configuration of the ^{12}C wave function is assumed to be $(0s_{1/2})^2(0p_{3/2})^4$ for both proton and neutron with the harmonic-oscillator (HO) single-particle wave functions whose length parameter is fixed to reproduce the measured charge radius. For ^{20}C and ^{22}C, single particle wave functions of ^{20}C and ^{22}C systems are generated from the phenomenological Woods–Saxon potential [8] with diffuseness and radius parameters being $a = 0.65$ fm and $R = 1.25 A_P^{1/3}$ fm, respectively. A proton configuration is assumed to be $(0s_{1/2})^2(0p_{3/2})^4$. The subshell closure of the neutron number 14 and 16 is assumed for neutron configurations of ^{20}C and ^{22}C and are taken respectively as $(0s_{1/2})^2(0p_{3/2})^4(0p_{1/2})^2(0d_{5/2})^6$ for ^{20}C and

$(0s_{1/2})^2(0p_{3/2})^4(0p_{1/2})^2(0d_{5/2})^6(1s_{1/2})^2$ for ^{22}C. These assumptions can be reasonable to describe ^{22}C as ^{20}C+n + n s-wave two-neutron halo structure [9]. To simulate the two neutron halo structure of ^{22}C, we firstly take the central mean-field potential strength commonly to all angular-momentum l states and fix it in such a way so as to reproduce the interaction reaction cross section of ^{20}C+^{12}C measured at ~900 MeV [4]. We vary the central potential strength only for $l = 0$ state as a free parameter that controls the radius of ^{22}C.

In general, the explicit evaluation of the Glauber amplitude of (36.2) is difficult because the expression involves $3(A_P + A_T)$-dimensional integration. Here we directly evaluate it by using Monte Carlo (MC) integration. To perform the MC integration accurately, we need to generate a large number of points, typically 10^{6-8}. We confirmed in [3] that 10^8 configurations for the MC integration are actually needed in order to ensure the accuracy of the total reaction cross sections of ^{22}C on ^{1}H target within 1% level, whereas an ordinary nuclear system, e.g., ^{12}C and ^{20}C, only needs typically 10^6 or 10^7 configurations. More MC configurations are needed to have sufficient statistics in the tail regions of the extended wave function of ^{22}C.

36.3 Results and Discussions

Figure 36.1 displays the total reaction cross sections of ^{12,20,22}C on ^{12}C and ^{1}H targets as a function of incident energies together with the available experimental data. It clearly shows that our results reasonably agree with the total reaction and interaction cross section data for both ^{12}C and ^{1}H targets in this consistent theoretical

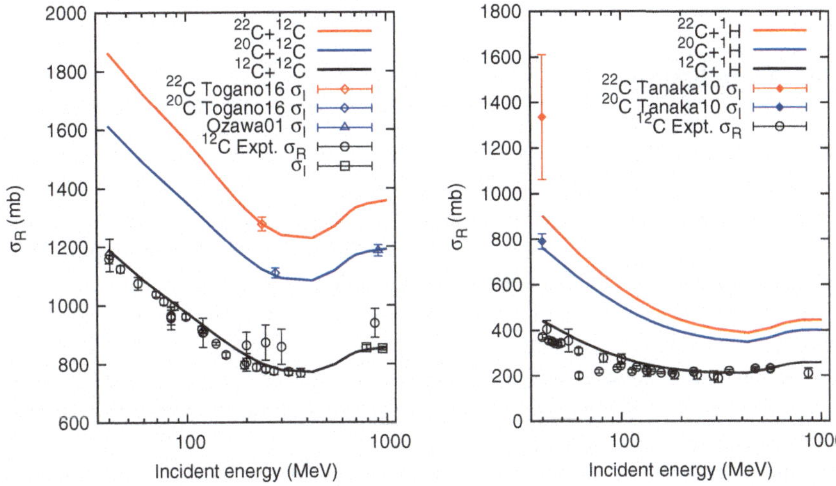

Fig. 36.1 Total reaction cross sections of ^{12,20,22}C on (left) ^{12}C and (right) ^{1}H targets as a function of incident energies. Experimental interaction cross section data of ^{20,22}C are taken from [1, 2, 4]. The figure is plotted based on the data of [3]

framework. The nuclear radius of ^{22}C is set to be 3.38 fm consistently with the recent interaction cross section data [2]. We find that the simultaneous reproduction of the cross section data by Tanaka et al. [1] is not feasible within 1σ, that is, for ^{1}H target, the experimental data is far from the theoretical values (However, it is consistent with 2σ as mentioned in [2]). We also extrapolate the rms radius using the data of [1] which is found to be huge $\gtrsim$5 fm even at the lower limit (1σ) of the experimental cross section, and never reach the central value of the experimental data 1338 mb [1] with the extrapolated function based on our theoretical cross sections.

At the end, we discuss the possible uncertainties in the radius determination:

- How good is the approximation used in the experimental analysis?
  We also calculate the total reaction cross section on ^{1}H target with the optical-limit approximation (OLA) which was employed in the analysis of [1]. The difference between the present complete Glauber and the OLA cross sections is small approximately 1%, being the situation unchanged.
- Is the incident energy of 40 MeV too low in the Glauber calculation?
  As shown in the figure, the theory reproduces fairly well the total reaction cross section of ^{20}C on ^{1}H target even at 40 MeV.
- What about the difference between the total reaction and interaction cross sections?
  Since any excited bound state of ^{22}C has not been observed so far, the total reaction and interaction cross sections are equal for ^{1}H target and its difference is expected to be small for ^{12}C target.
- What about the Coulomb breakup effect?
  The Coulomb breakup effect is expected to be small. For instance, the contribution is estimated less than 1% in the case of a one-neutron halo nucleus, ^{31}Ne on ^{12}C target [7]. It becomes even smaller in the case of ^{1}H target.

36.4 Conclusion

In order to resolve the radius puzzle in ^{22}C, we have investigated the total reaction cross sections of ^{22}C on ^{12}C and ^{1}H targets incident at medium- to high-incident energies within the framework of a microscopic high-energy reaction theory, the Glauber model. The complete optical phase-shift function or Glauber amplitude in the Glauber model is evaluated with use of a Monte Carlo technique.

The calculated total reaction cross sections on ^{12}C and ^{1}H targets reasonably reproduce the available experimental cross section data for ^{12}C and ^{20}C. The root-mean-square (rms) matter radius of ^{22}C deduced from our analysis is consistent with the radius given in [2] using the interaction cross section on ^{12}C target at 240 MeV, which corresponds to that of an $A \sim 40$ nucleus. We investigate possible uncertainties in the theoretical model and they are actually small. To conclude, the simultaneous reproduction of both the experimental interaction cross section data on ^{12}C and ^{1}H obtained by [1, 2] is not possible within the error bar (1σ). Since only two experimental cross section data are available, it is desired to have another data at different incident energy or target in order to clarify that the ^{22}C size is equivalent to a radius of medium- ($A \sim 40$) or heavy- ($A \sim 200$) mass nuclei. However, we already see theoretical consistency with the ^{20}C cross section data for both ^{1}H and ^{12}C target. It is unlikely to have a huge matter radius $\sim$5.4 fm ($A \sim 200$) of ^{22}C.

Acknowledgements This work was in part supported by JSPS KAKENHI Grant Numbers 18K03635 and 18H04569.

References

1. Tanaka, K., et al.: Phys. Rev. Lett. **104**, 062701 (2010)
2. Togano, Y., et al.: Phys. Lett. B **761**, 412–418 (2016)
3. Nagahisa, T., Horiuchi, W.: Phys. Rev. C **97**, 054614 (2018)
4. Ozawa, A., et al.: Nucl. Phys. A **691**, 599 (2001)
5. Glauber, R.J.: Brittin, W.E., Dunham, L.G. (eds.) Lectures in Theoretical Physics, vol. 1, p. 315. Interscience, New York (1959)
6. Abu-Ibrahim, B., et al.: Phys. Rev. C **77**, 034607 (2008)
7. Horiuchi, W., Suzuki, Y., Capel, P., Baye, D.: Phys. Rev. C **81**, 024606 (2010)
8. Horiuchi, W., et al.: Phys. Rev. C **75**, 044607 (2007)
9. Horiuchi, W., Suzuki, Y.: Phys. Rev. C **74**, 034311 (2006)

Chapter 37
Properties of Supersymmetric Transformed Alpha-Nucleus Potentials Studied with Electric-Multipole Transitions

T. Arai, W. Horiuchi and D. Baye

Abstract Towards applications to multi-cluster systems involving heavy clusters, we study properties of a shallow-singular potential generated by supersymmetric transformations from an original deep potential. Changes of the relative wave functions by the transformations are quantified with electric-multipole transitions which give a different radial sensitivity to the wave function depending on their multipolarity. Despite the fact that the original and transformed potentials give exactly the same phase shift, some observables are unfavorably modified. We propose another possible way to obtain a desired supersymmetric potential.

Cluster degrees-of-freedom is one of the most characteristic features of an atomic nucleus. Alpha(^{4}He or α)-cluster models have often been used to describe light nuclei. However, the computation becomes much involved in heavier cluster systems. In this contribution, we present our recent work [1] studying the relative wave functions of the α+^{16}O and α+^{40}Ca systems generated from phase-shift-equivalent potentials towards the application to multi-cluster systems involving heavy clusters such as ^{21}Ne (^{21}Na) = ^{16}O+$\alpha + n$ (p), ^{24}Mg = ^{16}O+$\alpha + \alpha$, ^{45}Ti (^{45}V) = ^{40}Ca+$\alpha + n$ (p), and ^{48}Cr = ^{40}Ca+$\alpha + \alpha$.

In general, a potential between clusters is deep accommodating several redundant bound states which should be removed in an appropriate way. To avoid such a complicated computation, we generate a shallow-singular potential by using supersymmetric transformations [2] from the original deep potential. For details, a reader is referred to [1].

Let us consider a two-spinless-cluster system that interacts only with a central, local, regular and deep potential which accommodates n forbidden (unphys-

T. Arai · W. Horiuchi (✉)
Department of Physics, Hokkaido University, Sapporo 060-0810, Japan
e-mail: whoriuchi@nucl.sci.hokudai.ac.jp

D. Baye
Physique Quantique, and Physique Nucléaire Thórique et Physique Mathématique, C.P. 229, Université Libre de Bruxelles (ULB), 1050 Brussels, Belgium

N. A. Orr et al. (eds.), *Recent Progress in Few-Body Physics*, Springer Proceedings in Physics 238,
https://doi.org/10.1007/978-3-030-32357-8_37

ical) bound-state solutions. The one-dimensional radial differential equation is written as $H\chi(E) = \left(-\frac{d^2}{dr^2} + V\right)\chi(E) = E\chi(E)$. The ground state is eliminated by supersymmetric transformations with a two-step procedure: The initial Hamiltonian is factorized into two first-order operators $L_0^{\pm} = \pm d/dr + d\ln\chi(E_0)/dr$ as $H = L_0^+ L_0^- + E_0$ where the ground-state energy E_0 is taken as factorization energy and $\chi(E_0)$ is the ground-state wave function. The SUSY partner, H_1, of H_0 is defined by $H_1 = L_0^- L_0^+ + E_0$. The lowest bound state is removed in H_1 but the phase shift is also modified. To recover the original phase shift, one performs another SUSY transformation by factorizing H_1 in the form $H_1 = L_1^+ L_1^- + E_0$ with $L_1^{\pm} = \pm d/dr + d\{\ln\int_0^r dt\,[\chi(E_0)]^2/\chi(E_0)\}/dr$ and defining its partner $H_2 = L_1^- L_1^+ + E_0$. The corresponding potential obtained by those two steps can be summarized in the form $V_2 = V - 2\frac{d^2}{dr^2}\ln\int_0^r dt\,[\chi(E_0)]^2$. This potential provides exactly the same phase shifts obtained with the original potential. The above procedure will be repeated until all the unphysical bound states are removed.

For a deep potential we assume a parity-dependent-single Gaussian form as $V(r) = (V_0 + V_r\hat{P}_r)\exp(-\mu_r r^2)$ where $\hat{P}_r$ is the parity operator that changes $\boldsymbol{r}$ into $-\boldsymbol{r}$. The potential strengths should be deep enough to accommodate a number of forbidden bound states with the total harmonic oscillator quanta of the cluster relative motion, $Q = 2n_r + l < 8$ and 12 for the α+^{16}O and α+^{40}Ca systems, respectively. The three parameters, V_0, V_r, and μ_r, are fixed so as to reproduce the ground state energy, its root-mean-square (rms) radius and the energy of the 1^- state. These potentials nicely reproduce the rotational spectra as well as the reduced electric-quadrupole transition probabilities, $B(E2)$, of ^{20}Ne and ^{44}Ti (Set C of [1]). See, also Fig. 37.1 of this contribution.

To help the understanding of a reader, here we formally define the modifications of observables caused by the SUSY transformation. For the calculation of matrix elements, we usually use a bare operator, i.e., we make the approximation that the operator is left unchanged by the SUSY transformation. If Ψ is the initial wave function, the transformed wave function is $\Phi = \mathcal{T}\Psi$, where $\mathcal{T}$ denotes an appropriate SUSY transformation. The ground-state wave function Ψ can be written formally

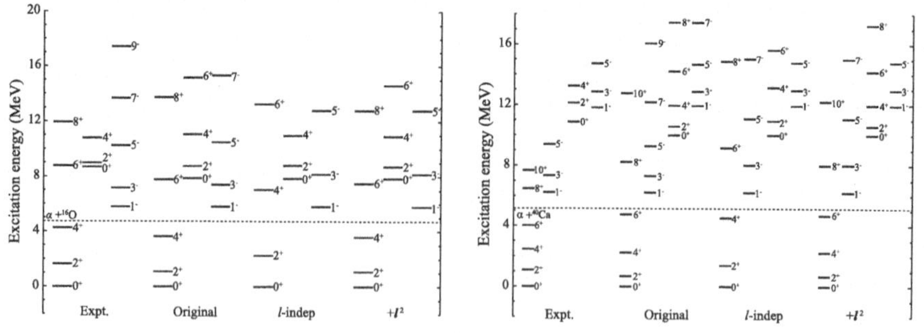

Fig. 37.1 Energy spectra of (left) ^{20}Ne and (right) ^{44}Ti with the original (Set C [1]), l-independent SUSY potentials, and the l-independent potential plus the l^2 term

as $\Psi = \mathcal{T}^{-1}\Phi$. The expectation value with an operator $\mathcal{O}$ can be evaluated with $\langle\Psi|\mathcal{O}|\Psi\rangle = \langle\Phi|(\mathcal{T}^{-1})^{\dagger}\mathcal{O}\mathcal{T}^{-1}|\Phi\rangle$. In principle, any expectation value is also identical if one evaluates this expectation value with the correlated operator $(\mathcal{T}^{-1})^{\dagger}\mathcal{O}\mathcal{T}^{-1}$ consistently derived from the transformation. However, it is not practical because the transformation employed in the present contribution, $\mathcal{T} = L_n^- \dots L_1^- L_0^-$, is too much involved. Hence, in practice, we actually evaluate $\langle\Phi|\mathcal{O}|\Phi\rangle$. Therefore, expectation values must change with the SUSY transformation, $\mathcal{T}$.

Let us briefly review the findings in [1]. With the deep potential, the relative wave functions exhibit several nodes due to the orthogonality to the forbidden states, whereas the ones with the SUSY potentials have no node at short distances. Despite the fact that the SUSY and deep potentials give exactly the same phase shift, the rms radii and the $B(E2)$ values are actually modified and increased by the standard SUSY transformation. Those observables are somewhat sensitive to the wave function at short distances. On the contrary, the changes in the reduced electric-hexadecapole transition probabilities, $B(E4)$, are relatively smaller than that of $B(E2)$. Since the operator is proportional to r^4, the $B(E4)$ values are less sensitive to the internal regions of the wave function, whereas the external regions contribute largely to the $B(E4)$ matrix element. Recalling that we fix the parameters of the deep potential in such a way so as to reproduce the binding energy and rms radius of the ground-state wave function, the standard SUSY transformation is no longer a desired potential. Therefore, we consider another SUSY transformation with one arbitrary parameter being fixed so as to reproduce the rms radius of the ground state. By this transformation, the $B(E2)$ values and rms radii of the bound 2^+ and 4^+ states are remedied. However, the $B(E4)$ value shows some deviation from the deep potential case. To conclude, though one always needs to check how observables of interest are modified by the transformation, as long as the rms radius and $B(E2)$ values are of interest, this SUSY transformation (SUSY-β in [1]) seems to provide a reasonable forbidden-state-free shallow potential which can be used for studying multi-cluster systems.

The SUSY potentials proposed in [1] are shallow and have no unphysical bound state but angular-momentum (l)-dependent which causes complications in practical computations. Here we propose another potential model which may be useful to reduce the computational cost. Figure 37.1 plots the original energy spectra and the one obtained by ignoring the l-dependence, i.e., a potential with $l = 0$ is used to compute all the l-states. Since the $l = 0$ state has the largest number of forbidden states, the excitation energies for higher l-state become higher. One may simulate the l-dependence by introducing an l^2 term $V_{ll}l^2\exp(-\mu_r r^2)$, whose parameter is fixed so as to reproduce the energy of the 2^+ state: $V_{ll} = -1.2732$ and -0.75965 MeV for the $\alpha+^{16}$O and $\alpha+^{40}$Ca systems, respectively. As shown in the figure, the energy spectra become almost identical to the original spectra by adding the l^2 term to the potential for $l = 0$. These potential sets are also useful for practical computations, although the phase-shift-equivalency to the original deep potential is lost for $l > 0$.

Acknowledgements This work was in part supported by JSPS KAKENHI Grant Numbers 18K03635 and 18H04569.

References

1. Arai, T., Horiuchi, W., Baye, W.: Nucl. Phys. A **977**, 82 (2018)
2. Baye, D.: Phys. Rev. Lett. **58**, 2738 (1987)

Chapter 38
Two-Neutron Correlations in ^{6}He Studied with Spin-Flip Charge-Exchange Transitions

N. Kawamura and W. Horiuchi

Abstract We investigate the spin-flip charge-exchange type nuclear excitations, that are, Gamow-Teller (GT) and isovector-spin-monopole (IVSM) transitions, to explore a possible probe to study the two-neutron correlations in the halo nuclei. A typical light halo nucleus, ^{6}He, is studied with ^{4}He (α) + two-nucleon (N) three-body model. We show that the IVSM transitions are superior to the GT ones to obtain the information on the structure of ^{6}He: The strong IVSM strengths, which are sensitive to the binding energy of ^{6}He, are found in the low-lying region below the α+deuteron (d) threshold, whereas the GT strengths have almost no peak due to small overlap between the wave functions of the ground state of ^{6}He and the $\alpha + d$ continuum of ^{6}Li.

The neutron-halo structure has been observed as one of the unique phenomena in neutron-rich unstable nuclei near the neutron dripline. Strong two-neutron correlations, the so-called dineutron correlations, are predicted in such halo nuclei [1] and have attracted much interest to observe the evidence of the correlations. However, its direct measurement is difficult partly because the two neutrons do not form a bound state.

Exploring such observables that reflects the dineutron correlations has gained interest. Here we study the spin-flip charge-exchange transitions from the ground state of ^{6}He to the continuum states of ^{6}Li and discuss the possibility of being observables to study the dineutron correlations. We consider two types of the operators, Gamow-Teller (GT) and isovector spin-monopole (IVSM), which give different sensitivity to the spatial overlap between the initial ground and the spin-flipped final states. Since the low-lying states of ^{6}Li are expected to have the $\alpha + d$ (dinucleon) structure, if the two valence neutrons in the ground state of ^{6}He are spatially localized, the transition strengths to these states could be enhanced.

^{6}He and ^{6}Li systems are described with the $\alpha + N + N$ three-body system [2]. We employ Minnesota (MN) potential [3] as an *NN* interaction, and the KKNN

N. Kawamura · W. Horiuchi (✉)
Department of Physics, Hokkaido University, Sapporo 060-0810, Japan
e-mail: whoriuchi@nucl.sci.hokudai.ac.jp

N. A. Orr et al. (eds.), *Recent Progress in Few-Body Physics*,
Springer Proceedings in Physics 238,
https://doi.org/10.1007/978-3-030-32357-8_38

potential [4] for the αN system. In order to see the binding energy dependence of the nuclear responses, we employ the modified MN potential (MMN) [2] which is set so as to reproduce the two-neutron separation energy of ^{6}He. The orthogonal condition to the $0s_{1/2}$ orbit in the α core is imposed to the two-nucleon motions.

The total wave functions of the $A = 6$ systems are expanded in many basis functions in the *LS* coupling scheme. The spatial part of the wave function is represented by the correlated Gaussian function with two global vectors [5]. The power of this approach is confirmed in many interesting examples, see recent review papers [6, 7]. The diagonal part of the Gaussian fall-off parameters are chosen by a geometrical progression that covers from 0.2 to about 20 fm. The correlations among the relative coordinates are taken into account by mixing of the rearrangement channels. All possible (LS) values are considered, i.e., $(LS) = (00), (11)$ for ^{6}He, $(LS) = (01), (10), (11),$ and (21) for ^{6}Li. We take intermediate angular momenta up to 4. In the end, the total number of basis is 2114 for ^{6}He and 4334 for ^{6}Li. The calculated ground state properties are all in consistent with those obtained in [2].

In this paper, we consider spin-flip charge-exchange transitions from the ground state of ^{6}He to ^{6}Li systems through GT and IVSM operators, $\mathcal{M}^{\mathrm{GT}-} = \sum_{i=1}^{2} \sigma_i \tau_i^-$ and $\mathcal{M}^{\mathrm{IVSM}-} = \sum_{i=1}^{2} \bar{\boldsymbol{r}}_i^2 \sigma_i \tau_i^-$, where σ_i, τ_i^-, and $\bar{\boldsymbol{r}}_i$ denote spin, particle exchange from neutron to proton, and single-particle coordinate from the center-of-mass (cm) operators, respectively.

We calculate the probabilities of finding the $\alpha + d$ component in the wave functions of ^{6}Li (S_d^2), which are evaluated by

$$S_d^2 = \left| \int d\boldsymbol{r} \langle \Psi_d(\boldsymbol{x}_1)\delta(\boldsymbol{x}_2 - \boldsymbol{r}) | \Psi_{^6\mathrm{Li}}(\boldsymbol{x}_1, \boldsymbol{x}_2) \rangle \right|^2, \tag{38.1}$$

where $\Psi_d(\boldsymbol{x}_1)$ and $\Psi_{^6\mathrm{Li}}(\boldsymbol{x}_1, \boldsymbol{x}_2)$ are the wave functions of d and $\alpha + n + p$ state with the MN potential. Figure 38.1a plots S_d^2 for the $\alpha + n + p$ states as a function of the excitation energy. In the ground state, the S_d^2 is 0.629. The d subsystem is distorted by the interaction as well as the Pauli principle from the α core. For several states, it appears that the S_d^2 is almost unity between the $\alpha + d$ and $\alpha + n + p$ threshold energies, which are of our interest.

Figure 38.1b, c display the reduced GT transition probabilities (B_{GT}) from the ground state of ^{6}He to the continuum states of ^{6}Li with the MN and MMN potentials. The validity of our model is confirmed by calculating the life time for the beta decay of ^{6}He to ^{6}Li, which can be evaluated by the formula [8] $(ft_{1/2})^{-1} = (2\ln 2)\pi^3(m_e c^2/\hbar)G_\beta^2\gamma^2 B_{\mathrm{GT}}$, where m_e is the electron mass, $G_\beta = 3.002 \times 10^{-12}$, and $\gamma = 1.226$. The calculated life times are 757.1 and 796.5 ms with the MMN and MN potentials, respectively, which are in reasonable agreement with the observation (806.7 ± 0.15 ms [9]). Since most of the transition strengths are exhausted by the ground-state to ground-state transition, in the figure, the ground state transition is omitted for the sake of visibility. The other states are fragmented in the energy regions beyond 5 MeV. No prominent peak appears between the two thresholds. Though the S_d^2 values are large for the states between the two thresholds, root-mean-square dis-

Fig. 38.1 **a** Probability of finding $\alpha + d$ component in the wave function of ^{6}Li, reduced, **b**, **c** Gamow-Teller and **d**, **e** isovector-spin-monopole transition probabilities as a function of excitation energy of ^{6}Li. The **b**, **d** MMN and **c**, **e** MN potentials are employed for the ground state of ^{6}He. Arrows indicate the $\alpha + d$ and $\alpha + p + n$ threshold energies

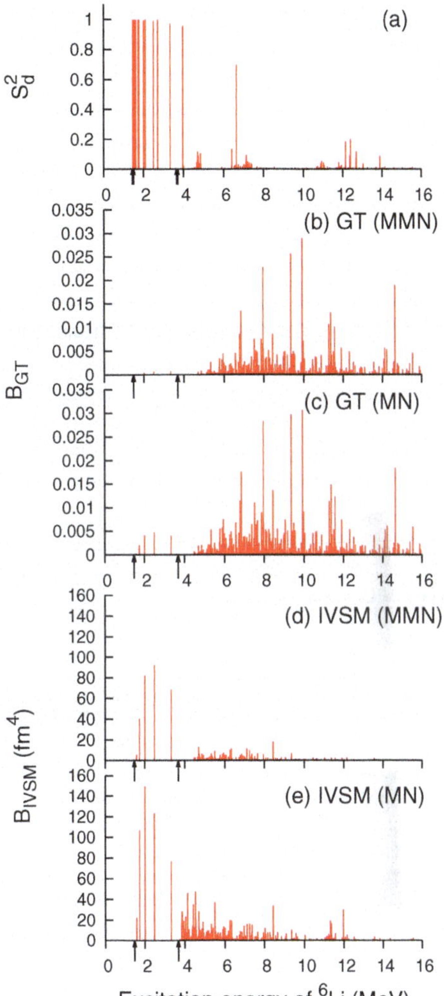

tances between α and the cm of the two nucleons of those continuum states are very large ~20–140 fm. Therefore, the spatial overlaps with the ground state of ^{6}He and the these continuum states become small. This can be understood in the early study of the beta-delayed deuteron emission of ^{6}He [10]. The GT strength to the $\alpha + d$ continuum states are strongly suppressed due to the orthogonality to the ground state. These oscillating $\alpha + d$ wave functions leads to the strong cancellation. In Fig. 38.1c, since the spatial distribution of the ground state of ^{6}He is extended with the MN interaction, the $B_{\rm GT}$ values become larger than those obtained with the MMN interaction, although the values are small. Fig. 38.1d, e display the reduced IVSM transition probabilities ($B_{\rm IVSM}$). Contrary to the $B_{\rm GT}$ case, prominent peaks appear between the two thresholds. As compared in Fig. 38.1d, e, the low-lying strengths are very sensitive to the spatial extension of the ground state. We also calculate the transition strength to the ground state of ^{6}Li are 472 and 482 (fm^4) for the MMN and MN interactions, respectively. They are not sensitive to the *NN* interaction employed.

In summary, to explore possible probes of the two-neutron correlations, we have investigated spin-flip-charge-exchange transitions of ^{6}He. The initial and final states are described with an $\alpha + N + N$ three-body model. Three-body dynamics with Pauli constraint are taken into account by superposing many correlated Gaussian basis functions. We find strong IVSM transition strengths in the low-lying energy region below the $\alpha + d$ threshold, whereas almost no GT strength is found due to small overlap between the ^{6}He and $\alpha + d$ continuum wave functions. We show the sensitivity of the GT and IVSM transition strengths to the binding energy of ^{6}He. We need more analysis to conclude this story. Especially, appropriate treatment of the three-body continuum states is necessary.

Acknowledgements We thank J. Singh for careful reading of the manuscript. This work was in part supported by JSPS KAKENHI Grant Numbers 18K03635 and 18H04569.

References

1. Zhukov, M.V., et al.: Phys. Rep. **231**, 151 (1993)
2. Horiuchi, W., Suzuki, Y.: Phys. Rev. C **76**, 024311 (2007)
3. Thompson, D.R., et al.: Nucl. Phys. **A 286**, 53 (1977)
4. Kanada, H., et al.: Prog. Theor. Phys. **61**, 1327 (1979)
5. Suzuki, Y., Horiuchi, W., Orabi, M., Arai, K.: Few-Body Syst. **42**, 33 (2008)
6. Mitroy, J., et al.: Rev. Mod. Phys. **85**, 693 (2013)
7. Suzuki, Y., Horiuchi, W.: Emergent phenomena in atomic nuclei from large-scale modeling: a symmetry-guided perspective, pp. 199–227 (2017)
8. Csótó, A., Baye, D.: Phys. Rev. C **49**, 818 (1994)
9. Ajzenberg-Selove, F.: Nucl. Phys. **A 490**, 1 (1998)
10. Baye, D., Suzuki, Y., Descouvemont, P.: Phys. Rev. C **89**, 064303 (2014)

Chapter 39
R-Mode Instability of Neutron Stars Using Nuclear EoS from DDM3Y

D. Atta, S. Mukhopadhyay and D. N. Basu

Abstract The r-mode instability windows and the gravitational wave signatures of neutron stars have been studied using β-equilibrated neutron-proton-electron neutron star matter at the core with a rigid crust. The fiducial gravitational and viscous timescales, the critical frequencies and the time evolutions of the frequencies and the rates of frequency change are calculated for a range of neutron star masses. It is found that the young and hot rotating massive neutron stars (NSs) are more susceptible to r-mode instability through gravitational radiation. We also find that NS for equation of state (EoS) with low L value lie in the r-mode instability region.

39.1 Introduction

The oscillations of rotating neutron stars have gained much importance because of the information they send about the properties of the high-density Equation of State (EoS) through electromagnetic and gravitational wave signals. In the present work the r-mode instability has been discussed with reference to the EoS obtained using the density dependent M3Y (DDM3Y) effective nucleon-nucleon (NN) interaction [1]. This EoS provides good descriptions for proton, α and cluster radioactivities, elastic and inelastic scattering, symmetric and isospin asymmetric nuclear matter, neutron star masses and radii, their core-crust transition and crustal fraction of moment of inertia [2, 3].

D. Atta (✉)
Government General Degree College, Kharagpur II,
Ambigeria 721149, West Bengal, India
e-mail: debasisa906@gmail.com

S. Mukhopadhyay · D. N. Basu
Variable Energy Cyclotron Centre, HBNI, 1/AF Bidhan Nagar,
Kolkata 700064, India

N. A. Orr et al. (eds.), *Recent Progress in Few-Body Physics*,
Springer Proceedings in Physics 238,
https://doi.org/10.1007/978-3-030-32357-8_39

39.2 Dissipative Time Scales

The dissipative time scale τ of an r-mode arises from various dissipative mechanisms such as gravitational wave emissions and viscosity (bulk and shear) and hence $\frac{1}{\tau(\Omega,T)} = \frac{1}{\tau_{GR}(\Omega,T)} + \frac{1}{\tau_{BV}(\Omega,T)} + \frac{1}{\tau_{SV}(\Omega,T)}$ where $1/\tau_{GR}$, $1/\tau_{BV}$ and $1/\tau_{SV}$ are the contributions from gravitational radiation, bulk viscosity and shear viscosity, respectively. The dissipative time scales τ_{GR} and τ_{SV} are given by

$$\frac{1}{\tau_{GR}} = -\frac{32\pi G \Omega^{2l+2}}{c^{2l+3}} \frac{(l-1)^{2l}}{[(2l+1)!!]^2} \left(\frac{l+2}{l+1}\right)^{(2l+2)} \int_0^{R_c} \rho(r) r^{2l+2} dr, \quad (39.1)$$

$$\frac{1}{\tau_{SV}} = \left[\frac{1}{2\Omega} \frac{2^{l+3/2}(l+1)!}{l(2l+1)!!I_l} \sqrt{\frac{2\Omega R_c^2 \rho_c}{\eta_c}}\right]^{-1} \left[\int_0^{R_c} \frac{\rho(r)}{\rho_c} \left(\frac{r}{R_c}\right)^{2l+2} \frac{dr}{R_c}\right]^{-1}, \quad (39.2)$$

where Ω is the angular velocity of the star in the inertial frame, G and c are the gravitational constant and velocity of light respectively; R_c, ρ_c, η_c in (39.1) are the radius, density and shear viscosity of the fluid at the outer edge of the core respectively and bulk viscosity contribution being small have been neglected. The quadrupole $l = 2$ r-mode is more strongly unstable to gravitational radiation than any other mode in neutron stars.

39.3 Theoretical Calculations and R-Mode Instability

The r-mode oscillation is analogous to Rossby waves in the ocean and results from perturbation in the velocity field of a star with little disturbance in the star's density (Fig. 39.1). The evolution of the angular velocity, as the angular momentum is radiated to infinity by the gravitational radiation is given by

$$\frac{d\Omega}{dt} = \frac{2\Omega}{\tau_{GR}} \frac{\alpha_r^2 Q}{1 - \alpha_r^2 Q}, \quad (39.3)$$

where α_r is the dimensionless r-mode amplitude and $Q = 3\tilde{J}/2\tilde{I}$ with,

$$\tilde{J} = \frac{1}{MR^4} \int_0^R \rho(r) r^6 dr, \quad \tilde{I} = \frac{8\pi}{3MR^2} \int_0^R \rho(r) r^4 dr. \quad (39.4)$$

α_r is treated as free parameter whose value varies within a wide range $1 - 10^{-8}$. Under the ideal consideration that the heat generated by the shear viscosity is same

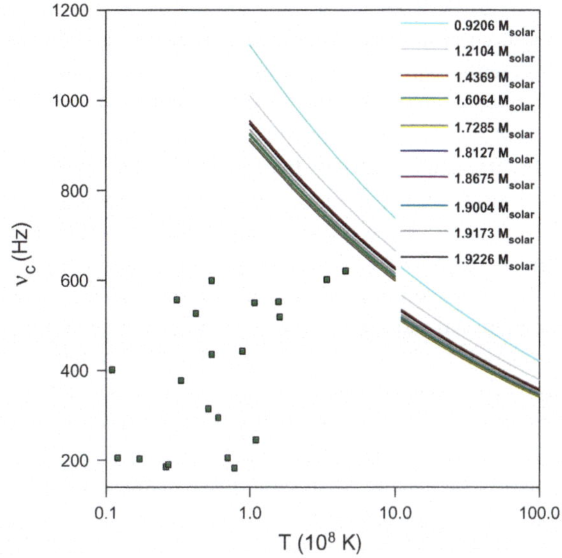

Fig. 39.1 Plots of critical frequency with temperature for different masses of neutron stars. The square dots represent observational data [6]

as that taken out by the emission of neutrinos [4, 5], (39.3) can be solved for the angular frequency $\Omega(t)$ as

$$\Omega(t) = \left(\Omega_{in}^{-6} - Ct\right)^{-1/6} \text{where } C = \frac{12\alpha_r^2 Q}{\widetilde{\tau}_{GR}\left(1 - \alpha_r^2 Q\right)} \frac{1}{\Omega_0^6}, \tag{39.5}$$

and Ω_{in} is considered as a free parameter whose value corresponds to be the initial angular velocity. The spin down rate can be obtained from (39.3) to be,

$$\frac{d\Omega}{dt} = \frac{C}{6}\left(\Omega_{in}^{-6} - Ct\right)^{-7/6}. \tag{39.6}$$

The neutron star spin shall decrease continually until it approaches its critical angular velocity Ω_c. The time t_c taken by neutron star to evolve from its initial value Ω_{in} to its minimum value Ω_c is given by (39.5)

$$t_c = \frac{1}{C}\left(\Omega_{in}^{-6} - \Omega_c^{-6}\right). \tag{39.7}$$

39.4 Results and Discussion

The r-mode damping mechanism is calculated using the shear viscosity timescale acting along the boundary layer between the rigid crust and fluid core. We have calculated the fiducial gravitational radiation and shear viscosity timescales within the DDM3Y framework for a wide range of neutron star masses. It is observed that the gravitational radiation timescale decreases rapidly with increasing neutron star mass while the viscous damping timescales exhibit an approximate linear increase. Next, we have studied the temperature dependence of the critical angular frequency for different neutron star masses. The implication is that for neutron stars rotating with frequencies greater than their corresponding critical frequencies have unstable r-modes leading to the emission of gravitational waves. Further, our study of the variation of the critical temperature as a function of mass shows that both the critical frequency and temperature decrease with increasing mass. The conclusion is that massive hot neutron stars are more susceptible to r-mode instability through gravitational radiation. Finally we have calculated the spin down rates and angular frequency evolution of the neutron stars through r-mode instability. We have also pointed out the fact that the critical frequency depends on the EoS through the radius and the symmetry energy slope parameter L. If the dissipation of r-modes from shear viscosity acts along the boundary layer of the crust-core interface then the r-mode instability region is enlarged to lower values of L [7].

39.5 Summary and Conclusions

In this work we study the r-mode instability windows and the gravitational wave signatures of neutron stars in the slow rotation approximation using the equation of state obtained from the density dependent M3Y effective interaction. The fiducial gravitational and viscous timescales, the critical frequencies and the time evolutions of the frequencies and the rates of frequency change are calculated for a range of neutron star masses. We show that the young and hot rotating neutron stars lie in the r-mode instability region. We also stress that if the dominant dissipative mechanism of the r-mode is the shear viscosity along the boundary layer of the crust-core interface, then the neutron stars with low L value lie in the r-mode instability region and hence emit gravitational radiation.

Acknowledgements The Author is thankful to DST, SERB, Govt. of India for sanctioning fund number ITS/2018/002339 for attending the FB22 conference.

References

1. Atta, D., Basu, D.N.: Phys. Rev. **C 90**, 035802 (2014)
2. Basu, D.N., Chowdhury, P.R., Mishra, A.: Eur. Phys. J. Plus **129**, 62 (2014)
3. Atta, D., Mukhopadhyay, S., Basu, D.N.: Indian J. Phys **91**, 235 (2017)
4. Bondarescu, R., Teukolsky, S.A., Wasserman, I.: Phys. Rev. **D 79**, 104003 (2009)
5. Moustakidis, Ch.: Phys. Rev. **C 91**, 035804 (2015)
6. Haskell, B., Degenaar, N., Ho, W.C.G.: Mon. Not. R. Astron. Soc. **424**, 93 (2012)
7. Mukhopadhyay, S., Lahiri, J., Atta, D., Imam, K., Basu, D.N.: Phys. Rev. **C 97**, 065804 (2018)

Chapter 40
Radiative ^{3}He(^{2}H,γ)^{5}Li Capture at Astrophysical Energy and Its Possible Role in Accumulation of ^{6}Li at the BBN

S. B. Dubovichenko, N. A. Burkova, A. V. Dzhazairov-Kakhramanov, A. S. Tkachenko and R. Ya. Kezerashvili

Abstract The Big Bang Nucleosynthesis (BBN) relevance reactions ^{3}He(^{2}H, γ)^{5}Li, ^{3}H(^{3}He, γ)^{6}Li, and ^{5}Li(n, γ)^{6}Li are treated as a source of ^{6}Li formation. Comparison of the numerical reaction rates of these processes, the prevalence of light elements suggest that the two-step process ^{2}H + ^{3}He → ^{5}Li + γ and n + ^{5}Li → ^{6}Li + γ can make a significant contribution, of the order of unity, to the formation of ^{6}Li nuclei at the BBN at least at temperatures T_9. Calculations of the total cross sections, astrophysical S-factor, and reaction rates have been performed for ^{3}He(^{2}H, γ)^{5}Li radiative capture reaction within the framework of the modified potential cluster model with forbidden states, which follow from the classification of the orbital cluster states according to Young diagrams. Numerical data and corresponding parametrizations cover the energy range 100 keV < $E_{\text{c.m.}}$ < 5 MeV and temperature ranges 0.01 < T_9 < 10. Detailed up-to-date experimental data for the reaction ^{3}He(^{2}H, γ)^{5}Li are analyzed.

The interest in radiative capture reactions in the isobar-analogue channels ^{3}He(^{2}H, γ)^{5}Li and ^{3}H(^{2}H, γ)^{5}He is due to two factors: i. The new data [1] on the diagnostics of nuclear fusion efficiencies of ^{2}H(^{3}H, n)^{4}He and ^{2}H(^{3}He, p)^{4}He reactions used for study of tokamak plasmas in experiments on Joint European Torus (JET) and

S. B. Dubovichenko (✉) · A. V. Dzhazairov-Kakhramanov · A. S. Tkachenko (✉)
Fesenkov Astrophysical Institute "NCSRT", Almaty 050020, Kazakhstan
e-mail: dubovichenko@gmail.com

A. S. Tkachenko
e-mail: tkachenko.alessya@gmail.com

N. A. Burkova · A. S. Tkachenko
al-Farabi Kazakh National University, Almaty 050040, Kazakhstan

R. Ya. Kezerashvili
New York City College of Technology, City University of New York, New York City, NY 11201, USA

Graduate School and University Center, City University of New York, New York City, NY 10016, USA

N. A. Orr et al. (eds.), *Recent Progress in Few-Body Physics*,
Springer Proceedings in Physics 238,
https://doi.org/10.1007/978-3-030-32357-8_40

International Thermonuclear Experimental Reactor (ITER); ii. The latest data on plasma diagnostics are presented in [2].

These reactions are also parts of the nucleosynthesis chain of processes occurring in the early stage of stable stars formation, as well as are identified as possible candidates for overcoming the well-known problem of the $A = 5$ gap in the synthesis of light elements in the primordial Universe [3].

There is an unambiguous opinion: due to the small cross section of the ^{3}He(^{2}H, γ)^{5}Li reaction, it does not contribute to astrophysical processes [4]. However, this statement is not entirely true, since the rate of this reaction is not negligible. To address this, we consider a possible scenario of ^{6}Li formation involving a short-lived ^{5}Li isotope in astrophysical processes.

The radiative ^{3}He(^{2}H, γ)^{5}Li capture is considered on the basis of the modified potential cluster model [5] and new results are obtained for dipole $E1$ and $M1$ transitions, taking into account the mixing of the doublet and quartet spin channels, both in scattering states and for the bound ground state. The potentials of the intercluster interaction were constructed on the basis of the description of the known scattering phase shifts and the main characteristics of the ground state (GS) of ^{5}Li. The total cross sections of ^{3}He(^{2}H, γ)^{5}Li capture on the GS of ^{5}Li at energies from 5 keV to 5 MeV in c.m. are calculated. For the astrophysical S-factor and the rate of this reaction, obtained in these calculations, simple analytic parametrizations are proposed.

In Fig. 40.1 one can see that at energies above 400 keV the results of calculation are much lower than the available experimental data [6], and at energies less than 2 MeV they lie below the results from [11]. This difference in the cross sections can be due to the following reasons: in our calculations, we considered 12 different electromagnetic transitions, but perhaps some additional valuable processes were not taken into account; in [6, 11] in the energy range from 0.4 to 2 MeV the effect of capture on the first excited state (FES) is not entirely excluded. In addition, the maximum of the calculated cross section in the 3 MeV region is lower than the data from [11], which can be explained by the inaccuracy of the spectra [14] known by the authors of [11]. We use this data anyway, as it was the most accurate measurement reported in [15]. In both cases, the results of [6, 11] require refinement on the basis of modern methods of experimental measurements.

The value of the calculated S-factor at an energy of 6–30 keV in c.m. turned out to be relatively stable and equal to 0.125(2) keV b, which is noticeably less than the experimental value reported in [6]. The error of the calculated S-factor shown here is obtained by averaging it over the indicated energy interval. The value of the calculated S-factor is 0.125 keV b at the energy of 6 keV. At a maximum energy of 230 keV in c.m. the S-factor is equal to 0.43 keV b.

To summarize, in this work the parametrization of the cross sections for the ^{3}H(^{3}He, γ)^{6}Li and ^{5}Li(n, γ)^{6}Li processes of radiative capture is carried out. The corresponding rates of these processes are calculated, their parametrization is performed, and a comparison with the ^{3}He(^{2}H, γ)^{5}Li and ^{4}He(^{2}H, γ)^{6}Li capture reactions rate is made.

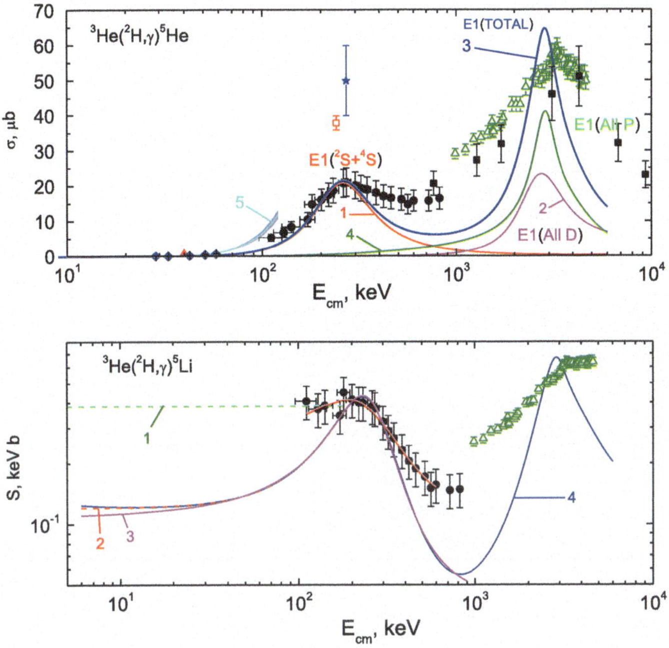

Fig. 40.1 Total cross section and astrophysical S-factor for ^{3}He(^{2}H,γ)^{5}Li below 5.0 MeV. Experimental data: ●—[6]; ▲—[7]; ♦—[8, 9]; ■—[10, 11]; Δ—[11]; ★—[12]; □—[13]; cyan band 5—results from [1]. Curves: 1—the cross section for the $E1$ transition from the ^2S and ^4S scattering states to the GS $^{2+4}P_{3/2}$: 2—the cross section for the $E1$ transition to the GS from the resonating $^{2+4}D_{5/2}$ waves; 3—the contribution of the $M1$ transitions from the resonating $^{2+4}P_{3/2}$ and $^{2+4}P_{5/2}$ waves, 4—the total cross section included all $E1$ and $M1$ transitions

On the basis of comparisons of the rates of these reactions and the prevalence of light elements, it is assumed that the two-step process ^{2}H + ^{3}He → ^{5}Li + γ and n + ^{5}Li → ^{6}Li + γ can make a definite contribution, of order unity, to the production of ^{6}Li at the BBN at least at temperatures T_9. In this temperature range the number of neutrons has not yet begun to decrease, and the number of ^{2}H and ^{3}He nuclei is already reaching its maximum, which leads to the increase in the reaction yield ^{2}H + ^{3}He → ^{5}Li + γ.

Acknowledgements This work was supported by the Grant of Ministry of Education and Science of the Republic of Kazakhstan through the program BR05236322. Participation in FB22 was sponsored by Foundation of the first president of the Republic of Kazakhstan—Elbasy.

References

1. Kiptily, V.G., et al.: Plasma Phys. Control. Fusion **48**, R59 (2006)
2. Sharapov, S.E., et al.: Nucl. Fusion **56**, 112021 (2016)
3. Barnes, C.A., et al.: Essays in Nuclear Astrophysics. Presented to William A. Fowler. Cambridge University Press, UK (1982)
4. Caughlan, G.R., et al.: At. Data Nucl. Data Tables **40**, 283 (1988)
5. Dubovichenko, S.B.: Thermonuclear Processes in Stars and Universe, 2nd edn. Scholar's Press, Germany (2015)
6. Buss, W., et al.: Nucl. Phys. A **112**, 47 (1968)
7. Cecil, F.E., et al.: Vacuum **44**, 181 (1993)
8. Cecil, F.E., et al.: Phys. Rev. C **32**, 690 (1985)
9. Aliotta, M.A., et al.: Nucl. Phys. A **690**, 790 (2001)
10. King, H.T., et al.: Nucl. Phys. A **178**, 337 (1972)
11. Schroder, H., et al.: Z. Phys. **235**, 234 (1970)
12. Blair, J.M., et al.: Phys. Rev. **96**, 1023 (1954)
13. Kraus, et al.: Nucl. Phys. A, **109**, 593 (1968)
14. Ajzenberg-Selove, F.: Nucl. Phys. A **490**, 1 (1988)
15. Tilley, D.R., et al.: Nucl. Phys. A **708**, 3 (2002)

Chapter 41
Experimental Studies of Unbound Neutron-Rich Nuclei

Yosuke Kondo

Abstract The unbound nuclei $^{25-28}$O have been studied by the invariant-mass method with SAMURAI at RIBF, RIKEN. The unbound nuclei are produced by one- and two-proton removal reactions from neutron-rich fluorine and neon isotopes at around 200 MeV/nucleon. The decay energies of the ground and first excited states of ^{26}O are determined for the first time. Preliminary results of the ^{27}O and ^{28}O measurements are presented.

41.1 Introduction

The structure of atomic nuclei near the limit of binding, called the drip-line, has attracted much attention since exotic phenomena such as neutron halos and neutron-neutron correlations have been found and discussed. Neutron-unbound oxygen isotopes located beyond the neutron drip line provide good examples of few-body systems, consisting of the subshell closed nucleus ^{24}O [1] and a few neutrons. Theoretical study of the unbound ^{26}O nucleus using a three-body model (an inert ^{24}O core and two valence neutrons) predicts an enhancement of back-to-back two neutron emission in the decay of the ^{26}O ground state. Such a correlation of neutrons can be interpreted as evidence for spatial neutron-neutron correlation, called the dineutron correlation [2]. In addition, a recent experiment [3] observed candidate events of a tetra neutron system. Therefore, it is interesting to investigate whether ^{28}O has a structure of a ^{24}O core and two dineutrons or a tetra neutron.

The unbound oxygen isotopes are also interesting in terms of the sudden drip-line gap, called the oxygen anomaly [4]. It is experimentally known that the neutron drip line for oxygen ($Z = 8$) is located at ^{24}O ($N = 16$), while it extends more 6 neutrons to ^{31}F ($N = 22$) for fluorine ($Z = 9$). Currently, the origin of the oxygen anomaly has not yet been understood. While the theoretical study [4] suggests that three-nucleon forces play an important role in the binding of the oxygen isotopes especially in

Y. Kondo (✉)
Tokyo Institute of Technology, O-okayama, Meguro, Tokyo 152-8551, Japan
e-mail: kondo@phys.titech.ac.jp

N. A. Orr et al. (eds.), *Recent Progress in Few-Body Physics*,
Springer Proceedings in Physics 238,
https://doi.org/10.1007/978-3-030-32357-8_41

the $N > 16$ region, little is known experimentally about ^{25}O and ^{26}O, and the more neutron-rich isotopes ^{27}O and ^{28}O have never been observed. The experimental study of the unbound oxygen isotopes towards ^{28}O is therefore strongly desired to examine the theoretical predictions.

Shell evolution is one of other important factors in understanding the oxygen anomaly. It has been established that the conventional magic number $N = 20$ disappears in the region of $Z = 10 - 12$. This is called the island of inversion. Since the disappearance of the shell closure emerges with decrease in atomic number along $N = 20$ from the β-stability line, the shell evolution in more proton-deficient nuclei towards the possible doubly magic nucleus ^{28}O at $Z = 8$ is an interesting topic. Recent in-beam γ-ray spectroscopy [5] has suggested that the island of inversion extends to $Z = 9$ from the comparison of the measured excitation energy (1.08 MeV) of the first excited state in the $N = 20$ nucleus ^{29}F with shell model calculations. It should be noted that the shell model calculations with the SDPF-M effective interaction [6, 7], which well reproduce the measured excitation energy of ^{29}F, gives small fraction of 0p-0h configuration in the ground and first excited states of ^{29}F as well as the ground state of ^{28}O. In this context, it is of great importance to study the doubly-magic candidate nucleus ^{28}O ($Z = 8$, $N = 20$) experimentally.

41.2 Experiment for ^{25}O and ^{26}O

The first experiment [8] was carried out for ^{25}O and ^{26}O at the RI Beam Factory (RIBF) operated by the RIKEN Nishina Center and the Center for Nuclear Study (CNS), University of Tokyo. The unbound nuclei ^{25}O and ^{26}O were produced by one-proton removal reactions from ^{26}F and ^{27}F respectively, at 201 MeV/nucleon. Unbound states were identified by the invariant mass method by measuring the momentum vectors of the ^{24}O and neutron(s) decay particles. The secondary beams of ^{26}F and ^{27}F were produced by projectile fragmentation of a primary ^{48}Ca beam at 345 MeV/nucleon. The beams were purified by the BigRIPS fragment separator [9] and transported to the experimental area of SAMURAI [10]. The incident beam particles were detected by two plastic scintillators, an ionization chamber, and two multi-wire drift chambers (MWDCs) before impinging on a carbon reaction target (1.8 g/cm^2). The outgoing ^{24}O and neutron(s) were separated by the superconducting dipole magnet. The ^{24}O ions were detected by two MWDCs before and after the magnet, and a plastic scintillator hodoscope. The decay neutron(s) were detected in coincidence by NEBULA. It consists of 120 plastic scintillation detectors. Each scintillator bar has a dimension of $12 \times 12 \times 180$ cm, equipped with two photo multiplier tubes (PMTs) and light guides at each end. The detectors are arranged in a two-wall configuration. Each wall is equipped with thin (1 cm thickness) plastic scintillators for a charged particle veto. Thanks to the two-wall configuration, it is possible to detect two neutrons emitted with zero relative energy by eliminating the crosstalk background using velocity between the two walls [11].

The ground state resonance of ^{25}O was observed at 749(10) keV in the one-proton removal reaction from ^{26}F. The observed energy is consistent with the previous measurements [12, 13]. In the one-proton removal reaction from ^{27}F, the ground state resonance of ^{26}O was observed near the neutron decay threshold. The statistics obtained in this experiment was about 5 times higher than the previous measurement [14]. The finite decay energy of 18±3(stat)±4(syst) keV for the ground state was determined. In addition, the first excited state was observed for the first time at $1.28^{+0.11}_{-0.08}$ MeV. This is most probably the first 2^+ state.

The fact that the excitation energy of the 2^+ state of ^{26}O is much lower than that of ^{24}O (4.7 MeV [15]) confirms the presence of the $N = 16$ subshell closure. Shell model calculations with the USDB effective interaction [16] reproduce the trend of the energies of the 2^+ states of the even-even oxygen isotopes, including the local maximum at ^{24}O. The calculations also reproduce the trend of the even-even $N = 18$ isotones in the range of $12 \leq Z \leq 18$. However, the calculations do not reproduce the gradual decrease in the 2^+ energies observed experimentally. Instead the model shows an increase in the 2^+ energies from $Z = 12$ to $Z = 8$. This discrepancy suggests the importance of effects not included in the calculation. The 2^+ energy of ^{26}O has also been compared to three-body model calculations [17, 18], shell model calculations including continuum effects [19, 20], and ab-initio type shell model calculations [21]. These calculations give closer energies to the experimental value than the USDB calculation as shown in Fig. 5 in [8]. To pin down the various effects quantitatively, further theoretical study is desired to disentangle the effects which are important to describe the nuclear structure of ^{26}O.

41.3 Experiment for ^{27}O and ^{28}O

The second experiment was also performed with the SAMURAI setup. The unbound nuclei ^{27}O and ^{28}O were produced by two- and one-proton removal reactions from ^{29}Ne and ^{29}F, respectively, at ~230 MeV/nucleon. The secondary beams were produced by the projectile fragmentation of a ^{48}Ca primary beam at 345 MeV/nucleon on a 15-mm thick beryllium target, and they were purified by the BigRIPS fragment separator with two aluminum energy degraders (15 and 7 mm thicknesses) at the F1 and F5 dispersive focal planes. A plastic scintillator with a thickness of 3 mm was installed at the F5 dispersive focal plane for measuring the magnetic rigidity. The typical intensities of the ^{29}Ne and ^{29}F beams were 8000 and 90 counts/s.

Figure 41.1 shows the experimental setup at SAMURAI. The field strength of the superconducting dipole magnet was 2.9 T at the center. To realize the detection of 3 and 4 neutrons from the decay of ^{27}O and ^{28}O, MINOS [22] and NeuLAND [23] were additionally installed to the experimental setup for ^{25}O and ^{26}O. MINOS consists of a liquid hydrogen target (thickness 15 cm) and time projection chamber (TPC), which were surrounded by the γ-ray detector array DALI2 [24]. The reaction vertex can be determined from proton track(s) measured in the MINOS TPC and the track of the incident beam determined by the MWDCs placed before the

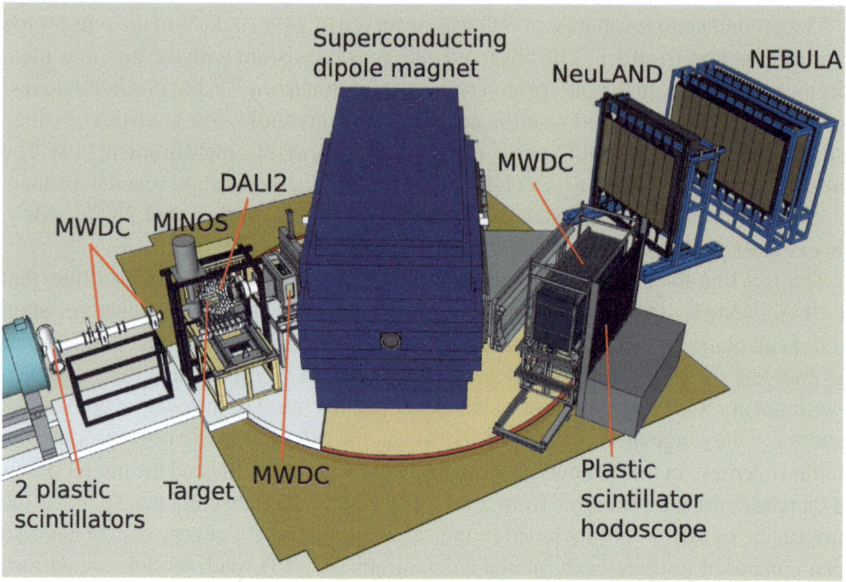

Fig. 41.1 Experimental setup for the ^{27}O and ^{28}O measurements. To show the target region, the top half of DALI2 is not shown

target. The MINOS device provides a high luminosity of the reaction without losing invariant mass resolution by calculating the energy loss in the target using the reconstructed reaction vertex. NeuLAND is a neutron detector array developed at GSI. In the present experiment, an array of prototype detectors, called the NeuLAND demonstrator, was installed in front of NEBULA. Four hundred detectors, consisting of $5 \times 5 \times 250$ cm plastic scintillators each with two PMTs, are arranged in 8 layers of alternating horizontal and vertical planes, providing 40-cm thickness in total. A layer of 1-cm thick plastic scintillators were installed in front of the NeuLAND demonstrator for charged particle veto. The combination of the NeuLAND demonstrator with NEBULA provides high neutron detection efficiency (about 50% for single neutron detection).

Figures 41.2 and 41.3 show the preliminary two-body decay energy (E_{fn}) spectra of ^{24}O and a neutron observed in the reactions from ^{29}Ne and ^{29}F. These plots are expected to include the ^{27}O and ^{28}O decay events as well as decays from ^{25}O and ^{26}O which are directly populated by multiple-nucleon removal reactions and/or by sequential decays from heavier oxygen isotopes. In both spectra, two prominent peaks are seen at $E_{fn} \sim 0$ and 0.8 MeV, corresponding to the decay of the ^{26}O and ^{25}O ground state resonances. The spectra show quite different intensities of the peaks.

Since the decay of the ^{26}O ground state resonance is identified by the events with $E_{fn} \sim 0$ obtained from the ^{24}O and single neutron coincidence, possible sequential decay through the resonance is investigated by a Dalitz-plot-like analysis of the ^{24}O and two neutron coincidence events. Candidate events of the sequential decay from

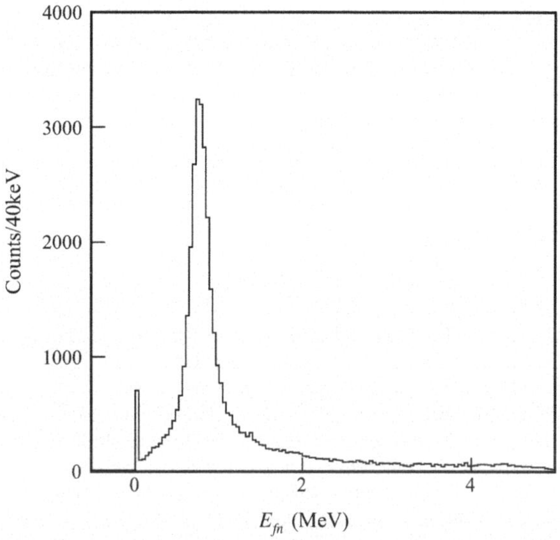

Fig. 41.2 Preliminary two-body decay energy spectrum of ^{24}O and neutron observed in the nucleon(s) removal reaction from ^{29}Ne

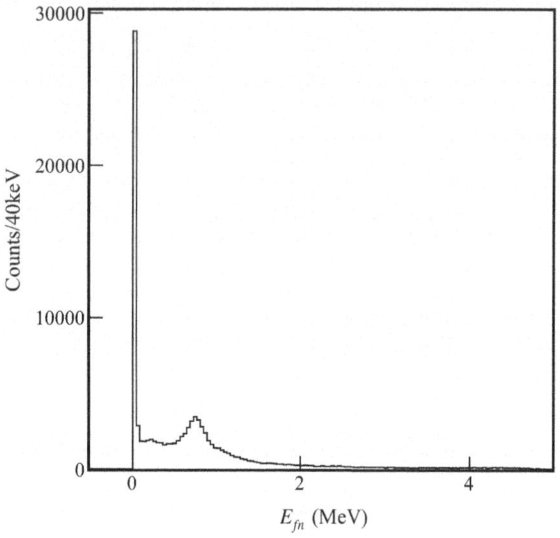

Fig. 41.3 Preliminary two-body decay energy spectrum of ^{24}O and neutron observed in the nucleon(s) removal reaction from ^{29}F

^{27}O and ^{28}O can be seen in the two-body decay energy spectra of ^{24}O and neutron by gating $E_{fn} \sim 0$ for ^{24}O and another neutron. Further analysis of 3- and 4-neutron coincidence events as well as simulation is now in progress to examine the candidate events.

Acknowledgements The author would like to thank all the collaborators in the studies described here. The author is also grateful to the R^3B collaboration for their efforts in setting up the NeuLAND demonstrator at SAMURAI, and the MINOS collaboration for their cooperation during the experiment. This work was supported in part by JSPS KAKENHI Grant No. 18K03672, 24740154, 16H02179, MEXT KAKENHI GRANT No. 24105005 and 18H05404.

References

1. Ozawa, A., et al.: Phys. Rev. Lett. **84**, 5493–5495 (2000)
2. Hagino, K., Sagawa, H.: Phys. Rev. C **89**, 014331 (2014)
3. Kisamori, K., et al.: Phys. Rev. Lett. **116**, 052501 (2016)
4. Otsuka, T., et al.: Phys. Rev. Lett. **105**, 032501 (2010)
5. Doornenbal, P., et al.: Phys. Rev. C **95**, 041301 (2017)
6. Utsuno, Y., Otsuka, T., Mizusaki, T., Honma, M.: Phys. Rev. C **60**, 054315 (1999)
7. Utsuno, Y., et al.: Phys. Rev. C **70**, 044307 (2004)
8. Kondo, Y., et al.: Phys. Rev. Lett. **116**, 102503 (2016)
9. Kubo, T.: Nucl. Instrum. Methods Phys. Res. Sect. B **204**, 97–113 (2003)
10. Kobayashi, T., et al.: Nucl. Instrum. Methods Phys. Res. Sect. B **317**, 294–304 (2013)
11. Nakamura, T., Kondo, Y.: Nucl. Instrum. Methods Phys. Res. Sect. B **376**, 156–161 (2016)
12. Hoffman, C.R., et al.: Phys. Rev. Lett. **100**, 152502 (2008)
13. Caesar, C., et al.: Phys. Rev. C **88**, 034313 (2013)
14. Lunderberg, E., et al.: Phys. Rev. Lett. **108**, 142503 (2012)
15. Hoffman, C.R., et al.: Phys. Lett. B **672**, 17–21 (2009)
16. Brown, B.A., Richter, W.A.: Phys. Rev. C **74**, 034315 (2006)
17. Hagino, K., Sagawa, H.: Phys. Rev. C **90**, 027303 (2014)
18. Grigorenko, L.V., Zhuko, M.V.: Phys. Rev. C **91**, 064617 (2015)
19. Volya, A., Zelevinsky, V.: Phys. Rev. C **74**, 064314 (2006)
20. Tsukiyama, K., Otsuka, T., Fujimoto, R.: Prog. Theor. Exp. Phys. **2015**, 093D01 (2015)
21. Bogner, S.K., et al.: Phys. Rev. Lett. **113**, 142501 (2014)
22. Obertelli, A., et al.: Eur. Phys. J. A **50**, 8 (2014)
23. NeuLAND technical design report. https://edms.cern.ch/document/1865739/1
24. Takeuchi, S., et al.: Nucl. Instrum. Methods Phys. Res. Sect. A **763**, 596–603 (2014)

Chapter 42
Extraction of Nucleon Polarisabilities from Light Nuclei

Judith A. McGovern, Harald W. Grießhammer and Daniel R. Phillips

Abstract We consider the response of nucleons and light nuclei to electromagnetic fields to test the ability of chiral effective field theory to describe such systems in a consistent and systematically-improvable manner.

42.1 Introduction

At sufficiently low energy, gauge and Lorentz invariance require that Compton scattering from a composite target reduces to Thomson scattering, depending only on the charge and mass. At somewhat higher energies, deviations become apparent, and the form of these deviations is a sensitive test of the important low-energy degrees of freedom determining the target's structure. The aims of Compton scattering experiments from light nuclei at photon energies around 50–150 MeV are therefore two-fold: an exploration both of nucleon and of nuclear structure. For a proton target, the principal such effects at low energy (after well-understood relativistic corrections, π^0 pole contributions and the anomalous magnetic moment) are the nucleon's electric and magnetic dipole polarisabilities, α_{E1} and β_{M1}. These reveal the extent to which charge and current distributions in the target shift under the influence of external electromagnetic fields and parametrise the strength of the induced radiation dipoles. Then, in the amplitudes that are sensitive to the target's spin, four "spin polarisabilities" γ_i govern the departure from point-like scattering and parametrise the response of the spin degrees of freedom. For light nuclei (d, ^{3}He, ^{4}He) the deviations from Thomson scattering set in much earlier, at energies around the break-up energy, and by 50 MeV the response involves an interplay of both nucleonic and nuclear excitations.

J. A. McGovern (✉)
University of Manchester, Manchester, UK
e-mail: judith.mcgovern@manchester.ac.uk

H. W. Grießhammer
George Washington University, Washington DC, USA

D. R. Phillips
Ohio University, Athens, USA

N. A. Orr et al. (eds.), *Recent Progress in Few-Body Physics*,
Springer Proceedings in Physics 238,
https://doi.org/10.1007/978-3-030-32357-8_42

A satisfactory description of such experiments both validates the underlying theory, and allows for an extraction of the otherwise-inaccessible neutron polarisabilities [1].

For reference we reproduce the low-energy non-relativistic effective Hamiltonian that indicates how polarisabilities affect the response of the nucleon to external electric and magnetic fields [2]:

$$H_{\text{eff}} = -\frac{1}{2}4\pi\Big(\alpha_{E1}\mathbf{E}^2 + \beta_{M1}\mathbf{H}^2 + \gamma_{E1E1}\sigma\cdot\mathbf{E}\times\dot{\mathbf{E}} + \gamma_{M1M1}\sigma\cdot\mathbf{H}\times\dot{\mathbf{H}} - 2\gamma_{M1E2}E_{ij}\sigma_i H_j + 2\gamma_{E1M2}H_{ij}\sigma_i E_j\Big), \tag{42.1}$$

where dots mean a time derivative and $X_{ij} = \frac{1}{2}(\nabla_i X_j + \nabla_j X_i)$. The scalar polarisabilities will be given throughout in units of 10^{-4} fm^3, and the spin polarisabilities in units of 10^{-4} fm^4.

In recent years there has been an upsurge in interest in the polarisabilities of the nucleon, both scalar and spin, with a number of new experiments planned, running or completed at MAX-Lab, MAMI and HIγS [3–7]. In addition the magnetic polarisability β^{p}_{M1} has been shown to be a crucial input in the determination of the two-photon-exchange contribution to the Lamb shift in muonic hydrogen and the isovector $\beta^{\text{p}}_{M1} - \beta^{\text{n}}_{M1}$ has been connected to the nucleon electromagnetic mass difference (see [8]). The calculation of nucleon polarisabilities is also an aim of lattice QCD, and several groups now have published results, albeit almost all at large pion masses (see [9]). In this contribution we report on the results of high-precision EFT fits of scalar polarisabilities to current data, and on prospects for the determination of spin polarisabilities. For more details and in particular for more complete references to earlier work the reader is referred to our review [1], and to subsequent papers [5, 8–10]. For related work see the contribution of H. W. G. and J. McG. in these proceedings [11].

42.2 Chiral Effective Field Theory

The framework we use in our calculations is based on heavy-baryon chiral perturbation theory (HBχPT) [12]. This is an effective field theory involving hadronic degrees of freedom rather than elementary quarks and gluons. It exploits the full symmetry of the QCD Lagrangian which arises from the fact that the up and down quarks are not only close in mass (hence isospin symmetry) but also very light compared to typical hadronic masses. Contrary to what one may read, the mass of ordinary matter does not arise primarily from coupling to the Higgs field. In fact if that were turned off, the proton and neutron would remain massive and would not change dramatically. In that limit the symmetry of isospin would be doubled, with one copy for each of the unmixed massless left- and right-handed quarks: the assumption of chiral perturbation theory is that we can treat the real world by expanding perturbatively about that limit. The symmetry is realised in the hidden mode, so the hadron spectrum does not contain degenerate parity doublets; instead pions, which are much lighter than other hadrons, can be interpreted as Goldstone bosons, massless in the chiral limit. The symmetry also requires their interactions with one another and with other hadrons to vanish in the low-momentum (soft) limit. Hence, somewhat surprisingly, a theory of the strong interaction turns into a theory of weakly interacting hadrons, provided we confine ourselves to energies below some breakdown scale, and in practice not too much greater than the pion mass. Being a field theory, the incorporation of coupling to photons (and indeed W and Z bosons) is straightforward, and gauge-invariance is built in. The theory is renormalisable order-by-order in the following sense. The Lagrangian consists of infinitely many terms but with a finite number at each order in a counting of derivatives and the pion mass, and predictions to a given order are generated from tree diagrams using vertices from the Lagrangian to that order, and from pion loops involving vertices of up to two orders lower. The divergences from the loops can be consistently absorbed into the parameters ("low-energy constants", LECs) at the relevant order. The field theory at its simplest contains only pions and nucleons; the effects of heavier mesons and nucleon excitations, being short-range, are encoded in the LECs.

A major advantage of a systematic, EFT-based approach is the ability to use the power-counting to estimate theoretical uncertainties. Though recognised from the earliest days, there has been a recent upsurge in interest in this point. A natural framework is the Bayesian one, see [13].

Of course in such a theory we give up any attempt to model quark substructure of hadrons. Any property which is dominated by such short-distance physics (the anomalous magnetic moments of the nucleons is an example) is encoded in low-energy constants in the Lagrangian. The electric polarisability on the other hand arises primarily from pionic fluctuations, and indeed would diverge in the chiral limit. It turns out to be very well described in leading-order HBχPT, as was famously demonstrated by Bernard et al. a number of years ago now [14]. The magnetic polarisability on the other hand seems to have comparable contributions from both longer and shorter distance physics. However working to NNLO in HBχPT allows for prediction of the full set of six energy- and scattering-angle-dependent amplitudes of Compton scattering from the nucleon in terms of only two free parameters, namely the fourth-order LECs which constitute the leading short-distance contribution to the electric and magnetic polarisabilities [15]. A fit of these parameters to experimental data is equivalent to an extraction of the polarisabilities, as we now describe.

42.3 Compton Scattering from the Nucleon

Chiral descriptions of Compton scattering [12, 16], and extractions of polarisabilities [17, 18], date back some years. Details of the calculations reported here are given in [10] including the specification of the chiral Lagrangian used (Fig. 42.1). One important point is that we included the $\Delta(1232)$ as an explicit degree of freedom; with an excitation energy of not much more than $2m_\pi$ it clearly stretches the definition

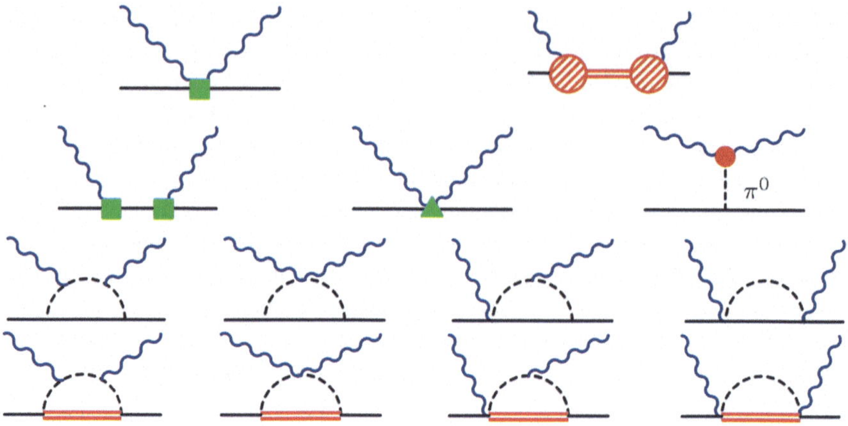

Fig. 42.1 Tree, $\Delta(1232)$ pole, leading $N\pi$, and $\Delta\pi$ loop diagrams that contribute to Compton scattering. The double line represents the $\Delta(1232)$. A number of fourth-order $N\pi$ loop diagrams are included but not depicted. Permuted and crossed diagrams not shown. Figure adapted from [10]

of "high-energy", and a cursory glance at Compton scattering data shows its overwhelming importance above about 200 MeV photon energy. The power counting that obtains in the low-energy region is no longer correct around the Delta peak [19] and our theory there is only good to NLO; hence we only use data from the resonance region to constrain the Delta parameters and not for our fits.

Grießhammer et al. [1] contains a critical evaluation of the data for Compton scattering from the proton below about 200 MeV, they are quite numerous but not all consistent, and some degree of selection was required. Our results for α_{E1}^{p} and β_{M1}^{p} were compatible with the Baldin sum rule $\alpha_{E1}^{\mathrm{p}} + \beta_{M1}^{\mathrm{p}} = 13.8 \pm 0.4$, so we reduced our statistical errors by imposing that constraint, giving

$$\begin{aligned} \alpha_{E1}^{\mathrm{p}} &= 10.65 \pm 0.35(\mathrm{stat}) \pm 0.2(\mathrm{Baldin}) \pm 0.3(\mathrm{theory}), \\ \beta_{M1}^{\mathrm{p}} &= 3.15 \mp 0.35(\mathrm{stat}) \pm 0.2(\mathrm{Baldin}) \mp 0.3(\mathrm{theory}), \end{aligned} \tag{42.2}$$

The predictions of the chiral theory with these fit parameters are shown together with world data in Fig. 42.2. Work on fitting polarisabilities to world data has also been done recently in dispersion-relation [20] and low-energy-expansion [21] frameworks.

42.4 Compton Scattering from the Deuteron

From a nuclear-physics point of view, a great merit of the chiral perturbation theory framework is that it allows for a description of nuclear forces from a starting-point which is better aligned with QCD than the traditional approaches, and very substantial progress has been made in describing light and even medium-mass nuclei starting

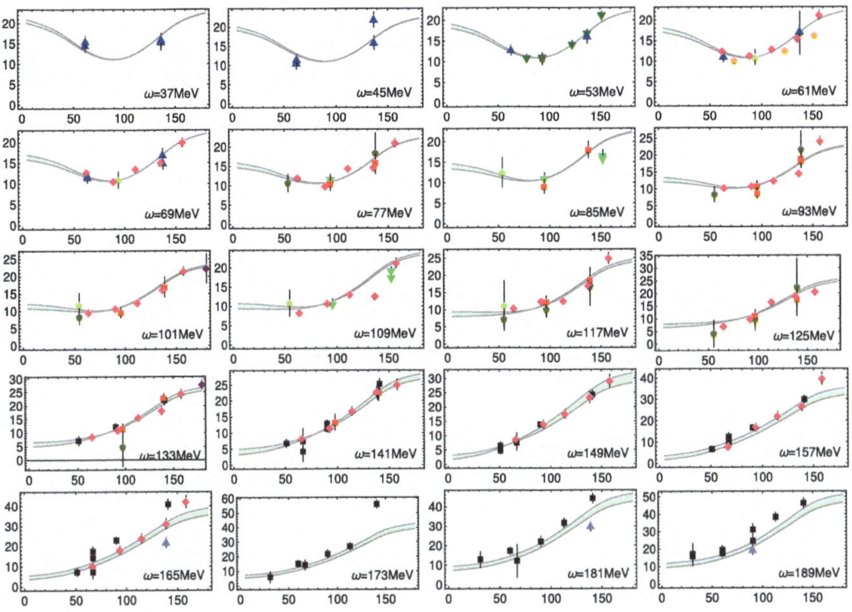

Fig. 42.2 World proton Compton scattering data and chiral EFT predictions. The labelled photon laboratory energy is the central value of 8 MeV bins, and the shaded bands span the same range (variation due to the errors on the extracted polarisabilities is small in comparison). The symbols are explained in Table 3.1 of [1]

with chiral forces that also do a good job of describing few-body systems (see for example [22]). In the current context the ability to describe the contribution of pions to both nucleon and nuclear Compton scattering on the same footing is of immense benefit. The extension of the work described above to light nuclei, therefore, has two aims: one is to extract the neutron polarisabilities in the absence of a free target, but the other, perhaps more relevant for this conference, is to test the efficacy of a chiral description of nuclear forces to describe a dynamical process, by comparing to experimental data and, if a good description can be given, comparing the neutron polarisabilities extracted from different targets.

The various diagrams which contribute are shown in Fig. 42.3. At chiral energies $E_\gamma \sim m_\pi$ the rescattering diagram (c) is higher-order. However at very low energies the same enhancement of the two-nucleon propagator that gives rise to the existence of a bound state in the first place requires this graph to be included. Only with the complete set is the correct Thomson limit reproduced; (a) and (b) alone give a result which is out by around a factor of two. As required by the power-counting the contribution of (c) rapidly diminishes with energy. However given that the NN force used in practice has regulators which introduce momentum-dependence beyond that arising from the Lagrangian, the inclusion of (c) even at higher energies allows consistency with Siegert's theorem and markedly reduces the spread between the results using different forces [23].

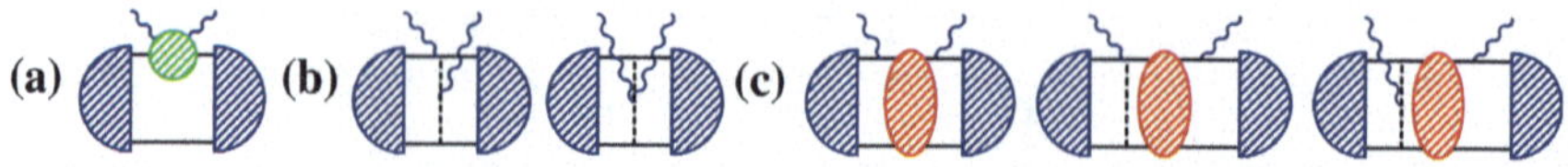

Fig. 42.3 Diagrams that contribute to Compton scattering on the deuteron. The green blob in the one-body diagram **a** represents the graphs of Fig. 42.1 while the red blob in **c** represents the full NN rescattering, including no interaction. The blue hemisphere represents the deuteron wave function. Figure adapted from [1]

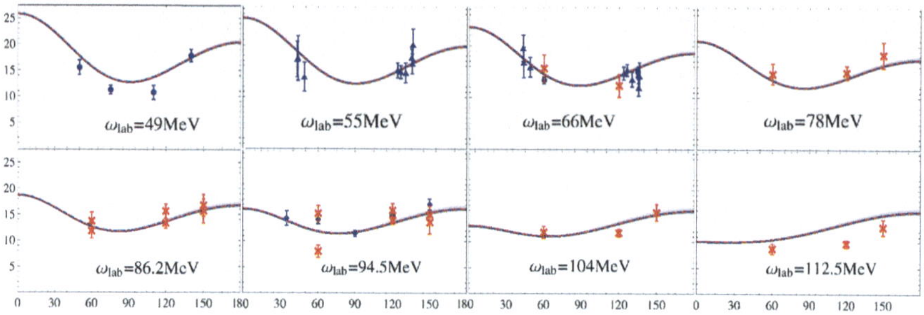

Fig. 42.4 World deuteron Compton scattering data and the chiral EFT fit. The new data of [5] are the red crosses. Figure adapted from [5]

In [1] we fitted $\alpha_{E1}^{(s)}$ and $\beta_{M1}^{(s)}$ to the pre-2014 world deuteron data, then in [5] the fit was updated to include the new MAX-Lab data which almost doubled the size of the database. The one-body diagrams were implemented to NNLO, rather than N^{3}LO as in the proton case, and the main consequence of this lower-order fit is a larger theory error than for the proton; however the statistical error still dominates. The isoscalar Baldin sum rule of $\alpha_{E1}^{(s)} + \beta_{M1}^{(s)} = 14.5 \pm 0.4$, [24] was used as a constraint. We obtained $\alpha_{E1}^{(s)} - \beta_{M1}^{(s)} = 7.8 \pm 1.2(\text{stat}) \pm 0.8(\text{th})$, with a χ^2 of 45.2 for 44 degrees of freedom.

This was then combined with the proton value to extract numbers for the neutron:

$$\begin{aligned} \alpha_{E1}^{n} &= 11.65 \pm 1.25(\text{stat}) \pm 0.2(\text{Baldin}) \pm 0.8(\text{theory}), \\ \beta_{M1}^{n} &= 3.55 \mp 1.25(\text{stat}) \pm 0.2(\text{Baldin}) \mp 0.8(\text{theory}), \end{aligned} \tag{42.3}$$

The world data and chiral EFT cross sections are shown in Fig. 42.4.

As can be seen from Fig. 42.4, the description of the data is good. Even more so than with the proton, the inclusion of the Delta is required to reproduce the backwards angle cross section, which otherwise falls too low. The results for the polarisabilities are very close to those of the proton (as expected in chiral perturbation theory, since the dominant effects are isoscalar) and indeed a significant isovector contribution is driven entirely by the Baldin sum rule and would be absent in a two-parameter fit.

42.5 Compton Scattering from ^{3}He

In principle the calculation for ^{3}He mirrors that of the deuteron, and the first studies were done by Shukla (née Choudhury) Nogga and Phillips [25, 26] and improved and extended by Margaryan et al. [27]. This is the subject of another contribution in these proceedings [11], so I will not show any results here. I will just comment that the equivalent of diagram (c) in Fig. 42.3, in which the three nucleons rescatter between absorbing and emitting the photons, is not yet included. Based on experience with the deuteron we expect that for energies above 50–60 MeV the results will still be qualitatively reliable, but work is in progress to correct this deficit.

42.6 Spin Polarisabilities

The effective Hamiltonian of (42.1) contains not only the easily-interpreted electric and magnetic polarisabilities, but also four spin-dependent polarisabilities. Because their contributions to the low-energy amplitudes go as ω^3, and also because they are not enhanced by interference with the leading Thomson amplitude, their influence on the cross section is rather small. They are predicted to the order at which we work in χEFT , and the existing data does not provide enough sensitivity to extract them without further constraints (see also [21]). A summary of our knowledge of these is given in [9, 28]. The natural observables for this purpose are those obtained with polarised targets and photons. Few such experiments have been done, and some are at energies sufficiently high (around 280 MeV lab energy) that realistic theory errors on polarisability extractions in a chiral framework would be very large. As an indication of this, it is helpful to consider the so-called "dynamical polarisabilities", which are linear combinations of Compton scattering multipoles, with the leading dependence on ω factored out, so that the $\omega \to 0$ limit coincides with the corresponding (static) polarisability (see [18, 28] and references therein). Figure 42.5 shows the the first 8 multipoles in the present theory (after fitting as described above), in the $\mathcal{O}(e^2\delta^3)$ covariant framework of Lensky et al. [28] (without fitting to data), and in the dispersion-relation framework of Pasquini et al. [18], based on integrals over pion-photoproduction multipoles (with $\alpha_{E1} - \beta_{M1}$ and γ_π fit to Compton scattering data). Varying a static polarisability is identical to simply shifting the corresponding dynamical polarisability up or down.

For low-energy scattering, the message of this plot is very positive. There is a substantial degree of agreement on the *shape* of the polarisabilities among the approaches up to around 250 MeV lab energy (200 MeV cm energy in the figure). Furthermore, after adjusting the static polarisabilities to a common value, the results generically lie very close to one another. Indeed, the same pion-loop and Delta-resonance physics is encoded in all three calculations. In the Delta-dominated multipoles, this agreement continues up to surprisingly high energies, but overall disagreement becomes quite pronounced above 250 MeV lab energy. Thus at these energies, a reliable link between the amplitudes and the static polarisability is lost.

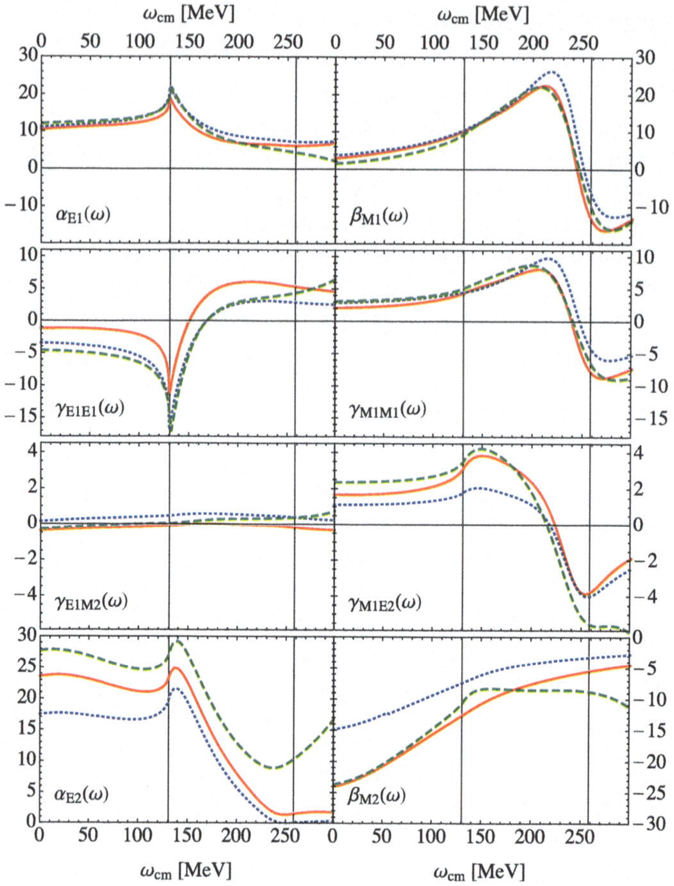

Fig. 42.5 Real parts of the dominant dynamical polarisabilities for low-energy Compton scattering from the proton, plotted as a function of cm photon energy. The units are 10^{-4} fm^n where $n = 3$ for α_{E1} and β_{M1}, $n = 4$ for the γ_i, and $n = 5$ for α_{E2} and β_{E2}. Red (solid): this work; green (dashed): DR-based by Pasquini et al. [29]; blue (dotted) 3rd-order covariant χPT by Lensky et al. [28]. Note that each row has its distinct plot scale. Figure reproduced from [8]

In our recent paper [8] we explored the sensitivity of various target-beam asymmetries to the spin polarisabilities. As an advertisement for our results I here show one of the sensitivity "heat-plots" from that paper in Fig. 42.6. It concerns the asymmetry Σ_{2x}, which involves circularly polarised photons and a target polarised perpendicularly to the scattering plane, and the colour indicates the derivative of the observable with respect to the polarisability (or combination), with deep red and deep blue indicating large positive and negative values respectively. A grey mist over the high-energy, right-hand end of the plots indicates the region where the different theoretical approaches considered no longer agree, as discussed above. This observable is particularly sensitive to the combination $\gamma_{E1E1} - \gamma_{E1M2}$, with little confounding sensitivity to other combinations if the Baldin Sum rule is used to fix $\alpha_{E1} + \beta_{M1}$. The plot also

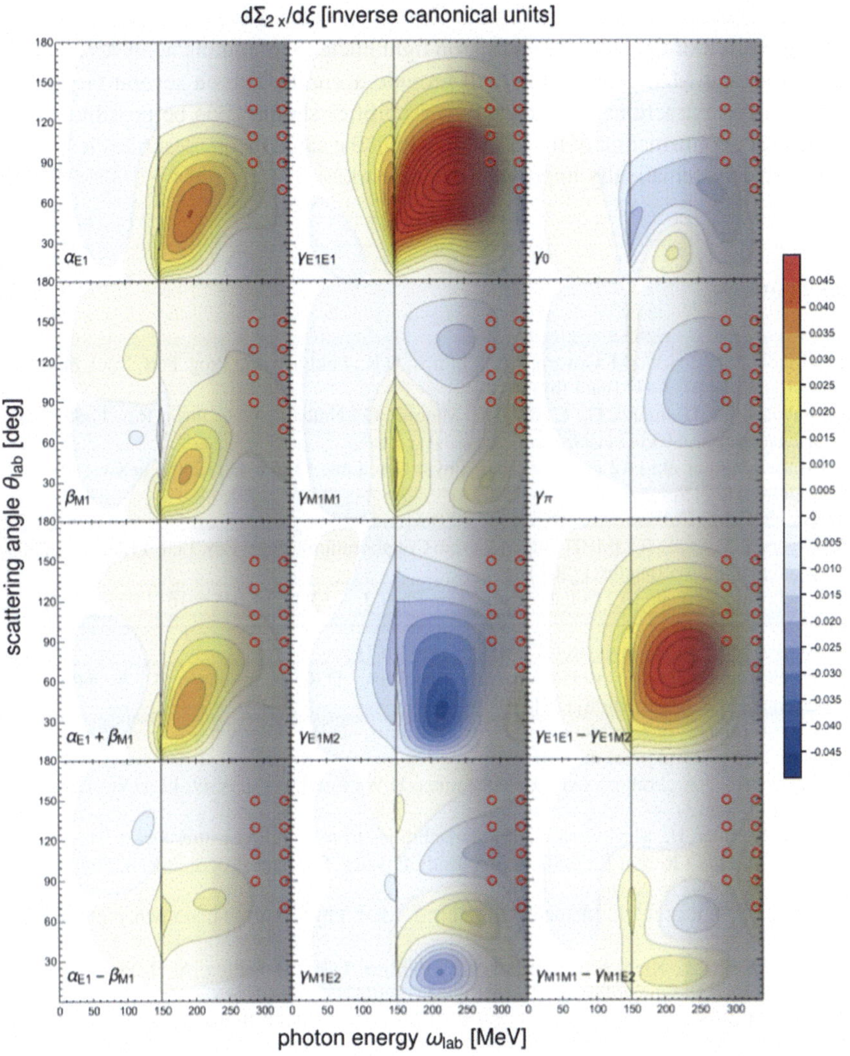

Fig. 42.6 Sensitivity of the double asymmetry Σ_{2x} (circularly polarised photons on a proton target polarised along the x axis) to varying the polarisabilities; see text for details. Circles indicate the approximate location of the data of [3]; their size does not reflect the error bars, nor the size of energy or angle bins. Figure reproduced from [8]

shows that the extant data (circles) is not at the best kinematics (even leaving aside the theory disagreement at such high energy). For an fuller explanation of the plot, the reader is referred to the original paper.

In conclusion, chiral effective field theory provides a highly effective tool for the exploration of Compton scattering on light nuclei. Up-coming analyses and new experiments at MAMI and HIγS will provide accurate date on several targets, so that multiple extractions of neutron polarisabilities should soon be possible, along with further confirmation of the extent to which the same theory describes few-body system in a systematically improvable framework.

References

1. Grießhammer, H.W., McGovern, J.A., Phillips, D.R., Feldman, G.: Prog. Part. Nucl. Phys. **67**, 841. arXiv:1203.6834 [nucl-th] (2012)
2. Babusci, D., Giordano, G., L'vov, A.I., Matone, G., Nathan, A.M.: Phys. Rev. C **58**, 1013. arXiv:hep-ph/9803347 (1998)
3. Martel, P.P., et al.: [A2 collaboration]. Phys. Rev. Lett. **114**/(11), 112501. arXiv:1408.1576 [nucl-ex] (2015)
4. Huber, G.M., Collicott, C.: arXiv:1508.07919 [nucl-ex]
5. Myers, L.S., et al.: [COMPTON@MAX-lab Collaboration]. Phys. Rev. Lett. **113**(26), 262506. arXiv:1409.3705 [nucl-ex] (2014)
6. Sokhoyan, V., et al.: [A2 Collaboration]. Eur. Phys. J. A **53**(1), 14. arXiv:1611.03769 [nucl-ex] (2017)
7. Sikora, M.H., et al.: Phys. Rev. C **96**, 055209 (2017)
8. Grießhammer, H.W., McGovern, J.A., Phillips, D.R.: Eur. Phys. J. A **54**(3), 37. arXiv:1711.11546 [nucl-th] (2018)
9. Grießhammer, H.W., McGovern, J.A., Phillips, D.R.: Eur. Phys. J. A **52**(5), 139. arXiv:1511.01952 [nucl-th] (2016)
10. McGovern, J.A., Phillips, D.R., Grießhammer, H.W.: Eur. Phys. J. A **49**, 12. arXiv:1210.4104 [nucl-th] (2013)
11. Grießhammer, H., McGovern, J.A.: Polarisabilities from compton scattering on ^{3}He. In: Orr, N., et al. (eds.) Recent Progress in Few-Body Physics, Springer Proceedings in Physics 238, arXiv:1812.04726. Springer
12. Bernard, V., Kaiser, N., Meißner, U.G.: Int. J. Mod. Phys. E **4**, 193. arXiv:hep-ph/9501384 (1995)
13. Wesolowski, S., Klco, N., Furnstahl, R.J., Phillips, D.R., Thapaliya, A.: J. Phys. G **43**(7), 074001 (2016). https://doi.org/10.1088/0954-3899/43/7/074001, arXiv:1511.03618 [nucl-th]
14. Bernard, V., Kaiser, N., Kambor, J., Meißner, U.G.: Nucl. Phys. B **388**, 315 (1992)
15. Bernard, V., Kaiser, N., Schmidt, A., Meißner, U.G.: Phys. Lett. B **319**, 269. arXiv:hep-ph/9309211 (1993)
16. McGovern, J.A.: Phys. Rev. C **63**, 064608 (2001). [Erratum-ibid. C **66** (2002) 039902] [nucl-th/0101057]
17. Beane, S.R., Malheiro, M., McGovern, J.A., Phillips, D.R., van Kolck, U.: Nucl. Phys. A **747**, 311. arXiv:nucl-th/0403088 (2005)
18. Hildebrandt, R.P., Grießhammer, H.W., Hemmert, T.R., Pasquini, B.: Eur. Phys. J. A **20**, 293. arXiv:nucl-th/0307070 (2004)
19. Pascalutsa, V., Phillips, D.R.: Phys. Rev. **C67**, 055202 (2003). [nucl-th/0212024]
20. Pasquini, B., Pedroni, P., Sconfietti, S.: arXiv:1711.07401 [hep-ph]
21. Krupina, N., Lensky, V., Pascalutsa, V.: arXiv:1712.05349 [nucl-th]
22. Lonardoni, D., Gandolfi, S., Lynn, J.E., Petrie, C., Carlson, J., Schmidt, K.E., Schwenk, A.: Phys. Rev. C **97**(4), 044318. arXiv:1802.08932 [nucl-th] (2018)
23. Hildebrandt, R.P., Grießhammer, H.W., Hemmert, T.R.: Eur. Phys. J. A **46**, 111. arXiv:nucl-th/0512063 (2010)

24. Levchuk, M.I., L'vov, A.I.: Nucl. Phys. A **674** 449. arXiv:nucl-th/9909066 (2000)
25. Choudhury, D., Nogga, A., Phillips, D.R.: Phys. Rev. Lett. **98**, 232303. arXiv:nucl-th/0701078 (2007)
26. Shukla, D., Nogga, A., Phillips, D.R.: Nucl. Phys. A **819**, 98. arXiv:0812.0138 [nucl-th] (2009)
27. Margaryan, A., Strandberg, B., Grießhammer, H.W., Mcgovern, J.A., Phillips, D.R., Shukla, D.: Eur. Phys. J. A **54**(7), 125. arXiv:1804.00956 [nucl-th] (2018)
28. Lensky, V., McGovern, J.A., Pascalutsa, V.: Eur. Phys. J. C **75**(12), 604. arXiv:1510.02794 [hep-ph] (2015)
29. Pasquini, B.: Private communication based on [18]

Chapter 43
The Quest for New Data on the Space Star Anomaly in Pd Breakup

A. Wilczek, N. Kalantar-Nayestanaki, St. Kistryn, A. Kozela, J. Messchendorp, I. Skwira-Chalot and E. Stephan

Abstract Even though the development of the theories providing a precise description of few-nucleon interactions is well advanced, certain inconsistencies between experimental data and theoretical predictions are still to be resolved, one of which is the Space Star Anomaly in deuteron-proton breakup. As the cross-sections for the star configurations are measured mainly for the energy range below 20 MeV, new measurements at higher energies could give an important hint for a possible source of the discrepancy between experimental data and the theoretical predictions. In this contribution, the very first preliminary 160 MeV deuteron on proton $p(d, pp)n$ breakup cross-sections for the star configuration measured with the BINA experimental setup are presented.

43.1 Introduction

The Space Star Anomaly, the most known discrepancy between the theoretical and the experimental cross-sections in the proton-deuteron breakup reaction, is named after the specific configuration for which it is observed [1].

The configuration is defined as a final state, for which the center-of-mass momenta form an equilateral triangle. Depending on the angle of inclination with respect to

A. Wilczek (✉) · E. Stephan
Institute of Physics, University of Silesia, 41500 Chorzow, Poland
e-mail: awilczek@us.edu.pl

N. Kalantar-Nayestanaki · J. Messchendorp
KVI-CART, University of Groningen, 9747 AA Groningen, The Netherlands

St. Kistryn
Institute of Physics, Jagiellonian University, 30348 Krakow, Poland

A. Kozela
Institute of Nuclear Physics, PAS, 31342 Krakow, Poland

I. Skwira-Chalot
Faculty of Physics, University of Warsaw, 02093 Warsaw, Poland

N. A. Orr et al. (eds.), *Recent Progress in Few-Body Physics*,
Springer Proceedings in Physics 238,
https://doi.org/10.1007/978-3-030-32357-8_43

the beam axis (α), one distinguishes between the Space Star, for $\alpha = 90°$, the Forward Plane Star $\alpha = 0°$, and the Backward Plane Star $\alpha = 180°$.

The highest ever energies, for which an analysis of the star configurations in pd-breakup was done, are 19 MeV [2] and 65 MeV [3]. Therefore due to a poor coverage of the energy range above 20 MeV it was not possible to draw clear conclusions about the source of the effect. The comparison of the measurement and the calculations at 65 MeV show lack of the Space Star Anomaly.

The systematic studies of cross-section as a function of energy for various orientations of the star relatively to the beam direction are important for better understanding of the process dynamics. The Big Instrument for Nuclear-polarization Analysis (BINA) [4], currently at the Cyclotron Centre Bronowice (CCB) in Krakow, is one of the detectors well suited for such studies [5]. The research program of the experiment aims i.a. at providing some additional data on the star cross-sections.

43.2 The BINA Experimental Setup

The detector consists of a scattering chamber equipped with a target system enabling to use liquid hydrogen or deuterium. The detection system is divided into two parts: the Ball surrounding the target, and the Wall. The Ball is build out of 149 detectors (polyhedrons of triangular cross-section) made of two kinds of scintillating polymers in the phoswich configuration. The polar angles $\theta \in (40°,\ 165°)$ are within the acceptance of the detector.

The track reconstruction for forward angles is based upon the charge detection in a MultiWire Proportional Chamber, which makes use of 3 detection layers. Optionally, it is possible to place the ΔE Wall just next to the MWPC. The ΔE detector works in combination with the Wall of energy detectors as a particle identification system. The Wall covers polar angles $\theta \in (10°,\ 40°)$

Such a combination of detectors covers a solid angle of almost 4π, what makes the system outstandingly fitted to the measurements of the Space Star geometry.

The experiments performed with BINA at KVI, Groningen, and the ongoing measurements at CCB, Krakow, make possible to extend the experimental data on Space Star cross-section with the measurements taken at 50, 80, 108, 135, and 160 MeV/nucleon. New measurements will be carried out in 2019.

43.3 Results

The very first preliminary results are obtained for $p(d(160\,\text{MeV}), pp)n$ reaction. The analysis [6] relies on reconstruction of proton energies, polar angles, and the relative azimuthal angle of the proton pair. Since the configurational correction for the star configuration (e.g. for loss of coincidences due to double-hits) was not fully implemented, the cross-section was normalized to the theoretical results

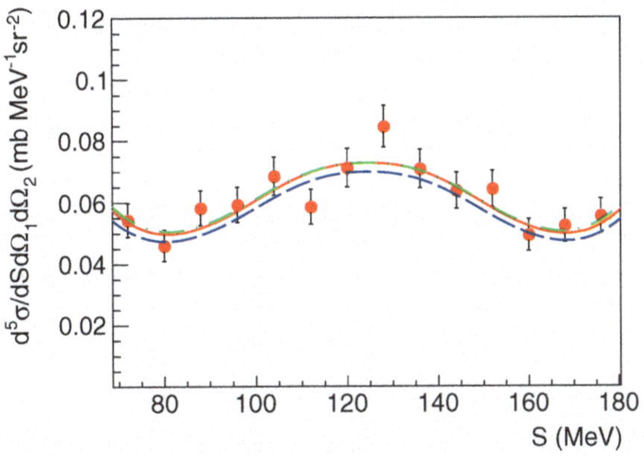

Fig. 43.1 Cross-section for $p(d(160\,\text{MeV}), pp)n$ breakup $\theta_1 = \theta_2 = 25.11°$, $\phi_{12} = 147.80°$ ($\alpha = 30°$) configuration, plotted against arc length measured anti-clockwise along the kinematics (S). The starting point ($S = 0$) is arbitrarily chosen as the point where energy of the second proton is minimal and starts to increase. The star configuration corresponds to the axis of symmetry of the plot. Theoretical predictions are plotted as lines (red solid: CD-Bonn+Δ+Coulomb, green chain: CD-Bonn+Δ, blue dashed: CD-Bonn+Coulomb) [7]. The normalization of the experimental data is arbitrary

(CD-Bonn+Δ+Coulomb) [7]. This makes possible to check the feasibility of providing new good-quality results on the star configurations.

The symmetry axis in the Fig. 43.1 corresponds to the star configuration at inclination angle $\alpha = 30°$. The statistics corresponds to about 1.5 days of continuous measurement, and the statistical error equals to 8%. Although the statistical errors do not allow distinguishing between the models based upon CD-Bonn, one has to take into account that the cross-section for the star configurations in deuteron on proton breakup increases with α, until it reaches the value corresponding to $\alpha = 180°$, which is about 3 times higher than for $\alpha = 30°$. On the other hand, the statistical error for $\alpha = 0°$ was still below 10%.

In order to reduce the statistical error for $\alpha = 30°$ to 5%, the next measurements with BINA should consist of about 4 days of continuous data-taking. One might consider 10 days of acquisition to reach the statistical error of about 2.5%.

Acknowledgements This project is supported by the National Science Centre, Poland (grant 2016/23/D/ST2/01703 and 2012/05/B/ST2/02556).

References

1. Strate, J., et al.: Differential cross section of the ^{2}H(n, nnp)-reaction at E_n = 13 MeV. Nucl. Phys. A **501**, 51 (1989)
2. Ley, J., et al.: Cross sections and tensor analyzing powers A_{yy} of the reaction $^1\text{H}(\vec{d}, pp)n$ in "symmetric constant relative energy" geometries at E_d = 19 MeV. Phys. Rev. C **73**, 064001 (2006)

3. Zejma, J., et al.: Cross sections and analyzing powers A_y in the breakup reaction $^2H(\mathbf{p}, pp)n$ at 65 MeV: star configurations. Phys. Rev. C **55**, 42 (1997)
4. Kistryn, St., Stephan, E.: Deuteron–proton breakup at medium energies, J. Phys. G **40**, 063101 (2013)
5. Wilczek, A.: Towards an explanation of the space star anomaly. Acta Phys. Pol. B **49**, 458 (2018)
6. Parol, W.: Investigation of three-nucleon force effects in the reaction of deuteron proton breakup at the energy of 80 MeV/nucleon (in Polish). Ph.D. thesis, Jagiellonian University, Krakow (2015)
7. Deltuva, A.: Momentum-space calculation of proton-deuteron scattering including Coulomb and irreducible three-nucleon forces. Phys. Rev. C **80**, 064002 (2009)

Chapter 44
New Ab Initio Approach to Nuclear Reactions Based on the Symmetry-Adapted No-Core Shell Model

Alexis Mercenne, Kristina D. Launey, Jutta E. Escher, Tomas Dytrych and Jerry P. Draayer

Abstract We present the current development of a new ab initio approach for nuclear reactions that takes advantage of SU(3) symmetry and its relevant dynamics combined with the resonating group method. In this model, the structure of the clusters is based on the ab initio symmetry-adapted no-core shell model, which enables the description of spatially enhanced nuclear configurations. We will present the formalism that involves the expression of the norm kernels in the SU(3) symmetry-adapted basis, in addition to first results for the p-α, p-^{16}O and p-^{20}Ne scattering reactions.

44.1 Introduction

The development of new experimental facilities has highlighted the need for new microscopic nuclear reaction models. In addition, recent progresses in ab initio nuclear theory using realistic, QCD inspired, interactions as well as many developments in high performance computing (HPC) have given the necessary tools to theoretical approaches such as the no-core shell model to provide an ab initio description of the structure of light nuclei [1, 2]. Its recent implementation within the resonating group method (RGM) [3] has allowed a microscopic study of nuclear reactions [4, 5], pursuing the long-lasting goal to unify the nuclear structure and reactions. However, even with the development of more advanced HPC techniques, the nuclear structure and reactions for certain mass region remain out of reach for ab initio approaches mainly due to the the size of the configuration space. Recently, the

A. Mercenne (✉) · K. D. Launey · T. Dytrych · J. P. Draayer
Department of Physics and Astronomy, Louisiana State University, Baton Rouge, LA 70803, USA
e-mail: amercenne1@lsu.edu

J. E. Escher
Lawrence Livermore National Laboratory, Livermore, CA 94550, USA

T. Dytrych
Nuclear Physics Institute, Academy of Sciences of the Czech Republic, 25068 Řež, Czech Republic

N. A. Orr et al. (eds.), *Recent Progress in Few-Body Physics*, Springer Proceedings in Physics 238,
https://doi.org/10.1007/978-3-030-32357-8_44

symmetry-adapted no-core shell model (SA-NCSM) [6, 7] which considers a more physically relevant basis, has proven its efficacy and has been successfully applied to the description of nuclear structure for nuclei up to medium mass. Motivated by the need for calculated nuclear cross sections in experimental research and astrophysics studies, and following the spirit of the NCSM/RGM, we combine the SA-NCSM with the RGM, with the view toward providing a complete description of structure and reactions for binary reactions in which the projectile is a nucleon.

44.2 Unified Ab Initio Approach for Medium-Mass Nuclei

In the RGM, the wave function is expanded within a cluster basis:

$$|\Psi^{J^\pi T}\rangle = \sum_\nu \int_r dr r^2 \frac{g_\nu^{J^\pi T}(r)}{r} \hat{\mathcal{A}} |\Phi_{\nu r}^{J^\pi T}\rangle \, , \tag{44.1}$$

where the index ν gathers all quantum numbers defining channels and partitions: $\nu = \{(A-a)\alpha_1 I_1 T_1;\, a\alpha_2 I_2 T_2;\, \ell s\}$, and the cluster states are defined as $|\Phi_{\nu r}^{J^\pi T}\rangle = \left[(|(A-a)\alpha_1 I_1 T_1\rangle \otimes |a\alpha_2 I_2 T_2\rangle)^{(sT)} \times Y_\ell(\hat{r}_{A-a,a})\right]^{(J^\pi T)} \frac{\delta(r-r_{A-a,a})}{r r_{A-a,a}}$. The wave functions $g_\nu^{J^\pi T}(r)$ in (44.1) are the quantities to be determined. They describe the relative motion between the target and the projectile for all channels ν, and the cross section can be extracted from their asymptotic behavior. The determination of $g_\nu^{J^\pi T}(r)$ is achieved by solving the Schrödinger equation:

$$\sum_\nu \int dr r^2 \left[H_{\nu'\nu}^{J^\pi T}(r, r') - E N_{\nu'\nu}^{J^\pi T}(r', r)\right] \frac{g_\nu^{J^\pi T}(r)}{r} = 0 \, . \tag{44.2}$$

Here, the Hamiltonian $H_{\nu'\nu}^{J^\pi T}(r', r)$ and norm $N_{\nu'\nu}^{J^\pi T}(r', r)$ kernels are expressed as: $\langle \Phi_{\nu' r'}^{J^\pi T} | \hat{\mathcal{A}} \hat{O} \hat{\mathcal{A}} | \Phi_{\nu r}^{J^\pi T}\rangle$, where $\hat{\mathcal{A}}$ is the antisymmetrizer ensuring the Pauli exclusion principle, and they are computed using the wave functions of the clusters. Once the kernels are computed within a given basis, (44.2) can then be solved using an R-matrix approach.

An ab initio application of this approach is the NCSM/RGM which uses ab initio NCSM wave functions generated using realistic interactions in order to compute the kernels. The NCSM/RGM has then been successfully applied to the description of several nuclear reactions involving light nuclei. However the method becomes numerically challenging for heavier systems due to the size and complexity of the configuration space. In addition, the inversion of the norm kernel as well as the treatment of the center-of-mass excitations become challenging tasks when the number of channels increases. In this context, the SA-NCSM combined with the RGM holds promise to obtain a unified ab initio description of structure and reaction for intermediate- up to medium-mass nuclei.

In the SA-NCSM, the microscopic many-body basis (Slater determinants) is based on the spherical harmonic oscillator single particle basis. In our case, we consider a basis made of the irreductible representations according to the group chain: $SU(3)_{(\lambda\mu)} \underset{\kappa}{\supset} SO(3)_L \supset SO(2)_{M_L}$. Consequently, for any given total spin and its projection JM, the wave function of a nucleus will be described within a basis $\{|\alpha_i(\lambda_i\mu_i)\kappa_i(L_iS_i)J_iM_i\rangle\}$ with each component weighted by a coefficient C_i, and where α_i gathers additional quantum numbers needed to enumerate the complete shell model space.

In the symmetry-adapted RGM (SA-RGM), the channels are defined by coupling each component f the SA-NCSM wave functions between the projectile and the target. Consequently, the channels with good SU(3), spin and isospin quantum numbers are given in the case of one nucleon projectile as:

$$|\Phi^{\rho(\lambda\mu)\kappa(LS)JMTM_T}_{\gamma n}\rangle = \left\{|\alpha_1(\lambda_1\mu_1)S_1T_1\rangle \otimes |(n0)\frac{1}{2}\frac{1}{2}\rangle\right\}^{\rho(\lambda\mu)\kappa(LS)JMTM_T}, \tag{44.3}$$

where the index: $\gamma \equiv \{(A-a)\alpha_1(\lambda_1\mu_1)S_1T_1; a\frac{1}{2}\frac{1}{2}\}$ label our channel basis, $(n0)$ represents the SU(3) relative motion of the projectile, and $(\frac{1}{2}\frac{1}{2})$ its spin and isospin respectively. In this basis, the exchange matrix, which ensures the antisymmetrization in the kernels, has the following form (in conventional notations [8]):

$$\begin{aligned}
&\langle\Phi^{\rho'(\lambda'\mu')\kappa'(L'S')JMTM_T}_{\gamma'n'}\rangle\hat{P}_{A,A-1}|\Phi^{\rho(\lambda\mu)\kappa(LS)JMTM_T}_{\gamma n}\rangle \\
&= \frac{1}{A-1}\delta_{\rho\rho'}\delta_{(\lambda\mu)(\lambda'\mu')}\delta_{\kappa\kappa'}\delta_{LL'}\delta_{SS'} \sum_{\substack{\tau\rho_o(\lambda_o\mu_o)\\ S_o\bar{\rho}}} \Pi_{\tau S_oS_1'T_1'}(-1)^{n+n'-(\lambda_o+\mu_o)} \\
&\times (-1)^{T_1+\frac{1}{2}+T'}(-1)^{S_1+\frac{1}{2}+S'} \begin{Bmatrix} S_1 & S_o & S_1' \\ \frac{1}{2} & S & \frac{1}{2} \end{Bmatrix} \begin{Bmatrix} T_1 & \tau & T_1' \\ \frac{1}{2} & T & \frac{1}{2} \end{Bmatrix} \\
&\times \sqrt{\frac{\dim(\lambda_o\mu_o)}{\dim(n0)}}\,\mathrm{U}\left[(\lambda_1\mu_1)(\lambda_o\mu_o)(\lambda'\mu')(n'0);\,(\lambda_1'\mu_1')\bar{\rho}\rho'(n0)\rho_o\rho''\right] \\
&\times \langle\alpha_1'(\lambda_1'\mu_1')S_1'T_1'\rangle|| \left\{a^\dagger_{(n0)\frac{1}{2}\frac{1}{2}} \otimes \tilde{a}_{(\tilde{0n'})\frac{1}{2}\frac{1}{2}}\right\}^{\rho_o(\lambda_o\mu_o)S_o\tau} |||\alpha_1(\lambda_1\mu_1)S_1T_1\rangle_{\bar{\rho}}.
\end{aligned} \tag{44.4}$$

An important advantage here is that the exchange matrix is diagonal within this SU(3) basis, and this allows one to overcome numerical inversion of the norm. So with such an approach, the dependence on angular momentum is deferred to the very last step in the calculations, and in turn, facilitates quick calculations [9]. Then, only at the end we transform back to the partial waves expansion, i.e $|\Phi^{\rho(\lambda\mu)K(LS)JMTM_T}_{\gamma n}\rangle \rightarrow |\Phi^{J^\pi T}_{\nu n}\rangle$ using the coefficients C_i, and calculate the norm $N^{J^\pi T}_{\nu'\nu}(r', r)$ using the conventional formula [4].

In order to demonstrate the efficacy of the approach, we present a benchmark calculation for p-^{4}He. We compare the exchange part of the norm (see in [4]) using the two NCSM/RGM and SA-RGM approaches Fig. 44.1a. The SA-RGM result has

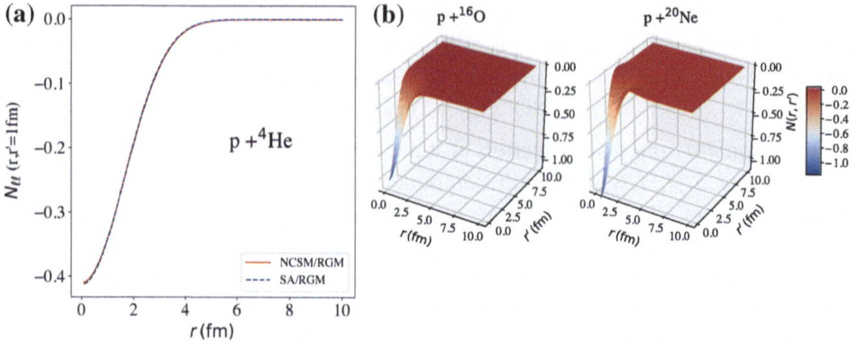

Fig. 44.1 **a** Exchange part of the norm kernel for p-^{4}He. The target wave function is calculated in a $N_{\max} = 4$ model space, and is truncated in the SA-RGM calculation by selecting only components greater than 1%. The calculation of ^{4}He was performed using the chiral N2LO$_{\text{opt}}$ NN interaction. **b** Exchange part of the norm calculated for two heavier system using SA-RGM. Target wave function has been generated using the chiral N2LO$_{\text{sat}}$ NN in 10 shells ($\hbar\Omega = 16$ MeV) for ^{16}O and the chiral N2LO$_{\text{opt}}$ NN in 13 shells ($\hbar\Omega = 15$ MeV) for ^{20}Ne, with selected SU(3) configurations that have a contribution greater than 2%. In all calculations, the spurious center-of-mass motion has been removed from the ab initio wave functions, but not from the cluster system

been obtained using a ^{4}He wave function truncated to only several SU(3) shapes. To illustrate the potential of applying this approach to heavier systems, we present calculations of this exchange part for heavier system in Fig. 44.1b, for p-^{16}O and p-^{20}Ne.

Hence those results show that the use of a physically relevant basis through the SA-RGM is a promising approach, where truncated target wave functions can be implemented to reach heavier system for nuclear reactions. So far, the spurious center-of-mass motion between the clusters has not been removed, but it is expected to be negligible for reactions involving one nucleon plus an $A > 16$ target. This work presents the method to implement the RGM within an SU(3) basis, more specifically how to take advantage of the SU(3) basis to calculate the norm kernel. The same procedure will be used for the Hamiltonian kernel.

Acknowledgements This work was supported by the U.S. National Science Foundation (OIA-1738287, ACI -1713690), the Czech Science Foundation (16-16772S) and under the auspices of the U.S. Department of Energy by Lawrence Livermore National Laboratory under Contract DE-AC52- 07NA27344, with support from LDRD project 19-ERD-017. In addition, this work benefitted from computing resources provided by LSU (www.hpc.lsu.edu) and Blue Waters.

References

1. Navrátil, P., Vary, J.P., Barrett, B.R.: Phys. Rev. Lett. **84**, 5728 (2000)
2. Barrett, B.R., Navrátil, P., Vary, J.P.: Prog. Part. Nucl. Phys. **69**, 131 (2013)
3. Tang, Y.C., LeMere, M., Thompson, D.R.: Phys. Rep. **47**, 167 (1978)

4. Quaglioni, S., Navrátil, P.: Phys. Rev. C **79**, 044606 (2009)
5. Baroni, S., Navrátil, P., Quaglioni, S.: Phys. Rev. Lett. **110**, 022505 (2013)
6. Dytrych, T., et al.: Phys. Rev. Lett. **111**, 252501 (2013)
7. Launey, K.D., Dytrych, T., Draayer, J.P.: Prog. Part. Nucl. Phys. **89**, 101 (2016)
8. Draayer, J.P., Akiyama, Y.: J. Math. Phys. **14**, 12 (1973)
9. Hecht, K.T.: Nucl. Phys. A **283**, 223 (1977)

Chapter 45
Cluster Structure of the Ground and Excited States of ^{9}Be and ^{10}B Nuclei

M. A. Zhusupov, K. A. Zhaksybekova, R. S. Kabatayeva and A. S. Kopenbayeva

Abstract Reactions of quasielastic (p, px), (e, ex) knockout and reactions of (γ, x) photodisintegration show that particles x—deuterons, tritons, α-particles and nucleons—escape with a comparable probability out of light nuclei. The significant values of spectroscopic S-factors in these channels are evidence of this. This situation is well interpreted by the wave functions of multiparticle shell model which allows considering from the unified positions both cluster and nucleonic degrees of freedom. The situation in ^{9}Be and ^{10}B nuclei is of peculiar interest. In these nuclei from analysis of symmetry of orbital part of wave functions of states at high energies it was suggested for instance in ^{9}Be nucleus to search three-cluster αtd-states in reactions with ^{6,7}Li ions on the same target nuclei. Our calculations showed that in excitation spectra of ^{9}Be nucleus there are three-cluster αtd-states situating in range of both 17–19 and 10–11 MeV. These states have large S-factors in triton, deuteron (and α-particle) channels simultaneously, moreover the last ones appear exactly due to [441] states. Similar conclusions were also obtained for ^{10}B nucleus where three-particle ατt-levels are connected not only to states with Young [433] scheme, but to [442].

45.1 Multicluster Structure of ^{9}Be Nucleus

A calculation of spectroscopic factors includes in wave functions of ^{9}Be levels the configurations with Young [441] and [432] schemes [1]. In transfer reactions the cross section σ of excitation of levels of final nucleus can be presented by an expression (in assumption of direct mechanism) $\sigma \sim (2J + 1)\Sigma S_L \cdot \Phi$, here S_L—respective spectroscopic factors and Φ—a factor depending on kinematic characteristics. If one considers the quantity Φ to be more or less smooth quantity depending on

M. A. Zhusupov · K. A. Zhaksybekova · R. S. Kabatayeva (✉) · A. S. Kopenbayeva
IETP, al-Farabi Kazakh National University, Almaty, Kazakhstan
e-mail: raushan.kabatayeva@gmail.com

R. S. Kabatayeva
International Information Technology University, Almaty, Kazakhstan

N. A. Orr et al. (eds.), *Recent Progress in Few-Body Physics*,
Springer Proceedings in Physics 238,
https://doi.org/10.1007/978-3-030-32357-8_45

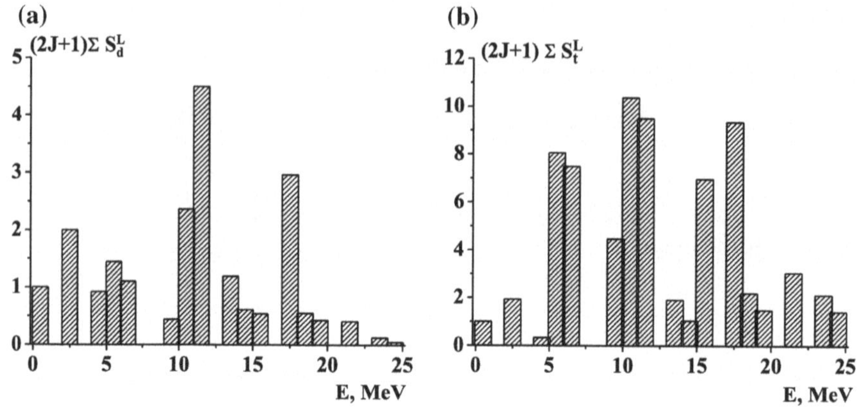

Fig. 45.1 Spectra of excitation of ^{9}Be nucleus in reactions: **a**–^{7}Li(^{6}Li, α)^{9}Be, **b** - ^{6}Li(^{7}Li, α)^{9}Be; adapted from [6]

energy, then the maxima observable in reactions should be connected with maxima of spectroscopic factors.

In Fig. 45.1 there are spectroscopic factors in form of histograms including the sums of values of S-factors over energy region of 1 MeV. A comparison with experimental data [2] shows, that the theory describes well the main maxima at energies E = 11.8, 15.2, 17.8 and 22 MeV. There are faint peaks at E = 0 and 3 MeV due to Young [441] scheme as well.

Thus three-cluster states having αtd-nature can respond to not only orbital Young [432] scheme but with no less weight to Young [441] scheme. That is why there is no surprise in successful description of photonuclear (γ, d) and (γ, t) processes on ^{9}Be nucleus achieved by the authors in ααn-model [3].

Our calculations showed that in spectrum of excitation of ^{9}Be nucleus there are three-cluster αtd-states in region both 17–19 and 10–11 MeV. These states have large S-factors in triton and deuteron channels simultaneously, and the last ones appear due to states [441] exactly. In [4] authors analyzed wave functions and suggested searching the three-cluster αtd-states in reactions with ^{6,7}Li ions on the same target nuclei.

45.2 Cluster Structure of ^{10}B Nucleus

The wave function of ^{10}B nucleus contains in the ground state two components with Young [442] and [433] schemes with contribution of the second one of slightly more than 3% [1]. Because of the smallness of binding energy of nuclei ^{7}Li in (α + t)-channel and ^{6}Li in (α + d)-channel the dominating mechanisms in both ^{6}Li(^{6}Li, d)^{10}B and ^{7}Li(^{7}Li, α)^{10}B cases are the transfer of alpha-particle and triton clusters respectively.

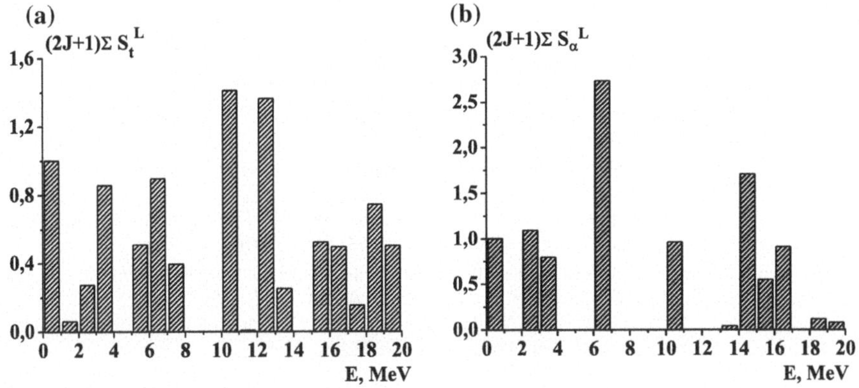

Fig. 45.2 Spectra of excitation of ^{10}B nucleus in reactions: **a**–$^{7}Be + t \rightarrow {}^{10}B^{*}$, **b**–$^{6}Li + \alpha \rightarrow {}^{10}B^{*}$

In Fig. 45.2 there are values of summarized spectroscopic factors. A comparison with experimental data shows that in whole the theory describes the main maxima observed at energies E = 7, 11 and 13 MeV for joining tritons and at energies E = 7, 11 and 16 MeV for alpha-particles.

45.3 Conclusion

With aim of study of cluster structure of ^{9}Be and ^{10}B nuclei the reactions of lithium isotopes interaction with each other were considered, such reactions lead to the ground and excited states of ^{9}Be and ^{10}B nuclei. The fact that the main mechanism in reaction with lithium ions is the mechanism of transfer of weakly bound deuteron, triton and alpha-particle was used. It is turned out that the energy dependence of excitation spectrum is well described by the summarized spectroscopic factors.

For calculations we used the wave functions of multiparticle shell model [1]. For ^{9}Be nucleus the ground state has a strongly pronounced ααn-structure [5]. And αdt-configuration in the ground state practically has no contribution. That is why some calculations for ^{9}Be nucleus on the base of αdt-model are simply mistaken. This is confirmed by the fact that in αdt-model the main transitions with neutron escape out of ^{9}Be nucleus with formation of ^{8}Be nucleus in the ground and the first excited states (with Young [44] scheme) are forbidden, and also the observable alpha-particle escape with formation of ^{5}He nucleus in the ground and the first excited states with Young [41] scheme. For ^{10}B nucleus the analogous calculations give the dominating ααd-structure.

Acknowledgements The work was supported by the Grant No. AP05132952.

References

1. Boyarkina, A.: Structure of 1p-shell Nuclei. MSU, Moscow (1973)
2. Glukhov, YuA, et al.: Study of Li + Li reactions and possible existence of three-cluster states. Phys. At. Nucl. **13**, 277 (1971)
3. Burkova, N.A., et al.: Potential theory of cluster photodisintegration of light nuclei. Phys. Part. Nucl. **36**, 427 (2005)
4. Neudatchin, V., et al.: Supermultiplet symmetry and near-threshold levels in systems featuring two or three extremely light clusters. Phys. At. Nucl. **63**, 195 (2000)
5. Voronchev, V.T., et al.: Analysis of the structure and properties of nuclei with A = 9 (Be-9 – B-9) in the dynamic multicluster 2-alpha + n model. Phys. At. Nucl. **57**, 1890 (1994)
6. Zhusupov, M.A., Kabatayeva, R.S.: Multicluster structure of the ground and excited states of ^{9}Be nucleus. Bull. Russ. Acad. Sci.: Phys. **76**, 429 (2012)

Chapter 46
Characteristics of ^{6}Li Nucleus Cluster Photodisintegration Reactions

M. A. Zhusupov, K. A. Zhaksybekova and R. S. Kabatayeva

Abstract On the base of potential theory the characteristics of Lithium nuclei cluster photodisintegration reactions are considered in the range of very low and intermediate energies. At low energies the important role of E1-multipole and its interference with E2-multipole were considered. The essential point is the different character of interference of E1- and E2-amplitude for the direct and inverse reactions. If for the direct reaction the interference at scattering in forward hemisphere has a constructive character, then in backward hemisphere the interference of E1- and E2-amplitudes is destructive. For the inverse reaction the interference has an opposite character: in forward hemisphere it is destructive, and in backward hemisphere it is constructive. In the energy range above several MeV the E2-multipole becomes dominating.

46.1 Introduction

The processes of two-particle photodisintegration of light self-conjugated ($N = Z$) nuclei with formation of particles with zero isotopic spin like $^4He(\gamma, d)d$ and $^6Li(\gamma, d)\alpha$ are of peculiar interest for the theory of photonuclear reactions. The cross sections of the reactions are unusually small because according to selection rules by isotopic spin the E1-transitions in case $\Delta T = 0$ are strongly suppressed and the E2-multipoles begin to play the determining role.

The reaction $\alpha d \rightarrow {}^6Li\gamma$ represents peculiar interest as a unique source of formation of ^6Li nuclei in the Big Bang [1]. Its study is important for thermonuclear applications as well. Despite the fact that the E1-transitions in the reaction $\alpha d \rightarrow$

M. A. Zhusupov · K. A. Zhaksybekova (✉) · R. S. Kabatayeva
IETP, al-Farabi Kazakh National University, Almaty, Kazakhstan
e-mail: zhaksybekova_kulyan_a@mail.ru

R. S. Kabatayeva
e-mail: raushan.kabatayeva@gmail.com

R. S. Kabatayeva
International Information Technology University, Almaty, Kazakhstan

N. A. Orr et al. (eds.), *Recent Progress in Few-Body Physics*,
Springer Proceedings in Physics 238,
https://doi.org/10.1007/978-3-030-32357-8_46

$^{6}Li\gamma$ are strongly suppressed due to selection rules by isospin, a violation of symmetry of angular distribution of γ-quanta with respect to angle $\theta = 90^0$, characteristic for the case of "pure" E2-transitions, shows the noticeable interference of these multipoles at low energies [2]. That is why a natural question about reasons leading to appearance of E1-multipole appears.

46.2 E1-Transitions at Low Energies

We think that in this case the appearance of E1-multipole is due to a clearly pronounced αd-structure of ^{6}Li nucleus, as a consequence of which the center of the charge does not coincide with the center of mass of the system. For a nucleus consisting of subsystems a and b the dipole operator can be presented in form of a sum dipole operators acting in each of the subsystems and $\vec{d}_\rho = e\vec{\rho} \cdot m_a m_b/(m_a + m_b) \cdot (Z_a/m_a - Z_b/m_b)$, here $\vec{\rho}$ – is a coordinate of relative motion of clusters a and b. Applying this formula for calculation of the reaction $\alpha + d \rightarrow {}^{6}Li + \gamma$ one finds that in αd-system because of the fact that $m_\alpha - 2m_d \neq 0$ the dipole moment $d_\rho = 4.3 \cdot 10^{-4} \cdot e\rho$ appears and it gives the appearance of E1-multipole [3]. The wave function of the ground state of ^{6}Li nucleus was chosen in αnp-model [4]. When constructing the wave function of αd-scattering a deep potential with forbidden states was used [3].

One can observe the E1-transition in angular distributions of processes $^{6}Li\gamma \leftrightarrow \alpha d$ in the interference with E2-multipole. In Fig. 46.1 there are our calculations in cluster model. As it is seen in Fig. 46.1b the theoretical calculation agrees qualitatively with experimental data [2]. Note the different character of interference of E1- and E2-amplitudes for direct $^{6}Li\gamma \rightarrow \alpha d$ and inverse $\alpha d \rightarrow {}^{6}Li\gamma$ reactions. If for the direct reaction the interference at scattering in the forward hemisphere (until $\pi/2$) has a constructive character, then in the backward hemisphere the interference of E1- and

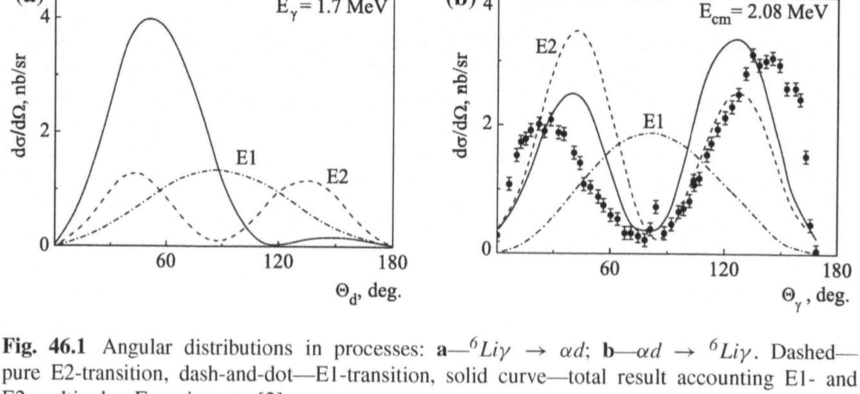

Fig. 46.1 Angular distributions in processes: **a**—$^{6}Li\gamma \rightarrow \alpha d$; **b**—$\alpha d \rightarrow {}^{6}Li\gamma$. Dashed—pure E2-transition, dash-and-dot—E1-transition, solid curve—total result accounting E1- and E2-multipoles. Experiment—[2]

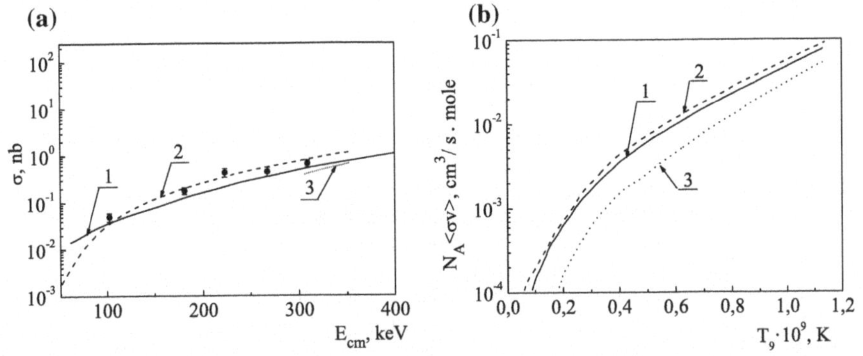

Fig. 46.2 Total cross sections (a) and astrophysical rates (b) of $\alpha d \rightarrow {}^6Li\gamma$ process. Experiment - [8]. Theory: **a** 1—calculation [6]; 2—[7]; 3—[8]; **b** 1—calculation [6]; 2 – [9]; 3—[8]

E2-amplitudes is destructive. For the inverse reaction the interference has opposite character. Concerning the reaction $\alpha d \rightarrow {}^6Li\gamma$ the region lower than 700 keV is still important. In this region only data obtained as a result of coulomb dissociation of lithium nuclei in the field of a heavy nucleus (*Pb*) is available that is using the method of virtual photons [5].

To answer the question about the role of E1-multipole it is necessary along with the angular distributions of deuterons to measure the total cross sections. Calculations in works [6, 7] reproduce the experimental data on total cross sections and astrophysical factor approximately in the same way (Fig. 46.2). The result in work [6] is a theoretical prediction since it had been obtained before the authors got acquainted with the experimental work [5].

46.3 Conclusion

The cluster E1-transition appears due to large difference in masses $2m_d - m_\alpha$, that is because of large binding energy of α-particle in *dd*-channel which is equal to 24.5 MeV. Because of the difference in penetrability of potential barrier the E1-multipole appears in astrophysical region where the interference effects of E1 and E2 multipoles in the angular distributions of particles are the strongest. In such a case the character of interference of E1- and E2-multipoles in the direct and inverse photonuclear reactions is different.

The work was financially supported by the Grant No. AP05132952.

References

1. Schramm, D.N., Wagoner, R.V.: Element production in early universe. Ann. Rev. Nucl. Sci. **27**, 37 (1997)
2. Robertson, R.G.H., et al.: Observation of the capture reaction H-2(Alpha-Gamma)Li-6 and its role in production of Li-6 in the Big-Bang. Phys. Rev. Lett. **47**, 1867 (1981)
3. Burkova, N.A., et al.: Is it possible to observe an isoscalar E1-multipole in Li-6-Gamma-Reversible-Alpha-D reactions. Phys. Lett. B **248**, 15 (1990)
4. Kukulin, V.I. et al.: Detailed study of the cluster structure of light-nuclei in a 3-body model. 1. Ground state of Li-6. Nucl. Phys. A **417**, 128 (1984)
5. Kiener, J., et al.: Measurements of the Coulomb dissociation cross-section of 156 MeV Li-6 projectiles at extremely low relative fragment energies of astrophysical interest. Phys. Rev. C **44**, 2195 (1991)
6. Zhusupov, M.A. et al.: One particle spectroscopic characteristics of Li-6 and Be-9 in the three particle models. Izv. Akad. Nauk Kaz. SSR. Ser. Fiz. **62**, 985 (1998)
7. Igamov, S.B., Yarmukhamedov, R.K.: Analysis of the nuclear astrophysical reaction Alpha-D—Li-6 + Gamma. Phys. At. Nucl. **58**, 1317 (1995)
8. Langanke, K.H.: Microscopic potential model studies of light nuclear capture reactions. Nucl. Phys. A **457**, 351 (1986)
9. Caughlan, G.R., Fowler, W.A.: Thermonuclear reaction rates. At. Data Nucl. Data Tables **40**, 283 (1988)

Chapter 47
Three-Body Approach to Deuteron-Alpha Scattering Using Realistic Forces in a Separable or Non-separable Representation

L. Hlophe, Jin Lei, Ch. Elster, A. Nogga and F. M. Nunes

Abstract Starting from an effective three-body Hamiltonian consisting of realistic two-body potentials, we solve the Faddeev-AGS equations for $d + \alpha$ scattering using the separable expansion method. First, we construct separable representations of the two-body interactions and solve one-dimensional integral equations to obtain angular distributions for elastic scattering. We find that the latter converge rapidly with the number of separable terms. Moreover, we observe that these cross sections are in excellent agreement with those calculated by directly solving the Faddeev-AGS equations using the original interactions. This result demonstrates that the separable expansion method can be used to obtain accurate solutions of the Faddeev-AGS equations.

47.1 Introduction

Deuteron-induced reactions are a powerful tool for probing nuclear structure as well as extracting neutron capture rates needed to determine r-process abundances. From a theoretical perspective, (d, p) reactions can be viewed as a three-body problem consisting of the neutron (n), proton (p), and nucleus (A). Exact solutions of the three-body scattering problem are given by the Faddeev equations. Since this work is carried out in momentum space, we adopt the Alt-Grassberger-Sandhas (AGS) [1]

L. Hlophe (✉) · F. M. Nunes
National Superconducting Cyclotron Laboratory and Department of Physics and Astronomy, Michigan State University, East Lansing, MI 48824, USA
e-mail: hlophe@nscl.msu.edu

J. Lei · Ch. Elster
Institute of Nuclear and Particle Physics, and Department of Physics and Astronomy, Ohio University, Athens, OH 45701, USA

A. Nogga
IAS-4, IKP-3, JHCP, and JARA-HPC, Forschungszentrum Jülich, 52428 Jülich, FRG, Germany

N. A. Orr et al. (eds.), *Recent Progress in Few-Body Physics*, Springer Proceedings in Physics 238,
https://doi.org/10.1007/978-3-030-32357-8_47

formulation of the Faddeev equations. Obtaining numerical solutions of these equations is complicated by the presence of the Coulomb potential, particularly for heavy systems and low beam energies. Those difficulties can be addressed by recasting the Faddeev-AGS equations in a Coulomb basis using separable effective two-body potentials [2]. In this work we demonstrate that using the Faddeev-AGS equations with separable two-body interactions based on the Ernst-Shakin-Thaler (EST) [3] scheme provides solutions for $d + \alpha$ scattering identical to those in which no separable representation is employed. We point out that an identical study [4] validated the method for the three-body bound state equations.

47.2 Faddeev-AGS Equations

The effective three-body Hamiltonian for the $n + p + \alpha$ system has the form

$$H_{3B} = H_0 + V_{np} + V_{p\alpha} + V_{n\alpha}, \tag{47.1}$$

where V_{np}, $V_{p\alpha}$, and $V_{n\alpha}$ represent the two-body subsystem interactions. To describe deuteron-alpha scattering we adopt the AGS formulation of the Faddeev equations

$$U^{ij}(z) = \bar{\delta}_{ij}\; G_0^{-1}(z) + \sum_{k=1}^{3} \bar{\delta}_{ik}\; t_k(z)\; G_0(z) U_{kj}(z), \tag{47.2}$$

where $\bar{\delta}_{ij} = 1 - \delta_{ij}$, $z = E_3 + i\varepsilon$, and E_3 the available three-body energy in the center-of-mass (c.m.) frame. Here $t_i(z)$ is the effective two-body t-matrix for the ith pair and $G_0(z)$ the free propagator. The AGS transition amplitudes $U^{ij}(z)$ describe all possible three-body processes which include elastic scattering, transfer reactions, and deuteron breakup. The momentum space representation of (47.2) results in a set of coupled integral equations that depend on the Jacobi momenta p_i and q_i. The former corresponds to the momentum of the pair, while the latter describes the momentum of the third particle relative to the pair. A great simplification occurs if the two-body t-matrix has a separable form

$$t_i^{\alpha_i\alpha_i'}(p_i, p_i; E_{q_i}) = \sum_{m,n=1}^{rank} h_m^{\alpha_i}(p_i)\; \tau_{mn}^{\alpha_i\alpha_i'}(z, q_k)\; h_n^{\alpha_i'}(p_i'), \tag{47.3}$$

where $m(n)$ is the rank index and $\alpha_i(\alpha_i')$ represents an angular momentum channel. The separability of the t-matrix allows to integrate out the pair momentum p_i leading to effective two-body transition amplitudes $X_{mn}^{ij}(z)$ which fulfill

$$X_{m\alpha_i,n\alpha_j}^{ij}(q_i, q_j; z) = Z_{m\alpha_i,n\alpha_j}^{ij}(q_i, q_j; z) + \sum_{km'n'} \sum_{\alpha_k\alpha_k'} \int dq_k\; q_k^2\; Z_{m\alpha_i,m'\alpha_k}^{ik}(q_i, q_k; z)$$

$$\times\ \tau^{\alpha_k,\alpha'_k}_{m'n'}(z,q_k)\, X^{kj}_{n'\alpha'_k,n\alpha_j}(q_k,q_j;z). \tag{47.4}$$

The three-body dynamics are contained in the so-called effective two-body transition potentials $Z^{(ij)}_{n\alpha'_i,\nu\alpha'_j}(q_i,q_j;z)$.

47.3 Results and Discussion

To evaluate $d+\alpha$ scattering observables, we need to specify the effective three-body Hamiltonian H_{3B}. For this purpose, the CD-Bonn [5] high precision potential is adopted in the np subsystem and the Bang [6] interaction in the $n/p-\alpha$ subsystem. The latter potential is fitted to low-energy phase shifts for $n/p-\alpha$ elastic scattering. The two-body model space is restricted to $l_i \le 2$ and $J_i \le 2$, with l_i and J_i being the orbital and total angular momenta for each subsystem respectively. The Coulomb potential is omitted in all calculations. Reference [7] showed that EST separable expansion of effective two-body nucleon-nucleus interactions works equally well in both the plane wave and Coulomb bases. We thus expect that the observations made in this study will hold when solving the Faddeev-AGS equations in the presence of the Coulomb potential.

The idea of the EST separable representation scheme is to perform an expansion of the two-body t-matrix (or equivalently the potential) using solutions of the Schrödinger equation as basis functions. Specifically, we proceed by fixing the asymptotic momentum p_{in}, and the two-body energy $E_{q_{in}}$, so that the form factors are given by $h^{\alpha'_i}_n(p'_i) \equiv t^{\alpha'_i,\alpha_{in}}_i(p'_i, p_{in}; E_{q_{in}})$, where α_{in} indicates the channels that are included in the expansion. The fixed momenta and energies constitute the so-called EST support points (ESPs). The index n represents the rank of the separable potential. The EST constraint ensures that at the ESPs the separable t-matrix is identical to the original one. Contrary to previous works which imposed the artificial constraint $p_{in} = \sqrt{2\mu_i|E_{q_{in}}|}$, we choose the values of p_{in} and $E_{q_{in}}$ independently as described in [4]. In that work, it was shown that this choice of ESPs improves the convergence of the separable expansion for negative energy support points. For positive energy support points, it is sufficient to choose the fixed momenta to be on-shell, i.e., $p_{in} = \sqrt{2\mu_i E_{q_{in}}}$.

To proceed, we construct rank-8 separable potentials for each two-body subsystem. The latter are specified by giving the fixed energies in MeV and momenta in fm^{-1}. For example, a support point with energy -60 MeV and momentum 1.0 fm^{-1} is represented as $(-60, 1.0)$. First, we define the separable potential EST8-1= $\{(-15, 0.2), (-15, 1.0), (-15, 1.2), (-15, 3.0), (-2, 0.2), (-2, 1.0), (-2, 1.2), (-2, 3.0)\}$, for the np system and EST8-1= $\{(-60, 0.2), (-60, 1.0), (-60, 1.2), (-60, 3.0), (-3, 1.0), (-3, 1.2), (-3.0), (-3, 3.0)\}$, for the $n/p-\alpha$ system. Next, we solve (47.4) for incident deuteron energies below ($E_d = 3$ MeV) and above ($E_d = 10,\ 20$ MeV) the deuteron breakup threshold using the aforementioned separable interactions. Figure 47.1 shows the differential cross sections for $d+\alpha$ elastic scattering as a

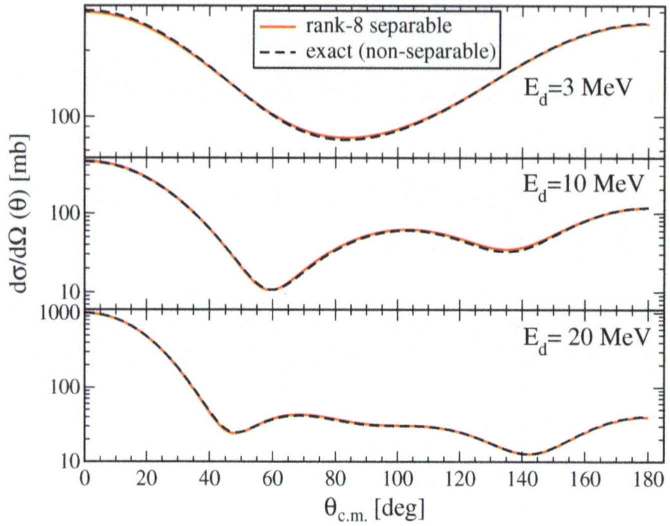

Fig. 47.1 The differential cross section for $d + \alpha$ elastic scattering as a function of the center-of-mass angle $\theta_{c.m.}$. The incident deuteron energies (E_d) are displayed in the figure. The solid curve shows cross sections computed using rank-8 separable potentials for both the np and $n/p - \alpha$ subsystems. The ESPs are given in the text. The exact results taken from [8] are indicated by the dashed curve

function of the center-of-mass angle $\theta_{c.m.}$. The solid curve indicates the angular distributions evaluated using the rank-8 separable potentials. Varying the values of the fixed energies by several MeV yields results that are indistinguishable in the plot. This provides an estimate of the precision of the separable expansion. The dashed curve shows results taken from [8] calculated by directly solving (47.2). This demonstrates that the separable expansion method works very well in obtaining solutions for the three-body scattering equations.

47.4 Summary and Outlook

Angular distributions for elastic $d + \alpha$ scattering at incident deuteron energies below ($E_d = 3$ MeV) and above ($E_d = 10,\ 20$ MeV) the deuteron breakup threshold have been computed using the separable expansion method. We find that the cross sections converge rapidly with the rank of the separable potential. Furthermore, the converged results are in excellent agreement with those exact obtained directly from the Faddeev-AGS equations without a separable expansion. In light of this successful benchmark, the method has been employed to study universal behavior in the $d + \alpha$ system without including the Coulomb interaction [9]. An effort to incorporate the Coulomb potential is underway.

Acknowledgements This work was in part supported by the U.S. NSF under contracts NSF-PHY-1520972, NSF-PHY-1520929, the U.S. DoE under contracts DE-FG02-93ER40756, DE-AC02-05CH11231, NSFC under Grant No. 11621131001, and DFG under Grant No. TRR110.

References

1. Alt, E.O., Grassberger, P., Sandhas, W.: Nucl. Phys. B **2**, 167 (1967)
2. Mukhamedzhanov, A.M., et al.: Phys. Rev. C **86**, 034001 (2012)
3. Ernst, D.J., Shakin, C.M., Thaler, R.M.: Phys. Rev. C **8**, 46 (1973)
4. Hlophe, L., Lei Jin, et al.: Phys. Rev. C **96**, 064003 (2017)
5. Machleidt, R.: Phys. Rev. C **63**, 024001 (2001)
6. Bang, J., et al.: Nucl. Phys. A **405**, 126 (1983)
7. Hlophe, L., et al.: Phys. Rev. C **90**, 061602 (2014)
8. Deltuva, A.: Private Communication
9. Lei Jin, Hlophe, L., et al.: arXiv:1809.06351 [nucl-th]

Chapter 48
A Fresh Look at Treatment of Radiative Capture in Nuclear Reactions: Applications to the $\alpha - \alpha$ Bremsstrahlung

Adam Arslanaliev and Aleksandr Shebeko

Abstract This contribution demonstrates how the gauge invariance (Eicheninvarianz) principle can be realized in the theory of electromagnetic interactions with nuclei.

48.1 Theoretical Background

Our departure point in describing electromagnetic (EM) interactions with nuclei (in general, bound systems of charged particles) is to use the Fock-Weyl criterion and a generalization of the Siegert theorem (see [1, 2], where this approach is compared with that by Friar and Fallieros [3]). It has been shown how one can meet the gauge invariance principle (GIP) in all orders in the charge e and construct the corresponding EM interaction operators in case of nuclear forces (in general, interparticle ones) arbitrarily dependent on velocity (see paper [4] and refs. therein). Along the guideline we have derived the conserved current density operator for a dicluster system (more precisely, the system of two finite-size clusters with many-body interaction effects included). To be more definite, for the photon emission with energy E_γ, momentum $\mathbf{k}_\gamma$ and polarization ε the reaction amplitude

$$T_{if} = \left[2(2\pi)^3 E_\gamma\right]^{-1/2} \left\langle \mathbf{P}_i - \mathbf{k}; f \left| \varepsilon^\mu \hat{J}_\mu(0) \right| \mathbf{P}_i; i \right\rangle$$

can be expressed [4] through electric $(\mathbf{E}(\mathbf{k}))$ and magnetic $(\mathbf{H}(\mathbf{k}))$ field strengths, these manifestly GI quantities, and matrix elements $\mathbf{D}_{if}(\mathbf{k}_\gamma)$ and $\mathbf{M}_{if}(\mathbf{k}_\gamma)$ of the so-called generalized electric and magnetic dipole moments of nucleus (system), viz.,

A. Arslanaliev
V.N. Karazin National University, Kharkiv, Ukraine

A. Shebeko (✉)
National Science Center "Kharkov Institute of Physics and Technology", Kharkiv, Ukraine
e-mail: shebeko@kipt.kharkov.ua

N. A. Orr et al. (eds.), *Recent Progress in Few-Body Physics*, Springer Proceedings in Physics 238,
https://doi.org/10.1007/978-3-030-32357-8_48

$$T_{if} = \mathbf{E}(\mathbf{k}_\gamma) \cdot \mathbf{D}_{if}(\mathbf{k}_\gamma) + \mathbf{H}(\mathbf{k}_\gamma) \cdot \mathbf{M}_{if}(\mathbf{k}_\gamma).$$

Henceforth, we will confine ourselves to nonrelativistic approach in which the nuclear Hamiltonian $H = \mathbf{P}^2/2M + H_{int} \equiv K_{CM} + H_{int}$, where M is the total mass of nuclear system, is divided into the kinetic energy operator K_{CM} of the center-of-mass (CM) motion and the intrinsic Hamiltonian H_{int} that depends on internal variables of interacting nucleons. Special attention is paid to the cluster structure of the T-matrix for radiative process $A + B \to \gamma + C$, in which a target-nucleus A captures a projectile-nucleus B that is followed by the single-photon emission and formation of a system $C = A + B$ in a bound or continuum state, e.g., as in case of $\alpha + \alpha \to \alpha' + \alpha' + \gamma$ bremsstrahlung.

48.2 Application to Alpha-Alpha Bremsstrahlung

In the Coulomb gauge (CG), where $\epsilon_0(\mathbf{k}_\gamma) = 0$, $\mathbf{k}_\gamma \cdot \varepsilon(\mathbf{k}_\gamma) = 0$, the T-matrix of interest is given by

$$T(\alpha + \alpha \to \alpha' + \alpha' + \gamma) = \langle \Psi^{int}_{\alpha'+\alpha'} | T | \Psi^{int}_{\alpha+\alpha} \rangle,$$

$$T = i[\varepsilon \cdot \mathbf{D}_{int}(\mathbf{k}_\gamma), H_{int}] - \frac{\varepsilon \cdot \mathbf{P}}{8m} \rho_{int}(\mathbf{k}_\gamma) + i[\mathbf{k}_\gamma \times \varepsilon] \cdot \mathbf{M}_{int}(\mathbf{k}_\gamma),$$

where we encounter Fourier transforms of the operators of charge density $\rho_{int}(\mathbf{k}_\gamma)$, electric dipole moment $\mathbf{D}_{int}(\mathbf{k}_\gamma)$ and magnetic moment $\mathbf{M}_{int}(\mathbf{k}_\gamma)$. For this short exposition we do not consider the last contribution. Explicit expressions of these quantities can be found in [4]. Here we show our calculations in a simple cluster picture with the initial $\Psi_{\alpha+\alpha} = \Phi_{\alpha_1} \Phi_{\alpha_2} \chi^{(+)}_{\mathbf{k}}$ and final $\Psi_{\alpha'+\alpha'} = \Phi_{\alpha'_1} \Phi_{\alpha'_2} \chi^{(-)}_{\mathbf{k}'}$ states. The ingoing and outgoing distorted waves $\chi^{(\pm)}$ meet the equation $(p^2/2\mu_\alpha + V)|\chi^{(\pm)}_k\rangle = E|\chi^{(\pm)}_k\rangle$ for the scattering of particle with mass μ_α by the potential $V = V_C + V_S$, viz., the repulsive! Coulomb potential $V_C = Z^2 e^2/\rho$ plus a short-range potential V_S, e.g. it may be a phenomenological nuclear potential modified by finite-size effects in Coulomb interaction between extended objects as in [5]. For simplicity, we assume that the $\Phi_\alpha = \Phi_{\alpha_1} = \Phi_{\alpha_2} = \Phi_{\alpha'_1} = \Phi_{\alpha'_2}$ describes the ^4He g.s.

After this the amplitude of interest $T(\alpha + \alpha \to \alpha + \alpha + \gamma) = \varepsilon \cdot \mathbf{A}$ with

$$\mathbf{A} = \left(E_\gamma + \frac{\mathbf{P}_i \cdot \mathbf{k}_\gamma}{2M_\alpha} + \frac{E_\gamma^2}{4M_\alpha} \right) \int_0^1 d\lambda F_{CH}(q) \nabla_{\mathbf{q}} \big[I(\mathbf{k}', \mathbf{k}; \mathbf{q}) + I(\mathbf{k}', \mathbf{k}; -\mathbf{q}) \big]$$
$$+ \frac{\mathbf{P}_i}{2M_\alpha} F_{CH}(k_\gamma) \big(I(\mathbf{k}', \mathbf{k}; \mathbf{k}_\gamma) + I(\mathbf{k}', \mathbf{k}; -\mathbf{k}_\gamma) \big)$$

where F_{CH} is the charge form factor of alpha-particle, $\mathbf{q} = \lambda \mathbf{k}_\gamma$ and the overlap integral

$$I(\mathbf{k}', \mathbf{k}; \pm\mathbf{q}) = \int \chi_{\mathbf{k}'}^{(-)*}(\rho) e^{\pm i \frac{1}{2}\mathbf{q}\cdot\rho} \chi_{\mathbf{k}}^{(+)}(\rho) d\rho = \langle \chi_{\mathbf{k}'}^{(-)} | e^{\pm i \frac{1}{2}\mathbf{q}\rho} | \chi_{\mathbf{k}'}^{(+)} \rangle,$$

Following a common practice one can split the overlap integral I into the purely Coulomb integral I_C and their difference $I_{CS} \equiv I - I_C$ (cf. decomposition (1) in [6]). It is the case, where the Nordsieck-type integral I_C is determined via the analytical expression (10) from [7].

48.3 Some Numerical Details

At this point, our calculations of the three dimensional integral I_{CS} are reduced to the summation of its partial -wave expansions with the radial integrals

$$I_{CS}(l', l, L) = \int_0^\infty dr j_L\left(\frac{1}{2}qr\right)\left[g_{l'l} w_{k'l'}(r) w_{kl}(r) + w_{k'l'}(r) w_{kl}(r) - F_{l'}(k'r) F_l(kr)\right],$$

where the factor $g_{l'l} = \frac{1}{4}(s_{l'} - 1)(s_l - 1) + \frac{1}{2}(s_{l'} - 1) + \frac{1}{2}(s_l - 1)$ is expressed through the S-matrix elements in the angular momentum representation $s_l = e^{2i\delta_l(k)}$ with the phase shift $\delta_l(k)$ for given values of the collision energy and angular momentum l. Every radial wave function $w_{kl}(r)$ that enters the partial-wave expansions of the solutions $\chi_{\mathbf{k}}^{(\pm)}(\mathbf{r})$ of the Schrödinger equation for the elastic $\alpha - \alpha$ scattering is normalized in such a way that starting from a value of $r = R_{eff}$ it can with a high precision be approximated by the superposition $F_l(kr) + \tan\delta_l G_l(kr)$ where $F_l(kr)$ and $G_l(kr)$ are the regular and irregular Coulomb functions, respectively. It is important to know since we calculate these integrals with the help of the so-called contour method [8]. As an illustration, in Fig. 48.1 we show the bremsstrahlung

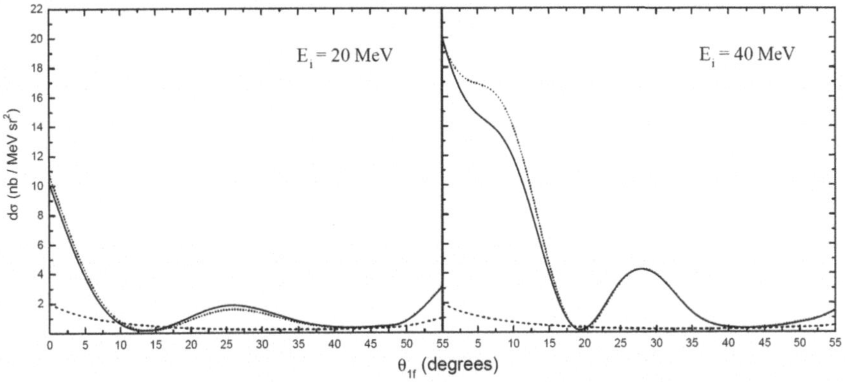

Fig. 48.1 Curves calculated for different incident energies E_i: dashed curve for the Coulomb bremsstrahlung, solid curve for potential model [5], dotted curve for potential model [9]

cross-section $d\sigma \equiv d^3\sigma/dE_\gamma d\Omega_{1'} d\Omega_{2'}$ calculated in the lab. system for the coplanar kinematics with the following disposal of the momenta: the photon momentum is directed along the Z- axis while the rest lie in the XZ - plane, viz., $\mathbf{k}_1 = k_1(\theta_i, 0)$, $\mathbf{k}_{1'} = k_{1'}(\theta_f, \pi)$. The calculation has been carried out for energies of the incident alpha-particle $E_i = 20$ MeV, 40 MeV and the emitted photon $E_\gamma = 1$ MeV. It is proved that under such conditions the partial contributions to the cross-section with $L = 0$ and $l'_{max} = l_{max} = 6$ are dominant. We believe that other details of our approach will be published somewhere else.

References

1. Shebeko, A.: Sov. J. Nucl. Phys. **49**, 30 (1989)
2. Levchuk, L.G., Shebeko, A.: Phys. At. Nucl. **56**, 227 (1993)
3. Friar, J., Fallieros, S.: Phys. Rev. C **34**, 2029 (1986)
4. Shebeko, A.: Phys. At. Nucl. **77**, 518 (2014)
5. Buck, et al.: Nucl. Phys. A **275**, (1977)
6. Baye, D., Sauwens, C., Descouvemont, P., Keller, S.: Nucl. Phys. A **529**, 467 (1991)
7. Gravielle, M., Miraglia, J.: Comp. Phys. Commun. **69**, 53 (1992)
8. Vincent, C., Fortune, H.: Phys. Rev. C **2**, 782 (1970)
9. Ali, S., Bodmer, A.R.: Nucl. Phys. **80**, 99 (1966)

Chapter 49
Direct Measurement of the $^{13}C(\alpha, n)^{16}O$ Reaction at LUNA

G. F. Ciani, L. Csedreki, J. Balibrea-Correa and A. Best

Abstract The $^{13}C(\alpha, n)^{16}O$ reaction is the main neutron source for the *s-process* in low mass AGB stars. Although several direct measurements have been performed, no dataset reaches the Gamow window (140–230 keV) due to the exponential drop of the cross section σ(E) with decreasing energy. The reaction rate becomes so low that the strong cosmic background would become predominant. In order to measure the $^{13}C(\alpha, n)^{16}O$ cross section at low energies, ancillary measurements to understand the behaviour of 99% enriched ^{13}C evaporated targets, under a high intensity alpha beam (100–200 μA). These measurements were carried out in deep underground laboratories of Laboratori Nazionali del Gran Sasso (LNGS) in the framework of the LUNA experiment. The preliminary results are reported in this contribution.

49.1 State of the Art

The $^{13}C(\alpha, n)^{16}O$ reaction (Q = 2.215 MeV) is the major neutron source for the main component of the *s*-process in low mass $(1 - 3M_{\odot})$ Asymptotic Giant Branch (AGB) stars, whose temperature of interest is about $1 - 2 \cdot 10^8$ K. This corresponds to a Gamow window between 140 and 230 keV, below the Coulomb potential energy of the reaction.

For the LUNA collaboration.

G. F. Ciani (✉) · L. Csedreki
Gran Sasso Science Institute, Viale F. Crispi 7, L'Aquila, Italy
e-mail: giovanni.ciani@gssi.it

INFN, Laboratori Nazionali del Gran Sasso, Via G. Acitelli, 22, 76100 Assergi, L'Aquila, Italy

J. Balibrea-Correa · A. Best
Università degli Studi di Napoli Federico II, Naples, Italy

INFN, Sezione di Napoli, 80126 Naples, Italy

N. A. Orr et al. (eds.), *Recent Progress in Few-Body Physics*,
Springer Proceedings in Physics 238,
https://doi.org/10.1007/978-3-030-32357-8_49

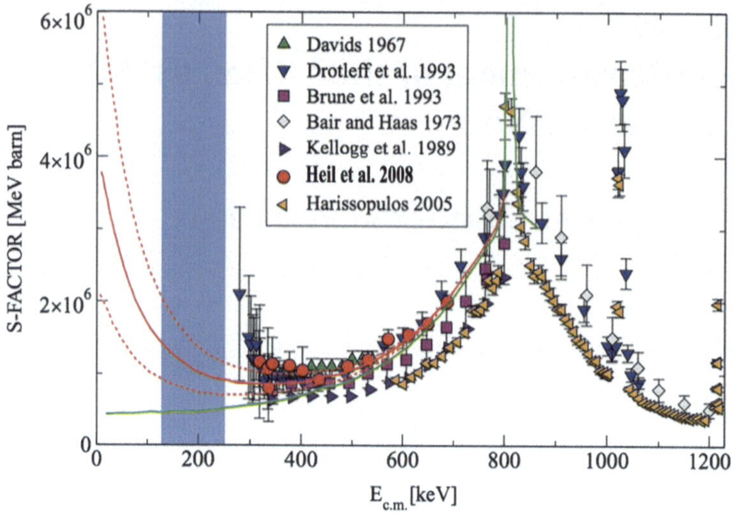

Fig. 49.1 (Colour online) State of the art of the direct measurements of the $^{13}C(\alpha, n)^{16}O$ reaction adapted from Heil's paper [1]. The violet band represents the Gamow window, the red solid line is the extrapolation of the astrophysical S-factor down to this region while its error bands are indicated by red dashed lines. The green line is an extrapolation where the presence of the threshold resonance is omitted

In the last 25 years, several direct measurements of this reaction cross section have been performed [1–4]. The astrophysical S(E)- factor, is shown in Fig. 49.1. The lowest energy point has been measured by Drotleff et al. [2] with an uncertainty of 50%.

In addition, the reaction mechanism at low energies includes also the contribution of the high energy the tail of a near-threshold resonance at $E_R = -3 \pm 8$ keV (the resonance energy in the center-of-mass system), corresponding to $E_x = 6.356$ MeV state in ^{17}O. The red and the green lines in Fig. 49.1, indicate the R Matrix extrapolation performed by Heil, considering and omitting the subthrehsold resonance, respectively. As one can see the two curves differ almost of one order of magnitude in the astrophysical energy region (violet bar).

The extrapolation at lower energies can be improved only by the extension of experimental cross section data towards the Gamow-window with moderated and fully controlled uncertainties. Another way to solve the problem are indirect measurements: the Trojan Horse Method (THM) [5] or the Asymptotic Normalization Coefficient (ANC) [6] have been used to measure the $^{13}C(\alpha, n)^{16}O$ reaction, but these techniques need a normalization with respect to direct data, so approach the Gamow Window with a direct measurement has a crucial importance.

A big effort in the nuclear astrophysics community was made to approach the Gamow window with uncertainties as low as possible to improve the stellar reaction rates calculation [7].

49.2 The $^{13}C(\alpha, n)^{16}O$ Measurement at LUNA

In this framework, the Laboratory for Underground Nuclear Astrophysics (LUNA) collaboration proposed a measurement taking advantage from the low background environment of Underground Laboratori Nazionali Del Gran Sasso (LNGS). Thanks to 1400 m of rocks (3500 m.w.e) the cosmic background is reduced by 6 orders of magnitude for muons and by three orders of magnitude for neutrons [8].

Furthermore, the LUNA accelerator provides an intense ($< I >= 200\,\mu A$) stable alpha beam in the energy range $50 < E_\alpha < 400$ keV [9] on a solid state target or on a windowless gas target system [10].

The experimental setup used for this measurement is based on 18 ^3He counters with low intrinsic background arranged in two rings (6 in the inner ring, 12 in the outer ring) concentric with respect to the target chamber. The counters are embedded in a polyethylene moderator.

In this campaign a vertical multi target chamber where three targets could be mounted was used: this guaranteed the target changing without break the vacuum and allowed the reproducibility devoted to deep target behaviour investigation. The schematic view of the setup is shown in Fig. 49.2a. A series of simulations based on the Monte Carlo code Geant4 has been performed in order to find the best counters

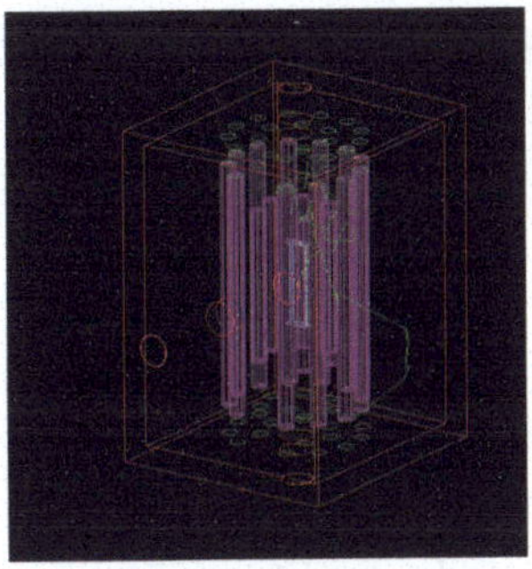

(a) Schematic design of the setup used for the measurement: the vertical violet volumes represent the active volumes of the ^3He counters concentric with respect the target holder (blue rectangle). Everything is embedded in the polyethylene moderator (red box).

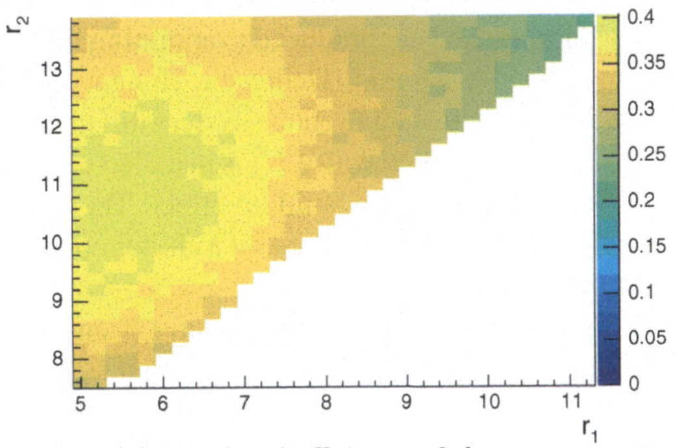

(b) Simulated efficiency of the setup as a function of the rings radii r_1 and r_2

Fig. 49.2 Some main features of the simulation GEANT4 code. In Fig. 49.2a the geometry implemented for the setup simulation. In Fig. 49.2b the efficiency as a function of the rings radii r_1 and r_2

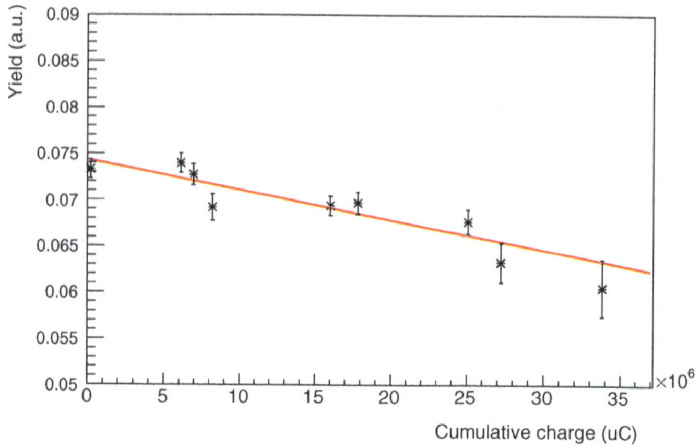

Fig. 49.3 Yield of the $^{13}C(p,\gamma)^{14}N$ $DC \rightarrow GS$ transition as a function of the cumulative charge

placement to maximize the efficiency. Figure 49.2b shows the detector efficiency as a function of the two rings' radii, r_1 and r_2: the optimal condition was found at $r_1 = 6$ cm and $r_2 = 11$ cm [11]. A validation of simulation will be performed through the $^{51}V(p,n)^{51}Cr$ reaction (Q $= -1535$ keV). The analysis is in progress.

The targets used are produced evaporating 99% enriched ^{13}C powder on tantalum backing at MTA Atomki (Debrecen, Hungary). In order to keep the target modification under control and reduce systematic uncertainties, several precautions were taken. A cold trap filled with LN_2 is installed along the beamline just before the target chamber to avoid carbon build up effect. The beam power dissipation is performed through a cooling system based on a chiller connected to the target chamber. It cools down deionised water at 5 °C.

Two complementary campaigns have been scheduled to investigate target modification under an intense beam. In the first step, immediately after the evaporation, the Nuclear Resonant Reaction Analysis with the $^{13}C(p,\gamma)^{14}N$ (Q = 7550 keV) reaction wide resonance at $E_{p,lab} = 1748$ keV was exploited at the 2 MV Tandetron accelerator installed at MTA Atomki. In this phase the uniformity of the evaporation was confirmed by means of the scan of the resonance in different points of the targets.

In the second step, the direct capture component at lower energies was used for the LUNA campaign and for testing the stability of targets under high intensity proton (I $= 150\ \mu A$) beam with $E_p = 310$ keV [12]. The $\gamma-$ray spectra were acquired with a HPGe 120% relative efficiency in close geometry. Estimating the yield in the region of interest of the direct capture (DC) peak to the ground state (GS) as a function of the proton charge on the target, we concluded that the degradation was 17% after 33 C (Fig. 49.3).

The negligible target damage after a short irradiation with a proton beam, permitted to use an alternation of proton irradiations and alpha ones on the ^{13}C targets.

An innovative gamma shape analysis has been elaborated to evaluate the target stoichiometry modification that causes the yield decrease (Ciani's PhD [13]). This allows to monitor the target conditions during intense irradiations and to correct $^{13}C(\alpha, n)^{16}O$ yield measured in each partial run for the loss due to degradation.

49.3 Preliminary Results and Conclusion

During the first campaign of measurement, held in winter 2018, five energy points have been measured between $360 < E_\alpha < 400$ keV in steps of 10 keV. A typical neutron spectrum acquired at $E_\alpha = 390$ keV is shown in Fig. 49.4. The neutron peak, together with the wall effect region, are delimited between the red bars. The contribution beyond the neutron peak comes from the alpha intrinsic background in the counters and can be considered negligible.

Figure 49.5 presents the preliminary result of the measured reaction yield as a function of alpha beam energy. The data are not yet corrected for the target degradation and error bars represent statistical uncertainties. The lowest point measured, corresponding to $E_\alpha = 360$ keV, is the lowest point ever measured with a direct measurement.

The final analysis is in progress: once the detector efficiency will be calculated, the cross section will be evaluated. Moreover the promising and encouraging quality of the data permits to schedule another measurement campaign to go lower in energy and approach the Gamow Window.

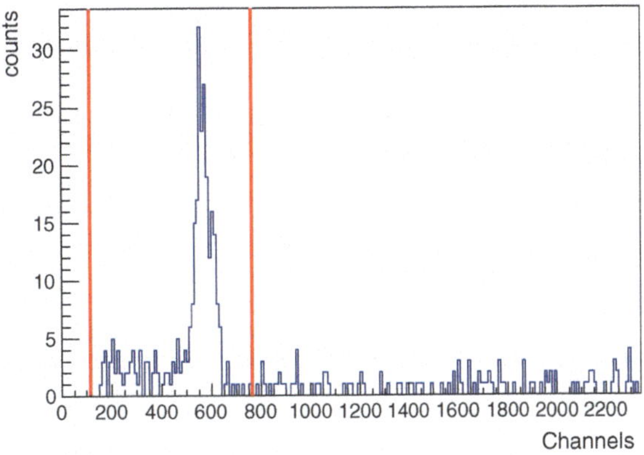

Fig. 49.4 Neutron spectrum acquired at $E_\alpha = 390$ keV. The red bars indicate the region of interest of the neutron peak in ^3He counters

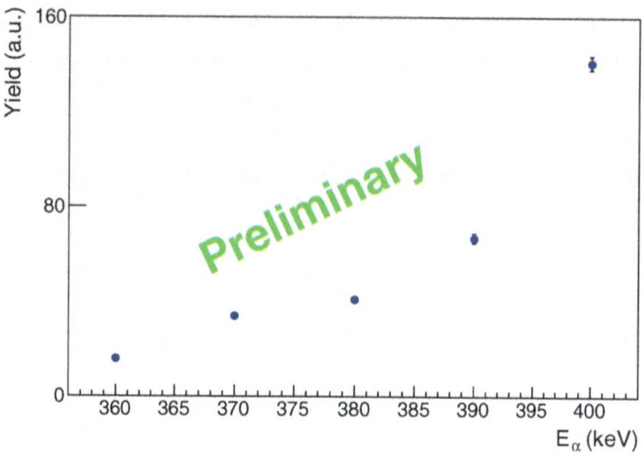

Fig. 49.5 Preliminary reaction yield as a function of alpha beam energy. Only statistical uncertainties are shown in the plot. Points are not corrected for the target degradation

References

1. Heil, M., et al.: The $^{13}C(\alpha, n)$ reaction and its role as a neutron source for the s process. Phys. Rev. C **78**, 2 (2008)
2. Drotleff, H.W., et al.: Reaction rates of the s-process neutron sources $Ne22(\alpha, n)Mg25$ and $C13(\alpha, n)O16$. Astrophys. J. **414**, 2 (1993)
3. Brune, C.R., et al.: Low-energy resonances in ^{13}C(α, n). Phys. Rev. C **48**, 6 (1993)
4. Harissopulos, S., et al.: Cross section of the $^{13}C(\alpha, n)^{16}O$ reaction: a background for the measurement of geo-neutrinos. Phys. Rev. C **72**, 6 (2005)
5. La Cognata, M., et al.: On the measurement of the $^{13}C(\alpha, n)^{16}O$ S-factor at negative energies and its influence on the s-process. Astrophys. J. **777**, 2 (2013)
6. Mukhamedzhanov, A., et al.: Subthreshold resonances and resonances in the R-matrix method for binary reactions and in the Trojan horse method. Phys. Rev. C **96**, 2 (2017)
7. Cristallo, S., et al.: The Importance of the $^{13}C(\alpha, n)^{16}O$ Reaction in asymptotic giant branch stars. Astrophys. J. **859**, 2 (2018)
8. Best, A., et al.: Low energy neutron background in deep underground laboratories. Nucl. Instrum. Methods Phys. Res. A **812**, (2016)
9. Formicola, A., et al.: The LUNA II 400kV accelerator. Nucl. Instrum. Methods Phys. Res. A **3**, 507 (2003)
10. Ferraro, F. et al. : A high-efficiency gas target setup for underground experiments, and redetermination of the branching ratio of the 189.5 keV Ne-22(p,γ)Na-23 resonance. Eur. Phys. J. A **54**, 44 (2018)
11. Csedreki, L., et al.: Introduction of the new LUNA experimental setup for high precision measurement of the $^{13}C(\alpha, n)^{16}O$ reaction for astrophysical purposes. In: EPJ Web Conference. NPA (2017)
12. Ciani, G.F., et al.: Target characterizations for direct measurement of the $^{13}C(\alpha, n)^{16}O$ reaction at LUNA 400. In: EPJ Web Conference. NPA (2017)
13. Ciani, G.F.: Cross section of the $^{13}C(\alpha, n)^{16}O$ reaction at low energies. GSSI & SISSA (2019)

Chapter 50
Validity of Quasi-classical Approaches to True Three-Body Decays

O. M. Sukhareva, L. V. Grigorenko, D. A. Kostyleva and M. V. Zhukov

Abstract Within the hyperspherical harmonics (HH) method the three-body problem is reduced to a motion of one effective particle in a "strongly deformed" field, which is described in the coupled-channel formalism. This method is well suited to studies of so-called true three-body decays. The reduction of the hyperspherical equations set to a single-channel approximation provides the basis of standard quasi-classical (QC) expression for width evaluation. We demonstrate that by itself the quasi-classical approach is quite precise in application to typical three-body effective potentials. However, the reduction to a single channel leads to a significant overestimation of the width.

Conventional methods of a width determination for resonant states, such as via an elastic phase shift energy dependence, or via S-matrix pole position in the complex energy plane could be technically complicated for very small widths $\Gamma \ll E$. Therefore studies of radioactive decays require specific methods for the decay width

O. M. Sukhareva (✉)
Omsk State Technical University, 644050 Omsk, Russia
e-mail: o.m.sukhareva@gmail.com

L. V. Grigorenko
Flerov Laboratory of Nuclear Reactions, JINR, 141980 Dubna, Russia

National Research Nuclear University "MEPhI", Kashirskoye shosse 31, 115409 Moscow, Russia

National Research Centre "Kurchatov Institute", Kurchatov sq. 1, 123182 Moscow, Russia

D. A. Kostyleva
II. Physikalisches Institut, Justus-Liebig-Universität, 35392 Giessen, Germany

GSI Helmholtzzentrum für Schwerionenforschung GmbH, 64291 Darmstadt, Germany

M. V. Zhukov
Department of Physics, Chalmers University of Technology, 41296 Göteborg, Sweden

N. A. Orr et al. (eds.), *Recent Progress in Few-Body Physics*, Springer Proceedings in Physics 238,
https://doi.org/10.1007/978-3-030-32357-8_50

determination. Among them are "natural" width definition via WF with pure outgoing asymptotics [1], Kadmensky-type integral formulas [2], and quasi-classical approach of Gamow type [3].

The standard QC approximation is easy to formally generalize for true three-body decays, where two protons are emitted simultaneously. The motion of such a system can be considered in a certain approximation as a single-channel motion in the hyperradius ρ value. In such an approximation evaluation of the decay width requires reduction of few-body problem to a single-channel formalism of some form, where the Gamow integral over the sub-barrier trajectory $\{\rho_2, \rho_3\}$ can be defined as

$$\Gamma = \nu \exp\left[2i \int_{\rho_2}^{\rho_3} p\, d\rho\right], \quad \nu = \left[2M \int_{\rho_1}^{\rho_2} \frac{d\rho}{p}\right]^{-1}, \quad p = \sqrt{2M[E_T - V_{\text{eff}}(\rho)]}. \tag{50.1}$$

Here ρ_1, ρ_2 are inner and ρ_3 is outer classical turning points of the potential $V_{\text{eff}}(\rho)$. Both the validity of the few-body problem reduction to a single channel formalism and the applicability of the QC approximation for barriers of specific shape can be questioned.

The Gamow-type formalism has been repeatedly used in the recent years for determination of three-body decay widths, e.g. [3, 4]. We examine the validity of this approximation by the example of the width of the first excited $3/2^-$ state of ^{17}Ne, which is known to decay via true two-proton decay mechanism. The width of this state is important for determination of the astrophysical capture rate for $^{15}\text{O}+p+p \rightarrow {}^{17}\text{Ne}+\gamma$ reaction [5]. Theoretical calculations of this width have produced so far a considerable controversy [2–4, 6]. Recently this issue was revisited experimentally [7] providing the improved upper limit for the two-proton width value $\Gamma < 8.8 \times 10^{-13}$ MeV.

The paper [4] is dedicated to finding "necessary conditions for accurate computations of three-body partial decay widths", so its results are expected to be especially accurate. For the decay of the ^{17}Ne $3/2^-$ state we have found that we can not reconcile the width values quoted in [4] with the potential curves provided in this work, see Fig. 50.1 and Table 50.1. The potential provided as a final result of these studies (solid curve in Fig. 50.1) gives for the ^{17}Ne $3/2^-$ state decay (according to our calculations) the width value $\Gamma = 8.3 \times 10^{-12}$ MeV. This value exceeds the recent experimental limit from [7] by an order of the magnitude.

The potentials obtained by a trivial single-channel reduction of the hyperspherical potentials of [2] (just diagonalization) look reasonably consistent with potential derived in [4], see Fig. 50.1. They also have analogous basis convergence trend. However, the QC results obtained with these potentials are more than an order of the magnitude larger than the results of the complete three-body calculations [2], see Table 50.2. So we find that the reduction to one-channel approximation leads to the significant width overestimation, compared to fully dynamical three-body calculations.

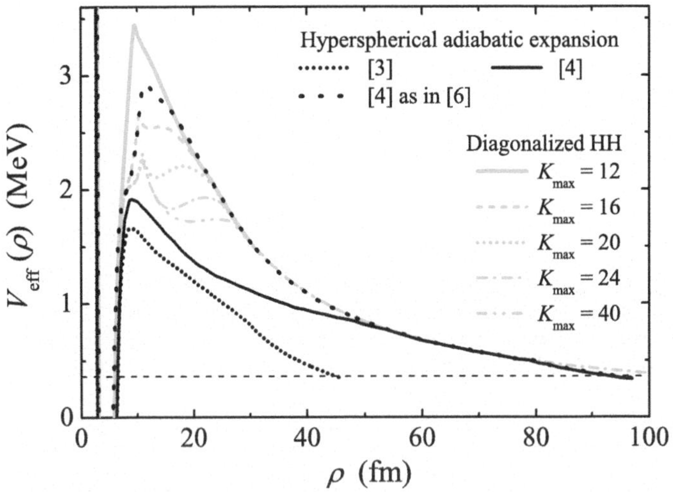

Fig. 50.1 Effective single channel potentials V_{eff} for quasi-classical calculations of two-proton width of the ^{17}Ne $3/2^-$ state adapted from [4]. The two-proton decay energy E_T is indicated by horizontal line. Our hyperspherical harmonics (HH) potentials (gray curves) are obtained by trivial diagonalization of the HH potential matrix from [2]

Table 50.1 The ^{17}Ne $3/2^-$ state widths (in MeV) obtained by (50.1) for hyperspherical adiabatic effective potentials V_{eff} from [4]. Column 1 corresponds to the non-converged result of [3], column 2 to the "accurate" converged result and column 3 aims to imitate the calculations of [6]. Our QC calculations with the same V_{eff} are in a strong contradiction with the results quoted in [4]

Case:	$K_{max} = 70,\ l_x, l_y \leq 2$	$K_{max} = 70,\ l_x, l_y \leq 9$	$K_{max} = 20,\ l_x, l_y \leq 10$
[4]	3.6×10^{-12}	1.7×10^{-14}	5.4×10^{-16}
QC, (50.1)	7.3×10^{-6}	8.3×10^{-12}	7.9×10^{-14}

Table 50.2 Widths of the ^{17}Ne $3/2^-$ state in 10^{-15} MeV units as a function of the hyperspherical basis size K_{max}. QC calculations with diagonalized HH potentials from [2], three-body calculations of [2] (see Fig. 15 of [2]), and their ratio. Column "Asympt." contains values obtained by exponential extrapolation to the infinite basis

K_{max}	12	16	20	24	28	32	36	40	Asympt.
QC, (50.1)	9.83	21.8	40.1	54.1	68.5	81.7	90.4	95.6	103
3-body, [2]	0.91	1.32	1.70	2.15	2.64	3.21	3.62	3.98	6.90
Ratio	10.8	16.5	23.5	26.1	25.9	25.5	25.0	24.0	14.9

To conclude, our results question a validity of the widths obtained in [4] for the ^{17}Ne $3/2^-$ state. The recent experimental data [7] also contradict the results of [4] recalculated by us. The QC method by itself is reasonably precise. However, the reduction of the few-body problem to a single channel problem leads to a significant overestimation of the width.

Acknowledgements O.M.S. and L.V.G. were partly supported by the Russian Science Foundation grant No. 17-12-01367.

References

1. Pfützner, M., Karny, M., Grigorenko, L.V., Riisager, K.: Rev. Mod. Phys. **84**, 567 (2012)
2. Grigorenko, L.V., Zhukov, M.V.: Phys. Rev. C **76**, 014008 (2007)
3. Garrido, E., Fedorov, D., Jensen, A.: Nucl. Phys. A **733**, 85 (2004)
4. Garrido, E., Jensen, A., Fedorov, D.: Phys. Rev. C **78**, 034004 (2008)
5. Parfenova, Y., et al.: Phys. Rev. C **98**, 034608 (2018)
6. Grigorenko, L.V., Mukha, I.G., Zhukov, M.V.: Nucl. Phys. A **713**, 372 (2003)
7. Sharov, P.G., et al.: Phys. Rev. C **96**, 025807 (2017)

Chapter 51
Complex-Range Gaussians as a Basis for Treatment of Charged Particle Scattering

D. A. Sailaubek, O. A. Rubtsova and V. I. Kukulin

Abstract An employment of the Complex-Range Gaussian basis for solving charged particle scattering is briefly described. The method is based on a discretization of the continuum and the Coulomb wave-packet formalism. As a result, the off-mass-shell Coulomb-nuclear t-matrix at any energy can be found from diagonalisation procedures for the total and pure Coulomb Hamiltonians. The approach is illustrated with the d-α partial phase shifts and pp scattering with the NN interaction containing non-nucleon degree of freedom.

51.1 Introduction

A problem of a choice of the optimal basis for quantum scattering calculations doesn't lose its importance, especially, for systems with a few charged particles. Recently [1] we suggested to use the so-called Complex-range Gaussian basis (CRGB) [2, 3] for that purpose. The functions of this basis are real-valued ones but they are constructed from two sets of Gaussians with complex scale parameters:

$$\psi(r) = r^l e^{-(1+ib)\alpha r^2}, \quad \psi^*(r) = r^l e^{-(1-ib)\alpha r^2} \tag{51.1}$$

where l is the relative orbital momentum and the parameter b is chosen arbitrarily. As a result, the basis functions have additional oscillating factors, so that, they are better suited for scattering calculations than ordinary real-valued Gaussians. At the same time, the usage of such oscillating basis functions keeps all the advantages

D. A. Sailaubek (✉)
Faculty of Physics and Technical Sciences, L.N. Gumilyov Eurasian National University, 010000 Nur-Sultan, Kazakhstan
e-mail: s.dinmuhamed@gmail.com

O. A. Rubtsova · V. I. Kukulin
Skobeltsyn Institute of Nuclear Physics, Moscow State University, 119991 Moscow, Russia

N. A. Orr et al. (eds.), *Recent Progress in Few-Body Physics*, Springer Proceedings in Physics 238,
https://doi.org/10.1007/978-3-030-32357-8_51

of the Gaussian expansion method such as analytical transformation between basis functions defined in different Jacobi sets in few-body case etc.

To apply the above basis for scattering calculations, we use the stationary wave-packet formalism [4] and our newly developed diagonalisation technique [5] which allows finding the off-mass-shell t-matrix at many energies from a spectral expansion for the total resolvent built from the pseudostates of the discretized continuum. Here we briefly explain the formalism and give numerical examples. More detailed explanation of the method can be found in [1].

51.2 The Formalism

Consider scattering of two charged particles with the Hamiltonian:

$$h = h_0 + \frac{z_1 z_2 e^2}{r} + v_S(r) \equiv h_C + v_S(r), \tag{51.2}$$

where h_0 is a kinetic energy operator for the relative motion of particles, r is a relative distance, v_S is a short-range interaction and h_C is the Coulomb Hamiltonian which includes the long-range Coulomb interaction between the charges z_1 and z_2 of the same sign.

As basis functions for practical calculations, we employ the cosine-type functions constructed from the complex Gaussians (51.1): $\phi_{nl}(r) = r^l e^{-\alpha_n r^2}\cos(b\alpha_n r^2)$ $(n = 1, \ldots, N)$, where the scale parameters α_n are defined on some grid.[1] Employing diagonalisation procedures for the total h and Coulomb h_C Hamiltonian matrices in this basis, one gets the sets of eigenfunctions which represent the so-called pseudostates of the discretized continua for the corresponding Hamiltonians and approximations for the possible bound-states in a case of the total Hamiltonian h.

Further, one should treat the corresponding pseudostates as approximations for the stationary wave-packets for h and h_C [4]. In particular, the *Coulomb wave-packets* $|x_k^C\rangle$ are constructed as integrals of the regular Coulomb wave functions $|F_l(E)\rangle$ over discretized energy intervals, while the scattering wave-packets $|z_k\rangle$ are constructed as similar integrals of the scattering wave-functions $|\psi(E)\rangle$ of the total Hamiltonian h. By using the wave-packet formalism [5], one obtains an explicit representation for the total resolvent and, finally, the off-mass-shell t-matrix in the following closed form:

$$t_{kk'}(E) = v_{kk'} + \sum_{n=1}^{N_b} \frac{\bar{v}_{kn}\bar{v}_{k'n}}{E - E_n} + \sum_{i=1}^{N-N_b} g_i(E)\tilde{v}_{ki}\tilde{v}_{k'i}, \quad k, k' = 1, \ldots, N, \tag{51.3}$$

[1]The properties of the basis were studied in detail in [1] where estimations for optimal values of b and a grid were given.

where $g_i(E)$ are known values, E_n are the energies of possible bound states of the total Hamiltonian h and we have introduced the matrix elements of the short-range potential between different normalized states, viz. Coulomb wave-packets $|x_k^C\rangle$, scattering wave-packets $|z_k\rangle$ and bound states $|\psi_n^b\rangle$ for the total Hamiltonian:

$$v_{kk'} = \langle x_k^C|v_S|x_{k'}^C\rangle, \quad \bar{v}_{kn} = \langle x_k^C|v_S|\psi_n^b\rangle, \quad \tilde{v}_{ki} = \langle x_k^C|v_S|z_i\rangle. \tag{51.4}$$

Single diagonalisation procedures for the Coulomb and total Hamiltonian matrices in the CRGB allow to get all three above types of the normalized states expanded over the CRGB basis functions (see details in [1]).

51.3 Numerical Illustrations

As the first illustration, the partial S- and D-wave phase shifts for deuteron scattering over ^{4}He target are represented in Fig. 51.1. The short-range nuclear potentials v_S have been taken in a simple Gaussian form [6] which allows, however, to reproduce the experimental data at low energies. The potential has one forbidden (by the Pauli principle) S-wave bound-state. However, we found that it should be included into the spectral expansion (51.3) of the t-matrix [1].

As the second example, we have calculated the proton-proton scattering phase shifts with the nucleon-nucleon interaction in which the intruder state $|a\rangle$ with energy E_0 in some inner channel exists (see Fig. 51.2). The Hamiltonian for such a system is written in a two-channel form:

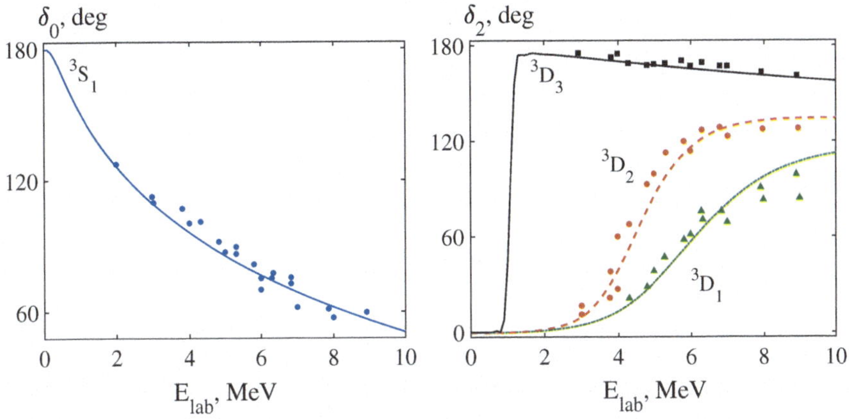

Fig. 51.1 S- and D-wave partial phase shifts for the elastic d-α scattering found in the CRGB (solid curves) in comparison with experimental data (filled circles) from [7]

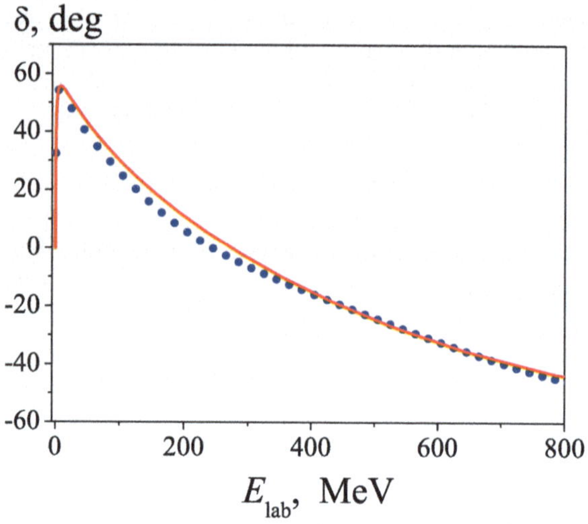

Fig. 51.2 $^1S_0$-partial phase shifts for *pp* scattering calculated in the CRGB (solid curve) in comparison with the SAID data (filled circles)

$$H = \begin{pmatrix} h_C + v_S & \sqrt{\lambda} E_0 |\chi\rangle\langle a| \\ \sqrt{\lambda} E_0 |a\rangle\langle\chi| & E_0 |a\rangle\langle a| \end{pmatrix}, \tag{51.5}$$

where $|\chi\rangle$ is a form factor which determines coupling of the scattering channel with the inner channel and the dimension of the inner channel is equal to 1. Such type an interaction is used the dibaryon model for NN interaction [8] where the intruder state has a meaning of the quark-meson component of a dibaryon resonance in the respective spin-angular NN channel.

As is seen from Fig. 51.2, the partial phase shifts for $^1S_0$ channel in the above model reproduce the corresponding SAID data [9] rather well. It should be emphasized that all the phase shift in a very wide energy region are found from single diagonalisation procedures for the total and Coulomb Hamiltonians.

51.4 Conclusions

The technique developed makes it possible to find rather easily the off-shell t-matrix both for short- and long-range interactions in a very broad energy range using only discretized L_2 representations for the total and Coulomb Hamiltonians in the CRGB. This technique being extended to few-body scattering will lead to some additional important features, such as a fully analytical form for all matrix elements of the total Hamiltonian and a convenient way of transformations between basis sets defined in different Jacobi coordinates.

Acknowledgements Authors appreciate a partial financial support from the RFBR grant 19-02-00014.

References

1. Sailaubek, D.A., Rubtsova, O.A., Kukulin, V.I.: Eur. Phys. J. A **54**, 126 (2018)
2. Egami, T., et al.: Phys. Rev. C **70**, 047604 (2004)
3. Hiyama, E. et al.: Prog. Theor. Exp. Phys. 01A204 (2012)
4. Rubtsova, O.A., Kukulin, V.I., Pomerantsev, V.N.: Ann. Phys. **360**, 613 (2015)
5. Müther, H., et al.: Phys. Rev. C **94**, 024328 (2016)
6. Dubovichenko, S.B.: Phys. At. Nucl. **73**, 9 (2010)
7. McIntyre, L.C., Haeberli, W.: Nucl. Phys. A **91**, 382 (1967); Keller, L.G., Haeberli, W.: Nucl. Phys. A **156**, 465 (1970); Gruebler, W. et al.: Nucl. Phys. A **242**, 265 (1975); Schmelzbach, P.A. et al.: Nucl. Phys. A **184**, 193 (1972)
8. Kukulin, V.I., et al.: J. Phys. G **27**, 1851 (2001)
9. Arndt, R.A., et al.: Phys. Rev. C **76**, 025209 (2007)

Chapter 52
Study of a 5-Alpha Cluster Candidate with the ^{22}Ne(p,t)^{20}Ne and ^{22}Ne(p,^{3}He)^{20}F Reactions

J. A. Swartz, R. Neveling, P. Papka, L. M. Donaldson, F. D. Smit, G. F. Steyn, P. Adsley, L. P. L. Baloyi, J. Carter, H. O. U. Fynbo, N. Hubbard, M. K. Mohamed, G. G. O'Neill, J. C. Nzobadila-Ondze, L. Pellegri, V. Pesudo, B. Rebeiro and S. Triambak

Abstract The isobaric nuclei ^{20}Ne and ^{20}F were investigated with the ^{22}Ne(p,t)^{20}Ne and ^{22}Ne(p,^{3}He)^{20}F reactions using a beam of protons at $E_p = 80$ MeV. The former reaction was employed to confirm a tentative candidate for the 5-alpha cluster state in ^{20}Ne, while the latter was used to look for T = 2 isobaric analogue partner states in ^{20}F to states which were already observed above $E_x = 20$ MeV in ^{20}Ne. The existence of a state which could be considered as a candidate for the 5-alpha cluster state is confirmed here as well as the T = 2 IAS states in ^{20}Ne.

52.1 Introduction

A deformed nucleus in its ground state [1, 2], ^{20}Ne is valuable for testing ab initio cluster models [3]. It offers an ideal arena for the occurrence of alpha clustering, and indeed many of its states are composed of ^{16}O + alpha or ^{12}C + 2-alpha structures

J. A. Swartz (✉) · H. O. U. Fynbo · N. Hubbard
Department of Physics and Astronomy, Aarhus University, Ny Munkegade 120, 8000 Aarhus, Denmark
e-mail: cobus.swartz@gmail.com

R. Neveling · P. Papka · L. M. Donaldson · F. D. Smit · G. F. Steyn · P. Adsley · G. G. O'Neill · J. C. Nzobadila-Ondze · L. Pellegri · V. Pesudo
Department of Subatomic Physics, iThemba LABS, Somerset West 7129, South Africa

P. Papka · P. Adsley
Department of Physics, University of Stellenbosch, Matieland 7600, Stellenbosch, South Africa

L. P. L. Baloyi · J. Carter · L. Pellegri
School of Physics, University of the Witwatersrand, Johannesburg 2050, South Africa

N. Hubbard
Department of Physics, University of York, York YO10 5DD, UK

M. K. Mohamed · G. G. O'Neill · J. C. Nzobadila-Ondze · V. Pesudo · B. Rebeiro · S. Triambak
Department of Physics, University of the Western Cape, Bellville 7535, South Africa

N. A. Orr et al. (eds.), *Recent Progress in Few-Body Physics*, Springer Proceedings in Physics 238,
https://doi.org/10.1007/978-3-030-32357-8_52

[4–7]. Above its 5-alpha breakup threshold at $E_x = 19.17$ MeV, it is postulated that a 5-alpha cluster state may exist in this nucleus [8]. This state would be analogous to the Hoyle state in ^{12}C which is well-known to have a strong alpha clustering character [9]. We may speculate that such a state could also be of alpha-condensate nature, although the feasibility of condensation of multi-alpha cluster states is still under investigation [10–12]. As with ^{16}O and ^{24}Mg, a multi-alpha cluster state in ^{20}Ne has in any case not been observed as yet.

A measurement performed with a magnetic spectrometer and a 60-MeV proton beam at iThemba LABS has, however, given an indication of what appears to be a 0^+ state at $E_x = 22.5$ MeV in ^{20}Ne which could not be reproduced by shell-model calculations. Said measurement also reported on the existence of three $T = 2$ isobaric analogue states (IAS) of ^{20}Ne at $E_x = 20.59$, 21.16 and 21.80 MeV [13]. The present measurement was proposed to reproduce the $^{22}Ne(p,t)^{20}Ne$ measurement at $E_x =$ 20–25 MeV, while investigating ^{20}F in the region where the $T = 2$ IAS partner states should exist in order to both confirm the $T = 2$ character of these states while proving, by its non-observation, that the 5-alpha candidate has the requisite $T = 0$ character.

52.2 Experimental Procedure

As with the measurements in [13], the K600 magnetic spectrometer at iThemba LABS [14] was employed with proton beams from the K200 Separated Sector Cyclotron facility. This time a beam energy of 80 MeV was used since the energy of 60 MeV used in [13] would be too low to allow for the ^3He ejectiles to reach the trigger detectors of the spectrometer's focal plane detection setup. The same energy was used for the (p,t) measurements to allow for comparison of the angular distribution spectra, which could be used to ascertain the spin-parities and isospins of these states. Both the $^{22}Ne(p,t)^{20}Ne$ and $^{22}Ne(p,^3He)^{20}F$ reactions were investigated with the same ^{22}Ne gas-cell target which was employed in [13]. This target, which is described in further detail in [15], was supplemented for this measurement with a gas control system which uses gas regulators and flow meters from Bronkhorst High-Tech B.V. to keep the gas pressure constant and to protect the gas target foils during gas filling [16, 17]. The regulation of the gas pressure is necessary to enable for more accurate cross section values than was possible with the 60 MeV data from [13] through accurate knowledge of the areal density of the gas target.

52.3 Preliminary Results and Discussion

The investigation of the $^{22}Ne(p,t)^{20}Ne$ reaction reproduced two out of the three $T =$ 2 IAS states from [13], as well as the 0^+ state which has been noted as a candidate for the 5-alpha cluster state. These are shown in Fig. 52.1 and listed in Table 52.1. The third $T = 2$ state, at $E_x = 20.59$ MeV, was probably not seen in this measurement

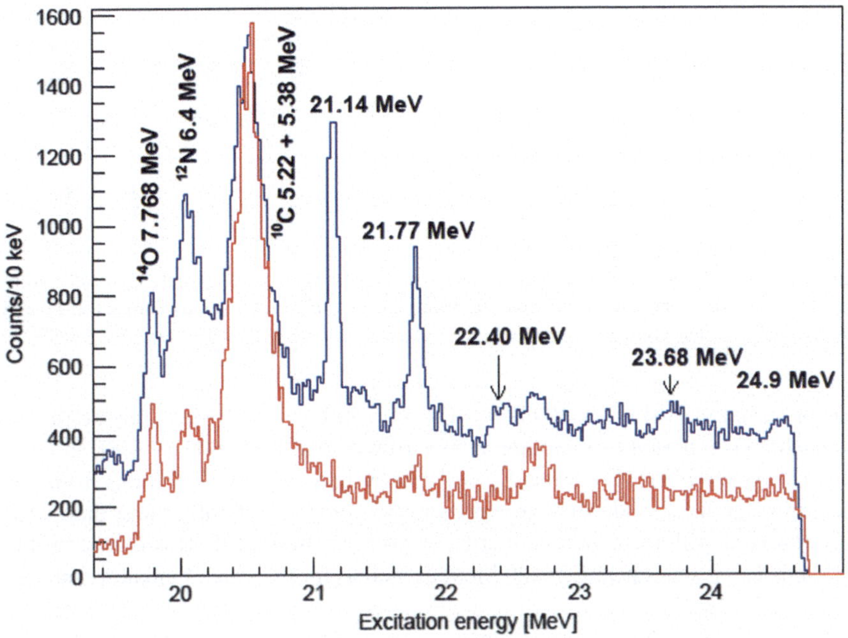

Fig. 52.1 The ^{22}Ne(p,t)^{20}Ne excitation energy spectrum at E_p = 80 MeV and Θ_{lab} = 0° (blue spectrum), and the aramid(p,t) background spectrum (red spectrum)

Table 52.1 States observed from ^{22}Ne(p,t)^{20}Ne at E_p = 80 MeV, compared to Ep = 60 MeV data from [11] with spins, parities and isospins inferred from [13]

E_x Present data [MeV]	E_x [11] [MeV]	Spin and parity J^π	Isospin T	Width FWHM [keV]	Cross section [μb.sr^{-1}]	Cross section [13] [μb.sr^{-1}]
16.73	16.73	0^+	2	72	298	370
21.14	21.16	0^+	2	66	32	43
21.77	21.80	1^-	2	97	35	38
22.40	22.50	(0^+)		192	19	22
23.68	23.70	(9^-)		153	9	Not observed

due to it being buried beneath foil-related contaminant peaks from the ^{12}C(p,t)^{10}C reaction. These peaks are also visible on the aramid foil spectrum in Fig. 52.1. The excitation energies observed for these states show very good agreement with those of [13]. These data were acquired over a period of over 6 h at Θ_{lab} = 0°, hence it may be that small shifts in the beam position or gas target thickness could have affected the observed resolution values noted in Table 52.1. The width observed for the 5-alpha cluster candidate at E_x = 22.4 MeV (Γ = 192 keV) is smaller than what

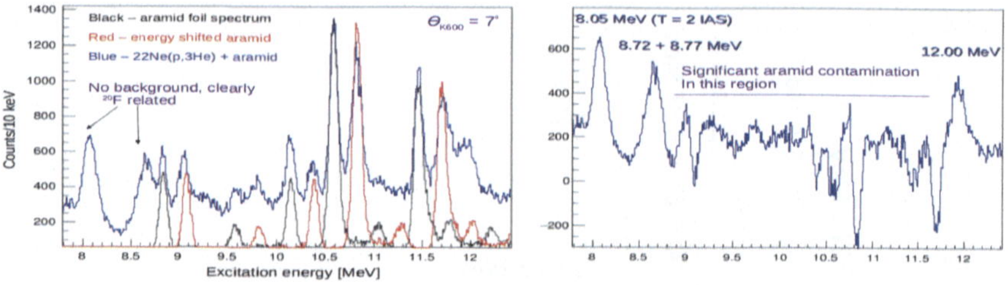

Fig. 52.2 The ^{22}Ne(p,^{3}He)^{20}F excitation energy spectrum at $E_p = 80$ MeV and $\Theta_{lab} = 7°$, before (left, blue spectrum) and after (right, blue spectrum) the attempted subtraction of target foil-related background (left, black and red spectra)

was observed in [13] ($\Gamma = 260$ keV). The state at $E_x = 23.7$ MeV may be the 9^- state which is reported by literature to be narrower than $\Gamma = 200$ keV [18], and was found to have a FWHM of around $\Gamma = 153$ keV here. The cross sections from this measurement at $E_p = 80$ MeV are very similar though all slightly lower than what was observed with the experiment at $E_p = 60$ MeV from [13]. This is true for the low-spin T = 2 states near $E_x = 20$ MeV, as well as for the state of unknown isospin at $E_x = 22.4$ MeV.

In Fig. 52.2 left, the black spectrum was measured with a single aramid foil, while the red spectrum is the same spectrum replotted after applying an energy shift which simulates the difference in energy between peaks from the reactions in the back and the front aramid foils. Both were subtracted from the gas-filled ^{22}Ne(p,^{3}He)^{20}F spectrum to produce the background subtracted spectrum in Fig. 52.2 right. The ^{22}Ne(p,^{3}He)^{20}F spectra shown in Fig. 52.2 are obscured by significant foil-related background which could not be effectively subtracted due to the higher energy loss of helions passing through the foils and the gas which makes the resultant double peaking of foil-related contaminant peaks far more severe and difficult to describe. We note here though that the second T = 2 state of ^{20}F at $E_x = 8.05$ MeV, which is thought to be an IAS to the $E_x = 18.43$ MeV 2^+ state in ^{20}Ne, is observed in the ^{22}Ne(p,^{3}He)^{20}F spectrum in Fig. 52.2.

To summarize and conclude, two out of the three T = 2 IAS candidates from [13] were reproduced as well as the 5-alpha cluster candidate, this time at $E_x =$ 22.4 MeV and with a narrower width than previously reported. The cross sections of the low-spin states near $E_x = 20$ MeV from the ^{22}Ne(p,t)^{20}Ne reaction are slightly suppressed by increasing the beam energy from $E_p = 60$ to 80 MeV, including the state at $E_x = 22.4$ MeV which may be considered as a 5-alpha cluster state candidate.

Acknowledgements We are grateful to the iThemba LABS accelerator team, led by Lowry Conradie and Dirk Fourie, for providing a high-quality proton beam to this experiment, and to our mechanical engineer, Zaid Dyers, for constructing the gas compressor unit.

References

1. Spear, R.H.: Phys. Rep. **73**, 369 (1981)
2. Ebran, J.-P., Khan, E., Niksic, T., Vretenar, D.: Letters to Nature **487**, 341 (2012)
3. Jansen, G.R., Schuster, M.D., Signoracci, A., Hagen, G., Navratil, P.: Phys. Rev. **94**, 011301(R) (2016)
4. Hindi, M.M., Thomas, J.H., Radford, D.C., Parker, P.D.: Phys. Rev. C **27**, 2902 (1983)
5. MacArthur, J.D., Evans, H.C., Leslie, J.R., Mak, H.-B.: Phys. Rev. C **22**, 356 (1980)
6. Fujiwara, Y., Horiuchi, H., Tamagaki, R.: Prog. Theor. Phys. **61**, 1629 (1979)
7. Nauruzbayev, D.K., Goldberg, V.Z., Nurmukhanbetova, A.K., Golovkov, M.S., Volya, A., Rogachev, G.V., Tribble, R.E.: Phys. Rev. **96**, 014322 (2017)
8. Ikeda, K., Takigawa, N., Horiuchi, H.: Prog. Theo. Phys. Suppl. **E68**, 464–475 (1968)
9. Tohsaki, A., Horiuchi, H., Schuck, P., Ropke, G.: Phys. Rev. Lett. **87**, 192501 (2001)
10. Zinner, N.T., Jensen, A.S.: Phys. Rev. C **78**, 041306(R) (2008)
11. Kokalova, Tz., Itagaki, N., Von Oertzen, W., Wheldon, C.: Phys. Rev. Lett. **96**, 192502 (2006)
12. Bishop, J., et al.: J. Phys: Conf. Ser. **863**, 012070 (2017)
13. Swartz, J.A., Brown, B.A. et al.: Phys. Rev. C **91** (3), (2015)
14. Neveling, R., et al.: Nucl. Inst. Meth. **654**(1), 29–39 (2011)
15. Swartz, J.A.: PhD thesis, http://scholar.sun.ac.za/handle/10019.1/86443, Stellenbosch University (2014)
16. Bronkhorst High-Tech B.V. (NL), https://www.bronkhorst.com/ (2018)
17. Baloyi, L.P.L.: MSc thesis, University of the Witwatersrand (2018)
18. National Nuclear Data Center, Brookhaven Laboratory, https://www.nndc.bnl.gov (2018)

Chapter 53
Unveiling New Features in Rare Isotopes with Direct Reactions

R. Kanungo

Abstract An overview is presented of how rare isotopes are ushering in a new era in nuclear science embodying exotic few-body correlations in these many-body nuclei at the edges of nuclear binding.

53.1 Introduction

Our Universe has a wide variety of isotopes that govern the characteristics of the various forms of visible matter. Most of the isotopes in the Universe are rare to find on earth though they serve as the breeding grounds for creation of majority of heavy elements that we find around us. The interest in these short lived rare isotopes stem from their pivotal role is serving as the bridge between stable nuclei found on earth and some of the exotic environments of our Universe such as neutron stars, X ray busts and supernovae. These rare isotopes span the vast majority of the nuclear landscape becoming rapidly weakly bound as they approach the edges of nuclear binding, referred to as the neutron- and proton- driplines. This rapid decrease in nucleon separation energy leads to formation of exotic structures in these drip-line nuclei whereby few body correlations manifest in these many-body systems and govern their characteristics.

Some of the key questions that rare isotopes address is how and what type of new features emerge with large neutron-proton asymmetry, how do they relate to the evolution of nuclear shells, what new characteristics of the nuclear force become visible in such extremes and how do the unconventional characteristics of the rare isotopes impact our understanding of extreme neutron-rich objects and heavy element synthesis in our Universe.

To explore these questions, rare isotopes are produced at facilities around the world with mainly two different methods, namely in-flight projectile fragmention/fission

R. Kanungo (✉)
Astronomy and Physics Department, Saint Mary's University, Halifax, NS B3H3C3, Canada
e-mail: ritu@triumf.ca

TRIUMF, Vancouver, BC V6T2A3, Canada

N. A. Orr et al. (eds.), *Recent Progress in Few-Body Physics*,
Springer Proceedings in Physics 238,
https://doi.org/10.1007/978-3-030-32357-8_53

and the Isotope Separator Online (ISOL) technique. In the following I will outline with a few selected experimental investigations from both types of production techniques, evidences of exotic structures in Sect. 53.2, new changes in nuclear shell structure in Sect. 53.3 and sensitivity to the nuclear force in Sect. 53.4.

53.2 Nuclear Halo and Skin

A new era in nuclear physics was triggered by the discovery of the nuclear halo [1, 2]. It is an unexpected form of a nucleus where one or two very weakly bound neutrons reside in the classically forbidden region far from the rest of the nucleus called the 'core'. There are only a handful of such nuclei spotted at the low-mass end of the nuclear chart till date. Of particular interest are the Borromean two-neutron halo nuclei that lie along the dripline. These unique quantum systems depict three-body correlation with two halo nucleons and the core forming a bound system while any two of these subcomponents taken together are not a bound nucleus. Therefore, the few- (three-) body core+n+n correlation is the driving mechanism behind the existence of these bound systems. The matter and proton radii offer the scope to derive the average three-body geometrical correlation of the Borromean nuclei [3]. The correlation derived considers smearing of the core radius due to center of mass motion of the core within the Borromean nucleus under the assumption that the core has the same radius as the bare core nucleus. Figure 53.1 summarizes the measured matter and proton radii from He to C isotopes. The matter radii are determined from measurements of interaction cross sections [4] at projectile fragmentation facilities such as GSI in Germany and RIKEN in Japan using high and intermediate energy rare isotope beams. The proton radii of ^{6,8}He [5, 6], $^{6-11}$Li [7] and $^{9-12}$Be [8, 9] are determined from isotope shift measurements at CERN-ISOLDE in Switzerland, Argonne National Laboratory in USA, TRIUMF in Canada and GANIL in France. Those for ^{14}Be [10], $^{10-17}$B [11], $^{12-19}$C [12] are determined from charge changing cross section measurements at GSI. The large increase in matter radii observed for ^{11}Li, ^{14}Be and ^{17}B indicate two-neutron halos. The evolution of the proton radii differs for these different chains of isotopes and that reflects the variation in the core+n+n correlation. The proton radius of ^{11}Li is considerably larger than that of ^{9}Li though being much smaller to its matter radius. This leads to a configuration with the center of mass of the two halo neutrons, R_{cn}, being ~ 6 fm from the center of the core and the relative distance between the two halo neutrons, R_{nn}, being ~ 6 fm as

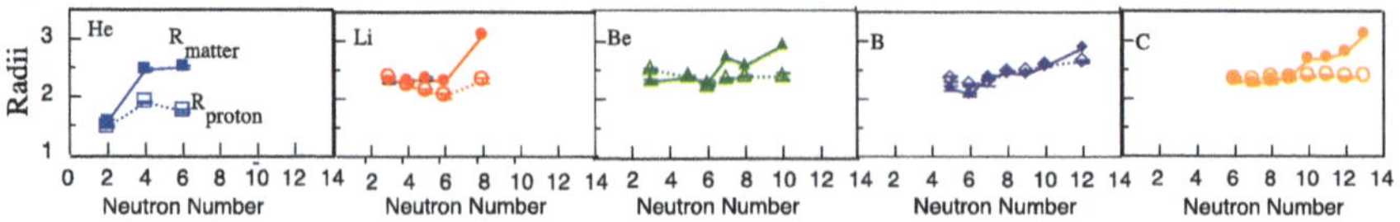

Fig. 53.1 Matter and proton radii of He to C isotopic chains

well. For ^{14}Be however, the proton radius is quite similar to that of ^{12}Be [10] that leads to R_{cn} being only ~3 fm while R_{nn} is ~8 fm showing presence of a cigar type two-neutron halo in this nucleus. The halo radius of ^{14}Be, R_h ~4 fm, is smaller than that of ^{11}Li, R_h ~6 fm. While ^{17}B has a similar halo radius (~4 fm) to ^{14}Be, its proton radius however is larger than that of the core nucleus ^{15}B [11]. This leads to quite a different core+n+n correlation where R_{cn} ~5 fm and R_{nn} is very small depicting a di-neutron type two neutron halo configuration.

A more deeper understanding of the two-neutron halo correlation can be obtained through investigations of two-neutron transfer reactions such as (p,t) or (t,p) reactions. The low beam intensities of the rare isotopes close to the driplines together with the small differential cross sections and wide angular distribution of the reaction ejectiles however have made such measurements challenging and hence rather limited. The ^{11}Li(p,t)^{9}Li reaction measurement was nonetheless accomplished using the high intensity re-accelerated beam of ^{11}Li at TRIUMF and the active target MAYA from GANIL. The angular distribution of the core ^{9}Li in its ground state when interpreted in a DWBA framework (Fig. 53.2a) with simultaneous and sequential transfer requires a strong mixing of the $1p_{1/2}$ and $2s_{1/2}$ orbitals. Such a mixed configuration drives the correlation of the halo neutrons. A new finding in the ^{11}Li(p,t) measurement [12] was the observation of the core ^{9}Li in its first excited state as well. In the framework of nuclear field theory (Fig. 53.2b) this has been interpreted to be the first signature of phonon mediated pairing in nuclei, where exchange of vibrational phonons between the core ^{9}Li and the halo neutrons are discussed to be essential for the binding of the halo [13].

The presence of a correlated pair of weakly bound neutrons in the Borrmean halo nuclei opens the possibility of a new mode of excitation. On one hand it was postulated by Ikeda that the oscillation of the halo neutrons against the core could give rise to low-energy dipole resonance states [14]. On the other hand Hansen and Jonson

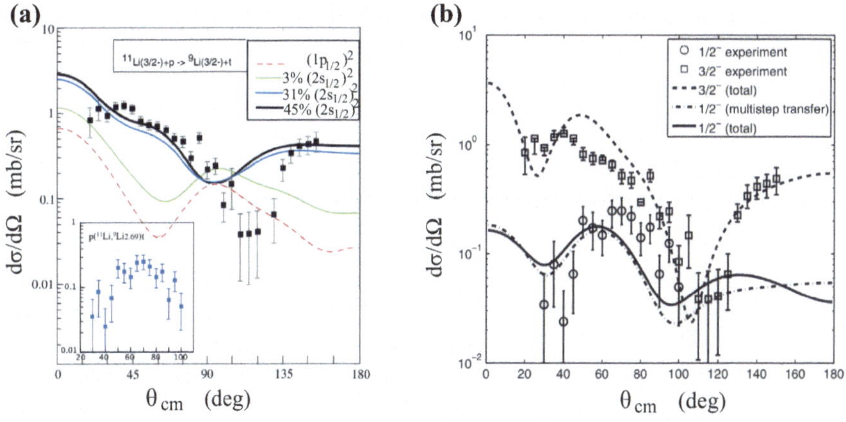

Fig. 53.2 ^{11}Li(p,t)^{9}Li differential cross sections **a** Curves show DWBA calculations, adapted from [12] **b** Curves show nuclear field theory calculations, adapted from [13]

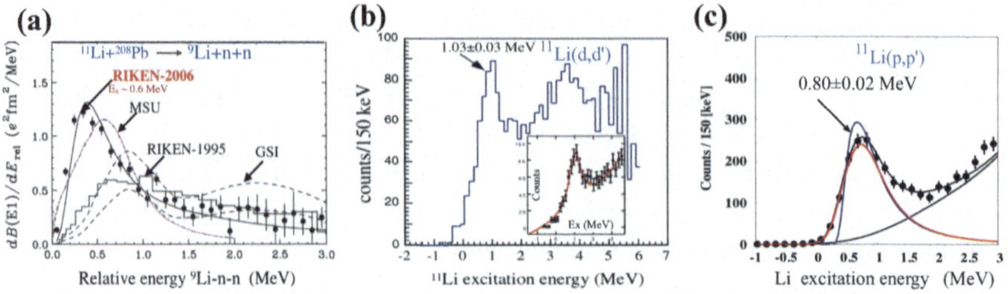

Fig. 53.3 Dipole peak in ^{11}Li observed from **a** Coulomb dissociation, adapted from [15] **b** ^{11}Li(d,d') adapted from [18] (c) ^{11}Li(p,p') adapted from [19]

proposed that the weak binding of the halo would give rise to soft electric dipole mode that could be excited in Coulomb collisions [2]. The most recent measurement of Coulomb dissociation of ^{11}Li observed a peak in the cross section at ~0.6 MeV excitation energy [15] (Fig. 53.3a). Early studies of proton inelastic scattering at intermediate energies of ~ E/A ~ 80 MeV reported a peak at 1.3±0.1 MeV with a width of $\Gamma = 0.75 \pm 0.6$ MeV [16]. The excitation energy resolution in this measurement was ~1.5 MeV (FWHM) which left open the question whether the reported structure was only a threshold effect or does a resonance state exist in ^{11}Li. Pion capture reaction found a structure at ~1.02 MeV [17]. However the nature of this state was not understood from this reaction. More recently, the availability of low-energy reaccelerated beam of ^{11}Li made it possible to perform high precision inelastic scattering measurements at TRIUMF. The ^{11}Li(*d,d'*) reaction at E/A = 5.5 MeV provided the first signature of a narrow isoscalar dipole resonance at 1.03 ± 0.03 MeV [18] (Fig. 53.3b). Subsequently the ^{11}Li(*p,p'*) measurement at E/A~6 MeV also found a resonance peak at 0.8±0.02 MeV [19] (Fig. 53.3c). The dipole nature of these resonances was ascertained from the shape of their angular distributions when compared to DWBA predictions. A dipole excitation in ^{11}Li can give rise to states with spins of $3/2^+$, $5/2^+$ or $1/2^+$. The relative population of these states could be different in the (*d,d'*) and (*p,p'*) reactions which potentially could explain the small difference in the resonance peak seen in the two reactions.

53.3 Shell Evolution Towards the Drip-Lines

The formation of the exotic structures is intricately related to metamorphosis of nuclear shells. In a conventional picture of nuclear shells ^{11}Li is an $N = 8$ closed shell nucleus where the two halo neutrons should reside in the $1p_{1/2}$ orbital with 100% occupancy. In such an arrangement, the ^{10}Li unbound sub-component of ^{11}Li would be in a *p*-wave resonance state with spin of 2^+- or 1^+. Predictions in the framework of the hyperspherical expansion of the Faddeev equations suggest 40% *p*-wave component for ^{11}Li with ^{10}Li being in *p*-wave resonance at ~0.5 MeV and requires a second p-wave resonance at a lower energy [20]. Neutron removal and

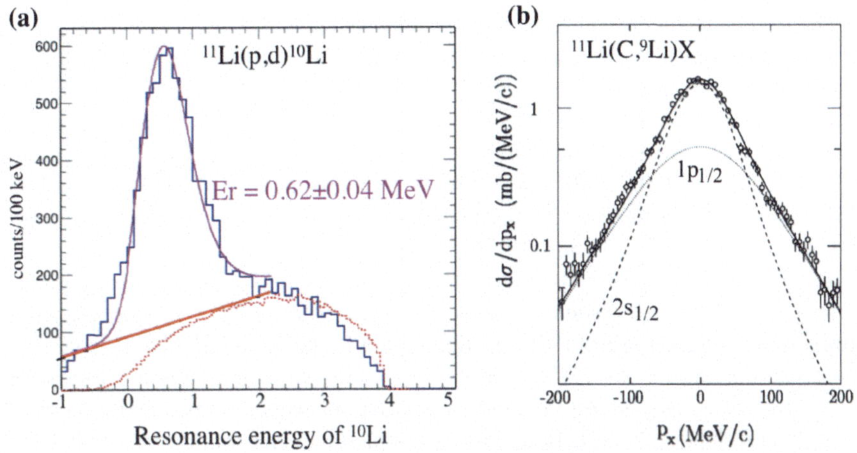

Fig. 53.4 **a** ^{10}Li p-wave resonance population in the ^{11}Li(p,d)^{10}Li reaction, adapted from [21] **b** Momentum distribution of two-neutron removal from ^{11}Li, adapted from [24]

neutron transfer reactions from ^{11}Li can provide experimental signature of the neutron occupancy and resonance state(s) of the ^{10}Li sub-component. The ^{11}Li(*p*,*d*)^{10}Li one-neutron transfer reaction measurement was found to populate a prominent resonance peak at E_r=0.62±0.04 MeV [21] (Fig. 53.4a). The measured differential cross section when analyzed in a DWBA framework confirmed this resonance to have p-wave character. The p-wave occupancy was found to be 33 ± 12%. This low probability of the halo neutrons filling the conventional $p_{1/2}$ orbital shows breakdown of the $N = 8$ shell closure. The large radius of ^{11}Li [22] and the neutron removal momentum distributions have shown significant occupation of the intruder $2s_{1/2}$ orbital [23–26] (Fig. 53.4b).

The ground states of neutron-rich Li–C isotopes exhibit configuration mixing with the $2s_{1/2}$ orbital coming lower in energy and crossing the $1d_{5/2}$ and in some cases even the $1p_{1/2}$ orbital [3]. There are many factors that drive the changes in conventional ordering of orbitals in the rare isotopes. While weak binding is only one of the aspects that tend to bring low angular momentum orbitals lower in energy, there are features of the nuclear force that show prominence in these neutron-proton asymmetric nuclei and affect the orbital arrangement. The monopole tensor interaction has been discussed to have a major contribution to shell changes [27]. The attractive p-n interaction between nucleons in the $j_> = l+1/2$ and $j_< = l-1/2$ orbitals and repulsive interaction between nucleons in the $j_>$–$j_>$ or $j_<$–$j_<$ orbitals can arise both from the tensor interaction as well as the spin-isospin central interaction [27, 28]. Changes of such interactions in moving from stable to rare isotopes cause migration of the conventional orbital scheme thereby causing shell gaps to change. In addition, the role of the three-nucleon force has been found be important in explaining the binding energy [29] and hence the drip-line location as well as excited states that reflect shell closures [30–32].

The isotope chain of oxygen isotopes is truncated at ^{24}O where a new magic N = 16 shows appearance at the neutron drip-line. Empirical signature of this shell closure was pointed out from neutron separation energy systematics [33]. The high excitation energy of the unbound 2^+ state in ^{24}O (Fig. 53.5a) affirms this [34]. The one-neutron removal momentum distribution finds the valence neutrons to exclusively occupy the $2s_{1/2}$ orbital (Fig. 53.5b) confirming ^{24}O to be a spherical closed shell at the drip-line [35]. This is further supported by the small deformation parameter that is found from inelastic scattering (Fig. 53.5c) [36] (Fig. 53.6).

The unbound nature of $^{26,28}O$ raise interest in understanding the nature of correlation and orbitals in their resonance states. The lowest state of ^{26}O was observed from proton removal reaction from ^{27}F reacting with a Be target at RIKEN. This narrow resonance observed at only 18(3 stat.)(4 syst.) keV above the ^{24}O+n+n threshold [37]. This was consistent with theoretical predictions where the two neutrons were strongly correlated [38, 39]. In a Gamow coupled channel framework predict the ground state wavefunction of ^{26}O to have three peaks associated with di-neutron, cigar shape and triangular shape n-n correlation [40].

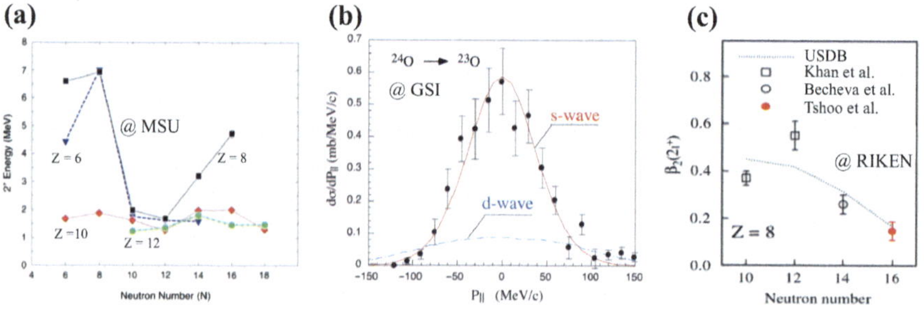

Fig. 53.5 **a** 2^+ state in ^{24}O, adapted from [34] **b** Momentum distribution from one neutron removal from ^{24}O adapted from [35] **c** Deformation parameter from ^{24}O(p,p'), adapted from [36]

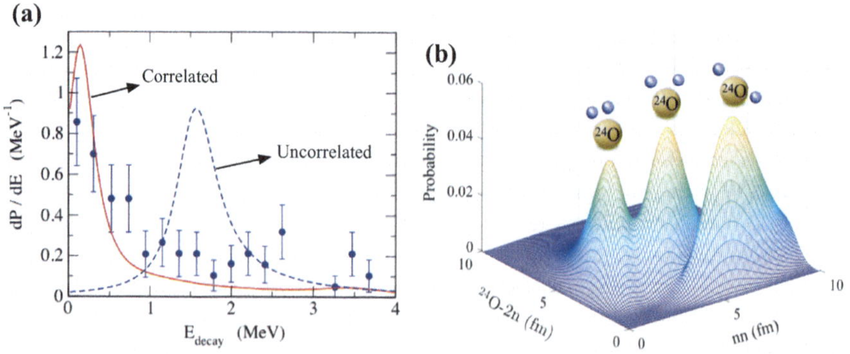

Fig. 53.6 **a** The two-neutron decay energy spectrum of ^{26}O [37] compared to calculations of [38], adapted from [38] **b** Wavefunction of ^{26}O in Gamow coupled channel framework adapted from [40]

53.4 Nuclear Force Sensitivity in Rare Isotopes

The unexpected nuclear properties that are emerging in rare isotopes point to our incomplete knowledge of the nuclear force. Describing the residual strong force that binds protons and neutrons into complex many-body systems has been a challenge for over a century. In recent times its description with a link to the fundamental theory of quantum chromodynamics (QCD) is enabled by the chiral effective field theory. This is applicable at low-energies where the energy of the probe has wavelength much larger than the size of nucleons. Here the Hamiltonian can be expressed as an expansion in powers of (Q/Λ) where Q is the pion mass, which is the soft scale, and Λ is the momentum cutoff, which is the hard scale. In the expansion the low-energy constants (LEC) containing the short-range contributions are found by various methods that fit to some know nuclear properties. This gives rise to different prescriptions of the nuclear force. An experimental guidance is therefore necessary to understand which of them are suitable to describe successfully the various characteristics of drip-line rare isotopes.

Most of the interactions are found to explain the binding energies pretty well. The three-nucleon force had been found to be important for explaining the nuclear binding. While mostly interactions are benchmarked against static properties of nuclei, dynamics observables can bring a greater sensitivity to constrain the description of the nuclear force. Recent development of *ab initio* reaction theory makes it possible to describe the structure and reactions of light nuclei in the vicinity of the drip-lines on the same footing. The proton elastic scattering angular distribution was measured using the re-accelerated beam of ^{10}C at $E_{cm} \sim 4.16$ MeV at TRIUMF [41]. The *ab initio* calculations were based on the no-core shell model with continuum [42, 43]. The forces from chiral effective field theory used were the two-nucleon NN force at N^{3}LO [44], the NN force at N^{3}LO + 3 N force at N^{2}LO with cutoff of L = 400 MeV namely NN + 3 N(400) [45] and the NN + 3 N force N^{2}LOsat [46] where the LECs are determined simultaneously for the NN and 3 N sectors. Figure 53.7a shows that the angular distribution using the NN force has a pronounced dip in cross section $\theta_{cm} \sim 85^\circ$ which is in contrast to the measured distribution. The inclusion of the 3 N force improves the chisquare fit to the data with a much more reduced minimum of cross section that is shifted to a higher angle of $\theta_{cm} \sim 120^\circ$. However, the shape of this angular distribution is still not in proper phase agreement with the data thereby showing that the NN + 3 N(400) force is an inadequate description of the nuclear force. The N^{2}LOsat force predicts an angular distribution which agrees well with the shape of the measured one as seen by the dashed curve in Fig. 53.7a. However the red solid curve in Fig. 53.7a shows that it overpredicts in magnitude. The dramatically different shapes of the angular distributions using the three different force prescriptions demonstrate the strong sensitivity of this dynamic observable to the nuclear force description. Furthermore, although the N^{2}LOsat force has been successful in explaining the bulk properties of many nuclei especially in the neutron-rich region its failure to explain the magnitude of the cross section suggests that it is still not an

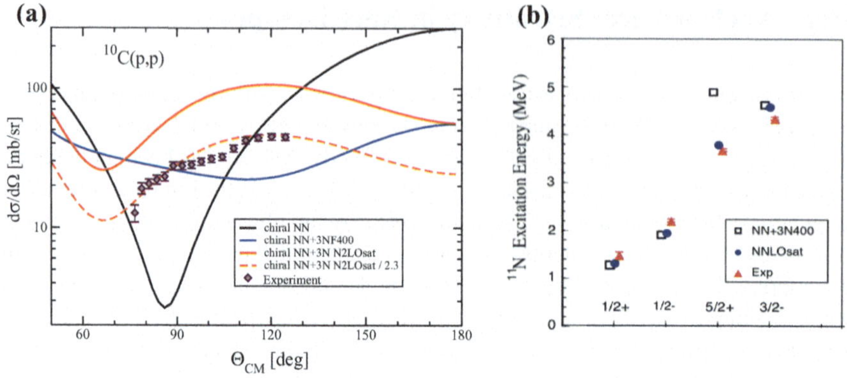

Fig. 53.7 **a** $^{10}C(p,p)$ differential cross section compared to no-core shell model calculations with different chiral forces [41] **b** Resonance energies of ^{11}N compared to *ab intio* calculations

appropriate description of the nuclear force and its shortcomings become visible at the proton drip-line.

Excitation spectra of drip-line nuclei can also serve as testing grounds for the nuclear force prescriptions. The p+^{10}C composite system is an unbound nucleus ^{11}N whose experimentally observed resonance states are compared with the NN + 3 N(400) and N^2LO_{sat} interactions in Fig. 53.7b. It is found that the two interactions predict very similar resonance energies except for the $5/2^+$ resonance state. Furthermore, the predictions with the N^2LOsat interactions agree fairly well with the data. This shows that one would have reached an incomplete conclusion on the nuclear forces with a study of the resonance energies alone, while the proton elastic scattering revealed a much stronger variance of the different forces.

Moving to ^{20}Mg, the next isotope in the proton drip-line with same mass/charge ratio as ^{10}C deuteron inelastic scattering reaction found presence of a new first resonance state at $3.70^{+0.02}_{-0.20}$ MeV above the proton emission threshold [47]. The observed resonance state is compared to *ab initio* predictions using two different many body frameworks namely the in-medium similarity renormalization group method (IMSRG) and the coupled cluster effective interaction (CCEI). In both these frameworks few different nuclear force prescriptions from the chiral effective field theory was used. The predicted excited states are shown in Fig. 53.8. The overall observation is that the predictions using the same force prescription is similar for the different many-body frameworks. This shows that the differences in the predictions arise largely due to the different force prescriptions. None of the forces seem to explain the observed resonance energy with all predictions being lower than the data. It is found that the prediction with the N^2LO_{sat} interaction has the largest deviation from the data thereby suggesting the shortcoming of this description at the proton drip-line despite its success in predicting bulk properties of some neutron-rich nuclei.

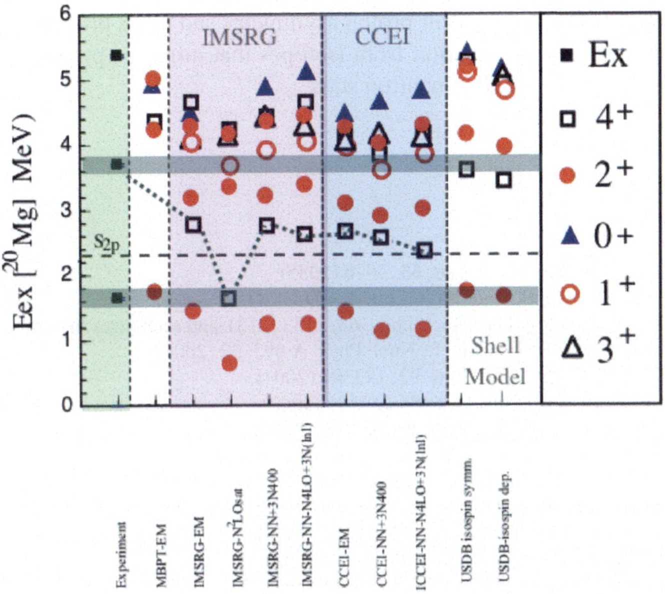

Fig. 53.8 Measured excited states in ^{20}Mg (black squares) compared to predictions in the framework of IMSRG, CCEI and shell model. The *ab initio* calculations use the chiral forces, while the shell model uses the USDB interaction as labeled in the figure

53.5 Summary

We have presented through selected examples how rare isotopes are rapidly changing our conventional concepts. The discussion has been largely focused on some reaction studies using both in-flight beams at high and intermediate energies as well as re-accelerated beams at low energies. We have found few-body correlations appearing in Borromean drip-line nuclei some of which form two-nucleon halos. The correlated halo neutron pair was postulated to induce a new soft dipole mode of oscillation whose existence has been now experimentally established. Reactions show strongly correlated halo neutrons in ground state as well as unbound states at and beyond the drip-lines. These correlations that were found to differ in different halo nuclei are strongly guided by the shell structure of the nuclei. Rare isotopes are revolutionizing the fundamentals pillars of nuclear physics through a metamorphosis of nuclear shells. The disappearance of conventional $N = 8$ and 20 shells are found to vanish through the halo features of ^{11}Li and unbound nature of ^{26,28}O, respectively. On the other hand a new shell closure making ^{24}O a doubly magic nucleus has emerged at the drip-line. The new features at the drip-lines calls for a renewed understanding of the nuclear force whose description in recent times through the chiral effective field theory builds a link to QCD. Dynamic observables of few-body drip-line nuclei ^{10}C and ^{20}Mg were shown to have strong sensitivity to constraining the nuclear force prescriptions. In the coming years new generation facilities will extend our reach to heavy isotopes with large neutron-proton asymmetry and holds promise to unfold a wealth of further new information from isotopes that might be produced in exotic environments such as colliding neutron stars.

References

1. Tanihata, I., et al.: Phys. Rev. Lett. **55**, 2676 (1985)
2. Hansen, P.G., Jonson, B.: Eur. Phys. Lett. **4**, 409 (1987)
3. Tanihata, I., et al.: Prog. Part. Nucl. Phys. **68**, 215 (2013). and references therein
4. Ozawa, A., Suzuki, T., Tanihata, I.: Nucl. Phys. A **693**, 32 (2001)
5. Wang, L.B., et al.: Phys. Rev. Lett. **92**, 142501 (2004)
6. Mueller, P., et al.: Phys. Rev. Lett. **99**, 252501 (2007)
7. Sanchez, R., et al.: Phys. Rev. Lett. **96**, 033002 (2006)
8. Nörtershäuser, W., et al.: Phys. Rev. Lett. **102**, 062503 (2009)
9. Krieger, A., et al.: Phys. Rev. Lett. **108**, 142501 (2012)
10. S. Terashima et al., Prog. Theor. Exp. Phys. 101D02 (2014)
11. Estrade, A., et al.: Phys. Rev. Lett. **113**, 132501 (2014)
12. Tanihata, I., et al.: Phys. Rev. Lett. **100**, 192502 (2008)
13. Potel, G., et al.: Phys. Rev. Lett. **105**, 172502 (2010)
14. Ikeda, K.: Nucl. Phys. A **538**, 355c (1992)
15. Nakamura, T., et al.: Phys. Rev. Lett. **96**, 252502 (2006)
16. Korshennikov, A.A., et al.: Phys. Rev. Lett. **78**, 2317 (1997)
17. Gornov, M.G., et al.: Phys. Rev. Lett. **81**, 4325 (1998)
18. Kanungo, R., et al.: Phys. Rev. Lett. **114**, 192502 (2015)
19. Tanaka, J., et al.: Phys. Lett. B **774**, 268 (2017)
20. Garrido, E., et al.: Nucl. Phys. A **700**, 117 (2012)
21. Sanetullaev, A., et al.: Phys. Lett. B **755**, 481 (2016)
22. Yabana, K., Ogawa, Y., Suzuki, Y.: Nucl. Phys. A **539**, 295 (1992)
23. Zinser, M., et al.: Nucl. Phys. A **619**, 151 (1997)
24. Simon, H., et al.: Phys. Rev. Lett. **83**, 496 (1999)
25. Aksyutina, Yu., et al.: Phys. Lett. B **666**, 430 (2008)
26. Simon, H., et al.: Nucl. Phys. A **791**, 267 (2007)
27. Otsuka, T., et al.: Phys. Rev. Lett. **95**, 232502 (2005)
28. Otsuka, T., et al.: Phys. Rev. Lett. **87**, 082502 (2001)
29. Otsuka, T., et al.: Phys. Rev. Lett. **105**, 232501 (2010)
30. Hagen, G., et al.: Phys. Rev. Lett. **108**, 242501 (2012)
31. Hagen, G., et al.: Phys. Rev. Lett. **109**, 032502 (2012)
32. Holt, J.D., et al.: Phys. Rev. Lett. **110**, 022502 (2013)
33. Ozawa, A., et al.: Phys. Rev. Lett. **84**, 5493 (2000)
34. Hofman, C.R., et al.: Phys. Lett. B **672**, 17 (2009)
35. Kanungo, R., et al.: Phys. Rev. Lett. **102**, 152501 (2009)
36. Tshoo, K., et al.: Phys. Rev. Lett. **109**, 022501 (2012)
37. Kondo, Y., et al.: Phys. Rev. Lett. **116**, 102503 (2016)
38. Hagino, K., et al.: Phys. Rev. C **89**, 014331 (2014)
39. Grigorenko, L.V., Zhukov, M.V.: Phys. Rev. C **91**, 064617 (2015)
40. Wang, S.M., et al.: Phys. Rev. C **96**, 044307 (2017)
41. Kumar, A., et al.: Phys. Rev. Lett. **118**, 262502 (2017)
42. Baroni, S., Navrátil, P., Quaglioni, S.: Phys. Rev. Lett. **110**, 022505 (2013)
43. Navrátil, P., et al.: Phys. Scr. T. **91**, 053002 (2016)
44. Entem, D.R., Machleidt, R.: Phys. Rev. C **68**, 041001 (2003)
45. Roth, R., et al.: Phys. Rev. Lett. **109**, 052501 (2012)

46. Ekström, A., et al.: Phys. Rev. C **91**, 051301R (2015)
47. Randhawa, J.S., et al.: Phys. Rev. C **99**, 021301R (2019)

Chapter 54
The Unitarity Expansion for Light Nuclei

Sebastian König

Abstract It is argued here that (at least light) nuclei may reside in a sweet spot: bound weakly enough to be insensitive to the details of the interaction, but dense enough to be insensitive to the exact values of the large two-body scattering lengths as well. In this scenario a systematic expansion of nuclear observables around the unitarity limit converges. In particular, the nuclear force in this scheme is constructed such that the gross features of states in the nuclear chart are determined by a very simple leading-order interaction, whereas—much like the fine structure of atomic spectra—observables are moved to their physical values by small *perturbative* corrections. Explicit evidence in favor of this conjecture is shown for the binding energies of three and four nucleons.

54.1 Introduction

Ever since the effective range expansion was developed as a theory to parameterize the low-energy two-nucleon system [1–4] it has been known that the nucleon-nucleon (NN) scattering lengths, $a_t \simeq 5.4$ fm and $a_s \simeq -23.7$ fm in the ${}^3S_1$ and ${}^1S_0$ channels, respectively, are large compared to the typical range of the nuclear interaction, $R \sim M_\pi^{-1} \simeq 1.4$ fm, set by the inverse pion mass. Considering quantum chromodynamics (QCD) as the underlying theory of the strong interaction, this particular feature of the low-energy two-nucleon ($2N$) system can be understood as an accidental "fine tuning" of the QCD parameters [5–9] (the quark masses) to be close to a critical point where the scattering lengths are infinite, the so-called "unitarity (or unitary) limit."

This curiosity of nature has profound consequences for the theoretical description of few-nucleon systems at low energies, placing them in the same universality class

S. König (✉)
Institut für Kernphysik, Technische Universität Darmstadt, 64289 Darmstadt, Germany
e-mail: sekoenig@theorie.ikp.physik.tu-darmstadt.de

ExtreMe Matter Institute EMMI, GSI Helmholtzzentrum für Schwerionenforschung GmbH, 64291 Darmstadt, Germany

N. A. Orr et al. (eds.), *Recent Progress in Few-Body Physics*,
Springer Proceedings in Physics 238,
https://doi.org/10.1007/978-3-030-32357-8_54

as other systems governed by large scattering lengths, such as cold atomic gases, where the scattering length can be tuned *via* Feshbach resonances [10], or certain mesons which can be interpreted as hadronic molecules [11]. Most notably, the triton is understood to be the single remaining bound state out of an infinite tower of Efimov states [12] that exists in the exact unitarity limit [13–15]. Recently it was shown in a model-independent way that a virtual state in the three-nucleon ($3N$) system, known to exist for a long time [16, 17], is an S-matrix pole that would be an excited Efimov state if nature were just slightly closer to the unitarity limit [18], confirming a relation previously observed in a separable potential model [19].

Following [20] it is argued here that nature is indeed close enough to unitarity such that it is possible to quantitatively describe the spectra of—at least light, and possible also heavier—nuclei by a perturbative expansion around the limit of infinite two-body scattering lengths. At leading order (LO) this yields an interaction which is parameter free in the $2N$ sector and determined by a single three-body parameter, adjusted to keep the triton binding energy fixed at its experimental value. This remarkably simple theory is shown to capture the gross features of nuclei up to ^{4}He, while corrections such as the actual finite values of the $2N$ scattering lengths and electromagnetic effects are accounted for in perturbation theory.

Quantitatively, this "unitarity expansion" is constructed as a variant of pionless effective field theory (pionless EFT). This theory, most recently reviewed in [21], describes low-energy nuclear systems in a model-independent way, guided only by the symmetries of QCD and the universal physics of systems governed by a large scattering length. As such, it is ideally suited to set up the unitarity expansion with a minimal set of assumptions. An important aspect of each EFT is the organizational principle called "power counting," which attributes the various terms to different orders in a systematic expansion. In the standard pionless theory, the expansion parameter is given by a typical low-momentum scale Q divided by the high scale $R^{-1} \sim M_\pi$. The unitarity expansion is obtained by assuming that $Q \sim Q_A = \sqrt{2M_N B_A/A}$, placing nuclei in a "sweet spot" $1/a_{s,t} < Q_A < 1/R$, where a combined expansion in $Q_A R$ and $1/(Q_A a_{s,t})$ converges.

In the following, the formalism is discussed by describing its application to calculate systems of up to four nucleons. Readers not interested in the more technical details are invited to skip ahead to Sect. 54.3, which presents the main results and provides a broader perspective that places the unitarity expansion in line with other recent results suggesting a fascinating simplification of nuclear physics.

54.2 Formalism

Following the notation of [20, 22], pionless EFT is defined in terms of a Lagrange density

$$\mathcal{L} = N^\dagger \left(\mathrm{i}\mathcal{D}_0 + \frac{\mathcal{D}^2}{2M_N} \right) N + \sum_{\mathbf{i}} C_{0,\mathbf{i}} \left(N^T P_{\mathbf{i}} N\right)^\dagger \left(N^T P_{\mathbf{i}} N\right) + D_0 \left(N^\dagger N\right)^3 + \cdots , \tag{54.1}$$

involving nonrelativistic nucleon isospin doublets $N = (p\ n)^T$ as well as photon fields A_μ which are coupled to the nucleons *via* the covariant derivative $\mathcal{D}_\mu = \partial_\mu + \mathrm{i}eA_\mu(1+\tau_3)/2$, where e is the proton charge and τ_a is used to label isospin Pauli matrices. Besides these electromagnetic interactions, of which only the static Coulomb potential is relevant to the order considered here, the theory involves only contact (zero range) interactions proportional to "low-energy constants" (LECs), such as the $C_{0,\mathbf{i}}$, D_0 shown in (54.1), plus other contributions (involving an increasing number of derivatives acting on the nucleon fields) contained in the ellipses. The $P_{\mathbf{i}}$ denote projectors onto the NN S waves, $\mathbf{i} = {}^1S_0, {}^3S_1$, corresponding to the short-hand labels used above for the singlet and triplet scattering lengths.

Leading-order terms are summed to all orders in a nonperturbative treatment to which higher-order corrections are added in perturbation theory. This procedure implies that the LECs of all operators are split into different orders, e.g., $C_{0,\mathbf{i}} = C_{0,\mathbf{i}}^{(0)} + C_{0,\mathbf{i}}^{(1)} + \cdots$. Typically, only the leading term in this expansion introduces a new parameter whereas the higher-order contributions are merely used to maintain lower-order renormalization conditions as additional corrections are included. The unitarity expansion departs from this scheme by moving the introduction of two-body parameters from $C_{0,\mathbf{i}}^{(0)}$ to $C_{0,\mathbf{i}}^{(1)}$.

The LO calculation can be carried out in closed form by solving the Lippmann-Schwinger equation for a separable potentials $V_{2,\mathbf{i}}^{(0)} = C_{0,\mathbf{i}}^{(0)} |g\rangle\langle g|$, where $\langle p|g\rangle = g(p^2) = \exp(-p^2/\Lambda^2)$ with a cutoff scale Λ is a Gaussian regulator and p is the NN center-of-mass momentum. This is shown diagrammatically in Fig. 54.1. Some results discussed in Sect. 54.3 are obtained using a slightly different implementation, employing so-called "dibaryon" fields to describe the two-body sector and using a sharp momentum cutoff, which is essentially equivalent to choosing $\langle p|g\rangle$ to be a step function. This approach, deriving two- and three-body equations directly from Feynman diagrams, has been discussed in detail in [22, 23], so that here only the potential formalism with Gaussian regulator is considered.

Starting from $t_{\mathbf{i}}^{(0)} = V_{2,\mathbf{i}}^{(0)} + V_{2,\mathbf{i}}^{(0)} G_0 t_{\mathbf{i}}^{(0)}$, where G_0 is the free two-body Green's function, the separable form of the interaction makes it possible to directly write down the solution as

Fig. 54.1 Diagrammatic version of the Lippmann-Schwinger equation for the two-nucleon T-matrix at LO, depicted by the circled zero. The solid lines represent nucleon fields, whereas the dot represents a contact interaction $C_{0,\mathbf{i}}^{(0)}$

$$t_{\mathrm{i}}^{(0)}(z;\mathbf{k},\mathbf{k}') = \langle\mathbf{k}|t_{\mathrm{i}}^{(0)}|\mathbf{k}'\rangle = g(k^2)\tau_{\mathrm{i}}(z)g(k'^2),\ \tau_{\mathrm{i}}(z) = \left[1/C_{0,\mathrm{i}}^{(0)} - \langle g|G_0|g\rangle\right]^{-1}, \tag{54.2}$$

where z denotes the energy. $C_{0,\mathrm{i}}^{(0)}$ can now be determined by matching this T-matrix to the effective range expansion at the on-shell point, $\mathbf{k} = \mathbf{k}'$ and $E = k^2/M_N$. The unitarity limit (infinite scattering length, $1/a_{\mathrm{i}} = 0$) is reproduced by setting $C_{0,\mathrm{i}}^{(0)} = \frac{-2\pi^2}{M_N \Lambda}\theta^{-1}$, where $\theta = 1/\sqrt{2\pi}$ for the Gaussian regulator used here. This means that the $C_{0,\mathrm{i}}^{(0)}$ do not introduce any physical parameters. At NLO, the correction to the T-matrix is

$$t_{\mathrm{i}}^{(1)} = V_{2,\mathrm{i}}^{(1)} + V_{2,\mathrm{i}}^{(1)}G_0 t_{\mathrm{i}}^{(0)} + t_{\mathrm{i}}^{(0)}G_0 V_{2,\mathrm{i}}^{(1)} + t_{\mathrm{i}}^{(0)}G_0 V_{2,\mathrm{i}}^{(1)}G_0 t_{\mathrm{i}}^{(0)}, \tag{54.3}$$

i.e., the sum of all possible terms linear in $V_{2,\mathrm{i}}^{(1)}$. The overall NLO T-matrices $t_{\mathrm{i}}^{(0)} + t_{\mathrm{i}}^{(1)}$ should reproduce the physical values of the *NN* S-wave scattering lengths, which leads to $C_{0,\mathrm{i}}^{(1)} = \frac{M_N}{4\pi a_{\mathrm{i}}} C_{0,\mathrm{i}}^{(0)2}$. Note that instead of using (54.3) it is possible to conveniently obtain $t_{\mathrm{i}}^{(1)}$ as well as higher-order corrections by solving integral equations similar to the one defining $t_{\mathrm{i}}^{(0)}$. Details about this procedure can be found in [23, 24].

With the two-body LECs thus determined it is possible to proceed with calculations for three and four nucleons. In the following the unified Faddeev + Faddeev-Yakubovsky (F+FY) framework that was used to obtain the four-nucleon results presented in [20] is summarized. The approach follows [25–28] (which, in turn, are based on the work of Kamada and Glöckle [29]), but uses an independently developed numerical implementation. Since a fully comprehensive description of the method is beyond the scope of this work, emphasis is put here primarily on details regarding the perturbative treatment of NLO contributions.

Three nucleons It is a distinct feature of pionless EFT that a three-nucleon interaction (3NI) enters at LO in the power counting, while naïvely it would be expected to contribute only much later. This promotion of the 3NI is a direct consequence of the unnaturally large *NN* S-wave scattering length, leading to the triton as an effective Efimov state [13–15]. In the separable potential formalism the LO 3NI can be implemented as

$$V_3^{(0)} = D_0^{(0)}\,|^3\mathrm{H}\rangle|\xi\rangle\langle\xi|\langle^3\mathrm{H}|\,, \tag{54.4}$$

where $|^3\mathrm{H}\rangle$ projects onto a $J = T = 1/2$ three-nucleon state and the regulator is now defined, for Jacobi momenta $\mathbf{u}_1 = \frac{1}{2}(\mathbf{k}_1 - \mathbf{k}_2)$ and $\mathbf{u}_2 = \frac{2}{3}[\mathbf{k}_3 - \frac{1}{2}(\mathbf{k}_1 + \mathbf{k}_2)]$, as $\langle\mathbf{u}_1\mathbf{u}_2|\xi\rangle = g\big(u_1^2 + \frac{3}{4}u_2^2\big)$. The $\mathbf{k}_i$ label the individual nucleon momenta. The NLO correction $V_3^{(1)}$ has the same form as (54.4), but involves the LEC $D_0^{(1)}$.

The Faddeev equations in an abstract operator notation take the form

$$|\psi\rangle = G_0\, t^{(0)}\, P|\psi\rangle + G_0\, t^{(0)}\, |\psi_3\rangle\,, \tag{54.5a}$$

$$|\psi_3\rangle = G_0\, t_3^{(0)}\, (1 + P)|\psi\rangle\,, \tag{54.5b}$$

where $|\psi\rangle = |\psi_{(12)3}\rangle$ is one of the (equivalent) two-body Faddeev components and $|\psi_3\rangle$ is an auxiliary three-body amplitude [27]. G_0 now denotes the free three-body Green's function and $P = P_{12}P_{23} + P_{13}P_{23}$ generates the non-explicit components through permutations. $t^{(0)}$ collectively denotes the two-body T-matrices $t_{\mathbf{i}}^{(0)}$, whereas $t_3^{(0)}$ is the solution of Lippmann-Schwinger like equation with $V_3^{(0)}$ as driving term. The equations are solved by projecting onto momentum states $|u_1 u_2; s\rangle$, where $s = |\left(\ell_2\left((\ell_1 s_1)\, j_1 \tfrac{1}{2}\right) s_2\right) J; \left(t_1 \tfrac{1}{2}\right) T\rangle$ collects the relevant angular momentum, spin, and isospin quantum numbers, coupled such that $(\ell_1 s_1)\, j_1$ describes the two-nucleon subsystem and ℓ_2 denotes the orbital angular momentum associated with the Jacobi momentum u_2. Since only S-wave interactions enter to the order considered here, all sums over s are naturally truncated to involve only states with $(\ell_1 s_1)\, j_1 = \mathbf{i} = {}^1S_0, {}^3S_1$. For details regarding the implementation and solution of (54.5), see [27, 30, 31], noting that the coupling scheme used here for $|s\rangle$ is a somewhat unusual choice for $3N$ calculations; it is chosen in order to be consistent with the four-nucleon states introduced below.

In order to calculate the NLO triton energy, the full LO wavefunction $|\Psi\rangle = (1+P)|\psi\rangle + |\psi_3\rangle$ is required. Assuming it to be normalized such that $\langle\Psi|\Psi\rangle = 1$, the NLO energy shift is given by

$$\Delta E = \langle\Psi|V_{\text{NLO}}|\Psi\rangle,\ V_{\text{NLO}} = \sum\nolimits_{\mathbf{i}} V_{2,\mathbf{i}}^{(1)} + V_3^{(1)}\,. \tag{54.6}$$

To check the calculation, one can verify that $\langle\Psi|H_{\text{LO}}|\Psi\rangle$ with $H_{\text{LO}} = H_0 + \sum_{\mathbf{i}} V_{2\mathbf{i}}^{(0)} + V_3^{(0)}$ gives the same energy as obtained directly from (54.5).

While for the Faddeev equations only the potential between a single pair of nucleons (chosen to be nucleons 1 and 2) is needed explicitly, evaluating matrix elements requires the *full* two-body potential including all pairwise interactions. Temporarily dropping sub- and superscripts for simplicity, this can be written as [31]

$$V_2 = V_{12} + (P_{12}P_{23})V_{12}(P_{23}P_{12}) + (P_{13}P_{23})V_{12}(P_{23}P_{13})\,. \tag{54.7}$$

Using the antisymmetry of the full wavefunction, $P_{ij}|\Psi\rangle = -|\Psi\rangle\ \forall\ P_{ij}$, one can write $V_2|\Psi\rangle = (1+P)V_{12}|\Psi\rangle$, and noting furthermore that $(1+P)^\dagger(1+P) = 3(1+P)$ gives $\langle\Psi|V_2|\Psi\rangle = 3\langle\Psi|V_{12}|\Psi\rangle$. Similar simplifications together with $(1+P)|\psi_3\rangle = 3|\psi_3\rangle$ can be applied to the norm and matrix elements of H_0.

Equation (54.5) are solved to tune $D_0^{(0)}$ at LO (with the two-body S-waves at unitarity) such that the triton bound state comes out at its physical energy. At NLO, where the finite physical scattering lengths are included *via* the $V_{2,\mathbf{i}}^{(1)}$, there is a corresponding shift in the triton energy. The LEC $D_0^{(1)}$ is adjusted such that this shift is compensated by $V_3^{(1)}$, thus keeping the triton at its physical position. Once this is done, all ingredients are in place to make predictions for four nucleons.

Four nucleons For the four-nucleon system, there are two distinct Faddeev-Yakubovsky components, $|\psi_A\rangle$ and $|\psi_B\rangle$, corresponding two 3+1 and 2+2 cluster configurations of the four-body system. For each of these components there is a natural set of Jacobi coordinates, $(\mathbf{u}_1, \mathbf{u}_2, \mathbf{u}_3)$ and $(\mathbf{v}_1, \mathbf{v}_2, \mathbf{v}_3)$, respectively, of which the former is a direct extension of the three-body Jacobi coordinates (defining $\mathbf{u}_3$ as the relative momentum of the fourth particle with respect to the center of mass of the other three). For the 2+2 setup, $\mathbf{v}_1 = \mathbf{u}_1$, $\mathbf{v}_3$ denotes the relative momentum in the (34) system, and $\mathbf{v}_2$ is defined at the relative momentum between the (12) and (34) subsystems. Using the formalism of [27, 29], the Faddeev-Yakubovsky equations are written as

$$|\psi_A\rangle = G_0 t^{(0)} P\big[(1 - P_{34})|\psi_A\rangle + |\psi_B\rangle\big] + \frac{1}{3}(1 + G_0 t^{(0)})G_0 V_3^{(0)}|\Psi\rangle \tag{54.8a}$$

$$|\psi_B\rangle = G_0 t^{(0)} \tilde{P}\big[(1 - P_{34})|\psi_A\rangle + |\psi_B\rangle\big]\,, \tag{54.8b}$$

where $|\Psi\rangle = (1 - P_{34} - P P_{34})(1 + P)|\psi_A\rangle + (1 + P)(1 + \tilde{P})|\psi_B\rangle$ is the full four-body wavefunction and G_0 now represents the free four-body Green's function. In addition to the permutation operators already encountered in the three-body system, (54.8) involve the operators P_{34} and $\tilde{P} = P_{13}P_{24}$.

As discussed above for the three-body system, the FY equations are solved in a partial-wave momentum basis, involving now two sets of Jacobi momenta defined and sums over channel states,

$$|a\rangle = |\,\big(\ell_2\big((\ell_1 s_1)\, j_1 \tfrac{1}{2}\big)\, s_2\big)\, j_2,\, \big(\ell_3 \tfrac{1}{2}\big)\, j_3,\, (j_2 j_3)\, J;\, \big(\big(t_1 \tfrac{1}{2}\big)\, t_2 \tfrac{1}{2}\big)\, T\rangle\,, \tag{54.9a}$$

$$|b\rangle = |\,(\lambda_2(\lambda_1\sigma_1)\,\iota_1)\,\iota_2,\, (\lambda_3\sigma_3)\,\iota_3,\, (\iota_2\iota_3)\,J;\, (\tau_2\tau_3)\,T\rangle\,, \tag{54.9b}$$

which refer, respectively, to the 3+1 and 2+2 cluster setups. The $|a\rangle$ are a natural extension of three-nucleon states $|s\rangle$, including the angular momentum ℓ_3 associated with $\mathbf{u}_3$ as well as spin and isospin $\frac{1}{2}$ for the fourth nucleon into the overall coupling scheme. For the b states, $(\lambda_1, \sigma_1, \tau_1)$ and $(\lambda_3, \sigma_3, \tau_3)$ are quantum numbers for the (12) and (34) two-body subsystems, respectively, where $\lambda_{1,3}$ are the angular momenta associated with the Jacobi momenta $v_{1,3}$. The separation between the clusters is described by the momentum v_2 and its associated angular momentum λ_2. The projection of (54.8) yields a set of coupled equations which, unlike the Faddeev equations, does not naturally truncate even if all interactions are pure S-wave. As a consequence it is necessary to truncate the sums in (54.9) (e.g., by choosing all total angular momenta j_i and ι_i less than some $j_{\max}$) and study the numerical convergence of results as $j_{\max}$ is increased. More details can be found in [29].

54.3 Results and Discussion

The convergence pattern of the unitarity expansion for the binding energies of light nuclei is summarized in Table 54.1. The deuteron remains a zero-energy bound state at NLO and only moves to $1/(M_N a_t^2)$ at N^{2}LO, see [23] for an explicit calculation. This is the case for both a pure expansion in $1/a_t$ (neglecting range correction) as well as for the paired unitarity expansion that includes effective ranges together with finite-a

Table 54.1 Unitarity expansion convergence pattern. Underlined values indicate energies which are used as input values to determine three-body LECs. An asterisk superscript indicates an incomplete NLO result which only includes the finite-scattering length but no contributions from effective ranges or electromagnetic interactions

State	E_B^{LO}/MeV	E_B^{NLO}/MeV	$E_B^{\text{N}^2\text{LO}}$/MeV	$E_B^{\text{exp.}}$/MeV
^{2}H	0	0	1.41 ± 1.12	2.22
^{3}H	<u>8.48</u>	<u>8.48</u>	<u>8.48</u>	8.48
^{3}He	8.5 ± 2.5	7.6 ± 0.2	<u>7.72</u>	7.72
^{4}He	39 ± 12	30 ± 9*		28.3

corrections. The dominant source of uncertainty for the deuteron energy comes from the $1/(Q_2 a_t)$ expansion, which still amounts to a 50% effect at N^{2}LO. Conservatively taking the experimental binding energy as reference for the uncertainty estimate gives $B_d^{\text{N}^2\text{LO}} = 1.41 \pm 1.12$ MeV.

At each order the triton binding energy remains fixed at its physical value because it is used as input to tune the 3NI. At LO, ^{3}H and ^{3}He are degenerate by construction, but the splitting between the two iso-doublet states is a prediction at NLO. As discussed in [22], range corrections cancel at this order because LO is isospin-symmetric. The dominant effects that determine the splitting are thus electromagnetic corrections as well as the difference between the np and pp (Coulomb-modified) scattering lengths. The unitarity expansion predicts the triton-helion energy splitting as (0.92 ± 0.18) MeV at NLO, in good agreement with the experimental value 0.764 MeV. At N^{2}LO the mixing between electromagnetic and range corrections introduces a divergence that requires an isospin-breaking 3NI to be promoted to this order [23]. For the unitarity expansion this means that a new input is required, which is most conveniently chosen to be the ^{3}He binding energy. Neglecting range corrections, however, [23] finds good convergence up to N^{2}LO for an expansion that only includes finite scattering lengths and electromagnetic corrections in perturbation theory.

In the unitarity limit, ^{4}He is formally equivalent to a system of four bosons. It is known that each three-boson Efimov state with binding energy B_3 is associated with two four-boson states (tetramers) [32] at energies $B_4/B_3 \simeq 4.611$ and $B_{4^*}/B_3 \simeq 1.002$ [33]. The experimental values for the ^{4}He ground and first excited states are, respectively, $B_\alpha/B_H \simeq 3.66$ and $B_{\alpha^*}/B_H \simeq 1.05$, where the ^{3}He binding energy $B_H \simeq 7.72$ MeV is used as reference to approximately account for electromagnetic corrections. The closeness of these values to what is found in the unitarity limit suggests that a perturbative expansion can be expected to work well. The numerical results shown in Fig. 54.2 confirm this expectation. The ^{4}He binding energy as a function of the momentum cutoff Λ is found to converge as Λ increases, indicating that the EFT calculation is properly renormalized. While any Λ above the breakdown scale (of order M_π) is a valid choice in principle, quadratic polynomials in $1/\Lambda$ are fitted at large Λ to quantitatively assess the convergence and conveniently extrapolate $\Lambda \to \infty$. Figure 54.2 also shows a standard pionless calculation that includes finite $a_{s,t}$ at LO and gives results consistent with [25–27]. In the unitarity limit $B_\alpha =$

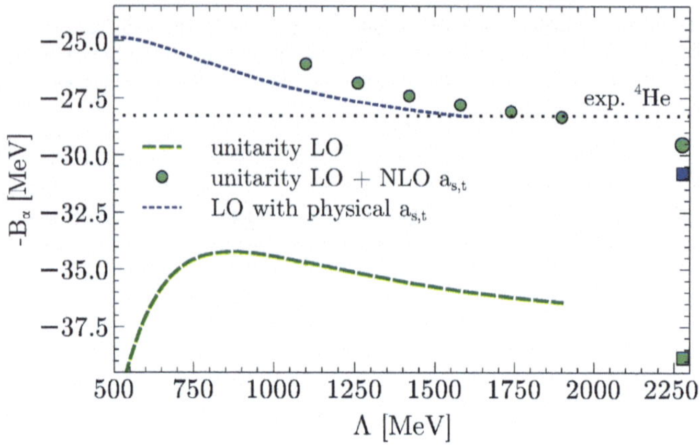

Fig. 54.2 ^{4}He binding energy as function of the Gaussian cutoff. (Blue) dotted and (green) dashed lines: standard Pionless EFT and full unitarity at LO, respectively. (Green) circles first-order corrections in $1/a_{s,t}$ added in perturbation theory. Large symbols on right edge: $\Lambda \to \infty$ extrapolation (see text)

39(12) MeV is found for the ^{4}He ground state. In addition, there is a bound excited state just below the proton-triton breakup threshold. Both these states are in excellent agreement with the universal unitarity expectation.

An incomplete NLO (neglecting effective ranges and electromagnetic contributions) is calculated here to study the effect of finite-scattering length corrections in ^{4}He. The result, 30(9) MeV for $\Lambda \to \infty$ comes out very close the standard pionless LO calculation, indicating that the $1/(Q_4 a_{s,t})$ expansion works remarkably well up to this order. The uncertainty of this value, as well as that of the LO result quoted above, is $\mathcal{O}(r_{s,t}/a_{s,t}) \simeq 30\%$ based on the expectation that range corrections are dominant in this case. Importantly, $(B_\alpha/B_T)^{\mathrm{NLO}(r=0)} \approx 3.48$ is also in good agreement with $(B_\alpha/B_T)^{\mathrm{exp}} = 3.34$. As shown in Fig. 54.3, the rapid convergence persists off the physical point: the correlation between $3N$ and $4N$ binding energies (Tjon line) is perturbatively close to the unitarity result over a significant range of energies. While a proper calculation of the excited state is computationally very demanding due to a slow convergence of the FY calculation for a state so close to a threshold, four-boson calculations with nuclear scales indicate that the $1/(Q_4 a_{s,t})$ corrections push the bound excited state into the continuum by about the amount expected from experiment [20].

Very recently it was found that a four-body force is required to renormalize the universal four-boson system once range corrections are included at NLO [34]. This result directly carries over to pionless EFT—and thus to the unitarity expansion considered here—and implies that a new observable, most obviously taken to be the ^{4}He binding energy, is required at this order to set the scale of the four-body force. Even with this additional required input the theory however remains predictive for

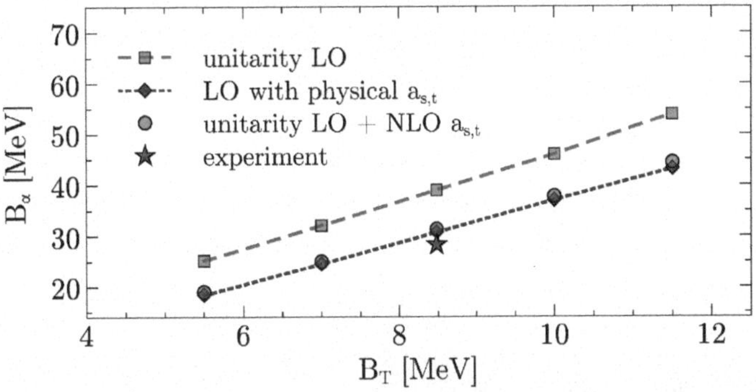

Fig. 54.3 Tjon line: correlation between the ^{4}He and ^{3}H binding energies. (Blue) dotted curve: standard pionless LO result; (green) dashed upper curve: unitarity limit at LO. Additional points nearly on top of the blue curve: inverse scattering lengths added in first-order perturbation theory. Star: experimental point

other four-body observables like ^{4}He charge radius and excited state energy, as well as for heavier systems, assuming the unitarity expansion converges for these.

The unitarity expansion constitutes a paradigm shift in the EFT-based description of light nuclei, deemphasizing the importance of two-body details in favor of using the three-body sector as "anchor point." As such, it is not unlike more phenomenological approaches using input from heavier nuclei to constrain few-nucleon forces. It is, however, more systematic by focusing on light nuclei and strives to answer the question what is essential to describe these systems. As discussed compellingly in [35], the idea can be boiled down to interpreting discrete scale invariance, the most striking manifestation of which is the Efimov effect, as a fundamental principle governing nuclear physics. In the bigger picture of things, the unitarity expansion furthermore stands in line with other recent results that suggest a fascinating simplification of nuclear physics. For examples, it has been observed that the isotopic chain of helium can be remarkably well described by adjusting a single parameter after tuning a model to reproduce α-n phase shifts [36]. More recently a correlation analogous to the Phillips line has been observed between the d-α scattering length and the ^{6}Li binding energy [37].

It is an exciting question how well the unitarity expansion works beyond what has been calculated so far. The observation that bosonic systems at unitarity exhibit saturation for large numbers of particles [38] and recent calculations of nuclear matter using interactions guided by unitarity [39] provide reasons to be optimistic. However, it remains to be seen to what extent lessons from universal bosonic systems carry over to nucleons, where beyond the four-body sector the influence of Fermi statistics is expected to become important. Concrete work looking at systems heavier than ^{4}He as well as observables beyond binding energies is currently in progress.

Acknowledgements I would like to thank Harald Grießhammer, Hans-Werner Hammer, and Bira van Kolck for their collaboration as well as for many insightful discussions and comments on this manuscript. This work was supported in part by the Deutsche Forschungsgemeinschaft (DFG, German Research Foundation) – Projektnummer 279384907 – SFB 1245 and by the ERC Grant No. 307986 STRONGINT.

References

1. Schwinger, J.: hectographed Notes on Nuclear Physics. Harvard University (1947)
2. Barker, F.C., Peierls, R.E.: Phys. Rev. **75**, 312 (1949)
3. Chew, G.F., Goldberger, M.L.: Phys. Rev. **75**, 1637 (1949)
4. Bethe, H.A.: Phys. Rev. **76**, 38 (1949)
5. Beane, S.R., Bedaque, P.F., Savage, M.J., van Kolck, U.: Nucl. Phys. A **700**, 377 (2002)
6. Beane, S.R., Savage, M.J.: Nucl. Phys. A **713**, 148 (2003)
7. Epelbaum, E., Meißner, U.-G., Glöckle, W.: Nucl. Phys. A **714**, 535 (2003)
8. Beane, S.R., Savage, M.J.: Nucl. Phys. A **717**, 91 (2003)
9. Braaten, E., Hammer, H.W.: Phys. Rev. Lett. **91**, 102002 (2003)
10. Chin, C., Grimm, R., Julienne, P., Tiesinga, E.: Rev. Mod. Phys. **82**, 1225 (2010)
11. Braaten, E., Kusunoki, M.: Phys. Rev. D **69**, 074005 (2004)
12. Efimov, V.: Phys. Lett. B **33**, 563 (1970)
13. Bedaque, P.F., Hammer, H.-W., van Kolck, U.: Phys. Rev. Lett. **82**, 463 (1999)
14. Bedaque, P.F., Hammer, H.-W., van Kolck, U.: Nucl. Phys. A **646**, 444 (1999)
15. Bedaque, P.F., Hammer, H.-W., van Kolck, U.: Nucl. Phys. A **676**, 35 (2000)
16. van Oers, W.T.H., Seagrave, J.D.: Phys. Lett. **24B**, 562 (1967)
17. Girard, B.A., Fuda, M.G.: Phys. Rev. C **19**, 579 (1979)
18. Rupak, G., Vaghani, A., Higa, R., van Kolck, U.: Phys. Lett. B **791**, 414 (2019)
19. Adhikari, S.K., Fonseca, A.C., Tomio, L.: Phys. Rev. C **26**, 77 (1982)
20. König, S., Grießhammer, H.W., Hammer, H.-W., van Kolck, U.: Phys. Rev. Lett. **118**, 202501 (2017)
21. Hammer, H.-W., König, S., van Kolck, U.: (submitted manuscript). arXiv:1906.12122 [nucl-th]
22. König, S., Grießhammer, H.W., Hammer, H.-W., van Kolck, U.: J. Phys. G **43**, 055106 (2016)
23. König, S.: J. Phys. G **44**, 064007 (2017)
24. Vanasse, J.: Phys. Rev. C **88**, 044001 (2013)
25. Platter, L., Hammer, H.-W., Meißner, U.-G.: Phys. Rev. A **70**, 052101 (2004)
26. Platter, L., Hammer, H.-W., Meißner, U.-G.: Phys. Lett. B **607**, 254 (2005)
27. Platter, L.: From Cold Atoms to Light Nuclei: The Four-Body Problem in an Effective Theory with Contact Interactions, Doctoral thesis (Dissertation), University of Bonn (2005)
28. Platter, L., Hammer, H.-W.: Nucl. Phys. A **766**, 132 (2006)
29. Kamada, H., Glöckle, W.: Nucl. Phys. A **548**, 205 (1992)
30. Glöckle, W.: The Quantum Mechanical Few-Body Problem. Springer, Berlin (1983)
31. Stadler, A., Glöckle, W., Sauer, P.U.: Phys. Rev. C **44**, 2319 (1991)
32. Platter, L., Hammer, H.-W.: Eur. Phys. J. A **32**, 113 (2007)
33. Deltuva, A.: Phys. Rev. A **82**, 040701 (2010)
34. Bazak, B., Kirscher, J., König, S., Pavón Valderrama, M., Barnea, N., van Kolck, U.: Phys. Rev. Lett. **122**, 143001 (2019)
35. van Kolck, U.: Few Body Syst. **58**, 112 (2017)
36. Fossez, K., Rotureau, J., Nazarewicz, W.: Phys. Rev. C **98**, 061302 (2018)
37. Lei, J., Hlophe, L., Elster, C., Nogga, A., Nunes, F.M., Phillips, D.R.: Phys. Rev. C **98**, 051001 (2018)
38. Carlson, J., Gandolfi, S., van Kolck, U., Vitiello, S.A.: Phys. Rev. Lett. **119**, 223002 (2017)
39. Kievsky, A., Viviani, M., Logoteta, D., Bombaci, I., Girlanda, L.: Phys. Rev. Lett. **121**, 072701 (2018)

Chapter 55
A New Measurement of the $^2H(p, \gamma)^3He$ Cross Section in the BBN Energy Range at LUNA

Francesca Cavanna

Abstract Deuterium is the first nucleus produced in the Universe whose accumulation marks the beginning of the so called Big Bang Nucleosynthesis (BBN). Its primordial abundance is very sensitive to some cosmological parameters like the baryon density and the number of the neutrino families. Presently the main obstacle to an accurate theoretical deuterium abundance evaluation is due to the poor knowledge of the $^2H(p, \gamma)^3He$ cross section at BBN energies. This paper reports on the measurements of the $^2H(p, \gamma)^3He$ reaction cross section performed at the Laboratory for Underground Nuclear Astrophysics (LUNA) facility in the Gran Sasso Laboratory (LNGS). After a general introduction on the astrophysical and cosmological relevance of $^2H(p, \gamma)^3He$ reaction cross section, the experimental setup used for the measurement campaign is described. The results obtained so far and the perspectives of the still on-going measurements are also discussed.

55.1 The Primordial Nucleosynthesis and the D/H Abundance Predictions

Nuclear physics plays a role in the very early life of the Universe: between 3 and 20 min after the Big Bang, a few light isotopes of H, He, Li and Be are formed through a net of reactions. This is known as Big Bang Nucleosynthesis and its importance is not just limited to the formation of the primordial material giving origin, about 10^9 y later, to the first pro-stars. Actually, the rate of the BBN reactions and the final abundances of the involved isotopes are strictly related to fundamental quantities like the barion density of the Universe. The barion to photon density is the sole free parameter to describe, according the Lambda Cold Dark Matter model (ΛCDM [1]),

For the LUNA collaboration

F. Cavanna (✉)
Istituto Nazionale di Fisica Nucleare sezionde di Genova,
Via Dodecaneso 33, 16146 Genova, Italy
e-mail: francesca.cavanna@ge.infn.it

N. A. Orr et al. (eds.), *Recent Progress in Few-Body Physics*,
Springer Proceedings in Physics 238,
https://doi.org/10.1007/978-3-030-32357-8_55

the Universe evolution. From the Cosmic Microwave background measured by the PLANK satellite [2] and the cross section of the BBN reactions, the abundances of the primordial isotopes can be calculated and compared with astronomical observations. This open the possibility to infer, from the accurate determination of the nuclear cross section, information widely beyond the limit of nuclear astrophysics. Among the relevant process for BBN nucleosynthesis there's the $^2H(p,\gamma)^3He$ currently under study at LUNA.

The primordial abundance of deuterium, $(D:H)_{obs}$, is presently known with good accuracy, $(D:H)_{obs}= (2.527 \pm 0.030)\ 10^{-5}$ [3], while the corresponding $(D:H)_{BBN}$ obtained from the BBN calculations, $(D:H)_{BBN} = (2.58 \pm 0.04)\ 10^{-5}$ [4], is affected by the insufficient knowledge of S_{12} in the relevant energy interval. Only a single dataset of S_{12} is available in the relevant energy range [5] and, according to the Authors, it is affected by a systematic error of 9%. The situation is even worst when considering a 20% discrepancy of that data with the theoretical previsions [6]. For all these reasons an experimental effort to measure the cross section with 3–5% accuracy is needed.

55.2 The $^2H(p,\gamma)^3He$ measurement

The measurement of the cross section of the $^2H(p,\gamma)^3He$ reaction at BBN energies is still ongoing at the LUNA 400 kV accelerator [7] installed in the underground Gran Sasso laboratory. Such machine is able to provide intense proton beams (up to 500 μA) with a precise absolute energy (~0.3 keV), low energy spread (0.1 keV) and long-term stability (5 eV/h). The experimental apparatus includes also a windowless deuterium gas target, consisting of three-stage pumping system able to increase the gas pressure from the accelerator high vacuum to the mbar level in the target chamber. The beam current is measured by a constant-gradient calorimeter characterized by two sides, a hot one heated to 70 °C by thermoresistors and a cold one cooled to −5 °C by a refrigerating system.

The experimental procedure consists of two main phases characterized by two different set-ups. The former foresees a deuterium gas target 10 cm long at 0.3 mbar of pressure and a cylindrical BGO detector [8]. Thanks to the high detection effciency, this set-up will provide a cross section measurement down to very low energies (~ 50 keV of proton energy). The efficiency calibration can be obtained at a few per cent level using Monte Carlo simulations of the set-up, tuned with radioactive sources (^{137}Cs, ^{60}Co and ^{88}Y) at low energies and with the well-known resonant reaction $^{14}N(p,\gamma)^{15}O$ at $E_r = 259$ keV, emitting γ-rays in the $^2H(p,\gamma)^3He$ energy range. In order to reduce systematic errors, the spatial position of the resonance has been identified inside the chamber using a well collimated NaI detector in close geometry with the target: keeping the scintillator fixed in the measurement position, the proton energy has been increased and decreased in 2–3 keV steps until the achievement of the maximum counting rate corresponding to the resonance peak. To determine the possible beam induced background, runs with proton beam impinging on evacuated

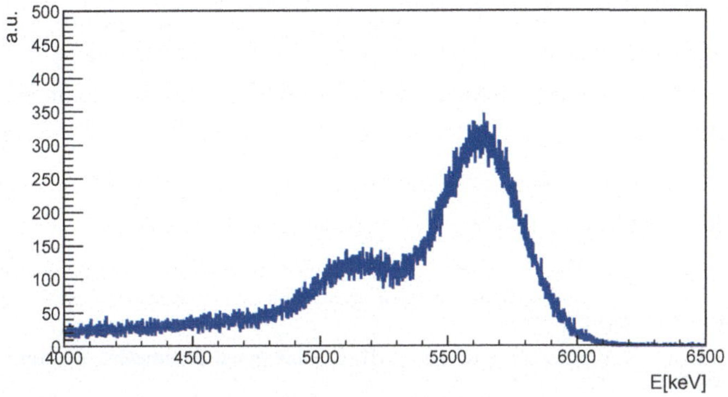

Fig. 55.1 BGO spectrum acquired at 150 keV proton energy

target chamber were performed. Figure 55.1 shows a spectrum acquired at 150 keV beam energy: the signal is well visible in the expected Region of Interest (RoI) 4000–6500 keV. The set up of the second phase consists of a 137% HPGe detector in close geometry with the interaction chamber [9, 10]. With such setup the angular distribution can be inferred by exploiting the high energy resolution of the detector and the Doppler effect responsible for the broad energy distribution of the detected γ-rays coming from different directions inside the extended gas target (33 cm long). The $^2H(p, \gamma)^3He$ photons have an energy of about 5.5 MeV, far away from the energy of the commonly used radioactive sources. Thus, for determining the setup efficiency a different technique based again on the well-known resonant reactions $^{14}N(p, \gamma)^{15}O$ and on ^{60}Co radioactive decay has been used. In order to reduce the systematic error due to the summing correction, the set-up efficiency has been measured exploiting the coincidence between two γ-rays emitted in cascade (from the source as well as from the reaction) and detected by two different germanium detectors, the main detector (Ge1) and a second one used as acquisition trigger (Ge2). Whenever Ge2 detects an event 1, it enables Ge1 that can thus detect the photon 2 emitted in cascade: the ratio of the observed photons and the number of triggers provides the Ge1 efficiency. In the case of the ^{60}Co source, for each radioactive decay process, two photons of energies E_1 = 1.17 MeV and E_2 = 1.33 MeV are produced. In the case of the resonant capture, several decay branches are able to provide two photons in cascade of energies up to 6.7 MeV, even higher than the $^2H(p,\gamma)^3He$ reaction. This method allows fixing precisely the detector energy response. To measure the cross section a scan has been performed in the energy range of interest (30 keV $< E_{cm} <$ 300keV) with 30–50 keV steps; two runs were performed for each energy: one with deuterium gas inside the scattering chamber, the other with ^4He in order to evaluate the beam induced background contribution and the eventual deuterium implantation. The data taking has been completed, the analysis is ongoing.

In conclusion a new measurement with a few percentage accuracy is very important to push down the BBN uncertainty on deuterium abundance to the same level as observations and to eventually constrain possible new physics effects. The D/H ratio is, in fact, very sensitive to the number of relativistic degrees of freedom and can contribute to setting tight bounds on the number of equivalent neutrino flavours.

References

1. Canuto, V., et al.: Scale-covariant theory of gravitation and astrophysical applications. Phys. Rev. D **16**, 1643–1663 (1977)
2. Ade, P.A.R. et al.: Planck 2015 results. XIII. Cosmological parameters. Astron. Astrophys. **594**(A13) (2016)
3. Cooke, R.J., et al.: One percent determination of the primordial deuterium abundance. Astrophys. J. **855**, 102 (2018)
4. Cyburt, R.H., et al.: Big bang nucleosynthesis: present status. Rev. Modern Phys. **88**, 015004 (2016)
5. Ma, L., et al.: Measurements of $^{1}H(d \rightarrow,)^{3}He$ and $^{2}H(p \rightarrow,)^{3}He$ at very low energies. Phys. Rev. C **55**, 588–596 (1997)
6. Marcucci, L.E., et al.: Implication of the proton-deuteron radiative capture for big bang nucleosynthesis. Phys. Rev. Lett. **116**, 102501 (2016)
7. Cavanna, F., et al.: Direct measurement of nuclear cross-section of astrophysical interest: results and perspectives. Int. J. Modern Phys. A **33**, 1843010–346 (2018)
8. Ferraro, F. et al.: A high-efficiency gas target setup for underground experiments, and redetermination of the branching ratio of the 189.5 keV $^{22}Ne(p,\gamma)^{23}Na$ resonance. Eur. Phys. J. A **54**, 44 (2018)
9. Cavanna, F., et al.: Three new low-energy resonances in the $^{22}Ne(p,\gamma)^{23}Na$ Reaction. Phys. Rev. Lett. **115**, 252501 (2015)
10. Depalo, R., et al.: Direct measurement of low-energy $^{22}Ne(p,\gamma\ )^{23}Na$ resonances. Phys. Rev. C **94**, 055804 (2016)

Chapter 56
Three-Nucleon Force Studies in Proton-Deuteron Break-Up Reaction with BINA at 190 MeV

M. Mohammadi-Dadkan, M. T. Bayat, N. Kalantar-Nayestanaki, St. Kistryn, A. Kozela, A. A. Mehmandoost-Khajeh-dad, J. G. Messchendorp, R. Ramazani-Sharifabadi, E. Stephan and H. Tavakoli-Zaniani

Abstract The present knowledge of nuclear forces is not sufficient to describe all experimental data for systems which consist of more than two nucleons. Recent three-nucleon scattering experiments have shown that the theoretical models based solely on nucleon-nucleon potentials fail to describe most of the experimental results. In this paper, we present data of the $\vec{p} + d \longrightarrow p + p + n$ break-up reaction that were obtained using a 190 MeV polarized-proton beam impinging on a liquid deuterium target. The experiment was performed by exploiting BINA (**B**ig **I**nstrument for **N**uclear-polarization **A**nalysis), a detector system with a large angular acceptance and a high energy resolution.

M. Mohammadi-Dadkan (✉) · M. T. Bayat · N. Kalantar-Nayestanaki · J. G. Messchendorp · R. Ramazani-Sharifabadi · H. Tavakoli-Zaniani
KVI-CART, University of Groningen, Groningen, The Netherlands
e-mail: m.mohammadi-dadkan@rug.nl

M. Mohammadi-Dadkan · A. A. Mehmandoost-Khajeh-dad
Department of Physics, University of Sistan and Baluchestan, Zahedan, Iran

St. Kistryn
Institute of Physics, Jagiellonian University, Kraków, Kraków, Poland

A. Kozela
Institute of Nuclear Physics, PAN, Kraków, Poland

R. Ramazani-Sharifabadi
Department of Physics, University of Tehran, Tehran, Iran

E. Stephan
Institute of Physics, University of Silesia, Chorzow, Poland

H. Tavakoli-Zaniani
Department of Physics, School of Science, University of Yazd, Yazd, Iran

N. A. Orr et al. (eds.), *Recent Progress in Few-Body Physics*,
Springer Proceedings in Physics 238,
https://doi.org/10.1007/978-3-030-32357-8_56

56.1 Introduction

The nature of the nuclear force is still not well understood. Two-nucleon force (2NF) models, most of which proposed based on Yukawa theory [1], deliver a superb description for nucleon-nucleon scattering data and the characteristics of the deuteron whereas, calculations based on the 2NF models underestimate the experimental results of the three-body system such as the binding energy of the triton [2] and show deviation when comparing their results with measured 3-body scattering cross sections. Therefore, 2NFs are not sufficient to describe three-nucleon systems and we need to take into account three-nucleon force (3NF) effects in the models. There are various 3NFs most of which are based on the model which is developed by Fujita-Miyazawa [3]. In these models, the 2π-exchange mechanism is used with an additional Δ excitation of one of the nucleons [4].

In the last two decades, a large number of scattering experiments have been performed to study 3NF effects by measuring differential cross sections and spin observables such as analyzing powers [5, 6]. The 3-body break-up reaction is one of the best tools to study 3NFs thanks to the rich kinematical phase space which allows us to check 3NF at different kinematical configurations.

A comprehensive study of 3NFs has been started at KVI using various beam energies and targets with the aim of measuring differential cross sections and analyzing powers [5–7]. The experimental setup (BINA), as it is shown in the left panel of Fig. 56.1, consists of two main parts: a forward-wall and a backward-ball. In this paper, results of measurements of the vector analyzing powers for the proton-deuteron break-up reaction using a 190 MeV proton beam are presented for a part

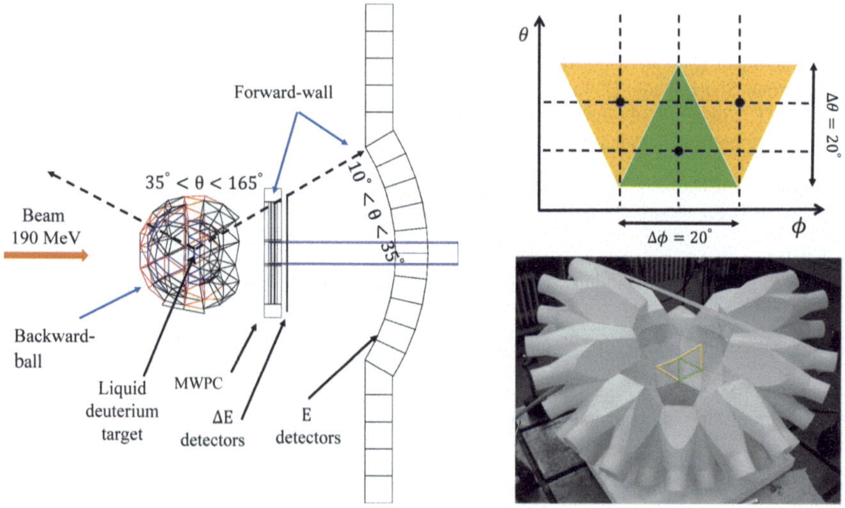

Fig. 56.1 The left panel shows the structure of BINA. On top of the right panel the angular position of each detector is marked with the center of gravity of each triangle

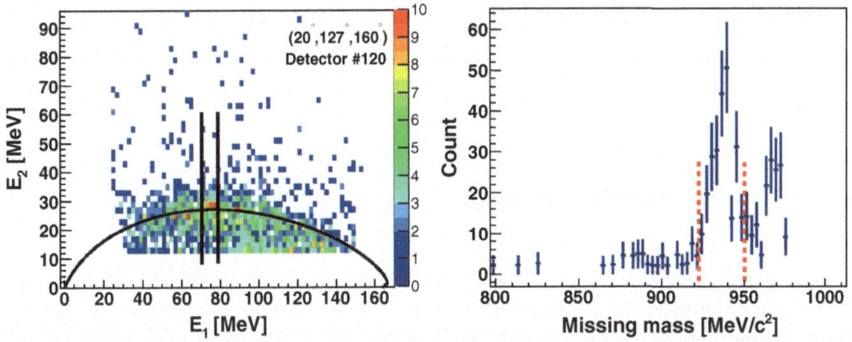

Fig. 56.2 The left panel shows the energy correlation of the two protons for kinematical configuration ($\theta_1 = 25° \pm 2°, \theta_2 = 127° \pm 10°, \phi_{12} = 160° \pm 10°$) and detector number 129, together with the expected kinematical *S*-curve. The right panel shows the missing-mass spectrum of the neutron for one of the gates alongside the *S*-curve (shown in the left panel)

of phase space which has not been explored before [8]. This part of the phase space represents kinematics in which one proton scatters to the forward-wall and the other to the backward-ball. As it is demonstrated in the right panel of Fig. 56.1, the position of each ball detector has been used to measure the scattering angles of protons in the backward-ball and therefore, the angular resolution is around $\pm 10°$.

56.2 Analysis Method

In the three-body breakup reaction, it is sufficient to measure 5 of the 9 kinematical variables in order to have all the information of the reaction. BINA is able to measure energy and scattering angles of two protons in coincidence, which provides an extra redundancy for extra check. Conventionally, in 3-body break-up reactions, kinematics are specified by the scattering angles of two protons ($\theta_1, \theta_2, \phi_{12} = \phi_2 - \phi_1$). The left panel of Fig. 56.2 shows the energy correlation of the two protons and the solid line represents the expected energy correlation based on the relativistic kinematics for a particular configuration. This curve is referred to as the *S*-curve. The value of the variable *S* is defined as the arc length of the *S*-curve which is starting from the minimum of E_2. Commonly, the breakup observables are presented as a function of the variable *S* for a given angle combination.

It is well known that the interaction of a polarized beam with an unpolarized target provides an azimuthal asymmetry in the scattering cross section which is given by [9]:

$$\frac{N^{\uparrow} - N^{\downarrow}}{N^{\downarrow} p_z^{\uparrow} - N^{\uparrow} p_z^{\downarrow}} = A_y \cos \phi - A_x \sin \phi, \tag{56.1}$$

where $N^{\uparrow}$ ($N^{\downarrow}$) and $p_z^{\uparrow}$ ($p_z^{\downarrow}$) are the cross section and beam polarization for up (down) mode, respectively, A_x and A_y are the vector analyzing powers, and ϕ is the azimuthal scattering angle of one the protons. The vector analyzing powers have been obtained for each configuration and ball detector. To get the final results, a weighted averaging has been applied over the ball detectors which are located at the same polar angle.

56.3 Results and Discussions

The analysis has been done for 48 kinematical configurations and in Fig. 56.3, as representative examples, the preliminary results of vector analyzing powers are shown for some kinematical configurations together with the results of Faddeev calculations using the CD-Bonn and AV18 two-nucleon potentials, with and without the inclusion of two different types of 3NFs. The calculations are performed for the kinematical configuration based on the center of each angular bins. To have a more accurate prediction and considering the fact that the angular resolution is around 10°, we should average the calculations over all covered kinematical configurations within an angular bin. This work is in progress. As observed in Fig. 56.3, the effects of the 3NF seem to be rather small for the configurations shown.

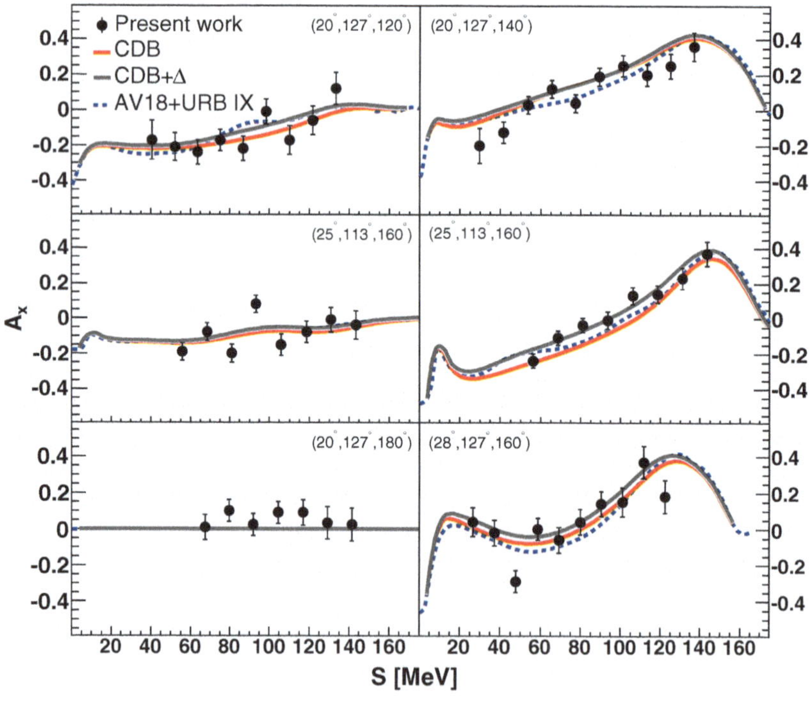

Fig. 56.3 A comparison between the results of the analyzing power measurements for a few selected break-up configurations with various theoretical predictions which are described in the legend. Errors are statistical only

References

1. Yukawa, H.: Proc. Phys. Math. Soc. Jpn. **17**, 48 (1935)
2. Pieper, S.C., Wiringa, R.B.: Ann. Rev. Nucl. Part. Sci. **51**, 53 (2001)
3. Fujita, J., Miyazawa, H.: Prog. Theor. Phys. **17**, 360 (1957)
4. Deltuva, A., Machleidt, R., Sauer, P.U.: Phys. Rev. C **68**, 024005 (2003)
5. Kalantar-Nayestanaki, N., et al.: Rep. Prog. Phys. **75**, 016301 (2012)
6. Kistryn, St., Stephan, E.: J. Phys. G: Nucl. Part. Phys. **40**, 063101 (2013)
7. Mehmandoost-Khajeh-Dad, A.A., et al.: Eur. Phys. J. A **47**, 59 (2011)
8. Mardanpour, H., et al.: Phys. Lett. B **149**, 687 (2010)
9. Ohlsen, G.G., et al.: Nucl. Instrum. Methods. **179**, 283 (1981)

Chapter 57
Structure of Beryllium Isotopes Beyond the Neutron Dripline

B. Monteagudo, J. Gibelin, F. M. Marqués, A. Corsi, Y. Kubota, N. A. Orr, G. Authelet, H. Baba, C. Caesar, D. Calvet, A. Delbart, M. Dozono, J. Feng, F. Flavigny, J. M. Gheller, A. Giganon, A. Gillibert, K. Hasegawa, T. Isobe, Y. Kanaya, S. Kawakami, D. Kim, Y. Kiyokawa, M. Kobayashi, N. Kobayashi, T. Kobayashi, Y. Kondo, Z. Korkulu, S. Koyama, V. Lapoux, Y. Maeda, T. Motobayashi, T. Miyazaki, T. Nakamura, N. Nakatsuka, Y. Nishio, A. Obertelli, A. Ohkura, S. Ota, H. Otsu, T. Ozaki, V. Panin, S. Paschalis, E. C. Pollacco, S. Reichert, J. Y. Rousse, A. T. Saito, S. Sakaguchi, M. Sako, C. Santamaria, M. Sasano, H. Sato, M. Shikata, Y. Shimizu, Y. Shindo, L. Stuhl, T. Sumikama, M. Tabata, Y. Togano, J. Tsubota, T. Uesaka, Z. H. Yang, J. Yasuda, K. Yoneda and J. Zenihiro

Abstract Dineutron decay is a forefront topic in nuclear structure that still lacks a firm experimental claim. The spontaneous emission of a dineutron should be favored in nuclei that are unbound with respect to two-neutron emission but bound with respect to single-neutron emission. A very interesting candidate can be found by adding two neutrons to the most neutron-rich Beryllium isotope, ^{14}Be, a well-known

---

B. Monteagudo (✉) · J. Gibelin · F. M. Marqués · N. A. Orr
LPC Caen, ENSICAEN, Université de Caen, CNRS/IN2P3, Caen, France
e-mail: monteagudo@lpccaen.in2p3.fr

A. Corsi · G. Authelet · D. Calvet · A. Delbart · J. M. Gheller · A. Giganon · A. Gillibert · V. Lapoux · A. Obertelli · E. C. Pollacco · J. Y. Rousse
Département de Physique Nucléaire, IRFU, CEA, Université Paris-Saclay, Saint-Aubin, France

Y. Kubota · H. Baba · T. Isobe · Z. Korkulu · T. Motobayashi · H. Otsu · V. Panin · M. Sako · C. Santamaria · M. Sasano · H. Sato · Y. Shimizu · L. Stuhl · T. Sumikama · T. Uesaka · Z. H. Yang · K. Yoneda · J. Zenihiro
RIKEN Nishina Center, Saitama, Japan

Y. Kubota · M. Dozono · Y. Kiyokawa · M. Kobayashi · S. Ota
Center for Nuclear Study, University of Tokyo, Tokyo, Japan

C. Caesar · A. Obertelli · S. Paschalis
Institut für Kernphysik, Technische Universität Darmstadt, Darmstadt, Germany

J. Feng
Department of Physics, Peking University, Beijing, China

F. Flavigny
IPNO, Université Paris-Sud, IN2P3-CNRS, Orsay, France

N. A. Orr et al. (eds.), *Recent Progress in Few-Body Physics*, Springer Proceedings in Physics 238,
https://doi.org/10.1007/978-3-030-32357-8_57

$2n$-halo nucleus. The intriguing nature of the phenomenon has motivated recent theoretical calculations and a new experimental campaign at RIKEN RIBF. In this work, the decay properties of ^{16}Be and the structure of ^{15}Be have been probed via the proton-knockout reaction from a ^{17}B beam.

57.1 Introduction

Exotic types of radioactivity are characteristic phenomena as we go far from stability, near the driplines. The already observed two-proton radioactivity appears beyond the proton dripline as a consequence of the evolution of the proton separation energy. In analogy to the proton dripline, due to the evolution of the n–n pairing correlations, the two-neutron separation energy can reach negative values while the one-neutron separation energy is still positive. When this is the case, spontaneous two-neutron emission is likely to happen since the sequential decay, via one neutron emission, would be energetically forbidden. In order to find such condition, we need to go beyond the neutron dripline. However, the limits for the neutron-rich side are experimentally difficult to explore and only probed for very light nuclei. As a result, few candidates are known.

One of the possible candidates is ^{16}Be. Experimental attempts to probe the structure of ^{15}Be found no evidence of the ground state but established a lower limit of 1.54 MeV above the ^{14}Be + n threshold [1]. Below this limit, a state in ^{16}Be was measured 1.35 MeV above the ^{14}Be + n + n threshold, which presented a very suitable scenario for the study of a spontaneous $2n$ emission. The study of n–n correlations showed a strong signal at very low relative energy and angle that was reported as a dineutron decay [2]. However, data were exclusively compared either to purely

K. Hasegawa · T. Kobayashi
Department of Physics, Tohoku University, Sendai, Japan

Y. Kanaya · S. Kawakami · Y. Maeda
Department of Applied Physics, University of Miyazaki, Miyazaki, Japan

D. Kim
Department of Physics, Ehwa Womans University, Seoul, South Korea

N. Kobayashi · S. Koyama · T. Miyazaki
Department of Physics, University of Tokyo, Tokyo, Japan

Y. Kondo · T. Nakamura · T. Ozaki · A. T. Saito · M. Shikata · Y. Togano · J. Tsubota
Department of Physics, Tokyo Institute of Technology, Tokyo, Japan

N. Nakatsuka
Department of Physics, Kyoto University, Kyoto, Japan

Y. Nishio · A. Ohkura · S. Sakaguchi · Y. Shindo · M. Tabata · J. Yasuda
Department of Physics, Kyushu University, Fukuoka, Japan

S. Reichert
Department of Physics, Technische Universitt Munchen, München, Germany

3-body phase space with no interaction between particles, either to the most extreme case, a dineutron case where the two neutrons are emitted as a single unit. The comparison with these two "opposed" cases led to the conclusion that only a dineutron decay was consistent with the data, neglecting a more standard case in which the attractive *n–n* interaction in the final state could be able to explain the observed signal [3].

57.2 Experimental Setup

A new experiment was performed at RIKEN RIBF facility as part of a campaign aiming to shed light over a wide range of phenomena related to neutron-rich nuclei. In particular, one of the main purposes was the study of the dineutron correlations, in unbound nuclei such as ^{16}Be but also in bound borromean nuclei such as ^{11}Li or ^{14}Be.

The ^{16}Be states were populated via one proton removal reaction from a high-intensity ^{17}B secondary beam produced via fragmentation of a primary beam of ^{48}Ca at an energy of 345 MeV/nucleon and separated using the BigRIPS fragment separator. The incoming beam was then identified using plastic scintillators located in the beam line ($TOF - \Delta E$ method) and by measurement of its magnetic rigidity. Two drift chambers (BDC1 and BDC2) were positioned along the beam direction to track the incident beam position onto the target. The ^{17}Be$(p, 2p)^{16}$Be reaction took place in MINOS [4], a 150 mm thick liquid hydrogen target. By combining the trajectories of the recoil proton and knocked-out protons, the vertex of the reaction was reconstructed with a resolution of about 5 mm. The full kinematics of the decay products was measured using the standard setup of SAMURAI spectrometer, as it is described in [5]. The SAMURAI superconducting dipole magnet in combination with the drift chambers at its entrance (FDC1) and at its exit (FDC2) was used to determine the $B\rho$ and momentum of the fragment. The selection and identification of the fragments was derived from the energy loss and time of flight provided by two plastic hodoscopes (HODOF, HODOP) placed right after the FDC2. The momenta of the neutrons were calculated from the time of flight between F13, a plastic start detector located a few meters before the target, and the detection position in NEBULA, a neutron detector array located at about 11 m from MINOS.

Due to the large amount of detectors that come into play, several reference reactions are used to check the validity of the calibrations and analysis methods. The already well known threshold state of ^{16}B at 40 keV [6] was the benchmark for the 1n detection case. The observation in our data of this narrow state confirmed the validity of our calibrations. Coincident detection of neutrons is, however, a more challenging but crucial task. The fake multi-neutron events, or cross-talk, were rejected according to a rejection algorithm based on causality conditions. The result was checked with the first excited state of ^{14}Be [7], located 250 keV above the ^{12}Be $+ n + n$ threshold. As we can see in Fig. 57.1, the 2^+ state appears only when the cross-talk rejection is applied.

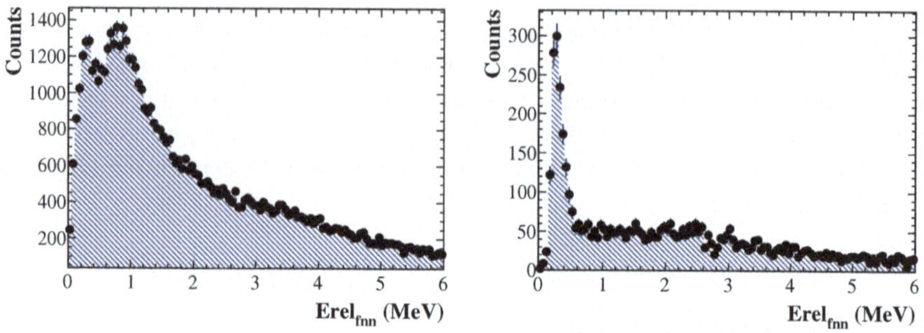

Fig. 57.1 Relative energy spectrum of ^{12}Be+n+n following a break-up reaction from ^{14}Be on a carbon target. *Left* Energy reconstruction if no cross-talk filter is used. *Right* Observation of the 2^+ state after the cross-talk rejection is applied

57.3 Preliminary Results

The relative energy E_{rel} of ^{16}Be was reconstructed by invariant mass method from the measured momenta of ^{14}Be and the two neutrons. The spectrum exhibits two different resonance-like structures below 5 MeV, identified as the ground state and the first excited state. Spin-parity were tentatively assigned for both states as 0^+ and 2^+, respectively. The best description of the spectrum was preliminary found for energies of $E_{0^+} = 0.83$ MeV and $E_{2^+} = 2.13$ MeV. The determination of the final values of energy and width, with their error bars, is in progress.

The decay mode of ^{16}Be depends on the structure of ^{15}Be. However, even if ^{15}Be was not directly populated during the campaign, the nature (sequential or direct) of the decay can be assessed by looking at the correlations of the decay particles. In particular, fragment-neutron relative energy distribution is very sensitive to the type of decay; and if sequential, to the energy of the intermediate state of ^{15}Be. Correlations between the fragment and one of the emitted neutrons were studied for both observed states in ^{16}Be by comparison with MC simulations, taking into account the detector response. A direct decay with no interaction between the two neutrons was assumed as the primary scenario for the ground state. Given the good agreement of the simulation with the $^{14}\text{Be} + n$ data, it was not necessary to suppose any other contribution coming from sequential decay. It was concluded then that the spontaneous two neutron emission to ground state of ^{14}Be was the most probable decay for the 0^+ state. With this result, the ground state of ^{15}Be was not expected to be below an energy of 830 keV.

In the case of the excited 2^+ state, a direct three-body decay was not consistent with the data. The fragment-neutron relative energy spectrum presented a peak-like structure that could not be described with a pure three body phase space simulation. A first attempt to find the energy of the ground state of ^{15}Be was carried out. The best description of data was achieved with a preliminary energy of 1.5 MeV as the intermediate state in ^{15}Be.

57.4 Summary and Perspectives

We investigated the decay of the unbound ^{16}Be using a proton knock-out reaction and a preliminary study of the fragment-neutron correlations was performed. The ground state, as well as the first excited state, were observed unambiguously for the first time. The best description of the spectrum was found for the following energies, $E_{0^+} = 0.83$ MeV and $E_{2^+} = 2.13$ MeV. The decay mode of the ground state seems to be, according to the fragment-neutron energy spectrum, purely direct by spontaneous emission of two neutrons. The decay of the excited state was found to be in better agreement with a sequential decay with an energy of 1.5 MeV for the ground state of ^{15}Be. Yet the description of the data for the 2^+ state could be improved by assuming both types of decay, direct and sequential, which is in progress.

Interesting theoretical calculations predict a very strong dineutron component in the structure of ^{16}Be [8, 9]. Whether the two neutron emission is in the form of a dineutron or not will be determined from the ongoing analysis of the n-n correlations.

References

1. Spyrou, A., et al.: Phys. Rev. C **84**, 044309 (2011)
2. Spyrou, A., et al.: Phys. Rev. Lett. **108**, 102501 (2012)
3. Marqués, F.M., et al.: Phys. Rev. Lett. **109**, 239201 (2012)
4. Obertelli, A., et al.: Eur. Phys. J. A **50**, 8 (2014)
5. Kobayashi, T. et al.: Nucl. Instrum. Methods Phys. Res., Sect. B **317**, 294 (2013)
6. Leblond, S.: Structure des isotopes de bore et de carbone riches en neutrons aux limites de la stabilité, Ph.D. thesis (2015)
7. Sugimoto, T., et al.: Phys. Rev. B **654**, 160 (2007)
8. Casal, J.: Phys. Rev. C **97**, 034613 (2018)
9. Casal, J., Gómez-Camacho, J.: Phys. Rev. C **99**, 014604 (2019)

Chapter 58
Approximate Sum Rule for the Electric Dipole Moment of Light Nuclei

Nodoka Yamanaka

Abstract The measurement of the electric dipole moment (EDM) is an excellent test of the standard model of particle physics, and the detection of a finite value is signal of a new source of CP violation beyond it. Among systems for which the EDM can be measured, light nuclei are particularly interesting due to their high sensitivity to new physics. In this proceedings contribution, we examine the sensitivity of the EDM of several light nuclei to the CP-odd one pion-exchange nucleon-nucleon interaction within the cluster model. We suggest an approximate sum rule for the nuclear EDM.

58.1 Motivation

The electric dipole moment (EDM) [1–4] is a good probe of CP violation. One of the most notable point is the almost negligible standard model contribution [5–9]. Recently, the experimental measurement of the EDM using storage rings is being developed [10]. Here we discuss the EDM of light nuclei [11, 12] as potentially interesting observables.

In the next section, we introduce the model and the interactions used in our work. In Sect. 58.3, we show our results of the calculations of the EDM of light nuclei, from which an interesting counting rule is suggested. The final section gives the summary.

58.2 The Model Setup

We consider the nucleons, the α (^{4}He), and triton (^{3}H) clusters as degrees of freedom, which are interacting themselves through phenomenological potentials [13–19]. The effect of the antisymmetrization is included using the orthogonality condition model [20].

N. Yamanaka (✉)
IPNO, Université Paris-Sud, CNRS/IN2P3, 91406 Orsay, France
e-mail: yamanaka@ipno.in2p3.fr

N. A. Orr et al. (eds.), *Recent Progress in Few-Body Physics*,
Springer Proceedings in Physics 238,
https://doi.org/10.1007/978-3-030-32357-8_58

The CP-odd nuclear force is modeled by the one-pion exchange [21]:

$$H^{\pi}_{\not P \not T} = \left\{ \bar{G}^{(0)}_{\pi} \boldsymbol{\tau}_1 \cdot \boldsymbol{\tau}_2 \boldsymbol{\sigma}_- + \frac{1}{2} \bar{G}^{(1)}_{\pi} (\tau^z_+ \boldsymbol{\sigma}_- + \tau^z_- \boldsymbol{\sigma}_+) + \bar{G}^{(2)}_{\pi} (3\tau^z_1 \tau^z_2 - \boldsymbol{\tau}_1 \cdot \boldsymbol{\tau}_2) \boldsymbol{\sigma}_- \right\} \cdot \frac{\boldsymbol{r}}{r} V(r), \quad (58.1)$$

where $\boldsymbol{r} \equiv \boldsymbol{r}_1 - \boldsymbol{r}_2$, $\boldsymbol{\sigma}_{\pm} \equiv \boldsymbol{\sigma}_1 \pm \boldsymbol{\sigma}_2$, and $\boldsymbol{\tau}_{\pm} \equiv \boldsymbol{\tau}_1 \pm \boldsymbol{\tau}_2$ denote the relative coordinate, spin, and isospin matrices, respectively, of the nucleons 1 and 2. The radial function is given by $V(r) = -\frac{m_\pi}{8\pi m_N} \frac{e^{-m_\pi r}}{r} \left(1 + \frac{1}{m_\pi r}\right)$. The CP-odd $\alpha - N$ and $\alpha - {}^3H$ potentials are obtained by folding [11] the CP-odd $N - N$ interaction (58.1) with the oscillator constant $b = 1.358$ fm $(\alpha - N)$ and $b = 1.482$ fm $(\alpha - {}^3H)$. In Fig. 58.1 we display the CP-odd potentials.

58.3 The Nuclear Electric Dipole Moment

The nuclear EDM generated by the CP-odd nuclear force is given by

$$\begin{aligned} d^{(\mathrm{pol})}_A &= \sum_{i=1}^{A} \frac{e}{2} \langle \Phi_J(A) \,|\, (1 + \tau^z_i) r_{iz} \,|\, \Phi_J(A) \rangle \\ &= \bar{G}^{(0)}_{\pi} a^{(0)}_{\pi} + \bar{G}^{(1)}_{\pi} a^{(1)}_{\pi} + \bar{G}^{(2)}_{\pi} a^{(2)}_{\pi} . \end{aligned} \quad (58.2)$$

where $|\, \Phi_J(A) \rangle$ is the polarized nuclear state. We show the results of our calculations in Table 58.1.

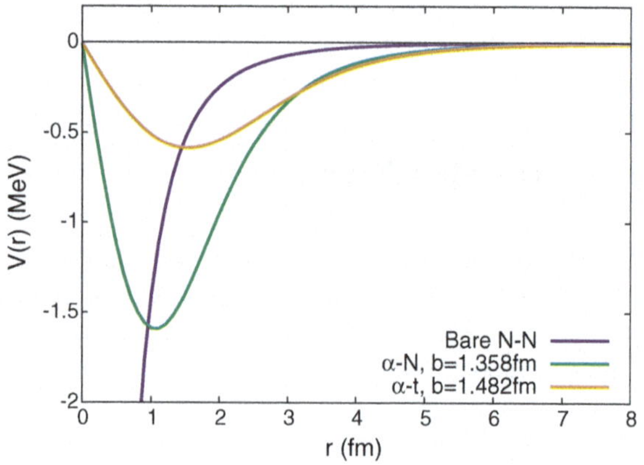

Fig. 58.1 The shape of the CP-odd potentials

Table 58.1 The linear coefficients of Eq. (58.2) in unit of e fm for several nuclei. The symbol "–" means zero within our framework

	$a_\pi^{(0)}$	$a_\pi^{(1)}$	$a_\pi^{(2)}$
^{2}H [22, 23]	–	0.0145	–
^{3}He [23, 24]	0.0059	0.0108	0.0168
^{3}H [23, 24]	−0.0059	0.0108	−0.0170
^{6}Li [23]	–	0.022	–
^{7}Li [25]	−0.006	0.016	−0.017
^{9}Be [23]	–	0.014	–
^{11}B [25]	−0.004	0.02	−0.01
^{13}C [26]	–	−0.0020	–
^{129}Xe [27]	7×10^{-5}	7×10^{-5}	4×10^{-4}

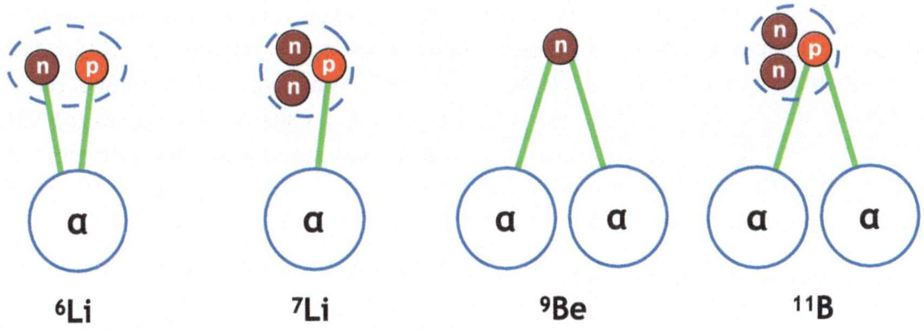

Fig. 58.2 Schematic picture of the counting rule for the EDMs of ^{6}Li, ^{7}Li, ^{9}Li, and ^{11}B

From this result, we can derive an approximate counting rule with the basic components the ^{2}H/^{3}H EDM and the CP-odd $\alpha - N$ polarization ~(0.005–0.007) $\bar{G}_\pi^{(1)} e$ fm. We indeed have

$$\begin{aligned} d_{^6\text{Li}} &= 2 \times (\alpha - N \text{ polarization}) + d_{^2\text{H}}, \\ d_{^7\text{Li}} &= 1 \times (\alpha - N \text{ polarization}) + d_{^3\text{H}}, \\ d_{^9\text{Be}} &= 2 \times (\alpha - N \text{ polarization}), \\ d_{^{11}\text{B}} &= 2 \times (\alpha - N \text{ polarization}) + d_{^3\text{H}}. \end{aligned} \tag{58.3}$$

We display in Fig. 58.2 the schematic picture of this rule. From it, we can predict

$$\begin{aligned} d_{^{10}\text{B}} &\sim 4 \times (\alpha - N \text{ polarization}) + d_{^2\text{H}} \sim 0.03\, \bar{G}_\pi^{(1)} e \text{ fm}, \\ d_{^{14}\text{N}} &\sim 6 \times (\alpha - N \text{ polarization}) + d_{^2\text{H}} \sim 0.04\, \bar{G}_\pi^{(1)} e \text{ fm}. \end{aligned} \tag{58.4}$$

We note that the EDM of ^{13}C does not respect the counting rule. This is due to the bad overlap between the structures of opposite parity states [26]. This suppression is certainly also relevant for ^{15}N which has a similar level structure.

Going along with the counting rule, we can naively predict that the EDM will increase if the the nucleon number grows thanks to the $\alpha - N$ polarization. The nuclear EDM will however be suppressed by the destructive interference due to the configuration mixing for heavy nuclei [27] (see the numerical value of the EDM of ^{129}Xe in Table 58.1).

58.4 Summary

In this proceedings contribution, we presented the results of the calculations of the nuclear EDM. The EDM of light nuclei seems to obey an approximate counting rule, if the nuclear structures of opposite parity states do not significantly differ. We could predict that the EDMs of ^{10}B or ^{14}N are more sensitive than the known ones. Increasing the number of nucleons will not give us a sensitive nucleus to the CP violation, since the destructive effect due to the configuration mixing will become important. Light nuclei seems to be the most suited for the EDM measurement using storage ring experiments.

References

1. Engel, J., Ramsey-Musolf, M.J., van Kolck, U.: Prog. Part. Nucl. Phys. **71**, 21 (2013)
2. Yamanaka, N.: Analysis of the Electric Dipole Moment in the R-parity Violating Supersymmetric Standard Model. Springer, Berlin (2014). https://doi.org/10.1007/978-4-431-54544-6
3. Yamanaka, N., Sahoo, B., Yoshinaga, N., Sato, T., Asahi, K., Das, B.: Eur. Phys. J. A **53**, 54 (2017)
4. Chupp, T.E., Fierlinger, P., Ramsey-Musolf, M.J., Singh, J.T.: Rev. Mod. Phys. **91**, 015001 (2019)
5. Pospelov, M., Ritz, A.: Phys. Rev. D **89**, 056006 (2014)
6. Seng, C.-Y.: Phys. Rev. C **91**, 025502 (2015)
7. Yamanaka, N., Hiyama, E.: JHEP **02**, 067 (2016)
8. Yamanaka, N.: Nucl. Phys. A **963**, 33 (2017)
9. Lee, J., Yamanaka, N., Hiyama, E.: Phys. Rev. C **99**, 055503 (2019)
10. Collaboration, J.E.D.I.: Phys. Rev. Lett. **117**, 054801 (2016)
11. JEDI Collaboration. Phys. Rev. Lett. **117**, 054801 (2016); Phys. Rev. Lett. **119**, 014801 (2017)
12. Yamanaka, N.: Int. J. Mod. Phys. E **26**, 1730002 (2017)
13. de Vries, J., Meißner, U.-G.: Int. J. Mod. Phys. E **25**, 1641008 (2016)
14. Wiringa, R.B., Stoks, V.G.J., Schiavilla, R.: Phys. Rev. C **51**, 38 (1995)
15. Kanada, H., Kaneko, T., Nagata, S., Morikazu, M.: Prog. Theor. Phys. **61**, 1327 (1979)
16. Hasegawa, A., Nagata, S.: Prog. Theor. Phys. **45**, 1786 (1971)
17. Nishioka, H., Saito, S., Yasuno, M.: Prog. Theor. Phys. **62**, 424 (1979)
18. Schmid, E.W., Wildermuth, K.: Nucl. Phys. **26**, 463 (1961)
19. Yamada, T., Funaki, Y.: Phys. Rev. C **82**, 064315 (2010)

20. Yamada, T., Funaki, Y.: Phys. Rev. C **92**, 034326 (2015)
21. Saito, S.: Prog. Theor. Phys. **40**, 893 (1968)
22. Liu, C.-P., Timmermans, R.G.E.: Phys. Rev. C **70**, 055501 (2004)
23. Towner, I.S., Hayes, A.C.: Phys. Rev. C **49**, 2391 (1994)
24. Liu, C.-P., Timmermans, R.G.E.: Phys. Rev. C **70**, 055501 (2004)
25. Yamanaka, N., Hiyama, E.: Phys. Rev. C **91**, 054005 (2015)
26. Bsaisou, J., de Vries, J., Hanhart, C., Liebig, S., Meißner, U.-G., Minossi, D., Nogga, A., Wirzba, A.: JHEP **03**, 104 (2015) [Erratum ibid. 05, 083 (2015)]
27. Yamanaka, N.: Hyperfine Interact. **239**, 35 (2018)

Chapter 59
Electron Scattering Experiments on Light Nuclei

Miha Mihovilovič

Abstract In this contribution we discuss double-polarised quasi-elastic electron scattering from ^{3}He and ^{12}C. For these nuclei precise data from recent experiments at Jefferson Lab and Mainz have become available, accompanied by a very strong theoretical effort dedicated to understanding these nuclei. The double polarisation experiments presented here offer insight into some of the details of the nuclear structure that could not be accessed by traditional cross-section measurements. The new experimental results show only rough agreement with the calculations and indicate that we do not yet fully understand the structure of these nuclei and nucleon dynamics inside them. Although the new data do not identify sources of the remaining discrepancies, they offer valuable clues that could lead to faster convergence of the theoretical models towards experimental results.

59.1 Introduction

The investigations of light nuclei are at the heart of nuclear physics. On the one hand light nuclei are interesting because they are small enough objects to be exactly calculable. On the other hand they are complex enough to exhibit all the important features that are usually present in the heavy nuclei. This combination makes them an ideal playground where the theoretical calculations can be compared to the data to an increasingly accurate degree.

In order to achieve a good agreement between the measurements and the theory, a precise knowledge on the nuclear ground-state wave function and the nuclear current that describes the transition from the initial to the final nuclear state are needed.

M. Mihovilovič (✉)
Jožef Stefan Institute, SI-1000 Ljubljana, Slovenia
e-mail: miha.mihovilovic@ijs.si

Faculty of Mathematics and Physics, University of Ljubljana,
SI-1000 Ljubljana, Slovenia

Institut für Kernphysik, Johannes Gutenberg-Universität Mainz,
DE-55128 Mainz, Germany

N. A. Orr et al. (eds.), *Recent Progress in Few-Body Physics*,
Springer Proceedings in Physics 238,
https://doi.org/10.1007/978-3-030-32357-8_59

They depend on the structure of the nuclei and the forces that bind nucleons together. In addition, effects that are accompanying the primary process, such as FSI and MEC also need to be well understood. Unfortunately, we can not measure different contributions individually, but observe a convolution of all the involved components. This means that if a discrepancy between the data and the theory is observed, it is not clear which part is responsible for it. This makes the interpretation of the observed results challenging.

In this paper recent studies of ^{3}He and ^{12}C will be discussed. The insight into the properties and structure of these nuclei is interesting not only in the context of fundamental nuclear physics, but is relevant also for other fields of research that use them as nuclear targets: ^{3}He is employed as an effective neutron target to investigate the spin properties of the nucleon, while ^{12}C is used in neutrino experiments to study their oscillation parameters.

59.2 Double-Polarised Electron Scattering Experiments

The traditional way of investigating the structure and dynamics of nuclei is by measuring electron-induced cross-sections. For quasi-elastic proton knockout the cross-section depends on six structure functions (R_L, R_T, R_{TL}, R_{TT}, $R_{T'}$ and $R_{TL'}$) which enclose all the information on the nucleus [1]:

$$\frac{d^6\sigma}{dE_{e'}d\Omega_{e'}dE_p d\Omega_p} = \sigma_0 \left[v_L R_L + v_T R_T + v_{TL} R_{TL} + v_{TT} R_{TT} \right.$$
$$\left. + \, h \left(v_{T'} R_{T'} + v_{TL'} R_{TL'} \right) \right] .$$

Unfortunately the measurements of this cross-section are suffering from systematic uncertainties which limit the precision of extracted structure functions and tarnish our insight into the structure of nuclei. These problems can be avoided by performing double-polarised experiments. With ^{3}He being a spin-1/2 nucleus, one can measure helicity ($h_\pm$) dependent asymmetries for different orientations of the target spin:

$$A(\vec{S}) = \frac{\sigma(h_+, \vec{S}) - \sigma(h_-, \vec{S})}{\sigma(h_+, \vec{S}) + \sigma(h_-, \vec{S})} \propto \left(v_{T'} R_{T'} + v_{TL'} R_{TL'} \right),$$

where $\vec{S}$ represents the spin of the nucleus. These asymmetries grant access to the suppressed responses ($R_{T'}$ and $R_{TL'}$) and allow for a validation of available ab-initio theories [2] explaining the structure of ^{3}He.

In a similar manner we can also improve our knowledge on the structure of the carbon nucleus. However, since ^{12}C can not be employed as a polarised target, we use different tactics and measure the polarisation of the ejected proton [3]:

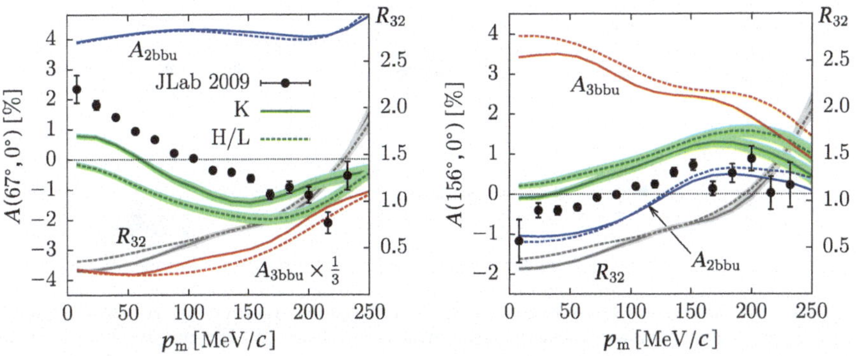

Fig. 59.1 The asymmetries $A(67°)$ (left) and $A(156°)$ (right) in the quasi-elastic $^3\vec{\mathrm{He}}(\vec{\mathrm{e}}, \mathrm{e'p})$ process (2bbu and 3bbu combined) as functions of missing momentum, compared to theoretical predictions (green) showing the 2bbu (blue) and 3bbu (red) contributions. All full lines correspond to Krakow-Bochum (K) calculations, while all dashed lines correspond to the Hannover-Lisbon (H/L) calculations [5]

$$\vec{P} = (P_x, P_y, P_z) = \frac{\sigma(h_+, \vec{s}_p) - \sigma(h_-, \vec{s}_p)}{\sigma_0},$$

where $\vec{s}_p$ represents the spin of the ejected proton in the system defined by the momentum of the virtual photon. The theory has shown that the polarisation components P_x and P_z are very sensitive to the details of the single-nucleon wave-function inside the nucleus [3]. Hence, the measurements of the polarisation components offer a unique opportunity to select the best description of the nuclear structure among different available models, which can not be accomplished with the cross-section data.

59.3 Investigating ^{3}He with $^3\vec{\mathrm{He}}(\vec{e}, e'p)$ Reactions

The most recent study of ^{3}He was performed at Jefferson Lab [4]. Using a high intensity electron beam in combination with a polarised ^{3}He target and spectrometers of Hall A we performed a high-precision measurement of the double-polarisation asymmetries for the $^3\vec{\mathrm{He}}(\vec{e}, e'p)$ reaction, as a function of the missing momentum in the range between 0 and 300 MeV/c. In particular, we investigated $A(67°)$ and $A(156°)$, representing the asymmetries when the target spin is oriented along and perpendicular to the direction of the incoming electron, respectively [5].

The measured asymmetries arise from an interplay of two reaction channels, that result in a proton in the final state: the two-body (2bbu) reaction $^3\vec{\mathrm{He}}(\vec{e}, e'p)d$, where a proton and a deuteron are produced, and a three body (3bbu) reaction $^3\vec{\mathrm{He}}(\vec{e}, e'p)pn$, where nucleus breaks into three separate nucleons. The 2bbu is interesting because it offers details on the wave function while the 3bbu channel is sensitive to the nuclear effects such as FSI. Since only one proton is detected in the final state, the two

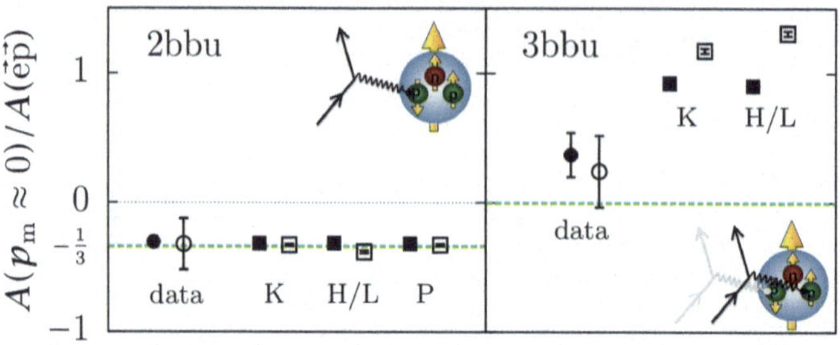

Fig. 59.2 The $A(67°)$ (full symbols) and $A(156°)$ (empty symbols) asymmetries for 2bbu (left) and 3bbu (right) divided by the corresponding asymmetries for elastic $\vec{e}\,\vec{p}$ scattering at the same value of Q^2 and for $E_m \leq 2.5$ MeV. In both panels the data (circles) are compared to the calculations (squares) [5]. Dashed green line denotes expected asymmetries within the PWIA

processes can be distinguished only though the examination of the missing energy spectrum. Such analysis requires apparatus with a very high momentum resolution, allowing a separation of two peaks just 2.2 MeV apart [6]. Unfortunately, due to the limited resolution of the spectrometers at hand, our resolving power was to small to be able to distinguish the two reaction channels. Consequently, we could not separate these two channels directly, but rather compared the measured asymmetries with theoretical calculations which correctly combine both reaction channels. For the experiment they were provided by theoretical groups from Krakow-Bochum [7, 8], Hannover-Lisbon [9–12] and Pisa [13].

The measured asymmetries as functions of missing momentum are shown in Fig. 59.1. The data are displayed together with the theoretical curves, which were obtained by averaging calculations for both contributing reactions over the acceptance and weighting them with the ratio of the cross-sections for the two channels. The theoretical calculations match the general trend of the measurements, but significantly underestimate the data. Since the theoretical 2bbu asymmetries obtained by all three theoretical groups are consistent with each other and agree with the elastic electron-proton asymmetry in the spectator picture, the discrepancy must be related to the 3bbu channel. We are most probably searching for a small effect which causes a tiny correction to the cross section but has significant influence on the 3bbu asymmetry. The Krakow group suspects that this effect could be the 3NF or the p-p Coulomb interaction which at present is not included in their calculations. However, calculations of the Hannover-Lisbon group already contain the 3NF but also suffer from the same discrepancy. Therefore, they suspect that the difference between the data and the theory is related to either unconsidered relativistic effects or uncertainties in the EM current operator, which needs to be improved in the future.

In order to gain more knowledge on the structure of ^{3}He despite inconsistencies with the theory, the data in the vicinity of $p_{\mathrm{miss}} = 0$ and with E_{miss} close to the 2bbu and 3bbu thresholds were examined. Within the PWIA, one expects that in the spectator limit the 2bbu asymmetry will be close to $-1/3$ of the elastic $\vec{e}\,\vec{p}$ asymmetry. On the other hand, for the 3bbu channel we expect an asymmetry close to zero, since the electron can scatter from either of the two protons with opposite spin directions. The analysis has shown that our data agree well with this hypothesis, see Fig. 59.2, and so do the calculations for the 2bbu channel. However, for the 3bbu channel the theories predict large asymmetries, which are close to the elastic ones. Since it is known that the 3bbu asymmetry arises from FSI [6], this has led us to the conclusion that the available theoretical calculations overestimate the contribution of the FSI.

59.4 Investigating ^{12}C with $^{12}C(\vec{e}, e'\vec{p})$ Reaction

The latest investigation of the double-polarised reaction $^{12}C(\vec{e}, e'\vec{p})$ has been performed in 2015 in Mainz. The experiment was initially motivated by the assumption that the nuclear environment changes the charge and magnetisation distributions of the constituent nucleons. In an attempt to observe this effect the polarisation components P_x and P_z were measured [14]. We know that for the free proton the ratio of the two polarisation components is proportional to the ratio of the electric and magnetic form-factors [15]. Hence, comparing the quasi-elastic measurements on the bound proton to the values for a free proton, one could get an idea about the influence of nuclear medium on the proton's elastic form-factors and its charge and magnetisation distributions.

The measured ratio $R_{^{12}C} = P_x/P_z$ relative to the corresponding ratio for a free proton is shown in Fig. 59.3. The obtained results are presented as a function of virtuality which, within the scope of the PWIA, tells us how far off the shell is the struck proton. The results show a clear deviation from those for a free proton and prove that the nuclear medium affects the interacting proton. However, the observed deviation can not be contributed solely to the modification of the proton's structure. This phenomenon is competing against other effects, like FSI and MEC, that also influence the result. In order to understand different contributions to the measured result, the data need to be studied in conjunction with the theoretical models. Therefore, we established a collaboration with the group from Pavia [3], and their calculations have demonstrated that the deviation of the measured ratio from that for a free proton is predominantly governed by the properties of the initial proton wave function, while is somewhat insensitive to the details of the optical potential and FSI. Furthermore, when comparing the results with the previous measurements on other nuclei [16–18], a surprising universal behaviour of the double polarisation ratios was observed. Within the measured uncertainties the double ratios for ^{12}C closely follow those for ^{4}He and ^{2}H, which are known to be governed by FSI [19]. Additionally, the virtuality dependence of the double polarisation ratio seems to be independent of Q^2 and beam energy. This observation is contra-intuitive and not understood. Therefore, further theoretical and experimental work is required in order to correctly interpret the measured data and explain the physics mechanisms behind them.

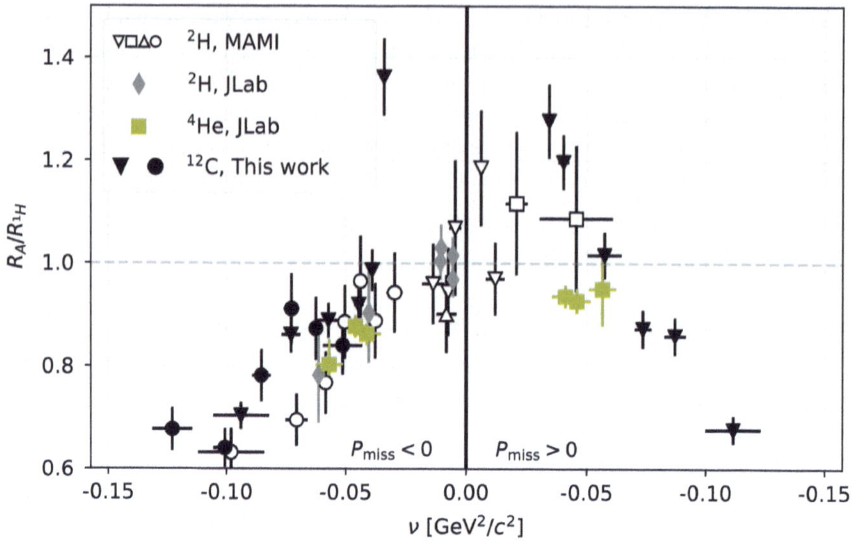

Fig. 59.3 The measured polarisation double-ratio of protons from ^{12}C compared with those obtained for ^{2}H and ^{4}He ($A = {}^{2}$H, ^{4}He, ^{12}C) as a function of the proton virtuality [14]. The ^{12}C results (black full symbols) represent combination data where protons are removed from the s- and p-shell. The open symbols are ^{2}H data measured at Mainz with essentially the same kinematics [16]. The light symbols are ^{2}H and ^{4}He data measured at JLab at $Q^2 = 0.8$ and $1\,\text{GeV}/c^2$ [17, 18]

59.5 Conclusions

Recent experiments at Jefferson Lab and Mainz provided new double polarisation data for ^{3}He and ^{12}C, which can be used for precise investigations of nuclear structure and dynamics. The comparison of these precise data to the available theoretical calculations revealed significant discrepencies for both ^{3}He and ^{12}C. At the moment it is not clear what causes the observed inconsistencies. However, in order to advance our knowledge on these nuclei and also to be able to use them as targets in other experiments, the discrepancies need to be addressed and understood. Therefore, dedicated theoretical efforts are being invested to further improve the models, and new experiments are planned at both Jefferson Lab and Mainz that could be used to advance our knowledge on these and also heavier nuclei like ^{40}Ar and ^{40}Ca.

References

1. Raskin, A.S., Donnelly, T.W.: Ann. Phys. **191**, 78 (1986)
2. Golak, J., et al.: Phys. Rep. **415**, 89 (2005)
3. Boffi, S., Giusti, C., Pacati, F.D.: Phys. Rep. A **226**, 1 (1993)

4. Širca, S., Gilad, S., Higinbotham, D.W., Korsch, W., Norum, B.E. (spokespersons): Measurement of A_x and A_z asymmetries in the quasi-elastic $^3\vec{\mathrm{He}}\left(\vec{e}, e'd\right)$. JLab experiment E05–102, June 2005
5. Mihovilovic, M., et al.: arXiv:1804.06043 [nucl-ex]
6. Achenbach, P., et al.: Eur. Phys. J. A **25**, 177 (2005)
7. Golak, J., et al.: Phys. Rev. C **65**, 064004 (2002)
8. Golak, J., et al.: Phys. Rev. C **72**, 054005 (2005)
9. Yuan, L.P., et al.: Phys. Rev. C **66**, 054004 (2002)
10. Deltuva, A., et al.: Phys. Rev. C **69**, 034004 (2004)
11. Deltuva, A., et al.: Phys. Rev. C **70**, 034004 (2004)
12. Deltuva, A., et al.: Phys. Rev. C **71**, 054005 (2005)
13. Marcucci, L.E., et al.: Phys. Rev. C **72**, 014001 (2005)
14. Izraeli, D., et al.: Phys. Lett. B **781**, 95 (2018)
15. Akhiezer, A.I., et al.: Sov. J. Part. Nucl. **4**, 277 (1974)
16. Yaron, I., et al.: Phys. Lett. B **769**, 21 (2017)
17. Hu, B., et al.: Phys. Rev. C **73**, 064004 (2006)
18. Strauch, S., et al.: Phys. Rev. Lett. **91**, 052301 (2003)
19. Arenhövel, H., et al.: Eur. Phys. J. A **23**, 147 (2005)

Chapter 60
The *Basic* Model of Nuclear Theory: From Atomic Nuclei to Infinite Matter

Maria Piarulli

Abstract A major goal of nuclear theory is to explain the wealth of data and peculiarities exhibited by nuclear systems in a fully microscopic way. In such an approach, which we refer to as the *basic model* of nuclear theory, the nucleons interact with each other via many-body (primarily, two- and three-body) effective interactions, and with external electroweak probes via effective current operators. These effective interactions and currents are the main inputs to ab-initio methods that are aimed at solving the many-body Schrödinger equation associated with the nuclear system under consideration. In this talk, I will review recent progress in Quantum Monte Carlo calculations of low-lying spectra and electroweak properties of light nuclei as well as nucleonic matter equation of state. Emphasis will be on calculations based on chiral effective field theory approach.

60.1 Introduction

The last few decades have witnessed the emergence of the *basic model* of nuclear theory in which nuclear systems—particularly atomic nuclei and infinite nucleonic matter—can be described as a collection of point-like nucleons interacting with each other in terms of many-body (primarily, two and three-body) effective interactions, and with external electroweak probes via effective current operators. Such an approach, in conjunction with a computational method of choice to solve the many-body Schrödinger equation, can thus be used to explain the structure and dynamics of nuclear systems in a fully microscopic way where the nucleons emerge as effective degrees of freedom, at sufficiently low-energy. This can be seen as the result of a more fundamental process involving the constituents of the nucleons, i.e., the quarks whose dynamics is described by Quantum Chromodynamics (QCD).

How the force between nucleons emerges from QCD has kept nuclear theorists occupied for many decades. One may try to solve such a problem with brute com-

M. Piarulli (✉)
Washington University, St. Louis, MO 63130, USA
e-mail: mpiarulli@physics.wustl.edu

N. A. Orr et al. (eds.), *Recent Progress in Few-Body Physics*,
Springer Proceedings in Physics 238,
https://doi.org/10.1007/978-3-030-32357-8_60

puting power on a discretized, Euclidean space-time lattice (known as lattice QCD). However, in spite of many advances [1–3], lattice QCD calculations are still in their infancy, and thus, at the present time, they can only be used to check few representative key-issues. Clearly, a different approach is necessary to address the full complexity of nuclear structure problems.

60.2 The Basic Model of Nuclear Theory

The basic model of nuclear theory assumes that a Hamiltonian consisting of non-relativistic kinetic energy, two- and three-body interactions provides a good approximation to the energy of interacting nucleons. Two-body interactions are characterized by a long-range component, for inter-nucleon separation $r \gtrsim 2$ fm, due to one-pion exchange (OPE) [4], and intermediate- and short-range components, for, respectively, $1\,\text{fm} \lesssim r \lesssim 2\,\text{fm}$ and $r \lesssim 1\,\text{fm}$. Up to the mid-1990's, such interactions were based almost exclusively on meson-exchange phenomenology. Those of the mid-1990's [5–7] were constrained by fitting nucleon-nucleon (*NN*) elastic scattering data up to lab energies of 350 MeV, with χ^2/datum $\simeq 1$ relative to the database available at the time [8]. Two well-known and still widely used examples in this class are the Argonne v_{18} (AV18) [6] and CD-Bonn [7]. These are so-called *phenomenological* interactions.

Already in the 1980's, accurate three-body calculations showed that contemporary *NN* interactions did not provide enough binding for the three-body nuclei, ^{3}H and ^{3}He [9]. In the late 1990's and early 2000's this realization was also extended to the spectra (ground and low-lying excited states) of light p-shell nuclei, for instance, in calculations based on quantum Monte Carlo (QMC) methods [10] and in no-core shell-model (NCSM) studies [11]. Consequently, the basic model with only *NN* interactions fit to scattering data, without the inclusion of a three-nucleon (3*N*) interaction, is definitlely incomplete.

Because of the composite nature of the nucleon and, in particular, the dominant role of the Δ resonance in pion-nucleon scattering, multi-nucleon interactions arise quite naturally in meson-exchange phenomenology. In particular, the Illinois 3*N* interactions [12] consist of a dominant two-pion exchange (TPE)—the Fujita-Miyazawa interaction [13]—and smaller multi-pion exchange components resulting from the excitation of intermediate Δ's. The most recent version, Illinois-7 (IL7) [14], also contains phenomenological isospin-dependent central terms. The parameters characterizing this 3*N* potential have been determined by fitting the low-lying spectra of nuclei in the mass range $A = 3$–10. The resulting AV18+IL7 Hamiltonian, generally utilized with QMC methods, then leads to predictions of 100 ground- and excited-state energies up to $A = 12$, including the ^{12}C ground- and Hoyle-state energies, in good agreement with the corresponding empirical values [15]. However, when used to compute the neutron star matter equation of state, these do not provide sufficient repulsion to garantee the stability of the observed starts against gravitational collapse [16]. Thus, in the context of the phenomenological nuclear interactions, we do not have a Hamiltonian that can predict the properties of all nuclear systems, from *NN* scattering to dense nuclear and neutron matter. Furthermore, high-precision phenomenological potentials suffer from several limitations, most notably the missing connection with the (approximate) chiral symmetry exhibited by QCD. For instance, they do not provide rigorous schemes to consistently derive two- and three-body forces and compatible electroweak currents.

To this end, a new phase in the evolution of the basic model, and renewed interest in its further development, have been spurred by the emergence in the early 1990's of chiral effective field theory (χEFT) [17–19]. Such a theory provides the most general scheme accommodating all possible interactions among nucleons, Δ isobars, and pions compatible with the relevant symmetries—in particular chiral symmetry—of low-energy QCD. By its own nature, χEFT is organized within a given power counting scheme and the resulting chiral potentials (and currents) are systematically expanded in powers of Q/Λ_χ with $Q \ll \Lambda_\chi$, where Q denotes generically a low-momentum scale entering the theory and $\Lambda_\chi \sim 1$ GeV specifies the chiral-symmetry breaking scale (see [20, 21] for recent review articles). Therefore, χEFT provides a rigourous scheme to systematically construct many-body forces and consistent electroweak currents, providing tools to estimate their uncertainties [22–27]. The Weinberg power counting of χEFT indicates that nuclear forces are dominated by *NN* interactions while many-body forces are suppressed by powers of Q. However, the inclusion of $3N$ interaction is mandatory at the level of accuracy now reached by few- and many-body calculations (see [28, 29] for a comprehensive review on this topic).

Nuclear forces in χEFT are separated into pions-exchange contributions and contact terms. Pions-exchange contributions represent the long- and intermediate-range of nuclear interactions while the contact terms encode the short-range physics and their strength is specified by unknown low-energy constants (LECs), which then need to be fixed by fitting experimental data. Within χEFT many studies have been carried out dealing with the construction and optimization of *NN* and $3N$ interactions [21, 23, 30–51]. These interactions were typically formulated in momentum space, and included cutoff functions to regularize their behavior at large momenta which, however, made them strongly non-local when Fourier-transformed in configuration space, and therefore unsuitable for use with QMC methods. Among these, Variational (VMC) and Green's Function Monte Carlo (GFMC) are the methods of choice to provide reliable solutions of the many-body Schrödinger equation—presently for up to $A = 12$ nucleons—with full account of the complexity of the many-body, spin- and isospin-dependent correlations induced by nuclear interactions. The sampling of configuration space in VMC and GFMC simulations gives access to many important properties of light nuclei as spectra, form factors, transitions, low-energy scattering and response. Auxiliary Field Diffusion Monte Carlo (AFDMC) uses Monte Carlo to also sample the spin-isospin degrees of freedom, enabling studies of, for example, neutron matter [52–56] that is so critical to determining the structure of neutron stars and properties of light- and medium-mass nuclei [57, 58]. QMC simulations have surely proved to be very valuable in attacking many nuclear structure issues over the last three decades. Therefore there was a need to develop local chiral interaction for QMC in order to test to what extent the chiral effective field theory framework impacts our knowledge of few- and many-body systems.

It is worth mentioning here that, besides QMC methods, there are several other sophisticated few- and many-methods [59–64], based on realistic Hamiltonians, that are addressing different interesting aspects of nuclear systems. In particular there is a big effort on developing microscopic models that include continuum couplings which are mandatory to describe, for istance, weakly bound nuclear systems [65].

60.3 Local Chiral Hamiltonians

A major thrust of our work is based on the theoretical derivation, optimization and implementation of chiral interactions suitable for QMC methods. In recent years local, configuration-space chiral *NN* interactions have been derived by two groups [66–68]. In this talk, we focus on the family of local chiral interactions constructed by our group in [68, 69]. They are written as the sum of an electromagnetic interaction component and a strong-interaction component, characterized by long- and short-range parts. The long-range part includes one-pion-exchange (OPE) and two-pion-exchange (TPE) terms up to next-to-next-to-leading order (N2LO) in the chiral expansion, derived in the static limit from leading and sub-leading πN and $\pi N\Delta$ chiral Lagrangians. The short-range part, however, is described by contact terms up to next-to-next-to-next-to-leading order (N3LO), characterized by 26 LECs.

We constructed two classes of interactions, which only differ in the range of laboratory energy over which the fits were carried out, either 0–125 MeV in class I or 0–200 MeV in class II (the fits used the 2013 *NN* database, including the deuteron ground-state energy and two-neutron scattering length, as assembled by the Granada group [70]). For each class, three different sets of cutoff radii (R_S, R_L) were considered (R_S, R_L) = (0.8, 1.2) fm in set a, (0.7, 1.0) fm in set b, and (0.6, 0.8) fm in set c, where R_S and R_L enter respectively the configuration-space cutoffs for the short- and long-range parts of the two-nucleon interaction [68]. The χ^2/datum achieved by the fits in class I (II) was $\lesssim 1.1 (\lesssim 1.4)$ for a total of about 2700 (3700) data points. We are refering to these high-quality *NN* interactions generically as the Norfolk potentials (NV2s), and designate those in class I as NV2-Ia, NV2-Ib, and NV2-Ic, and those in class II as NV2-IIa, NV2-IIb, and NV2-IIc.

The NV2s were found to provide insufficient attraction, in GFMC calculations, for the ground-state energies of nuclei with $A = 3$–6 [68], thus corroborating the insight realized in the early 2000's within the older (and less fundamental) meson-exchange phenomenology. To remedy this shortcoming, we constructed the leading $3N$ interaction in χEFT, including Δ intermediate states. It consists [34, 35] of a long-range piece mediated by TPE and a short-range piece parametrized in terms of two contact interactions. The two LECs—namely c_D and c_E—have been constrained either by fitting exclusively strong-interaction observables [52, 53, 71, 72] or by relying on a combination of strong- and weak-interaction ones [73–75]. This last approach is made possible by the relation, established in χEFT [76], between c_D in the three-nucleon contact interaction and the LEC in the *NN* contact axial current [73, 74], which allows one to use nuclear properties governed by either the strong or weak interactions to constrain simultaneously the 3*N* interaction and *NN* axial current.

In particular, in our recent work [72, 75], we constructed two different sets for the values of c_D and c_E, leading to two different parametrization of the 3*N* interaction. In the first, these LECs were determined by simultaneously reproducing the experimental trinucleon ground-state energies and *nd* doublet scattering length for each of the *NN* models considered [72]. In the second set, these c_D and c_E were constrained by fitting, in addition to the trinucleon energies, the empirical value of the Gamow-Teller matrix element in tritium β decay [75]. The resulting Hamiltonian models were designated as NV2+3-Ia/b and NV2+3-IIa/b (or Ia/b and IIa/b for short) in the first case, and as NV2+3-Ia*/b* and NV2+3-IIa*/b* (or Ia*/b* and IIa*/b*) in the second. These two different procedures for fixing c_D and c_E produced rather different values for these LECs, particularly for c_E which was found to be relatively large and negative in models Ia/b and IIa/b, but quite small, and not consistently negative, in models Ia* /b* and IIa* /b*. This in turn impacts predictions for the spectra of light nuclei and the equation of state of neutron matter, since a negative c_E leads to repulsion in light nuclei, but attraction in pure neutron matter.

60.4 Applications

In this section, we discuss some illustrative applications of nuclear χEFT to the few- and many-body systems.

In Fig. 60.1 are shown the GFMC energy results calculated with NV2+3-Ia model (described in the previous section) for 37 different nuclear states in $A = 4$–12 nuclei. They are compared to results from the older AV18+IL7 model [15] and experiment [77]. The agreement with experiment is impressive for both Hamiltonians, with absolute binding energies very close to experiment, and excited states reproducing the observed ordering and spacing, indicating reasonable one-body spin-orbit splittings. For both Hamiltonians, the inclusion of the 3*N* interactions is in many cases necessary to get ground states that are correctly bound against breakup, e.g., ^{6}He is not bound with just the *NN* interaction [68]. The lowest 3^+ and 1^+ states of ^{10}B are of particular interest. For both AV18 and Ia without 3*N* interactions, the 1^+ state is incorrectly predicted as the ground state (for NV2-Ia by 1.9 MeV) but including the 3*N* interactions gives the correct 3^+ ground state. However, it is important to emphasize that in the AV18+IL7 model the four parameters in the 3*N* interaction are fitted to the energies of many nuclear levels up to $A = 10$.

While model NV2+3-Ia provides an excellent description of the energy levels and level ordering of nuclei in the mass range $A = 4$–12 [72], it collapses neutron matter already at relatively low densities [56], and cannot sustain the existence of neutron stars of twice solar masses, in conflict with recent observations [78, 79]. Indeed, we find that the equation of state of neutron matter obtained with interactions that accurately describe $A \leq 12$ nuclei suffers from a strong regulator dependence, consistent

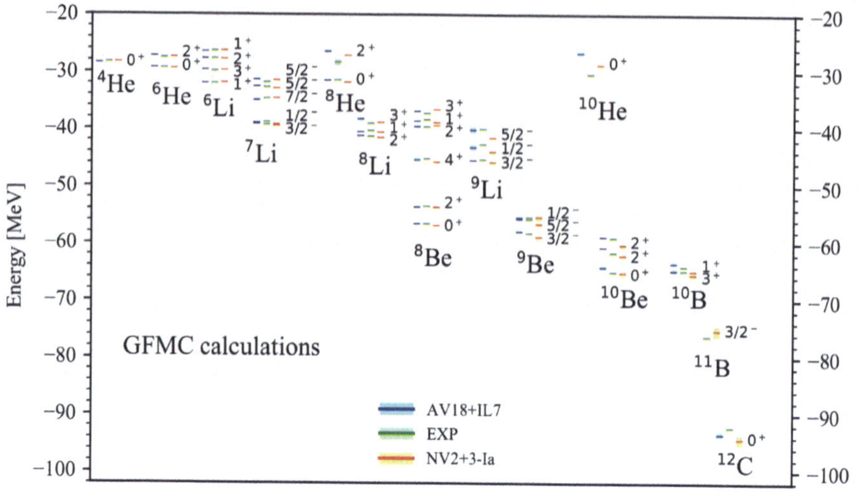

Fig. 60.1 The energy spectra of $A = 4$–12 nuclei obtained with the NV2+3-Ia chiral interactions are compared to experimental data [77]. Also shown are results obtained with the phenomenological AV18+IL7 interactions [15]

with [52]. More specifically, the potentials with the softest regulators do not provide sufficient repulsion to guarantee the stability of observed stars against gravitational collapse (this is analogous to what happens when employing the phenomenological AV18+IL7 interaction [16]), while those characterized by harder regulators are in much better agreement with astrophysical constraints. There are indications [56] that the smaller values of c_E characteristic of models Ia*/b* and IIa*/b*, mitigate, if not resolve, the collapse problem, however the prediction of light-nuclei spectra is not at the same accuracy level [56] of Fig. 60.1. Studies along this line are in progress.

Differential cross section, and vector and tensor polarization observables in proton-deuteron elastic scattering obtained with the present NV2+3 models have also been calculated using the Hyperspherical Harmonics (HH) [59] method and compared to experimental data [80]. The effect of the $3N$ interaction is small, marginally improving (appreciably worsening) the agreement between theory and experiment for the observables A_y, $i\,T_{11}$, and T_{22} (T_{20} and T_{21}). In particular, the well known discrepancy in the vector analyzing power—the "A_y puzzle" [81]—persists. It also appears to be unresolved when higher-order chiral loops are accounted for in the long-range component of the $3N$ interaction [82]. More details regarding this topic are provided in Girlanda's talk.

To have a full description of the properties of nuclear systems such has electroweak form factors, magnetic moments, radii, decays, etc, we also need to know how they interact with external probes, such as electroweak probes. Therefore we need to specify the charge and current operators which, in the contest of the basic model, are written in a series of many-body operators in analogy to the Hamiltonian.

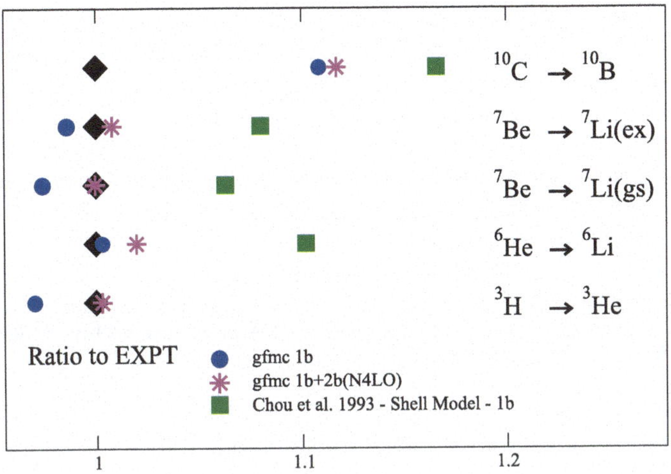

Fig. 60.2 Ratios of GFMC to experimental values of the GT RMEs in the ^{3}H, ^{6}He, ^{7}Be, and ^{10}C weak transitions. Theory predictions correspond to the χEFT axial current in LO (blue circles) and up to N4LO (magenta stars). Green squares indicate unquenched shell model calculations from [89] based on the LO axial current

More recently electroweak operators based on χEFT have been derived in both the electromagnetic and weak sectors [83–88]. For instance, in the weak sector, observables that are of great interest are β-decay and the Gamow-Teller (GT) matrix element contributing to it. Being well-known experimentally they have provided over the past several decades, a testing ground for models of the nuclear axial current and, in particular, for the role that many-body weak transition operators beyond the leading one-body GT operator play in this matrix element.

In Fig. 60.2 are displayed ratios of GFMC to experimental values performed by Pastore and collaborators for the GT in the ^{3}H, ^{6}He, ^{7}Be, and ^{10}C weak transitions. They find overall good agreement with data for the beta decay of ^{6}He and electron capture of ^{7}Be even if the first one is overpredicted by 2%. The experimental GT reduced matix element (RME) for the ^{10}C β-decay is overpredicted by 10%, with two-body currents giving a contribution that is comparable to the statistical GFMC error. The presence of a second (1+; 0) excited state at ∼2.15 MeV can potentially contaminate the wave function of the ^{10}B excited state at ∼0.72 MeV, making this the hardest transition to calculate reliably.

60.5 Conclusions

Where do we stand so far with the basic model of nuclear theory? From a theoretical point of view we went from a phenomenological approach of nuclear forces and currents to a framework rooted in low-energy QCD (χEFT) drawing a more deep connection with the fundamental theory of quarks and gluons.

Highly advanced few- and many-body techniques in conjunction with the rapid increase in computational resources is making possible to test the inputs of the basic model (nuclear interactions and corresponding electroweak currents) based on different theoretical approaches. Ab-initio calculations of light and medium-mass nuclei and infinite nucleonic matter surely provide important benchmarks for the computational methods that aim to understand heavy-mass nuclei where new physics might be hiding. More of these studies are needed in the context of χEFT to test to what extent such a theory impacts our knowledge of nuclear systems.

Acknowledgements I wish to thank my collaborators A. Baroni, L. Girlanda, A. Kievsky, A. Lovato, L.E. Marcucci, S. Pastore, S.C. Pieper, R. Schiavilla, M. Viviani, and R.B. Wiringa for their many contributions to the work presented in this talk.

References

1. Inoue, T., Aoki, S., Doi, T., Hatsuda, T., Ikeda, Y., Ishii, N., Murano, K., Nemura, H., Sasaki, K.: Phys. Rev. Lett. **111**(11), 112503 (2013)
2. Beane, S.R., Chang, E., Cohen, S., Detmold, W., Lin, H.W., Orginos, K., Parreno, A., Savage, M.J., Tiburzi, B.C.: Phys. Rev. Lett. **113**(25), 252001 (2014)
3. Savage, M.J., Shanahan, P.E., Tiburzi, B.C., Wagman, M.L., Winter, F., Beane, S.R., Chang, E., Davoudi, Z., Detmold, W., Orginos, K.: Phys. Rev. Lett. **119**(6), 062002 (2017)
4. Yukawa, H.: Proc. Phys. Math. Soc. Jpn. **17**, 48–57 (1935)
5. Stoks, V.G.J., Klomp, R.A.M., Terheggen, C.P.F., de Swart, J.J.: Phys. Rev. C **49**, 2950 (1994)
6. Wiringa, R.B., Stoks, V.G.J., Schiavilla, R.: Phys. Rev. C **51**, 38 (1995)
7. Machleidt, R.: Phys. Rev. C **63**, 024001 (2001)
8. Stoks, V.G.J., Klomp, R.A.M., Rentmeester, M.C.M., de Swart, J.J.: Phys. Rev. C **48**, 792 (1993)
9. Friar, J.L., Gibson, B.F., Payne, G.L.: Ann. Rev. Nucl. Part. Sci. **34**, 403 (1984)
10. Pudliner, B.S., Pandharipande, V.R., Carlson, J., Pieper, S.C., Wiringa, R.B.: Phys. Rev. C **56**, 1720 (1997)
11. Navratil, P., Vary, J.P., Barrett, B.R.: Phys. Rev. C **62**, 054311 (2000)
12. Pieper, S.C., Pandharipande, V.R., Wiringa, R.B., Carlson, J.: Phys. Rev. C **64**, 014001 (2001)
13. Fujita, J., Miyazawa, H.: Prog. Theor. Phys. **17**, 360 (1957)
14. Pieper, S.C.: AIP Conf. Proc. **1011**(1), 143 (2008)
15. Carlson, J., Gandolfi, S., Pederiva, F., Pieper, S.C., Schiavilla, R., Schmidt, K.E., Wiringa, R.B.: Rev. Mod. Phys. **87**, 1067 (2015)
16. Maris, P., Vary, J.P., Gandolfi, S., Carlson, J., Pieper, S.C.: Phys. Rev. C **87**(5), 054318 (2013)
17. Weinberg, S.: Phys. Lett. B **251**, 288 (1990)
18. Weinberg, S.: Nucl. Phys. B **363**, 3 (1991)
19. Weinberg, S.: Phys. Lett. B **295**, 114 (1992)
20. Epelbaum, E., Hammer, H.W., Meissner, U.G.: Rev. Mod. Phys. **81**, 1773 (2009)
21. Machleidt, R., Entem, D.R.: Phys. Rep. **503**, 1 (2011)
22. Furnstahl, R.J., Phillips, D.R., Wesolowski, S.: J. Phys. **G42**(3), 034028 (2015)
23. Epelbaum, E., Krebs, H., Meißner, U.G.: Eur. Phys. J. A **51**(5), 53 (2015)
24. Furnstahl, R.J., Klco, N., Phillips, D.R., Wesolowski, S.: Phys. Rev. C **92**(2), 024005 (2015)
25. Wesolowski, S., Klco, N., Furnstahl, R.J., Phillips, D.R., Thapaliya, A.: J. Phys. **G43**(7), 074001 (2016)
26. Melendez, J.A., Wesolowski, S., Furnstahl, R.J.: Phys. Rev. C **96**(2), 024003 (2017)
27. Wesolowski, S., Furnstahl, R.J., Melendez, J.A., Phillips, D.R.: arXiv:1808.08211 (2018)

28. Kalantar-Nayestanaki, N., Epelbaum, E., Messchendorp, J.G., Nogga, A.: Rep. Prog. Phys. **75**, 016301 (2012)
29. Hammer, H.W., Nogga, A., Schwenk, A.: Rev. Mod. Phys. **85**, 197 (2013)
30. Ordonez, C., Ray, L., van Kolck, U.: Phys. Rev. C **53**, 2086 (1996)
31. Epelbaum, E., Gloeckle, W., Meissner, U.G.: Nucl. Phys. A **637**, 107 (1998)
32. Entem, D.R., Machleidt, R.: Phys. Rev. C **68**, 041001 (2003)
33. Krebs, H., Epelbaum, E., Meissner, U.G.: Eur. Phys. J. A **32**, 127 (2007)
34. van Kolck, U.: Phys. Rev. C **49**, 2932 (1994)
35. Epelbaum, E., Nogga, A., Gloeckle, W., Kamada, H., Meissner, U.G., Witala, H.: Phys. Rev. C **66**, 064001 (2002)
36. Navratil, P.: Few Body Syst. **41**, 117 (2007)
37. Bernard, V., Epelbaum, E., Krebs, H., Meissner, U.G.: Phys. Rev. C **84**, 054001 (2011)
38. Girlanda, L., Kievsky, A., Viviani, M.: Phys. Rev. C **84**, 014001 (2011)
39. Krebs, H., Gasparyan, A., Epelbaum, E.: Phys. Rev. C **85**, 054006 (2012)
40. Epelbaum, E., Krebs, H., Meißner, U.G.: Phys. Rev. Lett. **115**(12), 122301 (2015)
41. Entem, D.R., Kaiser, N., Machleidt, R., Nosyk, Y.: Phys. Rev. C **91**(1), 014002 (2015)
42. Entem, D.R., Kaiser, N., Machleidt, R., Nosyk, Y.: Phys. Rev. C **92**(6), 064001 (2015)
43. Ekström, A., et al.: Phys. Rev. Lett. **110**(19), 192502 (2013)
44. Ekström, A., Jansen, G.R., Wendt, K.A., Hagen, G., Papenbrock, T., Carlsson, B.D., Forssén, C., Hjorth-Jensen, M., Navrátil, P., Nazarewicz, W.: Phys. Rev. C **91**(5), 051301 (2015)
45. Ekström, A., Hagen, G., Morris, T.D., Papenbrock, T., Schwartz, P.D.: Phys. Rev. C **97**(2), 024332 (2018)
46. Reinert, P., Krebs, H., Epelbaum, E.: Eur. Phys. J. A **54**(5), 86 (2018)
47. Binder, S., et al.: Phys. Rev. **C98**(1), 014002 (2018)
48. Krebs, H., Gasparyan, A.M., Epelbaum, E.: Phys. Rev. C **98**(1), 014003 (2018)
49. Friar, J.L., van Kolck, U.: Phys. Rev. C **60**, 034006 (1999)
50. Friar, J.L., van Kolck, U., Rentmeester, M.C.M., Timmermans, R.G.E.: Phys. Rev. C **70**, 044001 (2004)
51. Friar, J.L., Payne, G.L., van Kolck, U.: Phys. Rev. C **71**, 024003 (2005)
52. Lynn, J.E., Tews, I., Carlson, J., Gandolfi, S., Gezerlis, A., Schmidt, K.E., Schwenk, A.: Phys. Rev. Lett. **116**(6), 062501 (2016)
53. Tews, I., Gandolfi, S., Gezerlis, A., Schwenk, A.: Phys. Rev. C **93**(2), 024305 (2016)
54. Tews, I., Carlson, J., Gandolfi, S., Reddy, S.: Astrophys. J. **860**(2), 149 (2018)
55. Tews, I., Margueron, J., Reddy, S.: Phys. Rev. C **98**(4), 045804 (2018)
56. Piarulli, M.: https://indico.ectstar.eu/event/15/contributions/270/
57. Lonardoni, D., Gandolfi, S., Lynn, J.E., Petrie, C., Carlson, J., Schmidt, K.E., Schwenk, A.: Phys. Rev. C **97**(4), 044318 (2018)
58. Lonardoni, D., Gandolfi, S., Wang, X.B., Carlson, J.: Phys. Rev. C **98**(1), 014322 (2018)
59. Kievsky, A., Rosati, S., Viviani, M., Marcucci, L.E., Girlanda, L.: J. Phys. **G35**, 063101 (2008)
60. Barrett, B.R., Navratil, P., Vary, J.P.: Prog. Part. Nucl. Phys. **69**, 131 (2013)
61. Hagen, G., Papenbrock, T., Hjorth-Jensen, M., Dean, D.J.: Rep. Prog. Phys. **77**(9), 096302 (2014)
62. Hergert, H., Bogner, S.K., Binder, S., Calci, A., Langhammer, J., Roth, R., Schwenk, A.: Phys. Rev. C **87**(3), 034307 (2013)
63. Dickhoff, W.H., Barbieri, C.: Prog. Part. Nucl. Phys. **52**, 377 (2004)
64. Forssen, C., Hagen, G., Hjorth-Jensen, M., Nazarewicz, W., Rotureau, J.: Phys. Scripta T **152**, 014022 (2013)
65. Fossez, K., Rotureau, J., Nazarewicz, W.: Phys. Rev. C **98**(6), 061302 (2018)
66. Gezerlis, A., Tews, I., Epelbaum, E., Gandolfi, S., Hebeler, K., Nogga, A., Schwenk, A.: Phys. Rev. Lett. **111**(3), 032501 (2013)
67. Gezerlis, A., Tews, I., Epelbaum, E., Freunek, M., Gandolfi, S., Hebeler, K., Nogga, A., Schwenk, A.: Phys. Rev. C **90**(5), 054323 (2014)
68. Piarulli, M., Girlanda, L., Schiavilla, R., Kievsky, A., Lovato, A., Marcucci, L.E., Pieper, S.C., Viviani, M., Wiringa, R.B.: Phys. Rev. C **94**(5), 054007 (2016)

69. Piarulli, M., Girlanda, L., Schiavilla, R., Navarro Pérez, R., Amaro, J.E.: Ruiz ArriolaE. Phys. Rev. C **91**(2), 024003 (2015)
70. Navarro Pérez, R., Amaro, J.E., Ruiz Arriola, E.: Phys. Rev. **C88**(6), 064002 (2013) [Erratum: Phys. Rev. **C91**(2), 029901 (2015)]
71. Lynn, J.E., Tews, I., Carlson, J., Gandolfi, S., Gezerlis, A., Schmidt, K.E., Schwenk, A.: Phys. Rev. C **96**(5), 054007 (2017)
72. Piarulli, M., et al.: Phys. Rev. Lett. **120**(5), 052503 (2018)
73. Gazit, D., Quaglioni, S., Navratil, P.: Phys. Rev. Lett. **103**, 102502 (2009)
74. Marcucci, L.E., Kievsky, A., Rosati, S., Schiavilla, R., Viviani, M.: Phys. Rev. Lett. **108**, 052502 (2012) [Erratum: Phys. Rev. Lett. **121**(4), 049901 (2018)]
75. Baroni, A., et al.: Phys. Rev. **C98**(4), 044003 (2018)
76. Gardestig, A., Phillips, D.R.: Phys. Rev. Lett. **96**, 232301 (2006)
77. Audi, G., Wapstra, A.H., Thibault, C.: Nucl. Phys. A **729**, 337 (2002)
78. Demorest, P., Pennucci, T., Ransom, S., Roberts, M., Hessels, J.: Nature **467**, 1081 (2010)
79. Antoniadis, J., et al.: Science **340**, 6131 (2013)
80. Shimizu, S., Sagara, K., Nakamura, H., Maeda, K., Miwa, T., Nishimori, N., Ueno, S., Nakashima, T., Morinobu, S.: Phys. Rev. C **52**, 1193 (1995)
81. Gloeckle, W., Witala, H., Huber, D., Kamada, H., Golak, J.: Phys. Rep. **274**, 107 (1996)
82. Golak, J., et al.: Eur. Phys. J. A **50**, 177 (2014)
83. Pastore, S., Schiavilla, R., Goity, J.L.: Mod. Phys. Lett. A **24**, 931 (2009)
84. Pastore, S., Girlanda, L., Schiavilla, R., Viviani, M., Wiringa, R.B.: Phys. Rev. C **80**, 034004 (2009)
85. Kölling, S., Epelbaum, E., Krebs, H., Meissner, U.G.: Phys. Rev. C **80**, 045502 (2009)
86. Kölling, S., Epelbaum, E., Phillips, D.R.: Phys. Rev. C **86**, 047001 (2012)
87. Piarulli, M., Girlanda, L., Marcucci, L.E., Pastore, S., Schiavilla, R., Viviani, M.: Phys. Rev. C **87**, 014006 (2013)
88. Baroni, A., Girlanda, L., Pastore, S., Schiavilla, R., Viviani, M.: Phys. Rev. C **93**, 015501 (2016)
89. Chou, W.T., Warburton, E.K., Brown, B.A.: Phys. Rev. C **47**, 163 (1993)

Chapter 61
Effective Description of $^{5-10}$He and the Search for a Narrow ^{4}n Resonance

K. Fossez, J. Rotureau, W. Nazarewicz, N. Michel and M. Płoszajczak

Abstract Open quantum systems that are at or beyond the limit of particle-emission stability exhibit generic features stemming from their coupling to an environment of positive energy states and decay channels. In nuclear physics, the exotic helium isotopes $^{5-10}$He and the four-neutron system ^{4}n represent two prototypical cases of open quantum systems whose structures are highly impacted by the environment. In the first part, a practical approach inspired by halo effective field theory for the description of $^{5-10}$He within tens of keV uncertainties is presented and the parity inversion in ^{9}He, as well as the possible two-neutron decay of ^{10}He are discussed. The second part discusses the last ab initio results on the four-neutron system obtained using chiral forces and including continuum couplings.

The study of open quantum systems, *i.e.* quantum systems coupled to an environment with which they can exchange energy or particles, is becoming an important part of modern nuclear physics [1, 2]. Indeed, the exploration of the drip lines, *i.e.* the limits of nuclear stability with respect to proton and neutron emissions, revealed that weakly bound and unbound nuclei exhibit generic features of open quantum systems which can only be understood when acknowledging the importance of couplings to continuum states and decay channels.

Hereafter, two modern problems related to the study of nuclear open quantum systems are presented. In the first part, recent results on the neutron-rich helium isotopes and the controversies around the ground-states of ^{9,10}He are discussed, and in the second part, the last progress on the nature of the four-neutron system are briefly reviewed.

K. Fossez (✉) · J. Rotureau
NSCL/FRIB Laboratory, Michigan State University, East Lansing, MI 48824, USA
e-mail: fossez@nscl.msu.edu; kevin.fossez@gmail.com

W. Nazarewicz
Department of Physics and Astronomy and NSCL/FRIB Laboratory, Michigan State University, East Lansing, MI 48824, USA

N. Michel · M. Płoszajczak
Grand Accélérateur National d'Ions Lourds (GANIL), CEA/DSM - CNRS/IN2P3, BP 55027, 14076 Caen Cedex, France

N. A. Orr et al. (eds.), *Recent Progress in Few-Body Physics*,
Springer Proceedings in Physics 238,
https://doi.org/10.1007/978-3-030-32357-8_61

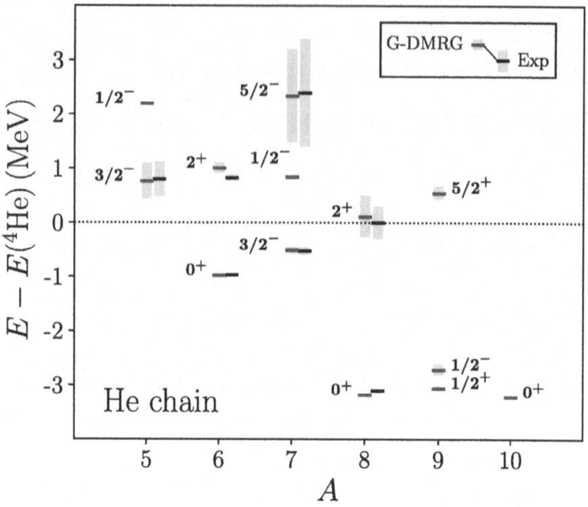

Fig. 61.1 Energy spectra of $^{5-10}$He with respect to the ^{4}He core. Experimental data [22] are compared to our Gamow-DMRG (G-DMRG) calculations. Decay widths are shown as shaded bars. The predicted $1/2^-$ resonant states in ^{5,7}He are so broad that their widths are not marked. For these states, as well as for states in ^{9,10}He, experimental information is not firm. Figure adapted from [23]

61.1 Effective Description of Neutron-Rich Helium Isotopes

The neutron-rich helium isotopes $^{5-10}$He present many of the open quantum system features found in exotic nuclei and hence, they represent an ideal playground to study the impact of continuum couplings in nuclear systems. For instance, the isotopes ^{6}He [3] and ^{8}He [4, 5] have two- and four-neutron Borromean halo structures in their ground states, respectively, whereas the isotopes ^{5}He [3, 6, 7] and ^{7}He [8, 9] are neutron-unbound and have broad excited states. However, the situation is still controversial concerning the parity of the ground-state of ^{9}He [10, 11] and little is known about the ground-state of ^{10}He [5, 6, 12–21]. The current established experimental information on the energy spectra of $^{5-10}$He is displayed in Fig. 61.1.

Many theoretical approaches have been applied on the neutron-rich helium isotopes, including ab initio methods [24–27] with some formulated in the open quantum system framework [28–34]. However, the shortcomings of those methods, such as the quality of the interaction or the approximations made in the many-body methods themselves, did not allow to reach the precision required to solve the controversies in ^{9,10}He.

In the work presented below, a precise description of the energy spectra of $^{5-10}$He was achieved using an effective approach that accounts for the emergence of effective scales and associated degrees of freedom in these nuclei [23]. The idea behind this approach is to use the fact that at low-energy, the tightly bound nature of ^{4}He makes it a natural core whose internal dynamics is largely decoupled from valence neutrons. This idea is also at the basis of halo effective field theories (halo-EFTs) [35, 36] which allow to systematically construct, order by order, effective interactions tailored to weakly bound systems [37–40]. The halo-EFT approach has however only be applied on a system-by-system basis and the Hamiltonians obtained at each order increase quickly in complexity, even requiring three-body forces at the leading order in the case of ^{6}He. Its application to heavier helium isotopes as it is formulated seems difficult in that regard.

Other phenomenological shell model approaches introduced the ^{4}He core and couplings to continuum states to describe the helium chain [41–45]. However, the in-medium two-body interactions used in those approaches were not built using effective scale arguments, and, except for the recent work in [45], no systematic study of the model parameter space has been carried out. Moreover, truncations were applied in the continuum space for the ^{9,10}He isotopes.

In this context, the development of an effective approach taking into account the emergent scale in neutron-rich helium isotopes, and formulated in the open quantum system framework, provides an alternative path to resolve the controversies in ^{9,10}He.

The strategy employed in the present work is based on a parameter reduction of the effective in-medium interaction defined in [46, 47] using effective scale arguments. First, the core-neutron interaction is represented by a Woods-Saxon (WS) potential containing central and spin-orbit terms optimized using the s and p phase shifts in the $\alpha - n$ scattering [48–50] as was done in [45, 51, 52]. The parameters of the WS potential are the depth $V_0 = 41.77$ MeV, the diffuseness $a = 0.618$ fm, the radius $R_0 = 2.162$ fm, and the spin-orbit strength $V_{so} = 6.991$ MeV. We note that the present model does not include effects beyond a static-core plus valence-neutron picture, as for instance the "core-swelling" effect (core polarization due to the valence neutrons) [44].

In a second step, the in-medium interaction between valence neutrons is reduced to a residual two-body force using insights from halo-EFT. Instead of considering the four terms in the isovector channel of the original interaction, *i.e.* two central terms in the spin-singlet and spin-triplet channels, and one spin-orbit term and one tensor term in the spin-triplet channel, we only consider the single central term in the spin-singlet channel as in halo-EFT. The leading-order of halo-EFT [37, 38] involves the $^1S_0$ channel only but here we also consider the $L > 0$ spin-singlet channels to be able to check a posteriori that the main contribution comes from the $^1S_0$ channel.

The form factor for the central term is a sum of three Gaussians with the ranges: ($r_0 = 0.160$ fm), ($r_1 = 1.127$ fm) and ($r_2 = 3.400$ fm). In halo-EFT at leading-order the interaction would be given by a regularized delta force in the $^1S_0$ channel [37, 38], which can be taken in a single Gaussian form, but we kept the original form factor with three Gaussians as it has proven to perform well in earlier studies [45, 53–55], and our objective is to show how a simple Hamiltonian based on effective scale arguments can capture the complex energy relations within the neutron-rich helium isotopes.

Couplings to continuum states are included in the many-body calculations using the single-particle (s.p.) Berggren basis [56, 57], which contains resonant (Gamow) states and nonresonant scattering states for each partial-wave channel $c = (\ell, j)$. It allows to naturally extend the configuration-interaction picture into the complex-energy plane [58]. We considered the spd one-body model space built on the s.p. poles $0p_{3/2}$ and $0p_{1/2}$ and associated continua, each made of three segments in the complex momentum plane defined by the points $(0.2, -0.1)$, $(0.4, 0.0)$ and $(6.0, 0.0)$ (all in fm^{-1}) for the $p_{3/2}$ partial wave, and the points $(0.25, -0.2)$, $(0.5, 0.0)$ and

(6.0, 0.0) (in fm^{-1}) for the $p_{1/2}$ partial wave. The continuum associated with the $s_{1/2}$ partial wave is real and defined by the points 0.1, 0.2, and 6.0 fm^{-1}. Each segment defining the s and p continua are discretized with 12 points. Additionally, the $1s_{1/2}$ state was added to the s.p. basis for the $^{8-10}$He calculations by increasing the depth of the basis-generating WS potential, as its absence would make the identification of many-body states difficult. Finally, the $d_{3/2}$ and $d_{5/2}$ continua are represented by six harmonic oscillator shells each. Adding higher partial waves only leads to an overall energy renormalization and hence does not change our results. Also, contrary to previous related approaches [41, 42, 45], no truncations on the number of particles in the continuum are imposed in the present work.

The many-body problem is solved using the density matrix renormalization group (DMRG) method for open quantum systems [43, 59] or Gamow-DMRG (G-DMRG), which has been shown to be a powerful technique to handle large many-body spaces. The method is also augmented using natural orbitals [60] to significantly speed-up the numerical convergence of the method [53, 61, 62].

Only one free parameter needs to be adjusted in the valence-space interaction, namely the strength of the spin-singlet central interaction V_c. We used the energies of the known states in $^{6-8}$He to fix the interaction strength at the optimal value $V_c^{(\mathrm{opt})} = -5.709$ MeV with a standard deviation of $\sigma = 0.008$ MeV. The small value of σ illustrates the ability of our model to describe the spectra of $^{6-8}$He. We wish to point out that the $J^\pi = 2_1^+$ state of ^{6}He requires an abnormally large interaction strength ($V_c \approx -6.8$ MeV) to reproduce the experimental value; hence, is not included in the calculation of $V_c^{(\mathrm{opt})}$. The reason for this discrepancy (around 180 keV) is the dominant $0p_{3/2} \to 0p_{1/2}$ structure of this state [44]. In fact, the deviation between the calculated and experimental values for the 2_1^+ state can be significantly reduced by slightly changing the strength of the spin-orbit term of the core-neutron potential, which we decided to keep fixed in this work.

The uncertainties on the energies associated with V_c are calculated as: $\Delta E = 0.5 \left| E(V_c^{(\mathrm{opt})} + \sigma) - E(V_c^{(\mathrm{opt})} - \sigma) \right|$ and are all between 10 and 30 keV. The G-DMRG results are shown in Fig. 61.1. In principle, there are also uncertainties coming from the core potential, but they were shown to be negligible as compared to the uncertainties coming from the valence-space interaction in [45]. Only a complete uncertainty quantification study (e.g., through a Bayesian analysis) could provide full theoretical uncertainties.

We predict very broad $1/2_1^-$ states in ^{5,7}He in agreement with [32, 33, 41, 42], which cannot be considered as genuine nuclear states because of their short lifetimes [62, 63]. The ground-state of ^{8}He is found to have a complex structure [64, 65], with $p_{3/2}$ and $p_{1/2}$ occupations being about 2.58 and 0.18, respectively, and the remaining occupations (0.24) shared between the s and d partial waves. For comparison, the first excited 2^+ state of ^{8}He has $p_{3/2}$ and $p_{1/2}$ occupations of almost 3.0 and 1.0, respectively, which is a situation reminiscent of ^{6}He, whose g.s. has a strong dineutron component and the excited state has predominantly a particle-hole structure [44].

Concerning the controversy around the ground-state of ^{9}He, we predict a narrow $J^\pi = 1/2^+$ g.s. and a close-lying $J^\pi = 1/2^-$ resonance with a larger width, as well as a $J^{\pi} = 5/2^{+}$ resonance at higher energy. The first two states could not be distinguished in [45] within statistical uncertainties. The uncertainty on the $J^{\pi} = 5/2^{+}$ state could not be estimated due to the instability of calculations for extreme values of V_c. These results are in relative agreement with experimental data from (d, p) reactions [11], and in disagreement with the study of isobaric analog states in ^{9}Li [66], as well as the no-core shell model with continuum calculations of [34] where the g.s. is predicted to have $J^{\pi} = 1/2^{-}$ and the first excited state to be a broader $J^{\pi} = 3/2^{-}$ resonance. A possible explanation for the latter is that in [34] only the two-body part of the normal-ordered three-body forces was considered, the 2_1^+ state of ^{8}He used to build the $^8\mathrm{He} + n$ channels was calculated as a bound state, and the only decay channel considered (for a fairly small number of channels) was the one-neutron emission. Earlier quantum Monte Carlo results [30] stated the possible existence of a virtual $J^{\pi} = 1/2^{+}$ state in ^{9}He, seen as a $\ell = 0$ single-particle state above ^{8}He, and a possible $J^{\pi} = 1/2^{-}$ state at higher energy (3–4 MeV). In the present work, the $J^{\pi} = 1/2^{+}$ state in ^{9}He is predicted to be a many-body resonance built almost entirely of excitations to the $s_{1/2}$ continuum, but not a virtual state, see below.

The most surprising prediction concerns the g.s. of ^{10}He, which is calculated at about the same energy than the g.s. of ^{8}He. Taking into account the uncertainty on the g.s. energies of $^{8-10}$He, and the decay width of the g.s. of ^{9}He, both one- and two-neutron decay channels are theoretically possible. Also, the ground states of ^{8}He, ^{9}He, and ^{10}He have almost identical partial-wave decompositions except for the $s_{1/2}$ occupations, which are almost exactly zero, one, and two, respectively. In comparison, the $1/2^{-}$ state of ^{9}He is mostly built of the $p_{1/2}$ component. The interplay between $s_{1/2}$ and $p_{1/2}$ continuum states is believed to be a determining factor for the phenomenon of parity inversion in ^{9}He [67].

The present results allow revisiting past conclusions drawn from three-body models about ^{9,10}He. Early on, it was proposed [17] that the ground state of ^{10}He might be a low-lying resonance at $E = 0.05$ MeV with a width of $\Gamma = 0.21$ MeV dominated by s waves, and that the experimental observations of higher-energy resonances at ~1.8 MeV [68]) might, in fact, correspond to the first excited state of ^{10}He. Later, other studies [18, 69] investigated the consistency between a possible narrow ground state in ^{9}He and a broad ground state at 1–3 MeV in ^{10}He, and concluded that either the ground state of ^{10}He has not been observed yet, or the s-wave scattering length in ^{9}He must be less attractive. There was also speculation about the fact that the observed state in ^{10}He might, in fact, corresponds to several states [16, 21]. This question could not be addressed in the present work, but it certainly goes beyond the limited three-body picture and supports the idea of a narrow ground state of ^{10}He dominated by s waves, built on the $1/2^{+}$ ground-state resonance of ^{9}He.

The almost identical energies and partial-wave occupations of the ground states of ^{8}He and ^{10}He support the $^8\mathrm{He} + 2n$ cluster picture of ^{10}He, in which an extended dineutron structure is present atop the four-neutron halo in ^{8}He. In other words, ^{10}He is predicted to be on the brink of forming a nuclear double-halo structure (^{4}He + $4n$ + $2n$) if not for a few tens of keV, similarly to the known $^3\mathrm{He}^4\mathrm{He}_2$ trimer [70].

In conclusion, we demonstrated in the case of the neutron-rich helium isotopes that a parameter reduction of a phenomenological interaction using effective scale arguments can yield a significant decrease in systematic uncertainties, allowing predictions within tens of keV for the energy spectra. The present results confirmed the parity inversion in ^{9}He and shed a new light on the ground state of ^{10}He. In future studies, we will explore more rigorous ways to build effective interactions with composite degrees of freedom while keeping the relative simplicity of the present approach.

61.2 The Search for a Narrow ^{4}n resonance

In this section, we briefly review the recent findings concerning the existence of a narrow ^{4}n resonance. Following the experimental hints of a ^{4}n resonance [71] at an energy $E = 0.83 \pm 0.65$ (stat) $\pm$ 1.25 (syst) MeV above the ^{4}n threshold and a maximal width $\Gamma = 2.6$ MeV, several ab initio works investigated the possibility to form a four-neutron system within the continuum using different formalisms [62, 72–74].

The first attempt at finding a narrow four-neutron resonance was done in [72], by solving the Faddeev-Yakubowsky equations and by including continuum couplings using the uniform complex-scaling method [75, 76]. This exact method offers the advantage of properly including all possible decay channels, however, the authors used a realistic phenomenological interaction and only focused on the effect of adding a $T = 3/2$ three-body force. Nevertheless, it was shown that unrealistic modifications to the nuclear interaction would be necessary to obtain a ^{4}n system at the experimental value and that most likely the four-neutron system would at most be a very broad resonance.

A subsequent work [73] based on the no-core shell model in the harmonic oscillator basis [77] and the J-matrix formalism [78] claimed, using a realistic phenomenological two-body interaction [79], that a resonance at an energy $E = 0.8$ MeV and with a width $\Gamma = 1.4$ MeV was indeed present in relative agreement with the experimental claim in [71]. However, it was assumed for the extraction of the energy and width from computed phase shifts that all four neutrons decay simultaneously.

In a similar spirit than in [80], where it was shown using a quantum Monte Carlo method that a bound four-neutron system would imply modifications of the nuclear interaction incompatible with what we know on other light systems, in [81] the authors demonstrated using a chiral interaction including two- and three-body terms that (i) three-body forces are not important in the three- and four-neutron systems, (ii) if a three-neutron resonance exists it must be lower than that of the four-neutron system and (iii) a four-neutron resonance might exist at about $E = 2.1$ MeV based on a simple extrapolation from an artificially bound four-neutron system in a Woods-Saxon potential.

In our work [62], we decided to answer whether or not a four-neutron resonance could form a narrow resonance, and hence be considered a genuine nucleus, using two-body chiral interactions (N3LO [82], N2LO$_{\text{opt}}$ [83] and N2LO$_{\text{sat}}$ [84]) and including continuum couplings.

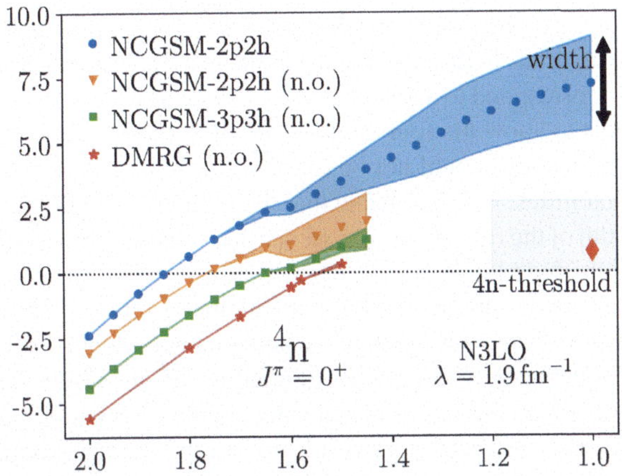

Fig. 61.2 Evolution of the energy and width (shaded area) of the four-neutron system with the scaling of the N3LO interaction from 2.0 to 1.0. The circles represent the NCGSM results with two neutrons in the continuum, which is used to generate the NCGSM results based on natural orbitals with two (triangles) and three (squares) neutrons in the continuum. The DMRG results without truncations are represented by stars. The experimental energy is indicated by a diamond and the gray area shows the maximal experimental uncertainties. This area is extended up to an interaction 20% more attractive to guide the reader. Figure adapted from [62]

We solved the many-body problem in the continuum using the no-core Gamow shell model (NCGSM) [31, 58] and the density matrix renormalization group method (DMRG) [85] for open quantum systems [43, 59], which both represent a generalization of the configuration interaction picture in the complex energy plane [58] using the Berggren basis [56]. One of the two major difficulties associated with the above-mentioned methods lies in the identification of the physical many-body solutions within the continuum space. This problem was solved by considering artificially bound four-neutron systems by increasing the strength of the nuclear interaction, where the identification issue is inexistent, and then follow the solution toward the physical region where the four-neutron system is unbound as the solution having a maximal overlap with the unambiguously identified solution in the unphysical region. The second issue was related to the size of the continuum space, which was made manageable by using the DMRG approach for the full calculations, augmented by the use of natural orbitals [53, 61].

The energy and width obtained within a sp continuum space augmented by a dfg harmonic oscillator space, for different interaction strengths or scaling factor (N3LO two-body chiral interaction with a renormalization cutoff of $\lambda = 1.9\,\text{fm}^{-1}$), are presented in Fig. 61.2. The different lines correspond to different levels of approximation, namely calculations with configurations having at most two and three neutrons in the continuum shells, without and with natural orbitals, and finally DMRG calculations without truncations.

The main finding is that removing truncations progressively, which is equivalent in the present case to increasing continuum couplings, lowers the energy but does not decrease the width. The hardest truncation considered, *i.e.* allowing configurations with at most two neutrons in the continuum shells, gives a large width $\Gamma \approx 3.7$ MeV which can hardly correspond to a genuine nuclear state as the corresponding lifetime would be too short to allow the four neutrons to all interact with each other. Calculations without truncations could not be extended into the physical region because of numerical instabilities, which indicate that the width is exploding, supporting the claim that the width of the four-neutron system is too large to correspond to a physical state. Details can be found in [62].

Our conclusion is in qualitative agreement with the results in [72] showing a rapid increase of the width when the strength of the phenomenological $T = 3/2$ three-body force decreases. Interestingly, a naive extrapolation of our results in the physical might be compatible with the experimental value, but there is still disagreement with the nature of the state observed. Our result is also in contrast with the prediction in [73] based on a simultaneous decay of the four-neutron system, as we explicitly show that having only two neutrons in the continuum as a first approximation (one- and two-neutron decay channel open only) already contributes significantly to the total width. In conclusion, we believe that the observation in [71] corresponds to a feature in scattering experiments and not a genuine nuclear state.

The last ab initio addressing the question of the existence of the four-neutron system [74] is based on a momentum space formulation of the Faddeev-Yakubowsky method. The main advantage of this exact method is that by design it does not suffer from a truncation of the continuum issue. Moreover, the results presented in this work were obtained using a two-body chiral interaction. It was found that no four-neutron system state can be formed at all within this approach, supporting the idea that the four-neutron system cannot exist as a genuine nuclear state (bound state or narrow resonance).

In summary, except for one approach [73] predicting a four-neutron state compatible with the experimental claim, and one approach [81] supporting the existence of a state above the experimental claim (without saying what the lifetime would be), the three remaining approaches are in agreement about the nonexistence of the four-neutron system as a genuine nuclear state. The situation is however not entirely clear as different interactions and truncations were used across different methods, and the nature of the four-neutron system seems to be rather delicate. Coming new experimental results might clear the question once for all.

Acknowledgements This work was supported by the U.S. Department of Energy, Office of Science, Office of Nuclear Physics under award numbers DE-SC0017887, DE-SC0013365 (Michigan State University) and DE-SC0018083 (NUCLEI SciDAC-4 collaboration), and by the National Science Foundation under award number PHY-1403906. An award of computer time was provided by the Institute for Cyber-Enabled Research at Michigan State University, and part of the computations was performed on local resources at the Chalmers University of Technology supported by the Swedish Foundation for International Cooperation in Research and Higher Education (STINT, IG2012-5158).

References

1. Forssén, C., Hagen, G., Hjorth-Jensen, M., Nazarewicz, W., Rotureau, J.: Living on the edge of stability, the limits of the nuclear landscape. Phys. Scr. T **2013**, 014022 (2013)
2. Nazarewicz, W.: Challenges in nuclear structure theory. J. Phys. G **43**, 044002 (2016)
3. Tilley, D.R., et al.: Energy levels of light nuclei $A = 5, 6, 7$. Nucl. Phys. A **708**, 3 (2002)
4. Tanihata, I., et al.: Measurements of interaction cross sections and radii of He isotopes. Phys. Lett. B **160**, 380 (1985)
5. Golovkov, M.S., et al.: The ^{8}He and ^{10}He spectra studied in the (t, p) reaction. Phys. Lett. B **672**, 22 (2009)
6. Kobayashi, T., et al.: Quasifree nucleon-knockout reactions from neutron-rich nuclei by a proton target. Nucl. Phys. A **616**, 223 (1997)
7. Aleksandrov, D., et al.: Invariant mass spectrum and $\alpha - n$ correlation function studied in the fragmentation of ^{6}He on a carbon target. Nucl. Phys. A **633**, 234 (1998)
8. Denby, D.H., et al.: Ground state energy and width of ^{7}He from ^{8}Li proton knockout. Phys. Rev. C **78**, 044303 (2008)
9. Aksyutina, Yu., et al.: Properties of the ^{7}He ground state from ^{8}He neutron knockout. Phys. Lett. B **679**, 191 (2009)
10. Tilley, D.R., et al.: Energy levels of light nuclei $A = 8, 9, 10$. Nucl. Phys. A **745**, 155 (2004)
11. Al Kalanee, T., et al.: Structure of unbound neutron-rich ^{9}He studied using single-neutron transfer. Phys. Rev. C **88**, 034301 (2013)
12. Korsheninnikov, A.A., et al.: Observation of ^{10}He. Phys. Lett. B **326**, 31 (1994)
13. Ostrowski, A.N., et al.: Spectroscopy of ^{10}He. Phys. Lett. B **338**, 13 (1994)
14. Johansson, H.T., et al.: The unbound isotopes ^{9,10}He. Nucl. Phys. A **842**, 15 (2010)
15. Sidorchuk, S.I., et al.: Structure of ^{10}He low-lying states uncovered by correlations. Phys. Rev. Lett. **108**, 202502 (2012)
16. Sharov, P.G., Egorova, I.A., Grigorenko, L.V.: Anomalous population of ^{10}He states in reactions with ^{11}Li. Phys. Rev. C **90**, 024610 (2014)
17. Aoyama, S.: Theoretical prediction for the ground state of ^{10}He with the method of analytic continuation in the coupling constant. Phys. Rev. Lett. **89**, 052501 (2002)
18. Grigorenko, L.V., Zhukov, M.V.: Problems with the interpretation of the ^{10}He ground state. Phys. Rev. C **77**, 034611 (2008)
19. Jones, M.D., et al.: Further insights into the reaction ^{14}Be(CH$_2$, X)^{10}He. Phys. Rev. C **91**, 044312 (2015)
20. Fortune, H.T.: Relative population of 0^{+} states in ^{10}He in various reactions. Phys. Rev. C **88**, 054623 (2013)
21. Fortune, H.T.: Constraints on energies of ^{10}He(0^{+}) and ^{9}He$(1/2^{+})$. Phys. Rev. C **91**, 034306 (2015)
22. http://www.nndc.bnl.gov/ensdf (2015)
23. Fossez, K., Rotureau, J., Nazarewicz, W.: Energy spectrum of neutron-rich helium isotopes: complex made simple. Phys. Rev. C **98**, 061302(R) (2018)
24. Pudliner, B.S., Pandharipande, V.R.: Quantum Monte Carlo calculations of nuclei with $A \leq 7$. Phys. Rev. C **56**, 1720 (1997)
25. Caurier, E., Navrátil, P.: Proton radii of ^{4,6,8}He isotopes from high-precision nucleon-nucleon interactions. Phys. Rev. C **73**, 021302(R) (2006)
26. Lisetskiy, A.F., Barrett, B.R., Kruse, M.K.G., Navrátil, P., Stetcu, I., Vary, J.P.: Ab-initio shell model with a core. Phys. Rev. C **78**, 044302 (2008)
27. Sääf, D., Forssén, C.: Microscopic description of translationally invariant core + N + N overlap functions. Phys. Rev. C **89**, 011303(R) (2014)
28. Nollett, K.M., Pieper, S.C., Wiringa, R.B.: Quantum Monte Carlo calculations of neutron-α scattering. Phys. Rev. Lett. **99**, 022502 (2007)
29. Hagen, G., Dean, D.J., Hjorth-Jensen, M., Papenbrock, T.: Complex coupled-cluster approach to an ab initio description of open quantum systems. Phys. Lett. B **656**, 169 (2007)

30. Nollett, K.M.: Ab initio calculations of nuclear widths via an integral relation. Phys. Rev. C **86**, 044330 (2012)
31. Papadimitriou, G., Rotureau, J., Michel, N., Płoszajczak, M., Barrett, B.R.: Ab-initio no-core Gamow shell model calculations with realistic interactions. Phys. Rev. C **88**, 044318 (2013)
32. Baroni, S., Navrátil, P., Quaglioni, S.: Ab initio description of the exotic unbound ^{7}He nucleus. Phys. Rev. Lett. **110**, 022505 (2013)
33. Baroni, S., Navrátil, P., Quaglioni, S.: Unified ab initio approach to bound and unbound states: no-core shell model with continuum and its application to ^{7}He. Phys. Rev. C **87**, 034326 (2013)
34. Vorabbi, M., Calci, A., Navrátil, P., Kruse, M.K.G., Quaglioni, S., Hupin, G.: Structure of the exotic ^{9}He nucleus from the no-core shell model with continuum. Phys. Rev. C **97**, 034314 (2018)
35. Bedaque, P.F., van Kolck, U.: Effective field theory for few-nucleon systems. Annu. Rev. Nucl. Part. Sci. **52**, 339 (2002)
36. Hammer, H.W., Ji, C., Phillips, D.R.: Effective field theory description of halo nuclei. J. Phys. G **44**, 103002 (2017)
37. Bertulani, C.A., Hammer, H.W., van Kolck, U.: Effective field theory for halo nuclei: shallow p-wave states. Nucl. Phys. A **712**, 37 (2002)
38. Bedaque, P.F., Hammer, H.W., van Kolck, U.: Narrow resonances in effective field theory. Phys. Lett. B **569**, 159 (2003)
39. Rotureau, J., van Kolck, U.: Effective field theory and the Gamow shell model—the ^{6}He halo nucleus. Few-Body Syst. **54**, 725 (2013)
40. Ji, C., Elster, Ch., Phillips, D.R.: ^{6}He nucleus in halo effective field theory. Phys. Rev. C **90**, 044004 (2014)
41. Michel, N., Nazarewicz, W., Płoszajczak, M., Rotureau, J.: Gamow shell-model description of weakly bound and unbound nuclear states. Revista Mexicana De Fisica **5**(Suplemento 2), 74 (2004)
42. Volya, A., Zelevinsky, V.: Discrete and continuum spectra in the unified shell model approach. Phys. Rev. Lett. **94**, 052501 (2005)
43. Rotureau, J., Michel, N., Nazarewicz, W., Płoszajczak, M., Dukelsky, J.: Density matrix renormalisation group approach for many-body open quantum systems. Phys. Rev. Lett. **97**, 110603 (2006)
44. Papadimitriou, G., Kruppa, A.T., Michel, N., Nazarewicz, W., Płoszajczak, M., Rotureau, J.: Charge radii and neutron correlation in helium halo nuclei. Phys. Rev. C **84**, 051304(R) (2011)
45. Jaganathen, Y., Id Betan, R.M., Michel, N., Nazarewicz, W., Płoszajczak, M.: Quantified Gamow shell model interaction for psd-shell nuclei. Phys. Rev. C **96**, 054316 (2017)
46. Furutani, H., Horiuchi, H., Tamagaki, R.: Structure of the second 0^+ state of ^{4}He. Prog. Theor. Phys. **60**, 307 (1978)
47. Furutani, H., Horiuchi, H., Tamagaki, R.: Cluster-model study of the $T = 1$ states in $A = 4$ system. Prog. Theor. Phys. **62**, 981 (1979)
48. Hoop Jr., B., Barschall, H.H.: Scattering of neutrons by α-particles. Nucl. Phys. **83**, 65 (1966)
49. Stammbach, Th, Walter, R.L.: R-matrix formulation and phase shifts for $n - {}^4$He and $p - {}^4$He scattering for energies up to 20 mev. Nucl. Phys. A **180**, 225 (1972)
50. Bond, J.E., Firk, F.W.K.: Determination of R-function and physical-state parameters for $n - {}^4$He elastic scattering below 21 mev. Nucl. Phys. A **287**, 317 (1977)
51. Varga, K., Suzuki, Y., Ohbayasi, Y.: Microscopic multicluster description of the neutron-rich helium isotopes. Phys. Rev. C **50**, 189 (1994)
52. Theeten, M., Baye, D., Descouvemont, P.: Comparison of local, semi-microscopic, and microscopic three-cluster models. Phys. Rev. C **74**, 044304 (2006)
53. Fossez, K., Rotureau, J., Michel, N., Liu, Q., Nazarewicz, W.: Single-particle and collective motion in unbound deformed ^{39}Mg. Phys. Rev. C **94**, 054302 (2016)
54. Fossez, K., Rotureau, J., Michel, N., Nazarewicz, W.: Continuum effects in neutron-drip-line oxygen isotopes. Phys. Rev. C **96**, 024308 (2017)
55. Jones, M.D., et al.: Search for excited states in ^{25}O. Phys. Rev. C **96**, 054322 (2017)

56. Berggren, T.: On the use of resonant states in eigenfunction expansions of scattering and reaction amplitudes. Nucl. Phys. A **109**, 265 (1968)
57. Berggren, T., Lind, P.: Resonant state expansion of the resolvent. Phys. Rev. C **47**, 768 (1993)
58. Michel, N., Nazarewicz, W., Płoszajczak, M., Vertse, T.: Shell model in the complex energy plane. J. Phys. G **36**, 013101 (2009)
59. Rotureau, J., Michel, N., Nazarewicz, W., Płoszajczak, M., Dukelsky, J.: Density matrix renormalization group approach to two-fluid open many-fermion systems. Phys. Rev. C **79**, 014304 (2009)
60. Brillouin, L.: Act. Sci. Ind. **71**, 159 (1933)
61. Shin, I.J., Kim, Y., Maris, P., Vary, J.P., Forssén, C., Rotureau, J., Michel, N.: *Ab initio* no-core solutions for ^{6}Li. J. Phys. G **44**, 075103 (2017)
62. Fossez, K., Rotureau, J., Michel, N., Płoszajczak, M.: Can tetraneutron be a narrow resonance? Phys. Rev. Lett. **119**, 032501 (2017)
63. Fossez, K., Nazarewicz, W., Jaganathen, Y., Michel, N., Płoszajczak, M.: Nuclear rotation in the continuum. Phys. Rev. C **93**, 011305(R) (2016)
64. Keeley, N., et al.: Probing the ^{8}He ground state via the ^{8}He$(p, t)^6$He reaction. Phys. Lett. B **646**, 222 (2007)
65. Skaza, F., et al.: Low-lying states and structure of the exotic ^{8}He via direct reactions on the proton. Nucl. Phys. A **788**, 260 (2007)
66. Uberseder, E., et al.: Nuclear structure beyond the neutron drip line: the lowest energy states in ^{9}He via their $T = 5/2$ isobaric analogs in ^{9}Li. Phys. Lett. B **754**, 323 (2016)
67. Hansen, P.G., Sherrill, B.M.: Reactions and single-particle structure of nuclei near the drip lines. Nucl. Phys. A **693**, 133 (2001)
68. Kohley, Z., et al.: Unresolved question of the ^{10}He ground state resonance. Phys. Rev. Lett. **109**, 232501 (2012)
69. Barker, F.C.: Level widths in ^{9}He and ^{10}He. Nucl. Phys. A **741**, 42 (2004)
70. Voigtsberger, J., et al.: Imaging the structure of the trimer systems ^{4}He$_3$ and ^{3}He^{4}He$_2$. Nat. Commun. **5**, 5765 (2014)
71. Kisamori, K., et al.: Candidate resonant tetraneutron state populated by the ^{4}He(^{8}He, ^{8}Be) reaction. Phys. Rev. Lett. **116**, 052501 (2016)
72. Hiyama, E., Lazauskas, R., Carbonell, J., Kamimura, M.: Possibility of generating a 4-neutron resonance with a $T = 3/2$ isospin 3-neutron force. Phys. Rev. C **93**, 044004 (2016)
73. Shirokov, A.M., Papadimitriou, G., Mazur, A.I., Mazur, I.A., Roth, R., Vary, J.P.: Prediction for a four-neutron resonance. Phys. Rev. Lett. **117**, 182502 (2016)
74. Deltuva, A.: Tetraneutron: rigorous continuum calculation. Phys. Lett. B **782**, 238 (2018)
75. Aoyama, S., Myo, T., Katō, K., Ikeda, K.: The complex scaling method for many-body resonances and applications to three-body resonances. Prog. Theor. Phys. **116**, 1 (2006)
76. Myo, T., Kikuchi, Y., Masui, H., Katō, K.: Recent development of complex scaling method for many-body resonances and continua in light nuclei. Prog. Part. Nucl. Phys. **79**, 1 (2014)
77. Barrett, B.R., Navrátil, P., Vary, J.P.: Ab initio no core shell model. Prog. Part. Nucl. Phys. **69**, 131 (2013)
78. Bang, J.M., Mazur, A.I., Shirokov, A.M., Smirnov, Y.F., Zaytsev, S.A.: P-matrix and J-matrix approaches: Coulomb asymptotics in the harmonic oscillator representation of scattering theory. Ann. Phys. 280, 299 (2000)
79. Shirokov, A.M., Vary, J.P., Mazur, A.I., Weber, T.A.: Realistic nuclear Hamiltonian: Ab exitu approach. Phys. Lett. B **644**, 33 (2007)
80. Pieper, S.C.: Can modern nuclear Hamiltonians tolerate a bound tetraneutron? Phys. Rev. Lett. **90**, 252501 (2003)
81. Gandolfi, S., Hammer, H.W., Klos, P., Lynn, J.E., Schwenk, A.: Is a trineutron resonance lower in energy than a tetraneutron resonance? Phys. Rev. Lett. **118**, 232501 (2017)
82. Entem, D.R., Machleidt, R.: Accurate charge-dependent nucleon-nucleon potential at fourth order of chiral perturbation theory. Phys. Rev. C **68**, 041001(R) (2003)
83. Ekström, A., Baardsen, G., Forssén, C., Hagen, G., Hjorth-Jensen, M., Jansen, G.R., Machleidt, R., Nazarewicz, W., Papenbrock, T., Sarich, J., Wild, S.M.: Optimized chiral nucleon-nucleon interaction at next-to-next-to-leading order. Phys. Rev. Lett. **110**, 192502 (2013)

84. Ekström, A., Jansen, G.R., Wendt, K.A., Hagen, G., Papenbrock, T., Carlsson, B.D., Forssén, C., Hjorth-Jensen, M., Navrátil, P., Nazarewicz, W.: Accurate nuclear radii and binding energies from a chiral interaction. Phys. Rev. C **91**, 051301(R) (2015)
85. White, S.R.: Density matrix formulation for quantum renormalization groups. Phys. Rev. Lett. **69**, 2863 (1992)

Chapter 62
Exploring Beyond the Proton Drip Line

Francois de Oliveira Santos

Abstract Several unbound nuclei: ^{19}Na, ^{18}Na, ^{16}F, ^{15}F, were studied at the GANIL SPIRAL 1 facility using the technique of resonant elastic scattering measured in inverse kinematics with thick targets and radioactive beams.

62.1 Introduction

The boundaries for nuclear stability against particle emission are called drip lines. Beyond the drip lines, the particle emission time τ is usually very short, shorter than typically 10^{-20} s. Unbound nuclei are often observed as broad resonances with an energy width $\Gamma = \hbar/\tau \sim 1$ MeV. Different nuclear reactions and experimental techniques can be used to study these broad resonances. Here, the experimental technique called Resonant Elastic Scattering was used, since it is an efficient method to study unbound nuclei. The technique is described shortly in Sect. 62.2. In this article, several results obtained at GANIL SPIRAL 1 facility are presented. Unbound nuclei are interesting for different reasons:

- Some unbound nuclei play an important role in nuclear astrophysics, that is the case of ^{8}Be in the triple alpha reaction, and ^{2}He in the $pp1$ reaction chain. The extreme case of ^{19}Na [1], involved in a two-proton capture reaction from ^{18}Ne, is presented in Sect. 62.3.
- The time reversal reaction of the two-proton capture is the two-proton emission or two-proton radioactivity. The same theoretical models can be used to calculate the two reactions. The study of ^{18}Na [2], the intermediate nucleus in the two-proton radioactivity of ^{20}Mg, is presented in Sect. 62.4.
- The symmetry of mirror nuclei is often used in nuclear astrophysics. The unknown properties of a neutron deficient nucleus can be deduced from the known properties of its mirror nucleus. Unbound nuclei are perfect cases to study the effect of the

F. de Oliveira Santos (✉)
GANIL, CEA/DSM-CNRS/IN2P3, Boulevard Henri Becquerel,
14076 Caen Cedex 5, France
e-mail: oliveira@ganil.fr

N. A. Orr et al. (eds.), *Recent Progress in Few-Body Physics*,
Springer Proceedings in Physics 238,
https://doi.org/10.1007/978-3-030-32357-8_62

coupling with continuum. The case of ^{16}F$-^{16}$N [3], described as a core plus one proton and one neutron, is presented in Sect. 62.5.

- The generalized conjecture of Ikeda says that the coupling to a nearby particle/cluster decay channel induces an increase of the particles/cluster correlations in the structure of the state. It is claimed in [4] this conjecture holds for all kinds of cluster states including unstable systems like dineutron, diproton, and ^{8}Be. We observed a narrow resonance in the unbound nucleus ^{15}F [5], located well above the one proton emission barrier, but just above the two-proton emission threshold. This state could be an example of increased two-proton correlation. Search for γ-transition in the unbound ^{15}F is also discussed in Sect. 62.6.

62.2 Resonant Elastic Scattering Technique

The Rutherford elastic scattering supposes that the two interacting atomic nuclei do not fuse together, they repel each other by the strong Coulomb force exerting itself between positively charged particles. The fusion of the two incident particles into one compound nucleus, followed by the separation into the two entrance channel particles, is another possibility of reaction mechanism. It is a resonant process. It has to be added to the Rutherford contribution in a coherent manner, giving rise to interferences effects. The exact shape of the resonance depends on the spin, parity, partial and total widths of the resonance in the compound nucleus. It is possible to study the structure of a nucleus by using this technique. The analyse of the resonances shape is relatively simple, a R-matrix code can be used. The cross section is generally high, several tens or hundred of millibarns per steradian, making resonant elastic scattering an ideal method to study unbound nuclei [6, 7]. Moreover, the thick target technique can be used efficiently, see Fig. 62.1.

62.3 The Case of ^{19}Na

In the astrophysical X-ray bursts, the flow of nuclear reactions goes from the light nuclei to the heaviest ones through proton captures and β-decays. These high temperature and high density environments could allow exotic reactions to happen, like the simultaneous capture of two protons [9, 10]. In the case of the radioactive nucleus ^{18}Ne, the capture of one proton gives ^{19}Na, which is a proton-unbound nucleus, decaying back to ^{18}Ne+p. The flow of reactions is stopped until ^{18}Ne β decays to ^{18}F. The double proton capture reaction ^{18}Ne($2p$, γ)^{20}Mg could be a breakout reaction. The rate of this reaction depends on the properties of the resonances in ^{19}Na. It is possible to show this two-proton capture reaction could be faster than the β-decay of ^{18}Ne [10]. One has to measure the properties of the resonances in ^{19}Na to determine the rate of this reaction. This was the objective of an experiment performed at GANIL. A beam of ^{18}Ne (the first radioactive beam produced at the

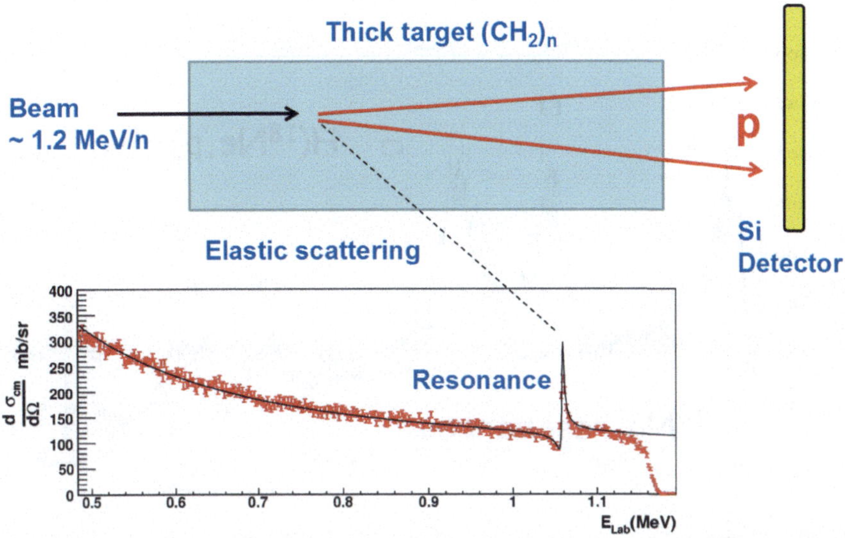

Fig. 62.1 Top: In the thick target technique, the incident ions slow down inside the target, the excitation function is measured at once from the incident to the exit energy of the beam. The number of scattered protons increases when the center of mass energy of the ions is equal to the energy of a resonance in the compound nucleus. Bottom: Differential cross section of the reaction $^{14}N(p, p)^{14}N$ [8], measured at $\theta_{CM} = 180°$ in inverse kinematics with a beam of ^{14}N accelerated to 1.2 MeV/nucleon, is shown as a function of the proton energy in lab. It was measured at once from 1.16 to 0.5 MeV using a thick target of polypropylene

SPIRAL 1 facility) was accelerated to 7.2 MeV/n and purified using a stripper foil located at the entrance of the LISE spectrometer. The 5×10^5 pps pure beam was incident upon a thick cryogenic solid hydrogen target. Despite the large thickness of the target, 1050 ± 20 μm, the resolution in the CM was only 30 keV FWHM, this was due to the quality of the target specially made for, and to the technique of the experiment (inverse kinematics). The obtained results, see Fig. 62.2, showed that the reaction $^{18}Ne(2p, \gamma)^{20}Mg$ plays a role only at extreme densities, much larger than those found in the X-ray bursters. More details can be obtained in [1, 10].

62.4 The Case of ^{18}Na

Two-proton radioactivity is a kind of time-reverse two-proton capture reaction. One way to study experimentally the astrophysical two-proton capture is, then, to study two-proton radioactivity. Both processes are calculated with the same nuclear models. A two-proton emitter is ^{19}Mg [11]. It is a good candidate for "true 2p-decay" since the intermediate nucleus ^{18}Na ($^{18}Na = {}^{19}Mg - p$) is probably a narrow resonance located well above the ^{19}Mg ground state. This two-proton radioactive nucleus ^{19}Mg is one

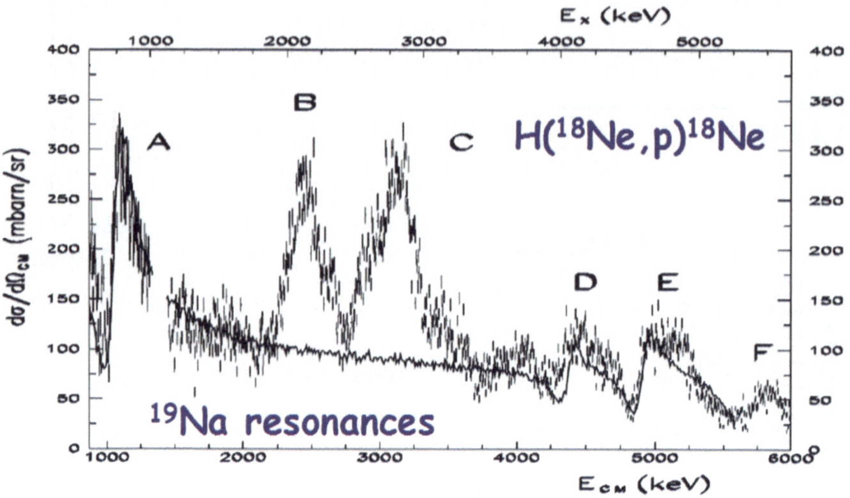

Fig. 62.2 Differential cross section of the reaction H(^{18}Ne, p)^{18}Ne is shown as a function of CM energy (bottom axis) and ^{19}Na excitation energy (top axis). Several resonances of ^{19}Na can be observed. The ground ($E_x = 0$) and the first excited states ($E_x \simeq 120$ keV) could not be observed. Peaks B and C correspond to inelastic resonances, see [1] for details. The line is a R-Matrix fit of the elastic excitation function taking into account the experimental conditions. Adapted from [1]

of the few cases, if not the only one, where the intermediate nucleus can be measured precisely, since the other cases are located too far from stability. Moreover, in the case of ^{6}Be, ^{12}O and ^{16}Ne, the decaying states have large widths and the intermediate state overlaps [12], these are sequential decays. We performed an experiment in order to measure, for the first time, the properties of the ^{18}Na low lying resonances.

These ^{18}Na resonances were studied through the measurement of the resonant elastic scattering reaction H(^{17}Ne, p)^{17}Ne. This experiment was performed at 4 A.MeV using a radioactive beam from the SPIRAL 1 facility. The measured excitation function is shown in Fig. 62.3. Several resonances are clearly visible, some are narrow, see [2] for details. Relatively to the other observed resonances, the ground state is expected to be at $E_{CM} \approx 1.35$ MeV. It is not visible in the data, we can deduce the width is narrower than 1 keV. This experiment confirmed ^{18}Na+p ground state resonance is narrow and located well above ^{19}Mg.

The measured half-life of ^{19}Mg ($t_{1/2} = 4.0(15)$ ps), the proton-proton angular correlations measured in the ^{19}Mg radioactivity, and the measured properties of the low-lying states of ^{18}Na are all in agreement with the theoretical three-body model predictions when assuming a dominant d-wave configuration for ^{19}Mg ground state [13, 14]. But this excellent agreement hides a big difference. While theoretical model of L. Grigorenko [14] predicts the two s-wave states of ^{18}Na 0_1^- and 1_2^- to be at $E_{CM} \geq 3.5$ MeV, these are measured at $E_{CM} = 1.842(40)$ and $2.030(20)$ MeV, i.e. $\approx$1.5 MeV down shifted in energy. The problem is that, with these s-wave resonances located at lower energy, the probability increases to get a sequential decay of two

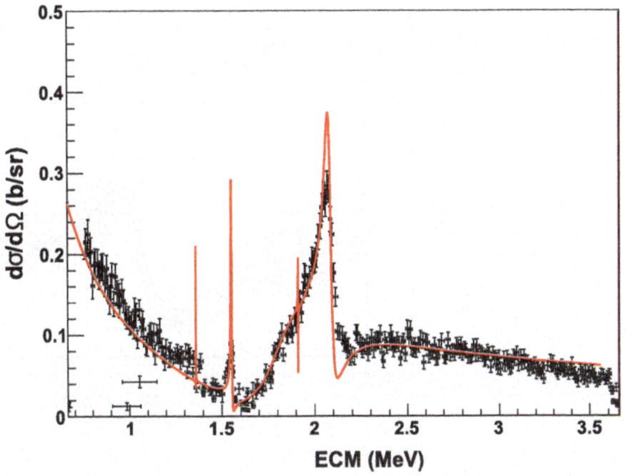

Fig. 62.3 Excitation function of the reaction H(^{17}Ne,p)^{17}Ne is shown as a function of CM energy. Experimental data were measured between 5° and 20° (LAB) and are reconstructed at 180° in the CM. Line shows the best R-matrix fit of the data, here it is not degraded by the energy resolution of the experiment

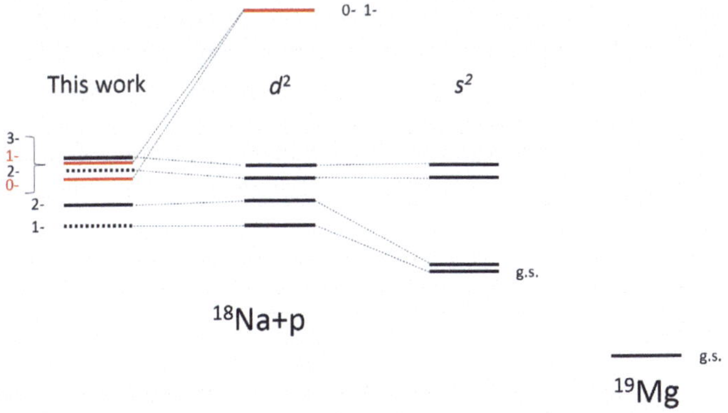

Fig. 62.4 The experimental level scheme is compared with predictions from [14]. In this study, two nuclear potential were used. With the first one the d^2 configuration (^{17}Ne+2p) is dominating the ^{19}Mg ground state structure, with the second the s^2 dominates. The agreement is better with the first one, but the experimental s-wave states (0_1^- and 1_2^-) are very shifted compared to the predicted ones

protons from ^{19}Mg through the low energy tails of these broad resonances. We calculated the partial width of this process using the quasi-classical R-matrix type model from [15] which is known to give results close to those obtained with a three-body model. It gives a ^{19}Mg lifetime shorter than the measured one. To be consistent with the experimental value, the s-wave component in the ^{19}Mg ground state has to be lower than $\approx 5\%$ (Fig. 62.4).

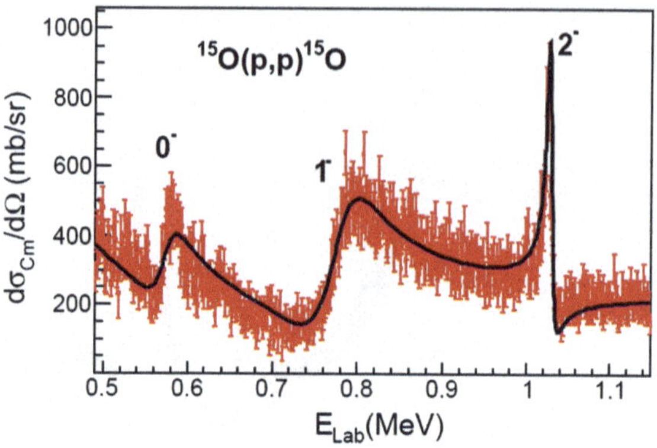

Fig. 62.5 Excitation function of the reaction $^{15}O(p, p)^{15}O$ measured in inverse kinematics. The line corresponds to an R-matrix fit. It shows the presence of the 0^- ground state, and two excited states 1^- and 2^-. The resolution of the experiment was $\sigma_{CM} \simeq 3$ keV. Adapted from [3]

62.5 The Case of ^{16}F

Understanding the role of specific parts of nuclear forces [16] in stabilising atomic nuclei and in inducing shell evolutions is a central theme of nuclear physics [17]. This understanding would bring a better predictive power for exotic nuclei such as those involved in the explosive r-process nucleosynthesis or rp-process in X-ray bursters. Most of the states involved in astrophysics are resonances, i.e. states coupled to continuum. An ideal case to study the effect of continuum was found in the mirror system: *unbound* $^{16}_{9}F_7$ and *bound* $^{16}_{7}N_9$ [18]. In the present work, we measured the energies and widths of the unbound states in ^{16}F using the resonant elastic scattering technique.

Radioactive beam of ^{15}O was used to populate the low lying resonant states 0^-, 1^-, 2^- in the unbound ^{16}F nucleus by means of proton elastic scattering reactions in inverse kinematics. The measured excitation function is shown in Fig. 62.5.

We used the measured properties of the low lying states of ^{16}F to derive the effective proton-neutron interaction energies. Based on their measured large proton spectroscopic factor values, the resonant states of ^{16}F can be viewed as a core of ^{14}O plus a proton in the $2s_{1/2}$ or $1d_{5/2}$ shell and a neutron in $1p_{1/2}$, see Fig. 62.6. Experimental energies were used to derive the strength of the $2s_{1/2}-1p_{1/2}$ and $1d_{5/2}-1p_{1/2}$ effective proton-neutron interactions. It is found that the former changes by 40% compared with the mirror nucleus ^{16}N, and the second by 10%. This apparent symmetry breaking of the nuclear force between mirror nuclei finds explanation in the large coupling to the continuum for the states built on an $\ell = 0$ proton configuration. Two effects were identified. Firstly, the $2s_{1/2}$ proton wave function in ^{16}F is different from the $2s_{1/2}$ proton in ^{15}F, since these nuclei have not the same binding energy, and so, not the same extension of wave functions, resulting in different Coulomb energies. This has to be taken into account. Secondly, the overlap between the $2s_{1/2}$ and the $1p_{1/2}$, proton and neutron wave functions, are different between ^{16}F and ^{16}N. In ^{16}F the $2s_{1/2}$ wave function is unbound, and in ^{16}N it is bound, the unbound wave

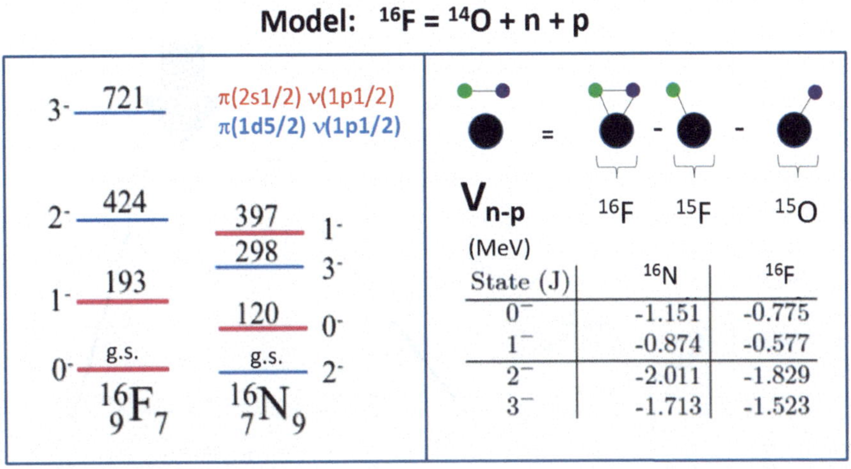

State (J)	^{16}N	^{16}F
0^-	-1.151	-0.775
1^-	-0.874	-0.577
2^-	-2.011	-1.829
3^-	-1.713	-1.523

Fig. 62.6 Left: The levels scheme of ^{16}F and ^{16}N mirror nuclei are compared. The states can be interpreted as a core nucleus plus a proton in the $2s_{1/2}$ or $1d_{5/2}$ shell and a neutron in $1p_{1/2}$, and vice versa. Right: Effective proton-neutron interaction energies are calculated from mass differences

function is more spread, resulting in different proton-neutron effective interactions. If the two contributions are taken into account, we check that the two nuclei have the same proton-neutron interaction energies. See [3] for more details.

62.6 The Case of ^{15}F

We also used the resonant elastic scattering technique to study the unbound ^{15}F nucleus. The excitation function of the elastic scattering reaction ^{14}O(p, p)^{14}O was measured at 180° (cm) in inverse kinematics with a thick target. It is shown in Fig. 62.7. This spectrum has high statistics, good energy resolution ($\sigma = 7(2)$ keV) and large energy covering from 0.4 to 5.6 MeV. An analysis of the excitation function using the R-Matrix method was performed with the code AZURE2 [19]. A deep minimum is observed at ≈ 1 MeV corresponding to the well known $J^{\pi} = 1/2^+_1$ ground state resonance of ^{15}F. It was fitted at an energy $E_R = 1270(10)(10)$ keV with $\Gamma = 376(70)(^{+200}_{0})$ keV, where the quoted uncertainties correspond to statistical and systematic uncertainties respectively. The measured width is lower than the values obtained in previous studies by at least 30%, but is still within the error bars. The best fit is shown with the continuous red line which leads to $\Gamma = 376$ keV. For comparison, the dashed-blue line shows the calculation made for the average value of the previous results, i.e. $\Gamma = 737$ keV. The peak observed at the resonance energy $E_R = 2763(9)(10)$ keV with $\Gamma = 305(9)(10)$ keV corresponds to the $J^{\pi} = 5/2^+_1$ first excited state. It is in good agreement with the previous measurements. In addition, for the first time in a resonant elastic scattering experiment, the second excited

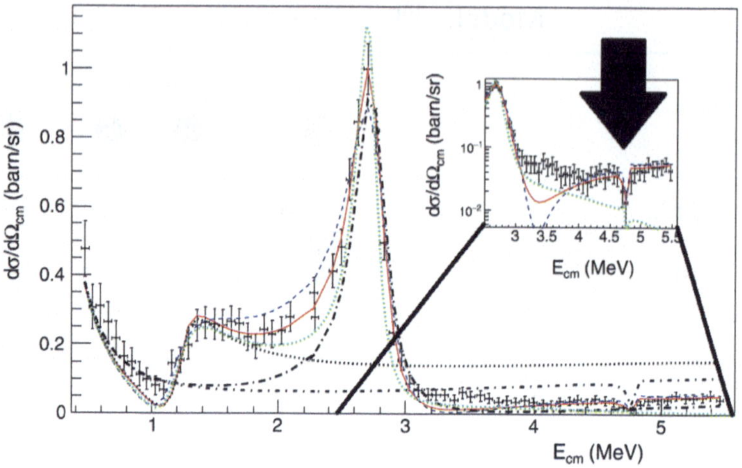

Fig. 62.7 Measured excitation function of the reaction ^{14}O(p, p)^{14}O, see text and [5] for details. Adapted from [5]

state is clearly observed as a narrow dip at a resonance energy of ≈4.8 MeV. In the corresponding mirror nucleus, the second excited state has spin $J^{\pi} = 1/2^{-}$. The shape of the dip corresponds very well with the R-matrix calculation using this spin and parity. It is the first time the spin of this state is assigned. The R-Matrix analysis of the excitation function was performed taking into account the experimental resolution. The resonance is measured to be $E_R = 4.757(6)(10)$ MeV with $\Gamma = 36(5)(14)$ keV. The observation of this narrow resonance in ^{15}F is surprising since this resonance is located well above the Coulomb plus centrifugal barrier ($B_C + B_\ell \approx 3.3$ MeV) for the proton emission, there is no barrier to retain the proton inside the nucleus.

The structure of this narrow above-barrier state in a nucleus located two neutrons beyond the proton drip line was investigated using the Gamow Shell Model in the coupled channel representation [20] with a ^{12}C core and three valence protons. Our calculations shows that the overlap with the $^{14}O_{g.s.}+p$ is very weak, this explains why the resonance is narrow. It was found that it is an almost pure wave function of two quasi-bound protons in the $2s_{1/2}$ shell. The state is located just above the two-proton emission threshold, as shown in Fig. 62.8. Is it another example of the so-called Ikeda [4] conjecture? The emission of two protons from this narrow state is energetically possible. Since there is no intermediate state accessible to ^{14}O, it should be a direct two-proton emission to the g.s. of ^{13}N. However, the available energy is only $Q_{2p} = 129$ keV, inducing a Wigner limit of $\Gamma_{^2He} = 4 \times 10^{-11}$ eV ($t_{1/2} = 16.5$ μs) for the emission of a ^{2}He cluster with $\ell = 0$. Therefore, the branching ratio for the emission of two protons is expected to be extremely small. This explains the resonance is narrow also for the two-proton emission. More details can be found in [5].

Electromagnetic transitions between resonances have been rarely observed, e.g. in the unbound nucleus ^{8}Be [21, 22], and in ^{56}Cu [23]. The second excited state

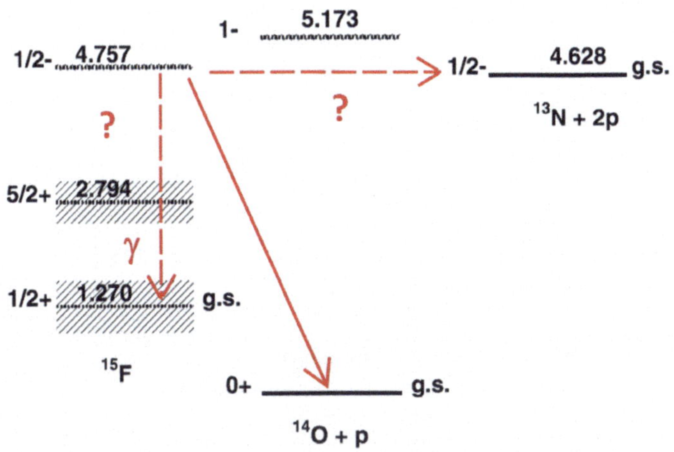

Fig. 62.8 Level scheme of ^{15}F relatively to ^{14}O+p and to ^{13}N+2p. The 1/2$^-$ resonance in ^{15}F is narrow, the γ-decay might compete with the one proton and two-proton emissions

of ^{15}F being a long-lived resonance, a $\gamma-$transition from this resonance to the g.s. resonance of ^{15}F is conceivable. $E1$ transitions occurring between $2s_{1/2} \rightarrow 1p_{1/2}$ single-particle states are expected to be extremely fast. The γ-width is larger since the electric transition is proportional to the radial integral $\int u_f(r)\ r\ u_i(r)dr$ where $u_{f,i}(r)$ are final and initial radial wave functions of the nucleon, the unbound states having an extended radial wave function. Taking this property into account in ^{15}F and the neutron/proton effective charge difference, we predict $\Gamma_\gamma \approx 50$ eV. It would be interesting to measure these γ-rays in order to elucidate the structure of this unique $1/2^-_1$ state and its $1s_{1/2}$ content. These γ-rays will be in coincidence with protons emitted from the ground state of ^{15}F.

Other narrow resonances were predicted in ^{15}F at higher excitation energies [24, 25] remain to be observed. The existence of relatively narrow resonances at high excitation energies may be actually more frequent than thought opening a possibility for the particle and gamma resonance spectroscopy in nuclei far beyond the drip lines.

Acknowledgements This article was written in the name of the E400S, E442S and E521S collaborations. I would like to thank all collaborators, and specially F. de Grancey, I. Stefan, M. Assié (GANIL France, present address: Institut de Physique Nucléaire Université Paris-Sud-11-CNRS/IN2P3 91406 Orsay, France), J. Mrazek (Nuclear Physics Institute ASCR CZ-25068 Rez Czech Republic), the GANIL accelerator staff for delivering the radioactive beams, and R.J. de Boer for his help. These studies were supported by OP RDE, MEYS Czech Republic under the project EF16-013/0001679, by LEA NuAG, by LIA COSMA.

References

1. de Oliveira Santos, F., et al.: Eur. Phys. J. A **24**, 237 (2005)
2. Assie, M., et al.: Phys. Lett. B **712**, 198 (2012)
3. Stefan, I., et al.: Phys. Rev. C **90**, 014307 (2014)
4. Okołowicz, J., Płoszajczak, M., Nazarewicz, W.: Prog. Theo. Phys. Suppl. **196**, 230 (2012)
5. De Grancey, F., et al.: Phys. Lett. B **758**, 26 (2016)
6. Axelsson, L., et al.: Phys. Rev. C **54**, R1511 (1996)
7. de Oliveira Santos, F., et al.: Eur. Phys. J. Web Conf. **184**, 01006 (2018)
8. Stefan, I.: Ph.D. Thesis, Université de Caen, France (2006)
9. Görres, J., Wiescher, M., Thielemann, F.-K.: Phys. Rev. C **51**, 392 (1995)
10. de Oliveira Santos, F.: Habilitation Thesis, Université de Caen, France (2017)
11. Mukha, I., et al.: Phys. Rev. Lett. **99**, 182501 (2007)
12. Blank, B., Płoszajczak, M.: Rep. Prog. Phys. **71**, 046301 (2008)
13. Mukha, I., et al.: Phys. Rev. C **77**, 061303(R) (2008)
14. Grigorenko, L.V., Mukha, I.G., Zhukov, M.V.: Nucl. Phys. A **713**, 372 (2003)
15. Grigorenko, L.V., Zhukov, M.V.: Phys. Rev. C **76**, 014009 (2007)
16. Dobaczewski, J., et al.: Prog. Part. Nucl. Phys. **59**, 432 (2007)
17. Forssén, C., et al.: Phys. Scr. T **152**, 014022 (2013)
18. Ogawa, K., et al.: Phys. Lett. B **464**, 157 (1999)
19. Azuma, R.E., et al.: Phys. Rev. C **81**, 045805 (2010)
20. Michel, N., et al.: Phys. G **36**, 013101 (2009)
21. Datar, V.M., et al.: Phys. Rev. Lett. **94**, 122502 (2005)
22. Datar, V.M., et al.: Phys. Rev. Lett. **111**, 062502 (2013)
23. Orrigo, S.E.A., et al.: Phys. Rev. lett. **112**, 222501 (2014)
24. Canton, L., et al.: Phys. Rev. Lett. **96**, 072502 (2006)
25. Fortune, H.T., Sherr, R.: Phys. Rev. Lett. **99**, 089201 (2007)

Chapter 63
Four-Body Effects in Nucleus-Nucleus Scattering

Pierre Descouvemont

Abstract We develop a four-body CDCC (continuum discretized coupled channel) model, aimed at describing reactions where both nuclei present a low breakup threshold. The only inputs are the interactions describing the colliding nuclei, and the four optical potentials between the fragments. The method is applied to the ^{11}Be + d and ^{7}Li + d systems. We show that, in general, breakup channels are crucial to reproduce the elastic cross sections. We suggest that the present wave functions could be used in future DWBA calculations.

63.1 Introduction

Since the 90's, the availability of radioactive beams provided a rich information on the structure and properties of exotic nuclei [1, 2]. A significant progress in scattering theory was provided by the CDCC (Coupled Channel Discretized Continuum) method [3, 4], where the many-body (in general two-body) structure of the projectile is taken into account. The CDCC formalism has been initially developed to investigate deuteron-induced reactions [3]. Although the deuteron is not considered as an exotic nucleus, its low binging energy ($B = 2.22$ MeV) makes breakup channels quite important, even for elastic scattering. In the CDCC model, the projectile breakup is simulated by a discrete number of approximate continuum states. This technique permits a strong improvement in the description of deuteron-nucleus cross sections (see, for example, [4–6] for reviews).

For many years, CDCC calculations involved three-body systems. Typical examples are deuteron + nucleus reactions, where the deuteron is described by a $p + n$ structure. Since more than 20 years, the three-body CDCC method was successfully applied to reactions involving weakly bound nuclei, such as ^{11}Be, which can be seen as a ^{10}Be + n system. The CDCC theory represents a natural framework for reac-

P. Descouvemont (✉)
Physique Nucléaire Théorique et Physique Mathématique, C.P. 229,
Université Libre de Bruxelles (ULB), 1050 Brussels, Belgium
e-mail: pdesc@ulb.ac.be

N. A. Orr et al. (eds.), *Recent Progress in Few-Body Physics*,
Springer Proceedings in Physics 238,
https://doi.org/10.1007/978-3-030-32357-8_63

tions involving exotic nuclei. The main property of exotic nuclei is their low breakup threshold, and couplings to the continuum are important. More recently, the CDCC method was extended to 4-body systems, i.e. to reactions involving three-body projectiles, such as ^{6}He [7], ^{9}Be [8] or ^{11}Li [9].

Until now, most scattering calculations assume that the target remains in its ground state. This is certainly a fair approximation for heavy targets, but is more questionable for light targets. A typical example is the deuteron, which is used as a target for the investigation of exotic nuclei by stripping reactions (see, for example, [10]). If the projectile also presents a two-body structure, current versions of the CDCC method are no longer sufficient. In this work, we present a new extension of CDCC, where the breakup of both colliding nuclei is included. We assume a two-body structure for the projectile and for the target, which leads to another variant of the four-body model (2 + 2, instead of 3 + 1).

A typical application of the 4-body model is the ^{11}Be + d reaction, which has been experimentally studied recently [11]. In this reaction, both ^{11}Be and the deuteron present a low breakup threshold, and four-body breakup effects are expected to play a role in the scattering cross section. Another natural application of a four-body model is the ^{7}Li + d reaction, since ^{7}Li is known to have a marked $\alpha + t$ cluster structure.

63.2 Overview of the Four-Body CDCC Method

We consider the scattering of two nuclei, each of them presenting a two-body cluster structure. The coordinates are shown in Fig. 63.1 for the ^{11}Be + d system: $\boldsymbol{r}_1$ and $\boldsymbol{r}_2$ are the internal coordinates, and $\boldsymbol{R}$ is the relative coordinate between the colliding nuclei. The Hamiltonian of this four-body system is given by

$$H = H_1(\boldsymbol{r}_1) + H_2(\boldsymbol{r}_2) + T_R + \sum_{i=1}^{2}\sum_{j=1}^{2} U_{ij}(\boldsymbol{R}, \boldsymbol{r}_1, \boldsymbol{r}_2), \tag{63.1}$$

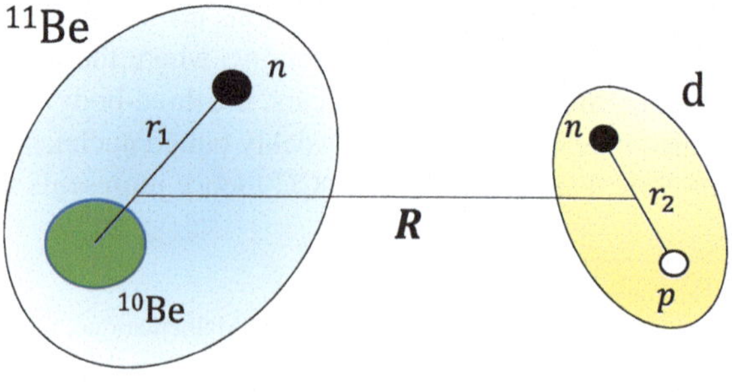

Fig. 63.1 Cluster configuration and coordinates used in the four-body model for the ^{11}Be + d system

where T_R is the relative kinetic energy, and U_{ij} are cluster-cluster optical potentials. The internal Hamiltonians H_1 and H_2 are chosen so as to reproduce the low-lying states of the nuclei.

Our goal is to solve the Schrödinger equation associated with (63.1) for scattering states. This four-body scattering problem can be approximately solved with the CDCC. We first define internal wave functions $\Phi_k^{I_i}(\boldsymbol{r}_i)$ from

$$H_i \Phi_k^{I_i}(\boldsymbol{r}_i) = E_k^{I_i} \Phi_k^{I_i}(\boldsymbol{r}_i), \tag{63.2}$$

where I_i is the angular momentum and k the level of excitation (the internal parity is implied in I_i). In the CDCC method, the radial part of the internal wave functions $\Phi_k^{I_i}(\boldsymbol{r}_i)$ is expanded over a set of N basis functions, such as Gaussian or Lagrange functions. In this way, (63.2) is converted to a standard eigenvalue problem. Negative energies $E_k^{I_i}$ correspond to physical states, and positive energies correspond to square-integrable approximations of the continuum [5]. These states, referred to as pseudostates, do not correspond to physical states, but are crucial to simulate the breakup of nuclei 1 and 2.

The total four-body wave function is then expanded over the internal states as

$$\Psi_\omega^{JM\pi}(\boldsymbol{R}, \boldsymbol{r}_1, \boldsymbol{r}_2) = \sum_c g_{\omega,c}^{J\pi}(R)\, \varphi_c^{JM\pi}(\Omega_R, \boldsymbol{r}_1, \boldsymbol{r}_2), \tag{63.3}$$

where ω is the entrance channel, and where the channel function with orbital angular momentum L is defined by

$$\varphi_c^{JM\pi}(\Omega_R, \boldsymbol{r}_1, \boldsymbol{r}_2) = \left[\left[\Phi_{k_1}^{I_1}(\boldsymbol{r}_1) \otimes \Phi_{k_2}^{I_2}(\boldsymbol{r}_2) \right]^I \otimes Y_L(\Omega_R) \right]^{JM}. \tag{63.4}$$

In these definitions, c stands for $c = (I_1, I_2, k_1, k_2, I, L)$, and I is the channel spin. This coupling mode is standard in scattering theory. In actual applications, the summation is truncated by a maximum energy and by a maximum angular momentum.

The radial functions $g_{\omega,c}^{J\pi}(R)$ are obtained from the coupled-channel system

$$\left(-\frac{\hbar^2}{2\mu} \frac{d^2}{dR^2} + \frac{\hbar^2}{2\mu} \frac{L(L+1)}{R^2} + E_c - E \right) g_{\omega,c}^{J\pi}(R) + \sum_{c'} V_{c,c'}^{J\pi}(R) g_{\omega,c'}^{J\pi}(R) = 0, \tag{63.5}$$

where μ is the reduced mass, and E_c is the energy of channel c, obtained from (63.2). The coupling potentials $V_{c,c'}^{J\pi}(R)$ are obtained from matrix elements of the potential between channel functions (63.4).

At large distance R, the radial functions tend to a combination of Coulomb functions as

$$g_{\omega,c}^{J\pi}(R) \to v_c^{-1/2} \Big(I_c(k_c R) \delta_{c\omega} - O_c(k_c R) U_{\omega,c}^{J\pi} \Big), \tag{63.6}$$

where $I_c(x)$ and $O_c(x)$ are the incoming and outgoing Coulomb functions, and k_c (v_c) is the wave number (velocity) in channel c. Scattering states associated with (63.5) are obtained within the R-matrix theory [12, 13] which provides the scattering matrix $U^{J\pi}_{\omega c}$. From scattering matrices in all partial waves $J\pi$, the various cross sections can be obtained by standard formulae [14].

63.3 Application to ^{11}Be + d Scattering

Data on ^{11}Be + d elastic scattering and breakup have been available recently at a ^{11}Be beam energy $E/A = 26.9$ MeV ($E_{\text{c.m.}} = 45.5$ MeV) [11]. The ^{11}Be + d system is an ideal candidate for the four-body CDCC method. Both colliding nuclei present a low separation energy (0.50 and 2.22 MeV, respectively). In d+nucleus and ^{11}Be+nucleus scattering, breakup effects of the deuteron and of ^{11}Be are known to be important.

For the ^{10}Be + n system, we use two different potentials: (i) the potential of [15] which neglects ^{10}Be excitation; (ii) the set-I potential of [16] which includes the ^{10}Be(2^+) + n channel. Both potentials reproduce the experimental energies of the $1/2^+$ ground state and of the $1/2^-$ excited state. More detail about the conditions of the calculation can be found in [17].

The 4-body CDCC elastic cross section is presented in Fig. 63.2, where we see that the breakup component of the wave function is crucial to reproduce the amplitude of the cross section. With the full CDCC calculation (black lines), the minimum near $\theta \approx 20°$ is well reproduced. The single-channel calculation (red lines), involving only the ground states of ^{11}Be and of the deuteron, overestimates the data by a factor two at the minimum. Including core excitation of ^{10}Be slightly enhances the cross section for $\theta > 30°$. This goes in the right direction but the calculation is still 10–20% lower than experiment.

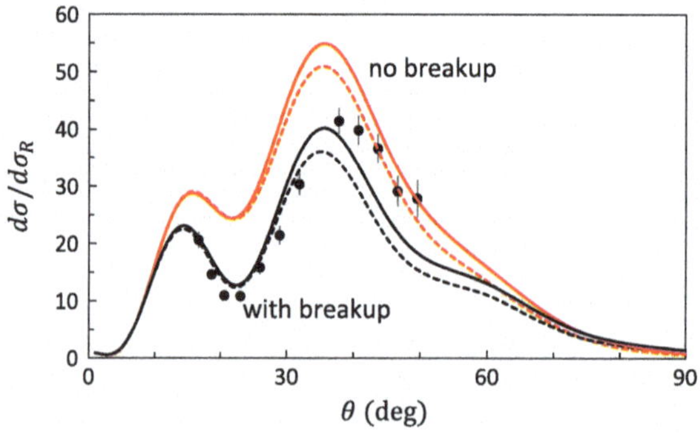

Fig. 63.2 ^{11}Be + d elastic cross section (divided by the Rutherford cross section) at $E(^{11}\text{Be}) = 26.9A$ MeV, with the single-channel approximation, and with all breakup channels. The solid lines are obtained with core excitation, and the dashed line with the ^{10}Be(0^+) + n configuration only. The data are taken from [11]

Notice that the full calculation involves a very large coupled-channel system. The number of channels reaches values around 600. The size of the coupled-channel system (63.5) is still larger since we have several channel spins I and several angular momenta L for a given physical channel. Such large systems can be solved by the R-matrix method, but they are extremely demanding in terms of computer time and memory.

63.4 Application to ^{7}Li + d Elastic Scattering

Although ^{7}Li is not a weakly bound nucleus, it presents a dominant $\alpha + t$ structure, and the ^{7}Li + d system is therefore a good candidate for a 4-body model. Data have been obtained, for example, in [18] at a deuteron energy $E_{\text{lab}} = 25$ MeV, which corresponds to $E_{\text{c.m.}} = 19.44$ MeV. This system is quite interesting in a theoretical point of view, since the four optical potentials between the fragments (α, t and p, n) are all real at low energies. Absorption is simulated by the breakup of ^{7}Li and of the deuteron. However, DWBA calculations [18] suggest that the ^{7}Li$(d, t)^6$Li, and probably ^{7}Li$(d, p)^8$Li, transfer channels play a role at large angles. The four-body model is therefore not expected to provide an excellent description of the cross section over a wide angular range.

The conditions of the calculation are given in [17]. The ^{7}Li + d elastic cross section at $E_d = 25$ MeV is shown in Fig. 63.3 with the data of [18]. At small angles, all calculations reproduce the data, which are dominated by the Coulomb interaction. Strong differences, however, are observed for $\theta > 30°$. Without any breakup channel, the model fails to reproduce the minimum near $\theta \simeq 35°$, and overestimates the data by 2 orders of magnitude at large angles. Introducing breakup channels improves the agreement between theory and experiment.

The behaviour of the cross section up to 60° is nicely reproduced when ^{7}Li and deuteron breakup channels are introduced simultaneously. At large angles, however,

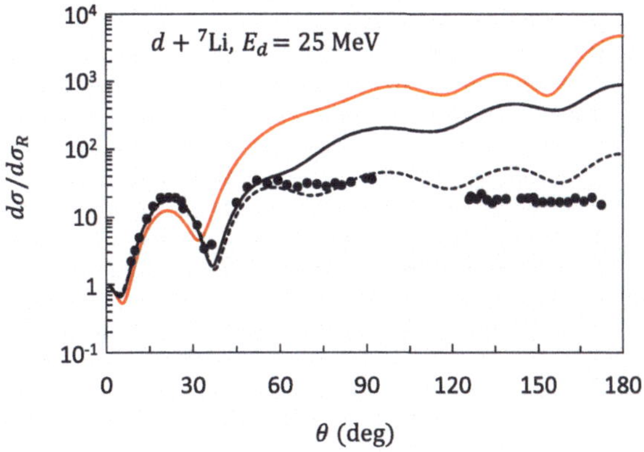

Fig. 63.3 ^{7}Li + d elastic cross section at $E_d = 25$ MeV with different conditions of calculation. The red and black lines correspond to the single-channel and to the full calculations, respectively. The dashed line is obtained by adding a phenomenological imaginary potential. The data are taken from [18]

the CDCC calculation still overestimates the data. Calculations have been performed by increasing the model space, but this has a minor effect on the theoretical curve. In [17], it is shown that the introduction of a phenomenological imaginary potential, which is justified by the lack of rearrangement channels in CDCC, improves the agreement with experiment.

63.5 Conclusion

The present work is a natural extension of the CDCC method to reactions involving two nuclei with low breakup thresholds. This situation is often met in deuteron induced reactions on exotic nuclei.

The main advantage of the present model is its predictive power. Optical potentials between the fragments are known in most cases. Then, the cross sections are computed without any parameter fitting. A possible extension of the model would be to use the wave functions in DWBA analyses. Nucleon transfer (d, p) or (d, n) reactions are often used to investigate the structure of exotic nuclei. Such calculations represent a challenge for future scattering studies.

References

1. Tanihata, I., Savajols, H., Kanungo, R.: Prog. Part. Nucl. Phys. **68**, 215 (2013)
2. Bonaccorso A.: Phys. Scr. **2013**(T152), 014019 (2013). http://stacks.iop.org/1402-4896/2013/i=T152/a=014019
3. Rawitscher, G.H.: Phys. Rev. C **9**, 2210 (1974)
4. Austern, N., Iseri, Y., Kamimura, M., Kawai, M., Rawitscher, G., Yahiro, M.: Phys. Rep. **154**, 125 (1987)
5. Kamimura, M., Yahiro, M., Iseri, Y., Sakuragi, S., Kameyama, H., Kawai, M.: Prog. Theor. Phys. Suppl. **89**, 1 (1986)
6. Yahiro, M., Ogata, K., Matsumoto, T., Minomo, K.: Prog. Theor. Exp. Phys. 01A206 (2012)
7. Matsumoto, T., Hiyama, E., Ogata, K., Iseri, Y., Kamimura, M., Chiba, S., Yahiro, M.: Phys. Rev. C **70**, 061601 (2004)
8. Descouvemont, P., Druet, T., Canto, L.F., Hussein, M.S.: Phys. Rev. C **91**, 024606 (2015). https://doi.org/10.1103/PhysRevC.91.024606. http://link.aps.org/doi/10.1103/PhysRevC.91.024606
9. Fernández-García, J.P., Cubero, M., Acosta, L., Alcorta, M., Alvarez, M.A.G., Borge, M.J.G., Buchmann, L., Diget, C.A., Falou, H.A., Fulton, B., Fynbo, H.O.U., Galaviz, D., Gómez-Camacho, J., Kanungo, R., Lay, J.A., Madurga, M., Martel, I., Moro, A.M., Mukha, I., Nilsson, T., Rodríguez-Gallardo, M., Sánchez-Benítez, A.M., Shotter, A., Tengblad, O., Walden, P.: Phys. Rev. C **92**, 044608 (2015)
10. Kanungo, R., Gallant, A., Uchida, M., Andreoiu, C., Austin, R., Bandyopadhyay, D., Ball, G., Becker, J., Boston, A., Boston, H., Brown, B., Buchmann, L., Colosimo, S., Clark, R., Cline, D., Cross, D., Dare, H., Davids, B., Drake, T., Djongolov, M., Finlay, P., Galinski, N., Garrett, P., Garnsworthy, A., Green, K., Grist, S., Hackman, G., Harkness, L., Hayes, A., Howell, D., Hurst, A., Jeppesen, H., Leach, K., Macchiavelli, A., Oxley, D., Pearson, C., Pietras, B., Phillips, A., Rigby, S., Ruiz, C., Ruprecht, G., Sarazin, F., Schumaker, M., Shotter, A., Sumitharachchi,

C., Svensson, C., Tanihata, I., Triambak, S., Unsworth, C., Williams, S., Walden, P., Wong, J., Wu, C.: Phys. Lett. B **682**, 391 (2010). https://doi.org/10.1016/j.physletb.2009.11.025. www.sciencedirect.com/science/article/pii/S0370269309013586
11. Chen, J., Lou, J.L., Ye, Y.L., Rangel, J., Moro, A.M., Pang, D.Y., Li, Z.H., Ge, Y.C., Li, Q.T., Li, J., Jiang, W., Sun, Y.L., Zang, Zhang, Y., Aoi, N., Ideguchi, E., Ong, H.J., Lee, J., Wu, J., Liu, H.N., Wen, C., Ayyad, Y., Hatanaka, K., Tran, T.D., Yamamoto, T., Tanaka, M., Suzuki, T., Nguyen, T.T.: Phys. Rev. C **94**, 064620 (2016). https://doi.org/10.1103/PhysRevC.94.064620. http://link.aps.org/doi/10.1103/PhysRevC.94.064620
12. Descouvemont, P., Baye, D.: Rep. Prog. Phys. **73**, 036301 (2010)
13. Descouvemont, P.: Comput. Phys. Commun. **200**, 199 (2016). https://doi.org/10.1016/j.cpc.2015.10.015. http://www.sciencedirect.com/science/article/pii/S0010465515003951
14. Satchler, G.R.: Direct Nuclear Reactions. Oxford University Press (1983)
15. Capel, P., Goldstein, G., Baye, D.: Phys. Rev. C **70**, 064605 (2004)
16. Summers, N.C., Pain, S.D., Orr, N.A., Catford, W.N., Angélique, J.C., Ashwood, N.I., Bouchat, V., Clarke, N.M., Curtis, N., Freer, M., Fulton, B.R., Hanappe, F., Labiche, M., Lecouey, J.L., Lemmon, R.C., Mahboub, D., Ninane, A., Normand, G., Nunes, F.M., Soić, N., Stuttge, L., Timis, C.N., Thompson, I.J., Winfield, J.S., Ziman, V.: Phys. Lett. B **650**(2), 124 (2007). https://doi.org/10.1016/j.physletb.2007.05.003. http://www.sciencedirect.com/science/article/pii/S0370269307005448
17. Descouvemont, P.: Phys. Rev. C **97**, 064607 (2018). https://doi.org/10.1103/PhysRevC.97.064607. https://link.aps.org/doi/10.1103/PhysRevC.97.064607
18. Burtebayev, N., Burtebayeva, J.T., Duisebayev, A., Kerimkulov, Z.K., Nassurlla, M., Zholdybayev, T., Artemov, S.V., Karakhodzhayev, A.A., Salikhbayev, U.S., Sakuta, S.B., Kliczewski, S., Piasecki, E., Rusek, K., Siudak, R., Trzcinska, A., Wolinska-Cichocka, M., Amar, A.: Acta Phys. Pol. B **46**, 1037 (2015)

Chapter 64
Few-Nucleon Reactions in Underground Laboratory

Alba Formicola, Giovanni Francesco Ciani, László Csedreki, Lucio Di Paolo and Matthias Junker

Abstract Accurate knowledge of thermonuclear reaction rates is important in understanding the generation of energy, the luminosity of neutrinos, and the synthesis of elements in stars. The LUNA Collaboration (Costantini et al. in Rep Prog Phys 72:086301, 2009; Broggini et al. in Annu Rev Nucl Part Sci 60:53–73, 2010) has shown how going underground and using the typical techniques of low background physics allows to measure nuclear cross sections at or close to energies relevant for the nucleosynthesis inside stars. This contribution will outline the general features of resonant and non resonant few nucleon reactions studied with stable beam including an overview of the experimental techniques adopted in underground nuclear astrophysics. Moreover, it will present a summary of the main recent results and achievements.

64.1 Introduction

How and where are stars born? Observational evidence points to the interstellar gas and dust clouds along the Galaxy's spiral arms as being the birthplace of stars. When stars like those responsible for planetary nebulae, novae and supernovae reach the end of their lives, they return some of their masses to the interstellar medium. New generations of stars are thus always forming from the "ashes" of previous generations. The abundances of the elements in the universe were studied since the beginning of the previous century by analyzing first Earth fragments and later meteorites. Later on it has been shown that the spectra of stars like the Sun, the interstellar matter and cosmic rays contain important information for the understanding of astrophysical scenarios, which have proven to be similar for most of these objects. Isotopic abundances are

A. Formicola (✉) · G. F. Ciani · L. Csedreki · L. Di Paolo · M. Junker
INFN, Laboratori Nazionali del Gran Sasso, via Acitelli 22 Assergi,
67100 L'Aquila, Italy
e-mail: formicola@lngs.infn.it

G. F. Ciani · L. Csedreki
GSSI, Gran Sasso Science Institute,viale F. Crispi 7, 67100 L'Aquila, Italy

N. A. Orr et al. (eds.), *Recent Progress in Few-Body Physics*,
Springer Proceedings in Physics 238,
https://doi.org/10.1007/978-3-030-32357-8_64

thus considered as "cosmic" or "universal", even though there are some exceptions for particular classes of stars [3].

The goals of nuclear astrophysics are to understand the energy generation of stars at all stages of their evolution and to explain the abundances of elements and their isotopes observed in nature. These aims are closely related since nuclear processes have been identified as the source of the enormous energy generated inside stars which stabilizes the stars and governs their evolution by transmuting nuclear species into other nuclear species through networks of nuclear reactions. It is therefore not surprising that the reaction rates of the nuclear processes are required for all studies of stellar evolution [3].

64.2 Thermonuclear Fusion: The pp-Chain and the CNO Cycle

A variety of different energy generating processes can take place in a star. As an example, in main sequence stars four hydrogen nuclei are transformed to helium by hydrogen burning.

$$4p \longrightarrow {}^4He + 2e^+ + 2\nu + 26.73\ \mathrm{MeV} \tag{64.1}$$

As in a stellar environment the probability of simultaneous interaction of four protons is essentially zero, a series of two body interactions is required to obtain the same end result. Two different processes are known: the pp-chain and the CNO cycle, both shown in Fig. 64.1.

The pp-chain starts with the reaction $p(p, e^+\nu)d$, which is the slowest reaction and therefore represents the rate-limiting step for the whole chain. As the cross section of the reaction at stellar energies is far too small to be measured in a laboratory, it is calculated from standard weak-interaction theory and results to be in the order 10^{-33} b at energies of a few keV and 10^{-23} b at energies of in the order of some MeV. The reaction d(p, γ)^{3}He immediately transforms all deuterium into ^{3}He and then into ^{4}He by the fusion of two ^{3}He nuclei (chain I):

$$^3He + {}^3He \rightarrow {}^4He + p + p \tag{64.2}$$

As an alternative ^{3}He can encounter an α particle generating ^{7}Be.

$$^3He + {}^4He \rightarrow {}^7Be + \gamma \tag{64.3}$$

The ^{7}Be generated this way decays via electron capture to ^{7}Li and a neutrino an energy of either 0.38 or 0.86 MeV. The ^{7}Li captures a proton producing ^{8}Be, which decays producing two helium nuclei (chain II). The ^{7}Be electron capture competes with proton capture producing ^{8}B. Subsequently, ^{8}Be decays generating a positron,

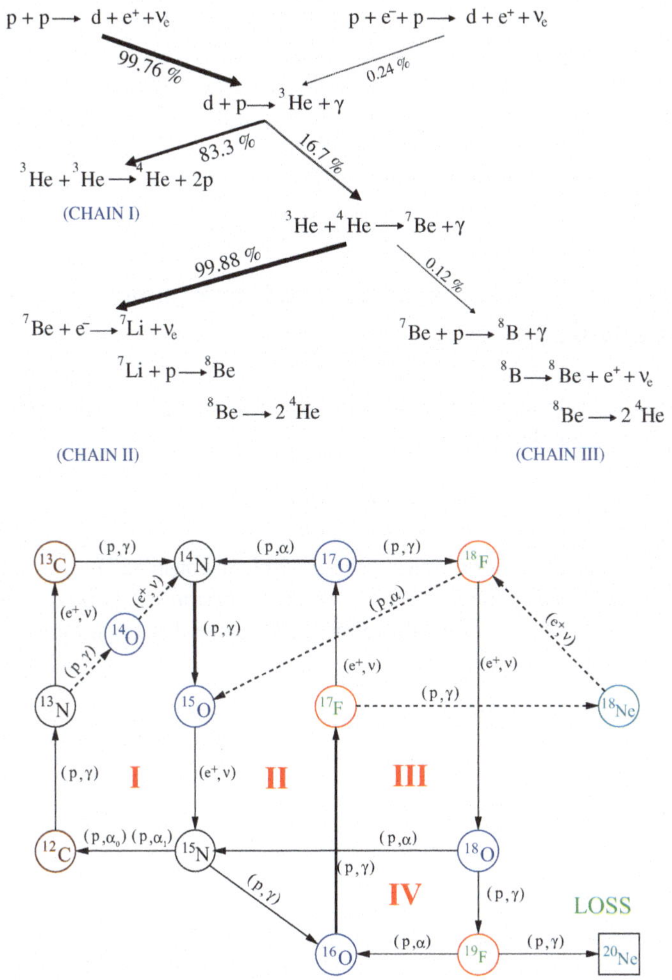

Fig. 64.1 Top panel: the pp-chain, the branching ratios shown are calculated for solar conditions in according to the Standard Solar Model (SSM) [4]. Bottom panel: the CNO-cycle. Dotted lines correspond to hot CNO

a high energy neutrino (8 MeV $\leq E_\nu \leq$15 MeV), and again two helium nuclei (chain III).

The second possibility for the conversion of hydrogen into helium is given by the CNO-cycle, a reaction network proposed by H. Bethe and C.F. von Weizäcker in 1938. Here, the elements carbon, nitrogen, and oxygen (C, N, O) act as "catalysts" in each single cycle (see lower panel in Fig. 64.1). Obviously this scenario requires the presence of heavier elements in addition to hydrogen and helium in the interior of a star, a condition which is fulfilled in all stars which can be observed today.

During hydrogen burning, pp-chain and CNO-cycle contribute to the energy production in different proportions, which are dependent on temperature and abundance of heavy nuclei inside the star. The pp-chain dominates the energy production at temperatures below approximately 0.02 GK. In solar conditions hydrogen burning occurs predominantly through the pp-chain and CNO-cycle contributes to the energy production by a very small amount (1.5%) [3].

64.3 Qualitative Features of the Nuclear Reaction Mechanism

The energy distribution of nuclei in hot stellar matter can be described by the Maxwell–Boltzmann distribution

$$\phi(E) \propto Eexp(-E/kT), \tag{64.4}$$

where T is the local temperature and k is the Boltzmann constant. Due to the Coulomb barrier of the entrance channel, the reaction cross section σ(E) drops nearly exponentially with decreasing energy. Therefore σ(E) can be described by:

$$\sigma(E) = \frac{1}{E} \exp(-2\pi\eta(E))\, S(E) \tag{64.5}$$

The function S(E) is called the Astrophysical S–Factor and contains all strictly nuclear effects of the reaction, while

$$P \simeq \exp(-2\pi\eta(E)) \tag{64.6}$$

is the penetration probability through the Coulomb barrier.

Assuming no centrifugal barrier in the entrance channel (i.e. only s-wave projectiles) and a Coulomb potential, which is much larger than the projectile energy and has a square-well-shape inside and a point source shape outside the nuclear radius, the term $\eta(E)$, also called Sommerfeld parameter, is:

$$\eta(E) = \frac{Z_1 Z_2 e^2}{\hbar} (\mu/2E)^{1/2}, \tag{64.7}$$

where E and μ are the center of mass energy the and reduced mass in the entrance channel and Z_1 and Z_2 denote the proton numbers of projectile and target nucleus, respectively.

All the previous approximations are valid in the case of a smooth energy dependence of the S(E)–factor, i.e. when no resonance is present and the cross section is dominated by direct reaction mechanisms. In case a virtually unbound state is present at the excitation energy of the compound system $E_x = E_{cm} + Q$, the cross section of the resonant excitation of this state in the entrance channel (a) and the subsequent decay into the exit channel (b) are described by the Breit-Wigner equation:

$$\sigma(E) = \pi\lambda^2 \cdot \frac{2J+1}{(2J_1+1)\cdot(2J_2+1)}(1+\delta_{12}) \cdot \frac{\Gamma_a\Gamma_b}{\left(E - E_{R,lab}\right)^2 + (\Gamma/2)^2}, \quad (64.8)$$

where J is the angular momentum of the compound nucleus, J_1 and J_2 those of the projectile and target, $E_{R,lab}$[1] is the resonance energy, and Γ_a, Γ_b and Γ are the partial and the total widths of the resonance.

The cross sections of reactions ^{3}He(^{4}He, γ)^{7}Be and the ^{14}N(p, γ)^{15}O are characterized by the direct capture process. At very low energies the latter one is also influenced by the contribution of the tail of a narrow resonance.

Experimental techniques have improved significantly, allowing to push the limit of measurements close to, or even inside, the thermal energy region in stars called Gamow window which frequently is much lower the height of the Coulomb barrier. Even in those cases, in which direct measurements prove to be feasible, uncertainties in low energy stopping power data or the insufficient understanding of the electron screening effect introduce significant systematic uncertainties. Consequently extrapolations of data from higher energies using e.g. R-matrix fits [6] are often required. In order to minimize the risks related to the extrapolation, the data sets used should cover a wide range of beam energies. Furthermore, it is important to aim for direct measurements with statistical and systematical uncertainties at energies which are as close as possible to the Gamow Peak. Scattering experiments or transfer reactions provide important additional information. Indirect methods like the Trojan Horse technique can be exploited as an alternative tool to determine σ(E) [7].

In order to cover all evolution phases of stars from the Main Sequence stars (T $\simeq 10^7$) to supernovae (T $\simeq 10^9$) reaction rates must be known over a wide range of temperatures, which requires the availability of σ(E) data over a wide range of energies from keV to MeV [3].

64.4 Experimental Approach in an Underground Laboratory

Due to the small cross sections involved in nuclear astrophysics, experiments must be designed carefully considering backgrounds from cosmic rays (CR) and environmental radioactivity. At earth surface CR are mainly composed of muons. As an effect of their high energy, CR can penetrate through also massive shielding even though the response of a specific detector is related to its material and shape. Also secondary particles generated by CR impacting on the surrounding materials can interact with the detector. While the direct interactions of CR can be efficiently rejected by active

[1] As a general notation in this work, $E_{R,lab}$ is the resonance energy in the laboratory system, while E_r is the resonance energy in the center of mass system.

shielding (e.g. with plastic scintillators), events related to secondary particles (e.g. neutrons) are difficult to suppress and limit the efficiency of active shielding. Only the very massive shielding of an underground laboratory can provide a significant reduction of the CR flux of several orders of magnitude [8]. This approach is commonly exploited by low event rate experiments in the field of Dark Matter research and Neutrino physics.

Environmental radioactivity is another source of background, which reaches gamma energies of up to 3.5 MeV. It is caused by radionuclides belonging to natural radioactive series (^{214}Bi, ^{226}Ra, ^{214}Pb, ^{208}Tl…), long-lived natural radionuclides (^{40}K, ^{87}Rb…), radionuclides of cosmogenic origin (^{3}H, ^{14}C, ^{22}Na, ^{7}Be…) and radionuclides of artificial origin (^{95}Nd, ^{137}Cs, ^{90}Sr, ^{144}Ce…). These nuclides can be found outside but also inside the detectors themselves. The related background can be reduced by careful material selection, by cleaning and by massive passive shielding with high purity lead and copper. As CR induce secondary particles when interacting with passive shielding, the reduction of the CR flux inside an underground laboratory makes massive Pb and Cu shielding more efficient underground than above ground. Furthermore, the higher the rock coverage of a site the higher the reduction of the underground CR flux and the higher the suppression of background achieved by lead and copper shielding in the gamma energy range up to 3.5 MeV [14].

The interaction of the ion beam with traces of material on the target can give rise to beam induced background, which is specific for each reaction. In particular materials with atomic numbers lower than or similar to the target material are candidates for beam induced backgrounds. Consequently each reaction must be treated individually considering the involved beam energy and particle spectrum as well as the Q-value and the resonance structure of the background reactions. The type of target (gaseous or solid) and the target production process are of fundamental importance. In some particular cases background can be induced also by target nuclei scattered by the impinging projectiles [15].

The excellent signal to background ratio in the high resolution spectra of HPGe detectors favors the identification of the background, but they provide a limited efficiency with respect to high density materials like BGO or suffer consistent intrinsic background (e.g. LaBr). On the other hand detectors with low resolution disfavor a reliable background recognition. In any case, the spectra obtained when operating underground significantly improve the identification and subsequent suppression of beam induced background.

The first accelerator-based nuclear astrophysics experiments in a deep environment are being performed by the LUNA collaboration[2] since 25 years at the Laboratori Nazionali del Gran Sasso (LNGS).[3] Located long side the motorway Teramo—Rome the underground site consists in three halls (100 m long, 20 m wide, 18 m high) covered by 1.400 m of rock corresponding to 3.800 m of water equivalent (m.w.e.). This rock overburden reduces the cosmic muon flux be six orders of magnitude. Accelerator activities started in 1994 using a home made 50 kV accelerator, which

[2]Laboratory for Underground Nuclear Astrophysics; http://luna.lngs.infn.it.

[3]http://www.lngs.infn.it.

has been decommissioned in 2003. Since 2001 a 400 kV Singletron machine constructed by High Voltage Engineering Europe, The Netherlands, is available. In brief, this machine can deliver hydrogen (alpha) ion beams with an intensity of up to 500 (200) μA over a continuous operating time of about 40 days. This facility is presently equipped with a gas target and a solid state target beam line and allows for 7/24 h operation with only minimal operators effort [10].

These activities of LUNA have proven that direct measurements in nuclear astrophysics benefit largely from the environmental background achievable underground [9, 10]. This success has stimulated the construction of new accelerators at LNGS (LUNA-MV [11]) and also in other underground laboratories, CASPAR in the US [12], and JUNA in China [13].

64.5 Overview of ^{3}He(^{4}He, γ)^{7}Be and ^{14}N(p, γ)^{15}O reactions

64.5.1 The ^{3}He(^{4}He, γ)^{7}Be reaction

The nuclear physics input from the ^{3}He(α, γ)^{7}Be cross section is a major uncertainty in the determination of solar ^{7}Be and ^{8}B neutrino fluxes predicted by solar models and in the ^{7}Li abundance obtained in big-bang nucleosynthesis calculations. The ^{3}He(α, γ)^{7}Be (Q-value: 1.586 MeV) reaction leads to the emission of prompt γ-rays, and the final ^{7}Be nucleus decays with a half-life of 53.22 days, emitting a 478 keV γ-ray in 10.44% of the cases. Using the 400kV accelerator [21] the LUNA collaboration performed an experiment exploiting the detection of prompt γs and of ^{7}Be (activation method) using the same experimental setup [18–20]. In preparing the experiment particular attention had been given to a good shielding of the used detectors as both, prompt and β delayed γ-rays, have energies which are lower than 2.6 MeV.

The α-beam entered the ^{3}He extended windowless gas target through a collimator with a diameter of 7 mm and was stopped on a detachable copper disk that was positioned at the hot side of the calorimeter used to measures the beam intensity (about 250 μA) and which served as the primary catcher for the produced ^{7}Be. Prompt γ-rays were counted with a 135% ultra-low-background HPGe detector shielded with 5 cm of OFHC copper and 25 cm of lead [17]. The detector and the shield were enclosed in a sealed plastic box flushed with dry N_2 to reduce ^{222}Rn background. Due to the significant cosmic muon reduction in the underground environment the shielding allowed to suppression the background by five orders of magnitude for γ-rays below 2 MeV.

In order to count the ^{7}Be nuclei produced inside the gas target and implanted into the removable calorimeter cap, the detachable disk was dismounted from the calorimeter and moved to the low-activity counting facility operating in the under-

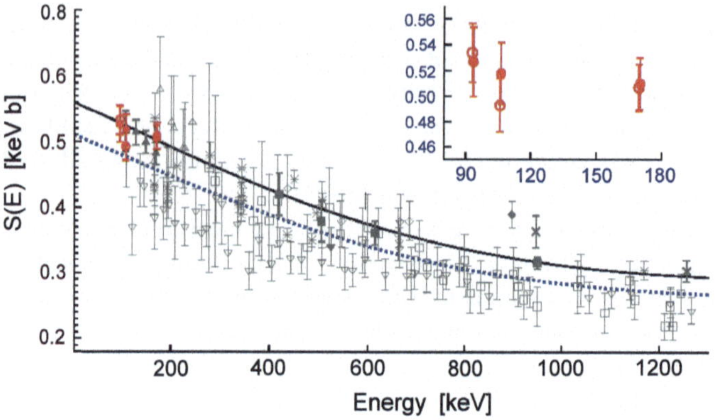

Fig. 64.2 Overview of all available S-factor values for the ^{3}He(α, γ)^{7}Be reaction. Filled and open circles: LUNA work. In the insert a zoom of prompt-γ (filled circles) and activation (open circles) data obtained by LUNA. For literature references see [20]

ground facilities of LNGS [10]. Overall, three couples of cross section values have been measured (prompt-γ and activation) at $E_\alpha = 220$, 250, and 400 keV.

A total (statistical and systematical) accuracy of about 3% for the S(0) value was obtained from LUNA data: $S(0) = 0.567 \pm 0.018 \pm 0.004$ keV b, where the last uncertainty term refers to the indetermination due to the theoretical model adopted for extrapolation to zero energy. The difference between data taken with the prompt-γ method and the activation method is smaller than $\pm 1.4\%$ (Fig. 64.2), so no discrepancy between LUNA prompt and activation data has been observed.

Concluding, the uncertainty contributed by the cross section determination of the reaction ^{3}He+^{4}He to the overall error budget of the calculated ^{8}B neutrino flux has been reduced from 7.5 to 2.4% while the total uncertainty of the calculated ^{8}B neutrino flux has been decreased from 12 to 10%, including astrophysical parameters. As a result, the uncertainty on the predicted ^{7}Be neutrino flux has been decreased from 9.4% down to 5.5% [17].

In 2014 deBoer et al. performed a global R-Matrix fit using ^{3}He(α, γ)^{7}Be data, including the higher energy data from the European recoil separator for nuclear astrophysics (ERNA)[23] and scattering data [22]. This effort resulted in $S(0) = 0.542 \pm 0.011(\text{MCfit}) \pm 0.006(\text{model})^{+0.019}_{-0.011}$ (phase shift) keVb. Nevertheless, a refined measurement of the slope of the S-factor in a wide energy range is desirable to verify theoretical calculations thus reducing the uncertainty on the extrapolated $S(0)$ [17].

64.5.2 The $^{14}N(p, \gamma)^{15}O$ reaction

The ^{14}N(p, γ)^{15}O reaction is the slowest process in the CNO cycle and, therefore, the key for understanding the timescale of the CNO cycles as well as the overall energy and neutrino production associated with CNO burning (see right panel in Fig. 64.1). The investigations at LUNA aimed to determine the ^{14}N(p,γ)^{15}O reaction rate at low energies in two phases. The first phase was a low efficiency, but high energy resolution experiment, which could be extended to energies down to E = 122 keV using a high purity Germanium detector and solid state set-up [24, 25]. The second phase consisted in a high efficiency but low energy resolution experiment, since a 4π BGO summing detector (80% efficiency) was used in combination with a windowless gas target system [26]. In this experiment a lower energy limit of E = 75 keV was reached with the disadvantage that no spectral information but only the total cross section of the reaction could be obtained. The reaction mechanism at low energies includes contributions from a resonance at E_r = 259 keV,[4] the direct capture process and the of the tail of a sub-threshold resonance at $E_r = -507$ keV, corresponding to the known E_x = 6793 keV state in ^{15}O. A series of new experiments using direct [24–27, 29, 29, 34, 35, 39] and indirect approaches [28, 32, 36–38] have been carried out over the last 18 years.

While an extensive discussion on different data sets and the relative R-Matrix approach is beyond the scope of this contribution, it is worth mentioning that in recent efforts [34, 35] the excitation function of ^{14}N(p, γ)^{15}O reaction has been measured by in-beam γ-ray spectroscopy for the ground state and the 6.79 MeV transition. In particular Li et al. [34] reported very accurate angular distribution data. Furthermore, as the analysis covered a wide energy range the contributions from broad resonances and direct capture have better been constrained.

The work of Wagner et al. [35], found some slight discrepancies with the R-matrix fit of the 6.79 transition by Li et al. [34]. In addition a lower value for S(0) of the ground state transition compared to [34] has been reported. In conclusion, these works show that a new data set connecting the low energy LUNA data and the extrapolation from higher-energy data is still required to further reduce the uncertainty of the total cross section at stellar energies.

64.6 Outlook and Conclusion

The results obtained at LUNA prove underground nuclear astrophysics as a powerful tool not only for pushing measurements to lower energies but also, and even more, for providing high quality data, which can be used to reduce the uncertainties

[4]Resonance energies are given in the center-of-mass system.

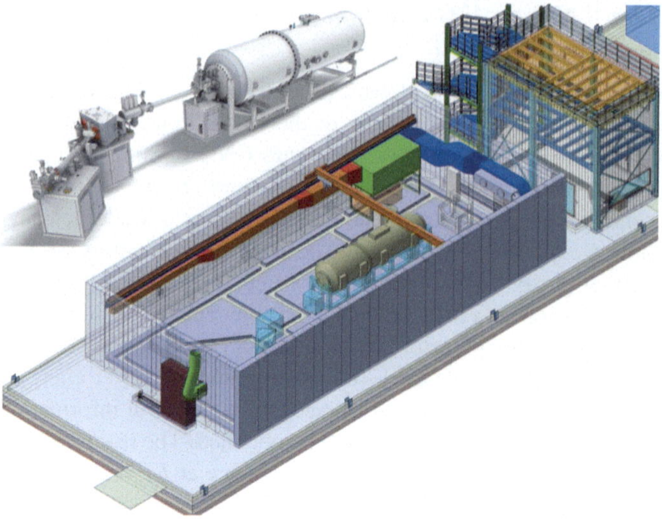

Fig. 64.3 Artists view of the LUNA-MV facility located in the underground laboratories of LNGS [40]

of unavoidable extrapolations. Stimulated by the success of the LUNA experiment several next generation facilities for underground nuclear astrophysics have been proposed in the last years.

In 2012 and 2013 the Italian Ministry of Research has granted INFN with financial support dedicated to the LUNA-MV project (Fig. 64.3). In March 2016 INFN assigned the contract for constructing and installing the LUNA-MV accelerator to High Voltage Engineering Europe (HVE) which is presently finalizing the construction of the machine at their factory in the Netherlands [40]. At the same time LNGS is setting up the necessary infrastructure including a shielded accelerator room which will reduce the neutron flux induced by LUNA-MV accelerator operations to a level which is lower than the natural background in the underground laboratories. Once in operation the LUNA-MV facility will deliver intense proton, alpha and Carbon beams of energies up to 3.5 MeV (7 MeV for double charged Carbon) providing thus unique possibilities to study process involved in AGB stars, Novae and other astrophysical scenarios.

Acknowledgements The experimental work described in this paper has been mainly developed in the framework of the LUNA experiments. The authors would thank all the members of this international collaboration.

References

1. Costantini, H., Formicola, A., Imbriani, G., et al.: LUNA: a laboratory for underground nuclear astrophysics. Rep. Prog. Phys. **72**, 086301 (2009). https://doi.org/10.1088/0034-4885/72/8/086301
2. Broggini, C., Bemmerer, D., Guglielmetti, A., Menegazzo, R.: LUNA: nuclear astrophysics deep underground. Annu. Rev. Nucl. Part. Sci. **60**, 53–73 (2010). https://doi.org/10.1146/annurev.nucl.012809.104526
3. Rolfs, C., Rodney, W.S.: Cauldrons in the Cosmos. University of Chicago Press, Chicago (1988)
4. Adelberger, E.G., et al.: Solar fusion cross sections II: the pp chain and CNO cycles. Rev. Mod. Phys. **83**, 195–245 (2010). https://doi.org/10.1103/RevModPhys.83.195
5. Iliadis, C.: Nuclear Physics in Stars. Wiley-VCH, Weinheim (2007)
6. Lane, A.M., Thomas, R.G.: R-matrix theory of nuclear reactions. Rev. Mod. Phys. **30**, 257–353 (1958)
7. Tumino, A., et al.: New advances in Trojan Horse Method as an indirect approach to nuclear astrophysics. Few-Body Sys. **54**, 745–753 (2013)
8. Bettini, A.: The world deep underground laboratories. Eur. Phys. J. Plus **127**, 114 (2012)
9. Best, A., et al.: Low energy neutron background in deep underground laboratories. Nucl. Instr. Meth. A **812**, 1–6 (2016). https://doi.org/10.1016/j.nima.2015.12.034
10. Best, A., et al.: Underground nuclear astrophysics. Eur. Phys. J. A **52**, 72–75 (2016)
11. Aliotta, M., et al.: Helium burning and neutron sources in the stars. Eur. Phys. J. A **52**, 76 (2016)
12. Strieder, F., et al.: Bulletin of the American Physical Society - 2015 Fall Meeting of the APS Division of Nuclear Physics, vol. 60 (2015)
13. Yu-Cheng, Wu, et al.: Measurement of cosmic-ray flux in the China JinPing underground laboratory. Chin. Phys. C **37**, 086001 (2013). https://doi.org/10.1088/1674-1137/37/8/086001
14. Heusser, G.: Low-radioactivity background techniques. Ann. Rev. Nucl. Part. Sci. **45**, 543–590 (1995). https://doi.org/10.1146/annurev.ns.45.120195.002551
15. Anders, M., Trezzi, D., Bellini, A., et al.: Neutron-induced background by an α-beam incident on a deuterium gas target and its implications for the study of the $^2H(\alpha,\gamma)^6Li$ reaction at LUNA. Eur. Phys. J. A **49**, 28 (2013)
16. Arpesella, C., et al.: Measurement of the $^3H(^3H,2p)^4He$ cross section within the solar Gamow peak. Phys. Lett. B **389**, 452–456 (1996)
17. Formicola, A., et al.: The nuclear physics of the hydrogen burning in the sun. Eur. Phys. J. A **52**, 73 (2016)
18. Bemmerer, D., et al.: Activation measurement of the $^3He(\alpha,\gamma)^7Be$ cross section at low energy. Phys. Rev. Lett. **97**, 122502 (2006). https://doi.org/10.1103/PhysRevLett.97.122502
19. Gyurky, G., et al.: $^3He(\alpha,\gamma)^7Be$ cross section at low energies. Phys. Rev. C. **75**, 035805 (2007). https://doi.org/10.1103/PhysRevC.75.035805
20. Confortola, F., et al.: Astrophysical S factor of the $^3He(\alpha,\gamma)^7Be$ reaction measured at low energy via detection of prompt and delayed γ rays. Phys. Rev. C **75**, 065803 (2007). https://doi.org/10.1103/PhysRevC.75.065803
21. Formicola, A., et al.: The LUNA II 400 kV accelerator. Nucl. Instr. Meth. A **507**, 609–615 (2003). https://doi.org/10.1016/S0168-9002(03)01435-9
22. deBoer, R.J., et al.: Monte Carlo uncertainty of the $^3He(\alpha,\gamma)^7Li$ reaction rate. Phys. Rev. C **90**, 035804 (2014). https://doi.org/10.1103/PhysRevC.90.035804
23. Di Leva, A., et al.: Stellar and Primordial Nucleosynthesis of the ^7Be: Measurement of $^3He(\alpha,\gamma)^7Li$. Phys. Rev. Lett. **102**(23), 232502 (2009). https://doi.org/10.1103/PhysRevLett.102.232502
24. Formicola, A., et al.: Astrophysical S-factor of $^{14}N(p,\gamma)^{15}O$. Phys. Lett. B **591**, 61–68 (2004). https://doi.org/10.1016/j.physletb.2004.03.092
25. Imbriani, G., et al.: S-factor of $^{14}N(p,\gamma)^{15}O$ at astrophysical energies. Eur. Phys. J. A **25**, 455–466 (2005). https://doi.org/10.1140/epja/i2005-10138-7

26. Lemut, A., et al.: First measurement of the $^{14}N(p,\gamma)^{15}O$ cross section down to 70 keV. Phys. Lett. B **634**, 483–487 (2006). https://doi.org/10.1016/j.physletb.2006.02.021
27. Marta, M., et al.: Precision study of ground state capture in the $^{14}N(p,\gamma)^{15}O$ reaction. Phys. Rev. C **78**, 022802(R) (2008). https://doi.org/10.1103/PhysRevC.78.022802
28. Mukhamedzhanov, A.M., et al.: Asymptotic normalization coefficients for $^{14}N+p\rightarrow^{15}O$ and the astrophysical S factor for $^{14}N(p,\gamma)^{15}O$. Phys. Rev. C **67**, 065804 (2003). https://doi.org/10.1103/PhysRevC.67.065804
29. Runkle, R.C., et al.: Direct measurement of the $^{14}N(p,\gamma)^{15}O$ S factor. Phys. Rev. Lett. **94**, 082503 (2005). https://doi.org/10.1103/PhysRevLett.94.082503
30. Schröder, U., et al.: Stellar reaction rate of $^{14}N(p,\gamma)^{15}O$ and hydrogen burning in massive stars*. Nucl. Phys. A **467**, 240–260 (1987)
31. Serenelli, A., et al.: Using the standard solar model to constrain composition and S-factors. arXiv:1211.6740
32. Bertone, P.F., et al.: Lifetime of the 6793-keV State in ^{15}O. Phys. Rev. Lett **87**, 152501 (2001). https://doi.org/10.1103/PhysRevLett.87.152501
33. Angulo, C., Descouvemont, P.: The $^{14}N(p,\gamma)^{15}O$ low-energy S-factor. Nucl. Phys. A **690**, 755–768 (2001)
34. Li, Q., et al.: Cross section measurement of $^{14}N(p,\gamma)^{15}O$ in the CNO cycle. Phys. Rev. C **93**, 055806 (2016). https://doi.org/10.1103/PhysRevC.93.055806
35. Wagner, L., et al.: Astrophysical S factor of the $^{14}N(p,\gamma)^{15}O$ reaction at 0.41.3 MeV. Phys. Rev. C 97, 015801 (2017). https://doi.org/10.1103/PhysRevC.97.015801
36. Yamada, K., et al.: E1 strength of the subthreshold 3/2+ state in ^{15}O studied by Coulomb excitation. Phys. Lett. B **579**, 265–270 (2004). https://doi.org/10.1016/j.physletb.2003.11.024
37. Schrmann, D., et al.: Lifetime measurement of the 6792 keV state in^{15}O, important for the astrophysical S factor extrapolation in $^{14}N(p,\gamma)^{15}O$. Phys. Rev. C **77**, 055803 (2008). https://doi.org/10.1103/PhysRevC.77.055803
38. Nelson, S.O., et al.: Analyzing power measurement for the $^{14}N(\vec{p},\gamma)^{15}O$ reaction at astrophysically relevant energies. Phys. Rev. C **68**, 065804 (2003). https://doi.org/10.1103/PhysRevC.68.065804
39. Bemmerer, D., et al.: Low energy measurement of the $^{14}N(p,\gamma)^{15}O$ total cross section at the LUNA underground facility. Nucl. Phys. A **779**, 297–317 (2006). https://doi.org/10.1016/j.nuclphysa.2006.09.001
40. Sen, A., et al.: High intensity, high stability 3.5 MV Singletron accelerator. Nucl. Instr. Meth. B (in press). https://doi.org/10.1016/j.nimb.2018.09.016

Chapter 65
A Comprehensive Measurement of Analyzing Powers in the Proton-Deuteron Break-Up Channel at 135 MeV

H. Tavakoli-Zaniani, M. T. Bayat, M. Eslami-Kalantari, N. Kalantar-Nayestanaki, St. Kistryn, A. Kozela, J. G. Messchendorp, M. Mohammadi-Dadkan, R. Ramazani-Sharifabadi and E. Stephan

Abstract In this contribution, vector analyzing powers are presented for the proton-deuteron break-up reaction studied using a polarised-proton beam at 135 MeV. For the experiment, we used the Big Instrument for Nuclear-polarisation Analysis (BINA) at KVI, the Netherlands. With this setup, we determined, for the first time, A_x for a large range in the kinematical S, polar and azimuthal angles of the two outgoing protons. Our data are reasonably well described by Faddeev calculations for kinematical configurations at which the three-nucleon force effect is predicted to be small. However, striking discrepancies are observed at specific configurations, for which the relative azimuthal angle between the two protons becomes small. In this contribution, some of these configurations along with the analysis techniques are presented.

H. Tavakoli-Zaniani (✉) · M. T. Bayat · N. Kalantar-Nayestanaki · J. G. Messchendorp · M. Mohammadi-Dadkan · R. Ramazani-Sharifabadi
KVI-CART, University of Groningen, Groningen, The Netherlands
e-mail: h.tavakolizaniani@rug.nl

H. Tavakoli-Zaniani · M. Eslami-Kalantari
Department of Physics, School of Science, Yazd University, Yazd, Iran

St. Kistryn
Institute of Physics, Jagiellonian University, Kraków, Poland

A. Kozela
Institute of Nuclear Physics, PAN, Kraków, Poland

M. Mohammadi-Dadkan
Department of Physics, University of Sistan and Baluchestan, Zahedan, Iran

R. Ramazani-Sharifabadi
Department of Physics, University of Tehran, Tehran, Iran

E. Stephan
Institute of Physics, University of Silesia, Chorzow, Poland

N. A. Orr et al. (eds.), *Recent Progress in Few-Body Physics*,
Springer Proceedings in Physics 238,
https://doi.org/10.1007/978-3-030-32357-8_65

65.1 Introduction and Experimental Setup

A detailed description of nuclear forces is essential for understanding the properties of nuclei and the dynamics in few-nucleon scattering processes. The need for an additional three-nucleon potential became evident when comparing three-body scattering observables and light-nuclei binding energies with state-of-the-art calculations [1]. In this work, we measured polarisation observables in the proton-deuteron break-up channel with the BINA setup. The detector system BINA is composed of two main parts: the forward wall which can measure the energy, the position, and the type of the particle at scattering angles between 10°–35°. It has three parts, namely E-scintillators, ΔE-scintillators, and a Multi-Wire Proportional Chamber (MWPC). Scattered particles with enough energy traverse from the target to the scintillators. They pass through the MWPC and as a result, their coordinates are recorded. Subsequently, particles pass through the ΔE-scintillators in which a small fraction of their energy is deposited. Finally, the particle is stopped (for protons with an energy less than 140 MeV) inside the E-scintillators and its deposited energy is measured. A combination of E and ΔE allows us to identify the type of particle detected, e.g. proton, deuteron, etc. The backward ball has 149 detectors. It is the scattering chamber and a detector at the same time and particles scattering to angles larger than 35° up to 165° are detected by the ball [2, 3]. The two parts together, therefore, cover almost the entire kinematical phase space of the elastic and break-up reactions.

65.2 Data Analysis

To obtain the analyzing power as a function of the proton energy or, equivalently, as a function of the arc-length, S, along the kinematical-correlation curve, an energy calibration is required between two forward scattered protons in the range where the reaction is being investigated. For a given kinematical configuration the analyzing powers can be obtained by:

$$N(\xi, \phi) = N^0(\xi, \phi)(1 + p_z A_y \cos\phi - p_z A_x \sin\phi), \tag{65.1}$$

where N^s and N^0 are the number of events for a polarized beam and the number of events for an unpolarized beam respectively. The vector polarization of the beam is given by p_z and the vector analyzing powers are indicated by A_y and A_x. Here ϕ is the angle between quantization axis for the polarization and the normal to the scattering plane in the laboratory frame of reference. ξ defines a kinematical point $(\theta_1, \theta_2, \phi_{12}, S)$.

65.3 Results

The analysis of the **p**+d break-up reaction was measured with BINA using a polarized proton beam with an energy of 135 MeV on a liquid deuterium target. Analyzing powers have successfully been measured and are presented as a function of S for different combinations of $(\theta_1, \theta_2, \phi_{12})$. In Fig. 65.1 the filled circles show the vector analyzing powers, A_y and A_x, measured for small, intermediate and large azimuthal opening angles between the two protons. The predictions of the Faddeev calculations based on a nucleon-nucleon (NN) potential and including a three-nucleon potential (3NF) are added to every panel with line colors and styles. The black (solid), the blue (dotted) and red (dash) lines correspond to calculations based on CDB (NN), CDB+Δ (3NF), and CDB+Δ+Coulomb calculations from the Hannover-Lisbon group [4], respectively. The effects of the Coulomb force are generally predicted to be very small for the analyzing power of these configurations as shown in Fig. 65.1.

At small azimuthal opening angles, the predictions based on a NN potential are closest to the data specially for the analyzing powers of A_y, although, the disagreement is still significant. Therefore, the origin of this discrepancy must lie in the treatment of the 3NF. Strikingly, the calculation with an additional 3NF results in an even larger discrepancy between data and theory. At intermediate azimuthal opening angles for A_x there is a better agreement between data and a 3NF calculation. For $\phi_{12} = 180°$, A_x is zero as expected due to parity arguments and the agreement between calculations and data for A_y is reasonably good.

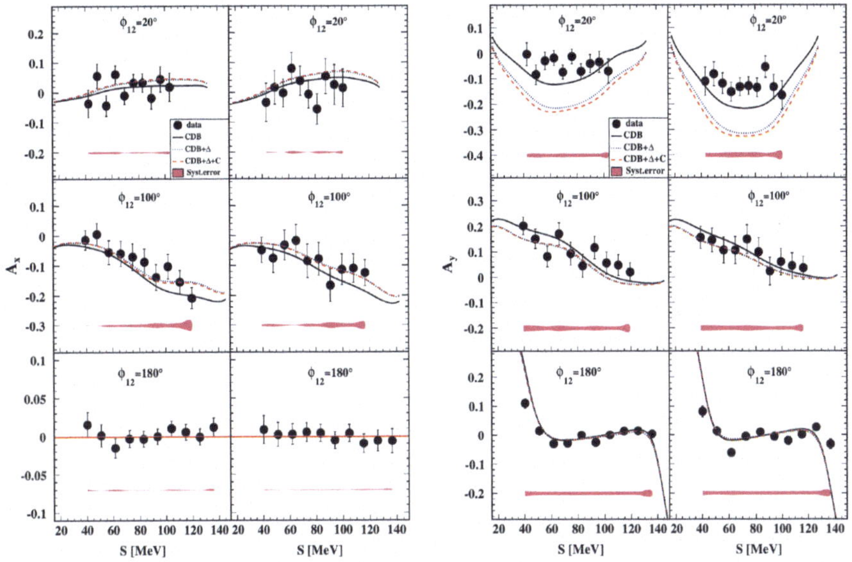

Fig. 65.1 Analyzing powers A_x and A_y as a function of S. For each panel, the left column corresponds to $(\theta_1, \theta_2) = (28°, 24°)$ and the right column corresponds to $(\theta_1, \theta_2) = (28°, 28°)$

In general, the state-of-the-art calculations describe reasonably well the experimental break-up data. Possibly, the modeling of short-range 3NF can be significantly improved. The present data would serve as a good test bench for fine tuning the details of the 3NF.

References

1. Friar, J.L., Payne, G.L., Stoks, V.G.J., de Swart, J.J.: Phys. Lett. B **311** (1993)
2. Witala, H., Glockle, W., Huber, D., Golak, J., Kamada, H.: Phys. Rev. Lett. **81**, 1183 (1993)
3. Shimizu, H., et al.: Nuci. Phys. A **382**, 242 (1993)
4. Deltuva, A., et al.: Phys. Rev. C **71** (2005)

Part III
Few-Nucleon Systems and Qcd Inspired Approaches

Chapter 66
Low-Energy QCD Research at TUNL

Calvin R. Howell, Mohammad W. Ahmed and Werner Tornow

Abstract Experiments are underway at the Triangle Universities Nuclear Laboratory (TUNL) that provide data for evaluating calculations based on low-energy QCD theories, e.g., effective field theory formulations of two-nucleon and three-nucleon interactions, chiral-perturbation theory, and Lattice QCD. Few-nucleon reaction measurements are carried out using the monoenergetic polarized gamma-ray beam at the High Intensity Gamma-ray Source (HIγS) and the neutron beams at the tandem laboratory. Experiments in nucleon structure are performed at HIγS. In this paper we present the status of recent few-nucleon experiments that measure the cross sections for neutron-neutron quasifree scattering in neutron-deuteron breakup and for photodisintegration of ^{3}He.

66.1 Introduction

The mechanisms by which macroscopic properties of nucleon structure and the residual strong nuclear force emerge from color interactions in QCD are not well understood. These phenomena are consequences of quark and gluon interaction dynamics at distances where color forces are strong, i.e., in the non-perturbative regime of QCD. Effective field theories (EFT) and Lattice QCD calculations provide theoretical frames that connect low-energy nuclear phenomena to QCD. Experiments at TUNL are providing data that through analysis using low-energy theories provide insights about nucleon structure, few-nucleon reaction dynamics, and two-nucleon and three-nucleon interactions within the context of few-nucleon reactions. The measurements at TUNL are in two broad categories: (1) Compton scattering on the proton

C. R. Howell (✉) · W. Tornow
Duke University, Durham, NC 27705, USA
e-mail: howell@tunl.duke.edu

M. W. Ahmed
North Carolina Central University, Durham, NC 27707, USA

C. R. Howell · M. W. Ahmed · W. Tornow
Triangle Universities Nuclear Laboratory, Durham, NC 27708, USA

N. A. Orr et al. (eds.), *Recent Progress in Few-Body Physics*,
Springer Proceedings in Physics 238,
https://doi.org/10.1007/978-3-030-32357-8_66

and few-nucleon systems that probe the long-range properties of nucleons, and (2) few-nucleon reactions induced by photons and neutrons.

The aim of the Compton-scattering experiments at HIγS is to determine the dipole electric and magnetic polarizabilities of the proton and neutron to high precision. The few-nucleon measurements performed at HIγS and in the tandem lab at TUNL provide complementary information in evaluating ab-initio theory calculations. The current focus is on three-nucleon (3N) reactions that have sufficient complexity to exhibit influences of 3 N interactions. The first exclusive differential cross sections for photodisintegration of ^{3}He at low energy have been measured at HIγS at an incident beam energy of 15 MeV. The focus of the analysis is on determining the cross section for the $^1S_0$ neutron-proton (*np*) final-state interaction for the purpose of extracting a value for the $^1S_0$ *np* scattering length. The high sensitivity of the scattering length to the strength of the interaction between two nucleons enables using this reaction to probe for 3 N interactions that are not included in the calculations. Also, neutron-induced 3 N reactions in the tandem laboratory at TUNL are used to search for 3 N force effects using *s-wave* neutron-neutron (*nn*) quasi-free scattering (QFS) in neutron-deuteron (*nd*) breakup. These measurements are carried out at incident neutron beam energies of 10 and 16 MeV. This paper will present the status of the experiments at TUNL on photodisintegration of ^{3}He and *nn* QFS in *nd* breakup.

66.2 Exclusive Photodisintegration of ^{3}He

Photonuclear reactions on few-nucleon systems are sensitive to meson exchange currents (MEC) that contribute to pairwise nucleon-nucleon interactions and to 3 N current operators, which are associated with 3 N interactions. HIγS experiment E-13-16 will produce the first cross-section measurements of kinematically complete, i.e., exclusive, three-body photodisintegration of ^{3}He at low energies. A 3D rendering of the experiment setup is shown in Fig. 66.1. An example of the signal-to-background quality of the charged-particle coincidence spectra is illustrated in Fig. 66.2. The data have been collected and analysis is underway to compare experiment to ab-initio 3N calculations by Witala [1] and Deltuva [2].

66.3 Cross-Section Measurements of *nn* QFS in *nd* Breakup

An analysis by Witała and Glöckle [3] showed that the discrepancy between calculations and the cross-section data for *nn* QFS in *nd* breakup can be resolved by increasing the strength of the CD-Bonn *nn s-wave* potential [4] by about 8%, thereby introducing large charge-symmetry breaking in the 2 N interaction. Other possible explanations are that 3 N interactions that are not in current models are important or that the systematic uncertainties of the published data are underestimated. We are performing new *nd* breakup measurements. The calculated cross section at 10

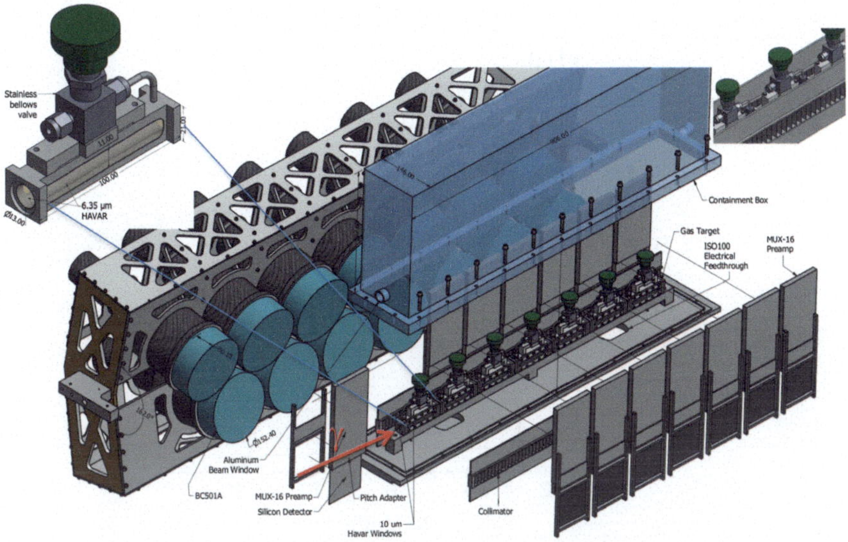

Fig. 66.1 Three-dimensional rendering of the experimental setup for HIγS P-13-16. The setup consists of an array of BC-501 liquid scintillators (12.5 cm dia. × 5 cm thick) for neutron detection, 14 collimated silcon-strip detectors for proton and deuteron detection, and seven ^{3}He gas targets with 10-μm thick Havar windows. Each target is 10 cm long and pressurized with 5 atm of ^{3}He gas. The area of each silicon strip is 5.9 mm × 91.6 mm, and each panel has 16 strips. The neutron detectors are shown only on one side of the beam axis. The average neutron flight path is 42 cm

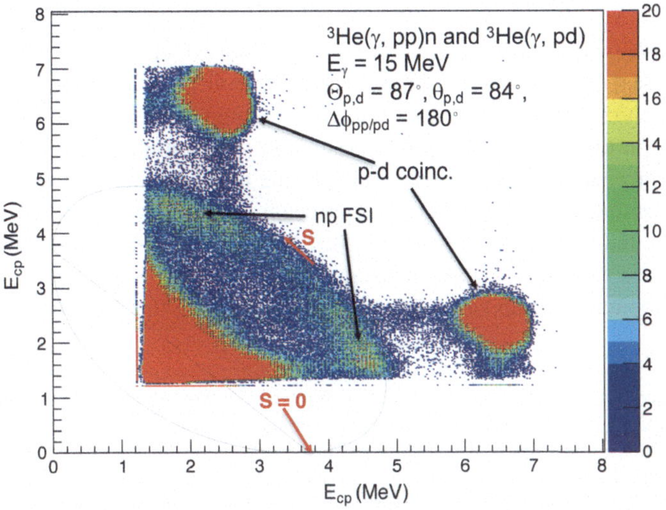

Fig. 66.2 Two-dimensional histogram of the energies for the two charged particles detected in coincidence in photodisintegration of ^{3}He in HIγS experiment P-13-16. The particle scattering angles are given in the plot. The peaks for 2-body photodisintegration and the locus for 3-body photodisintegration are indicated. Also, the cross-section enhancements for the *np* final-state interactions (FSI) are identified

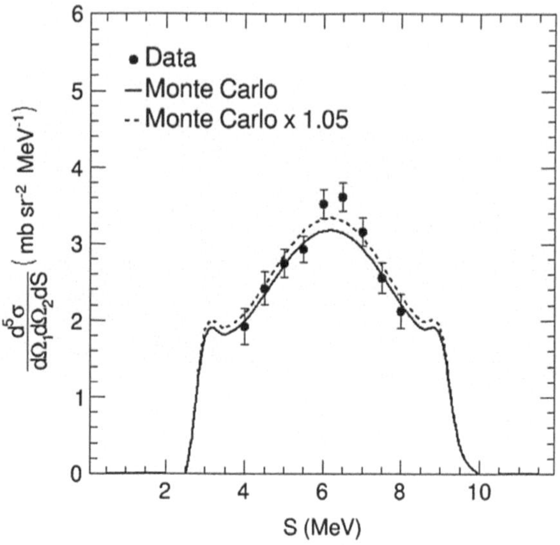

Fig. 66.3 A plot of the preliminary TUNL cross-section data for *nn* QFS in *nd* breakup as a function of the kinematic S curve for incident neutron beam energy of 10 MeV and symmetric *nn* coincidence angles of 36.7°. The error bars represent statistical uncertainties only. The solid curve is a Monte-Carlo simulation based on ab-initio *nd* breakup calculations of Witała that use the CD-Bonn nucleon-nucleon potential model [4]. The dashed curve is the solid curve multiplied by a factor of 1.05 to obtain the minimum chi-square fit to the data with the simulated cross section

MeV is relatively insensitive to details of the *nn* interaction; the measurements at this energy are used to validate our techniques. The data collection at 10 MeV is complete and data analysis is underway. A plot of our preliminary data is shown in Fig. 66.3 in comparison to Monte-Carlo simulations based on ab-initio *nd* calculations. The systematic error in the data is ±6%, which is due mostly to the uncertainties in the neutron detection efficiency and the target-beam luminosity. The good agreement between the data and calculations indicates that our experimental techniques are solid. The next step is to perform measurements at 16 MeV, where the sensitivity to the *nn* interaction strength should be adequate to validate the data around 26 MeV [5, 6].

Acknowledgements This work is supported in part by the U.S. Department of Energy under grant Nos. DE-FG02-97ER41033 and DE-SC0005367. The contributions of the following people are appreciated: L.C. Cumberbatch, B.A. Fallin, F.Q.L. Friesen, C. Malone, and R.C. Malone, Duke University; D.R. Ticehurst, University of NC at Chapel Hill; D. Markoff and B. Crowe, NC Central University; H. Witała, Jagiellonian University; and A. Deltuva, Vilnius University.

References

1. Skibinski, R., Golak, J., Witała, H., Gloeckle, W., Kamada, H., Nogga, A.: Phys. Rev. C **67**, 054002 (2003)
2. Deltuva, A., Fonseca, A.C., Sauer, P.U.: Phys. Rev. C **80**, 064004 (2009)
3. Witała, H., Glöckle, W.: Phys. Rev. C **83**, 034004 (2011)
4. Machleidt, R.: Phys. Rev. C **63**, 024001 (2001)
5. Siepe, A., et al.: Phys. Rev. C **65**, 034010 (2002)
6. Ruan, X., et al.: Phys. Rev. C **75**, 057001 (2007)

Chapter 67
How to Use Renormalization Group Analysis in Lattice Nuclear Effective Field Theory

Koji Harada, Satoru Sasabe and Masanobu Yahiro

Abstract We propose a new approach to Nuclear Effective Field Theory (NEFT) on a lattice on the basis of Renormalization Group (RG) analysis. In order to perform Markov-chain Monte Carlo lattice simulation of NEFT, we introduce auxiliary fields to integrate nucleon field so that its effects are represented as a determinant. The problem is that the determinant becomes complex and cannot be considered as a part of probability distribution function. We introduce a reweighting method, in which the reference determinant is chosen to be optimal in the RG analysis sense: the reference determinant contains only the relevant interactions and the closest to the original determinant. We calculate the standard deviation of the absolute value of the reweighting factor in a simple model, isospin-symmetric S-wave NLO NEFT without pions, and explain why our choice is optimal.

67.1 Introduction: Sign Problem and the Reweighting Method

Nuclear Effective Field Theory (NEFT) is a low-energy effective field theory in which nucleons (and optionally pions, hyperons, etc.) are fundamental degrees of freedom instead of quarks and gluons. See [1] for a review. In recent years, several lattice simulations have been done. See [2] for an early review, and [3] for a recent review. Note that most of recent calculations are based on Quantum Monte Carlo algorithms. In this contribution we concentrate on the Markov-chain Monte Carlo (MCMC) algorithm, which is the most common in lattice QCD simulations.

In order to perform MCMC simulations, fermion fields must be integrated over. In NEFT, nucleons interact themselves through contact interactions (as well as pion exchanges when pions are included). By introducing auxiliary fields and making

K. Harada (✉)
Faculty of Arts and Science, Kyushu University, Fukuoka 819-0395, Japan
e-mail: harada@artsci.kyushu-u.ac.jp

S. Sasabe · M. Yahiro
Department of Physics, Kyushu University, Fukuoka 819-0395, Japan

N. A. Orr et al. (eds.), *Recent Progress in Few-Body Physics*,
Springer Proceedings in Physics 238,
https://doi.org/10.1007/978-3-030-32357-8_67

Hubbard-Stratonovich transformations, the nucleon part of the action is written as a bilinear form, so that nucleon field can be integrated to produce a determinant. In the MCMC simulations, the determinant is regarded as a part of the probability distribution function, which must be positive definite.

The reality and the positivity of the determinant depend on the coupling constants of the contact interactions of the original action. For the realistic case, the reality of the determinant turns out to be lost. This is a kind of "sign" problem.

To be concrete, let us consider a simple model: the isospin-symmetric S-wave NLO NEFT without pions. Here "NLO" means that the Lagrangian contains operators with up to including two derivatives. The Euclidean Lagrangian is given by

$$\begin{aligned}\mathcal{L}_{\mathrm{E}} &= N^{\dagger}\left(-\partial_4 + \frac{\nabla^2}{2M} + \mu\right)N - c_0(N^{\dagger}\tilde{\tau}N)^2 \\ &+ c_2\Bigg[\left\{N^{\dagger}\tilde{\tau}\left(\overleftarrow{\nabla}^2 + \overrightarrow{\nabla}^2\right)N\right\}\cdot(N^{\dagger}\tilde{\tau}N) + (N^{\dagger}\tilde{\tau}N)\cdot\left\{N^{\dagger}\tilde{\tau}\left(\overleftarrow{\nabla}^2 + \overrightarrow{\nabla}^2\right)N\right\} \\ &- \left\{(N^{\dagger}\tilde{\tau}\overleftarrow{\nabla}N) - (N^{\dagger}\tilde{\tau}\overrightarrow{\nabla}N)\right\}^2 + (N^{\dagger}\tilde{\tau}N)\cdot\nabla^2(N^{\dagger}\tilde{\tau}N)\Bigg], \end{aligned} \tag{67.1}$$

where $\tilde{\tau} = (I_{2\times 2}, i\tau^1, i\tau^2, i\tau^3)$. By introducing auxiliary fields ϕ_0 and $\overrightarrow{\phi}_2$, the contact interactions can be written in nucleon bilinear forms. After integrating the nucleon field, we end up with the following determinant:

$$\begin{aligned}\det\left[M(\phi_0, \overrightarrow{\phi}_2)\right] = \det\Bigg[&\left(-\partial_4 + \frac{\nabla^2}{2M} + \mu\right) - 2i\sqrt{c_0}\phi_0\cdot\tilde{\tau} + 2i\frac{c_2}{\sqrt{c_0}}\phi_0\cdot\tilde{\tau}\left(\overleftarrow{\nabla}^2 + \overrightarrow{\nabla}^2\right) \\ &+ i\frac{c_2}{\sqrt{c_0}}\phi_0\cdot\tilde{\tau}\nabla^2 - 2i\sqrt{c_2}\overrightarrow{\phi}_2\cdot\left(\tilde{\tau}\overleftarrow{\nabla} - \tilde{\tau}\overrightarrow{\nabla}\right)\Bigg]. \end{aligned} \tag{67.2}$$

It is now clear that the determinant is not real in the physical case where $c_0 < 0$ and $c_2 > 0$. See [7] and Fig. 67.1.

A well-known strategy to circumvent this problem is the reweighting method, in which a positive definite reference determinant det M_{R} is introduced and the expectation value of an operator $\mathcal{O}(x)$ is rewritten (schematically) as

$$\langle\mathcal{O}(x)\rangle = \frac{\int d\Phi\ \mathcal{O}(x)\left(\frac{\det M(\Phi,\mu)}{\det M_{\mathrm{R}}(\Phi,\nu)}\right)\det M_{\mathrm{R}}(\Phi,\nu)e^{-\int\Phi^2}}{\int d\Phi\left(\frac{\det M(\Phi,\mu)}{\det M_{\mathrm{R}}(\Phi,\nu)}\right)\det M_{\mathrm{R}}(\Phi,\nu)e^{-\int\Phi^2}}, \tag{67.3}$$

where Φ (collectively) denotes the auxiliary fields and $(\det M_{\mathrm{R}})e^{-\int\Phi^2}$ is regarded as the probability distribution function in the MCMC algorithm.

The question now is: what is the optimal choice of the reference determinant? The answer is provided by RG analysis.

67.2 RG Analysis and the Choice of the Reference Determinant

It is important to notice that NEFT for the realistic case is close to the nontrivial fixed point (FP) of the RG transformations [4–6]. The closeness is related to the unnaturally large scattering lengths in the S-waves.

Operators around a FP are classified according to the behaviors under RG transformations. Relevant operators (ROs) get more important in the IR, while irrelevant operators (IOs) get less. In the S-wave NEFT, there is only one RO. In the effective range expansion (ERE) language, it corresponds to the scattering length. Other operators are irrelevant, and correspond to other ERE parameters.

In a lattice formulation, the inverse of the lattice constant plays a role of (floating) momentum cutoff. If one changes the lattice constant, one needs to change the coupling constants on the lattice so that the physical quantities do not change. This change of coupling constants is a RG transformation.

Even though the existence of FPs and the scaling dimensions are independent of the regularization, the location of the FPs and the explicit forms of ROs and IOs depend on the specific regularization one employs. We have performed RG analysis for the S-wave NEFT without pions in [7]. We find that the physical system corresponds to the case $c_0^{\rm phys} < 0$ and $c_2^{\rm phys} > 0$ with the lattice constant $a = 5$ fm, which corresponds to the momentum cutoff $\Lambda = \pi/a \approx 124$ MeV.

With lattice regularization, the RO is *not* the term without derivatives (c_0-term), but a certain linear combination of operators with and without derivatives. It is shown graphically as the direction of the RG flow in the coupling constant space flown out of the nontrivial FP.

Since the RO is important in the IR, we propose to use the reference determinant that contains the RO only.

67.3 Results

We perform MCMC simulations with the number of sites $N^4 = 4^4$ and the lattice constant $a = 5$ fm. Totally 250,000 configurations are generated and 500 configurations are used for thermalization. Measurements are performed for every 100 configurations.

We calculate the standard deviation (SD) of the absolute value of the reweighting factor $\left|\frac{\det M}{\det M_{\rm R}}\right|$ as a measure of the "similarity" of the two determinants. The measurements are done for seven points (I1,. . .,I7) in the irrelevant direction, and four points (R1,. . .,R4) in the relevant direction, as shown in Fig. 67.1. The point I7 corresponds to the physical situation. In Fig. 67.1, the red thick line shows the critical line on which the scattering length is infinite and the blue bullet on it is the nontrivial FP where the effective range is zero. The green dashed line indicate the relevant direction. The magenta thin dotted line is the RG flow of the physical system. (It flows

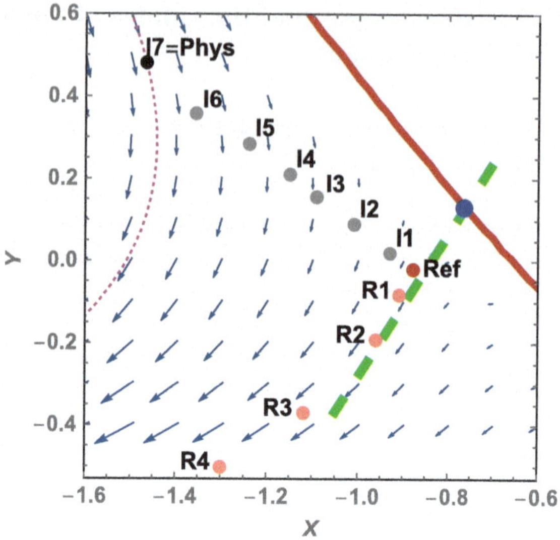

Fig. 67.1 RG flow in the strong-coupling phase near the nontrivial FP. X and Y are dimensionless coupling constants defined in [7]

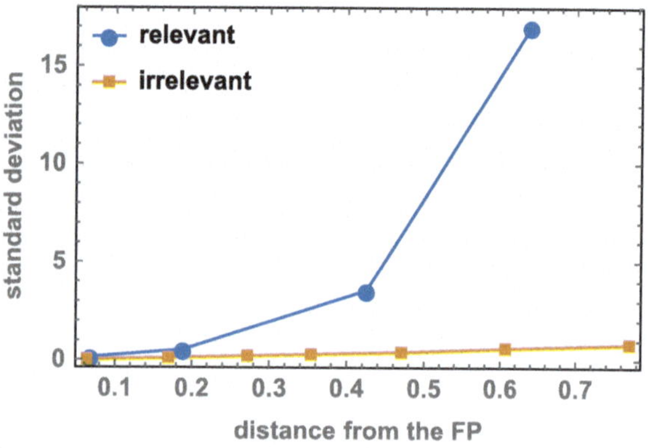

Fig. 67.2 The SD of the absolute value of the reweighting factor when the chemical potential $\nu \approx -16$ MeV. The horizontal axis shows the (Euclidean) distance from the nontrivial FP in the X-Y plane

downward as the lattice constant gets larger.) The reference point is shown as a red bullet.

Figure 67.2 shows the SD of the absolute value of the reweighting factor when the chemical potential $\nu \approx -16$ MeV. The horizontal axis shows the (Euclidean) distance from the nontrivial FP in the X-Y plane. It is easily seen that the SD grows rapidly as one goes in the relevant direction, while it grows much slower in the irrelevant direction. It clearly illustrates the difference between the RO and the IO: the RO changes the theory strongly while the IO does not change much. It is the IO that can be regarded as perturbation.

It is observed that the growth of the SD is faster as the chemical potential increases.

67.4 Summary

We have shown that the RO changes the theory very much while the IO does not. It conforms with the general concept of the relevant and the irrelevant operators in the RG analysis. It is therefore legitimate and optimal to choose the reference determinant that contains only the RO part of the original determinant in the reweighting method.

Acknowledgements This work was supported by JSPS KAKENHI Grants No. 15K05082 (K.H), No. 26-5861 (S.S), and No. 26400278 (M.Y.).

References

1. Epelbaum, E., Hammer, H.-W., Meissner, U.-G.: Rev. Mod. Phys. **81**, 1773 (2009)
2. Lee, D.: Prog. Part. Nucl. Phys. **63**, 117 (2009)
3. Elhatisari, S.: Plenary talk in this proceedings
4. Weinberg, S.: Phys. Lett. B **251**, 288 (1990)
5. Birse, M.C., McGovern, J.A., Richardson, K.G.: Phys. Lett. B **464**, 169 (1999)
6. Harada, K., Kubo, H.: Nucl. Phys. B **758**, 304 (2006)
7. Harada, K., Sasabe, S., Yahiro, M.: Phys. Rev. C **94**, 024004 (2016)

Chapter 68
Polarisabilities from Compton Scattering on ^{3}He

Harald W. Grießhammer and Judith A. McGovern

Abstract This executive summary of recent theory progress in Compton scattering off ^{3}He focuses on determining neutron polarisabilities; see Margaryan et al (Eur Phys J A 54:125, 2018) and references therein for details and a better bibliography.

Two plenaries discuss the large-scale international effort, gains and goals of a new generation of high-precision facilities to extract nucleon polarisabilities from Compton scattering experiments, and show that determining them by experiments takes years of planning, execution and analysis—and commensurate theory support. Others highlight the importance of electromagnetic polarisabilities in many contexts. We thus refer to all these contributions [2] for motivation and context, and concentrate on theory progress for one target nucleus: ^{3}He.

Setting the Stage Low-energy Compton scattering $\gamma X \to \gamma X$ probes a target's internal degrees of freedom in the electric and magnetic fields of a real photon. These fields induce radiation multipoles by displacing the target constituents. The angular and energy dependence of the emitted radiation encodes information from the symmetries and strengths which govern the interactions of the constituents with each other and with photons. After subtracting the "Born contributions" (known from one-photon data like form factors), its multipoles parametrise the stiffness of a nucleon N (spin $\frac{\sigma}{2}$) against transitions $Xl \to Yl'$ at frequency ω ($l' = l \pm \{0; 1\}$; $X, Y = E, M$; $T_{ij} = \frac{1}{2}(\partial_i T_j + \partial_j T_i)$; $T = E, B$):

$$2\pi N^\dagger \big[\alpha_{E1}(\omega)\mathbf{E}^2 + \beta_{M1}(\omega)\mathbf{B}^2 + \gamma_{E1E1}(\omega)\boldsymbol{\sigma}\cdot(\mathbf{E}\times\dot{\mathbf{E}}) + \gamma_{M1M1}(\omega)\boldsymbol{\sigma}\cdot(\mathbf{B}\times\dot{\mathbf{B}}) \\ - 2\gamma_{M1E2}(\omega)\sigma^i B^j E_{ij} + 2\gamma_{E1M2}(\omega)\sigma^i E^j B_{ij} + \text{(higher multipoles)}\big] N.$$

H. W. Grießhammer (✉)
Department of Physics, The George Washington University, Washington, DC, USA
e-mail: hgrie@gwu.edu

J. A. McGovern
School of Physics and Astronomy, The University of Manchester, Manchester, UK

N. A. Orr et al. (eds.), *Recent Progress in Few-Body Physics*,
Springer Proceedings in Physics 238,
https://doi.org/10.1007/978-3-030-32357-8_68

Six two-photon response functions suffice up to about 400 MeV: two scalar polarisabilities $\alpha_{E1}(\omega)$ and $\beta_{M1}(\omega)$ for electric and magnetic dipole transitions; and the four dipole spin-polarisabilities $\gamma_{E1E1}(\omega)$, $\gamma_{M1M1}(\omega)$, $\gamma_{E1M2}(\omega)$, $\gamma_{M1E2}(\omega)$. These test the nucleon-spin structure and complement information from Jefferson Lab's spin programme. Intuitively, the electromagnetic field of the spin degrees causes bi-refringence in the nucleon, like in the classical Faraday-effect.

The static values, $\alpha_{E1} \equiv \alpha_{E1}(\omega = 0)$ etc., are often just called *"the" polarisabilities* and condense the rich information on the pion cloud, on the $\Delta(1232)$ excitation, and on the interplay between chiral symmetry breaking and short-distance interactions. These fundamental quantities provide stringent tests for theoretical descriptions of hadron structure. Moreover, they are ingredients to the neutron-proton mass difference, the proton charge-radius puzzle, and the Lamb shift of muonic hydrogen. To extract them, one must reliably extrapolate from data to $\omega = 0$. Since pure neutron targets are unfeasible, nuclear binding and meson-exchange effects must also be subtracted with reliable theory uncertainties. Fortunately, Chiral Effective Field Theory (χEFT) provides model-independent estimates of higher-order corrections and encodes the correct low-energy dynamics of QCD. For few-nucleon systems, it consistently incorporates hadronic and nuclear currents, rescattering effects and wave functions. The photon's interaction with the charged pion-exchange between nucleons also probes few-nucleon binding. Even if scattering on a free neutron were feasible, cross sections and signals for coherent scattering from nuclei are markedly larger.

Elastic Compton scattering from ^3He is a promising means to access neutron polarisabilities. In [3] and subsequent publications, Shukla *et al.* showed that the differential cross section between 50 and 120 MeV is sensitive to the electric and magnetic dipole polarisabilities of the neutron, $\alpha_{E1}^{(n)}$ and $\beta_{M1}^{(n)}$, and that scattering on polarised ^{3}He provides good sensitivity to the neutron spin polarisabilities. This triggered several approved proposals at MAMI and HIγS.

We recently extended these χEFT predictions by one order to N^{3}LO [$\mathcal{O}(e^2\delta^3)$] by adding a dynamical Delta degree of freedom, and provided results for photon lab energies between 50 and 120 MeV for the differential cross section, for the beam asymmetry Σ_3, and for the two double asymmetries with circularly polarised photons and transversely or longitudinally polarised targets, Σ_{2x} and Σ_{2z}. These are the only non-zero observables below pion-production threshold in our formulation. We also found that the pioneering results were obtained from a computer code which contained mistakes, triggering an erratum to [3].

At such energies, the complete photonuclear operator at N^{3}LO [$\mathcal{O}(e^2\delta^3)$] is: the Thomson and other minimal-substitution terms; magnetic-moment couplings; dynamical single-nucleon effects such as virtual pion loops and the Delta excitation; and couplings of photons to the charged-pion exchange. All terms are evaluated with ^{3}He wave functions found from the same χEFT expansion.

Results The dynamical Delta effects are obvious in all observables for $\omega_{\text{lab}} \gtrsim$ 100 MeV; see Fig. 68.1. They markedly invert the fore-aft asymmetry of the cross section and increase the magnitude of double asymmetries and their sensitivity to

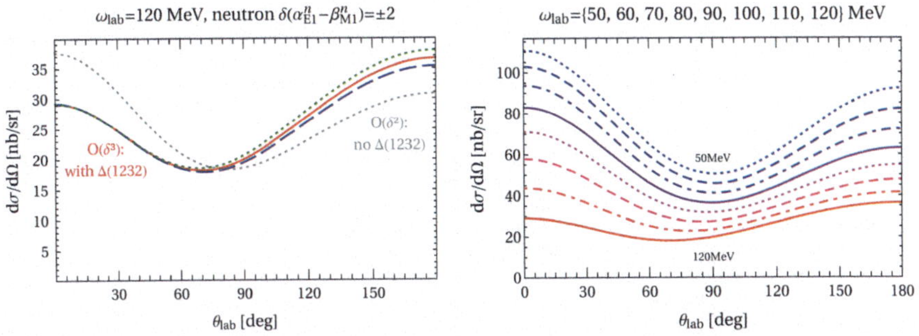

Fig. 68.1 Differential cross section, adapted from [1]. Left: at $\mathcal{O}(e^2\delta^2)$ [no Delta] and $\mathcal{O}(e^2\delta^3)$ [with Delta], and sensitivity to the neutron's scalar polarisabilities. Right: Energy dependence

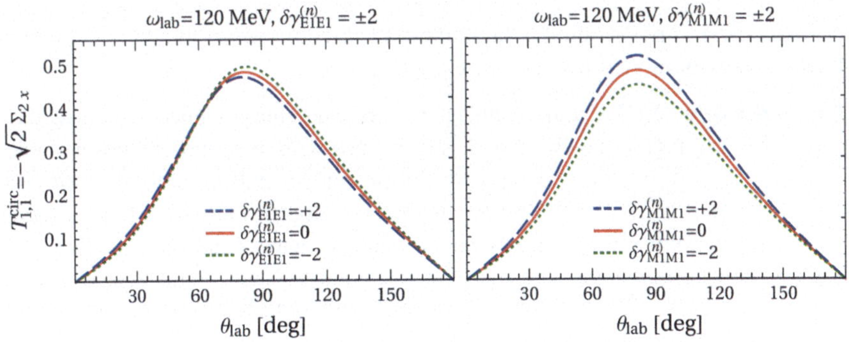

Fig. 68.2 The sensitivity of Σ_{2x} on the two spin polarisabilities with the biggest impact, adapted from [1]

spin polarisabilities, echoing similar findings for the deuteron. The chiral expansion converges in this energy range quite well; see e.g. Fig. 68.1. The dependence on the choice of the ^{3}He wave function is small and can usually be distinguished from the effects of polarisabilities by a different angular dependence.

We found that $\alpha^{(n)}_{E1} - \beta^{(n)}_{M1}$ can be extracted from the cross section; Σ_{2x} has a non-degenerate sensitivity to $\gamma^{(n)}_{M1M1}$ around 90°; and Σ_{2z} to $\gamma^{(n)}_{E1E1}$ and $\gamma^{(n)}_{E1M2}$; see Fig. 68.2. The beam asymmetry Σ_3 is dominated by the single-nucleon Thomson term and not very useful to directly determine polarisabilities. Ultimately, the most accurate polarisabilities will be inferred from data of all four observables. For the spin polarisabilities, data at $\omega_{lab} \gtrsim 100$ MeV will be crucial.

This exploration is part of an ongoing dialogue with our experimental colleagues on the best kinematics and observables to extract neutron polarisabilities. An interactive *Mathematica* notebook is available from `hgrie@gwu.edu`. Results are quite robust. Varying the single-nucleon amplitudes of complementary approaches like dispersion relations will lead to sensitivities which are hardly discernible from ours.

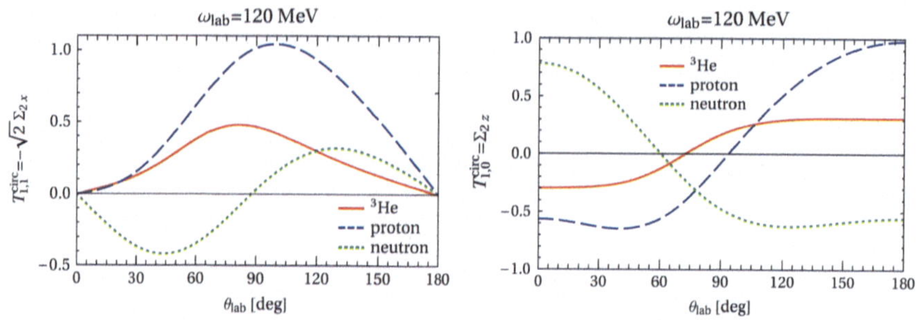

Fig. 68.3 Double asymmetries with circularly polarised beam and transversely (left) or longitudinally (right) polarised target, for proton, neutron and ^{3}He, adapted from [1]

Once data exist, a polarisability extraction will of course need to address residual uncertainties with more diligence; see e.g. [4].

Nuclear Binding χEFT also quantifies the angle- and energy-dependent corrective to the naïve ^{3}He picture as the sum of two protons with antiparallel spins and one neutron. Sensitivity to the scalar polarisabilities enters indeed approximately via $2\alpha_{E1}^{(p)} + \alpha_{E1}^{(n)}$ and $2\beta_{M1}^{(p)} + \beta_{M1}^{(n)}$, and the double-asymmetries are 10–20 times more sensitive to the spin polarisabilities of the neutron than of the proton. However, Fig. 68.3 confirms that there is no energy where polarised ^{3}He simply acts as a "free neutron-spin target". The *sensitivities* to neutron spin polarisabilities closely mimic those of free-neutron observables. But their *magnitudes* do not.

An impulse approximation would thus omit a key mechanism: charged pion-exchange currents. Without their large interference with the polarisabilities, results are severely distorted. The χEFT expansion provides quantitative predictions of the two-body currents, with reliable theory uncertainties. Detailed checks of the convergence of the expansion for exchange currents and for the other pieces of the ^{3}He-Compton amplitude by performing a N^{4}LO [$\mathcal{O}(e^2\delta^4)$] calculation and extending the applicable energy range are under way. They will allow for even more accurate extractions of polarisabilities from upcoming data.

Acknowledgements HWG is particularly indebted to JMcG for filling in at the oral presentation on very short notice. We are also grateful to the other co-authors of [1]. This work was supported in part by the US Department of Energy under contract DE-SC0015393 (HWG), and by UK Science and Technology Facilities Council grants ST/L005794/1 and ST/P004423/1 (JMcG).

References

1. Margaryan, A., et al.: Eur. Phys. J. A **54**, 125 (2018). https://doi.org/10.1140/epja/i2018-12554-x. arXiv:1804.00956 [nucl-th]
2. See the contributions to these proceedings by J. McGovern and C. Howell, as well as by M. Elihau, R. Higa, R. Pohl and M. S. Safronova

3. Choudhury, D., et al.: Phys. Rev. Lett. **98**, 232303 (2007). https://doi.org/10.1103/PhysRevLett.120.249901[nucl-th/0701078]; erratum ibid. **120**, 249901 (2018). https://doi.org/10.1103/PhysRevLett.98.232303. arXiv:1804.01206 [nucl-th]
4. Grießhammer, H.W., et al.: Eur. Phys. J. A **52**, 139 (2016). https://doi.org/10.1140/epja/i2016-16139-5. arXiv:1511.01952 [nucl-th]

Chapter 69
Relation Between Scattering Amplitude and Bethe-Salpeter Wave Function Inside Interaction Range

Takeshi Yamazaki

Abstract We discuss an exact relation between the two-particle scattering amplitude and the Bethe-Salpeter (BS) wave function inside the interaction range in quantum field theory. The reduced BS wave function, which is defined by the BS wave function inside the interaction range, plays an essential role in the relation. The half off-shell scattering amplitude can be obtained from the relation as well as the on-shell scattering amplitude. The relation can be regarded as the LSZ reduction formula using the BS wave function. The corresponding relation on the lattice can be derived with the condition that the interaction range is less than the half of the spatial extent. In order to apply the relation on the lattice to the S-wave isospin $I = 2$ two-pion scattering channel, we calculate the BS wave function in the quenched lattice QCD. It is found that the scattering length obtained from the on-shell scattering amplitude agrees with that obtained from the finite volume method. We also calculate the half off-shell amplitude for the first time, and observe a clear signal of the amplitude in a wide range of the momentum. From the slope of the half off-shell amplitude, the effective range is estimated using two assumptions. Although the half off-shell amplitude is not an observable in experiment, it could be useful to understand property of hadron scatterings.

69.1 Introduction

The scattering phase shift $\delta(k)$ has been calculated from the finite volume method [1] in various lattice QCD calculations. The finite volume formula in the method is a relation between $\delta(k)$ and the two-particle energy on finite volume of L^3. The formula was derived from a two-particle wave function outside the two-particle interaction

T. Yamazaki (✉)
Faculty of Pure and Applied Sciences, University of Tsukuba, Tsukuba, Ibaraki 305-8571, Japan
e-mail: yamazaki@het.ph.tsukuba.ac.jp

Center for Computational Sciences, University of Tsukuba, Tsukuba, Ibaraki 305-8577, Japan

N. A. Orr et al. (eds.), *Recent Progress in Few-Body Physics*, Springer Proceedings in Physics 238,
https://doi.org/10.1007/978-3-030-32357-8_69

range R in quantum mechanics. In quantum field theory the same formula was obtained by a similar discussion with the Bethe-Salpeter (BS) wave function outside R [2, 3]. It means that relation between the BS wave function outside R and the scattering amplitude is well understood in the finite volume method.

On the other hand, in quantum field theory relation between the BS wave function inside R and the scattering amplitude is poorly known. Only a method using an effective potential to obtain $\delta(k)$ was proposed [4] based on a discussion in quantum mechanics. In this report, we discuss an exact relation between the inside BS wave function in the infinite volume and the scattering amplitude in quantum field theory. This relation gives both the on-shell and half off-shell amplitudes. Furthermore, we extend the relation to that on the lattice and perform an exploratory calculation of the scattering amplitude with the relation in the S-wave isospin $I = 2$ two-pion scattering channel. Almost all the results in this report have been already presented in the two papers [5, 6].

69.2 Relation in the Infinite Volume

The BS wave function in the infinite volume is defined through the LSZ reduction formula [2, 3] as,

$$\phi(x;k) = e^{i\mathbf{k}\cdot\mathbf{x}} + \int \frac{d^3p}{(2\pi)^3} \frac{H(p;k)}{p^2 - k^2 - i\varepsilon} e^{i\mathbf{p}\cdot\mathbf{x}}, \tag{69.1}$$

where $\mathbf{x}$ is the relative coordinate of two particles, k is the relative momentum determined from the two-particle energy $E_k = 2\sqrt{m^2 + k^2}$ with m being the mass of the particle. $H(p;k)$ is the half off-shell amplitude. Here we assume the S-wave scattering of spinless two particles and neglect inelastic scattering contributions. At the on-shell $p = k$, the on-shell amplitude $H(k;k)$ is written by $\delta(k)$,

$$H(k;k) = \frac{4\pi}{k} e^{i\delta(k)} \sin\delta(k). \tag{69.2}$$

The interaction range R is determined from the reduce BS wave function given by

$$h(x;k) = (\Delta + k^2)\phi(x;k). \tag{69.3}$$

It is assumed that $h(x;k)$ is non-zero in the interacting region of $x < R$, while $h(x;k)$ is zero in $x > R$. This property is similar to a potential in quantum mechanics. Substituting (69.1) to (69.3) and performing the Fourier transformation, an exact relation of $H(p;k)$ written by $h(x;k)$ is obtained as

$$H(p;k) = -\int d^3x\ h(x;k) e^{-i\mathbf{p}\cdot\mathbf{x}}. \tag{69.4}$$

The integrand gives non-zero value only in the interacting region. Therefore, it is regarded as a relation between $\phi(x;k)$ inside R and $H(p;k)$, since the integrand is given by $\phi(x;k)$ through (69.3). It is noted that this relation is useful to discuss uncertainties in the effective potential method as presented in [5, 7].

Using the partial integration and the definition of $h(x;k)$ in (69.3), the exact relation (69.4) can be expressed by the BS wave function in the momentum space $\phi(p;k) = \int d^3x\, \phi(x;k)e^{-i\mathbf{x}\cdot\mathbf{p}}$ as,

$$H(p;k) = (p^2 - k^2)\phi(p;k). \tag{69.5}$$

This formula can be regarded as the LSZ reduction formula in the relative coordinate [8], since it gives the on-shell amplitude in the on-shell limit, $p \to k$. A similar discussion with the four-dimensional BS wave function in the momentum space was reported in [9].

69.3 Relation on the Lattice

We will discuss an extension of (69.4) to the one on the lattice of L^3. The integration range in (69.4) can be changed from $[-\infty,\infty]^3$ to $[-X, X]^3$, if $X > R$ where $h(X;k) = 0$. Thus, the following summation gives the S-wave half off-shell amplitude on the lattice $H_L(p;k)$, which is the same as $H(p;k)$ except for the overall factor,

$$H_L(p;k) = -\sum_{\mathbf{x}\in L^3} h_L(x;k)\, j_0(px), \tag{69.6}$$

where $j_0(px)$ is the spherical Bessel function of $l = 0$. The reduced BS wave function on the lattice $h_L(x;k)$ is defined by the BS wave function on the lattice $\phi_L(\mathbf{x};k)$ in a similar way to the one in the infinite volume (69.3). A sufficient condition of (69.6) is $L/2 > R$ as in the finite volume method [1]. The overall constant is removed by taking appropriate ratios such as $H_L(k;k)/\phi_L(\mathbf{x};k)$ and $H_L(p;k)/H_L(k;k)$. We cannot obtain the phase of the scattering amplitude from the ratios. This is a similar situation to the finite volume method.

The corresponding relation to (69.5) on the lattice is obtained from (69.6) by the partial integration on the lattice. In general $\phi_L(\mathbf{x};k)$ is non-zero on the boundary of the lattice, so that a surface term appears as

$$H_L(p;k) = (p^2 - k^2)\sum_{\mathbf{x}\in L^3} \phi_L(\mathbf{x};k)\, j_0(px) + (\text{surface term}), \tag{69.7}$$

where the surface term depends on p, k, and L. It is noted that the on-shell amplitude is given by only the surface term, because the first term vanishes at the on-shell. The explicit form of the surface term will be reported in other paper [8].

Using the on-shell amplitude $H_L(k;k)$, $\tan\delta(k)$ is given by

$$\tan\delta(k) = \frac{\sin(kx_{\text{ref}})}{4\pi x_{\text{ref}}\frac{\phi_L(\mathbf{x}_{\text{ref}};k)}{H_L(k;k)} - \cos(kx_{\text{ref}})}, \tag{69.8}$$

where $\mathbf{x}_{\text{ref}}$ is a reference position which satisfies $x_{\text{ref}} > R$. In the equation we use the relation

$$\frac{H_L(p;k)}{\phi_L(\mathbf{x}_{\text{ref}};k)} = \frac{4\pi x_{\text{ref}}\sin\delta(k)}{\sin(kx_{\text{ref}} + \delta(k))}, \tag{69.9}$$

where we assume that $\phi_L(\mathbf{x}_{\text{ref}};k)$ has only the S-wave contribution.

69.4 Lattice QCD Calculation

As a pilot study to apply the formula (69.6) to a hadron scattering, we calculate the two-pion BS wave function of the ground state in the S-wave $I = 2$ scattering with the quenched approximation at $m_\pi = 0.86$ GeV [6]. The lattice size is $L^3 \times T = 24^3 \times 64$. From $\mathbf{x}$ dependence of $h_L(x;k)$, we estimate $R \sim 10$ which is less than the half of the spatial extent. It suggests that the sufficient condition to use (69.6) is satisfied in our data.

We evaluate $\tan\delta(k)$ from the on-shell amplitude using (69.8) with $x_{\text{ref}} = (12, 7, 2)$. The scattering length a_0 is estimated from $\tan\delta(k)$ with an approximation $a_0 \sim \tan\delta(k)/k$, since the ground state of the two-pion scattering has tiny momentum $k^2 = 0.001549(45)$ GeV^2 determined from the two-pion energy E_k. Figure 69.1 compares our result of a_0/m_π with the one from the finite volume method and also the one in [3]. The consistency of our result with those results suggests that the formula (69.6) could be an alternative method to calculate $\delta(k)$ to the finite volume method.

Similarly to $H(k;k)$, we also calculate the half off-shell amplitude $H(p;k)$ with (69.6). The result of $H(p;k)/H(k;k)$ is plotted as a function of p^2 in Fig. 69.2.

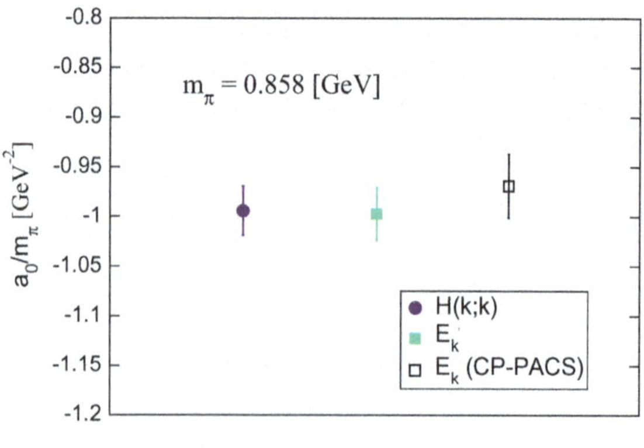

Fig. 69.1 Comparison of a_0/m_π adapted from [6]. Filled circle and square symbols represent the results evaluated from the on-shell amplitude $H(k;k)$ and the finite volume method with E_k, respectively. Open symbol denotes the result from the finite volume method by CP-PACS Collaboration [3]

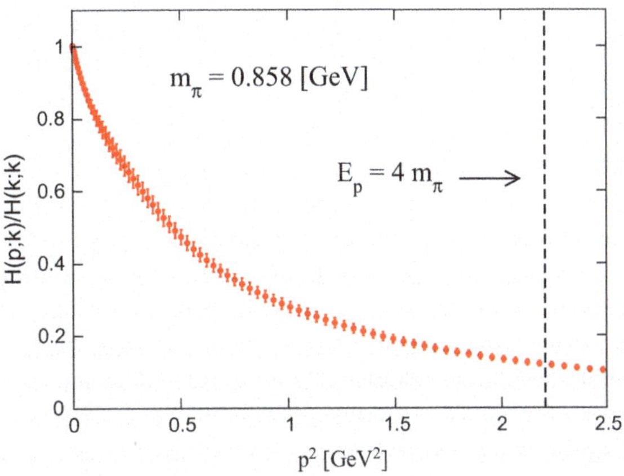

Fig. 69.2 Ratio of half off-shell to on-shell amplitudes $H(p;k)/H(k;k)$ as a function of p^2 adapted from [6]. The horizontal dotted line represents the momentum at the inelastic threshold

A clear signal of the ratio is obtained even in a higher momentum region than the inelastic threshold, $E_p = 4m_\pi$.

Although the half off-shell amplitude is not an observable in experiment, we try to estimate the effective range r of the expansion, $k \cot\delta(k) = a_0^{-1} + rk^2 + O(k^4)$, from the slope of $H(p;k)/H(k;k)$ at $p = k$. In this estimation, we use two assumptions: 1) the phase of $H(p;k)$ is $e^{i\delta(k)}$ near $p = k$, and 2) $(H(p;k)e^{-i\delta(k)})' = (H(p;p)e^{-i\delta(p)})'$ at $p = k$ with the prime (′) denoting the derivative with respect to p^2. These assumptions need to validate in future. Our result $r = -2.64(41)\ \text{GeV}^{-1}$ is compatible with an estimation $r = -0.3(8.4)\ \text{GeV}^{-1}$ using the data of $\tan\delta(k)$ in [3].

69.5 Summary

We have discussed an exact relation between the half off-shell scattering amplitude and the BS wave function inside the interaction range. In order to derive the relation in the infinite volume, we have followed the discussion in [3] using the LSZ reduction formula. The reduced BS wave function defined by the BS wave function inside the interaction range plays an essential role in this relation. The relation we derived is useful to discuss uncertainties employing in the effective potential method as reported in [5, 7].

The corresponding relation on the lattice is derived with a condition that the interaction range is less than the half of the spatial extent of the lattice. It gives the same scattering amplitude in the infinite volume except for the overall factor. In order to calculate the scattering amplitude in a hadron scattering channel through the relation, we have performed a quenched lattice QCD simulation and calculated the S-wave $I = 2$ two-pion BS wave function at $m_\pi = 0.86$ GeV. The scattering length obtained from the on-shell amplitude agrees with the one from the finite volume method. For the half off-shell amplitude, we have observed a clear signal even in a large momentum region beyond the inelastic threshold. The effective range is estimated from the slope of the half off-shell amplitude under two assumptions. The estimated effective range is compatible with the one obtained from the data in the previous calculation. In future we will need a more realistic comparison of the effective ranges obtained from the half off-shell amplitude and the finite volume method, and also a study to validate the two assumptions.

The half off-shell amplitude cannot be directly measured in experiment, while it could be useful to understand hadron scatterings. For example, it can be used to constrain parameters in effective theories of hadron scatterings by comparing half off-shell amplitudes calculated from lattice QCD and an effective theory. Furthermore, the half off-shell amplitude in more than two-particle systems could be calculated in a similar way, because the exact relation of the scattering amplitude was essentially derived from the LSZ reduction formula. Its extension could be straightforward.

Acknowledgements The author thanks J. Carbonell and V. A. Karmnov for fruitful discussions in this conference. This work is supported in part by JSPS KAKENHI Grant Numbers 16H06002 and 18K03638.

References

1. Lüscher, M.: Nucl. Phys. B **354**, 531 (1991)
2. Lin, C.J.D., Martinelli, G., Sachrajda, C.T., Testa, M.: Nucl. Phys. B **619**, 467 (2001)
3. Aoki, S., Fukugita, M., Ishikawa, K.I., Ishizuka, N., Iwasaki, Y., Kaneko, T., Kuramashi, Y., Okawa, M., Ukawa, A., Yamazaki, T., Yoshié, T.: Phys. Rev. D **71**, 094504 (2005)
4. Aoki, S., Hatsuda, T., Ishii, N.: Prog. Theor. Phys. **123**, 89 (2010)
5. Yamazaki, T., Kuramashi, Y.: Phys. Rev. D **96**(11), 114511 (2017)
6. Namekawa, Y., Yamazaki, T.: Phys. Rev. D **98**(1), 011501 (2018)
7. Yamazaki, T., Kuramashi, Y.: Phys. Rev. D **98**(3), 038502 (2018)
8. Namekawa, Y., Yamazaki, T.: Phys. Rev. D **99**(11), 114508 (2019)
9. Carbonell, J., Karmanov, V.A.: Phys. Lett. B **754**, 270 (2016)

Chapter 70
3N Continuum Reactions with Semilocal Coordinate-Space Regularized Chiral Forces

H. Witała, J. Golak, R. Skibiński and K. Topolnicki

Abstract We applied the semilocal coordinate-space (SCS) regularized chiral N^4LO nucleon-nucleon (NN) potential supplemented by the chiral N^2LO three-nucleon force (3NF), regularized in the same way, to study the nucleon-deuteron (Nd) elastic scattering and deuteron breakup reaction. We checked that the elastic Nd scattering cross section can be used to fix, together with the ^3H binding energy, the strengths of the contact terms of the N^2LO 3NF. We found that the N^2LO 3NF provides effects comparable to those of (semi)phenomenological 2π-exchange 3NF's.

70.1 Introduction

The question about the importance of a 3NF is one of the main topics of 3 N system studies. Comparison of theoretical predictions based on high precission semi(phenomenological) nucleon-nucleon interactions, such as AV18 [1], CDBonn [2], Nijm1 and Nijm2 [3], with precise Nd elastic scattering and breakup data in a wide range of incoming nucleon energies enabled to find clear-cut discreppancies between theory and data, thus helping to reveal regions of large 3NF's effects. However, the models of 3NF's used in these studies [4], mainly of the 2π-exchange character, such as Tucson-Melbourne (TM99) [5] and Urbana IX [6], which are derived independently from the applied NN interactions, prevented the formulation of unambiguous conclusions. For numerous observables these two models of 3NF's combined with the same NN potential led to different predictions, not supported by existing data [4], indicating that these models contained only a part of contributions to a 3 N interaction.

With the advent of chiral perturbation theory (ChPT) approach it became possible to construct consistent two- and many-body nuclear forces which incorporate all possible contributions up to a given order of chiral expansion. High precision NN interactions up to the fifths order of chiral expansion (N^4LO) appeared which very

H. Witała (✉) · J. Golak · R. Skibiński · K. Topolnicki
M. Smoluchowski Institute of Physics, Jagiellonian University, Kraków, Poland
e-mail: henryk.witala@uj.edu.pl

N. A. Orr et al. (eds.), *Recent Progress in Few-Body Physics*,
Springer Proceedings in Physics 238,
https://doi.org/10.1007/978-3-030-32357-8_70

precisely describe NN data below the pion production threshold [7, 8]. In these first versions of chiral forces a nonlocal regularization in momentum space was applied to remove large momenta contributions by employing a multiplicative factor $e^{-(p^2/\Lambda^2)^n}$ with some value of a cut-off regulator parameter Λ [7, 8]. Such a regularization led, for small values of Λ, to a distortion of long-range behaviour of the NN potential and to problems when applying these forces to Nd elastic scattering at higher energies. To avoid such unwanted behaviour a new method of regularization was proposed in [9, 10]. It made use of a local regularization of the pion exchange contributions. The resulting semilocal coordinate-space regularized (SCS) chiral potentials contain the long-range part of the NN force predicted in a parameter-free way. First applications of these NN potentials to elastic Nd scattering and to the nucleon induced deuteron breakup reaction were described in [11].

Starting from N^2LO order of the chiral expansion a 3NF appears in a 3N Hamiltonian [12]. Before applying such a Hamiltonian to structure and reaction calculations one must fix two free parameters of this N^2LO 3NF, namely the strength parameters c_D and c_E of the leading 1π-contact and the three-nucleon-contact terms, respectively. In Sect. 70.2 we shortly describe how we fixed c_D and c_E parameters and in Sect. 70.3 present some results for the Nd elastic scattering and complete deuteron breakup reaction, obtained when applying that N^2LO 3NF. We summarize and conclude in Sect. 70.4.

70.2 Theoretical Formalism and Determination of c_D and c_E

Neutron-deuteron (nd) scattering with nucleons interacting through a two- and three-body interaction is described in terms of a breakup operator T satisfying the Faddeev-type integral equation [13, 14]. The amplitude for elastic scattering leading to the two-body final state and for the breakup reaction with a final state of three free nucleons can be obtained from that operator [13, 14]. We refer the reader to [13, 14] for a general overview of 3N scattering and for more details on the practical implementation of the Faddeev equations. We solved 3N Faddeev equations taking as a NN interaction the semi-locally regularized N^4LO chiral potential of [9, 10] with the regulator $R = 0.9$ fm alone or combined with the chiral N^2LO 3NF, regularized in the same SCS scheme [12, 15].

The nuclear Hamiltonian with a 3NF at N^2LO is fixed by specifying the values of LECs c_D and c_E, which parametrize the strengths of the leading 1π-contact and the three-nucleon-contact terms [12]. To fix them we used the experimental triton binding energy $E(^3H)$ to first determine the dependence of $E(^3H)$ on c_E for a given value of c_D. The requirement to reproduce the experimental value of the triton binding energy yields a set of pairs (c_D, c_E). These values (c_D, c_E) were then used in the calculations of the elastic Nd scattering cross section, what allowed us to find out the (c_D, c_E) pair describing both observables simultaneously.

The low-energy elastic nd and proton-deuteron (pd) scattering cross section is an observable which is well described by standard theory without the inclusion of 3NF's [4]. It turns out that when increasing the incoming nucleon laboratory energy, starting from ≈ 60 MeV, clear discrepancies between theory and data develop and pure NN potential predicitions underestimate the data in the region of the cross section minimum up to backward scattering angles [4]. Therefore it seems reasonable that the cross section in that region of angles could be used to determine the constants c_D and c_E. Since at larger energies higher orders of chiral expansion become important and a 3NF gets additional contributions, the procedure should be applied at energies where the disrepancies start to appear. Therefore we used precise pd data sets taken by different groups, one at $E = 65$ MeV from [16] and second at $E = 70$ MeV from [17], and performed χ^2 fits of theory to data using (c_D, c_E) pairs which reproduce ^{3}H binding energy. We have got for pairs of (c_D, c_E) the following values and errors: $(5.79 \pm 0.33, -1.00 \pm 0.14)$ at 65 MeV and $(6.09 \pm 0.23, -1.13 \pm 0.10)$ at 70 MeV. The values of c_D and c_E found at 65 MeV and 70 MeV are compatible and agree within their error bars.

70.3 Results

The good description of the Nd elastic scattering cross section data up to about 130 MeV when the N^{2}LO 3NF is included in calculations resembles effects found with standard 2 N and 3 N forces [4]. Namely adding the TM99 or Urbana IX 3NF to standard NN potentials removed in that energy region discrepancies between theory based on NN potentials only and the data, as exemplified in the left part of Fig. 70.1 for elastic Nd scattering at 135 MeV. At larger energies it turned out, however, that the inclusion of the 2π-exchange 3 N force models is unable to explain growing discrepancies between data and theory [4]. A similar pattern of agreement and discrepancies is found when instead of standard interactions chiral SCS forces are used, as shown in the right part of Fig. 70.1 in case of the agreement to data when the N^{2}LO 3NF is included in the calculations. The astonishing similarity of standard and chiral predictions can be traced back to the fact that the basic mechanism underlying these 3NF's is the 2π-exchange mechanism.

With increasing incoming nucleon energy large discrepnacies between pure 2 N theory and data start to develop also for numerous spin observables. Again the pattern of discrepancies found at these energies with chiral forces qualitatively resembles that obtained with standard interactions [4]. It is the region of higher energies, where higher chiral order terms start to play an important role and short-range components of a 3 N force contribute, which presents a challenge for future applications of a N^{3}LO 3NF [18, 19].

Among numerous kinematically complete configurations of the Nd breakup reaction the so-called symmetric space star (SST) configuration has attracted special attention. The cross section for that geometry is very stable with respect to the underlying dynamics and is dominated by the S-waves [22].

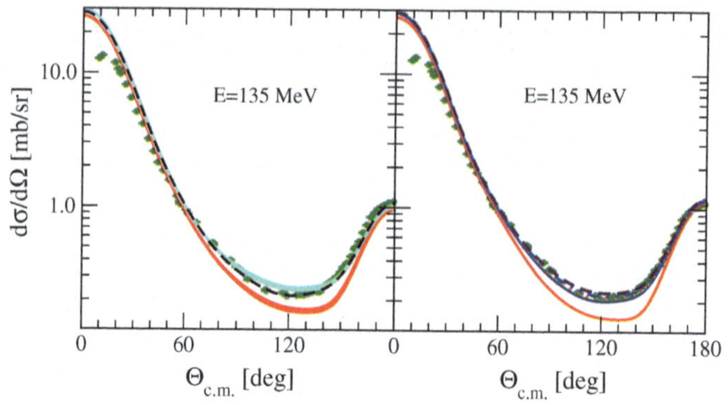

Fig. 70.1 (Color online) The nd elastic scattering cross section at the incoming neutron laboratory energy $E = 135$ MeV. In the left part predictions of (semi-)phenomenological NN potentials (AV18, CD Bonn, Nijm1 and Nijm2) are shown by dark (red) band. Results obtained when they are combined with the TM99 3NF are shown by a light (cyan) band. The dashed (black) line is the result of combination of the AV18 potential with the Urbana IX 3NF. In the right part the solid (red) line is prediction of the N^4LO SCS NN potential with the regulator $R = 0.9$ fm. Combining it with the N^2LO 3NF with four different strengths (c_D, c_E) of the contact terms, which reproduce ^3H binding energy, leads to results shown by different lines: solid (blue) (4.0, −0.270), dotted (red) (6.0, −1.094), double-dotted-dashed (blue) (8.0, −2.032), and dashed (maroon) (10.0, −3.108). In left and right parts the (green) dots are pd data from [17]

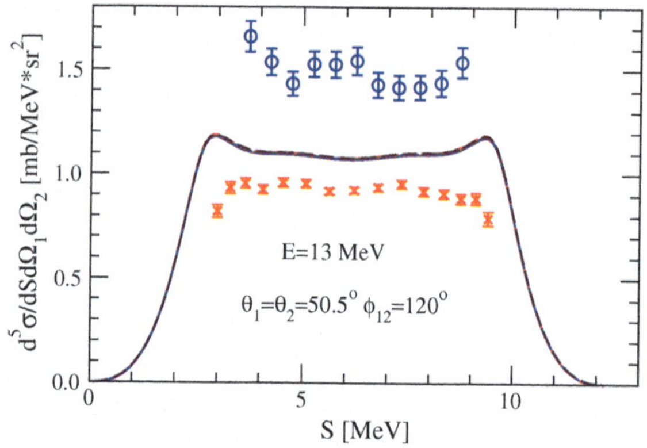

Fig. 70.2 (Color online) The nd breakup five-fold differential cross section in the SST complete geometry at the incoming neutron laboratory energy $E = 13$ MeV, shown as a function of arc-length of the S-curve. The dashed (red) line is prediction of the N^4LO SCS NN potential with the regulator $R = 0.9$ fm. Combining it with the N^2LO 3NF with four different strengths (c_D, c_E) of the contact terms leads to results shown by different lines, for explanation of which see Fig. 70.1. The (blue) circles are nd data from [20] and the (red) x-ces are pd data from [21]

In Fig. 70.2, we compare predictions of the N^{4}LO SCS chiral potential to the SST cross section data at the incoming nucleon lab. energy $E = 13$ MeV. The nd and pd breakup data are far away from the theory. The pd data set shown in Fig. 70.2 is supported by other SST pd breakup measurements [23]. The inclusion of the chiral N^{2}LO 3NF does not affect the results for the SST configuration cross section at this energy at all, leaving a big discrepancy to both nd and pd data. The calculations of the pd breakup with inclusion of the pp Coulomb force [24] revealed only very small Coulomb force effects for this configuration, what makes the discrepancy between the theory and pd data puzzling. Since at that energy the SST configuration is practically dominated by the S-wave NN force components, the big difference between pd and nd data seems to indicate significant charge-symmetry breaking in the $^1S_0$ NN partial wave. We anticipate difficulties with explaining the large difference between the nd and pd data sets by the inclusion of a 3NF without introducing large charge symmetry breaking interactions.

70.4 Summary and Conclusions

We applied the semilocal coordinate-space regularized N^{4}LO chiral NN potential combined with the N^{2}LO 3NF, regularized in the same way, to reactions in 3N continuum. We checked that using elastic Nd scattering cross section data, as an additional (to the ^{3}H binding energy) 3N observable, provides an efficient tool to fix the strengths parameters c_D and c_E of the N^{2}LO 3NF contact terms. The application of this tool should be restricted to the region of the incoming nucleon energies where 3NF effects start to appear in the cross section.

The application of that particular combination of NN and 3N forces provides at higher energies effects for the elastic scattering cross section which are very similar to those obtained with standard interactions. The N^{2}LO 3NF is incapable of explaining discrepancies between data and theory for 13 MeV SST complete breakup configuration. At higher energies and for numerous elastic scattering spin observables a complex pattern of discrepancies to data was found, similar to that revealed in previous investigations for standard NN and 3N forces. This points to the growing importance of short-range components of a 3NF at higher energies and calls for an application of the N^{3}LO 3NF, in which numerous new components contribute [18, 19].

Acknowledgements This study has been performed within Low Energy Nuclear Physics International Collaboration (LENPIC) project and was supported by the Polish National Science Centre under Grants No. 2016/22/M/ST2/00173 and 2016/21/D/ST2/01120. The numerical calculations were performed on the supercomputer cluster of the JSC, Jülich, Germany.

References

1. Wiringa, R.B., et al.: Accurate nucleon-nucleon potential with charge-independence breaking. Phys. Rev. C **51**, 38–51 (1995)
2. Machleidt, R.: High-precision, charge-dependent Bonn nucleon-nucleon potential. Phys. Rev. C **63**, 024001-1–32 (2001)
3. Stoks, V.G.J., et al.: Construction of high-quality NN potential models. Phys. Rev. C **49**, 2950–2963 (1994)
4. Witała, H., et al.: Nd elastic scattering as a tool to probe properties of 3N forces. Phys. Rev. C **63**, 024007-1–12 (2001), and references therein
5. Coon, S.A., Han, H.K.: Reworking the Tucson-Melbourne three-nucleon potential. Few Body Syst. **30**, 131–141 (2001)
6. Pudliner, B.S., et al.: Quantum Monte Carlo calculations of nuclei with $A < 7$. Phys. Rev. C **56**, 1720–1750 (1997)
7. Epelbaum, E.: Few nucleon forces and systems in chiral effective field theory. Prog. Part. Nucl. Phys. **57**, 654–741 (2006)
8. Machleidt, R., Entem, D.R.: Chiral effective field theory and nuclear forces. Phys. Rep. **503**, 1–75 (2011)
9. Epelbaum, E., et al.: Improved chiral nucleon-nucleon potential up to next-to-next-to-next-to-leading order. Eur. Phys. J. A **51**, 53–81 (2015)
10. Epelbaum, E., et al.: Precision nucleon-nucleon potential at fifth order in the chiral expansion. Phys. Rev. Lett. **115**, 122301-1–5 (2015)
11. Binder, S., et al.: Few-nucleon systems with state-of-the-art chiral nucleon-nucleon forces. Phys. Rev. C **93**, 044002-1–6 (2016)
12. Epelbaum, E., et al.: Three-nucleon forces from chiral effective field theory. Phys. Rev. C **66**, 064001-1–17 (2002)
13. Glöckle, W., et al.: The three-nucleon continuum: achievements, challenges and applications. Phys. Rep. **274**, 107–285 (1996)
14. Hüber, D., et al.: How to include a three-nucleon force into Faddeev equations for the 3N continuum: a new form. Acta Phys. Pol. B **28**, 1677–1685 (1997)
15. Hebeler, K., et al.: Efficient calculation of chiral three-nucleon forces up to N^3LO for ab initio studies. Phys. Rev. C**91**, 044001-1–9 (2015)
16. Shimizu, H., et al.: Analyzing powers and cross sections in elastic scattering at 65 MeV. Nucl. Phys. A **382**, 242–254 (1982)
17. Sekiguchi, K., et al.: Complete set of precise deuteron analyzing powers at intermediate energies: comparison with modern nuclear force predictions. Phys. Rev. C **65**, 034003-1–16 (2002)
18. Bernard, V., et al.: Subleading contributions to the chiral three-nucleon force: long-range terms. Phys. Rev. C **77**, 064004-1–13 (2008)
19. Bernard, V., et al.: Subleading contributions to the chiral three-nucleon force. II. short-range terms and relativistic corrections. Phys. Rev. C **84**, 054001-1–12 (2011)
20. Setze, H.R., et al.: Verification of the space-star anomaly in nd breakup. Phys. Lett. B **388**, 229–234 (1996)
21. Rauprich, G., et al.: Study of the kinematically complete breakup reaction $^2H(\vec{p}, -pp)n$ at $E_p = 3$ MeV with polarized protons. Nucl. Phys. A **535**, 313–330 (1991)
22. Witała, H., Glöckle, W.: On the discrepancies in the low-energy neutron-deuteron breakup. J. Phys. G: Nucl. Part. Phys. **37**, 064003-1–11 (2010)
23. Sagara, K.: Experimental investigations of discrepancies in three-nucleon reactions. Few-Body Syst. **48**, 59–108 (2010)
24. Deltuva, A., et al.: Momentum-space description of three-nucleon breakup reactions including the Coulomb interaction. Phys. Rev. C**72**, 054004-1–12 (2005)

Chapter 71
Momentum Distributions in ^{3}He with Chiral Potentials

L. E. Marcucci, F. Sammarruca, M. Viviani and R. Machleidt

Abstract We present predictions for the np momentum distributions in ^{3}He, obtained by applying the most recent high-quality chiral nucleon-nucleon potentials up to fifth order in the chiral expansion, together with the leading chiral three-nucleon force. Three-nucleon interaction contributions as well as chiral convergence are briefly discussed.

71.1 Introduction

The study of high-momentum distributions in nuclei can provide information about the short-range few-nucleon dynamics, and, ultimately, about the nature of the strong force which governs it. In fact, the picture of a nucleus as a dilute system, in which hard objects (the nucleons) are collected in a mean field, has strong limitations: the wave functions of nucleons in a medium strongly overlap, giving rise to the so-called short-range correlations (SRCs). Momentum distributions and SRCs in ^{3}He are the subject of this contribution. In particular, we focus on np momentum distributions, defined as

$$n^{np}(k_{rel}, K_{c.m.}) = \int d\hat{\mathbf{k}}_{rel} \int d\hat{\mathbf{K}}_{c.m.} \Psi^{\dagger}(\mathbf{k}_{rel}, \mathbf{K}_{c.m.}) P_{np} \Psi(\mathbf{k}_{rel}, \mathbf{K}_{c.m.}) \quad (71.1)$$

$$n^{np}(k_{rel}) = 4\pi \int_0^{K^+_{c.m.}} K^2_{c.m.}\, dK_{c.m.} n^{np}(k_{rel}, K_{c.m.})\,, \quad (71.2)$$

L. E. Marcucci (✉)
Dipartimento di Fisica "E. Fermi", Pisa University, 56127 Pisa, Italy
e-mail: laura.elisa.marcucci@unipi.it

L. E. Marcucci · M. Viviani
Istituto Nazionale di Fisica Nucleare, Sezione di Pisa, 56127 Pisa, Italy

F. Sammarruca · R. Machleidt
Department of Physics, University of Idaho, Moscow, ID 83844, USA

N. A. Orr et al. (eds.), *Recent Progress in Few-Body Physics*,
Springer Proceedings in Physics 238,
https://doi.org/10.1007/978-3-030-32357-8_71

where k_{rel} and $K_{c.m.}$ are the np relative and center-of-mass momenta, P_{np} is the projection operator on the np pair, and $\Psi(\mathbf{k}_{rel}, \mathbf{K}_{c.m.})$ is the ^{3}He nuclear wave function. This is calculated using the Hyperspherical Harmonics (HH) method, as explained in [1] (see also [2] and references therein). Note that the HH wave functions can be written in coordinate- as well as in momentum-space [2], and therefore there is no restriction on the choice of the nuclear potential model, which can be either local or non-local. In (71.2) the upper limit of the integral $K^+_{c.m.}$ is taken to be 1.5 fm^{-1}, since a highly correlated np pair is expected to have small center-of-mass momentum [3, 4]. Another quantity of interest is the so-called back-to-back np momentum distribution, i.e. $n^{np}(k_{rel}, K_{c.m.} = 0)$. We will consider this quantity as well. Due to lack of space, we will not discuss the results for the np momentum distributions integrated over k_{rel}. They can be found in [1].

This contribution is organized as follows: in Sect. 71.2 we present the results for the above mentioned quantities obtained with the most recent chiral two-nucleon potentials of [5], without and with the inclusion of the corresponding three-nucleon interactions, as constructed in [1]. Note that the np momentum distributions have been first studied in [1] using the older Argonne v_{18} (AV18) [6] purely phenomenological potential model, without or with three-nucleon interaction, i.e. the Urbana IX (UIX) [7] model. This has allowed us to compare with previous results obtained by other groups [3, 4, 8], finding a good agreement. Furthermore, using also the meson-theoretic charge dependent Bonn (CDBonn) potential model [9], augmented or not by an appropriate three-nucleon force, namely the Tucson-Melbourne (TM) [10] model, it has been found in [1] that (i) the three-nucleon interaction contributions are very small; (ii) the model dependence is very large, as it was already found in [11] for the case of the deuteron. This model dependence can be traced back to locality vs. non-locality of the adopted models, especially in their tensor components, and, together with the small contributions of the three-nucleon interactions, will be a recurrent feature of this study. Finally, in Sect. 71.3 we summarize our results and present our conclusions, with an outlook on future work.

71.2 Results

We present results for the np momentum distributions $n^{np}(k_{rel}, K_{c.m.} = 0)$ and $n^{np}(k_{rel})$, obtained using the most recent chiral two-nucleon potentials of [5], from leading (LO) to fifth chiral order (N4LO), with three different values of the cutoff Λ, i.e. 450, 500 and 550 MeV. In addition, we include the chiral three-nucleon forces, as constructed in [1], although we expect and indeed find them to give extremely small contributions.

The np back-to-back momentum distributions, $n^{np}(k_{rel}, K_{c.m.} = 0)$, calculated with the two-nucleon chiral potentials of [5] for a fixed cutoff value ($\Lambda = 500$ MeV), at different chiral orders is shown in Fig. 71.1. By inspection of the figure we can conclude that the LO results at small k_{rel} values are much smaller than those obtained at the other orders. This suggests that the asymptotic part of the wave function at LO

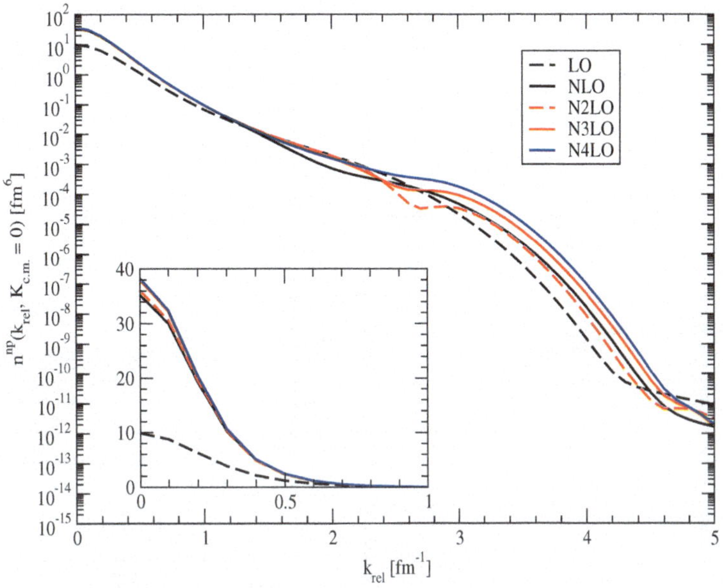

Fig. 71.1 The np back-to-back momentum distributions $n^{np}(k_{rel}, K_{c.m.} = 0)$, calculated using the two-nucleon chiral potentials of [5], with $\Lambda = 500$ MeV. Adapted from [1]

is significantly different than at the higher orders. Furthermore, the N3LO and N4LO curves are very similar up to $k_{rel} \simeq 2.2$ fm^{-1}, indicating satisfactory order-by-order convergence at least in the region where the distributions still have non-negligible size. Similar features can be found for the other values of the cutoff Λ.

The np momentum distributions $n^{np}(k_{rel})$ calculated with and without three-nucleon interaction, at different chiral order and for different values of the cutoff Λ, are shown in Fig. 71.2. By inspection of the figure we can conclude that we have essentially no cutoff dependence below $k_{rel} \simeq 2.2 - 2.5$ fm^{-1}, and increasingly strong cutoff dependence above it. Furthermore, the contributions of three-nucleon interaction are visible only for $k_{rel} \geq 3.0 - 3.5$ fm^{-1}. Note, however, that above $k_{rel} \simeq 2.5$ fm^{-1} all momentum distributions are so small that the differences are of no practical relevance.

71.3 Summary and Conclusions

We have presented predictions for the np momentum distributions in ^{3}He, obtained using state-of-the-art chiral two-nucleon potentials (with or without the leading chiral three-nucleon force). A more comprehensive study of these quantities and related ones can be found in [1]. Here we only remark that we have found the contribution from three-nucleon interactions to be essentially negligible, whereas dependence on

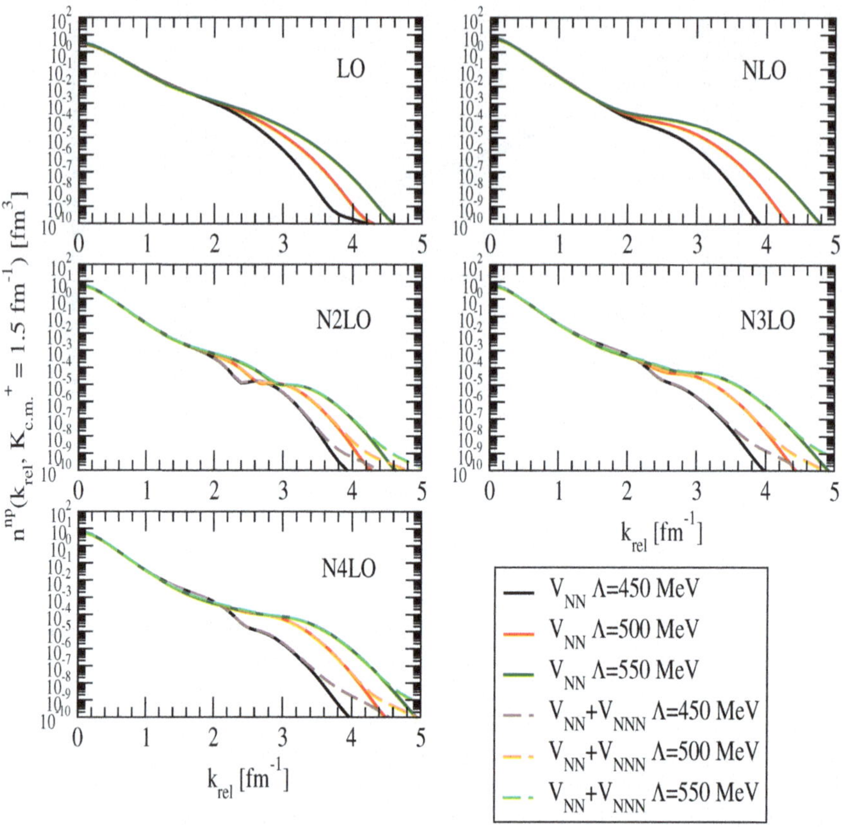

Fig. 71.2 (Color online) The np momentum distributions $n^{np}(k_{rel})$ calculated using only two-nucleon (solid lines) and two- and three-nucleon (dashed lines) chiral interactions, at different chiral order and for three values of the cutoff $\Lambda = 450, 500, 550$ MeV. Adapted from [1]

the nucleon-nucleon potential model is very strong, especially in the high-momentum tail of the distributions. This must be taken into account when extracting SRCs "empirical" information from high momentum transfer electron scattering experiments (see for instance [12]).

References

1. Marcucci, L.E., et al.: Momentum distributions and short-range correlations in the deuteron and ^{3}He with modern chiral potentials. Phys. Rev. C **99**, 034003 (2019)
2. Kievsky, A., et al.: A high-precision variational approach to three- and four-nucleon bound and zero-energy scattering states. J. Phys. G **35**, 063101 (2008)
3. Alvioli, M., et al.: Nucleon momentum distributions, their spin-isospin dependence, and short-range correlations. Phys. Rev. C **87**, 034603 (2013)

4. Alvioli, M., et al.: Universality of nucleon-nucleon short-range correlations: the factorization property of the nuclear wave function, the relative and center-of-mass momentum distributions, and the nuclear contacts. Phys. Rev. C **94**, 044309 (2016)
5. Entem, D.R., Machleidt, R., Nosyk, Y.: High-quality two-nucleon potentials up to fifth order of the chiral expansion. Phys. Rev. C **96**, 024004 (2017)
6. Wiringa, R.B., Stoks, V.G.J., Schiavilla, R.: Accurate nucleon-nucleon potential with charge-independence breaking. Phys. Rev. C **51**, 38 (1995)
7. Pudliner, B.S., et al.: Quantum Monte Carlo calculations of nuclei with $A < 7$. Phys. Rev. C **56**, 1720 (1997)
8. Wiringa, R.B.: https://www.phy.anl.gov/theory/research/momenta2/
9. Machleidt, R.: High-precision, charge-dependent Bonn nucleon-nucleon potential. Phys. Rev. C **63**, 024001 (2001)
10. Coon, S.A., Han, H.K.: Reworking the Tucson-Melbourne three-nucleon potential. Few-Body Syst. **30**, 131 (2001)
11. Sammarruca, F.: Short-range correlations in the deuteron: chiral effective field theory, meson-exchange, and phenomenology. Phys. Rev. C **92**, 044003 (2015)
12. Korover, I., et al. (Jefferson Lab Hall A Collaboration): Probing the repulsive core of the nucleon-nucleon interaction via the ^{4}He$(e, e'pN)$ triple-coincidence reaction. Phys. Rev. Lett. **113**, 022501 (2014)

Chapter 72
Reaction Mechanisms in Deuteron-Proton Elastic Scattering at Intermediate Energies

Nadezhda Ladygina

Abstract Deuteron-proton elastic scattering is considered in the relativistic multiple scattering expansion framework. The four reaction mechanisms are included into consideration: one-nucleon exchange, single scattering, double scattering, and the term corresponding to the delta excitation in the intermediate state. The theoretical results are compared with the experimental data.

72.1 Introduction

Elastic deuteron-proton scattering is the simplest example of the hadron nucleus collision. Nowadays, a significant amount of the experimental data has been accumulated in a wide energy range both with unpolarised and polarised beams. However, we do not have any theory to describe the data for the energies above a few hundred MeV, especially, at backward scattering angles.

A good theoretical description of the deuteron-nucleon process was obtained at low energies, where the multiple scattering formalism based on the solution of the Faddeev equations, has been applied to this problem [1]. However, at the nucleon energies above 130 MeV there is some discrepancy between the experimental data and theoretical predictions in the minimum of the differential cross section [2].

The Glauber theory taking into account both single and double nucleon-nucleon interaction successfully describes the differential cross sections of the dp-elastic scattering at small angles [3, 4]. But it does not properly work at larger scattering angles.

We have previously proposed to use a model based on the multiple expansion of the reaction amplitude in powers of the nucleon-nucleon t-matrix [5–7]. We consider a multiple-scattering series up to the second-order terms of the NN t-matrix. Also Δ-isobar excitation in the intermediate state is included into consideration.

N. Ladygina (✉)
Joint Institute for Nuclear Research, Dubna 141980, Russia
e-mail: nladygina@jinr.ru

N. A. Orr et al. (eds.), *Recent Progress in Few-Body Physics*,
Springer Proceedings in Physics 238,
https://doi.org/10.1007/978-3-030-32357-8_72

72.2 General Formalism

According to the three-body collision theory, the amplitude of the deuteron-proton elastic scattering $\mathcal{J}$ is defined by the matrix element of the transition operator U_{11}:

$$U_{dp\to dp} = \delta(E_d + E_p - E'_d - E'_p)\mathcal{J} = \\ < 1(23)|[1 - P_{12} - P_{13}]U_{11}|1(23) > . \tag{72.1}$$

Here, the state $|1(23)>$ corresponds to the configuration, when nucleons 2 and 3 form the deuteron state and nucleon 1 is free. The permutation operators for two nucleons P_{ij} reflects the fact that the initial and final states are antisymmetric due to the two particles exchange.

The transition operators for rearrangement scattering are defined by the Alt–Grassberger–Sandhas equations:

$$\begin{aligned} U_{11} &= \qquad t_2 g_0 U_{21} + t_3 g_0 U_{31}, \\ U_{21} &= g_0^{-1} + t_1 g_0 U_{11} + t_3 g_0 U_{31}, \\ U_{31} &= g_0^{-1} + t_1 g_0 U_{11} + t_2 g_0 U_{21}, \end{aligned} \tag{72.2}$$

where $t_1 = t(2, 3)$, etc., is the t-matrix of the two-nucleon interaction and g_0 is the free three-particle propagator. The indices ij for the transition operators U_{ij} denote free particles i and j in the final and initial states, respectively.

Iterating these equations up to the t_i-second-order terms, we can present the reaction amplitude as a sum of the four contributions:

$$\mathcal{J}_{dp\to dp} = \mathcal{J}_{\mathrm{ONE}} + \mathcal{J}_{\mathrm{SS}} + \mathcal{J}_{\mathrm{DS}} + \mathcal{J}_{\Delta}, \tag{72.3}$$

one-nucleon exchange, single scattering, double scattering, and rescattering with Δ -excitation in the intermediate state.

72.3 Discussion and Conclusion

The results of the calculations for the differential cross sections at deuteron energies of 880 and 1200 MeV are presented in Figs. 72.1 and 72.2. Three different theoretical curves in the figures correspond to calculations taking into account only one-nucleon-exchange (ONE) and single-scattering (SS) terms (green), ONE+SS and double-scattering (DS) terms (blue), and ONE+SS+DS and Δ- isobar excitation in an intermediate state (red). All these curves practically coincide up to the scattering angles of about 60°. It shows that the main contribution in this angular range gives SS term. We can get a good description of the differential cross sections at $\Theta^* \leq 60°$ taking into account only ONE+SS mechanisms. But at larger angles differ-

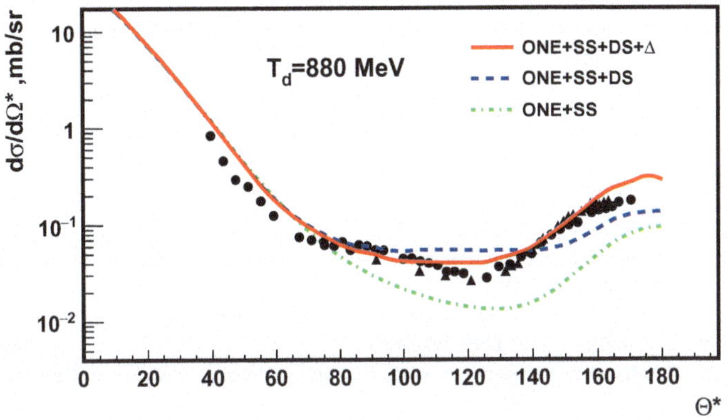

Fig. 72.1 The differential cross section at the deuteron kinetic energy of 880 MeV as a function of the c.m. scattering angle. The data are taken from [8] (•), and [10] (▲)

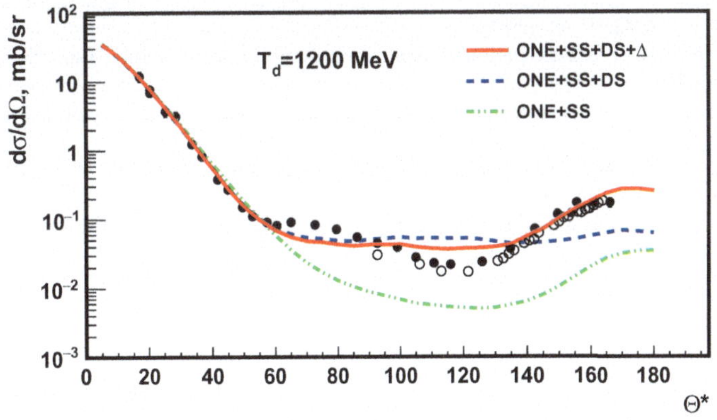

Fig. 72.2 The differential cross section at the deuteron kinetic energy of 1200 MeV as a function of the c.m. scattering angle. The data are taken from [9] (•), and [10] (∘)

ence between the experimental data and ONE+SS-curve is significant and increases with the energy. Adding DS-term lets improve the agreement between the data and theory in the angular range about 60°–140°. Inclusion of Δ- isobar term helps to describe a rise of the differential cross sections at backward angles.

The model gives rather good agreement between the experimental data and theoretical predictions at energies where Faddeev calculations do not work. However, the data in some angular ranges cannot be properly described in the framework without further improvement of the model. The expansion of knowledge about nucleon-nucleon interactions is very important for it, especially for higher energies. The approach deals with reaction amplitudes what gives the opportunity to get both differential cross sections and polarisation observables.

Acknowledgements This work has been supported by the Russian Foundation for Basic Research under grant No.16-02-00203a.

References

1. Glöckle, W., et al.: The tree-nucleon continuum: achievements, challenges and applications. Phys. Rep. **274**, 107–284 (1996)
2. Sekiguchi, K., et al.: Complete set of deuteron analyzing powers from $\mathbf{d}p$ elastic scattering at 190 MeV/nucleon. Phys. Rev. **C96**, 064001 (2017)
3. Franco, V.: Small-angle high-energy scattering by deuterons. Phys. Rev. Lett. **16**, 944–947 (1966)
4. Franco, V., Coleman, E.: Double scattering in high-energy elastic collisions with deuterons. Phys. Rev. Lett. **17**, 827–830 (1966)
5. Ladygina, N.B.: Deuteron-proton elastic scattering at intermediate energies. Phys. Atom. Nucl. **71**, 2039–2051 (2008)
6. Ladygina, N.B.: Differential cross section of DP-elastic scattering at intermediate energies. Eur. Phys. J. A **42**, 91–96 (2009)
7. Ladygina, N.B.: Delta excitation in deuteron-proton elastic scattering. Eur. Phys. J. A **52**, 199–206 (2016)
8. Booth, N.E., et al.: Proton-deuteron elastic scattering at 1.0 GeV/c. Phys. Rev. **D4**, 1261–1267 (1971)
9. Boschitz, E.T., et al.: Elastic scattering of 600-MeV protons from H, D, ^3He, and ^4He. Phys. Rev. **C6**, 457–466 (1972)
10. Alder, J.C., et al.: Elastic pd-scattering at 316, 364, 470, and 590 MeV in the backward hemisphere. Phys. Rev. **C6**, 2010–2019 (1972)

Chapter 73
Relativistic Faddeev Calculation for Nucleon-Deuteron Scattering with the Kharkov Potential

Hiroyuki Kamada, Oleksandr Shebeko, Adam Arslanaliev, Henryk Witała, Jacek Golak, Roman Skibiński, Margarita Stepanova and Sergey Yakovlev

Abstract The Kharkov potential is a recent field theoretical model of the nucleon-nucleon (NN) interaction that has been built up in the framework of the instant form of relativistic dynamics starting with the total Hamiltonian of interacting meson and nucleon fields and using the method of unitary clothing transformations (UCTs). It is the first time that the relativistic UCT potential is used in the Faddeev equation to study nucleon-deuteron (Nd) elastic scattering. Our results of the relativistic Faddeev calculations with UCT describe elastic observables well up to 65 MeV, just like the corresponding nonrelativistic predictions. The theoretical predictions based on the Kharkov and CDBonn potentials are compared with recent precise experimental data (iT_{11}, T_{20}, T_{21} and T_{22}) taken at the deuteron beam of energy 186.6 MeV/nucleon.

73.1 Introduction

Relativity in the three-nucleon (3N) system has been studied [1] under the Bakamjian-Thomas frame, which belongs to the relativistic quantum mechanics and is dictated by the Poincaré algebra. Since a relativistic NN potential is not easily provided, we need

H. Kamada (✉)
Department of Physics, Faculty of Engineering, Kyushu Institute of Technology, Kitakyushu 804-8550, Japan
e-mail: kamada@mns.kyutech.ac.jp

O. Shebeko
National Science Center "Kharkov Institute of Physics and Technology", Kharkov, Ukraine
e-mail: shebeko@kipt.kharkov.ua

A. Arslanaliev
V.N. Karazin Kharkiv National University, Kharkov, Ukraine

H. Witała · J. Golak · R. Skibiński
M. Smoluchowski Institute of Physics, Jagiellonian University, Kraków, Poland

M. Stepanova · S. Yakovlev
Saint-Petersburg State University, Saint Petersburg, Russia

N. A. Orr et al. (eds.), *Recent Progress in Few-Body Physics*, Springer Proceedings in Physics 238,
https://doi.org/10.1007/978-3-030-32357-8_73

some schemes which would allow us to transform a nonrelativistic potential into the corresponding relativistic one. Such schemes are required to fulfill the condition that the generated relativistic potential yields the same observables in the NN system as the original nonrelativistic potential. There are two schemes that satisfy that condition. One has been proposed by Coester et al. [2] and we call it the CPS scheme. It requires a solution of a nonlinear integral equation, which can be achieved numerically by an iteration method [3]. The other scheme is a momentum scaling method (MSM) [4]. It is our aim to compare the original nonrelativistic and the modified relativistic potential predictions for the 3N system. In the case of the ^{3}H binding energy and the elastic Nd scattering, we have already demonstrated such a comparison [5–7] by using the CPS and MSM schemes. In view of higher energy measurements (see e.g. [8]), relativistic treatment of the 3N dynamics has become even more topical.

The important feature of the Kharkov model [9] is that it provides directly the relativistic NN potential, so neither CPS nor MSM procedure is required. The calculations of the triton binding energy with the Kharkov potential were reported in [10]. In this context we present the Nd elastic scattering observables. Previous relativistic calculations [6, 7] employing the CDBonn potential [11] showed only small relativistic effects for elastic scattering cross sections and practically no effects for spin observables at energies below 100 MeV, however, clear effects were found for the nucleon-induced deuteron breakup [12] reaction.

73.2 Kharkov Potential

At the beginning the original Hamiltonian H can be expressed through the so-called bare creation (annihilation) operators $a^\dagger(a)$, $b^\dagger(b)$ and $d^\dagger(d)$, respectively, for bosons, fermions and antifermions. Since in such a bare-particles representation (BPR) the interaction part of H is given by primary Yukawa-type trilinear couplings between π-, η-, ρ-, ω-, σ-mesons and nucleons (antinucleons), the Hamiltonian has nonzero matrix elements between the states of different sectors in Fock space (e.g., between the bare vacuum state $|0\rangle$ and the bare one-meson states $a^\dagger|0\rangle$). Therefore, the sectors πN, NN, NNN etc. are not separated from each other (see [9] and references therein).

By introducing the so-called "clothed" particle operators $a_c \equiv W^\dagger a W$, $b_c \equiv W^\dagger b W$ and etc. with a unitary operator W, one can express our Hamiltonian in terms of these new creation and annihilation operators. Algebraically, such a transition is given via the similarity transformation $H \equiv H(\alpha) = W H(\alpha_c) W^\dagger$ where a symbol α (α_c) denotes the set of all bare particle creation and annihilation operators (their clothed particle counterparts). Being fulfilled some requirements imposed upon the clothing transformation W the UCT method allows one to rewrite $H(\alpha)$ in the form, $H(\alpha) = K_F(\alpha_c) + K_I(\alpha_c) \equiv K(\alpha_c)$. The free part of the new decomposition is determined by $K_F(\alpha_c) = \int d\mathbf{k}\,\omega_{\mathbf{k}} a_c^\dagger(\mathbf{k}) a_c(\mathbf{k}) + \int d\mathbf{p}\,E_{\mathbf{p}} \sum_\mu \left[b_c^\dagger(\mathbf{p},\mu) b_c(\mathbf{p},\mu) + d_c^\dagger(\mathbf{p},\mu) d_c(\mathbf{p},\mu)\right]$, while K_I contains only interactions responsible for physical processes, these quasipotentials between the clothed particles, e.g., $K_I^{(2)}(\alpha_c) = K(NN \to NN) + K(\bar{N}\bar{N} \to \bar{N}\bar{N}) + K(N\bar{N} \to N\bar{N}) + K(bN \to N) + K(b\bar{N} \to b\bar{N}) + K(bb' \to N\bar{N}) + K(N\bar{N} \to bb')$. One of such interactions known as Kharkov potential is generated by the nucleon - nucleon interaction operator $K(NN \to NN) \sim b_c^\dagger b_c^\dagger b_c b_c$. The meson propagators and the cutoff functions in the potential depend on the Lorentz invariants constructed of four-momentum vectors for the closed mesons and nucleons, i.e., on their mass shells.

The Kharkov potential enters the relativistic Lippmann-Schwinger equation. Therefore, the boosted Kharkov potential is easily converted by our method [3].

73.3 Relativistic Calculations of the Nd Scattering at $E = 186.6$ MeV/nucleon

When the energy of the incident particle exceeds approximately 65 MeV/nucleon, it gradually becomes impossible to explain the differential cross section at backward scattering angles only with the two-body forces. The Fujita-Miyazawa (FM) type of the three-nucleon (3NF), for example the Tucson-Melbourne 3NF or Urbana IX 3NF, have been included to explain the discrepancy. However, for energies above 200 MeV it is again not realizable to describe experimental data, even including 3NF of the FM type. There are some possible ways to solve this problem. One of them is to incorporate additional structures in 3NF and this path is being pursued mainly in the framework of chiral effective field theory, now at higher orders of the chiral expansion [13, 14]. The other possibility, discussed here, is solving the relativistic Faddeev equation. Such solutions were obtained in [6, 7], starting from the nonrelativistic CDBonn and Argonne V18 potentials and helped reduce slightly the discrepancies between theory and data.

It is, however, plausible that the CDBonn, Argonne V18 and other realistic forces might be not appropriate starting points for relativistic calculations, even if the SCP or MSM transformations are applied to them. That is why we use Kharkov potential which can be directly used in the relativistic LS equation. We compare our theoretical predictions with the results of a recent precise measurement of all deuteron analyzing powers for elastic dp scattering with polarized deuteron beam of energy 186.6 MeV/nucleon, performed at the RIKEN RI Beam Factory [15]. In Fig. 73.1 the measured polarization observables (iT_{11}, T_{20}, T_{21} and T_{22}) are compared with our theoretical predictions based on the Kharkov and the CDBonn potentials. Both theoretical predictions describe the deuteron vector polarization iT_{11} and one of the deuteron tensor analyzing powers T_{21} rather well. In contrast, predictions for T_{20} and T_{22} strongly differ from the data [15].

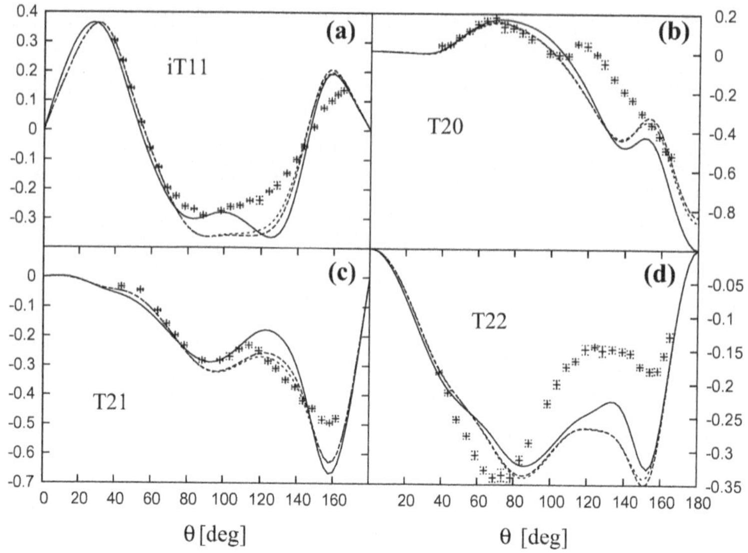

Fig. 73.1 Polarization observables in the elastic Nd scattering: iT_{11} (**a**), T_{20} (**b**), T_{21} (**c**) and T_{22} (**d**), at $E = 186.6$ MeV/nucleon. The solid (long-dashed) line represents the theoretical predictions obtained with the Kharkov UCT1 [9] (CDBonn [11]) potential. The additional short-dashed line is used for the nonrelativistic results obtained with the CDBonn potential. The experimental data are from [15]

73.4 Remark and Outlook

The Kharkov potential can be employed directly in the relativistic LS equation and its Lorentz-boosted version can be later generated, which makes the application of this NN potential in the relativistic Faddeev equation more straightforward. The calculation of the elastic Nd scattering polarization observables with the Kharkov and CDBonn potentials lead to quite similar results. At the considered energy a clear discrepancy between the calculations and the data [15] persists. As shown in [15] for T_{22} the difference can be only partially reduced by using 3NF. In the future we plan to use a 3NF of the FM type in the calculations with the Kharkov potential. Also a new version [16] of the Kharkov potential will be prepared based on recent NN data.

Acknowledgements This work was supported by the Japan Society for the Promotion of Science (JSPS) for financial support under Grant-in-Aid for Scientific Research (B) No. 16H04377, and the Polish National Science Centre under Grant No. 2016/22/M/ST2/00173. The numerical calculations were performed on the supercomputer cluster of the JSC, Jülich, Germany, and the interactive server at RCNP, Osaka University, Japan.

References

1. Glöckle, W., Lee, T.-S.H., Coester, F.: Phys. Rev. C **33**, 709 (1986)
2. Coester, F., Pieper, S.C., Serduke, F.J.D.: Phys. Rev. C **11**, 1 (1975)
3. Kamada, H., Glöckle, W.: Phys. Lett. B **655**, 119 (2007)
4. Kamada, H., Glöckle, W.: Phys. Rev. Lett. **80**, 2457 (1998)
5. Kamada, H., et al.: EPJ Web Conf. **3**, 05025 (2010)
6. Witała, H., et al.: Phys. Rev. C **71**, 054001 (2005)
7. Witała, H., Golak, J., Skibiński, R., Glöckle, W., Kamada, H., Polyzou, W.N.: Phys. Rev. C **83**, 044001 (2011); Erratum, Phys. Rev. C **88**, 069904(E) (2013)
8. Sekiguchi, K., et al.: Phys. Rev. Lett. **95**, (2005) 162301; Maeda, Y., et al.: Phys. Rev. C **76**, 014004 (2007)
9. Dubovyk, I., Shebeko, O.: Few-Body Syst. **48**, 109 (2010)
10. Kamada, H., Shebeko, O., Arslanaliev, A.: Few-Body Syst. **58**, 70 (2017)
11. Machleidt, R., Sammarruca, F., Song, Y.: Phys. Rev. C **53**, R1483 (1996)
12. Witała, H., Golak, J., Skibiński, R.: Phys. Lett. B **634**, 374 (2006)
13. Epelbaum, E., et al.: Phys. Rev. C**66**, 064001 (2002); Epelbaum, E.: Nucl. Phys. **57**, 654 (2006)
14. Hebeler, K., et al.: Phys. Rev. C **91**, 044001 (2015)
15. Sekiguchi, K., et al.: Phys. Rev. C **96**, 064001 (2017)
16. Arslanaliev, A., et al.: Ukr. J. Probl. At. Sci. Technol. **115**(3), 3 (2018)

Chapter 74
Study of Three-Nucleon Dynamics in the dp Breakup Collisions Using the WASA Detector

B. Kłos, I. Ciepał, A. Kozela, St. Kistryn, A. Magiera, W. Parol, I. Skwira-Chalot and E. Stephan

Abstract An experiment to investigate the $^1H(d, pp)n$ breakup reaction using a deuteron beam of 300, 340, 380 and 400 MeV and the WASA detector has been performed at the Cooler Synchrotron COSY-Jülich. As a first step the data collected at the beam energy of 340 MeV are analysed, with a focus on the proton-proton coincidences registered in the Forward Detector. The differential cross section is determined for 189 configurations on a dense angular grid defined by the emission angles of the two outgoing protons: two polar angles θ_1 and θ_2 (in the range between 5° and 15° with the step size of 2°) and the relative azimuthal angle φ_{12} (in the range from 20° to 180°, with the step size of 20°).The cross section data are compared to theoretical predictions based on the state-of-the-art nucleon-nucleon potentials, combined with three-nucleon force, Coulomb interaction or carried out in a relativistic regime.

74.1 Introduction

Differential cross section for the 1H(d,pp)n breakup reaction at intermediate energies is sensitive to details of nuclear interactions, in particular to the so-called three nucleon force (3NF). The modern theoretical calculations include 3NF [1] but also the long range Coulomb interactions [2] or, separately, the relativistic effects [3].

The WASA-at-COSY Collaboration

B. Kłos (✉) · E. Stephan
Institute of Physics, University of Silesia, 41500 Chorzów, Poland
e-mail: barbara.klos@us.edu.pl

I. Ciepał · A. Kozela · W. Parol
Institute of Nuclear Physics PAN, 31342 Kraków, Poland

St. Kistryn · A. Magiera
Institute of Physics, Jagiellonian University, 30059 Kraków, Poland

I. Skwira-Chalot
Faculty of Physics, University of Warsaw, Warsaw, Poland

N. A. Orr et al. (eds.), *Recent Progress in Few-Body Physics*,
Springer Proceedings in Physics 238,
https://doi.org/10.1007/978-3-030-32357-8_74

All these ingredients of the dynamics influence significantly the results of calculations for the breakup cross section, in particular at certain parts of the phase space. With the lack of complete calculations performed in relativistic regime, including 3NF and Coulomb interaction, there is a need of the systematic (in the beam energy) data set collected in a large phase space. This may allow us to trace the effects in the kinematic regions where they play locally very important role. It was a motivation to perform new measurement of the $^{1}H(d, pp)n$ breakup cross section using a deuteron beam of 300, 340, 380 and 400 MeV in order to check the theoretical predictions and to unambiguously fix a relevance of the 3NF.

74.2 Data Analysis

The experiment studying the $^{1}H(\mathbf{d}, pp)n$ breakup reaction at beam energies 300, 340, 380 and 400 MeV has been performed at the Cooler Synchrotron COSY-Jülich with the WASA (Wide Angle Shower Apparatus) detector [4, 5]. The detection system consists of four main components: Central Detector (CD), Forward Detector (FD), a pellet target device and a scattering chamber. The 4π detector facility WASA allowed to study production and decays of light mesons in proton and deuteron reactions [6–8].

The data analysis is focused on the proton-proton coincidences registered in the Forward Detector with the aim to determine the differential cross section on dense angular grid of kinematical configurations defined by the emission angles of the two outgoing protons: two polar angles θ_1 and θ_2 (in the range between 4° and 18°) and the relative azimuthal angle ϕ_{12}. The first step of data analysis is the identification of interesting events, *i.e.* proton pairs from the breakup process and deuterons from the elastic scattering channel registered in the Forward Detector. The particle identification is based on the ΔE-E technique. Energy calibration of FD is based on measurements of dp elastic scattering at energies corresponding to minimum ionization and on Monte Carlo simulations based on the Geant package, as described in detail in [9].

The missing mass spectrum is a tool to control the proton energy calibration and the procedure of selection of proton-proton coincidences. The missing mass of the neutron is calculated according to the formula (in which $c = 1$):

$$MM = \sqrt{(E_{in} - E_{p_1} - E_{p_2})^2 - (\mathbf{P}_{in} - \mathbf{P}_{p_1} - \mathbf{P}_{p_2})^2}, \tag{74.1}$$

where E_{in} and $\mathbf{P}_{in}$ are the sum of energy and momenta of the incident deuteron and target proton and E_{p_i} and $\mathbf{P}_{p_i}$ $(i = 1, 2)$ are the total energies and momenta of the two outgoing protons registered in coincidence. Figure 74.1 presents the missing mass spectrum, built for all pairs of coincident protons registered in FD. The tail of the distribution at energies exceeding a neutron mass is mainly due to hadronic interactions of the outgoing protons. Contribution of accidental coincidences is expected to

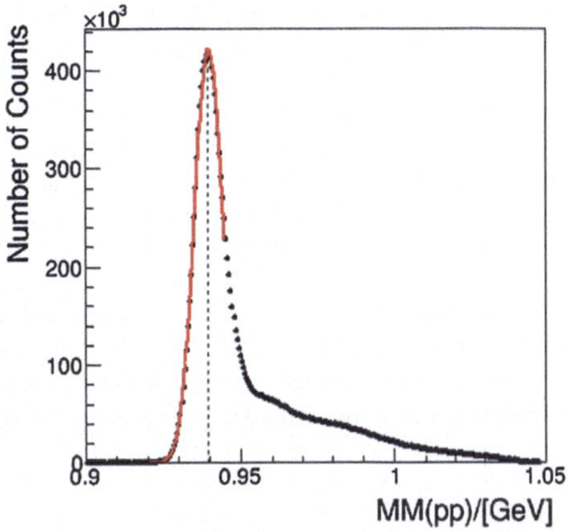

Fig. 74.1 Missing mass reconstructed from momenta of two outgoing protons detected in coincidence. Clearly, the peak corresponding to the neutron mass dominates. Red line represents Gaussian fit with mean value of 0.94 GeV. The right tail of the distribution originates from proton energy loss due to hadronic interactions and from accidental coincidences

be low. The data are compared to Monte Carlo simulations for the same phase space region. Obtained agreement indicates correct simulation of the hadronic interactions and validates the efficiency corrections calculated on the basis of simulations.

To obtain differential cross section, the luminosity is determined on the basis of the number of the elastically-scattered deuterons at a given θ angle and the known cross section for elastic scattering at the studied energy [10].

The normalized experimental breakup event rate obtained for the chosen kinematical configurations at the energy of 340 MeV have been compared to calculations performed by two theory groups (H. Witała et al., A. Deltuva) with the state-of-the-art 2N potentials, combined with 3NF, Coulomb interaction or carried out in a relativistic regime. The results indicate an interplay of 3NF effects, Coulomb force and relativistic effects [11].

Quantitative analysis of the description of the cross section data (σ_{exp}) provided by various calculations (σ_{teor}) is performed in terms of χ^2-like variables. Due to dominating contribution of systematic uncertainties, the following definition has been applied:

$$\chi^2 = \frac{1}{n_{d.o.f.}} \sum \frac{(\sigma_{teor}(\xi) - \sigma_{exp}(\xi))^2}{(\Delta\sigma_{st}(\xi) + \Delta\sigma_{sys}(\xi))^2}, \tag{74.2}$$

where ξ represents a set of kinematic variables ξ=(θ_1, θ_2, φ_{12},S), $\Delta\sigma_{st}(\xi)$ and $\Delta\sigma_{sys}(\xi)$ denote statistical and systematic uncertainties, respectively, summing goes over certain set of kinematic variables and $n_{d.o.f.}$ is a number of data points included in this sum. So defined quantity has no precise statistical meaning, however, it is still a measure of description provided by different models. When its value reaches roughly 2 or more, it can be treated as a signal of inconsistency between the model predictions and the measured data.

The analysis of data sorted in function of combination of polar angles θ_1, θ_2 (Fig. 74.2) provides examples of the Coulomb force dominance in the FSI region,

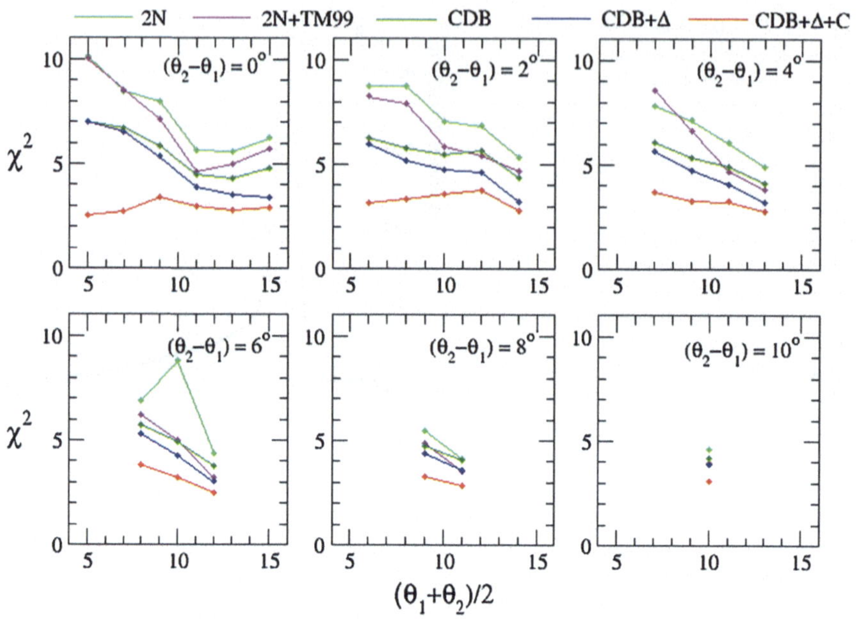

Fig. 74.2 Quality of description of the differential cross section for the breakup reaction at 170 MeV/nucleon beam energy at forward angles given by various models (listed in the legend), presented as χ^2 per degree of freedom calculated for each set of data characterised with the given combination of proton polar angles. The results are ordered according to difference of polar angles $\theta_{12} = \theta_1 - \theta_2$; in each panel results for one value of θ_{12} are shown

characterised by the lowest difference of polar angles. Dominant influence of Coulomb interaction at forward proton emission angles (in laboratory system of the ^{1}H(d, pp)n reaction) is in agreement with studies at other beam energies, see for example [12].

74.3 Summary and Outlook

The analysis is continued with the aim to determine the differential cross sections for the the deuteron breakup process in the d+p system for a large set of kinematic configurations covering a significant part of the reaction phase space. It will be also extended to data sets collected at 300, 380 and 400 MeV.The data will be compared to the theoretical predictions for three nucleon systems. The calculations including relativistic effects with 3NFs and studies of Coulomb effects are very important to draw definitive conclusions.

References

1. Glöckle, W., Witała, H., Hüber, D., Kamada, H., Golak, J.: The three-nucleon continuum: achievements, challenges and applications. Phys. Rep. **274**, 107 (1996)
2. Deltuva, A., et al.: Benchmark calculation for proton-deuteron elastic scattering observables including the Coulomb interaction. Phys. Rev. C **71**, 064003 (2005)
3. Witała, H., et al.: Three-nucleon force in relativistic three-nucleon Faddeev calculations. Phys. Rev. C **83**, 044001 (2011)
4. Bargholtz, Chr., et al.: The WASA detector facility at CELSIUS. Nucl. Instr. Meth. A **594**, 339–350 (2008)
5. Adam, H.H., et al.: arXiv:nucl-ex/0411038 (2004)
6. Adlarson, P., et al.: (WASA-at-COSY collaboration): isotensor dibaryon in the $pp \rightarrow pp\pi^+\pi^-$ reaction? Phys. Rev. Lett. **121**, 052001 (2018)
7. Adlarson, P., et al.: (WASA-at-COSY collaboration): total and differential cross sections of η-production in proton-deuteron fusion for excess energies between $Q\eta = 13$ MeV and $Q\eta = 81$ MeV. Phys. Lett. B **782**, 297–304 (2018)
8. Adlarson, P., et al.: (WASA-at-COSY collaboration): measurement of the $\omega \rightarrow \pi^+\pi^-\pi^0$ Dalitz plot distribution. Phys. Lett. B **770**, 418–425 (2017)
9. Vlasov, P.: Analysis of the $\eta \rightarrow 3\pi^o$ decay in the pp interaction. Ph.D. thesis, Ruhr-Universität Bochum, Germany (2008)
10. Ermisch, K., et al.: Systematic investigation of the elastic proton-deuteron cross section-at intermidate energies. Phys. Rev. C **68**, 051001(R) (2003)
11. Kłos, B., Jamróz, B., Ermisch, K., et al.: Experimental study of three-nucleon dynamics in the dp breakup collisions using the WASA detector. Few-Body Syst. **58**, 38 (2017)
12. Ciepał, I., et al.: Investigation of the deuteron breakup on proton target in the forward angular region at 130 MeV. Few-Body Syst. **56**, 665 (2015)

Chapter 75
Few Nucleon Experiments in the Hadronic Weak Interaction

Jason Fry

Abstract The Hadronic Weak Interaction (HWI) is of great interest as it involves the mixing between weak and strong interaction between quarks. To describe the HWI, meson exchange and effective field theories are utilized. Theoretical and experimental study has remained steadfast since the 1980s, yet the HWI is not constrained. We present an introduction to the theory and recent results from few-body experiments needed to further constrain the HWI.

75.1 Introduction

In General, the weak interaction can be separated into three sectors: leptonic, semi-leptonic, and hadronic weak interactions (HWI). The theoretical development in the leptonic and semi-leptonic sectors has been of great study and the underlying mechanisms are mostly understood. In contrast, the HWI between nucleons is not as well understood. The weak interaction between quarks and leptons is understood, however the HWI competes directly with the strong interaction at the quark level. Since the range of the weak interaction is 100 times smaller than the strong interaction and the strong interaction at short distances is repulsive, HWI dynamics must involve both the quark-quark weak interaction and the low energy, non-perturbative limit of QCD. Thus, we think of the HWI as an "inside out" probe of QCD [1].

At low energies, the HWI between quarks takes on the form of a Hamiltonian in terms of the charged J_μ^W and neutral J_μ^Z weak currents involving only the lightest u, d, and s quarks. The effective strangeness conserving Hamiltonian is [2]

$$\mathcal{H}_W^{\Delta S=0} = \frac{G_F}{2\sqrt{2}}[\cos^2\theta_W J_\mu^{W,0\dagger} J^{W,0\mu} + \sin^2\theta_W J_\mu^{W,1\dagger} J^{W,1\mu} + J_\mu^{Z\dagger} J^{Z\mu}] \quad (75.1)$$

J. Fry (✉)
Institute of Nuclear and Particle Physics, University of Virginia, Charlottesville, VA 22904, USA
e-mail: jaafry@virginia.edu

N. A. Orr et al. (eds.), *Recent Progress in Few-Body Physics*, Springer Proceedings in Physics 238,
https://doi.org/10.1007/978-3-030-32357-8_75

where θ_W is the weak mixing angle and G_F is the fermi constant. The first term consists of $\Delta I = 0$ and 2 contributions from charged currents, the second term consists of suppressed $\Delta I = 1$ contributions from charged currents, and the last term consists of is $\Delta I = 1$ contributions sensitive to weak neutral currents. The latter fact adds additional motivation to study the $\Delta I = 1$ channel.

The mixing of the strong and weak amplitudes between nucleons produces a small parity-violating (PV) signal. The natural scale of this PV interaction is about $\sim 10^{-7}$, but some enhancements in heavy nuclei due to small nuclear energy spacings or resonances can amplify this signal by five orders of magnitude. While heavy nuclei experiments offer large PV admixtures from small level spacings, the theoretical interpretations are not as clean as few nucleon experiments. Few nucleon experiments have large level spacings and small PV asymmetries, which make these experiments difficult, but can provide clean or exact theoretical interpretations. Section 75.3 discusses recent few-body experiments which hope to determine or constrain the different isospin contributions to the HWI.

75.2 Theoretical Landscape

The two main theoretical approaches used study the HWI include a meson exchange model devised by Desplanque, Donoghue, and Holstein (DDH) [3] and various effective field theories (EFT) (see for example, [4–6]). The DDH model describes the low energy HWI with a weak PV potential in terms of meson exchange of π, ρ, and ω mesons which couple a weak vertex to a strong vertex. This forms seven PV couplings, h_π^1, $h_\rho^{0,1,1',2}$, and $h_\omega^{0,1}$ to describe the theory in terms of the meson exchange (subscript) and isospin (superscript). In the DDH benchmark calculation, a reasonable range of the couplings was calculated with error $\sim 100\%$. Additionally, the DDH model contains assumptions of short-distance interactions as well as many-body physics, compounding the theoretical error. Nonetheless, the DDH model has remained an important framework for analyzing experiments for close to 40 years, yet the couplings remain unconstrained.

In order to study the HWI without the model dependence of DDH, vigorous development of EFTs began in the late 1990s and early 2000s. While different theories were developed using slightly different phenomenology [4–7], EFTs describe the HWI in terms of five low energy constants (LEC), as there are only five S–P wave transitions between two nucleons at low energy [8]. These five LECs are to be determined from experiment. While treatments of EFT with and without pions as dynamical degrees of freedom have been analyzed in detail [4], pionless EFT has been successfully used to describe the parity-conserving (PC) processes for few-body systems [9].

Most recently, the $1/N_c$ expansion of QCD has been treated in the HWI framework. This technique was applied to the DDH model [10] and pionless EFT [11] for two-body systems, giving a hierarchy of the couplings or LECs with $\sim 30\%$ error. It was shown that at leading order (LO), the two isoscalar terms are not independent, but they are related by a factor of 3 [11]. This reduces the number of experiments needed to describe the HWI for few-body systems. Finally, the phenomenology of the HWI in the EFT and $1/N_c$ framework was analyzed to find a new LO basis in order to predict the LO expectation of future experiments and provide an assessment of the field [12]. The $1/N_c$ expansion has been a crucial step forward towards clearing up the landscape in the HWI. Below we outline previous results, recent results, and future possible experiments.

75.3 Low Energy Experiments

Prior to the few-body results and outlook in Sect. 75.4, only four experiments bear a reasonable theoretical interpretation and enough precision to place meaningful constraints in the HWI:

1. The longitudinal analyzing power A_L of polarized $\mathbf{p} + p$ scattering performed at 13.6, 15, 45, and 221 MeV gives the most precise constraints to date
2. $\mathbf{p} + \alpha$ elastic scattering at 46 MeV gives a necessary and different linear combination of isospin channels
3. The circular polarization P_γ from the 0^-0 excited state of ^{18}F to the 1^+0 ground state is sensitive to $\Delta I = 1$ couplings, including the long range pion coupling h_π^1. See the results of NPDGamma for a more detailed discussion
4. The angular asymmetry A_γ from the $1/2^-$ excited state to the $1/2^+$ ground state in ^{19}F yields a similar and consistent linear combination with $\mathbf{p} + \alpha$

See the references within [2] for more details. To visualize how these experimental results fit within the 5-6D coupling space, a 2D projection was created in a basis that accommodates most observables [2]. In a recent $1/N_c$ analysis, a new LO hierarchy and basis as been proposed [12]. Both conclude that more few-body experiments are still needed to fully explain the phenomenology.

75.4 Recent Results in Few-Body Systems and Outlook

A set of few-body experiments where the theoretical error can be treated properly and produce further constraints in the HWI include: the spin-angular asymmetry A_γ of γs in $\mathbf{n} + p \to d + \gamma$, circular polarization P_γ of γs emitted in $n + p \to d + \gamma$, the spin-angular asymmetry A_γ of γs in $\mathbf{n} + d \to t + \gamma$, the longitudinal asymmetry A_L in the capture ^{3}He($\mathbf{n}$,p)^{3}H, neutron spin rotation (NSR) $d\phi/dz$ in the p, d, and α systems, the analyzing power A_L of $\mathbf{p} + d$ scattering, and circular polarization of $\gamma + d \to n + p$. See for example [2, 13] for more details.

A program of studying few-body systems in the HWI with neutrons is in progress. At the Spallation Neutron Source (SNS) in Oak Ridge National Lab (ORNL), the NPDGamma collaboration and the n3He collaboration have measured A_γ in

$\mathbf{n} + p \rightarrow d + \gamma$ and A_L in ^{3}He($\mathbf{n}$,p)^{3}H, respectively. The observable A_γ in the NPDGamma experiment is directly related to the long range pion coupling h_π^1 (and analogous LEC), similar to ^{18}F. After about 20 years, the systematics and statistics of the NPDGamma experiment reached its goal uncertainty. The final result from the experiment is $A_\gamma = -3.0 \pm 1.4 \times 10^{-8}$ [14], yielding a slightly higher value of h_π^1 than that extracted from ^{18}F. Since the observable P_γ in ^{18}F also contains contributions from the other $\Delta I = 1$ channels with smaller coefficients, the results agree if the other two isovector couplings give a slightly higher value than previously thought. These results also follow the $1/N_c$ hierarchy. In contrast to NPDGamma, the n3He observable is most sensitive to the $\Delta I = 0$ and 2 channels. The n3He collaboration recently announced the final asymmetry of the PV longitudinal asymmetry, $A_{PV} = 1.2 \pm 1.0 \times 10^{-8}$ [15]. Together, these results tighten the constraints in the HWI, but at least one more few-body experiment is needed to completely constrain the theory.

With the results from NPDGamma and n3He, we are close to a full set of experiments (many- and few-body) needed to completely determine the couplings or LECs in the HWI. The NSR experiment in ^{4}He is planned at NIST in the near future and the $\mathbf{n} + d \rightarrow t + \gamma$ experiment is in development. An $n - p$ spin rotation experiment is appealing as it is sensitive to LO isoscalar and isotensor channels. Recent lattice QCD calculations [16] encourage future calculations of the $\Delta I = 2$ channel as it contains no disconnected quark-loop diagrams.

References

1. Snow, W.: Neutron measurements and the weak nucleon–nucleon interaction. Eur. Phys. J. A **24** (2005)
2. Haxton, W.C., Holstein, B.R.: Hadronic parity violation. Prog. Part. Nucl. Phys. **71** (2013)
3. Desplanques, B., Donoghue, J.F., Holstein, B.R.: Unified treatment of the parity violating nuclear force. Ann. Phys. **124** (1980)
4. Zhu, S.L., Maekawa, C., Holstein, B., Ramsey-Musolf, M., van Kolck, U.: Nuclear parity violation in effective field theory. Nucl. Phys. A **748** (2005)
5. Liu, C.-P.: Parity-violating observables of two-nucleon systems in effective field theory. Phys. Rev. C **75** (2007)
6. Phillips, D.R., Schindler, M.R., Springer, R.P.: An effective-field-theory analysis of low-energy parity-violation in nucleon-nucleon scattering. Nucl. Phys. A **822** (2009)
7. Schindler, M., Springer, R.: The theory of parity violation in few-nucleon systems. Prog. Part. Nucl. Phys. **72** (2013)
8. Danilov, G.: Circular polarization of gamma quanta in absorption of neutrons by protons and isotopic structure of weak interactions. Phys. Lett. **18** (1965)
9. Bedaque, P.F., van Kolck, U.: Effective field theory for few-nucleon systems. Annu. Rev. Nucl. Part. Sci. **52** (2002)
10. Phillips, D.R., Samart, D., Schat, C.: Parity-violating nucleon-nucleon force in the $1/N_c$ expansion. Phys. Rev. Lett. **114** (2015)
11. Schindler, M.R., Springer, R.P., Vanasse, J.: Large-n_c limit reduces the number of independent few-body parity-violating low-energy constants in pionless effective field theory. Phys. Rev. C **93** (2016)

12. Gardner, S., Haxton, W., Holstein, B.R.: A new paradigm for hadronic parity nonconservation and its experimental implications. Annu. Rev. Nucl. Part. Sci. **67** (2017)
13. Fry, J.: Exploring the hadronic weak interaction with cold neutrons in the NPDGamma and NSR experiments. Ph.D. dissertation, Indiana University (2015)
14. Blyth, D., Fry, J., Fomin, N., et al.: First observation of p-odd γ asymmetry in polarized neutron capture on hydrogen. Phys. Rev. Lett. **121**, 242002 (2018)
15. Gericke, M.: Final results from the n^{3}he experiment: parity violation in n-^{3}he capture. https://conferences.lbl.gov/event/137/session/18/contribution/178/material/slides/0.pdf (2018)
16. Wasem, J.: Lattice QCD calculation of nuclear parity violation. Phys. Rev. C **85** (2012)

Chapter 76
Resonance Production of Two Pions in the Reaction $pd \to pd\pi\pi$ at 1–2 GeV

Nurbek Tursunbayev and Yuriy Uzikov

Abstract The recent ANKE@COSY data on the reaction $pd \to pd\pi\pi$ demonstrate a peak at invariant mass of the final $d\pi\pi$ system $M_{d\pi\pi} = 2.38$ GeV that corresponds to the isocalar $J^P = 3^+$ dibaryon D_{03}. The data are analyzed within a theoretical model assuming excitation of the D_{03} dibaryon via the σ-meson exchange. An upper limit on the partial width $D_{03} \to d\sigma$ is estimated from comparison to the data.

Physics of dibaryon resonances in two-nucleon and two-hyperon systems had got a great attention after observation by the WASA@COSY collaboration of the $D_{IJ} = D_{03}$ resonance in the total cross section of the reaction of two-pion production $pn \to d\pi^0\pi^0$ [1], here I is the isospin and J is the total angular momentum of the resonance. The mass of this resonance 2.380 GeV is close to the $\Delta\Delta$-threshold, but its width $\Gamma = 70$ MeV is twice lower as compared to the width of the free Δ-isobar. This narrow width is considered as the most serious indication to quark content of the observed resonance state [2], although the hadronic structure is also possible [3, 4]. One possible mechanism of the reaction $pn \to d\pi^0\pi^0$ suggested in paper [5] involves sequential excitation and decay of two dibaryon resonances, $D_{03}(2380)$ and $D_{12}(2150)$. The spin-parity of this resonance $J^P = 3^+$ were established by polarized measurements, however information about its partial widths is non-complete. Here we are focusing on some production/decay channels of the D_{03} resonance.

N. Tursunbayev · Y. Uzikov
Joint Institute for Nuclear Researches, Dubna 141980, Russia

Dubna State University, Dubna 141980, Russia

Y. Uzikov (✉)
Department of Physics, M.V. Lomonosov Moscow State University, 11991 Moscow, Russia
e-mail: uzikov@jinr.ru

N. Tursunbayev
Institute of Nuclear Physics, Almaty, Kazakhstan

N. A. Orr et al. (eds.), *Recent Progress in Few-Body Physics*, Springer Proceedings in Physics 238,
https://doi.org/10.1007/978-3-030-32357-8_76

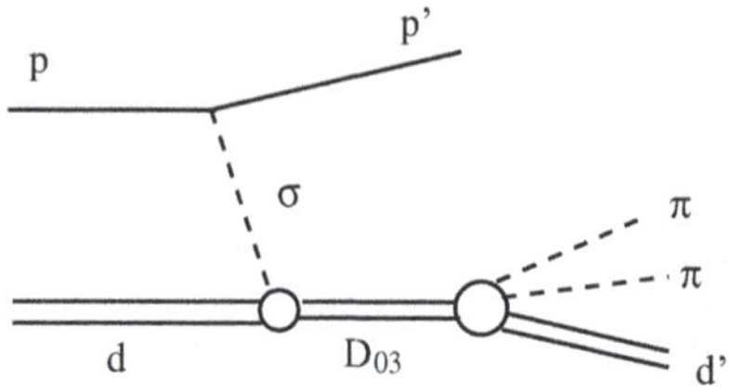

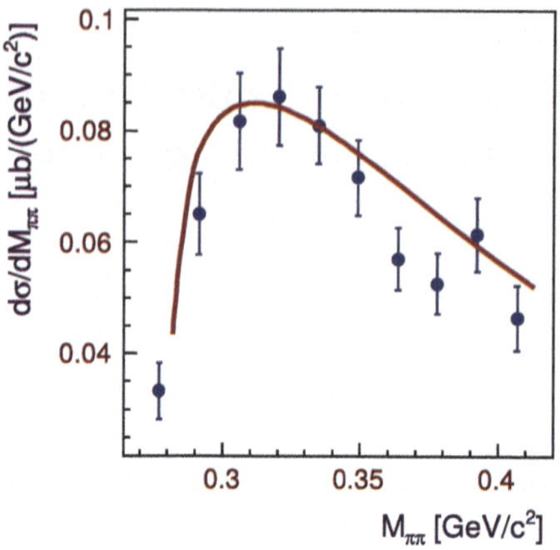

Fig. 76.1 Left: Mechanism of the D_{03} resonance excitation via the σ-meson exchange in the reaction $pd \to pd\pi\pi$. Right: Results of our calculations (line) normalized to the data (•) on the distribution over the invariant mass of two final pions $M_{\pi\pi}$ in the region $M_{d\pi\pi} = 2.357$–2.380 GeV [6] (see text)

Recently ANKE@COSY collaboration have measured the differential cross section of the two-pion production reaction $pd \to pd\pi\pi$ at proton beam energies 0.8-2.0 GeV and small scattering angles of the final proton and deuteron [6]. The resonance peaks were observed in distribution over the invariant mass $M_{d\pi\pi}$ of the final $d\pi\pi$ system at $M_{d\pi\pi} \approx 2.380$ GeV with the averaged width of 115 MeV. Kinematics in [6] considerably differs from that in the quasi-free reaction studied in [1]. Since transferred momentum to the deuteron is large in [6], the mechanisms involving the deuteron form factor strongly underestimate magnitude of the cross section of this reaction. Thus, in order to describe the observed behaviour of the reaction $pd \to pd\pi\pi$ we apply the σ-meson exchange between the proton and deuteron with excitation of the D_{03} resonance (Fig. 76.1) which then decays via the $D_{12}(2150)$ resonance as $D_{03} \to D_{12}\pi \to d\pi\pi$. This mechanism [5] was considered on qualitative level in [6].

Within this model the absolute value of the cross section of the reaction $pd \to pd\pi\pi$ is proportional to the product of the partial widths $\Gamma^{(l=2)}_{D_{03}\to d\sigma}$, $\Gamma^{(l=1)}_{D_{03}\to D_{12}\pi}$, $\Gamma^{(l=1)}_{D_{12}\to d\pi}$. The experimental data on some partial widths are given in [7]. However, the partial widths $\Gamma^{(l=2)}_{D_{03}\to d\sigma}$ and $\Gamma^{(l=1)}_{D_{03}\to D_{12}\pi}$ are not known at present. Here we normalize $\Gamma^{(l=1)}_{D_{03}\to D_{12}\pi}$ in such a way to reproduce the width $\Gamma_{D_{03}\to d\pi^0\pi^0} = 10$ MeV, and after that we can find the width $\Gamma^{(l=2)}_{D_{03}\to d\sigma}$ by normalization the calculated differential cross section to the data [6]. When doing so we obtain $\Gamma^{(l=2)}_{D_{03}\to d\sigma} = 8.5$ MeV for the $M_{\pi\pi}$ distribution in the intervals $M_{d\pi\pi} = 2.357 - 2.380$ GeV (Fig. 76.1) and $M_{d\pi\pi} = 2.380 - 2.412$ GeV. The same result we obtain using $M_{d\pi\pi}$ distribution at beam energy 1.1 GeV, however at 1.4 GeV the result is twice smaller.

The partial widths $\Gamma_{D_{12}\to d\pi}$ and $\Gamma_{D_{12}\to NN}$ were estimated in [8] from the analysis of the cross section of the reaction $pp \to d\pi^+$. The partial width $\Gamma_{D_{03}\to d\pi^0\pi^0}$ is about 10 MeV [7]. Within the two-step model $D_{03} \to D_{12}\pi \to d\pi\pi$ the partial width $\Gamma_{D_{03}\to d\pi^0\pi^0}$ can be estimated as

$$\Gamma(D_{03} \to d\pi^0\pi^0) = (2\pi)^{-3}(2M_{D_{03}})^{-2} \int_{m_d+m_\pi}^{M_{D_{03}}-m_\pi} (p_1k_1)^3|f(M_{d\pi})|^2 dM_{d\pi},$$
$$f(M_{d\pi}) = F_{D_{03}\to D_{12}\pi}\, F_{D_{12}\to d\pi}\, \Pi_{D_{12}}, \quad (76.1)$$

where k_1 is the D_{12} momentum in the D_{03} cms and p_1 is the momentum of the final deuteron in the D_{12} cms, $M_{d\pi}$ is the mass of the final system in the decay $D_{12} \to d\pi$; m_d (m_π) is the mass of the deuteron (pion), $M_{D_{03}}$ is the invariant mass of the D_{03}. In (76.1) $\Pi_{D_{12}}$ is the propagator of the D_{12}, and F are vertex factors defined in [5] as

$$F_{R\to ab} = M_{ab}\sqrt{\frac{8\pi\Gamma^l_{R\to ab}(q_{ab})}{(q_{ab})^{2l+1}}};\ \Gamma^l_{R\to ab}(q) = \Gamma^l_{R\to ab}\left(\frac{q}{q_0}\right)^{2l+1}\left(\frac{q_0^2+\lambda_{ab}^2}{q^2+\lambda_{ab}^2}\right)^{l+1}. \quad (76.2)$$

Here M_{ab} is the mass of the system $a + b$, q_{ab} is the relative momentum in the $a + b$ cms, q_0 is the same momentum at the resonance point $M_{ab} = M_R$, l is the orbital momentum in the decay channel $R \to a + b$.

In our calculations we assume P-waves in the decays $D_{03} \to D_{12}\pi$, $D_{12} \to d\pi$, and D-wave in $\sigma d \to D_{03}$, $D_{12} \to NN$ and use the following parameters $\lambda_{\pi D_{12}} = 0.12$ GeV, $\lambda_{d\sigma} = 0.18$ GeV and $\lambda_{d\pi} = 0.25$ GeV obtained in [5]. The mass and width of the σ-meson are 0.5 GeV and 0.55 GeV, respectively. If one put in (76.1) $\Gamma^{l=1}_{D_{12}\to d\pi} = 10$ MeV [8], then in order to get agreement with the experimental value $\Gamma_{D_{03}\to d\pi^0\pi^0} \approx 10$ MeV one need to put $\Gamma^{(l=1)}_{D_{03}\to D_{12}\pi} = 140$ MeV. Formally this result seems to be in strong contradiction with the total width $\Gamma_{D_{03}} = 70$ MeV. However, we can show that the partial width $\Gamma_{D_{03}\to\pi NN}$ calculated with this value for $\Gamma^{(l=1)}_{D_{03}\to D_{12}\pi}$ via the decays $D_{03} \to D_{12}\pi$ and $D_{12} \to NN$ similarly to (76.1), is $\Gamma_{D_{03}\to\pi NN} = 10$ MeV. This result should be compared with the upper limit to the branching fraction $BF(D_{03} \to \pi NN) \leq 9\%$ [9] that means this width is less than 6.3 MeV.[1]

Conclusion. When normalizing the two-resonance decay width $D_{03} \to D_{12}\pi^0 \to d\pi^0\pi^0$ to the experimental value $\Gamma_{D_{03}\to d\pi^0\pi^0}$, we have got $\Gamma_{D_{03}\to d\sigma} = 8.5$ MeV from further normalization of the calculated cross section to the data on the reaction $pd \to pd\pi\pi$. The D_{03} partial width for the decay $D_{03} \to D_{12}\pi \to NN\pi$ is found to be $\Gamma_{D_{03}\to\pi NN} = 10$ MeV.

[1] In the $\pi N\Delta$ model [10] is found $\Gamma_{D_{03}\to\pi NN} = 5.8$ MeV whereas in the quark model [11] $\Gamma_{D_{03}\to\pi NN}$ is about several hundred keV.

References

1. Adlarson, P., et al.: Phys. Rev. Lett. **106**, 242302 (2011)
2. Dong, Y., Huang, F., Shen, P., Zhang, Z.: Phys. Rev. C **94**, 014003 (2016)
3. Gal, A., Garcilazo, H.: Phys. Rev. Lett. **111**, 172301 (2013)
4. Niskanen, J.A.: Phys. Rev. C **95**, 054002 (2017)
5. Platonova, M.N., Kukulin, V.I.: Phys. Rev. C **87**, 025202 (2013)
6. Komarov, V., et al.: Eur. Phys. J. A **54**, 206 (2018)
7. Bashkanov, M., Clement, H., Skorodko, T.: Eur. Phys. J. A **51**, 87 (2015)
8. Platonova, M.N., Kukulin, V.I.: Nucl. Phys. A **946**, 117 (2016)
9. Adlarson, P., et al.: Phys. Lett. B **774**, 559 (2017)
10. Gal, A.: arxiv:1803.08788 [nucl-th]
11. Dong, Y., Huang, F., Shen, P., Zhang, Z.: Phys. Lett. B **769**, 223 (2017)

Chapter 77
Four-Body Continuum with Three-Nucleon Forces

Michele Viviani, Luca Girlanda, Alejandro Kievsky and Laura E. Marcucci

Abstract We present a study of the effect of three-nucleon forces (3NFs) in $p + {}^3\mathrm{He}$, $p + {}^3\mathrm{H}$ and $n + {}^3\mathrm{He}$ scattering at low energies. The used 3NF is derived from effective field theory at next-to-next-to-leading order. The four-nucleon scattering observables are calculated using the Kohn variational principle and the hyperspherical harmonics technique and the results are compared with available experimental data. We have found that the effect of introducing this particular 3NF is in general tiny except for $p + {}^3\mathrm{H}$ scattering below the opening of the $n + {}^3\mathrm{He}$ channel. In such a case, the effect of 3NF is evident and a clear dependence on the cutoff used to regularize the high-momentum tail of the interactions is observed. Such a dependence is related to the presence of a poorly known sharp 0^+ resonance, considered to be the first excited state of ${}^4\mathrm{He}$.

77.1 Introduction

Four nucleon systems have been object of intense studies in recent years. First of all, these systems are particularly interesting as a "theoretical laboratory" to test the validity of our present knowledge of the nucleon-nucleon (NN) and three nucleon forces (3NFs), in particular after the advent of the theoretical framework of chiral effective field theory (χEFT) (see, for example, [1–4]). The test of these new potentials in few-nucleon scattering will give very stringent and critical information [5].

Moreover, there is a number of reactions involving four nucleons which are of extreme importance for astrophysics, energy production, and studies of fundamental

M. Viviani (✉) · A. Kievsky · L. E. Marcucci
INFN-Pisa, Largo B. Pontecorvo 3, 56127 Pisa, Italy
e-mail: michele.viviani@pi.infn.it

L. Girlanda
Department of Mathematics and Physics, University of Salento,
and INFN-Lecce, Via Arnesano, 73100 Lecce, Italy

L. E. Marcucci
Department of Physics, University of Pisa, Largo B. Pontecorvo 3, 56127 Pisa, Italy

N. A. Orr et al. (eds.), *Recent Progress in Few-Body Physics*,
Springer Proceedings in Physics 238,
https://doi.org/10.1007/978-3-030-32357-8_77

symmetries. As an example, the reactions $d + d \to p + {}^3\mathrm{H}$ and $d + d \to n + {}^3\mathrm{He}$ play a key role in the theory of big-bang nucleosynthesis.

In this work, we present the results of some recent studies of the inclusion of 3NF in low energy $p + {}^3\mathrm{He}$, $p + {}^3\mathrm{H}$ and $n + {}^3\mathrm{He}$ scattering observables. The NN potential used in this paper is the N3LO model by Entem and Machleidt [4], an interaction based on the χEFT approach and derived up to next-to-next-to-next-to-leading order (N3LO) of chiral perturbation theory (χPT). To this NN force, we have added 3NFs derived within the same approach at next-to-next-to-leading order (N2LO) [6, 7], with parameters c_D and c_E fixed in order to reproduce the three-nucleon binding energy and the Gamow-Teller (GT) matrix element in the tritium β-decay [8, 9]. Note that the parameters c_D and c_E have been recently redetermined [9, 10] after finding and correcting an inconsistency between the 3NF and the axial current used so far [11].

In this contribution, we present also a study of the dependence of the results on the cutoff parameter Λ used to regularize the high-momentum tail of these interactions. Specifically, we present the results obtained for two values of the cutoff, $\Lambda = 500$ and 600 MeV. The interactions including only the NN force has been correspondingly denoted as N3LO500 and N3LO600, while those including also the 3NF as N3LO500/N2LO500 and N3LO600/N2LO600, respectively.

The paper is organized as follows. In Sect. 77.2 we report the results of the theoretical calculations, and compare them with the available experimental data. Finally, in Sect. 77.3 we present our conclusions.

77.2 Results

The results presented in this contribution have been obtained using the Kohn variational principle and expanding the wave functions in terms of the hyperspherical harmonic functions (for more details see, for example, [17]). Benchmark papers [18, 19] have been completed where the results obtained with this method have been compared with the analogous results obtained in the framework of the Alt-Sandhas-Grassberger (AGS) equations [20–22], solved in momentum space, and of the Faddeev-Yakubovsky (FY) equations, solved in configuration space [23, 24]. An overall good agreement among the results of the three methods has been found.

Some of the results obtained for $p + {}^3\mathrm{He}$ elastic scattering are reported in Fig. 77.1. The results obtained using NN interaction only or including also the 3NF are shown by bands, whose widths represent the dependence on the cutoff parameter Λ between 500 and 600 MeV. As can be seen inspecting the figure, the widths of the bands are always small, showing that the dependence on Λ is not critical. The effect of the inclusion of 3NF is particularly important for the A_{y0} observable, as already reported some years ago [5].

The results for some $n + {}^3\mathrm{He}$ elastic scattering and $p + {}^3\mathrm{H} \to n + {}^3\mathrm{He}$ charge exchange reaction observables are reported in Figs. 77.2 and 77.3, together with the

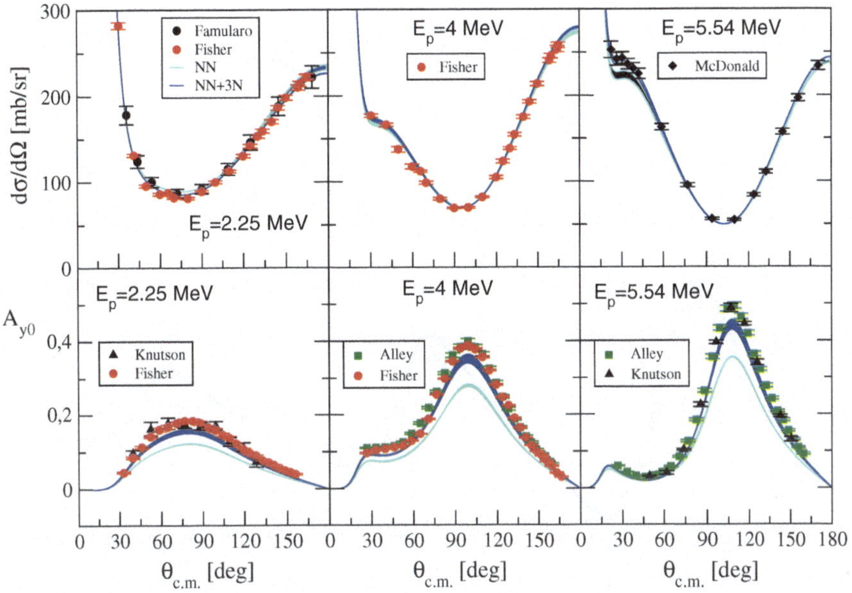

Fig. 77.1 $p + {}^3$He differential cross section and proton analyzing power A_{y0} at different proton laboratory beam energy calculated with the adopted interactions. The cyan bands are the results obtained using NN interaction only, while the blue bands collect the results obtained including the 3NF. The widths of the bands is connected to the dependence of the results on the variation of the cutoff parameter Λ between 500 and 600 MeV. The experimental data are from [12–16]

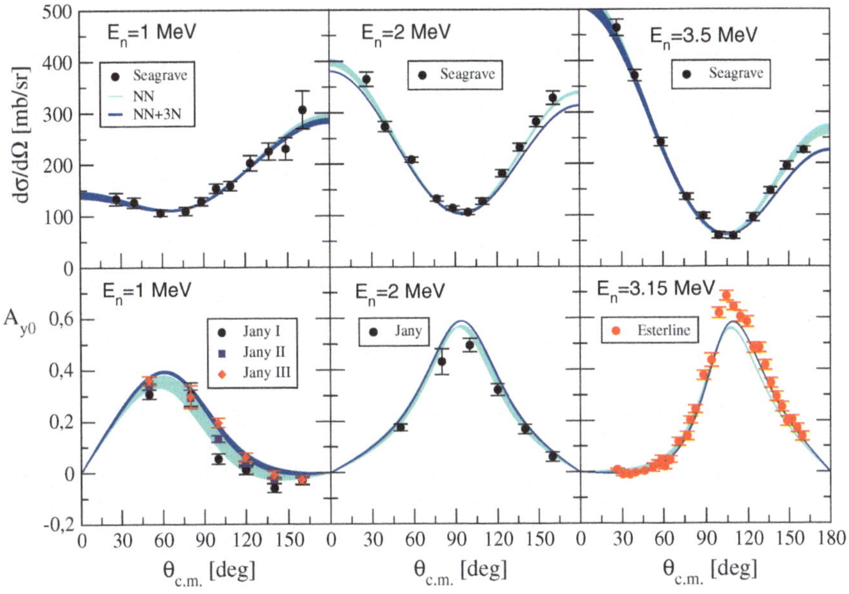

Fig. 77.2 The same as Fig. 77.1 but for $n + {}^3$He scattering. The experimental data are from [25–27]

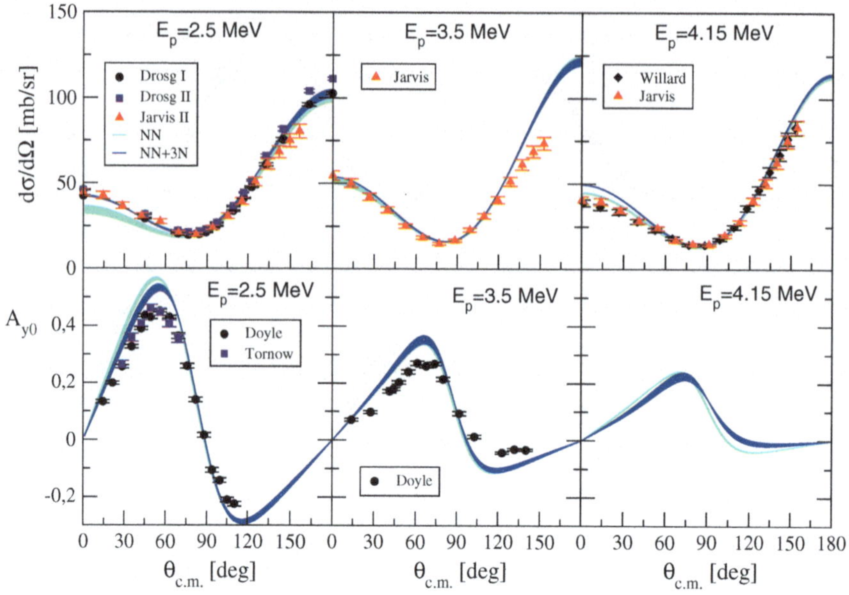

Fig. 77.3 The same as in Fig. 77.1 but for the $p + {}^3\mathrm{H} \rightarrow n + {}^3\mathrm{He}$ reaction. The experimental data are from [28–32]

available experimental data. We see that the contribution of the 3NF is small for these observables. Again, the dependence on the cutoff is not critical.

More interesting are the results for the $p + {}^3\mathrm{H}$ elastic scattering differential cross section, reported in Fig. 77.4. Here we observe that at the two lowest energies the effect of 3NF is sizable. We note also that including the 3NF the observable becomes very dependent on the cutoff used to regularize the interaction. For those energies the $n + {}^3\mathrm{He}$ channel is closed, and the cross section sizeably depends on the position of the first excited state of ${}^4\mathrm{He}$. In fact, the differences in the cross section originates mainly from the 0^+ phase-shift. This quantity is shown in Fig. 77.5. From the phase-shift it is possible to extract the position E_r of the resonance as the maximum of $\delta'(E)$, while the width of the resonance from $\Gamma = 2/\delta'(E_r)$, $\delta'(E)$ being the first derivative of the phase-shift with respect to the energy [39]. The calculated values for these quantities are reported in Table 77.1. As can be seen, E_r is not very dependent on the interaction. However, the width calculated with N3LO500/N2LO500 is rather close to the experimental one (extracted from an R-matrix analysis [40]), while for N3LO600/N2LO600 it is more than twice larger. This study suggests a critical dependence of the position and width of the resonance on the 3NF and the adopted cutoff. The origin of this strong dependence is not clear and it is currently under study.

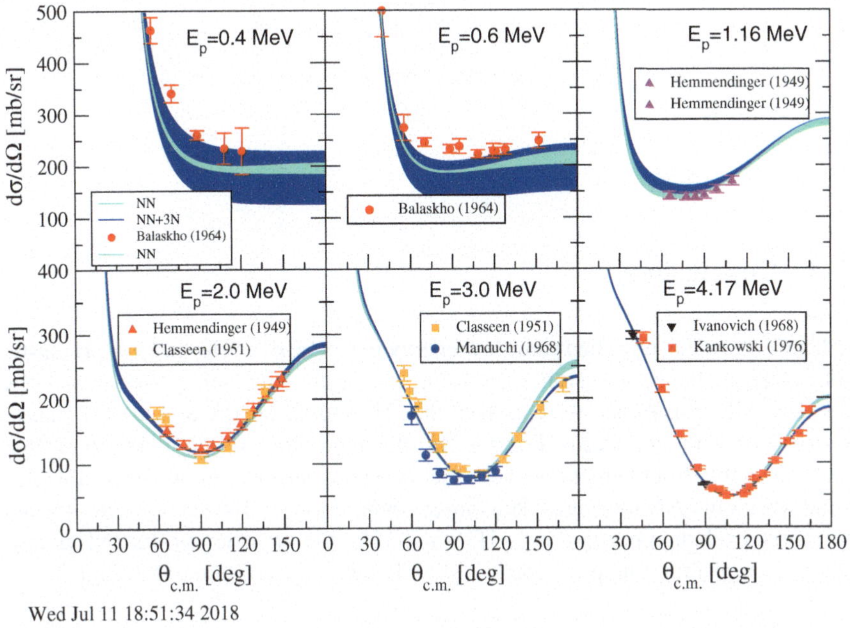

Fig. 77.4 The same as in Fig. 77.1 but for $p + {}^3\mathrm{H}$ scattering. The experimental data are from [33–38]

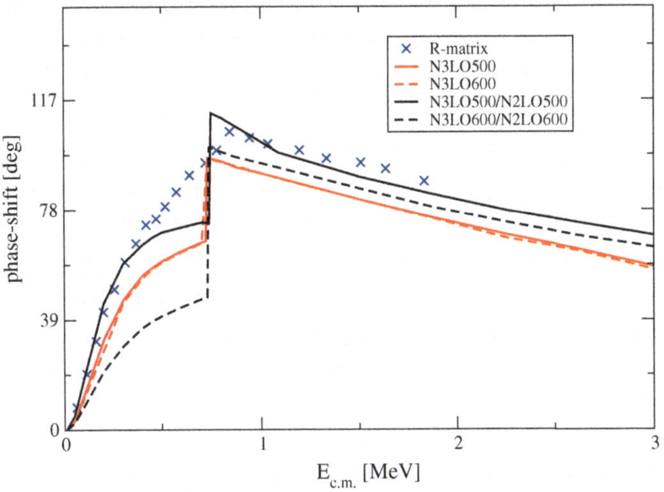

Fig. 77.5 $p + {}^3\mathrm{H}$ phase-shift for the 0^+ wave calculated with the different interactions considered in this contribution. The crosses are the phase-shift extracted from the data using an R-matrix analysis [40]

Table 77.1 Energy of the resonance and its width as extracted from the phase-shifts reported in Fig. 77.5. The experimental values are extracted from the R-matrix analysis reported in [40]

	N3LO500	N3LO500/N2LO500	N3LO600/N2LO600	Expt.
E_r [MeV]	0.09	0.07	0.09	0.39
Γ [MeV]	0.57	0.40	0.91	0.50

77.3 Conclusions

In this work, we have studied some low energy $p + {}^3$He, $n + {}^3$He, and $p + {}^3$H elastic and charge-exchange observables by using interactions including NN forces only, and also 3NF constrained to reproduce the ^{3}H binding energy and the GT matrix element in tritium beta decay. For $p + {}^3$He the main effect of the inclusion of 3NF is to reduce the disagreement between theory and experiment in the observable A_{y0} found when only NN force are taken into account. For $n + {}^3$He elastic scattering and the charge exchange reaction $p + {}^3\text{H} \to n + {}^3$He, the inclusion of the 3NF is tiny, although in general it helps to obtain a slightly better description of the data.

On the other hand, for $p + {}^3$H elastic scattering at energies below the opening of the $n + {}^3$He channel some sizable effects are observed, in particular a rather strong cutoff dependence when the 3NF is included in the calculations. We have found that this is related to a critical dependence on the 3NF of the position and width of the resonance representing the first excited state of ^{4}He. Further studies of this resonance are currently in progress. Moreover, it would be rather important to have new and more accurate measurements of $p + {}^3$H elastic scattering at these low energies, as the available experimental data are rather old and of limited angular range.

References

1. Weinberg, S.: Phys. Lett. B **251**, 288 (1990); Nucl. Phys. B **363**, 3 (1991); Phys. Lett. B **295**, 114 (1992)
2. Ordonez, C., Ray, L., van Kolck, U.: Phys. Rev. C **53**, 2086 (1996)
3. Epelbaum, E., Hammer, H.W., Meissner, U.-G.: Rev. Mod. Phys. **81**, 1773 (2009)
4. Machleidt, R., Entem, D.R.: Phys. Rep. **503**, 1 (2011)
5. Viviani, M., Girlanda, L., Kievsky, A., Marcucci, L.E.: Phys. Rev. Lett. **111**, 172302 (2013)
6. van Kolck, U.: Phys. Rev. C **49**, 2932 (1994)
7. Epelbaum, E., et al.: Phys. Rev. C **66**, 064001 (2002)
8. Gardestig, A., Phillips, D.R.: Phys. Rev. Lett. **96**, 232301 (2006); Gazit, D., Quaglioni, S., Navrátil, P.: ibid. **103**, 102502 (2009)
9. Marcucci, L.E., et al.: Phys. Rev. Lett. **108**, 052502 (2012); Erratum, Phys. Rev. Lett. **121**, 049901 (2018)
10. Baroni, A., et al.: Phys. Rev. C **98**, 044003 (2018)
11. Schiavilla, R.: Unpublished
12. Famularo, K.F., et al.: Phys. Rev. **93**, 928 (1954)
13. McDonald, D.G., Haberli, W., Morrow, L.W.: Phys. Rev. **133**, B1178 (1964)

14. Alley, M.T., Knutson, L.D.: Phys. Rev. C **48**, 1890 (1993)
15. Viviani, M., et al.: Phys. Rev. Lett. **86**, 3739 (2001)
16. Fisher, B.M., et al.: Phys. Rev. C **74**, 034001 (2006)
17. Kievsky, A., Rosati, S., Viviani, M., Marcucci, L.E., Girlanda, L.: J. Phys. G: Nucl. Part. Phys. **35**, 063101 (2008)
18. Viviani, M., et al.: Phys. Rev. C **84**, 054010 (2011)
19. Viviani, M., et al.: Phys. Rev. C **95**, 034003 (2017)
20. Deltuva, A., Fonseca, A.C.: Phys. Rev. C **75**, 014005 (2007)
21. Deltuva, A., Fonseca, A.C.: Phys. Rev. C **76**, 021001 (2007)
22. Deltuva, A., Fonseca, A.C.: Phys. Rev. Lett. **113**, 102502 (2014); Phys. Rev. C **90**, 044002 (2014)
23. Lazauskas, R., Carbonell, J.: Nucl. Phys. A **737**, S79 (2004)
24. Lazauskas, R.: Phys. Rev. C **79**, 054007 (2009)
25. Seagrave, J.D., Cranberg, L., Simmons, J.E.: Phys. Rev. **119**, 1981 (1960)
26. Jany, P., Heeringa, W., Klages, H.O., Zeitnitz, B., Garret, R.: Nucl. Phys. A **483**, 269 (1988)
27. Esterline, J., Tornow, W., Deltuva, A., Fonseca, A.C.: Phys. Rev. Lett. **110**, 152503 (2013)
28. Willard, H.B., Bair, J.K., Kington, J.D.: Phys. Rev. **90**, 865 (1953)
29. Jarvis, G.A.: Los Alamos Scientific Report No. 2014, p. 35 (1956)
30. Drosg, M.: Los Alamos Scientific Report No. 8215 (1980)
31. Doyle Sr., M.A., et al.: Nucl. Phys. A **371**, 225 (1981)
32. Tornow, W., Byrd, R.C., Lisowski, P.W., Walter, R.L., Donoghue, T.R.: Nucl. Phys. A **371**, 235 (1981)
33. Hemmendinger, A., Jarvis, G.A., Taschek, R.F.: Phys. Rev. **76**, 1137 (1949)
34. Claassen, R.S., Brown, R.J.S., Freier, G.D., Stratton, W.R.: Phys. Rev. **82**, 589 (1951)
35. Balashko, Y.G., Barit, I.Y., Dulkova, L.S., Kurepin, A.B.: Bull. Russ. Acad. Sci. Phys. **28**, 1028 (1965)
36. Manduchi, C., Moschini, G., Tornielli, G., Zannoni, G.: Nuovo Cim. B **57**, 340 (1968)
37. Ivanovich, M., Young, P.G., Ohlsen, G.G.: Nucl. Phys. A **110**, 441 (1968)
38. Kankowsky, R., Fritz, J.C., Kilian, K., Neufert, A., Fick, D.: Nucl. Phys. A **263**, 29 (1976)
39. Thompson, I.J., Nunes, F.: Nuclear Reactions for Astrophysics, p. 301. Cambridge University Press, 2009
40. Tilley, D.R., Weller, H.R., Hale, G.M.: Nucl. Phys. A **541**, 1 (1992)

Chapter 78
4-Nucleon System Dynamics in Proton Helium-3 Scattering

A. Kozela, I. Ciepał, I. Skwira-Chalot, B. Głowacz, P. Kulessa, T. Pałasz, W. Parol, E. Stephan, B. Włoch and J. Zejma

Abstract We propose kinematically complete measurement of the differential cross section and analyzing power in proton polarized helium-3 elastic scattering and breakup. The proposed method, involving tracking of charged reaction products and subsequent reconstruction of the reaction vertex, should be regarded as a novel experimental approach in this low energy nuclear physics experiment. The details of the proposed experimental setup and prospects of the first measurement at the new Cyclotron Center Bronowice in Kraków are also given.

78.1 Introduction

While the best available nucleon-nucleon potentials reached their mature state, being able to describe systems composed of two nucleons close to perfectly [1, 2], addition of the third nucleon spoils the situation considerably. Accounting for a new dynamical component, appearing in such systems, referred to as a three-nucleon force (3NF), greatly decreases the differences between the experiment and theoretical calculations, still leaving a lot of space for improvement. The same holds true even after recent, technically very demanding, inclusion of Coulomb force and relativistic effects [3, 4]. One still observes discrepancies in differential cross section and observables involving nucleon's polarization [5], which change smoothly with relevant dynamical variables. It is a general expectation that these deficiencies will increase after transition to the systems involving four nucleons. Though even more

A. Kozela (✉) · I. Ciepał · P. Kulessa · W. Parol · B. Włoch
Henryk Niewidniczański Institute of Nuclear Physics PAS, Kraków, Poland
e-mail: adam.kozela@ifj.edu.pl

E. Stephan
Marian Smoluchowski Institute of Physics, Jagiellonian University, Kraków, Poland

B. Głowacz · T. Pałasz · J. Zejma
August Chełkowski Institute of Physics, University of Silesia, Chorzow, Poland

I. Skwira-Chalot
Faculty of Physics, University of Warsaw, Warsaw, Poland

N. A. Orr et al. (eds.), *Recent Progress in Few-Body Physics*,
Springer Proceedings in Physics 238,
https://doi.org/10.1007/978-3-030-32357-8_78

difficult, such systems offer quite new possibilities for 3NF studies (as e.g. accessibility of new isospin channels, new combinations of beam and target system) and appearing here 4-nucleon force. Different theoretical approaches, like e.g. single-scattering approximation [6], call for measurements precise enough to validate or fine tune applied models.

78.2 Experimental Method

A schematic view of the proposed experimental setup is shown in Fig. 78.1. Isotopically pure polarized ^{3}He gas is contained in a cylindrical glass target cell under near atmospheric pressure. Its axis matches that of the proton beam line. Thin entrance and exit windows on the beam axis minimize beam induced background, competitive with events of interest originating from proton – ^{3}He reactions. Additional windows in the side walls of the cell allow to reduce energy loss and angular straggling of charged particles produced in the reactions inside the cell. This is especially important in the case of double-charged ^{3}He nuclei.

Two identical detector modules, placed directly in front of the target cell side windows will be used for particle identification and tracking. Each module will consist of a multiwire drift chamber (MWDC) and a scintillator detector. The main role of the MWDC detector will be precise reconstruction of charged particles trajectories. Application of at least six measuring planes with wires at three different directions will allow for unique and efficient reconstruction of events with even two trajectories in one module. Combined pulse height information from all wires belonging to

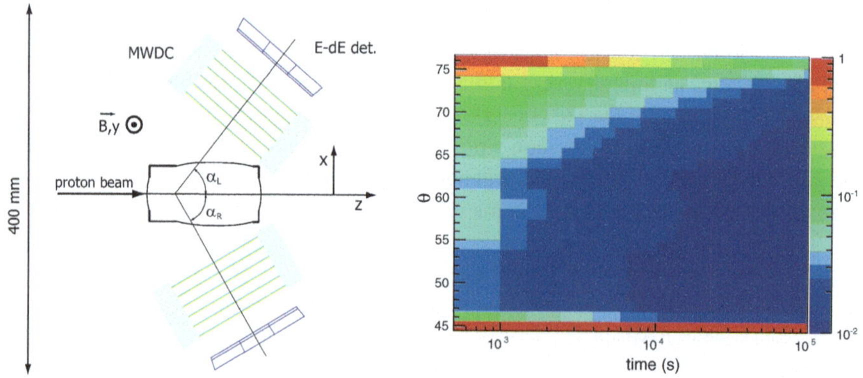

Fig. 78.1 Left panel: schematic top view of the experimental setup. The arrangement of the left and right detector modules with respect to the beam is described by polar angles α_L, α_R, adjustable independently for each module within a range 45°–135°. The whole system is immersed in uniform spin-holding vertical magnetic field $B \approx 2$ mT. Right panel: estimated uncertainty of vector analyzing power as a function of proton emission polar angle θ in laboratory frame and the time of the measurement for optimal, symmetric detector configuration ($\alpha_L = \alpha_R \approx 56°$)

one track, proportional to specific energy losses of a particle, will contribute to the particle identification. In order to accomplish this goal dedicated electronics has to be used for all MWDC channels.

Directly behind each MWDC a scintillator hodoscope for the energy loss and the total kinetic energy measurement will be positioned, see Fig. 78.1. Depending on proton beam energy scintillating material of 30–150 mm thickness, characterized by a long decay time constant, can be used. For the energy loss measurement 1–2 mm thick and much faster plastic scintillator (e.g. BC-408) will be used, allowing for the application of so called phoswich mode of operation in which thin and thick scintillators are read out by the same photomultiplier device. Based on the obtained in this way information particle identification will be applied, which in the case of the proposed measurement can be reduced to distinguishing between protons, deuterons and ^{3}He nuclei. In order to minimize effects resulting from high background count rates, and to allow for two charged particles reconstruction in a given detector module, the calorimeter will be divided into seven, optically isolated segments.

The polarization of ^{3}He will be produced with the use of Spin-Exchange Optical Pumping method applied in a separate polarizing cell. This method allows one to obtain substantial polarization for relatively high pressure of the gas sample. The polarization will be measured with a standard NMR technique. Two pairs of standard Helmholtz coils, producing field perpendicular to each other and to the beam axis will be used to create uniform spin holding magnetic field. Smooth control of the current flow in these coils will allow one to rotate the magnetic field (and consequently ^{3}He spins) around the target axis. A dedicated set of coils to compensate for the Earth magnetic field can be used if required.

Figure 78.1, right shows simulated dependence of relative uncertainty of vector analyzing power as a function of measurement time and proton polar emission angle in LAB. The calculations assumed conservatively rather small beam current of 100 pA, fine angular binning ($\Delta\theta = 1°$) and the uncertainty of helium polarization equal to 1%, which seams reasonable in the proposed method of polarization measurement. Obtained results allow one to conclude, that in the symmetric configuration of proposed experiment the contribution of statistical precision to the analyzing power error can reach the level of the contribution of polarization uncertainty already after about an hour of measurement. Due to the quick drop of the differential cross section with the polar angle θ, the situation changes rapidly with the change of the proton emission angle and for utterly disadvantageous conditions of asymmetric configuration, ($\theta \approx 120°$) about two days of measurement will be necessary, to attain similar precision. Concerning the statistical uncertainty of the differential cross section measurement the situation is correspondingly better, however good control of the beam current is mandatory in this case.

The experiment received positive recommendation from International Advisory Committee working at the Institute of Nuclear Physics PAS in Kraków and is scheduled for implementation at Cyclotron Center Bronowice in this institute.

References

1. Sauer, P.U., et al.: Phys. Rev. C **68**, 024005 (2003)
2. Viviani, M., et al.: Phys. Rev. Lett. **86**, 3739 (2001)
3. Deltuva, A., et al.: Phys. Rev. C **73**, 057001 (2006)
4. Witala, H., et al.: Phys. Rev. C **83**, 044001 (2011)
5. Kistryn, St., Stephan, E.: J. Phys. G: Nucl. Part. Phys. **40**, 063101 (2013)
6. Deltuva, A., et al.: Phys. Rev. C **93**, 044001 (2016)

Chapter 79
Measurement of ^{3}He Analyzing Power for $p-^3$He Elastic Scattering at 70 MeV

A. Watanabe, K. Sekiguchi, T. Akieda, D. Etoh, Y. Inoue, K. Kawahara, H. Kon, K. Miki, T. Mukai, S. Nakai, D. Sakai, S. Shibuya, Y. Shiokawa, T. Taguchi, Y. Wada, M. Watanabe, M. Itoh, T. Ino and T. Wakui

Abstract The ^{3}He analyzing powers for $p-^3$He elastic scattering with the polarized ^{3}He target have been measured at 70 MeV. ^{3}He polarization which was calibrated by the electron paramagnetic resonance method was achieved to 50%. In the conference the data were compared with the theoretical calculations based on the modern nucleon–nucleon potentials. Large discrepancies between the theoretical calculations and the experimental data were found at the angles where the ^{3}He analyzing power takes minimum and maximum.

79.1 Introduction

The three nucleon forces (3NFs) are essentially important to clarify various nuclear properties such as few-nucleon scattering [1], binding energies of light mass nuclei [2] and equation of state of nuclear matter [3]. With the aim of studying 3NFs, we have performed the experimental studies of nucleon–deuteron scattering at intermediate energies (65–300 MeV/nucleon) [4]. In the case of the cross section for the elastic deuteron–proton scattering at 135 MeV/nucleon, large discrepancies between the

A. Watanabe (✉) · K. Sekiguchi · T. Akieda · D. Etoh · Y. Inoue · K. Kawahara · H. Kon · K. Miki · T. Mukai · S. Nakai · D. Sakai · S. Shibuya · Y. Shiokawa · T. Taguchi · Y. Wada · M. Watanabe
Department of Physics, Tohoku University, Sendai, Miyagi 980-8578, Japan
e-mail: watanabe@lambda.phys.tohoku.ac.jp

K. Sekiguchi · T. Akieda · D. Etoh · Y. Inoue · K. Kawahara · H. Kon · K. Miki · T. Mukai · S. Nakai · D. Sakai · S. Shibuya · Y. Shiokawa · T. Taguchi · Y. Wada · M. Watanabe · M. Itoh
Cyclotron and Radioisotope Center (CYRIC), Tohoku University, Sendai, Miyagi 980-8578, Japan

K. Sekiguchi · T. Akieda · D. Etoh · Y. Inoue · K. Kawahara · H. Kon · K. Miki · T. Mukai · S. Nakai · D. Sakai · S. Shibuya · Y. Shiokawa · T. Taguchi · Y. Wada · M. Watanabe · T. Ino
High Energy Accelerator Research Organization (KEK), Tsukuba, Ibaraki 305-0801, Japan

K. Sekiguchi · T. Akieda · D. Etoh · Y. Inoue · K. Kawahara · H. Kon · K. Miki · T. Mukai · S. Nakai · D. Sakai · S. Shibuya · Y. Shiokawa · T. Taguchi · Y. Wada · M. Watanabe · T. Wakui
National Institute of Radiological Science, Chiba 263-8555, Japan

N. A. Orr et al. (eds.), *Recent Progress in Few-Body Physics*,
Springer Proceedings in Physics 238,
https://doi.org/10.1007/978-3-030-32357-8_79

experimental data and the rigorous numerical calculations based on the modern nucleon–nucleon (NN) potentials (such as AV18 [5], CD-Bonn [6] and Nijmegen I, II, 93 [7]) are found. The theoretical calculations with the Tucson-Melbourne'99 [8] or Urbana IX [9] 3NF reproduce well the experimental data. This is taken as the first signature of the 3NF effects in three-nucleon scattering. However, the total isospin of Nd scattering system is limited to $T = 1/2$. Recently, the importance of the isospin dependence of 3NFs has been pronounced for the understanding of asymmetric nuclear matter (e.g., neutron-rich nuclei, neutron matter and neutron stars) [2, 3]. In order to explore the properties of the 3NFs in $A \geq 3$ nuclear systems and to approach the isospin dependence of 3NFs, we are planning the measurement of proton–^{3}He elastic scattering at intermediate energies. It should also be noted that rigorous numerical calculations of this system are becoming available up to 35 MeV [10, 11]. As the first step, we have performed the measurement of ^{3}He analyzing powers at 70 MeV.

79.2 The Polarized ^{3}He Target

We developed the polarized ^{3}He target system for the measurement of ^{3}He analyzing power. A schematic view of the polarized ^{3}He target system is shown in Fig. 79.1. In order to polarize ^{3}He nucleus, we adopted the alkali-hybrid spin-exchange optical pumping (AH-SEOP) method. This method consisted of two step processes. First, Rb vapor was polarized by optical pumping with circularly polarized light in the presence of a static magnetic field. Then the polarization of Rb atoms was transferred to K atoms via spin-exchange collision. Second, the alkali metal polarization was transferred to ^{3}He nuclei by hyper-fine interaction. We applied a spectrally narrowed laser with the optics to produce circularly polarized light. The output power of the laser was 60 W and FWHM was 0.2 nm. The main coils were the Helmholtz type providing a static magnetic field with high homogeneity. The target cell consisted of a cylindrical pumping chamber with a length of 4.5 cm and a cylindrical target chamber with a length of 15.0 cm. It contained the ^{3}He gas with pressure of 3 atm at room temperature together with a small amount of Rb and K as well as N_2 gas. The target cell was placed at the center of the main coils, and the pumping chamber was placed in the oven to obtain a sufficient Rb and K vapor. We measured the target polarization by using the adiabatic fast passage-NMR (AFP-NMR) method in combination with the electron paramagnetic resonance (EPR) method [12]. The AFP-NMR system consisted of the main coils, the drive coils and the pick-up coil. The spin axis of ^{3}He was flipped by sweeping a static magnetic field and applying a RF field by the drive coils under the AFP condition. The induced NMR signals were detected by the pick-up coil. We have obtained the absolute values of the ^{3}He polarization by using the EPR method. The maximum ^{3}He polarization was $\sim$50% in our system.

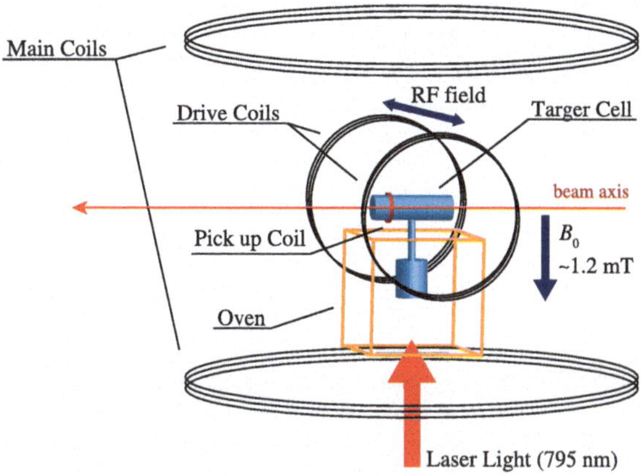

Fig. 79.1 Schematic view of the polarized ^{3}He target

79.3 The Measurement of ^{3}He Analyzing Power

We performed the measurement of ^{3}He analyzing powers for p–^{3}He elastic scattering at Cyclotron and Radioisotope Center (CYRIC), Tohoku University. The proton beams were accelerated up to 70 MeV by the AVF cyclotron. A typical beam intensity was 5–10 nA. Charge collection of the beams was performed by using the Faraday cup placed downstream of the target. We also monitored relative values of the beam intensities by using the beam monitor installed in the vacuum chamber placed upstream of the target. The scattered protons from the polyethylene film with thickness of 20 μm were detected by the beam monitor detectors. The scattered protons from the polarized ^{3}He target were detected by $\Delta E - E$ detectors placed on a left and right side of the target. The ΔE detectors were plastic scintillators with thickness of 0.2, 0.5 or 1 mm coupled with photomultiplier tubes (PMTs). The E detectors were a NaI(Tl) scintillator with thickness of 50 mm. The measured angles were $\theta_{\mathrm{c.m.}} = 46°$–$141°$ in the center of mass system. During the experiment, we measured the ^{3}He polarization and flipped ^{3}He nuclear spin direction by using the AFP-NMR method every hour.

79.4 Results

The ^{3}He analyzing powers for p–^{3}He elastic scattering are formulated as

$$\frac{d\sigma}{d\Omega} = \left(\frac{d\sigma}{d\Omega}\right)_0 (1 + p_y A_y), \tag{79.1}$$

where $d\sigma/d\Omega$ ($(d\sigma/d\Omega)_0$) is the polarized (un-polarized) differential cross section. The p_y denotes the vector polarizations of the target. The polarization axis is per-

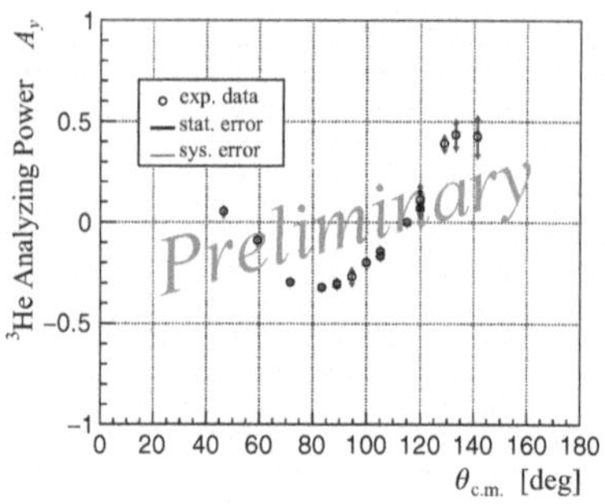

Fig. 79.2 ^{3}He analyzing power A_y for $p-^3$He elastic scattering at 70 MeV. The theoretical calculations are not shown here

pendicular to the reaction plane in this experiment. The A_y denotes the analyzing powers for $p-^3$He elastic scattering.

Figure 79.2 shows the preliminary results of the experimental data. In the conference the data were compared with the rigorous numerical four-nucleon calculations based on the modern NN potentials [13]. The angular distribution of the experimental data had a moderate agreement with the theoretical calculation. However, the large discrepancies are found at the angles $\theta_{c.m.} \simeq 80°$ and $130°$.

79.5 Summary and Outlook

We have performed the measurement of ^{3}He analyzing powers for $p-^3$He elastic scattering at 70 MeV in a wide angular range ($\theta_{c.m.} = 46°$–$141°$). In the conference the experimental data were compared with the calculations based on the modern NN potentials. Large discrepancies between the data and the calculations were found at the angles $\theta_{c.m.} \simeq 80°$ and $130°$. In order to perform quantitative discussions of the 3NF effects, study is in progress both from experiment and theory. As the next step, we are planning the measurement of spin observables for $p-^3$He at higher energies at Research Center for Nuclear Physics (RCNP), Osaka university.

References

1. Kalantar-Nayestanaki, N., et al.: Rep. Prog. Phys. **75**, 016301 (2012)
2. Pieper, S.C., et al.: Phys. Rev. C **64**, 014001 (2001)
3. Akmal, A., Pandharipande, V.R., Ravenhall, D.G.: Phys. Rev. C **58**, 1804 (1998)
4. See e.g., Sekiguchi, K., et al.: Phys. Rev. C **65**, 034003 (2002); ibid. **89**, 064007 (2014)
5. Wiringa, R.B., Stoks, V.G.J., Schiavilla, R.: Phys. Rev. C **51**, 38 (1995)
6. Machleidt, R.: Phys. Rev. C **63**, 024001 (2001)
7. Stoks, V.G.J., et al.: Phys. Rev. C **49**, 2950 (1994)
8. Coon, S.A., Han, H.K.: Few-Body Syst. **30**, 131 (2001)
9. Pudliner, B.S., et al.: Phys. Rev. C **56**, 1720 (1997)
10. Viviani, M., et al.: Phys. Rev. Lett. **111**, 172302 (2013)

11. Deltuva, A., Fonseca, A.C.: Phys. Rev. C **87**, 054002 (2013)
12. Romalis, M.V., Cates, G.D.: Phys. Rev. A **58**, 3004 (1998)
13. Deltuva, A.: Private communications

Chapter 80
Exploring High Quality Chiral Forces

E. F. Batista, S. Szpigel and V. S. Timóteo

Abstract We investigate the interplay between the pion exchanges and the contact interactions in the $^1S_0$ channel using a N4LO chiral potential. We compute the pairing gap without the pions and show that, although the contact interactions dominate in this channel, the gap is strongly enhanced without the attraction that comes from the pions.

80.1 Introduction

The nuclear interaction has been subject of intense research in the last decades with the advent of the Effective Field Theory (EFT) approach [1, 2]. Many nucleon-nucleon interactions were constructed based on the chiral expansion and currently chiral forces have reached the same level of precision than phenomenological potentials [3–7]. The EFT approach is fundamental and systematic but the organization of the processes into a hierarchy of interactions, known as Power Counting, and different renormalization schemes for the theory have also been investigated [8–10].

E. F. Batista
Departamento de Ciências Exatas e Naturais - DCEN,
Universidade Estadual do Sudoeste da Bahia - UESB, Itapetinga, BA, Brazil

S. Szpigel
Centro de Rádio-Astronomia e Astrofísica Mackenzie - CRAAM,
Universidade Presbiteriana Mackenzie, São Paulo, SP, Brazil

V. S. Timóteo (✉)
Grupo de Óptica e Modelagem Numérica - GOMNI, Faculdade de Tecnologia - FT,
Universidade Estadual de Campinas - UNICAMP, Limeira, SP, Brazil
e-mail: varese@uincamp.br

N. A. Orr et al. (eds.), *Recent Progress in Few-Body Physics*,
Springer Proceedings in Physics 238,
https://doi.org/10.1007/978-3-030-32357-8_80

80.2 Pion Exchanges Versus Contact Interactions at N4LO

The nucleon-nucleon interaction in chiral perturbation theory can be separated in two parts: the pion exchanges (1π, 2π, 3π) and the contact terms, whose strengths are determined by matching the full interaction to scattering observables for two nucleons. For a given partial wave, we have $V^{\rm ful}_{\rm N4LO} = V^{\rm pions}_{\rm N4LO} + V^{\rm contacts}_{\rm N4LO}$.

Figure 80.1 shows the contributions from the pion exchanges and the contact terms to the N4LO potential for the case of the $^1S_0$ partial wave. It is clear from the pion contributions (in blue), the contact contributions (in red) and the full interaction (in black) that the "unknown" part of the interaction dominates in this channel. The contact terms and the full interaction are very similar while the pions have the opposite behavior (see lower right panel).

80.3 Phase-Shift and Pairing Gap in the $^1S_0$ Channel

Both the phase-shifts for two nucleons and the pairing gap in nuclear matter are calculated directly from the matrix elements of the NN potential. The phase-shifts can be obtained from the T-matrix, which is a solution of the Lippman-Schwinger equation

$$T(p, p') = V(p, p') + \frac{2}{\pi}\int dq\, q^2\, V(p, q)\frac{1}{p^2 - q^2 + i\epsilon}T(q, p')\,, \qquad (80.1)$$

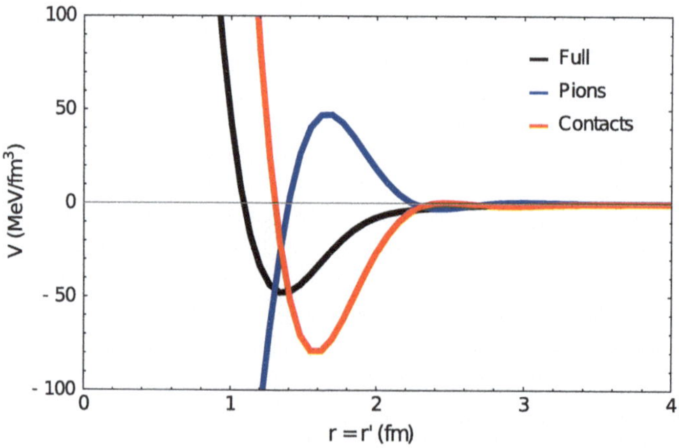

Fig. 80.1 Nuclear force in the $^1S_0$ channel at N4LO in configuration space. Only the diagonal part of the potential matrix is shown: Pion exchanges (blue), Contact interactions (red) and Full interaction (black)

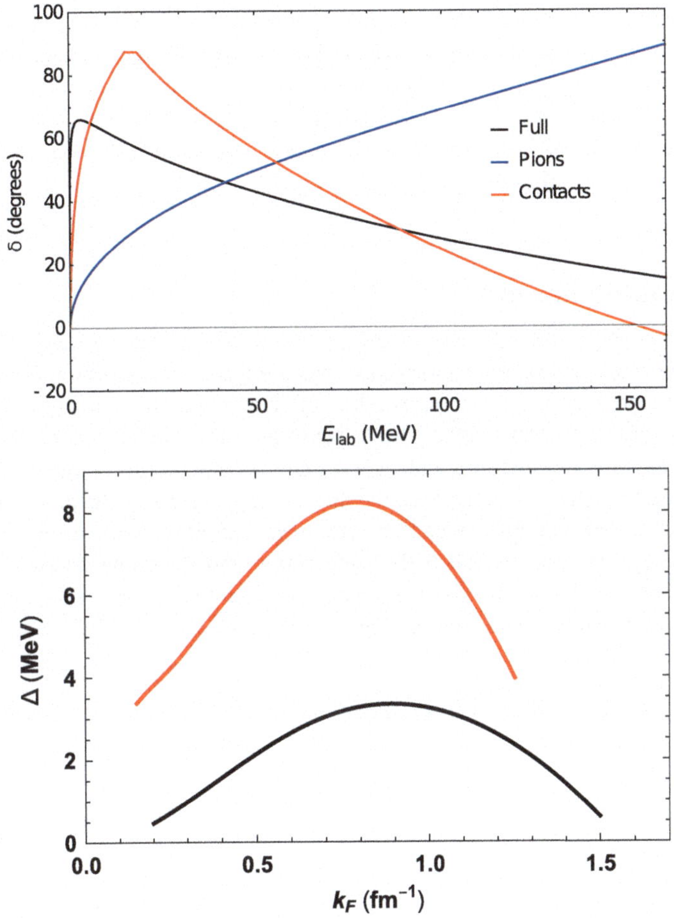

Fig. 80.2 Contributions to the phase-shift (upper panel) and to the pairing gap (lower panel). Pion exchanges are shown in blue, contact terms in red and full interaction in black

and the pairing gap is a solution of the BCS equation

$$\Delta(k_F) = -\frac{2}{\pi}\int dp\, p^2\, V(k_F, p)\frac{1}{2ME(p)}\Delta(p)\,, \qquad (80.2)$$

where $2ME(p) = \sqrt{(p^2 - k_{\rm F}^2)^2 + 4M^2\Delta^2(p)}$ (M is the nucleon mass and k_F is the Fermi momentum).

The phase shifts and the pairing gap corresponding to $V^{\rm pions}_{\rm N4LO}$ and $V^{\rm contacts}_{\rm N4LO}$ can then be calculated using (80.1) and (80.2). In the upper panel of Fig. 80.2 we show the calculations for the phase-shift indicating that the dominance of contact terms over pions observed in Fig. 80.1 makes the shape of the phase-shifts to be mostly given by the contact interactions and removing pions essentially enhances the phases at low energies but makes them repulsive at about 150 MeV. Such enhancement also applies to the pairing gap, although in this case the size of the gap increases much more than the values of the phases when the pions are removed. This is shown in the lower panel of Fig. 80.2, where the pairing gap is computed with pions (black line) and without pions (red line).

80.4 Final Remarks

We study the interplay between pions and contact interactions in a chiral NN interaction at N4LO and discuss the corresponding effects in two-nucleon phase-shifts and the pairing gap in nuclear matter. The S-waves are the channels where the nuclear force is stronger at low energies and, at N4LO, the contact terms dominate the interaction in the $^1S_0$ wave. Nevertheless, the attraction provided by the pion exchanges is important at low energies as can be seen from the phase-shifts at low energies. The similarity between the full N4LO interaction and the contact terms suggests that it may be possible to absorb the pion exchanges into the contact interactions by re-tunning their strengths and adding higher order terms.

Acknowledgements We would like to thank Prof. R. Machleidt for the special version of the chiral force. This work was supported by FAPESP (grant 2016/07061-3) and CNPq (grant 306195/2015-1).

References

1. Weinberg, S.: Effective chiral lagrangians for nucleon-pion interactions and nuclear forces. Nucl. Phys. B **363**, 3–18 (1991)
2. Weinberg, S.: Three-body interactions among nucleons and pions. Phys. Lett. B **295**, 114–121 (1992)
3. Ordóñez, C., Ray, L., van Kolck, U.: Two-nucleon potential from chiral Lagrangians. Phys. Rev. C **53**, 2086–2105 (1996)
4. Entem, D.R., Machleidt, R.: Accurate charge-dependent nucleon-nucleon potential at fourth order of chiral perturbation theory. Phys. Rev. C **68**, 041001 (2003)
5. Epelbaum, E., Glöckle, W., Meissner, U.-G.: The two-nucleon system at next-to-next-to-next-to-leading order. Nucl. Phys. A **747**, 362–424 (2005)
6. Entem, D.R., Kaiser, N., Machleidt, R., Nosyk, Y.: Peripheral NN scattering at fifth order of chiral perturbation theory. Phys. Rev. C **91**, 014002 (2015)
7. Epelbaum, E., Krebs, H., Meissner, U.-G.: Precision nucleon-nucleon potential at fifth order in the chiral expansion. Phys. Rev. Lett. **115**, 122301 (2015)
8. Nogga, A., Timmermans, R.G.E., van Kolck, U.: Renormalization of the one-pion exchange and power counting. Phys. Rev. C **72**, 054006 (2005)
9. Szpigel, S., Timóteo, V.S.: Power counting and renormalization group invariance in the subtracted kernel method. J. Phys. G: Nucl. Part. Phys. **39**, 105102 (2012)
10. Batista, E.F., Szpigel, S., Timóteo, V.S.: Renormalization of chiral forces with multiple subtractions. Adv. High Energy Phys. **2017**, 2316247 (2017)

Chapter 81
Experimental Study of Few Nucleon Correlations Using Deuteron Beam at Nuclotron

Marian Janek, Vladimir Petrovich Ladygin, Olena Mezhenska, Alexandr Vladimirovich Averyanov, Evgenyi Vasilievich Chernykh, Dan Enache, Yuriy Vitalievich Gurchin, Alexandr Yurievich Isupov, Julia-Tatiana Karachuk, Anatolyi Nikolaevich Khrenov, Dmitryi Olegovich Krivenkov, Pavel Konstantievich Kurilkin, Nadezhda Borisovna Ladygina, Alexei Nikolaevich Livanov, Semen Mikhailovich Piyadin, Sergei Grigorievich Reznikov, Yaroslav Tarasovich Skhomenko, Arkadiy Arkadievich Terekhin, Aleksei Viktorovich Tishevsky, Tomohiro Uesaka and Jozef Urban

Abstract Polarized and unpolarized data for the dp breakup reaction at deuteron energies of 270, 400 MeV and 300–500 MeV using $\Delta E - E$ technique are obtained. Dp elastic scattering has been investigated using polarized deuteron beam under various kinematic configurations in the angular range (65°–135°) in c.m. at deuteron energies of 400, 700, 800, 1000, 1100, 1300, 1500 and 1800 MeV. Preliminary results of analyzing powers obtained in dp elastic scattering at 800 MeV of deuteron energy are compared with the calculations based on relativistic multiple scattering model.

M. Janek (✉)
Zilina University, 01001 Žilina, Slovakia
e-mail: janek.marian@gmail.com

V. P. Ladygin · A. V. Averyanov · E. V. Chernykh · Y. V. Gurchin · A. Y. Isupov · A. N. Khrenov · D. O. Krivenkov · P. K. Kurilkin · N. B. Ladygina · A. N. Livanov · S. M. Piyadin · S. G. Reznikov · Y. T. Skhomenko · A. A. Terekhin · A. V. Tishevsky
Joint Institute for Nuclear Research, 141980 Dubna, Moscow region, Russia

O. Mezhenska · J. Urban
Pavol Jozef Safarik University, 04001 Košice, Slovakia

D. Enache · J.-T. Karachuk
Advanced Research Institute for Electrical Engineering, Bucharest, Romania

Y. T. Skhomenko
Belgorod State University, 308015 Belgorod, Russia

T. Uesaka
Nishina Center for Accelerator-Based Science, RIKEN, Wako, Japan

N. A. Orr et al. (eds.), *Recent Progress in Few-Body Physics*,
Springer Proceedings in Physics 238,
https://doi.org/10.1007/978-3-030-32357-8_81

81.1 Short Introduction

Processes in which the lightest nuclei are participating, such as *dp* elastic scattering and *dp* breakup reaction, have great advantage that can be studied in almost independent way with relatively small number of parameters.

Short range deuteron structure can be obtained from *dp* elastic scattering. *Dp* breakup reaction has richer phase space to be investigated which allows us to study reaction mechanisms in different kinematic conditions. Three nucleon scattering can be solved with modern two and three nucleon forces up to 200 MeV/nucleon [1]. Experiments performed at RIKEN [2] were aimed to study three nucleon forces (3NFs) contribution and to test modern models of 3NFs.

81.2 Results

Few nucleon correlations are studied using polarized and unpolarized deuteron beams and polyethylene and carbon targets at Nuclotron, JINR. More detailed information about performed experiments can be found in [3]. The *dp* breakup reaction is investigated at the deuteron energies from 300 to 500 MeV in the region where non-nucleonic degrees of freedom and relativistic effects can play a significant role. Analyzing powers of the *dp* breakup reaction using polarized beam were investigated at Internal Target Station (ITS) of Nuclotron at deuteron energies of 270 and 400 MeV using $\Delta E - E$ technique [4]. Recent studies of *dp* breakup reaction are aimed to study 3NFs and relativistic effects behaviour under special kinematic conditions. In order to obtain polarization observables calibration procedure has been performed using unpolarized deuteron beam data [5]. Recently, *dp* elastic scattering is investigated using polarized deuteron beam at ITS under various kinematic configurations in the angular range (65°–135°) in c.m. at deuteron energies of 400, 700, 800, 1000, 1100, 1300, 1500 and 1800 MeV. The A_y, A_{yy} and A_{xx} analyzing powers for the *dp* elastic scattering have been obtained for the first time at the ITS of the Nuclotron at deuteron energy of 800 MeV in angular range between 60° and 135° in the center of mass system. Theoretical calculations based on relativistic multiple scattering model are compared with the data containing statistical errors only, see Fig. 81.1. Dot-dashed, dashed and solid curves represent relativistic multiple scattering model calculations including only single scattering term, single + double scattering term [6] and calculations including single + double scattering term and also Δ—isobar excitation [7], respectively. Reasonably good agreement is found in description of vector analyzing power A_y in the middle of the angular range (85°–115°). The contribution which comes from double scattering and Δ isobar excitation ones is relatively small in this range. Behaviour of tensor analyzing power A_{yy} is reproduced qualitatively using double scattering and Δ isobar terms. It shows on importance of double scattering term in the range the data were obtained. Quantitative description is obtained around the angle of ~115° only. Tensor analyzing power A_{xx} shows linear growth in the angular range from 65° to 90° and then the data have constant behaviour which is in agreement with calculation based on single scattering term.

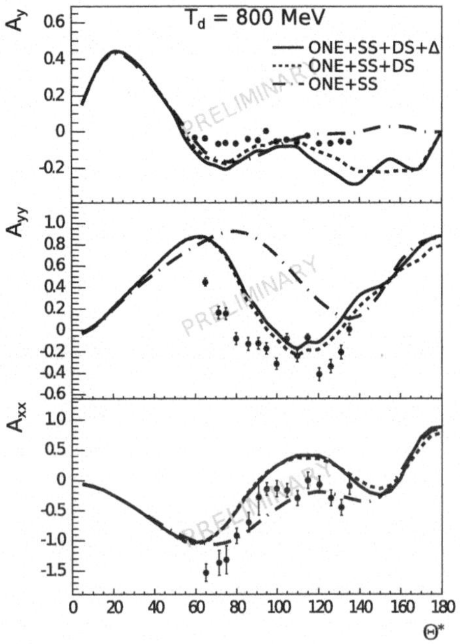

Fig. 81.1 The angular dependencies of the vector A_y and tensor A_{yy} and A_{xx} analyzing powers in dp elastic scattering measured at 800 MeV at ITS of Nuclotron. Theoretical calculations based on relativistic multiple scattering model are presented, see text for the curve explanation

81.3 Conclusion

Dp elastic and *dp* breakup reactions are investigated at intermediate energies using polarized and unpolarized deuteron beams of Nuclotron. *Dp* elastic data along with theoretical calculations based on multiple scattering model obtained at 800 MeV of deuteron energy are discussed.

The work has been supported in part by the Ministry of Education, Science, Research, and Sport of the Slovak Republic (VEGA Grant No. 1/0113/18), by the RFBR under grant No. 16-02-00203a and by JINR- Slovak Republic cooperation programs in 2018.

References

1. Glockle, W., Witala, H., Huber, D., Kamada, H., Golak, J.: The three-nucleon continuum: achievements, challenges and applications. Phys. Rep. **274**, 107 (1996)
2. Sekiguchi, K., et al.: Complete set of precise deuteron analyzing powers at intermediate energies: comparison with modern nuclear force predictions. Phys. Rev. C **65**, 034003 (2002)
3. Janek, M., et al.: Investigation of reactions using polarized and unpolarized deuteron beam at Nuclotron: Current status and perspectives. EPJ WoC **138**, 01028 (2017)
4. Janek, M., et al.: Analyzing powers iT_{11} and T_{20} of $dp \rightarrow ppn$ reaction at 400 MeV investigated at Nuclotron. Sci. Lett. Univ. Zilina. **19**(3), 62 (2017)

5. Janek, M., et al.: Calibration procedure of the $\Delta E - E$ detectors for dp breakup investigation at Nuclotron. Phys. Part. Nucl. Lett. **15**(1), 76 (2018)
6. Ladygina, N.B.: Differential cross-section of dp elastic scattering at intermediate energies. Eur. Phys. J. A **42**, 91 (2009)
7. Ladygina, N.B.: Delta excitation in deuteron-proton elastic scattering. Eur. Phys. J. A **52**, 199 (2016)

Chapter 82
High-Precision Nucleon-Nucleon Potentials from Chiral EFT

Patrick Reinert, Hermann Krebs and Evgeny Epelbaum

Abstract We present new momentum-space regularized nucleon-nucleon (NN) potentials up to fifth order in chiral effective field theory, whose adjustable parameters have been determined from experimental NN scattering data. The resulting N^4LO^+ potential achieves the same precision as high-quality phenomenological forces while the new momentum-space regulator paves the way for the implementation of consistently regularized three-nucleon forces and current operators.

82.1 Introduction

Over the last few years, nuclear forces derived from chiral effective field theory (EFT) have seen widespread use in few-body calculations. In the framework of chiral EFT, the two-nucleon force, many-body forces and current operators are consistently derived from an effective Lagrangian which encodes the symmetries of QCD with pions and nucleons as explicit degrees of freedom. Nuclear forces and currents are then derived as an expansion in a low-momentum variable Q up to some given order. This makes quantities in chiral EFT systematically improvable (by going to higher orders in the expansion) and allows one to access the uncertainty of chiral predictions (which, in a convergent expansion, are dominated by the next-higher, omitted order).

In addition to these conceptual advantages, chiral forces are also expected to yield good quantitative results. With the rapid progress in the field of ab initio few- and many-body methods, a reliable input in form of precise and consistent many-body forces becomes crucial.

Recently, a new generation of chiral two-nucleon forces [1–3] has been published where the chiral expansion has been pushed to the fifth order (N^4LO) and which fix their adjustable parameters from the two-nucleon continuum and bound state but

P. Reinert (✉) · H. Krebs · E. Epelbaum
Fakultät für Physik und Astronomie, Institut für Theoretische Physik II,
Ruhr-Universität Bochum, 44780 Bochum, Germany
e-mail: Patrick.Reinert@rub.de

N. A. Orr et al. (eds.), *Recent Progress in Few-Body Physics*,
Springer Proceedings in Physics 238,
https://doi.org/10.1007/978-3-030-32357-8_82

differ in regularization. Parallel developments are e.g. chiral potentials which include the $\Delta(1232)$ resonance as an explicit degree of freedom [4, 5] as well as potentials like [6], which have explicitly been tuned to the properties of light nuclei.

82.2 The Semi-local Momentum-Space (SMS) Regularized Potential

In this talk, the semi-local momentum-space (SMS) regularized potentials [3] up to fifth order ($\mathrm{N^4LO}$) in the chiral expansion are presented. The long-range part of the interaction consists of the one- and two-pion exchange potentials whose low-energy constants (LECs) are taken from the recent Roy-Steiner equation analysis [7] of pion-nucleon scattering. The long-range interaction is thus parameter-free in the NN system. On the other hand, the short-range contact interaction part of the potential has been fixed by a fit to the 2013 Granada database [8] of nucleon-nucleon scattering data. The deuteron binding energy and the np coherent scattering length enter as additional constraints in this fit.

We have also added the four leading $\mathrm{N^5LO}$ contact interactions in F-waves to the $\mathrm{N^4LO}$ potential ($\mathrm{N^4LO^+}$) to probe the uncertainty stemming from the still unparameterized F-waves at $\mathrm{N^4LO}$. In the following, we will discuss the new momentum-space regularization and the removal of redundant contact interactions in more detail.

82.2.1 *Regularization*

The application of nuclear forces in few- and many-body problems requires a non-perturbative resummation of the potential. For iterations of the potential to be UV-convergent, a regulator has to be applied. Following previous works [1, 9], we multiply the momentum-space matrix elements of the short-range contact interaction part of the potential with a non-local gaussian regulator $\exp\left(-(p'^2 + p^2)/\Lambda^2\right)$ while using a local regulator for the long-range pion-exchange part, hence the name semi-local regularization. In [1, 9], the regularization of the long-range potential was performed in coordinate space according to:

$$V_{\pi,\mathrm{reg.}}(\mathbf{r}) = V_\pi(\mathbf{r}) \left[1 - e^{-\frac{r^2}{R^2}}\right]^6, \tag{82.1}$$

where $R = 0.8$–1.2 fm. While the local coordinate-space regulator helped to significantly reduce long-range cutoff artifacts, its application is inconvenient for consistently regularized higher-order three-nucleon forces (3NF) and currents. For a systematic extension to those quantities, a momentum-space regulator is preferable. Inspired by [10], we introduce a new local regularization of the long-range potential in momentum space. Its main idea is to regularize static pion propagators with a gaussian form factor in the following way:

$$\frac{1}{\mathbf{l}^2 + M_\pi^2} \rightarrow \frac{1}{\mathbf{l}^2 + M_\pi^2} e^{-\frac{\mathbf{l}^2 + M_\pi^2}{\Lambda^2}}, \tag{82.2}$$

with $\mathbf{l}$ being the three-momentum of the exchanged pion. The regularization manifestly induces only short-range terms and does not affect the long-range behavior of the pion-exchange potentials. Additionally, the freedom in the polynomial short-range contributions of the regularized potentials has been fixed in such a way, that its coordinate space version and as many derivatives thereof as allowed by power counting vanish at the origin, leading to a qualitatively similar behavior at short distances than the coordinate-space regulator of [1, 9].

82.2.2 Redundant Contact Interactions at Order Q^4

Another important feature of the SMS potential is the removal of redundant contact interactions in S-waves starting at N^{3}LO (Q^4). The full partial-wave decomposed Q^4 contact potential in $^1S_0$, $^3S_1$ and $^3S_1 - {}^3D_1$ can be written as:

$$\begin{aligned}
\langle {}^1S_0, p'|V_{\text{cont}}|{}^1S_0, p\rangle &= \tilde{C}_{1S0} + C_{1S0}\left(p^2 + p'^2\right) + D_{1S0}\, p^2 p'^2 + D^{\text{off}}_{1S0}\left(p'^2 - p^2\right)^2 \\
\langle {}^3S_1, p'|V_{\text{cont}}|{}^3S_1, p\rangle &= \tilde{C}_{3S1} + C_{3S1}\left(p^2 + p'^2\right) + D_{3S1}\, p^2 p'^2 + D^{\text{off}}_{3S1}\left(p'^2 - p^2\right)^2 \\
\langle {}^3S_1, p'|V_{\text{cont}}|{}^3D_1, p\rangle &= C_{\epsilon 1}\, p^2 + D_{\epsilon 1}\, p^2 p'^2 + D^{\text{off}}_{\epsilon 1}\, p^2\left(p'^2 - p^2\right)
\end{aligned} \tag{82.3}$$

Here, the D_i^{off} terms contribute only off-shell and can be changed by suitably chosen unitary transformations, establishing them as redundant at the order Q^4 (see [3] for details). This redundancy manifests itself as strong correlations among the LECs in the corresponding partial-wave channels, which was first noted in the framework of bayesian statistical analysis [11, 12]. In order to arrive at a well-conditioned optimization problem when fitting the contact LECs, we make use of the unitary freedom and fix those off-shell contacts at $D^{\text{off}}_{1S0} = D^{\text{off}}_{3S1} = D^{\text{off}}_{\epsilon 1} = 0$. This choice leads to softer, more perturbative interactions which is a welcome feature regarding application in many-body methods.

82.3 Results

We performed the fits of the NN contact LECs to the two-nucleon experimental scattering data for all chiral orders and all cutoffs $\Lambda = 350$–500 MeV. Table 82.1 shows χ^2/datum values for the cutoff $\Lambda = 450$ MeV at different chiral orders.

Table 82.1 χ^2/datum for the regulator $\Lambda = 450\,\text{MeV}$ at various orders of the chiral expansion. The subscript of each order indicates the number of fitted short-range LECs

E_{lab} bin	$\text{LO}_{(3)}$	$\text{NLO}_{(10)}$	$\text{N}^2\text{LO}_{(10)}$	$\text{N}^3\text{LO}_{(22)}$	$\text{N}^4\text{LO}_{(23)}$	$\text{N}^4\text{LO}^+{}_{(27)}$
Neutron-proton scattering data						
0–100	73	2.2	1.2	1.08	1.07	1.08
0–200	62	5.4	1.7	1.10	1.08	1.07
0–300	75	14	4.2	2.01	1.16	1.06
Proton-proton scattering data						
0–100	2290	10	2.2	0.90	0.88	0.86
0–200	1770	90	37	1.99	1.42	0.95
0–300	1380	90	41	3.43	1.67	1.00

Going from LO to N^4LO^+, a clear order-by-order improvement in the description of NN scattering data is seen. It should be noted, that the improvement from NLO to N^2LO is entirely due to the parameter-free chiral two-pion exchange. A similar effect is also seen in the transition from N^3LO to N^4LO, where apart from parameter-free contributions, only one isospin-breaking LEC is added to the contact potential in the ${}^1S_0$ phase. Finally, the addition of the leading F-wave contact interactions in the N^4LO^+ potential allows the accurate description of very precise proton-proton observables at intermediate energies, the most notable example being the data of [13]. One is then able to describe the entire scattering database with χ^2/datum $\sim$1, which is on par with high-quality phenomenological forces in the considered energy range of $E_{\text{lab}} = 0$–300 MeV. In comparison to e.g. the CD-Bonn potential [14], which has 43 adjustable parameters, this precision is achieved with only 27 adjustable contact interactions, showing the importance of the chiral two-pion exchange.

To summarize, the presented chiral two-nucleon potentials offer precision comparable to high-quality phenomenological forces and simultaneously pave the way for the inclusion of consistent higher-order three-nucleon forces and current operators. Work along these lines is in progress by LENPIC [15].

Acknowledgements This work was supported by BMBF (contract No. 05P2015—NUSTAR R&D) and by DFG through funds provided to the Sino-German CRC 110 "Symmetries and the Emergence of Structure in QCD" (Grant No. TRR110).

References

1. Epelbaum, E., Krebs, H., Meißner, U.G.: Phys. Rev. Lett. **115**, 122301 (2015)
2. Entem, D.R., Machleidt, R., Nosyk, Y.: Phys. Rev. C **96**, 024004 (2017)
3. Reinert, P., Krebs, H., Epelbaum, E.: Eur. Phys. J. A **54**, 86 (2018)
4. Piarulli, M., Girlanda, L., Schiavilla, R., Navarro Pérez, R., Amaro, J.E., Ruiz Arriola, E.: Phys. Rev. C **91**, 024003 (2015)

5. Ekström, A., Hagen, G., Morris, T.D., Papenbrock, T., Schwartz, P.D.: Phys. Rev. C **97**, 024332 (2018)
6. Ekström, A., et al.: Phys. Rev. C **91**, 051301 (2015)
7. Hoferichter, M., Ruiz de Elvira, J., Kubis, B., Meißner, U.G.: Phys. Rev. Lett. **115**, 192301 (2015)
8. Navarro Pérez, R., Amaro, J.E., Ruiz Arriola, E.: Phys. Rev. C **88**, 064002 (2013)
9. Epelbaum, E., Krebs, H., Meißner, U.G.: Eur. Phys. J. A **51**, 53 (2015)
10. Rijken, T.A.: Ann. Phys. **208**, 253 (1991)
11. Wesolowski, S.: Bayesian methods for effective field theories, talk at the INT Program INT-16-2a on Bayesian Methods in Nuclear Physics, INT Seattle, USA, 13 June–8 July 2016
12. Wesolowski, S., Furnstahl, R.J., Melendez, J.A., Phillips, D.R.: J. Phys. G **46**(4), 045102 (2019)
13. Cox, G.F., Eaton, G.H., Van Zyl, C.P., Jarvis, O.N., Rose, B.: Nucl. Phys. B **4**, 353 (1968)
14. Machleidt, R.: Phys. Rev. C **63**, 024001 (2001)
15. Low Energy Nuclear Physics International Collaboration (LENPIC). www.lenpic.org

Chapter 83
Configurational Efficiency with Reconstructed Neutron for Deuteron Breakup Reaction

Bogusław Włoch, Izabela Ciepał, Adam Kozela, Joanna Kuboś and Wiktor Parol

Abstract Number of experiments devoted to the study of few nucleon system dynamics were carried out at KVI with the use of the BINA detector and 160 MeV deuteron beam on deuteron or proton targets. One of the crucial steps of the analysis is to uniquely identify real breakup reactions. While in a standard approach only charged particles were used for this, it also seams to be interesting to exploit relatively high sensitivity of the thick scintillators to neutron detection. In this paper we investigate the influence of the scintillator granularity to the detection efficiency of proton-neutron coincidences.

83.1 Motivation

Deuteron breakup leading to the final state of three nucleon is one of the simplest systems, providing however an excellent testing ground for various models of few nucleon system dynamics. One of the most sensitive observables to investigate a complicated interplay of different components of this dynamics is the differential cross section measured in a broad range of the phase space. Experiments at Kernfysisch Versneller Instituut in Groningen with the use of the 160 MeV deuteron beam provided by the AGOR cyclotron and with BINA detector were performed, and a large set of high precision data on ^{2}H(d, dp)n and ^{1}H(d, pp)n deuteron breakup was obtained [1]. This and most of previously preformed analysis of the data obtained with the BINA detector were focused on reactions with two charged particles detected, proton-deuteron and proton-proton respectively. Development of techniques aimed at direct neutron identification in BINA experimental setup gives us possibility to examine reaction channels with neutron instead of one of the charged particles. Having determined the differential cross sections for three-body ^{2}H(d, dn)p breakup reaction

B. Włoch (✉) · I. Ciepał · A. Kozela · J. Kuboś · W. Parol
Institute of Nuclear Physics Polish Academy of Sciences, 31342 Kraków, Poland
e-mail: boguslaw.wloch@ifj.edu.pl

N. A. Orr et al. (eds.), *Recent Progress in Few-Body Physics*,
Springer Proceedings in Physics 238,
https://doi.org/10.1007/978-3-030-32357-8_83

with the neutron momentum reconstructed one can compare it with already analysed ^{2}H(d, dp)n channel [2] at the analogous kinematic conditions and directly study the Coulomb effects and possible charge symmetry breaking, like it was suggested in [3].

83.2 Experimental Setup

The BINA detection system is characterized by large angular acceptance and low energy threshold for charged particles detection. It has been specially designed for investigation of a few-nucleon systems in the intermediate energy range. It is composed of two main parts, Forward Wall for angles from 10° to 40° and Backward Ball for 40° to 165°, covering together almost 4π solid angle. The Backward Ball consists of 149 phoswich detectors arranged in fulleren-like structure. It plays also a role of vacuum scattering chamber with liquid deuterium or hydrogen target inside. Forward Wall is composed of a multi-wire proportional chamber (MWPC), 12 vertical, 2 mm thick plastic scintillator stripes (ΔE) and 10 horizontal, 12 cm thick plastic scintillator slabs (E). Together, both detectors form an array of virtual ΔE-E telescopes used for charged particles identification. Detailed information about the detector can be found in [4]. The highest sensitivity for neutron detection has the thick E scintillator, with an average efficiency 10% as obtained from experimental data [5]. Both ΔE detector and MWPC can be used as an active veto, rejecting events caused by charged particles. Position of detected neutron can be determined independently by the time and pulse height asymmetries of signals obtained at both ends of 2 m long E slabs, while its energy can be calculated from the time of flight information. This method allows to reconstruct neutron emission angle in BINA with a reasonable angular resolution of around 5°, as it was shown in [6].

83.3 Configurational Efficiency

Whenever two (or more) particles hit the same scintillator element neither positions nor energies can be properly reconstructed and in consequence the event is lost. In order to reconstruct the actual number of coincidences and in consequence to obtain the absolute values of cross section it is crucial to account for this effect, which is referred to as the configurational efficiency. Total efficiency for registration a pair of particles from the breakup reaction is given by:

$$\epsilon^{Total}(\theta_1, \theta_2, \Delta\phi_{12}, S) = \epsilon_1^{det}(\theta_1, \phi_1) \cdot \epsilon_2^{det}(\theta_2, \phi_2) \cdot \epsilon^{conf}(\theta_1, \theta_2, \Delta\phi_{12}, S), \quad (83.1)$$

where: $\epsilon_i^{det}(\theta, \phi)$ is the detector efficiency for a given particle, and $\epsilon^{conf}(\theta_1, \theta_2, \Delta\phi_{12}, S)$ is the configurational efficiency and S is defined as the arc-length along three-body breakup kinematics. The detector efficiency calculations were described

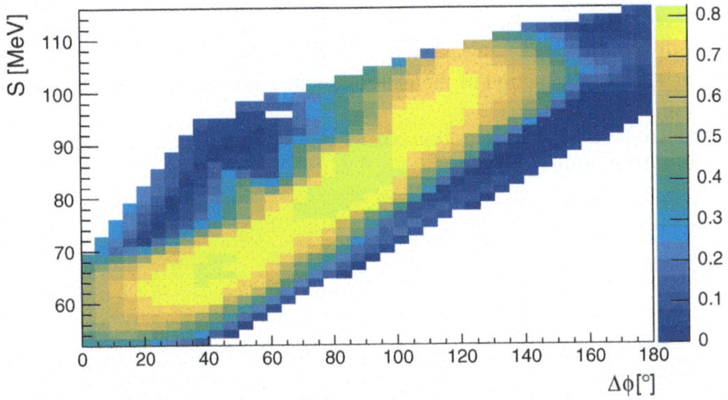

Fig. 83.1 An example of configurational efficiency for chosen angular configurational of $\theta_1 = 29°$, $\theta_2 = 25°$ as a function of relative azimuthal angle $\Delta\phi$ and S variable

in details in previous papers [1]. Configurational efficiency depends on the detector granularity and geometry, as well as on the geometrical configuration of the velocity vectors of both particles, which in the case of breakup events is defined by two polar angles (θ_1, θ_2) and relative azimuthal angle of emission ($\Delta\phi_{12}$). In the case of neutron detection, the additional component of active veto has to be taken into account. Unlike in the case of configurational efficiency for two charged particles (see [7] for details), where unobserved neutron is of almost no influence, clean identification of neutron impose stringent conditions on the emission angle of the third particle-proton. This introduces the dependence of the efficiency to the energies of detected particles, which for a given angular configuration can be parameterized by the S variable. A gap of at least one not activated scintillator between detected particles was required to avoid neutron misidentification due to cross-talks between neighbouring elements. Figure 83.1 presents a sample of configurational efficiency as a function of $\Delta\phi_{12}$ angle and S variable, calculated from the ^{1}H(d, pn)p breakup events simulated with the use of BINA detector geometry and GEANT4 framework. Described above configurational efficiency for events with neutron and accompanying charged particle detected in exit channel is one of the crucial steps in the analysis of deuteron breakup. Similar maps as in Fig. 83.1 for all interesting configurations will be used in further analysis of differential cross-sections in ^{1}H(d, pn)p and ^{2}H(d, dn)p reactions.

References

1. Khatri, G., et al.: Deuteron-deuteron collision at 160 MeV. Acta Phys. Pol. B **47** (2016)
2. Khatri, G.: Investigation of Deuteron Disintegration. Ph.D. thesis Jagiellonian University, Kraków (2015)
3. Howell, C.R., et al.: Novel probe of charge symmetry breaking: deuteron-induced deuteron breakup. Phys. Rev. C **48** (1993)
4. Stephan, E., et al.: Three-nucleon interaction dynamics studied via the deuteron-proton breakup. Int. J. Mod. Phys. A **24** (2009)

5. Kuboś, J.: Wyznaczenie wydajnosci detektora BINA na rejestracje neutronow w reakcji breakupu dp przy energii 160 MeV. B.Sc. thesis, AGH Kraków (2014)
6. Włoch, B., et al.: Reconstruction of the neutron momentum in the deuteron breakup reaction. Acta Phys. Pol. B **49** (2018)
7. Parol, W., Włoch, B., et al.: Configuration efficiency for deuteron breakup reaction investigation. Acta Phys. Pol. Proc. Suppl. B **10** (2017)

Chapter 84
Compelete Set of Deuteron Analyzing Powers for *dp* Elastic Scattering at 70–300 MeV/nucleon and Three-Nucleon Forces

K. Sekiguchi, Y. Wada, A. Watanabe, D. Eto, T. Akieda, H. Kon, K. Miki, N. Sakamoto, H. Sakai, M. Sasano, Y. Shimizu, H. Suzuki, T. Uesaka, Y. Yanagisawa, M. Dozono, S. Kawase, Y. Kubota, K. Yako, Y. Maeda, S. Kawakami, T. Yamamoto, S. Sakaguchi, T. Wakasa, J. Yasuda, A. Ohkura, Y. Shindo, M. Tabata, E. Milman, S. Chebotaryov, H. Okamura and T. L. Tang

Abstract In order to study three-nucleon force (3NF) effects in three-nucleon scattering we performed the measurements of the cross sections and the spin observables for the deuteron–proton (*dp*) scattering with the polarized deuteron beams at the incident energies up to 135 MeV/nucleon. Recently we have extended the measurements at the RIKEN RI Beam Factory (RIBF) with the polarized deuteron beams at 190, 250, and 300 MeV/nucleon.

K. Sekiguchi (✉) · Y. Wada · A. Watanabe · D. Eto · T. Akieda · H. Kon · K. Miki
Department of Physics, Tohoku University, Sendai 980-8578, Japan
e-mail: kimiko@lambda.phys.tohoku.ac.jp

N. Sakamoto · H. Sakai · M. Sasano · Y. Shimizu · H. Suzuki · T. Uesaka · Y. Yanagisawa
RIKEN Nishina Center, Wako, Saitama 351-0198, Japan

M. Dozono · S. Kawase · Y. Kubota · K. Yako
Center for Nuclear Study, University of Tokyo, Bunkyo, Tokyo 113-0033, Japan

Y. Maeda · S. Kawakami · T. Yamamoto
Faculty of Engineering, University of Miyazaki, Miyazaki 889-2192, Japan

S. Sakaguchi · T. Wakasa · J. Yasuda · A. Ohkura · Y. Shindo · M. Tabata
Department of Physics, Kyushu University, Higashi, Fukuoka 812-8581, Japan

E. Milman · S. Chebotaryov
Department of Physics, Kyungpook National University, Daegu 702-701, Korea

H. Okamura · T. L. Tang
RCNP, Osaka University, Osaka, Ibaraki 567-0047, Japan

N. A. Orr et al. (eds.), *Recent Progress in Few-Body Physics*,
Springer Proceedings in Physics 238,
https://doi.org/10.1007/978-3-030-32357-8_84

84.1 Introduction

One of the main interests of nuclear physics is to understand the forces acting between nuclear constituents. Importance of the three-nucleon force ($3NF$) in the nuclear Hamiltonian has been studied in few-nucleon systems as well as in many-nucleon systems [1–3]. Three-nucleon ($3N$) systems, where numerically exact solutions of the corresponding Faddeev equations for any two- and three-nucleon forces are feasible, play an especially important role in these investigations. Nucleon–deuteron (Nd) scattering offers a good opportunity to study dynamical aspects of $3N$Fs, that are momentum, spin and isospin dependences, since it provides not only cross sections but also a variety of spin observables at different incident nucleon energies. The last two decades have witnessed the extensive experimental and theoretical investigations of the Nd scattering performed in a wide range of incoming nucleon energies up to $E \sim 300$ MeV/nucleon.

84.2 Experimental Results of *dp* Scattering

The experiments of the *dp* scattering were performed at the RIKEN Accelerator Facility using the polarized deuteron beams at the incident energies up to 135 MeV/nucleon. Measured observables are the cross section, the all deuteron analyzing powers (iT_{11}, T_{20}, T_{21}, T_{22}), and the deuteron to proton polarization transfer coefficients [4]. Direct comparison between the data and the Faddeev calculations based on the realistic nuclear forces have shown the first clear signatures of 3NFs in the cross section [4]. Later the measurements have been extended to the RIKEN RI Beam Factory (RIBF). All deuteron analyzing powers were obtained at 190, 250, 294 MeV/nucleon [5–7].

In Fig. 84.1 some representative data of the deuteron analyzing powers are shown with open circles. Statistical errors are only shown. The data are compared with the Faddeev calculations based on the modern nucleon–nucleon forces combined with the three nucleon forces. The red (blue) bands in the figure are the Faddeev calculations with (w/o) Tucson–Melbourne'99 (TM99) 3NF [8] based on the modern NN potentials, namely CDBonn [9], AV18 [10], Nijmegen I and II [11]. The dashed lines are the calculations with including Urbana IX 3NF [12] based on the AV18 potential.

For the vector analyzing power iT_{11} the discrepancies between the data and the predictions based on 2NFs (blue bands) are seen at the angles $\theta_{\text{c.m.}} \sim 120°$. At 135 and 190 MeV/nucleon the data have good agreements to the predictions with the 3NFs while at 250 MeV/nucleon discrepancy exists at backward angle $\theta_{\text{c.m.}} \gtrsim 120°$. Tensor analyzing power T_{22} reveals different energy dependence from that of iT_{11}. At 135 MeV/nucleon adding 3NFs worsens the description of data in a large angular region. It is contrary to what happens at 190 and 250 MeV/nucleon, where large 3NF effects are supported by the measured data. The results of comparison shows that the 3NF is definitely needed in Nd elastic scattering. However the spin dependent parts of the 3NF may be deficient. It is interesting to see how the potential of the

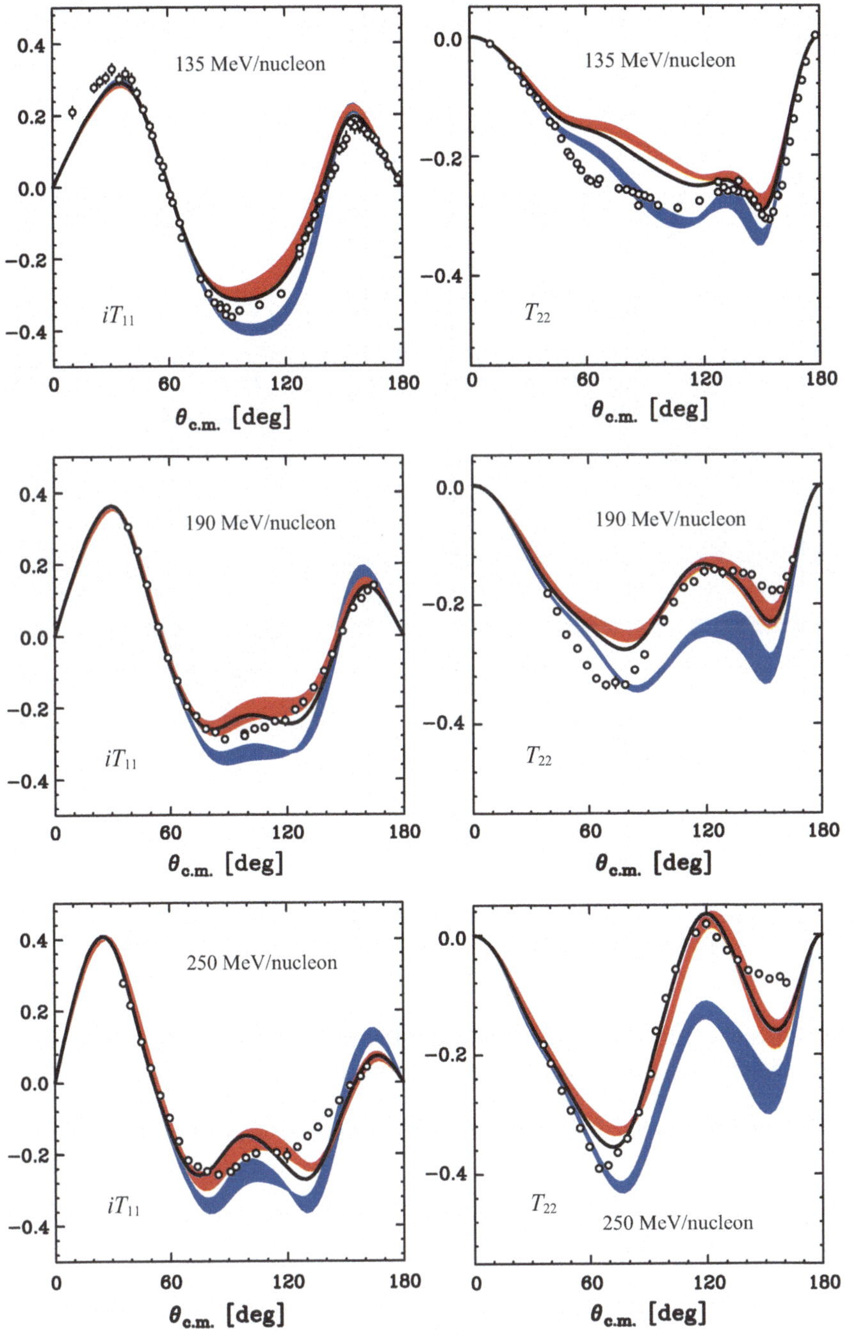

Fig. 84.1 Deuteron analyzing powers iT_{11}, T_{22} for elastic *Nd* scattering at 135, 190 and 250 MeV/nucleon. For descriptions of the calculations see the text

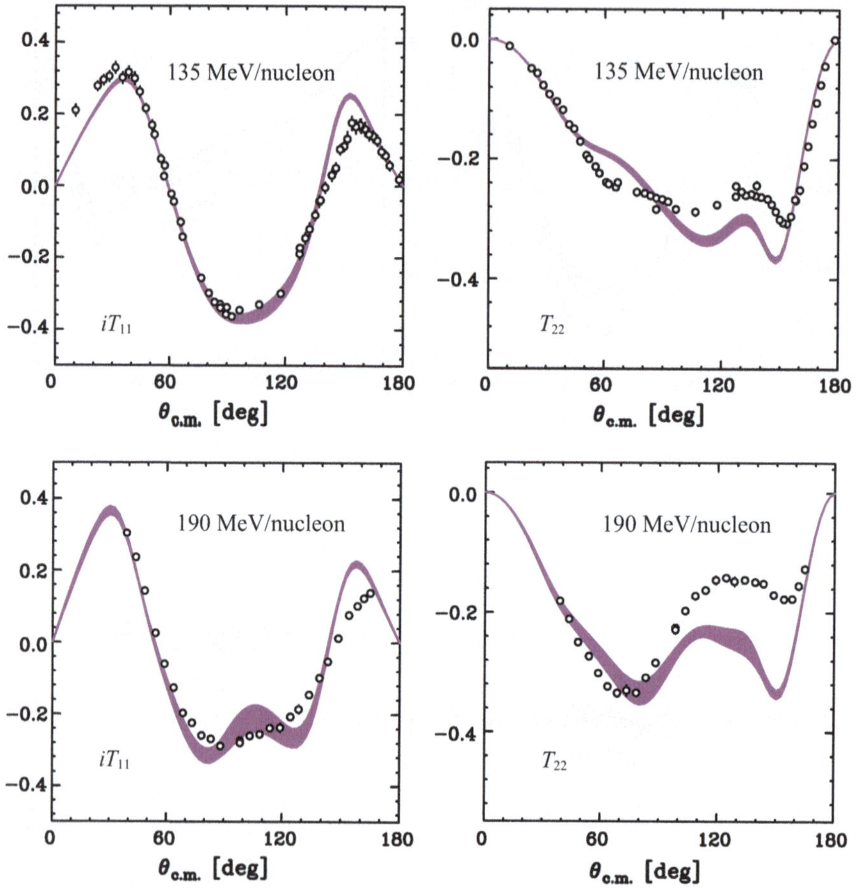

Fig. 84.2 Deuteron analyzing powers iT_{11}, T_{22} for elastic *Nd* scattering at 135 and 190 MeV/nucleon. For descriptions of the calculations see the text

chiral effective field theory (χEFT) describe deuteron analyzing powers for *dp* elastic scattering. In Fig. 84.2 the data are compared with the calculations based on the χEFT N4LO NN potentials [13]. The vector analyzing power iT_{11} data are well described by the χEFT N4LO NN potentials, while large discrepancies are found for the tensor analyzing power T_{22}. In order to see how χEFT 3NFs describe the data the theoretical treatments are now in progress [14].

84.3 Summary

The *Nd* scattering data provide rich sources to explore the properties of 3NFs such as momentum and spin dependence. In this contribution the experiments performed with polarized deuteron beams at RIKEN are presented and the recent achievements of study of 3NFs via *dp* scattering are discussed. The energy and angular dependent results of the cross sections as well as the polarization observables show that clear signatures of the 3NF effects are found in the cross section. Meanwhile the spin dependent parts of the 3NFs may be deficient. In order to obtain consistent understanding the effects of three nucleon forces in the 3N scattering further investigation should be necessary. It would be interesting to see how well the theoretical approaches, e.g. addition of 3NFs other than 2π exchange types, and the potentials based on the chiral effective field theory describe these obtained data.

References

1. Glöckle, W., Witała, H., Hüber, D., Kamada, H., Golak, J.: Phys. Rep. **274**, 107 (1996)
2. Hammer, H.-W., Nogga, A., Schwenk, A.: Rev. Mod. Phys. **85**, 197 (2013), and references therein
3. Kalantar-Nayestanaki, N., et al.: Rep. Prog. Phys. **75**, 016301 (2012)
4. Sakamoto, N., et al.: Phys. Lett. B **367**, 60 (1996); Sakai, H., et al.: Phys. Rev. Lett. **84**, 5288 (2000); Sekiguchi, K., et al.: Phys. Rev. C **65**, 034003 (2002); Sekiguchi, K., et al.: Phys. Rev. C **70**, 014001 (2004); Sekiguchi, K., et al.: Phys. Rev. Lett. **95**, 162301 (2005)
5. Sekiguchi, K., et al.: Phys. Rev. C **83**, 061001(R) (2011)
6. Sekiguchi, K., et al.: Phys. Rev. C **89**, 064007 (2014)
7. Sekiguchi, K., et al.: Phys. Rev. C **96**, 064001 (2017)
8. Coon, S.A., Han, H.K.: Few Body Syst. **30**, 131 (2001)
9. Machleidt, R.: High-precision, charge-dependent Bonn nucleon-nucleon potential. Phys. Rev. C **63**, 024001 (2001)
10. Wiringa, R.B., et al.: Accurate nucleon-nucleon potential with charge-independence breaking. Phys. Rev. C **51**, 38 (1995)
11. Stoks, V.G.J., et al.: Construction of high-quality NN potential models. Phys. Rev. C **49**, 2950 (1994)
12. Pudliner, B.S., et al.: Phys. Rev. C **56**, 1720 (1997)
13. Binder, S., et al.: Few-nucleon systems with state-of-the-art chiral nucleon-nucleon forces. Phys. Rev. C **93**, 044002 (2016)
14. Epelbaum, E.: Private communications

Chapter 85
Measurement for p–^{3}He Elastic Scattering with a 65 MeV Polarized Proton Beam

S. Nakai, K. Sekiguchi, K. Miki, A. Watanabe, T. Mukai, S. Shibuya, M. Watanabe, K. Kawahara, D. Sakai, Y. Wada, Y. Shiokawa, T. Taguchi, D. Eto, T. Akieda, H. Kon, M. Ito, T. Ino, K. Hatanaka, A. Tamii, H. J. Ong, N. Kobayashi, A. Inoue, S. Nakamura, T. Wakasa, S. Mitsumoto, H. Ohshiro, S. Goto, Y. Maeda, H. Sakai, T. Uesaka and T. Wakui

Abstract We performed the measurement of the cross section and the proton analyzing power A_y for p–^{3}He elastic scattering with a 65 MeV polarized proton beam at Research Center for Nuclear Physics (RCNP), Osaka University. The proton analyzing power A_y data are compared with the theoretical calculations based on the nucleon–nucleon potential (INOY04).

S. Nakai (✉) · K. Sekiguchi · K. Miki · A. Watanabe · T. Mukai · S. Shibuya · M. Watanabe · K. Kawahara · D. Sakai · Y. Wada · Y. Shiokawa · T. Taguchi · D. Eto · T. Akieda · H. Kon
Department of Physics, Tohoku University, Sendai, Miyagi 980-8578, Japan
e-mail: nakai@lambda.phys.tohoku.ac.jp

M. Ito
Cyclotron and Radioisotope Center (CYRIC), Tohoku University, Sendai, Miyagi 980-8578, Japan

T. Ino
KEK, High Energy Accelerator Research Organization, Tsukuba, Ibaraki 305-0801, Japan

K. Hatanaka · A. Tamii · H. J. Ong · N. Kobayashi · A. Inoue · S. Nakamura
Research Center for Nuclear Physics, Osaka University, Osaka, Ibaraki 567-0047, Japan

T. Wakasa · S. Mitsumoto · H. Ohshiro · S. Goto
Department of Physics, Kyushu University, Higashi, Fukuoka 812-8581, Japan

Y. Maeda
Faculty of Engineering, University of Miyazaki, Miyazaki 889-2192, Japan

H. Sakai · T. Uesaka
RIKEN Nishina Center, Wako, Saitama 351-0198, Japan

T. Wakui
National Institute of Radiological Sciences, Chiba, Chiba 263-8555, Japan

N. A. Orr et al. (eds.), *Recent Progress in Few-Body Physics*,
Springer Proceedings in Physics 238,
https://doi.org/10.1007/978-3-030-32357-8_85

85.1 Introduction

One of the most important topics of nuclear physics is to describe various nuclear phenomena based on the nucleon–nucleon (NN) interactions combined with the three-nucleon forces (3NFs). 3NFs are key elements to understand various nuclear phenomena, e.g. binding energies of light mass nuclei [1] and equation of state of nuclear matter [2]. In order to study the dynamical aspects of 3NFs, such as momentum, spin, and iso-spin dependencies, few-nucleon scattering is a good probe. The first indication of the 3NF effects in the few-nucleon scattering was found in the cross section minimum for deuteron–proton (dp) elastic scattering at intermediate energies ($E/A \gtrsim 65$ MeV) [3]. As an extension of the study of 3NF effects in nucleon–deuteron scattering, we performed the measurement for the p–^{3}He scattering at 65 MeV. The motivation of this experiment is to explore the 3NF effects in four-nucleon scattering as well as to approach to the 3NFs with the channels of the total iso-spin $T = 3/2$.

85.2 Experiment

The measurement of p–^{3}He elastic scattering was performed in the west experimental hall of the RCNP cyclotron facility. Figure 85.1 shows the schematic view of the experimental setup. The polarized proton beams were provided by the High Intensity Polarized Ion Source and they were accelerated by the AVF cyclotron up to 65 MeV. After bombarding the ^{3}He gaseous target in the scattering chamber, the beams were stopped in the Faraday cup. The beam intensity was 20–100 nA. The polarization of the beam was measured by using the beam line polarimeter. The polarimetry

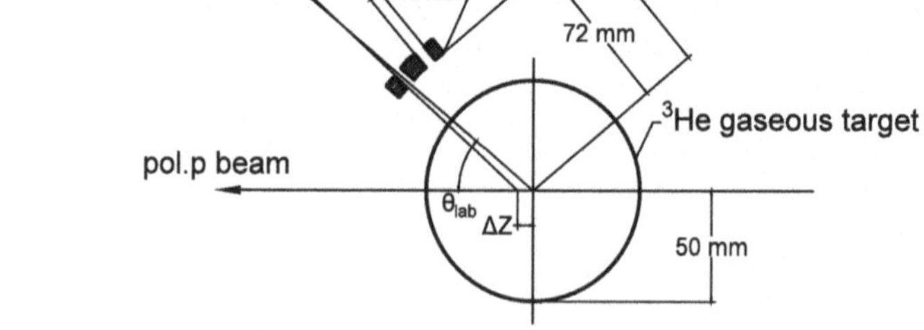

Fig. 85.1 Schematic view of the experimental setup

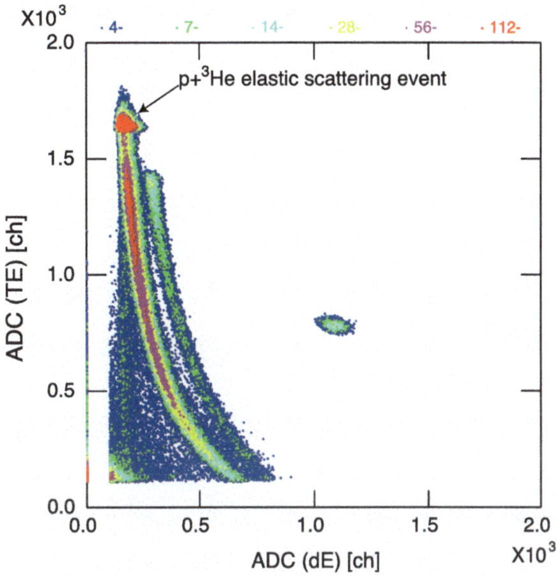

Fig. 85.2 Two-dimensional plot of the light outputs of the dE and E detectors

was made by p–^{12}C elastic scattering. The typical beam polarizations were 45–55 % of the theoretical maximum values. In the experiment, the ^{3}He gaseous target was operated at the room temperature under the atmospheric pressure. The scattered particles were detected by the dE–E detectors which consisted of plastic and NaI(Tl) scintillators. The measured angles were 26.9°–170.1° in the center of mass system.

85.3 Results

Particle identification was made by using the correlation between the dE and E detectors. Figure 85.2 shows a two-dimensional plot of the light outputs of the dE and E detectors.

The events from the p–^{3}He elastic scattering are clearly seen around the highest ADC channels of the E detector. Time of flight information was also used for event selection.

As expressed in (85.1), the proton analyzing power A_y was extracted by using the beam polarization (p_y) and the difference of the yields (N) between the spin up and down modes (the subscripts "u" and "d" denote the spin-up and spin-down, respectively).

$$A_y^p = \frac{N^u - N^d}{N^d p_y^u + N^u p_y^d} \tag{85.1}$$

Figure 85.3 shows the experimental results of the proton analyzing power A_y as a function of the scattering angle in the center of mass system. Solid circles are the

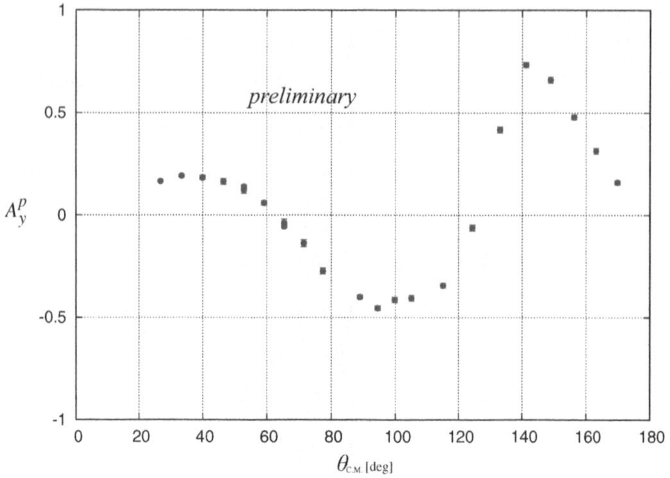

Fig. 85.3 Proton analyzing power A_y for the p–^{3}He elastic scattering at 65 MeV

experimental data. Only the statistical errors are shown. Statistical errors are less than 0.02 for all the measured data. Comparing the data to the calculations based on the realistic NN potential (INOY04 [4]) [5], the angular distribution of the experimental data has a moderate agreement with the theoretical calculation. However, the clear discrepancies are seen at the angles $\theta_{\text{C.M.}} \sim 80°$ and $\theta_{\text{C.M.}} \sim 140°$.

85.4 Summary

We performed the measurement of the cross section and proton analyzing power A_y for p–^{3}He elastic scattering using 65 MeV polarized proton beams. We obtained the experimental data in a wide angular range ($\theta_{\text{C.M.}} = 26.9°$–$170.1°$). The experimental data are compared with the theoretical calculations based on the INOY04 NN potential. Clear discrepancies between the data and the calculations are found at the angles where the proton analyzing power A_y takes minimum and maximum. The analysis of the cross section is in progress now.

References

1. Pieper, S.C., et al.: Phys. Rev. C **64**, 014001 (2001)
2. Akmal, A., et al.: Phys. Rev. C **58**, 1804 (1998)
3. Sekiguchi, K., et al.: Phys. Rev. C **65**, 034003 (2002)
4. Doleschall, P.: Phys. Rev. C **69**, 054001 (2004)
5. Deltuva, A.: Private communications

Chapter 86
Effective Field Theory Descriptions of Few-Nucleon Systems

L. Girlanda, M. Gattobigio, A. Kievsky, L. E. Marcucci and M. Viviani

Abstract The understanding of nuclear systems as composed of interacting nucleons has been considerably sharpened by the effective field theory (EFT) framework. The latter provides a link between the nuclear interaction and the underlying quantum chromodynamics, as the relevant degrees of freedom result, at least ideally, from a decimation process starting from fundamental quarks and gluons. Owing to chiral symmetry and the Goldstone bosons' characters of the interchanged pions among nucleons, the properties of heavier nuclei can in principle be traced back to a restricted set of low-energy constants (LECs) to be determined in lighter systems in the framework of a systematic low-energy expansion. At smaller energy scales, in pionless EFT, the interactions simplify becoming of contact type. The low-energy expansion is organized differently, relying on the emergence of universal properties, characteristic of systems with large two-body scattering lengths. We will examine the two above schemes and discuss their relation, with the aim of devising viable power counting schemes for applications in nuclear physics.

86.1 Generalities on Effective Field Theories

Since the birth of physical science, the key to the successful mathematical description of phenomena has been the capability of identifying their truly relevant features, and of isolating them from all disturbances and complications, that could eventually

L. Girlanda (✉)
Dipartimento di Matematica e Fisica, University of Salento, Lecce, Italy
e-mail: girlanda@le.infn.it

INFN, Sez. di Lecce, Lecce, Italy

M. Gattobigio
CNRS, Institut de Physique de Nice, Université Côte d'Azur, Nice, France

A. Kievsky · L. E. Marcucci · M. Viviani
INFN, Sez. di Pisa, Pisa, Italy

L. E. Marcucci
Dipartimento di Fisica, University of Pisa, Pisa, Italy

N. A. Orr et al. (eds.), *Recent Progress in Few-Body Physics*,
Springer Proceedings in Physics 238,
https://doi.org/10.1007/978-3-030-32357-8_86

be treated as small perturbations. It would indeed be impractical to face the full complexity of physical phenomena in their finest details, even in the fortunate case in which those finest details are actually understood. The quest towards such ultimate understanding has led to the formulation of quantum field theories (QFTs), forming the basis of the currently accepted Standard Model of elementary particles. One of the guiding principles in such an accomplishment has been the requirement of renormalizability for the relevant QFTs, i.e. the requirement that they be valid down to the tiniest distances.

After the completion of this program a new understanding of the significance of this requirement began to emerge: renormalizability was henceforth considered not as a requirement of principle, but rather as the consequence of a separation of scales [1–3]. The QFTs describing the interactions of elementary particles are no longer to be considered as valid at all scales, they will break down at an unknown scale at which new unknown physics sets in. The satisfactory description in terms of renormalizable QFTs, and their associated high predictive power, is just a manifestation of the large separation between us and the scale of new physics, since the contributions from non-renormalizable terms (with the corresponding proliferation of unknown coupling constants) is suppressed by inverse powers of the scale ratio. It would only be the sign of a pretentious ingenuity to ask that Nature follows the rules of renormalizable, infinitely predictive, QFTs. Instead, the Standard Model is only valid below a cutoff, a high-energy scale Λ, thus it qualifies as an effective field theory (EFT).

The cutoff Λ in a given EFT is not just a regulator, a mathematical artifact that one can dispose of by sending it to infinity, after the renormalization procedure. It is instead a physical parameter that refers to the very definition of the EFT [4]. It encodes the scale characterizing those fine details that are unimportant at the energies where the EFT is applicable, suppressed by inverse powers of Λ.

The EFTs reflect the procedure of choosing the appropriate degrees of freedom (the interacting fields) for a given physical problem, and identifying the relevant symmetries. The interactions among the fields follow from these two steps and they are typically of arbitrary complexity. Each interaction term comes with a coupling constant, a low-energy constant (LEC), and there is an infinite number of them: indeed, according to Weinberg [5], the EFT Lagrangian is just the most general one respecting the assumed symmetries. Therefore, changing the cutoff Λ, the theory will remain the same, only with changed values for the LECs, that define their renormalization group flow. As the cutoff is moved towards the infrared the LECs multiplying higher dimension operators (e.g. with higher derivatives) become less and less important. Under the hypothesis of naturalness, i.e. that the magnitudes of the LECs conform with their renormalization group running, higher dimension operators become effectively irrelevant. Thus, an expansion scheme emerges in powers of p/Λ, p denoting the typical momentum involved in a given process. The interactions are ordered according to this expansion, so that, at a given order, only a finite number of operators contributes. The values of the corresponding LECs represent physical effects beyond the cutoff scale. The larger the scale separation, the faster the convergence of this expansion. If the scales are not widely separated then, in order to achieve a satisfactory description, high orders in the expansion have to be taken into account, with a consequently large number of LECs involved. This amounts to less predictive power, since the LECs have to be fitted from data.

It is important to stress the sharp contrast of the EFT framework with models: the latter give definite predictions to be confronted with data, while the EFT is driven from data, through the fitting of the LECs. The falsification of an EFT involves either the discovery of new interactions violating the symmetry principles, or the failure of the low-energy expansion scheme. The true underlying theory is just one realization of the EFT, among many other alternative theories sharing the same symmetries on which the EFT is built. Ideally, the LECs could even be computed. Being unable to do so, one fixes them from data, with no guarantee that the extracted values of the LECs are the same as their true values. In the best case the series is well convergent and the determined LECs are close to their true values: as one proceeds in the expansion, the LECs get small corrections.

A particularly important feature of the EFT framework, compared to models, is the capability of estimating the theoretical uncertainty of a given calculation, as determined e.g. from the truncation of the low-energy expansion and by propagating the error propagation from the LECs [6]. In this ideal situation suppose we compute a given set of observables at the leading order (LO) of the perturbative scheme. Probably the χ^2 will not be satisfactory but we can equip the theoretical predictions with a theoretical uncertainty δ, and determine δ as the minimum for which the χ^2 per degree of freedom is 1. We can have thus a measure of the theoretical uncertainty δ_{LO} of the LO description. By going to the next order, usually the number of intervening LECs increases, allowing for a better χ^2. Consequently also δ_{NLO}, the theoretical uncertainty of the next-to-leading order description will be less than δ_{LO}. The δ_X so determined can be confronted with the convergence pattern which is expected on the basis of the separation of scales [3]. A slow convergence is a sign of insufficient separation of scales, which might be due to the emergence of relevant physics at a lower energy scale than expected.

86.2 Chiral Versus Pionless EFT

Various types of EFT are relevant for the description of few-nucleon systems, that have different domains of applicability. For the discussion to follow, let us focus on two of them, the chiral effective field theory (ChEFT) and the pionless effective field theory ($\not\pi$EFT) [7–14].

The ChEFT is the more microscopic one, its domain of applicability extends to shorter distances, and it encompasses more processes. It is based on the chiral symmetry of QCD, an approximate symmetry valid in the limit of vanishing light quark masses. Dynamical chiral symmetry breaking leads to the appearance of massless Goldstone bosons, identified with the pions, the lightest among the hadrons. The other hadrons, whose mass is unprotected by chiral symmetry, define a hadronic scale Λ_{H} of the same order as the chiral symmetry breaking scale $4\pi F_\pi$. As a consequence of explicit chiral symmetry breaking by the light quark masses the pions acquire a mass, which nevertheless can be considered as much smaller than Λ_{H}. This provides the sought after separation of scales, the mass gap between the pions and the heavier hadrons that mediate the nuclear interaction. The applicability domain is determined by typical momenta of order of M_π or less, which are much smaller than the hadronic scale $\Lambda_H \sim 1$ GeV. The proximity of the Δ resonance in πN scattering suggests to include it as an explicit degree of freedom, thus extending the domain of applicability. Among the distinctive features of this EFT in its different variants, following distinct power counting rules, there is the predicted hierarchy of few-nucleon forces, according to which the three-nucleon force (3NF) is a small perturbation on the top of the dominant 2N interaction. Furthermore, this theory encodes all constraints from chiral symmetry, that relate different processes, like for instance nuclear beta decays and the long-range component of the 2N interaction, as encoded in the Goldberger-Treiman relation. Chiral 2N potentials that provide excellent fit to the 2N database have been developed up to the fifth order of the chiral expansion [15–20], while the 3NF has so far been developed only up to the third order [21–23].

$\not\pi$EFT applies to a more restricted energy domain: for the resolution of this less microscopic EFT, the pion mass is a heavy scale and the pion exchanges are contracted to multinucleon contact interactions. So the pions are integrated out and their effects are encoded in the values of the LECs parametrizing contact interactions. On the basis of this reasoning, the 2N scattering lengths, denoted generically by a, should be of the order $1/M_\pi$, but instead they are much larger. Without a pion, this cannot be due to accidental cancellations and calls for a non-perturbative resummation. A new soft scale emerges $\sim 1/a$ (as was M_π in the ChEFT) and the theory is applicable in a much narrower domain of $p \sim 1/a \ll M_\pi$. As shown by Bedaque, Hammer and van Kolck [24, 25] this resummation calls for the introduction of a 3-body force, otherwise the 3-body system remains completely arbitrary. This arbitrariness can in turn be traced back to the Efimov effect [26], characterized by a discrete scale invariance, reminiscent of the absence of scales in the unitary limit $a \to \infty$ [27–29]. As a result, the 3NF needs to be part of the LO EFT, contrary to ChEFT. This is a direct consequence of the large scattering length scale in the absence of the pion. It is therefore interesting to examine more closely the role of the pion.

To this end let us recall that, in ChEFT, the expected order of magnitude of the kinetic energy in a two-nucleon (2N) system like the deuteron is $O(Q^2/\Lambda_{\mathrm{H}}) \sim$ 20 MeV and the same for the potential energy $O[Q^3/(4\pi F_\pi^2)] \sim 20$ MeV [30]. There is no expectation nor explanation for an almost exact compensation of these contributions. It seems that the LO suffers from an accidental cancellation. Even without spoiling the overall convergence of the low-energy expansion of the two-body binding energy B, this might induce instabilities in the expansion of observables which are non-linearly related to B. Indeed, considering the low-energy expansion $B = B_{\mathrm{LO}} + B_{\mathrm{NLO}} + \delta B$, if B_{LO} happens to be smaller than expected then the following terms take more relative importance: the perturbative series of a quantity like $1/B$ will consequently be flawed, despite the overall convergence of B, i.e. $\delta B \ll B$. And the two-body dynamics enters in the 3-body systems quite non-linearly. This is indeed manifested in the binding energy of the three-nucleon (3N) system: it is known that the 3NF produces ~ 1 MeV more attraction compared to the two-body force. Thus the 2N force produces about 2 MeV binding per pair, while the 3NF contributes about 1 MeV. It is not a small perturbation at any rate.

86.3 Impact of the Pion-Exchange in the LO $\not\pi$EFT

In the two-nucleon sector, the only difference between chiral and pionless EFT at LO is the one-pion exchange potential V^π, whose strength is fixed by the Goldberger-Treiman relation. In both schemes two LECs parametrize the S-wave short-range interaction $V_{\rm sr}$ in the singlet and triplet channels. Thus they lead to equivalent descriptions in the low-energy domain, since the two scattering lengths can be reproduced. On the contrary, in the 3N sector a marked difference appears, due to the presence of a leading 3N interaction in $\not\pi$EFT, which is absent in ChEFT. It can be thought that in the latter case the mass scale in the 3N system is determined by the pion-range. If this happens then the necessity of a leading 3NF would be lifted by the inclusion of V^π. In this respect it is instructive to consider a LO 2N potential inspired to ChEFT [31]

$$V = C_S\delta_{r_0}(\mathbf{r}) + C_T\delta_{r_0}(\mathbf{r})\boldsymbol{\sigma}_1\cdot\boldsymbol{\sigma}_2 + V^\pi_\beta, \tag{86.1}$$

where $C_{S/T}$ are the contact interaction LECs, the delta functions are regularized and become Gaussian functions of range r_0 and the one-pion exchange potential V^π is also regularized with a range β, suppressing the short distance behaviour by appropriate powers of the factor $1-\exp(-r^2/\beta^2)$. The regulator β is changed from 1 fm, corresponding approximately to the short-distance scale of the theory, to infinity, in which case the theory reduces to the pionless one. For each value of the regulators we fix the short-distance LECs to the singlet and triplet S-wave scattering lengths. With these values we compute the triton binding energy, which is shown in Fig. 86.1, and observe a pronounced dependence on the short-distance cutoff r_0 between 1 and 2 fm. On general grounds, this is a symptom of a missing contribution represented by the 3NF, whose LEC has to absorb this dependence [32]. It is true that, by including the pion (smaller β), the spread is somewhat reduced, but for the reasonable cutoffs $\beta \gtrsim 1$ fm that we explored it remains quite unstable, hinting at the necessity of including a 3NF also in the LO of the ChEFT. In the α particle binding energy this cutoff dependence is still more dramatic, without the inclusion of 3NF, as shown in the light band of Fig. 86.2. Whereas the results are much more stable, within 10% of the experimental value, by including a 3N contact interaction

$$V^{(0)}_{3N}(i,j,k) = W_0 e^{-r^2_{ij}/r^2_0} e^{-r^2_{ik}/r^2_0}, \tag{86.2}$$

regularized by Gaussian functions of the same range r_0, whose strength W_0 is fixed to the triton binding energy. According to our previous reasoning, the accidental emergence of a soft scale destabilizes the chiral expansion: important contributions will eventually compensate the cutoff dependence, but at the cost of going to very high orders.

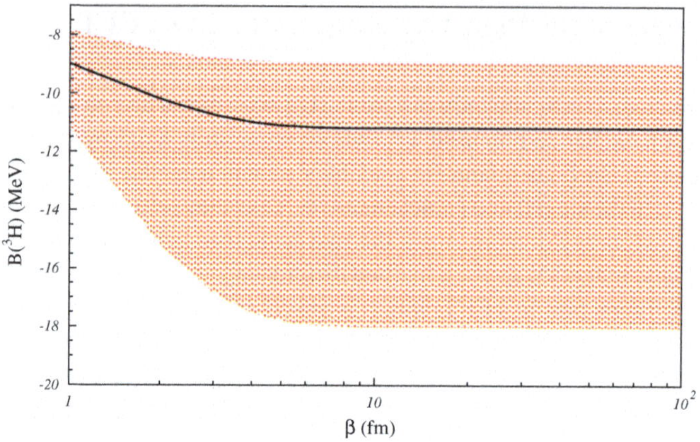

Fig. 86.1 Cutoff dependence of the triton binding energy. The band represents variation with the short-distance cutoff r_0 between 1 and 2 fm (adapted from [31])

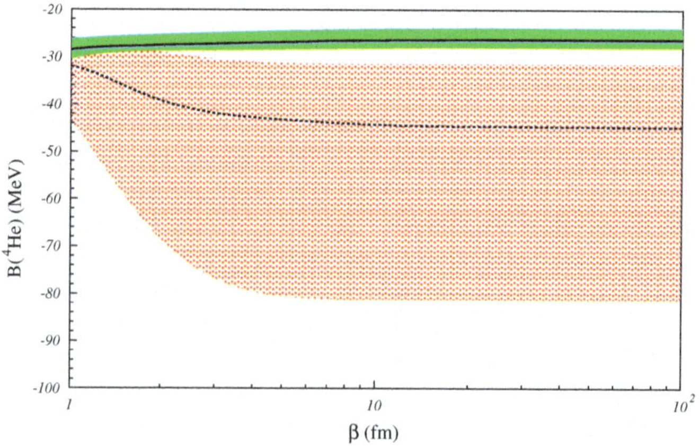

Fig. 86.2 Cutoff dependence of the α particle binding energy with (solid line and green band) and without (dashed line and light red band) a leading three-nucleon interaction adjusted to the triton binding energy (adapted from [31])

86.4 Cutoff Optimization at LO

As already stated, in EFT the cutoff is not only a mathematical artifact to manipulate the equations, it represents a physical scale, unresolved by the EFT. As the cutoff is changed within this region, the EFT description should not change, within the accuracy associated to the truncation of the low-energy expansion. This implies that the cutoff dependence should be reduced as one proceeds in the perturbative scheme. Thus, at a given order, the residual cutoff dependence reflects the effect of neglected higher orders. There are choices of the cutoff that minimize the effect of individual higher order terms. One can therefore optimize the choice of the cutoff to improve the description at a given order. This of course is only legitimate provided the low-energy expansion is well behaved, which can only be checked by actually going to higher orders. Taking such optimistic perspective, in [33], in order to improve the LO description of the previous section, the short distance cutoffs in the singlet and triplet channels, respectively $r_0^{S/T}$, have been chosen so as to reproduce the corresponding effective ranges. Analogously, with all cutoffs and couplings fixed in the two-body sector, and the 3NF strength adjusted to the triton, one can use the 3N short-distance cutoff r_3 as a further optimization parameter, replacing r_0 in (86.2). It is remarkable that such cutoff optimization not only can lead to the correct value of the α-particle binding energy, but also, in the framework of Brueckner-Hartree-Fock calculations, to satisfactory many-body properties like the energy per particle of symmetric nuclear matter and the corresponding saturation point and incompressibility, the nuclear symmetry energy and its slope parameter, an equation of state of β-stable nuclear matter compatible with presently observed neutron star masses [33].

As a general feature, such a satisfactory description would be impossible at LO if r_3 was not allowed to vary, leading to the idea of A-dependent cutoffs. More precisely, values of r_3 smaller than the two-body cutoffs r_0^X are preferred, This may also be understood from the way the contact interactions are regularized in momentum space: if the momentum transfers $\mathbf{k}_1$ and $\mathbf{k}_2$ of two particles are limited to a range Λ, the third one, $\mathbf{k}_3 = -\mathbf{k}_1 - \mathbf{k}_2$, may take larger values; with Gaussian cutoffs, the $\mathbf{k}_3$ distribution would have a width larger by a factor $\sqrt{2}$ compared to $\mathbf{k}_1$ and $\mathbf{k}_2$, implying the necessity of shorter ranges for the regularized three-body contact interaction.

86.5 Accurate N2LO Description of the A = 3 Continuum

If the contact 3NF is promoted to leading order in the ChEFT, due to the instabilities induced in the low-energy expansion by the proximity of the unitary limit, then the subleading 3N contact interaction, which merely specifies finer details of the underlying theory, will get the same relative promotion. It would therefore be part of the N2LO chiral potential[1].

This component of the 3N interaction has been studied in [38], where it was found to depend on 10 LECs, denoted E_i ($i = 1, \ldots, 10$), leading to a local potential in coordinate space,

[1]By this we mean an interaction suppressed by two orders relative to the leading one in the low-energy expansion, i.e. $O(p^2/\Lambda^2)$. Due to the vanishing of the $O(p/\Lambda)$ chiral potential, it is customarily denoted as NLO.

$$
\begin{aligned}
V_{3N}^{(2)}(i,j,k) = \sum_{\text{perm}} & (E_1 + E_2 \boldsymbol{\tau}_i \cdot \boldsymbol{\tau}_j + E_3 \boldsymbol{\sigma}_i \cdot \boldsymbol{\sigma}_j + E_4 \boldsymbol{\tau}_i \cdot \boldsymbol{\tau}_j \boldsymbol{\sigma}_i \cdot \boldsymbol{\sigma}_j) \\
& \times \left[Z_0''(r_{ij}) + 2\frac{Z_0'(r_{ij})}{r_{ij}} \right] Z_0(r_{ik}) \\
& + (E_5 + E_6 \boldsymbol{\tau}_i \cdot \boldsymbol{\tau}_j) S_{ij} \left[Z_0''(r_{ij}) - \frac{Z_0'(r_{ij})}{r_{ij}} \right] Z_0(r_{ik}) \\
& + (E_7 + E_8 \boldsymbol{\tau}_i \cdot \boldsymbol{\tau}_k)(\mathbf{L} \cdot \mathbf{S})_{ij} \frac{Z_0'(r_{ij})}{r_{ij}} Z_0(r_{ik}) \\
& + (E_9 + E_{10} \boldsymbol{\tau}_j \cdot \boldsymbol{\tau}_k) \boldsymbol{\sigma}_j \cdot \hat{\mathbf{r}}_{ij} \boldsymbol{\sigma}_k \cdot \hat{\mathbf{r}}_{ik} Z_0'(r_{ij}) Z_0'(r_{ik}) \quad (86.3)
\end{aligned}
$$

where $\boldsymbol{\sigma}_i$ ($\boldsymbol{\tau}_i$) are the Pauli spin (isospin) matrices, $\mathbf{r}_{ij}$ is the ij relative distance, S_{ij} and $(\mathbf{L} \cdot S)_{ij}$ are respectively the tensor and spin-orbit operators for particles i and j, and the function $Z_0(r)$ is the Fourier transform of the cutoff function $F(\mathbf{p}^2; \Lambda)$,

$$
Z_0(r; \Lambda) = \int \frac{d\mathbf{p}}{(2\pi)^3} e^{i\mathbf{p} \cdot \mathbf{r}} F(\mathbf{p}^2; \Lambda). \quad (86.4)
$$

The appearance of 10 new LECs is a very welcome feature from a phenomenological point of view since, contrary to the two-nucleon sector, none of the currently used 3N interaction models lead to a satisfactory description of the 3N scattering observables. In particular a long-standing discrepancy exists in polarization observables of low-energy $p-d$ scattering, most notably the A_y [39, 40]. As shown in [37, 41, 42], the subleading 3N contact potential (86.3) has enough flexibility to correctly address this problem leading to reasonable fits of these 3N observables below the deuteron breakup threshold. For the sake of illustration, we show in Fig. 86.3 fit results to very precise $p-d$ scattering data at 3 MeV proton energy [34]. Also fitted are the triton binding energy and the doublet and quartet $N-d$ scattering length. The resulting χ^2/d.o.f. is ~ 1.6 for cutoff values in the range $\Lambda = 200 - 500$ MeV. It will be interesting to see whether the same is true in a more extended kinematical region, i.e. beyond the breakup threshold.

86.6 Conclusions

Either due to basic phenomena deeply rooted in the underlying QCD, like spontaneous chiral symmetry breaking, or to accidental unexplained facts, like the large values of the 2N scattering lengths, small ratios of scales can be identified which allow to describe nuclear systems in an EFT framework. In addition to a more extended domain of applicability as compared to $\not\pi$EFT, a particularly attractive feature of ChEFT is the incorporation of all constraints from chiral symmetry, which is specially important when one considers nuclear interactions with external probes of quark bilinears within and beyond the Standard Model. However, the proximity of

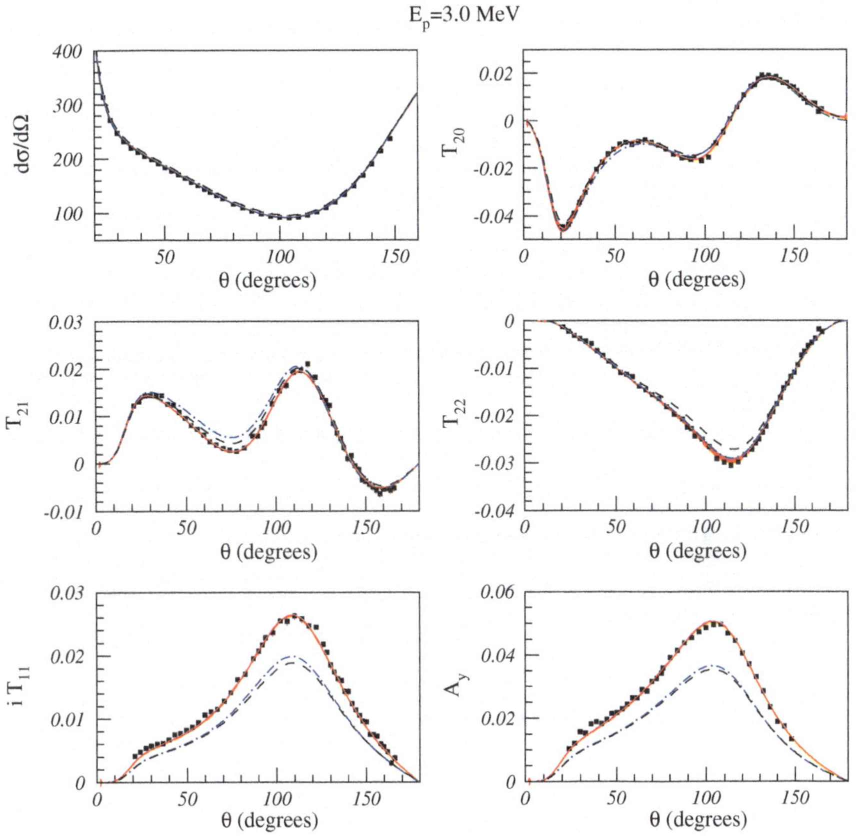

Fig. 86.3 Fit to cross section and polarization observables for $p - d$ scattering at 3 MeV proton energy. Data are from [34]. The adopted potential is the phenomenological AV18 2N interaction [35], supplemented by the leading and subleading contact 3N interaction, regularized with a cutoff Λ in the range 200–500 MeV (red bands). Also shown for reference are the predictions of the purely 2-body interaction (dashed lines) and with the addition of the Urbana IX 3N interaction [36] (adapted from [37])

nuclear physics to the unitary limit might destabilize the chiral expansion, requiring to reach very high orders for a satisfactory description. In order to cope with this possibility, we advocate a hybrid power counting for ChEFT in which the contact 3N interaction is promoted to the leading order. Remarkably, this simple interaction, with duly optimized A-dependent cutoffs, seems to bridge successfully few-nucleon to many-nucleon systems [33, 43]. In addition, thanks to the appearance of new unknown LECs, there is the possibility of obtaining an accurate description of the 3N continuum already at the next-to-next-to leading order. It would be interesting to explore the possibility of tailoring this interaction on the spectral properties of medium-light nuclei using Green Function Montecarlo techniques [44, 45].

References

1. Georgi, H.: Effective field theory. Ann. Rev. Nucl. Part. Sci. **43**, 209 (1993)
2. Weinberg, S.: The Quantum Theory of Fields. Cambridge University Press, Cambridge (1995)
3. Lepage, G.P.: What is renormalization? arXiv:hep-ph/0506330
4. Lepage, G.P.: How to renormalize the Schrodinger equation. arXiv:nucl-th/9706029
5. Weinberg, S.: Phenomenological Lagrangians. Phys. A **96**, 327 (1979)
6. Furnstahl, R.J., Phillips, D.R., Wesolowski, S.: A recipe for EFT uncertainty quantification in nuclear physics. J. Phys. G **42**, 034028 (2015)
7. Weinberg, S.: Nuclear forces from chiral Lagrangians. Phys. Lett. B **251**, 288 (1990)
8. Weinberg, S.: Effective chiral Lagrangians for nucleon–pion interactions and nuclear forces. Nucl. Phys. B **363**, 3 (1991)
9. van Kolck, U.: Effective field theory of nuclear forces. Prog. Part. Nucl. Phys. **43**, 337 (1999)
10. Epelbaum, E.: Few-nucleon forces and systems in chiral effective field theory. Prog. Part. Nucl. Phys. **57**, 654 (2006)
11. Machleidt, R., Entem, D.R.: Chiral effective field theory and nuclear forces. Phys. Rep. **503**, 1 (2011)
12. Epelbaum, E., Hammer, H.W., Meissner, U.G.: Modern theory of nuclear forces. Rev. Mod. Phys. **81**, 1773 (2009)
13. Braaten, E., Hammer, H.W.: Universality in few-body systems with large scattering length. Phys. Rep. **428**, 259 (2006)
14. Bedaque, P.F., van Kolck, U.: Effective field theory for few nucleon systems. Ann. Rev. Nucl. Part. Sci. **52**, 339 (2002)
15. Ordonez, C., Ray, L., van Kolck, U.: The two nucleon potential from chiral Lagrangians. Phys. Rev. C **53**, 2086 (1996)
16. Epelbaum, E., Gloeckle, W., Meissner, U.G.: The two-nucleon system at next-to-next-to-next-to-leading order. Nucl. Phys. A **747**, 362 (2005)
17. Entem, D.R., Machleidt, R.: Accurate charge dependent nucleon nucleon potential at fourth order of chiral perturbation theory. Phys. Rev. C **68**, 041001 (2003)
18. Epelbaum, E., Krebs, H., Meissner, U.G.: Improved chiral nucleon-nucleon potential up to next-to-next-to-next-to-leading order. Eur. Phys. J. A **51**, 53 (2015)
19. Epelbaum, E., Krebs, H., Meissner, U.G.: Precision nucleon-nucleon potential at fifth order in the chiral expansion. Phys. Rev. Lett. **115**, 122301 (2015)
20. Entem, D.R., Kaiser, N., Machleidt, R., Nosyk, Y.: Peripheral nucleon-nucleon scattering at fifth order of chiral perturbation theory. Phys. Rev. C **91**, 014002 (2015)
21. Epelbaum, E., Nogga, A., Gloeckle, W., Kamada, H., Meissner, U.G., Witala, H.: Three nucleon forces from chiral effective field theory. Phys. Rev. C **66**, 064001 (2002)
22. Bernard, V., Epelbaum, E., Krebs, H., Meissner, U.G.: Subleading contributions to the chiral three-nucleon force. I. Long-range terms. Phys. Rev. C **77**, 064004 (2008)
23. Bernard, V., Epelbaum, E., Krebs, H., Meissner, U.G.: Subleading contributions to the chiral three-nucleon force II: short-range terms and relativistic corrections. Phys. Rev. C **84**, 054001 (2011)
24. Bedaque, P.F., Hammer, H.W., van Kolck, U.: Effective theory of the triton. Nucl. Phys. A **676**, 357 (2000)
25. Bedaque, P.F., Hammer, H.W., van Kolck, U.: Renormalization of the three-body system with short range interactions. Phys. Rev. Lett. **82**, 463 (1999)
26. Efimov, V.: Energy levels arising form the resonant two-body forces in a three-body system. Phys. Lett. **33B**, 563 (1970)
27. König, S., Griesshammer, H.W., Hammer, H.W., van Kolck, U.: Nuclear physics around the unitarity limit. Phys. Rev. Lett. **118**, 202501 (2017)
28. Hammer, H.W.: Nuclei and the unitary limit. Few Body Syst. **59**, 58 (2018)
29. König, S.: The unitarity expansion for light nuclei. These proceedings, arXiv:1812.05327 [nucl-th]

30. Friar, J.L.: Dimensional power counting in nuclei. Few Body Syst. **22**, 161 (1997)
31. Kievsky, A., Viviani, M., Gattobigio, M., Girlanda, L.: Implications of Efimov physics for the description of three and four nucleons in chiral effective field theory. Phys. Rev. C **95**, 024001 (2017)
32. Nogga, A., Timmermans, R.G.E., van Kolck, U.: Renormalization of one-pion exchange and power counting. Phys. Rev. C **72**, 054006 (2005)
33. Kievsky, A., Viviani, M., Logoteta, D., Bombaci, I., Girlanda, L.: Correlations imposed by the unitary limit between few-nucleon systems, nuclear matter and neutron stars. Phys. Rev. Lett. **121**, 072701 (2018)
34. Shimizu, S., et al.: Analyzing powers of p+d scattering below the deuteron breakup threshold. Phys. Rev. C **52**, 1193 (1995)
35. Wiringa, R.B., Stoks, V.G.J., Schiavilla, R.: An Accurate nucleon-nucleon potential with charge independence breaking. Phys. Rev. C **51**, 38 (1995)
36. Pudliner, B.S., Pandharipande, V.R., Carlson, J., Pieper, S.C., Wiringa, R.B.: Quantum Monte Carlo calculations of nuclei with A <= 7. Phys. Rev. C **56**, 1720 (1997)
37. Girlanda, L., Kievsky, A., Viviani, M., Marcucci, L.E.: Short-range three-nucleon interaction from A = 3 data and its hierarchical structure. arXiv:1811.09398 [nucl-th]
38. Girlanda, L., Kievsky, A., Viviani, M.: Subleading contributions to the three-nucleon contact interaction. Phys. Rev. C **84**, 014001 (2011)
39. Viviani, M., Girlanda, L., Kievsky, A., Marcucci, L.E.: Effect of three-nucleon interaction in p-3He elastic scattering. Phys. Rev. Lett. **111**, 172302 (2013)
40. Golak, J., et al.: Low-energy neutron-deuteron reactions with N3LO chiral forces. Eur. Phys. J. A **50**, 177 (2014)
41. Girlanda, L., Kievsky, A., Viviani, M., Marcucci, L.: Progress in the quest for a realistic three-nucleon force. PoS CD **15**, 103 (2016)
42. Girlanda, L., Kievsky, A., Viviani, M., Marcucci, L.E.: Tuning the 3N force from 3N scattering data. EPJ Web Conf. **113**, 04009 (2016)
43. Kievsky, A., Polls, A., Juliá-Daz, B., Timofeyuk, N.K.: Saturation properties of helium drops from a leading order description. Phys. Rev. A **96**, 040501 (2017)
44. Piarulli, M., et al.: Light-nuclei spectra from chiral dynamics. Phys. Rev. Lett. **120**, 052503 (2018)
45. Piarulli, M.: The basic model of nuclear theory: from atomic nuclei to infinite nuclear matter. These proceedings

Chapter 87
Study of Multi-neutron Systems with SAMURAI Spectrometer

Z. H. Yang, F. M. Marqués, N. L. Achouri, D. S. Ahn, T. Aumann, H. Baba, D. Beaumel, M. Böhmer, K. Boretzky, M. Caamaño, S. Chen, N. Chiga, M. L. Cortés, D. Cortina, P. Doornenbal, C. A. Douma, F. Dufter, J. Feng, B. Fernández-Domínguez, Z. Elekes, U. Forsberg, T. Fujino, N. Fukuda, I. Gašparić, Z. Ge, R. Gernhäuser, J. M. Gheller, J. Gibelin, A. Gillibert, B. M. Godoy, Z. Halász, T. Harada, M. N. Harakeh, A. Hirayama, S. W. Huang, N. Inabe, T. Isobe, J. Kahlbow, N. Kalantar-Nayestanaki, D. Kim, S. Kim, M. A. Knösel, T. Kobayashi, Y. Kondo, P. Koseoglou, Y. Kubota, I. Kuti, C. Lehr, P. J. Li, Y. Liu, Y. Maeda, S. Masuoka, M. Matsumoto, J. Mayer, H. Miki, M. Miwa, I. Murray, T. Nakamura, A. Obertelli, N. A. Orr, H. Otsu, V. Panin, S. Park, M. Parlog, S. Paschalis, M. Potlog, S. Reichert, A. Revel, D. Rossi, A. Saito, M. Sasano, H. Sato, H. Scheit, F. Schindler, T. Shimada, Y. Shimizu, S. Shimoura, I. Stefan, S. Storck, L. Stuhl, H. Suzuki, D. Symochko, H. Takeda, S. Takeuchi, J. Tanaka, Y. Togano, T. Tomai, H. T. Törnqvsit, J. Tscheuschner, T. Uesaka, V. Wagner, K. Wimmer, H. Yamada, B. Yang, L. Yang, Y. Yasuda, K. Yoneda, L. Zanetti and J. Zenihiro

Abstract The tetraneutron has been drawing the attention of the nuclear physics community for decades, but a firm conclusion on its existence and properties is still far from being reached despite many experimental and theoretical efforts. New

---

Z. H. Yang (✉)
Research Center for Nuclear Physics, Osaka University, Ibaraki, Osaka 567-0047, Japan
e-mail: zhyang@ribf.riken.jp

Z. H. Yang · D. S. Ahn · H. Baba · K. Boretzky · N. Chiga · M. L. Cortés · P. Doornenbal · J. Feng · Z. Elekes · N. Fukuda · I. Gašparić · Z. Ge · T. Harada · A. Hirayama · S. W. Huang · N. Inabe · T. Isobe · J. Kahlbow · Y. Kubota · C. Lehr · Y. Liu · M. Matsumoto · M. Miwa · I. Murray · H. Otsu · V. Panin · M. Sasano · H. Sato · Y. Shimizu · H. Suzuki · H. Takeda · Y. Togano · T. Tomai · H. T. Törnqvsit · T. Uesaka · B. Yang · Y. Yasuda · K. Yoneda · L. Zanetti · J. Zenihiro
RIKEN Nishina Center, 2-1 Hirosawa, Wako, Saitama 351-0198, Japan

F. M. Marqués · N. L. Achouri · J. Gibelin · B. M. Godoy · N. A. Orr · M. Parlog
LPC Caen, ENSICAEN, Université de Caen, CNRS/IN2P3, 14050 Caen Cedex, France

T. Aumann · J. Kahlbow · M. A. Knösel · P. Koseoglou · C. Lehr · A. Obertelli · S. Paschalis · D. Rossi · H. Scheit · S. Storck · D. Symochko · J. Tanaka · H. T. Törnqvsit · J. Tscheuschner · V. Wagner · L. Zanetti
Institut für Kernphysik, Technische Universität Darmstadt, 64289 Darmstadt, Germany

N. A. Orr et al. (eds.), *Recent Progress in Few-Body Physics*,
Springer Proceedings in Physics 238,
https://doi.org/10.1007/978-3-030-32357-8_87

measurements have recently been performed at RIBF with the SAMURAI spectrometer by applying complementary reaction probes, which will help to pin down the properties of this four-neutron system.

T. Aumann · K. Boretzky · P. Koseoglou · J. Tanaka
GSI Helmholtzzentrum für Schwerionenforschung, 64291 Darmstadt, Germany

D. Beaumel · I. Stefan
Institut de Physique Nucléaire Orsay, IN2P3-CNRS, 91406 Orsay Cedex, France

M. Böhmer · F. Dufter · R. Gernhäuser · S. Reichert · F. Schindler
Technische Universität München, 85748 Garching, Germany

M. Caamaño · D. Cortina · B. Fernández-Domínguez
Departamento de Física de Partículas and IGFAE, Universidade de Santiago de Compostela, 15782 Santiago de Compostela, Spain

S. Chen · P. J. Li
Department of Physics, The University of Hong Kong, Pokfulam Road, Hong Kong, China

C. A. Douma · M. N. Harakeh · N. Kalantar-Nayestanaki
KVI-CART, University of Groningen, Zernikelaan 25, 9747 AA Groningen, The Netherlands

J. Feng · S. W. Huang · Y. Liu · B. Yang
School of Physics and State Key Laboratory of Nuclear Physics and Technology, Peking University, Beijing 100871, China

Z. Elekes · Z. Halász · I. Kuti
MTA ATOMKI, Debrecen 4001, Hungary

U. Forsberg · S. Paschalis
Department of Physics, University of York, York YO10 5DD, UK

T. Fujino · Y. Togano
Department of Physics, Rikkyo University, Toshima, Tokyo 172-8501, Japan

I. Gašparić
Ruđer Bošković Institut (RBI), Zagreb, Croatia

J. M. Gheller · A. Gillibert
CEA, Centre de Saclay, IRFU, 91191 Gif-sur-Yvette, France

T. Harada · M. Miwa
Toho University, Tokyo 143-8540, Japan

A. Hirayama · Y. Kondo · M. Matsumoto · H. Miki · T. Nakamura · A. Saito · T. Shimada · S. Takeuchi · T. Tomai · H. Yamada · Y. Yasuda
Department of Physics, Tokyo Institute of Technology, Meguro, Tokyo 152-8551, Japan

D. Kim · S. Kim · S. Park
Department of Physics, Ehwa Womans University, Seoul, Korea

T. Kobayashi
Department of Physics, Tohoku University, Miyagi 980-8578, Japan

Y. Maeda
Faculty of Engineering, University of Miyazaki, Miyazaki 889-2192, Japan

87.1 Introduction

There has been a long-standing question among the nuclear physics community whether a nucleus made purely of neutrons without any protons can exist or not. These so-called "*Neutral nuclei*" have attracted in particular a lot of attention over the past decades. These multi-neutron systems, whether existing as bound or quasi-bound (resonant) states, have fundamental importance in nuclear physics. They provide the possibility to investigate "purely" the nuclear forces free from Coulomb interaction, which is essential for developing the nuclear theory, and critical for our understanding of neutron-rich nuclear matter and neutron stars. The two-neutron system, dineutron (^2n), has been well known to be unbound. For the tetraneutron (^4n), however, no firm conclusion has been drawn yet despite many experimental and theoretical efforts.

Earlier experimental attempts to search for a bound ^4n with a wide variety of methods all failed to find positive evidence. The existence of a bound state of four neutrons has been ruled out by calculations based on standard nuclear forces. But the possibility for ^4n existing as a resonant state is supported by some theoretical calculations. In 2002 Marqués et al. reported the possible existence of a bound or low-lying resonant ^4n state [1]. The resurgence of interest on this topic in recent years was triggered by the report on the observation of a low-lying ^4n resonance (although the resolution made the result also compatible with a bound state within error bars) released by Kisamori et al. in 2016. This concerned a background-clean measurement with the SHARAQ spectrometer, but only four events were identified indicative of a "candidate" ^4n resonant state [2]. This observation was qualitatively reproduced by recent Quantum Monte Carlo (QMC) [3] and No-Core Shell Model calculations [4], but was not supported by some other *ab initio-type* calculations [5, 6].

New experimental measurements on ^4n have been recently performed at RIKEN Radioactive Isotope Beam Factory (RIBF) facility: updated double-charge exchange reaction ^{4}He(^{8}He, $\alpha\alpha$)^4n by Shimoura et al. (for details see report by Shimoura in the present proceedings), ^{8}He(p, $p\alpha$)^4n with inverse kinematics by Rossi and Paschalis et al. and ^{8}He(p, $2p$)^{7}H$\{t + {}^4n\}$ with inverse kinematics by Yang and Marqués et al. These experiments will provide independent and complementary

S. Masuoka · S. Shimoura · L. Stuhl · K. Wimmer · L. Yang
Center for Nuclear Study, University of Tokyo, 2-1 Hirosawa, Wako, Saitama 351-0198, Japan

J. Mayer
Institut für Kernphysik, Universität zu Köln, Cologne, Germany

U. Forsberg
Department of Physics, Lund University, 22100 Lund, Sweden

M. Potlog
Institute of Space Sciences, Magurele, Romania

A. Revel
Grand Accélérateur National d'Ions Lourds (GANIL), CEA/DRF, CNRS/IN2P3, Bvd Henri Becquerel, 14076 Caen, France

information about ^4n, considering the different population processes, resolutions, production yields and analysis methods, which will help to pin down the properties of this four-neutron system. The latter two experiments, both requiring high luminosity from the thick liquid hydrogen target MINOS [7] and the high neutron detection efficiency of the combined neutron detection array NeuLAND + NEBULA, were arranged in a campaign with the SAMURAI spectrometer in 2017. In the present report we will focus on the ^{8}He(p, 2p) experiment, and a brief introduction of the ^{8}He(p, $p\alpha$) experiment will also be presented.

87.2 Description of the ^{8}He(P, 2p)^{7}H{T + 4n} Experiment

The ^{8}He(p, 2p)^{7}H{t + 4n} experiment was carried out at the RIBF, which is operated by the RIKEN Nishina Center and the Center for Nuclear Study (CNS), University of Tokyo. The secondary ^{8}He beam with an energy of ~150 MeV/nucleon and an intensity of ~10^5 pps was produced in the fragmentation of ^{18}O on the ^{9}Be primary target, and then purified and transported by the BigRIPS beam line. The incident ^{8}He particles were identified event-by-event with TOF-ΔE method by using plastic scintillators on the beam line and tracked with two multi-wire drift chambers (BDC1 and BDC2) onto the 150-mm-thick vertex-tracking liquid hydrogen target MINOS. ^{7}H was then produced from the (p, 2p) reaction off ^{8}He. A schematic view of the experimental setup is presented in Fig. 87.1.

The recoil protons from the (p, 2p) reaction were tracked by the TPC of MINOS and then recorded by a compact NaI array surrounding the target, constructed with 36 crystals from the DALI2 in-beam gamma-ray spectrometer [8] and arranged into two symmetric rings. The energy resolution of the NaI scintillators was determined to be

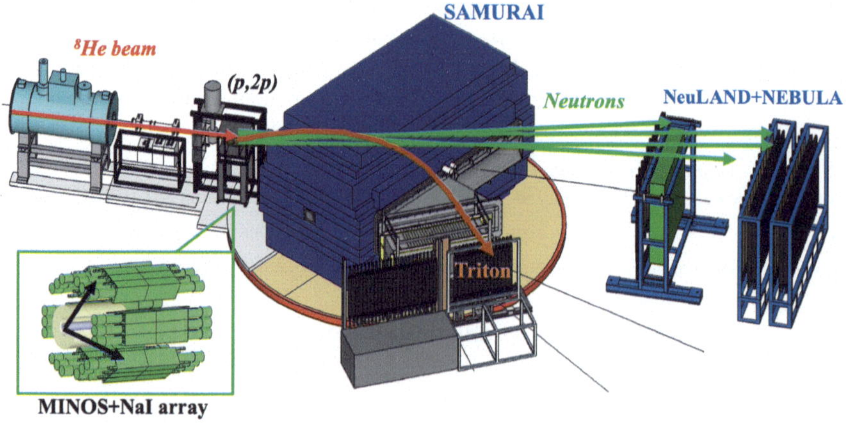

Fig. 87.1 Schematic view of the setup for the ^{8}He(p, 2p)^{7}H{t + 4n} experiment with inverse kinematics

around 1% (FWHM) in measurements with 80-MeV protons at CYRIC of Tohoku University, and an energy calibration was performed by measuring the proton-proton elastic scattering at 175 MeV with the same setup. The charged fragments were analyzed by the SAMURAI spectrometer [9]. Coincident detection of the multiple decay neutrons is critical but extremely challenging for studies of these multi-neutron systems. The NeuLAND demonstrator with four double-planes from GSI [10] was added to the existing NEBULA array, which provides the highest 4-neutron detection efficiency ($\varepsilon_{4n} \sim 1\%$) achievable at present. It is worthwhile to note that all the reaction products were recorded in the present experiment, providing the complete kinematics of the reaction and therefore eliminating possible ambiguity from identification of multi-neutron events. Another advantage of the present kinematically complete measurement is the applicability of the so-called "*Missing-Invariant-Mass method*", which requires the detection of only three of the four neutrons by reconstructing the kinematics of the missing neutron from the remaining particles and therefore largely enhances the sensitivity in the vicinity of the threshold (by a factor of ~20).

The charged fragments are identified from the TOF and ΔE signals from the HODO plastic scintillator array at the exit of SAMURAI. As shown in Fig. 87.2a, tritons are well recorded and clearly separated from ^{6}He fragments. Angular information of the recoil protons is provided by the TPC surrounding the target. In Fig. 87.2b and c, the correlations of polar angles and azimuthal angles of the reoil proton pair in coincidence with outgoing triton fragments are presented, respectively. An opening angle of ~70° and the coplanarity are evidently observed, in agreement with the expected correlation pattern for the quasi-free (p, $2p$) reaction off ^{8}He. We have also checked the registered neutron multiplicity in the NeuLAND + NEBULA array by applying causality cuts on the space-time separation of neutron hits in order to remove fake multi-neutron events (due to the so called cross-talk). In this way, we estimate that the recorded full-kinematics events will be of the order of 50 k, which will permit a detailed investigation of the tetraneutron system.

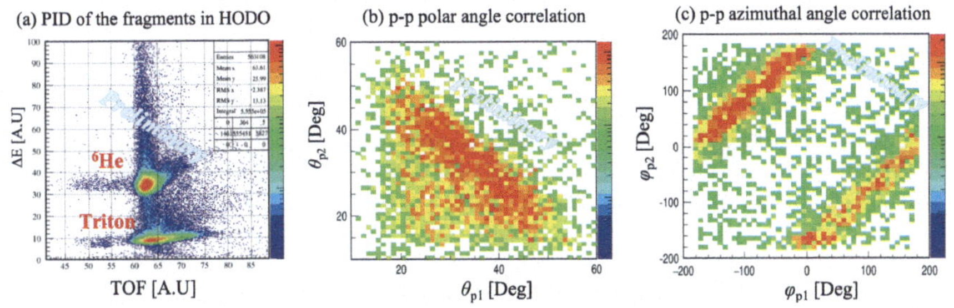

Fig. 87.2 **a** PID of fragments by measuring ToF and ΔE with HODO. Polar angle, **b** and azimuthal angle, **c** correlations of the two recoil protons populated in the ^{8}He(p, $2p$) reaction

87.3 Perspectives

Within the same campaign, ^{8}He(p, $p\alpha$)4n was measured by Rossi and Paschalis et al., taking advantage of the well developed $\alpha + 4n$ cluster structure of ^{8}He. The final-state interaction with the charged fragments is minimized by carrying out the measurement at very backward scattering angles in the center of mass. The experimental setup is similar to the ^{8}He(p, 2p)^{7}H{t + 4n} experiment (Fig. 87.1). Instead of the TPC + NaI detectors, a silicon-tracker system was introduced at the target region for the tracking of protons and alpha particles [11]. The data analysis is still in progress, but protons and alpha particles are clearly identified in coincidence, and the four-neutron system can then be reconstructed from the missing-mass method.

In 2018, the first direct experimental investigation of the 6-neutron system ("hexaneutron") was made by Beaumel et al. by measuring the ^{14}Be(p, $p\alpha\alpha$) reaction with SAMURAI spectrometer. Furthermore, some new proposals on multi-neutron systems based on (p, 2p) and (p, p) reactions are also in preparation.

There are in general two key yet challenging factors for experimental studies of these exotic multi-neutron systems. One is the population process, and the other is the detection and identification of the multi-neutron events. It is not clear at present which reaction will selectively populate the multi-neutron systems we are interested in, and complementary measurements with different reaction probes would in this sense help to reach a definite conclusion. The high-quality secondary beams provided by RIBF together with the large-acceptance SAMURAI spectrometer and the associated large neutron detector arrays (including the EXPAND upgrade project for NEBULA) are providing new opportunities for sophisticated studies of these multi-neutron systems.

We acknowledge the support of the RIBF accelerator staff and the BigRIPS team for providing the high-quality beam. Z. H. Yang acknowledges the financial support from the Foreign Postdoctoral Researcher program of RIKEN. I. G. was supported by HIC for FAIR and Croatian Science Foundation Project No. 7194.

References

1. Marqués, F.M., et al.: Phys. Rev. C **65**, 044006 (2002). arXiv:nucl-ex/0504009
2. Kisamori, K., et al.: Phys. Rev. Lett. **116**, 052501 (2016)
3. Pieper, S.C.: Phys. Rev. Lett. **90**, 252501 (2003)
4. Shirokov, A.M., et al.: Phys. Rev. Lett. **117**, 182502 (2016)
5. Lazauskas, R., Carbonell, J.: Phys. Rev. C **72**, 034003 (2005)
6. Hiyama, E., et al.: Phys. Rev. C **93**, 044004 (2016)
7. Obertelli, A., et al.: Eur. Phys. J. A **50**, 8 (2014)
8. Takeuchi, S., et al.: Nucl. Instrum. Methods A **763**, 596 (2014)
9. Kobayashi, T., et al.: Nucl. Instrum. Methods B **317**, 294 (2013)
10. Kahlbow, J., et al.: RIKEN Accel. Prog. Rep **49**, 17 (2016)
11. Dufter, F., et al.: RIKEN Accel. Prog. Rep **51**, 149 (2018)

Chapter 88
Nucleon Structure from Lattice QCD

Kyriakos Hadjiyiannakou

Abstract We present selected recent results on the nucleon structure using lattice QCD simulations with physical or close to physical pion masses. Results include, the nucleon axial charge, tensor charge, σ-terms, average momentum fraction and the decomposition of the nucleon spin. Results about the proton electric form factors are also discussed.

88.1 Introduction

Lattice QCD has entered a new and exciting era. Progress in algorithms and the increased availability of powerful computers have enabled simulations at physical values of the parameters of theory. The so-called physical ensembles eliminate chiral extrapolations, which especially for baryons, introduce uncontrolled systematic errors. Other sources of systematic errors include discretization effects from finite lattice spacings and finite volume effects. Therefore simulations for at least three different lattice spacings and at least three different volumes are needed to reliably extrapolate to the continuum and infinite volume limit. An additional source of systematic error that needs special attention is the pollution of the ground-state from contributions from excited states.

Understanding the complex structure of the nucleon is among the frontiers of Nuclear and Particle Physics. Lattice QCD provides the framework to compute nucleon observables from first principles providing significant input for experimental and phenomenological studies. In this proceedings we summarize important nucleon quantities such as the nucleon axial charge the average momentum fraction allowing us to decompose the nucleon spin, the tensor charge an important quantity for searches beyond the standard model, σ-terms involving in dark matter searches and

K. Hadjiyiannakou (✉)
The Cyprus Institute, 20 Kavafi Str., 2121 Nicosia, Cyprus
e-mail: k.hadjiyiannakou@cyi.ac.cy

N. A. Orr et al. (eds.), *Recent Progress in Few-Body Physics*,
Springer Proceedings in Physics 238,
https://doi.org/10.1007/978-3-030-32357-8_88

the electric structure of the proton. We note that the inclusion of disconnected contributions is crucial to avoid uncontrolled systematic errors in the nucleon matrix elements.

88.2 Nucleon Structure

On the lattice in order to extract the nucleon matrix elements one has to compute the nucleon two-point and three-point Euclidean correlation functions given respectively as

$$C^{2pt}(\Gamma_0, \mathbf{p}; t_s, t_0) = \sum_{\mathbf{x}_s} \mathrm{Tr}\left[\Gamma_0 \langle J(t_s, \mathbf{x}_s) \bar{J}(t_0, \mathbf{x}_0)\rangle\right] e^{-i(\mathbf{x}_s - \mathbf{x}_0)\cdot \mathbf{p}}, \tag{88.1}$$

and

$$C^{3pt}_{\mu}(\Gamma_\nu, \mathbf{q}, \mathbf{p}'; t_s, t_{\mathrm{ins}}, t_0) = \sum_{\mathbf{x}_{\mathrm{ins}}, \mathbf{x}_s} e^{i(\mathbf{x}_{\mathrm{ins}} - \mathbf{x}_0)\cdot \mathbf{q}} e^{-i(\mathbf{x}_s - \mathbf{x}_0)\cdot \mathbf{p}'} \times$$
$$\mathrm{Tr}\left[\Gamma_\nu \langle J(t_s, \mathbf{x}_s) \mathcal{O}_\mu(t_{\mathrm{ins}}, \mathbf{x}_{\mathrm{ins}}) \bar{J}(t_0, \mathbf{x}_0)\rangle\right]. \tag{88.2}$$

The standard nucleon interpolating field is $J(x) = \epsilon^{abc}\left(u^{aT}(x) C\gamma_5 d^b(x)\right) u^c(x)$ with the creation interpolator at x_0 and the annihilation operator at x_s and the current operator is $\mathcal{O}(x_{\mathrm{ins}})$ inserted at x_{ins}. The matrix element can be extracted by taking appropriate combinations of three- to two-point functions. Since the nucleon interpolating field creates any state with the quantum numbers of the nucleon, the ground-state should be isolated. In the limit of large time separation states with higher than the lowest energy die out and one can isolate the ground state. Since arbitrarily large separations are practically not allowed due to the exponential increase in the statistical error with the time separation, alternative approaches should be followed. One way to increase the overlap with the nucleon is to utilize smearing techniques to increase the overlap with the ground state. Common methods to extract the nucleon ground state are the so-called plateau method, summation method, multi-state fits and variational approaches.

88.2.1 Nucleon Axial and Tensor Charges and σ-Terms

The isovector nucleon axial charge is considered a benchmark quantity for lattice QCD since it has been measured in high precision in neutron β-decay. It can be extracted from the axial-vector operator $\mathcal{O}_A = \bar{\psi}(x)\gamma_\mu\gamma_5\frac{\tau^3}{2}\psi(x)$ at zero momentum transfer. Due to its isovector structure disconnected contributions vanish allowing for a relatively easy extraction. In Fig. 88.1 we show recent lattice QCD results for g_A. As one can see lattice results are in general in agreement with the experimental

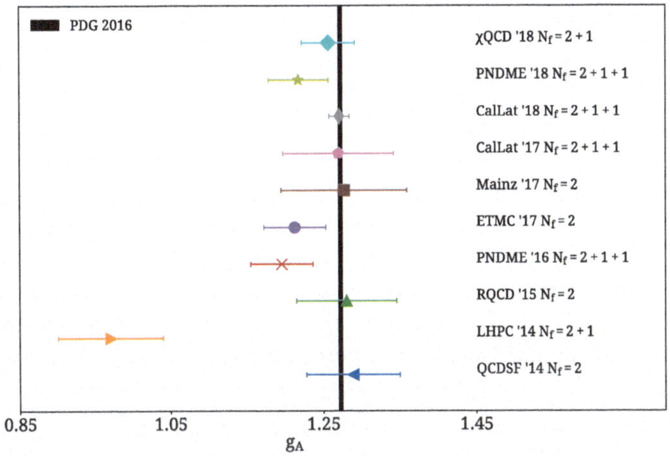

Fig. 88.1 Comparison of lattice QCD results for g_A from [2, 5–8, 10, 11, 13]. The black band indicates the PDG value from [16]

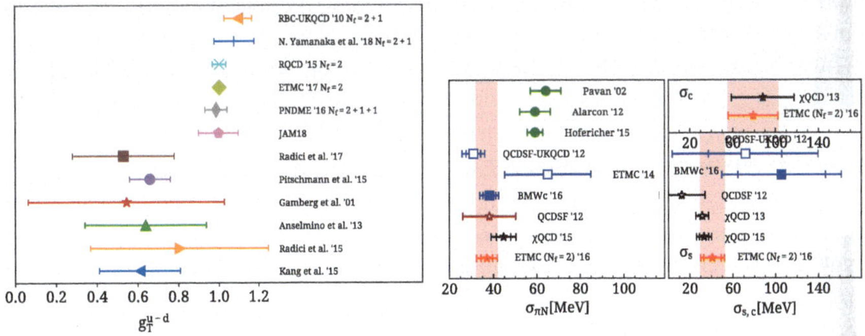

Fig. 88.2 Left panel: Comparison of results for the isovector tensor charge. Reading from down to up the first six points correspond to results from phenomenological studies while the rest are results from lattice QCD studies. Right panel: Comparison of results for the $\sigma_{\pi N}$, σ_s and σ_c. Green circles [1, 12, 17] are results from phenomenological studies while the rest are lattice QCD results [4, 9, 14, 18]

value. The tensor charge can be extracted from the operator $\mathcal{O}_T = \bar{\psi}(x)\sigma^{\mu\nu}\frac{\tau^3}{2}\psi(x)$ while the σ-term is $\mathcal{O}_S = m_q\bar{q}(x)q(x)$. For the σ-terms disconnected contributions should be included.

Results for the g_T^{u-d} is shown in Fig. 88.2. Phenomenological studies have significantly smaller values compared to lattice QCD results with the latter having a trend to higher values. Results for the σ-terms are presented also in Fig. 88.2. For $\sigma_{\pi N}$ lattice results are in good agreement with values lower than phenomenology. For σ_s and σ_c there are no phenomenological studies and lattice results are well compatible.

88.2.2 Spin Decomposition

One gauge invariant way to decompose the nucleon spin was introduced by X. Ji [15], namely $J_N = \sum_q J_q + J_g = \sum_q \left(\frac{1}{2}\Delta\Sigma_q + L_q\right) + J_g$. The intrinsic quark spin is $\frac{1}{2}\Delta\Sigma_q$ and L_q is the quark orbital angular momentum. The total contribution of the gluons to the nucleon is J_g where J_g cannot decompose further, in contrast to the Jaffe-Manohar decomposition. The intrinsic quark spin can be computed from the first Mellin moment of the polarized PDF and is the nucleon matrix element of the axial-vector operator at zero momentum transfer. The total quark contribution J_q, can be extracted from the second Mellin moments of the unpolarized PDF or the Generalized Form Factors (GFFs) at zero momentum transfer, namely $J_q = \frac{1}{2}\left[A_{20}^q(0) + B_{20}^q(0)\right]$. The GFFs can be extracted from the nucleon matrix element of the vector one-derivative operator $\mathcal{O}_V^{\mu\nu} = \bar{q}\gamma^{\{\mu}\overleftrightarrow{D}^{\nu\}}q$, namely $\langle N(p', s')|\mathcal{O}_V^{\mu\nu}|N(p, s)\rangle = \bar{u}_N(p', s')\Lambda^{\mu\nu}(Q^2)u_N(p, s)$ where $\Lambda^{\mu\nu}(Q^2) = A_{20}^q(Q^2)\gamma^{\{\mu}P^{\nu\}} + B_{20}^q(Q^2)\frac{\sigma^{\{\mu\alpha}q_\alpha P^{\nu\}}}{2m} + C_{20}^q(Q^2)\frac{1}{m}Q^{\{\mu}Q^{\nu\}}$ with $Q^2 = (p' - p)^2$ is the momentum transfer square and $P = (p' + p)/2$. The $A_{20}(0)$ is directly accessible from the lattice data while $B_{20}(0)$ needs to be extrapolated at zero momentum transfer from finite Q^2 values.

In order to compute the J_g term we construct the gluon operator [3], $\mathcal{O}_g^{\mu\nu} = 2\mathrm{Tr}\left[G_{\mu\sigma}G_{\nu\sigma}\right]$ where $G_{\mu\nu}$ is the field strength tensor. We use the scalar operator $\mathcal{O}_B = \mathcal{O}_{44} - \frac{1}{3}O_{jj}$ and from the matrix element $\langle N|\mathcal{O}_B|N\rangle = -2m_N\langle x\rangle_g$ one can extract $A_{20}^g(0) = \langle x\rangle_g$. To compute $J_g = \frac{1}{2}\left[A_{20}^g(0) + B_{20}^g(0)\right]$, evaluation of $B_{20}^g(0)$ is also needed. Assuming that spin and momentum sums are satisfied one can relate B_{20} of gluons to quarks, such as $\sum_q B_{20}^q(0) = -B_{20}^g(0)$. In Fig. 88.3 we present our results for the complete nucleon spin and momentum decomposition [3]. We find

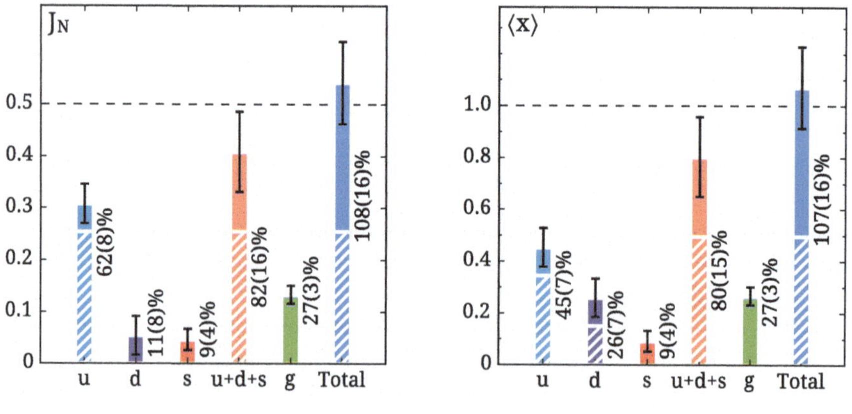

Fig. 88.3 Left: Nucleon spin decomposition. Right: Nucleon momentum decomposition. The striped segments show valence quark contributions (connected) and the solid segments the sea quark and gluon contributions (disconnected). Results are given in $\overline{\text{MS}}$-scheme at 2 GeV

that disconnected diagrams contribute significantly and that the spin and momentum sums are satisfied within the current statistical accuracy.

88.2.3 Electromagnetic Form Factors

The proton electric form factor is an important quantity probing the charge density distribution inside the proton. The vector nucleon matrix element is

$$\langle N(p',s')|j_\mu|N(p,s)\rangle = \sqrt{\frac{m_N^2}{E_N(\mathbf{p}')E_N(\mathbf{p})}} \times \\ \bar{u}_N(p',s')\left[\gamma_\mu F_1(q^2) + \frac{i\sigma_{\mu\nu}q^\nu}{2m_N}F_2(q^2)\right]u_N(p,s)\,. \tag{88.3}$$

and the electric form factor can be expressed in terms of the Dirac $F_1(Q^2)$ and Pauli $F_2(Q^2)$ form factors as

$$G_E(Q^2) = F_1(Q^2) - \frac{Q^2}{2m_N^2}F_2(Q^2). \tag{88.4}$$

In Fig. 88.4 we show our results extracted from a physical ensemble for the proton electric form factor. The disconenected contributions have also been evaluated and are found to have a negligible contribution for this quantity. Our results have been found to lie one to two standard deviations higher that experiment.

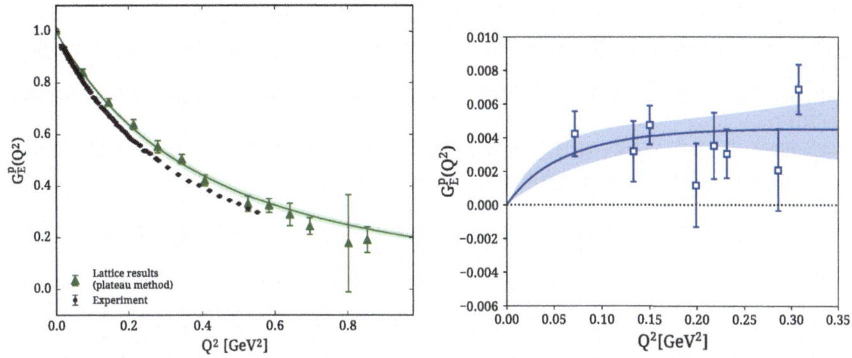

Fig. 88.4 Left panel: Proton electric form factor including disconnected contributions in comparison with experimental results. Right panel: Disconnected contributions to the proton electric form factor

88.3 Conclusions

Lattice QCD can deliver high quality results for several nucleon observables directly at the physical point. Recent results for the nucleon axial charge are compatible with the experimental, for the tensor charge lattice results are significantly more accurate as well as for the nucleon σ-terms. With state-of-the-art simulations we can now shed light to the proton spin decomposition and also compute in high precision crucial form factors such as the electric form factor allowing us to access the proton charge radius.

Acknowledgements I would like to thank my collaborators C. Alexandrou, M. Constantinou, K. Jansen, C. Kallidonis, G. Koutsou, A. Vaquero. Results were obtained using Piz Daint at CSCS, via projects with s540 and s702. We also gratefully acknowledge the Gauss Centre for Supercomputing e.V. (www.gauss-centre.eu) for funding the project pr74yo by providing computing time on the GCS Supercomputer SuperMUC at Leibniz Supercomputing Centre (www.lrz.de).

References

1. Alarcon, J.M., Martin Camalich, J., Oller, J.A.: The chiral representation of the πN scattering amplitude and the pion-nucleon sigma term. Phys. Rev. D **85**, 051503 (2012)
2. Alexandrou, C., Constantinou, M., Hadjiyiannakou, K., Jansen, K., Kallidonis, C., Koutsou, G., Vaquero Aviles-Casco, A.: Nucleon axial form factors using $N_f = 2$ twisted mass fermions with a physical value of the pion mass. Phys. Rev. D **96**(5), 054507 (2017)
3. Alexandrou, C., Constantinou, M., Hadjiyiannakou, K., Jansen, K., Panagopoulos, H., Wiese, C.: Gluon momentum fraction of the nucleon from lattice QCD. Phys. Rev. D **96**(5), 054503 (2017)
4. Alexandrou, C., Drach, V., Jansen, K., Kallidonis, C., Koutsou, G.: Baryon spectrum with $N_f = 2 + 1 + 1$ twisted mass fermions. Phys. Rev. D **90**(7), 074501 (2014)
5. Bali, G.S., Collins, S., Gläßle, B., Göckeler, M., Najjar, J., Rödl, R.H., Schäfer, A., Schiel, R.W., Söldner, W., Sternbeck, A.: Nucleon isovector couplings from $N_f = 2$ lattice QCD. Phys. Rev. D **91**(5), 054501 (2015)
6. Berkowitz, E., et al.: An accurate calculation of the nucleon axial charge with lattice QCD (2017)
7. Capitani, S., Della Morte, M., Djukanovic, D., von Hippel, G.M., Hua, J., Jäger, B., Junnarkar, P.M., Meyer, H.B., Rae, T.D., Wittig, H.: Iso-vector axial form factors of the nucleon in two-flavour lattice QCD (2017)
8. Chang, C.C., et al.: A per-cent-level determination of the nucleon axial coupling from quantum chromodynamics. Nature **558**(7708), 91–94 (2018)
9. Gong, M., et al.: Strangeness and charmness content of the nucleon from overlap fermions on $2 + 1$-flavor domain-wall fermion configurations. Phys. Rev. D **88**, 014503 (2013)
10. Green, J.R., Engelhardt, M., Krieg, S., Negele, J.W., Pochinsky, A.V., Syritsyn, S.N.: Nucleon structure from lattice QCD using a nearly physical pion mass. Phys. Lett. B **734**, 290–295 (2014)
11. Gupta, R., Jang, Y.-C., Yoon, B., Lin, H.-W., Cirigliano, V., Bhattacharya, T.: Isovector charges of the nucleon from $2 + 1 + 1$-flavor lattice QCD. Phys. Rev. D **98**, 034503 (2018)
12. Hoferichter, M., Ditsche, C., Kubis, B., Meissner, U.G.: Dispersive analysis of the scalar form factor of the nucleon. JHEP **06**, 063 (2012)

13. Horsley, R., Nakamura, Y., Nobile, A., Rakow, P.E.L., Schierholz, G., Zanotti, J.M.: Nucleon axial charge and pion decay constant from two-flavor lattice QCD. Phys. Lett. B **732**, 41–48 (2014)
14. Horsley, R., Nakamura, Y., Perlt, H., Pleiter, D., Rakow, P.E.L., Schierholz, G., Schiller, A., Stuben, H., Winter, F., Zanotti, J.M.: Hyperon sigma terms for 2 + 1 quark flavours. Phys. Rev. D **85**, 034506 (2012)
15. Ji, X.-D.: Gauge-invariant decomposition of nucleon spin. Phys. Rev. Lett. **78**, 610–613 (1997)
16. Patrignani, C., et al.: Review of particle physics. Chin. Phys. C **40**(10), 100001 (2016)
17. Pavan, M.M., Strakovsky, I.I., Workman, R.L., Arndt, R.A.: The Pion nucleon Sigma term is definitely large: results from a G.W.U. analysis of pi nucleon scattering data. PiN Newslett. **16**, 110–115 (2002)
18. Yang, Y.-B., Alexandru, A., Draper, T., Liang, J., Liu, K.-F.: πN and strangeness sigma terms at the physical point with chiral fermions. Phys. Rev. D **94**(5), 054503 (2016)

Chapter 89
Experimental Analysis of Few-Body Physics

Yukie Maeda

Abstract The differential cross sections of the elastic *Nd* scattering at the energy below 150 MeV can be well reproduced by introducing 3NF in the Faddeev calculations based on modern nucleon-nucleon (NN) interactions. On the other hand, some spin-observables, and the differential cross sections of *Nd* elastic and inelastic scatterings at over 250 MeV show large discrepancies between the data and the Faddeev calculations with 3NF. It indicates the presence of the missing features of the three nucleon systems. For the systematic study about these large discrepancies at higher energy regions, we measured the differential cross sections and the vector analyzing power for the inclusive and exclusive breakup reaction at 170 and 250 MeV. The experiments were carried out at RCNP. The data were compared with the results of the Faddeev calculations with and without the 3NF.

89.1 Introduction

The study of three-nucleon force (3NF) effects is attracting attentions not only in the few-body studies but also in the studies of heavier system like neutron-rich nuclei and neutron star. Historically, the need for many-body interaction was suggested just after Yukawa's meson theory in 1939 [1]. Based on the meson exchange picture, the main component of the 3NF is considered to be Fujita-Miyazawa type, a 2π-exchange between three nucleons with a Δ isobar excitation as an intermediate state [2]. Firstly, the study of the 3NF effects was started through the bound states of few nucleon systems. The correct binding energies of 3N bound systems were reproduced by introducing the Tucson-Melbourne (TM) or Urbana-IX 3NFs, which showed the clear effects of 3NF in the nucleus.

The study of the 3N continuum states through the *Nd* scattering allows us to measure not only the cross sections but also the spin observables, which offer more details of 3NF properties. The recent progress in computational resources has made

Y. Maeda (✉)
University of Miyazaki, 1-1 Gakuenkibanadai-Nishi, Miyazaki 889-2192, Japan
e-mail: yukie@cc.miyazaki-u.ac.jp

N. A. Orr et al. (eds.), *Recent Progress in Few-Body Physics*,
Springer Proceedings in Physics 238,
https://doi.org/10.1007/978-3-030-32357-8_89

it possible to obtain rigorous numerical results of Faddeev calculations for the 3N scattering at intermediate energy region with using the realistic NN forces and 3N forces. In the experimental sides, highly precise measurements of the *dp* elastic scattering for the deuteron energies of E_d^{lab} = 140, 200 and 270 MeV have been performed [3, 4]. The calculations with NN forces only fail to reproduce the data of cross sections and vector analyzing powers at the angular range where the cross section takes minimum. These discrepancies are filled by adding 3NFs. These results mean that the *Nd* elastic scattering at intermediate energy is a good probe to study the 3NF effects.

Theoretically, 3NF effects are expected to be significant at higher energy. However, the Faddeev calculations with 3NF still underestimate the data for *nd* [5] and *pd* [6] elastic scattering at 250 MeV. The discrepancies between the data and the theory can not be explained by the relativistic corrections [7]. This is quite different to the results of the comparisons between the data and the theoretical predictions at 135 MeV/nucleon.

For the next step of the study of 3NF effects, the breakup reactions are expected to give us more information because the total cross sections of the breakup are predicted to become larger than elastic at higher energy region. So we performed measurements of both inclusive and exclusive breakup reactions at 170 and 250 MeV at Research Center for Nuclear Physics (RCNP), Osaka university.

89.2 Inclusive Breakup Reaction

We performed two different types of measurements of inclusive breakup reactions at RCNP. One was $D(p, p)pn$ inclusive breakup reaction at $E_p = 250$ MeV [8]. In this experiment, polarized proton beam of 250 MeV was bombarded on the liquid deuterium target and scattered proton was analyzed by LAS spectrometer at θ_{LAB} = 10, 15, 20, 25, and 30°. In [8], the differential cross sections and vector analyzing powers are compared with the Faddeev calculations with and without 3NF. Concerning about the differential cross sections, only the half value of the data in the high energy transfer region were explained by the theory. This result is similar to that of elastic scattering at 250 MeV.

Another was $D(p, n)pp$ inclusive breakup reaction at 170 MeV. Because this energy is lower than the π-threshold of 210 MeV, it would allows us explicit comparison between the data and the theory. In this experiment, we accelerated the polarized proton beam up to E_p = 170 MeV and transported it to the N0 experimental hall. We used the deuterated polyethylene (CD_2) sheets [9] as the deuteron targets. The thickness of them were 44 and 110 mg/cm^2. To subtract the contributions from the carbons in CD_2 targets, we also used a graphite target of 140 mg/cm^2.

The proton beam was injected to the targets in the vacuum scattering chamber and bent to the beam dump by the swinger magnet. The scattered neutrons passed through the vacuum window and ran through the 100 m time-of-flight (TOF) tunnel in the air then be detected by NPOL3 [10]. The efficiency of NPOL3 was determined

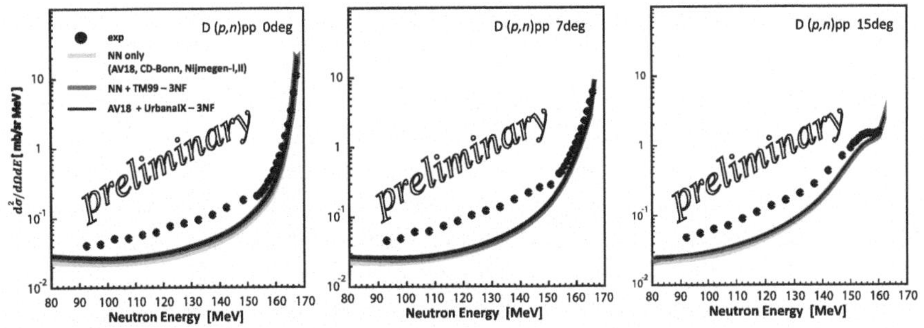

Fig. 89.1 The double differential cross sections for the D(p, n) breakup reactions at $\theta_{lab} = 0, 7, 15°$ as functions of the scattered neutron energy. The solid circles show the data with statistical errors only. The calculations including various NN forces only (light shaded band), various NN forces with TM99-3NF (dark shaded band) and AV18+UrbanaIX-3NF (solid line) are also shown

by measuring the ^{7}Li(p, n)^{7}Be(g.s. + 0.4 MeV) reaction, of which cross sections were measured over the wide energy range. In this work, the efficiency was deduced to be about 1.5%. The energy of a detected neutron was deduce by TOF method.

Preliminary results for the double differential cross sections at θ_{lab} = 0, 7, 15° are shown in Fig. 89.1 as functions of the scattered neutron energy. The data are compared with the Faddeev calculations [11]. The dark (light) shaded bands show the calculations based on the NN forces (CD-Bonn, AV18, Nijmegen-I&II) with (without) TM99-3NF. The solid lines represent the calculations based on AV18 and UrbanaIX-3NF.

We can see large discrepancies between the data and the calculations in the low neutron energy regions. In this regions, the calculations including 3NF can explain only about 50% of the data, which is similar to the results of the D(p, p) inclusive breakup reaction at 250 MeV. It may indicates the presence of the missing features of the three nucleon system at intermediate energy region, for instance $\pi - \rho$ or $\rho - \rho$ exchange type 3NF effects, which are not included in the present calculations. On the other hand, the energy dependence of discrepancies between the data and the theoretical calculations are rather small, which is different from the case of elastic scatterings. It suggests that more study of the inclusive breakup reactions in the different beam energies are important to explore the energy dependence of the 3NF effects.

Concerning about the analyzing powers of D(p, n)pp reacrions, the statistical errors of the data are large to discuss about the effects of 3NF. The data are almost consistent with the calculations within the error bars.

89.3 Exclusive Breakup Reactions

In the studies by using the exclusive breakup reactions, it is important to select kinematical configurations to measure because the three-nucleon final states are kinematically much more complicated than elastic and inclusive breakup reactions. To investigate the origin of large discrepancies in the differential cross sections of the elastic and inelastic scattering which were mentioned above, we performed the two different sets of measurements of D(p, pp)n exclusive breakup reactions at 250 MeV at RCNP west experimental hall. The polarized proton beam of 250 MeV was bombarded to liquid deuterium target and scattered two protons (p_1 and p_2) were analyzed by double arm magnetic spectrometer LAS and Grand Raiden (GR), respectively.

In one set of measurements, we focused on the kinematical configurations that one proton p_1 was scattered into $\theta_{LAS} = 15°$, because the data of D(p, p)pn inclusive measurement showed large disagreement with the theories at forward angular region [12]. Another proton p_2 was measured at $\theta_{GR} = 35$, 50, 65, and 80°. In [12], the comparison between the data of differential cross sections and the theoretical calculations show angular distribution: the data are well reproduced only in configurations with large θ_2.

In another set of measurements, we focused on the kinematical configurations, so-called final state interaction (FSI) geometry, because FSI configurations are predicted to be a good probe for not only the 3NF effects but also the difference between the 3NF models [13]. According to this theoretical prediction, we measured A_y and the differential cross sections of the following reaction $\vec{p} + d \rightarrow (p_1 n)_{FSI} + p_2$ at E_p = 250 MeV. We chose the configuration that the np pair is around FSI, because it is difficult to include the Coulomb force in the Faddeev calculations for the pp pair in FSI. The recoiled proton p_1 in FSI is detected by using LAS spectrometer at θ_{LAB} = 40°. Another proton p_2 which scattered in the direction around θ_{LAB} = 67°, which depends of the relative energy between proton p_1 and neutron, is detected by using GR spectrometer in coincidence with p_1. By using the momentum vectors of p_1 and p_2, we obtained the missing mass spectrum of neutrons with the energy resolutions of 1.2 MeV.

We obtained the preliminary results of the differential cross sections and the vector analyzing powers with the S-curve energy bin of the 7 MeV. The results are plotted in Fig. 89.2 by solid circles as a function of the energy of p_1. The statistical errors of the data are less than 1% and 0.02 for the differential cross sections and the vector analyzing powers, respectively. The peak corresponds to the FSI can be seen around $E_p = 65$ MeV.

The solid (dashed) lines represent the results of the Faddeev calculations [14] with (without) Tucson-Melbourne99 (TM99) 3NF based on the CDBonn potential. The dot-dashed (dotted) lines are the same predictions with solid (dashed) lines but integrated with the experimental solid angles. From Fig. 89.2, we can see that the experimental results are well reproduced by the calculations which integrated with the angular acceptance of the measurements. The data around FSI are well reproduced

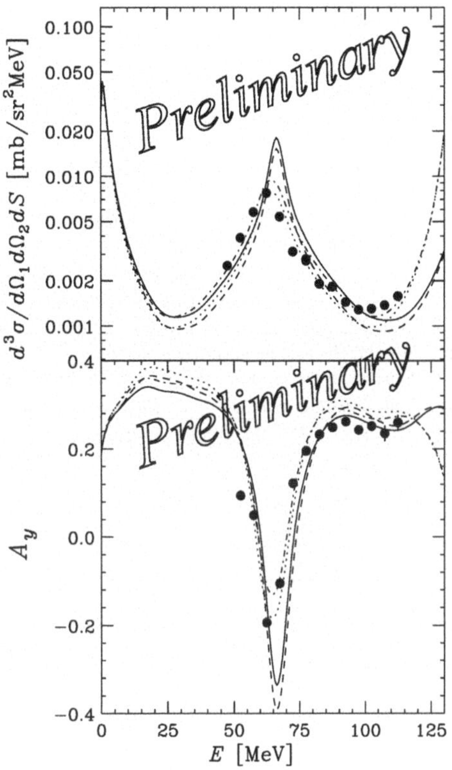

Fig. 89.2 The upper and the lower panel show the results of the differential cross sections and the vector analyzing powers for the *pd* break reactions at 250 MeV for $\theta_p = \theta_n = 40°$ in the laboratory system (solid circles). Only the statistical errors are represented. The predictions with CDBonn and TM99-3NF (solid line) and that with CDBonn only (dashed line) are shown simultaneously. The dot-dashed (dotted) lines are the theoretical results with (without) TM99-3NF which integrated with the angular acceptance of the measurements

by the calculations with CDBonn only but the data around the cross section minimum are well reproduced by that including TM99-3NF. This result is different from the case of elastic and inclusive breakup reactions. It would be interesting to compare the data with the theoretical predictions which include the relativistic corrections [7] or the Coulomb interaction effects [15].

Acknowledgements The author would like to thank the collaborators of the experiments E270 and E336, T. Saito, H. Miyasako, R. Chen, M. Dozono, K. Hatanaka, E. Ihara, T. Kawabata, S. Kawase, T. Kikuchi, T. Matsubara, S. Noji, S. Ota, K. Sagara, H. Sakaguchi, S. Sakaguchi, H. Sakai, Y. Sasamoto, M. Sasano, K. Sekiguchi, T. Shima, Y. Shimizu, T. Suzuki, K. Suda, M. Takechi, Y. Tameshige, A. Tamii, H. Tokieda, T. Uesaka, T. Wakasa, and K. Yako. We acknowledge the RCNP accelerator group for providing the excellent beam during the beam time. And we would like to thank Prof. Witała and Prof. H. Kamada for their calculations and discussions. This work was supported by Grant-in-Aid for JSPS Research Fellow, and JSPS KAKENHI Grant Number 24740168, and 15K17655.

References

1. Primakoff, H., Holstein, T.: Phys. Rev. **55**, 1218 (1939)
2. Fujita, J., Miyazawa, H.: Prog. Theor. Phys. **17**, 360 (1957)
3. Sakai, H., et al.: Phys. Rev. Lett. **84**, 5288 (2000)
4. Sekiguchi, K., et al.: Phys. Rev. C **65**, 034003 (2002)
5. Maeda, Y., et al.: Phys. Rev. C **76**, 014004 (2007)
6. Hatanaka, K., et al.: Phys. Rev. C **66**, 044002 (2002)
7. Witała, H., Golak, J., Glöckle, W., Kamada, H.: Phys. Rev. C **71**, 054001 (2005)
8. Kuroita, S., et al.: AIP Conf. Proc. **1011**, 98 (2008)
9. Maeda, Y., et al.: Nucl. Instrum. Methods A **490**, 518 (2002)
10. Wakasa, T., et al.: Nucl. Instrum. Methods A **547**, 569 (2005)
11. Witała, H.: Private communication
12. Kuroita, S., et al.: Few-Body Syst. **50**, 287 (2011)
13. Kuroś-Żołnierczuk, J., et al.: Phys. Rev. C **66**, 024003 (2002)
14. Kamada, H. Private communication
15. Deltuva, A., Fonseca, A.C., Sauer, P.U.: Phys. Rev. C **71**, 054005 (2005)

Chapter 90
Strong Interactions for Precision Nuclear Physics

Andreas Ekström

Abstract One of the key challenges in ab initio nuclear theory is to understand the emergence of nuclear structure from quantum chromodynamics. I will address this challenge and focus on the statistical aspects of uncertainty quantification and parameter estimation in chiral effective field theory.

It is well-known that quantum chromodynamics (QCD) is non-perturbative in the low-energy region where atomic nuclei exist. This feature prevents us from direct application of perturbation theory. To make progress, two complementary approaches are presently employed; lattice QCD (LQCD) [1] and chiral effective field theory (χEFT) [2]. The former amounts to numerical evaluation of the QCD path integral on a space-time lattice, while the latter is aimed at exploiting the decoupling principles of the renormalization group (RG) to systematically formulate a potential description of the nuclear interaction rooted in QCD. LQCD is a computationally expensive approach that requires at least exascale resources for a realistic analysis of multi-nucleon systems, and will most likely not be the most economical choice for analyzing nuclear systems. Nevertheless, in cases where numerically converged results can be obtained, LQCD offers a unique computational laboratory for theoretical studies of QCD in a low-energy setting [3].

The derivation of a nuclear potential in χEFT proceeds via the construction of an effective Lagrangian consisting of pions, nucleons, sometimes also the Δ isobar, endowed with all possible interactions compatible with the symmetries of low-energy QCD. The details can be found in extensive reviews [4–6]. All short-distance physics, normally associated with quarks and gluons, reside beyond a hard momentum scale $\Lambda_b \sim 1$ GeV, that remains unresolved in χEFT. Such high-momentum dynamics is instead encoded in a set of low-energy constants (LECs) that must be determined from experimental data, or in a future scenario hopefully computed directly from LQCD. χEFT is the theoretical framework to calculate observables in an expan-

A. Ekström (✉)
Department of Physics, Chalmers University of Technology, SE-412 96 Gothenburg, Sweden
e-mail: andreas.ekstrom@chalmers.se

N. A. Orr et al. (eds.), *Recent Progress in Few-Body Physics*,
Springer Proceedings in Physics 238,
https://doi.org/10.1007/978-3-030-32357-8_90

sion expressed in powers of the small ratio Q/Λ_b, where Q is a soft momentum scale $\sim m_\pi$. If done right, this approach allows for a systematically improvable description of low-energy nuclear properties in harmony with the symmetries of low-energy QCD.

The promise of being systematically improvable is a unique selling point of χEFT, or any EFT for that matter. Indeed, although the order-by-order expansion contains an infinite number of terms and must be truncated, the omitted terms represent neglected physics and contribute to the systematic uncertainty. The upshot is that higher-order corrections should be less important and follow a pattern determined by the EFT expansion ratio. The organization of this expansion, such that increasingly unimportant physics appear at consecutively higher orders, is called power counting (PC).

The leading-order (LO) in this expansion consists of the well-known one-pion exchange potential (Yukawa term) accompanied by a contact potential to describe any unresolved short-ranged physics at this order. The potentials at higher orders, i.e. next-to-leading order (NLO) etc., systematically introduce multiple-pion exchanges, accompanied additional zero-ranged contact potentials, possibly Δ perturbations, and irreducible many-nucleon interactions, see Fig. 90.1.

To achieve an accurate theoretical description of the nuclear interaction, with quantified statistical and systematic uncertainties of the theoretical predictions, can

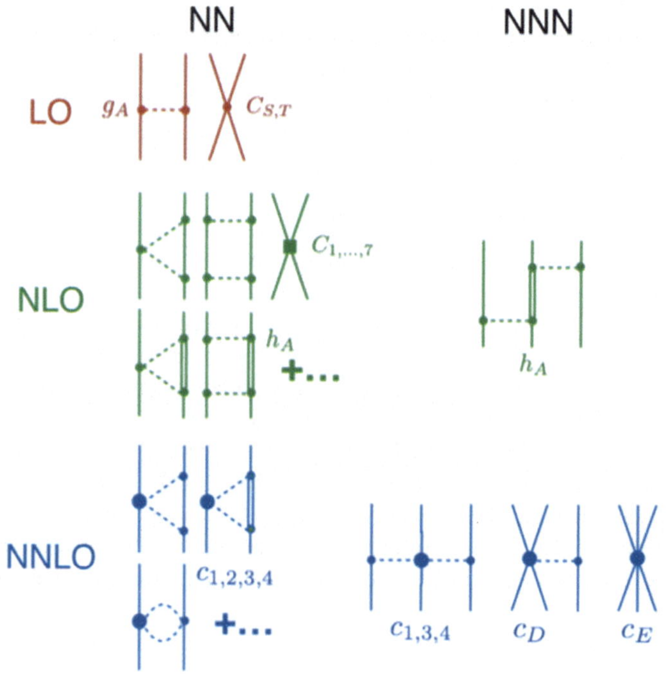

Fig. 90.1 Diagrammatic order-by-order representation of the Δ-full two-nucleon (NN) and three-nucleon (NNN) nuclear interaction up to NNLO in χEFT based on so-called Weinberg power counting (WPC)

be referred to as achieving a state of *precision nuclear physics*. There are several interesting facets of this ongoing endeavor:

- On a fundamental physics level, it is well-known that the nuclear potentials from χEFT that are based on Weinberg power counting (WPC) do not generate observables that respect RG invariance, see e.g. [7] and references therein. At the same time, there is an ongoing debate regarding the need or validity for probing large momenta in a potential description from EFT that is only valid at low-energies to begin with, see e.g. [8–10] for a selection of viewpoints. Presently, most ab initio calculations of atomic nuclei, including the calculations presented here, employ potentials based on WPC. There exist potentials with alternative PCs that fulfill the fundamental tests of RG invariance for observables in two- and three-nucleon systems, see e.g. [11]. Unfortunately, such potentials have not yet been employed in nuclear many-body calculations.
- The numerical values of the LECs in χEFT must be determined from data before any quantitative analysis can proceed. From a frequentist perspective, parameter estimation often amounts to maximizing a likelihood. For χEFT, this turns into a non-linear optimization problem over a high-dimensional parameter domain [12–14]. Bayesian parameter estimation is explored more and more in ab initio nuclear theory [15, 16]. This approach captures the entire probability distribution of the relevant parameters, and not just the values at the maximum of the probability mode. However, the computational demands are substantially higher compared to most of the frequentist methods, mainly due to repeated sampling of the model across the parameter domain.
- There are several sources of uncertainty in model calibration. For instance, the calibration data itself come with uncertainties. Thus, any parameter estimation process will contain covariances that must be quantified and propagated. There exist well-known methods, frequentist as well as Bayesian, for quantifying the statistical uncertainties at any level of the calculation, see e.g. [17, 18]. However, it remains a challenge to achieve full uncertainty quantification in complex models that require substantial high-performance resources for a *single* evaluation at one point in the parameter domain. Well-designed surrogate models can hopefully provide some leverage, see e.g. [19–21].
- A theoretical model will never represent nature fully. Consequently, there are theory errors (sometimes referred to as systematic uncertainties or model discrepancies). The statistical uncertainties stemming from the calibration data discussed above are typically not the main source of error in χEFT predictions [22, 23]. It is therefore of key importance to identify and quantify the sources of systematic errors in χEFT. At the moment, such analyses are rarely performed in ab initio nuclear theory. χEFT models combined with ab initio methods are often computationally complex and require substantial computational resources. As such, Markov Chain Monte Carlo with long mixing times can be prohibitively expensive. Furthermore, it is not clear how to identify and exploit the relevant momentum scales in descriptions of atomic nuclei.

90.1 Ab initio Nuclear Theory with χEFT

Ab initio methods, such as the no-core shell-model (NCSM) [24], the coupled cluster method (CC) [25], in-medium similarity renormalization (IM-SRG) [26], or lattice EFT [27], for solving the many-nucleon Schrödinger equation

$$\left(\sum_{i=1}^{A} \frac{p_i^2}{2m_N} + \sum_{i<j=1}^{A} \mathcal{V}_{ij}^{NN}(\boldsymbol{\alpha}) + \sum_{i<j<k=1}^{A} \mathcal{V}_{ijk}^{NNN}(\boldsymbol{\alpha}) \right) |\Psi\rangle = E|\Psi\rangle \tag{90.1}$$

with two-nucleon (NN) and three-nucleon (NNN) potentials derived from χEFT with a set of LECs $\boldsymbol{\alpha}$, make use of controlled mathematical approximations. Such many-body approaches can provide numerically exact nuclear wave functions for several bound, resonant, and scattering states in isotopes well into the region of medium-mass nuclei [28–31]. This development has drastically changed the agenda in development of nuclear interactions for atomic nuclei.

In the beginning of the previous decade, a lot of effort was spent on constructing so-called high-precision nuclear interactions, most prominently Idaho-N3LO [32], AV18 [33], and CD-Bonn [34], that could reproduce the collected data on NN scattering below the pion-production threshold with nearly surgical precision. We now know that such interactions often fail to reproduce important bulk properties of atomic nuclei [31, 35–37]. However, fifteen years ago, it was unclear how to gauge the quality of the many-nucleon wave functions since they relied on a series of involved approximations. Although it is still a challenge to quantify the theoretical uncertainty in many-body calculations, modern ab initio methods are tremendously refined. Indeed, their fidelity, and domain of applicability have been dramatically extended during the recent decade. This development has led to an increased focus on designing improved microscopic nuclear potentials that are based on novel fitting protocols. To ensure steady progress, we need to critically examine and systematically compare the quality of different sets of interaction models and their predictive power.

90.1.1 Optimization of LECs and Uncertainty Quantification of Predictions from χEFT

The canonical approach to estimate the numerical values of the LECs $\boldsymbol{\alpha}$ in χEFT is to minimize some weighted sum of squared residuals

$$\chi^2(\boldsymbol{\alpha}) = \sum_{i \in \mathcal{D}} \left(\frac{\mathcal{O}_i^{\text{theo}}(\boldsymbol{\alpha}) - \mathcal{O}_i^{\text{exp}}}{\sigma_i} \right)^2, \tag{90.2}$$

where $\mathcal{D}$ represents the calibration dataset, and $\mathcal{O}_i$ denotes experimental and theoretical values for observable i in $\mathcal{D}$ with an obvious notation. The theoretical description of each observable depends explicitly on the LECs $\boldsymbol{\alpha}$. In the limit of independent data, the uncertainty associated with each observable is represented by σ_i. Known, or estimated, correlations across the data can also be incorporated [38, 39]. Using well-known methods from statistical regression analysis, often assuming normally distributed residuals, it is possible to extract the covariance matrix of the parameters that minimize the objective.

An order-by-order uncertainty analysis of chiral interactions up to NNLO was undertaken in [13]. The objective function in that work incorporated an estimate of the theory uncertainty from χEFT, and the dataset $\mathcal{D}$ comprised πN and NN scattering data as well as bound-state observables in $A = 2, 3$ nuclei. The total covariance matrix for the LECs was determined for each analyzed interaction model. Additional components of the systematic uncertainty were probed by varying the regulator cutoff $\Lambda \in [450, 600]$ MeV as well as the maximally allowed scattering energy in the employed database of measured scattering cross sections. This effort resulted in a family of 42 chiral interactions at NNLO. Together, they furnish a valuable tool for probing uncertainties in ab initio few-nucleon predictions, see [40, 41] for representative examples of their use.

90.1.2 With an EFT, We Can Do Better

One way to estimate the effect of the first excluded term in an EFT expansion was suggested in [42]. Building on the work in [43], this was given a Bayesian interpretation in [18]. In brief, if we write the order-by-order expansion of some observable $\mathcal{O}$ as

$$\mathcal{O} = \mathcal{O}_0(a_0 q^0 + a_1 q^1 + a_2 q^2 + a_3 q^3 + \ldots), \tag{90.3}$$

where $\mathcal{O}_0$ is the overall scale, e.g. the leading order contribution, and we know the expansion parameter, e.g. $q = (Q/\Lambda_b)$, then we can compute the probability distribution of the expansion coefficient a_i provided that we know the values of the lower-order coefficients $a_0, \ldots, a_{i-1}$. The application of Bayes theorem with identically distributed, independent, boundless, and uniform prior distributions of the expansion coefficients a_i, leads to a simple expression for the estimate of a_i, with $(100 \times i/(i+1))\%$ confidence, given by

$$a_i = \max\{|a_n|\}_{n<i}. \tag{90.4}$$

Although the above expression only provides an estimate, theoretical predictions equipped with truncation errors provide important guidance and demonstrate one of the main advantages of using an EFT. Refined methods for quantifying EFT truncation errors in nuclear physics is of key importance.

90.2 Muon-Capture on the Deuteron

An excellent example of where theoretical uncertainty quantification plays an important role is in the theoretical analysis of the muon-deuteron $\mu - d$ (doublet) capture rate Γ_D, i.e. the rate of

$$\mu^- + d \rightarrow \nu_\mu + n + n. \tag{90.5}$$

Experimentally, this will be determined with 1.5% precision in the MuSun experiment. Such precision, if attained, corresponds to a tenfold improvement over previous experiments. The centerpiece of the MuSun experiment is to extract the two-body weak LEC d_R from a two-nucleon process. This LEC is of central importance in several other low-energy processes that are currently studied. It is proportional to the proton-proton (*pp*) fusion cross-section, an important low-energy process that generates energy in the Sun. Given its extremely low cross-section, this cannot be measured on earth. The LEC d_R also appears in neutrino-deuteron scattering, and once the πN couplings c_3 and c_4 are fixed, it determines the LEC c_D which governs the strength of the one-pion exchange plus contact piece of the leading *NNN* interaction. A thorough analysis of the uncertainties in theoretical descriptions of $\mu - d$ capture was carried out in [44], using the covariance matrices from [13], yielding for the $S-$wave contribution $\Gamma_D^{^1S_0} = 252.4^{+1.5}_{-2.1}\,\mathrm{s}^{-1}$.

Exploiting the Roy-Steiner analysis from [45], it was also possible to quantify the correlation between the $\mu - d$ capture rate and the *pp*-fusion low-energy cross section in terms of the LEC c_D. Furthermore, assuming an EFT expansion ratio $q = \frac{m_\pi}{\Lambda_b} \sim 0.28$, i.e. estimating $\Lambda_b \sim 500$ MeV, allowed for an order-by-order estimate of the EFT truncation error of the capture rate along the lines presented in Sect. 90.1.2. The LO-NLO-NNLO predictions of the capture rate are $\Gamma_D^{^1S_0} = 186.3 + 61.0 + 5.5$ s^{-1}, where the second and third term indicate the NLO and NNLO contributions, respectively, to the LO result (first term). This information leads to an estimated EFT truncation error of 4.6 s^{-1}, with 75%-confidence. Clearly, the dominating source of uncertainty.

90.3 From Few to Many

Increasing the number of nucleons in the system under study introduces several new challenges. The presence of multiple scales, emergence of many-body effects such as collectivity, clusterization, and saturation are not trivial to understand from first principles, nor particularly easy to handle when solving the Schrödinger equation and therefore not straightforward to incorporate when calibrating the interaction. In [36], the LECs of a chiral NNLO interaction was optimized to reproduce few-nucleon data as well as binding energies and radii in ^{14}C and selected oxygen isotopes. This approach to parameter estimation, resulting in the NNLO$_{\mathrm{sat}}$ interaction was facilitated by a novel application the POUNDERs optimization algorithm [46] coupled to jacobi-

NCSM and CC methods. NNLO_{sat} has enabled accurate predictions of radii and ground-state energies in selected medium-mass nuclei [47].

It should be pointed out that the NNLO_{sat} interaction does not provide an accurate description of *NN* scattering cross-sections, in particular for *pp* scattering, at relative momenta beyond $\sim m_\pi$. At the same time, it is not obvious how to determine the domains of applicability of an interaction model and exploit this information such that the risk of overfitting is minimized. This, and other challenges are intimately related to quantifying truncation errors in χEFT and predictions from ab initio nuclear theory.

90.3.1 Delta Isobars and Nuclear Saturation

It turns out that the inclusion of the Δ isobar as an explicit low-energy degree of freedom in the effective Lagrangian, in addition to pions and nucleons, play an important role for accurately reproducing the saturation properties of the nuclear interaction. See [48] for additional details. Figure 90.2 demonstrates the effect of incorporating the Δ up to NNLO in CC calculations of symmetric nuclear matter. Additional advantages of including the Δ were observed in [49–51]. Such results are not surprising from an EFT perspective, given that the $\Delta - N$ mass splitting is only twice the pion mass and therefore below the expected breakdown scale of χEFT potentials [52]. Thus, the Δ-full chiral interaction provides a valuable starting point for constructing more refined χEFT interactions with improved uncertainty estimates.

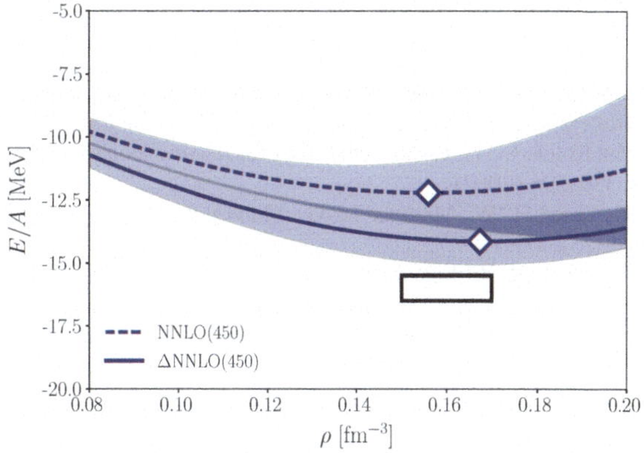

Fig. 90.2 CC calculations of the energy per nucleon (in MeV) in symmetric nuclear matter at NNLO in χEFT with (solid line) and without (dashed line) the Δ isobar. Both interactions employ a momentum regulator-cutoff $\Lambda = 450$ MeV. The shaded areas indicate the estimated EFT-truncation errors following the prescription presented in Sect. 90.1.2. The diamonds mark the saturation point and the black rectangle indicates the region $E/A = 16 \pm 0.5$ MeV and $\rho = 0.16 \pm 0.01$ fm^3

90.4 Discussion and Outlook

It is clear that the computational capabilities in ab initio nuclear physics exceed the accuracy of available chiral interactions. To make further progress requires improved statistical analysis and evaluation of interaction models. Hopefully, such efforts will bring us closer to a well-founded and microscopically rooted formulation of the nuclear interaction. There are several interesting challenges ahead of us. We must push the frontier of accurate ab initio methods further towards exotic systems and decays; systematically exploit information from *NNN* scattering data, decay probabilities, and saturation properties of infinite matter when optimizing the LECs of chiral interactions; demonstrate a connection between EFT(s) applied to nuclei and low-energy QCD, e.g. test PCs for RG invariance; and quantify systematic and statistical uncertainties in theoretical predictions. Continuous development of efficient computer codes to harness high-performance computing resources will hopefully enable detailed Bayesian analyses of ab initio calculations in the near future.

Acknowledgements I would like to thank all my collaborators for sharing their insights during our joint work on the range of topics presented here. This work has received funding from the European Research Council (ERC) under the European Unions Horizon 2020 research and innovation programme (Grant Agreement No. 758027) and the Swedish Research Council under Grant No. 2015-00225 and Marie Sklodowska Curie Actions, Cofund, Project INCA 600398.

References

1. Beane, S.R., Detmold, W., Orginos, K., Savage, M.J.: Nuclear physics from lattice QCD. Prog. Part. Nucl. Phys. **66**(1), 1–40 (2011)
2. Weinberg, Steven: Effective chiral Lagrangians for nucleon—pion interactions and nuclear forces. Nucl. Phys. B **363**, 3–18 (1991)
3. Chang, C.C., Nicholson, A.N., Rinaldi, E., et al.: A per-cent-level determination of the nucleon axial coupling from quantum chromodynamics. Nat. Publ. Group **558**(7708), 91–94 (2018)
4. Bedaque, P.F., Van Kolck, U.: Effective field theory for few-nucleon systems. Annu. Rev. Nucl. Part. Sci. **52**(1), 339–396 (2002)
5. Epelbaum, E., Hammer, H.-W., Meißner, U.-G.: Modern theory of nuclear forces. Rev. Mod. Phys. **81**, 1773–1825 (2009)
6. Machleidt, R., Entem, D.R.: Chiral effective field theory and nuclear forces. Phys. Rep. **503**, 1–75 (2011)
7. Nogga, A., Timmermans, R.G.E., van Kolck, U.: Renormalization of one-pion exchange and power counting. Phys. Rev. C **72**, 054006 (2005)
8. Phillips, D.R. Recent results in chiral effective field theory for the NN system. PoS **CD12**, 172 (2013)
9. Griehammer, H.W.: Assessing theory uncertainties in EFT power countings from residual cutoff dependence. PoS **CD15**, 104 (2016)
10. Epelbaum, E., Meissner, U.G.: on the renormalization of the one-pion exchange potential and the consistency of Weinberg's power counting. Few Body Syst. **54**, 2175–2190 (2013)
11. Song, Y.-H., Lazauskas, R., van Kolck, U.: Triton binding energy and neutron-deuteron scattering up to next-to-leading order in chiral effective field theory. Phys. Rev. C **96**, 024002 (2017)

12. Ekström, A., Baardsen, G., Forssén, C., et al.: Optimized chiral nucleon-nucleon interaction at next-to-next-to-leading order. Phys. Rev. Lett. **110**(19), 192502 (2013)
13. Carlsson, B.D., Ekström, A., Forssén, C., et al.: Uncertainty analysis and order-by-order optimization of chiral nuclear interactions. Phys. Rev. X **6**(1), 011019 (2016)
14. Reinert, P., Krebs, H., Epelbaum, E.: Semilocal momentum-space regularized chiral two-nucleon potentials up to fifth order. Eur. Phys. J. A **54**(5), 86 (2018)
15. Schindler, M.R., Phillips, D.R.: Bayesian methods for parameter estimation in effective field theories. Ann. Phys. **324**(3), 682–708 (2009)
16. Wesolowski, S., Klco, N., Furnstahl, R.J., Phillips, D.R., Thapaliya, A.: Bayesian parameter estimation for effective field theories. J. Phys. G: Nucl. Part. Phys. **43**(7), 074001 (2016)
17. Dobaczewski, J., Nazarewicz, W., Reinhard, P.-G.: Error estimates of theoretical models: a guide. J. Phys. G: Nucl. Part. Phys. **41**(7), 074001 (2014)
18. Furnstahl, R.J., Klco, N., Phillips, D.R., et al.: Quantifying truncation errors in effective field theory. Phys. Rev. C **92**(2), 024005 (2015)
19. McDonnell, J.D., Schunck, N., Higdon, D., et al.: Uncertainty quantification for nuclear density functional theory and information content of new measurements. Phys. Rev. Lett. **114**, 122501 (2015)
20. Vernon, I., Goldstein, M., Bower, R.G.: Galaxy formation: a Bayesian uncertainty analysis. Bayesian Anal. **5**(4), 619–669 (2010)
21. Neufcourt, L., Cao, Y., Nazarewicz, W., et al.: Neutron drip line in the ca region from bayesian model averaging. Phys. Rev. Lett. **122**, 062502 (2019)
22. Ekström, A., Carlsson, B.D., Wendt, K.A., et al.: Statistical uncertainties of a chiral interaction at next-to-next-to leading order. J. Phys. G: Nucl. Part. Phys. **42**(3), 034003 (2015)
23. Pérez, RN., Amaro, J.E., Arriola, ER., Maris, P., Vary, J.P.: Statistical error propagation in ab initio no-core full configuration calculations of light nuclei. Phys. Rev. C **92**, 064003 (2015)
24. Barrett, B.R., Navrátil, P., Vary, J.P.: Ab initio no core shell model. Prog. Part. Nucl. Phys. **69**, 131–181 (2013)
25. Hagen, G., Papenbrock, T., Hjorth-Jensen, M., et al.: Coupled-cluster computations of atomic nuclei. Rept. Prog. Phys. **77**(9), 096302 (2014)
26. Hergert, H., Bogner, S.K., Morris, T.D., et al.: The in-medium similarity renormalization group: a novel ab initio method for nuclei. Phys. Rept. **621**, 165–222 (2016)
27. Lee, Dean: Lattice simulations for few and many-body systems. Prog. Part. Nucl. Phys. **63**, 117–154 (2009)
28. Hagen, G., Ekström, A., Forssén, C., et al.: Neutron and weak-charge distributions of the ^{48}ca nucleus. Nat. Phys. **12**(2), 186–190 (2016)
29. Hagen, G., Jansen, G.R., Papenbrock, T.: Structure of ^{78}Ni from first principles computations. Phys. Rev. Lett. **117**(17), 172501 (2016)
30. Morris, T.D., Simonis, J., Stroberg, S.R., et al.: Structure of the lightest tin isotopes. Phys. Rev. Lett. **120**, 152503 (2018)
31. Lapoux, V., Somà, V., Barbieri, C., et al.: Radii and binding energies in oxygen isotopes: achallenge for nuclear forces. Phys. Rev. Lett. **117**, 052501 (2016)
32. Entem, D.R., Machleidt, R.: Accurate charge dependent nucleon nucleon potential at fourth order of chiral perturbation theory. Phys. Rev. C **68**, 041001 (2003)
33. Wiringa, R.B., Stoks, V.G.J., Schiavilla, R.: An accurate nucleon-nucleon potential with charge independence breaking. Phys. Rev. C **51**, 38–51 (1995)
34. Machleidt, R.: The high precision, charge dependent Bonn nucleon-nucleon potential (CD-Bonn). Phys. Rev. C **63**, 024001 (2001)
35. Binder, S., Langhammer, J., Calci, A., et al.: Ab initio path to heavy nuclei. Phys. Lett. B **736**(C), 119–123 (2014)
36. Ekström, A., Jansen, G.R., Wendt, K.A., et al.: Accurate nuclear radii and binding energies from a chiral interaction. Phys. Rev. C **91**(5), 051301 (2015)
37. Drischler, C., Hebeler, K., Schwenk, A.: Chiral interactions up to next-to-next-to-next-to-leading order and nuclear saturation. Phys. Rev. Lett. **122**, 042501 (2019)

38. Stump, D., Pumplin, J., Brock, R., et al.: Uncertainties of predictions from parton distribution functions. I. the lagrange multiplier method. Phys. Rev. D, **65**(1), 014012 (2001)
39. Wesolowski, S., Furnstahl, R., Melendez, J.A., et al.: Exploring Bayesian parameter estimation for chiral effective field theory using nucleon-nucleon phase shifts. J. Phys. G: Nucl. Part. Phys. (2018)
40. Hernandez, O.J. Ekström, A., Dinur, N.N., et al.: The deuteron-radius puzzle is alive: a new analysis of nuclear structure uncertainties. Phys. Lett. B **778**, 377–383 (2018)
41. Gazda, D., Catena, R., Forssén, C.: Ab initio nuclear response functions for dark matter searches. Phys. Rev. D **95**, 103011 (2017)
42. Epelbaum, E., Krebs, H., Meißner, U.-G.: Improved chiral nucleon-nucleon potential up to next-to-next-to-next-to-leading order. Eur. Phys. J. A **51**(5), 53 (2015)
43. Cacciari, M., Houdeau, N.: Meaningful characterisation of perturbative theoretical uncertainties. J. High Energy Phys. **2011**(9), 39 (2011)
44. Acharya, B., Ekström, A., Platter, Lucas: Effective-field-theory predictions of the muon-deuteron capture rate. Phys. Rev. C **98**, 065506 (2018)
45. Hoferichter, M., de Elvira, J.R., Kubis, B., Meiner, U.-G.: Roysteiner-equation analysis of pionnucleon scattering. Phys. Rep. **625**, 1–88 (2016)
46. Wild, S.M.: Solving derivative-free nonlinear least squares problems with POUNDERS. In: Terlaky, T., Anjos, M.F., Ahmed, S. (eds.) Advances and trends in optimization with engineering applications, pp. 529–540. SIAM (2017)
47. Hagen, G., et al.: Neutron and weak-charge distributions of the ^{48}Ca nucleus. Nature Phys. **12**(2), 186–190 (2015)
48. Ekström, A., Hagen, G., Morris, T.D., et al.: Δ isobars and nuclear saturation. Phys. Rev. C **97**(2), 024332 (2018)
49. Piarulli, M., Girlanda, L., Schiavilla, R., et al.: Minimally nonlocal nucleon-nucleon potentials with chiral two-pion exchange including Δ resonances. Phys. Rev. C **91**(2), 024003 (2015)
50. Piarulli, M., et al.: Light-nuclei spectra from chiral dynamics. Phys. Rev. Lett. **120**(5), 052503 (2017)
51. Logoteta, D., Bombaci, I., Kievsky, A.: Nuclear matter properties from local chiral interactions with Δ isobar intermediate states. Phys. Rev. C **94**, 064001 (2016)
52. van Kolck, U.: Few nucleon forces from chiral Lagrangians. Phys. Rev. C **49**, 2932–2941 (1994)

Chapter 91
Recent Developments in Solving the Few-Particle Scattering Problem by the Solution of The Faddeev-Yakubovsky Equations

Rimantas Lazauskas

Abstract In this article a short overview of the Faddeev-Yakubovsky (FY) equation formalism is provided. The progress in solving the few-particle problem based on solution of the FY equations is briefly discussed. The first numerical solution of the 5-body FY equations is undertaken, presenting the formalism how to include the three-nucleon interactions. Finally the results related with low energy neutron scattering on ^{4}He are presented and discussed.

91.1 Introduction

The few-body problem plays a special role in nuclear physics. Since the first days of nuclear physics one hoped to construct reliable models to describe nucleon-nucleon (NN) interaction and required accurate theoretical tools to test them in solving non-relativistic Schrödinger equation. The pioneering estimates were based on variational method and allowed, in particular, to demonstrate that the NN forces should be of the finite range to avoid collapse of the $A > 2$ nuclei. Even more important but also much more troublesome turned to be the route of solving the few-body scattering problem. Soon after the Lippmann-Schwinger equations have been formulated [19] it has been realized that these equations do not provide a unique solution for the multiparticle scattering problems. Furthermore these equations are inappropriate to treat the problems, where interactions are defined by boundary conditions (solid spheres or zero-range interactions) and not by the potentials. The partial solution for the second problem has been proposed in [27], however this approach has not allowed to handle the interactions of the finite range. The well founded integral equations to describe the three-particle scattering problem governed by the short range interactions have been derived by Faddeev in [6] and proved to provide a unique solution. Few years later Faddeev equations have been generalized to arbitrary number of

R. Lazauskas (✉)
Université de Strasbourg, CNRS, IPHC UMR 7178,
F-67000 Strasbourg, France
e-mail: rimantas.lazauskas@iphc.cnrs.fr

N. A. Orr et al. (eds.), *Recent Progress in Few-Body Physics*,
Springer Proceedings in Physics 238,
https://doi.org/10.1007/978-3-030-32357-8_91

particles by Yakubovsky [32]. An additional step has been required to formulate Faddeev-Yakubovsky (FY) equations in their diffferential form and work out appropriate wave-function boundary conditions. This feat has been realized in [20] for a three-body and in [21] for a four-body cases.

From the day of their formulation the actual applications of the Faddeev-Yakubovsky equations pose severe numerical challenges being due to the limited numerical resources. The first solutions of the three-body equations have been performed for the separable potentials [1], which allow to spare wave functions dependence on one vector variable. The first solution for the nonseparable potentials have appeared in early seventies using momentum space techniques by Utrecht group [9] and in configuration space by Grenoble group [11]. The first solution of the three-nucleon problem based on realistic interactions appeared more than one decade later [10]. Around that time the first solutions of the four-nucleon problem based on Faddeev-Yakubovsky equations for the simplistic nucleon-nucleon interaction models have appeared [21]. The fully realistic four-nucleon bound state problem for alpha-particle has been solved one decade later [13], whereas one more decade has been required to reach breakthrough for elastic [30] and non-elastic [4, 13] four-nucleon scattering problems. Finally the first attempt to solve the five-nucleon scattering problem has been undertaken just a few years ago [15] and extended to handle realistic nuclear interaction models in the very recent study [16].

91.2 Five-Body FY Equations

The five-body FY equations have been carefully detailed in the work [26]. Their derivation starts by decomposing systems total wave function Ψ into binary components ϕ_{ij}, similar to the three-body Faddeev components:

$$\phi_{ij} = G_0 V_{ij} \Psi, \tag{91.1}$$

here and in the following by letters $(ijklm)$ I denote indexes of the five-particles, without supposing any preferential ordering. For a five body system there exist 10 two-particle interactions V_{ij} and thus one may construct 10 different binary components ϕ_{ij}. Next step is to introduce four-body like FY components. These components are defined as

$$\begin{cases} \psi_{ij}^{ijk} = G_{ij} V_{ij} (\phi_{jk} + \phi_{ik}), \\ \psi_{ij}^{ij,kl} = G_{ij} V_{ij} \phi_{kl}. \end{cases} \tag{91.2}$$

In the last set of equations five-body Green's function G_{ij} includes single interaction term V_{ij}, i.e. $G_{ij} = (E - H_0 - V_{ij})^{-1}$. For a five-body system there exist 30 different four-body components of each type ψ_{ij}^{ijk} and $\psi_{ij}^{ij,kl}$. Using Yakubovsky's scheme the last components are further decomposed into the five-body FY components, schematically depicted in Fig. 91.1, and assorted into the FY equations as:

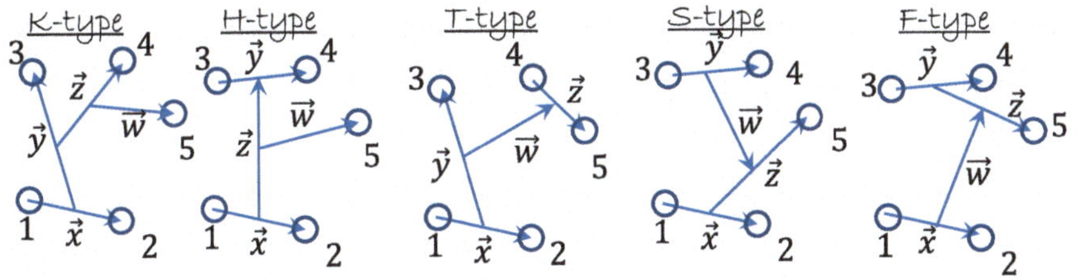

Fig. 91.1 Jacobi coordinate sets proper to 5-body Faddeev-Yakubovsky components. From left to right components $\mathcal{K}_{12,3}^4$, $\mathcal{H}_{12}^{34}$, $\mathcal{T}_{12,3}$, $\mathcal{S}_{12}^{34}$ and $\mathcal{F}_{12}^{34}$ are represented

$$\begin{cases} \mathcal{K}_{ij,k}^l = G_{ij}V_{ij}\left(\mathcal{K}_{ik,j}^l + K_{jk,i}^l + \psi_{ik}^{ikl} + \psi_{jk}^{jkl} + \psi_{ik}^{ik,jl} + \psi_{jk}^{jk,il}\right), \\ \mathcal{H}_{ij}^{kl} = G_{ij}V_{ij}\left(\mathcal{H}_{kl}^{ij} + \psi_{kl}^{jkl} + \psi_{kl}^{ikl}\right), \\ \mathcal{T}_{ij,k} = G_{ij}V_{ij}\left(\mathcal{T}_{ik,j} + T_{jk,i} + \psi_{ik}^{lm} + \psi_{jk}^{lm}\right), \\ \mathcal{S}_{ij}^{lm} = G_{ij}V_{ij}\left(\mathcal{F}_{lm}^{ij} + \psi_{lm}^{jk} + \psi_{lm}^{ik}\right), \\ \mathcal{F}_{ij}^{lm} = G_{ij}V_{ij}\left(\mathcal{S}_{lm}^{ij} + \psi_{lm}^{klm}\right). \end{cases} \tag{91.3}$$

The five-body components are related to the four-body ones, via:

$$\psi_{ij}^{ijk} = \mathcal{K}_{ij,k}^l + K_{ij,k}^m + \mathcal{T}_{ij,k}, \tag{91.4}$$

$$\psi_{ij}^{ij,kl} = \mathcal{H}_{ij,kl} + \mathcal{S}_{ij,kl} + \mathcal{F}_{ij,kl}. \tag{91.5}$$

The FY equations do not provide direct framework to implement the many-body forces. The framework to implement three-body forces in three- and four-body FY equations has been proposed in [7] and has been slightly modified in [12]. After some relatively simple combinatorial analysis three-body forces can also be implemented into the five-body FY equations. It is required only to modify equations for $\mathcal{K}$ and $\mathcal{T}$ components appearing in the left-hand side of the equation set (91.3):

$$\begin{aligned} (E - \widehat{H}_0 - V_{12})\mathcal{K}_{12,3}^4 = {} & V_{12}(\mathcal{K}_{13,2}^4 + \mathcal{K}_{23,1}^4 + \psi_{13}^{134} + \psi_{23}^{234} + \psi_{13}^{13,24} + \psi_{23}^{23,14}) \\ & + \mathcal{W}_{12}^3(\mathcal{K}_{12,3}^4 + \mathcal{K}_{13,2}^4 + \mathcal{K}_{23,1}^4 + \psi_{13}^{134} + \psi_{23}^{234} + \psi_{13}^{13,24} + \psi_{23}^{23,14} \\ & + \psi_{12}^{12,34} + \psi_{12}^{124} + \phi_{14} + \phi_{24} + \phi_{34}), \end{aligned} \tag{91.6}$$

$$\begin{aligned} (E - \widehat{H}_0 - V_{12})\mathcal{T}_{12,3} = {} & V_{12}(\mathcal{T}_{13,2} + \mathcal{T}_{23,1} + \psi_{13}^{13,45} + \psi_{23}^{23,45}) \\ & + \mathcal{W}_{12}^3(\mathcal{T}_{12,3} + \mathcal{T}_{13,2} + \mathcal{T}_{23,1} \\ & + \psi_{13}^{13,45} + \psi_{23}^{23,45} + \psi_{12}^{12,45} + \phi_{45}) \end{aligned} \tag{91.7}$$

In the last set of equations the terms $\mathcal{W}_{12}^3$ denote symmetrized components of a three-body force $\mathcal{W}_{123}$ acting between the particles (123) and being decomposed as:

$$\mathcal{W}_{ijk} = \mathcal{W}_{ij}^k + \mathcal{W}_{jk}^i + \mathcal{W}_{ki}^j \tag{91.8}$$

I proceed solution of the 5-body FY equations using the techniques developed in our previous studies. Spatial, spin and isospin dependence of the FY components is expressed by means of the partial wave expansion:

$$F^{JM}(\vec{x}, \vec{y}, \vec{z}, \vec{w}) = \sum \frac{f_\alpha(x, y, z, w)}{xyzw} \left| \left\{ \{l_x l_y\}_{l_{xy}} \{l_z l_w\}_{l_{zw}} \right\}_L \{S\} \right\rangle_{JM} \{T\}_{TT_z}, \tag{91.9}$$

here $\alpha \equiv (l_x, l_y, l_z, l_w, l_{xy}, l_{zw}, L, \{S\}, \{T\})$ is an index representing set of intermediate quantum numbers, coupled to total angular momentum J and total isospin T with its projection T_z (for a n-^{4}He scattering, considered in this work, total isospin and its projection are fixed to T=1/2 and T$_z$=1/2$^-$). The partial wave amplitudes $f_\alpha(x, y, z, w)$ depend only on the radial parts of the coordinates. These amplitudes are further expanded using Lagrange-Laguerre basis functions by means of Lagrange-mesh method [2]. Number of basis functions is adjusted to the partial angular momentum state they represent, by systematically reducing basis size describing partial amplitudes with large partial angular momentum values. For a realistic interaction models employed in this work—partial basis has been limited by setting $max(l_x, l_y, l_z) < 5$ and $l_w < 3$. Nucleons have been considered to be of the equal mass m_N by imposing $\hbar^2/m_N$ =41.471 MeV·fm^2.

For the scattering calculations asymptotic behavior of the FY components is supplemented with a long range w-dependent functional, composed of a regular and irregular spherical Bessel functions, adapting similar procedure as one used in [8]. This functional carries additional unknowns related to the K-matrix elements to be calculated. In order to balance resulting linear algebra problem additional equations are added, requiring the FY components combined with the incoming wave to satisfy Wronskian relation, see [14].

91.3 Results

In Fig. 91.2 my results of the calculated n-^{4}He phaseshifts are presented for the Hamiltonian based on the χEFT approach, derived up to next-to-next-to-next-to-leading order in chiral perturbation theory [5], denoted by I-N3LO. These results are compared with the ones computed by No Core Shell Model with Continuum (NCSMC) technique [24] as well as with the phaseshifts extracted from the experimental data performing R-matrix analysis [3]. Keeping in mind that my calculations as well as the ones performed using NCSMC techniques are estimated to be accurate at a few percent level—one may conclude nice agreement between the two theoretical approaches. Comparison with the phaseshifts extracted from the experimental data employing R-matrix analysis [3] reveals nice agreement for the repulsive S-wave, whereas theoretical NN interaction models fail to provide sufficient splitting of the resonant P-waves.

The accurate description of the n-^{4}He P-waves is quite a challenging problem, as pointed out in [24, 25] it requires presence of a three-nucleon force containing strong

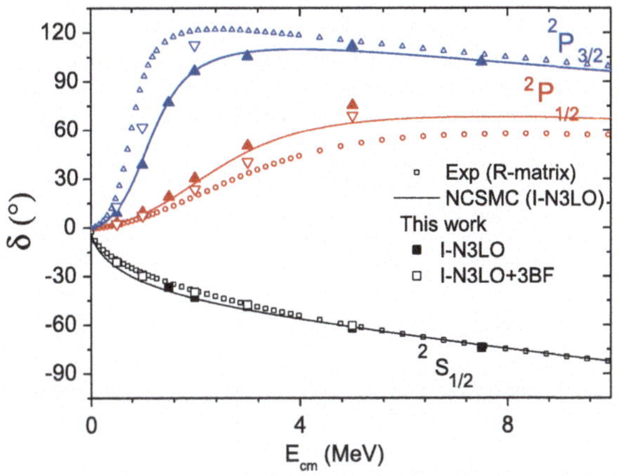

Fig. 91.2 Elastic phaseshifts for neutron scattering on ^{4}He nucleus at low energies calculated using I-N3LO NN potential without three-nucleon force (hollow symbols) and with the N2LO three-nucleon force parameterized by Marccuci et al. (full symbols). These results are compared with the calculations performed within NCSMC framework (continuous line) using I-N3LO model [24] and with the phaseshifts extracted from the experimental data, employing R-matrix analysis (small hollow symbols) [3]

spin-orbit components. The three-nucleon force (3NF) proposed by Navratil [23], being constructed using local regulators and derived up to next-to-next-to-leading order, once employed with I-N3LO NN interaction is able to a large extent reproduce the required P-wave splitting [24]. A similar feat has been achieved in Greens Function Monte Carlo calculations [25] employing AV18 NN interaction combined with IL7 three-nucleon force. I have performed calculations combining the I-N3LO NN interaction in conjunction with the newly parameterized next-to-next-to-leading order 3NF of [23] by Marcucci et al. It is expected, from the calculations in neighboring nuclei, that the two different parameterizations of 3NF should not differ in predictions of the binding energies or the low energy cross sections. Inclusion of the last 3NF delivers almost excellent description of the phaseshifts of doublet P-wave ($J^\pi = 1/2^-$). Description of the quartet P-wave ($J^\pi = 3/2^-$) is also significantly improved. Nevertheless in the resonance region my phaseshifts are still slightly smaller than the 'experimental' ones, as a result of less pronounced (situated deeper in the continuum) ^{5}He $J^\pi = 3/2^-$ resonant state. This issue requires more detailed analysis, as it could be also related with a slower than usual convergence of the PW decomposition in the region dominated by a narrow resonance, where a small variation of the resonance position can have dramatic effects on the calculated scattering observables.

It makes certain interest to concentrate on the n-^{4}He S-wave. As pointed out in my previous study this channel is dominated by strong Pauli repulsion between the projectile neutron and the ones, present within ^{4}He target. This repulsion produce phaseshifts, which are not very sensitive to the details of the nuclear interaction. Sim-

Table 91.1 Calculated n-^{4}He scattering lengths a_0 for two different NN interaction models presented together with ^{4}He binding energies $B(^4$He) obtained within the same model space as used in the scattering calculations (third column) and fully converged model space results (fourth column). My results are also compared with the theoretical and the recommended experimental values found in the literature

Model	a_0 [fm]	$B(^4He)$ [MeV] scatt.	$B(^4He)$ [MeV] full	a_0 [fm]	Reference
I-N3LO	2.72(3)	25.24	25.39	3.19	[24]
AV18	2.96(6)	24.08	24.23(1)	2.4	[25]
Experiment	2.456(16)	28.29567			TUNL [28]
	2.608 (24)				NIST [29]
	2.507 (32)				Atlas n-res.[22]

ilar feature is observed in the repulsive S-waves of neutron-deuteron and neutron-triton scattering [17, 31]. Nevertheless study of the thermal neutron scattering on ^{4}He may reveal interesting correlations with the binding energy of a ^{4}He ground state. Furthermore accurate description of the thermal neutron scattering is required in relation with the ongoing Parity violation studies, aiming to measure slow neutron spin rotation for n-^{4}He elastic scattering [18]. Curiously experimental situation with a measured n-^{4}He scattering length is quite unsettled, see Table 91.1. I have calculated n-^{4}He scattering lengths for several nuclear Hamiltonians and compared them in Table 91.1 with existing values in the literature. On may see that discrepancy between the theoretical predictions is even larger than for the experimental values. Deviation of the NCSMC result is well understood and is apparently due to the effect of induced three-nucleons forces, which have not been estimated in the former calculations with sufficient accuracy leading to overestimated S-wave phaseshifts at energies below 2 MeV. More accurate evaluation of the induced three-nucleons forces matrix element should allow to reduce the predicted scattering length, providing better low energy behavior of the S-wave phaseshifts. Situation with GFMC result is more troublesome—disagreement with my calculated values using AV18 potential is dramatic. To shed some light on this situation I have presented in Fig. 91.3 my calculated scattering lengths as a function of the model predicted ^{4}He binding energy. One may see clear signs for these two observables to be correlated, which reflects a well known property of strongly repulsive systems—the scattering lengths being correlated with the target's size. Worths noting that a similar tendency is observed for the n-^{3}H scattering lengths [17]. In the GFMC calculation such correlation pattern is absent, as their predicted scattering lengths for AV18 and AV18+IL7 models coincide, regardless the presence of $\sim$4 MeV difference in calculated ^{4}He binding energies and mean square radii. This indicates that GFMC might be not sufficiently accurate at very low energies. Finally correlation pattern presented in Fig. 91.3 suggests that between three recommended scattering length values the NIST result [29] might be the most reliable.

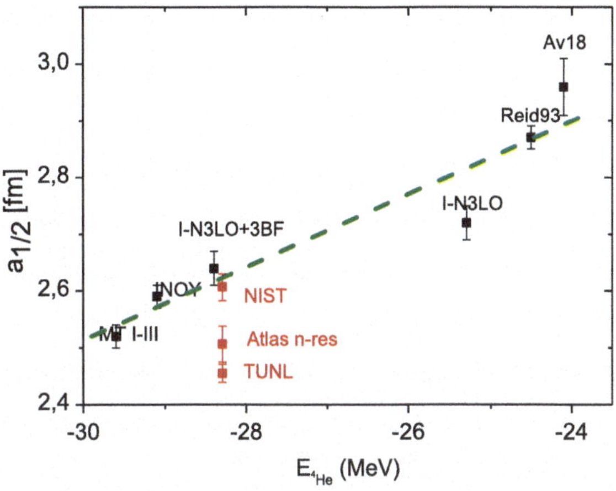

Fig. 91.3 Dependence of the model calculated n-^{4}He scattering length a_0 on the binding energy of ^{4}He. Recommended experimental values from NIST [29], Atlas n-res. [22] and TUNL [28] are also provided. A green dashed line is added just to guide an eye

Acknowledgements His work was granted access to the HPC resources of TGCC/IDRIS under the allocation 2018-A0030506006 made by GENCI (Grand Equipement National de Calcul Intensif).

References

1. Aaron, R., Amado, R.D., Yam, Y.Y.: Phys. Rev. **140**, B1291 (1965)
2. Baye, D.: The lagrange-mesh method. Phys. Rep. **565**, 1–107 (2015)
3. Csótó, A., Hale, G.M.: *S*. Phys. Rev. C **55**, 536–539 (1997). https://doi.org/10.1103/PhysRevC.55.536
4. Deltuva, A., Fonseca, A.C.: Neutron-^{3}h scattering above the four-nucleon breakup threshold. Phys. Rev. C **86**, 011,001 (2012). https://doi.org/10.1103/PhysRevC.86.011001
5. Entem, D.R., Machleidt, R.: Accurate charge-dependent nucleon-nucleon potential at fourth order of chiral perturbation theory. Phys. Rev. C **68**, 041,001 (2003). https://doi.org/10.1103/PhysRevC.68.041001
6. Faddeev, L.D.: Scattering theory for a three particle system. Sov. Phys. JETP **12**, 1014–1019 (1961)
7. Hüber, D., Witała, H., Nogga, A., Glöckle, W., Kamada, H.: A new look into the partial-wave decomposition of three-nucleon forces. Few-Body Syst. **22**(3), 107–135 (1997)
8. Kievsky, A., Viviani, M., Barletta, P., Romero-Redondo, C., Garrido, E.: Variational description of continuum states in terms of integral relations. Phys. Rev. C **81**, 034,002 (2010). https://doi.org/10.1103/PhysRevC.81.034002
9. Kloet, W., Tjon, J.: A study of break-up processes in neutron-deuteron scattering. Nucl. Phys. A **210**(2), 380–396 (1973)
10. Koike, Y., Plessas, W., Zankel, H.: Nucleon-deuteron scattering based on meson-exchange nucleon-nucleon dynamics. Phys. Rev. C **32**(5), 1796 (1985)
11. Laverne, A.: Nucl. Phys. **203**, 597 (1973)
12. Lazauskas, R.: Scattering of heavy charged particles in atomic and nuclear systems. Ph.D. thesis, Université Joseph Fourier, Grenoble (2003). https://hal.archives-ouvertes.fr/tel-00004178
13. Lazauskas, R.: Modern nuclear force predictions for $n -^3$ H scattering above the three- and four-nucleon breakup thresholds. Phys. Rev. C **91**, 041,001 (2015). https://doi.org/10.1103/PhysRevC.91.041001

14. Lazauskas, R.: Solution of n-^{4}He elastic scattering problem using Faddeev-Yakubovsky equations. arXiv:1711.04716 (2017)
15. Lazauskas, R.: Description of five-nucleon systems using faddeev-yakubovsky equations. Few-Body Syst. **59**(2), 13 (2018)
16. Lazauskas, R.: Solution of the n-^{4}He elastic scattering problem using the faddeev-yakubovsky equations. Phys. Rev. C **97**(4), 044,002 (2018)
17. Lazauskas, R., Song, Y.H., Park, T.S.: Heavy-baryon chiral perturbation theory approach to thermal neutron capture on He^3. Phys. Rev. C **83**(3), 034,006 (2011)
18. Lehnert, R., Snow, W., Yan, H.: A first experimental limit on in-matter torsion from neutron spin rotation in liquid ^{4}He. Phys. Lett. B **730**, 353–356 (2014)
19. Lippmann, B.A., Schwinger, J.: Variational principles for scattering processes. Phys. Rev. **79**(3), 469 (1950)
20. Merkuriev, S., Gignoux, C., Laverne, A.: Three-body scattering in configuration space. Ann. Phys. **99**(1), 30–71 (1976)
21. Merkuriev, S., Yakovlev, S., Gignoux, C.: Four-body yakubovsky differential equations for identical particles. Nucl. Phys. A **431**, 125–138 (1984)
22. Mughabghab, S.F.: Atlas of Neutron Resonances: resonance Parameters and Thermal Cross Sections. Z = 1–100. Elsevier (2006)
23. Navrátil, P.: Local three-nucleon interaction from chiral effective field theory. Few-Body Syst. **41**(3), 117–140 (2007). https://doi.org/10.1007/s00601-007-0193-3
24. Navrátil, P., Quaglioni, S., Hupin, G., Romero-Redondo, C., Calci, A.: Unified ab initio approaches to nuclear structure and reactions. Phys. Scr. **91**(5), 053,002 (2016). http://stacks.iop.org/1402-4896/91/i=5/a=053002
25. Nollett, K.M., Pieper, S.C., Wiringa, R.B., Carlson, J., Hale, G.M.: Quantum monte carlo calculations of neutron-α scattering. Phys. Rev. Lett. **99**, 022,502 (2007). https://doi.org/10.1103/PhysRevLett.99.022502. https://link.aps.org/doi/10.1103/PhysRevLett.99.022502
26. Sasakawa, T.: Five-body equation. Prog. Theor. Phys. Suppl. **61**, 149–160 (1977)
27. Skornyakov, G., Ter-Martirosian, K.: Three-body problem with short-range forces. Neutron Scatt. Deutrons Small Energy (Zh. Eksp. Teor. Fiz) **31**, 775 (1956)
28. Tilley, D., et al.: Nucl. Phys. A **708**, 3 (2002)
29. Varley, F.S.: Neutron scattering lengths and cross sectioirn. Neutron News **3**(3), 29–37 (1992)
30. Viviani, M., Deltuva, A., Lazauskas, R., Carbonell, J., Fonseca, A., Kievsky, A., Marcucci, L.E., Rosati, S.: Benchmark calculation of n_^{3}H and p_^{3}He scattering. Phys. Rev. C **84**(5), 054,010 (2011)
31. Witała, H., Nogga, A., Kamada, H., Glöckle, W., Golak, J., Skibiński, R.: Modern nuclear force predictions for the neutron-deuteron scattering lengths. Phys. Rev. C **68**(3), 034,002 (2003)
32. Yakubovsky, O.A.: On the integral equations in the theory of n particle scattering. Sov. J. Nucl. Phys. **5**, 937 (1967)

Chapter 92
Few-Body Systems in Minkowski Space: The Bethe-Salpeter Equation Challenge

Giovanni Salmè

Abstract Solving the homogeneous Bethe-Salpeter equation directly in Minkowski space is becoming a very alive field, since, in recent years, a new approach has been introduced, and the reachable results can be potentially useful in various areas of research, as soon as the relativistic description of few-body bound systems is relevant. A brief review of the status of the novel approach, which benefits from the consistent efforts of different groups, will be presented, together with some recent results.

92.1 Introduction

A fully covariant and non perturbative description of a bound system, with and without spin degrees of freedom (dofs), represents a Holy Grail in different areas of research, as soon as relativistic effects become urgent for obtaining a reliable control of the dynamical evolution of the state. Indeed, a bound system has a unavoidably non perturbative nature, and if it is necessary a relativistic description, one keenly quests a suitable approach with the above mentioned features for tackling the issue. In view of this, the Bethe-Salpeter equation (BSE) in Minkowski space can play a role, since it was proposed by Salpeter and Bethe [1] within a fully field-theoretical and non perturbative framework (see [2] for a first review). In particular, the capability to offer a non perturbative description has to be ascribed to its-own structure, since the BSE for bound states is a homogeneous integral equation, and hence it is able to take into account all the infinite exchanges, needed to reproduce the bound-state pole (in the relevant Green's function). Given such an attractive feature, the BSE has been actively studied, but the presence of many challenges related to, e.g., the need of kernels beyond the ladder one (see the next Section), the presence of self-energy and vertex corrections and, last but not least, the non trivial analytic structures of the BS amplitude (i.e. the solution of the BSE), has urged the introduction of wise mathematical trick, like the Wick rotation [3] and/or sharp approximations. As is

G. Salmè (✉)
INFN, Sezione di Roma, P.le A. Moro 2, I-00185 Rome, Italy
e-mail: salmeg@roma1.infn.it

N. A. Orr et al. (eds.), *Recent Progress in Few-Body Physics*,
Springer Proceedings in Physics 238,
https://doi.org/10.1007/978-3-030-32357-8_92

well-known, very popular approaches to the BS formalism are based on the rotation to imaginary relative energy (see, e.g. [4] for an application to the deuteron) or on productive 3D-reductions of the BSE (see e.g. [5], for some introductory details), and they are devised for circumventing the issue of the cumbersome analytic structure in Minkowski momentum-space, related to the role of the relative energy (as well as the meaning of the relative time). There exist also covariant approximation of the BSE (see, e.g. [6], illustrating the covariant spectator model approach and [7] for the approach based on separable kernels), where some simplifying assumptions are introduced but still keeping safe the covariant nature of BSE. Till nineties, only one case was known to be exactly solved in Minkowski momentum-space, though with the interaction kernel in ladder approximation, namely the Wick-Cutkosky model [3, 8], which is composed by two massive scalars interacting through a *massless* scalar. In 1997, a new approach, based on (i) the so-called Nakanishi Integral Representation (NIR) of the BS amplitude[9] and (ii) its uniqueness, was introduced in order to successfully solve the ladder BSE governing a two-scalar system interacting through a *massive* scalar [10]. In particular, the emphasis of the work was on the calculations of both coupling constants needed for binding the system with assigned masses, and the so-called Nakanishi weight functions (see Sect. 92.3), rather than on the evaluation of the momentum distributions, so important in the phenomenological investigation of the inner dynamics of the bound systems. Nonetheless, [10] opened a new and attractive path, though the approach involved a lengthy algebraic manipulations. A substantial improvement was reached through the clever introduction [11] of the light-front (LF) framework (see e.g. [12, 13], for detailed reviews). Indeed, by adopting LF variables, $x^{\pm} = x^0 \pm x^3$ and $\mathbf{x}_\perp \equiv \{x^1, x^2\}$, it is possible to simplify the treatment of the needed analytical integrations, extending the realm of application even including bound system with spin dofs [14–16], and, more important, a probabilistic framework can be established.

In Sects. 92.2 and 92.3 some snapshots on the Bethe-Salpeter equation and the Nakanishi integral representation of the BS amplitude will be presented, while Sect. 92.4 is devoted to the LF framework. Finally Sects. 92.5 and 92.6 contain some numerical results and perspectives, respectively.

92.2 The Bethe-Salpeter Equation in a Nutshell

To briefly illustrate the main steps leading to the BSE, let us consider a simple two-scalar case (without antiparticles). The 4-point Green's Function (ϕ_i are the scalar fields) is given by

$$G(x_1, x_2; y_1, y_2) =< 0 \,|\, T\{\phi_1(x_1)\phi_2(x_2)\phi_1^+(y_1)\phi_2^+(y_2)\} \,|\, 0 >, \qquad (92.1)$$

and it fulfills the following inhomogeneous integral equation, sketched in Fig. 92.1,

$$G = G_0 \, + \, G_0 \, \mathcal{I} \, G \qquad (92.2)$$

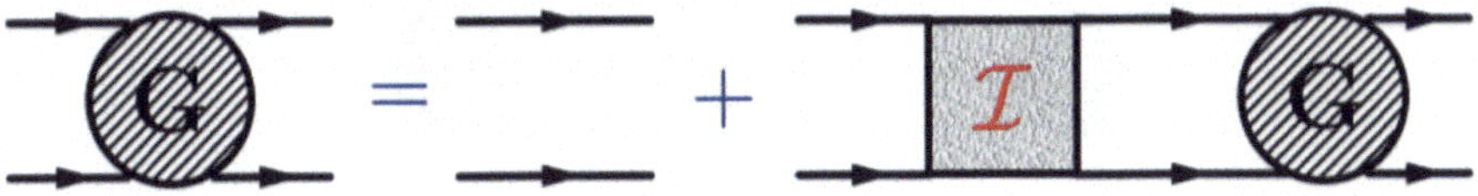

Fig. 92.1 Pictorial representation of the inhomogeneous integral equation fulfilled by the four-leg Green's function. The first irreducible diagrams contributing to the interaction kernel $\mathcal{I}$ are shown in Fig. 92.2

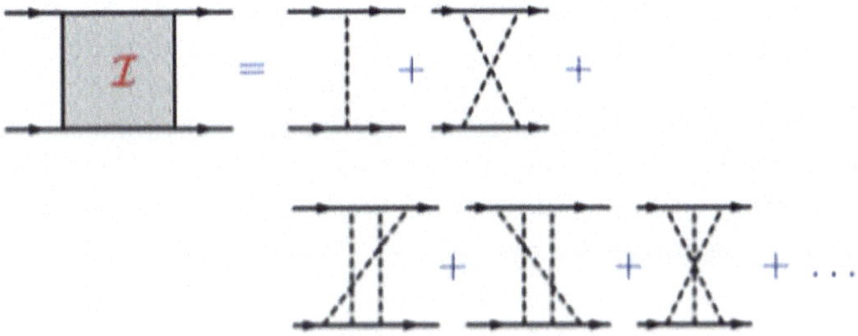

Fig. 92.2 The lowest-order irreducible (see text) Feynman diagrams contributing to the interaction kernel of the integral equation for the four-leg Green's function, (92.2). In a full theory, both interaction vertexes and propagators have to be properly dressed

where G_0 is the product of four free propagators and $\mathcal{I}$ is the interaction kernel. It is given by the infinite sum of *irreducible Feynman graphs*, i.e. the ones that cannot be split into two lower order diagrams after cutting two internal particle lines and without touching the quantum lines [1]. The set of the irreducible diagrams up to the 6th power of the coupling constant are shown in Fig. 92.2. The first diagram, though it contains two interaction vertexes (hence, two coupling constants), generates the infinite *ladder* series, after iterating the integral equation in (92.2). The second diagram, that has four coupling constants, is the simplest crossed diagram. Finally, the last set of diagrams contain crossed diagram with 6 interaction vertexes. It is important to stress that each class of irreducible diagrams, contributing to the interaction kernel, in turn gives rise to an infinite series of diagrams by iteration. Such an observation allows us to anticipate that the solutions of the BSE corresponding to a truncated interaction kernel $\mathcal{I}$ contains however effects generated by an infinite set of powers of the coupling constant. After properly inserting a complete 2-body Fock basis in $G(x_1, x_2; y_1, y_2)$, one can isolate the bound state contribution (assuming only one non-degenerate bound state, for the sake of simplicity) that manifests itself as a pole, in momentum space, i.e.

$$G(k, q; p) \Rightarrow G_B(k, q; p) \simeq \frac{i}{(2\pi)^{-4}} \frac{\chi(k; p_B)\, \bar{\chi}(q; p_B)}{2\omega_B(p^0 - \omega_B + i\epsilon)} \tag{92.3}$$

where $\omega_B = \sqrt{M_B^2 + |\mathbf{p}|^2}$, $p_B^\mu \equiv \{\omega_B, \mathbf{p}\}$ with M_B the mass of the bound state (the adopted metric is $\{1, -1, -1, -1\}$), k is the relative four-momentum of the pair and

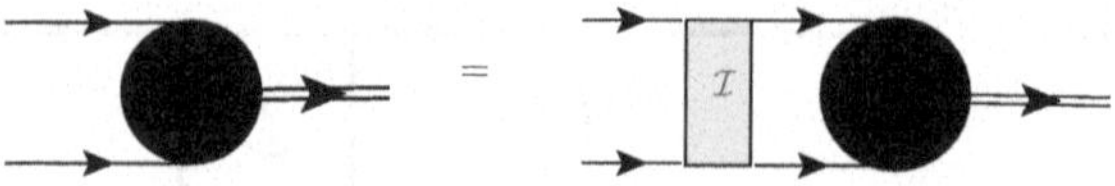

Fig. 92.3 Pictorial representation of the Bethe-Salpeter integral equation for a bound system

$\beta \equiv$ further quantum numbers. The function $\chi(k; p_B)$ is called the BS amplitude, and describes the residue together with its conjugate (notice that the conjugate $\bar{\chi}$ is actually defined through (92.3) and involves the chronological product, to be carefully considered [2]). Unfortunately, the BS amplitude does not have a probabilistic interpretation, but one can profitably retrieve a probabilistic environment once the LF framework and the Fock expansion of the interacting two-body state is introduced (see Sect. 92.4). In conclusion, close to the bound-state pole, i.e. $p^0 \to \omega_B =$ the Green's function is approximated by

$$G \simeq \frac{i}{(2\pi)^{-4}} \frac{\chi(k; p_B)\, \bar{\chi}(q; p_B)}{2\omega_B(p^0 - \omega_B + i\epsilon)} + \textit{regular terms} \tag{92.4}$$

By going through the following steps: (i) insert the approximation (92.4) in both sides of $G = G_0 + G_0\, \mathcal{I}\, G$, (ii) multiply by $(p^0 - \omega_B)$ and (iii) look close to the pole, one eventually gets the BSE, viz (see Fig. 92.3)

$$\chi(k; p_B, \beta) = G_0(k; p_B, \beta) \int d^4k'\, \mathcal{I}(k, k'; p_B)\, \chi(k'; p_B, \beta) \tag{92.5}$$

The normalization of the BS amplitude is given by (see [2, 17–19] for further interesting issues on the BS norm)

$$\int \frac{d^4q}{(2\pi)^4} \int \frac{d^4k}{(2\pi)^4}\, \bar{\chi}(q, p_B)\, \frac{\partial}{\partial p_\mu}\Big[G_0^{-1}(k, p)(2\pi)^4\delta^4(q-k) - i\mathcal{I}(q, k, p)\Big]\Big|_{p^2=M_B^2} \times \chi(k, p_B) = i2p_B^\mu \tag{92.6}$$

92.3 The Nakanishi Integral Representation and the BS Amplitude

In the sixties, Noburo Nakanishi [9, 20] proposed an integral representation of transition amplitudes (see Fig. 92.4), based on the parametric formula for the Feynman diagrams. Within the perturbation theory for scalars, the transition amplitude with N external legs, f_N, gets an infinite number of contributions, whose generic expression is given by

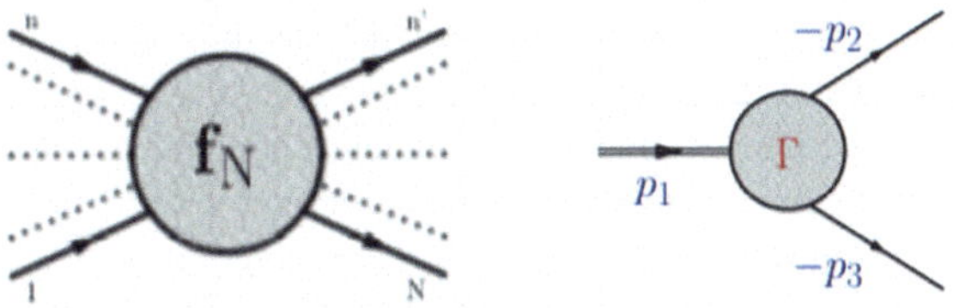

Fig. 92.4 Left panel: transition amplitude with N external legs. The bubble represents the infinite possible contributions, and the legs are represented just for enumerating the external momenta, but they do not indicate any external propagation. Right panel: the 3-leg transition amplitude, relevant for the application to the two-body Bethe-Salpeter amplitude (see text), the same previous caveat has to be applied to the external legs

$$f_{\mathcal{G}}(p_1, p_2, \ldots, p_N) \propto \prod_{i=1}^{k} \int d^4 q_i \; \frac{1}{(\ell_1^2 - m_1^2)(\ell_2^2 - m_2^2) \; \ldots \; (\ell_n^2 - m_n^2)} \tag{92.7}$$

where n indicates the total number of propagators, k is the total number of loops ($\equiv$ number of integration variables), ℓ_j are the internal momenta, combinations of external momenta p_r and loop ones q_i. The label $\mathcal{G}$ is a shorthand notation replacing $\{n, k\}$. By standard manipulations one can write

$$f_{\mathcal{G}}(\{p_i\}) \propto \prod_{j=1}^{n} \int_0^1 d\alpha_j \; \frac{\delta(1 - \sum_r \alpha_r)}{U(\{\alpha_r\}) \left[\sum_r \alpha_r m_r^2 - \sum_h \eta_h(\{\alpha_i\}) \; s_h + i\epsilon\right]^{n-2k}} \tag{92.8}$$

where $\{\alpha_i\}$ are n Feynman parameters, $\{s_h\}$ all the *independent scalar products* one can construct from the N external legs with $1 \leq h \leq N' < N$, $\eta_h(\{\alpha_r\})$ and $U(\{\alpha_r\})$ are well-defined combinations of the Feynman parameters [20]. It is worth noticing that the dependence upon $\{n, k\}$ affects the denominator, and therefore each contribution to the total transition amplitude has its-own analytic structure. Nakanishi introduced a compact and elegant expression of the full N-leg amplitude $f_N(s) = \sum_{\mathcal{G}} f_{\mathcal{G}}(s)$, by inserting in the generic contribution $f_{\mathcal{G}}$ the following identity

$$1 \doteq \prod_{j=1}^{N'} \int_0^1 dz_j \; \delta\left(z_j - \frac{\eta_j(\{\alpha_i\})}{\beta}\right) \int_0^\infty d\gamma \; \delta\left(\gamma - \sum_i \frac{\alpha_i m_i^2}{\beta}\right) \tag{92.9}$$

where $\beta = \sum_h \eta_h(\{\alpha_i\})$ (see [20] for details). After integrating by parts $n - 2k - 1$ times the expression of $f_{\mathcal{G}}$, one gets from (92.8)

$$f_{\mathcal{G}}(\tilde{s}) \propto \prod_{j=1}^{N'} \int_0^1 dz_j \int_0^\infty d\gamma \frac{\delta(1 - \sum_j z_j) \; \tilde{\phi}_{\mathcal{G}}(\mathbf{z}, \gamma)}{(\gamma - \sum_j z_j \; s_j)} \tag{92.10}$$

where $\tilde{\phi}_{\mathcal{G}}(\mathbf{z},\gamma)$ is a proper weight function (that in this perturbative framework lives in the realm of the distribution functions), with $\mathbf{z} \equiv \{z_1, z_2, \ \dots \ , z_{N'}\}$ and $\tilde{s} \equiv \{s_1, s_2, \ \dots \ , s_{N'}\}$. Summarizing, the dependence upon the details of the diagram, i.e. $\{n, k\}$, moves from the denominator to the numerator. Hence, one has exactly the same formal expression for the denominator of any diagram $\mathcal{G}$.

The full N-leg transition amplitude is the sum of infinite diagrams $\mathcal{G}$ and it can be formally written as

$$f_N(\tilde{s}) = \sum_{\mathcal{G}} f_{\mathcal{G}}(\tilde{s}) \propto \prod_{j=1}^{N'} \int_0^1 dz_j \int_0^\infty d\gamma \frac{\delta(1-\sum_j z_j)\, \phi_N(\mathbf{z},\gamma)}{(\gamma - \sum_i z_i\, s_i)} \tag{92.11}$$

where $\phi_N(\mathbf{z},\gamma) = \sum_{\mathcal{G}} \tilde{\phi}_{\mathcal{G}}(\mathbf{z},\gamma)$ is called a Nakanishi weight function (NWF). It is a real function, depending upon N' compact variables, $\mathbf{z}$, and one non compact, γ.

For the 3-leg transition amplitude, (cf. the right panel in Fig. 92.4) after eliminating one compact variable and redefining the NWF, one reduces to

$$f_3(p^2, k^2, k\cdot p) = \int_{-1}^1 dz \int_0^\infty d\gamma \frac{\pi_3(z,\gamma)}{\gamma + m^2 - \frac{p^2}{4} - k^2 - zk\cdot p - i\epsilon} \tag{92.12}$$

with $p = p_1 + p_2$ and $k = (p_1 - p_2)/2$. Notice that ϕ_3 is also known as the vertex function, Γ, for a scalar theory (N.B. for fermions one has to add spinor indexes). The expression holds at *any* order in perturbation-theory.

Once we put one leg on the mass shell, and we assume $\phi_3(z,\gamma)$ as an unknown functions, (92.12) becomes a natural choice as a trial function for obtaining actual solution of BSE for a two-scalar interacting system [9–11, 21]. The validation of this assumption, namely translating an expression formally elaborated within a perturbative framework to a non perturbative one, has been successfully obtained numerically, as discussed in what follows. This result should be not too much surprising if one takes into account the freedom associated to the NWF, that is the unknown to be determined.

A vertex function $f_3(k,p)$ with one leg on mass-shell is related to the BS amplitude χ by attaching propagators to the two external virtual legs, i.e. schematically: $\chi = G_1 \otimes G_2 \otimes f_3(\tilde{s})$ where G_i are the propagators of the two particles. Finally one writes

$$\chi(k,p) = -i \int_{-1}^1 dz' \int_{-\infty}^\infty d\gamma' \frac{g(z',\gamma';\kappa^2)}{\left[\gamma' + \kappa^2 - k^2 - z'k\cdot p - i\epsilon\right]^3} \tag{92.13}$$

where $\kappa^2 = m^2 - M^2/4$, with m the mass of the constituent scalars and $g(z',\gamma';\kappa^2)$ indicates a NWF corresponding to a given mass M of the interacting system. Indeed the lower extremum of γ' has to be put equal to zero, in order to avoid the spontaneous decay of the bound state.

92.4 Projecting BSE onto the Hyper-Plane $x^+ = 0$

As mentioned above, NIR yields the *needed freedom* for exploring non perturbative problems, once the NWFs are taken as unknown real quantities. Unfortunately, even adopting NIR, BSE still remains a highly singular integral equation in the Minkowski momentum space. The strategy adopted in [10] for obtaining actual solutions of the two-scalar homogeneous BSE, in ladder approximation, relied on the uniqueness of the NWF. Profitably, a more general approach was introduced by Carbonell and Karmanov [11, 14], that exploited the known analytic structure of the BS amplitude, once NIR is applied. Since the target is to determine the NWF, one can perform analytic integrations, to formally obtain a new integral equation, more suitable for the numerical treatment. The approach can be seen as a variant of the 3D reduction, but in this case the assumed expression of the BS amplitude in terms of NIR allows one the formally exact reconstruction of the full BS amplitude. In particular, in [11] an explicitly-covariant LF framework [13] was adopted, while the non-explicitly covariant version can be found in [21, 22]. It should be pointed out that the latter approach appears more suitable for isolating and mathematically treating further singularities one meets, particularly when the spin dofs are involved (see [15, 16]). In general, the LF approach allows one a simpler treatment of the analytic integration, since, e.g., double poles in the k^0 complex plane splits in pairs of single poles in the k^- and k^+ complex planes.

In order to obtain an integral equation for the NWF more tractable from the numerical point of view (spin dofs lead to a *system* of integral equations), there is an illuminating step toward the final goal. One can recognize that within the LF framework a probabilistic interpretation is recovered by (i) expanding the BS amplitude on a LF Fock basis, and (ii) singling out the *valence component*. This is the amplitude of the Fock state with the lowest number of constituents, and it has (together with all the other amplitudes in the Fock expansion) the property to be invariant under LF-boost transformations (see [12]). In particular, the integral of the square modulus of the valence component gives the probability to find only two constituents in the interacting state. In the case of the non-explicitly covariant LF framework, the valence component is formally obtained by integrating the BS amplitude on $k^- = k^0 - k^3$. Namely

$$\text{Valence w.f.} = \psi_{n=2}(\xi, k_\perp) = \frac{p^+}{\sqrt{2}}\, \xi\,(1-\xi) \int \frac{dk^-}{2\pi} \chi(k,p) =$$
$$= \frac{1}{\sqrt{2}} \xi\,(1-\xi) \int_0^\infty d\gamma' \, \frac{g(\gamma', 1-2\xi; \kappa^2)}{[\gamma' + k_\perp^2 + \kappa^2 + (2\xi-1)^2 \frac{M^2}{4} - i\epsilon]^2} \tag{92.14}$$

where $\xi = (1-z)/2, M = 2m - B$. with B the binding energy. Notice that the above mathematical step is equivalent to make vanishing the relative *LF-time* x^+ in the BS amplitude in coordinate space (see, e.g.,[21, 22]), i.e. projecting the BS amplitude onto the hyperplane $x^+ = 0$. The above observations make attractive to study what happens when the k^- integration is applied to both sides of BSE. As a matter of fact, on the lhs one has (a part trivial factors) the valence component that lives in a 3D space, and contains the NWF, while on the rhs one remains with a folding of the NWF with the so-called Nakanishi kernel, $V^{LF}(\alpha;\gamma,z;\gamma',z')$, viz

$$\text{valence w.f.} \propto \int_0^\infty d\gamma' \frac{g(\gamma',z;\kappa^2)}{[\gamma'+\gamma+z^2m^2+(1-z^2)\kappa^2-i\epsilon]^2} =$$
$$= \int_0^\infty d\gamma' \int_{-1}^{1} dz'\; V^{LF}(\alpha;\gamma,z;\gamma',z')\; g(\gamma',z';\kappa^2). \tag{92.15}$$

where α is the coupling constant. N.B. $V^{LF}(\alpha;\gamma,z;\gamma',z')$, is determined by the irreducible kernel $\mathcal{I}(k,k',p)$ (see, e.g. [21, 22], for details). In presence of spin dofs, the evaluation of the Nakanishi kernel, has been carried out in ladder approximation, but it was shown that it is plagued by further singularities, as recognized for the first time in [23]. Fortunately, within the non-explicitly covariant LF framework, those singularities can be isolated and mathematically integrated [15, 16] by using the standard approach introduced in [24].

In the numerical calculations of [15, 16, 21, 25], an orthonormal basis given by the Cartesian product of Laguerre and Gegenbauer polynomials (the first ones depends upon the non compact variable γ and the second ones upon the compact variable z) is adopted for expanding $g(\gamma,z;\kappa^2)$. In ladder approximation, once we assign the mass of the system, i.e. κ^2, the integral equation in (92.15) becomes a generalized eigen-equation, with eigenvalue α, and the eigenvector composed by the coefficients of the expansion of $g(\gamma,z;\kappa^2)$. If the eigen-equation admits a solution, for a given mass M (it is a non linear parameter) of the system, then we know how to reconstruct the BS amplitude, and eventually we validate the whole procedure from the 4D BSE to its projection onto the $x^+=0$ hyperplane, or better its integration on the k^- component of the relative four-momentum.

92.5 Excerpt of the Numerical Results

Many calculations have been performed for the two-scalar system by using the ladder approximation, i.e. a kernel with the following massive scalar exchange $\mathcal{K}_S = g^2/[(k-k')^2-\mu^2+i\epsilon]$, where the coupling g is related to the coupling α in (92.15) by $\alpha = g^2/(16\pi m^2)$. In particular a successful comparison between the results in [11] (explicitly-covariant LF approach) and [21] (non-explicitly covariant LF framework) has been obtained for eigenvalues and eigenvectors. Encouraged by this achievement, both excited states [26] and even scattering lengths [27] have been studied. There have been also investigations of the cross-diagram contribution to the kernel (second diagram in Fig. 92.2) [14], and its effect on the calculation of the electromagnetic form factor [28].

In the two-fermion case, the ladder kernel has been generalized to take into account (i) the pseudo-scalar exchange and (ii) the vector one (massless and massive). As

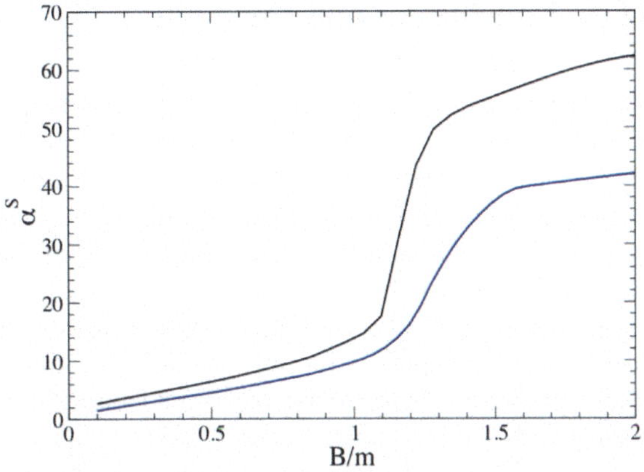

Fig. 92.5 The coupling constant $\alpha = \lambda_F \lambda_S/(8\pi m)$, with $\lambda_{F(S)}$ the fermion (scalar) coupling in the interaction Lagrangian, versus the binding energy per unit mass. Blue solid line: exchanged mass, $\mu/m = 0.15$. Black solid line: exchanged mass, $\mu/m = 0.5$

mentioned above, in this case one has to solve a system of four integral equations for determining the four NWFs needed to describe an s-wave interacting system [15, 16, 23]. A part the fixing of the issue of the LF singularities generated by the presence of spin dofs, in [15, 16] an important feature has been pointed out: the theoretically expected behavior of the high-momentum tail of the valence component of a fermion-antifermion s-wave state, interacting through the exchange of a massless vector, has been recovered. This means that dynamical quantities can be addressed, teven with a simple interaction kernel.

The extensive collection of results and the agreement achieved between different groups leads to extend the application of NIR to other cases. In particular, some preliminary results for the fermion-scalar system can be presented [25]. In this case, the BS amplitude contains two unknown scalar functions ϕ_i, viz

$$\Phi_{BS}(k,p) = \left[\phi_1(k,p) + \frac{\not{k}}{M}\,\phi_2(k,p)\right] U(p,s) \tag{92.16}$$

with $(\not{p} - M)\;U(p,s) = 0$. By applying NIR to each ϕ_i, analogously to the two fermion case, one obtains a system of two coupled integral equations, and one can apply the same tools for exactly transforming the initial ladder BSE in a generalized eigenvalue problem. In Fig. 92.5, preliminary results for the coupling constants needed to bind a fermion-scalar system, with a given mass $M = 2m - B$ and interacting through a scalar exchange, are shown. The interesting fast growing that appears for $B/m \geq 1$ is a signature of the increasing relevance of the repulsion produced by the small component in the Dirac spinor of the constituent fermion. The attraction of the scalar interaction is softened for increasing binding energy, due to the competition between large and small components in $(\bar{u}\;u)$, at the fermionic vertex. Large B/m values mean large kinetic energy, and in turn big effects from the small components.

92.6 Conclusions and Perspectives

The technique based on the Nakanishi integral representation of the BS amplitude has to be considered a viable and effective tool for solving BSE Indeed, the cross-check of results obtained by different groups, for different interacting systems, with kernels in ladder and cross-ladder contributions as well as with and without spin dofs, has produced a clear numerical evidence of the validity of NIR for obtaining actual solutions. Noteworthy, the possibility to obtain the valence component straightforwardly leads to evaluate relevant quantities, like the the LF momentum distributions, so important for phenomenological investigations.of bound systems in relativistic regime. In particular, one should remind the second ingredient of the technique, i.e. the LF framework. It has well-known advantages in performing analytical integrations, and in the fermionic case it shows its effectiveness in full glory.

The numerical validation of NIR strongly encourages to face with the next challenges represented by including self-energies and vertex corrections evaluated within the same framework (works in progress on the Dyson-Schwinger Equations) as well as by the possible construction of interaction kernels beyond the sum of the first two contributions in Fig. 92.2, e.g., moving from the fully off-shell ladder series, that fulfills an integral equation, to a closed form for the cross-ladder T-matrix.

References

1. Salpeter, E.E., Bethe, H.A.: A relativistic equation for bound-state problems. Phys. Rev. **84**, 1232 (1951). https://doi.org/10.1103/PhysRev.84.1232
2. Nakanishi, N.: A general survey of the theory of the Bethe-Salpeter equation. Prog. Theor. Phys. Suppl. **43**, 1 (1969). https://doi.org/10.1143/PTPS.43.1
3. Wick, G.C.: Properties of Bethe-Salpeter wave functions. Phys. Rev. **96**, 1124 (1954). https://doi.org/10.1103/PhysRev.96.1124
4. Zuilhof, M.J., Tjon, J.A.: electromagnetic properties of the deuteron and the Bethe-Salpeter equation with one boson exchange. Phys. Rev. C **22**, 2369 (1980). https://doi.org/10.1103/PhysRevC.22.2369
5. Lucha, W., Schoberl, F.F.: Instantaneous Bethe-Salpeter equation with exact propagators. J. Phys. G **31**, 1133 (2005). https://doi.org/10.1088/0954-3899/31/11/001
6. Gross, F., The, C.S.T.: Its achievements and its connection to the light cone. Few Body Syst. **58**(2), 39 (2017). https://doi.org/10.1007/s00601-017-1215-4
7. Bondarenko, S.G., Burov, V.V., Molochkov, A.V., Smirnov, G.I., Toki, H.: Bethe-Salpeter approach with the separable interaction for the deuteron. Prog. Part. Nucl. Phys. **48**, 449 (2002). https://doi.org/10.1016/S0146-6410(02)00142-4
8. Cutkosky, R.E.: Solutions of a Bethe-Salpeter equations. Phys. Rev. **96**, 1135 (1954). https://doi.org/10.1103/PhysRev.96.1135
9. Nakanishi, N.: Partial-wave Bethe-Salpeter equation. Phys. Rev. **130**(3), 1230 (1963)
10. Kusaka, K., Simpson, K., Williams, A.G.: Solving the Bethe-Salpeter equation for bound states of scalar theories in Minkowski space. Phys. Rev. D **56**, 5071 (1997). https://doi.org/10.1103/PhysRevD.56.5071
11. Karmanov, V.A., Carbonell, J.: Solving Bethe-Salpeter equation in Minkowski space. Eur. Phys. J. A **27**, 1 (2006). https://doi.org/10.1140/epja/i2005-10193-0

12. Brodsky, S.J., Pauli, H.C., Pinsky, S.S.: Quantum chromodynamics and other field theories on the light cone. Phys. Rept. **301**, 299 (1998). https://doi.org/10.1016/S0370-1573(97)00089-6
13. Carbonell, J., Desplanques, B., Karmanov, V.A., Mathiot, J.F.: Explicitly covariant light front dynamics and relativistic few body systems. Phys. Rept. **300**, 215 (1998). https://doi.org/10.1016/S0370-1573(97)00090-2
14. Carbonell, J., Karmanov, V.A.: Cross-ladder effects in Bethe-Salpeter and light-front equations. Eur. Phys. J. A **27**, 11 (2006). https://doi.org/10.1140/epja/i2005-10194-y
15. de Paula, W., Frederico, T., Salmè, G., Viviani, M.: Advances in solving the two-fermion homogeneous Bethe-Salpeter equation in Minkowski space. Phys. Rev. D **94**, 071901 (2016). https://doi.org/10.1103/PhysRevD.94.071901
16. de Paula, W., Frederico, T., Salmè, G., Viviani, M., Pimentel, R.: Fermionic bound states in Minkowski-space: light-cone singularities and structure. Eur. Phys. J. C **77**(11), 764 (2017). https://doi.org/10.1140/epjc/s10052-017-5351-2
17. Lurié, D., Macfarlane, A.J., Takahashi, Y.: Normalization of Bethe-Salpeter wave functions. Phys. Rev. **40**, B1091 (1965). https://doi.org/10.1103/PhysRev.140.B1091
18. Nakanishi, N.: Normalization condition and normal and abnormal solutions of the Bethe-Salpeter equation. Phys. Rev. **138**, B1182 (1965). https://doi.org/10.1103/PhysRev.138.B1182
19. Ahlig, S., Alkofer, R.: (In)consistencies in the relativistic description of excited states in the Bethe-Salpeter equation. Ann. Phys. **275**, 113 (1999). https://doi.org/10.1006/aphy.1999.5922
20. Nakanishi, N.: Graph Theory and Feynman Integrals. Gordon and Breach, New York (1971)
21. Frederico, T., Salmè, G., Viviani, M.: Quantitative studies of the homogeneous Bethe-Salpeter equation in Minkowski space. Phys. Rev. D **89**, 016010 (2014). https://doi.org/10.1103/PhysRevD.89.016010
22. Frederico, T., Salmè, G., Viviani, M.: Two-body scattering states in Minkowski space and the Nakanishi integral representation onto the null plane. Phys. Rev. D **85**, 036009 (2012). https://doi.org/10.1103/PhysRevD.85.036009
23. Carbonell, J., Karmanov, V.A.: Solving Bethe-Salpeter equation for two fermions in Minkowski space. Eur. Phys. J. A **46**, 387 (2010). https://doi.org/10.1140/epja/i2010-11055-4
24. Yan, T.M.: Quantum field theories in the infinite-momentum frame. IV. scattering matrix of vector and dirac fields and perturbation theory, Phys. Rev. D **7**, 1780 (1973). https://doi.org/10.1103/PhysRevD.7.1780
25. Alvarenga Nogueira, J.H., Gherardi, V., Frederico, T., Salmè, G., Colasante, D., Pace, E.: Solving the Bethe-Salpeter equation in Minkowski space for a fermion-scalar system. Phys. Rev. D **100**, 0160021 (2019)
26. Gutierrez, C., Gigante, V., Frederico, T., Salmè, G., Viviani, M., Tomio, L.: Bethe-Salpeter bound-state structure in Minkowski space. Phys. Lett. B **759**, 131 (2016). https://doi.org/10.1016/j.physletb.2016.05.066
27. Frederico, T., Salmè, G., Viviani, M.: Solving the inhomogeneous Bethe-Salpeter equation in Minkowski space: the zero-energy limit. Eur. Phys. Jou. C **75**(8), 398 (2015). https://doi.org/10.1140/epjc/s10052-015-3616-1
28. Gigante, V., Nogueira, J.H.A., Ydrefors, E., Gutierrez, C., Karmanov, V.A., Frederico, T.: Bound state structure and electromagnetic form factor beyond the ladder approximation. Phys. Rev. D **95**(5), 056012 (2017). https://doi.org/10.1103/PhysRevD.95.056012

Chapter 93
Time Reversal Violation in Two and Three Nucleon Systems

Alex Gnech and Michele Viviani

Abstract Time reversal violation sources in fundamental theories induce interactions between nucleons which can be revealed by looking at the presence of permanent electric dipole moments of light nuclei. In this work we derive the time reversal violation potential up to the N2LO in the chiral effective field theory. Using this potential we compute the electric dipole moments of ^{2}H and ^{3}H. Our results, combined with experimental data, can be used to test various possible sources of time reversal violation.

93.1 Introduction

Time Reversal Violation (TRV) and Parity Violation (PV) are key ingredients in the explanation of the observed baryon-antibaryon asymmetry in the Universe (BAU) [1]. The Standard Model (SM) has a natural source of CP-violation in the Cabibbo-Kobayashi- Maskawa (CKM) quark mixing matrix, however this mechanism is not sufficient to explain the observed value of BAU [2]. This discrepancy opens a window in possible TRV effect in extension of SM, such as the θ-term in the Quantum Chromodynamics (QCD) sector [3], or in beyond-SM (BSM) theories [4].

The measurement of Electric Dipole Moments (EDMs) of particles is the most promising observable for studying TRV effects beyond CKM mixing matrix. The present experimental upper bounds on the EDMs are $|d_n| < 2.9 \times 10^{-13}$ e fm for neutron [5], $|d_p| < 7.9 \times 10^{-12}$ e fm for proton [6, 7], and $|d_e| < 8.7 \times 10^{-16}$ e fm for electron [8]. In this context, there are proposals of direct measurements of EDMs of charged particles in dedicated storage rings [9–13]. However, a single measurement would not be sufficient to identify the source of TRV. For this reason, the measurement of EDMs of various light nuclei such as ^{2}H, ^{3}H and ^{3}He can help to impose constrains on the TRV sources.

A. Gnech (✉)
Gran Sasso Science Institute, I-67100 L'Aquila, Italy
e-mail: alex.gnech@gssi.it

A. Gnech · M. Viviani
Istituto Nazionale di Fisica Nucleare, sezione di Pisa, I-56100 Pisa, Italy

N. A. Orr et al. (eds.), *Recent Progress in Few-Body Physics*,
Springer Proceedings in Physics 238,
https://doi.org/10.1007/978-3-030-32357-8_93

The use of light nuclei as probes for TRV results to be advantageous because the nuclear physics of the systems is theoretically under control. In particular, the chiral effective field theory (χEFT) provides a practical scheme to study TRV nuclear effects treating all the possible sources [14–16]. Each Lagrangian term is associated to a low-energy constants (LECs) which must be determined fitting the experimental data. The χEFT approach permits not only to determine the TRV interactions but also to determine the chiral order of the LECs and their values as function of the fundamental parameters providing a direct connection between the fundamental theories and the nuclear observables [14–16].

Starting from the Lagrangian of [14–16] adding only the isotensor interactions we derived the chiral potential up to next-to-next-leading order (N2LO). We use this potential to study the EDM of ^{2}H, ^{3}H and ^{3}He focusing in particular in the error related to the truncation of the chiral expansion. Such an approach provides a suitable framework for the future determination of the LECs.

93.2 The Nuclear TRV Potential

The relevant terms of the Lagrangian which can give contribution to the nuclear potential up to N2LO are [14–16],

$$
\begin{aligned}
\mathcal{L}_{\text{TRV}} = {} & g_0 \overline{\psi} \mathbf{ß} \cdot \boldsymbol{ø} \psi + g_1 \overline{\psi} \pi_3 \psi + g_2 \overline{\psi} \pi_3 \tau_3 \psi + M \Delta \pi_3 \pi^2 \\
& -2i \overline{\psi} (d_0 + d_1 \tau_z) \gamma_5 \sigma^{\mu\nu} \psi F_{\mu\nu} + \mathcal{L}_{\text{TRV}}^{NN} ,
\end{aligned}
\tag{93.1}
$$

where $\mathcal{L}_{\text{TRV}}^{NN}$ includes five contact interactions which permits us to take care of the exchange of heavier mesons and reabsorb the divergences in the potential. Further terms of the Lagrangian involving derivatives would give rise to potential terms which can be reabsorbed in those generated by the Lagrangian given in (93.1) [17]. Moreover, $d_p = (d_0 + d_1)/2$ and $d_n = (d_0 - d_1)/2$ are the proton and neutron EDM.

We evaluate the T matrix in terms of time-ordered perturbation theory amplitudes whose associated diagrams are shown in Fig. 93.1. The nuclear potential was then derived from the T matrix by inverting order by order in the power counting the Lippman-Schwinger equation. he leading order (LO) of the potential is given by the one pion exchange (OPE) associated to the LECs g_0, g_1 and g_2. The two pion exchange contribution coming from these LECs appears only for g_0 and g_2 as already observed in [18]. The three pion vertex (TPV) gives rise to contribution at order Q^0 but also of order Q^1 when the LECs c_1, c_2 and c_3 of the parity conserving (PC) Lagrangian, are taken into account. The TPV vertex generates also three body forces at order Q^0 while at order Q^1 all the time ordered diagrams cancel. The contribution of the contact terms appear at N2LO while the single nucleon contribution at LO. The final expression of the potential contains 11 LECs which must be determine from the experiment. The potential is then regularized introducing a regularization function $C_\Lambda(k) = \mathrm{e}^{-(k/\Lambda)^4}$. Three cutoff values are considered $\Lambda = 450, 500, 550$ MeV.

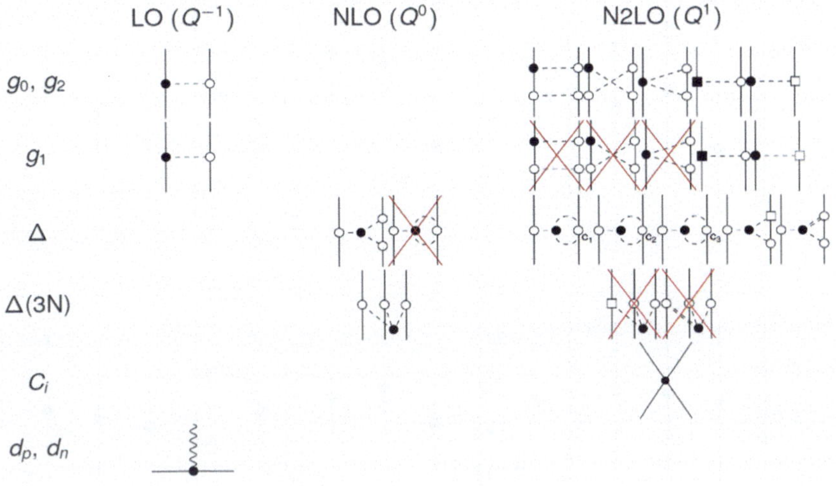

Fig. 93.1 Time-ordered diagrams contributing to the TRV potential. Nucleons and pions are denoted by solid and dashed lines, respectively. The open (solid) circle represents a PC (TRV) vertex. The squares represent NLO vertices, while the double dashed lines recoil corrections coming from the energy denominators

93.3 The EDM of ^{2}H and ^{3}H

In this section we present the results for the EDM of ^{2}H and ^{3}H. The calculation for ^{3}He is in progress. The EDM of each nucleus can be expressed as the product of the LECs with numerical coefficients which contains all the dynamics given by the potential, namely,

$$d^A = g_0 a_0 + g_1 a_1 + g_2 a_2 + \Delta\, a_\Delta + d_p a_p + d_n a_n + \sum_{i=1,5} C_i A_i\,. \qquad (93.2)$$

where the C_i are the LECs which comes from the contact terms. In the calculation of the coefficients we use the various chiral order of the NN PC potential of [19] and for ^{3}H the NNN PC potential at N2LO [20]. In particular we focus on the errors associated to the truncation of the chiral expansion as,

$$(\delta a_i)^2 = (\delta a_i^{\rm PC})^2 + (\delta a_1^{\rm TRV})^2 + (\delta a_i^{\psi})^2 \qquad (93.3)$$

where $\delta a_i^{\rm PC}$ ($\delta a_i^{\rm TRV}$) are the error associated PC (TRV) potential evaluated using the prescriptions of [21]. Moreover for ^{3}H we introduce an error related to the uncertainties on the wave function δa_i^{ψ}.

The results for some of the coefficients of ^{3}H are shown in Fig. 93.2 while for the ^{2}H in Fig. 93.3. The EDM of ^{2}H receives contribution only from the isovector term and

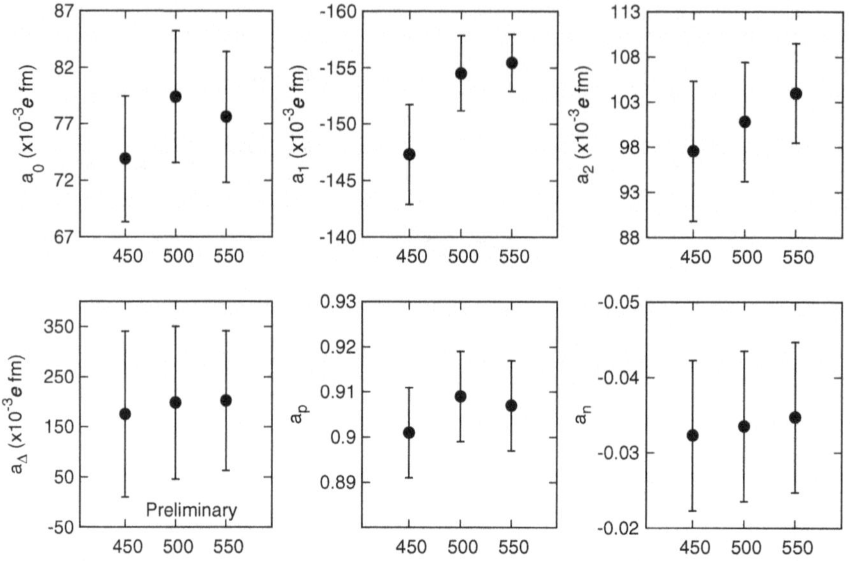

Fig. 93.2 Values of selected coefficients a_i for ^{3}H. The error bars are evaluated following [21]

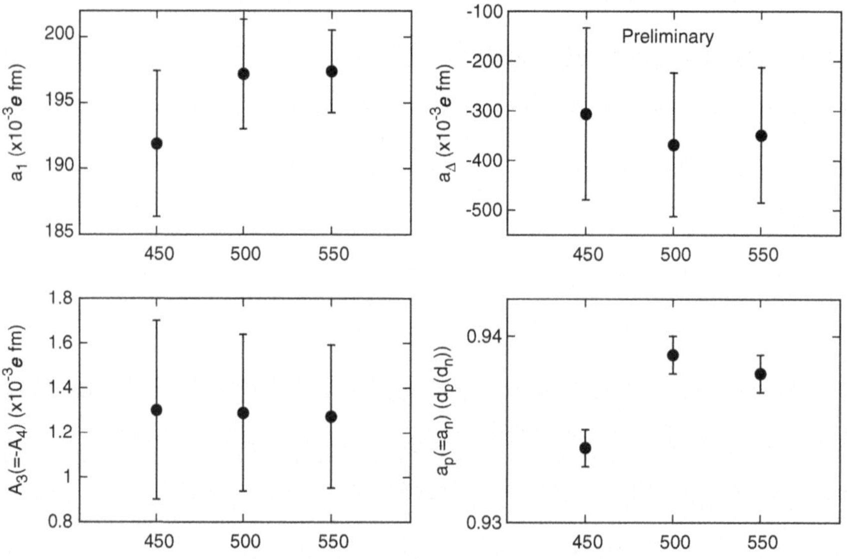

Fig. 93.3 The same as Fig. 93.2 for ^{2}H

the single nucleon EDM while the EDM of ^{3}H from all the LECs. Note the huge errors reported for the coefficients a_Δ which are due to the fact that the N2LO diagrams give a correction of $\sim 70\%$. This effect is mainly due to the large contribution of the c_1, c_2, and c_3 LECs [17]. For ^{3}H, the calculation of the coefficients a_Δ does not include yet the TRV three-body forces. As regarding the cutoff dependence, in both the Figs. 93.2 and 93.3 we observe a nice agreement for the results obtained with $\Lambda = 500$ and 550 MeV while the results obtained with $\Lambda = 450$ MeV seems always at variance compared to the other two cutoff. However the results are consistent within the theoretical uncertainties. More detailed calculations of these observables are in progress.

References

1. Sakarhov, A.D.: Pisma. Zh. Eksp. Teor. Fiz. **5**, 32 (1967)
2. Cohen, A.G., Kaplan, D.B., Nelson, A.E.: Ann. Rev. Nucl. Part. Sci. **43**, 27 (1993)
3. Hooft, G.T.: Phys. Rev. Lett. **37**, 8 (1976)
4. Pospelov, M., Ritz, A.: Ann. Phys. **318**, 119 (2005)
5. Baker, C.A., et al.: Phys. Rev. Lett. **97**, 131801 (2006)
6. Griffith, W.C., Swallows, M.D., Loftus, T.H., Romalis, M.V., Heckel, B.R., Fortson, E.N.: Phys. Rev. Lett. **102**, 101601 (2009)
7. Dmitriev, V.F., Senkov, R.A.: Phys. Rev. Lett. **91**, 212303 (2003)
8. Baron, J., et al.: Science **343**(6168), 269 (2014)
9. Orlov, Y.F., Morse, W.M., Semertzidis, Y.K.: Phys. Rev. Lett. **96**, 214802 (2006)
10. Semertzidis, Y.K.: Proceedings of the DPF 2011 Conference (2011)
11. Lehrach, A., Lorentz, B., Morse, W., Nikolaev, N., Rathmann, F.: arXiv:1201.5773 (2013)
12. Pretz, J.: Hyperfine Interact. **214**(1–3), 111 (2013)
13. Rathmann, F., Saleev, A., Nikolaev, N.N.: J. Phys. Conf. Ser. **447**, 012011 (2013)
14. Mereghetti, E., Hockings, W.H., van Kolck, U.: Ann. Phys. **325**, 2363 (2010)
15. de Vries, J., Mereghetti, E., Timmermans, R.G.E., van Kolck, U.: Ann. Phys. **338**, 50 (2013)
16. Bsaisou, J., Meissner, U.-G., Nogga, A., Wirzba, A.: Ann. Phys. **359**, 317 (2015)
17. Gnech, A., Viviani, M.: In preparation
18. Maekawa, C.M., Mereghetti, E., de Vries, J., van Kolck, U.: Nucl. Phys. A **872**, 117 (2011)
19. Entem, D.R., Kaiser, N., Machleidt, R., Nosyk, Y.: Phys. Rev. C **91**, 014002 (2015)
20. Marcucci, L.E., Sammarruca, F., Viviani, M., Machleidt, R.: arXiv:1809.01849
21. Epelbaum, E., Krebs, H., Meissner, U.-G.: Phys. Rev. Lett. **115**, 122301 (2015)

Chapter 94
Lattice Simulations with Chiral Effective Field Theory for Light and Medium-Mass Nuclei

Serdar Elhatisari

Abstract In this proceedings we present recent results from lattice simulations with chiral effective field theory up to next-to-next-to-next-to-leading order. We discuss our investigation on the degree of locality of the short-range nucleon-nucleon interactions. We also discuss ground state energies of light and medium-mass nuclei as well as new algorithms for the proton and neutron density distributions and other properties.

94.1 Introduction

Lattice effective field theory is a powerful numerical method formulated in the framework of chiral effective field theory which organizes the nuclear interactions as an expansion in powers of low energy scales, Q, such as the momenta, the pion mass etc. Chiral effective field theory gives a modern description for the nuclear forces in the chiral limit where the light quarks are massless. A relevant recent review can be found in [1]. In the chiral expansion the first term dominates and is called the leading order (LO or Q^0) interaction. The first correction to the LO is the next-to-leading order (NLO or Q^2) interaction, the second correction is called the next-to-next-to-leading order (NNLO or Q^3) and so on. These interactions contain sets of coupling constants (or low-energy constants (LECs)) to be determined by fitting to the experimental data.

In lattice effective field theory these interactions are formulated in a periodic cubic lattice, and the LECs on the lattice are determined by fitting to experimental data. Reference [2] discusses the details of lattice interactions and a new lattice formulation

Nuclear Lattice Effective Field Theory Collaboration

S. Elhatisari (✉)
Karamanoglu Mehmetbey University, 70100 Karaman, Turkey
e-mail: elhatisari@hiskp.uni-bonn.de

Helmholtz-Institut für Strahlen- und Kernphysik (Theorie) and Bethe Center for Theoretical Physics, Universität Bonn, D-53115 Bonn, Germany

N. A. Orr et al. (eds.), *Recent Progress in Few-Body Physics*,
Springer Proceedings in Physics 238,
https://doi.org/10.1007/978-3-030-32357-8_94

of short-range chiral effective field theory interactions with a simpler decomposition into spin channels.

94.2 Euclidean Time Projection Monte Carlo

In lattice simulations we study the low-lying states of nuclei using the normal ordered transfer-matrix formalism. The transfer matrix is defined in Euclidean time as in the following,

$$M_{\rm LO} = : \exp(-\alpha_t\, H_{\rm LO}) : \tag{94.1}$$

where H is the lattice Hamiltonian, and α_t is the ratio of the temporal lattice spacing a_t to the spatial lattice spacing a. The symbol :: signifies normal ordering, which moves all annihilation operators to the right and creation operators to the left with the appropriate number of anticommutation minus signs. In our simulation we employ projection Monte Carlo with auxiliary fields for nucleon-nucleon interactions. In auxiliary-field Monte Carlo simulations, the interactions are recast as single particle interacting with fluctuating auxiliary fields. See [3] for a detailed discussion on auxiliary-field Monte Carlo calculations.

Nuclear structure on the lattice: To compute the ground state energies or the properties of nuclei, we consider some initial and final states, respectively $|\Psi_i\rangle$ and $|\Psi_f\rangle$, as Slater determinants of free-particle standing waves on the lattice. These states are projected in Euclidean time using the transfer matrix to form the Euclidean time projection amplitude at LO,

$$Z_{\rm LO}(L_t) = \langle \Psi_f |\, M_{\rm LO}^{L_t}\, |\Psi_i\rangle\,. \tag{94.2}$$

We perform the auxiliary-field Monte Carlo simulations to compute the quantum amplitude $Z_{\rm LO}(L_t)$, and the ground state energy at LO is determined from the ratio $Z_{\rm LO}(L_t+1)/Z_{\rm LO}(L_t)$ in the limit $L_t \to \infty$. In our simulation the higher order calculations are computed using perturbation theory, and we compute the Euclidean time projection amplitude at higher order,

$$Z_{\rm ho}(L_t+1) = \langle \Psi_f |\, M_{\rm LO}^{(L_t-1)/2}\, M_{\rm ho}\, M_{\rm LO}^{(L_t-1)/2}\, |\Psi_i\rangle\,, \tag{94.3}$$

where

$$M_{\rm ho} = : \exp[-\alpha_t\,(\,H_{\rm LO} + \,H_{\rm ho})]\,. \tag{94.4}$$

Therefore, the energy correction to the LO energy is computed from the ratio $Z_{\rm ho}(L_t+1)/Z_{\rm LO}(L_t+1)$.

Nuclear scattering on the lattice: Scattering and reactions involving clusters can be studied on the lattice using the adiabatic projection method, which is a general framework that constructs a low energy effective field theory for the clusters. See [4–6] for the details of the method. The method uses initial states,

$$|\mathbf{R}\rangle = \sum_{\mathbf{r}} |\mathbf{r}+\mathbf{R}\rangle_1 \otimes |\mathbf{r}\rangle_2 \,, \tag{94.5}$$

which is parameterized by the relative separation between clusters, $\mathbf{R}$. Furthermore, we project these initial states onto spherical harmonics Y_{ℓ,ℓ_z} with angular momentum numbers ℓ, ℓ_z,

$$|R\rangle^{\ell,\ell_z} = \sum_{\mathbf{R}'} Y_{\ell,\ell_z}(\hat{R}')\, \delta_{R,|\mathbf{R}'|}\, |\mathbf{R}\rangle \,. \tag{94.6}$$

Then we evolve these states using the LO transfer matrix in Euclidean time to form dressed cluster states,

$$|R\rangle^{\ell,\ell_z}_{n_t} = M^{n_t}_{\mathrm{LO}}\, |R\rangle^{\ell,\ell_z} \,. \tag{94.7}$$

These dressed cluster states are used to construct the transfer matrix of the cluster-cluster system,

$$[M_{n_t}]^{\ell,\ell_z}_{R',R} = {}^{\ell,\ell_z}_{n_t}\langle R' \,| M_{\mathrm{LO}} |\, R\rangle^{\ell,\ell_z}_{n_t} , \tag{94.8}$$

and the norm matrix,

$$[N_{n_t}]^{\ell,\ell_z}_{R',R} = {}^{\ell,\ell_z}_{n_t}\langle R' \mid R\rangle^{\ell,\ell_z}_{n_t} . \tag{94.9}$$

The adiabatic projection in Euclidean time gives a systematically improvable description of the low-lying scattering states of clusters, and in the limit of large Euclidean time the description becomes exact. We use the auxiliary-field Monte Carlo simulations to compute the amplitude matrices in (94.8) and (94.9). In addition, we perform metropolis sampling of the cluster positions. Then we use (94.8) and (94.9) to construct the adiabatic transfer matrix,

$$[M^a_{n_t}]^{\ell,\ell_z}_{R',R} = \left[N_{n_t}^{-\frac{1}{2}} M_{n_t} N_{n_t}^{-\frac{1}{2}} \right] , \tag{94.10}$$

and by employing the spherical wall method [4, 7] the adiabatic transfer matrix is used to compute the scattering phase shifts for two-cluster systems. The computational scaling of lattice calculations consisting of A_1-body and A_2-body clusters is roughly $(A_1 + A_2)^2$, and this makes *ab initio* calculations involving a heavier projectile accessible and practical.

94.3 Results

Alpha-alpha scattering: In this section, we present the recent results from the lattice simulations. We start with the first *ab initio* calculation of the elastic ^{4}He+^{4}He scattering using the adiabatic projection method. As described above, thanks to mild computational scaling, *ab initio* calculations involving ^{4}He cluster as a projectile are possible with the adiabatic projection method. We used the lattice action developed and used in [8], and performed the first *ab initio* calculation of ^{4}He+^{4}He scattering up to next-to-next-to-leading order in chiral effective field theory [5]. In these calculations the spatial lattice spacing is $a = 1.97$ fm and the temporal lattice spacing is $a_t = 1.32$ fm.

Figure 94.1 shows the S-wave (left) and the D-wave (right) scattering phase shifts versus laboratory energy up to NNLO in chiral effective field theory comparison with experimental data [9]. We found that for the S-wave the NNLO result is in good agreement with experiment, and we found a fairly good agreement between the D-wave the NNLO result and experiment.

The adiabatic projection method is of significant importance not only because the ^{4}He nuclei has been the heaviest projectile used in the ab initio calculations of scattering and reactions, but also has opened the door towards using experimental data from collisions of heavier nuclei as input to improve ab initio nuclear structure theory.

Degree of locality of nuclear forces: In [10] we used the ^{4}He+^{4}He scattering as a tool for probing the degree of locality of the short-range nuclear interactions and the nuclear structure of alpha-conjugate nuclei which are nuclei with equal and even numbers of protons and neutrons. We started with two leading order lattice interactions $V_{\rm A}(\mathbf{r}', \mathbf{r})$ and $V_{\rm B}(\mathbf{r}', \mathbf{r})$ where $\mathbf{r}$ is the spatial separation between the two incoming nucleons and $\mathbf{r}'$ is the spatial separation between the two outgoing

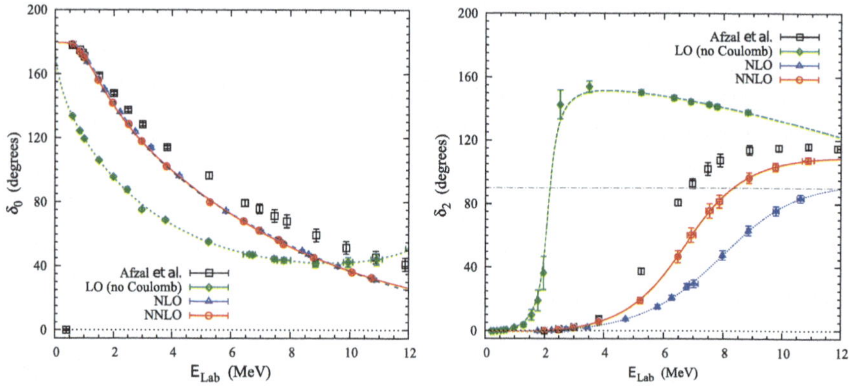

Fig. 94.1 Left: S-wave phase shifts up to NNLO and comparison with experimental data [9]. Right: D-wave phase shifts up to NNLO and comparison with experimental data [9]

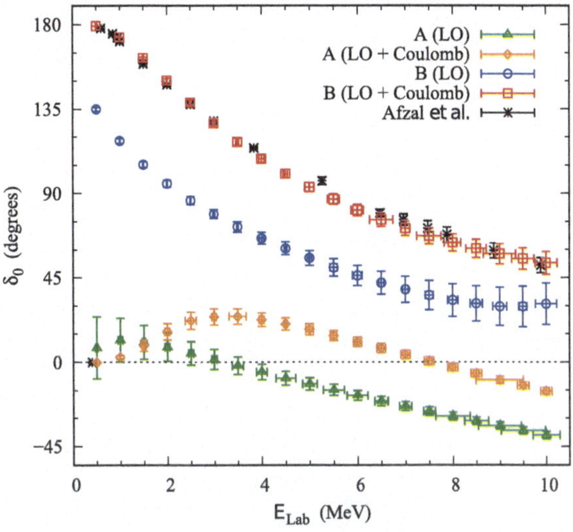

Fig. 94.2 ^{4}He+^{4}He S-wave scattering. We plot S-wave phase shifts δ_0 for alpha-alpha scattering for interactions $V_A(\mathbf{r}', \mathbf{r})$ and $V_B(\mathbf{r}', \mathbf{r})$ versus laboratory energy. We show LO for A (green tiangle), LO + Coulomb for A (orange diamonds), LO for B (blue circles), and LO + Coulomb results for B (red squares) comparison with experimental data [9]

nucleons. The interaction $V_A(\mathbf{r}', \mathbf{r})$ includes non-local short-range interactions, while the interaction $V_B(\mathbf{r}', \mathbf{r})$ consists of both non-local and local short-range interactions. We tuned the interactions $V_A(\mathbf{r}', \mathbf{r})$ and $V_B(\mathbf{r}', \mathbf{r})$ to produce the experimental low-energy nucleon-nucleon scattering phase shifts, while the extra parameters of the interaction $V_B(\mathbf{r}', \mathbf{r})$ due to the local terms were tuned to give the alpha-alpha S-wave scattering phase shifts. Figure 94.2 shows the ^{4}He+^{4}He S-wave scattering phase shifts as a function of laboratory energy for interactions $V_A(\mathbf{r}', \mathbf{r})$ and $V_B(\mathbf{r}', \mathbf{r})$. The results from Fig. 94.2 clearly shows that the ^{4}He+^{4}He scattering phase shifts are highly sensitive to the degree of locality of the short-range nuclear interactions. Reference [10] explains these results in detail closely looking at the structure of the ^{4}He wave function.

For interactions $V_A(\mathbf{r}', \mathbf{r})$ and $V_B(\mathbf{r}', \mathbf{r})$ we computed the ground state energies of ^{3}H, ^{3}He, ^{4}He as well as alpha-conjugate nuclei ^{8}Be, ^{12}C, ^{16}O, ^{20}Ne given in Table 94.1. We found that nuclei up to ^{8}Be are equally well described by both interactions. For the interaction $V_B(\mathbf{r}', \mathbf{r})$ the results of nuclei heavier than ^{8}Be are in agreement with the experimental data, while they are underbound for the interaction $V_A(\mathbf{r}', \mathbf{r})$. Also the ^{4}He+^{4}He scattering phase shifts for interaction $V_A(\mathbf{r}', \mathbf{r})$ is very weak as given in Fig. 94.2. To illuminate what is going on with the interaction $V_A(\mathbf{r}', \mathbf{r})$ we found it useful to look at the ratio of the LO energy for each of the alpha-conjugate nuclei to that of the ^{4}He particle. The results for the ratios are 1.997(6), 3.00(1), 4.00(2), and 5.03(3) for ^{8}Be, ^{12}C, ^{16}O, and ^{20}Ne, respectively. The important result revealed here is that in each case the interaction $V_A(\mathbf{r}', \mathbf{r})$ forms a weakly-interacting Bose gas of alpha particles.

In this study we found that the correct description of the ^{4}He+^{4}He scattering phase shifts play a crucial role to describe alpha-conjugate nuclei well. In order to understand the many-body limit in details, we switched off the Coulomb interactions

Table 94.1 The lattice results for the ground state energies of ^{3}H, ^{3}He, ^{4}He, ^{8}Be, ^{12}C, ^{16}O, ^{20}Ne from the interactions $V_{\mathrm{A}}(\mathbf{r}', \mathbf{r})$ and $V_{\mathrm{B}}(\mathbf{r}', \mathbf{r})$. All energies are in units of MeV

Nucleus	A (LO)	B (LO)	A (LO + Coulomb)	B (LO + Coulomb)	Experiment
^{3}H	−7.82(5)	−7.78(12)	−7.82(5)	−7.78(12)	−8.482
^{3}He	−7.82(5)	−7.78(12)	−7.08(5)	−7.09(12)	−7.718
^{4}He	−29.36(4)	−29.19(6)	−28.62(4)	−28.45(6)	−28.296
^{8}Be	−58.61(14)	−59.73(6)	−56.51(14)	−57.29(7)	−56.591
^{12}C	−88.2(3)	−95.0(5)	−84.0(3)	−89.9(5)	−92.162
^{16}O	−117.5(6)	−135.4(7)	−110.5(6)	−126.0(7)	−127.619
^{20}Ne	−148(1)	−178(1)	−137(1)	−164(1)	−160.645

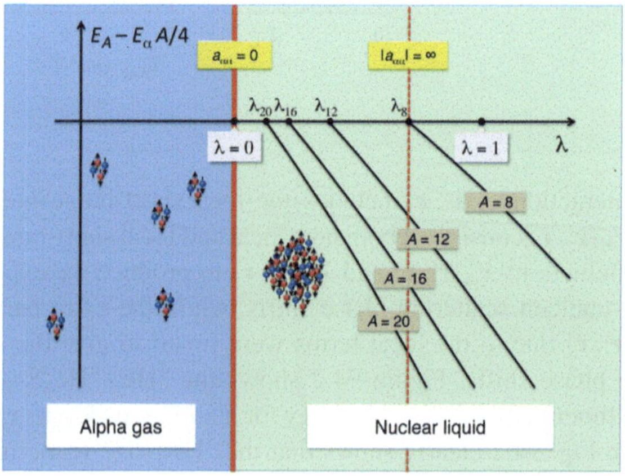

Fig. 94.3 Zero-temperature phase diagram as a function of the parameter λ in the nuclear interaction $V_\lambda = (1 - \lambda)V_A + \lambda V_B$. A first-order quantum phase transition from a Bose gas to nuclear liquid at the point appears where the alpha-alpha scattering length crosses zero. This is very close to the value $\lambda = 0$. Also shown are the alpha-conjugate nuclear ground state energies E_A for A nucleons up to $A = 20$ relative to the corresponding multi-alpha threshold $E_\alpha A/4$. The last alpha-conjugate nucleus to be bound is ^{8}Be at the unitarity point

and connect the interaction $V_{\mathrm{A}}(\mathbf{r}', \mathbf{r})$ to the interaction $V_{\mathrm{B}}(\mathbf{r}', \mathbf{r})$ by a simple interpolation, $V_\lambda = (1 - \lambda)V_A + \lambda V_B$. We did not observe any significant change in the properties of the two-, three-, and four-nucleon systems with λ, while the many-body ground state of the interpolated interaction V_λ undergoes a quantum phase transition from a Bose-condensed gas to a nuclear liquid. A schematic view of the zero temperature phase diagram is shown in Fig. 94.3.

Density profiles for nuclei: The simulations with auxiliary-field Monte Carlo methods involve quantum states that are superposition of many different center-of-mass

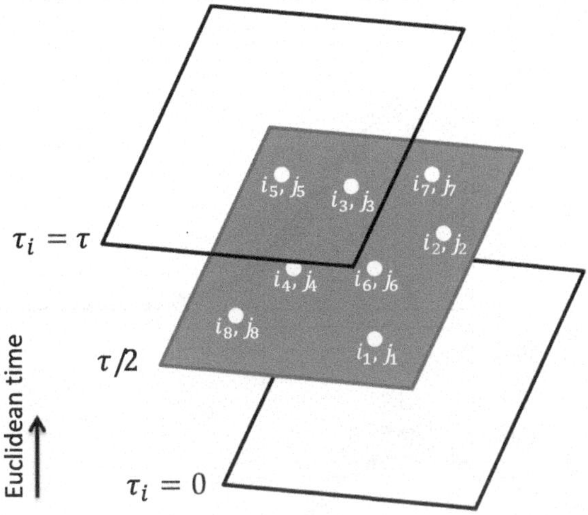

Fig. 94.4 A sketch of the pinhole locations and spin-isospin indices at time $\tau/2 = L_t a_t/2$

positions. Therefore, the density distrubitions of the nucleons cannot be computed direclty. To solve this problem we developed and introduced a new computational approach called the pinhole algorithm [11], which solves a long-standing deficiency of auxiliary-field Monte Carlo simulations in computing density correlations relative to the center of mass.

In this algorithm we consider a screen placed at the middle time step having pinholes with spin and isospin labels that allow nucleons with the corresponding spin and isospin to pass. This screen corresponds to the insertion of the normal-ordered A-body density operator at the middle time step,

$$\rho_{i_1,j_1,\dots,i_A,j_A}(\mathbf{n}_1,\dots,\mathbf{n}_A) =: \rho_{i_1,j_1}(\mathbf{n}_1)\ \dots\ \rho_{i_A,j_A}(\mathbf{n}_A) : \qquad (94.11)$$

where $\rho_{i,j}(\mathbf{n}) = a^{\dagger}_{i,j}(\mathbf{n})a_{i,j}(\mathbf{n})$ is the density operator for nucleon with spin i and isospin j. Figure 94.4 shows a sketch of the pinhole locations and spin-isospin indices for the operator $\rho_{i_1,j_1,\dots,i_A,j_A}(\mathbf{n}_1,\dots,\mathbf{n}_A)$ inserted at time $\tau/2$. The screen has A pinholes for a simulation consist of A nucleons, and we perform Metropolis sampling for the locations as well as the spin and isospin labels of the pinholes. Using the pinhole algorithm, we have computed the proton and neutron densities for the ground states of ^{12}C, ^{14}C, and ^{16}C given in Fig. 94.5.

Ground state energies for light and medium-mass nuclei: Recently we have constructed a set of short-range chiral effective field theory interactions on the lattice with a simpler decomposition into spin channels. Li et al. [2] presents the full details of these lattice interactions and the results for the neutron-proton scattering on the lattice with various lattice spacings comparision with the empirical phase shifts.

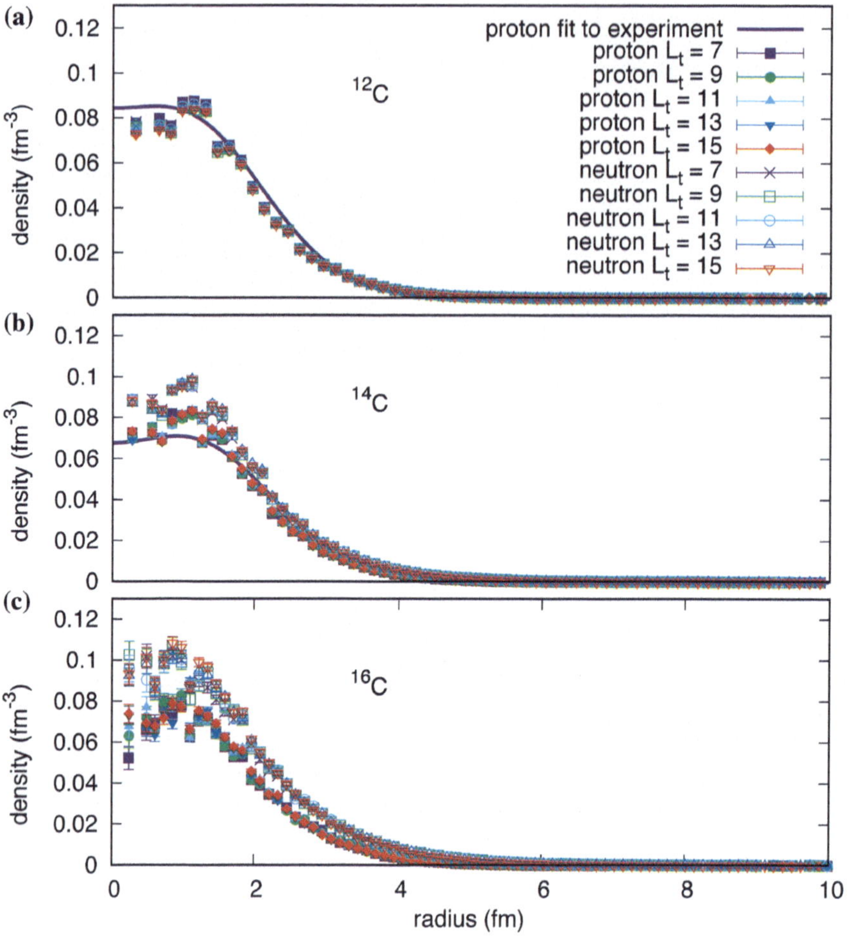

Fig. 94.5 Plots of the proton and neutron densities as function of radial distance for the ground states of ^{12}C, ^{14}C, and ^{16}C comparison with the experimentally observed proton densities for 12C and 14C [12]. We show data for various L_t time steps

Using the lattice action developed in [2], we have studied the neutron-proton scattering to determine the LECs and computed the ground states energies of light and medium-mass nuclei at lattice spacing $a = 1.97$ fm. Figures. 94.6 and 94.7 show the neutron-proton scattering phase shifts comparision with the empirical phase shifts. Lattice results for ground state energies of light and medium-mass nuclei up to N3LO in chiral effective field theory are given in Fig. 94.8 [13]. We stress that these results are preliminary. Also these results do not include any three-body force, and the relevant work is in progress.

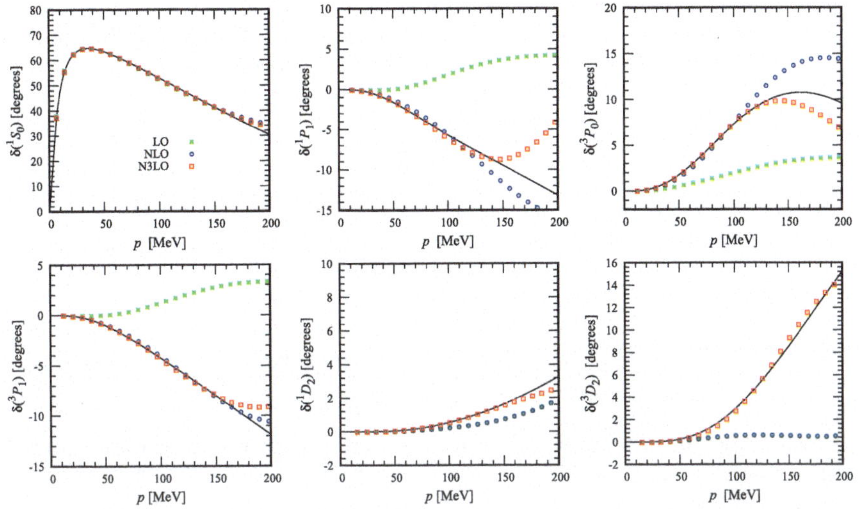

Fig. 94.6 Neutron-proton scattering uncoupled channel phase shifts as function of relative momenta at lattice spacing $a = 1.97$ fm

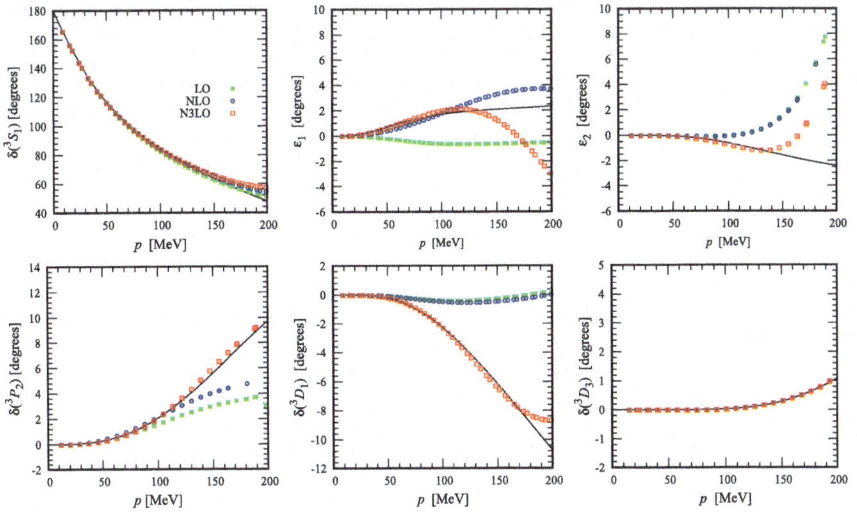

Fig. 94.7 Neutron-proton scattering coupled channel phase shifts and mixing angles as function of relative momenta at lattice spacing $a = 1.97$ fm

Summary We have reported recent results from lattice simulations with chiral effective field theory up to next-to-next-to-next-to-leading order. We have also discussed our investigation on the degree of locality of the short-range nucleon-nucleon interactions, and algorithm that we developed recently to study the proton and neutron density distributions.

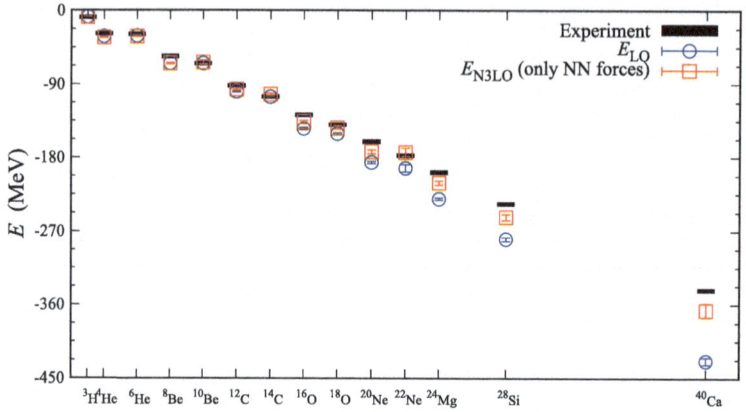

Fig. 94.8 PRELIMINARY RESULTS: Lattice results for ground state energies of light and medium-mass nuclei up to N3LO in chiral effective field theory without the three-body forces

Acknowledgements The author is grateful to his collaborators, Jose Manuel Alarcón, Dechuan Du, Evgeny Epelbaum, Nico Klein, Hermann Krebs, Timo A. Lähde, Dean Lee, Ning Li, Bing-nan Lu, Thomas Luu, Ulf-G. Meißner, Gautam Rupak, and Alexander Rokash, for their contributions to the work presented in this proceedings. We acknowledge partial financial support from the Deutsche Forschungsgemeinschaft (SFB/TR 110, "Symmetries and the Emergence of Structure in QCD"), the U.S. Department of Energy (DE-SC0018638), and the Scientific and Technological Research Council of Turkey (TUBITAK project no. 116F400). Further support was provided by the Chinese Academy of Sciences (CAS) Presidents International Fellowship Initiative (PIFI) (Grant No. 2018DM0034) and by VolkswagenStiftung (Grant No. 93562). The computational resources were provided by the Julich Supercomputing Centre at Forschungszentrum Julich, Oak Ridge Leadership Computing Facility, RWTH Aachen, North Carolina State University, and Michigan State University.

References

1. Epelbaum, E., Hammer, H.W., Meißner, U.G.: Modern theory of nuclear forces. Rev. Mod. Phys. **81**, 1773 (2009). https://doi.org/10.1103/RevModPhys.81.1773
2. Li, N., Elhatisari, S., Epelbaum, E., Lee, D., Lu, B.N., Meißner, U.G.: Neutron-proton scattering with lattice chiral effective field theory at next-to-next-to-next-to-leading order. Phys. Rev. C **98**(4), 044002 (2018) https://doi.org/10.1103/PhysRevC.98.044002
3. Lee, D.: Lattice simulations for few- and many-body systems. Prog. Part. Nucl. Phys. **63**, 117 (2009). https://doi.org/10.1016/j.ppnp.2008.12.001
4. Rokash, A., Pine, M., Elhatisari, S., Lee, D., Epelbaum, E., Krebs, H.: Scattering cluster wave functions on the lattice using the adiabatic projection method. Phys. Rev. C **92**(5), 054612 (2015) https://doi.org/10.1103/PhysRevC.92.054612
5. Elhatisari, S., Lee, D., Rupak, G., Epelbaum, E., Krebs, H., Lähde, T.A., Luu, T., Meißner, U.G.: Ab initio alpha-alpha scattering. Nature **528**, 111 (2015). https://doi.org/10.1038/nature16067
6. Elhatisari, S., Lee, D., Meißner, U.G., Rupak, G.: Nucleon-deuteron scattering using the adiabatic projection method. Eur. Phys. J. A **52**(6), 174 (2016). https://doi.org/10.1140/epja/i2016-16174-2

7. Borasoy, B., Epelbaum, E., Krebs, H., Lee, D., Meißner, U.G.: Two-particle scattering on the lattice: phase shifts, spin-orbit coupling, and mixing angles. Eur. Phys. J. A **34**, 185 (2007). https://doi.org/10.1140/epja/i2007-10500-98
8. Epelbaum, E., Krebs, H., Lahde, T.A., Lee, D., Meißner, U.G.: Structure and rotations of the Hoyle state. Phys. Rev. Lett. **109**, 252501 (2012). https://doi.org/10.1103/PhysRevLett.109.252501
9. Afzal, S.A., Ahmad, A.A.Z., Ali, S.: Ystematic survey of the α-α interaction. Rev. Mod. Phys. **41**, 247 (1969). https://doi.org/10.1103/RevModPhys.41.247
10. Elhatisari, S., Li, N., Rokash, A., Alarcon, J.M., Du, D., Klein, N., Lu, B.N., Meißner, U.-G., Epelbaum, E., Krebs, H., Lähde, T.A., Lee, D., Rupak, G.: Nuclear binding near a quantum phase transition. Phys. Rev. Lett. **117**(13), 132501 (2016). https://doi.org/10.1103/PhysRevLett.117.132501
11. Elhatisari, S., Epelbaum, E., Krebs, H., Lähde, T.A., Lee, D., Li, N., Lu, B.N., Meißner, U.-G., Rupak, G.: Ab initio calculations of the isotopic dependence of nuclear clustering. Phys. Rev. Lett. **119**(22), 222505 (2017). https://doi.org/10.1103/PhysRevLett.119.222505
12. Kline, F.J., Crannell, H., O'Brien, J.T., Mccarthy, J., Whitney, R.R.: Elastic electron scattering from C-14. Nucl. Phys. A **209**, 381 (1973). https://doi.org/10.1016/0375-9474(73)90585-X
13. Nuclear lattice effective field theory collaboration: work in progress

Chapter 95
Light Nuclei from Lattice QCD: Spectrum, Structure and Reactions

Zohreh Davoudi

Abstract Lattice Quantum Chromodynamics (LQCD) studies of light nuclei have entered an era when first results on structure and reaction properties of light nuclei have emerged in recent years, complementing existing results on their lowest-lying spectra. Although in these preliminary studies the quark masses are still set to larger than the physical values, a few results at the physical point can still be deduced from simple extrapolations in the quark masses. The progress paves the road towards obtaining several important quantities in nuclear physics, such as nuclear forces and nuclear matrix elements relevant for pp fusion, single and double-β decay processes, neutrino-nucleus scattering, searches for CP violation, nuclear response in direct dark-matter detection experiments, as well as gluonic structure of nuclei for an Electron-Ion Collider (EIC) program. Some of the recent developments, the results obtained, and the outlook of the field will be briefly reviewed in this talk, with a focus on results obtained by the Nuclear Physics From LQCD (NPLQCD) collaboration.

95.1 One the Goals and Impact of a LQCD Program for Nuclear Physics

The standard approach in nuclear structure and reaction theory has shifted from relying on phenomenological nuclear potentials to studies based on nuclear effective field theories (EFTs), hence providing a systematic way to assess uncertainties of a calculation. In order for this program to succeed, not only the nuclear EFTs must offer a valid power counting with convergent and renormalization-scale independent results, but also their multitude of low-energy coefficients (LECs) must

For the NPLQCD collaboration

Z. Davoudi (✉)
Maryland Center for Fundamental Physics and Department of Physics, University of Maryland, College Park, MD 20742, USA
e-mail: davoudi@umd.edu

RIKEN Center for Accelerator-based Sciences, Wako 351-0198, Japan

N. A. Orr et al. (eds.), *Recent Progress in Few-Body Physics*, Springer Proceedings in Physics 238,
https://doi.org/10.1007/978-3-030-32357-8_95

be fit to experiment, so the EFTs can acquire a predictive power. In situations where experimental data are scarce or nonexistent, such as multi-neutron and hyperon-nucleon interactions, nuclear effects in the response of a nucleus to external probes, or nuclear matrix elements for the neutrinoless double-β decay of a nucleus, studies based on the underlying theory of quantum chromodynamics (QCD) are essential. Reliable predictions for a number of grand-challenge problems in nuclear physics (a few examples of which are enumerated in the chart in Fig. 95.1) will benefit from a coherent program in nuclear theory, in which the input from the underlying theory of QCD in the few-body sector provides the stepping stone for a nuclear many-body study based upon the constrained EFTs. A roadmap of this program is depicted in Fig. 95.1.

The only reliable method that enables QCD determination of observables in nuclear physics is LQCD, a method that relies on Monte Carlo sampling of the quantum fields in QCD, and provides n-point correlation functions obtained in a finite discretized Euclidean spacetime. Physical observables can be obtained in a

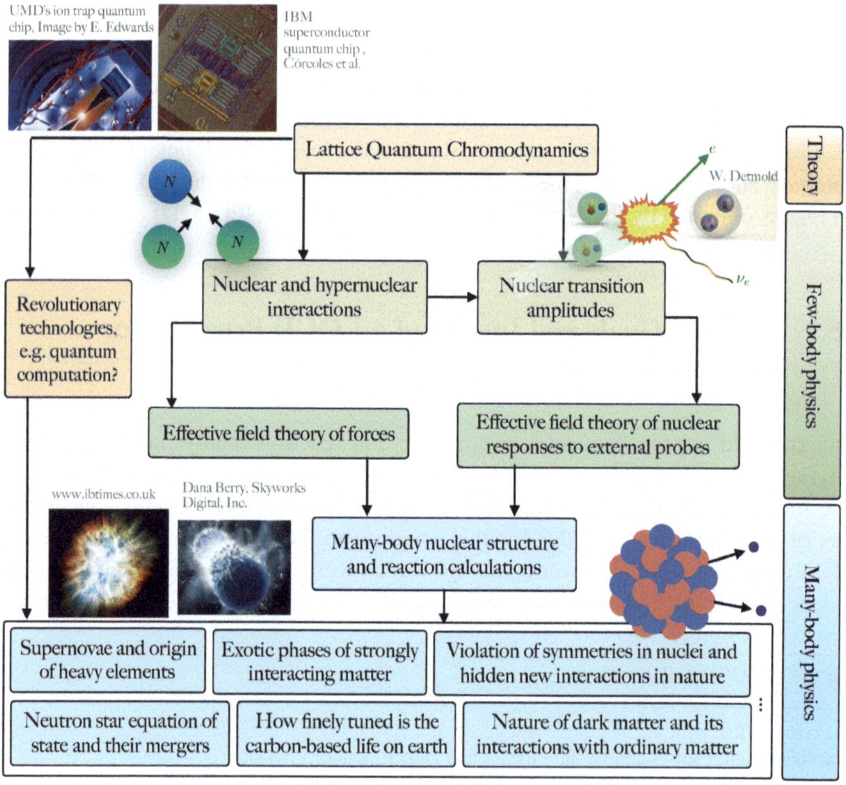

Fig. 95.1 A roadmap illustrating a systematic path from QCD to addressing grand-challenge problems in nuclear physics

systematic way using various extrapolations, or in the case of scattering amplitudes and transition rates, through mappings between the finite and infinite-volume theory. Heroic effort has been devoted in recent years to studies of multi-nucleon systems, considering the great computational complexity of these studies, and impressive progress has been made. While this short review can not do justice to the wealth of the results obtained in this area, I will focus on selected results by the NPLQCD collaboration on hadronic interactions, nuclear structure and nuclear reactions from LQCD. For a recent review of multi-nucleon physics from LQCD, see [1].

95.2 Hadronic Interactions

Constraining nuclear and hypernuclear forces remains a central component of research in nuclear physics. This effort complements experiments on neutron-rich isotopes, and provides the input to research on the nature of dense matter in astrophysical environments. A milestone for nuclear physics from LQCD was reached in 2012 when the emergence of light nuclei and hypernuclei from QCD was demonstrated in [2], albeit at larger-than-physical values of the quark masses, Fig. 95.2. This was enabled by algorithmic advances in forming nuclear correlation functions based on [3, 4] and the availability of computational resources. Such spectral studies at closer-to-physical values of the quark masses have since been con-

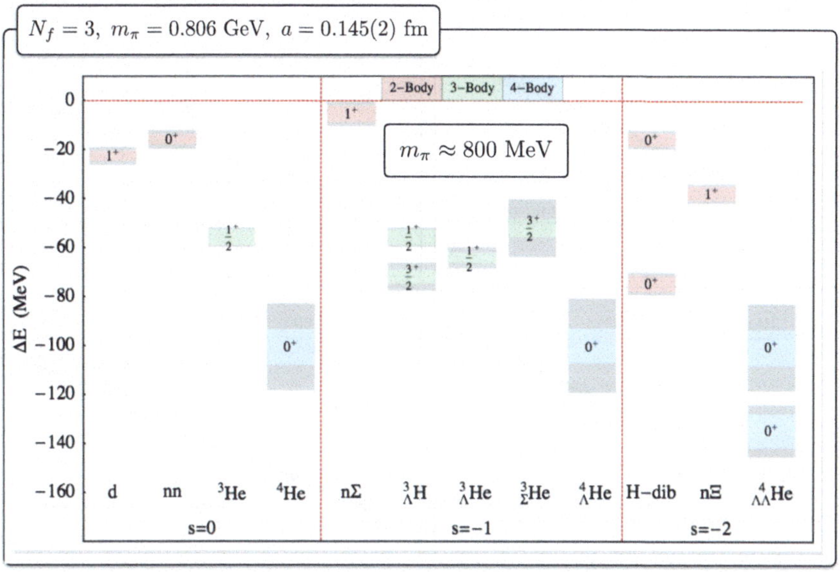

Fig. 95.2 The lowest-lying spectra of light nuclei and hypernuclei from LQCD obtained at larger-than-physical values of the quark masses [2]. The figure and all subsequent figures courtesy of the NPLQCD collaboration

ducted and appear promising. Further, a leap in the application of LQCD to nuclear physics was the realization [5] that large unnatural scattering lengths in two-nucleon systems is not an impediment in applying the powerful Lušcher's method [6]—a method that turns the finite-volume spectra obtained from LQCD to scattering amplitudes in the two- (and in recent extensions of the method to three-) body scattering amplitudes.

A recent example of such application is shown in Fig. 95.3, in which the SU(3) flavor-symmetric s-wave scattering phase shifts and the ground-state binding energies in four different scattering channels, corresponding to representative flavor channels $NN(^1S_0)$, $NN(^3S_1)$, $N\Sigma(^3S_1)$ and $N\Xi(^3S_1)$, were constrained, albeit at larger-than-physical values of the quark masses [7]. An interesting finding, arrived at by the observation of nearly identical scattering lengths and effective ranges in the four SU(3) flavor-symmetric channels, is that the low-energy interactions among two octet baryons exhibit not only a SU(6) spin-flavor symmetry that is argued to exist in QCD in the limit of a large number of colors [8], but also an extended SU(16) symmetry, which is now conjectured to be in place to minimize the entangling power of the S-matrix at low energies [9]. Further, this study demonstrates the matching between LQCD output and the EFT LECs, a program that can enable studies of larger systems of nucleons currently not accessible directly with LQCD.

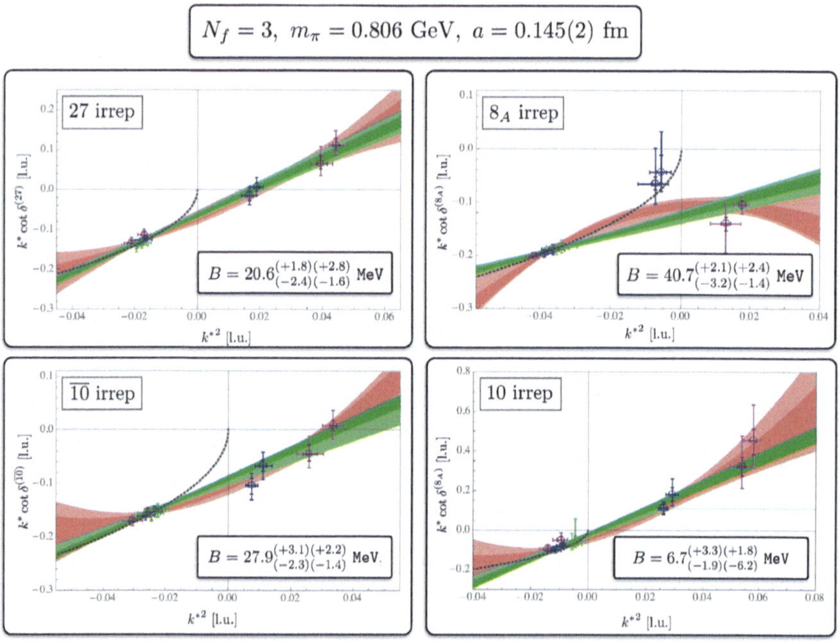

Fig. 95.3 Low-energy scattering phase shifts of various two octet-baryon channels along with the binding energy of the lowest-lying states, obtained with LQCD at larger-than-physical values of the quark masses [7]

95.3 Nuclear Structure

Investigations into the structure of hadrons and nuclei aim to provide further insight into the nature of strong dynamics. They are further essential in interpreting the outcome of high-energy collider experiments by providing a more accurate picture of the internal structure of the colliding protons or heavy ions. Like in experiment, certain electromagnetic (EM) properties of hadrons, such as magnetic moment, electric and magnetic polarizabilities and charge radii, can be deduced in a LQCD calculation from the response of the hadron to external EM fields. Such studies have been extended to light nuclei in recent years. As is shown in Fig. 95.4, the shift in the lowest finite-volume energy of proton, neutron, deuteron, ^{3}He and ^{3}H in an external magnetic field are used to deduce their magnetic moment, albeit at a larger-than-physical value of the quark masses [10]. When expressed in units of natural nuclear Magneton defined with the mass of the nucleon/nuclei at the corresponding value of the quark masses, they are surprisingly close to their values in nature, suggesting that much of the quark-mass dependence of the magnetic moment is captured by the quark-mass dependence of the mass. Additionally, it is observed that as in nature, nuclei at such large values of the quark masses still appear to behave as a collection of the nucleons, i.e., they can be described by a shell-model picture.

A further motivation for a nuclear structure program from LQCD is in supporting experiments in Fundamental Symmetries and Searches for New Physics. For example, LQCD studies of the matrix elements of scalar, axial and tensor currents in light

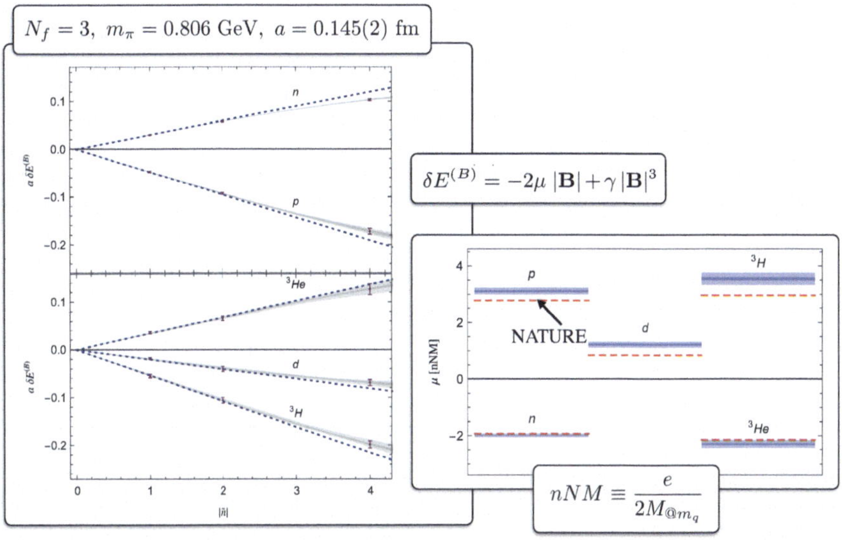

Fig. 95.4 The shift in the energy of select light nuclei, $\delta E^{(B)}$, in a background magnetic field **B** (left) and the extracted magnetic moments, μ, from LQCD at larger-than-physical values of the quark masses (right) [10]

nuclei can determine how significant nuclear effects (those arising from the fact that a nucleus is more than a collection of nearly noninteracting nucleons) are for current and future searches for CP violation in nuclei, in the single and double-β decay of a large nucleus and in the direct searches for dark matter candidates using heavy isotope as targets. This is enabled through matching the LQCD results in the few-body sector to the corresponding EFT description of these processes, a process that can constrain unknown two and multi-nucleon short-distance effective couplings of the EFT, see Fig. 95.1. A first LQCD study of scalar, axial and tensor quark-bilinear currents in light nuclei was performed in [11] at a large value of the quark masses (see the left panel of Fig. 95.5), and found nonnegligible nuclear effects in the scalar matrix element. If this conclusion persists at the physical values of the quark masses, significant nuclear effects may need to be accounted for in obtaining the cross section of nuclear targets with dark-matter candidates in scalar portals (Fig. 95.6).

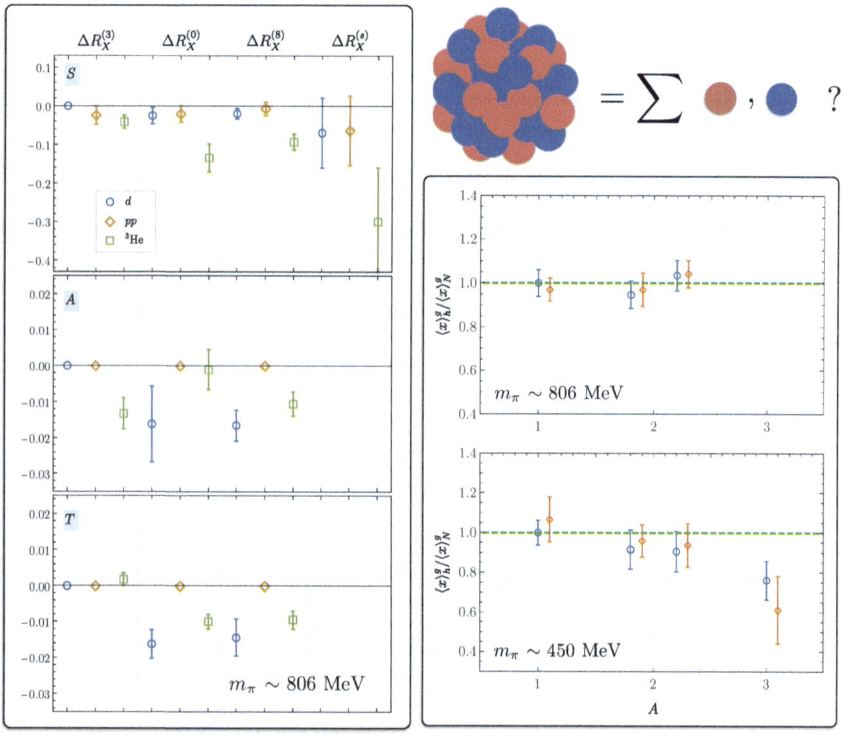

Fig. 95.5 Depicted in left is the deviation of the matrix element $\langle h|\bar{q}\Gamma q|h\rangle$ from that in a non-interacting model of nucleons. $\bar{q}\Gamma q$ denotes scalar, axial and tensor quark bilinear currents and $h = \{d, pp, {}^3\text{He}\}$. Depicted in right is the ratio of the gluon momentum fraction in select light nuclei to that of the single nucleon, obtained from LQCD at larger-than-physical values of the quark masses [11]. In the right panel, blue and orange colors correspond to two different sink operators in forming the correlation functions. For a detailed description of quantities, see [11, 12]

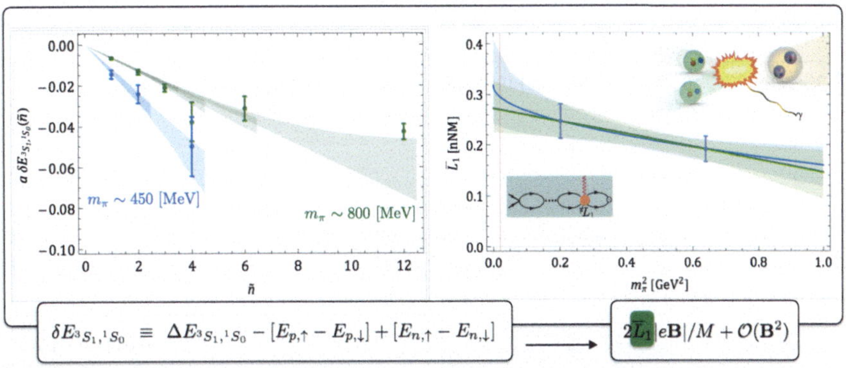

Fig. 95.6 The background magnetic field breaks the (near) degeneracy of two-nucleon systems in a LQCD calculation (left). This gives access to the nuclear matrix element for the M1 magnetic transition in the radiative capture process $np \to d\gamma$, and constrains the two-nucleon short-distance coupling of the pionless EFT, L_1, (right) [13]

Finally, LQCD enables gluonic probes of nuclear structure, promising a growing program that can provide the theoretical support for an EIC in upcoming years. Investigations into the role of the gluons in the structure of single hadrons have reached significant milestones in recent years, while in the case of the nuclei, this effort has just been started. The first LQCD determination of select gluonic observables in light nuclei was reported in [12], albeit at two unphysically large values of the quark masses, in search for a gluonic analog of the EMC effect, see the right panel of Fig. 95.5. These studies demand significant computational resources but are planned to be improved, in scope and precision, in upcoming years.

95.4 Nuclear Reactions

While constraining nuclear reaction cross sections appeared to be a distant goal in the early stages of the field, some phenomenologically important reactions in the two-nucleon sector have now been computed from LQCD, albeit at unphysically large values of the quark masses. The first reaction studied is the M1 transition rate in the radiative capture process $np \to d\gamma$, a primary reaction in big-bang nucleosynthesis, and responsible for forming most of the light nuclei in the cosmos. An important quantity is the two-nucleon short-distance coupling of the pionless EFT, namely L_1, which quantifies the size of contributions to the rate beyond that induced by the magnetic moment of each of the nucleons. This coupling was constrained by applying an external constant magnetic field in a LQCD calculation to induce a transition between the two-nucleon isosinglet and isotriplet channels, at two unphysically large values of the quark masses, and was extrapolated to the physical values of the quark masses, giving rise to a cross section consistent with the experimental value [13].

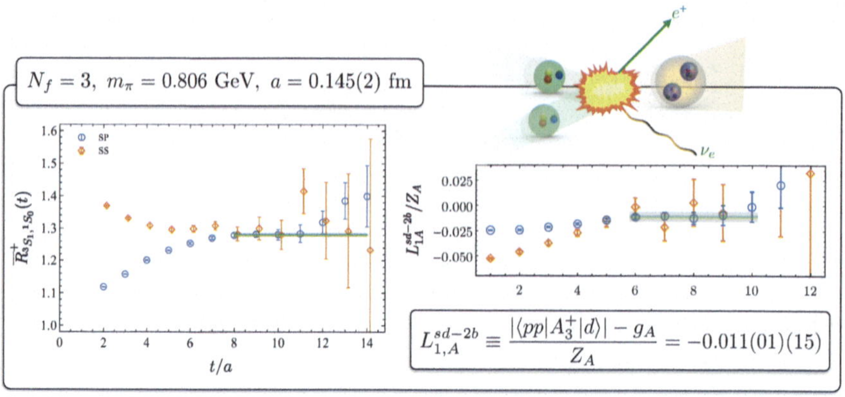

Fig. 95.7 The plot in the left obtains the nuclear matrix element corresponding to the pp fusion cross section obtained from LQCD with two (blue and orange) different sink operators in forming the correlation functions. The plot in the right obtains the short-distance solely two-body coupling of the pionless EFT, $L^{sd-2b}_{1,A}$. For a detailed description of quantities, see [14]

Perhaps a more phenomenologically interesting and less-known cross section is that of the weak counterpart of the process above. This is the pp fusion process at low incident velocities, which is relevant for the energy production in sun. Here, there remains a large uncertainty on the value of the two-nucleon short-distance coupling of the pionless EFT, namely $L_{1,A}$, which is planned to be reduced to the few-percent level using the MuSun experiment. This coupling was calculated in a recent LQCD calculation from a direct evaluation of the corresponding matrix element at a large value of the quark masses, see Fig. 95.7, and under the assumption of a mild dependence on the quark masses, was extrapolated to the physical point [14]. The obtained value was found consistent with the current phenomenological value. This study will be improved in the upcoming years towards the physical values of the quark masses.

Another milestone in the area of nuclear reactions from LQCD is the study of a doubly-weak process, namely the neutrinoifull double-β decay. In a recent study, the matrix element for the $nn \to pp$ transition was calculated, and both the long-distance contribution (arising from a deuteron pole in the intermediate state) and the short-distance contribution to the process were isolated, albeit at a large value of the quark masses, see Fig. 95.8. A new short-distance two-nucleon doubly-weak coupling of the pionless EFT was identified for the first time and its value was constrained in this study [15, 16]. It was concluded that the contribution from this coupling to the matrix element was comparable to that from the $L_{1,A}$ coupling, introducing a potential source of modification of the value of an "effective" axial charge in a heavy isotope. It is important to realize that such new short-distance couplings, both in the context of neutrinofull and neutrinoless double-β decays are unknown, and in particular in the latter case, they can only be deduced from a direct LQCD calculation. This strongly motivates a continuation of this program in the upcoming years.

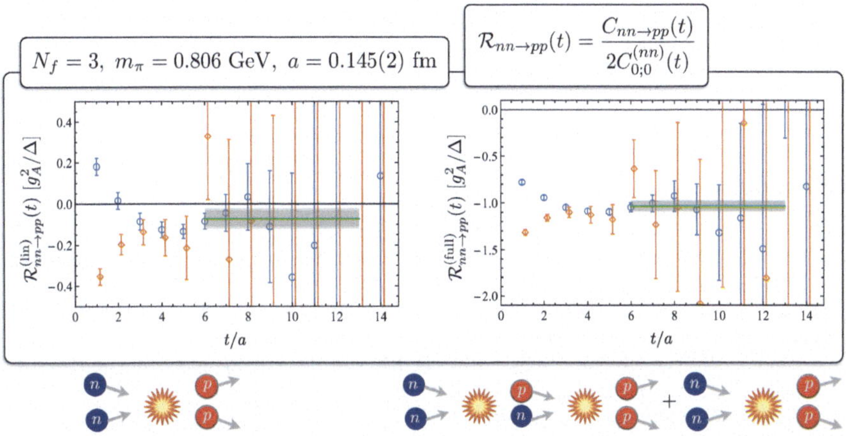

Fig. 95.8 The short-distance contribution to the nuclear matrix element for $nn \to pp$ (left) and the full contribution including that from the intermediate deuteron pole (right) obtained from LQCD at larger-than-physical values of the quark masses, with two (blue and orange) different sink operators. For a detailed description of quantities, see [15, 16]

95.5 Summary and Outlook

This review was aimed at showcasing a few accomplishments of the field of LQCD for nuclear physics. While most of the studies are still limited to unphysical values of the quark masses (given the significant computational cost of computations at the physical values), formal, numerical and algorithmic advancements have placed the field in a position that spectra, structure and scattering and reaction properties of few-nucleon systems, in a bound or unbound form, can be studies directly from the underlying theory. Given the emergence of exceedingly fine energy scales in the spectrum of larger nuclei and limited precision of LQCD calculations into the future, the complexity of the correlation functions when formed out of quark-level interpolating operators, and the signal-to-noise degradation in the nuclear correlation function due to the presence of lighter states in the theory, the computational complexity of LQCD-based calculations of nuclei increases dramatically with increasing the system's size. As a result, a coordinated effort by the larger community needs to be in place to systematically match experiment and LQCD results in the few-body sector to nuclear EFTs, hence enable studies of systems in the realm of nuclear many-body physics. In the meantime, with the fast and exciting progress in harnessing quantum entanglement to perform highly parallelized computations, the prospect of this alternative approach to classical computation will be investigated by nuclear physicists in the upcoming years [17], with the ultimate goal of overcoming the impeding sign and signal-to-noise problems inherent in Monte Carlo-based studies of finite-density systems and their real-time dynamics.

Acknowledgements The NPLQCD collaboration acknowledges support from Nuclear Physics, High Energy Physics and Advanced Scientific Computing Research programs at the Office of Science of the U.S. Department of Energy.

References

1. Davoudi, Z.: EPJ Web Conf. **175**, 01022 (2018). https://doi.org/10.1051/epjconf/201817501022. arXiv:1711.02020 [hep-lat]
2. Beane, S.R., et al.: [NPLQCD Collaboration], Phys. Rev. D **87**,(3), 034506 (2013). https://doi.org/10.1103/PhysRevD.87.034506. arXiv:1206.5219 [hep-lat]
3. Detmold, W., Savage, M.J.: Phys. Rev. D **82**, 014511 (2010). https://doi.org/10.1103/PhysRevD.82.014511. arXiv:1001.2768 [hep-lat]
4. Detmold, W., Orginos, K.: Phys. Rev. D **87**(11), 114512 (2013). https://doi.org/10.1103/PhysRevD.87.114512. arXiv:1207.1452 [hep-lat]
5. Beane, S.R., Bedaque, P.F., Parreno, A., Savage, M.J.: Phys. Lett. B **585**(106) (2004). https://doi.org/10.1016/j.physletb.2004.02.007 [hep-lat/0312004]
6. Luscher, M.: Commun. Math. Phys. **105**, 153 (1986). https://doi.org/10.1007/BF01211097
7. Wagman, M.L., Winter, F., Chang, E., Davoudi, Z., Detmold, W., Orginos, K., Savage, M.J., Shanahan, P.E.: Phys. Rev. D **96**(11), 114510 (2017). https://doi.org/10.1103/PhysRevD.96.114510. arXiv:1706.06550 [hep-lat]
8. Kaplan, D.B., Savage, M.J.: Phys. Lett. B **365**, 244 (1996). https://doi.org/10.1016/0370-2693(95)01277-X [hep-ph/9509371]
9. Beane, S.R., Kaplan, D.B., Klco, N., Savage, M.J.: arXiv:1812.03138 [nucl-th]
10. Beane, S.R., et al.: Phys. Rev. Lett. **113**(25), 252001 (2014). https://doi.org/10.1103/PhysRevLett.113.252001. arXiv:1409.3556 [hep-lat]
11. Chang, E., et al.: [NPLQCD Collaboration], Phys. Rev. Lett. **120**(15), 152002 (2018). https://doi.org/10.1103/PhysRevLett.120.152002. arXiv:1712.03221 [hep-lat]
12. Winter, F., Detmold, W., Gambhir, A.S., Orginos, K., Savage, M.J., Shanahan, P.E., Wagman, M.L.: Phys. Rev. D **96**(9), 094512 (2017). https://doi.org/10.1103/PhysRevD.96.094512. arXiv:1709.00395 [hep-lat]
13. Beane, S.R., et al.: [NPLQCD Collaboration], Phys. Rev. Lett. **115**(13), 132001 (2015). https://doi.org/10.1103/PhysRevLett.115.132001. arXiv:1505.02422 [hep-lat]
14. Savage, M.J., et al.: Phys. Rev. Lett. **119**(6), 062002 (2017). https://doi.org/10.1103/PhysRevLett.119.062002. arXiv:1610.04545 [hep-lat]
15. Shanahan, P.E., et al.: Phys. Rev. Lett. **119**(6), 062003 (2017). https://doi.org/10.1103/PhysRevLett.119.062003. arXiv:1701.03456 [hep-lat]
16. Tiburzi, B.C., et al.: Phys. Rev. D **96**(5), 054505 (2017). https://doi.org/10.1103/PhysRevD.96.054505. arXiv:1702.02929 [hep-lat]
17. Carlson, J., et al.: A white paper on quantum computing for theoretical nuclear physics. http://www.int.washington.edu/PROGRAMS/17-66W/

Part IV
Hadron and High-Energy Physics

Chapter 96
Study of Non-strange Dibaryon Resonances Via Coherent Double Neutral-Meson Photoproduction from the Deuteron

T. Ishikawa, H. Fujimura, H. Fukasawa, R. Hashimoto, Q. He, Y. Honda, T. Iwata, S. Kaida, H. Kanda, A. Kawano, S. Kuwasaki, K. Maeda, S. Masumoto, M. Miyabe, F. Miyahara, K. Mochizuki, N. Muramatsu, A. Nakamura, K. Nawa, S. Ogushi, Y. Okada, K. Okamura, Y. Onodera, K. Ozawa, Y. Sakamoto, M. Sato, H. Shimizu, H. Sugai, K. Suzuki, Y. Tajima, S. Takahashi, Y. Taniguchi, Y. Tsuchikawa, H. Yamazaki, R. Yamazaki and H. Y. Yoshida

Abstract The non-strange $B = 2$ system (dibaryon) has been studied using coherent neutral-meson photoproduction from the deuteron, $\gamma d \to \pi^0\pi^0 d$ and $\gamma d \to \pi^0\eta d$, measured with the FOREST detector at the ELPH facility. A rather flat angular distribution of deuteron emission and a peak in the $\pi^0 d$ invariant-mass distribution give evidence for the 2.15-GeV isovector dibaryon. The excitation function of the total cross section as a function of the γd center-of-mass energy for the former reaction shows the evidence for the existence of 2.47- and 2.63-GeV isoscalar dibaryons.

T. Ishikawa (✉) · H. Fujimura · H. Fukasawa · R. Hashimoto · Q. He · Y. Honda · S. Kaida · S. Kuwasaki · M. Miyabe · F. Miyahara · K. Mochizuki · N. Muramatsu · A. Nakamura · K. Nawa · S. Ogushi · Y. Okada · K. Okamura · Y. Onodera · M. Sato · H. Shimizu · H. Sugai · K. Suzuki · S. Takahashi · Y. Taniguchi · Y. Tsuchikawa · H. Yamazaki · R. Yamazaki
Research Center for Electron Photon Science (ELPH), Tohoku University, Sendai 982-0826, Japan
e-mail: ishikawa@lns.tohoku.ac.jp

T. Iwata · Y. Tajima · H. Y. Yoshida
Department of Physics, Yamagata University, Yamagata 990-8560, Japan

H. Kanda · K. Maeda
Department of Physics, Tohoku University, Sendai 980-8578, Japan

A. Kawano · Y. Sakamoto
Department of Information Science, Tohoku Gakuin University, Sendai 981-3193, Japan

S. Masumoto
Department of Physics, University of Tokyo, Tokyo 113-0033, Japan

K. Ozawa
Institute of Particle and Nuclear Studies, High Energy Accelerator Research Organization (KEK), Tsukuba 305-0801, Japan

N. A. Orr et al. (eds.), *Recent Progress in Few-Body Physics*,
Springer Proceedings in Physics 238,
https://doi.org/10.1007/978-3-030-32357-8_96

96.1 Dibaryons

The study of the $B = 2$ systems (dibaryons) has a long history [1]. An early work by Dyson and Xuong [2] predicts the sextet of non-strange dibaryons $\mathcal{D}_{IJ}$ with isospin I, spin J, and $I + J = 1, 3$ based on the deuteron ($\mathcal{D}_{01}$)—the first known dibaryon, and $^3S_1$-NN state—, the $^1S_0$-NN virtual states ($\mathcal{D}_{10}$), and a resonance-like structure at $M = 2.16$ GeV ($\mathcal{D}_{12}$) for the $^3P_2$-πd ($^1D_2$-pp) amplitude of the partial-wave analysis for $\pi^+ d \rightarrow pp$. The recent observation of $d^*(2380)$ ($\mathcal{D}_{03}$) by the CELSIUS/WASA and WASA-at-COSY collaborations [3, 4] has made us pay attention to the dibaryon sextet. It is important to establish the excitation spectrum of dibaryons in order to understand their internal structures. A dibaryon is of particular interest to study a phase transition of its basic configuration from a molecule-like state (such as the deuteron) to a spatially-compact hexaquark hadron state.

Although the πd system has a resonance-like structure at around 2.15 GeV, it can be also understood as a quasi-free (QF) Δ excitation from a nucleon in the deuteron in the existing experimental data ($\pi^+ d \rightarrow pp$ and $\pi^\pm d \rightarrow \pi^\pm d$). In photoinduced reactions, $\gamma d \rightarrow \pi^0 d$ is a convenient approach to study $d^*(2150)$. It is, however, difficult to exclude the effect of a QF Δ excitation in this reaction. An alternative reaction $\gamma d \rightarrow \pi^0 \pi^0 d$ is more advantageous to find $d^*(2150)$ because the QF Δ excitation is kinematically separable. Thus, we study $\gamma d \rightarrow \pi^0 \pi^0 d$, aiming to observe $d^*(2150)$ in the $\pi^0 d$ system through π^0 decay from a possible higher-mass dibaryon in the $\pi^0 \pi^0 d$ system.

Mechanisms for $\gamma d \rightarrow \pi^0 \pi^0 d$ can be classified as:

1 QF $\pi^0 \pi^0$ production on a nucleon followed by deuteron coalescence (QFC),
2 the first π^0 is emitted from the QF nucleon, the intermediate $d^*(2150)$ state is generated from the NN or ΔN reaction with the spectator nucleon, finally leaving the second π^0 and deuteron (semi-QF), and
3 both the two π^0s and deuteron are emitted from a dibaryon resonance.

In the mechanism 1 (2), the second π^0 should be emitted to compensate for the momentum given to the QF participant nucleon by the first emitted π^0 to coalesce into a deuteron (the intermediate $d^*(2150)$ state). Thus, the angular distribution of deuteron emission shows a strongly backward-peaking (sideway-peaking) structure for the mechanism 1 (2), while the mechanism 3 makes a rather flat distribution. Since the three mechanisms give completely different angular distributions, the three mechanisms can be separated (distinguished).

96.2 Experiments and Analysis

A series of meson photoproduction experiments [5] have been conducted with an energy-tagged bremsstrahlung-photon beam [6–9] at the ELPH facility. The tagging energy ranges from 0.75 to 1.15 GeV. The target used in the experiments is liquid

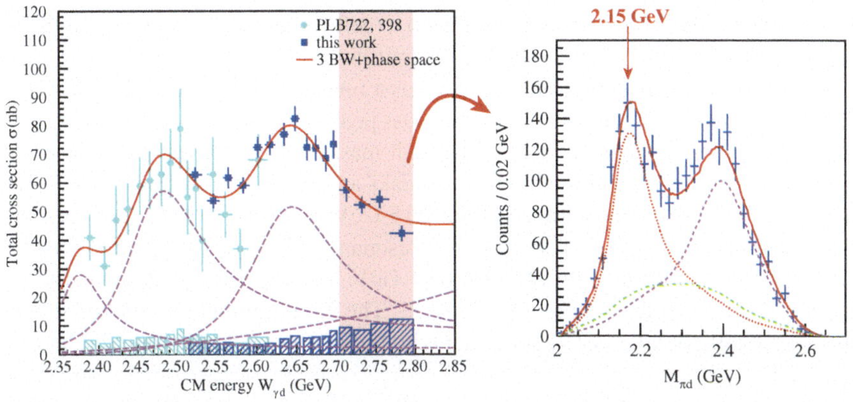

Fig. 96.1 Total cross section σ as a function of $W_{\gamma d}$ (left). The squares show σ obtained in this work, while the circles show that presented in [11]. The horizontal error of each data point corresponds to the coverage of the incident photon energy, and the vertical error shows the statistical error of σ. The solid curve shows the fitted function expressed by a sum of three BW-peaks and phase-space contributions. The lower hatched histograms show the systematic errors of σ. Typical $\pi^0 d$ invariant-mass $M_{\pi d}$ distribution (right). The solid curve shows the fitted function expressed as a sum of a Breit-Wigner peak, its reflection, and phase-space contributions, to the data. This figure is taken from [12]

deuterium with a thickness of 45.9 mm. All the final-state particles in $\gamma d \to \pi^0 \pi^0 d$ are measured with the FOREST detector [10], which consists of three different electromagnetic-calorimeter (EMC) systems. Each system has a plastic-scintillator hodoscope (PSH) placed in front of each EMC to identify charged particles.

Event selection is made for $\gamma d \to \pi^0 \pi^0 d \to \gamma\gamma\gamma\gamma d$, where the charged particles (d candidates) are detected with the forward PSH. A kinematic fit with six constraints is applied to select the reaction of interest: energy and three-momentum conservation, and every $\gamma\gamma$ invariant mass being the π^0 mass. Here, the momentum of the charged particle is obtained from the time delay assuming that the particle has the deuteron mass. Events for which the χ^2 probability is higher than 0.4 are selected to reduce those from other background processes. The details of the analysis are described in [11, 12] (Fig. 96.1).

96.3 Results and Discussion

The total and differential cross sections are deduced for $\gamma d \to \pi^0 \pi^0 d$ at the incident energy ranging from 0.75 to 1.15 GeV. The excitation function as a function of γd center-of-mass system $W_{\gamma d}$ is not monotonically increasing but shows resonance-like behavior peaked at around 2.47 and 2.63 GeV. The angular distribution of deuteron emission is rather flat, and the kinematic condition for the obtained data completely differs from the QFC and semi-QF processes. Thus, we consider the possibility that the resonance-like structure might be due to a manifestation of dibaryons. Both the two nucleons obviously participate before emitting π^0s.

The differential cross section $d\sigma/dM_{\pi d}$ as a function of πd invariant-mass $M_{\pi d}$ shows two peaks. The centroid of the low-mass peak is ~2.15 GeV independently of the incident energy. However, that of the high-mass peak decreases with a decrease of the incident energy, and finally the two peaks are merged into a bump. The high-mass peak reflects the appearance of the 2.15-GeV peak in $d\sigma/dM_{\pi d}$ between the other pion and deuteron (reflection). The resonance-like structure give a mass of 2.14±0.01 GeV with a width of 0.09±0.01 GeV. The angular distributions for the two π^0s limit J^π of the state to 1^+, 2^+, or 3^-. The 2^+ assignment is consistent with the theoretically predicted $\mathcal{D}_{12}$ state, and with the resonance structure of the $^3P_2$-πd amplitude.

The cross sections for $\gamma d \to \pi^0 \eta d$ are also deduced in a similar analysis. The angular distribution of deuteron emission shows an almost flat behavior, suggesting a non-QFC process. The 2.15-GeV isovector resonance in the $\pi^0 d$ subsystem also appears in the $d\sigma/dM_{\pi d}$ distribution.

96.4 Summary and Acknowledgements

The total and differential cross sections have been measured for $\gamma d \to \pi^0 \pi^0 d$ and $\gamma d \to \pi^0 \eta d$ at incident energies from 0.75 to 1.15 GeV. The total cross section as a function of $W_{\gamma d}$ shows resonance-like behavior peaked at around 2.47 and 2.63 GeV for the former reaction. The angular distributions of deuteron emission can never be understood in the QFC mechanism for these reactions. The present work shows evidence for the 2.15-GeV dibaryon in the $\pi^0 d$ channel. The details including figures can be found elsewhere [12].

Acknowledgements The authors express their gratitude to the ELPH accelerator staff. This work was supported in part by JSPS KAKENHI Grant Nos. 17340063, 19002003, 24244022, 26400287, and 16H02188.

References

1. Clement, H.: On the history of dibaryons and their final observation. Prog. Part. Nucl. Phys. 93, 195–242 (2017). https://doi.org/10.1016/j.ppnp.2016.12.004
2. Dyson, F.J., Xuong, N.-H.: $Y = 2$ states in SU(6) theory. Phys. Rev. Lett. **13**, 815 (1964)
3. Bashkanov, M., et al.: [CELSIUS/WASA collaboration]: double-pionic fusion of nuclear systems and the "ABC" effect: approaching a puzzle by exclusive and kinematically complete measurements. Phys. Rev. Lett. **102**, 052301 (2009)
4. Adlarson, P., et al.: [WASA-at-COSY collaboration]: abashian-booth-Crowe effect in basic double-pionic fusion: a new resonance? Phys. Rev. Lett. **106**, 242302 (2011)
5. Ishikawa, T., et al.: Meson Photoproduction Experiments at ELPH. In: JPS Conference on Proceedings, vol. 10, p. 031001. Tohoku University (2016)

6. Ishikawa, T., et al.: The second GeV tagged photon beamline at ELPH. Nucl. Instrum. Meth. A **622**, 1 (2010)
7. Ishikawa, T., et al.: A fast profile monitor with scintillating fiber hodoscopes for high-intensity photon beams. Nucl. Instrum. Meth. A **811**, 124 (2016)
8. Matsumura, Y., Ishikawa, T., et al.: Development of a transmittance monitor for high-intensity photon beams. Nucl. Instrum. Meth. A **902**, 103 (2018)
9. Obara, Y., Ishikawa, T., et al.: Profile measurement of circulating electrons in a synchrotron by inserting a carbon wire. Nucl. Instrum. Meth. A **922**, 103 (2018)
10. Ishikawa, T., et al.: The FOREST detector for meson photoproduction experiments at ELPH. Nucl. Instrum. Meth. A **832**, 108 (2016)
11. Ishikawa, T., et al.: First measurement of coherent double neutral-pion photoproduction on the deuteron at incident energies below 0.9 GeV. Phys. Lett. B **772**, 398 (2017)
12. Ishikawa, T., et al.: Non-strange dibaryons studied in the $\gamma d \to \pi^0\pi^0 d$ reaction. Phys. Lett. B **789**, 413 (2019)

Chapter 97
ηn Photoproduction and Nucleon Resonances

Jung-Min Suh, Sang-Ho Kim and Hyun-Chul Kim

Abstract We present results of a recent work on the reaction mechanism of ηn photoproduction off the neutron in the range of $\sqrt{s} \approx 1.5 - 1.9$ GeV and discuss the role of various nucleon resonances listed in the Particle Data Group (PDG). We make use of an effective Lagrangian approach combining with a Regge method. The total and helicity-dependent cross sections $\sigma_{1/2}$, $\sigma_{3/2}$ are computed and the numerical results are in good agreement with the A2 experimental data.

97.1 Introduction

ηn photoproduction is one of the most practical and useful reaction processes to investigate various N^* resonances. More interesting is that the narrow bump-like structure near $\sqrt{s} \sim 1.68$ GeV is seen only at this neutron channel, which was coined as the neutron anomaly [1–8]. A number of theoretical studies have been performed whether this phenomena is due to the existence of the narrow resonance $N(1685, 1/2^+)$ or not. In this talk, we revisit the $\gamma n \to \eta n$ process, taking account of a total of fifteen N^* resonances given in the PDG and the $N(1685, 1/2^+)$ additionally. All the

J.-M. Suh (✉) · H.-C. Kim
Department of Physics, Inha University, Incheon 22212, Republic of Korea
e-mail: suhjungmin@inha.edu

S.-H. Kim
Center for Extreme Nuclear Matters (CENuM), Korea University, Seoul 02841, Republic of Korea
e-mail: sangho_kim@korea.ac.kr

Department of Physics, Pukyong National University (PKNU), Busan 48513, Republic of Korea

H.-C. Kim
Advanced Science Research Center, Japan Atomic Energy Agency, Shirakata, Tokai, Ibaraki 319-1195, Japan
e-mail: hchkim@inha.ac.kr

School of Physics, Korea Institute for Advanced Study (KIAS), Seoul 02455, Republic of Korea

N. A. Orr et al. (eds.), *Recent Progress in Few-Body Physics*,
Springer Proceedings in Physics 238,
https://doi.org/10.1007/978-3-030-32357-8_97

resonance parameters were extracted from the experimental data on the PDG [9] and quark model predictions [10]. The numerical results show that the $N(1685, 1/2^+)$ is an essential part of describing the ηn photoproduction together with the dominant $N(1535, 1/2^-)$ resonance.

97.2 Formalism

The Feynman diagrams for the $\gamma n \to \eta n$ reaction are drawn in Fig. 97.1. As a background contribution, we take into account ρ and ω Reggeon exchanges in the t channel, and N exchanges in both the s and u channels. We can construct the effective Lagrangians for the photon (meson) and nucleon vertices as follows

$$\begin{aligned}
\mathcal{L}_{\gamma NN} &= -\bar{N}\left[e_N\gamma_\mu - \frac{e\kappa_N}{2M_N}\sigma_{\mu\nu}\partial^\nu\right]A^\mu N, \\
\mathcal{L}_{VNN} &= -g_{VNN}\bar{N}\left[\gamma_\mu N - \frac{\kappa_{VNN}}{2M_N}\sigma_{\mu\nu}N\partial^\nu\right]V^\mu + \text{h.c.}, \\
\mathcal{L}_{\eta NN} &= \frac{g_{\eta NN}}{2M_N}\bar{N}\gamma_\mu\gamma_5 N\partial^\mu\phi_\eta,
\end{aligned} \tag{97.1}$$

where the values of coupling constants are fixed to be $\kappa_N = -1.91$ [9], $g_{\rho NN} = 2.6$, $\kappa_{\rho NN} = 3.7$, $g_{\omega NN} = 10.4$, $\kappa_{\omega NN} = 0.41$, $g_{\eta NN} = 6.34$ [11, 12], respectively.

In addition to these Born terms, we include the fifteen N^* resonances [9] and the narrow $N(1685, 1/2^+)$ in the s-channel diagram. We can calculate the corresponding transition amplitudes from the following the effective Lagrangians

$$\begin{aligned}
\mathcal{L}^{1/2^\pm}_{\gamma NN^*} &= \frac{eh_1}{2M_N}\bar{N}\Gamma^{(\mp)}\sigma_{\mu\nu}\partial^\nu A^\mu N^* + \text{h.c.}, \\
\mathcal{L}^{3/2^\pm}_{\gamma NN^*} &= -ie\left[\frac{h_1}{2M_N}\bar{N}\Gamma^{(\pm)}_\nu - \frac{ih_2}{(2M_N)^2}\partial_\nu\bar{N}\Gamma^{(\pm)}\right]F^{\mu\nu}N^*_\mu + \text{h.c.}, \\
\mathcal{L}^{5/2^\pm}_{\gamma NN^*} &= e\left[\frac{h_1}{(2M_N)^2}\bar{N}\Gamma^{(\mp)}_\nu - \frac{ih_2}{(2M_N)^3}\partial_\nu\bar{N}\Gamma^{(\mp)}\right]\partial^\alpha F^{\mu\nu}N^*_{\mu\alpha} + \text{h.c.}, \\
\mathcal{L}^{7/2^\pm}_{\gamma NN^*} &= ie\left[\frac{h_1}{(2M_N)^3}\bar{N}\Gamma^{(\pm)}_\nu - \frac{ih_2}{(2M_N)^4}\partial_\nu\bar{N}\Gamma^{(\pm)}\right]\partial^\alpha\partial^\beta F^{\mu\nu}N^*_{\mu\alpha\beta} + \text{h.c.},
\end{aligned} \tag{97.2}$$

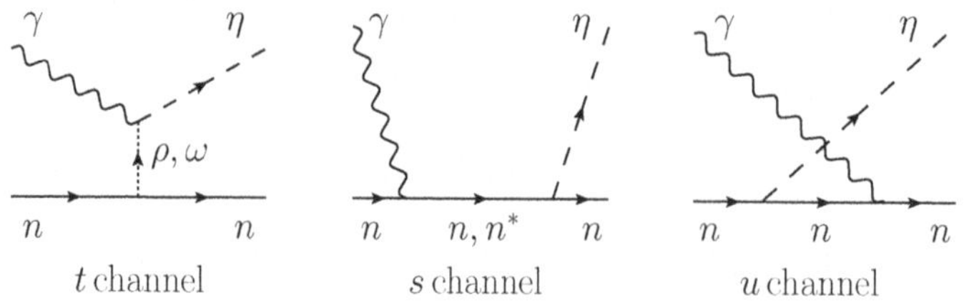

Fig. 97.1 Feynman diagrams for the $\gamma n \to \eta n$ reaction

for the electromagnetic interactions and

$$
\begin{aligned}
\mathcal{L}^{1/2^\pm}_{\eta NN^*} &= -ig_{\eta NN^*}\eta\bar{N}\Gamma^{(\pm)}N^* + \text{h.c.},\\
\mathcal{L}^{3/2^\pm}_{\eta NN^*} &= \frac{g_{\eta NN^*}}{M_\eta}\partial^\mu\eta\bar{N}\Gamma^{(\mp)}N^*_\mu + \text{h.c.},\\
\mathcal{L}^{5/2^\pm}_{\eta NN^*} &= \frac{ig_{\eta NN^*}}{M_\eta^2}\partial^\mu\partial^\nu\eta\bar{N}\Gamma^{(\pm)}N^*_{\mu\nu} + \text{h.c.},\\
\mathcal{L}^{7/2^\pm}_{\eta NN^*} &= -\frac{g_{\eta NN^*}}{M_\eta^3}\partial^\mu\partial^\nu\partial^\alpha\eta\bar{N}\Gamma^{(\mp)}N^*_{\mu\nu\alpha} + \text{h.c.},
\end{aligned}
\tag{97.3}
$$

for the strong interactions with the notations

$$
\Gamma^{(\pm)} = \begin{pmatrix} \gamma_5 \\ I_{4\times 4} \end{pmatrix}, \quad \Gamma^{(\pm)}_\nu = \begin{pmatrix} \gamma_\nu\gamma_5 \\ \gamma_\nu \end{pmatrix}. \tag{97.4}
$$

Note that N^*_μ, $N^*_{\mu\alpha}$, and $N^*_{\mu\alpha\beta}$ stand for the Rarita-Schwinger fields of spin $-3/2$, $-5/2$, and $-7/2$, respectively. We refer to [13] for more details of how the resonance couplings are determined.

97.3 Results and Discussion

Figure 97.2 shows the numerical results of the total cross section with various contributions. The total result describes the A2 data well [6]. The contribution of ρ Reggeon exchange becomes large as W increases. Meanwhile, the four N^* resonances, i.e., $N(1520, 3/2^-)$, $N(1535, 1/2^-)$, $N(1685, 1/2^+)$, and $N(1710, 1/2^+)$ are dominant near threshold region.

To clarify the role of spin $J = 1/2$ and $J \geq 3/2$ N^* resonances separately, we draw the helicity-dependent cross sections in Fig. 97.3. Both the $\sigma_{1/2}$ and $\sigma_{3/2}$ are in good agreement with the A2 data [7]. It turns out from the $\sigma_{1/2}$ that the

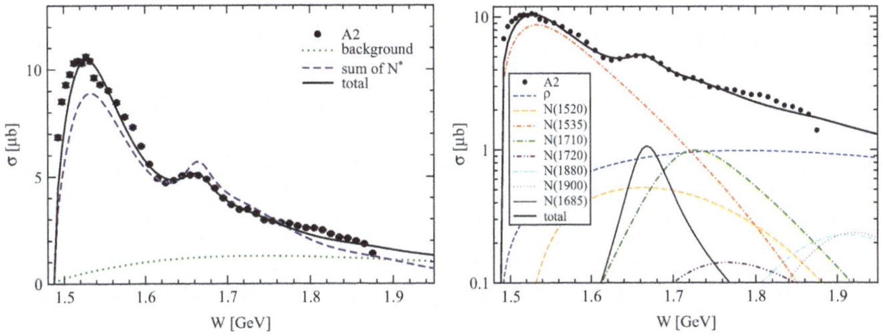

Fig. 97.2 Total cross section for $\gamma n \to \eta n$ adapted from [13]. The circles denote the A2 data [6]

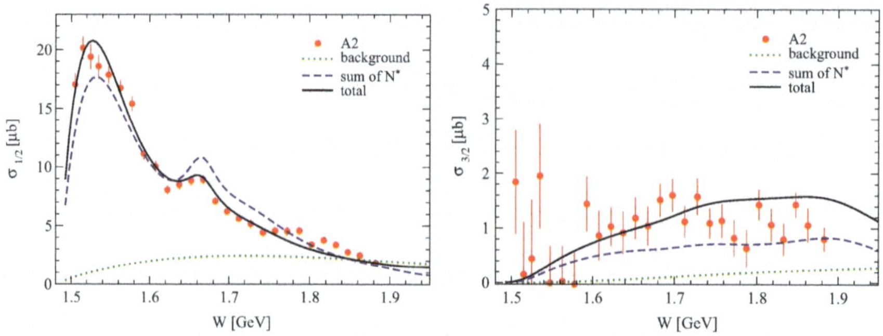

Fig. 97.3 Helicity-dependent cross sections for $\gamma n \to \eta n$ adaped from [13]. The circles denote the A2 data [7]

background term and $N(1535, 1/2^-)$ interfere constructively but destructively with $N(1685, 1/2^+)$ and $N(1710, 1/2^+)$. The N^* resonances with higher spin $J \geq 3/2$ also have certain effects as found from the $\sigma_{3/2}$. The constructive interference effect between the backgound and higher-spin N^* contributions plays an important role in explaining the data. It implies that the N^* resonances with spin $J = 1/2$ and $J \geq 3/2$ are all correctly treated in the present model. That is, the N^* resonances with spin $J = 3/2$, i.e., $N(1520, 3/2^-)$, $N(1720, 3/2^+)$, and $N(1900, 3/2^+)$ give important contributions to the $\gamma n \to \eta n$ reaction besides the dominant spin $-1/2$ N^* resonances $N(1535, 1/2^-)$, $N(1685, 1/2^+)$, and $N(1710, 1/2^+)$.

97.4 Conclusion and Outlook

In this talk, we presented a recent investigation of ηn photoproduction, employing an effective Lagrangian approach combining with a Regge model. The total and helicity-dependent cross sections were discussed in comparison with the A2 data. It was found that the $N(1685, 1/2^+)$ plays a crucial role for the description of the $\gamma n \to \eta n$ reaction. This present results favor the existence of the narrow $N(1685, 1/2^+)$ together with our recent work on the $\gamma n \to K^0\Lambda$ reaction where a clue on the evidence of the $N(1685, 1/2^+)$ was also given [14].

Finally, we want to mention that a detailed analysis on the helicity amplitudes in the partial-wave expansion is required in order to elucidate the existence of the narrow $N^*(1685)$ resonance. The corresponding work is under way.

Acknowledgements H.-Ch.K is grateful to M. V. Polyakov for critical comments and criticism. This work was supported by the National Research Foundation of Korea (NRF) grant funded by the Korea government (MSIT) (No. 2018R1A5A1025563).

References

1. Kuznetsov, V., et al.: (GRAAL Collaboration): Phys. Lett. B **647**, 23 (2007)
2. Jaegle, I., et al.: (CBELSA and TAPS Collaborations): Phys. Rev. Lett. **100**, 252002 (2008)
3. Witthauer, L., et al.: (CBELSA/TAPS Collaboration): Eur. Phys. J. A **53**, 58 (2017)
4. Werthmüller, D., et al.: (A2 Collaboration): Phys. Rev. Lett. **111**, 232001 (2013)
5. Witthauer, L., et al.: (A2 Collaboration): Eur. Phys. J. A **49**, 154 (2013)
6. Werthmüller, D., et al.: (A2 Collaboration): Phys. Rev. C **90**, 015205 (2014)
7. Witthauer, L., et al.: (A2 Collaboration): Phys. Rev. Lett. **117**, 132502 (2016)
8. Witthauer, L., et al.: (A2 Collaboration): Phys. Rev. C **95**, 055201 (2017)
9. Tanabashi, M., et al.: (Particle Data Group): Phys. Rev. D **98**, 030001 (2018)
10. Capstick, S., Roberts, W.: Phys. Rev. D **58**, 074011 (1998)
11. Stoks, V.G.J., Rijken, T.A.: Phys. Rev. C **59**, 3009 (1999); Rijken, T.A., Stoks, V.G.J., Yamamoto, Y.: ibid **59**, 21 (1999)
12. Hohler, G., Pietarinen, E.: Nucl. Phys. B **95**, 210 (1975)
13. Suh, J.M., Kim, S.H., Kim, H.C.: arXiv:1810.05056 [hep-ph]
14. Kim, S.H., Kim, H-Ch.: Phys. Lett. B **786**, 156 (2018)

Chapter 98
Hidden-Charm and Bottom Meson-Baryon Molecules Coupled with Five-Quark States

Alessandro Giachino, Atsushi Hosaka, Elena Santopinto, Sachiko Takeuchi, Makoto Takizawa and Yasuhiro Yamaguchi

Abstract We investigate the hidden-charm and hidden-bottom pentaquarks as meson-baryon molecules coupled to the five-quark states. As a result of our calculation it emerges that in the charm sector, one needs to add the five-quark potential to the pion exchange potential in order to produce bound and resonant states, whereas, in the bottom sector, the pion exchange interaction alone is strong enough to produce

A. Giachino (✉) · E. Santopinto
Istituto Nazionale di Fisica Nucleare (INFN), Sezione di Genova, via Dodecaneso 33, 16146 Genova, Italy
e-mail: agiachino@ge.infn.it

E. Santopinto
e-mail: elena.santopinto@ge.infn.it

A. Hosaka · S. Takeuchi
Research Center for Nuclear Physics (RCNP), Osaka University, Ibaraki, Osaka 567-0047, Japan
e-mail: hosaka@rcnp.osaka-u.ac.jp

S. Takeuchi
e-mail: s.takeuchi@jcsw.ac.jp

A. Hosaka
Advanced Science Research Center, Japan Atomic Energy Agency, Tokai, Ibaraki 319-1195, Japan

S. Takeuchi · M. Takizawa · Y. Yamaguchi
Theoretical Research Division, Nishina Center, RIKEN, Hirosawa, Wako, Saitama 351-0198, Japan
e-mail: takizawa@ac.shoyaku.ac.jp

Y. Yamaguchi
e-mail: yasuhiro.yamaguchi@ge.infn.it

S. Takeuchi
Japan College of Social Work, Kiyose, Tokyo 204-8555, Japan

M. Takizawa
Showa Pharmaceutical University, Machida, Tokyo 194-8543, Japan

J-PARC Branch, KEK Theory Center, Institute for Particle and Nuclear Studies, KEK, Tokai, Ibaraki 319-1106, Japan

N. A. Orr et al. (eds.), *Recent Progress in Few-Body Physics*,
Springer Proceedings in Physics 238,
https://doi.org/10.1007/978-3-030-32357-8_98

states. Thus, from this investigation, it comes out that the hidden-bottom pentaquarks are more likely to be produced than their hidden-charm counterparts, and, for this reason, we suggest that the experimentalists should look for the hidden-bottom pentaquarks.

98.1 Introduction

In 2015, the Large Hadron Collider beauty experiment (LHCb) collaboration announced the observation of two hidden-charm pentaquarks, $P_c^+(4380)$ and $P_c^+(4450)$, in $\Lambda_b^0 \to J/\psi K^- p$ decay [1–3]. These two pentaquark states are found to have masses of $4380 \pm 8 \pm 28$ MeV and $4449.8 \pm 1.7 \pm 2.5$ MeV, with corresponding widths of $205 \pm 18 \pm 86$ MeV and $39 \pm 5 \pm 19$ MeV. The spin-parity J^P of these states has not yet been determined. The parities of these states are preferred to be opposite, in particular one state has $J = 3/2$ and the other one has $J = 5/2$. With $(J^P_{P_c^+(4380)}, J^P_{P_c^+(4450)}) = (3/2^-, 5/2^+)$ one obtains the best fit solution, but also $(3/2^+, 5/2^-)$ and $(5/2^-, 3/2^+)$ provide acceptable results.

Hidden-charm pentaquark states, such as $uudc\bar{c}$ and $udsc\bar{c}$ compact structures, have been studied so far. Before the LHCb observation Yuan et al. in [4] studied the $uudc\bar{c}$ and $udsc\bar{c}$ systems by the non-relativistic harmonic oscillator Hamiltonian with three kinds of the schematic interactions: a chromomagnetic-, a flavor-spin- and instanton-induced- interactions. In [5] Santopinto et al. investigated the hidden-charm pentaquark states as S-wave five-quark compact states with a constituent quark model approach. The hidden-charm and hidden-bottom pentaquark masses have been evaluated by Wu et al. in [6], by means of a color-magnetic interaction between the three light quarks and the $c\bar{c}$ ($b\bar{b}$) pair in a color octet state. Takeuchi et al. [7] has also investigated the hidden-charm pentaquark states with a quark cluster model. The diquark model has also been applied in [8–12] to investigate the compact five-quark state, but these authors do not provide any information about the pentaquark widths. Unfortunately, there is so far no clear evidence of such compact multiquark states while it is widely accepted that there are candidates for hadronic molecular states. Now, the P_c^+ pentaquarks have been found just below the $\bar{D}\Sigma_c^*$ and $\bar{D}^*\Sigma_c$ thresholds. Thus, the $\bar{D}\Sigma_c^*$ and $\bar{D}^*\Sigma_c$ molecular components are expected to be dominant [13–27]. In general, if more than one state is allowed for a given set of quantum numbers, the hadronic resonant states are mixtures of all the possible states. For this reason, an important issue is to clarify how these components are mixed in the physical hadrons. This problem prompted us to investigate the hidden-charm pentaquarks and the hidden bottom pentaquarks as meson-baryon molecules coupled to the five-quark states [28].

98.2 Model and Results

In [28] we study pentaquark states as quantum superpositions of the meson-baryon channels and the five-quark compact core. The inclusion of the five-quark state is inspired by the recent work by Santopinto et al. in [5] and Takeuchi et al. in [7]. In this work we model the coupling to the five-quark states as a short-range interaction between the meson and the baryon, while the long-range force is provided by the one-pion exchange potential. Table 98.1 summarizes the model space that we considered for open charm hadrons. In the bottom sector, the model space can be easily obtained by replacing the c quark with with b. As we discussed previously the interactions between all the relevant channels is modelled with the one-pion exchange potential, which is a well established interaction based on chiral symmetry and its spontaneous breaking. The five-quark part describes the short distance dynamics is considered to be consider to be in the order of 1 fm or less. Inspired by the recent discussion [7], we consider $5q$ compact states formed by color-octet light quarks ($3q$) and color octet $c\bar{c}$. The relevant channels are summarized in Table 98.2.

The bound and resonant states are obtained by solving the coupled-channel Schrödinger equation for ψ^{MB} which contains the one pion exchange potential, OPEP, $V^{\pi}(r)$, and the five-quark $5q$ potential, $V^{5q}(r)$:

$$H = \begin{pmatrix} H^{MB} & V \\ V^{\dagger} & H^{5q} \end{pmatrix} \tag{98.1}$$

where H^{MB} describes the meson-baryon part and contains K_i; the kinetic energy of each MB channel i and V^{π}_{ij} is the OPEP potential, while H^{5q} is referred to as the $5q$ channels. Figure 98.1 shows the Jacobi coordinates of "$\bar{D}$ meson" and "Y_c baryon" in the $5q$ configuration. The numerical results for the hidden-charm meson-baryon molecules are displayed in Fig. 98.2. The bound and resonant state energies of the hidden-charm molecules are presented as a function of different values of a

Table 98.1 Various channels of open-charm meson-baryons of total spin parity J^P with ${}^{2S+1}L$

Channels	$\bar{D}\Lambda_c$	$\bar{D}^*\Lambda_c$	$\bar{D}\Sigma_c$	$\bar{D}\Sigma_c^*$	$\bar{D}^*\Sigma_c$	$\bar{D}^*\Sigma_c^*$
J^P						
$1/2^-$	${}^2S$	${}^2S$, ${}^4D$	${}^2S$	${}^4D$	${}^2S$, ${}^4D$	${}^2S$, ${}^4D$, ${}^6D$
$3/2^-$	${}^2D$	${}^4S$, ${}^2D$, ${}^4D$	${}^2D$	${}^4S$, ${}^4D$	${}^4S$, ${}^2D$, ${}^4D$	${}^4S$, ${}^2D$, ${}^4D$, ${}^6D$, ${}^6G$
$5/2^-$	${}^2D$	${}^2D$, ${}^4D$, ${}^4G$	${}^2D$	${}^4D$, ${}^4G$	${}^2D$, ${}^4D$, ${}^4G$	${}^6S$, ${}^2D$, ${}^4D$, ${}^6D$, ${}^4G$, ${}^6G$

Table 98.2 Channels of $5q$'s with color octet qqq and $c\bar{c}$ ($b\bar{b}$) with possible total spin J. Notations are $[q^3 D_C, S_{3q}]S_{c\bar{c}(b\bar{b})}$ where $D_C = 8$ indicates that qqq form the color octet, S_{3q} is the spin of the light quarks $qqq = uud$, and $S_{c\bar{c}(b\bar{b})}$ the spin of $c\bar{c}$ ($b\bar{b}$)

Channel	$[q^3 8, \frac{1}{2}]0$	$[q^3 8, \frac{1}{2}]1$	$[q^3 8, \frac{3}{2}]0$	$[q^3 8, \frac{3}{2}]1$
J	1/2	1/2, 3/2	3/2	1/2, 3/2, 5/2

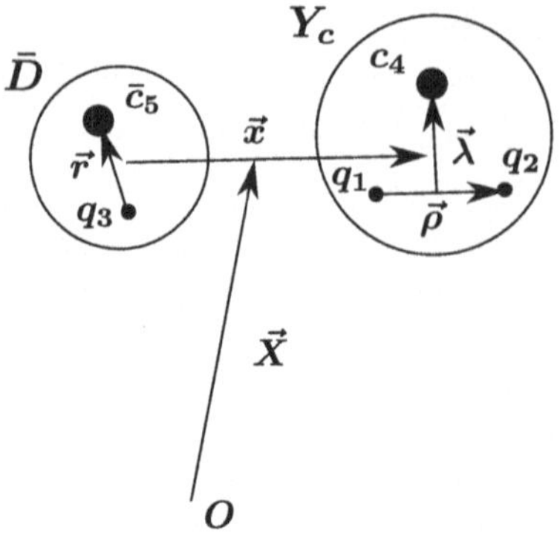

Fig. 98.1 Jacobi coordinates of "$\bar{D}$ meson" and "Y_c baryon" in the $5q$ configuration. q_i ($i = 1, 2, 3$) stands for the light quark, and c_4 ($\bar{c}_5$) stands for the (anti)charm quark. The coordinate $\boldsymbol{\rho}$ is the relative coordinate of q_1q_2, $\boldsymbol{\lambda}$ is the relative coordinate between the center of mass of q_1q_2 and c_4, $\mathbf{r}$ is the relative coordinate of $q_3\bar{c}_5$, while $\mathbf{x}$ is the relative coordinate between the centers of mass of $q_1q_2c_4$ and $\bar{c}_5q_3$. (APS copyright [28])

free parameter $\frac{f}{f_0}$. This parameter is introduced to parametrise the coupling strength between the meson-baryon and the compact five-quark channels. The red solid line are the eigenvalues of the coupled-channel equation 98.1 while the filled circles gives the minimum value of the coupling, $\frac{f}{f_0}$, which is needed for the formation of a resonant or bound state. Let us consider, for example, the solution obtained for quantum number $J^P = \frac{5}{2}^-$. As one can see from Fig. 98.2 when the coupling constant $\frac{f}{f_0}$ is larger than 25 one resonance appears below the $\bar{D}^*\Sigma_c^*$ threshold. When the energy of a resonant state is lower than the lowest threshold, $\bar{D}\Lambda_c$, (4150 MeV), this state becomes a bound state, and the energy gap between its energy and the lowest energy threshold, $\bar{D}\Lambda_c$, is the binding energy for that state.

As one can see from Fig. 98.2, no states are produced when one eliminates the coupling between the compact and the meson-baryon channels, i.e. when the coupling constant $\frac{f}{f_0}$ is set to zero, $\frac{f}{f_0} = 0$. This means that, in the hidden-charm sector, the one pion exchange potential (OPEP) is not enough strong to produce bound and resonant pentaquark states.

Figure 98.3 is similar as Fig. 98.2 but for it describes the situation in the bottom sector. As one can see from Fig. 98.3 some bound states are produced without the necessity to add the five-quark potential, i.e., even when $\frac{f}{f_0} = 0$. The only possibility to explain this behaviour is that the attractive interaction provided by OPEP is strong enough to bind the meson and baryon inside the pentaquark states. The main reason is due to the fact that, in the hidden bottom sector, the kinetic energy of the meson-baryon system is suppressed with respect to the charm sector because of the higher reduced mass of the system. A secondary reason is the fact that, in the bottom sector, the mixing effect are enhanced by the smaller mass splitting between the heavy mesons, B and B^*, and between the heavy baryons Σ_b and Σ_b^*. Finally, we observe that the number of resonant and bound states increases when one switches on the $5q$ potential.

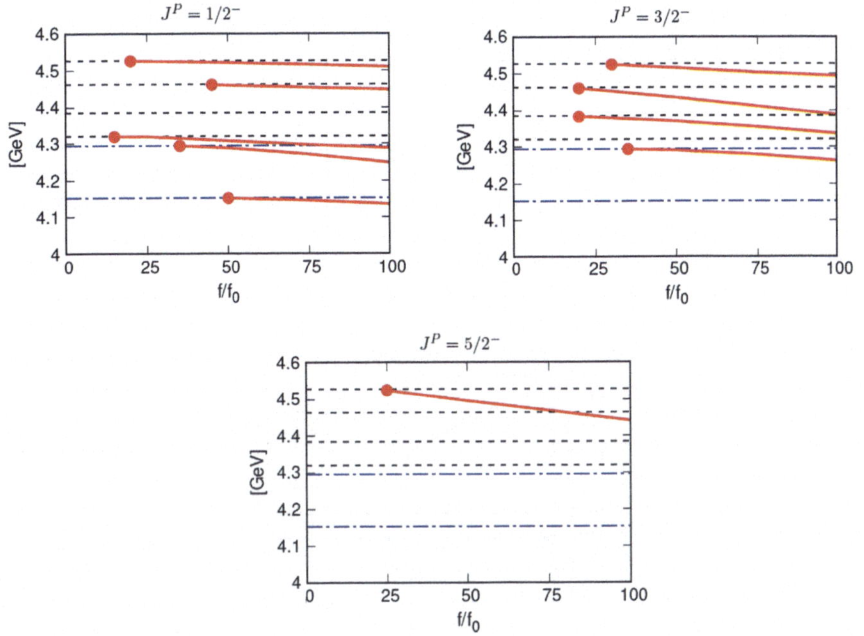

Fig. 98.2 Bound and resonant state energies of the hidden-charm molecules (solid lines) with various coupling constants f. Dot-dashed lines are the $\bar{D}\Lambda_c$ and $\bar{D}^*\Lambda_c$ thresholds. Dashed lines are the $\bar{D}\Sigma_c$, $\bar{D}\Sigma_c^*$, $\bar{D}^*\Sigma_c$ and $\bar{D}^*\Sigma_c^*$ thresholds. The lowest threshold, $\bar{D}\Lambda_c$, is at 4150 MeV and the state whose energy is lower than the threshold is a bound state (Adapted from [28])

98.3 Conclusions

We showed how the negative parity hidden-charm (hidden-bottom) pentaquark states can arise from the interplay between the meson-baryon channels, and the compact five-quark configurations. As a result we found that, in the hidden charm sector, the one pion exchange alone is not enough strong to produce resonant or bound states, while, in the hidden bottom sector, it pentaquark states are produced even without adding the coupling interaction between the compact five-quark configuration and the meson-baryon molecular configuration. As the hidden bottom pentaquark states are more stable than their hidden charm partners we encourage experimentalists look for new resonant states in the hidden bottom sector. Finally, we observe that another possibility to gain insight into the nature of the pentaquark states consists of producing these states in a prompt production reaction. In fact, an observation pentaquark states in prompt production would indicate that the pentaquark has a compact nature, while on the contrary, a non-observation would not exclude the the possibility that they are molecular states.

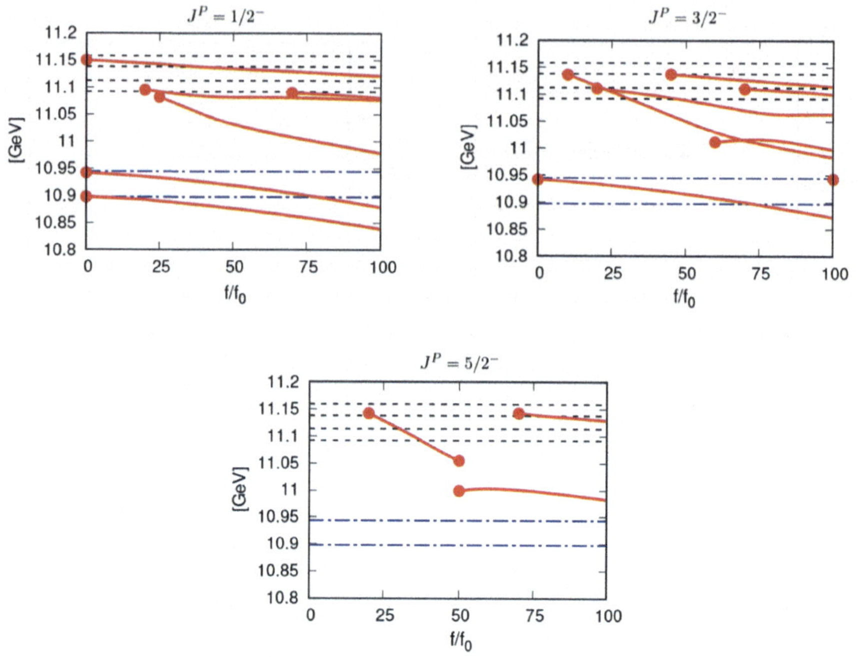

Fig. 98.3 As Fig. 98.2, but for the bottom sector. The lowest threshold, $\bar{B}\Lambda_b$, is at about 10,900 MeV and the state whose energy is lower than the threshold is a bound state (Adapted from [28])

References

1. Aaij, R., et al.: [LHCb Collaboration]: Observation of $J/\psi p$ resonances consistent with pentaquark states in $\Lambda_b^0 \to J/\psi K^- p$ decays. Phys. Rev. Lett. **115**, 072001 (2015)
2. Aaij, R., et al.: [LHCb Collaboration]: Model-independent evidence for $J/\psi p$ contributions to ${}^0_b \to J/\psi p K^-$ decays. Phys. Rev. Lett. **117**, 082002 (2016)
3. Aaij, R., et al.: [LHCb Collaboration]: Evidence for exotic hadron contributions to ${}^0_b \to J/\psi p \pi^-$ decays. Phys. Rev. Lett. **117**, 082003 (2016)
4. Yuan, S.G., Wei, K.W., He, J., Xu, H.S., Zou, B.S.: Study of $qqqc\bar{c}$ five quark system with three kinds of quark-quark hyperfine interaction. Eur. Phys. J. A **48**, 61 (2012)
5. Santopinto, E., Giachino, A.: Compact pentaquark structures. Phys. Rev. D **96**, 014014 (2017)
6. Wu, J., Liu, Y.R., Chen, K., Liu, X., Zhu, S.L.: Hidden-charm pentaquarks and their hidden-bottom and B_c-like partner states. Phys. Rev. D **95**, 034002 (2017)
7. Takeuchi, S., Takizawa, M.: The hidden charm pentaquarks are the hidden color-octet uud baryons. Phys. Lett. B **764**, 254 (2017)
8. Maiani, L., Polosa, A.D., Riquer, V.: The new pentaquarks in the diquark model. Phys. Lett. B **749**, 289 (2015)
9. Li, G.N., He, X.G., He, M.: Some predictions of diquark model for hidden charm pentaquark discovered at the LHCb. J. High Energy Phys. **1512**, 128 (2015)
10. Ali, A., Ahmed, I., Aslam, M.J., Rehman, A.: Heavy quark symmetry and weak decays of the b-baryons in pentaquarks with a $c\bar{c}$ component. Phys. Rev. D **94**, 054001 (2016)
11. Lebed, R.F.: The pentaquark candidates in the dynamical diquark picture. Phys. Lett. B **749**, 454 (2015)

12. Zhu, R., Qiao, C.F.: Pentaquark states in a diquarktriquark model. Phys. Lett. B **756**, 259 (2016)
13. Wu, J.J., Molina, R., Oset, E., Zou, B.S.: Prediction of narrow N^* and Λ^* resonances with hidden charm above 4 GeV. Phys. Rev. Lett. **105**, 232001 (2010)
14. Wu, J.J., Molina, R., Oset, E., Zou, B.S.: Dynamically generated N^* and Λ^* resonances in the hidden charm sector around 4.3 GeV, Phys. Rev. C **84**, 015202 (2011)
15. Garcia-Recio, C., Nieves, J., Romanets, O., Salcedo, L.L., Tolos, L.: Hidden charm N and Delta resonances with heavy-quark symmetry. Phys. Rev. D **87**, 074034 (2013)
16. Karliner, M., Rosner, J.L.: New exotic meson and baryon resonances from doubly-heavy hadronic molecules. Phys. Rev. Lett. **115**, 122001 (2015)
17. Chen, R., Liu, X., Li, X.Q., Zhu, S.L.: Identifying exotic hidden-charm pentaquarks. Phys. Rev. Lett. **115**, 132002 (2015)
18. Roca, L., Nieves, J., Oset, E.: LHCb pentaquark as a $\bar{D}^*\Sigma_c - \bar{D}^*\Sigma_c^*$ molecular state. Phys. Rev. D **92**, 094003 (2015)
19. He, J.: $\bar{D}\Sigma_c^*$ and $\bar{D}^*$ interactions and the LHCb hidden-charmed pentaquarks. Phys. Lett. B **753**, 547 (2016)
20. Meißner, U.G., Oller, J.A.: Testing the $\chi_{c1}\,p$ composite nature of the $P(4450)$. Phys. Lett. B **751**, 59 (2015)
21. Chen, H.X., Chen, W., Liu, X., Steele, T.G., Zhu, S.L.: Towards exotic hidden-charm pentaquarks in QCD. Phys. Rev. Lett. **115**, 172001 (2015)
22. Uchino, T., Liang, W.H., Oset, E.: Baryon states with hidden charm in the extended local hidden gauge approach. Eur. Phys. J. A **52**, 43 (2016)
23. Burns, T.J.: Phenomenology of $P_c(4380)^+$, $P_c(4450)^+$ and related states. Eur. Phys. J. A **51**, 152 (2015)
24. Shimizu, Y., Suenaga, D., Harada, M.: Coupled channel analysis of molecule picture of $P_c(4380)$. Phys. Rev. D **93**, 114003 (2016)
25. Yamaguchi, Y., Santopinto, E.: Hidden-charm pentaquarks as a meson-baryon molecule with coupled channels for $\bar{D}^{(*)}$c and $\bar{D}^{(*)}\Sigma_c^{(*)}$. Phys. Rev. D **96**, 014018 (2017)
26. Shimizu, Y., Harada, M.: Hidden Charm Pentaquark $P_c(4380)$ and doubly charmed baryon $\Xi_{cc}^*(4380)$ as hadronic molecule states. Phys. Rev. D **96**, 094012 (2017)
27. Yamaguchi, Y., Santopinto, E.: Hidden-charm pentaquarks as a meson-baryon molecule with coupled channels for $\overline{D}^{(*)}\Lambda_c$ and $\overline{D}^{(*)}\Sigma_c^{(*)}$. Phys. Rev. D **96**, 014018 (2017)
28. Yamaguchi, Y., Giachino, A., Hosaka, A., Santopinto, E., Takeuchi, S., Takizawa, M.: Hidden-charm and bottom meson-baryon molecules coupled with five-quark states Phys. Rev. D **96**, 114031 (2017)

Chapter 99
Short Range $\pi J/\psi - D\bar{D}^*$ Potential Described by the Quark Exchange Diagram

Yasuhiro Yamaguchi, Yukihiro Abe, Kenji Fukukawa and Atsushi Hosaka

Abstract Recently, unexpected Exotic hadrons have been reported in the heavy flavor sector. Especially near the thresholds, these exotic states would be produced by the hadron-hadron interaction dynamically. In this talk, we study the short range $\pi J/\psi - D\bar{D}^*$ potential whose importance is indicated by the Lattice QCD study. We compare the results from the $D^{(*)}$ meson exchange, and the Born-order quark exchange models, and find the large difference between two models.

99.1 Introduction

Exotic hadrons reported in the heavy flavor sector have been one of the interesting topics in hadron and nuclear physics [1, 2]. Especially, the charged charmonium $Z_c(3900)$ [3–6] is interesting. The ordinary charmonium being $c\bar{c}$ does not have electric charge, and hence $Z_c(3900)$ is a genuine exotic state which must have a multiquark component such as $c\bar{c}u\bar{d}$. There have been a lot of theoretical studies of $Z_c(3900)$, such as the multiquark picture, hadronic molecule, kinematical effect and so on, which is summarized in [2].

The exotic states have also been studied by the Lattice QCD approach. Recently HALQCD collaboration [7] study $Z_c(3900)$ as the hadron-hadron scattering including the $\pi J/\psi - \rho\eta_c - D\bar{D}^*$ channels at the non physical pion mass $m_\pi = 410 - 700$ MeV. In their simulation, the interesting result is obtained, which indicates that $Z_c(3900)$ is a virtual state induced by the strong $\pi J/\psi - D\bar{D}^*$ potential. The

Y. Yamaguchi (✉)
RIKEN Nishina Center, Hirosawa, Wako, Saitama 351-0198, Japan
e-mail: yasuhiro.yamaguchi@riken.jp

Y. Abe · A. Hosaka
Research Center for Nuclear Physics (RCNP), Osaka University, Ibaraki, Osaka 567-0047, Japan

K. Fukukawa
Suma Gakuen, Kobe, Hyogo 654-0009, Japan

N. A. Orr et al. (eds.), *Recent Progress in Few-Body Physics*, Springer Proceedings in Physics 238,
https://doi.org/10.1007/978-3-030-32357-8_99

$\pi J/\psi - D\bar{D}^*$ potential needs the charm quark (hadron) exchange process, therefore this interaction had been considered to suppress due to the large charm quark mass (>1 GeV).

In this talk, we study the short range $\pi J/\psi - D\bar{D}^*$ interaction by the theoretical model approaches. We focus on the interaction for the quantum number $J^{PC} = 1^{+-}$ with the total angular momentum J, parity P and C−parity C. The models employing here are (a) Born-order quark exchange models [8, 9], and (b) the $D^{(*)}$ meson exchange model. In order to understand the mechanism of the $\pi J/\psi - D\bar{D}^*$ interaction, the results obtained by these models are compared.

99.2 Meson Exchange Model

The D meson exchange potential is obtained as the Born-term of the t−channel Feynman diagram. The t−channel diagram is described by the effective Lagrangians respecting to the heavy quark and chiral symmetries [10, 11]. The Lagrangians of the $\pi D^{(*)} D^{(*)}$ and $J/\psi D^{(*)} D^{(*)}$ couplings are given by

$$\mathcal{L}_{\pi HH} = -\frac{g_\pi}{2f_\pi}\mathrm{Tr}\left[H_1\gamma_\mu\gamma_5\partial^\mu\hat{\pi}\bar{H}_1\right], \tag{99.1}$$

$$\mathcal{L}_{\psi HH} = g'\mathrm{Tr}\left[\mathcal{J}\bar{H}_2\overleftrightarrow{\partial}_\mu\gamma^\mu\bar{H}_1\right] + \mathrm{H.c.}, \tag{99.2}$$

with $g_\pi = 0.59$, $g' = 4/\sqrt{m_\psi m_D^2}$, and the pion decay constant $f_\pi = 93$ MeV. The heavy meson fields H and $\mathcal{J}$ are

$$H_1 = \frac{1+\not{v}}{2}\left[D^{*\mu}\gamma_\mu + iD\gamma_5\right], \quad H_2 = \left[\bar{D}^{*\mu}\gamma_\mu + i\bar{D}\gamma_5\right]\frac{1-\not{v}}{2}, \tag{99.3}$$

$$\mathcal{J} = \frac{1+\not{v}}{2}\left[\psi^\mu\gamma_\mu + i\eta_c\gamma_5\right]\frac{1-\not{v}}{2}. \tag{99.4}$$

From these Lagrangians, the meson exchange potentials are given by

$$V^\pi = -\frac{1}{2}\left(\frac{g_\pi}{f_\pi}\right)^2\left[\mathbf{S}_1\cdot\mathbf{S}_2 C(r;m_\pi) + S_{12}(\hat{r})T(r;m_\pi)\right]\boldsymbol{\tau}_1\cdot\boldsymbol{\tau}_2, \tag{99.5}$$

$$V^D = \frac{2}{3}\frac{g_\psi g_\pi}{f_\pi\sqrt{E_\pi}}\left[\mathbf{S}_1\cdot\mathbf{S}_2 C(r;m_D) + S_{12}(\hat{r})T_D(r;m_D)\right], \tag{99.6}$$

$$V^{D^*} = \frac{2}{3}\frac{g_\psi g_\pi}{f_\pi\sqrt{E_\pi}}\left[2\mathbf{S}_1\cdot\mathbf{S}_2 C(r;m_{D^*}) - S_{12}(\hat{r})T(r;m_{D^*})\right]. \tag{99.7}$$

The spin operator $\mathbf{S}_i$ $(i = 1, 2)$ is given by the polarization vector ε or the spin-one operator T. The tensor operator is $S_{12}(\hat{r}) = 3(\mathbf{S}_1 \cdot \hat{r})(\mathbf{S}_2 \cdot \hat{r}) - \mathbf{S}_1 \cdot \mathbf{S}_2$. The operator $\boldsymbol{\tau}_1 \cdot \boldsymbol{\tau}_2$ is for the isospin. The functions $C(r; m)$ and $T(r; m)$ are introduced in [12–16].

99.3 Quark Exchange Model

The Quark exchange model is described by the Born-order quark exchange diagram introduced in [8, 9]. The amplitude is given by $\mathcal{M} \propto \langle \phi_C, \phi_D | H_q | \phi_A, \phi_B \rangle$ with the quark Hamiltonian H_q and the meson wave function ϕ_i $(i = A, B, C, D)$.

The constituent quark model Hamiltonian H^q is given by [8, 9]

$$H^q_{ij} = K_q + \left(-\frac{3}{4} br + \frac{\alpha_s}{r} - C \right) \mathbf{F}_i \cdot \mathbf{F}_j - \frac{8\pi\alpha_h}{3m_i m_j} \left(\frac{\sigma^3}{\pi^{3/2}} e^{-\sigma^2 r_{ij}^2} \right) \mathbf{S}_i \cdot \mathbf{S}_j \mathbf{F}_i \cdot \mathbf{F}_j. \tag{99.8}$$

These parameters are determined to reproduce the meson masses, summarized in Table 99.1. K_q is the kinetic term, and $\mathbf{F}_i = \boldsymbol{\lambda}_i/2$ with the Gell-Mann matrix λ_i.

For the meson wave function ϕ, we employ the single Gaussian, $\phi(r) = (4\pi b_G)^{-3/4} \exp\left(-r^2/8b_G\right)$, with the Gaussian parameter b_G determined to minimize $E(b_G) = \langle \phi | H^q | \phi \rangle$. The meson masses and Gaussian parameter are summarized in Table 99.2.

Table 99.1 Parameters of the quark Hamiltonian determined to reproduce the meson masses in [8, 9]. The charm quark mass m_c is also determined to fit the charmed meson masses

m_q [GeV]	m_c [GeV]	α_s	α_h	b [GeV^{-2}]	C [GeV]	σ [GeV]
0.375	1.9	0.857	0.840	0.154	−0.4358	0.70

Table 99.2 The meson masses m and Gaussian parameter b_G obtained in the single Gaussian Approximation

	(m [GeV], b_G [GeV^{-2}])		(m [GeV], b_G [GeV^{-2}])		(m [GeV], b_G [GeV^{-2}])
π	(0.258, 0.854)	D	(1.876, 0.965)	η_c	(2.826, 0.261)
ρ	(0.782, 2.549)	D^*	(2.016, 1.298)	J/ψ	(2.910, 0.290)

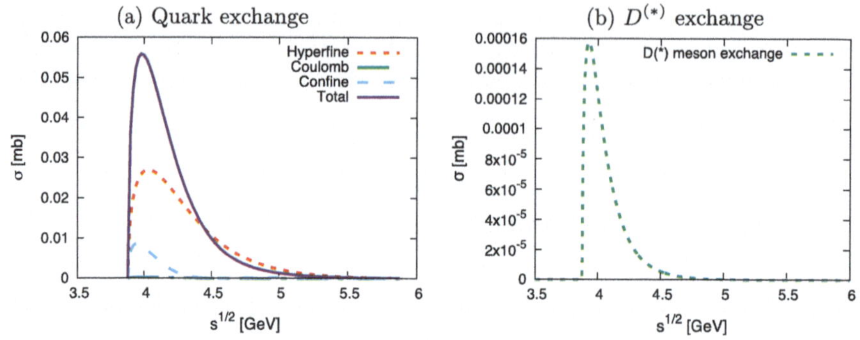

Fig. 99.1 The cross section of the $\pi J/\psi - D\bar{D}^*$ transition by **a** the quark exchange model and **b** $D^{(*)}$ exchange model. In (a), the solid line shows the cross section including all terms, spin-spin (hyperfine), coulomb, and confinement, while the short-dashed line is only for the spin-spin (hyperfine) term, the dashed-dot for coulomb term, and the long-dashed line for the confinement term

99.4 Numerical Results

From the interaction models introduced in Sects. 99.2 and 99.3, the cross section of the $\pi J/\psi - D\bar{D}^*$ transition is obtained numerically as shown in Fig. 99.1. Figure 99.1a is for the cross section by the quark exchange model. We find the large contribution from the spin-spin term, while the coulomb term has a minor role. Since the spin-spin term with the charm quark is suppressed, this strong spin-spin contribution found here is given by the light quark dynamics

In comparison with Fig. 99.1a, the cross section by the meson exchange model in Fig. 99.1b is much smaller than that by the quark exchange one. This observation would be important to understand the short range $\pi J/\psi - D\bar{D}^*$ interaction. The $D^{(*)}$ exchange potential plays a minor role, while the very short-range quark exchange contribution has the important role in the $\pi J/\psi - D\bar{D}^*$ scattering.

99.5 Summary

The understanding hadron-hadron interaction is important to investigate the exotic structure produced dynamically near the thresholds. We study the short range $\pi J/\psi - D\bar{D}^*$ potential which is described by the quark exchange and the meson exchange models. We find that the cross section in the quark exchange model is dominated by the spin-spin term which is contributed by the light quark dynamics. The cross section obtained by the $D^{(*)}$ meson exchange model is much smaller than that by the quark exchange model. The large difference between two models is obtained, and it would be useful to understand the short range $\pi J/\psi - D\bar{D}^*$ interaction.

References

1. Brambilla, N., et al.: Eur. Phys. J. C **71**, 1534 (2011)
2. Hosaka, A., Iijima, T., Miyabayashi, K., Sakai, Y., Yasui, S.: Prog. Theor. Exp. Phys. **2016**, 062C01 (2016)
3. Ablikim, M., et al.: BESIII collaboration. Phys. Rev. Lett. **110**, 252001 (2013)
4. Liu, Z.Q., et al.: Belle collaboration. Phys. Rev. Lett. **110**, 252002 (2013)
5. Xiao, T., Dobbs, S., Tomaradze, A., Seth, K.K.: Phys. Lett. B **727**, 366 (2013)
6. Abazov, V.M., et al.: D0 collaboration. Phys. Rev. D **98**, 052010 (2018)
7. Ikeda, Y., et al.: HAL QCD collaboration. Phys. Rev. Lett. **117**, 242001 (2016)
8. Barnes, T., Swanson, E.S.: Phys. Rev. D **46**, 131 (1992)
9. Swanson, E.S.: Ann. Phys. **220**, 73 (1992)
10. Casalbuoni, R., Deandrea, A., Di Bartolomeo, N., Gatto, R., Feruglio, F., Nardulli, G.: Phys. Reports **281**, 145 (1997)
11. Manohar, A.V., Wise, M.B.: Heavy quark physics, Cambridge monographs on particle physics, nuclear physics and cosmology, p. 191. Cambridge University Press, Cambridge, England (2000)
12. Yamaguchi, Y., Ohkoda, S., Yasui, S., Hosaka, A.: Phys. Rev. D **84**, 014032 (2011)
13. Ohkoda, S., Yamaguchi, Y., Yasui, S., Sudoh, K., Hosaka, A.: Phys. Rev. D **86**, 014004 (2012)
14. Hosaka, A., Hyodo, T., Sudoh, K., Yamaguchi, Y., Yasui, S.: Prog. Part. Nucl. Phys. **96**, 88 (2017)
15. Yamaguchi, Y., Santopinto, E.: Phys. Rev. D **96**, 014018 (2017)
16. Yamaguchi, Y., Giachino, A., Hosaka, A., Santopinto, E., Takeuchi, S., Takizawa, M.: Phys. Rev. D **96**, 114031 (2017)

Chapter 100
The Origin of the Nucleon Mass

Cédric Lorcé

Abstract It is often claimed that 98% of the nucleon mass is generated by quantum chromodynamics. The decomposition of the nucleon mass based on the trace of the energy-momentum tensor suggests that gluons play by far a dominant role. About 25 years ago, Ji proposed another decomposition based on the energy component of the energy-momentum tensor, leading to a quite different picture. Recently, we critically revisited these decompositions and argued that both overlooked pressure effects. In particular we showed that Ji's decomposition, although mathematically correct, makes little sense from a physical point of view. We identify the proper mass decomposition along with a balance equation for the pressure forces.

100.1 Introduction

While the Brout-Englert-Higgs mechanism generates the current quark masses, it accounts only for about 2% of the nucleon mass. The remaining 98% comes from the relativistic kinetic energy and the strong interactions confining quarks and gluons inside hadrons [1, 2], described by quantum chromodynamics (QCD).

The question of the origin of the nucleon mass can be addressed based on the QCD energy-momentum tensor (EMT)

$$T^{\mu\nu} = \overline{\psi}\gamma^\mu \tfrac{i}{2}\overleftrightarrow{D}^\nu\psi - G^{a\mu\lambda}G^{a\nu}{}_\lambda + \tfrac{1}{4}\,\eta^{\mu\nu}G^2, \tag{100.1}$$

where ψ is the quark field, $\overleftrightarrow{D}^\mu = \overrightarrow{\partial}^\mu - \overleftarrow{\partial}^\mu - 2igA^{a\mu}t^a$ is the symmetric non-abelian covariant derivative, $G^{a\mu\nu}$ is the gluon field strength, and $\eta_{\mu\nu}$ is the Minkowski metric. The sum over the quark flavors is implicit. Note that in Particle Physics there is no fundamental reason for requiring the EMT to be symmetric [3, 4]. Beside the classical

C. Lorcé (✉)
Centre de Physique Théorique, École polytechnique, CNRS, Université Paris-Saclay, F-91128 Palaiseau, France
e-mail: cedric.lorce@polytechnique.edu

N. A. Orr et al. (eds.), *Recent Progress in Few-Body Physics*,
Springer Proceedings in Physics 238,
https://doi.org/10.1007/978-3-030-32357-8_100

term $\overline{\psi}m\psi$, the trace of the renormalized EMT

$$T^{\mu}_{\ \mu} = \frac{\beta(g)}{2g}\,G^2 + (1+\gamma_m)\,\overline{\psi}m\psi, \tag{100.2}$$

includes anomalous quantum contributions involving the β function and the anomalous quark mass dimension γ_m, see e.g. [5–8].

In the following, we critically review the two standard mass decompositions found in the literature, and propose a new one free of the problems associated with the former [9, 10].

100.2 Standard Decompositions

100.2.1 Trace Decomposition

Using the covariant normalization $\langle P'|P\rangle = 2P^0\,(2\pi)^3\,\delta^{(3)}(\boldsymbol{P}'-\boldsymbol{P})$, Poincaré invariance imposes that the forward matrix elements of the total EMT in a nucleon state with momentum P take the form [11]

$$\langle P|T^{\mu\nu}(0)|P\rangle = 2P^{\mu}P^{\nu}. \tag{100.3}$$

Considering the trace of this expression and using (100.2), one finds that

$$2M^2 = \langle P|\tfrac{\beta(g)}{2g}\,G^2|P\rangle + \langle P|(1+\gamma_m)\,\overline{\psi}m\psi|P\rangle. \tag{100.4}$$

Since the second term is known to give a rather small contribution, this picture suggests that most of the nucleon mass comes from gluons [12–14].

Although manifestly covariant, we find that the physical interpretation of this decomposition is somewhat misleading for the following reasons:

1. We know from Quantum Mechanics that we are in principle free to choose the normalization of states since physical quantities associated with an operator O are expressed as $\langle\Psi|O|\Psi\rangle/\langle\Psi|\Psi\rangle$. The standard trace decomposition does not involve the normalization factor $1/\langle P|P\rangle$ and appears to be manifestly covariant only because of the particular choice for $\langle P'|P\rangle$.
2. The relation between the trace of the EMT and the nucleon mass holds only at the level of the matrix elements and for the total EMT. At the operator level, it is not known how to relate the individual operators $\frac{\beta(g)}{2g}\,G^2$ and $(1+\gamma_m)\,\overline{\psi}m\psi$ to actual gluon and quark contributions to the nucleon mass.

Note that the forward matrix elements of any scalar operator are necessarily proportional to some power of the nucleon mass, since the latter is the only natural scale at our disposal. If we followed the same logic as with the trace operator, many of these scalar operators would lead to quite different "decompositions" of the nucleon mass. In summary, although the decomposition (100.4) is mathematically correct, one has to be very careful with the physical interpretation of the individual terms.

100.2.2 Ji's Decomposition

A decomposition of the nucleon mass analogous to the virial theorem for a harmonic oscillator and the hydrogen atom has been proposed by Ji [15, 16]. The idea is to decompose first the renormalized QCD EMT into traceless and trace parts

$$T^{\mu\nu} = \bar{T}^{\mu\nu} + \hat{T}^{\mu\nu} \qquad \text{with} \qquad \hat{T}^{\mu\nu} = \tfrac{1}{4}\,\eta^{\mu\nu} T^{\alpha}_{\ \alpha}. \tag{100.5}$$

The traceless part is then further decomposed into quark and gluon contributions $\bar{T}^{\mu\nu} = \bar{T}_q^{\mu\nu} + \bar{T}_g^{\mu\nu}$, whereas the trace part is further decomposed into quark mass and trace anomaly contributions $\hat{T}^{\mu\nu} = \hat{T}_m^{\mu\nu} + \hat{T}_a^{\mu\nu}$. The corresponding forward matrix elements can be written as

$$\langle P|\bar{T}_q^{\mu\nu}(0)|P\rangle = 2\,a(\mu^2)\left(P^\mu P^\nu - \tfrac{1}{4}\,\eta^{\mu\nu} M^2\right), \tag{100.6}$$
$$\langle P|\bar{T}_g^{\mu\nu}(0)|P\rangle = 2\,[1 - a(\mu^2)]\left(P^\mu P^\nu - \tfrac{1}{4}\,\eta^{\mu\nu} M^2\right), \tag{100.7}$$
$$\langle P|\hat{T}_m^{\mu\nu}(0)|P\rangle = \tfrac{1}{2}\,b(\mu^2)\,\eta^{\mu\nu} M^2, \tag{100.8}$$
$$\langle P|\hat{T}_a^{\mu\nu}(0)|P\rangle = \tfrac{1}{2}\,[1 - b(\mu^2)]\,\eta^{\mu\nu} M^2, \tag{100.9}$$

where the coefficients $a(\mu^2)$ and $b(\mu^2)$ depend generally on the renormalization scheme and scale μ.

According to the Hamiltonian formalism, the mass of a system is identified with the total energy defined in the rest frame. Ji then proposed to decompose the nucleon mass as

$$M = M_q + M_g + M_m + M_a, \tag{100.10}$$

where the various contributions on the right-hand side are defined as $M_i = \langle P|H_i|P\rangle/\langle P|P\rangle|_{\boldsymbol{P}=\boldsymbol{0}}$ with

$$H_q = \int \mathrm{d}^3r\, \psi^\dagger (i\boldsymbol{D}\cdot\boldsymbol{\alpha})\psi, \tag{100.11}$$

$$H_g = \int \mathrm{d}^3r\, \bar{T}_g^{00}(r), \tag{100.12}$$

$$H_m = \int \mathrm{d}^3r\, \left(1 + \tfrac{1}{4}\,\gamma_m\right) \overline{\psi} m \psi, \tag{100.13}$$

$$H_a = \int \mathrm{d}^3r\, \hat{T}_a^{00}(r). \tag{100.14}$$

Note that the QCD equations of motion have been used to rearrange quark mass contributions between $\bar{T}_q^{00}$ and $\hat{T}_m^{00}$. Using the parametrization in Eqs. (100.6)–(100.9), one finds that $M_q = \frac{3}{4}\left(a - \frac{b}{1+\gamma_m}\right) M$, $M_g = \frac{3}{4}\,(1-a)\,M$, $M_m = \frac{1}{4}\,\frac{4+\gamma_m}{1+\gamma_m}\,b\,M$, and $M_a = \frac{1}{4}\,(1-b)\,M$.

Sometimes, Ji's decomposition is criticized because it is obtained in the nucleon rest frame [13]. We do not consider this as an actual problem since most of the physical quantities, like e.g. energy and momentum, are known to be frame-dependent. For massive systems like the nucleon, the rest frame plays a special role and is commonly chosen as the preferred frame for a decomposition. Note also that if one insists on manifest Lorentz invariance, it is always possible to trade a frame-dependent quantity defined in a particular frame for a frame-independent quantity with simple interpretation in that particular frame only [3, 17]. The archetypical example is the four-momentum squared $p^2 = (p^0)^2 - \boldsymbol{p}^2 = m^2$, where $m = p^0|_{\boldsymbol{p}=\boldsymbol{0}}$ represents the rest-frame energy. In the case of Ji's decomposition, a covariant form can formally be obtained by trading $\langle T^{00}\rangle|_{\boldsymbol{P}=\boldsymbol{0}}$ for the Lorentz-invariant quantity $\langle T^{\mu\nu}u_\mu u_\nu\rangle$, where $u^\mu = P^\mu/M$ is the nucleon four-velocity.

The actual problem with Ji's decomposition is that although T^{00}, $\bar{T}^{00}$ and $\hat{T}^{00}$ all have the dimension of energy densities, they correspond to *different* thermodynamic potentials. This is not apparent because the non-covariant treatment, focused on the $\mu = \nu = 0$ component in the rest frame, is unable to distinguish pressure-volume work from other forms of energy. In summary, although Ji's decomposition (100.10) is also mathematically correct, it amounts to adding apples and oranges since all four individual contributions correspond to four different combinations of internal energy and pressure-volume work.

100.3 New Decomposition

Lorentz covariance implies that the mass decomposition follows directly from a decomposition of the EMT. Since the various components of the EMT correspond to different mechanical properties, one should not consider a decomposition based on the tensor structure like in (100.5), but rather a decomposition based on the sole nature of the constituents. The QCD EMT can naturally be decomposed into quark ($i = q$) and gluon ($i = g$) contributions $T^{\mu\nu} = \sum_i T_i^{\mu\nu}$. The corresponding forward matrix elements in a nucleon state read [3, 18].

$$\langle P|T_i^{\mu\nu}(0)|P\rangle = 2P^\mu P^\nu A_i(0) + 2M^2\eta^{\mu\nu}\bar{C}_i(0), \tag{100.15}$$

where $A_i(0)$ and $\bar{C}_i(0)$ are two gravitational form factors (GFFs) evaluated at zero momentum transfer. Poincaré invariance (100.3) implies that $\sum_i A_i(0) = 1$ and $\sum_i \bar{C}_i(0) = 0$. Defining the quark and gluon contributions as

$$T_q^{\mu\nu} \equiv \bar{T}_q^{\mu\nu} + \hat{T}_m^{\mu\nu} \quad \text{and} \quad T_g^{\mu\nu} \equiv \bar{T}_g^{\mu\nu} + \hat{T}_a^{\mu\nu}, \tag{100.16}$$

these GFFs can easily be related to Ji's coefficients as follows $a_i = A_i(0)$, $b_i = A_i(0) + 4\bar{C}_i(0)$.

If we average the expectation value of $T_i^{\mu\nu}$ over the nucleon volume $\mathcal{V} = V\,M/P^0$

$$\langle\langle T_i^{\mu\nu}\rangle\rangle \equiv \frac{1}{\mathcal{V}}\frac{\langle P|\int d^3r\, T_i^{\mu\nu}(r)|P\rangle}{\langle P|P\rangle} = \frac{\langle P|T_i^{\mu\nu}(0)|P\rangle}{2M^2}\frac{M}{V}, \tag{100.17}$$

we find using $u^\mu = P^\mu/M$ [9, 10]

$$\langle\langle T_i^{\mu\nu}\rangle\rangle = (\varepsilon_i + \bar{p}_i)\, u^\mu u^\nu - p_i\, \eta^{\mu\nu} \tag{100.18}$$

with $\varepsilon_i = [A_i(0) + \bar{C}_i(0)]\,\frac{M}{V}$ and $p_i = -\bar{C}_i(0)\,\frac{M}{V}$. The structure in (100.18) is similar to that of an element of perfect fluid in relativistic hydrodynamics and allows us to interpret ε_i and p_i as partial proper internal energy density and isotropic pressure averaged over the nucleon, respectively.

Multiplying ε_i and p_i by the nucleon proper volume V, we obtain the partial internal energy $U_i = \varepsilon_i V = [A_i(0) + \bar{C}_i(0)]\, M$ and pressure-volume work $W_i = p_i V = -\bar{C}_i(0)\, M$ which satisfy the sum rules

$$M = \sum_i U_i, \qquad 0 = \sum_i W_i \tag{100.19}$$

derived from Poincaré invariance (100.3). The first sum rule is nothing but the mass decomposition we were looking for. The second sum rule expresses the stability of the nucleon by imposing that the total pressure forces must balance between the various parts of the system.

100.4 Discussion

Now that internal energy and pressure-volume work contributions are well identified, we can unravel the meaning of the old decompositions. Starting with the trace decomposition (100.4) divided by $2M$, we see that it does not correspond to a decomposition of the total energy of the system, but rather to a decomposition of the interaction measure $I = \sum_i I_i$ with $I_i = U_i - 3W_i$. Since the total pressure-volume work vanishes, the total interaction measure coincides with the nucleon mass $I = M$. The fact that the gluon contribution dominates $I_g \gg I_q$ does not mean that it is responsible for most of the nucleon mass as claimed e.g. in [12–14]. It turns out in fact that the nucleon mass is more or less equally shared between quarks and gluons $U_q \approx U_g$ [9, 10]. What the dominance of the gluon contribution to the interaction measure indicates is that the gluon average pressure $p_g = -p_q$ is large and negative. In average, gluons are therefore responsible for the net attractive forces inside the nucleon, balanced by the net repulsive forces associated with quarks.

Turning now to Ji's decomposition, we find that using the definition of quark and gluon EMT given in (100.16), the four terms in (100.10) correspond to the following combinations of average internal energy and pressure-volume work

$$M_q = \tfrac{3}{4}\,\tfrac{1}{1+\gamma_m}\left[\gamma_m U_q + (4+\gamma_m)W_q\right], \tag{100.20}$$

$$M_m = \tfrac{1}{4}\,\tfrac{4+\gamma_m}{1+\gamma_m}\left(U_q - 3W_q\right), \tag{100.21}$$

$$M_g = \tfrac{3}{4}\left(U_g + W_g\right), \tag{100.22}$$

$$M_a = \tfrac{1}{4}\left(U_g - 3W_g\right). \tag{100.23}$$

Clearly, each term M_i represents a different physical quantity. One is adding apples and oranges, which is not something we would like for a genuine mass decomposition. In order for Ji's decomposition to make sense, one has to give up something. If one sacrifices covariance, one can impose from the onset the further decompositions $U_q = M_q + M_m$ and $U_g = M_g + M_a$, keeping the pressure-volume works unchanged. If one wants to preserve covariance, one then has to treat $\bar{T}_i^{\mu\nu}$ and $\hat{T}_i^{\mu\nu}$ as actual separate EMTs. In that case, one is fixing arbitrarily the equation of state of the individual contributions because of the restriction on the Lorentz structure. For example, the gluon contribution in Ji's decomposition is divided into kinetic+potential energy treated as a pure radiation $\bar{p}_g = \frac{1}{3}\,\bar{\varepsilon}_g$, and trace anomaly treated as a cosmological constant $\hat{p}_a = -\hat{\varepsilon}_a$. In other words, a covariant decomposition into four terms can only be achieved by combining in an arbitrary way the two a priori independent sum rules (100.19) into a single one.

Acknowledgements This work is a result of discussions held at the workshop "The Proton Mass: At the Heart of Most Visible Matter" at the ECT* Trento, on 3-7 April 2017. It has been supported by the Agence Nationale de la Recherche under the project ANR-16-CE31-0019.

References

1. Schumacher, M.: Nambu's Nobel Prize, the σ meson and the mass of visible matter. Ann. Phys. **526**(5–6), 215–226 (2014). https://doi.org/10.1002/andp.201400077
2. Gao, H., Liu, T., Peng, C., Ye, Z., Zhao, Z.: Proton remains puzzling. Universe **3**(2), 18–30 (2015)
3. Leader, E., Lorcé, C.: The angular momentum controversy: whats it all about and does it matter? Phys. Reptorts **541**(3), 163–248 (2014). https://doi.org/10.1016/j.physrep.2014.02.010
4. Lorcé, C.: The light-front gauge-invariant energy-momentum tensor. JHEP **1508**, 045 (2015). https://doi.org/10.1007/JHEP08(2015)045
5. Crewther, R.J.: Nonperturbative evaluation of the anomalies in low-energy theorems. Phys. Rev. Lett. **28**, 1421–1424 (1972). https://doi.org/10.1103/PhysRevLett.28.1421
6. Chanowitz, M.S., Ellis, J.R.: Canonical anomalies and broken scale invariance. Phys. Lett. **40B**, 397–400 (1972). https://doi.org/10.1016/0370-2693(72)90829-5
7. Adler, S.L., Collins, J.C., Duncan, A.: Energy-momentum-tensor trace anomaly in spin 1/2 quantum electrodynamics. Phys. Rev. D **15**, 1712–1721 (1977). https://doi.org/10.1103/PhysRevD.15.1712
8. Nielsen, N.K.: The energy momentum tensor in a nonabelian quark gluon theory. Nucl. Phys. B **120**, 212–220 (1977). https://doi.org/10.1016/0550-3213(77)90040-2
9. Lorcé, C.: On the hadron mass decomposition. Eur. Phys. J. C **78**(2), 120 (2018). https://doi.org/10.1140/epjc/s10052-018-5561-2
10. Lorcé, C., Moutarde, H., Trawiński, A.P.: Revisiting the mechanical properties of the nucleon. Eur. Phys. J. C **79**(1), 89 (2019). https://doi.org/10.1140/epjc/s10052-019-6572-3

11. Jaffe, R.L., Manohar, A.: The G(1) problem: fact and fantasy on the spin of the proton. Nucl. Phys. B **337**, 509–546 (1990). https://doi.org/10.1016/0550-3213(90)90506-9
12. Shifman, M.A., Vainshtein, A.I., Zakharov, V.I.: Remarks on higgs boson interactions with nucleons. Phys. Lett. **78B**, 443–446 (1978). https://doi.org/10.1016/0370-2693(78)90481-1
13. Roberts, C.D.: Perspective on the origin of hadron masses. Few Body Syst. **58**(1), 5 (2017). https://doi.org/10.1007/s00601-016-1168-z
14. Krein, G., Thomas, A.W., Tsushima, K.: Nuclear-bound quarkonia and heavy-flavor hadrons. Prog. Part. Nucl. Phys. **100**, 161–210 (2018). https://doi.org/10.1016/j.ppnp.2018.02.001
15. Ji, X.D.: A QCD analysis of the mass structure of the nucleon. Phys. Rev. Lett. **74**, 1071–1074 (1995). https://doi.org/10.1103/PhysRevLett.74.1071
16. Ji, X.D.: Breakup of hadron masses and energy-momentum tensor of QCD. Phys. Rev. D **52**, 271–281 (1995). https://doi.org/10.1103/PhysRevD.52.271
17. Hoodbhoy, P., Ji, X.D., Lu, W.: Implications of color gauge symmetry for nucleon spin structure. Phys. Rev. D **59**, 074010 (1999). https://doi.org/10.1103/PhysRevD.59.074010
18. Ji, X.D.: Gauge-invariant decomposition of nucleon spin. Phys. Rev. Lett. **78**, 610–613 (1997). https://doi.org/10.1103/PhysRevLett.78.610

Chapter 101
Pion-Cloud Contribution to the $N \to \Delta$ Transition Form Factors

Ju-Hyun Jung and Wolfgang Schweiger

Abstract We examine the contribution of the pion cloud to the electromagnetic $N \to \Delta$ transition form factors within a relativistic hybrid constituent-quark model. In this model baryons consist not only of the $3q$ valence component, but contain, in addition, a $3q\pi$ non-valence component. We start with constituent quarks which are subject to a scalar, isoscalar confining force. This leads to an $SU(6)$ spin-flavor symmetric spectrum with degenerate nucleon and Delta masses. Mass splitting is caused by pions which are assumed to couple directly to the quarks. The point-form of relativistic quantum mechanics is employed to achieve a relativistically invariant description of this system. The $N \to \Delta$ transition current is then determined from the one-photon exchange contribution to the Δ electroproduction amplitude. We will give predictions for the ratios R_{EM} and R_{SM} of electric to magnetic and Coulomb to magnetic form factors, which are supposed to be most sensitive to pion-cloud effects.

101.1 Formalism

For a proper relativistic description of the $N \to \Delta$ transition form factors we make use of point-form relativistic quantum mechanics in connection with the Bakamjian-Thomas construction [1]. Like in previous work [2, 3] we use this framework to determine the one-photon-exchange amplitude for $e^- p \to e^- \Delta^+$ scattering. From this scattering amplitude we extract the electromagnetic $p \to \Delta^+$ transition current and determine the form factors by means of a covariant analysis of the transition current. Thereby both, the nucleon and the Delta are assumed to consist of a $3q$ and a $3q+\pi$ component and, in addition to the dynamics of electron and quarks, the dynamics of the photon and the pion are fully taken into account. This is accomplished by means of a multichannel formulation that comprises all states which can occur during the scattering process (i.e. $|3q, e\rangle$, $|3q, \pi, e\rangle$, $|3q, e, \gamma\rangle$, $|3q, \pi, e, \gamma\rangle$). After reducing the mass eigenvalue equation for this system of coupled states to the

J.-H. Jung (✉) · W. Schweiger
Institute of Physics, University of Graz, 8010 Graz, Austria
e-mail: ju.jung@uni-graz.at

N. A. Orr et al. (eds.), *Recent Progress in Few-Body Physics*,
Springer Proceedings in Physics 238,
https://doi.org/10.1007/978-3-030-32357-8_101

$3qe$-component, one ends up with an eigenvalue equation of the form

$$\left[\hat{M}^{\text{conf}}_{3qe} + \hat{K}_\pi(\sqrt{s} - \hat{M}^{\text{conf}}_{3q\pi e})^{-1}\hat{K}^\dagger_\pi + \hat{V}^{\text{opt}}_{1\gamma}(\sqrt{s})\right]|\psi_{3qe}\rangle = \sqrt{s}\,|\psi_{3qe}\rangle\,, \qquad (101.1)$$

where $\hat{V}^{\text{opt}}_{1\gamma}(\sqrt{s})$ is the 1γ-exchange optical potential, $\sqrt{s}$ the invariant mass of the scattering system and $\hat{K}_\pi$ the $qq\pi$ vertex operator. We assume an instantaneous scalar and isoscalar confining force between the quarks, which enters $\hat{M}^{\text{conf}}_{3q(\pi)e}$. The invariant 1γ-exchange amplitude for electroproduction of the Delta is now obtained by sandwiching $\hat{V}^{\text{opt}}_{1\gamma}(\sqrt{s})$ between (the valence component of) physical electron-nucleon $|eN\rangle$ and electron-Delta $|e\Delta\rangle$ states , i.e. eigenstates of $[\hat{M}^{\text{conf}}_{3qe} + \hat{K}_\pi(\sqrt{s} - \hat{M}^{\text{conf}}_{3q\pi e})^{-1}\hat{K}^\dagger_\pi]$. The crucial point is now to observe that, due to instantaneous confinement, propagating intermediate states do not contain free quarks, they rather contain bare nucleons N_0 or bare Deltas Δ_0 (or corresponding excitations, which are neglected in our calculations). The bare particles are eigenstates of the pure confinement problem. This allows us to rewrite the scattering amplitude in terms of pure hadronic degrees of freedom with the quark substructure being hidden in strong and electromagnetic vertex form factors of the bare baryons. This is graphically represented in Fig. 101.1.

For scalar, isoscalar confinement the masses of the bare nucleon and Delta are the same, $m_{N_0} = m_{\Delta_0} =: m_0$, and also the three-quark wave functions coincide due to $SU(6)$ spin-flavor symmetry. Instead of choosing a particular confining interaction we therefore rather parameterize the three-quark wave function of N_0 and Δ_0 by means of a Gaussian. Knowing the bare mass m_0, the (pseudovector) pion-quark coupling $f_{\pi qq}$ and the constituent-quark masses $m_u = m_d =: m_q$, one can first calculate the strong couplings and form factors at the $\pi N_0 N_0$, $\pi N_0 \Delta_0$ and $\pi N_0 \Delta_0$ vertices and in the sequel the renormalization effect of pion loops on the nucleon and Delta mass.

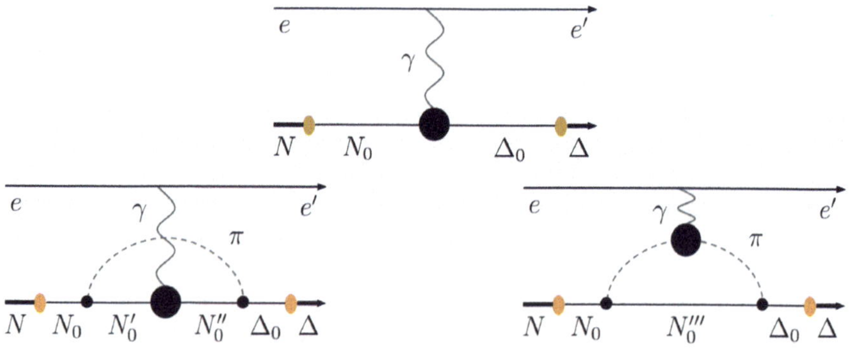

Fig. 101.1 The three graphs contributing to electroexcitation of the Δ resonance in the presence of a pion cloud. The big blobs represent electromagnetic (transition) form factors involving the bare nucleon N_0 and the bare Delta Δ_0. The small black blobs represent strong form factors at the $\pi N_0 N_0$, $\pi N_0 \Delta_0$ vertices. All these form factors are determined by the valence-quark wave functions of the bare baryons. A vertex form factor, calculated within a constituent-quark model [2] and the same approach as used here, is also assumed at the pion-photon vertex

Table 101.1 Model parameters for two common choices of the constituent-quark mass m_q

	m_q [GeV]	$f_{qq\pi}$	m_0 [GeV]	α
Model I	0.263	0.8067	1.380	2.660
Model II	0.340	0.7565	1.390	2.585

Fixing the constituent-quark mass m_q in advance, we have varied the remaining three parameters (m_0, $f_{qq\pi}$ and α) by means of a self-consistent procedure such that the solution of a mass-eigenvalue problem analogous to (101.1) (just without electron and photon) gives the physical nucleon and Delta masses. Our resulting parameters for two common choices of the constituent-quark mass are given in Table 101.1. A more detailed account of the parameter fixing can be found in [4].

What is still necessary to calculate the leading-order electroproduction amplitude, as depicted in Fig. 101.1, are the electromagnetic (transition) form factors of the bare baryons. These are obtained from the first graph in Fig. 101.1 by identifying the bare and the physical baryons. As one would expect, the one-photon exchange amplitude for $eB_0 \to eB_0'$ scattering can be written as (covariant) photon propagator times electron current contracted with the baryonic current, $\mathcal{M}_{1\gamma}^{eB_0 \to eB_0'} \propto j_{e\mu} I^{\mu}_{B_0 \to B_0'}/Q^2$. This allows to extract a microscopic expression for the baryonic current $I^{\mu}_{B_0 \to B_0'}$. The form factors are then obtained by means of a general covariant decomposition of $I^{\mu}_{B_0 \to B_0'}$. The resulting model current $I^{\mu}_{B_0 \to B_0'}$, however,can be afflicted by unphysical contributions (depending on the electron momentum) which are partly eliminated by extracting the form factors in the infinite momentum frame. But problems with the "angular condition" may still persist. This deficiency can be traced back to problems with cluster separability inherent in the Bakajian-Thomas construction [1]. A more detailed account of how we deal with these problems in case of the $N \to \Delta$ transition current can be found in [5].

101.2 Results and Discussion

With the strong and electromagnetic form factors of the bare baryons we are now able to calculate the pion-loop contributions to the electromagnetic $p \to \Delta^+$ transition form factors. We account only for the $N_0\pi$ component in the physical nucleon and Delta, but neglect the $\Delta_0\pi$ component. There is some evidence from phenomenological hadronic models that an $SU(6)$ spin-flavor symmetric model like ours would overestimates the $\Delta_0\pi$ component considerably. A common choice for electromagnetic $p \to \Delta^+$ transition form factors is the one suggested by Jones and Scadron [6]. Pion-cloud effects are most visible in the small form factors G_E^* and G_C^*. What is often plotted are the ratios $R_{EM} = -G_E^*/G_M^*$ and $R_{SM} = -(Q_+Q_-/(4m_\Delta^2))\, G_C^*/G_M^*$, where $Q_\pm = \sqrt{(m_\Delta \pm m_N)^2 + Q^2}$. These are shown in Fig. 101.2 for the two parameterizations of our model given in Table 101.1. Our results compare with the outcome of other theoretical predictions coming from constituent-quark models [10–12]. For

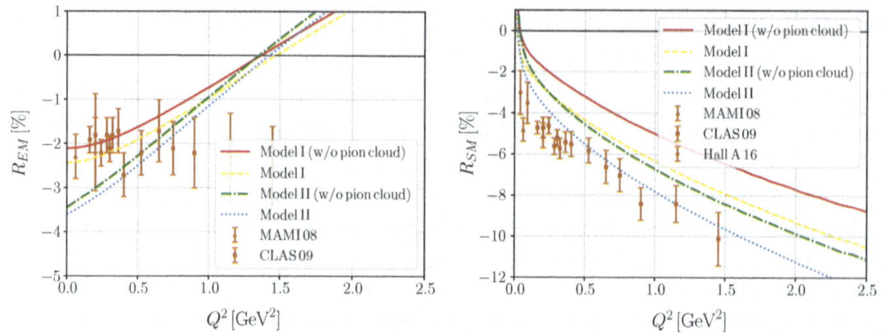

Fig. 101.2 Our model predictions for the ratios R_{EM} and R_{SM} (for the definition, see text) as compared with experimental data from MAMI [7], CLAS [8] and JLab [9]. Model I and model II refer to the two parameterizations given in Table 101.1. Curves labelled "w/o pion cloud" refer to the conventional constituent-quark model without pion cloud, whereas the other results are those for the full calculation, including the pion cloud

$Q^2 \gtrsim 0.5$ GeV^2 our predictions for G_M^* agree well with the data, for vanishing Q^2, however, we underestimate the data by about 15% for model I and about 25% for model II. Model I works also better for R_{EM}, whereas a better reproduction of R_{SM} is achieved with model II. The pion-cloud contribution is clearly visible in both ratios and it goes into the right direction.

There is, of course, room left for improvement. One should keep in mind that our starting point was $SU(6)$ spin-flavor symmetry for the bare baryons. One could, e.g., think of introducing $SU(6)$ symmetry-breaking effects right from the beginning, which lead to different masses and wave functions for the bare nucleon and Delta. This perhaps will also lead to a more reasonable probability for finding the $\pi\Delta_0$ component in the physical nucleon and Delta. Contributions from $\pi\Delta_0$ intermediate states could then also help to improve agreement with data. It is the topic of future work to find out, whether $SU(6)$-symmetry breaking effects on the bare baryon level (in addition to pion-cloud effects) suffice to improve the agreement with data, or whether, e.g., an explicit d-wave contribution to the Δ wave function, as it is asserted by several authors (see, e.g., [11]), will be necessary to achieve a satisfactory reproduction of data.

Acknowledgements J.-H. Jung acknowledges the support of the Fonds zur Förderung der wissenschaftlichen Forschung in Österreich (Grant No. FWF DK W1203-N16).

References

1. Klink, W.H., Schweiger, W.: Relativity, Symmetry, and the Structure of Quantum Theory: Volume 2 (Point Form Relativistic Quantum Mechanics), Morgan & Claypool Publishers, Bristol (2018)
2. Biernat, E.P., Schweiger, W., Fuchsberger, K., Klink, W.H.: Phys. Rev. C **79**, 055203 (2009)

3. Biernat, E.P., Schweiger, W.: Phys. Rev. C **89**, 055205 (2014)
4. Jung, J.-H., Schweiger, W.: Few Body Syst. **58**, 73 (2017)
5. Jung, J.-H., Schweiger, W., Biernat, E.P.: Few Body Syst. **60**, 13 (2019)
6. Jones, H.F., Scadron, M.D.: Annals Phys. **81**, 1 (1973)
7. Sparveris, N.F., et al.: Phys. Rev. C **78**, 018201 (2008)
8. Aznauryan, I.G., et al.: [CLAS Collaboration]: Phys. Rev. C **80**, 055203 (2009)
9. Blomberg, A., et al.: Phys. Lett. B **760**, 267 (2016)
10. Cardarelli, F., Pace, E., Salme, G., Simula, S.: Phys. Lett. B **357**, 267 (1995)
11. Ramalho, G., Pena, M.T., Gross, F.: D **78**, 114017 (2008)
12. Sanchis-Alepuz, H., Alkofer, R., Fischer, C.S.: Eur. Phys. J. A **54**, 41 (2018)

Chapter 102
Pasta Phases Within the QMC Model

G. Grams, A. M. Santos, P. K. Panda, C. Providência and D. P. Menezes

Abstract In this work the low density regions of nuclear and neutron star matter are studied. The search for the existence of pasta phases in this region is performed within the context of the quark-meson coupling (QMC) model, which incorporates quark degrees of freedom. Fixed proton fractions are considered, as well as nuclear matter in beta equilibrium at zero temperature. We analyze the influence of the nuclear pasta on some neutron star properties.

102.1 Introduction

At very low nuclear matter density, a competition between the strong and the electromagnetic interactions takes place [1], leading to a configuration in which its free energy per particle may be lower than the corresponding to the homogeneous phase at the same density. The so-called *pasta phases* are therefore the preferred shapes of some systems at these densities [2, 3]. These structures look like droplets, bubbles, rods, tubes and slabs [3], and are expected to exist [4] both in the crust of neutron stars (zero temperature, very low proton fraction, matter in β- equilibrium) and in supernova (finite temperature, proton fraction around 0.3). The existence of the pasta phase in the neutron star crust was shown to considerably alter the neutrino mean-free paths and its diffusion coefficients as compared with the homogeneous matter results. The consequent differences in neutrino opacities certainly influence the Kelvin-Helmholtz phase of the star evolution [5, 6].

In the present work, we study the possible existence of the pasta structures within the QMC model at zero temperature and its dependence on the neutron star properties.

G. Grams (✉) · A. M. Santos · D. P. Menezes
Depto de Física, CFM, Universidade Federal de Santa Catarina, Florianópolis, Brazil
e-mail: grams.guilherme@gmail.com

P. K. Panda
Department of Physics, Utkal University, Bhubaneswar, India

C. Providência
CFisUC, Department of Physics, University of Coimbra, Coimbra, Portugal

N. A. Orr et al. (eds.), *Recent Progress in Few-Body Physics*,
Springer Proceedings in Physics 238,
https://doi.org/10.1007/978-3-030-32357-8_102

102.2 The Quark-Meson Coupling Model

In the QMC model, the nucleon in nuclear medium is assumed to be a static spherical MIT bag in which quarks interact with the scalar (σ) and vector (ω, ρ) fields, and those are treated as classical fields in the mean field approximation (MFA) [7]. The quark field, ψ_{q_N}, inside the bag then satisfies the equation of motion:

$$\left[i\partial\!\!\!/ - (m_q^0 - g_\sigma^q) - g_\omega^q \omega \gamma^0 + \frac{1}{2} g_\rho^q \tau_z \rho_{03} \gamma^0 \right] \psi_{q_B}(x) = 0 \, , \quad q = u, d \tag{102.1}$$

where m_q^0 is the current quark mass, and g_σ^q, g_ω^q and g_ρ^q denote the quark-meson coupling constants. The energy of a static bag describing nucleon N consisting of three quarks in ground state is expressed as

$$E_N^{\mathrm{bag}} = \sum_q n_q \frac{\Omega_{q_N}}{R_N} - \frac{Z_N}{R_N} + \frac{4}{3} \pi R_N^3 B_N \, , \tag{102.2}$$

where Z_N is a parameter which accounts for zero-point motion of nucleon N and B_N is the bag constant. The set of parameters used in the present work is determined by enforcing stability of the nucleon (here, the "bag"), much like in [8], so there is a single value for proton and neutron masses. The effective mass of a nucleon bag at rest is taken to be $M_N^* = E_N^{\mathrm{bag}}$.

The equilibrium condition for the bag is obtained by minimizing the effective mass, M_N^* with respect to the bag radius $d M_N^*/d R_N^* = 0$, where $N = p, n$. By fixing the bag radius $R_N = 0.6$ fm and the bare nucleon mass $M = 939$ MeV the unknowns $Z_N = 4.0050668$ and $B_N^{1/4} = 210.85$MeV are then obtained. Furthermore, the desired values of $B/A \equiv \epsilon/\rho - M = -16.45$ MeV at saturation $n = n_0 = 0.15\ \mathrm{fm}^{-3}$, are achieved by setting $g_\sigma^q = 5.9810$, $g_\omega = 8.9817$, $g_\rho = 8.6510$, where $g_\omega = 3 g_\omega^q$ and $g_\rho = g_\rho^q$. The meson masses are $m_\sigma = 550$ MeV, $m_\omega = 783$ MeV and $m_\rho = 770$ MeV.

The total energy density of the nuclear matter reads

$$\begin{aligned} \varepsilon = & \frac{1}{2} m_\sigma^2 \sigma + \frac{1}{2} m_\omega^2 \omega_0^2 + \frac{1}{2} m_\rho^2 \rho_{03}^2 + 3 \Lambda_v g_\omega^2 g_\rho^2 \omega_0^2 \rho_{03}^2 \\ & + \sum_N \frac{1}{\pi^2} \int_0^{k_N} k^2 dk [k^2 + M_N^{*2}]^{1/2}, \end{aligned} \tag{102.3}$$

The vector mean field ω_0 and ρ_{03} are determined through $\omega_0 = g_\omega (n_p + n_n)/m_\omega^{*^2}$, and $\rho_{03} = g_\rho (n_p - n_n)/2 m_\rho^{*^2}$, where $n_B = n_p + n_n = \sum_N (2 k_N^3)/(3\pi^2)$, is the baryon density, and the effective masses of the meson fields are $m_\omega^{*^2} = m_\omega^2 + 2 \Lambda_v g_\omega^2 g_\rho^2 \rho_{03}^2$ and $m_\rho^{*^2} = m_\rho^2 + 2 \Lambda_v g_\omega^2 g_\rho^2 \omega_0^2$.

Finally, the mean field σ is fixed by imposing that $\partial \varepsilon / \partial \sigma = 0$.

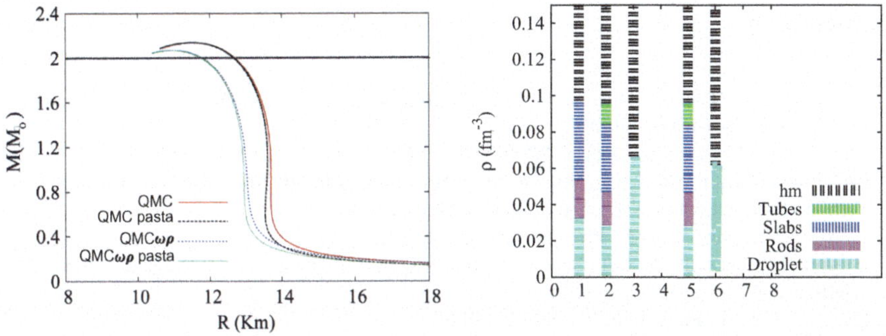

Fig. 102.1 Left: Mass-radius relation for a family of neutron stars described with the QMC and QMC$\omega\rho$ models with (QMC pasta and QMC$\omega\rho$ pasta) and without (QMC and QMC$\omega\rho$) the pasta phases. Right: Phase diagrams at T = 0 obtained with CP approximation for QMC$\omega\rho$ with $L = 69$ MeV (1):$Y_p = 0.5$, (2):$Y_p = 0.3$, (3):β equilibrium, and for QMC with $L = 90$ MeV, (5):$Y_p = 0.3$, (6): β equilibrium, hm stands for homogeneous matter. Figures adapted from [9]

The bulk nuclear matter properties for the QMC are: $B/A = -16.4$ MeV, $n_0 = 0.15$ MeV, $\Lambda = 0.00$, $M*/M = 0.77$, $\varepsilon_{sym} = 34.5$ MeV, $L_0 = 90$ MeV and $K = 295$ MeV. And for QMC$\omega\rho$: $B/A = -15.7$ MeV, $n_0 = 0.15$ MeV, $\Lambda = 0.03$, $M*/M = 0.77$, $\varepsilon_{sym} = 30.9$ MeV, $L_0 = 69$ MeV and $K = 295$ MeV

102.3 Results

In this section we present our results for the pasta phases obtained with the QMC model at zero temperature, within the coexisting phases approximation, where matter is organized in regions of lower density, generally with a neutron gas in the background and regions of higher density. We would like to remark that pasta is only predicted when its free energy per baryon is lower than the homogeneous *npe* (neutron-proton-electron) matter (Fig. 102.1).

102.4 Conclusions

In the present work we have revisited the calculation of the pasta phases now using a model with quark degrees of freedom, the QMC model. The general conclusions related to the size of the pasta phases, its internal structure and the transition density from the pasta to homogeneous matter go in line with the ones obtained in previous works [1]. All EoSs satisfies the $2M_\odot$ constraint for the neutron star maximum mass.

The work presented here was published on the Physical Review C journal [9] where you can find more details of our calculations and results.

References

1. Avancini, S.S., Menezes, D.P., Alloy, M.D., Marinelli, J.R., Moraes, M.M.W., Providência, C.: Phys. Rev. C **78**, 015802 (2008)
2. Maruyama, T., et al.: Phys. Rev. C **72**, 015802 (2005)
3. Ravenhall, D.G., Pethick, C.J., Wilson, J.R.: Phys. Rev. Lett. **50**, 2066 (1983)
4. Pethick, C.J., Potekhin, A.Y.: Phys. Lett. B **427**, 7 (1998)
5. Alloy, M.D., Menezes, D.P.: Phys. Rev. C **83**, 035803 (2011)
6. Furtado, U., Avancini, S., Marinelli, J., Martarello, W., Providencia, C.: Eur. Phys. J. A **52**, 290 (2016)
7. Guichon, P.A.M.: Phys. Lett. B **200**, 235 (1988)
8. Santos, A.M., Providência, C.: Phys. Rev. C **79**, 045805 (2009)
9. Grams, G., Santos, A.M., Panda, P.K., Providência, C., Menezes, D.P.: Phys. Rev. C **95**, 055807 (2017)

Chapter 103
Elastic Form Factor of Pseudoscalar Mesons

Muyang Chen and Lei Chang

Abstract We calculated the elastic form factor of pion-like mesons with masses $m_{0^-}/\text{GeV} = 0.14, 0.47, 0.69, 0.83$ on the space-like domain, upto $Q^2/\text{GeV}^2 = 4, 5, 6, 7$ respectively. Comparing the preliminary lattice QCD results with ours, we find that the form factors on the domain $Q^2 < 1.0\ \text{GeV}^2$ are consistent, while the lattice QCD results can not all fit our pattern of the mass denpedence of the form factor on larger Q^2 region.

103.1 Introduction

Perturbation theory in quantum chromodynamics [QCD] is applicable to hard exclusive processes; and for almost forty years the leading-order factorised result for the elastic form factor of a pseudoscalar meson has excited experimental and theoretical interest. Namely, $\exists Q_0 > \Lambda_{\text{QCD}}$, such that

$$Q^2 F_{0^-}(Q^2) \overset{Q^2 > Q_0^2}{\approx} 16\pi\alpha_s(Q^2) f_{0^-}^2 \omega_{0^-}^2(Q^2), \tag{103.1}$$

where f_{0^-} is the mesons leptonic decay constant; $\alpha_s(Q^2)$ is the leading-order strong running-coupling $\alpha_s(Q^2) = 4\pi/[\beta_0 \ln(Q^2/\Lambda_{\text{QCD}}^2)]$, with $\beta_0 = 11 - (2/3)n_f$, n_f is the number of active quark flavours; and $\omega_{0^-}(Q^2) = \frac{1}{3}\int_0^1 dx \frac{1}{x} \varphi_{0^-}(x; Q^2)$, where $\varphi_{0^-}(x; Q^2)$ is the meson's dressed-valence-quark parton distribution amplitude (PDA).

However, the value of Q_0 is not predicted by perturbative QCD. In the domain of low Q^2, people have to resort to non-perturbative method and experiment. The elastic form factor of the pion have been investaged experimentally for more than 30 years,

M. Chen (✉) · L. Chang
Nankai University, Tianjin, China
e-mail: muyang@nankai.edu.cn

L. Chang
e-mail: leichang@nankai.edu.cn

N. A. Orr et al. (eds.), *Recent Progress in Few-Body Physics*,
Springer Proceedings in Physics 238,
https://doi.org/10.1007/978-3-030-32357-8_103

restricted in the domain $Q^2 < 2.5$ GeV^2. The upgraded Jefferson Lab [JLab 12] aim for precision measurements of $F_\pi(Q^2)$ to $Q^2 \approx 6$ GeV^2. Nowdays, no lQCD predictions at m_π are available on the full domain accessible to JLab 12, but new results exist on $Q^2 < 6$ GeV^2 at bound-state mass-squared values $m_{0^-}^2 \approx 10m_\pi^2,\ 25m_\pi^2$ [1, 2]. We employ the continuum approach to the QCD bound-state problem to calculate the pion elastic form factor and have a compare with the lQCD results.

103.2 Formulations

In the rainbow-ladder approximation, the pion elastic form factor is given by

$$2P_\nu F(Q^2) = N_c \int_{\mathrm{d}k}^{\Lambda} \frac{\mathrm{d}^4k}{(2\pi)^4} \mathrm{tr}\Big[S^b(k_3)\,\bar{\Gamma}^{a\bar{b}}(k_1, k_3; -P_+) \\ \times S^a(k_1)\, i\Gamma_\nu^a(k_2, k_1; Q) S^a(k_2)\, \Gamma^{a\bar{b}}(k_3, k_2; P_-)\Big], \quad (103.2)$$

where Q is the photon momentum, $P_\pm$ is the momentum of the final and initial pion, satistying the on shell condition $P_\pm^2 = -M^2$, M being the meson mass, S^f is the dressed quark propagator with flavor f, $\Gamma^{a\bar{b}}$ is the Bethe-Salpeter amplitude(BSA) of the meson with valence quark a and antiquark $\bar{b}$, $\bar{\Gamma}^{a\bar{b}}$ is its charge conjugate, Γ_ν^f is the quark-photon vertex with quark flavor f and Lorentz index ν. $k_1,\ k_2,\ k_3$ are momenta of the the internal quarks. The notation $\int_{dk}^{\Lambda} = \int^{\Lambda} d^4k/(2\pi)^4$ stands for a Poincaré invariant regularized integration, with Λ the regularization mass-scale. We truncate the equations in the rainbow-ladder(RL) approximation. The rainbow truncated Dyson-Schwinger equation(DSE) for the quark propagator in Euclidean space is

$$S(p)^{-1} = Z_2 i\gamma\cdot p + Z_4 m_q(\mu) + \frac{4}{3}Z_2^2 \int_{dq}^{\Lambda} \mathcal{G}((p-q)^2) D_{\alpha\beta}^f(p-q)\gamma_\alpha S(q)\gamma_\beta, \quad (103.3)$$

where $D_{\alpha\beta}^f(k) = \left(\delta_{\alpha\beta} - \frac{k_\alpha k_\beta}{k^2}\right)\frac{1}{k^2}$ represent the free gluon propagator and the effective interaction is denoted by $\mathcal{G}(k^2)$. Z_2 is the wave functions and $Z_4 = Z_2 Z_m$, where Z_m is the mass renormalization constant, $m_q(\mu)$ is the current quark mass at spacelike renormalization point μ.

The equation for the quark-photon interaction vertex, $\hat{\Gamma}_\mu^q = \hat{Q}^q \Gamma_\mu^q$, and the meson amplitude are

$$\Gamma_\mu^q(k;P) = Z_2\gamma_\mu - \frac{4}{3}Z_2^2 \int_{dq}^{\Lambda} \Big[\mathcal{G}((k-q)^2) D_{\alpha\beta}^f(k-q)\gamma_\alpha S^q(q_+)\Gamma_\mu^q(q;P)S^q(q_-)\gamma_\beta\Big],$$

$$\Gamma(k;P) = -\frac{4}{3}Z_2^2 \int_{dq}^{\Lambda} \Big[\mathcal{G}((k-q)^2) D_{\alpha\beta}^f(k-q)\gamma_\alpha S(q_+)\Gamma(q;P)S(q_-)\gamma_\beta\Big]. \quad (103.4)$$

where k and P are the $q\bar{q}$ state's relative and total momenta, $q_\pm = q \pm P/2$. We normalize the amplitude $\Gamma(k; P)$ by the Nakanishi normalization condition [3]. To solve the equations (103.3) and (103.4), we need to know the coupling function $\mathcal{G}(s)$, and we employ the infrared constant model [4].

103.3 Results and Conclusion

Our results of the decay constants, charge radii and 2rd moment of PDA of the pion-like mesons at $\omega = 0.4, 0.5, 0.6$ GeV are listed in Table 103.1, and the elastic form factors are shown in Fig. 103.1. The main points of our results are:

Table 103.1 Input current-quark masses [one-loop evolved from an associated value of $\hat{m}$] for four pion-like mesons and related results computed with interaction width $\omega = 0.5 \pm 0.1$ GeV. M_{0^-}, f_{0^-} and r_{0^-} are the mass, decay constant and radius of the meson, and $\langle \varepsilon^2 \rangle$ is the second momentum of the meson PDA. [In the table, all dimensioned quantities listed in GeV, except r_{0^-}, in fm.]

m_{ζ^2}	M_{0^-}	$\omega = 0.4$			$\omega = 0.5$			$\omega = 0.6$		
		f_{0^-}	r_{0^-}	$\langle \varepsilon^2 \rangle$	f_{0^-}	r_{0^-}	$\langle \varepsilon^2 \rangle$	f_{0^-}	r_{0^-}	$\langle \varepsilon^2 \rangle$
0.0046	0.14	0.092	0.63	0.255	0.094	0.66	0.265	0.097	0.68	0.273
0.053	0.47	0.115	0.53	0.217	0.115	0.55	0.226	0.115	0.56	0.229
0.107	0.69	0.135	0.47	0.196	0.133	0.49	0.207	0.133	0.49	0.211
0.152	0.83	0.147	0.43	0.180	0.145	0.45	0.193	0.145	0.45	0.200

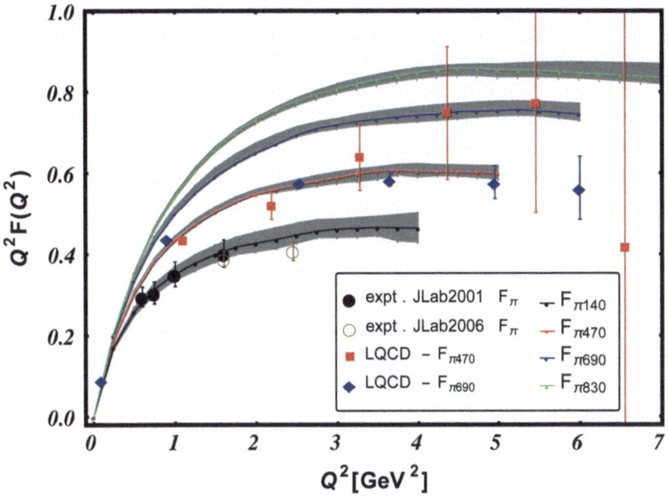

Fig. 103.1 The elastic form factor of the pion-like mesons: lines $F_{\pi 140}$, $F_{\pi 470}$, $F_{\pi 690}$, $F_{\pi 830}$, our calculations with meson masses m_{0^-}/GeV $= 0.14, 0.47, 0.69, 0.83$, the gray shadows correspond to $\omega \in [0.4, 0.6]$ GeV; expt. JLab2001 (see [5]) and expt. JLab2006 (see [6]) are the experimental results; LQCD-$F_{\pi 470}$ (see [1]) and LQCD-$F_{\pi 690}$ (see [2]) are the lattice QCD results

1. The charge radius decreases with increasing mass, i.e. the bound-states become more pointlike; and $r_{0^-} \propto 1/f_{0^-}$, up to $ln(m_{0^-})$ corrections. r_{0^-} is an intrinsic length-scale in these systems. The meson becomes a more highly correlated state as it diminishes. Hence, steadily increasing values of Q^2 are required to reach the domain upon which (103.1) provides a useful guide to $F_{0^-}(Q^2)$.
2. $f_{0^-} w_{0^-}$, where w_{0^-} is the inverse moment of the PDA, is roughly constant on the domain of meson masses considered. Consequently, the prediction of the hard-scattering formula is weakly varying on $_{0^-} \in [0.1, 0.9]$ GeV, whereas the form factor itself rises steadily with m_{0^-}, owing primarily to the decreasing radius of the system.
3. The lQCD results in [1, 2] are mutually inconsistent: the lighter meson mass in [1] is associated with a form factor which is larger in magnitude than that describing the internal structure of the heavier 0^{-+}−meson in [2]. Whilst the lowscale results from both studies match our predictions, only [1] is consistent with our calculations on the domain of larger Q^2.

References

1. Chambers, A.J.: Electromagnetic form factors at large momenta from lattice QCD. Phys. Rev. D **96**, 114509 (2017)
2. Koponen, J., et al.: Pseudoscalar meson electromagnetic form factor at high Q^2 from full lattice QCD. Phys. Rev. D **96**, 054501 (2017)
3. Nakanishi, N.: Normalization condition and normal and abnormal solutions of the Bethe–Salpeter equation. Phys. Rev. B1182–B1192, 138 (1965)
4. Qin, S., et al.: Interaction model for the gap equation. Phys. Rev. C **84**, 042202 (2011)
5. Volmer, J., et al.: Measurement of the Charged Pion Electromagnetic Form-Factor: . and others. Phys. Rev. Lett. **86**, 1713–1716 (2001)
6. Horn, T., et al.: Determination of the charged Pion form factor at Q^2 = 1.60 and 2.45 $(GeV/c)^2$. Phys. Rev. Lett. **97**, 192001 (2006)

Chapter 104
Light Hadron Spectroscopy at BESIII

Francesca De Mori

Abstract The BESIII experiment at the BEPCII electron-positron collider is successfully operating since 2008 and has collected large data samples in the tau-charm mass region, including the world's largest data samples at the J/ψ and $\psi(2S)$ resonances. In particular their decays provide a rich and clean environment to study hadrons consisting of light quarks and to search for exotics. In this report, recent BESIII results of the light hadron physics program will be highlighted.

104.1 Introduction

Light hadron spectroscopy (LHS) is a unique laboratory to test Quantum Chromo Dynamics (QCD) in low energy regime, where the perturbative approach is not holding anymore. Beside the conventional hadrons, predicted by quark model, QCD allows also exotic hadrons like glueballs, multi-quark states, molecules and hybrids. One of the main aims of LHS is the search for exotic states. Many candidates have been reported, but none of them has been unambiguously established yet.

The Beijing Electron Spectrometer III (BESIII) detector [1] is a general purpose magnetic spectrometer, installed at the Beijing Electron Positron Collider II (BEPCII). BEPCII is a double ring e^+e^- collider, hosted at the Institute of High Energy Physics (IHEP) of Beijing (P. R. C.) with a tunable beam energy from 1.0 to 2.3 GeV. It reached its design luminosity of 10^{33} cm^{-2} s^{-1} in April 2016. Since the beginning of its successfully operation in 2008, BESIII has collected the world's largest samples of J/ψ and $\psi(3686)$: 2.25×10^8 and 1.06×10^8 in 2009 plus 1.09×10^9 and 3.41×10^8 in 2012, respectively.

F. De Mori (✉)
Università degli Studi di Torino and INFN sez. Torino, via P. Giuria 1, 10126 Torino, Italy
e-mail: francesca.demori@unito.it

N. A. Orr et al. (eds.), *Recent Progress in Few-Body Physics*,
Springer Proceedings in Physics 238,
https://doi.org/10.1007/978-3-030-32357-8_104

J/ψ and $\psi(2S)$ radiative decays are gluon rich processes, with favourable kinematics and they provide a clean and powerful environment to study light hadron spectroscopy, without combinatorial background. Furthermore lattice QCD calculations [2] foresee, e.g., the 0^+ glueball ground state in the range 2.3–2.6 GeV/c^2, that can be easily explored in charmonia radiative decays at BESIII.

104.1.1 X($p\bar{p}$) and X(1835)

A strong enhancement at $p\bar{p}$ threshold, called X($p\bar{p}$), was observed for the first time, more than 10 years ago, by BESII [3] in $J/\psi \rightarrow \gamma p\bar{p}$, being confirmed later by BESIII [4] and CLEO-c [5] in the same channel with the J/ψ from $\psi(2S) \rightarrow \pi^+\pi^- J/\psi$. Based on 2009 BESIII data, a Partial Wave Analysis of J/ψ radiative decay in $p\bar{p}$ has been performed, including Final State Interaction [6], determining its spin-parity as 0^{-+}. Similar structures have not been observed in related channels, like $\psi(2S)$ decays, $\Upsilon(1S) \rightarrow \gamma p\bar{p}$ or $J/\psi \rightarrow \omega p\bar{p}$ [7]. The most appealing interpretation of this structure describes it as a $p\bar{p}$ bound state, the so-called baryonium [8]. The analysis of $J/\psi \rightarrow \gamma n\bar{n}$, on-going in BESIII, would help to clarify this issue.

Studying the $J/\psi \rightarrow \gamma\eta'\pi^+\pi^-$ decay, BESII observed a structure in the $\eta'\pi^+\pi^-$ invariant mass [9] in the same mass-region, close to the $p\bar{p}$ threshold, named X(1835). Afterwards the X(1835) was confirmed by BESIII in the same process [10] with a sample of 225 $\times 10^6$ J/ψ. Its angular distribution was found to be consistent with $J^P = 0^-$. The nature of this state is not fully understood yet and several hypotheses have been proposed, explaining it as a $p\bar{p}$ bound state or a glueball candidate, e.g.

Based on 1.09×10^9 J/ψ collected in 2012, BESIII was able to study the $\eta'\pi^+\pi^-$ invariant mass to investigate the coupling to $p\bar{p}$ with larger statistics. An abrupt in the line-shape has been found near the $p\bar{p}$ mass threshold [11]. The distorted line-shape has been described equally well by both the Flatté formula [12], taking into account the opening of the $p\bar{p}$ decay channel in the mass spectrum, with an additional narrow resonance, named X("1920")(M=1918.6 $\pm$ 3.0 MeV/c^2 and $\Gamma = 50.6 \pm 20.9$ MeV/c^2), as shown in Fig. 104.1 on the left, and the coherent sum of X(1835) plus a narrow resonance near the $p\bar{p}$ threshold (X(1870) with mass and width 1870.2 $\pm$ 2.2 MeV/c^2 and 13.0 $\pm$ 6.1 MeV/c^2). The large coupling to $p\bar{p}$ indicates most likely a connection between X($p\bar{p}$) and X(1835), but more investigations and more data are needed to clarify the nature of these states.

For this purpose, it is important to search for other decay modes. The X(1835) was also observed by BESIII [13] in the $J/\psi \rightarrow \gamma\eta K^0_S K^0_S$ decay, based on 1.3×10^9 J/ψ events, i.e. the full statistics of 2009 and 2012. A clear structure has been observed in the $\eta K^0_S K^0_S$ mass spectrum, peaking around 1.85 GeV/c^2 and strongly correlated to $f_0(980)$. A Partial Wave Analysis has been performed for $M(K^0_S K^0_S) < 1.1$GeV/c^2 and $M(K^0_S K^0_S \eta) < 2.8$GeV/c^2. The results are shown in Fig. 104.1 on the right. The final best fit includes: a X(1835) resonance in the $f_0(980)\eta$ channel, a X(1560)$\rightarrow$ $f_0(980)\eta$ and a non-resonant $f_0(1500)\eta$ component. In the nominal solution the

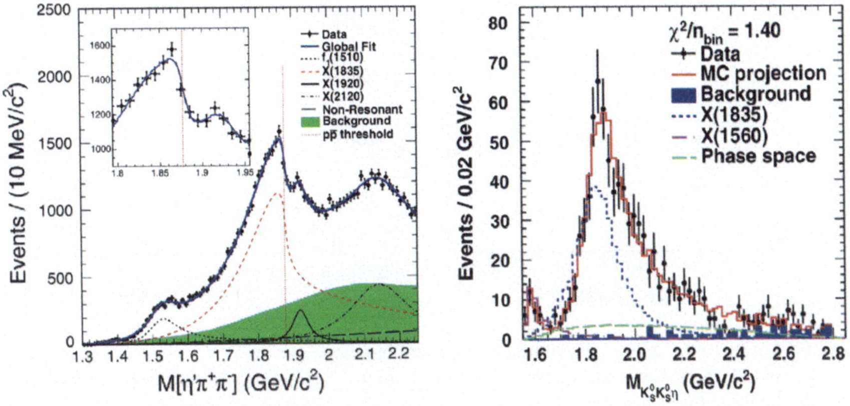

Fig. 104.1 On the left the fit of M($\eta'\pi^+\pi^-$) with Flatté formula is shown overlapped with the data points, adapted from [11]. On the right it is the reported the comparison between data and PWA fit projections for the $K_s K_s \eta$ invariant mass distribution (two entries/event), adapted from [13]

X(1835) is required with a significance larger than 12.9 σ and the mass and width have been determined to be $(1844 \pm 9^{+16}_{-25})$ MeV/c^2 and $(192^{+20}_{-17}{}^{+70}_{-43}$)MeV/c^2, respectively. The spin-parity of X(1835) turns out to be $J^P = 0^{+-}$ [13]. The resonance parameters are consistent with those obtained for X(1835) from $J/\psi \rightarrow \gamma\eta'\pi^+\pi^-$ by BESIII, while this structure results to be wider than the X($p\bar{p}$).

104.1.2 $\eta(1405)/\eta(1475)$

A puzzling state, the $\eta(1440)$, was first observed more than 50 years ago in $p\overline{p}$ collisions. Further studies, by different experiments, reported evidences for the existence of two pseudoscalar mesons in this region, the $\eta(1405)$ and the $\eta(1475)$. Theoretical interpretations favour the $\eta(1475)$ as the first radial excitation of the η_0 while the $\eta(1405)$ is a candidate for a 0^+ glueball in the flux-tube model but not in lattice gauge calculations, that predict for this state a mass larger than 2 GeV/c^2. Based on a 1.31$\times$ 10^9 J/ψ sample, BESIII studied its radiative decay to $\gamma\phi$ [16].

Two structures were observed in the $\gamma\phi$ invariant mass spectrum for the first time. The fit of the M($\gamma\phi$) distribution, taking into account their possible interference, is shown in Fig. 104.2 with their resonance parameters consistent with $\eta(1475)$ and X(1835). Their statistical significances are 13.5 σ and 6.3 σ, respectively. Further analysis of the polar angle of the radiative photon in the J/ψ rest system is used to investigate their spin-parity and it favoured their assignment to $\eta(1475)$ and X(1835).

According to Zhao et al. [14, 15] the property of $\eta(1405)/\eta(1475)$ in the J/ψ radiative decays can be explained by the so-called Triangle Singularity Mechanism (TSM). The spectrum could consist of a single state, i.e. the $\eta(1440)$, but the TSM

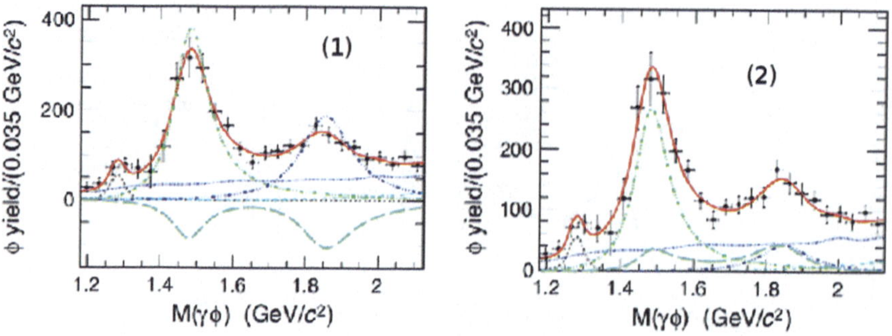

Fig. 104.2 Fits to the M($\gamma\phi$) distributions (two combinations per event) for the case of (1) constructive and (2) destructive interference. Together with the data points, the fit results are shown in red with the three fitted structures in green, cyan and black. The background is represented with a blue dotted line while the interference contribution with long dashed line, adapted from [16]

can produce obvious shifts of peak position in different channels. Under the one-state assumption, the partial width ratio between its $\gamma\rho$ and $\gamma\phi$ decay modes is predicted to be $\frac{\Gamma_{\gamma\rho}}{\Gamma_{\gamma\phi}} = 3.8$ in [15]. In our analysis [16] we found that this ratio is 11.10 ± 3.50 for destructive interference and 7.53 ± 2.49 for constructive interference, where the branching fraction of $J/\psi \to \gamma\eta(1405/1475) \to \gamma\gamma\rho$ was taken from BES measurement [17]. The ratio is slightly larger than the prediction.

104.1.3 Conclusions

Thanks to the world's largest J/ψ and $\psi(3686)$ data samples, BESIII can play an important role in the light hadron spectroscopy sector. Two "puzzles" have been pointed out in this proceedings even if the current statistics does not allow BESIII to find a conclusive answer. With the future increased statistics, towards 9 Billions J/ψ, we expect new important results.

References

1. Ablikim, M., et al.: BESIII collaboration. Nucl. Instrum. Meth. A **614**, 345 (2010)
2. Chen, Y., et al.: Phys. Rev. 73, 014516 (2006) (1981). https://doi.org/10.1103/PhysRevD.73.014516
3. Bai, J.Z., et al.: BESII collaboration. Phys. Rev. Lett. **91**, 022001 (2003)
4. Ablikim, M., et al.: BESIII collaboration. Chin. Phys. C **34**, 421 (2010)
5. Alexander, J.P., et al.: CLEO collaboration. Phys. Rev. D **82**, 092002 (2010)
6. Ablikim, M., et al.: BESIII collaboration. Phys. Rev. Lett. **108**, 112003 (2012)
7. Ablikim, M., et al.: BESIII collaboration. Eur. Phys. J. C **53**, 15 (2008)
8. Datta, A., O'Donnel, P.J.: Phys. Lett. B **567**, 273 (2003); Yan M. L. et al.: Phys. Rev. D **72**, 034027 (2005); Loiseau, B., Wycech, S., Phys. Rev. C **72**, 011001 (2005)

9. Bai, J.Z., et al.: BESII collaboration. Phys. Rev. Lett. **95**, 262001 (2005)
10. Ablikim, M., et al.: BESIII collaboration. Phys. Rev. Lett. **106**, 072002 (2011)
11. Ablikim, M., et al.: BESIII collaboration. Phys. Rev. Lett. **117**, 042002 (2016)
12. Flatté, S.M.: Phys. Lett. B **63**, 224 (1976)
13. Ablikim, M., et al.: BESIII collaboration. Phys. Rev. Lett. **115**, 091803 (2015)
14. Wu, X.G., et al.: Phys. Rev. Lett. **108**, 081803 (2012)
15. Zhao, Q., Zou, B.S.: Phys. Rev. D **87**, 014023 (2013)
16. Ablikim, M., et al.: BESIII collaboration Phys. Rev. D 97, 051101(R) (2018). https://doi.org/10.1103/PhysRevD.97.051101
17. Bai, J.Z., et al.: BES collaboration. Phys. Lett. B 594, 47 (2004). https://doi.org/10.1016/j.physletb.2004.04.085

Chapter 105
Relativistic Effects in Non-relativistic Calculations of Electroweak Cross Sections

G. Orlandini, N. Rocco, A. Lovato and W. Leidemann

Abstract In view of the growing interest in neutrino scattering with nuclei we show that it is possible to extend to relatively high energies and momentum transfers the accurate ab initio results obtained by the integral transform (IT) approach for quasi-elastic (q.e) kinematics. This is made possible by performing the calculation in a specific frame and by employing a q.e. kinematical model that helps in reducing the frame dependence. The results are validated comparing theoretical and experimental (e, e′) cross sections.

105.1 Introduction

In recent years we have seen a growing interest in understanding how neutrinos scatter on nuclei. Several experiments are underway and others are planned. However, they all share the difficulties of interpreting the results of the measurements, since it is necessary to know the neutrino cross sections on the nuclei present in the detectors. Resolving this problem by dedicated experiments is very difficult, so that one has to rely on nuclear theory. However, in this case one has to solve the non perturbative many-body problem of *A* strongly interacting nucleons, both in their initial (ground) and final excited state.

G. Orlandini (✉) · W. Leidemann
Department of Physics, University of Trento, Trento, Italy
e-mail: giuseppina.orlandini@unitn.it

G. Orlandini · A. Lovato · W. Leidemann
INFN-TIFPA Trento Institute of Fundamental Physics and Applications, Via Sommarive, 14, 38123 Trento, Italy

N. Rocco · A. Lovato
Physics Division, Argonne National Laboratory, Argonne, IL 60439, USA

N. Rocco
Theoretical Physics Department, Fermi National Accelerator Laboratory, P.O. Box 500, Batavia, IL 60510, USA

N. A. Orr et al. (eds.), *Recent Progress in Few-Body Physics*, Springer Proceedings in Physics 238,
https://doi.org/10.1007/978-3-030-32357-8_105

In the last decades theoretical nuclear physics has made remarkable progress in producing accurate ab initio results for describing nuclear ground state properties. An analogous accuracy in describing the nuclear dynamics at positive energies (many-body scattering problem) represents a tremendous challenge. However, it has been shown that, at least for not too heavy systems, one can aim at comparable accuracies, by extending ground state techniques to excited states. Such approaches make use of IT's and are summarized in Sect. 105.2.

An additional problem is the rather high energy and momentum transfers involved in neutrino scattering. The accurate ab initio approaches mentioned above are based on a nonrelativistic (n.r) quantum mechanical description of the A-nucleon system. Hence, the range of reliability of those results has to be investigated. This can be done testing the frame independence of the results at increasing energy/momentum transfers. In Sect. 105.3 this aspect is discussed.

Since electron and neutrino scattering are very similar under the nuclear many-body point of view, it seems appropriate to test the precision of theoretical results on electron scattering cross sections. This is done in Sect. 105.4.

105.2 The Integral Transform Approach

At first order perturbation theory, the inclusive lepton scattering cross section is described by so-called "inelastic structure functions" of form

$$R_{\mu\nu}(\mathbf{q},\omega) = \sum_n \langle 0|\Theta^\dagger_\mu|n\rangle\langle n|\Theta_\nu|0\rangle\delta(\omega - E_{rec} - E_n + E_0)\,, \qquad (105.1)$$

where $|0\rangle$, $|n\rangle$ and E_0, E_n are the eigenstates and eigenenergies of the initial and final nuclear states, respectively, while ω and $\mathbf{q}$ are the energy and momentum transferred by the electroweak probe. The operators $\Theta_{\mu\nu}$ are the components of the proper 4-currents and E_{rec} represents the target recoil energy. Because of the difficulties in calculating $|n\rangle$ one turns to calculate an IT of $R(\omega)$

$$\Phi_{\mu\nu}(\sigma) = \int d\omega R_{\mu\nu}(\omega) K(\omega,\sigma)\,, \qquad (105.2)$$

where $K(\omega,\sigma)$ is a suitable kernel that allows the direct calculation of $\Phi_{\mu\nu}(\sigma)$. The functions $R_{\mu\nu}$ are recovered by inverting the transform. The inversion may be delicate, since, due to inaccuracies in the calculation of Φ, instabilities may arise. Two particular kernels are known in the literature. One is $K_{\mathrm{E}}(\omega,\sigma) = e^{-\sigma\omega}$, largely used in many fields of physics. In condensed matter and nuclear physics the real parameter σ coincides with the *imaginary time* τ, which describes the evolution of the many-body system and allows the application of diffusion Monte Carlo techniques for the evaluation of Φ_{E}. The inversion of Φ_{E} can be problematic and requires a very high accuracy in its calculation. The other kernel is a Lorentzian, $K_{\mathrm{L}}(\omega,\sigma) =$

$[(\omega - \sigma_R)^2 + \sigma_I^2]^{-1}$, where σ is complex. The transform Φ_{L} can be accessed by diagonalization methods. Due to the width σ_I, and different from Φ_{E}, the transform Φ_{L} filters information only from a restricted energy range. This makes the inversion of Φ_{L} much less problematic than in the previous case.

105.3 Different Frames and the Role of Relativity

The importance of relativistic effects can be tested by studying the frame dependence of results obtained within a n.r. framework. This can be achieved by performing the n.r. calculation in different reference frames with a subsequent boost of the obtained results to the laboratory (LAB) frame. If relativistic effects are not important the results should all be very similar.

Besides the LAB frame, one can consider the Breit frame (B), where the nucleus moves with $-\mathbf{q}/2$ and the Antilab frame (AL) where the nucleus moves with $-\mathbf{q}$. In the q.e. region it is appropriate to consider also another frame, where the nucleus moves with $-A\mathbf{q}/2$. This is a sort of Breit frame, but for the nucleon that in the q.e. picture is the *active* one (ANB frame). This would have an average initial momentum of $-\mathbf{q}/2$ and a final one of $+\mathbf{q}/2$.

In [1, 2] a study of the frame dependence of the longitudinal and transverse inclusive (e,e') structure functions R_L and R_T has been performed and a noticeable frame dependence has been found, both for ^{3}He and ^{4}He. At this point one should remember that the n.r. calculation is always performed in the center of mass (c.m.) system. Relativistically the separation of c.m. and internal energy cannot be performed. However, one might expect that among all frames there should also be some where the error introduced by such a separation is minimized. To find out the best frame in that respect, one can apply the q.e. kinematics, namely that of a two-body 1-$(A-1)$ final state. It is well known that in this case one cannot at the same time calculate internal (relative) momentum and energy in a relativistically correct way. However, one can choose to calculate correctly one of them, say the relative momentum p, and then use $p^2/2\mu$ as the final internal energy (for further details of the q.e. two-fragment model see [1, 2]).

105.4 Results

First we concentrate on the longitudinal structure function $R_L(\mathbf{q}, \omega)$ of ^{4}He (nuclear force AV18+IL7). In this case Θ of (105.1) is the density operator ρ. In Fig. 105.1 the result obtained via the IT approach with the kernel K_{E} is compared with that of [3], where the kernel K_{L} has been used. The agreement is rather satisfactory, considering that the ingredients of the two calculations are slightly different. In fact in the present case ρ contains spin orbit corrections, moreover the three-body force is IL7, while in [3] UIX has been used.

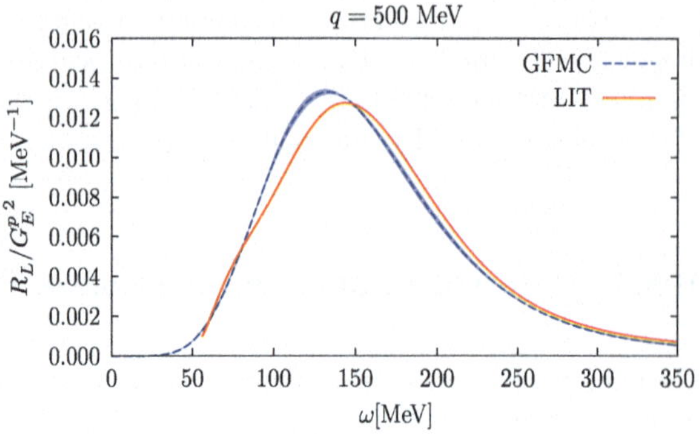

Fig. 105.1 R_L at $q = 500$ MeV with kernels K_E as in Green Function Monte Carlo (GFMC) method and K_L as in Lorentz Integral Transform (LIT) method. (Adapted from [2])

The frame dependence (LAB, ANB) of the results can be observed in the left panel of Fig. 105.2. One sees that the q.e. two-fragment model reduces the frame dependence (LAB rel, ANB rel). One also notes that results obtained in the ANB frame are little affected by the use of the two-fragment model. This indicates that in this frame relativistic effects are minimized. In particular one notes that the peak position does not change when the two fragment model is employed. This is understandable in the q.e. picture, where the peak energy in the ANB system is found at $\omega_{\rm ANB} = 0$, a correct value both relativistically and non-relativistically.

In the right panel of Fig. 105.2 we report one of the results of [2] for the total cross section. Besides the enhancement brought about by the two-body currents, it has to be noticed the importance of the ANB frame to reproduce the peak position.

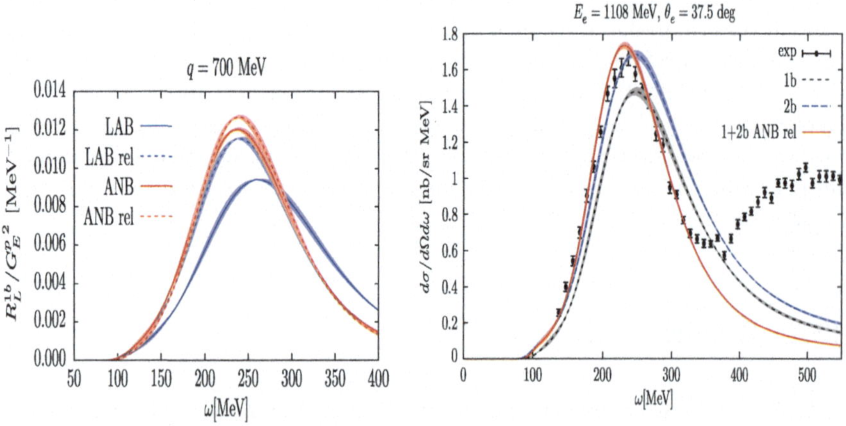

Fig. 105.2 Left: R_L in ANB and LAB frame, with and without relativistic kinematics (two-fragments). Right: (e, e′) total cross section from ANB frame calculation. The red line is the same as the blue line with the relativistic kinematics. (Adapted from [2])

Here we should mention that one has to work with interpolations to boost the results from a given frame to the LAB frame. In order to do so a novel algorithm has been developed, based on the concept of first-kind scaling (for further details see [2]).

References

1. Efros, V.D., Leidemann, W., Orlandini, G., Tomusiak, E.D.: Phys. Rev. C **72**, 011002(R) (2005)
2. Rocco, N., Leidemann, W., Lovato, A., Orlandini, G.: Phys. Rev. C **81**, 034001 (2010)
3. Rocco, N., Leidemann, W., Lovato, A., Orlandini, G.: Phys. Rev. C **97**, 055501 (2018)

Chapter 106
Bethe-Salpeter Approach to Three-Body Bound States with Zero-Range Interaction

E. Ydrefors, J. H. Alvarenga Nogueira, V. A. Karmanov and T. Frederico

Abstract The relativistic three-boson system with zero-range interaction has recently been investigated by our group by using the valence Light-Front and the four-dimensional Euclidean Bethe-Salpeter equations. We review some results from those studies, focusing on the impact of higher-Fock contributions on the computed physical quantities. Furthermore, we have also solved the Bethe-Salpeter equation by direct integration in Minkowski space and some first results from these calculations are presented. A full solution in Minkowski space is important since it gives direct access to dynamical observables such as electromagnetic form factors and momentum distributions. The obtained results are in fair agreement with the ones computed in Euclidean when the comparison is suitable.

106.1 Introduction

The Bethe-Salpeter (BS) equation [1, 2] constitutes a reliable framework for the treatment of relativistic few-body systems in the non-perturbative regime. From the numerical point of view, the simplest way to solve the integral equation is to deal with its analytic continuation to Euclidean space through theWick rotation [3]. Some

---

E. Ydrefors (✉) · J. A. Nogueira · T. Frederico
Instituto Tecnológico de Aeronáutica, DCTA, 12228-900,
São José dos Campos, Brazil
e-mail: ydrefors@kth.se

T. Frederico
e-mail: tobias@ita.br

J. H. A. Nogueira
Dipartimento di Fisica, Università di Roma "La Sapienza" & INFN,
Sezione di Roma Piazzale A. Moro 5, 00187 Roma, Italy
e-mail: dealvare@roma1.infn.it

V. A. Karmanov
Lebedev Physical Institute, Leninsky Prospekt 53, 119991 Moscow, Russia
e-mail: karmanovva@lebedev.ru

N. A. Orr et al. (eds.), *Recent Progress in Few-Body Physics*,
Springer Proceedings in Physics 238,
https://doi.org/10.1007/978-3-030-32357-8_106

quantities, such as binding energies, are unchanged under this transformation. However, as shown in [4] the Euclidean BS amplitude cannot be used naively to compute some dynamical observables. For such applications one needs the BS amplitude solution in Minkowski space.

The BS equation for the system of two scalars with one-boson-exchange interaction has been successfully solved in Minkowski space by several research groups, see e.g. [5–8]. In those works the Nakanishi integral representation [9, 10] was adopted to transform BS equation into a non-singular solvable form.

Understanding the structure of relativistic three-body systems is important for applications in nuclear- and particle physics, but it is also more challenging compared to two-body systems. The BS and Light-Front (LF) equations for the three-boson system with zero-range interaction were introduced in [11]. The LF equation, obtained by the integration over k^- of the BS amplitude, only retains the valence component of the BS amplitude, and was solved by Frederico in a limited range and later re-analyzed in [12]. Recently, Ydrefors et al. [13] solved the BS equation, derived in [11], in Euclidean space and it was found that many-body components beyond the valence affect dramatically the structure of the three-body system. In this contribution, we review some of the results calculated in [13], with a special emphasis on the contributions coming from higher-Fock states. As already mentioned, it is essential to obtain the BS amplitude directly in Minkowski space. For this purpose, we have also solved the BS equation by direct integration in Minkowski space and some of the results are presented.

106.2 Bethe-Salpeter Equation

We consider in this work the system of three bosons, having equal mass m, with zero-range interaction. The BS equation for the Faddeev component of the vertex function reads [11]

$$v(p,q) = 2iF(M_{12}) \int \frac{d^4k}{(2\pi)^4} \frac{i}{[k^2 - m^2 + i\epsilon]} \frac{i}{[(p-q-k)^2 - m^2 + i\epsilon]}, \quad (106.1)$$

where p is the total four momentum of the three-body system and $F(M_{12})$ denotes the two-body scattering amplitude. Here the squared mass of the two-body subsystem is defined as $M_{12}^2 = (p-q)^2$.

The BS equation (106.1) constitutes a highly-singular integral equation, being thus difficult to solve numerically. If the purpose is to compute binding energies (or other well-defined quantities), one can transform (106.1) to the complex plane by means of Wick rotation. In the rest frame, the Euclidean BS equation takes the form [13]

$$v_E(q_4', q_v') = 2F(-M_{12}'^2) \int_{-\infty}^{\infty} dk_4' \int_0^{\infty} \frac{dk_v'}{(2\pi)^3} \frac{\Pi_E(q_4', q_v', k_4', k_v')}{(k_4' - \frac{i}{3}M_3)^2 + k_v'^2 + m^2}, \quad (106.2)$$

where Π_E is a non-singular kernel (fully given in [13]) and we performed the change of variables $k = k' + \frac{p}{3}$ and $q = q' + \frac{p}{3}$, so that the Wick rotation could be done without crossing any singularities of the integrand in (106.1).

In the Light-Front formalism, the equation for the three-body vertex function is given by [11, 12]

$$\Gamma(k_\perp, x) = \frac{F(\bar{M}_{12})}{(2\pi)^3} \int_0^{1-x} \frac{dx'}{x'(1-x-x')} \int_0^\infty \frac{d^2k'_\perp}{M_0^2 - M_3^2} \Gamma(k'_\perp, x') \tag{106.3}$$

where the kernel of the integral equation is, again, smooth and is explicitly written in [13]. The LF equation (106.3) is fully defined in Minkowski space, but it only gives access to the valence component. For realistic calculations of dynamical observables, it is thus important to study also the complete solution of (106.1) directly in Minkowski space. By following the method introduced in [14], we thus re-write (106.1) in the partially non-singular form [15]

$$\begin{aligned} v(q_0, q_v) = {} & \frac{\mathcal{F}(M_{12})}{(2\pi)^4} \int_0^\infty k_v^2 dk_v \left\{ \frac{\pi i}{\varepsilon_k} \left[\Pi(q_0, q_v; \varepsilon_k, k_v) v(\varepsilon_k, k_v) + \Pi(q_0, q_v; -\varepsilon_k, k_v) v(-\varepsilon_k, k_v) \right] \right. \\ & - 2 \int_{-\infty}^0 dk_0 \left[\frac{\Pi(q_0, q_v; k_0, k_v) v(k_0, k_v) - \Pi(q_0, q_v; -\varepsilon_k, k_v) v(-\varepsilon_k, k_v)}{k_0^2 - \varepsilon_k^2} \right] \\ & \left. -2 \int_0^\infty dk_0 \left[\frac{\Pi(q_0, q_v; k_0, k_v) v(k_0, k_v) - \Pi(q_0, q; \varepsilon_k, k_v) v(\varepsilon_k, k_v)}{k_0^2 - \varepsilon_k^2} \right] \right\}, \end{aligned} \tag{106.4}$$

where the propagator singularities have been eliminated by using subtractions. The kernel Π has now weak (logarithmic) singularities to be treated numerically, and is explicitly given in [15].

106.3 Results

The Euclidean BS and LF equations were solved in [13] for various values of the scattering length a, a parameter introduced to renormalize the two-body zero-range interaction. In Fig. 106.1, we display the computed squared three-body mass, M_3^2, versus the inverse scattering length for the two lowest states obtained within these approaches. The results clearly show the large impact of many-body contributions on the structure of the three-body system. These additional contributions can be interpreted as an effective three-body force of relativistic origin. Furthermore, it is seen that for $a > 0$, the lowest state has a negative squared three-body mass and is unphysical. However, it exists a range of negative of values of a, for which that state has $M_3^2 > 0$ and is physical. However, it should be noted that the M_3^2 always remains finite and a Thomas collapse in the non-relativistic sense is thus avoided.

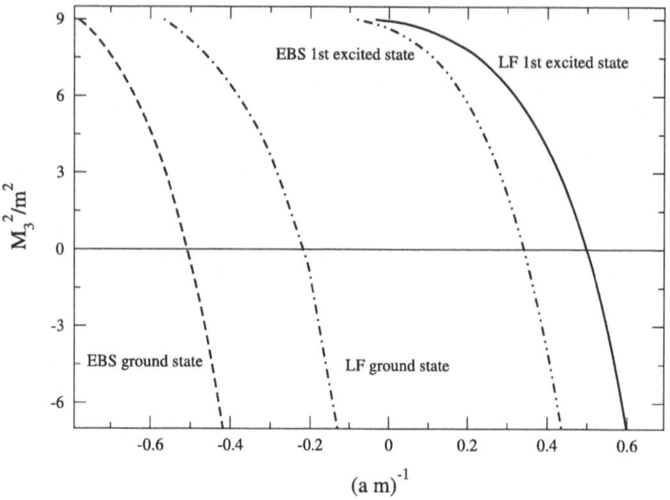

Fig. 106.1 Computed M_3^2 versus the inverse scattering length for the Euclidean BS ground state, LF ground state, Euclidean BS first excited state and the LF first excited state

Table 106.1 Eigenvalues of the three-body ground state for three scattering lengths, a, computed by using the Euclidean three-body binding energies

am	B_3/m	λ
-1.280	0.006	$0.999 - 0.054i$
-1.500	0.395	$1.000 + 0.002i$
-1.705	1.001	$0.997 + 0.106i$

We have solved the Minkowski-space BS equation (106.4) by multiplying the right-hand side by a factor λ, using as inputs the two-body scattering length and the corresponding three-body energy computed from the Euclidean BS equation (106.2). Therefore, Minkowski and Euclidean space computations are consistent if the eigenvalue $\lambda = 1.0$ is obtained. In the Table 106.1, we show the computed λ for three different values of the binding energy, i.e. $B_3/m = 0.006, 0.395, 1.001$. It is seen that for all three cases the real part is close to one. At this point, it should be noticed that in the Minkowski-space calculations were used finite cut-offs on the variables q_0 and q_v, so that a stable solution could be obtained. Differently from this, in the solution of the Euclidean BS equation the infinite limits were retained by using a mapping procedure. Consequently, the calculations are not fully comparable and this is one of the reasons for the small imaginary parts and error in the real parts.

Our results show that the binding energies obtained by direct integration agree with the Euclidean ones and more details on this work were explored in [15]. In that paper, we also computed the three-body transverse amplitudes, that depend on the momenta $k_{1\perp}$ and $k_{2\perp}$, and good agreement with the results calculated in Euclidean space was obtained. However, the method is quite challenging from the numerical point of view, because of the many singularities which have to be taken into account analytically and numerically. One way to improve the numerical accuracy and also to be able to treat more realistic kernels is to adopt the Nakanishi integral representation. This is a work in progress.

Acknowledgements We are grateful to Jaume Carbonell for stimulating discussions. E.Y. thanks for the financial support of the grant #2016/25143-7 from Fundação de Amparo à Pesquisa do Estado de São Paulo (FAPESP). J.H.A.N. acknowledges the support of the grants #2014/19094-8 and #2017/14695-1, and V.A.K. of the grant #2015/22701-6, from FAPESP. T.F. thanks the Brazilian funding agencies CNPq, CAPES, FAPESP (grant #17/05660-0) and Project INCT-FNA Proc. No. 464898/2014-5.

References

1. Salpeter, E.E., Bethe, H.: Phys. Rev. **84**, 1232 (1951)
2. Gell-Man, M., Low, F.: Phys. Rev. **84**, 350 (1951)
3. Wick, G.C.: Phys. Rev. **96**, 1135 (1954)
4. Carbonell, J., Karmanov, V.A.: Few-Body Syst. **49**, 205 (2011)
5. Kusaka, K., Williams, A.G.: Phys. Rev. D **51**, 7026 (1995)
6. Kusaka, K., Simpson, K., Williams, A.G.: Phys. Rev. D **56**, 5071 (1997)
7. Karmanov, V.A., Carbonell, J.: Eur. Phys. J. A **27**, 1 (2006)
8. Frederico, T., Salmè, G., Viviani, M.: Phys. Rev. D **89**, 016010 (2014)
9. Nakanishi, N.: Phys. Rev. **130**, 1230 (1963)
10. Nakanishi, N.: Prog. Theor. Phys. Suppl. **43**, 1 (1969)
11. Frederico, T.: Phys. Lett. B **282**, 409 (1992)
12. Carbonell, J., Karmanov, V.A.: Phys. Rev. C **67**, 037001 (2003)
13. Ydrefors, E., et al.: Phys. Lett. B **770**, 131 (2017)
14. Carbonell, J., Karmanov, V.A.: Phys. Rev. D **90**, 056002 (2014)
15. Ydrefors, E., et al.: Phys. Lett. B **791**, 276 (2019)

Chapter 107
Meson-Baryon Scattering in Extended-on-Mass-Shell Scheme Up to NNLO

Junxu Lu, Lisheng Geng, Xiulei Ren and Menglin Du

Abstract In this present work, we study the scattering of a pseudoscalar meson off one ground state octet baryon in covariant baryon chiral perturbation theory up to the next-to-next-to-leading order. We remove the power counting breaking terms with the extended-on-mass-shell scheme. We perform the first combined study of the pion-nucleon and kaon-nucleon scattering data and show that the covariant baryon Chiral perturbation theory can provide a reasonable description of the experimental data for both channels.

107.1 Introduction

Chiral perturbation theory (ChPT), as a low-energy effective field theory of QCD, plays an important role in our understanding of the non-perturbative strong interactions [1]. Elastic meson-baryon scattering is a fundamental process that not only can test our understanding of the strong interaction but also plays a relevant role in the studies of the properties of single and multi baryons. Because of the large nonzero baryon masses m_0 in the chiral limit, lower order analytical terms appear in nominal higher order loop calculations, and therefore a consistent power counting is lost [2]. In the past three decades, several solutions have been proposed. The most studied

J. Lu · L. Geng (✉)
School of Physics, Beihang University, Beijing 100191, China
e-mail: lisheng.geng@buaa.edu.cn

J. Lu
Groupe de Physique Théorique, IPN (UMR8608), Université Paris-Sud 11, 91406 Orsay, France

X. Ren
Institut für Theoretische Physik II, Ruhr-Universität Bochum, 44780 Bochum, Germany

M. Du
Helmholtz-Institut für Strahlen- und Kernphysik and Bethe Center for Theoretical Physics, Universität Bonn, 53115 Bonn, Germany

N. A. Orr et al. (eds.), *Recent Progress in Few-Body Physics*,
Springer Proceedings in Physics 238,
https://doi.org/10.1007/978-3-030-32357-8_107

ones are the heavy baryon ChPT [3, 4], the infrared (IR) baryon ChPT [5], and the extended-on-mass-shell (EOMS) BChPT [6, 7]. Among them, the EOMS BChPT turns out to be more appealing because it satisfies all the symmetry and analyticity constraints and converges relatively faster [8, 9].

In this talk, we present our work on meson-baryon scattering up to next-to-next-to-leading order in the EOMS scheme.

107.2 Theoretical Formalism

In the isospin limit, the standard decomposition of the meson-baryon scattering amplitude reads [2],

$$T_{MB} = \bar{u}(p', s')\left[D + \frac{i}{m_i + m_f}\sigma^{\mu\nu} q'_\mu q_\nu B\right] u(p, s), \tag{107.1}$$

where the $p(p')$ and $q(q')$ are the momentum for initial (final) baryons and mesons, respectively.

In order to calculate the meson-baryon scattering amplitudes up to the leading one-loop order, i.e., $\mathcal{O}(p^3)$, we need the following meson-meson and meson-baryon Lagrangians:

$$\mathcal{L}_{\rm eff} = \mathcal{L}^{(2)}_{MM} + \mathcal{L}^{(4)}_{MM} + \mathcal{L}^{(1)}_{MB} + \mathcal{L}^{(2)}_{MB} + \mathcal{L}^{(3)}_{MB}, \tag{107.2}$$

where the explicit form of each term can be found in [10–12].

It should be noted that not all of the remaining $\mathcal{O}(p^2)$ and $\mathcal{O}(p^3)$ terms contribute to a specific process. Particularly, for pion-nucleon and kaon-nucleon scattering, only 24 out of the total 37 LECs contribute. They are tabulated in Table. 107.1.

Because the baryon mass in the chiral limit does not vanish, the power counting rules are violated [2]. Within the EOMS scheme, as shown in [7], these power counting breaking terms are all analytic and can be absorbed into low energy constants. Thus in the present work, we apply the $\overline{MS} - 1$ first to absorb the ultraviolet divergence and then the EOMS scheme to remove the PCB terms to restore the power counting rules.

107.3 Fitting and Results

In the present work we focus on the πN and KN channels, because only for these channels partial wave phase shifts are available.

With the amplitudes properly renormalized, we determined the LECs by fitting to the partial wave phase shifts from the analysis of WI08 [13] for πN. And correspondingly, the phase-shift analysis of the SP92 solution [14] are used for KN. The

Table 107.1 Independent (combinations of) LECs contributing to πN and KN scattering. Note that in the fitting process, we have neglected the contributions from tree level at $\mathcal{O}(p^3)$ in the KN channels

πN	FIT	$KN_{I=0}$	FIT	$KN_{I=1}$	FIT	
$b_1+b_2+b_3+2b_4$	−7.64(6)	b_3-b_4	−0.767(1)	$b_1+b_2+b_4$	−0.419(2)	[GeV^{-1}]
$b_5+b_6+b_7+b_8$	1.42(2)	b_6-b_8	0.126(1)	$2b_5+2b_7+b_8$	0.429(2)	[GeV^{-2}]
c_1+c_2	1.34(1)	$4c_1+c_3$	0.604(3)	$4c_2+c_3$	0.616(1)	[GeV^{-1}]
$2b_0+b_D+b_F$	−1.36(6)	b_0-b_F	0.093(1)	b_0+b_D	−0.090(3)	[GeV^{-1}]
d_2	0.61(2)	$d_1+d_2+d_3$	0	$d_1-d_2-d_3$	0	[GeV^{-4}]
d_4	3.25(6)	$d_4+d_5+d_6$	0	$d_4-d_5+d_6$	0	[GeV^{-2}]
d_8+d_{10}	1.45(3)	$d_7-d_8+d_{10}$	0	$d_7+d_8+d_{10}$	0	[GeV^{-3}]
d_{49}	−0.32(12)	$d_{48}+d_{49}+d_{50}$	0	$d_{48}+d_{49}-d_{50}$	0	[GeV^{-2}]
$\chi^2/d.o.f$	0.154	$\chi^2/d.o.f$	0.971	$\chi^2/d.o.f$	0.471	

fitting results are shown in Figs. 107.1 and 107.2. The corresponding fit results are compared with the empirical data in both figures. For the sake of comparison, we show as well the $\mathcal{O}(p^3)$ results of the SU(3) heavy baryon(HB) [16] and the SU(2) EOMS BChPT [15].

Clearly, the EOMS results can describe the phase shifts quite well. While the data are only fitted up to $\sqrt{s} = 1.13$GeV, the phase shifts are described very well even up to $\sqrt{s} = 1.16$GeV. Besides, our calculation in SU(3) shows a compatible description compared to that in SU(2), which implies that the inclusion of strangeness has small effects on the fitting results. For the KN scattering, a quite good description of the phase shifts can already be achieved at NLO. We found that in the EOMS ChPT, the phase shifts alone cannot uniquely fix the eight LECs up to $\mathcal{O}(p^3)$. For the sake of comparison, we show as well the HB results of [16]. It is clear that the EOMS descriptions are slightly better that the HB results when extended to higher energy region.

107.4 Summary

In this talk, we present our results on meson-baryon scattering up to next-to-next-to-leading order in the framework of covariant ChPT with the EOMS scheme. We determined the LECs by fitting to the partial wave phase shifts from GWU group. We

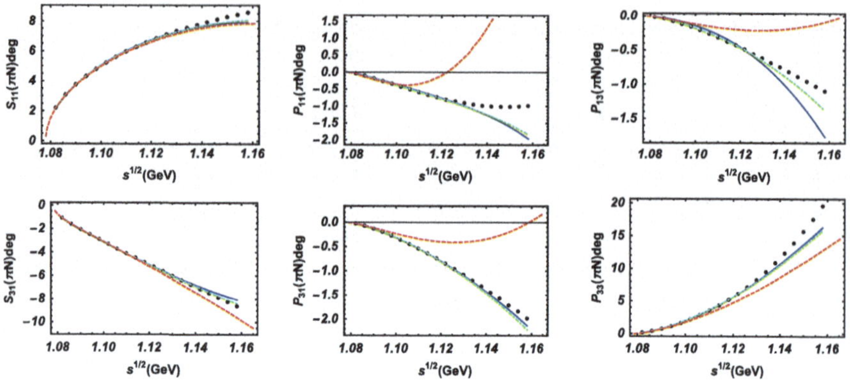

Fig. 107.1 Pion-nucleon phase shifts adapted from [17]. The blue lines denote our results and the dots the WI08 solutions. For the sake of comparison, we show as well the EOMS SU(2) results [15] (green dot-dashed lines) and the HB SU(3) results [16] (red dashed lines)

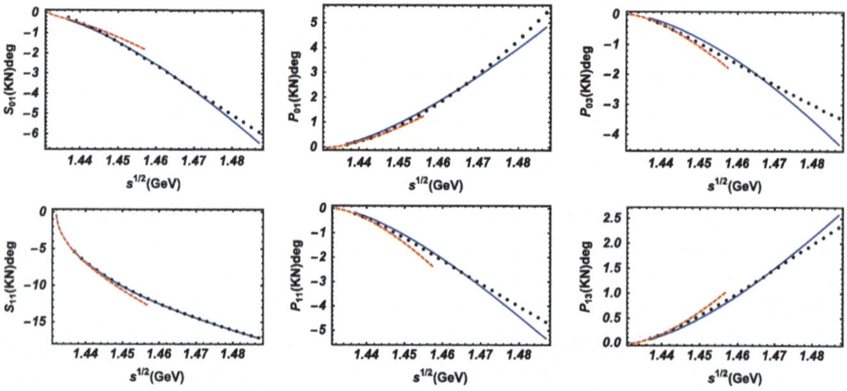

Fig. 107.2 $I = 0$ (upper panel) and $I = 1$ (lower panel) KN phase shifts adapted from [17]. The blue lines represent our results while the red dashed lines denote those of the HB ChPT [16]. The dots donate the WI08 solutions

achieved a pretty good description for πN and KN channels simulatively. Compared with those in the HB scheme, our results in πN channels are much better, while in KN channels, the improvements are not significant as in πN cases.

Acknowledgements This work is partly supported by the National Natural Science Foundation of China under Grants No.11522539, 11735003 and the fundamental Research Funds for the Central Universities.

References

1. Weinberg, S.: Physica A **96**(1–2), 327 (1979)
2. Gasser, J., Sainio, M.E., Svarc, A.: Nucl. Phys. B **307**, 779 (1988)
3. Jenkins, E.E., Manohar, A.V.: Phys. Lett. B **255**, 558 (1991)
4. Bernard, V., Kaiser, N., Meissner, U.G.: Int. J. Mod. Phys. E **4**, 193 (1995)
5. Becher, T., Leutwyler, H.: Eur. Phys. J. C **9**, 643 (1999)
6. Gegelia, J., Japaridze, G.: Phys. Rev. D **60**, 114038 (1999)
7. Fuchs, T., Gegelia, J., Japaridze, G., Scherer, S.: Phys. Rev. D **68**, 056005 (2003)
8. L. S. Geng, J. Martin Camalich, L. Alvarez-Ruso and M. J. Vicente Vacas, Phys. Rev. Lett. **101** (2008) 222002
9. Geng, L.: Front. Phys. (Beijing) **8**, 328 (2013)
10. Frink, M., Meissner, U.G.: Eur. Phys. J. A **29**, 255 (2006)
11. J. A. Oller, M. Verbeni and J. Prades, hep-ph/0701096
12. Oller, J.A., Verbeni, M., Prades, J.: JHEP **0609**, 079 (2006)
13. Arndt, R.A., Briscoe, W.J., Strakovsky, I.I., Workman, R.L.: Phys. Rev. C **74**, 045205 (2006)
14. Hyslop, J.S., Arndt, R.A., Roper, L.D., Workman, R.L.: Phys. Rev. D **46**, 961 (1992)
15. Chen, Y.H., Yao, D.L., Zheng, H.Q.: Phys. Rev. D **87**, 054019 (2013)
16. Huang, B.L., Zhang, J.S., Li, Y.D., Kaiser, N.: Phys. Rev. D **96**, 016021 (2017)
17. Lu, J.X., Geng, L.S., Ren, X.L., Du, M.L.: Phys. Rev. D **99**, 054024 (2019)

Chapter 108
Electromagnetic Transitions of Doubly Charmed Baryons Within Light-Cone Sum Rules

Er-Liang Cui, Hua-Xing Chen, Wei Chen, Xiang Liu and Shi-Lin Zhu

Abstract We study the electromagnetic transition of the Ξ_{cc}^{*++} into $\Xi_{cc}^{++}\gamma$ by employing the method of light-cone sum rules. The radiative decay width of the Ξ_{cc}^{*++} is estimated to be $13.7^{+17.7}_{-7.9}$ keV, which is large enough for it to be observed in the $\Xi_{cc}^{++}\gamma$ channel. We propose to continually search for it in future LHCb and BelleII experiments.

108.1 Introduction

The doubly heavy baryons, which provide an ideal platform to study the heavy quark symmetry, have been investigated in various experimental and theoretical studies during the past three decades [1]. Very recently, the doubly charmed baryon $\Xi_{cc}^{++}(3621)$ was observed by the LHCb collaboration in the mass spectrum of $\Lambda_c^+ K^- \pi^+ \pi^+$ [2].

The discovery of the $\Xi_{cc}^{++}(3621)$ quickly attracted much attention from the hadron physics community, and lots of theoretical methods were employed to study it [3].

E.-L. Cui (✉) · H.-X. Chen
School of Physics and Beijing Key Laboratory of Advanced Nuclear Materials and Physics, Beihang University, Beijing 100191, China
e-mail: erliangcui.phy@buaa.edu.cn

W. Chen
School of Physics, Sun Yat-Sen University, Guangzhou 510275, China

X. Liu
School of Physical Science and Technology, Lanzhou University, Lanzhou 730000, China

Research Center for Hadron and CSR Physics, Lanzhou University and Institute of Modern Physics of CAS, Lanzhou 730000, China

S.-L. Zhu
School of Physics and State Key Laboratory of Nuclear Physics and Technology, Peking University, Beijing 100871, China

Collaborative Innovation Center of Quantum Matter, Beijing 100871, China

Center of High Energy Physics, Peking University, Beijing 100871, China

N. A. Orr et al. (eds.), *Recent Progress in Few-Body Physics*,
Springer Proceedings in Physics 238,
https://doi.org/10.1007/978-3-030-32357-8_108

Since the LHCb experiment preferentially retains longer-lived Ξ_{cc}^{++} candidates and favors the $J^P = 1/2^+$ [2], it is natural to continually search for the doubly charmed baryon Ξ_{cc}^{*++} of $J^P = 3/2^+$, which is possible to be observed in its radiative decay $\Xi_{cc}^{*++} \to \Xi_{cc}^{++}\gamma$. This has been investigated in various phenomenological models [4–9]. We have also used the method of light-cone sum rules to study the electromagnetic transition of the Ξ_{cc}^{*++} into $\Xi_{cc}^{++}\gamma$ [10]. In this paper we briefly introduce our method and results.

108.2 Light-Cone Sum Rules and Numerical Analyses

In this section we use the currents $J_{\Xi_{cc}^{++}}$ and $J_{\Xi_{cc}^{*++},\alpha}$ to perform the light-cone sum rules. The two currents have been given in [11]. To use this method we first investigate the following three-point correlation function:

$$\Pi_\alpha(p,\,k,\,q,\,\epsilon) = \int d^4x e^{-ik\cdot x}\langle 0|J_{\Xi_{cc}^{*++},\alpha}(0)\bar{J}_{\Xi_{cc}^{++}}(x)|\gamma(q,\epsilon)\rangle\,, \qquad (108.1)$$

where $p(=k+q), k, q$ are the momenta of the Ξ_{cc}^{*++}, Ξ_{cc}^{++}, and γ, respectively; ϵ is the polarization vector of the γ. Then we assume that the currents $J_{\Xi_{cc}^{++}}$ and $J_{\Xi_{cc}^{*++},\alpha}$ respectively couple to the two physical states through

$$\langle 0|J_{\Xi_{cc}^{++}}|\Xi_{cc}^{++}\rangle = f_{\Xi_{cc}^{++}}u_{\Xi_{cc}^{++}}(p)\,, \qquad (108.2)$$

$$\langle 0|J_{\Xi_{cc}^{*++},\alpha}|\Xi_{cc}^{*++}\rangle = f_{\Xi_{cc}^{*++}}u_{\Xi_{cc}^{*++},\alpha}(p)\,. \qquad (108.3)$$

At the hadronic level, the amplitude of the $\Xi_{cc}^{*++} \to \Xi_{cc}^{++}\gamma$ is written as

$$\mathcal{M}_{\Xi_{cc}^{*++}\to\Xi_{cc}^{++}\gamma} = e\;g\;\epsilon^{\mu\nu\rho\sigma}\;\overline{u}_{\Xi_{cc}^{++}}\;u_{\Xi_{cc}^{*++},\rho}\;p_\mu q_\nu \epsilon_\sigma^*\,, \qquad (108.4)$$

where g is the coupling constant and e is the charge of the proton. By inserting (108.2–108.4) into (108.1), we find $\Pi_\alpha(p,\,k,\,q,\,\epsilon)$ has the following pole terms:

$$\Pi_\alpha(p,\,k,\,q,\,\epsilon) \approx \frac{2e}{3}\;g\;\epsilon^{\alpha\nu\rho\sigma}p_\nu q_\rho \epsilon_\sigma \times \frac{f_{\Xi_{cc}^{*++}}f_{\Xi_{cc}^{++}}}{(p^2-\widetilde{M}^2)^2} \times (p^2+\widetilde{M}^2)+\cdots\,, \qquad (108.5)$$

where we assume that $p \approx k \gg q$, $M_{\Xi_{cc}^{++}} \approx M_{\Xi_{cc}^{*++}}$, and $\widetilde{M} \equiv (M_{\Xi_{cc}^{++}} + M_{\Xi_{cc}^{*++}})/2$ to obtain it. Note that we have kept only the double-pole term but omitted the single-pole terms, which gives some but not large uncertainties.

At the quark-gluon level, we calculate $\Pi_\alpha(p,\,k,\,q,\,\epsilon)$ using the method of operator product expansion (OPE). The detailed results are shown in the [10].

Inserting all values of parameters [10] into the sum rules we obtained that the coupling constant $g_{\Xi_{cc}^{*++}\to\Xi_{cc}^{++}\gamma}$ is evaluated to be $0.30^{+0.16}_{-0.11}$ GeV^{-2}. Finally, we use the following decay formula

Table 108.1 Electromagnetic transitions of doubly heavy baryons, in units of keV

Process	Our results	[4]	[5]	[6]	[7]	[8]	[9]
$\Xi_{cc}^{*++} \to \Xi_{cc}^{++}\gamma$	$13.7^{+17.7}_{-7.9}$	1.43	4.35	7.21	16.7	22.0	23.46 ± 3.33
$\Xi_{cc}^{*+} \to \Xi_{cc}^{+}\gamma$	$8.1^{+11.1}_{-4.9}$	2.08	3.96	3.90	14.6	9.57	28.79 ± 2.51
$\Omega_{cc}^{*+} \to \Omega_{cc}^{+}\gamma$	$5.4^{+6.9}_{-3.1}$	0.95	1.35	0.82	6.93	9.45	2.11 ± 0.11
$\Xi_{bb}^{*0} \to \Xi_{bb}^{0}\gamma$	$0.11^{+0.13}_{-0.07}$	–	–	0.98	1.19	–	0.31 ± 0.06
$\Xi_{bb}^{*-} \to \Xi_{bb}^{-}\gamma$	$0.28^{+0.24}_{-0.13}$	–	–	0.21	0.24	–	0.0587 ± 0.0142
$\Omega_{bb}^{*-} \to \Omega_{bb}^{-}\gamma$	$0.08^{+0.05}_{-0.04}$	–	–	0.04	0.08	–	0.0226 ± 0.0045

$$\Gamma_{\Xi_{cc}^{*++} \to \Xi_{cc}^{++}\gamma} = \frac{|\mathbf{q}|}{8\pi M^2_{\Xi_{cc}^{*++}}} \times |\frac{1}{4}\sum_{\text{spin}} \mathcal{M}_{\Xi_{cc}^{*++} \to \Xi_{cc}^{++}\gamma}|^2 , \qquad (108.6)$$

to obtain $\Gamma_{\Xi_{cc}^{*++} \to \Xi_{cc}^{++}\gamma} = 13.7^{+17.7}_{-7.9}$ keV, where $|\mathbf{q}|$ is the momentum of the photon in the rest frame of the Ξ_{cc}^{*++}. We also apply the same approach to investigate the radiative decay processes including $\Xi_{cc}^{*+} \to \Xi_{cc}^{+}\gamma$, $\Omega_{cc}^{*+} \to \Omega_{cc}^{+}\gamma$, $\Xi_{bb}^{*0} \to \Xi_{bb}^{0}\gamma$, $\Xi_{bb}^{*-} \to \Xi_{bb}^{-}\gamma$, and $\Omega_{bb}^{*-} \to \Omega_{bb}^{-}\gamma$. All the results are listed in Table 108.1.

108.3 Summary and Discussions

In this paper we have applied the method of light-cone sum rules to study the electromagnetic transition of the Ξ_{cc}^{*++} into $\Xi_{cc}^{++}\gamma$, whose coupling constant is evaluated to be $0.30^{+0.16}_{-0.11}$ GeV^{-2} and decay width to be $13.7^{+17.7}_{-7.9}$ keV. We have also investigated electromagnetic transitions of some other doubly heavy baryons, whose decay widths are all summarized in Table 108.1 together with those obtained in [4–9] for comparison. We propose to continually search for the state Ξ_{cc}^{*++} of $J^P = 3/2^+$ in the future LHCb and BelleII experiments.

References

1. Patrignani, C., et al.: [Particle data group], Chin. Phys. C **40**(10), 100001 (2016). https://doi.org/10.1088/1674-1137/40/10/100001
2. Aaij, R., et al.: [LHCb Collaboration], Phys. Rev. Lett. **119**(11), 112001 (2017). https://doi.org/10.1103/PhysRevLett.119.112001
3. Guo, Z.H.: Phys. Rev. D **96**(7), 074004 (2017). https://doi.org/10.1103/PhysRevD.96.074004

4. Bernotas, A., Simonis, V.: Phys. Rev. D **87**(7), 074016 (2013). https://doi.org/10.1103/PhysRevD.87.074016
5. Hackman, R.H., Deshpande, N.G., Dicus, D.A., Teplitz, V.L.: Phys. Rev. D **18**, 2537 (1978). https://doi.org/10.1103/PhysRevD.18.2537
6. Lv, Q.F., Wang, K.L., Xiao, L.Y., Zhong, X.H.: Phys. Rev. D **96**(11), 114006 (2017). https://doi.org/10.1103/PhysRevD.96.114006
7. Branz, T., Faessler, A., Gutsche, T., Ivanov, M.A., Korner, J.G., Lyubovitskij, V.E., Oexl, B.: Phys. Rev. D **81**, 114036 (2010). https://doi.org/10.1103/PhysRevD.81.114036
8. Li, H.S., Meng, L., Liu, Z.W., Zhu, S.L.: Phys. Lett. B **777**, 169 (2018). https://doi.org/10.1016/j.physletb.2017.12.031
9. Xiao, L.Y., Wang, K.L., Lu, Q.F., Zhong, X.H., Zhu, S.L.: Phys. Rev. D **96**(9), 094005 (2017). https://doi.org/10.1103/PhysRevD.96.094005
10. Cui, E.L., Chen, H.X., Chen, W., Liu, X., Zhu, S.L.: Phys. Rev. D **97**(3), 034018 (2018). https://doi.org/10.1103/PhysRevD.97.034018
11. Chen, H.X., Mao, Q., Chen, W., Liu, X., Zhu, S.L.: Phys. Rev. D **96**(3), 031501 (2017). Erratum: [Phys. Rev. D **96**(11), 119902 (2017)] https://doi.org/10.1103/PhysRevD.96.031501. https://doi.org/10.1103/PhysRevD.96.119902

Chapter 109
Pion Effects in N and Δ Masses and Strong Form Factors

Willibald Plessas and Regina A. Schmidt

Abstract Baryon resonances cannot be described realistically along $\{QQQ\}$ configurations only. Here we present results for N and Δ masses with explicit pion contributions. We adhere to a coupled-channels formalism and construct a relativistic constituent-quark model coupling a (bare) $\{QQQ\}$ state to an additional $\{QQQ\pi\}$ channel. This approach allows to consistently derive strong form factors for the πNN, $\pi N\Delta$, $\pi\Delta N$, and $\pi\Delta\Delta$ interaction vertices and to calculate at the same time one-pion effects on the N and Δ masses.

109.1 Motivation

Baryon ground states can to a large extent be described reasonably well along $\{QQQ\}$ configurations. This is especially true for the nucleons with regard to their mass as well as their electromagnetic, weak, hadronic, and gravitational form factors [1]. The same concept based on a relativistic constituent-quark model (RCQM) works also for the baryons with flavors u, d, and s [2–5], and it can even be extended to the heavy-flavor baryons containing charm and beauty [6–8].

However, as far as resonances are concerned, one obviously meets shortcomings with regard to the hadronic decay properties of baryons. In general, the decay widths turn out to be too small as compared to phenomenological values, specifically for the single-π, η, and K decay modes [9–12]. The reasons are suspected to lie in an inadequate description of the resonances as excited bound states (poles on the negative real axis in the complex energy plane, thus with real eigenvalues) rather than as proper resonant states (poles in the complex energy plane). The latter are evidently connected with mesonic channels on top of the $\{QQQ\}$ configurations.

In [13] we have already investigated the effects of one-π loops on the N and Δ masses by following a coupled-channels (CC) approach on the macroscopic

W. Plessas (✉) · R. A. Schmidt
University of Graz, Institute of Physics, Universitätsplatz 5, 8010 Graz, Austria
e-mail: willibald.plessas@uni-graz.at

N. A. Orr et al. (eds.), *Recent Progress in Few-Body Physics*,
Springer Proceedings in Physics 238,
https://doi.org/10.1007/978-3-030-32357-8_109

(hadronic) level. In summary it has there been found that the corresponding pionic effects

1. reduce the mass of a bare $\widetilde{N}$ by about 100 MeV to the realistic N mass of 939 MeV and
2. make the Δ mass complex with a similar reduction of its real part (resonance energy) and produce a finite π-decay width Γ, where the value of the latter remains still too small (by about 50%) in comparison to experiment.

The detailed results varied according to the models employed for the πNN and $\pi N\Delta$ interaction vertices, which were then taken from the literature.

Here we report first results from an analogous CC study on the microscopic quark level. It will allow to generate consistent results for effects due to explicit π contributions both for the interaction vertices and the baryon masses.

109.2 Coupled-Channels Constituent Quark Model

We start out by constructing a relativistic constituent-quark model based on a CC mass operator of the form

$$\begin{pmatrix} M_{\widetilde{QQQ}} & K_{\pi\widetilde{QQQ}} \\ K^{\dagger}_{\pi\widetilde{QQQ}} & M_{\widetilde{QQQ}+\pi} \end{pmatrix} \begin{pmatrix} |\psi_{QQQ}\rangle \\ |\psi_{QQQ+\pi}\rangle \end{pmatrix} = m \begin{pmatrix} |\psi_{QQQ}\rangle \\ |\psi_{QQQ+\pi}\rangle \end{pmatrix} . \tag{109.1}$$

It describes a bare cluster $\widetilde{QQQ}$ of three confined quarks that is coupled to a channel containing an additional π. It gives access to a π-dressed baryon QQQ and, of course, the QQQ+π scattering channel. The coupling interaction $K_{\pi\widetilde{QQQ}}$ is furnished by a π-Q potential derived from a pseudovector-type Lagrangian density as in [14]

$$\mathcal{L}^{PV}_{\pi QQ}(x) = -\frac{f_{\pi QQ}}{m_\pi}\bar{\psi}(x)\gamma^{\mu}\gamma_5\boldsymbol{\tau}\psi(x)\cdot\partial_\mu\boldsymbol{\phi}(x) \tag{109.2}$$

connecting the Q fields ψ with the fundamental π fields $\boldsymbol{\phi}$. After a Feshbach reduction eliminating the QQQ+π channel one arrives at the eigenvalue equation for a dressed baryon state

$$[M_{\widetilde{QQQ}} + \underbrace{K_{\pi\widetilde{QQQ}}(m - M_{\widetilde{QQQ}+\pi})^{-1}K^{\dagger}_{\pi\widetilde{QQQ}}}_{V_{opt}}]|\psi_{QQQ}\rangle = m|\psi_{QQQ}\rangle , \tag{109.3}$$

where the eigenvalue m occurring both on the right- and left-hand sides is complex-valued for decaying resonances. For a $\widetilde{QQQ}$ cluster representing an arbitrary bare baryon $\widetilde{B}$ the corresponding optical potential in (109.3) can pictorially be viewed as in Fig. 109.1.

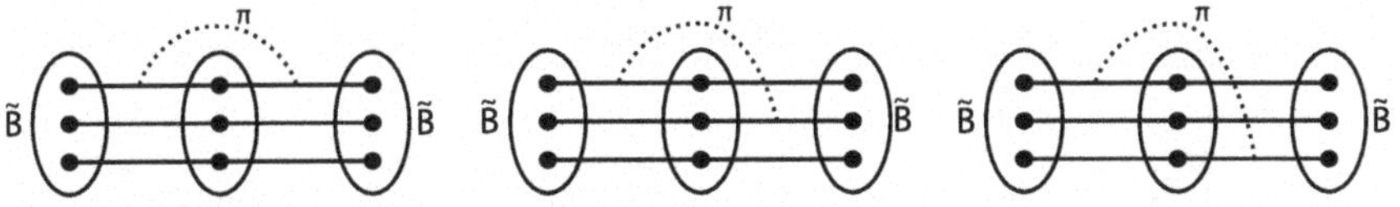

Fig. 109.1 Optical-potential terms for the π-Q interaction in (109.3)

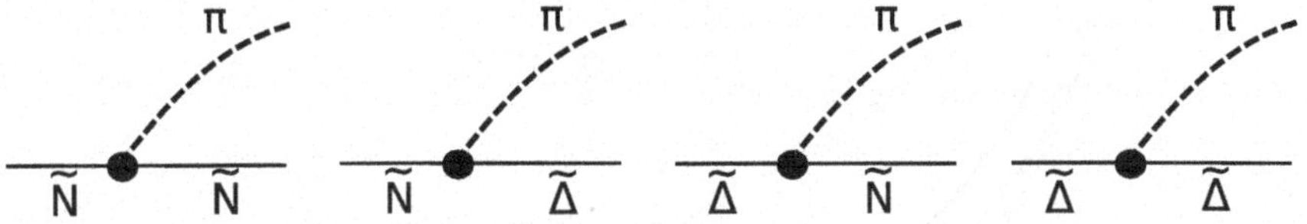

Fig. 109.2 $\pi\tilde{N}\tilde{N}$, $\pi\tilde{N}\tilde{\Delta}$, $\pi\tilde{\Delta}\tilde{N}$, and $\pi\tilde{\Delta}\tilde{\Delta}$ interaction vertices

Further details on the construction of the CC relativistic constituent-quark model, in particular the parameter ingredients, and on the method to solve the eigenvalue equation (109.3) can be found in [15] and will be published in a forthcoming paper.

The big advantage of the present microscopic CC theory is that it allows to produce consistently the vertex form factors and simultaneously predicts the pionic effects in the mass eigenvalues and eigenstates of a dressed baryon. For the vertex form factors and corresponding coupling strengths we only need to equate the microscopic optical-potential term in (109.3) with the analogous one on the macroscopic level as in [13]:

$$\int K_{\pi\widetilde{QQQ}}(m - M_{\widetilde{QQQ}+\pi})^{-1} K^{\dagger}_{\pi\widetilde{QQQ}} \\ \sim \int \mathcal{F}_{\pi\tilde{B}\tilde{B}}(\mathbf{k}_{\pi}^2) K_{\pi\tilde{B}\tilde{B}}\,(m - M_{\tilde{B}+\pi})^{-1} K^{\dagger}_{\pi\tilde{B}\tilde{B}} \mathcal{F}^{*}_{\pi\tilde{B}\tilde{B}}(\mathbf{k}_{\pi}^2)\,. \quad (109.4)$$

In this way we obtain the various form factors $\mathcal{F}_{\pi\tilde{N}\tilde{N}}$, $\mathcal{F}_{\pi\tilde{N}\tilde{\Delta}}$, $\mathcal{F}_{\pi\tilde{\Delta}\tilde{N}}$, and $\mathcal{F}_{\pi\tilde{\Delta}\tilde{\Delta}}$ at the interaction vertices shown in Fig. 109.2.

Due to limitations in space, we show here only the resulting $\pi\tilde{N}\tilde{N}$ and $\pi\tilde{N}\tilde{\Delta}$ form factors, where we can also compare with phenomenological meson-baryon models and a previous microscopic prediction from a RCQM. As can be seen from Fig. 109.3 the momentum dependences of the form factors produced by the CC relativistic constituent-quark model are very similar to the ones of the RCQM relying on $\{QQQ\}$ degrees of freedom only [14]. The same is true for the resulting $\pi\tilde{N}\tilde{N}$ coupling constant, whereas the $\pi\tilde{N}\tilde{\Delta}$ coupling constant differs to some extent (see Tables 109.1 and 109.2). For $\pi\tilde{N}\tilde{N}$ both microscopic predictions, the one of the RCQM and the one of the CC relativistic constituent-quark model, are also very close to the meson-nucleon/baryon models of SL [16] and KNLS [18], whereas they differ from the two versions of PR [17]. For $\pi\tilde{N}\tilde{\Delta}$ the results fall more apart both with regard to the momentum dependences of the form factors and the sizes of the coupling strengths.

The pionic effects on the N mass turn out biggest for the CC relativistic constituent-quark model followed by the RCQM of [14] and the phenomenolog-

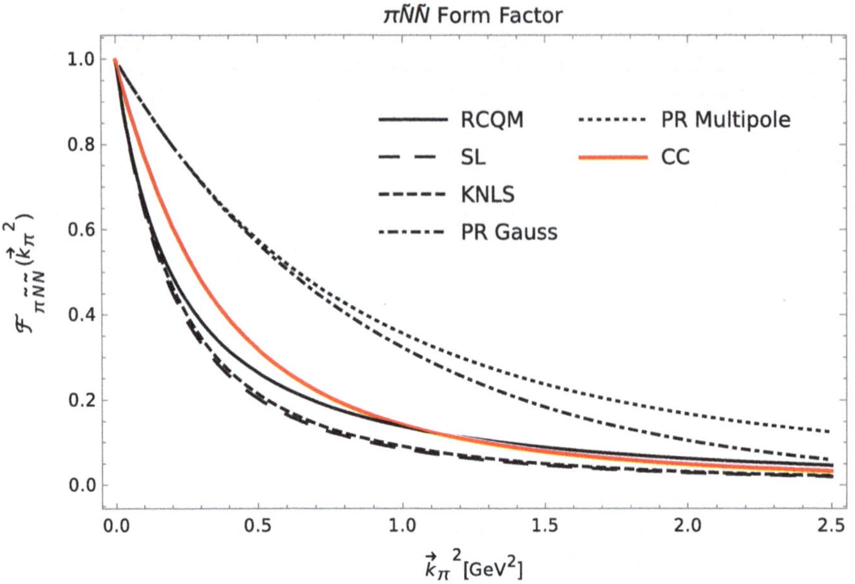

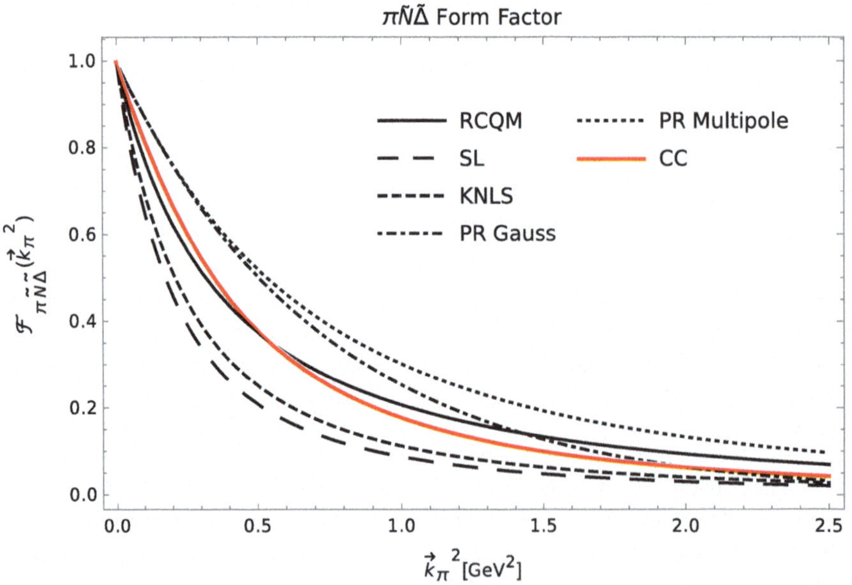

Fig. 109.3 Strong form factors for the $\pi\widetilde{N}\widetilde{N}$ and $\pi\widetilde{N}\widetilde{\Delta}$ vertices as predicted by the present CC relativistic constituent quark model in comparison to a previous RCQM [14] and phenomenological meson-baryon models by Sato and Lee (SL) [16], by Polinder and Rijken (PR Gauss as well as PR Multipole) [17] and by Kamano, Nakamura, Lee, and Sato (KNLS) [18]

Table 109.1 Predictions of the present CC relativistic constituent-quark model for the $\pi\widetilde{N}\widetilde{N}$ coupling constant and the pionic effects in the N mass in comparison to a previous RCQM [14], and meson-baryon phenomenological models by SL [16], PR [17], and KNLS [18]. For further explanations see the text

	CC	RCQM	SL	KNLS	PR Gauss	PR multipole
$f^2_{\pi\widetilde{N}\widetilde{N}}/4\pi$	0.071	0.0691	0.08	0.08	0.013	0.013
m_N	939	939	939	939	939	939
$m_{\widetilde{N}}$	1096	1067	1031	1037	1025	1051
$m_N - m_{\widetilde{N}}$	−157	−128	−92	−98	−86	−112

Table 109.2 Same as in Table 109.1 but for the $\pi\widetilde{N}\widetilde{\Delta}$ coupling constant and the Δ mass

	CC	RCQM	SL	KNLS	PR Gauss	PR multipole
$f^2_{\pi\widetilde{N}\widetilde{\Delta}}/4\pi$	0.239	0.188	0.334	0.126	0.167	0.167
$Re[m_\Delta]$	1232	1232	1232	1232	1232	1232
$m_{\widetilde{\Delta}}$	1327	1309	1288	1261	1329	1347
$Re[m_\Delta] - m_{\widetilde{\Delta}}$	−95	−77	−56	−29	−96	−115
$2\,Im[m_\Delta] = \Gamma$	67	47	64	27	52	52
$\Gamma_{\text{exp}}(\Delta \to \pi N)$	~ 117					

ical meson-baryon models. The effect on the real part of the Δ mass (resonance energy) produced by the CC relativistic constituent-quark model is quite similar to the meson-baryon model of PR. However, the effect on the imaginary part of the Δ mass, i.e. the π-decay width is largest, now comparable to the case of SL. Still it remains much too small as compared to the experimental value, even though we have in the intermediate state already assumed a realistic N with dressed mass $m =$ 939 MeV.

We note that contrary to the N case the coupling constant of the CC relativistic constituent-quark model at the $\pi\widetilde{N}\widetilde{\Delta}$ vertex is relatively smaller than in the case of SL. Also we must consider the coupling constants produced so far as bare coupling constants. In the literature it is claimed that the dressed coupling constants would be bigger by a factor of ~1.3 [16]. If we enlarge the $\pi\widetilde{N}\widetilde{\Delta}$ coupling constant of the CC relativistic constituent-quark model by such an amount, the π-decay width of the Δ is enhanced to 118 MeV, now in good agreement with experiment. Further studies are necessary to explore this issue.

So far we have succeeded to set up a microscopic CC framework that is able to describe the N and the Δ with explicit pionic contributions. Here we have reported the results due to one-π loops. We have already checked the influences of higher pionic contributions and found only negligible effects. These findings appear to be reasonable for the Δ, which decays to almost 100% to πN. In case of other N resonant states it will presumably be necessary to include further mesonic channels, such as ρN, ωN and so forth.

Acknowledgements The authors have profited from discussions with W. Schweiger and J.-H. Jung. The work was supported by the Austrian Science Fund, FWF, through the Doctoral Program on *Hadrons in Vacuum, Nuclei, and Stars* (FWF DK W1203-N16).

References

1. Plessas, W.: The constituent-quark model - Nowadays. Int. J. Mod. Phys. A **30**, 1530013 (2015). https://doi.org/10.1142/S0217751X15300136
2. Choi, K.-S.: Electromagnetic and axial structures of baryon ground and resonant states. PhD Thesis, University of Graz (2011)
3. Choi, K.-S., Plessas, W., Wagenbrunn, R.F.: Axial charges of the nucleon and N* resonances. Phys. Rev. C **81**, 028201 (2010). https://doi.org/10.1103/PhysRevC.81.028201
4. Choi, K.-S., Plessas, W., Wagenbrunn, R.F.: Axial charges of octet and decuplet baryons. Phys. Rev. D **82**, 014007 (2010). https://doi.org/10.1103/PhysRevD.82.014007
5. Choi, K.-S., Plessas, W.: Electromagnetic and axial form factors of octet and decuplet baryons. Few-Body Syst. **54**, 1055 (2013). https://doi.org/10.1007/s00601-012-0577-x
6. Day, J.P.: Approaches to non-perturbative problems in hadron physics. PhD Thesis, University of Graz (2013)
7. Day, J.P., Choi, K.-S., Plessas, W.: Spectroscopy of baryons as relativistic three-quark systems. Few-Body Syst. **54**, 329 (2013). https://doi.org/10.1007/s00601-012-0386-2
8. Day, J.P., Choi, K.-S., Plessas, W.: Relativistic quark-model spectroscopy of light and heavy baryons. PoS QNP 2012, 078 (2012). https://doi.org/10.22323/1.157.0078
9. Melde, T., Plessas, W., Wagenbrunn, R.F.: Covariant calculation of mesonic baryon decays. Phys. Rev. C 72, 015207 (2005). Erratum: [Phys. Rev. C 74, 069901 (2006)] https://doi.org/10.1103/PhysRevC.72.015207, https://doi.org/10.1103/PhysRevC.74.069901
10. Melde, T., Plessas, W., Sengl, B.: Covariant calculation of nonstrange decays of strange baryon resonances. Phys. Rev. C **76**, 025204 (2007). https://doi.org/10.1103/PhysRevC.76.025204
11. Sengl, B., Melde, T., Plessas, W.: Covariant calculation of strange decays of baryon resonances. Phys. Rev. D **76**, 054008 (2007). https://doi.org/10.1103/PhysRevD.76.054008
12. Melde, T., Plessas, W., Sengl, B.: Quark-model identification of baryon ground and resonant states. Phys. Rev. D **77**, 114002 (2008). https://doi.org/10.1103/PhysRevD.77.114002
13. Schmidt, R.A., Canton, L., Plessas, W., Schweiger, W.: Baryon masses and hadronic decay widths with explicit pionic contributions. Few-Body Syst. **58**, 34 (2017). https://doi.org/10.1007/s00601-016-1195-9
14. Melde, T., Canton, L., Plessas, W.: Structure of meson-baryon interaction vertices. Phys. Rev. Lett. **102**, 132002 (2009). https://doi.org/10.1103/PhysRevLett.102.132002
15. Schmidt, R.A.: Masses and decay widths of hadron resonances with explicit pionic contributions. PhD Thesis, University of Graz (2018)
16. Sato, T., Lee, T.-S.H.: Meson-exchange model for πN scattering and γN $\rightarrow$ πN reaction. Phys. Rev. C **54**, 2660 (1996). https://doi.org/10.1103/PhysRevC.54.2660
17. Polinder, H., Rijken, T.A.: Soft-core meson-baryon interactions. I. One-hadron-exchange potentials. Phys. Rev. C 72, 065210 (2005). https://doi.org/10.1103/PhysRevC.72.065210; Soft-core meson-baryon interactions. II. πN and K^+N scattering. Ibid. 065211 (2005). https://doi.org/10.1103/PhysRevC.72.065211
18. Kamano, H., Nakamura, S.X., Lee, T.-S.H., Sato, T.: Nucleon resonances within a dynamical coupled-channels model of πN and γN reactions. Phys. Rev. C **88**, 035209 (2013). https://doi.org/10.1103/PhysRevC.88.035209

Chapter 110
Poincaré Covariant Light-Front Spectral Function and Transverse Momentum Distributions

Emanuele Pace, Giovanni Salmè and Sergio Scopetta

Abstract In valence approximation the fermion correlator is simply related to the light-front spectral function. Then the leading twist time-reversal even transverse momentum distributions can be explicitly obtained from the light-front wave function of the system and the twist-three distributions are linear combinations of the transverse distributions at leading twist.

110.1 Introduction

Transverse momentum distributions (TMDs) are a powerful tool to study hadron structure [1]. Light-cone models have been used to study the three-dimensional hadron structure, to disentangle contributions from different angular momentum components and to investigate possible relations among the TMDs, with the aim to offer a guide for the extraction of TMDs from experimental data [2, 3].

In this paper a Poincaré covariant, light-front (LF) spin-dependent spectral function is considered to investigate hadrons within the LF Hamiltonian dynamics in valence approximation. We present both the most general expression for the spin-dependent momentum distribution in terms of six scalar functions and a linear relation between the LF spectral function and the fermion correlator. This link implies approximate relations between the six time-reversal even (T-even) TMDs, as well relations between the leading twist and the twist-three TMDs [4].

E. Pace (✉)
Università di Roma "Tor Vergata" and INFN, Sezione di Roma Tor Vergata,
Via Della Ricerca Scientifica 1, 00133 Rome, Italy
e-mail: pace@roma2.infn.it

G. Salmè
INFN, Sezione di Roma, P.le A. Moro 2, 00185 Rome, Italy

S. Scopetta
Università di Perugia and INFN, Sezione di Perugia, Via Alessandro Pascoli, 06123
Perugia, Italy

N. A. Orr et al. (eds.), *Recent Progress in Few-Body Physics*,
Springer Proceedings in Physics 238,
https://doi.org/10.1007/978-3-030-32357-8_110

The LF spectral function, $\mathcal{P}^{\tau}_{\mathcal{M},\sigma'\sigma}(\tilde{\boldsymbol{\kappa}}, \epsilon, S)$, was defined in [5], starting from the LF wave function for a three-body system with spin 1/2, third component $\mathcal{M}$ and polarization vector **S**. The energy ϵ is the energy of a fully interacting two-particle [23] subsystem and the variable $\tilde{\boldsymbol{\kappa}} = (\kappa^+, \boldsymbol{\kappa}_\perp)$ is the LF momentum for particle 1 in the intrinsic reference frame of the cluster [1,(23)].

The spectral function is defined through the overlaps between the LF wave function of the system and the tensor product of a plane wave of momentum $\tilde{\boldsymbol{\kappa}}$ and the state which describes the intrinsic motion of the two-particle spectator subsystem. The mentioned tensor product allows one to take care of macrocausality and to introduce a new effect of binding in the spectral function. The LF spectral function, through the Bakamjian-Thomas construction of the Poincaré generators [6], allows one to embed the successful phenomenology for few-nucleon systems in a Poincaré covariant framework and to satisfy at the same time normalization and momentum sum rule. As a first test of our approach the EMC effect for ^3He is being evaluated. Preliminary results show encouraging improvements with respect to a convolution approach with a momentum distribution [7].

110.2 Light-Front Spin-Dependent Spectral Function and Transverse Momentum Distributions

110.2.1 Spin-Dependent Momentum Distribution

Integration of the LF spectral function on the intrinsic energy ϵ of the $(A-1)$ system, gives the LF spin-dependent momentum distribution [5]

$$n^{\tau}_{\sigma'\sigma}(x, \mathbf{k}_\perp; \mathcal{M}, \mathbf{S}) = \sum\!\!\!\!\!\!\int d\epsilon \, \frac{1}{2\,(2\pi)^3} \, \frac{1}{1-x} \, \frac{E_S}{\kappa^+} \, \mathcal{P}^{\tau}_{\mathcal{M},\sigma'\sigma}(\tilde{\boldsymbol{\kappa}}, \epsilon, S) \tag{110.1}$$

where $\kappa^+ = x\,\mathcal{M}_0[1,(23)]$, with $\mathcal{M}_0[1,(23)]$ the free mass of the cluster [1,(23)], and $E_S = \sqrt{4m^2 + 4m\epsilon + |\boldsymbol{\kappa}|^2}$. Within the LF approach, the momentum distribution can be expressed through the three available independent vectors : (i) the polarization vector **S**; (ii) the unit vector $\hat{n}$ (identified with $\hat{z}$) which defines the $\pm$ LF components, $v^{\pm} = v^0 \pm \hat{n} \cdot \mathbf{v}$, and (iii) the transverse (with respect to the z axis) momentum component $\mathbf{k}_\perp = \mathbf{p}_\perp = \boldsymbol{\kappa}_\perp$ of the momentum **p**

$$n^{\tau}_{\sigma'\sigma}(x, \mathbf{k}_\perp; \mathcal{M}, \mathbf{S}) = \frac{1}{2}\left\{b_{0,\mathcal{M}} + \boldsymbol{\sigma} \cdot \boldsymbol{f}_{\mathcal{M}}(x, \mathbf{k}_\perp; \mathbf{S})\right\}_{\sigma'\sigma} \tag{110.2}$$

where $\boldsymbol{f}_{\mathcal{M}}(x, \mathbf{k}_\perp; \mathbf{S})$ is a pseudovector

$$\begin{aligned} \boldsymbol{f}_{\mathcal{M}}(x, \mathbf{k}_\perp; \mathbf{S}) = \mathbf{S}\, b_{1,\mathcal{M}} &+ \hat{\mathbf{k}}_\perp\,(\mathbf{S}\cdot\hat{\mathbf{k}}_\perp)\, b_{2,\mathcal{M}} + \hat{\mathbf{k}}_\perp\,(\mathbf{S}\cdot\hat{z})\, b_{3,\mathcal{M}} \\ &+ \hat{z}\,(\mathbf{S}\cdot\hat{\mathbf{k}}_\perp)\, b_{4,\mathcal{M}} + \hat{z}\,(\mathbf{S}\cdot\hat{z})\, b_{5,\mathcal{M}}. \end{aligned} \tag{110.3}$$

The functions $b_{i,\mathcal{M}}(x, \mathbf{k}_\perp)$ $(i = 0, 1, ..., 5)$ and then $n^\tau_{\sigma'\sigma}(x, \mathbf{k}_\perp; \mathcal{M}, \mathbf{S})$ can be obtained from the LF wave function of the system in momentum space. It results that the spin-dependent momentum distribution is an integral on the relative intrinsic momentum k_{23} of the interacting spectator pair [4]

$$n^\tau_{\sigma\sigma'}(x, \mathbf{k}_\perp; \mathcal{M}, \mathbf{S}) = \frac{2(-1)^{\mathcal{M}+1/2}}{(1-x)} \int dk_{23} \sum_L \mathcal{Z}^\tau_{\sigma\sigma'}(x, \mathbf{p}_\perp, k_{23}, L, \mathbf{S}) \quad (110.4)$$

where L is the orbital angular momentum of the contributions to the one-body off-diagonal density matrix (only the values $L = 0$ or $L = 2$ are allowed).

110.2.2 Fermion Correlator and LF Spectral Function

The fermion correlator in terms of the LF coordinates is [1]

$$\Phi^\tau_{\alpha,\beta}(p, P, S) = \frac{1}{2} \int d\xi^- d\xi^+ d\boldsymbol{\xi}_T \, e^{ip\xi} \left\langle P, S, A | \bar{\psi}^\tau_\beta(0) \psi^\tau_\alpha(\xi) | A, S, P \right\rangle \quad (110.5)$$

where $|A, S, P\rangle$ is the A-particle state and $\psi^\tau_\alpha(\xi)$ the particle field (e.g. a nucleon of isospin τ in a nucleus, or a quark in a nucleon). A linear relation exists between the correlator in valence approximation, $\Phi^{\tau p}$, and the spectral function [4]

$$\Phi^{\tau p}_{\alpha,\beta}(p, P, S) = \frac{D}{2p^+} \sum_{\sigma\sigma'} \left\{ u_\alpha(\tilde{\mathbf{p}}, \sigma') \, \mathcal{P}^\tau_{\mathcal{M},\sigma'\sigma}(\tilde{\boldsymbol{\kappa}}, \epsilon, S) \, \bar{u}_\beta(\tilde{\mathbf{p}}, \sigma) \right\}, \quad (110.6)$$

where $D = [(P^+)^2 \pi \, E_S] / \{p^+ \, m \, \mathcal{M}_0[1, (23)]\}$. Then, traces of Φ^p can be expressed by traces of the spectral function:

$$\mathrm{Tr}(\gamma^+ \Phi^\mathrm{p}) = D \, Tr \left[\hat{\mathcal{P}}_\mathcal{M}(\tilde{\boldsymbol{\kappa}}, \epsilon, S) \right] \quad (110.7)$$

$$\mathrm{Tr}(\gamma^+ \gamma_5 \, \Phi^\mathrm{p}) = D \, Tr \left[\sigma_z \, \hat{\mathcal{P}}_\mathcal{M}(\tilde{\boldsymbol{\kappa}}, \epsilon, S) \right] \quad (110.8)$$

$$\mathrm{Tr}(\not{\mathbf{p}}_\perp \, \gamma^+ \, \gamma_5 \, \Phi^\mathrm{p}) = D \, Tr \left[\mathbf{p}_\perp \cdot \boldsymbol{\sigma} \, \hat{\mathcal{P}}_\mathcal{M}(\tilde{\boldsymbol{\kappa}}, \epsilon, S) \right] \quad (110.9)$$

The proper integration on p^- of (110.7–110.9) and taking $p^+ = xP^+$ gives relations between the TMDs at leading twist and the functions $b_{i,\mathcal{M}}$ in (110.2, 110.3), viz

$$f(x, |\mathbf{p}_\perp|^2) = b_0 \qquad \Delta f(x, |\mathbf{p}_\perp|^2) = b_{1,\mathcal{M}} + b_{5,\mathcal{M}} \quad (110.10)$$

$$g_{1T}(x, |\mathbf{p}_\perp|^2) = \frac{M}{|\mathbf{p}_\perp|} b_{4,\mathcal{M}} \qquad \Delta'_T f(x, |\mathbf{p}_\perp|^2) = \frac{1}{2} \left\{ 2 \, b_{1,\mathcal{M}} + b_{2,\mathcal{M}} \right\} \quad (110.11)$$

$$h^\perp_{1L}(x, |\mathbf{p}_\perp|^2) = \frac{M}{|\mathbf{p}_\perp|} b_{3,\mathcal{M}} \qquad h^\perp_{1T}(x, |\mathbf{p}_\perp|^2) = \frac{M^2}{|\mathbf{p}_\perp|^2} b_{2,\mathcal{M}} \quad (110.12)$$

Linear equations between transverse parton distributions were discussed in [2]

$$\Delta f = \Delta_T' f + \frac{|\mathbf{p}_\perp|^2}{2M^2} h_{1T}^\perp \qquad g_{1T} = -h_{1L}^\perp \tag{110.13}$$

From the explicit expressions of the functions $b_{i,\mathcal{M}}$ in terms of the wave function of the system, one finds that these equalities hold exactly in valence approximation when the contribution to the transverse distributions from the angular momentum L = 2 is absent. This implies a vanishing value of the orbital angular momentum of the particle in the system wave function [4].

On the contrary the quadratic relation presented in [2] does not hold in our approach, even if the contribution from the angular momentum $L = 2$ is absent, because of the presence of $\int dk_{23}$ in the expressions (110.4) of the transverse momentum distributions.

By evaluating proper traces of both the correlator and the spectral function, one can also obtain the twist-three TMDs in terms of the functions $b_{i,\mathcal{M}}$ and relate twist-three and twist-two TMDs. In our approximation the same linear relations found in [8] hold, once the gluon contributions are disregarded. Obviously the T-odd TMDs vanish in valence approximation [4].

110.3 Conclusions and Perspectives

The LF spin-dependent momentum distribution for a spin 1/2 system composed of three fermions (as ^3He or a nucleon in valence approximation) can be expressed through six functions $b_{i,\mathcal{M}}$, that can be written in terms of the LF wave function of the system in momentum space. A simple relation exists between the fermion correlator in valence approximation and the LF spectral function. Then it follows that the TMDs are combinations of the functions $b_{i,\mathcal{M}}$.

As a result we found that the linear relations proposed between the T-even twist-two TMDs hold in valence approximation whenever the contribution from the L = 2 orbital angular momentum term in the one-body off-diagonal density matrix is absent, while the quadratic relation does not hold even in this case. Furthermore, in valence approximation the proper relations between the twist-three and the twist-two TMDs hold, once the gluon contributions are disregarded.

In the close future we will evaluate the transverse momentum distributions for a nucleon in ^3He, that could be extracted from measurements of appropriate spin asymmetries in $^3\overrightarrow{He}(\overrightarrow{e}, e'p)$ experiments at high momentum transfer.

References

1. Barone, V., Drago, A., Ratcliffe, P.G.: Transverse polarization of quarks in hadrons. Phys. Rep. **359**, 1 (2002)
2. Jacob R., Mulders P.J., Rodrigues J.: Modelling quark distribution and fragmentation functions. Nucl. Phys. A 626, 937 (1997); Pasquini B., Cazzaniga S., Boffi S.: Transverse momentum dependent parton distributions in a light-cone quark model. Phys. Rev. D 78, 034025 (2008); Lorcé C., Pasquini B.: Origin of model relations among transverse-momentum dependent parton distributions. Phys. Rev. D 84, 034039 (2011)
3. Burkardt, M., Pasquini, B.: Modelling the nucleon structure. Eur. Phys. J. A **52**, 161 (2016)
4. Del Dotto A., Pace E., Salmè G., Scopetta S., to be published
5. Del Dotto, A., Pace, E., Salmè, G., Scopetta, S.: Light-Front spin-dependent Spectral Function and Nucleon Momentum Distributions for a Three-Body System. Phys. Rev. C **95**, 014001 (2017)
6. Bakamjian, B., Thomas, L.H.: Relativistic particle dynamics. II. Phys. Rev. **92**, 1300 (1953)
7. Pace E., Del Dotto A., Kaptari L., Rinaldi M. , Salmè G., Scopetta S. : Light-Front Dynamics and the 3He Spectral Function. Few-Body Systems 57, 601 (2016); PoS INPC2016, 313 (2017), arXiv:1705.00966v1
8. Bacchetta, A., et al.: Semi-inclusive deep inelastic scattering at small transverse momentum. J. High Energy Phys. **02**, 093 (2007)

Chapter 111
Description of the Z_c Exotics States in a Quark Model Coupled Channel Calculation

Francisco Fernández, Pablo G. Ortega, Jorge Segovia and David R. Entem

Abstract The nature of the $Z_c(3900)^{\pm}$ and $Z_c(4020)^{\pm}$ is analyzed in a coupled-channels calculation, including $D^*\bar{D}^{(*)} + h.c.$, $\pi J/\psi$ and $\rho\eta_c$ channels, performed in the framework of a constituent quark model and the Resonating Group Method. The interactions among the different channels are dominated by the non-diagonal terms, which indicates the $Z_c(3900)^{\pm}$ and the $Z_c(4020)^{\pm}$ are unusual structures. The study of the analytic structure of the S-matrix on the complex energy plane leads us to conclude that the behaviour of the line shapes in the $\pi J/\psi$ and the $D^*\bar{D}^{(*)}$ invariant mass distributions is due to the presence of virtual states.

111.1 Introduction

The discovery in 2011 by the Belle Collaboration in the bottom sector of the $Z_b(10610)$ and $Z_b(10650)$ charged structures in the $\Upsilon(5S) \to \pi^+\pi^-\Upsilon(nS)$ reaction showed the existence of structures with exotic nature wich cannot be described by quark-antiquark pairs. The later discovery in 2013 by the BESIII and Belle Collaboration of the charged structure $Z_c(3900)$ in the $\pi^+\pi^- J/\psi$ invariant mass spectrum of the $e^+e^- \to \pi^+\pi^- J/\psi$ process at $\sqrt{s} = 4.26$ GeV extended this kind of structures to the charm sector. Soon after all these experimental activities, the BESIII Collaboration reported the discovery of another charged state, the $Z_c(4020)$ resonance, in the $e^+e^- \to \pi^+\pi^- h_c$ channel with a mass of $M = (4022.9 \pm 0.8 \pm 2.7)$ MeV/c^2 and a width of $\Gamma = (7.9 \pm 2.7 \pm 2.6)$ MeV. The corresponding experimental references for all these states can be found in [4].

F. Fernández (✉) · P. G. Ortega · D. R. Entem
Grupo de Física Nuclear and Instituto de Física Fundamental y Matemáticas (IUFFyM)
Universidad de Salamanca, 37008 Salamanca, Spain
e-mail: fdz@usal.es

J. Segovia
Departamento de Sistemas Físicos, Químicos y Naturales. Universidad Pablo de Olavide, 41013 Sevilla, Spain

N. A. Orr et al. (eds.), *Recent Progress in Few-Body Physics*,
Springer Proceedings in Physics 238,
https://doi.org/10.1007/978-3-030-32357-8_111

In the case of the $Z_c(3900)$ and $Z_c(4020)$, which are the objects of this study, the closeness of their masses to $D\bar{D}^*$ and $D^*\bar{D}^*$ thresholds, respectively suggest that $D^{(*)}\bar{D}^*$ components are significant in their wave functions. However, the fact that the Z_c structures have $I = 1$ makes the $D^{(*)}\bar{D}^{(*)}$ interaction too weak, because in these channels the one-pion-interaction is weaker than in the $I = 0$ sector. This leaves the coupled channel calculations as the most promising option to produce such structure.

This hypothesis has been confirmed by the work of Ikeda et al. [5], which performed a lattice coupled-channels calculation taking into account the $\pi J/\psi$, $\rho\eta_c$ and $D\bar{D}^*$ channels and found that the interactions among them are dominated by off-diagonal couplings which are not strong enough to produce a bound state or a resonance, which may indicate that the $Z_c(3900)^{\pm}$ can be explained as a threshold cusp. However, there are technicalities involving LQCD computations, such as large pion masses, small volumes and a set of interpolators not large enough for having overlap with the physical state, which still prevent to make a definitive statement.

111.2 Theory

In this work we perform a coupled-channels calculation for the $I^G(J^{PC}) = 1^+(1^{+-})$ sector, including the closest thresholds to the experimental masses of the $Z_c(3900)$ and $Z_c(4020)$ hadrons, that is: $\pi J/\psi$ (3234.19 MeV/c^2), $\rho\eta_c$ (3755.79 MeV/c^2), $D\bar{D}^*$ (3875.85 MeV/c^2), $D^*\bar{D}^*$ (4017.24 MeV/c^2), where the threshold masses are shown in parenthesis. The system is described in the framework of a constituent quark model successfully employed to explain the meson and baryon phenomenology from the light to the heavy quark sector. The interested reader is referred to [8–11] for detailed reviews about the naive quark model in which this work is based. The model parameters and explicit expressions for the potentials can be also found therein.

The aforementioned CQM specifies the microscopic interaction among constituent quarks. In order to describe the interaction at the meson level we employ the Resonating Group Method, where mesons are considered as quark-antiquark clusters and an effective cluster-cluster interaction emerges from the underlying $q\bar{q}$ dynamics. The solution of the coupled-channels RGM equations is performed deriving from a set of coupled Lippmann-Schwinger equations of the form

$$\begin{aligned} T_{\alpha}^{\alpha'}(E; p', p) = V_{\alpha}^{\alpha'}(p', p) + \textstyle\sum_{\alpha''} \int dp''\, p''^2\, V_{\alpha''}^{\alpha'}(p', p'') \\ \times \frac{1}{E - \mathcal{E}_{\alpha''}(p'')} T_{\alpha}^{\alpha''}(E; p'', p)\,, \end{aligned} \tag{111.1}$$

where α labels the set of quantum numbers needed to uniquely define a certain partial wave, $V_{\alpha}^{\alpha'}(p', p)$ is the projected potential that contains the direct and rearrangement potentials, and $\mathcal{E}_{\alpha''}(p'')$ is the energy corresponding to a momentum p''. Once the T-matrix is calculated, we determine the on-shell part which is directly related to the S-matrix. Details of the procedure can be found in [6].

111.3 Results

Our results for the invariant mass distribution of the $D\bar{D}^*$, $\pi J/\psi$ and $D^*\bar{D}^*$channels are shown in, respectively, the left and right panels of Fig. 111.1 and in left panel of Fig. 111.2. Note that all channels mentioned in the above paragraph are included in the calculation of the theoretical line shapes. To translate the decay rates into events, we use a normalization factor $\mathcal{N}_{AB}$ fitted for each channel. The shaded area around the theoretical curve shows the statistical 68%-confident level (CL) of the fit.

The contribution of the different channels to the line shape are shown in the right panel Fig. 111.2. When the $D\bar{D}^*$ scattering channel is considered alone, a small enhancement, basically due to the pion tensor interaction between the S and D waves, appears but the generated peak is too wide. The inclusion of the $\rho\eta_c$ channel narrows that peak, making it more compatible with the experimental situation. On the other hand, adding to the $D\bar{D}^*$ channel the $D^*\bar{D}^*$ one generates a second structure at its threshold opening associated to the $Z_c(4020)$ peak. One can see that the second

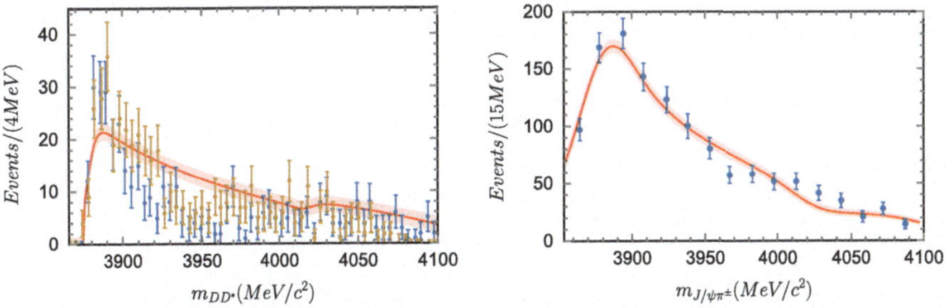

Fig. 111.1 Line shapes for $D\bar{D}^*$ (left panel) and $\pi J/\psi$ (right panel) at $\sqrt{s} = 4.26$ GeV. Experimental data are from [2, 3]. The theoretical line shapes have been convoluted with the experimental resolution. The line-shape's error is shown as a shadowed area

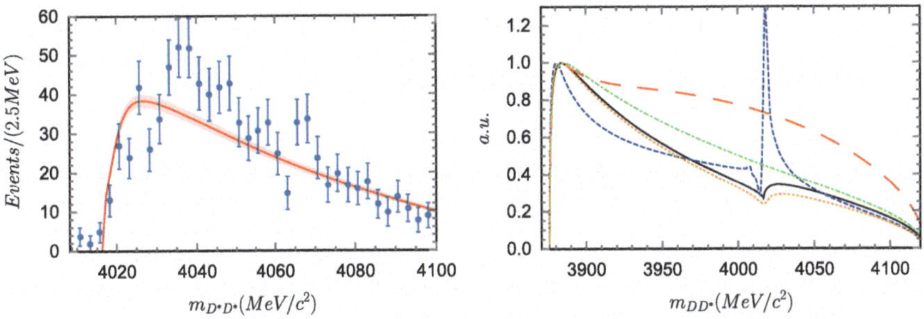

Fig. 111.2 Left panel: Line shapes for $D^*\bar{D}^*$ at $\sqrt{s} = 4.26$ GeV. Experimental data are from [1]. The theoretical line shapes have been convoluted with the experimental resolution. The line-shape's error is shown as a shadowed area. Right panel: Line shapes for different coupled-channels calculations. Only $D\bar{D}^*$ channel (red, long-dashed line), $D\bar{D}^* + D^*\bar{D}^*$ (blue, dashed line), $\rho\eta_c + D\bar{D}^*$ (green, dot-dashed line), $\rho\eta_c + D\bar{D}^* + D^*\bar{D}^*$ (orange, dotted line) and $\pi J/\psi + \rho\eta_c + D\bar{D}^* + D^*\bar{D}^*$ (black, solid line)

enhancement is much higher than the experimental data if we only include $D^{(*)}\bar{D}^*$ channels. The line shape moves closer to the experimental situation when the $\rho\eta_c$ and $\pi J/\psi$ channels are considered in the calculation, showing that they play an important role in building the observed enhancements.

To deepen into the nature of the $Z_c(3900)$ and the $Z_c(4020)$, we have examined the analytic structure on the complex energy plane of the S-matrix for the different coupled-channels calculations. For the $Z_c(3900)$, one can see that even for a one-channel $D\bar{D}^*$ calculation the S-matrix shows a pole $(3871.37 - 2.17i)$ below threshold in the second Riemann sheet. When the $\rho\eta_c$ is included the pole remains in the second Riemann sheet but moves to the real axis $(3871.74 - 0.00i)$ corresponding with a virtual state. The inclusion of the rest of the channels does not change drastically the pole position. However, it is necessary to reproduce the experimental line shape. The situation is similar in the case of the $Z_c(4020)$, which is interpreted as a virtual state located below the $D^*\bar{D}^*$ threshold. A extended version of the work can be found in [7].

Acknowledgements This work has been partially funded by Ministerio de Economía, Industria y Competitividad under Contracts No. FPA2016-77177-C2-2-P, FPA2014-55613-P, FPA2017-86989-P, SEV-2016-0588, IJCI-2016-28525 and IJCI-2016-30028.

References

1. Ablikim, M., et al.: Observation of a charged charmoniumlike structure in $e^+e^- \to (D^*\bar{D}^*)^{\pm}\pi^{\mp}$ at $\sqrt{s} = 4.26$GeV. Phys. Rev. Lett. **112**(13), 132,001 (2014)
2. Ablikim, M., et al.: Confirmation of a charged charmoniumlike state $Z_c(3885)^{\mp}$ in $e^+e^- \to \pi^{\pm}(D\bar{D}^*)^{\mp}$ with double D tag. Phys. Rev. **D92**(9), 092,006 (2015)
3. Ablikim, M., et al.: Determination of the spin and parity of the $Z_c(3900)$. Phys. Rev. Lett. **119**(7), 072,001 (2017). https://doi.org/10.1103/PhysRevD.92.092006
4. Guo, F.K., Hanhart, C., Meiner, U.G., Wang, Q., Zhao, Q., Zou, B.S.: Hadronic molecules. Rev. Mod. Phys. **90**(1), 015,004 (2018)
5. Ikeda, Y., Aoki, S., Doi, T., Gongyo, S., Hatsuda, T., Inoue, T., Iritani, T., Ishii, N., Murano, K., Sasaki, K.: Fate of the tetraquark candidate $Z_c(3900)$ from Lattice QCD. Phys. Rev. Lett. **117**(24), 242,001 (2016). https://doi.org/10.1103/PhysRevLett.117.242001
6. Ortega, P.G., Entem, D.R., Fernandez, F.: Molecular structures in charmonium spectrum: The XYZ puzzle. J. Phys. **G40**, 065,107 (2013). https://doi.org/10.1088/0954-3899/40/6/065107
7. Ortega, P.G., Segovia, J., Entem, D.R., Fernández, F.: The Z_c structures in a coupled-channels model. Eur Phys J C 79(1), **78** (2019). https://doi.org/10.1140/epjc/s10052-019-6552-7
8. Segovia, J., et al.: Constituent quark model description of charmonium phenomenology. Int. J. Mod. Phys. **E22**, 1330,026 (2013). https://doi.org/10.1142/S0218301313300269
9. Segovia, J., Yasser, A.M., Entem, D.R., Fernandez, F.: JPC=1– hidden charm resonances. Phys. Rev. **D78**, 114,033 (2008). https://doi.org/10.1103/PhysRevD.78.114033
10. Valcarce, A., Garcilazo, H., Fernandez, F., Gonzalez, P.: Quark-model study of few-baryon systems. Rept. Prog. Phys. **68**, 965–1042 (2005). https://doi.org/10.1088/0034-4885/68/5/R01
11. Vijande, J., Fernandez, F., Valcarce, A.: Constituent quark model study of the meson spectra. J. Phys. **G31**, 481 (2005). https://doi.org/10.1088/0954-3899/31/5/017

Chapter 112
Electromagnetic Properties of Singly Heavy Baryons

June-Young Kim and Hyun-Chul Kim

Abstract In this presentation, we summarize selectively recent results of the electromagnetic form factors of singly heavy baryons with spin 1/2 and related quantities, which were derived within a framework of the self-consistent chiral quark-soliton model. The results are compared with those from the lattice QCD and their physical implications are discussed.

112.1 Introduction

The chiral quark-soliton model (χQSM) was developed as a pion mean-field approach to describe the structures of the nucleon [1, 2]. The model was extended to describe also hyperons [3–5]. Recently, the model was successfully applied to the description of singly heavy baryons [6–10] in the limit of the infinitely heavy quark mass ($m_Q \to \infty$). In this presentation, we summarize selectively a recent work on the electromagnetic form factors of singly heavy baryons with spin 1/2 [11] and compare the results with those from a lattice QCD [12].

J.-Y. Kim (✉) · H.-C. Kim (✉)
Department of Physics, Inha University, Incheon 22212, Republic of Korea
e-mail: juneyoung.ghim@gmail.com

H.-C. Kim
e-mail: hchkim@inha.ac.kr

H.-C. Kim
School of Physics, Korea Institute for Advanced Study (KIAS),
Seoul 02455, Republic of Korea

Advanced Science Research Center, Japan Atomic Energy Agency,
Shirakata, Tokai, Ibaraki 319–1195, Japan

N. A. Orr et al. (eds.), *Recent Progress in Few-Body Physics*,
Springer Proceedings in Physics 238,
https://doi.org/10.1007/978-3-030-32357-8_112

112.2 Results and Discussion

In Fig. 112.1, we show the results of the electric form factors of the charmed heavy baryons Σ_c^{++} and Ω_c^0, comparing them with those from the lattice QCD [12]. The present results seem to fall off faster than those from the lattice QCD. This can be easily understood: When the value of the pion mass employed by the lattice QCD calculation is larger than the experimental one, the nucleon EM form factors decrease less slowly than the experimental data. This tendency can be also found in the magnetic form factors of the heavy baryons as drawn in Fig. 112.2.

In Fig. 112.1, we depict the light-quark contributions of the valence and sea parts to the electric form factors separately. In the present pion mean-field approach and in the limit of $m_Q \to \infty$, the heavy quark can be considered as a point-like particle, so that it plays only a part in making the concerned baryon charge complete. Note that the heavy-quark contribution to the magnetic form factors also vanishes in the limit of $m_Q \to \infty$. So, the light quarks govern all dynamics in a heavy baryon.

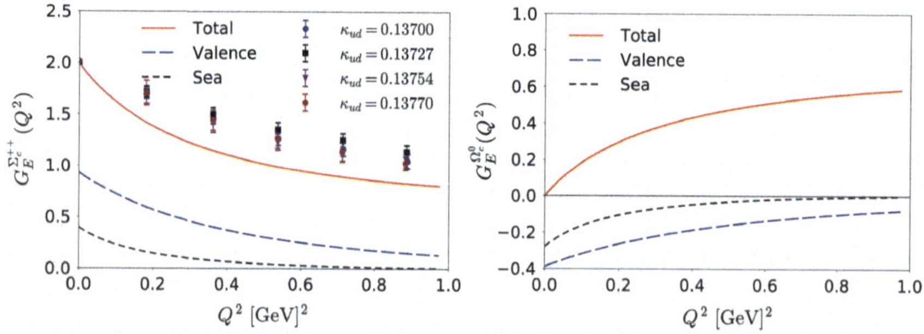

Fig. 112.1 Decomposition of the valence and sea contributions of the electric form factors for the charmed baryons Σ_c^{++} and Ω_c^0, in the left panel, and the right panel, respectively. The results are compared with those from the lattice QCD

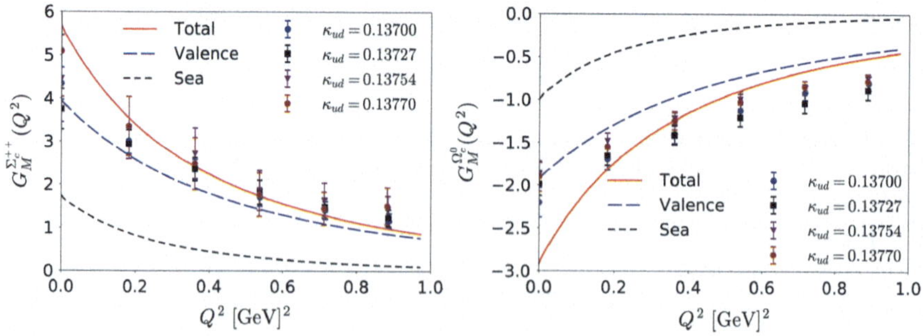

Fig. 112.2 Decomposition of the valence and sea contributions of the magnetic form factors for the charmed baryons Σ_c^{++} and Ω_c^0, in the left panel, and the right panel, respectively. The results are compared with those from the lattice QCD

Table 112.1 Dipole magnetic moments of the singly heavy baryons. The second and third columns list the present results without and with the effects of SU(3) symmetry breaking, respectively

B_c	$\mu_{B_c}^{(m_s=0)}$	$\mu_{B_c}^{(m_s=180)}$	[12, 13]	[10]
Σ_c^{++}	2.147	2.176	2.220(505)	2.15(10)
Σ_c^{+}	0.537	0.445	–	0.46(3)
Σ_c^{0}	−1.073	−1.286	−1.073(269)	−1.24(5)
Ξ_c^{+}	0.537	0.603	0.315(141)	0.60(2)
Ξ_c^{0}	−1.073	−1.147	−0.599(71)	−1.05(4)
Ω_c^{0}	−1.073	−1.014	−0.639(88)	−0.85(5)

In Table 112.1, we list the results of the magnetic moments of the charmed baryons, comparing them with those from the lattice QCD and from a "model-independent" approach of the χQSM. We want to mention that the nuclear magneton in the model is evaluated not by the experimental value of the nucleon mass but by the model value, which is larger than the experimental one. Keeping this in mind, we find that except for $\mu_{\Sigma^{++}}$ the present results turn out larger than those from the lattice QCD.

Acknowledgements The work is supported by Basic Science Research Program through the National Research Foundation (NRF) of Korea funded by the Korean government (Ministry of Education, Science and Technology(MEST)): Grant No. NRF-2018R1A2B2001752.

References

1. Diakonov, D., Petrov, V.Y., Pobylitsa, P.V.: Nucl. Phys. B **306**, 809 (1988)
2. Diakonov, D.: hep-ph/9802298
3. Blotz, A., Diakonov, D., Goeke, K., Park, N.W., Petrov, V., Pobylitsa, P.V.: Nucl. Phys. A **555**, 765 (1993)
4. Praszalowicz, M., Watabe, T., Goeke, K.: Nucl. Phys. A **647**, 49 (1999)
5. Christov, C.V., Blotz, A., Kim, H.-Ch., Pobylitsa, P., Watabe, T., Meissner, T., Ruiz Arriola E., Goeke, K.: Prog. Part. Nucl. Phys. **37**, 91 (1996)
6. Yang, Gh-S, Kim, H-Ch., Polyakov, M.V., Praszałowicz, M.: Phys. Rev. D **94**, 071502 (2016)
7. Kim, H.-Ch., Polyakov, M.V., Praszałowicz, M.: Phys. Rev. D **96**, 014009 (2017) Addendum: [Phys. Rev. D **96**, 039902 (2017)]
8. Kim, H.-Ch., Polyakov, M.V., Praszalowicz, M., Yang, G.S.: Phys. Rev. D **96**, 094021 (2017) Erratum: [Phys. Rev. D **97**, 039901 (2018)]
9. Kim, J.Y., Kim, H-Ch., Yang, G.S.: Phys. Rev. D **98**, 054004 (2018)
10. Yang, G.S., Kim, H-Ch.: Phys. Lett. B **781**, 601 (2018)
11. Kim, J.Y., Kim, H-Ch.: Phys. Rev. D **97**, 114009 (2018)
12. Can, K.U., Erkol, G., Isildak, B., Oka, M., Takahashi, T.T.: JHEP **1405**, 125 (2014)
13. Bahtiyar, H., Can, K.U., Erkol, G., Oka, M., Takahashi, T.T.: Phys. Lett. B **772**, 121 (2017). https://doi.org/10.1016/j.physletb.2017.06.022. arXiv:hep-lat/1612.05722

Chapter 113
Performance of the FOREST/BLC Spectrometer for Study of the η-nucleon Interaction via the $\gamma d \to p\eta n$ Reaction

S. Miyata, K. Aoki, H. Fujioka, Y. Honda, T. Hotta, T. Ishikawa, K. Itahashi, H. Kanda, H. Kawai, K. Maeda, M. Miyabe, Y. Matsumura, N. Muramatsu, H. Ohnishi, K. Ozawa, H. Shimizu, M. Tabata, A. O. Tokiyasu, Y. Tsuchikawa, T. Ueda and C. Yoshida

Abstract The current experimental information about interactions between the η meson and nucleon (N) is not enough to understand the nature of the η meson in a nucleus. We installed a new spectrometer and started an experiment to determine the ηN scattering parameters at the Research Center for Electron Photon Science, Tohoku University. We present the current status of the experiment.

S. Miyata (✉)
Department of Physics, University of Tokyo, Bunkyo 113-8654, Japan
e-mail: smiyata@post.kek.jp

K. Aoki · K. Ozawa
IPNS, KEK, Tsukuba 305-0801, Japan

H. Fujioka
Department of Physics, Tokyo Institute of Technology, Meguro 152-8551, Japan

Y. Honda · T. Ishikawa · M. Miyabe · N. Muramatsu · H. Ohnishi · H. Shimizu · A. O. Tokiyasu · T. Ueda · C. Yoshida
ELPH, Tohoku University, Sendai 982-0826, Japan

T. Hotta · H. Kanda · Y. Matsumura
RCNP, Osaka University, Ibaraki 567-0047, Japan

K. Itahashi
RIKEN Nishina Center, Wako 351-0198, Japan

H. Kawai · M. Tabata
Department of Physics, Chiba University, Chiba 263-8522, Japan

K. Maeda
Department of Physics, Tohoku University, Sendai 980-8572, Japan

Y. Tsuchikawa
Department of Physics, Nagoya University, Nagoya 464-8602, Japan

N. A. Orr et al. (eds.), *Recent Progress in Few-Body Physics*,
Springer Proceedings in Physics 238,
https://doi.org/10.1007/978-3-030-32357-8_113

113.1 Introduction

The interaction between the η meson and nucleon (N) is known to be attractive due to the existence of the nucleon resonance $N^*(1535)$. However, there are large uncertainties in the values of the scattering parameters especially the real part of the scattering length (Re $a_{\eta N}$ =0.2–1.1 fm [1]), since it is difficult to realize ηn scattering. An accurate determination of Re $a_{\eta N}$ is important to provide strong constraints on the existence of an η-mesic nucleus [2], a bound state of an η meson in a nucleus. The η bound nucleus has been searched for three decades. A new experiment is prepared to measure Re $a_{\eta N}$ using a special kinematics. As for η photoproduction on the deuteron, an η is likely to be produced almost at rest when the incident photon has a certain momentum ($\sim$ 0.94 GeV/c) and when protons are emitted in the forward region ($\sim 0°$). Then the final-state $p\eta$ and pn interaction will be suppressed due to the large $p\eta$ and pn relative momentum. This kinematics is ideal for determination of the low-energy ηn scattering parameters and the differential cross section as a function of the ηn center-of-mass energy (relative ηn momentum) is sensitive to Re $a_{\eta N}$.

113.2 Experimental Setup

We perform the experiment using an energy-tagged bremsstrahlung-photon beam (0.80–1.25 GeV) [3] from a 1.3 GeV electron synchrotron at the Research Center for Electron Photon Science (ELPH), Tohoku University. The η mesons are detected using the FOREST detector [4] via the $\eta \to \gamma\gamma$ decay. The energy and momentum of the forward-going protons are measured with the BLC spectrometer consisting of a bending magnet, two drift chambers (DCs), plastic-scintillator (PS) hodoscope, and additional SF5 lead-glass (LG) counters for e/π separation [5]. The DCs are placed 4–4.5 m downstream from the target and rotated by 15.7° against the beam axis so that protons with a specific momentum ($\sim$ 0.94 GeV/c) enter the front face of each DC at the incident angle θ of 0°.

113.3 Forward Proton Measurements

In this section, the measurement of forward emitted protons are described. As shown in Fig. 113.1, we placed DC1 and DC2 to determine the horizontal and vertical positions. Because the charged particles emitted in the forward region go through the bending magnet, a hit in the X direction can be interpreted as the momentum of the particle. The wire number increases with the distance from the beamline. This is because the magnet has a small coverage for forward emitted particles and the particles enter the magnet with incident angles of approximately 0°. For forward

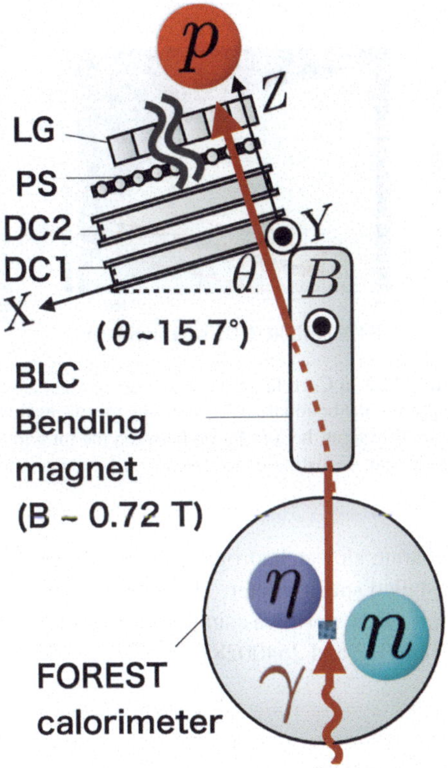

Fig. 113.1 Detector setup. Protons emitted at the forward angles (~ 0°) are deflected by the magnetic flux and are incident on the front face of each DC. The definition of the local coordinates (X, Y, and Z) in the two DC system is described

charged particles, there is a strong correlation between the hit wire number (DC1) and θ, incident angle to DC1, as shown in Fig. 113.2a. We reconstruct incident angles to the DCs using the hit positions in DC1 and DC2. The condition that both DC1 and DC2 have hits in the X direction limits the acceptance and the correlation is in the range of 15–100. By selecting the events in the region between the lines in Fig. 113.2a, we evaluate the arrival time at the PS hodoscope for charged particles. Here, the arrival time is given by the difference between the hit time of a photon-tagging counter set in the upstream and that of a PS hodoscope. Figure 113.2b shows the arrival time as a function of the hit X position in DC1. The zero point of the arrival timing corresponds to the speed of light so that the particles with the speed of light (mainly pions) and slower particles (protons) can be separated clearly.

113.4 Summary and Acknowledgements

In the ELPH facility, a new experiment for accurate determination of the ηN interaction is performed. In the experiment, the 1.3 GeV bremsstrahlung photon beam is used. The experimental setup consists of the electromagnetic calorimeter used to identify η production and spectrometer to detect forward charged particles. We can

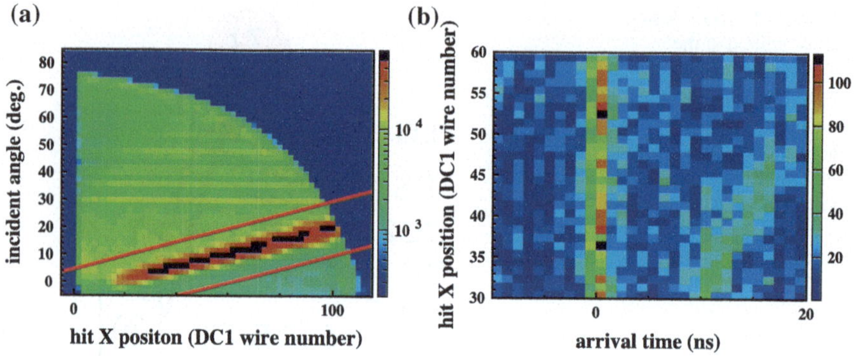

Fig. 113.2 **a** Correlation between the incident angles and hit wire number (DC1). The lines (red) represent the boundaries for selecting the events that contain the forward-emitted charged particles from the target. **b** Correlation between the hit wire number (DC1) and the arrival timing. Protons (right) can be distinguished from $\beta = 1$ particles (left)

separate slower hadrons from $\beta = 1$ particles in the forward region by the newly installed spectrometer. The authors express their gratitude to the ELPH accelerator staff. This work was supported in part by JSPS KAKENHI Grant Nos. 20684008, 24244022 and 26400287.

References

1. Nakamura, S.X., Kamano, H., Ishikawa, T.: Low-energy η-nucleon interaction studied with η photoproduction off the deuteron. Phys. Rev. C 96, 042201 (2017)
2. Haider, Q., Liu, L.C.: Formation of an eta-mesic nucleus. Phys. Lett. B **172**, 257 (1986)
3. Ishikawa, T., et al.: The second GeV tagged photon beamline at ELPH. Nucl. Instrum. Meth. A **622**, 1 (2010)
4. Ishikawa, T., et al.: The FOREST detector for meson photoproduction experiments at ELPH. Nucl. Instrum. Meth. A **832**, 108 (2016)
5. Ishikawa, T., et al.: Low-energy scattering parameters between the eta meson and nucleon from eta photoproduction on the deuteron. Acta Phys. Polon. B **48**, 1801 (2017)

Chapter 114
Strong Decays of p Wave Heavy Mesons in $HH\chi PT$

Jin-Yun Wu, Yong-Lu Liu, Jian-Rong Zhang, Chen Dong and Ming-Qiu Huang

Abstract In this work, we studied the strong decays of the p wave heavy mesons within the framework of heavy hadron chiral perturbation theory ($HH\chi PT$). Up to next-to-leading order in $1/\Lambda_\chi$, the chiral symmetry-breaking coupling constants are extracted. The single-pion decay widths for $D^*_{s0}(2317)$ and $D^{'}_{s1}(2460)$ are 9.2 ± 2.3 KeV and 9.0 ± 2.1 KeV, respectively. Meanwhile, the widths for their beauty partners are also given under the heavy quark symmetry. The predicted widths are consistent with the experimental measurements and comparable with other theoretical predictions.

114.1 Introduction

The measured masses and widths of $D^*_{s0}(2317)$ and $D^{'}_{s1}(2460)$ citeexp1 do not match the predictions from potential-based quark models citeGodfrey1991, unexpectedly, i.e. they lie below the DK and D^*K thresholds respectively and their widths are extremely narrow. Moreover, there is a big progress in 2015 on $D_0^{*\pm}$ by LHCb collaboration citeLHCb15X,LHCb15Y, and the gap between the masses of $D_0^{*\pm}$ and D_0^{*0} is greatly suppressed:

$$M_{D_0^{*\pm}} - M_{D_0^{*0}} \sim 90\text{MeV(before 2015)} \rightarrow \sim 30\text{MeV(after 2015)} \qquad (114.1)$$

J.-Y. Wu (✉) · C. Dong
College of Information and Communication, National University of Defense Technology, Xi'an 710006, China
e-mail: jinyunwu@nudt.edu.cn

Y.-L. Liu · J.-R. Zhang · M.-Q. Huang
College of Liberal Arts and Sciences, National University of Defense Technology, Changsha 410073, China

C. Dong
State Key Laboratory of Cryptology, Beijing 100878, China

N. A. Orr et al. (eds.), *Recent Progress in Few-Body Physics*, Springer Proceedings in Physics 238,
https://doi.org/10.1007/978-3-030-32357-8_114

This suppression will do a great favor to the determination of the chiral symmetry-breaking coupling constants and finally the decay rates of $D^*_{s0}(2317)$ and $D'_{s1}(2460)$.

114.2 The Chiral Lagrangian

Considering heavy quark spin-flavor symmetry and light quark chiral symmetry, an effective Lagrangian responsible for the strong decay $S \to HM$ (M is a light pseudoscalar meson) can be written with these superfields. The leading order contribution in $1/\Lambda_\chi$ and $1/m_Q$ is

$$\mathcal{L}_{mix} = hTr[\bar{H}_b S_a \mathcal{A}_{ab} \gamma_5] + h.c.. \tag{114.2}$$

According to [5, 6], the corresponding chiral symmetry breaking corrections to next-to-leading order in $1/\Lambda_\chi$ read

$$\begin{aligned}\mathcal{L}^{sb}_{mix} &= 1/\Lambda_\chi\{\kappa_1 Tr[(\bar{H}S\mathcal{A}\gamma_5)_{ab}(m^\xi_q)_{ba}] + \kappa_2 Tr[(\bar{H}S\mathcal{A}\gamma_5)_{aa}(m^\xi_q)_{bb}]\\ &+ \kappa_3 Tr[\bar{H}_a S_a \mathcal{A}_{bc}\gamma_5(m^\xi_q)_{cb}] + \kappa_4 Tr[\bar{H}_c S_a \mathcal{A}_{bc}\gamma_5(m^\xi_q)_{ab}]\\ &+ \kappa_5 Tr[\bar{H}_a S_b iv\cdot \mathcal{D}_{bc}\mathcal{A}_{ca}\gamma_5] + \kappa_6 Tr[\bar{H}_a S_b i\mathcal{D}\!\!\!/_{bc} v\cdot \mathcal{A}_{ca}\gamma_5]\} + h.c..\end{aligned} \tag{114.3}$$

Meanwhile, the effective Lagrangian responsible for $\eta - \pi^0$ mixing, through which the pionic decays of $D^*_{s0}(2317)$ and $D'_{s1}(2460)$ occur, can be described by the isospin violating piece in the chiral Lagrangian

$$\begin{aligned}\mathcal{L}_{\eta-\pi^0} &= \frac{m^2_\pi f^2_\pi}{4(m_u + m_d)} Tr[m^\dagger_q \Sigma + \Sigma^\dagger m_q]\\ &= \frac{m^2_\pi (m_u - m_d)}{\sqrt{3}(m_u + m_d)}\pi^0\eta + \cdots .\end{aligned} \tag{114.4}$$

114.3 Coefficients

For simplify, as mentioned in [7], the following transformations of the parameters are made:

(i) As κ_2 can be absorbed into the definition of h, we set $h' = h + \frac{2(m_u+m_d+m_s)}{\Lambda_\chi}$.
(ii) As κ_5 and κ_6 always enter in a fixed combination, they are properly represented by a united parameter $\kappa'_5 = \kappa_5 + \kappa_6$.
(iii) For κ_1 and κ_4, if we define $\kappa'_1 = \frac{\kappa_1+\kappa_4}{2}$ and $\kappa'_4 = \frac{\kappa_1-\kappa_4}{2}$, they will be distinguishable that κ'_1 concerns only the isospin conserving contributions, while κ'_4 involves only the isospin violating contributions.

Table 114.1 Results of the six $1/\Lambda_\chi$ parameters by minimizing χ^2

h'	κ_1'	κ_2'	κ_4'	κ_5'	$\chi^2/2$
0.56 ± 0.01	0.86 ± 0.11	0.52 ± 0.26	0.40 ± 0.31	0.28 ± 0.03	0.66 ± 0.01

By minimizing the χ^2, the unknown chiral-symmetry breaking coupling constants are extracted, as shown in Table 114.1, together with the value of corresponding χ^2. The main contribution to the value of χ^2 comes from the relatively large discrepancy between the decay rates of D_0^* and D_1' mesons.

114.4 Decay Widths and Conclusion

The results are shown in Tables 114.2 and 114.3. It shows that our results are consistent with the experimental constraints and comparable with the other theoretical works in the literature. And the chiral symmetry-breaking corrections of $c\bar{q}$ ($b\bar{q}$) are small in comparison with the leading order contributions, while those of $c\bar{s}$ ($b\bar{s}$) are significant due to relatively large mass of the strange quark. The confirmation of such predictions is expected in the near future by experiments at the LHCb and the hadron B factories.

Table 114.2 Strong decay widths of $D_{s0}^*(2317)$ to $D_s\pi^0$ and $D_{s1}'(2460)$ to $D_s^*\pi^0$ (in KeV)

Approach	$\Gamma(D_{s0}^* \to D_s\pi^0)$	$\Gamma(D_{s1}' \to D_s^*\pi^0)$
PDG [8]	< 3.8 MeV	< 3.6 MeV
[9]	21.5	21.5
[10]	≈ 10	≈ 10
[11]	7 ± 1	7 ± 1
[12]	150 ± 70	150 ± 70
[13]	46.7–111.9	50.1–79.2
[14]	96 ± 19	78 ± 14
[15]	10–100	–
[16]	6 ± 2	–
Γ(leading)	5.0 ± 1.0	4.9 ± 1.0
Γ(full)	9.2 ± 2.3	9.0 ± 2.1

Table 114.3 Predicted masses of the S doublet beauty mesons (in MeV). And strong decay widths of B_0^* and $B_1^{'}$ (in MeV), B_{s0}^* and $B_{s1}^{'}$ (in KeV)

	$B_0^*(0^+)$	$B_1^{'}(1^+)$	$B_{s0}^*(0^+)$	$B_{s1}^{'}(1^+)$
Mass [17]	5708.2 ± 22	5753.3 ± 31	5706.6 ± 1.2	5765.6 ± 1.2
Γ [17]	269 ± 58	268 ± 70	–	–
Γ [9]	–	–	21.5	21.5
Γ [14]	–	–	0.8 ± 0.8	1.8 ± 1.8
Γ [18]	–	–	13.6 ± 5.6	13.8 ± 3.6
Γ [19]	87	93	1.6	1.9
Γ [20]	–	–	55.2–89.9	57.0–94.0
Γ(leading)	284 ± 47	286 ± 52	6.5 ± 0.1	7.1 ± 0.1
Γ(full)	313 ± 53	314 ± 67	11.6 ± 1.6	12.3 ± 1.7

Acknowledgements This work was supported in part by the National Natural Science Foundation of China under Contract Nos. 11704412, 11675263, 11475257 and 11475258. YLL was supported by the Chinese Scholarship Council (CSC), JRZ was supported by the project in NUDT for excellent youth talents, and CD was supported by the Research Fund of National University of Defense and Technology (Grant No. ZK17-02-09) and Research Fund of State Key Laboratory of Cryptology (Grant No. MMKFKT201823).

References

1. B. Aubert *et al.* (BABAR Collaboration), Phys. Rev. Lett., **90**: 242–001 (2003); D. Besson *et al.* (CLEO Collaboration), Phys. Rev. D, **68**: 032002 (2003).
2. S. Godfrey and R. Kokoski, Phys. Rev. D, **43**: 1679 (1991); S. Godfrey and N. Isgur, Phys. Rev. D, **2**: 189 (1985); M. Di. Pierro and E. Eichten, Phys. Rev. D, **64**: 114004 (2001).
3. Aaij, R., et al.: LHCb Collaboration. Phys. Rev. D **92**, 012012 (2015)
4. Aaij, R., et al.: LHCb Collaboration. Phys. Rev. D **92**, 032002 (2015)
5. C. G. Boyd and B. Grinstein, Nucl. Phys. B, **442**: 205 (1995); M. Q. Huang, Y. B. Dai, C. S. Huang, Phys. Rev. D, **52**: 3986 (1995); I. W. Stewart, Nucl. Phys. B, **529**: 62 (1998).
6. Fajfer, S., Kamenik, J.: Phys. Rev. D **74**, 074023 (2006)
7. Stewart, I.W.: Nucl. Phys. B **529**, 62 (1998)
8. Patrignani, C., et al.: Particle Data Group. Chin. Phys. C **40**, 100001 (2016)
9. Bardeen, W.A., Eichten, E.J., Hill, C.T.: Phys. Rev. D **68**, 054024 (2003)
10. Godfrey, S.: Phys. Lett. B **568**, 254 (2003)
11. Colangelo, P., De, F.: Fazio. Phys. Lett. B **570**, 180 (2003)
12. Ishida, S., Ishida, M., Komada, T., Maeda, T., Oda, M., Yamada, K., Yamauchi, I., Conf, A.I.P.: Proc. **717**, 716 (2004)
13. A. Faessler, T. Gutsche, V. E. Lyubovitskij, and Y. L. Ma, Phys. Rev. D, **76**: 014005 (2007); A. Faessler, T. Gutsche, V. E. Lyubovitskij, and Y. L. Ma, Phys. Rev. D, **76**: 014008 (2007).
14. M. Cleven, H. W. Grießhammer, F. K. Guo, C. Hanhart and U. G. Meißner, arXiv:hep-ph/1405.2242
15. Cheng, H.Y., Hou, W.S.: Phys. Lett. B **566**, 193 (2003)
16. Nielsen, M.: Phys. Lett. B **634**, 35 (2006)
17. Colangelo, P., De Fazio, F., Giannuzzi, F., Nicotri, S.: Phys. Rev. D **86**, 054024 (2012)

18. Wang, Z.H., Wang, G.L., Fu, H.F., Jiang, Y.: Phys. Lett. B **706**, 389 (2012)
19. Matsuki, T., Seo, K.: Phys. Rev. D **85**, 014036 (2012)
20. Faessler, A., Gutsche, T., Lyubovitskij, V.E., Ma, Y.L.: Phys. Rev. D **77**, 114013 (2008)

Chapter 115
Relativity in Few-Hadron Systems: Analysis of Baryon Electromagnetic Transition Form Factors in the Covariant Spectator Theory

M. T. Peña and G. Ramalho

Abstract We discuss the role of valence quarks and meson cloud excitations for the Transition Form Factors (TFF) of the electromagnetic transition of the nucleon (N) to nucleonic excitations (N^*), calculated within the Covariant Spectator model. We extend TFF from the space-like to the time-like regime and calculate Dalitz di-electron decay widths for the $\Delta(1232)$ and the $N^*(1520)$.

115.1 Introduction

Nucleonic excitations play a crucial role in the transition of hadron matter to the Quark Gluon Plasma and the properties, as the critical end point, of the phase diagram of QCD. On the other hand, diquark clusters seem to have a role not only in tetraquark structures but also in baryons [1].

How N^* excited states decay and radiate in the highly dense medium is probed experimentally in dilepton production rates from relativistic proton-proton and heavy-ion reactions [2], and on the other-hand the microscopic structure of the nucleon and its excited states is revealed by electron scattering measurements [3]. These methods appear unrelated. However they are connected since they explore the same physical systems in different space-like and time-like ranges of photon virtualities.

Baryon transition form factors (TFFs). TFFs and the evolution of the photon-quark coupling with 4-momentum transfer squared, Q^2, constitute the common ground of the two methods. Form factors in the time-like region are measured from in-medium dilepton production and probe particle production channels at small Q^2, i.e. the spectroscopy and formation of vector mesons. Form factors in the space-like region are measured in electron scattering experiments and probe structure, as

M. T. Peña (✉)
CFTP, Instituto Superior Técnico, Universidade de Lisboa, Lisboa, Portugal
e-mail: teresa.pena@tecnico.ulisboa.pt

G. Ramalho
LFTC, Universidade Cruzeiro do Sul, São Paulo, Brazil

N. A. Orr et al. (eds.), *Recent Progress in Few-Body Physics*,
Springer Proceedings in Physics 238,
https://doi.org/10.1007/978-3-030-32357-8_115

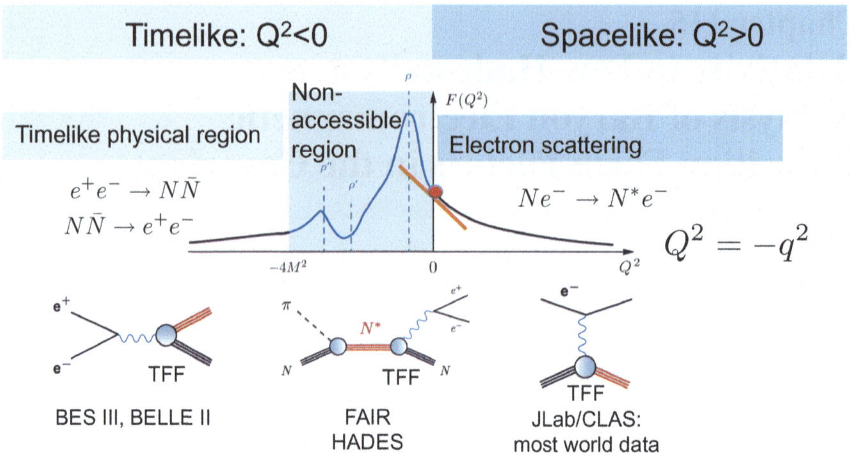

Fig. 115.1 CLAS/JLab electron scattering data and HADES dilepton production data: results have to match at the photon point. TFF sketch from [1]

shape and qqq excitations versus hybrid components. Spectroscopy and structure influence both space-like and time-like regimes, in electron-scattering and dilepton production reactions, respectively. Both regimes have to be matched smoothly at the photon point $Q^2 = 0$, and are illustrated in Fig. 115.1. This link is explored in this work, that uses the constraints from electron scattering to calculate TFFs in the time-like region.

Definition of degrees of freedom. What makes the information contained in baryon TFFs so challenging is the interplay between the microscopic quark-gluon structure and the hadronic states to which the baryonic resonances couple in their decays. A prime example that illustrates this interplay is the missing strength of the dominant magnetic form factor G_M of the $\gamma^* N \to \Delta(1232)$ transition in the region near $Q^2 = 0$ [4]. Theory, including state-of-the-art dynamical quark calculations of the three-quark system based on the Dyson-Schwinger formalism [1], systematically fails to describe the data in the region of small Q^2.

This seems to support that for baryons three quark degrees of freedom are complemented by structure effects beyond three constituents, and this is not surprising: for small momentum transfer the photon does not have enough resolution for the baryon structure, and therefore one expects that it couples to the baryon as a whole, and not to each quark. In addition, in the region around $Q^2 = 0$ chiral symmetry is crucial. Chiral symmetry breaking in QCD originates a Goldstone boson, the pion, with an almost vanishing mass. This is the reason why in the region around $Q^2 = 0$, where time-like and space-like form factors have to be in harmony, there are significant contributions of pion-loops to the electromagnetic transition form factors. This means that the photon sees the peripheral pion cloud instead of the inner three-quark core of the baryon. Pions were indeed seen as crucial for the description of

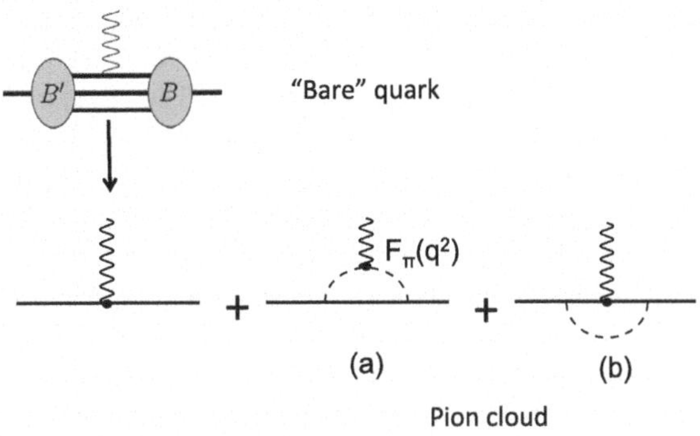

Fig. 115.2 Diagrams to be added to the quark degrees of freedom (top panel): the photon couples with a pion in flight dressing the full baryon (**a**), or with the fully dressed baryon while the pion is in flight (**b**)

the neutron electric form factor in constituent quark models [4]. At high momentum transfer, however, the direct coupling of the photon to an individual dressed quark is expected to be more important.

In Fig. 115.2 we represent our convention for the separation of quark and baryon degrees of freedom. We define here the "bare" quark contribution as the impulse diagram on the top panel where the photon couples with only one dressed quark, and the baryon degrees of freedom or "pion cloud" as the processes of diagrams (a) and (b) on the lower panel, where the photon couples with the baryon as a system of three interacting quarks.

115.2 The CST Model and the Electromagnetic Quark Current

To access the importance of the pion cloud in electromagnetic transitions from first principles, calculations for three-quark dynamics have to be extended to five quarks. This was not yet done in the framework of QCD or any constituent quark model. Here we take a first step into that direction, at a phenomenological level only. Our model is rooted in the Covariant Spectator Theory (CST), a field theoretic based framework, discussed in more detail in [5] where its successful application to the meson spectrum is presented.

The CST model. Within CST the covariant 4D integration over the momenta of the two non-active quarks (not interacting with the photon) in the upper panel of Fig. 115.2—forming what we call in what follows a "diquark", just for simplicity and not meaning that it corresponds to a pole of the scattering matrix—is reduced to a 3D integration by a prescription of the energy integration that keeps the dominant energy pole contribution. Within impulse approximation, i.e. for the contribution defined in the upper panel of Fig. 115.2, the electromagnetic coupling does not depend on the relative motion variables of the quarks of the "diquark", and we can integrate over those variables, ending up with only the integration on the quark-"diquark" relative coordinate. The implication is that in the energy regime where the impulse approximation is valid, one can use the effective quark-"diquark" result from integrating out the internal "diquark" coordinates. The "diquark" here is not a "point diquark" but has a scalar and/or pseudo-vector structure. The spin part of the wave functions for several baryon systems is presented in [6–11] in terms of Dirac structures. It was then possible to construct wave functions of several baryons that are explicitly covariant, have the correct nonrelativistic limit [6, 7], and can include components with angular momentum $L > 0$, in particular P and D waves [10].

In short, our model has three general features: (i) it is covariant; (ii) it assumes the impulse approximation for the coupling of a baryon with a photon for reasonably high momentum transfer where the photon couples with only one of the three constituent quarks at a time (Fig. 115.2 top panel); (iii) it uses Vector Meson Dominance (VMD) for the quark-photon current [6]. VMD allows the model to be extrapolated from space-like to the time-like kinematics. Also, it is VMD that allowed, where lattice QCD data was available, contact with lattice QCD input for determining parameters of the model. Aspects (ii) and (iii) are discussed next.

Vector meson dominance. Our starting point is that the quark-photon vertex Γ_μ [12] is given by the inhomogeneous Bethe-Salpeter equation,

$$\Gamma_\mu(p, Q) = \gamma_\mu + \int \frac{d^4q}{(2\pi)^4} K(p, q, Q) S(q + \eta Q) \Gamma_\mu(q, Q) S(q - \eta Q) \quad (115.1)$$

where η gives the photon 4-momentum fraction shared by the initial and final quark, K is the quark-antiquark interaction, S is the quark propagator. The inhomogeneous term is the coupling of the photon to a bare quark. The iterations of the quark-antiquark interaction kernel K generate loops of quark-antiquark excitations summed by the integral equation. It becomes clear from (115.1) how the meson spectrum ties with the behavior of the quark-photon coupling: the infinite iteration of the quark-antiquark interaction K generates poles at the vector meson masses for time-like kinematics, and the first pole is the ρ meson pole. This justifies using a VMD phenomenological parameterization for the quark current.

Because the quarks are effectively dressed and have anomalous magnetic moments, the single constituent quark current j_q^μ contains two terms

$$j_q^\mu = j_1 \hat{\gamma}^\mu + j_2 \frac{i\sigma^{\mu\nu} q_\nu}{2M}, \quad \hat{\gamma}^\mu = \gamma^\mu - \frac{\not{q} q^\mu}{q^2}, \quad (115.2)$$

where j_1 and j_2 are the Dirac and Pauli quark form factors, and in our parametrization M is taken as the nucleon mass. Each form factor $j_i (i = 1, 2)$ has an isoscalar and an isovector component, respectively f_{i+} and f_{i-} (functions of Q^2), $j_i = \frac{1}{6} f_{i+} + \frac{1}{2} f_{i-} \tau_3$. The second term in the second equation in (115.2) is equivalent to using the Landau prescription, and guarantees the conservation of the baryon current.

At the scales considered here contact quark-antiquark interactions can be assumed, and the series of the integral equation (115.1) analytically summed with the meson

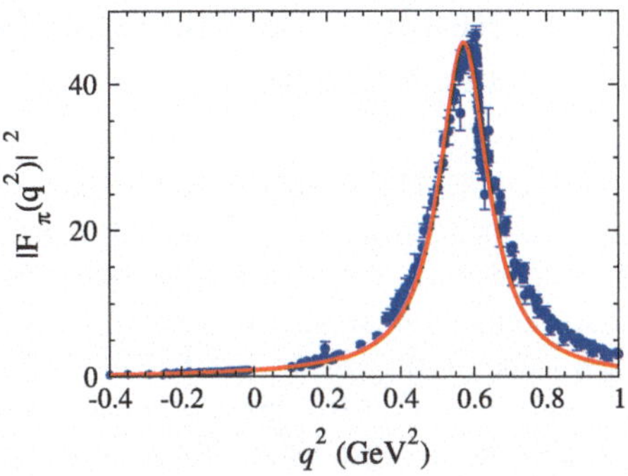

Fig. 115.3 The pion form factor in the timelike and spacelike regions around $q^2 = 0$

poles made explicit. The forms of the Dirac and Pauli quark form factors, $f_{1\pm}$ and $f_{2\pm}$, are then chosen to be consistent with this result from vector meson dominance [6, 7]. They involve mass parameters $m_v = m_\rho$ or m_ω of the light vector meson masses, and a $M_h = 2M$ term that takes into account meson resonances with a larger mass, as well as $\kappa_\pm$, the quark anomalous magnetic moments. The parameter for the high energy photon-quark vertex is fixed by deep inelastic scattering data. The functions $f_{1\pm}$ and $f_{2\pm}$ are normalized to reproduce the charge and anomalous magnetic moment of the u and d quarks. The mixture coefficients for the different vector mesons contributions are phenomenologically fixed by the proton and neutron elastic electromagnetic form factors and then used in the predictions of resonance TFFs.

Finally, for the diagram (a) of the baryon structure effects the coupling between the photon and the pion is taken to be momentum dependent and fitted to the experimentally well known the pion form, since the residue of the pion from factor F_π at the time-like pole is proportional to the $\rho \to \pi\pi$ decay. Figure 115.3 shows the fit that we used in the calculations. Diagram (b) is more suppressed for the $N^*(1520)$ than for the $\Delta(1232)$.

Constraints from LQCD. Because (i) the pion cloud effects are negligible for large unphysical pion masses and (ii) the electromagnetic quark current is based on vector meson dominance, the vector meson mass can be taken as a function of the running pion mass, and the electromagnetic current bare quark core model in the upper panel of Fig. 115.2 can be calibrated by the space-like LQCD data for large pion masses. Afterwards, taking the limit of the model in the physical pion mass, the experimental space-like data is well described in the high momentum transfer Q^2 region. This procedure was checked first for the $\gamma^* N \to \Delta(1232)$ reaction [13, 14]. The $\Delta(1232)$ wave function was fixed by the LQCD results for the three $\gamma^* N \to \Delta(1232)$ electromagnetic form factors, making use of a running pion mass to vary the ρ meson mass. Then by subtracting the experimental data from the CST constituent quark model, we inferred the pion cloud, and obtained that indeed it was non-zero in the vicinity of $Q^2 \approx 0$. The extracted information on the pion cloud contribution to the $\Delta(1232)$ electro-excitation is consistent with the results from QCD in the large N_c limit [15], recently revisited in [16].

115.3 Results in the Time-Like Region for the $\Delta(1232)$ and the $N^*(1520)$

To extend the TFF from space-like (in electron-scattering) to time-like (in dilepton production) regime, we generalize the physical Δ mass M_Δ to a running Δ mass W, making $M_\Delta \to W$ in the intermediate states away from the resonance pole. Four-momentum conservation implies that there is an upper limit for $q^2 = -Q^2$ for each value of W, and the excitation reaction becomes kinematically forbidden for $q^2 > (W - M)^2$. Also, the VMD parametrization is again particularly convenient for the time-like extension. To avoid the pole at $q^2 = m_\rho^2$ we include a non-zero width for the ρ. This extension reproduced the space-like results at the physical point point $W = M_\Delta$ [17], and enabled the first extraction of the $\Delta(1232)$ Dalitz decay branching ratio by the HADES collaboration [2].

Our results for the $\gamma^* N \to N^*(1520)$ $(J_\pi = 3/2^+)$ transition [18] gives, as in the $\Delta(1232)$ case, a good description of the helicity amplitudes and form factors for large Q^2. The time-like $N^*(1520)$ transition form factors in the region $Q^2 \approx 0$ are dominated by the meson cloud contributions due to the orthogonality of the initial and final state wave functions. At low Q^2 the transverse $A_{3/2}$ amplitude is found to be negligible. The inferred meson cloud has predominantely an iso-vector character, in agreement with the PDG data for this reaction, $A^V_{3/2} = 0.13\ \text{GeV}^{-1/2}$ for the iso-vector component of the $A_{3/2}$ amplitude at $Q^2 = 0$, versus $A^S_{3/2} = 0.01\ \text{GeV}^{-1/2}$ for the iso-scalar component. Once the form factors are calculated we obtain the Dalitz differential light dilepton decay rate from

$$\Gamma'_{e^+e^-N}(q, W) \equiv \frac{d\Gamma}{dq}(q, W) \frac{2\alpha}{3\pi q^3}(2\mu^2 + q^2)\sqrt{1 - \frac{4\mu^2}{q^2}}\,\Gamma_{\gamma^* N}(q, W), \qquad (115.3)$$

with

$$\Gamma_{\gamma^* N}(q, W) = \frac{3\alpha}{32} \frac{(W - M)^2}{M^2 W^3} \sqrt{y_+ y_-}\, y_+ |G_T(q^2, W)|^2, \qquad (115.4)$$

where $q = \sqrt{q^2}$, α is the fine-structure constant, $y_\pm = (W \pm M)^2 - q^2$, and $|G_T(q^2, W)|^2$ is the combination of the electromagnetic transition form factors $|G_T(q^2, W)|^2 = 3|G_M(q^2, W)|^2 + |G_E(q^2, W)|^2 + \frac{q^2}{2W^2}|G_C(q^2, W)|^2$ (G_M, G_E and G_C are respectively the magnetic dipole, electric and Coulomb quadrupole form factors). The Dalitz decay width is determined by the integral of $\Gamma'_{e^+e^-N}(q, W)$ in the region $4\mu^2 \leq q^2 \leq (W - M)^2$.

Figure 115.4 gives the first CST results for the $\Gamma_{\gamma N}$ decay width of the $N^*(1520)$, where the neutron and proton cases are separated. Our result $G_T(0, 1520) = 0.73$ is consistent with the PDG data. Our neutron and proton results do not differ much due to the iso-vector dominance of the meson cloud current in the model. Figure 115.5 shows in the left panel the differential di-electron production rate for the same resonance.

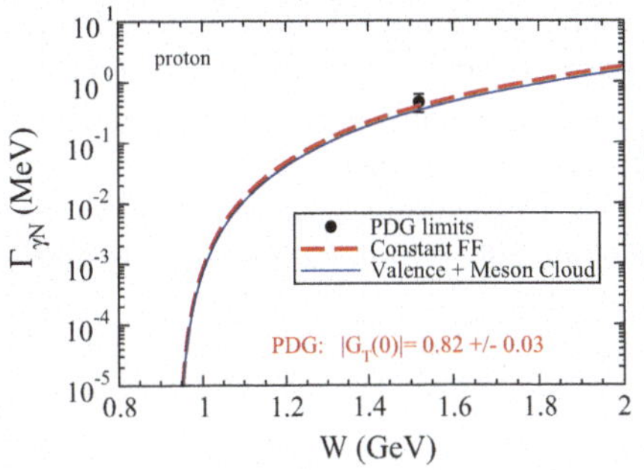

Fig. 115.4 Proton (left) and neutron (right) results for the $\Gamma_{N\gamma}$ decay width of the $N^*(1520)$. The dashed-line shows the approximation of taking the photon coupling constant and equal to its value at $q^2 = 0$

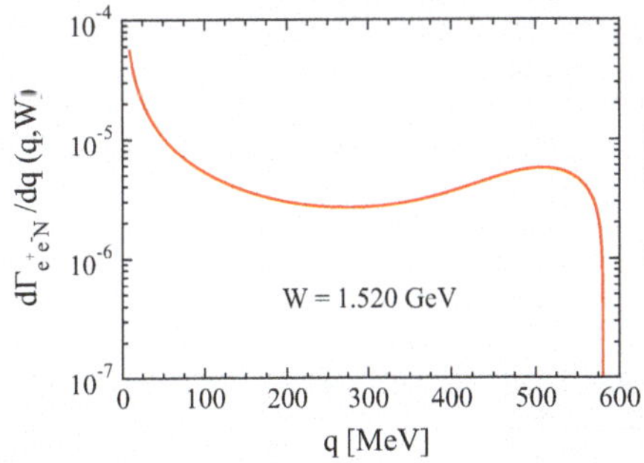

Fig. 115.5 Proton and neutron di-electron differential and total decay width of the $N^*(1520)$. CFF labels results obtained with constant photon-baryon coupling

Preliminary results from the HADES collaboration [19] confirm our results. The right panel of Fig. 115.5 gives the results for the integrated decay width and shows that for high energies the evolution of the photon coupling strength $G_T(q^2, W)$ with q^2 becomes important.

115.4 Conclusions and Outlook

CST calculations are consistent with LQCD in the large pion mass regime. When combined with LQCD they allow the indirect extraction of meson cloud effects. Our results show that the CST approach describes well the behavior of several $\gamma^* N \to N^*$ hadronic vertices in an efficient way, at least in the high Q^2 region. The constant baryon-photon coupling approximation, usually taken in the literature, can underestimate the $N^*(1520)$ di-electron differential Dalitz decay width. More LQCD simulations below the N* threshold will help us to refine the interpretation of data provided by theoretical quark models. In particular, LQCD data for the electromagnetic form factors of the baryon octet are a precious source of information, due to scarcity of experimental information. New experimental data at large Q^2 and even more precise data in all ranges can as well improve the interpretation of empirical results. Dynamical calculations of "diquark" vertices within CST is to be done, to support the quark-"diquark" dominance in baryons, seen within the Dyson-Schwinger approach [1].

References

1. Eichmann, G., Fischer, C.S.: Phys. Rev. D **85**, 034015 (2012); Eichmann, G., Sanchis-Alepuz, H., Williams, R., Alkofer, R., Fischer, C.S.: Prog. Part. Nucl. Phys. **91** (2016)
2. Adamczewski-Musch, J., et al. (HADES Collaboration): Phys. Rev. D **95**, 065205 (2017)
3. Aznauryan, I.G., et al.: Int. J. Mod. Phys. E **22**, 1330015 (2013)
4. Lu, D.H., Yang, S.N., Thomas, A.W.: Nucl. Phys. A **684**, 296 (2001); Cloet, I.C., Eichmann, G., Flambaum, V.V., Roberts, C.D., Bhagwat, M.S., Holl, A.: Few Body Syst. **42**, 91–113 (2008)
5. Stadler, A., Leitão, S., Peña, M.T., Biernat, E.P.: Submitted Proceedings to this Conference
6. Gross, F., Ramalho, G., Peña, M.T.: Phys. Rev. C **77**, 015202 (2008)
7. Ramalho, G., Peña, M.T., Gross, F.: Eur. Phys. J. A **36**, 329 (2008)
8. Gross, F., Ramalho, G., Peña, M.T.: Phys. Rev. D **85**, 093005 (2012)
9. Ramalho, G., Peña, M.T., Gross, F.: Phys. Rev. D **78**, 114017 (2008)
10. Gross, F., Ramalho, G., Peña, M.T.: Phys. Rev. D **85**, 093006 (2012)
11. Ramalho, G., Tsushima, K., Gross, F.: Phys. Rev. D **80**, 033004 (2009)
12. Maris, P., Tandy, P.C.: Phys. Rev. C **61**, 045202 (2000)
13. Ramalho, G., Peña, M.T.: J. Phys. G **36**, 115011 (2009)
14. Ramalho, G., Peña, M.T.: Phys. Rev. D **80**, 013008 (2009)
15. Pascalutsa, V., Vanderhaeghen, M., Yang, S.N.: Phys. Rept. **437**, 125 (2007)
16. Ramalho, G.: Eur. Phys. J. A **54**, 75 (2018)
17. Ramalho, G., Peña, M.T., Weil, J., van Hees, H., Mosel, U.: Phys. Rev. D **93**, 033004 (2016)
18. Ramalho, G., Peña, M.T.: Phys. Rev. D **95**, 014003 (2017)
19. Ramstein, B., Scozzi, F.: Private Communication (2018)
20. Ramalho, G., Peña, M.T., Gross, F.: Phys. Rev. D **81**, 113011 (2010)

Chapter 116
Masses and Structure of Heavy Quarkonia and Heavy-Light Mesons in a Relativistic Quark Model

Alfred Stadler, Sofia Leitão, M. T. Peña and Elmar P. Biernat

Abstract Using the framework of the Covariant Spectator Theory, we calculated the masses and vertex functions of heavy and heavy-light mesons, described as quark-antiquark bound states. Our interaction kernel consists of an adjustable mixture of Lorentz scalar, pseudoscalar, and vector linear confining interactions, together with a one-gluon-exchange kernel. We performed a series of fits to the heavy and heavy-light meson spectrum, and we discuss what conclusions can be drawn from it, especially about the Lorentz structure of the kernel.

116.1 Theoretical Framework

We present results of calculations of the mass spectrum and structure of heavy and heavy-light mesons in the framework of the Covariant Spectator Theory (CST) [1, 2]. Among the features that distinguish our CST treatment of $q\bar{q}$ bound states from other covariant approaches, such as Lattice QCD (e.g., [3–6]) and Dyson-Schwinger-Bethe-Salpeter equations [7–12], the more prominent are that we implement confinement through an effective linear confining interaction, and that we work in the physical Minkowski space. An advantage is that we can calculate highly excited meson states rather easily, which is not the case for the Euclidean approaches mentioned above.

The so-called one-channel covariant spectator equation (1CSE) for $q\bar{q}$ bound states is represented graphically in Fig. 116.1. It is obtained from the Bethe-Salpeter equation (BSE) by carrying out the integration over the energy component of the internal loop momentum in the complex plane and using Cauchy's integral formula, but keeping only the residue of the positive-energy propagator pole of the quark (in

A. Stadler (✉)
Departamento de Física, Universidade de Évora, 7000-671 Évora, Portugal
e-mail: stadler@uevora.pt

A. Stadler · S. Leitão · M. T. Peña · E. P. Biernat
CFTP, Instituto Superior Técnico, Universidade de Lisboa, Av. Rovisco Pais 1, 1049-001 Lisbon, Portugal

N. A. Orr et al. (eds.), *Recent Progress in Few-Body Physics*, Springer Proceedings in Physics 238,
https://doi.org/10.1007/978-3-030-32357-8_116

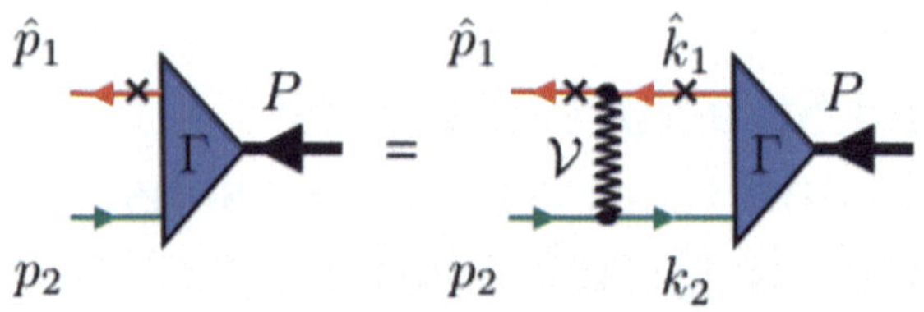

Fig. 116.1 The one-channel spectator equation (1CSE) for the bound-state vertex function Γ of a quark (particle 1) and an antiquark (particle 2), interacting through a kernel $\mathcal{V}$. On-shell four-momenta are characterized through a " ˆ ", and the corresponding lines in the diagram are marked with a "×"

our convention the mass of the antiquark is lower or equal to the mass of the quark). Neglecting the residues of the kernel's poles can lead to faster convergence to the full BS result—with all ladder and crossed ladder diagrams—than the BSE in ladder approximation, because these residues have a tendency to cancel [1].

The 1CSE has the correct one-body limit: when one quark becomes infinitely heavy, the 1CSE becomes an effective one-body (Dirac) equation for the lighter quark. It also has a smooth nonrelativistic limit. The more complicated full four-channel CST equation (4CSE) includes all quark propagator poles, but the 1CSE is a very good approximation for heavy and heavy-light systems. However, in contrast to the 4CSE it is not charge-conjugation symmetric, which means that we cannot assign a C-parity to its solutions for heavy quarkonia. On the other hand, the mass difference between axialvector $C = +$ and $C = -$ pairs is only 5–6 MeV in bottomonium and 14 MeV in charmonium, so this disadvantage is of no practical importance for our purposes.

The dynamic of the $q\bar{q}$ pair is determined by the interaction kernel $\mathcal{V}$. We adopt a kernel consisting of a covariant generalization of the linear (L) confining potential used in [13], a one-gluon exchange (OGE), and a covariantized constant (C) interaction,

$$\mathcal{V} = \left[(1-y)\left(\mathbf{1}_1 \otimes \mathbf{1}_2 + \gamma_1^5 \otimes \gamma_2^5\right) - y\,\gamma_1^\mu \otimes \gamma_{2\mu}\right] V_{\rm L}(p,k) + \gamma_1^\mu \otimes \gamma_{2\mu}\left[V_{\rm OGE}(p,k) + V_{\rm C}(p,k)\right], \tag{116.1}$$

where $p = \hat{p}_1 - P/2$ and $k = \hat{k}_1 - P/2$. The explicit expressions for the functions $V_{\rm L}(p,k)$, $V_{\rm OGE}(p,k)$, and $V_{\rm C}(p,k)$ can be found in [14].

Because the Lorentz structure of the confining interaction is not precisely known, we allow for a mixture of scalar-plus-pseudoscalar and vector structure, controlled by a mixing parameter y that we determine by fitting to the data. Although in principle scalar and pseudoscalar interactions break chiral symmetry, we have shown in [15] that our equal-weight scalar and pseudoscalar linear confining interaction satisfies the axial-vector Ward-Takahashi identity.

Whatever value of y is used, the signs in (116.1) make sure that the nonrelativistic limit of the kernel always yields the same Fourier transform of the Cornell potential, $V(r) = \sigma r - \alpha_s/r - C$. The three coupling strengths, σ, α_s, and C, and in some cases the quark masses, are free parameters of our models. We use Pauli-Villars regularization for the linear and the OGE kernels. The corresponding cut-off parameter Λ is not fitted, but scaled with the heavy quark mass, in the form $\Lambda = 2m_1$.

The 1CSE for the CST vertex function with a kernel of the general form $\mathcal{V} \equiv \sum_K V_K(p,k)\Theta_1^K \otimes \Theta_2^K$, where in our case $\Theta_i^K = \mathbf{1}_i, \gamma_i^5$, or γ_i^μ, is

$$\Gamma(\hat{p}_1, p_2) = -\int \frac{d^3k}{(2\pi)^3}\frac{m_1}{E_{1k}} \sum_K V_K(p,k)\Theta_1^K \frac{m_1 + \hat{\not{k}}_1}{2m_1}\Gamma(\hat{k}_1, k_2)\frac{m_2 + \not{k}_2}{m_2^2 - k_2^2 - i\epsilon}\Theta_2^K . \tag{116.2}$$

We then reformulate it as an equation for relativistic CST wave functions, which are expanded in a set of basis functions with definite orbital angular momentum and total quark-antiquark spin. We solve the equation for the wave functions numerically by applying techniques developed in [13].

116.2 Numerical Results and Conclusions

Our *global* model parameters (Table 116.1), i.e., taken equal for *all* described mesons, were determined through least square fits to different sets of experimental masses for $J^P = 0^\pm$ and $1^\pm$ mesons. The smallest set consisted of only 9 pseudoscalar (PS) states, whereas the largest set contained a total of 39 states of all kinds (a detailed list of these data sets can be found in [14]).

The mass spectra for the models of Table 116.1 are shown in Fig. 116.2. Model M0_{S1} was fitted with a fixed $y = 0$ (no Lorentz vector coupling in the confining kernel) and fixed quark masses. It is remarkable that—although it was fitted to a set of 9 PS states alone—it also predicts the full spectrum of $J^P = 0^\pm$ and $1^\pm$ mesons almost as well as the more extensive fits. The results of M0_{S1} remain almost the same once the PS component in the confining kernel is turned off (the largest difference is about 40 MeV in PS $c\bar{q}$), confirming the expectation that PS coupling does not play an important role in mesons with heavy quarks.

In the fit of M1_{S3}, the quark masses and y were allowed to vary, yielding $y = 0.20$. This seems to indicate that a 20% contribution of vector coupling in the confining kernel is preferred. However, a systematic variation of y showed that the minimum of the least-square-difference at $y = 0.20$ is very shallow, and fixing y anywhere between 0 and 0.3 produces fits of essentially the same quality. As an example, model M0_{S3} was fitted under the same conditions as M1_{S3}, except that $y = 0$ was imposed. The resulting rms difference to the data is almost as good as the one of M1_{S3}.

We conclude that our models provide a very good description of the heavy and heavy-light meson masses. However, the mass spectrum alone does not constrain the

Table 116.1 Kernel parameters of the models discussed in the text (we use $m_u = m_d \equiv m_q$). N_{st} is the number of states in the data set used in fitting the model, Δ_{rms} is the rms difference to the data for the largest set. The values in boldface were held fixed. The units for the quark masses, Δ_{rms}, and C are GeV, and σ is in GeV^2

Model	σ	α_s	C	y	m_b	m_c	m_s	m_q	N_{st}	Δ_{rms}
$M0_{S1}$	0.2493	0.3643	0.3491	**0.0000**	**4.892**	**1.600**	**0.4478**	**0.3455**	9	0.037
$M1_{S3}$	0.2022	0.4129	0.2145	0.2002	4.875	1.553	0.3679	0.2493	39	0.030
$M0_{S3}$	0.2058	0.4172	0.2821	**0.0000**	4.917	1.624	0.4616	0.3514	39	0.031

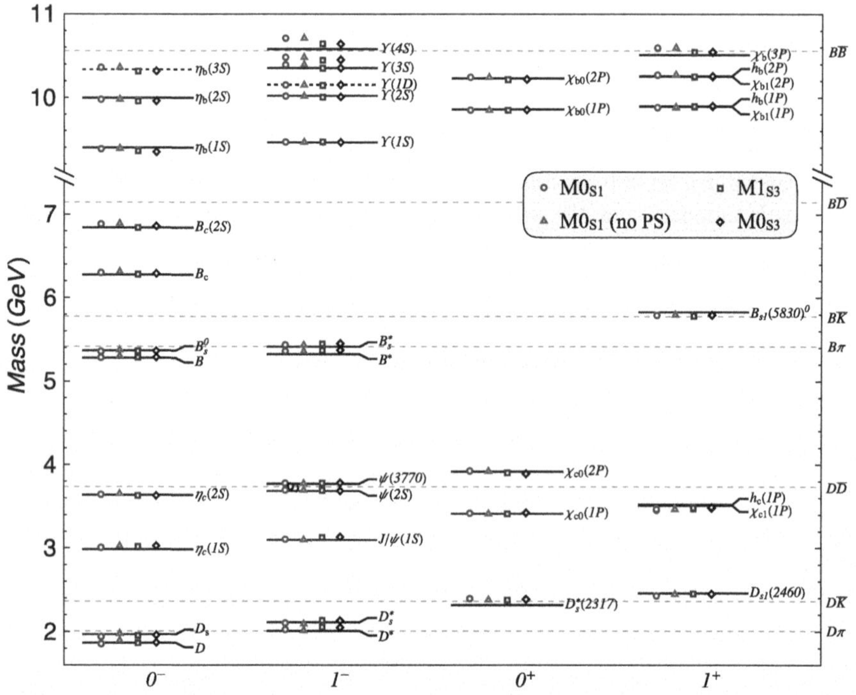

Fig. 116.2 Spectrum of heavy and heavy-light mesons with $J^P = 0^{\pm}$ and $1^{\pm}$. The symbols represent results of calculations with models $M0_{S1}$ (circles), $M0_{S1}$ with PS coupling turned off (triangles), $M1_{S3}$ (squares), and $M0_{S3}$ (diamonds). The solid lines are experimental data from the PDG [16]

Lorentz mixing parameter y very much, and we have to look for other observables in order to obtain more detailed information on the Lorentz structure of the confining interaction. In first exploratory calculations of quarkonia decay constants we found strong sensitivity to details of the wave functions, but a systematic study of the y dependence has not yet been done.

Acknowledgements We thank Fundação para a Ciência e a Tecnologia (FCT) for support under grants CFTP-FCT (UID/FIS/00777/2013), SFRH/BPD/100578/2014, and SFRH/BD/92637 /2013.

References

1. Gross, F.: Phys. Rev. **186**, 1448 (1969)
2. Stadler, A., Gross, F.: Few-Body Syst. **49**, 91 (2011)
3. UKQCD Collaboration, McNeile, C., Michael, C.: Phys. Rev. D **74**, 014508 (2006)
4. BGR [Bern-Graz-Regensburg] Collaboration, Burch, T., et al.: Phys. Rev. D **73**, 094505 (2006)
5. Dudek, J.J., Edwards, R.G., Mathur, N., Richards, D.G.: Phys. Rev. D **77**, 034501 (2008)
6. Briceno, R.A., Dudek, J.J., Young, R.D.: (2017). 1706.06223
7. Burden, C.J., Qian, L., Roberts, C.D., Tandy, P.C., Thomson, M.J.: Phys. Rev. C **55**, 2649 (1997)
8. Maris, P., Roberts, C.D.: Phys. Rev. **C56**, 3369 (1997). nucl-th/9708029
9. Fischer, C.S., Alkofer, R.: Phys. Rev. D **67**, 094020 (2003)
10. Eichmann, G., Cloët, I.C., Alkofer, R., Krassnigg, A., Roberts, C.D.: Phys. Rev. C **79**, 012202 (2009)
11. Hilger, T., Popovici, C., Gómez-Rocha, M., Krassnigg, A.: Phys. Rev. D **91**, 034013 (2015)
12. Eichmann, G., Sanchis-Alepuz, H., Williams, R., Alkofer, R., Fischer, C.S.: Prog. Particle Nucl. Phys. **91**, 1 (2016)
13. Leitão, S., Stadler, A., Peña, M.T., Biernat, E.P.: Phys. Rev. D **90**, 096003 (2014)
14. Leitão, S., Stadler, A., Peña, M.T., Biernat, E.P.: Phys. Rev. D **96**, 074007 (2017)
15. Biernat, E.P., Peña, M.T., Ribeiro, J.E., Stadler, A., Gross, F.: Phys. Rev. D **90**, 096008 (2014)
16. Particle Data Group, Olive, K.A., et al.: Chin. Phys. C **38**, 090001 (2014)

Chapter 117
The Molecular Nature of Some Ω_c^0 States

Glòria Montaña, Àngels Ramos and Albert Feijoo

Abstract A vector meson exchange model based on effective Lagrangians is used to build the meson–baryon interaction in the charm +1, strangeness −2 and isospin 0 sector. The s-wave scattering amplitudes resulting from the unitarization in coupled-channels show two resonances with masses and widths that are in very good agreement with those of the experimental $\Omega_c(3050)^0$ and $\Omega_c(3090)^0$ states observed by the LHCb collaboration. The interpretation of these resonances as pseudoscalar meson–baryon molecules would mean the assignment $J^P = 1/2^-$ to their spin–parity.

117.1 Introduction

A lot of theoretical effort in the field of baryon spectroscopy has lately arisen with the aim of explaining the inner structure of the five narrow Ω_c^0 excited resonances observed by the LHCb Collaboration [1] and possibly establishing their unknown values of spin–parity. Some works suggest a css quark description within revisited quark models [2–9] while others propose a pentaquark interpretation [10–12]. Models that can describe some resonances as quasi-bound states of an interacting meson–baryon pair [13–15] offer a complementary scenario, an approach that we have re-examined in [16] in view of the new experimental data. It is plausible to expect that some excitations in the $C = 1$, $S = -2$ sector can be obtained by adding a $u\bar{u}$ pair to the natural css content of the Ω_c^0, just as a pentaquark structure with a $c\bar{c}$ pair is more natural to explain the $P_c(4380)$ and $P_c(4450)$ excited nucleon resonances than an

G. Montaña (✉) · À. Ramos
Departament de Física Quàntica i Astrofísica, Institut de Ciències del Cosmos (ICCUB), Universitat de Barcelona, Martí i Franquès 1, 08028 Barcelona, Spain
e-mail: gmontana@fqa.ub.edu

À. Ramos
e-mail: ramos@fqa.ub.edu

A. Feijoo
Nuclear Physics Institute, 25068 Řež, Czech Republic
e-mail: feijoo@ujf.cas.cz

N. A. Orr et al. (eds.), *Recent Progress in Few-Body Physics*, Springer Proceedings in Physics 238,
https://doi.org/10.1007/978-3-030-32357-8_117

extremely high energy excitation of the three quark system. The hadronization of the five quarks can then lead to bound states, generated by the meson–baryon interaction in coupled channels. This possibility is supported by the fact that the masses of the excited Ω_c^0 baryons under study lie near the $\bar{K}\Xi_c$ and $\bar{K}\Xi_c'$ thresholds and that they have been observed in the $K^-\Xi_c^+$ invariant mass spectrum.

117.2 Formalism

The sought resonances are dynamically generated as poles of the scattering amplitude T_{ij}, unitarized by means of the on-shell Bethe-Salpeter equation in coupled channels, which implements the resummation of loop diagrams to infinite order:

$$T_{ij} = V_{ij} + V_{il} G_l T_{lj} . \qquad (117.1)$$

The G_l function for the meson–baryon loop is regularized using the *dimensional regularization* approach, which introduces the dependence on a subtraction constant $a_l(\mu)$ for each intermediate channel l at a given regularization scale μ (see (18) in [16]).

The s-wave interaction kernel V_{ij} is obtained from a t-channel vector meson exchange amplitude [13], that has the same structure as the contact Weinberg-Tomozawa term in the $t \ll m_V$ limit:

$$V_{ij}(\sqrt{s}) = -C_{ij} \frac{1}{4f^2} \left(2\sqrt{s} - M_i - M_j\right) N_i N_j , \qquad (117.2)$$

with M_i, M_j and E_i, E_j being the masses and the energies of the baryons, and N_i, N_j the normalization factors $N = \sqrt{(E+M)/2M}$.

The coefficients C_{ij} are obtained from the evaluation of the t-channel interaction diagram, with the effective Lagrangians of the hidden gauge formalism:

$$\mathcal{L}_{VPP} = ig\langle\left[\partial_\mu \phi, \phi\right] V^\mu\rangle , \qquad (117.3)$$

$$\mathcal{L}_{VBB} = \frac{g}{2} \sum_{i,j,k,l=1}^{4} \bar{B}_{ijk} \gamma^\mu \left(V^k_{\mu,l} B^{ijl} + 2 V^j_{\mu,l} B^{ilk}\right) , \qquad (117.4)$$

describing the vertices coupling the vector meson to pseudoscalars (VPP) and baryons (VBB), respectively, in the pseudoscalar meson–baryon (PB) scattering, and assuming $SU(4)$ symmetry [13].

The interaction of vector mesons with baryons (VB) is built in a similar way and involves the three-vector VVV vertex, which is obtained from:

$$\mathcal{L}_{VVV} = ig\langle\left[V^\mu, \partial_\nu V_\mu\right] V^\nu\rangle . \qquad (117.5)$$

Table 117.1 The C_{ij} coefficients for the $(I, S, C) = (0, -2, 1)$ sector of the PB interaction

	$\bar{K}\Xi_c$	$\bar{K}\Xi_c'$	$D\Xi$	$\eta\Omega_c^0$	$\eta'\Omega_c^0$
$\bar{K}\Xi_c$	1	0	$\sqrt{3/2}\,\kappa_c$	0	0
$\bar{K}\Xi_c'$		1	$\sqrt{1/2}\,\kappa_c$	$-\sqrt{6}$	0
$D\Xi$			2	$-\sqrt{1/3}\,\kappa_c$	$-\sqrt{2/3}\,\kappa_c$
$\eta\Omega_c^0$				0	0
$\eta'\Omega_c^0$					0

The resulting interaction is that of (117.2) with the addition of the product of polarization vectors, $\epsilon_i \cdot \epsilon_j$.

The interaction potential is not $SU(4)$ symmetric even though this symmetry is encoded in the Lagrangians. It is broken with the use of the physical masses of the mesons and baryons involved, and a factor $\kappa_c = 1/4$ that accounts for the higher mass of the charmed mesons exchanged in some of the non-diagonal transitions. In fact, the transitions mediated by the exchange of light vector mesons like the dominant diagonal ones do not make explicit use of $SU(4)$ symmetry since they are effectively projected into their $SU(3)$ content.

The available PB channels in the $(I, S, C) = (0, -2, 1)$ sector are $\bar{K}\Xi_c(2964)$, $\bar{K}\Xi_c'(3070)$, $D\Xi(3189)$, $\eta\Omega_c(3246)$, $\eta'\Omega_c(3656)$, $\bar{D}_s\Omega_{cc}(5528)$, and $\eta_c\Omega_c(5678)$, with the corresponding thresholds in parenthesis. The doubly charmed $\bar{D}_s\Omega_{cc}$ and $\eta_c\Omega_c$ channels are neglected as their energy is much larger than that of the other channels. The matrix of C_{ij} coefficients for the resulting 5-channel interaction is given in Table 117.1.

In the VB case, the allowed states are $D^*\Xi(3326)$, $\bar{K}^*\Xi_c(3363)$, $\bar{K}^*\Xi_c'(3470)$, $\omega\Omega_c(3480)$, $\phi\Omega_c(3717)$, $\bar{D}_s^*\Omega_{cc}(5662)$ and $J/\psi\Omega_c(5794)$, where, again, we neglect the doubly charmed states. The coefficients C_{ij} can be straightforwardly obtained from those in Table 117.1 with: $\pi \to \rho$, $K \to K^*$, $\bar{K} \to \bar{K}^*$, $D \to D^*$, $\bar{D} \to \bar{D}^*$, $1/\sqrt{3}\eta + \sqrt{2/3}\eta' \to \omega$ and $-\sqrt{2/3}\eta + 1/\sqrt{3}\eta' \to \phi$.

Resonance poles of the scattering amplitude appear in the *second Riemann sheet* of the complex energy plane. The residues at the pole position z_p give the coupling constants g_i of the resonance to the various channels and the real part of $-g_i^2(\partial G/\partial\sqrt{s})|_{z_p}$ corresponds to the compositeness, i.e., the amount of ith-channel meson–baryon component.

117.3 Results

The values of the subtraction constant, $a_l(\mu = 1\,\text{GeV})$, used when solving (117.1) are chosen so as the loop function in dimensional regularization coincides with the loop function regularized with a cut-off $\Lambda = 800\,\text{MeV}$ at the channel threshold ("Model 1"). The resulting PB scattering amplitude shows two poles,

Table 117.2 Position ($\sqrt{s} = M - i\Gamma/2$), couplings and compositeness of the Ω_c^0 states generated employing "Model 2"

$0^- \otimes 1/2^+$ interaction in the $(I, S, C) = (0, -2, 1)$ sector				
M (MeV)	3050.3		3090.8	
Γ (MeV)	0.44		12	
	$\lvert g_i \rvert$	$-g_i^2 dG/dE$	$\lvert g_i \rvert$	$-g_i^2 dG/dE$
$\bar{K}\Xi_c(2964)$	0.11	$0.00 + i\,0.00$	0.49	$-0.02 + i\,0.01$
$\bar{K}\Xi_c'(3070)$	1.80	$0.61 + i\,0.01$	0.35	$0.02 - i\,0.02$
$D\Xi(3189)$	1.36	$0.07 - i\,0.01$	4.28	$0.91 - i\,0.01$
$\eta\Omega_c(3246)$	1.63	$0.14 + i\,0.00$	0.39	$0.01 + i\,0.01$
$\eta'\Omega_c(3656)$	0.06	$0.00 + i\,0.00$	0.28	$0.00 + i\,0.00$

$$
\begin{aligned}
M_1 &= \mathrm{Re} z_1 = 3051.6\,\mathrm{MeV}, & \Gamma_1 &= -2\mathrm{Im} z_1 = 0.45\,\mathrm{MeV} \\
M_2 &= \mathrm{Re} z_2 = 3103.3\,\mathrm{MeV}, & \Gamma_2 &= -2\mathrm{Im} z_2 = 17\,\mathrm{MeV},
\end{aligned}
\tag{117.6}
$$

corresponding to resonances with spin–parity $J^P = 1/2^-$. Their energies are very similar to the second and fourth Ω_c^0 states discovered by LHCb [1].

These results clearly show that the meson–baryon dynamical models are able to generate states in the energy range of interest, although the mass of our heavier state is larger by 10 MeV and its width is about twice the experimental one. In an attempt to explore the possibilities of our model, we let the values of the five subtraction constants vary freely within a reasonably constrained range and look for a combination that reproduces the characteristics of the two observed states, $\Omega_c(3050)^0$ and $\Omega_c(3090)^0$, within 2σ of the experimental errors. Table 117.2 displays the new properties of the poles for a representative set of $a_l(\mu = 1\,\mathrm{GeV})$ with equivalent cut-off values in the 320–950 MeV range (referred as "Model 2" in [16]). We note that the strongest change corresponds to $a_{\bar{K}\Xi_c}$, needed to decrease the width of the $\Omega_c(3090)^0$. Its equivalent cut-off value of 320 MeV is on the low side of the usually employed values but it is still naturally sized.

We also show in Table 117.2 the couplings of each resonance to the various meson–baryon channels and the corresponding compositeness. The lowest energy state at 3050 MeV has a strong coupling to $\bar{K}\Xi_c'$ and has a high compositeness in this channel but also couples appreciably to $D\Xi$ and $\eta\Omega_c$ channels. The higher energy resonance at 3090 MeV couples strongly to $D\Xi$ and clearly qualifies as a $D\Xi$ bound state with a compositeness in this channel of 0.91.

The five Ω_c^0 states were observed from the $K^-\Xi_c^+$ invariant mass spectrum obtained in high energy pp collision data by the LHCb [1], which is tremendously difficult to model. In Fig. 117.1 we display a merely illustrative plot of the spectrum that our model would predict that retains certain similarities with the spectrum of Fig. 2 in [1] in the energy regions of the 3050 and 3090 MeV states.

Next we discuss the dependence of these results on the assumed value of the cut-off, as well as the influence of a certain amount of $SU(4)$ symmetry violation

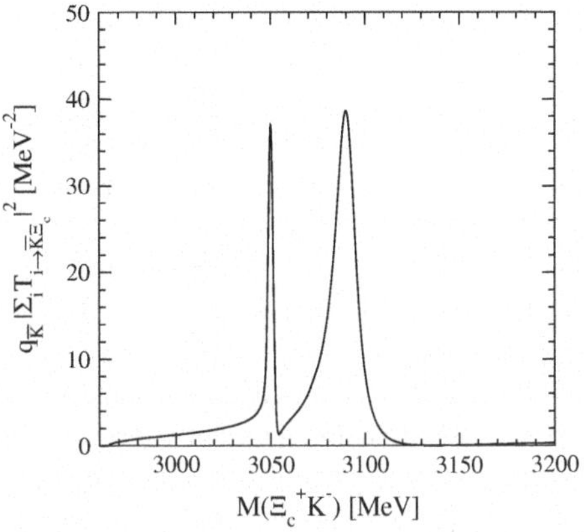

Fig. 117.1 Sum of amplitudes squared times the momentum of the K^- versus the $\bar{K}\Xi_c$ energy in the centre-of-mass frame, where $T_{i\to\bar{K}\Xi_c}$ is the amplitude for the $i \to \bar{K}\Xi_c$ transition obtained here with "Model 2", with i being any of the five coupled channels. The $q_{\bar{K}}$ acts as a phase-space modulator. The calculated spectrum has been convoluted with the energy dependent resolution of the experiment. (Adapted from [16])

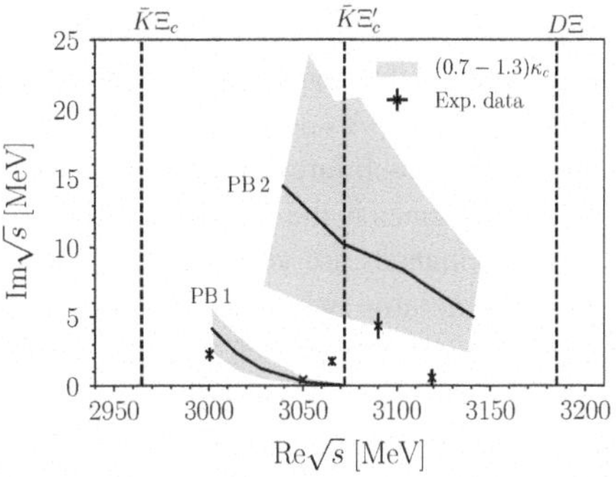

Fig. 117.2 Evolution of the position of the resonance poles with the cut-off, from $\Lambda = 700$ MeV (PB1) and $\Lambda = 650$ MeV (PB2) at the right end of the solid lines to $\Lambda = 1000$ MeV at the left end. The grey areas indicate the regions of results covered when a variation of 30% in the $SU(4)$ breaking is assumed in the transitions mediated by heavy-meson exchange. (Adapted from [16])

associated to the fact that the charm quark is substantially heavier than the light quarks. The solid lines in Fig. 117.2 indicate the evolution of the poles as the value of the cut-off is increased from 700 MeV for the low-energy pole (PB1) or 650 MeV for the high-energy one (PB2) to 1000 MeV. On the other hand, we note that the violation of $SU(4)$ symmetry is already partly implemented by the use of the physical meson and baryon masses in the interaction kernel. Moreover, only $SU(3)$ is effectively acting in the transitions mediated by light vector mesons and thus these will be left untouched. Therefore, up to an additional 30% of $SU(4)$ breaking is implemented only in the matrix elements that connect states via the t-channel exchange of a charmed vector meson, and this is achieved by allowing the factor κ_c to vary in the range $(0.7-1.3)\kappa_c$. The grey areas in Fig. 117.2 correspond to the regions in the complex plane where the resonances can be found varying both the cut-off and the amount of $SU(4)$ violation. The fact that these bands of uncertainties include the 3050 MeV and the 3090 MeV resonances measured at LHCb reinforces their interpretation as meson–baryon molecules.

In the case of *VB* scattering we have followed a similar procedure to look for resonances, which are degenerate in spin, $J^P = 1/2^-, 3/2^-$. Employing subtraction constants mapped onto a cut-off of $\Lambda = 800$ MeV, we see a similar pattern as that found for the PB case. A lower energy resonance mainly classifying as a $D^*\Xi$ molecule appears at 3231 MeV and a higher energy resonance is generated at 3419 MeV and corresponds to a $\bar{K}^*\Xi_c'$ composite state with some admixture of $\omega\Omega_c^0$ and $\phi\Omega_c$ components. There is an additional pole in between these two, coupling strongly to $\bar{K}^*\Xi_c$ states. These three resonances are located in an energy region above the states reported by the LHCb collaboration where no narrow structures have been seen [1]. We note, however, that the states found here from the VB interaction are artificially narrow as they do not couple to, and hence cannot decay into, the PB states that lie at lower energy.

Finally, we show the results of extending our model to the bottom sector by employing the meson–baryon interaction kernels obtained from the Lagrangians of (117.3)–(117.5), but replacing the charm mesons and baryons by their bottom counterparts (see the details in [16]). A coefficient $\kappa_b = 0.1$ in certain non-diagonal transitions that accounts for the much larger mass of the exchanged bottom vector mesons with respect to the light ones is the analogous to κ_c.

Our results for the Ω_b^- resonances are very similar to those found in the charm sector. The PB interaction generates two states at 6418 and 6519 MeV with spin $J^P = 1/2^-$, the former couples strongly to $\bar{K}\Xi_c'$ and $\eta\Omega_b$ while the later is essentially a $B\Xi$ bound state. In the VB interaction we find $J = 1/2^-, 3/2^-$ spin degenerate Ω_b^- states at 6560 MeV, coupling strongly to $\bar{B}^*\Xi$, 6665 MeV, coupling to $K^*\Xi_b$, and 6797 MeV, being a mixture of $\omega\Omega_b$, $\bar{K}^*\Xi_b'$ and $\Phi\Omega_b$.

117.4 Conclusions

Employing a t-channel vector meson exchange model with effective Lagrangians, we have studied the interaction of the low-lying pseudoscalar and vector mesons with the ground-state baryons in the charm +1, strangeness −2 and isospin 0 sector. Two resonances with energies and widths very similar to some of the Ω_c^0 states discovered recently at LHCb are found in the unitarized scattering amplitudes of the interaction of pseudoscalar mesons with baryons. We have extended the model to the bottom sector and predicted several Ω_b^- resonances in the energy region 6400–6800 MeV with a molecular meson–baryon structure.

Several other works [17–20] have also addressed the possibility of interpreting some of the Ω_c states seen at CERN as quasi-bound meson–baryon systems, as well as the prediction of analogous states in the bottom sector [21], finding results which are similar to those of our work [16] and hinting that the meson–baryon description cannot be ignored when trying to understand the nature of these excited heavy baryons.

Acknowledgements This work is supported by the Spanish MINECO under the contracts MDM-2014-0369, FIS2017-87534-P and the doctoral grant FPU17/04910 and by the Generalitat de Catalunya under the doctoral grant 2018 FI_B 00234.

References

1. Aaij, R., et al.: Phys. Rev. Lett. **118**(18), 182001 (2017)
2. Agaev, S.S., Azizi, K., Sundu, H.: EPL **118**(6), 61001 (2017)
3. Chen, H.X., Mao, Q., Chen, W., Hosaka, A., Liu, X., Zhu, S.L.: Phys. Rev. D **95**(9), 094008 (2017)
4. Karliner, M., Rosner, J.L.: Phys. Rev. D **95**(11), 114012 (2017)
5. Wang, K.L., Xiao, L.Y., Zhong, X.H., Zhao, Q.: Phys. Rev. D **95**(11), 116010 (2017)
6. Wang, W., Zhu, R.L.: Phys. Rev. D **96**(1), 014024 (2017)
7. Cheng, H.Y., Chiang, C.W.: Phys. Rev. D **95**(9), 094018 (2017)
8. Wang, Z.G.: Eur. Phys. J. C **77**(5), 325 (2017)
9. Chen, B., Liu, X.: Phys. Rev. D **96**(9), 094015 (2017)
10. Huang, H., Ping, J., Wang, F.: Phys. Rev. D **97**(3), 034027 (2018)
11. An, C.S., Chen, H.: Phys. Rev. D **96**(3), 034012 (2017)
12. Kim, H.C., Polyakov, M.V., Praszaowicz, M.: Phys. Rev. D **96**(1), 014009 (2017)
13. Hofmann, J., Lutz, M.F.M.: Nucl. Phys. A **763**, 90 (2005)
14. Jimenez-Tejero, C.E., Ramos, A., Vidana, I.: Phys. Rev. C **80**, 055206 (2009)
15. Romanets, O., Tolos, L., Garcia-Recio, C., Nieves, J., Salcedo, L.L., Timmermans, R.G.E.: Phys. Rev. D **85**, 114032 (2012)
16. Montaña, G., Feijoo, A., Ramos, A.: Eur. Phys. J. A **54**(4), 64 (2018)
17. Debastiani, V.R., Dias, J.M., Liang, W.H., Oset, E.: Phys. Rev. D **97**(9), 094035 (2018)
18. Wang, C., Liu, L.L., Kang, X.W., Guo, X.H., Wang, R.W.: Eur. Phys. J. C **78**(5), 407 (2018)
19. Chen, R., Hosaka, A., Liu, X.: Phys. Rev. D **97**(3), 036016 (2018)
20. Nieves, J., Pavao, R., Tolos, L.: Eur. Phys. J. C **78**(2), 114 (2018)
21. Liang, W.H., Dias, J.M., Debastiani, V.R., Oset, E.: Nucl. Phys. B **930**, 524 (2018)

Chapter 118
Meson Studies with a Contact Interaction

Marco A. Bedolla and Elena Santopinto

Abstract We present the spectrum of ground state heavy-light mesons. Our analysis is provided by a symmetry preserving Schwinger-Dyson Bethe-Salpeter Equation (SDBSE) approach of a vector-vector contact interaction model. This model offers a simple-to-implement alternative to perform exploratory studies of QCD. Despite the simplicity of this model, our results for D are in agreement within 9%, and for B within 2% when we compare them with experimental data.

118.1 Introduction

Heavy-light mesons are still a theoretical challenge within the Schwinger-Dyson Bethe-Salpeter Equation (SDBSE) approach [1–3]. This approach has proven to be a reliable instrument to explain a wide range of QCD phenomena. However, to solve fully covariant kernels, we require brute force numerical computation when evaluating at large momentum transfer regions [4, 5]. Additionaly, studies of unequal mass systems becomes a more difficult task because we have to deal with a multiscale problem: the heavy and the light quarks regimes [6–9].

The contact interaction (CI) model appeared as an alternative to full QCD studies. In this model, quarks interact not via mass-less vector-boson exchanges, but instead through a symmetry preserving vector-vector contact interaction [10–14]. This interaction is embedded within the SDBSE approach in the rainbow-ladder approximation, and it implements confinement through a proper time regularization scheme. A fully consistent treatment of the CI model is simple to execute, and it produces

M. A. Bedolla (✉) · E. Santopinto
INFN, Sezione di Genova, via Dodecaneso 33, 16146 Genova, Italy
e-mail: mbedolla@ge.infn.it

E. Santopinto
e-mail: santopinto@ge.infn.it

M. A. Bedolla
Instituto de Física y Matemáticas, Universidad Michoacana de San Nicolás de Hidalgo, Edificio C-3, Ciudad Universitaria, 58040 Morelia, Michoacán, Mexico

N. A. Orr et al. (eds.), *Recent Progress in Few-Body Physics*,
Springer Proceedings in Physics 238,
https://doi.org/10.1007/978-3-030-32357-8_118

useful results that can be compared and contrasted with full QCD calculations and experimental data.

In this work, we present the spectrum of heavy-light mesons using the CI model, we compare with experimental data when these are avalaible. Our results are a direct application of our ideas developed in previos works [15–17].

118.2 SDBSE Approach and CI Model

Since this work is a direct application of the unified CI model presented in [17], we only present the model and the extension needed to make our calculations. For complete description see [14, 15].

118.2.1 Contact Interaction Model

The SDBSE formalism solves the bound-state problem in terms of their building blocks (quarks), and their interactions with gluons. In order to solve a meson bound state equation, we need to know the quark propagator, the gluon propagator and the quark-gluon interaction. In the CI model, we assume that the quark-gluon interaction is led by symmetry-preserving vector × vector contact interaction; here, quarks are attached through an interaction defined as

$$g^2 D_{\mu\nu}(k) = \frac{4\pi\alpha_{\rm IR}}{m_g^2}\delta_{\mu\nu} \equiv \frac{1}{m_G^2}\delta_{\mu\nu}, \tag{118.1}$$

$$\Gamma_\mu^a(p,q) = \frac{\lambda^a}{2}\gamma_\mu, \tag{118.2}$$

where $m_g = 800$ MeV is a gluon mass scale generated dynamically in QCD [18], and $\alpha_{\rm IR}$ is the CI model parameter, which can be interpreted as the interaction strength in the infrared region [19, 20].

With this interaction, we obtain a constant mass function. Since we need to solve divergent integrals, we must adopt a regularization procedure. We employ the proper time regularization scheme [21], and we obtain an expression for the mass function

$$M_f = m_f + \frac{M_f^3}{3\pi^2 m_G^2}\Gamma(-1, \tau_{\rm UV} M_f^2, \tau_{\rm IR} M_f^2), \tag{118.3}$$

where $\Gamma(a, z_1, z_2)$ is the generalized incomplete Gamma function:

$$\Gamma(a, z_1, z_2) = \Gamma(a, z_1) - \Gamma(a, z_2). \tag{118.4}$$

The parameters $\tau_{\rm IR}$ and $\tau_{\rm UV}$ are infrared and ultraviolet regulators, respectively. A nonzero value for $\tau_{\rm IR} \equiv 1/\Lambda_{\rm IR}$ implements confinement [22]. On the other hand, since the CI model represents a nonrenormalizable theory, $\tau_{\rm UV} \equiv 1/\Lambda_{\rm UV}$ becomes part of the model and therefore sets the scale for all dimensional quantities.

118.2.2 CI model running coupling

In a previous work [17], we explained how the CI can be used to study light and heavy mesons. When studying the heavy sector, a change in the model parameters has to be done: an increase in the ultraviolet regulator, and a reduction in the coupling strength. Subsequently, we found out that different set of parameters are needed in order to study each sector: light, charm and bottom.
We define a dimensionless coupling in terms of the ultraviolet cutoff:

$$\alpha = \alpha(\Lambda_{\rm UV}) = \frac{\alpha_{\rm IR}}{m_g^2}\Lambda_{\rm UV}^2 \,. \tag{118.5}$$

It is interesting that this dimensional coupling can be fitted by an inverse logarithmic curve, as a reminiscent of the running coupling QCD with the momentum scale at which it is measured:

$$\alpha(\Lambda_{\rm UV}) = a \ln^{-1}\left(\Lambda_{UV}/\Lambda_0\right) \,, \tag{118.6}$$

where $a = 0.923$ and $\Lambda_0 = 0.357$ [17]. With this fit, we can recover the value of the strength coupling α once given a value of Λ_{UV}. We maintain the infrared cutoff Λ_{UV} and the gluon mass m_g.

118.3 Results

In order to calculate the mass spectrum, we follow the expressions found in [15]. In our approach, we retain each quark dressed mass computed in previous works [17]. We find the $\Lambda_{\rm IR}$ and α parameters through (118.5–118.6) to obtain the pseudoscalar meson experimental mass value. Subsequently, we calculate the other mesons masses with those parameters. We present our results in plots comparing our results on heavy-light mesons with experimental data.

118.3.1 Charmed Mesons

We display the spectrum of D mesons in Fig. 118.1. Our spectrum is in good agreement with experimental results. Additionally, in Fig. 118.2 we display the spectrum of D_s mesons with the same parameters of D mesons. Our results are within 9% of agreement with experiment for both set of mesons.

118.3.2 Bottom Mesons

In Fig. 118.3 we show the B-meson mass spectrum. Similar to D mesons, the spectrum agrees perfectly with experimental data. In addition, we present a mass value

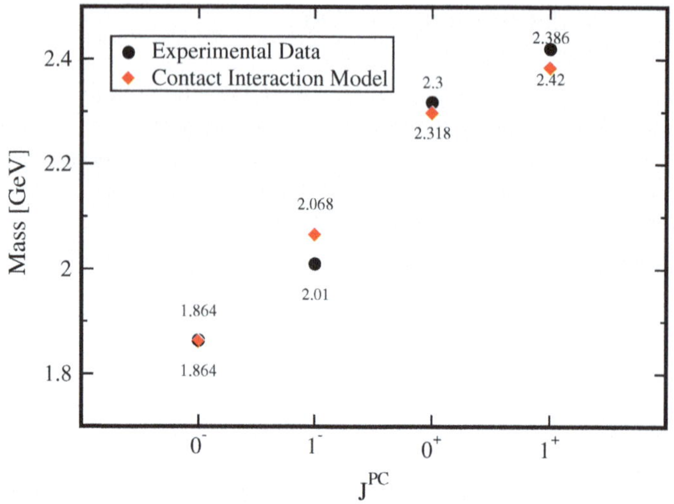

Fig. 118.1 Mass spectrum of ground-state D mesons in GeV. Our results were obtained with the parameter set: $\alpha_{IR} = 0.93\pi/4.528$ and $\Lambda_{\rm UV} = 1.532$ GeV. The dynamically generated constituent-like masses are $M_n = 0.367$ GeV and $M_c = 1.482$ GeV. Experimental (below) data are from [23]

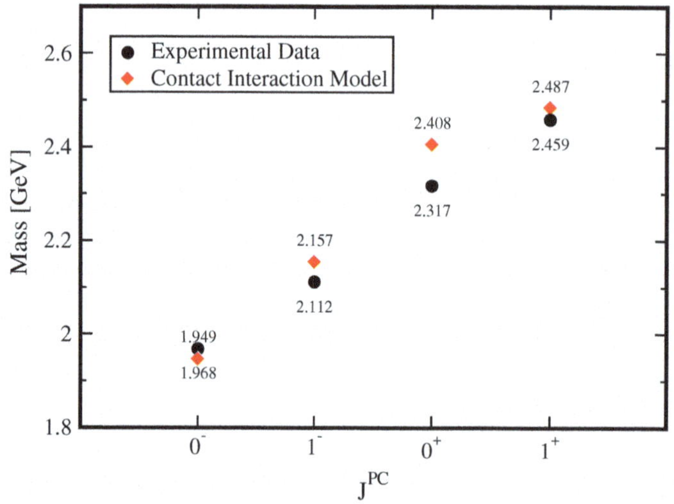

Fig. 118.2 Mass spectrum of ground-state D_s mesons in GeV. Our results were obtained with the parameter set: $\alpha_{IR} = 0.93\pi/4.528$ and $\Lambda_{\rm UV} = 1.532$ GeV. The dynamically generated constituent-like masses are $M_s = 0.533$ GeV and $M_c = 1.482$ GeV. Experimental data (below) are from [23]

for the scalar (0^+) meson which we expect to be aproximately 60 MeV below a future experimental result.

In a similar fashion, Figs. 118.4 and 118.5 present our predictions for B_s and B_c mesons, respectively. For B_s mesons, we expect that the 0^+ channel to be around

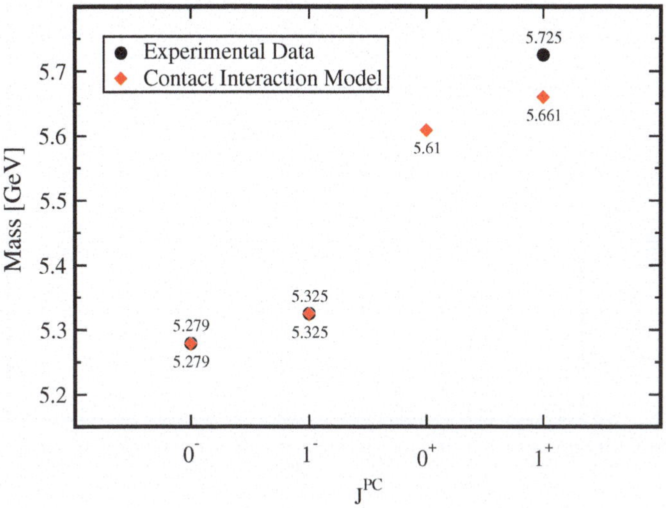

Fig. 118.3 Mass spectrum of ground-state of B mesons in GeV. Our results were obtained with the parameter set: $\alpha_{IR} = 0.93\pi/11.969$ and $\Lambda_{\text{UV}} = 2.223$ GeV. The dynamically generated constituent-like mass are $M_n = 0.367$ GeV and $M_b = 4.710$ GeV. Experimental data are from [23]

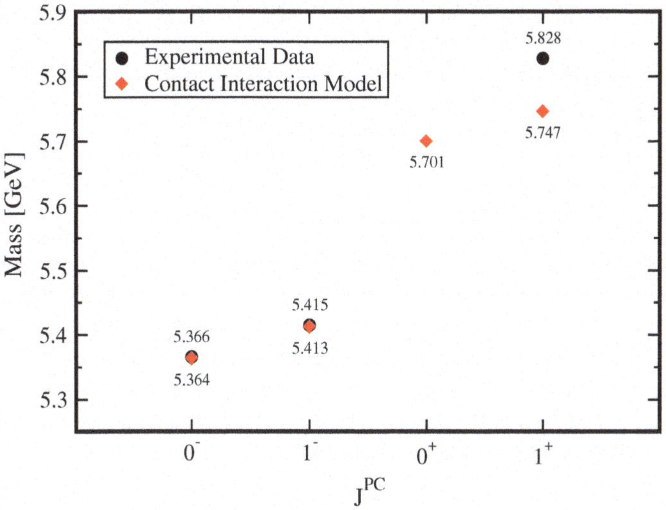

Fig. 118.4 Mass spectrum of ground-state B_s mesons in GeV. Our results were obtained with the parameter set: $\alpha_{IR} = 0.93\pi/11.969$ and $\Lambda_{\text{UV}} = 2.223$ GeV. The dynamically generated constituent-like masses are $M_s = 0.533$ GeV and $M_b = 4.710$ GeV. Experimental data (above) are from [23]

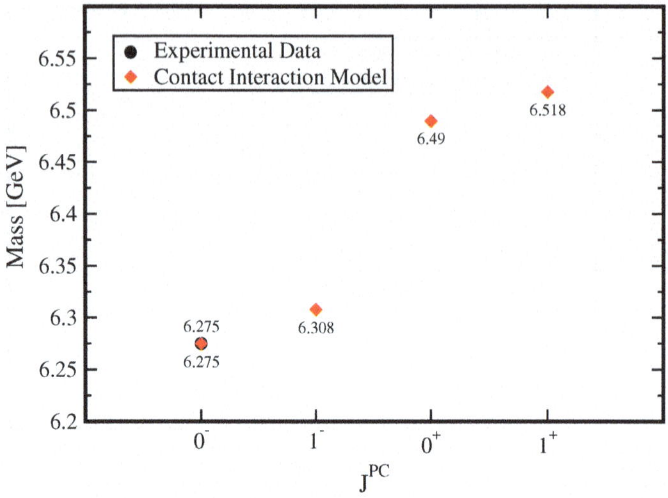

Fig. 118.5 Mass spectrum of ground-state B_c mesons in GeV. Our results were obtained with the parameter set: $\alpha_{IR} = 0.93\pi/59.046$ and $\Lambda_{\rm UV} = 4.244\,$GeV. The dynamically generated constituent-like masses are $M_c = 1.482\,$GeV and $M_b = 4.710\,$GeV. Experimental data (above) are from [23]

80 MeV below a future experimental value. Similarly, for B_c mesons, we expect that the 1^- meson to be in perfect agreement with the experimental value. While for the 0^+ and 1^+ we expect to be within to 2% of agreement with experimental results.

118.4 Conclusions

We calculated the spectrum of heavy-light mesons through (118.6) to find the coupling as a function of the ultraviolet cutoff fitted to 0^{-+} D, B and B_c mesons, while the other masses are predictions of these parameters. When we study D and D_s mesons, light coupling gets reduced by a factor of ≈4, by a factor of ≈12 for B and B_s mesons, and a factor of ≈60 for B_c mesons. Our mass spectrum is in a good agreement with experimental data when it is avalaible. More details about our approach will be described in a future publication [24].

References

1. Jain, P., Munczek, H.J.: q anti-q bound states in the Bethe-Salpeter formalism. Phys. Rev. D **48**, 5403 (1993)
2. Roberts, C.D., Williams, A.G.: Dyson-Schwinger equations and their application to hadronic physics. Prog. Part. Nucl. Phys. **33**, 477 (1994)

3. Bashir, A., Chang, L., Cloet, I.C., El-Bennich, B., Liu, Y.X., Roberts, C.D., Tandy, P.C.: Collective perspective on advances in Dyson-Schwinger Equation QCD. Commun. Theor. Phys. **58**, 79 (2012)
4. Maris, P., Tandy, P.C.: The pi, K+, and K0 electromagnetic form-factors. Phys. Rev. C **62**, 055204 (2000)
5. Chen J., Ding, M., Chang, L. Liu, Y.X.: Two photon transition form factor of $\bar{c}c$ Quarkonia. Phys. Rev. D **95**(1), 016010 (2017)
6. Nguyen, T., Souchlas, N.A., Tandy, P.C.: Soft and Hard scale QCD dynamics in mesons. AIP Conf. Proc. **1361**, 142 (2011)
7. Fischer, C.S., Kubrak, S., Williams, R.: Spectra of heavy mesons in the Bethe-Salpeter approach. Eur. Phys. J. A **51**, 10 (2015)
8. Gómez, M., Rocha, T., Hilger, Krassnigg, A.: Effects of a dressed quark-gluon vertex in vector heavy-light mesons and theory average of the B_c^* meson mass. Phys. Rev. D **93**(7), 074010 (2016)
9. Hilger, T., Gómez-Rocha, M., Krassnigg A., Lucha, W.: Aspects of open-flavour mesons in a comprehensive DSBSE study. Eur. Phys. J. A **53**(10), 213 (2017)
10. Gutierrez-Guerrero, L.X., Bashir, A., Cloet, I.C., Roberts, C.D.: Pion form factor from a contact interaction. Phys. Rev. C **81**, 065202 (2010)
11. Roberts, H.L.L., Roberts, C.D., Bashir, A., Gutierrez-Guerrero, L.X., Tandy, P.C.: Abelian anomaly and neutral pion production. Phys. Rev. C **82**, 065202 (2010)
12. Roberts, H.L.L., Chang, L., Cloet, I.C., Roberts, C.D.: Masses of ground and excited-state hadrons. Few Body Syst. **51**, 1 (2011)
13. Roberts, H.L.L., Bashir, A., Gutierrez-Guerrero, L.X., Roberts, C.D., Wilson, D.J.: pi- and rho-mesons, and their diquark partners, from a contact interaction. Phys. Rev. C **83**, 065206 (2011)
14. Chen, C., Chang, L., Roberts, C.D., Wan, S., Wilson, D.J.: Spectrum of hadrons with strangeness. Few Body Syst. **53**, 293 (2012)
15. Bedolla, M.A., Cobos-Martínez J.J., Bashir, A.: Charmonia in a contact interaction. Phys. Rev. D **92**(5), 054031 (2015)
16. Bedolla, M.A., Raya, K., Cobos-Martínez J.J., Bashir, A.: η_c elastic and transition form factors: contact interaction and algebraic model. Phys. Rev. D **93**(9), 094025 (2016)
17. Raya, K., Bedolla, M.A., Cobos-Martínez, J.J., Bashir, A.: Heavy quarkonia in a contact interaction and an algebraic model: mass spectrum, decay constants, charge radii and elastic and transition form factors. Few Body Syst. **59**(6), 133 (2018)
18. Boucaud, P., Leroy, J.P., Yaouanc, A.L., Micheli, J., Pene, O., Rodriguez-Quintero, J.: The Infrared Behaviour of the pure Yang-Mills green functions. Few Body Syst. **53**, 387 (2012)
19. Binosi, D., Mezrag, C., Papavassiliou, J., Roberts, C.D., Rodriguez-Quintero, J.: Process-independent strong running coupling. Phys. Rev. D **96**(5), 054026 (2017)
20. Deur, A., Brodsky, S.J., de Teramond, G.F.: The QCD running coupling. Prog. Part. Nucl. Phys. **90**, 1 (2016)
21. Ebert, D., Feldmann, T., Reinhardt, H.: Extended NJL model for light and heavy mesons without q - anti-q thresholds. Phys. Lett. B **388**, 154 (1996)
22. Roberts, C.D.: Hadron properties and Dyson-Schwinger equations. Prog. Part. Nucl. Phys. **61**, 50 (2008)
23. Patrignani, C., et al. [Particle Data Group].: Review of particle physics. Chin. Phys. C **40**(10), 100001 (2016)
24. Bedolla, M.A., Santopinto, E.: Heavy-light mesons spectra in a contact interaction. arXiv:1810.04391 [hep-ph]

Chapter 119
XYZ Mesons at BESIII

Chang-Zheng Yuan

Abstract The BESIII is one of the leading experiments in the study of the hadron spectroscopy. We report the progress in the study of the exotic hadrons, also called XYZ states, at BESIII experiment with its unique data samples in e^+e^- annihilation at center-of-mass energies of 3.8–4.6 GeV.

119.1 Introduction

In the conventional quark model, mesons are composed of one quark and one antiquark, while baryons are composed of three quarks. Although this picture is very simple, it describes almost all the hadrons observed so far [1]. However, many charmonium-like states were discovered at two B-factories BaBar and Belle in the first decade of this century [2]. Whereas some of these are good candidates of charmonium states, many other states have exotic properties, which may indicate that exotic states, such as multi-quark state, hadronic molecule, or hybrid, have been observed [3]. Experimentally, these states are also called XYZ states, to indicate their nature is still unclear.

BaBar and Belle experiments finished their data taking in 2008 and 2010, respectively, and the data are still used for various physics analyses. In 2008, the BESIII [4], a τ-charm factory experiment at the BEPCII e^+e^- collider, started data taking, and contributed to the study of the XYZ particles ever since. In this article, we focus on the measurements of the three mostly studied XYZ states, i.e., the $X(3872)$, the $Y(4260)$, and the $Z_c(3900)$, at BESIII [5].

C.-Z. Yuan (✉)
Institute of High Energy Physics, Chinese Academy of Sciences, Beijing, China
e-mail: yuancz@ihep.ac.cn

University of Chinese Academy of Sciences, Beijing, China

N. A. Orr et al. (eds.), *Recent Progress in Few-Body Physics*,
Springer Proceedings in Physics 238,
https://doi.org/10.1007/978-3-030-32357-8_119

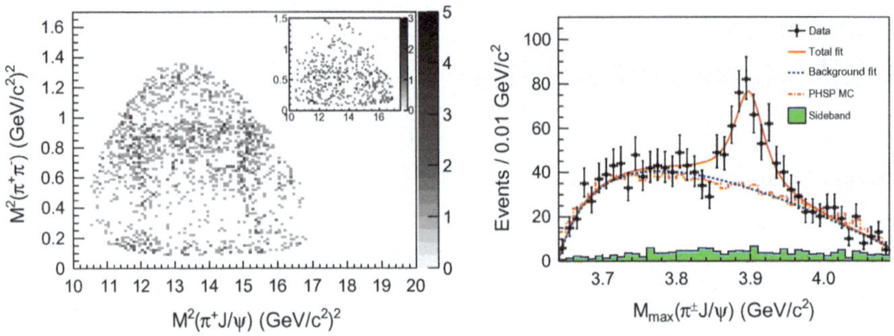

Fig. 119.1 Dalitz plot for selected $e^+e^- \to \pi^+\pi^- J/\psi$ events in the J/ψ signal region (left panel), and unbinned maximum likelihood fit to the distribution of the $M_{\max}(\pi J/\psi)$ (right panel). Adapted from [6]

119.2 The Charged Charmonium-Like State $Z_c(3900)$

119.2.1 Discovery of the $Z_c(3900)$

The BESIII experiment studied the $e^+e^- \to \pi^+\pi^- J/\psi$ process at a center-of-mass (c.m.) energy of 4.26 GeV using a 525 pb^{-1} data sample [6], with J/ψ decays into a pair of e^+e^- or $\mu^+\mu^-$. The J/ψ signal is selected by requiring the invariant mass of the lepton pair is consistent with the J/ψ, and a sample of 1595 $\pi^+\pi^- J/\psi$ events with a purity of 90% is obtained. The intermediate states are studied by examining the Dalitz plot (shown in Fig. 119.1) of the selected candidate events.

A structure at around 3.9 GeV/c^2 was observed in the $\pi^\pm J/\psi$ invariant mass distribution with a statistical significance larger than 8σ, which is referred to as the $Z_c(3900)$. A fit to the $\pi^\pm J/\psi$ invariant mass spectrum with a constant width Breit-Wigner (BW) function (Fig. 119.1), neglecting interference with other amplitudes, results in a mass of $(3899.0 \pm 3.6 \pm 4.9)$ MeV/c^2 and a width of $(46 \pm 10 \pm 20)$ MeV. The $Z_c(3900)$ state was reported shortly after at Belle [7] with initial state radiation (ISR) data and with CLEO-c data at a c.m. energy of 4.17 GeV [8], and the mass and width agreed very well with the BESIII measurement.

119.2.2 Spin-Parity of the $Z_c(3900)$

BESIII determines the spin-parity of the $Z_c(3900)$ based on a partial wave analysis (PWA) of $e^+e^- \to \pi^+\pi^- J/\psi$ events at $\sqrt{s} = 4.23$ and 4.26 GeV [9]. Following the event selection reported in [6], the numbers of selected candidate events are 4154 at $\sqrt{s} = 4.23$ GeV and 2447 at $\sqrt{s} = 4.26$ GeV, with 365 and 272 background events, respectively, estimated by using the J/ψ mass sidebands.

The fit indicates that the spin-parity $J^P = 1^+$ of the $Z_c(3900)$ are favored by more than 7σ over other quantum numbers (0^-, 1^-, 2^-, and 2^+). The pole mass of the $Z_c(3900)$ is measured as $(3881.2 \pm 4.2 \pm 52.7)$ MeV/c^2 and pole width $(51.8 \pm 4.6 \pm 36.0)$ MeV. The Born cross sections for $e^+e^- \rightarrow \pi^+ Z_c(3900)^- + c.c. \rightarrow \pi^+\pi^- J/\psi$ are measured to be $(21.8 \pm 1.0 \pm 4.4)$ pb at $\sqrt{s} = 4.23$ GeV and $(11.0 \pm 1.2 \pm 5.4)$ pb at $\sqrt{s} = 4.26$ GeV.

119.2.3 Observation of $Z_c(3900)^\mp \rightarrow (D\bar{D}^*)^\mp$

With the data sample at $\sqrt{s} = 4.26$ GeV, the BESIII experiment studied $e^+e^- \rightarrow \pi^\pm(D\bar{D}^*)^\mp$ with single-tag method (only one of the two D mesons is reconstructed) and observed the open-charm decay $Z_c(3900)^\pm \rightarrow (D\bar{D}^* + c.c.)^\pm$ [10]. The analysis is refined with double-tag method (both D mesons are reconstructed) and more luminosity [11]. In this analysis, both $e^+e^- \rightarrow \pi^+ D^0 D^{*-} + c.c.$ and $\pi^+ D^- D^{*0} + c.c.$ are measured with data samples at $\sqrt{s} = 4.23$ and 4.26 GeV. The double D tag technique allows the use of more D decay modes and the background level is greatly suppressed.

The double-tag analysis only has ~9% events in common with the single-tag analysis, so the two analyses are almost statistically independent and the results can be combined to have a better measurement. The combined pole mass and width are $(3882.2 \pm 1.1 \pm 1.5)$ MeV/c^2 and $(26.5 \pm 1.7 \pm 2.1)$ MeV, respectively. The combined production rate $\sigma(e^+e^- \rightarrow \pi^\mp Z_c(3900)^\pm) \times \mathcal{B}(Z_c(3900)^\pm \rightarrow (DD^*)^\pm)$ is $(104.4 \pm 4.8 \pm 8.4)$ pb at $\sqrt{s} = 4.26$ GeV. The production rate $\sigma(e^+e^- \rightarrow \pi^\mp Z_c(3900)^\pm) \times \mathcal{B}(Z_c(3900)^\pm \rightarrow (DD^*)^\pm) = (141.6 \pm 7.9 \pm 12.3)$ pb at $\sqrt{s} = 4.23$ GeV which is from double-tag method only.

119.2.4 Evidence for $Z_c(3900) \rightarrow \rho\eta_c$

BESIII searches for $e^+e^- \rightarrow \pi^+\pi^-\pi^0\eta_c$ and intermediate states decay into $\rho\eta_c$ with data collected at 4.23, 4.26, and 4.36 GeV [12]. The recoil mass of one charged pion (equivalent to the invariant mass of $\rho^\pm\eta_c$) is shown in Fig. 119.2 for the data at $\sqrt{s} = 4.23$ GeV, the $Z_c(3900)^\pm$ signal is found while there is no significant $Z_c(4020)^\pm$ signal. The $\rho^\pm\eta_c$ invariant mass distribution is fitted with the contributions from $Z_c(3900)$ and $Z_c(4020)$ together with a smooth background. 240 ± 56 $Z_c(3900)^\pm$ events is observed with a statistical significance of 4.3σ (3.9σ including the systematical uncertainty). The $Z_c(3900)$ signals at other c.m. energies and the $Z_c(4020)$ signals at all the c.m. energies are not statistically significant.

Using the results from [9], the ratio of the branching fractions of different $Z_c(3900)^\pm$ decays is calculated as $R_{Z_c(3900)} = \frac{\mathcal{B}(Z_c(3900)\rightarrow\rho\eta_c)}{\mathcal{B}(Z_c(3900)\rightarrow\pi J/\psi)} = 2.1 \pm 0.8$ at $\sqrt{s} = 4.23$ GeV and less than 6.4 at $\sqrt{s} = 4.26$ GeV at the 90% C.L. The theoretical predictions for this ratio varies depending on model assumptions and ranges from a few per cent to a few hundreds [13–18].

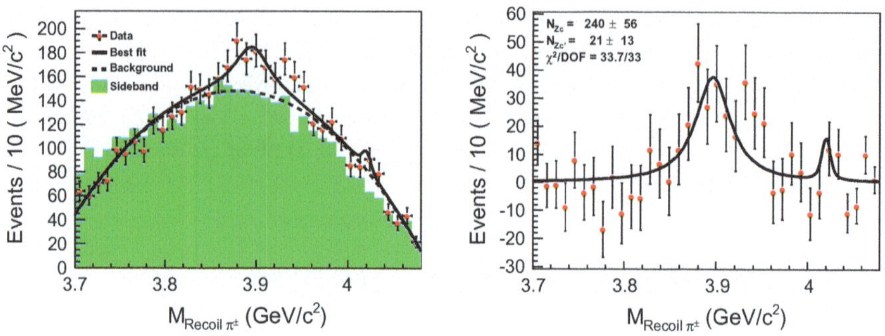

Fig. 119.2 The $\pi^{\pm}$ recoil mass distribution in $e^+e^- \to \pi^{\pm}\rho^{\mp}\eta_c$ at $\sqrt{s} = 4.23$ GeV and the fit with $Z_c(3900/4020)^{\pm}$ signals (left panel); and the same plot with background subtracted (right panel). Dots with error bars are data, shaded histogram is from η_c sidebands, normalized to the number of backgrounds from the fit, the solid lines are total fit and the dotted line is background

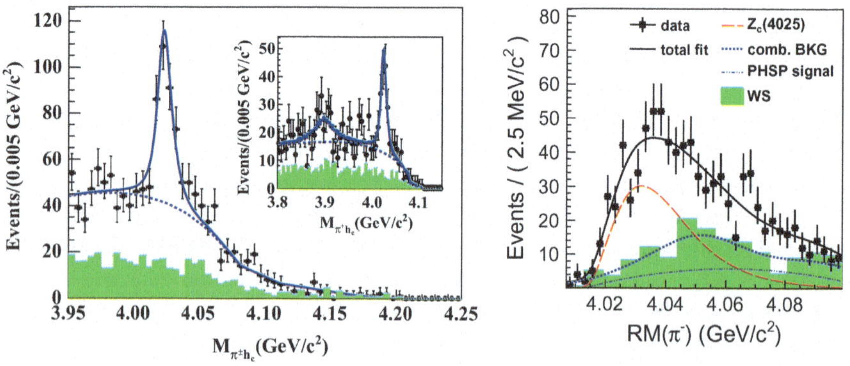

Fig. 119.3 The fit to the $M(\pi^{\pm}h_c)$ distribution (left panel, adapted from [19]) for selected $e^+e^- \to \pi^+\pi^-h_c$ events, summed over all energy points; and the fit to the $\pi^{\mp}$ recoil mass spectrum (right panel, adapted from [20]) in $e^+e^- \to (D^*\bar{D}^*)^{\pm}\pi^{\mp}$ at $\sqrt{s} = 4.26$ GeV

119.2.5 Observation of the $Z_c(4020)$

The partner state of the $Z_c(3900)$ close to the $D^*\bar{D}^*$ is searched in $e^+e^- \to \pi^+\pi^-h_c$ and $e^+e^- \to (D^*\bar{D}^*)^{\pm}\pi^{\mp}$. BESIII measures cross sections of $e^+e^- \to \pi^+\pi^-h_c$ at c.m. energies of 3.90–4.42 GeV [19]. Intermediate states are studied by examining the Dalitz plot of the selected $\pi^+\pi^-h_c$ candidate events. There is distinct signal for an exotic charmonium-like structure in the $\pi^{\pm}h_c$ system (Fig. 119.3). The mass and width of the $Z_c(4020)$ are measured to be $(4022.9 \pm 0.8 \pm 2.7)$ MeV$/c^2$ and $(7.9 \pm 2.7 \pm 2.6)$ MeV, respectively. The statistical significance of the $Z_c(4020)$ signal is found to be greater than 8.9σ.

The study of $e^+e^- \to (D^*\bar{D}^*)^{\pm}\pi^{\mp}$ process using data at $\sqrt{s} = 4.26$ GeV [20] indicates the open-charm decay $Z_c(4020) \to (D^*\bar{D}^*)^{\pm}$ in the $\pi^{\mp}$ recoil mass

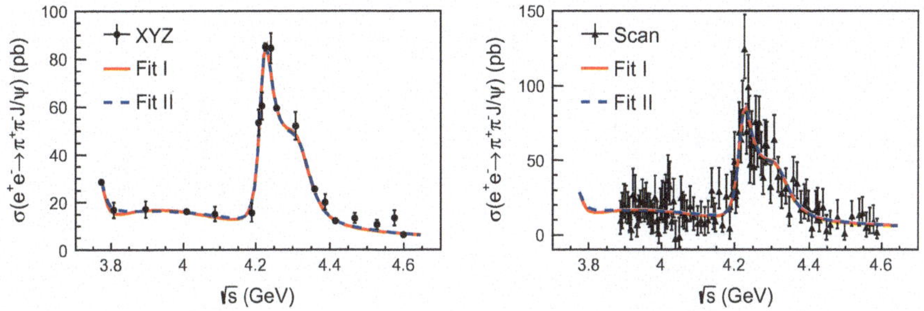

Fig. 119.4 Measured cross section $\sigma(e^+e^- \to \pi^+\pi^- J/\psi)$ and the fit to the data. Adapted from [24]

spectrum (see Fig. 119.3). The measured mass and width of the structure are $(4026.3 \pm 2.6 \pm 3.7)$ MeV/c^2 and $(24.8 \pm 5.6 \pm 7.7)$ MeV, respectively, and the statistical significance is 13σ.

119.3 The Vector Charmonium-Like State $Y(4260)$

The vector charmonium-like states with quantum numbers $J^{PC} = 1^{--}$ that are usually called Y states, like the $Y(4260)$ [21], the $Y(4360)$ [22, 23], and the $Y(4660)$ [23]. These states show strong coupling to hidden-charm final states in contrast to the vector charmonium states in the same energy region ($\psi(4040)$, $\psi(4160)$, $\psi(4415)$) which couple dominantly to open-charm meson pairs [1].

119.3.1 $e^+e^- \to \pi^+\pi^- J/\psi$

The cross sections of $e^+e^- \to \pi^+\pi^- J/\psi$ are measured precisely at c.m. energies from 3.77 to 4.60 GeV using 9 fb^{-1} of BESIII data [24]. Figure 119.4 shows the measured cross sections, one can see clearly the $Y(4260)$ structure observed by BaBar and Belle experiments, but it is peaked at around 4.22 GeV rather than 4.26 GeV from the previous fits [7, 25].

Two resonant structures in the $Y(4260)$ peak region are needed in a fit to the cross sections. The first one has a mass of $(4222.0 \pm 3.1 \pm 1.4)$ MeV/c^2 and a width of $(44.1 \pm 4.3 \pm 2.0)$ MeV, while the second one has a mass of $(4320.0 \pm 10.4 \pm 7.0)$ MeV/c^2 and a width of $(101.4^{+25.3}_{-19.7} \pm 10.2)$ MeV. The mass of first resonance is lower than that of the $Y(4260)$ and it is much narrower. The second resonance is observed in $e^+e^- \to \pi^+\pi^- J/\psi$ for the first time, with a statistical significance larger than 7.6σ.

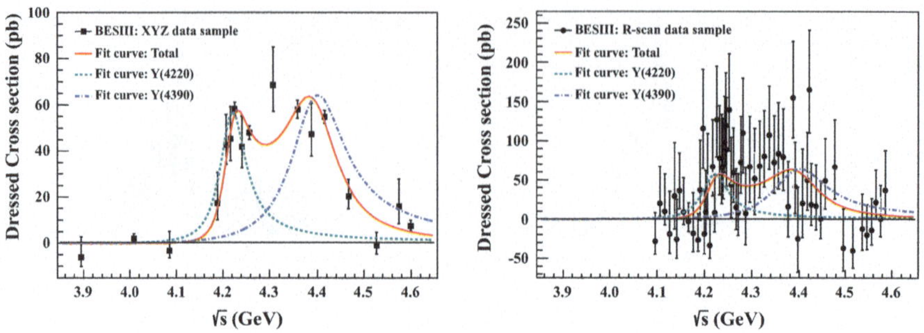

Fig. 119.5 Fit to the cross section of $e^+e^- \to \pi^+\pi^- h_c$ with the coherent sum of two BW functions (solid curve). Adapted from [26]

It is worth pointing out that the lower mass structure (called $Y(4220)$ hereafter) is the main component of the $Y(4260)$ structure but with improved measurement of the resonant parameters thanks to the high luminosity data from BESIII.

119.3.2 $e^+e^- \to \pi^+\pi^- h_c$

In 2013, BESIII reported the cross section measurement of $e^+e^- \to \pi^+\pi^- h_c$ at 13 c.m. energies from 3.9 to 4.4 GeV [19]. In 2017, BESIII presented a follow-up study at c.m. energies from 3.9 to 4.6 GeV [26]. The cross sections are shown in Fig. 119.5.

Assuming the $\pi^+\pi^- h_c$ events come from two resonances, BESIII obtains $M = (4218.4^{+5.5}_{-4.5} \pm 0.9)$ MeV/c^2, $\Gamma = (66.0^{+12.3}_{-8.3} \pm 0.4)$ MeV for $Y(4220)$, and $M = (4391.5^{+6.3}_{-6.8} \pm 1.0)$ MeV/c^2, $\Gamma = (139.5^{+16.2}_{-20.6} \pm 0.6)$ MeV for $Y(4390)$. The parameters of the low mass structure are consistent with those of the resonance observed in $e^+e^- \to \omega\chi_{c0}$ [27] and in $e^+e^- \to \pi^+\pi^- J/\psi$ [24]. The high mass structure is different from the $Y(4360)$ [28, 29] and $\psi(4415)$ [1].

119.3.3 $e^+e^- \to \pi^+\pi^-\psi(2S)$

BESIII measures the cross sections of $e^+e^- \to \pi^+\pi^-\psi(2S)$ using 5.1 fb^{-1} of data collected from 4.0 to 4.6 GeV [30]. The measurements are in good consistency with previous BaBar and Belle results [28, 29], but with much improved precision, as shown in Fig. 119.6.

As the BESIII data can only reach 4.6 GeV, the parameters of the $Y(4660)$ are fixed to Belle measurement [29] in the fit to the $e^+e^- \to \pi^+\pi^-\psi(2S)$ cross sections. The data require a lower-mass resonance with a mass $M = (4209.5 \pm 7.4 \pm 1.4)$ MeV/c^2 and a width $\Gamma = (80.1 \pm 24.6 \pm 2.9)$ MeV with a statistical significance of 5.8σ, this is the first observation of the new decay mode $Y(4220) \to \pi^+\pi^-\psi(2S)$. The fit results are also presented in Fig. 119.6.

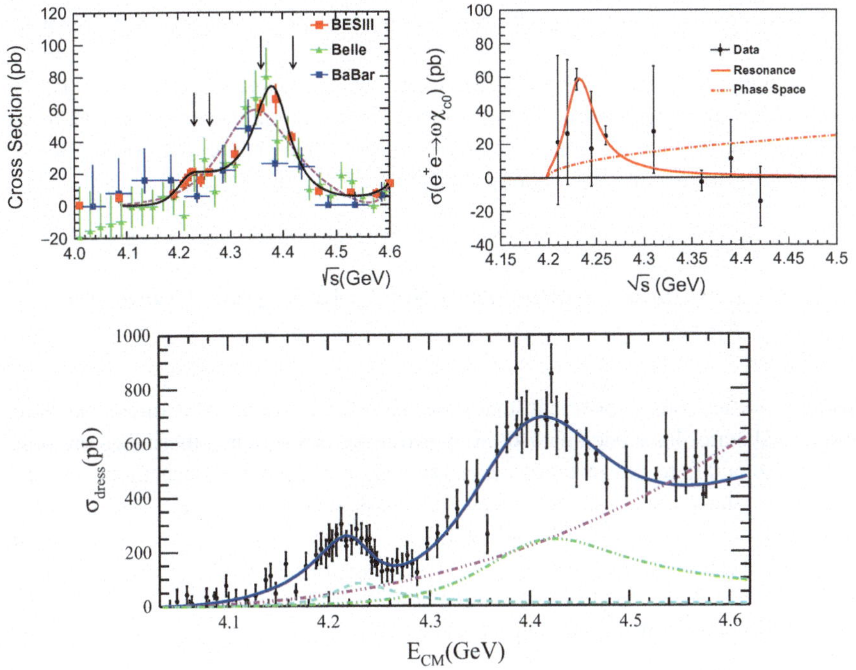

Fig. 119.6 Cross sections of $e^+e^- \to \pi^+\pi^-\psi(2S)$ (top left panel, adapted from [30]), $\omega\chi_{c0}$ (top right panel, adapted from [27]), and $D^0D^{*-}\pi^+ + c.c.$ (bottom panel, adapted from [33]), as a function of c.m. energy

119.3.4 $e^+e^- \to \omega\chi_{c0}$

The process $e^+e^- \to \omega\chi_{c0}$ is observed for the first time and the cross sections or the upper limits on the cross sections (when the signal is not significant) at the 90% C.L. are determined [27, 31]. By assuming the $\omega\chi_{c0}$ signals come from a single resonance, the mass is determined to be $(4226 \pm 8 \pm 6)$ MeV/c^2, and width $(39 \pm 12 \pm 2)$ MeV (shown in Fig. 119.6). In fact, it is in the $e^+e^- \to \omega\chi_{c0}$ mode that the $Y(4220)$ (or $Y(4230)$) structure was first reported [27]. The analysis was updated with more data recently in [32].

119.3.5 $e^+e^- \to D^0D^{*-}\pi^+ + c.c.$

BESIII reported measurements of the cross section of $e^+e^- \to D^0D^{*-}\pi^+ + c.c.$ at c.m. energies from 4.05 to 4.60 GeV [33] (shown in Fig. 119.6), which is a significant improvement over the Belle measurement [34] with ISR technique.

A fit to the cross section is performed to determine the parameters of the resonant structures. The fit yields a mass of $(4228.6 \pm 4.1 \pm 6.3)$ MeV/c^2 and a width of $(77.0 \pm 6.8 \pm 6.3)$ MeV for the lower mass structure, and a mass of (4404.7 ± 7.4) MeV/c^2 and a width of (191.9 ± 13.0) MeV for the higher mass one. Here for the higher mass state, the errors are statistical only. This is the first observation of the $Y(4220)$ decays into open-charm final state $D^0 D^{*-} \pi^+ + c.c.$.

119.3.6 Resonant Parameters of the Y(4260) from Combined Fit

As the cross sections of different final states have some common features, and some of the final states have been measured by different experiments, these data are used to do combined fit to extract more information about the resonant structures [35, 36].

In [35], the authors use the measured cross sections of $e^+e^- \to \omega\chi_{c0}$ [31], $\pi^+\pi^- h_c$ [26], $\pi^+\pi^- J/\psi$ [24], and $D^0 D^{*-}\pi^+ + c.c.$ [33] processes to determine the resonant parameters of the $Y(4220)$. The fit determines the mass of the $Y(4220)$ as $(4219.6 \pm 3.3 \pm 5.1)$ MeV/c^2 and the width is $(56.0 \pm 3.6 \pm 6.9)$ MeV. The coupling of this state to the lepton pair is also extracted.

119.4 The $X(3872)$

The $X(3872)$ was observed in $B^\pm \to K^\pm \pi^+\pi^- J/\psi$ decays 15 years ago at the Belle experiment [37]. It was confirmed subsequently by several other experiments [38–40]. BESIII observed a new process of the $X(3872)$ production: $e^+e^- \to \gamma X(3872) \to \gamma \pi^+\pi^- J/\psi$ [41].

The $X(3872)$ signal is shown in Fig. 119.7 (left), the mass is measured as $(3871.9 \pm 0.7 \pm 0.2)$ MeV/c^2 and the width is less than 2.4 MeV at the 90% C.L. The statistical significance of $X(3872)$ is 6.3σ. The energy-dependent cross sections (Fig. 119.7 (right)) can be described with the $Y(4260)$ resonance well, which supports the existence of radiative transition $Y(4260) \to \gamma X(3872)$. The analysis was updated with more data recently in [42].

Together with the hadronic transition of the $Y(4260) \to \pi Z_c(3900)$ [6–8], these suggest that there might be some commonality in the nature of $X(3872)$, $Y(4260)$, and $Z_c(3900)$, and so the model developed to interpret any one of them should also consider the other two.

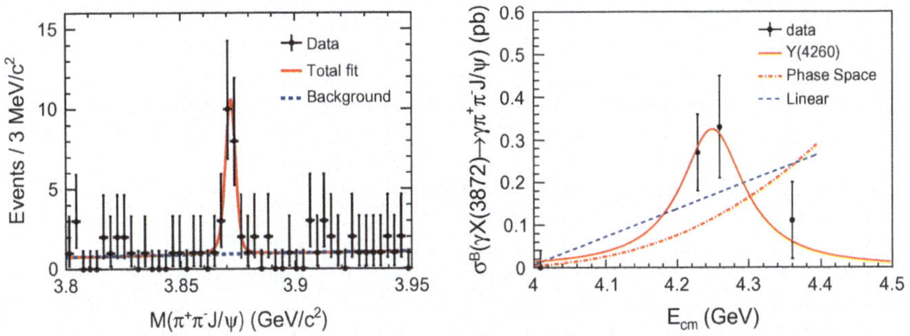

Fig. 119.7 Fit the $M(\pi^+\pi^- J/\psi)$ distribution observed at BESIII (left panel), and fit to $\sigma^B[e^+e^- \to \gamma X(3872)] \times \mathcal{B}[X(3872) \to \pi^+\pi^- J/\psi]$ (right panel). Adapted from [41]

119.5 Summary and Perspectives

BESIII has achieved a lot in the study of the XYZ states. There are still some data at BESIII not analyzed and data at more energy points will be taken [43]. More analyses with these data samples will allow many improved understanding of the XYZ states, especially the $X(3872)$, $Y(4260)$, $Z_c(3900)$, and $Z_c(4020)$. BEPCII is upgrading the maximum c.m. energy from 4.6 to 4.9 GeV in two years, this will enable a full coverage of the $Y(4660)$ [23] and $Y(4630)$ [44] resonances, improved measurements of their properties are expected.

Acknowledgements Many thanks to the organizers for the invitation to give this talk and for the perfectly organized conference. This work is supported in part by National Natural Science Foundation of China (NSFC) under contract Nos. 11835012, 11521505, and 11475187; the Ministry of Science and Technology of China under Contract No. 2015CB856701; and the CAS Center for Excellence in Particle Physics (CCEPP).

References

1. Patrignani, C., et al. [Particle Data Group]: Chin. Phys. C **40**, 100001 (2016); Tanabashi, M. et al. [Particle Data Group]: Phys. Rev. D **98**, 030001 (2018)
2. Bevan, A.J., et al. [BaBar and Belle Collaborations]: Eur. Phys. J. C **74**, 3026 (2014)
3. For recent reviews, see Guo, F.K., Hanhart, C. Meißner, U.G. Wang, Q., Zhao, Q., Zou, B.S.: Rev. Mod. Phys. **90**(1), 015004 (2018); Chen, H.X., Chen, W., Liu, X., Zhu, S.L.: Phys. Rept. **639**, 1 (2016); Brambilla, N., et al.: Eur. Phys. J. C **71**, 1534 (2011)
4. Ablikim, M., et al. [BESIII Collaboration]: Nucl. Instrum. Methods Phys. Res. Sect. A **614**, 345 (2010)
5. More results from BESIII can be found in Yuan, C.Z.: The XYZ states revisited. Int. J. Mod. Phys. A **33**(21), 1830018 (2018). arXiv:1808.01570 [hep-ex]
6. Ablikim, M., et al. [BESIII Collaboration]: Phys. Rev. Lett. **110**, 252001 (2013)
7. Liu, Z.Q., et al. [Belle Collaboration]: Phys. Rev. Lett. **110**, 252002 (2013)
8. Xiao, T., Dobbs, S., Tomaradze, A., Seth, K.K.: Phys. Lett. B **727**, 366 (2013)

9. Ablikim, M., et al. [BESIII Collaboration]: Phys. Rev. Lett. **119**, 072001 (2017)
10. Ablikim, M., et al. [BESIII Collaboration]: Phys. Rev. Lett. **112**, 022001 (2014)
11. Ablikim, M., et al. [BESIII Collaboration]: Phys. Rev. D **92**, 092006 (2015)
12. Yuan, C.-Z.: The 9th International Workshop on Charm Physics, 21–25 May 2018, Novosibirsk, Russia. https://indico.inp.nsk.su/event/10/session/2/contribution/41/material/slides/0.pdf
13. Esposito, A., Guerrieri, A.L., Pilloni, A.: Phys. Lett. B **746**, 194 (2015)
14. Li, G., Liu, X.H., Zhou, Z.: Phys. Rev. D **90**, 054006 (2014)
15. Ke, H.W., Wei, Z.T., Li, X.Q.: Eur. Phys. J. C **73**, 2561 (2013)
16. Ma, L., Liu, X.H., Liu, X., Zhu, S.L.: Phys. Rev. D **91**, 034032 (2015)
17. Ma, L., Deng, W.Z., Chen, X.L., Zhu, S.L.: arXiv:1512.01938 [hep-ph]
18. Voloshin, M.B.: arXiv:1806.05651 [hep-ph]
19. Ablikim, M., et al. [BESIII Collaboration]: Phys. Rev. Lett. **111**, 242001 (2013)
20. Ablikim, M., et al. [BESIII Collaboration]: Phys. Rev. Lett. **112**, 132001 (2014)
21. Aubert, B., et al. [BaBar Collaboration]: Phys. Rev. Lett. **95**, 142001 (2005)
22. Aubert, B., et al. [BaBar Collaboration]: Phys. Rev. Lett. **98**, 212001 (2007)
23. Wang, X.L., et al. [Belle Collaboration]: Phys. Rev. Lett. **99**, 142002 (2007)
24. Ablikim, M., et al. [BESIII Collaboration]: Phys. Rev. Lett. **118**, 092001 (2017)
25. Lees, J.P., et al. [BaBar Collaboration]: Phys. Rev. D **86**, 051102(R) (2012)
26. Ablikim, M., et al. [BESIII Collaboration]: Phys. Rev. Lett. **118**, 092002 (2017)
27. Ablikim, M., et al. [BESIII Collaboration]: Phys. Rev. Lett. **114**, 092003 (2015)
28. Lees, J.P., et al. [BaBar Collaboration]: Phys. Rev. D **89**, 111103 (2014)
29. Wang, X.L., et al. [Belle Collaboration]: Phys. Rev. D **91**, 112007 (2015)
30. Ablikim, M., et al. [BESIII Collaboration]: Phys. Rev. D **96**, 032004 (2017)
31. Ablikim, M., et al. [BESIII Collaboration]: Phys. Rev. D **93**, 011102 (2016)
32. Ablikim, M., et al. [BESIII Collaboration]: arXiv:1903.02359 [hep-ex]
33. Ablikim, M., et al. [BESIII Collaboration]: Phys. Rev. Lett. **122**, 102002 (2019)
34. Pakhlova, G., et al. [Belle Collaboration]: Phys. Rev. D **80**, 091101 (2009)
35. Gao, X.Y., Shen, C.P., Yuan, C.Z.: Phys. Rev. D **95**, 92007 (2017)
36. Zhang, J., Yuan, L., Wang, R.: arXiv:1805.03565 [hep-ph]
37. Choi, S.K., et al. [Belle Collaboration]: Phys. Rev. Lett. **91**, 262001 (2003)
38. Acosta, D., et al. [CDF Collaboration]: Phys. Rev. Lett. **93**, 072001 (2004)
39. Abazov, V.M., et al. [D0 Collaboration]: Phys. Rev. Lett. **93**, 162002 (2004)
40. Aubert, B., et al. [BaBar Collaboration]: Phys. Rev. D **71**, 071103 (2005)
41. Ablikim, M., et al. [BESIII Collaboration]: Phys. Rev. Lett. **112**, 092001 (2014)
42. Ablikim, M., et al.: arXiv:1903.04695 [hep-ex]
43. Yuan, C.-Z.: Front. Phys. **10**, 101401 (2015)
44. Pakhlova, G., et al. [Belle Collaboration]: Phys. Rev. Lett. **101**, 172001 (2008)

Chapter 120
Few-Body Insights of Multiquark Exotic Hadrons

Javier Vijande, Jean-Marc Richard and Alfredo Valcarce

Abstract In this contribution we discuss the adequate treatment of the 4- and 5-body dynamics within a constituent quark framework. We stress that the variational and Born-Oppenheimer approximations give energies rather close to the exact ones, while the diquark approximation might be rather misleading. Hall-Post inequalities provide very useful lower bounds that exclude possible stable states for some mass ratios and color wave functions.

120.1 Introduction

Recent contributions on multiquarks are stimulated by the discovery of a double-charm baryon [1], which is interesting by itself and also triggers speculations about exotic double-charm mesons $QQ\bar{q}\bar{q}$. For years, the sector of flavor-exotic tetraquarks has been somewhat forgotten, and even omitted from some reviews on exotic hadrons, as much attention was paid to hidden-flavor states $Q\bar{Q}q\bar{q}$. However the flavor-exotic multiquarks have been investigated already some decades ago [2] and has motivated an abundant literature (see [3] and references therein) that has been unfortunately ignored in some recent papers.

In this contribution, we stress that a careful treatment of the few-body problem is required before drawing any conclusion about the existence of stable states in

J. Vijande (✉)
Departamento de Física Atómica, Molecular y Nuclear, IFIC (UV-CSIC) and IRIMED Joint Research Unit (IIS Hospital La Fe-UV), Universidad de Valencia (UV), Valencia, Spain
e-mail: javier.vijande@uv.es

J.-M. Richard
Institut de Physique Nucléaire de Lyon, IN2P3-CNRS–UCBL, Université de Lyon, 4 rue Enrico Fermi, 69622 Villeurbanne, France

A. Valcarce
Departamento de Física Fundamental and IUFFyM, Universidad de Salamanca, 37008 Salamanca, Spain

N. A. Orr et al. (eds.), *Recent Progress in Few-Body Physics*,
Springer Proceedings in Physics 238,
https://doi.org/10.1007/978-3-030-32357-8_120

a particular model. Not surprisingly, the main difficulties are encountered when a multiquark state is found near its lowest dissociation threshold. The question of whether or not there is a bound state requires a lot of care. We consider that it is important to clarify the somewhat contradictory results in the literature. In particular, some authors who use similar ingredients obtain either stability or instability for the all-heavy configuration $QQ\bar{Q}\bar{Q}$, and in our opinion, this is due to an erroneous handling of the four-body problem.

120.2 Diquark Approximation

A few decades ago, the main concern in baryon spectroscopy was the problem of missing resonances predicted by the quark model and not observed experimentally. Many states of the symmetric quark model disappear if baryons are constructed out of a frozen diquark and a quark. However, the missing resonances are not very much coupled to the typical investigation channels πN or γN, which privilege states with one pair of quarks shared with the target nucleon N. In recent photoproduction experiments with improved statistics, some of the missing states have been identified, which cannot be accommodated as made of a ground-state diquark and a third quark [4].

The diquark model is nevertheless regularly revisited, to accommodate firmly established exotics such as the $X(3872)$, or even candidates awaiting confirmation. Unfortunately, some unwanted multiquarks are also predicted in this approach, though this is not always explicitly stated or even realized. The issue of unwanted multiquarks within the diquark model was raised many years ago by Fredriksson and Jandel [5], and is sometimes rediscovered, without any reference to the 1982 paper. The paradox is perhaps that the diquark model, that produces fewer baryon states, produces too many multiquarks!

There are many variants of the so-called diquark model. An extreme point of view is that diquarks are almost-elementary objects, with their specific interaction with quarks and between them. A whole baryon phenomenology can be built starting from well-defined assumptions about the diquark constituent masses and the potential linking a quark to a diquark. Then, a diquark-diquark interaction has to be introduced as a new ingredient for the multiquark sector.

Another extreme is to estimate the energy and wave function of, say $(a_1a_2a_3)$ with masses m_i and interaction $v_{ij}(r)$, first solving for (a_1a_2) with v_{12} alone with energy η_{12}, and then estimating the bound state of a point-like system (a_1a_2) of mass m_1+m_2 located at $\mathbf{R}_{12}$ interacting with a_3 through the potential $v_{13}(\mathbf{r}_3-\mathbf{R}_{12})+v_{23}(\mathbf{r}_3-\mathbf{R}_{12})$, resulting in binding energy $\eta_{12,3}$. Thus, the whole energy is given by $\eta_{12}+\eta_{12,3}$.

This strategy is of course fully justified for the deuterium atom considered as a pne^- system, as the inter-nuclear motion is not significantly modified by the electron. On the other hand, this approach ruins some subtle collective binding, for instance, that of Borromean states [6]. Also, one cannot see either how $\mathrm{H}^-(pe^-e^-)$

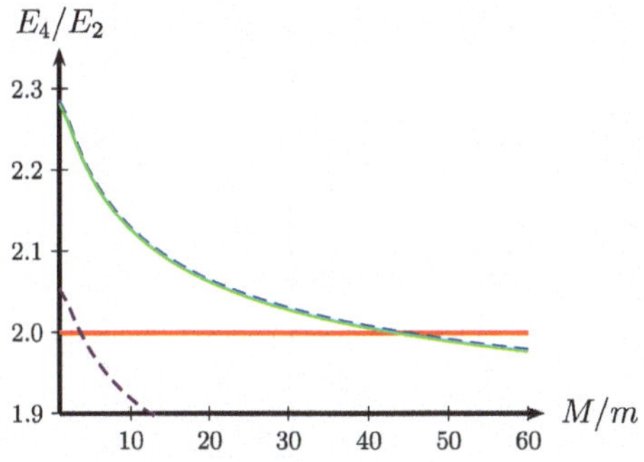

Fig. 120.1 Comparison of the variational upper bound (green curve) and Hall-Post lower bound (dotted blue curve), hardly distinguishable from the variational estimate at this scale, for the tetraquark Hamiltonian (120.1) with a linear interaction. Also shown is the naive diquark-antidiquark approximation (dashed violet curve). Adapted from [3]

could become bound in this approach, or the hydrogen molecule be described as a "diproton" linked to a "dielectron"! In some other cases, like in the case of constituent quark models, the method just overestimates the binding, since the *ad-hoc* clustering lowers significantly the energy.

For simplicity, we consider for doubly-heavy tetraquarks the case of a frozen $\bar{3}3$ color wave function. Color mixing has to be introduced to have the proper threshold in the model, and it has been seen in explicit calculations that the mixing with meson-meson configurations is crucial for states at the edge of stability. Nevertheless the comparison of various approximations is instructive for the toy model,

$$H_{33} = \frac{\mathbf{p}_1^2 + \mathbf{p}_2^2}{2\,M} + \frac{\mathbf{p}_3^2 + \mathbf{p}_4^2}{2\,m} + \frac{v(r_{12}) + v(r_{34})}{2} + \frac{v(r_{13}) + v(r_{14}) + v(r_{23}) + v(r_{24})}{4}. \tag{120.1}$$

In Fig. 120.1, we compare the exact solution of (120.1) with the approximation consisting of first computing the QQ diquark with $r_{12}/2$ alone and qq with r_{34} alone, and then $(QQ)(\bar{q}\bar{q})$ as a meson with a potential $r_{12,34}$ and constituent masses $2M$ and $2m$.

120.3 Relating Mesons, Baryons and Tetraquarks

In a recent paper, Eichten and Quigg [7] use heavy-quark symmetry to relate meson, baryon and tetraquark energies. In a simplified version without spin effects, it reads

$$QQ\bar{q}\bar{q} = QQq + Qqq - Q\bar{q}, \tag{120.2}$$

where the configuration stands for the ground-state energy. For fixed m and $M \to \infty$, the identity is exact. For finite M, there is some departure. If one treats the tetraquark

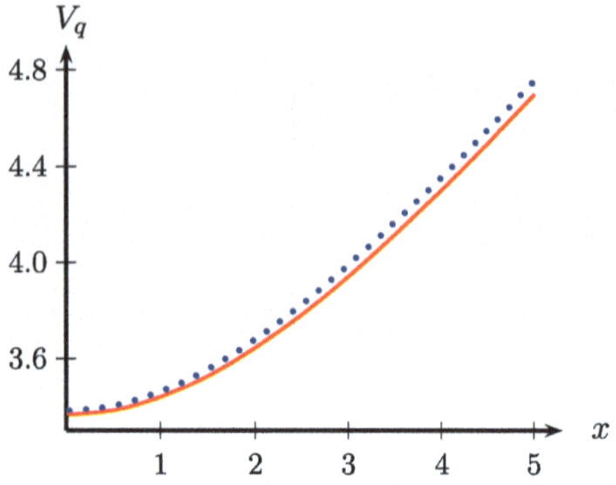

Fig. 120.2 Comparison of the light quark energies for $QQ\bar{q}\bar{q}$ (solid red line) and QQq (dotted blue line) as a function of the QQ separation x. The second curve is shifted by the difference of energies $Qqq - Q\bar{q}$. The units are such $m = 1$, $M = 5$ and $v_{ij} = r_{ij}$. Adapted from [3]

$QQ\bar{q}\bar{q}$ and the doubly-heavy baryon QQq in the Born-Oppenheimer approximation, one can compare the two effective potentials as a function of the QQ separation x, the baryon one being shifted by $Qqq - Q\bar{q}$ which is independent of x. Without recoil correction, the two potentials are identical at $x = 0$. For finite M, there is slight difference, as the single q recoils against either M or $2M$, and similarly qq recoils against one or two heavy quarks.

The comparison is shown in Fig. 120.2. Clearly the two effective potentials are very similar, and thus give almost identical energies, up to an additive constant that corresponds to the last two terms in (120.2).

120.4 Hall-Post Inequalities

The Hall-Post inequalities have been derived in the 50s to relate the binding energies of light nuclei with different number of nucleons [8]. They have been rediscovered in the course of studies on the stability of matter [9], or to link meson and baryon masses in the quark model [2, 10].

For their application to tetraquarks we consider the toy Hamiltonian (120.1). It can be rewritten as,

$$\left(\sum \mathbf{p}_i\right) \cdot (A(\mathbf{p}_1 + \mathbf{p}_2) + B(\mathbf{p}_3 + \mathbf{p}_4)) + \frac{\tilde{h}_{12}(x_{12}) + \tilde{h}_{34}(x_{34})}{2} + \sum{}' \frac{\tilde{\tilde{h}}_{ij}(x, a, b)}{4},$$
$$\tilde{\tilde{h}}_{13}(x, a, b) = \frac{1}{x}\left(\frac{\mathbf{p}_1 - \mathbf{p}_3 + a\,\mathbf{p}_2 + b\,\mathbf{p}_4}{2}\right)^2 + v_{ij}, \tag{120.3}$$

where the masses x_{12}, x_{34} and x are readily calculated from the parameters A, B and a, and b. This results into

$$E_4(M, m) \geq \max_{A,B,a,b} \left[E_2(x_{12}) + E_2(x_{34}) + E_2(x)\right]. \tag{120.4}$$

Hence a rigorous lower bound is obtained from simple algebraic manipulations and the knowledge of the 2-body energy as a function of the reduced mass. For a linear interaction the results for $E_4/E_2(1)$ as a function of M/m are shown in Fig. 120.1. The sum $1/M + 1/m$ is kept equal to 2 to fix the threshold energy at $2E_2(1)$.

120.5 Color Mixing and Spin-Dependent Corrections

Any model with a pairwise potential, due to color-octet exchange, induces mixing between $\bar{3}3$ and $6\bar{6}$ states in the $QQ - \bar{q}\bar{q}$ basis. Perhaps the true dynamics inhibits the call for higher color representations such as sextet, octet, etc., for the subsystems of a multiquark, but for the time being, let us adopt the color-additive model. If one starts from a $\bar{3}3$ state with QQ in a spin triplet, and, for instance $\bar{q}\bar{q} = \bar{u}\bar{d}$ with spin and isospin $S = I = 0$, then its orbital wave function is mainly made of an s-wave in all coordinates. It can mix with a color $6\bar{6}$ with orbital excitations in the $\mathbf{x}$ and $\mathbf{y}$ linking QQ and $\bar{q}\bar{q}$, respectively.

To illustrate the role of color-mixing for the AL1 potential we use the potential AL1 by Semay and Silvestre-Brac [11]. Its central part is a Coulomb-plus-linear term, while its spin-spin part is a regularized Breit-Fermi interaction with a smearing parameter that depends on the reduced mass.

The energy, normalized to the lowest threshold, as a function of M/m without and with color-mixing is shown in Fig. 120.3 (left). The ground state of the $QQ\bar{u}\bar{d}$ with $J^P = 1^+$, which is a candidate for stability, has its main component with color $\bar{3}3$, and spin $\{1, 0\}$ in the $QQ - \bar{u}\bar{d}$ basis. The main admixture consists of $6\bar{6}$ with spin $\{1, 0\}$ and an antisymmetric orbital wavefunction, and of $6\bar{6}$ with spin $\{0, 1\}$ with a symmetric orbital wavefunction.

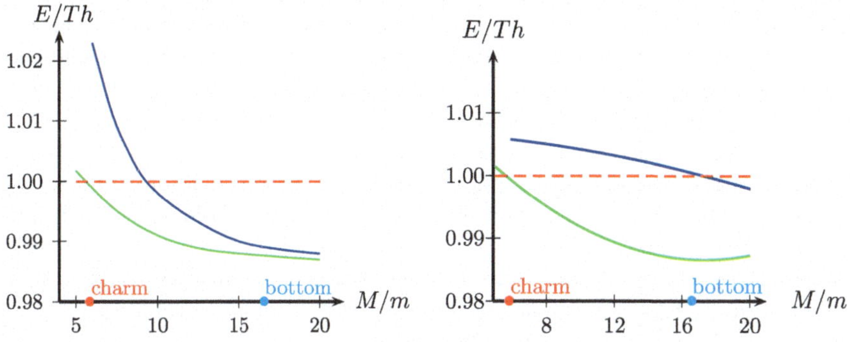

Fig. 120.3 (left) Effect of color-mixing on the binding of $QQ\bar{u}\bar{d}$, within the AL1 model. The tetraquark energy is calculated with only the color $\bar{3}3$ configurations (blue curve) and with the $6\bar{6}$ components (green curve). (right) Effect of the spin-spin interaction of the binding of $QQ\bar{u}\bar{d}$, within the AL1 model. The tetraquark energy is calculated with (green line) and without (blue line) the chromomagnetic term. Adapted from [3]

It has been acknowledged in the literature that a pure additive interaction such as (120.1) will not bind $cc\bar{q}\bar{q}$, on the sole basis that this tetraquark configuration benefits from the strong cc chromoelectric attraction that is absent in the $Q\bar{q} + Q\bar{q}$ threshold. In the case where $qq = ud$ in a spin and isospin singlet, however, there is in addition a favorable chromomagnetic interaction in the tetraquark, while the threshold experiences only heavy-light spin-spin interaction, whose strength is suppressed by a factor m/M.

To study the spin-dependent corrections, we make use once again of the AL1 potential. The results are shown in Fig. 120.3 (right) for $QQ\bar{u}\bar{d}$, as a function of the mass ratio M/m.

The system $bb\bar{u}\bar{d}$ is barely bound without the spin-spin term, though the mass ratio $m_b/(m_u \simeq m_d)$ is very large. It acquires its binding energy of the order of 150 MeV [3] when the spin-spin is restored. The system $cc\bar{u}\bar{d}$ is clearly unbound when the spin-spin interaction is switched off. This is shown here for the AL1 model, but this is true for any realistic interaction, including an early model by Bhaduri et al. [12]. The case of $cc\bar{u}\bar{d}$ is actually remarkable. Semay and Silvestre-Brac, who used their AL1 potential, missed the binding because their method of systematic expansion on the eigenstates of an harmonic oscillator is not very efficient to account for the short-range correlations, and it was abandoned in the latest quark model calculations. Janc and Rosina were the first to obtain binding with such potentials, and their calculation was checked by Barnea et al. (see [3] for references).

The role played by color mixing and spin-dependent corrections has also been addressed for hidden-charm pentaquarks ($\bar{c}cqqq$) [13]. The main difficulties faced when attempting to solve the 5-body problem including both color-mixing and spin-dependent corrections is the larger numbers of color vectors and the increased complexity of the radial equation. There are three independent color states for the pentaquarks: (i) ($\bar{c}c$) singlet coupled to (qqq) singlet, (ii) ($\bar{c}c$) octet coupled to the first (qqq) octet, in which the quarks 3 and 4 are in a $\bar{3}$ state, and (iii) ($\bar{c}c$) octet associated to the second (qqq) octet, in which the quarks 3 and 4 form a sextet. Besides this, the larger number of coordinate ensembles and its complexity makes even more important than in the 4-quark case to use a reliable numerical technique to solve the radial part. Few states were found to be bound. $(J, I) = (1/2, 3/2)$ and $(3/2, 3/2)$ are found below their lowest S- and D-wave thresholds: $\Delta\eta_c$ and $D\Sigma_c$. In fact they are substantially lower, so that they remain stable or metastable even if one accounts for the width of the Δ and considers that the actual lowest threshold is $N\pi\eta_c$. $(J, I) = (5/2, 1/2)$ is above its lowest D-wave threshold while it is below the lowest S-wave threshold. In this case, as highlighted long ago in [14], the contribution of color vectors different from the singlet-singlet combination would prevent by the centrifugal barrier the tunneling of the quarks to combine in a colorless object, enhancing in this way the stability of this state.

The role of the spin-spin interaction versus the spin-independent one is shown in Fig. 120.4 by modifying the strength of the chromomagnetic interaction using a multiplicative factor. It is seen that the binding starts already with a small fraction of the spin-spin interaction. This means that there is a favorable interplay of chromoelectric and chromomagnetic effects, although the binding disappears in the pure chromoelectric limit.

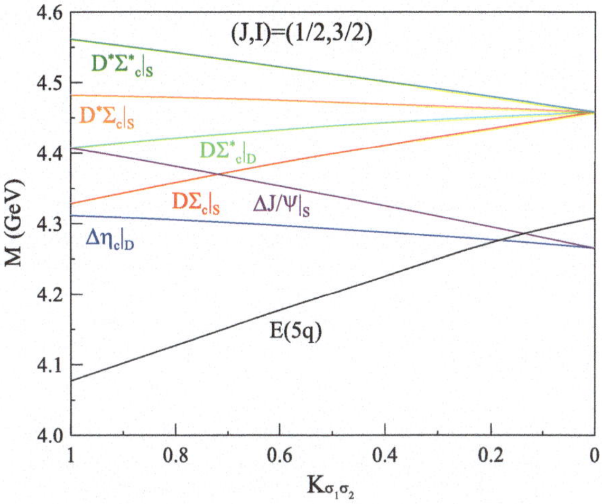

Fig. 120.4 Mass of the $(\bar{c}cqqq)$ $(J, I) = (1/2, 3/2)$ state and its thresholds as a function of the strength of the chromomagnetic interaction, decreased by a multiplicative factor $K_{\sigma_1\sigma_2}$. Adapted from [13]

120.6 Conclusions

Let us summarize. The few-body problem is rather delicate, especially for systems at the edge of stability. In the case of four-quark states, the analogy with atomic physics is a good guidance to indicate the most favorable configurations in the limit of dominant chromoelectric interaction. However, unlike the positronium molecule, the all-heavy configuration $QQ\bar{Q}\bar{Q}$ is not stable if one adopts a standard quark model and solves the four-body problem correctly.

The mixing of the $\bar{3}3$ and $6\bar{6}$ color configurations is important, especially for states very near threshold. This mixing occurs by both the spin-independent and the spin-dependent parts of the potential.

Approximations are welcome, especially if they shed some light on the four-body dynamics. The diquark-antidiquark approximation is not supported by a rigorous solution of the 4-body problem, but benefits of a stroke of luck, as the erroneous extra attraction introduced in the color $\bar{3}3$ channel is somewhat compensated by the neglect of the coupling to the color $6\bar{6}$ channel. The equality relating $QQ\bar{q}\bar{q}$, QQq, Qqq and $Q\bar{q}$ works surprisingly well as long as one is restricted to color $\bar{3}3$, but does not account for the attraction provided by color mixing. On the other hand, for asymmetric configurations ($QQ\bar{q}\bar{q}$), the Born-Oppenheimer method provides a very good approximation, and an interesting insight into the dynamics.

In short, $cc\bar{u}\bar{d}$ with $J^P = 1^+$ is at the edge of binding within current quark models. For this state, all contributions to the binding should be added, in particular the mixing of states with different internal spin and color structure, and in addition, the four-body problem should be solved with extreme accuracy. In comparison, achieving the binding of $bb\bar{u}\bar{d}$ looks easier. Still, with a typical quark model, the stability of the ground state below the threshold cannot be reached if spin-effects and color mixing are both neglected. The crucial role of spin effects explains why one does not expect too many states besides 1^+ [15].

These effects become even more acute in the five-body case, where color-mixing and a proper balance between chromoelectric and chromomagnetic terms are basic to get binding.

Acknowledgements This work has been partially funded by Ministerio de Economía, Industria y Competitividad and EU FEDER under Contract No. FPA2016-77177.

References

1. Aaij, R., et al.: Observation of the doubly charmed baryon Ξ_{cc}^{++}. Phys. Rev. Lett. **119**, 112001 (2017)
2. Ader, J. P., Richard, J.M., Taxil, P.: Do narrow heavy multiquark states exist? Phys. Rev. D **25**, 2370 (1982)
3. Richard, J.M., Valcarce, A., Vijande, J.: Few-body quark dynamics for doubly-heavy baryons and tetraquarks. Phys. Rev. C **97**, 035211 (2018)
4. Klempt, E., Sarantsev, A.V., Thoma, U.: Partial wave analysis. EPJ Web Conf. **134**, 02002 (2017)
5. Fredriksson, S., Jandel, M.: The Diquark Deuteron. Phys. Rev. Lett. **48**, 14 (1982)
6. Frederico, T., et al.: Structure of exotic three-body systems. Few-Body Syst. **38**, 57 (2006)
7. Eichten, E.J., Quigg, C.: Heavy-quark symmetry implies stable heavy tetraquark mesons $Q_i Q_j \bar{q}_k \bar{q}_l$. Phys. Rev. Lett. **119**, 202002 (2017)
8. Hall, R.L., Post, H.R.: Many-particle systems: IV. Short-range interactions. Proc. Phys. Soc. **90**, 381 (1967). For a recent review, see, Richard J.-M., Valcarce A., Vijande J., "Hall-post inequalities: review and application to molecules and tetraquarks", Annals of Physics (in press), arXiv:1910.08295 [nucl-th]
9. Fisher, M.E., Ruelle, D.: The stability of many-particle systems. J. Math. Phys. **7**, 260 (1966); Lavy-Leblond, J.-M.: Nonsaturation of gravitational forces. J. Math. Phys. **10**, 806 (1969)
10. Nussinov, S., Lampert, M.A.: QCD inequalities. Phys. Rept. **362**, 193 (2002)
11. Semay, C., Silvestre-Brac, B.: Diquonia and potential models. Z. Phys. C **61**, 271 (1994)
12. Bhaduri, R.K., Cohler, L.E., Nogami, Y.: A unified potential for mesons and baryons. Nuovo Cim. A **65**, 376 (1981)
13. Richard, J.M., Valcarce, A., Vijande, J.: Stable heavy pentaquarks in constituent models. Phys. Lett. B **774**, 710 (2018)
14. Hogaasen, H., Sorba, P.: The systematics of possibly narrow quark states with baryon number one. Nucl. Phys. B **145**, 119 (1978)
15. Vijande, J., Valcarce, A., Barnea, N.: Exotic meson-meson molecules and compact four-quark states. Phys. Rev. D **79**, 074010 (2009)

Chapter 121
Imaging the Partonic Structure of the Nucleon

Barbara Pasquini

Abstract We discuss the main properties of different types of parton distribution functions, which provide complementary multidimensional images of the partonic structure of the nucleon. These distributions are the generalized parton distributions, the transverse-momentum dependent parton distributions and the Wigner distributions. They have attracted increasing attention in the last years as they represent new tools to study how the composite structure of the proton results from the underlying quark-gluon dynamics.

121.1 Introduction

For a long time, the parton distributions have been explored in collinear processes such as fully inclusive deep inelastic scattering processes. This corresponds to a one-dimensional imaging of the proton as a set of partons moving collinearly with the direction of motion of the parent hadron, identified as the longitudinal direction. Recently, it has become clear that exclusive processes or semi-inclusive deep inelastic processes open the way to study the partonic structure of hadrons in more dimensions, using, respectively, generalized parton distributions (GPDs) and transverse-momentum dependent parton distributions (TMDs). GPDs can be used to map the partons in a mixed space of transverse spatial coordinates and longitudinal momentum. TMDs generalize the concept of collinear parton distributions by including also the dependence on the transverse momentum of the partons. GPDs and TMDs are truly independent functions, but can be seen as different projections of a larger class of distributions, known as generalized transverse-momentum dependent parton distributions (GTMDs). GTMDs have a direct connection with the Wigner functions,

B. Pasquini (✉)
Dipartimento di Fisica, Università degli Studi di Pavia, 27100 Pavia, Italy
e-mail: barbara.pasquini@unipv.it

Istituto Nazionale di Fisica Nucleare, Sezione di Pavia, 27100 Pavia, Italy

N. A. Orr et al. (eds.), *Recent Progress in Few-Body Physics*,
Springer Proceedings in Physics 238,
https://doi.org/10.1007/978-3-030-32357-8_121

which represent the quantum mechanical analogues of the classical phase-space distributions. In the following sections, we will discuss key features of these distributions and our present understanding from available experimental data.

121.2 Generalized Parton Distributions

Exclusive electroproduction of a real photon or a meson off a nucleon target at high momentum transfer is theoretically the cleanest way to access GPDs. Owing to the factorization property of QCD, these processes give access to different flavor combination of GPDs and are complementary to disentangle the various GPDs (see, e.g., [26, 46] for recent reviews of the GPD phenomenology). GPDs depend on three variables (considering the dependence on the factorization scale Q^2 to be known): x, that is the fraction of average longitudinal momentum of the active quark, referred to the average target momentum; the skewness variable ξ and t, that are, respectively, the fraction of longitudinal momentum and four-momentum transferred to the hadron target. However, one does not have direct access to this multidimensional structure, since the dependence on the three variables enters observables in nontrivial convolution with coefficient functions. Therefore, fits of GPDs require educated assumptions for the choice of the fitting functions such to incorporate the known theoretical constraints of GPDs, i.e. polinomiality, sum rules, and positivity (see, e.g., [9, 13, 21, 30, 31, 38] for a detailed preface to the GPD formalism and properties). Fits existing in the present literature are reasonably successful and it looks like there are no major problems with the theoretical framework.

Impact parameter distributions can be reconstructed by taking a Fourier transform of the GPDs in the variable t at $\xi = 0$. These distributions represent densities of partons with a given fraction of x as function of the position $\mathbf{b}_\perp$ from the centre of momentum of the nucleon in the plane perpendicular to the longitudinal direction [16]. A first attempt to obtain this information directly from photon electroproduction measurements was illustrated in [22, 23], using a model-dependent extrapolation to the point $\xi = 0$ that is not accessible experimentally. Recently, dispersion relation techniques have been used in [59] to constraint the GPDs at $\xi = 0$ from data. Both these analysis confirm that the width of the impact-parameter distribution for unpolarized quarks in unpolarized protons has a vey peaked transverse profile in the limit of $x \to 1$, as shown in Fig. 121.1. This behaviour is expected, since, in this limit, the active quark is always very close to the transverse center of momentum [16, 61].

The GPDs can also be viewed as the generating functions for the form factors of the twist-two operators governing the interaction mechanisms of hard processes in the deep inelastic regime. The most peculiar example are the form factors of the energy momentum tensor (EMT), which can be studied indirectly looking at moments of the GPDs. This is a unique and practical opportunity to access through electromagnetic processes the EMT form factors, which are canonically probed through gravity. We have four independent EMT form factors for the separate quark and gluon con-

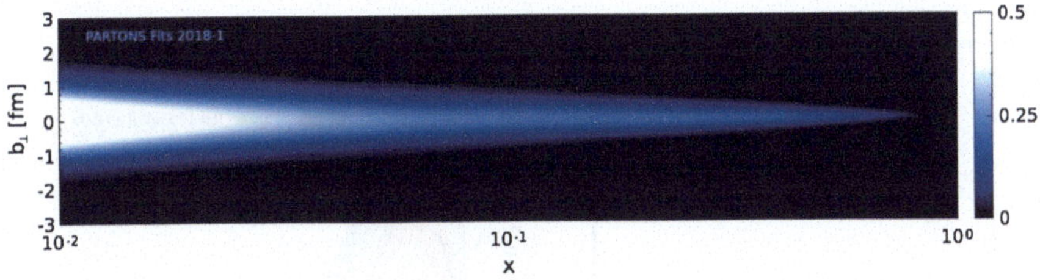

Fig. 121.1 The transverse position $b_\perp$ of up quarks in an unpolarized proton as a function of the longitudinal momentum fraction x. (Adapted from [59])

tribution, which reduce to three for the sum of quark and gluon terms due to the conservation of the EMT [66]. They are usually referred as $A(t)$, $J(t)$ and $D(t)$. At $t = 0$, the corresponding "charges" for quarks and gluons give, respectively, the fraction of nucleon momentum carried by the partons, the Ji's relation [41] for the quark and gluon contribution to the total angular momentum of the nucleon, and the D-term which is sometimes referred as the "last unknown global property" [66].

The physical content of the information encoded in the EMT form factors is revealed in the so-called Breit frame [65, 66] and has been recently discussed in other frames in [48, 56]. Working in the Breit frame, the D-term form factor can be related to the spatial distribution of shear forces $s(r)$ and pressure $p(r)$ as

$$p(r) = \frac{1}{3}\int \frac{d^3\boldsymbol{\Delta}}{2m(2\pi)^3} e^{-i\boldsymbol{\Delta}\cdot\mathbf{r}} P_0(\cos\theta)\,[tD(t)]\,, \tag{121.1}$$

$$s(r) = \frac{3}{4}\int \frac{d^3\boldsymbol{\Delta}}{2m(2\pi)^3} e^{-i\boldsymbol{\Delta}\cdot\mathbf{r}} P_2(\cos\theta)\,[tD(t)]\,, \tag{121.2}$$

where $t = \boldsymbol{\Delta}^2$. The relation for the shear forces holds also for quark and gluon separately, while it is defined only for the total system in the case of the pressure. Thanks to (121.1) and (121.2), the form factor $D(t)$ provides the key to introduce mechanical properties of the nucleon and reflects the internal dynamics of the system through the distribution of forces. Requiring that for the mechanical stability of the system the corresponding force must be directed outwards, one expects the local criterion $2s(r) + p(r) > 0$, which implies that the D-term for any stable system must be negative, $D < 0$, as confirmed in models [20, 29, 43], calculations from dispersion relations [63] and lattice QCD [33] for the nucleon. Another consequence of the EMT conservation is the von Laue condition, which shows how the internal forces balance inside a composed particle, i.e.,

$$\int_0^\infty p(r) r^2 dr = 0. \tag{121.3}$$

This relation implies that the pressure must have at least one node. In all model studies so far it was found that the pressure is positive in the inner region, and negative

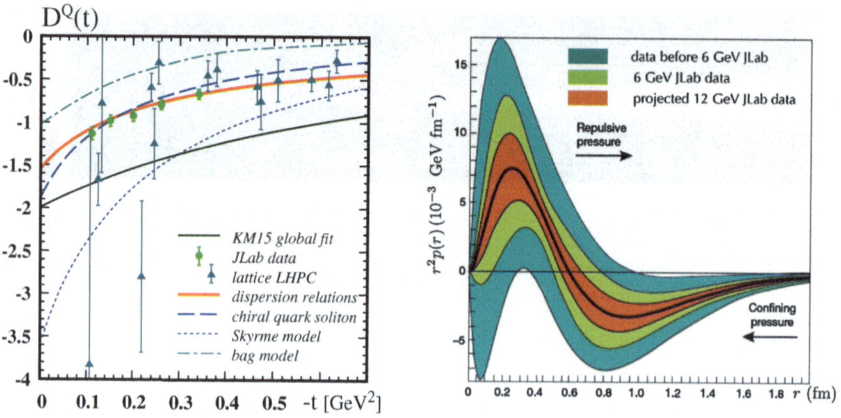

Fig. 121.2 Left panel (adapted from [66]): The $D^Q(t)$ form factor from the JLab analysis [19], in comparison with the KM15 fit [47] and calculations from dispersion relations [63] and lattice QCD [33], and results from the bag [43], chiral quark soliton [29] and Skyrme [20] model. Right panel (adapted from [19]): quark contribution to the pressure distributions $r^2 p(r)$ as function of the radial distance r from the centre of the proton

in the outer region, with the positive sign meaning repulsion towards outside and the negative sign meaning attraction directed towards inside. Recently, an analysis of the published JLab data measured at 6 GeV [28, 44] lead to experimental information on the quark contribution to the D-term form factor [19], as shown in Fig. 121.2 (squared in the left panel, referring to a scale of $\mu^2 = 1.5\,\text{GeV}^2$), in comparison with the KM15 fit [47] and calculations from dispersion relations [63] and lattice QCD [33] (all to the scale of 4 GeV^2) and scale independent results from the bag [43], chiral quark soliton [29] and Skyrme [20] models. The D-term parameters fitted to the JLab data, were used to plot the radial pressure distribution shown by the black solid curve in the right panel of Fig. 121.2, by assuming that the gluon and quark contributions are equal. The corresponding estimated uncertainties are displayed as the light-green shaded area. The blue area represents the uncertainties from all the data that were available before the 6 GeV experiment, and the red shaded area shows projected results from future JLab experiments at 12 GeV. Within the uncertainties of the analysis, the distribution satisfies the stability condition (121.3), with a zero crossing near $r = 0.6$ fm separating the inner region of a positive repulsive distribution from the outer region with a negative pressure responsible for the binding.

121.3 Transverse-Momentum Dependent Parton Distributions

In order to be sensitive to intrinsic transverse parton momenta it is necessary to measure transverse momenta of the produced hadrons in the final state, e.g., in processes like semi-inclusive lepton-nucleon deep inelastic scattering (SIDIS), hadron production in e^+e^- annihilation or Drell-Yan (DY) processes in hadron-hadron collisions.

In all these cases, factorization has been proved at leading twist enabling to access information on TMDs as well as on fragmentation functions (FFs), which describe the hadronization process of the hit quark decaying into the detected hadrons. The study of factorization and evolution properties of TMDs and FFs has rapidly evolved in the last years thanks to the contribution of several groups (see, e.g., the review in [67] and references therein). While DY and e^+e^- annihilation processes provide independent information on either TMDs or FFs, the SIDIS cross section involves the convolution of one TMD and one FF. Therefore, for a clean extraction of TMDs it is essential to have an independent and unbiased knowledge of FFs. At present, the challenge of the phenomenological extractions is to develop a consistent procedure to analyze all three kind of processes for a simultaneous fit of TMDs and FFs (for a recent review on the phenomenology of TMD see, e.g., [14] and for a review on TMD and FF measurements see, respectively, [4, 27]).

At leading twist, there are eight independent TMDs, three of them surviving when integrated over the transverse momentum and giving rise to the familiar parton density, helicity and transversity distributions. These three TMDs involve matrix elements which are all diagonal in the orbital angular momentum (OAM), but probe different transverse momentum and helicity correlations of the quarks inside the nucleon. Viceversa, the other TMDs vanish without the contribution of quark OAM and contain non-trivial information on the spin-spin and spin-orbit correlations of the partons and nucleon. The OAM content of the different quark TMDs becomes evident when using the representation of TMDs in terms of overlap of light-front wave functions that are eigenstates of the OAM operator [12, 62, 64]. However, we still do not have a direct and rigorous relation between TMDs and OAM, such, for example, in the case of the Ji's relation for GPDs. A relation between OAM and the so called pretzelosity TMD has been suggested [5, 70], but this is valid only within a particular class of models [49, 51] and it is not a rigorous prediction of QCD.

In the following, I will focus on the unpolarized TMD $f_1(x, k_\perp)$ and the Sivers function $f_{1T}^{\perp}(x, k_\perp)$, which have seen recently a considerable work from the phenomenological point of view. The unpolarized TMD is present in all measurements to construct the spin asymmetries which give access to the polarized TMDs, and therefore represents a fundamental building block. Several extractions of the unpolarized TMD exist in literature (see, e.g., [14] and references therein). Among them, the recent work by Scimemi and Vladimirov [69] reached the highest currently available level of perturbative accuracy, but it takes into account a limited number of experimental points, coming from DY and Z production. On the other side, the extraction in [7] applies the TMD evolution at one order lower, but it includes the largest set of data points available from experiments of different kinds, i.e. SIDIS, DY and Z production, providing an important test for the universality property of TMDs. The present extractions of f_1 give indications that the width of the TMDs at low scales, 1–2 GeV, is around 300−500 MeV and increases to more than 1 GeV at the Z mass, due to TMD evolution. Data indicate also that the width is probably increasing as x decreases and there is room for a flavor dependence, even though also a flavor-independent scenario is not ruled out. There are still important questions that need to be answered with more precision in future analyses [6], also thanks to the avail-

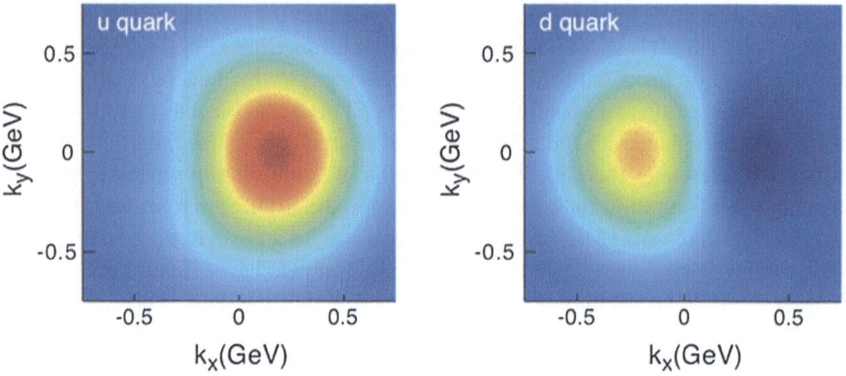

Fig. 121.3 Distributions in the transverse-momentum plane, at fixed $x = 0.1$, of unpolarized up and down quarks inside a proton transversely polarized in the $+y$ direction, as described by the Sivers function. The deep red (blue) indicates large negative (positive) values for the Sivers function. (Adapted from [1])

ability of new data. For example, we aim to reduce the uncertainties and eventual bias in the choice of the functional form adopted as input in the fit as well as in the prescription to deal with the non-perturbative part of the TMD evolution. The x dependence along with the flavor dependence are practically unknown. Calculations in different models [5, 18, 62, 68] and on lattice [60] indicate that the dependence of the transverse-momentum distributions on the quark polarisation and flavor may be significant at low scale, although it may be reduced under TMD evolution. A thorough study in these directions is driving the upgrades of several existing facilities (JLab, COMPASS and RHIC), and plays an important role also in the design and construction of new facilities worldwide (EIC, FAIR, NICA and JPARC).

The Sivers function describes the difference between the probability to find a quark with longitudinal momentum fraction x and transverse momentum $\mathbf{k}_\perp$ inside a hadron polarised transversely to its momentum direction, and the one where the polarisation points in the opposite direction. As such it encodes the correlation between the partonic intrinsic motion and the transverse spin of the nucleon, and it generates a dipole deformation in momentum space, as shown in Fig. 121.3 for the density distribution in the transverse momentum space at $x = 0.1$ of unpolarized up and down quarks in a nucleon transversely polarized in the $+y$ direction. The Sivers function has been phenomenologically extracted by several groups, mainly from analysing the azimuthal distributions of single hadrons in SIDIS (see, e.g., the review of the different works in [14]). The existence of the Sivers function as well as of the homolog Boer-Mulders function, describing transversely polarized quarks in an unpolarized nucleon, is ultimately related to initial-state and final-state collinear gluon interactions, which are summed into two different gauge links. This difference in the gauge link leads to a modified universality properties of the Sivers and Boer-Mulders functions, consisting in a sign flip in going from SIDIS to DY processes. First experimental evidences of this sign change of the Sivers function come from the transverse spin asymmetries measured at RHIC in transversely polarized proton-proton collision [2] and the pion-induced DY measurements at COMPASS [3].

121.4 Wigner Distributions

The concept of Wigner distributions in QCD for quarks and gluons was first explored in [8, 42]. In these works, the standard three-dimensional Fourier transform in the Breit frame was used, leading to six-dimensional Wigner distributions (three position and three momentum coordinates), which are valid only for infinitely massive target in order to get rid of relativistic corrections. A five-dimensional phase-space distribution free of relativistic corrections was introduced within the light-front formalism in [50]. In this case, the Wigner distributions appear to be the Fourier transform of GTMDs [24, 52, 57, 58], which reduce in particular limits to GPDs and TMDs. However, the Wigner distributions contain richer physics than TMDs and GPDs combined, as they carry information about the correlations between the quark momentum $(x, \mathbf{k}_\perp)$ and transverse space position $\mathbf{b}_\perp$, which cannot be accessed by separately studying TMDs or GPDs. Because of the uncertainty principle which prevents knowing simultaneously the position and momentum of a quantum-mechanical system, these phase-space distributions do not have a simple probabilistic interpretation. For this reason, they are often called quasi distributions, and only in the classical limit they become positive definite. Nonetheless, the physics of phase-space distributions is very rich and one can try to select certain situations where a semiclassical interpretation is still possible. Currently, the best hope to access the GTMDs is in the low-x regime for the gluon contribution [10, 15, 32, 35, 36, 40], while a single process has been identified sofar for the quark sector [11].

At leading twist, there are 32 quark phase-space distributions, half of them being associated to naive T-odd GTMDs and hence encoding initial and/or final-state interactions. A detailed study of these distributions has been presented in [50, 53]. All the contributions can be understood as encoding all the possible correlations between target and quark angular momenta, see Table 121.1. In particular, the function ρ_{LU} for unpolarized quarks in a longitudinally polarized nucleon has attracted consider-

Table 121.1 Correlations between target polarization $(S_L, \mathbf{S}_T)$, quark polarization $(S_L^q, \mathbf{S}_T^q)$ and quark OAM $(\ell_L^q, \boldsymbol{\ell}_T^q)$ encoded in the phase-space distributions ρ_X. U, L, T stand, respectively, for unpolarized, longitudinal and transverse polarizations

ρ_X	U	L	T_x	T_y
U	$\langle 1 \rangle$	$\langle S_L^q \ell_L^q \rangle$	$\langle S_x^q \ell_x^q \rangle$	$\langle S_y^q \ell_y^q \rangle$
L	$\langle S_L \ell_L^q \rangle$	$\langle S_L S_L^q \rangle$	$\langle S_L \ell_L^q S_x^q \ell_x^q \rangle$	$\langle S_L \ell_L^q S_y^q \ell_y^q \rangle$
T_x	$\langle S_x \ell_x^q \rangle$	$\langle S_x \ell_x^q S_L^q \ell_L^q \rangle$	$\langle S_x S_x^q \rangle$	$\langle S_x \ell_x^q S_y^q \ell_y^q \rangle$
T_y	$\langle S_y \ell_y^q \rangle$	$\langle S_y \ell_y^q S_L^q \ell_L^q \rangle$	$\langle S_y \ell_y^q S_x^q \ell_x^q \rangle$	$\langle S_y S_y^q \rangle$

able attention because it gives direct access to the quark OAM. One has just to take the phase-space average of the classical expression ($\mathbf{b}_\perp \times \mathbf{k}_\perp$) of the OAM as if the Wigner distributions were classical distributions [50, 55], i.e.

$$L_z^{q,g} = \int \mathrm{d}x\mathrm{d}^2\mathbf{b}_\perp\mathrm{d}^2\mathbf{k}_\perp^2\,(\mathbf{b}\times\mathbf{k}_\perp)\rho_{LU}^{q,g}(x,\mathbf{b}_\perp,\mathbf{k}_\perp;\mathcal{W}) = \int \mathrm{d}^2\mathbf{b}_\perp\mathbf{b}_\perp \times \langle\mathbf{k}_\perp\rangle^{q,g}, \quad (121.4)$$

where $\langle\mathbf{k}_\perp\rangle^{q,g}$ is the distribution in impact-parameter space of the quark/gluon mean transverse momentum, i.e.

$$\langle\mathbf{k}_\perp\rangle^{q,g} = \int \mathrm{d}x\mathrm{d}^2\mathbf{k}_\perp\mathbf{k}_\perp\rho_{LU}^{q,g}(x,\mathbf{b}_\perp,\mathbf{k}_\perp;\mathcal{W}). \quad (121.5)$$

In (121.4), the shape of the Wilson line $\mathcal{W}$ determines the type of OAM [17, 34]. For a staple-like Wilson line, like e.g. the one involved in the description of SIDIS scattering and DY processes, (121.4) leads to the canonical quark/gluon OAM, corresponding to the Jaffe-Manohar definition [37], irrespective of whether the staple is future or past-pointing. For a straight Wilson line, it leads to the kinetic quark OAM, corresponding to the Ji's definition [41], and to the Ji-Xiong-Yuan [39] definition of the gauge-invariant gluon OAM. Under a Fourier transform, the phase-space distribution is related to the GTMD $F_{1,4}$ [34, 45, 50], providing a relation between the OAM and GTMDs that gives access to the so far elusive canonical OAM in lattice QCD [25]. As an example, in Fig. 121.4 we show the distribution of the quark mean transverse momentum obtained within a light-front constituent quark model [54, 55]. We clearly see that in a longitudinally polarized nucleon, the quarks have on average nonzero OAM. The u quarks tend to orbit anti-clockwise inside the nucleon, corresponding to $L_z^u > 0$ since the proton is represented with its spin pointing out of the figure. For the d quarks, we see two regions. In the central region of the

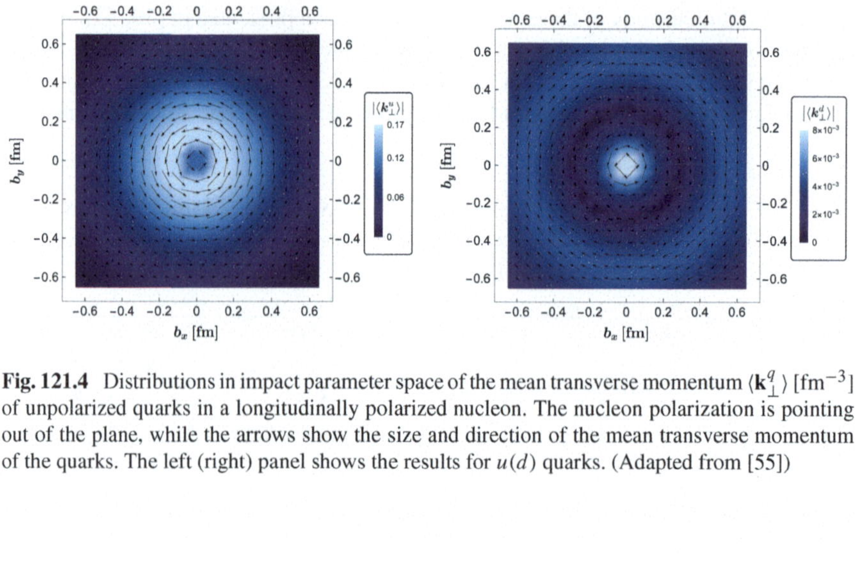

Fig. 121.4 Distributions in impact parameter space of the mean transverse momentum $\langle\mathbf{k}_\perp^q\rangle$ [fm^{-3}] of unpolarized quarks in a longitudinally polarized nucleon. The nucleon polarization is pointing out of the plane, while the arrows show the size and direction of the mean transverse momentum of the quarks. The left (right) panel shows the results for $u(d)$ quarks. (Adapted from [55])

nucleon, $b_\perp < 0.3$ fm, the d quarks tend to orbit anti-clockwise like the u quarks. In the peripheral region, $b_\perp > 0.3$ fm, the d quarks tend to orbit clockwise. The mean transverse momentum $\langle \mathbf{k}_\perp \rangle^q$ is always orthogonal to the impact-parameter vector $\mathbf{b}_\perp$. In principle, there can also be a radial component, representing the effect of initial and final state interactions. Such interactions being absent in the model, we do not see any radial contribution.

Acknowledgements This work is partially supported by the European Research Council (ERC) under the European Union's Horizon 2020 research and innovation programme (grant agreement No. 647981, 3DSPIN).

References

1. Accardi, A., et al.: Eur. Phys. J. A **52**(9), 268 (2016)
2. Adamczyk, L., et al.: Phys. Rev. Lett. **116**(13), 132301 (2016)
3. Aghasyan, M., et al.: Phys. Rev. Lett. **119**(11), 112002 (2017)
4. Avakian, H., Bressan, A., Contalbrigo, M.: Eur. Phys. J. **A52**(6), 150 (2016). [Erratum: Eur. Phys. J. **A52**(6), 165(2016)]
5. Avakian, H., Efremov, A.V., Schweitzer, P., Yuan, F.: Phys. Rev. D **81**, 074035 (2010)
6. Bacchetta, A.: Eur. Phys. J. A **52**(6), 163 (2016)
7. Bacchetta, A., Delcarro, F., Pisano, C., Radici, M., Signori, A.: JHEP **06**, 081 (2017)
8. Belitsky, A.V., Ji, X.d., Yuan, F.: Phys. Rev. **D69**, 074014 (2004)
9. Belitsky, A.V., Radyushkin, A.V.: Phys. Rept. **418**, 1 (2005)
10. Bhattacharya, S., Metz, A., Ojha, V.K., Tsai, J.Y., Zhou, J.: arXiv:1802.10550 [hep-ph]
11. Bhattacharya, S., Metz, A., Zhou, J.: Phys. Lett. B **771**, 396 (2017)
12. Boffi, S., Efremov, A.V., Pasquini, B., Schweitzer, P.: Phys. Rev. D **79**, 094012 (2009)
13. Boffi, S., Pasquini, B.: Riv. Nuovo Cim. **30**, 387 (2007)
14. Boglione, M., Prokudin, A.: Eur. Phys. J. A **52**(6), 154 (2016)
15. Boussarie, R., Hatta, Y., Xiao, B.W., Yuan, F.: Phys. Rev. D **98**(7), 074015 (2018)
16. Burkardt, M.: Int. J. Mod. Phys. A **18**, 173 (2003)
17. Burkardt, M.: Phys. Rev. D **88**(1), 014014 (2013)
18. Burkardt, M., Pasquini, B.: Eur. Phys. J. A **52**(6), 161 (2016)
19. Burkert, V.D., Elouadrhiri, L., Girod, F.X.: Nature **557**(7705), 396 (2018)
20. Cebulla, C., Goeke, K., Ossmann, J., Schweitzer, P.: Nucl. Phys. A **794**, 87 (2007)
21. Diehl, M.: Phys. Rept. **388**, 41 (2003)
22. Dupré, R., Guidal, M., Niccolai, S., Vanderhaeghen, M.: Eur. Phys. J. A **53**(8), 171 (2017)
23. Dupré, R., Guidal, M., Vanderhaeghen, M.: Phys. Rev. D **95**(1), 011501 (2017)
24. Echevarria, M.G., Idilbi, A., Kanazawa, K., Lorcé, C., Metz, A., Pasquini, B., Schlegel, M.: Phys. Lett. B **759**, 336 (2016)
25. Engelhardt, M.: Phys. Rev. D **95**(9), 094505 (2017)
26. Favart, L., Guidal, M., Horn, T., Kroll, P.: Eur. Phys. J. A **52**(6), 158 (2016)
27. Garzia, I., Giordano, F.: Eur. Phys. J. A **52**(6), 152 (2016)
28. Girod, F.X., et al.: Phys. Rev. Lett. **100**, 162002 (2008)
29. Goeke, K., Grabis, J., Ossmann, J., Polyakov, M.V., Schweitzer, P., Silva, A., Urbano, D.: Phys. Rev. D **75**, 094021 (2007)
30. Goeke, K., Polyakov, M.V., Vanderhaeghen, M.: Prog. Part. Nucl. Phys. **47**, 401 (2001)
31. Guidal, M., Moutarde, H., Vanderhaeghen, M.: Rept. Prog. Phys. **76**, 066202 (2013)
32. Hagiwara, Y., Hatta, Y., Pasechnik, R., Tasevsky, M., Teryaev, O.: Phys. Rev. D **96**(3), 034009 (2017)
33. Hagler, P., et al.: Phys. Rev. D **77**, 094502 (2008)

34. Hatta, Y.: Phys. Lett. B **708**, 186 (2012)
35. Hatta, Y., Nakagawa, Y., Yuan, F., Zhao, Y., Xiao, B.: Phys. Rev. D **95**(11), 114032 (2017)
36. Hatta, Y., Xiao, B.W., Yuan, F.: Phys. Rev. Lett. **116**(20), 202301 (2016)
37. Jaffe, R.L., Manohar, A.: Nucl. Phys. B **337**, 509 (1990)
38. Ji, X.: Ann. Rev. Nucl. Part. Sci. **54**, 413 (2004)
39. Ji, X., Xiong, X., Yuan, F.: Phys. Rev. D **88**(1), 014041 (2013)
40. Ji, X., Yuan, F., Zhao, Y.: Phys. Rev. Lett. **118**(19), 192004 (2017)
41. Ji, X.D.: Phys. Rev. Lett. **78**, 610 (1997)
42. Ji, X.d.: Phys. Rev. Lett. **91**, 062001 (2003)
43. Ji, X.D., Melnitchouk, W., Song, X.: Phys. Rev. D **56**, 5511 (1997)
44. Jo, H.S., et al.: Phys. Rev. Lett. **115**(21), 212003 (2015)
45. Kanazawa, K., Lorcé, C., Metz, A., Pasquini, B., Schlegel, M.: Phys. Rev. D **90**(1), 014028 (2014)
46. Kumericki, K., Liuti, S., Moutarde, H.: Eur. Phys. J. A **52**(6), 157 (2016)
47. Kumericki, K., Müller, D.: EPJ Web Conf. **112**, 01012 (2016)
48. Lorcé, C., Mantovani, L., Pasquini, B.: Phys. Lett. B **776**, 38 (2018)
49. Lorcé, C., Pasquini, B.: Phys. Rev. D **84**, 034039 (2011)
50. Lorcé, C., Pasquini, B.: Phys. Rev. D **84**, 014015 (2011)
51. Lorcé, C., Pasquini, B.: Phys. Lett. B **710**, 486 (2012)
52. Lorcé, C., Pasquini, B.: JHEP **09**, 138 (2013)
53. Lorcé, C., Pasquini, B.: Phys. Rev. D **93**(3), 034040 (2016)
54. Lorcé, C., Pasquini, B., Vanderhaeghen, M.: JHEP **05**, 041 (2011)
55. Lorcé, C., Pasquini, B., Xiong, X., Yuan, F.: Phys. Rev. D **85**, 114006 (2012)
56. Lorcé, C., Moutarde, H., Trawiski, A.P.: Eur. Phys. J. C **79**(1), 89 (2019)
57. Meissner, S., Metz, A., Schlegel, M.: JHEP **08**, 056 (2009)
58. Meissner, S., Metz, A., Schlegel, M., Goeke, K.: JHEP **08**, 038 (2008)
59. Moutarde, H., Sznajder, P., Wagner, J.: Eur. Phys. J. C **78**(11), 890 (2018)
60. Musch, B.U., Hagler, P., Negele, J.W., Schafer, A.: Phys. Rev. D **83**, 094507 (2011)
61. Pasquini, B., Boffi, S.: Phys. Lett. B **653**, 23 (2007)
62. Pasquini, B., Cazzaniga, S., Boffi, S.: Phys. Rev. D **78**, 034025 (2008)
63. Pasquini, B., Polyakov, M.V., Vanderhaeghen, M.: Phys. Lett. B **739**, 133 (2014)
64. Pasquini, B., Yuan, F.: Phys. Rev. D **81**, 114013 (2010)
65. Polyakov, M.V.: Phys. Lett. B **555**, 57 (2003)
66. Polyakov, M.V., Schweitzer, P.: Int. J. Mod. Phys. A **33**(26), 1830025 (2018)
67. Rogers, T.C.: Eur. Phys. J. A **52**(6), 153 (2016)
68. Schweitzer, P., Strikman, M., Weiss, C.: JHEP **01**, 163 (2013)
69. Scimemi, I., Vladimirov, A.: Eur. Phys. J. C **78**(2), 89 (2018)
70. She, J., Zhu, J., Ma, B.Q.: Phys. Rev. D **79**, 054008 (2009)

Chapter 122
Nucleon Parton Distribution Amplitude: A Scalar Diquark Picture

Cédric Mezrag, Jorge Segovia, Minghui Ding, Lei Chang and Craig D. Roberts

Abstract We report progress on the development of Perturbative Integral Representation Ansätze to compute the nucleon Faddeev Wave function, using an explicit quark-diquark picture. Our formalism is able to handle non-pointlike diquark and to mimic the strong, dynamical correlations observed when solving the Faddeev Equation. We then project the wave function in order to compute the leading-twist Parton Distribution Amplitude.

122.1 Introduction

One of the main objective of modern physics is to manage to describe hadron structure in terms of quarks and gluons, the fundamental degrees of freedom of Quantum Chromodynamics (QCD). A major effort is undertaken, both on the experimental

C. Mezrag (✉)
Istituto Nazionale di Fisica Nucleare, Sezione di Roma, P. le A. Moro 2, 00185 Rome, Italy
e-mail: cedric.mezrag@roma1.infn.it; cedric.mezrag@cea.fr

IRFU, CEA, Université Paris-Saclay, 91191 Gif-sur-Yvette, France

J. Segovia
Departamento de Sistemas Físicos, Químicos y Naturales, Universidad Pablo de Olavide, 41013 Seville, Spain

M. Ding
European Centre for Theoretical Studies in Nuclear Physics and Related Areas (ECT*) and Fondazione Bruno Kessler, Villa Tambosi, Strada delle Tabarelle 286, 38123 Villazzano (TN), Italy

L. Chang
School of Physics, Nankai University, Tianjin 300071, China

C. D. Roberts
School of Physics, Nanjing University, Nanjing, Jiangsu 210093, China

Institute for Nonperturbative Physics, Nanjing University, Nanjing, Jiangsu 210093, China

N. A. Orr et al. (eds.), *Recent Progress in Few-Body Physics*, Springer Proceedings in Physics 238,
https://doi.org/10.1007/978-3-030-32357-8_122

and theoretical sides. Insight of hadron structure can be gained experimentally thanks to the so-called factorisation theorem, which allows us to split cross sections or amplitudes of various inclusive or exclusive processes into a "hard part", expandable in perturbation theory, and "soft part" encoding the non-perturbative information on hadron structure. Within this framework, parton distribution amplitudes (PDAs) play an important role in our understanding of exclusive processes at large momentum transfer [1–4].

The PDAs can be computed from the Lightfront Wave Functions (LFWFs) by integrating out, to a certain scale ζ, the transverse momentum degrees of freedom of every parton belonging to the considered Fock state. If the energy involved in the scattering event is high enough, the so-called leading twist PDA, coming from one of the projection of the lowest Fock state is expected to dominate the description of exclusive processes. Contrary to the light-quark meson sector, where a significant effort has been performed in the recent years both using continuum techniques [5–7] and lattice-QCD [8–11], the nucleon leading-twist PDA remains to be computed at experimentally available energy, as only few modern attempts have partially tackled the issue [12–14]. However, its limit when the typical scale, ζ, goes to infinity is known and called the asymptotic PDA.

In this paper, we report the results (and the improvements made since [14]) of our approach to compute an insightful PDA for the nucleon. It relies on two main ingredients. First, we take advantage of the emerging picture coming from decades of work on solving the Faddeev equation [15–19] yielding a borromean picture of the nucleon [20]. This picture, which implies the constant breaking and formation of dynamical diquark correlations within baryons, has scored phenomenological successes (see e.g. [21–25]). Then, the Perturbation-Theory Integral Representation (PTIR) introduced by Nakanishi [26, 27] plays the second essential role in our approach. It allows us to write Bethe-Salpeter and Faddeev amplitudes in terms of a known momentum-dependent kernel and a momentum-independent Nakanishi weight function. Interestingly, PTIR is proved to be valid at all order of perturbation theory, and has already been used successfully in the meson sector to compute PDAs and beyond, Parton Distribution Functions (PDFs) and Generalised Parton Distributions (GPDs) (see [5, 28–38]).

122.2 Generalities

In order to define the nucleon PDA, we need to introduce two light-like vectors n and p, related to the nucleon momentum P such that $p^\mu = P^\mu - n^\mu P^2/(2P \cdot n)$. Introducing a three-quarks matrix element, we can define the nucleon leading twist PDA in Euclidean space as:

$$\langle 0|\, \epsilon^{ijk} \left(\tilde{u}^i_\uparrow(z_1 n) C^\dagger \not{n} u^j_\downarrow(z_2 n)\right) \not{n} d^k_\uparrow(z_3 n)\, |P, \lambda\rangle$$
$$= i\frac{1}{2}(p\cdot n) f_N \not{n} B^\uparrow \int \mathcal{D}x \varphi([x]) e^{-ip\cdot n \sum_i x_i z_i}, \qquad (122.1)$$

where $\tilde{u}$ is the quark field u transposed, B is the baryon Euclidean Dirac spinor, f_N is the normalisation constant of the PDA, or the value of the wave function at the origin, $[x] = (x_1, x_2, x_3)$, and

$$\mathcal{D}x = \mathrm{d}x_1 \mathrm{d}x_2 \mathrm{d}x_3 \delta\left(1 - \sum_{i=1}^{3} x_i\right) \quad \text{and} \quad q_{\uparrow\downarrow} = L^{\uparrow\downarrow} q = \frac{1 \pm \gamma_5}{2} q. \qquad (122.2)$$

The PDA can be readily computed once the nucleon Faddeev wave function χ is known. The latter is the non-amputated version of the Faddeev amplitude Ψ. As already emphasised in the introduction, modern studies of the Faddeev amplitude strongly suggest the existence of dynamical diquark correlations,[1] and we therefore describe the amplitude as:

$$\Psi = \psi_1 + \psi_2 + \psi_3, \qquad (122.3)$$

where the labels refer to the quark bystander. $\psi_{1,2}$ can be deduced from ψ_3 by cyclic permutations of the indices. One can decompose:

$$\psi_3 = \mathcal{N}_3^0 + \mathcal{N}_3^1 \qquad (122.4)$$
$$\mathcal{N}_3^0 = \left[\Gamma^0(k, K)\right]^{\alpha_1\alpha_2}_{\tau_1\tau_2} \Delta^0(K) \left[\mathcal{S}(\ell; P) B(P)\right]^{\alpha_3}_{\tau_3}, \qquad (122.5)$$
$$\mathcal{N}_3^1 = \left[\Gamma^{j;1}_\mu(k, K)\right]^{\alpha_1\alpha_2}_{\tau_1\tau_2} \Delta^1_{\mu\nu}(K) \left[\mathcal{A}^j_\nu(\ell; P) B(P)\right]^{\alpha_3}_{\tau_3} \qquad (122.6)$$

where $\mathcal{N}_3^0$ and $\mathcal{N}_3^1$ are respectively the scalar and axial-vector diquark contributions to ψ_3, $(\{p\}, \{\alpha\}, \{\sigma\})$ are respectively the momenta, Dirac and isospin labels of the nucleon Faddeev Amplitude. We have $P = p_1 + p_2 + p_3$, $K = p_1 + p_2$, $\ell = p_3 - 1/3P$, $k = \frac{p_1 - p_2}{2}$. The j sum runs over the isospin projections. The functions Γ are the diquark correlations amplitudes, Δ_0 and $\Delta^1_{\mu\nu}$ are the diquark propagators. Finally, the functions $\mathcal{S}$ and $\mathcal{A}^j_\nu$ are the quark-diquark Faddeev amplitudes. We left the colour structure implicit as it generates only an overall prefactor absorbed in the normalisation constant.

We do not tackle the computation of the PDA directly. Instead, we compute the general form of its Mellin moments, and deduce the PDA from this general expression. They are defined through:

$$\langle x_1^l x_2^m \rangle = \int \mathcal{D}x\, x_1^l\, x_2^m\, \varphi(x_i), \qquad (122.7)$$

[1]We stress that these diquarks are *not* the elementary ones introduced 50 years ago.

where, due to momentum conservation, only two indices are necessary to obtain their entire set. Putting (122.7) in perspective with (122.1), it is straightforward to realised that the Mellin moments are expectation values of local operators. Consequently, we do not need to handle Euclidean, complex valued "light-like" combinations of the Faddeev amplitude internal momenta. The moments are given through the projection of the Faddeev Amplitude with:

$$\frac{i}{2} f_B p \cdot n \not{n} B^{\uparrow} \langle x_1^l x_2^m \rangle = \int \mathcal{D}x x_1^l x_2^m \int \frac{d^4\ell}{(2\pi)^4} \times \int \frac{d^4k}{(2\pi)^4} \prod_{i=1}^{3} \delta\left(x_i - \frac{p_i \cdot n}{P \cdot n}\right) \chi(p_1, p_2, p_3) O_{21}^{\varphi} O_3^{\varphi} \tag{122.8}$$

where the projection operators are:

$$O_{21}^{\varphi} = L^{\downarrow} C^{\dagger} \not{n} L^{\uparrow}, \quad O_3^{\varphi} = \not{n} L^{\uparrow}. \tag{122.9}$$

122.3 Examples for a Scalar Diquark

We exemplify the computation of the PDA based on the scalar diquark component of the Faddeev Wave Function. The leading twist projection of (122.8) can be written in terms of $\gamma \cdot \mathcal{L}^0$ with:

$$\begin{aligned} \mathcal{L}^{0\nu} &= \frac{1}{4}\mathrm{Tr}\left[\gamma^{\nu} \not{n} L^{\uparrow} S(p_3) \tilde{\Gamma}^0 \tilde{S}(p_2) L^{\downarrow} C^{\dagger} \not{n} L^{\uparrow} S(p_1) \mathcal{S}\right] \Delta^0(K) \\ &= \frac{1}{4}\mathrm{Tr}\left[S(p_3) \tilde{\Gamma}^0 \tilde{S}(p_2) L^{\downarrow} C^{\dagger} \not{n} L^{\uparrow}\right] \mathrm{Tr}\left[\gamma^{\nu} \not{n} L^{\uparrow} S(p_1) \mathcal{S}\right] \Delta^0(K), \end{aligned} \tag{122.10}$$

where S is the quark propagator. The scalar diquark contribution to the leading twist PDA can therefore been split into two parts, one being the PDA of the scalar diquark itself, the second the projected quark-diquark amplitude. These two pieces are unsurprisingly convoluted through the momentum of the diquark.

A rigorous evaluation of the PDA certainly requires the computation of $\mathcal{L}^0$ using propagators and amplitudes computed within a QCD-connected, symmetry preserving framework. However, multiple example have highlighted the fact that, the use of algebraic models based on Perturbative Integral Representation (PTIR) reveals itself already insightful [7, 32–35, 39]. We will therefore proceed with developing such a model for the nucleon Faddeev Amplitude.

122.3.1 Scalar Diquark Model and Structure

We start by modelling and evaluating the structure of the scalar diquark itself, following (122.10). For this, we model the quark propagator and the diquark correlation vertex through:

$$S(p) = (-i \not{p} + M)\sigma_M(p^2), \quad \eta_0 \Gamma(k, K)C^\dagger = i\gamma_5 \int \mathrm{d}z(1 - z^2)\sigma_{\Lambda_\Gamma}(q_+) \tag{122.11}$$

where $\sigma_M(p^2) = (p^2 + M^2)^{-1}$ and $q_+ = q + (z/2)K$. We then used the methods developed in [5, 35] in order to compute the Mellin moments of the Scalar diquark PDA. We therefore introduce:

$$\mathcal{D}_0^m(K^2) = \frac{1}{P \cdot n} \int \frac{\mathrm{d}^4 k}{(2\pi)^4} \left(\frac{k \cdot n}{P \cdot n}\right)^m \mathrm{Tr}\left[S(p_3)\tilde{\Gamma}^0 \tilde{S}(p_2) L^\downarrow C^\dagger \not{n} L^\uparrow\right], \tag{122.12}$$

use a Feynman parametrisation to rearrange the denominator and introduce a carefully thought change of variable, in such a way that we can obtain:

$$\mathcal{D}_0^m(K^2) = \eta_0' \left(\frac{K \cdot n}{P \cdot n}\right)^{m+1} \int \frac{\mathrm{d}v \mathrm{d}u \mathrm{d}\beta (1 - z^2(\beta, u, v))\beta^m}{[(\beta(v(\beta - 2) + \beta) + u(v - \beta^2)] [\mathcal{M}^2 + K^2]}, \tag{122.13}$$

with $z(\beta, u, v) = -1 + 2(u - \beta)/(u - v)$, $0 \leq v \leq \beta \leq u \leq 1$ and:

$$\mathcal{M} = 4 \frac{(1 - u + v)M^2 + (u - v)\Lambda_\Gamma^2}{(\beta(v(\beta - 2) + \beta) + u(v - \beta^2))} (u - v) \,. \tag{122.14}$$

Equation (122.13) allows us to directly extract a pointwise expression for the DA, as the expansion of a continuous function in terms of Mellin moments is unique. In the case where $M = \Lambda_\Gamma$, it is actually even possible to obtain a simple, algebraic result for the structure of the diquark:

$$\eta_0''(y)\varphi_0(\hat{x}_2, \hat{x}_3) = 12y \left(1 - \frac{y}{\hat{x}_2 \hat{x}_3} \ln\left[1 + \frac{\hat{x}_2 \hat{x}_3}{y}\right]\right), \tag{122.15}$$

with $y = M^2/K^2$, and $\eta_0''(y)$ ensuring the PDA is normalised to 1 for every y. Interestingly, this result is compatible with previous continuum studies of the pion PDA [5, 40]; namely the limit when $y \gg 1$, i.e. $K^2 \ll \Lambda_\Gamma^2$ the PDA goes to the asymptotic one: $\varphi_0(\hat{x}_2\hat{x}_3) \to 6\hat{x}_2\hat{x}_3$. This can be understood as the correlation amplitudes momentum-space extent is far larger than the bound-state's mass-scale, yielding an effectively scale-free system. On the other hand, when $y \ll 1$, i.e. $K^2 \gg \Lambda_\Gamma^2$ the system tends to look like a pointlike particle and $\varphi(\hat{x}_2, \hat{x}_3) \to 1$. One should note that the end-point behaviour of our PDA is linear, independently of y.

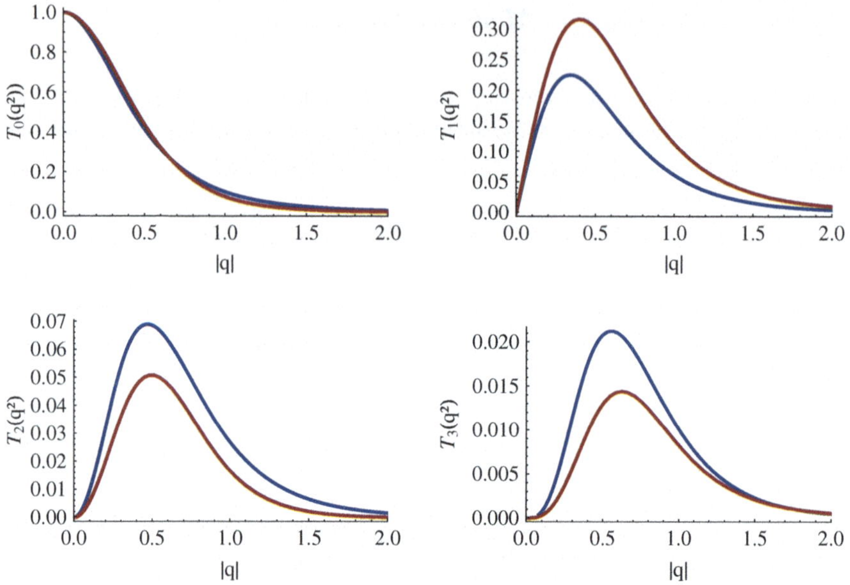

Fig. 122.1 Comparison of the 4 first Chebychev moments of the nucleon Faddeev Amplitude. Blue curves, results of the present model; red curves, realistic solution of the Faddeev Amplitude from [24]

122.3.2 Scalar Quark-Diquark Amplitude

In order to obtain the scalar diquark contribution to the nucleon PDA, it is necessary to perform the convolution of our scalar diquark structure with the quark-diquark Faddeev wave function. The latter is modelled using an effective diquark propagator and a Faddeev amplitude also computed thanks to PTIR:

$$\Delta_0(K^2) = \sigma_{M_0}(K^2), \quad \mathcal{S}(\ell, P) = i\eta \int \mathrm{d}z(1-z^2)\rho(z)\sigma^3_{\Lambda_0}(\ell_-^2), \qquad (122.16)$$

with $\ell_- = \ell - (1+3z)/6P$. In the purpose of determining a realistic weight, we expand the ρ function on the 3/2-Gegenbauer polynomial basis and we tune the coefficients in order to reproduce the first Chebychev moments coming from realistic numerical solution of the Faddeev equation [23, 24]. The results, shown in Fig. 122.1, are in fair but not perfect agreement with the realistic computations. From that point, we apply the same computing strategy than before in order to integrate the system over ℓ. The results are shown on Fig. 122.2. The nucleon PDA appears to be skewed with respect to the asymptotic one, emphasising that the bystander quark is more likely to carry the baryon lightfront momentum than the two quarks forming the strong diquark correlation.

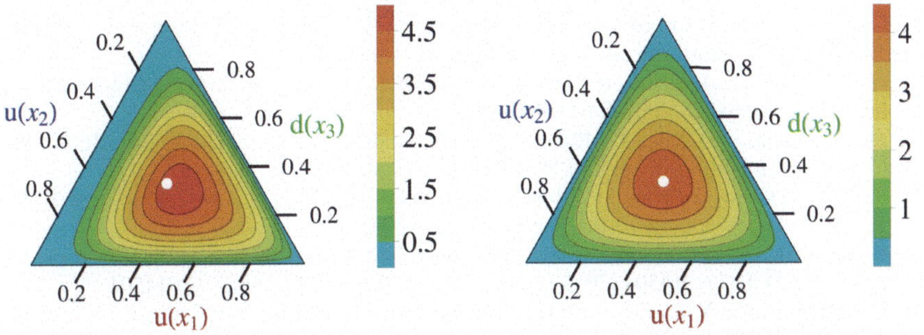

Fig. 122.2 Left: Result obtained for the computation of the Scalar diquark contribution using the parameters (in unit of the nucleon mass): $\{M, M_0, \Lambda_\Gamma, \Lambda_0\} = \{2/5, 9/10, 3/5, 6/5\}$; Right: asymptotic nucleon PDA

122.4 Conclusion

We present here a new step toward realistic models of the baryon PDA, by improving our previous work and presenting the preliminary results we obtain in the scalar case. The results are qualitatively identical to the previous ones, the scalar distribution is skewed, emphasising the bystander quark, but quantitatively different as they are now less skewed. We now look at extending our improved description of the Nakanishi weight functions to the Axial-vector diquark contributions, and beyond to the first radial excitation of the nucleon, namely the Roper resonance.

Acknowledgements We are grateful for insightful comments from F. Gao, V. Mokeev, H. Moutarde, S.-X. Qin, J. Rodríguez-Quintero, G. Salmè and S.-S. Xu. Work supported by: European Unions Horizon 2020 research and innovation programme under the Marie Sklodowska-Curie Grant Agreement No. 665919; Spanish MINECOs Juan de la Cierva-Incorporaci on programme, Grant Agreement No. IJCI-2016-30028; and MINECO Contract Nos. FPA2014-55613-P, FPA2017-86989-P and SEV-2016-0588; the Chinese Government's Thousand Talents Plan for Young Professionals; the Chinese Ministry of Education, under the International Distinguished Professor programme; and U.S. Department of Energy, Office of Science, Office of Nuclear Physics, under contract no. DE-AC02-06CH11357.

References

1. Collins, J.C., Soper, D.E., Sterman, G.F.: Adv. Ser. Direct. High Energy Phys. **5**, 1 (1989). https://doi.org/10.1142/9789814503266_0001
2. Efremov, A., Radyushkin, A.: Phys. Lett. B **94**, 245 (1980). https://doi.org/10.1016/0370-2693(80)90869-2
3. Lepage, G.P., Brodsky, S.J.: Phys. Rev. D **22**, 2157 (1980). https://doi.org/10.1103/PhysRevD.22.2157
4. Chernyak, V.L., Zhitnitsky, I.R.: Nucl. Phys. B **246**, 52 (1984). https://doi.org/10.1016/0550-3213(84)90114-7

5. Chang, L., Cloet, I., Cobos-Martinez, J., Roberts, C., Schmidt, S., et al.: Phys. Rev. Lett. **110**, 132001 (2013). https://doi.org/10.1103/PhysRevLett.110.132001
6. Gao, F., Chang, L., Liu, Y.X., Roberts, C.D., Schmidt, S.M.: Phys. Rev. D **90**(1), 014011 (2014). https://doi.org/10.1103/PhysRevD.90.014011
7. Shi, C., Chen, C., Chang, L., Roberts, C.D., Schmidt, S.M., Zong, H.S.: Phys. Rev. D **92**, 014035 (2015). https://doi.org/10.1103/PhysRevD.92.014035
8. Segovia, J., Chang, L., Clot, I.C., Roberts, C.D., Schmidt, S.M., Zong, H.S.: Phys. Lett. **B731**, 13 (2014). https://doi.org/10.1016/j.physletb.2014.02.006
9. Braun, V.M., Collins, S., Gckeler, M., Prez-Rubio, P., Schfer, A., Schiel, R.W., Sternbeck, A.: Phys. Rev. D **92**(1), 014504 (2015). https://doi.org/10.1103/PhysRevD.92.014504
10. Bali, G.S., Braun, V.M., Gckeler, M., Gruber, M., Hutzler, F., Korcyl, P., Lang, B., Schfer, A.: Phys. Lett. B **774**, 91 (2017). https://doi.org/10.1016/j.physletb.2017.08.077
11. Zhang, J.H., Chen, J.W., Ji, X., Jin, L., Lin, H.W.: Phys. Rev. D **95**(9), 094514 (2017). https://doi.org/10.1103/PhysRevD.95.094514
12. Braun, V.M., Collins, S., Glle, B., Gckeler, M., Schfer, A., Schiel, R.W., Sldner, W., Sternbeck, A., Wein, P.: Phys. Rev. D **89**, 094511 (2014). https://doi.org/10.1103/PhysRevD.89.094511
13. Bali, G.S., et al.: JHEP **02**, 070 (2016). https://doi.org/10.1007/JHEP02(2016)070
14. Mezrag, C., Segovia, J., Chang, L., Roberts, C.D.: Phys. Lett. B **783**, 263 (2018). https://doi.org/10.1016/j.physletb.2018.06.062
15. Cahill, R.T., Roberts, C.D., Praschifka, J.: Austral. J. Phys. **42**, 129 (1989). https://doi.org/10.1071/PH890129
16. Burden, C.J., Cahill, R.T., Praschifka, J.: Austral. J. Phys. **42**, 147 (1989). https://doi.org/10.1071/PH890147
17. Cahill, R.T.: Austral. J. Phys. **42**, 171 (1989). https://doi.org/10.1071/PH890171
18. Reinhardt, H.: Phys. Lett. B **244**, 316 (1990). https://doi.org/10.1016/0370-2693(90)90078-K
19. Efimov, G.V., Ivanov, M.A., Lyubovitskij, V.E.: Z. Phys. C **47**, 583 (1990). https://doi.org/10.1007/BF01552323
20. Segovia, J., Roberts, C.D., Schmidt, S.M.: Phys. Lett. B **750**, 100 (2015). https://doi.org/10.1016/j.physletb.2015.08.042
21. Segovia, J., Chen, C., Clot, I.C., Roberts, C.D., Schmidt, S.M., Wan, S.: Few Body Syst. **55**, 1 (2014). https://doi.org/10.1007/s00601-013-0734-x
22. Segovia, J., Chen, C., Roberts, C.D., Wan, S.: Phys. Rev. C **88**(3), 032201 (2013). https://doi.org/10.1103/PhysRevC.88.032201
23. Segovia, J., Cloet, I.C., Roberts, C.D., Schmidt, S.M.: Few Body Syst. **55**, 1185 (2014). https://doi.org/10.1007/s00601-014-0908-1
24. Segovia, J., El-Bennich, B., Rojas, E., Cloet, I.C., Roberts, C.D., Xu, S.S., Zong, H.S.: Phys. Rev. Lett. **115**(17), 171801 (2015). https://doi.org/10.1103/PhysRevLett.115.171801
25. Segovia, J., Roberts, C.D.: Phys. Rev. C **94**(4), 042201 (2016). https://doi.org/10.1103/PhysRevC.94.042201
26. Nakanishi, N.: Phys. Rev. **130**, 1230 (1963). https://doi.org/10.1103/PhysRev.130.1230
27. Nakanishi, N.: Prog. Theor. Phys. Suppl. **43**, 1 (1969). https://doi.org/10.1143/PTPS.43.1
28. Frederico, T., Salmè, G., Viviani, M.: Phys. Rev. D **85**, 036009 (2012). https://doi.org/10.1103/PhysRevD.85.036009
29. Carbonell, J., Frederico, T., Karmanov, V.A.: Phys. Lett. B **769**, 418 (2017). https://doi.org/10.1016/j.physletb.2017.04.016
30. de Paula, W., Frederico, T., Salmè, G., Viviani, M., Pimentel, R.: Eur. Phys. J. C **77**(11), 764 (2017). https://doi.org/10.1140/epjc/s10052-017-5351-2
31. Salmè, G., de Paula, W., Frederico, T., Viviani, M.: Few Body Syst. **58**(3), 118 (2017). https://doi.org/10.1007/s00601-017-1286-2
32. Chang, L., Mezrag, C., Moutarde, H., Roberts, C.D., Rodriguez-Quintero, J., et al.: Phys. Lett. B **737**, 23 (2014). https://doi.org/10.1016/j.physletb.2014.08.009
33. Mezrag, C., Chang, L., Moutarde, H., Roberts, C., Rodrguez-Quintero, J., et al.: Phys. Lett. B **741**, 190 (2014). https://doi.org/10.1016/j.physletb.2014.12.027

34. Mezrag, C., Moutarde, H., Rodríguez-Quintero, J., Sabatié, F. (2014). arXiv:1406.7425 [hep-ph]
35. Mezrag, C., Moutarde, H., Rodriguez-Quintero, J.: Few Body Syst. **57**(9), 729 (2016). https://doi.org/10.1007/s00601-016-1119-8
36. Chouika, N., Mezrag, C., Moutarde, H., Rodrguez-Quintero, J.: Eur. Phys. J. C **77**, 906 (2017). https://doi.org/10.1140/epjc/s10052-017-5465-6
37. Chouika, N., Mezrag, C., Moutarde, H., Rodrguez-Quintero, J.: Phys. Lett. B **780**, 287 (2018). https://doi.org/10.1016/j.physletb.2018.02.070
38. Shi, C., Mezrag, C., Zong, H.S.: Phys. Rev. **D98**(5), 054029 (2018). https://doi.org/10.1103/PhysRevD.98.054029
39. Chen, C., Chang, L., Roberts, C.D., Wan, S., Zong, H.S.: Phys. Rev. D **93**(7), 074021 (2016). https://doi.org/10.1103/PhysRevD.93.074021
40. Qin, S.X., Chen, C., Mezrag, C., Roberts, C.D.: Phys. Rev. C **97**(1), 015203 (2018). https://doi.org/10.1103/PhysRevC.97.015203

Chapter 123
Hadron Spectroscopy and Structure in the Dyson-Schwinger Approach

Gernot Eichmann

Abstract The Dyson-Schwinger/Bethe-Salpeter approach and its application to hadron spectroscopy and structure calculations are briefly summarized. The method allows one to calculate meson and baryon spectra, form factors, scattering amplitudes and other quantities from QCD's correlation functions. The spectrum of excited baryons is discussed along with advances towards understanding their transition form factors.

123.1 Motivation

The nucleon and its excitation spectrum have traditionally been at the heart of strong interaction studies. The proton is the only truly stable hadron and an ubiquitous ingredient in hadron structure experiments: from elastic and deep inelastic *ep* scattering to *pp* and $p\bar{p}$ reactions, $N\pi$ scattering, pion photo- and electroproduction, nucleon Compton scattering and more; also searches for physics beyond the Standard Model are typically performed on protons and nuclei. In addition, these experiments create meson and baryon resonances too and thereby allow us to extract their properties.

The data collected at Jefferson Lab, CERN and other facilities around the world show that hadrons are more complicated than the naive quark model suggests. For example, ostensibly simple baryon resonances may be mixtures of three-quark and multihadron states, and even our understanding of the nucleon is far from complete. Hadrons are complicated objects made of quarks and gluons, and it is the complexity of their interaction described by Quantum Chromodynamics (QCD) that encodes phenomena such as confinement and spontaneous chiral symmetry breaking. With quarks close to being massless, the dynamics of gluons plays a key role: for all we know today, the major fraction of the mass of the proton and other light hadrons is produced by glue.

G. Eichmann (✉)
CFTP, Instituto Superior Técnico, Universidade de Lisboa, 1049-001 Lisboa, Portugal
e-mail: gernot.eichmann@tecnico.ulisboa.pt

N. A. Orr et al. (eds.), *Recent Progress in Few-Body Physics*,
Springer Proceedings in Physics 238,
https://doi.org/10.1007/978-3-030-32357-8_123

A theoretical description of hadrons is tied to a thorough understanding of nonperturbative QCD, which requires combined efforts from lattice QCD, functional methods, amplitude analyses, phenomenological models and other theoretical approaches. Here we give a brief account of progress with functional methods, in particular the combination of Dyson-Schwinger equations (DSEs) and Bethe-Salpeter equations (BSEs), in calculating hadron spectroscopy and structure properties from QCD. For details we refer to the review [1].

123.2 From QCD's Correlation Functions to Hadrons

The basic starting point are QCD's correlation functions or n-point functions, some of which are collected in Fig. 123.1: the 'dressed' quark and gluon propagators, the quark-gluon vertex, three-gluon vertex and so on. In contrast to the few tree-level propagators and interactions that define the classical Lagrangian, the full information on the quantum field theory is encoded in its (infinitely many) correlation functions. These can be calculated from QCD's partition function, either directly using lattice QCD, or by deriving coupled equations for them, namely the DSEs which are the quantum equations of motion.

The DSEs are nonperturbative, self-consistent, exact equations which form an infinitely coupled system. At large momenta the coupling becomes small and they reproduce perturbation theory, as illustrated in Fig. 123.1 for the quark propagator. At small momenta, on the other hand, they encode effects which cannot happen at any order in perturbation theory, such as the nonperturbative generation of a quark mass scale (a 'constituent-quark mass') due to spontaneous chiral symmetry breaking, or the disappearance of the massless gluon pole and thus a gluon 'mass gap'; see [2–5] and references therein.

Because the correlation functions are gauge-dependent, one has to choose a gauge and the most convenient one in practice is Landau gauge. Landau-gauge lattice results

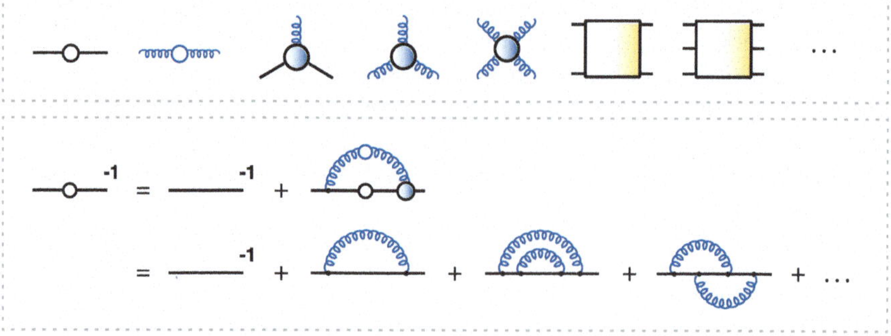

Fig. 123.1 *Top*: Some of QCD's elementary n-point functions. *Bottom*: Nonperturbative quark DSE together with its perturbative expansion

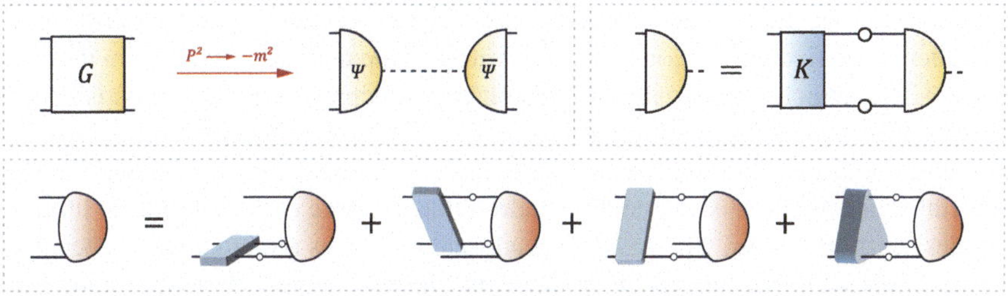

Fig. 123.2 *Top*: Four-point $q\bar{q}$ correlator, its pole behavior and the corresponding meson Bethe-Salpeter equation. *Bottom*: Covariant Faddeev equation for a baryon

for some of the elementary two- and three-point functions are available, see e.g. [6–10] and references therein. A different approach is the functional renormalization group (FRG), which leads to a similar tower of equations [11, 12]. To arrive at closed equations which can be systematically improved, the functional methods require truncations, either by neglecting higher n-point functions or higher-order terms in the quantum effective action. For the two- and three-point functions investigated so far [13–17] the three methods—lattice QCD, FRG equations and DSEs—provide qualitatively similar results, which suggests that a quantitative agreement is indeed within reach.

The properties of higher n-point functions become progressively more complicated since they depend on an increasing number of kinematic variables and Lorentz/Dirac tensors. General principles such as Lorentz invariance and gauge invariance pose constraints on them, which should be worked out before calculating their dynamical properties. So far, apart from the two- and three-point correlators, the structure of higher n-point functions is still largely unknown territory but progress is underway [12, 17–20].

Also the properties of hadrons are encoded in the correlation functions, namely in higher n-point functions which permit a spectral representation in terms of gauge-invariant hadron bound states and multiparticle states. For example, the $q\bar{q}$ four-point function contains all meson poles, and so does any other n-point function that creates meson quantum numbers ($q\bar{q}g$, $q\bar{q}q\bar{q}$ etc.). Likewise, the qqq six-point function contains all baryon poles. The residue at a pole defines the Bethe-Salpeter (BS) wave function Ψ shown in Fig. 123.2, which encodes the properties of the respective hadron.

The BS wave function can be calculated from its BSE, which at the same time determines the mass of the state. The BSEs for mesons and baryons are shown in Fig. 123.2. They are homogeneous integral equations in momentum space, which become eigenvalue equations for the respective $q\bar{q}$ and qqq kernels. The three-body BSE is also known as the covariant Faddeev equation; in this case the kernel is the sum of the irreducible two- and three-quark kernels.

In practice, the general strategy is to keep the full relativistic structure of the BS wave functions intact and make approximations only at the level of the kernels. A popular truncation is the so-called rainbow-ladder interaction, where the two-body kernels are approximated by a vector-vector interaction with a momentum-dependent effective interaction $\alpha(k^2)$. The quark propagator is solved from its DSE in Fig. 123.1, which dynamically breaks chiral symmetry: if the interaction is strong enough, it produces a running quark mass function which becomes large at small momenta and transforms the current quark into a dynamical 'constituent quark'. The resulting mass function is what makes hadron masses large even in the chiral limit of massless current quarks, whereas the pseudoscalar mesons remain massless because they are the Goldstone bosons of spontaneous chiral symmetry breaking. Rainbow-ladder has been frequently used in hadron spectrum and structure calculations [21, 22]. It provides a good overall description of heavy mesons but also light mesons in the pseudoscalar and vector channels [23–28], their decays, form factors, scattering amplitudes and, as it turns out, also a range of baryon properties including masses, elastic and transition form factors and more [1]. Moreover, the analogous four-quark equation for tetraquarks reproduces the mass pattern of the light scalar mesons [29].

To make a step forward to a quantitative understanding of hadron properties, one must improve the description of the underlying n-point functions. For example, in a recent beyond rainbow-ladder calculation all two- and three point functions in the system were solved, so that the BS kernel is no longer an input but a dynamical result [34]. This significantly improves the light meson spectrum: while the pseudoscalar and vector-meson ground states are less sensitive, the scalar and axialvector mesons (which are too strongly bound in rainbow-ladder) acquire large repulsive shifts, which puts them in the ballpark of experimental results, see Fig. 3.21 in [1].

123.3 Excited Baryons

Most baryon spectroscopy calculations so far have been performed in rainbow-ladder and attempts to go beyond rainbow-ladder are progressing [1, 30, 35]. In fact, already the rainbow-ladder kernels reproduce the masses of the nucleon, the $\Delta(1232)$ and the Roper resonance $N(1440)$, cf. Fig. 123.3. The remaining channels, however,

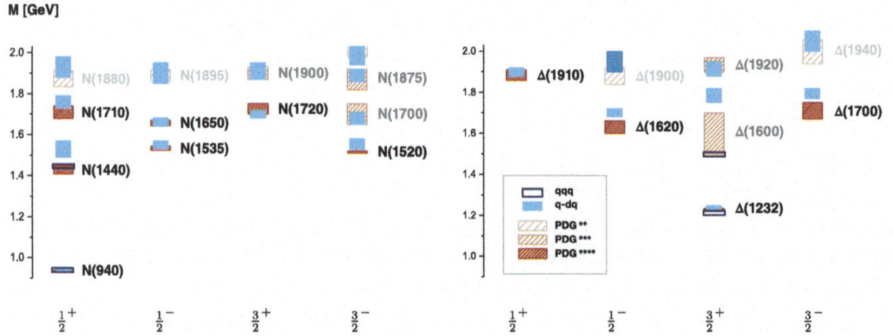

Fig. 123.3 Nucleon and Δ spectrum for $J^P = 1/2^{\pm}$ and $3/2^{\pm}$ states. The three-body (open boxes [30, 31]) and quark-diquark results (filled boxes [32]) are compared to the PDG values with their experimental uncertainties [33], see [32] for details

come out too low, which is similar to the situation in the meson sector beyond pseudoscalar and vector mesons. The three-body calculations are numerically expensive especially for excited states, because with the full structure of the relativistic Faddeev amplitudes the equations become eigenvalue problems for matrices of the size $10^6 \times 10^6 \dots 10^9 \times 10^9$ depending on the numerics.

At this point a quark-diquark interpretation can provide further insight. Instead of solving the three-body Faddeev equation directly, one can approximate it to a quark-diquark BSE where the baryon is treated as a quark-diquark system that interacts by quark exchange. To minimize the model input in such approaches [38–41], the diquark ingredients in [32] were calculated self-consistently from their DSEs and BSEs, so that also the quark-diquark system can be traced back to the same underlying quark-gluon interaction.

As a result, the three-quark and quark-diquark results essentially agree with each other; the $N(1/2^+)$ and $\Delta(3/2^+)$ channels are described well but the remaining ones show deficiencies. Those are due to the higher-lying diquarks: whereas N and Δ are dominated by scalar and axialvector diquarks, the remaining channels are also sensitive to pseudoscalar and vector diquarks [42] which are 'too strongly bound' just like their scalar and axialvector meson partners. By reducing the strength of these diquarks through one parameter, which is fixed by the $\rho - a_1$ splitting in the meson sector, one arrives at the spectrum in Fig. 123.3 [32]. One observes a 1:1 correspondence between the number of levels with PDG states and the masses are in almost quantitative agreement. An extension of the approach to the hyperon spectrum is underway [43].

The Faddeev amplitudes carry a rich tensorial structure which can be organized into eigenstates of spin and orbital angular momentum (OAM) in the rest frame [1, 38]. The resulting s, p, d and f-wave components ($L = 0, \dots, 3$) in Fig. 123.4 are different from the nonrelativistic quark model, where each baryon has definite L. Relativistically, these components can mix: the nucleon and $\Delta(1232)$ have p-wave components, the $N(1535)$ has s waves, and p waves are even dominant for the Roper. The subleading partial waves have consequences for form factors, for example in the $N\gamma \to \Delta$ transition [44], which demonstrates that relativity is important in the description of light baryons. Although these features should appear in other relativistic approaches as well, they are rarely discussed; exceptions are the covariant spectator theory [45] and light-front holographic QCD [46]. In the relativistic quark model of [47] there is no reference to non-traditional OAM contributions in baryons, but it is conceivable that they can appear after boosting the wave functions, e.g. in form factor calculations [48].

In addition to spectroscopy, more detailed properties of excited baryons can be extracted from their electromagnetic transition form factors $\gamma^* N \to N^*$. Figure 123.5 shows the generic properties of transition form factors. In the spacelike region ($Q^2 > 0$) they are accessible in meson photo- and electroproduction experiments at Jefferson Lab, ELSA and MAMI [49–51]. These have been the main experimental sources for the discovery of new nucleon resonances and the combination of precision data with multichannel partial-wave analyses has led to the addition of several new

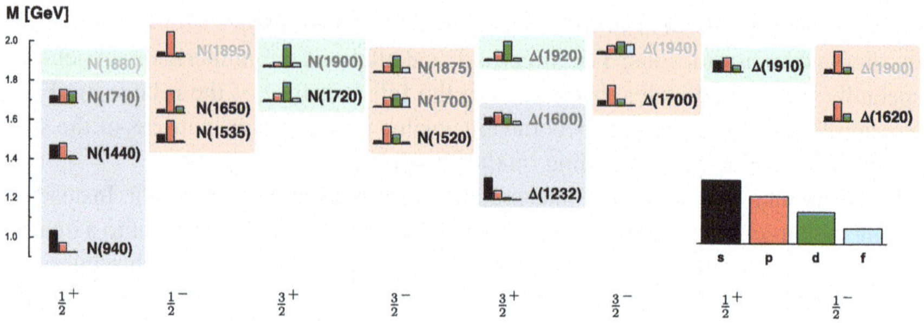

Fig. 123.4 Orbital angular-momentum contributions in the Faddeev amplitude of each baryon (in %); all bars sum up to 100% [36]. The rectangular backgrounds are the orbital angular-momentum assignments in the non-relativistic quark model [37]

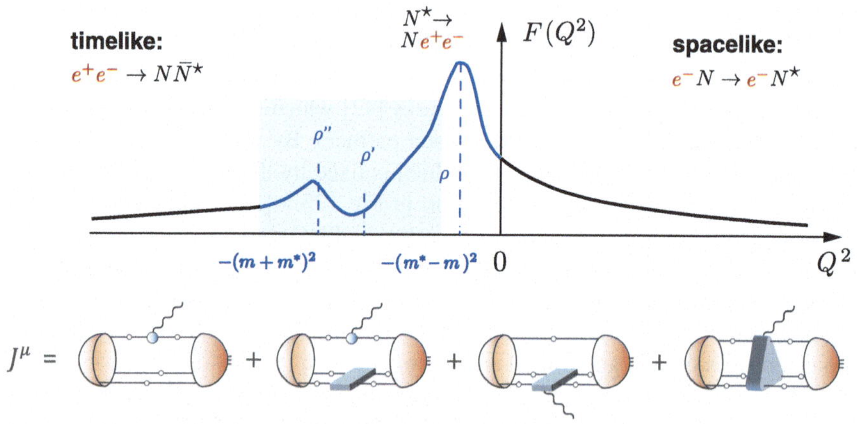

Fig. 123.5 *Top*: Generic behavior of an electromagnetic transition form factor in the spacelike and timelike regions. *Bottom*: Form factor diagram in the Faddeev approach

states to the PDG [33]. The timelike region ($Q^2 < 0$) above $N\bar{N}^*$ threshold is accessible in $e^+e^- \leftrightarrow N\bar{N}^*$ reactions, whereas the near timelike region $Q^2 > -(m^* - m)^2$ can be measured through the Dalitz decays $N^* \to Ne^+e^-$ at HADES/GSI [52].

In the Faddeev approach, whose form factor decomposition is shown in the bottom panel of Fig. 123.5, these different kinematical regions are tightly connected. Spacelike results for elastic and transition form factors are available from both three-body [53–55] and quark-diquark calculations [41, 44], see [1] for a review. With the exception of missing meson-cloud effects at low Q^2, they describe the existing data relatively well. The timelike structure, on the other hand, should be dominated by $\rho, \omega, \ldots$ bumps because a photon can fluctuate into vector mesons. The bumps originate in the quark-photon vertex, which is the contraction of the $q\bar{q}$ four-point function in Fig. 123.2 with γ^μ and thus inherits its vector-meson poles. The mesons in rainbow-ladder are stable hadrons and produce poles on the real axis instead of poles in the complex plane on higher Riemann sheets. Thus, to access the time-like properties of form factors one must implement the proper resonance mechanism beyond rainbow-ladder but also develop the necessary numerical tools in terms of residue calculus and contour deformations. Both strategies are currently being explored [56, 57].

123.4 Nucleon Resonances from Compton Scattering

In the last decade meson photo- and electroproduction experiments have become the main tools for gathering information on the baryon excitation spectrum. Understanding the structure and dynamics of scattering amplitudes is clearly important: dynamical reaction models and amplitude analyses based on general principles such as unitarity, analyticity and crossing symmetry are necessary to organize the experimental data and disentangle the various partial-wave contributions to extract resonance properties. Moreover, scattering amplitudes contain an abundance of information in addition to spectroscopy and thus their study also serves a purpose beyond the extraction of resonances.

An example is nucleon Compton scattering (CS), which encodes a broad range of applications from nucleon polarizabilities, structure functions, two-photon corrections to form factors and the proton radius puzzle to generalized parton distributions, see e.g. [58–61] for reviews. Our experimental knowledge of the CS amplitude is restricted to a few kinematic limits where direct measurements are possible, such as real and virtual CS and the forward limit.

Here we only focus on the nucleon resonances that appear in CS through the process $\gamma^* N \to N^* \to \gamma^* N$. The absence of spurious singularities in the CS amplitude poses constraints on the transition current matrix elements $\gamma^* N \to N^*$, which must satisfy electromagnetic gauge invariance and spin-3/2 gauge invariance [62]. The most general tensors according to these principles can be found in [63], along with a structure analysis of CS including all measured nucleon resonances with $J^P = 1/2^\pm$ and $J^P = 3/2^\pm$. As a consequence, the corresponding transition form factors are free of kinematic constraints, so their only singularities are physical poles and cuts such as in Fig. 123.5.

An example is the Roper resonance, the first excitation of the nucleon. For timelike Q^2 its transition form factors should resemble Fig. 123.5, so that their rise at low spacelike Q^2 is due to the first ρ-meson pole. However, to see such a behavior one must first cast the experimental data into the constraint-free form factors $F_1(Q^2)$ and $F_2(Q^2)$ corresponding to the tensors in [63]. These are shown in the top panels of Fig. 123.6 for spacelike Q^2. The bands are fits to the data and analogous fits for all nucleon resonances with $J^P = 1/2^\pm$ and $J^P = 3/2^\pm$ can be found in [63].

The existing reaction models usually do not extract the form factors but instead their helicity amplitudes (bottom panels), which are linear combinations of the form factors but not free of kinematic constraints. In particular, in the case of the Roper both $A_{1/2}$ and $S_{1/2}$ must vanish at the pseudothreshold $Q^2 = -(m^* - m)^2$, where m and m^* are the nucleon and Roper masses. For example, the MAID curve [67] in

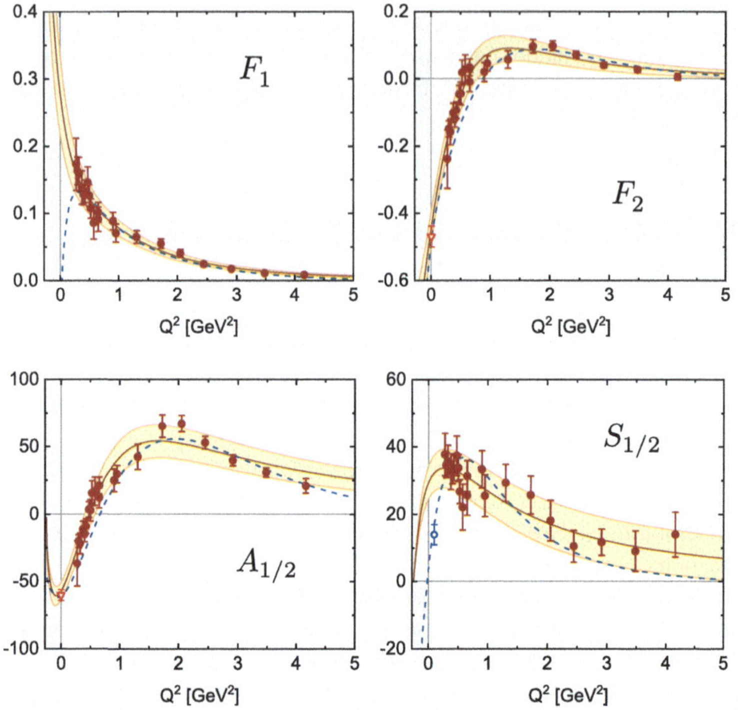

Fig. 123.6 $\gamma^* N \to N(1440)$ transition form factors and helicity amplitudes. The PDG [33] and CLAS data [50, 64, 65] are shown together with the A1/MAMI point for $S_{1/2}$ [66] and the MAID parametrization (dashed) [67]. The bands are fits [63]. The form factors are dimensionless and the helicity amplitudes carry units of 10^{-3} GeV$^{-1/2}$

Fig. 123.6 is compatible with the recent A1/MAMI measurement for $S_{1/2}$ at very low Q^2 [66] but does not reproduce the behavior at the pseudothreshold. This translates into a turnover of $F_1(Q^2)$ at very low Q^2, which would be difficult to explain from the analytic structure in the timelike region if the first ρ-meson pole has a positive residue. While such kinematic constraints must be implemented explicitly in the helicity amplitudes, they follow automatically when using constraint-free form factors.

The Q^2-dependence of several transition form factors is still poorly known, especially at low Q^2: even the best known resonances such as the $N(1440)$, $N(1520)$ and $N(1535)$ do not have any data below $Q^2 \lesssim 0.3$ GeV^2. In view of connecting the properties of form factors across the spacelike and timelike regions, this clearly motivates the need for future measurements at low Q^2.

Acknowledgements This work has been supported by the Portuguese Science Foundation FCT under FCT Investigator Grant IF/00898/2015.

References

1. Eichmann, G., Sanchis-Alepuz, H., Williams, R., Alkofer, R., Fischer, C.S.: Prog. Part. Nucl. Phys. **91**, 1–100 (2016)
2. Roberts, C.D., Williams, A.G.: Prog. Part. Nucl. Phys. **33**, 477–575 (1994)
3. Alkofer, R., von Smekal, L.: Phys. Rept. **353**, 281 (2001)
4. Fischer, C.S.: J. Phys. **G32**, R253–R291 (2006)
5. Binosi, D., Papavassiliou, J.: Phys. Rept. **479**, 1–152 (2009)
6. Cucchieri, A., Dudal, D., Mendes, T., Vandersickel, N.: Phys. Rev. **D85**, 094,513 (2012)
7. Aouane, R., Burger, F., Ilgenfritz, E.M., Mueller-Preussker, M., Sternbeck, A.: Phys. Rev. **D87**(11), 114,502 (2013)
8. Sternbeck, A., Balduf, P.H., Kizilersu, A., Oliveira, O., Silva, P.J., Skullerud, J.I., Williams, A.G.: PoS LATTICE **2016**, 349 (2017)
9. Boucaud, P., De Soto, F., Rodriguez-Quintero, J., Zafeiropoulos, S.: Phys. Rev. **D95**(11), 114,503 (2017)
10. Oliveira, O., Silva, P.J., Skullerud, J.I., Sternbeck, A. (2018)
11. Mitter, M., Pawlowski, J.M., Strodthoff, N.: Phys. Rev. **D91**, 054,035 (2015)
12. Cyrol, A.K., Mitter, M., Pawlowski, J.M., Strodthoff, N.: Phys. Rev. **D97**(5), 054,006 (2018)
13. Blum, A., Huber, M.Q., Mitter, M., von Smekal, L.: Phys. Rev. **D89**, 061,703 (2014)
14. Eichmann, G., Williams, R., Alkofer, R., Vujinovic, M.: Phys. Rev. **D89**(10), 105,014 (2014)
15. Williams, R.: Eur. Phys. J. A **51**(5), 57 (2015)
16. Aguilar, A.C., Binosi, D., Papavassiliou, J.: Front. Phys. China **11**(2), 111,203 (2016)
17. Huber, M.Q. (2018). arXiv:1808.05227 [hep-ph]
18. Binosi, D., Ibanez, D., Papavassiliou, J.: JHEP **09**, 059 (2014)
19. Eichmann, G., Fischer, C.S., Heupel, W.: Phys. Rev. **D92**(5), 056,006 (2015)
20. Ahmadiniaz, N., Schubert, C.: EPJ Web Conf. **182**, 02,114 (2018)
21. Maris, P., Roberts, C.D.: Phys. Rev. C **56**, 3369–3383 (1997)
22. Maris, P., Tandy, P.C.: Phys. Rev. **C60**, 055,214 (1999)
23. Maris, P., Tandy, P.C.: Nucl. Phys. Proc. Suppl. **161**, 136–152 (2006). [,136(2005)]
24. Krassnigg, A.: Phys. Rev. **D80**, 114,010 (2009)
25. Fischer, C.S., Kubrak, S., Williams, R.: Eur. Phys. J. A **50**, 126 (2014)
26. Rojas, E., El-Bennich, B., de Melo, J.P.B.C.: Phys. Rev. **D90**, 074,025 (2014)
27. Hilger, T., Popovici, C., Gomez-Rocha, M., Krassnigg, A.: Phys. Rev. **D91**(3), 034,013 (2015)
28. Fischer, C.S., Kubrak, S., Williams, R.: Eur. Phys. J. A **51**, 10 (2015)
29. Eichmann, G., Fischer, C.S., Heupel, W.: Phys. Lett. B **753**, 282–287 (2016)
30. Eichmann, G., Alkofer, R., Krassnigg, A., Nicmorus, D.: Phys. Rev. Lett. **104**, 201,601 (2010)
31. Sanchis-Alepuz, H., Eichmann, G., Villalba-Chavez, S., Alkofer, R.: Phys. Rev. **D84**, 096,003 (2011)
32. Eichmann, G., Fischer, C.S., Sanchis-Alepuz, H.: Phys. Rev. **D94**(9), 094,033 (2016)
33. Patrignani, C., et al.: Chin. Phys. **C40**(10), 100,001 (2016)
34. Williams, R., Fischer, C.S., Heupel, W.: Phys. Rev. **D93**(3), 034,026 (2016)
35. Sanchis-Alepuz, H., Williams, R.: Comput. Phys. Commun. **232**, 1–21 (2018)
36. Eichmann, G.: Few Body Syst. **58**(2), 81 (2017)
37. Crede, V., Roberts, W.: Rept. Prog. Phys. **76**, 076,301 (2013)
38. Oettel, M., Hellstern, G., Alkofer, R., Reinhardt, H.: Phys. Rev. C **58**, 2459–2477 (1998)
39. Oettel, M., Alkofer, R., von Smekal, L.: Eur. Phys. J. A **8**, 553–566 (2000)
40. Roberts, H.L.L., Chang, L., Cloet, I.C., Roberts, C.D.: Few Body Syst. **51**, 1–25 (2011)
41. Segovia, J., El-Bennich, B., Rojas, E., Cloet, I.C., Roberts, C.D., Xu, S.S., Zong, H.S.: Phys. Rev. Lett. **115**(17), 171,801 (2015)
42. Eichmann, G.: Few Body Syst. **57**(10), 965–973 (2016)
43. Eichmann, G., Fischer, C.S.: Few Body Syst. **60**(1), 2 (2019)
44. Eichmann, G., Nicmorus, D.: Phys. Rev. **D85**, 093,004 (2012)
45. Leitao, S., Stadler, A., Pena, M.T., Biernat, E.P.: Phys. Rev. **D96**(7), 074,007 (2017)

46. Brodsky, S.J., de Teramond, G.F., Dosch, H.G., Erlich, J.: Phys. Rept. **584**, 1–105 (2015)
47. Melde, T., Plessas, W., Sengl, B.: Phys. Rev. **D77**, 114,002 (2008)
48. Plessas, W.: Private communication
49. Sarantsev, A.V., et al.: Phys. Lett. B **659**, 94–100 (2008)
50. Aznauryan, I.G., et al.: Phys. Rev. **C80**, 055,203 (2009)
51. Aznauryan, I.G., et al.: Int. J. Mod. Phys. **E22**, 1330,015 (2013)
52. Ramstein, B.: Few Body Syst. **59**(6), 143 (2018)
53. Eichmann, G.: Phys. Rev. **D84**, 014,014 (2011)
54. Sanchis-Alepuz, H., Fischer, C.S.: Eur. Phys. J. A **52**(2), 34 (2016)
55. Sanchis-Alepuz, H., Alkofer, R., Fischer, C.S.: Eur. Phys. J. A **54**(3), 41 (2018)
56. Weil, E., Eichmann, G., Fischer, C.S., Williams, R.: Phys. Rev. **D96**(1), 014,021 (2017)
57. Williams, R. (2018). arXiv:1804.11161 [hep-ph]. https://doi.org/10.1016/j.physletb.2019.134943
58. Drechsel, D., Pasquini, B., Vanderhaeghen, M.: Phys. Rept. **378**, 99–205 (2003)
59. Griesshammer, H.W., McGovern, J.A., Phillips, D.R., Feldman, G.: Prog. Part. Nucl. Phys. **67**, 841–897 (2012)
60. Guidal, M., Moutarde, H., Vanderhaeghen, M.: Rept. Prog. Phys. **76**, 066,202 (2013)
61. Hagelstein, F., Miskimen, R., Pascalutsa, V.: Prog. Part. Nucl. Phys. **88**, 29–97 (2016)
62. Pascalutsa, V., Timmermans, R.: Phys. Rev. **C60**, 042,201 (1999)
63. Eichmann, G., Ramalho, G.: Phys. Rev. **D98**(9), 093,007 (2018)
64. Mokeev, V.I., et al.: Phys. Rev. **C86**, 035,203 (2012)
65. Mokeev, V.I., et al.: Phys. Rev. **C93**(2), 025,206 (2016)
66. Stajner, S., et al.: Phys. Rev. Lett. **119**(2), 022,001 (2017)
67. Tiator, L., Drechsel, D., Kamalov, S.S., Vanderhaeghen, M.: Eur. Phys. J. ST **198**, 141–170 (2011)

Part V
Interdisciplinary Aspects of Few-Body Physics and Techniques

Chapter 124
A Possibility of Nuclear Reaction Near the Three-Body Break-Up Threshold

Shinsho Oryu, Takashi Watanabe, Yasuhisa Hiratsuka, Masayuki Takeda and Yoshio Togawa

Abstract We investigate a possibility of nuclear reaction near the three-body break-up threshold (3BT) by a viewpoint of the three-body Hamiltonian. We study a virtual molecule: CsD_2 which is well developed in a cuboctahedron CsD_2Pd_{12}-cluster, where CsD_2 is surrounded by the Pd_{12}-cage. We found that the energy levels, which are obtained without the Coulomb repulsive potential, could be boosted up by the repulsive Coulomb correction, however the Coulomb excited resonant-level does not decay by the barrier of Pd_{12}-cage. The *boost up* effect resolves a difficulty of the penetration problem. If the molecular level $E_{\rm mol}$ and the boosted up nuclear levels $E_{\rm nucl}(<E_{\rm mol})$ can not close each other at the top of the Coulomb barrier or the 3BT, then their wave functions do not overlap, and the electromagnetic multipole (EMM) moment becomes zero. Thereby, the molecular level is stable without such an EMM-transition. However, if both of the molecular state and the excited nuclear state approach each other by a *long range* hadron potential, then the molecular state transfers to the nuclear La states by the EMM-transition: Cs(2d,γ)La with proper selection rules, step by step.

Recently, Watari investigated a cuboctahedron: Pd_{13}/D_n and also CsD_2Pd_{12} by using the well-known electron-density functional method, and found that a quasi-molecule CsD_2 is well developed inside of the Pd-cluster [1].

In this paper, we would like to investigate the d-Cs-d three-body problem for the reaction Cs(2d,γ)La which is surrounded by the Pd-cage from a viewpoint of the three-body Hamiltonian. The total Hamiltonian is given by

$$\begin{aligned} H &= (K_{3-ion} + V^{had}_{3-ion} + V^{C}_{3-ion}) + (K_{Z-el} + V^{M}_{ion-el} + V^{M}_{el-el}) \\ &= H_{had} + h^{M}_{c} \end{aligned} \tag{124.1}$$

S. Oryu (✉) · T. Watanabe · M. Takeda · Y. Togawa
Tokyo University of Science, Noda, Chiba 278-8510, Japan
e-mail: oryu@rs.noda.tus.ac.jp

Y. Hiratsuka
Preparation School of HLF Ltd., Kiryu, Gunma 376-0021, Japan

N. A. Orr et al. (eds.), *Recent Progress in Few-Body Physics*,
Springer Proceedings in Physics 238,
https://doi.org/10.1007/978-3-030-32357-8_124

with the Hamiltonian for three-hadron: H_{had}, and electrons: h_c^M. They are defined by

$$H_{had} = K_{3-ion} + V_{3-ion}^{had} + V_{3-ion}^{C}, \tag{124.2}$$

$$\approx K_{3-ion} + V_{3-ion}^{had} + V_{3-ion}^{M}, \tag{124.3}$$

$$h_c^M = K_{Z-el} + V_{ion-el}^{M} + V_{el-el}^{M}, \tag{124.4}$$

and also we define the molecular Hamiltonian,

$$H_M = \left(K_{3-ion} + V_{3-ion}^{C}\right) + h_c^M, \tag{124.5}$$

where in (124.2), K_{3-ion}, V_{3-ion}^{had} and V_{3-ion}^{C} are the kinetic energy terms for the three hadrons, the corresponding three hadron potentials, and the three Coulomb repulsive potentials. In (124.2), (124.3), V_{3-ion}^{M} differs from V_{3-ion}^{C} where V_{3-ion}^{M} is modified by electrons which break in the ion-chain. However, the difference is negligible, because the ratio of inner electrons for the large Z is very small, and also the effect for V_{3-ion}^{M} is calculated over the electron region: $0 \le r < \infty$. In the electron Hamiltonian h_c^M for the Z-electrons system, K_{Z-el}, V_{ion-el}^{M}, and V_{el-el}^{M} are the kinetic energy for Z-electrons, and the attractive ion-electrons potentials, and the repulsive electron-electron potentials, respectively.

The eigen-equation for (124.1): $H|\Psi> = E|\Psi>$ should be solved in the region $0 \le r < \infty$, however, it is technically rather hard. Therefore, we adopt the second best way to find a solution where the Hamiltonian is divided for two parts H_{had} with the eigen function $|\psi>$, and h_c^M with the electron wave function $|\psi_M>$. Therefore, the total wave function $|\Psi>$ could be given by $|\Psi> = |\psi \otimes \psi_M>$.

On the other hand, the molecular Hamiltonian satisfies

$$\begin{aligned} <\Psi|H_M|\Psi> &= <\Psi|K_{3-ion}|\Psi> + <\Psi|V_{3-ion}^{M}|\Psi> \\ &+ <\Psi|h_c^M|\Psi> \equiv \varepsilon_M, \end{aligned} \tag{124.6}$$

where the first two terms of the right hand side (r.h.s.) belong to the hadron part, and the last term belongs to the electron system.

Therefore, the expectation value for the electron Hamiltonian: h_c^M for the three fixed ions becomes

$$<\Psi|h_c^M|\Psi> = \varepsilon_M - <\Psi|V_{3-ion}^{M}|\Psi> = \varepsilon_M - \varepsilon_c^M, \tag{124.7}$$

with the static approximation: $<\Psi|K_{3-ion}|\Psi> = 0$, and also

$$<\Psi|V_{3-ion}^{C}|\Psi> = <\Psi|V_{\mathrm{Csd}_1}^{C} + V_{\mathrm{Csd}_2}^{C} + V_{\mathrm{d}_1\mathrm{d}_2}^{C}|\Psi> \equiv \varepsilon_c, \tag{124.8}$$

$$<\Psi|V_{3-ion}^{M}|\Psi> = <\Psi|V_{\mathrm{Csd}_1}^{M} + V_{\mathrm{Csd}_2}^{M} + V_{\mathrm{d}_1\mathrm{d}_2}^{M}|\Psi> \equiv \varepsilon_c^M, \tag{124.9}$$

for the d_2-Cs-d_1 three-ion system. The difference between ε_c and ε_c^M is very small as mentioned above. In (124.7), the molecular energy: ε_M is obtained from $< \Psi|h_c^M|\Psi >$ which is *boosted up* by the value ε_c^M,

$$\varepsilon_M =< \Psi|h_c^M|\Psi > +\varepsilon_c^M. \tag{124.10}$$

For the eigenvalue of the hadron Hamiltonian, let us adopt a *highest nuclear* (HN)-level: $E_{n_{max}} = \eta(< 0)$, we obtain by using (124.2), (124.3), and also (124.8), (124.9),

$$\begin{aligned} < \Psi|H_{had}|\Psi > &= < \Psi|(K_{had} + V_{3-ion}^{had})|\Psi > + < \Psi|V_{3-ion}^C|\Psi > \\ &= \eta + < \Psi|V_{3-ion}^C|\Psi >= \eta + \varepsilon_c \qquad (124.11) \\ &\approx \eta + < \Psi|V_{3-ion}^M|\Psi >= \eta + \varepsilon_c^M. \qquad (124.12) \end{aligned}$$

In (124.11) and (124.12), the HN-level is *boosted up* to near the 3BT or the top of the Coulomb barrier. Therefore, if $\eta \approx< \Psi|h_c^M|\Psi >$ is satisfied, the equilibrium between the hadron state and the molecular state could be realized. By the boost-up effect, the values $\eta + \varepsilon_c^M$ and $< \Psi|h_c^M|\Psi > +\varepsilon_c^M$ are close to the top of the Coulomb barrier or the 3BT. Therefore the molecular wave function of the CsD_2 state could easily penetrate the V_{3-ion}^C Coulomb barrier.

In conclusion, (124.1), (124.10), and (124.11), (124.12) give the total energy by,

$$\begin{aligned} < \Psi|H|\Psi > &= < \Psi|H_{had}|\Psi > + < \Psi|h_c^M|\Psi > \\ &= (\eta + \varepsilon_c) + < \Psi|h_c^M|\Psi >= (\eta + \varepsilon_c^M) + < \Psi|h_c^M|\Psi > \qquad (124.13) \\ &= (\eta + \varepsilon_c) + (\varepsilon_M - \varepsilon_c) = (\eta + \varepsilon_c^M) + (\varepsilon_M - \varepsilon_c^M) \\ &= \eta + \varepsilon_M = E, \qquad (124.14) \end{aligned}$$

where $< \Psi|H_{had}|\Psi >\approx< \psi|H_{had}|\psi >\equiv E_{n_{max}} = \eta$ is calculated only in the hadron system, while $< \Psi|h_c^M|\Psi >\approx< \psi_M|h_c^M|\psi_M >$ can be obtained in the electron system corresponding to the three-ion coordination. This means that the numerical calculations should be done separately in the nuclear region, and in the molecular region. Thus, we can obtain both results accurately.

If the $|\eta|$ is too large in the hadron system, the HN-level of the d-Cs-d nuclear binding energy will be smaller than the top of the Coulomb barrier, where the molecular wave function and the hadron wave function do not overlap, therefore the EMM-moment becomes very small, and the transfer cannot occur. If $\eta + \varepsilon_c^M$ comes very close to the molecular binding energy: ε_M by a long range hadron potential, such a state could transfer to the HN-level [2–7], and to the deeper nuclear levels step by step by the EM-transitions, until the La (see Fig. 124.1).

There are some experimental reports about the low temperature nuclear synthesis which are mainly observed in the Pd-crystal [8–10]. If these experiments are trustworthy, our theory may be helpful to investigate them. On the contrary, the experimental

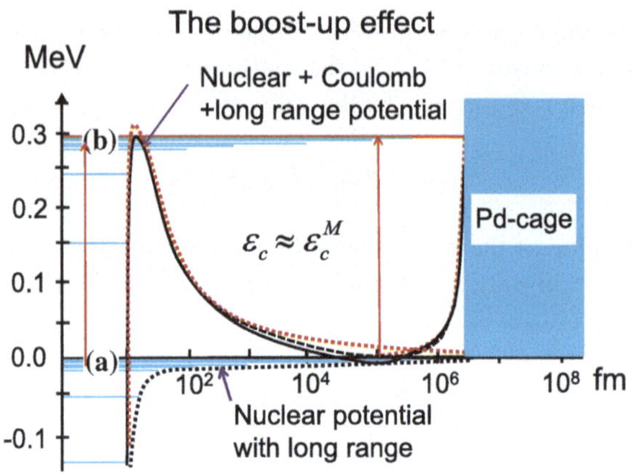

Fig. 124.1 An image of the one-dimensional effective Cs-d potential in the Pd_{12}-cage on the d-Cs-d linear bond. The right-half of wine-bottle-like potential for central Cs-ion is illustrated. The boost-up effect for the energy levels by the Cs-d Coulomb barrier is sketched. The vertical axis is the energy with the MeV scale, and the horizontal axis is the distance from Cs to d with the log-scale by fm. The Cs-d repulsive Coulomb plus the nuclear potential, and a repulsive Pd_{12}-cage (dotted red line) are illustrated. The black dashed line represents the correction by electrons with V^M_{ion-el} and V^M_{el-el}, and the black dotted line shows the nuclear potential with a long range effect. The black solid line is an image of the total effective two-body potential. Energy levels are indicated by blue solid lines under the zero Coulomb level (a), and also under the top of the Cs-d Coulomb barrier (b), where some of them belong to the molecular states. The boost-up effect raises the energy levels of (a) to (b). Therefore, the penetration difficulty is resolved

success could approve not only the present theory, but also our prediction for the *supper long range* hadron potential [2–5].

The authors would like to express our deep gratitude to Drs. N. Watari, I. Toyoda, S. Tsuruga, A. Kodama, and I. Lagaris for valuable discussions and comments. The financial support is provided by MHI Co. Ltd.

References

1. Watari, N., Ohnishi, S., Ishii, Y.: Hydrogen storage in Pd clusters. J. Phys. Condens. Matter **12**, 6799–6823 (2000)
2. Oryu, S.: Universal structure of the three-body system. Phys. Rev. C **86**, 044001 (2012)
3. Oryu, S.: Few-Body Syst. **54**, 283–286 (2013)
4. Oryu, S., et al.: J. Phys. Conf. Ser. **915**, 012001 (2017)
5. Oryu, S.: A New feature of the Efimov-like structure in the Hadron system—Long range force as a Recoil Effect. Few-Body Syst. **59**, 51 (2018). https://doi.org/10.1007/s00601-018-1373-z
6. Efimov, V.: Energy levels arising from resonant two-body forces in a three-body system. Phys. Lett. B **33**, 563 (1970)
7. Efimov, V.: Energy levels of three resonantly interacting particles. Nucl. Phys. A **210**, 157 (1973)

8. Iwamura, Y., Sakano, M., Itoh, T.: Elemental analysis of Pd complexes: Effects of D2 gas permeation. Jpn. J. Appl. Phys. **41**(4642-4650), (2002); ibid. J. Condens. Matter Nucl. Sci. **13** (242–252) (2014)
9. Hioki, T., Takahashi, N., Kosaka, S., Nishi, T., Azuma, H., Hibi, S., Higuchi, Y., Murase, A., Motohiro, T.: Inductively coupled plasma mass spectrometry study on the increase in the amount of Pr atoms for Cs-Ion-Implanted Pd/CaO multilayer complex with deuterium permeation. Jpn. J. Appl. Phys. **52**, 107301 (2013)
10. Tsuruga, S., Muta, K., Tanaka, Y., Shimazu, T., Fujimori, K., Nishida, T.: Transmutation tractions induced by deuterium permeation through nano-structured multilayer thin film. Mitsubishi Heavy Ind. Tech.-Rep. **52**(4), 104–108 (2015)

Chapter 125
The Problem of Cluster Separability in Relativistic Few-Body Systems

Wolfgang Schweiger, Nikita Reichelt and William H. Klink

Abstract An appropriate framework for dealing with hadron structure and hadronic physics in the few-GeV energy range is relativistic quantum mechanics. The Bakamjian-Thomas construction provides a systematic procedure for implementing interactions in a relativistic invariant way. It leads, however, to problems with cluster separability. It has been known for some time, due to Sokolov's pioneering work, that mass operators with correct cluster properties can be obtained through a series of unitary transformations making use of so-called *packing operators*. In the present contribution we sketch an explicit construction of packing operators for three-particle systems consisting of distinguishable, spinless particles.

125.1 Relativistic Quantum Mechanics

By a *relativistic quantum mechanics* we mean a quantum theory for a finite number of particles invariant under Poincaré transformations. Speaking more formally, one has to find a representation of all the Poincaré generators in terms of self-adjoint operators on an appropriate Hilbert space such that these operators satisfy the Poincaré algebra. For an interacting theory it is well known that at least three of the Poincaré generators have to contain interaction terms. Depending on which of the Poincaré generators become interaction dependent, one distinguishes different forms of relativistic dynamics [1]. For our purposes it turns out to be most convenient to use the point-form of relativistic dynamics in which all the components of the four-momentum operator P^{μ} contain interactions, whereas the generators of Lorentz-transformations stay interaction free [2].

The only systematic procedure for implementing interactions in the Poincaré generators of an N-particle system such that the Poincaré algebra is preserved was

W. Schweiger (✉) · N. Reichelt
Institute of Physics, University of Graz, 8010 Graz, Austria
e-mail: wolfgang.schweiger@uni-graz.at

W. H. Klink
Department of Physics and Astronomy, University of Iowa, Iowa City, USA

N. A. Orr et al. (eds.), *Recent Progress in Few-Body Physics*,
Springer Proceedings in Physics 238,
https://doi.org/10.1007/978-3-030-32357-8_125

suggested long ago by Bakamjian and Thomas [3]. In the point form this procedure amounts to factorizing the four-momentum operator into a four-velocity operator and a mass operator and putting interaction terms into the mass operator:

$$P^{\mu} = MV^{\mu}_{\text{free}} = (M_{\text{free}} + M_{\text{int}})V^{\mu}_{\text{free}} = P^{\mu}_{\text{free}} + P^{\mu}_{\text{int}} \,. \tag{125.1}$$

Since the mass operator is a Casimir operator of the Poincaré group, the constraints on the interaction terms that guarantee Poincaré invariance become simply that M_{int} should be a Lorentz scalar and that it should commute with V^{μ}_{free}, i.e. $[M_{\text{int}}, V^{\mu}_{\text{free}}] = 0$.

A very convenient basis for representing Bakajian-Thomas (BT) type mass operators consists of velocity states [2], $|\mathbf{v}; \mathbf{k}_1, \mu_1; \mathbf{k}_2, \mu_2; \ldots; \mathbf{k}_N, \mu_N\rangle$. These specify the state of an N-particle system by its overall velocity $\mathbf{v}$, the particle momenta $\mathbf{k}_i$ in the rest frame of the system ($\sum_i \mathbf{k}_i = 0$) and the (canonical) spin projections μ_i of the individual particles. The overall velocity factors out in velocity-state matrix elements of BT-type mass operators, leading to the separation of overall and internal motion of the system.

125.2 Cluster Separability and Packing Operators

An important requirement for a quantum mechanical system, in addition to relativistic invariance, is *cluster separability*. Cluster separability roughly means that subsystems of a quantum mechanical system should behave independently, if they are sufficiently space-like separated. There are weaker and stronger notions of cluster separability depending, e.g., on whether it is demanded for the S-operator or the Poincaré generators [4]. Also the kind of convergence of these quantities when letting the separation distance go to infinity plays an important role. We are interested in cluster separability of the Poincaré generators and, in particular, of the invariant mass operator. For our purposes it is useful to introduce a purely algebraic notion of cluster separability which we call *coupling constant separability*. Coupling constant separability means that, for a particular clustering, after separation of the clusters, operators behave as if there were no interaction between the clusters. It does not say much about the range of the interactions, but for the construction we are going to make it is a useful and reasonable concept.

Problems with cluster separability start to show up for interacting three-particle systems and are closely connected with how the two-particle subsystems are implemented in the three-particle Hilbert space. Let us assume for simplicity spinless, distinguishable particles and start with BT-type four-momentum operators for the two-particle subsystems, $P^{\mu}_{ij} = M_{ij}\, V^{\mu}_{ij\,\text{free}}$, $i, j = 1, 2, 3$, $i \neq j$. The third particle is then added by means of the usual tensor-product construction,

$$\tilde{P}^{\mu}_{ij|k} := P^{\mu}_{ij} \otimes I_k + I_{ij} \otimes P^{\mu}_{k} \,. \tag{125.2}$$

The individual four-momentum operators $\tilde{P}^{\mu}_{ij|k}$ describe 2+1-body systems in a Poincaré invariant way and also exhibit coupling constant separability. One may now think of adding the four-momentum operators for the different clusterings, as in the non-relativistic case, to end up with a four momentum operator for a three-particle system with pairwise interactions, $\tilde{P}^{\mu}_{123} = \tilde{P}^{\mu}_{12|3} + \tilde{P}^{\mu}_{23|1} + \tilde{P}^{\mu}_{31|2} - 2P^{\mu}_{123\,\mathrm{free}}$. But the components of $\tilde{P}_{123}$ do not commute, $[\tilde{P}^{\mu}_{123}, \tilde{P}^{\nu}_{123}] \neq 0$ since $[M_{ij\,\mathrm{int}}, V^{\mu}_{j}] \neq 0$. The three-particle system, described by the four-momentum operator $\tilde{P}^{\mu}_{123}$, is thus not relativistically invariant. The first step to overcome this problem is to factorize the four-momentum operators for the different clusterings again into a mass operator and a four-velocity operator:

$$\tilde{P}^{\mu}_{ij|k} = \tilde{M}_{ij|k}\, \tilde{V}^{\mu}_{ij|k} \quad \text{with} \quad \tilde{M}^{2}_{ij|k} = \tilde{P}_{ij|k} \cdot \tilde{P}_{ij|k} \,. \tag{125.3}$$

The four-velocities $\tilde{V}^{\mu}_{ij|k}$ contain interactions and differ for different clusterings. The key observation is now that there exist unitary operators which relate the four-velocity operators, since they all have the same spectrum ($\mathbb{R}^3$). One can find, in particular, unitary operators $U_{ij|k}$ such that

$$\tilde{V}^{\mu}_{ij|k} = U_{ij|k}\, V^{\mu}_{123\,\mathrm{free}}\, U^{\dagger}_{ij|k} \,. \tag{125.4}$$

Applied to the four-momentum operators $\tilde{P}^{\mu}_{ij|k}$ these unitary operators pack the interaction dependence of the four-velocity operators $\tilde{V}^{\mu}_{ij|k}$ into the mass operator $M_{ij|k} = U^{\dagger}_{ij|k}\tilde{M}_{ij|k}U_{ij|k}$ such that $P^{\mu}_{ij|k} = U^{\dagger}_{ij|k}\tilde{P}^{\mu}_{ij|k}U_{ij|k} = M_{ij|k}V^{\mu}_{123\,\mathrm{free}}$ is of BT-type. Therefore $U_{ij|k}$ were called *packing operators* by Sokolov in his seminal paper on the formal solution of the cluster problem [5]. The new four-momentum operator $P^{\mu}_{123\,\mathrm{BT}} = P^{\mu}_{ij|k} + P^{\mu}_{ij|k} + P^{\mu}_{ij|k} - 2P^{\mu}_{123\,\mathrm{free}} = M_{123\,\mathrm{BT}}V^{\mu}_{123\,\mathrm{free}}$ is now also of BT-type and thus provides a relativistic invariant description of a three particle system with pairwise two-particle interactions; but it still misses coupling-constant separability, as one can easily check. The way out is a further unitary transformation which involves the packing operators we have already introduced:

$$P^{\mu}_{123} = \left(\prod U_{ij|k}\right) P^{\mu}_{123\,\mathrm{BT}} \left(\prod U_{ij|k}\right)^{\dagger} . \tag{125.5}$$

If $U_{ij|k} \to 1$ for separations $ki|j$, $jk|i$ and $i|j|k$ and if $(\prod U_{ij|k})$ commutes with the generators of Lorentz transformations, it can be shown that such a *generalized BT construction* leads to relativistic invariance and coupling-constant separability of the resulting three-body model [5].

The procedure just outlined solves the cluster problem for three-body systems formally, but its practical applicability depends strongly on the capability to calculate the packing operators. The solution to this problem can also be found in Sokolov's paper [5]. The trick is to split the packing operator further into a product of unitary operators which depend on the corresponding two-particle mass operators in a way to be determined, $U_{ij|k} = W^\dagger(M_{ij})W(M_{ij\,\mathrm{free}})$. With this splitting one can rewrite Eq. (4) in the form $W(M_{ij\,\mathrm{free}})V^\mu_{123\,\mathrm{free}}W^\dagger(M_{ij\,\mathrm{free}}) = W(M_{ij})\tilde{V}^\mu_{ij|k}W^\dagger(M_{ij})$. Since this equation should hold for any interaction, the right- and left-hand sides can be chosen to equal some simple four-velocity operator, for which $V^\mu_{ij\,\mathrm{free}} \otimes I_k$ is a good choice. In order to compute the action of W it is then convenient to take bases in which matrix elements of $V^\mu_{123\,\mathrm{free}}$, $V^\mu_{ij\,\mathrm{free}} \otimes I_k$ and $\tilde{V}^\mu_{ij|k}$ can be calculated. This is the basis of (mixed) velocity eigenstates $|\mathbf{v}_{ij}; \tilde{\mathbf{k}}_i, \tilde{\mathbf{k}}_j, \mathbf{p}_k\rangle = |\mathbf{v}_{ij}; \tilde{\mathbf{k}}_i, \tilde{\mathbf{k}}_j\rangle \otimes |\mathbf{p}_k\rangle$ of $M_{ij|k\,\mathrm{free}}$ if one wants to calculate the action of $W(M_{ij\,\mathrm{free}})$ and corresponding eigenstates of $M_{ij|k}$ if one wants to calculate the action of $W(M_{ij})$. It turns out that the effect of these operators is mainly to give the two-particle subsystem ij the velocity $\mathbf{v}_{ij|k}$ of the whole three-particle system. After some calculations one finds that the whole effect of the packing operator $U_{ij|k}$ on the mass operator $\tilde{M}_{ij|k}$ is just the replacement of kinematical factors in the mixed velocity-state matrix elements, so that the transformed mass operator $U_{ij|k}\,\tilde{M}_{ij|k}\,U_{ij|k}$ attains BT-type structure [2]:

$$
\begin{gathered}
\frac{1}{m'^{3/2}_{ij}\, m^{3/2}_{ij}}\, v^0_{ij}\, \delta^3(\mathbf{v}'_{ij} - \mathbf{v}_{ij}) \\
\downarrow \\
\frac{\sqrt{v'_{ij}\cdot v_{123}}}{m'^{3/2}_{123}} \frac{\sqrt{v_{ij}\cdot v_{123}}}{m^{3/2}_{123}}\, v^0_{123}\, \delta^3(\mathbf{v}'_{123} - \mathbf{v}_{123}) .
\end{gathered}
\tag{125.6}
$$

Here m_{ij} and m_{123} are the invariant masses of the non-interacting two-particle subsystem ij and the non-interacting three-particle system, respectively, v_{ij} and v_{123} are the corresponding four-velocities.

125.3 Concluding Remarks

Knowing the packing operators and their action on the mass operators $\tilde{M}_{ij|k}$, we can construct a three-body mass operator $M_{123} = (\prod U_{ij|k})\, M_{123\,BT}\, (\prod U_{ij|k})^\dagger$ for distinguishable, spinless particles that has correct cluster properties. The inclusion of spin should be straightforward. The treatment of identical particles is, however, more intricate, because the simple product $(\prod U_{ij|k})$ has to be replaced by some kind of symmetrized product to preserve the symmetry under exchange of particles. Although the formal solution of the cluster problem and explicit expressions for the packing operators have been known for quite some time [4, 5], cluster separability has always been neglected in practical applications. This may well be justified for weakly bound nuclear systems [6], but has to be thoroughly investigated for strong binding. With our results we are now, at least in the simple case of spinless, distinguishable particles, in the position to perform numerical studies for clarifying the precise role of cluster separability in relativistic three-body systems.

References

1. Dirac, P.A.M.: Rev. Mod. Phys. **21**, 392 (1949)
2. Klink, W.H., Schweiger, W.: Relativity, Symmetry, and the Structure of Quantum Theory, Volume 2: Point Form Relativistic Quantum Mechanics. Morgan & Claypool Publishers, Bristol (2018)
3. Bakamjian, B., Thomas, L.H.: Phys. Rev. **92**, 1300 (1953)
4. Keister, B.D., Polyzou, W.N.: Adv. Nucl. Phys. **20**, 225 (1991)
5. Sokolov, S.N.: Theor. Math. Phys. **36**, 682 (1978)
6. Keister, B.D., Polyzou, W.N.: Phys. Rev. C **86**, 014002 (2012)

Chapter 126
Scattering with Real-Time Path Integrals

W. N. Polyzou and E. S. Nathanson

Abstract Sharp-momentum transition matrix elements for scattering from a short-range Gaussian potential are computed using a real-time path integral. The computation is based on a numerical implementation of a new interpretation of the path integral as the expectation of a potential functional with respect to a complex probability distribution on cylinder sets of paths. The method is closely related to a unitary transfer matrix computation.

126.1 Path Integrals and Complex Probabilities

Path integrals [1, 2] provide a means for treating problems in quantum mechanics that are often difficult to treat by other means. Normally path integral calculations are limited to quadratic interactions or generating perturbation theory. When the time can be made imaginary they can also be approximated using Monte Carlo [3] integration.

Most applications are naturally formulated using real time. Scattering is one such application. While the computation of real-time path integrals is more challenging, it is interesting to explore how far real-time applications can be pushed. One motivation is because real-time path integrals represent unitary time evolution, direct treatments of real-time path integrals have the potential to be candidates for applications of quantum computing algorithms.

The problem with the interpretation of the real-time path integral as an integral is that the measure has to be additive on a countable union of disjoint measurable sets. When the measure is not positive, the evaluation of a countable sum of non-positive numbers can be both infinite or finite, depending on how it is evaluated.

W. N. Polyzou (✉)
The University of Iowa, Iowa City, IA 52242, USA
e-mail: polyzou@uiowa.edu

E. S. Nathanson
Georgia Gwinnett College, Lawerenceville, GA 30043, USA
e-mail: enathanson@ggc.edu

N. A. Orr et al. (eds.), *Recent Progress in Few-Body Physics*,
Springer Proceedings in Physics 238,
https://doi.org/10.1007/978-3-030-32357-8_126

In [4–6] the Feynman path integral is reinterpreted as the expectation of a potential functional

$$E[F] \qquad F[\gamma] := e^{-i\int V(\gamma(t))dt} \tag{126.1}$$

with respect to a complex probability on a space of paths. It was shown in [5, 6] that this interpretation results in a global solution of the Schrödinger equation.

Classical probabilities on the real line are defined on measurable sets that are generated by intervals under complements and countable unions. These sets are the Lebesgue measurable sets. A Probability is a non-negative Lebesgue measurable function that integrates to 1.

The Henstock-Kurzweil integral [7, 8] provides an extension of the Lebesgue integral that is defined similar to a Riemann integral. Hesntock developed an alternative probability theory based on this integral. Henstock's probability agrees with classical probability theory when the probability is non-negative, however he realized that it could be extended to non-positive and complex probabilities if countable additivity was replaced by finite additivity. Muldowney [4] suggested that this interpretation could also be extended to reinterpret path integrals not as integrals, but as the expectation of a potential functional with respect to a complex probability distribution on a space of paths. Jørgensen and Nathanson verified that this interpretation leads to approximations that converge to global solutions of the Schrödinger equation.

In one dimension the path-integral representations of the unitary time evolution operator for a particle of mass μ in a potential V can be expressed as:

$$\langle x_0 | e^{-iHt} | \psi \rangle =$$

$$\lim_{N\to\infty} \left(\frac{\mu}{2\pi i \Delta t}\right)^{N/2} \int \prod_{i=n}^{N} dx_n e^{i\frac{\mu}{2\Delta t}(x_{n-1}-x_n)^2 - iV(x_n)\Delta t} \langle x_N | \psi \rangle . \tag{126.2}$$

where $\Delta t := t/N$. This follows from the Trotter product formula [9] and is the standard form of the path integral. It is expressed as the limit of N-dimensional integrals as $N \to \infty$.

The complex probability interpretation arises by expressing the real line as the union of a finite number of disjoint intervals

$$\mathbb{R} = \underbrace{(-\infty, x_{1n})}_{I_{0n}} \underbrace{[x_{1n}, x_{2n})}_{I_{1n}}, \cdots, \underbrace{[x_{M-1,n}, x_{M,n})}_{I_{M-1,n}}, \underbrace{[x_{M,n}, \infty)}_{I_{M,n}} . \tag{126.3}$$

and replacing the integral over each time slice by a sum of integrals over each of these intervals

$$\langle x_0 | e^{-iHt} | \psi \rangle =$$

$$\lim_{N\to\infty} \left(\frac{\mu}{2\pi i \Delta t}\right)^{N/2} \sum_{m_1 \cdots m_N} \prod_{n=1}^{N} \int_{I_{mn}} dx_n e^{i\frac{\mu}{2\Delta t}(x_{n-1}-x_n)^2 - iV(x_n)\Delta t} \langle x_N | \psi \rangle . \tag{126.4}$$

Computationally the intervals should be chosen so $e^{-iV(x)}$ is approximately constant on each interval. This also applies to the half-infinite intervals, where on these intervals, for a scattering problem with a short-range interaction, $e^{-iV(x)} \approx 1$. When these conditions hold the potential terms can be factored out of the integral, and evaluated at any point $y_{mn} \in I_{mn}$:

$$\langle x_0|e^{-iHt}|\psi\rangle \approx$$

$$\lim_{N\to\infty}(\frac{\mu}{2\pi i\Delta t})^{N/2}\sum_{m_1\cdots m_N}\prod_{n=1}^{N} e^{-iV(y_{mn})\Delta t}\int_{I_{mn}} dx_n e^{i\frac{\mu}{2\Delta t}(x_{n-1}-x_n)^2}\langle x_N|\psi\rangle. \tag{126.5}$$

What remains has the form

$$\langle x_0|e^{-iHt}|\psi\rangle \approx$$

$$\lim_{N\to\infty}\sum_{m_1\cdots m_N} P(x_0, I_{m1}, \cdots, I_{mN})e^{-i\sum_n V(y_{mn})\Delta t}\langle y_{m_N}|\psi\rangle \tag{126.6}$$

where

$$P(x_0, I_{m1}, \cdots, I_{mN}) = (\frac{\mu}{2\pi i\Delta t})^{N/2}\prod_{n=1}^{N}\int_{I_{mn}} dx_n e^{i\frac{\mu}{2\Delta t}(x_{n-1}-x_n)^2} \tag{126.7}$$

represents the complex probability that a path passes through I_{mN} at time t_1, I_{mN-1} at time t_2, ... I_{m1} at time t_N. It has all of the properties of a complex probability. The approximation converges in the limit that N gets large, the size of the intervals gets small, and the boundary of the half-infinite intervals approaches $\pm\infty$. In principle the intervals and evaluation points needed for a given accuracy are determined by the Henstock theory of integration.

This representation has the advantage that, if the intervals are chosen the same way on each time slice, the potential only needs to be evaluated at a small number of points where the potential is not zero.

The computational problem is that the number of cylinder sets grows like M^N in the limit that both M and N become infinite. In order to make this computable the complex probability is approximately factored [10] into a sum of products of one-step complex probabilities. They represent the complex probability of starting at a point in a given interval and coming out in another interval. This has the property that sum over all final intervals is 1, since the particle has to come out in some interval. The factorization has the form

$$P(x_0, I_{m1}, \cdots, I_{mN}) \approx P(x_0, I_{m1})\prod_{n=2}^{N} P(y_{m(n-1)}, I_{nm}) \tag{126.8}$$

Table 126.1 Sums of complex probabilities (5000 intervals)

y	$\sum_n \Re(P(y, I_n))$	$\sum_n \Im(P(y, I_n))$
−25.005001	p = 1.000000e+00	+i(5.273559e−16)
−20.204041	p = 1.000000e+00	−i(1.387779e−16)
−15.003001	p = 1.000000e+00	+i(2.775558e−17)
−10.202040	p = 1.000000e+00	+i(2.775558e−16)
−5.001000	p = 1.000000e+00	+i(7.771561e−16)
0.200040	p = 1.000000e+00	+i(3.608225e−16)
5.001000	p = 1.000000e+00	+i(1.665335e−16)
10.202040	p = 1.000000e+00	+i(5.273559e−16)
15.003001	p = 1.000000e+00	+i(8.049117e−16)
20.204041	p = 1.000000e+00	+i(1.137979e−15)
25.005001	p = 1.000000e+00	−i(2.164935e−15)

where

$$P(y_m, I_k) = \sqrt{\frac{1}{i\pi}} \int_{\sqrt{\frac{\mu}{2\Delta t}}(x_k - y_m)}^{\sqrt{\frac{\mu}{2\Delta t}}(x_{k+1} - y_m)} e^{i\beta^2} d\beta \qquad (126.9)$$

can be evaluated analytically [11]. With this factorization the "path integral" can be approximated by

$$\langle x_0 | e^{-iHt} | \psi \rangle \approx$$

$$\sum_{m1 \cdots mN} P(x_0, I_{m1}) e^{-iV(y_{m1})\Delta t} \prod_{n=2}^{N} P(y_{m(n-1)}, I_{nm}) e^{-iV(y_{mn})\Delta t} \langle y_{mN} | \psi \rangle . \qquad (126.10)$$

This expresses the path integral as the N-th power of a complex matrix, applied to a vector. Powers of matrices can be efficiently computed. In addition, because the matrix can be computed analytically, the matrix elements can be evaluated on the fly, so it is not necessary to store large matrices.

The table shows the sum of the one-step probabilities over 5000 intervals for various values of y. The imaginary part sums to zero with no visible round-off error (Table 126.1).

The goal of this exercise is to calculate scattering observables. The basic observables are functions of on-shell transition matrix elements

$$\langle \mathbf{p}_f | T(E_i 0^+) | \mathbf{p}_i \rangle \approx \langle \psi_{f0}(0) | V e^{-iHt} | \psi_{i0}(-t) \rangle \qquad (126.11)$$

where $\langle \mathbf{p} | \psi_{f0}(0) \rangle$ and $\langle \mathbf{p} | \psi_{i0}(0) \rangle$ are narrow Gaussian wave packets centered at $\mathbf{p}_f$ and $\mathbf{p}_i$ with delta-function normalizations (i.e. they integrate to 1). The initial state is evolved to $-t$ using the free dynamics. In one dimension the on-shell transition matrix elements are related to phase shifts and transmission and reflection coefficients

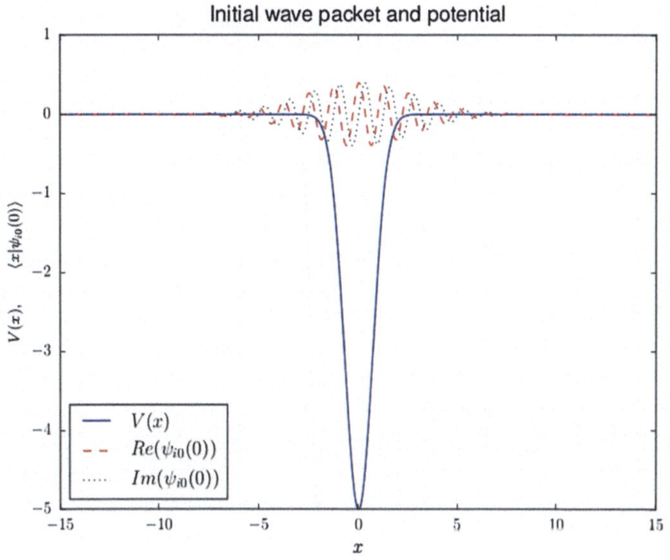

Fig. 126.1 V(x),$\langle x|\psi_{0i}(0)\rangle$

by

$$e^{2i\delta_\pm(E)} = 1 - 2\pi i \frac{\mu}{|p|}\langle \pm p|T(E+i0^+)|p\rangle \tag{126.12}$$

$$T = 1 - \frac{2\pi i p}{\mu}\langle p|T(E+i0^+)|p\rangle \qquad R = -\frac{2\pi i p}{\mu}\langle -p|T(E+i0^+)|p\rangle \tag{126.13}$$

This method was tested by computing the half-shell scattering matrix elements for a particle of mass μ off of a repulsive Gaussian potential of $-\lambda e^{-(r/r_0)^2}$. Because this is a time-dependent computation it is necessary to choose the parameters carefully. The time has to be chosen so, after accounting for wave packet spreading, the wave packet should be outside of the range of the potential. The exhibited calculations are for $\mu = 1$, $r_0 = 1$, $\lambda = 5$, $p_i = 5$ and $\Delta p = 0.25$. The calculations used 120 time steps and 10,000 intervals between ± 15. The results are compared to an exact numerical calculation.

Figures 126.1 and 126.2 show the width of the wave packet and potential. The second figure shows that at $t = -3$ the wave packet is out of the range of the potential, which means that the time limit can be replaced by an evaluation at $t = 3$. Figures 126.3 and 126.4 show the real and imaginary parts of the half-shell transition matrix elements using the path integral (short dashes) compared to the exact calculation (long dashes) of the same quantity.

While this is not the most efficient method to solve this problem, it demonstrates that real-time path integrals can be used to calculate scattering observables. The

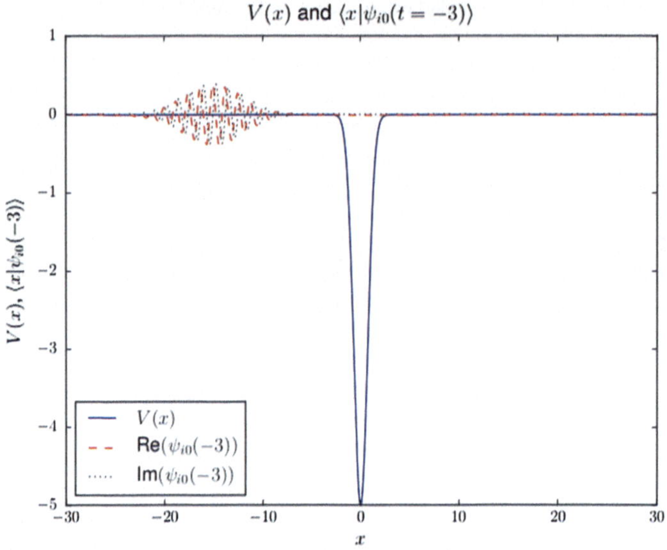

Fig. 126.2 V(x),$\langle x|\psi_{0i}(-3)\rangle$

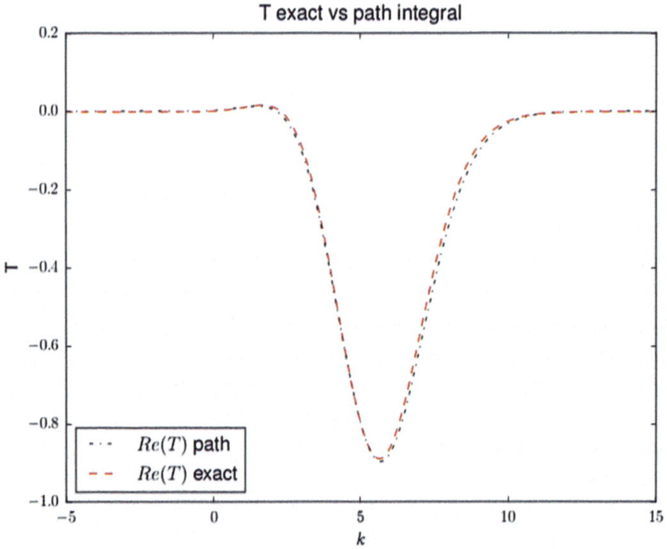

Fig. 126.3 Re(T) exact vs path integral

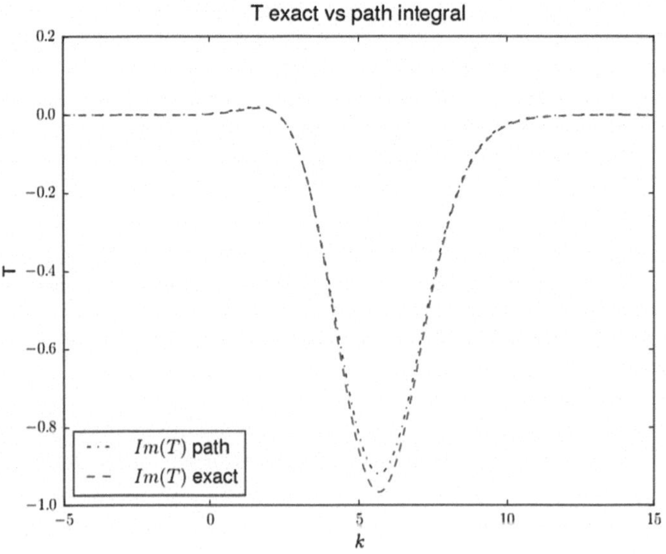

Fig. 126.4 Im(T) exact vs. path integral

exhibited calculations used fixed time steps and interval widths. There are many possible improvements in computational efficiency.

The most important observation is that after the one-step factorization, the method is numerically equivalent to a unitary transfer-matrix calculation. The factorization that arises from the Trotter product formula allows the transfer matrix to be factored into a product of a potential term and a one step probability matrix. The potential term is exactly unitary and the complex one-step probability matrix is approximately unitary. Because the calculated quantity is a matrix element of the product of the potential with a wave operator, it is only necessary to evolve the system for a finite time in a finite volume. These observations are the key to understanding both the strength and limitations of this method. A reasonable expectation is that, with some refinement, this method could be applied to the same class of problems that can be treated using transfer matrix methods.

Acknowledgements This work supported by the U.S. Department of Energy, Office of Science, Grant number: DE-SC0016457.

References

1. Feynman, R.: Space-time approach to non-relativistic quantum mechanics. Rev. Mod. Phys. **2**, 367 (1948)
2. Feynman, R., Hibbs, A.R.: Quantum Mechanics and Path Integrals. McGraw-Hill, USA (1965)
3. Metropolis, N., Ulam, S.: The Monte Carlo method. J. Am. Stat. Assoc. **44**, 335 (1949)

4. Muldowney, P.: A Modern Theory of Random Variation. Wiley, NJ (2012)
5. Nathanson, Ekaterina, S., Jørgensen, Palle, E.T.: A global solution to the Schrø:dinger equation: From Henstock to Feynman. J. Math. Phys. 56, 092102 (2015)
6. Nathanson, E.S.: Path integration with non-positive distributions and applications to the Schrödinger equation. University of Iowa Thesis, (2015)
7. Henstock, R.: Theory of Integration. Butterworths, London (1963)
8. Bartle, R.G.: A Modern Theory of Integration. American Mathematical Society, Providence, RI (2001)
9. Reed, M., Simon, B.: Methods of Modern Mathematical Physics. Academic Press, San Diego, p. 295 (1980)
10. Polyzou, W.N., Nathanson, E.S.: Scattering using real-time path integrals (2018). arxiv:1712.00046,
11. Abramowitz, M., Stegun, I.: Handbook of Mathematical Functions. National Bureau of Standards, Washington, DC (1964)

Chapter 127
A Simple Tool to Study Many-Body Forces

Claude Semay and Guillaume Sicorello

Abstract The envelope theory, also known as the auxiliary field method, is a technique to compute approximate solutions of a quantum system with N identical particles. The basic idea is to replace the Hamiltonian H under study by an auxiliary Hamiltonian $\tilde{H}$ which is solvable, the eigenvalues of $\tilde{H}$ being optimized to be as close as possible to those of H. The method is easy to implement since it reduces to find the solution of a transcendental equation. We show here that the envelope theory can be extended to compute the eigensolutions of a system of identical particles with a type of many-body forces often used in phenomenological models.

127.1 Introduction

Generally, two-body forces are the only type of interaction considered in many-body quantum systems. But three-body forces (and more generally many-body forces) are sometimes a crucial ingredient in atomic physics [1], nuclear physics [2], or hadronic physics [3]. Many-body forces have deep theoretical foundations, but their structure can be very difficult to compute. Effective forms can then be used to take into account at best possible these complicated many-body contributions. Among the possible structures for a K-body force in a N-body system, the one chosen in this work is given by

$$\sum_{\{i_1,\ldots,i_K\}}^{N} V\left(r_{\{i_1,\ldots,i_K\}}\right) \quad \text{with} \quad r^2_{\{i_1,\ldots,i_K\}} = \sum_{i<j}^{\{i_1,\ldots,i_K\}} r^2_{ij}, \tag{127.1}$$

C. Semay (✉) · G. Sicorello
Service de Physique Nucléaire et Subnucléaire, Université de Mons, UMONS Research Institute for Complex Systems, Place du Parc 20, 7000 Mons, Belgium
e-mail: claude.semay@umons.ac.be

N. A. Orr et al. (eds.), *Recent Progress in Few-Body Physics*, Springer Proceedings in Physics 238,
https://doi.org/10.1007/978-3-030-32357-8_127

where $r_{ij}^2 = (\boldsymbol{r}_i - \boldsymbol{r}_j)^2$ and $\{i_1, \ldots, i_K\}$ is a set of K particles among the N possible ones, with $i_1 < \ldots < i_K$. The sum $\sum_{\{i_1,\ldots,i_K\}}^N$ runs over the C_N^K different sets $\{i_1, \ldots, i_K\}$, while the sum $\sum_{i<j}^{\{i_1,\ldots,i_K\}}$ runs over the C_K^2 different pairs in a particular set $\{i_1, \ldots, i_K\}$, where C_A^B is a usual binomial coefficient.

The envelope theory (ET) [4–6], independently rediscovered under the name of auxiliary field method [7], is a simple technique to compute approximate solutions, eigenvalues and eigenvectors, of many-body systems with arbitrary kinematics in D dimensions [8, 9]. The basic idea is to replace the Hamiltonian H under study by an auxiliary Hamiltonian $\tilde{H}$ which is solvable, the eigenvalues of $\tilde{H}$ being optimised to be as close as possible to those of H. Quite good approximations can be obtained for various systems containing up to 10 bosons [10]. The accuracy can be improved, but to the detriment of the possible variational character [11]. The ET can yield interesting results for systems of N identical particles, whose Hamiltonians are given by [8]

$$H = \sum_{i=1}^N T(p_i) + \sum_{i=1}^N U(s_i) + \sum_{i<j}^N V(r_{ij}), \tag{127.2}$$

with $p_i = |\boldsymbol{p}_i|$ and $s_i = |\boldsymbol{r}_i - \boldsymbol{R}|$, where $\boldsymbol{R} = \frac{1}{N}\sum_{i=1}^N \boldsymbol{r}_i$ is the centre of mass position. T is the kinetic energy, and U and V are potentials ($\hbar = c = 1$). As only the internal motion is relevant, $\sum_{i=1}^N \boldsymbol{p}_i = \boldsymbol{0}$. The momentum $\boldsymbol{p}_i$ and position $\boldsymbol{r}_i$ of the particle i are conjugate variables.

The purpose of this work is to show that ET can treat Hamiltonians with K-body forces of type

$$H = \sum_{i=1}^N T(p_i) + \sum_{i=1}^N U(s_i) + \sum_{\{i_1,\ldots,i_K\}}^N V\left(r_{\{i_1,\ldots,i_K\}}\right). \tag{127.3}$$

Let us note that the Hamiltonian can contain several many-body potentials with various values of K. We keep here only one many-body contribution to lighten the notations. The one-body term U is kept, because its treatment is a little bit different.

In Sec. 127.2, the exact solution for the non-relativistic system of N identical harmonic oscillators with K-body forces is given, with the ET equations for general Hamiltonians. An analytical example is presented in Sec. 127.3.

127.2 Envelope Theory Equations

The Hamiltonian $\tilde{H}$ mentioned in the previous section is

$$H_{\text{ho}} = \frac{1}{2\mu}\sum_{i=1}^N p_i^2 + \nu \sum_{i=1}^N s_i^2 + \rho \sum_{\{i_1,\ldots,i_K\}}^N r_{\{i_1,\ldots,i_K\}}^2, \tag{127.4}$$

whose exact solutions are given by [7]

$$E_{\text{ho}} = Q\sqrt{\frac{2}{\mu}\left(\nu + N\, C_{N-2}^{K-2}\, \rho\right)} \quad \text{with} \quad Q = \sum_{i=1}^{N-1}\left(2\,n_i + l_i + \frac{D}{2}\right), \tag{127.5}$$

where D is the dimension of the space. The set of equations giving an approximate energy of Hamiltonian (127.3) is given by

$$E = N\,T(p_0) + N\,U\left(\frac{r_0}{N}\right) + C_N^K\,V\left(\sqrt{\frac{C_K^2}{C_N^2}}r_0\right), \tag{127.6}$$

$$p_0 = \frac{Q}{r_0}, \tag{127.7}$$

$$N\,p_0\,T'\left(p_0\right) = r_0\,U'\left(\frac{r_0}{N}\right) + C_N^K\sqrt{\frac{C_K^2}{C_N^2}}\,r_0\,V'\left(\sqrt{\frac{C_K^2}{C_N^2}}r_0\right), \tag{127.8}$$

where the particular state considered is fixed by the value of Q. Detailed calculations to obtain these equations are given in [12]: p_0 (which can be interpreted as a mean momentum for a particle) and r_0 (which is linked to the radius of the system) are intermediate variables which are determinate by (127.7) and (127.8). If $K = 2$, the equations in [8] are recovered. Let us note that (127.8) is also obtained by setting $dE/dr_0 = 0$ with the constraint (127.7), which shows the extremum character of E. As expected, the exact solution (127.5) is recovered for $T(x)$, $U(x)$ and $V(x)$ proportional to x^2.

The approximate energy E can be an upper or a lower bound. Using the comparison theorem [13], a simple procedure exists to verify if such a situation happens. It is necessary to define three functions b_T, b_U and b_V such that

$$T(x) = b_T(x^2),\ \ U(x) = b_U(x^2) \text{ and } V(x) = b_V(x^2). \tag{127.9}$$

It can be shown that, if $b_T''(x)$, $b_U''(x)$ and $b_V''(x)$ are all concave functions, E is an upper bound [5]. Conversely, if all these second derivatives are convex functions, E is a lower bound. If the second derivative is vanishing for one or two of these functions, the variational character is solely ruled by the convexity of the other(s). In the other cases, the variational character of the solution cannot be guaranteed.

127.3 An Analytical Example

Let us consider the generic kinetic energy

$$T(p) = D_\alpha\, p^\alpha, \tag{127.10}$$

with $D_\alpha > 0$ and $\alpha > 0$, in order that T be positive and growing with the modulus of the momentum p, and the power-law K-body potential

$$V(r) = a\,\mathrm{sgn}(b)\,r^b \quad \text{with} \quad a > 0. \tag{127.11}$$

Then, (127.7) and (127.8) implies that

$$r_0 = \left[\frac{\alpha\, N\, D_\alpha\, Q^\alpha}{C_N^K\, a\, |b|}\left(\frac{C_N^2}{C_K^2}\right)^{b/2}\right]^{1/(b+\alpha)}. \tag{127.12}$$

After some simple algebra, the approximate energy is given by

$$E = \mathrm{sgn}(b)\,(b+\alpha)\left[\left(\frac{N\, D_\alpha}{|b|}\right)^b \left(\frac{a\, C_N^K}{\alpha}\right)^\alpha \left(\frac{C_K^2}{C_N^2}\right)^{\alpha\, b/2} Q^{\alpha\, b}\right]^{1/(b+\alpha)}. \tag{127.13}$$

The sign of E must be given by the sign of b, so the constraint $b > -\alpha$ appears. With $\alpha = b = 2$, the exact solution is found. For $K = \alpha = 2$ and $b = -1$, this result coincides with the one in [10], where the numerical accuracy has been tested. For $K = 2$ and $D_\alpha = \alpha = 1$, it coincides with a calculation in [7]. An upper bound is obtained if $\alpha \leq 2$ and $b \leq 2$. Another analytical example is given in [12], where results about critical coupling constants [14] are also presented.

References

1. Gattobigio, M., Kievsky, A., Viviani, M.: Spectra of helium clusters with up to six atoms using soft-core potentials. Phys. Rev. A **84**, 052503 (2011)
2. Ishikawa, S.: Three-body potentials in α-particle model of light nuclei. Few-Body Syst. **58**, 37 (2017)
3. Pepin, S., Stancu, Fl.: Three-body confinement force in hadron spectroscopy. Phys. Rev. D **65**, 054032 (2002)
4. Hall, R.L.: Energy trajectories for the N-boson problem by the method of potential envelopes. Phys. Rev. D **22**, 2062 (1980)
5. Hall, R.L.: A geometrical theory of energy trajectories in quantum mechanics. J. Math. Phys. **24**, 324 (1983)
6. Hall, R.L., Lucha, W., Schöberl, F.F.: Relativistic N-boson systems bound by pair potentials $V(r_{ij}) = g(r_{ij}^2)$. J. Math. Phys. **45**, 3086 (2004)
7. Silvestre-Brac, B., Semay, C., Buisseret, F., Brau, F.: The quantum $\mathcal{N}$-body problem and the auxiliary field method. J. Math. Phys. **51**, 032104 (2010)
8. Semay, C., Roland, C.: Approximate solutions for N-body Hamiltonians with identical particles in D dimensions. Res. Phys. **3**, 231 (2013)
9. Semay, C., Buisseret, F.: Bound cyclic systems with the envelope theory. Few-Body Syst. **58**, 151 (2017)
10. Semay, C.: Numerical tests of the envelope theory for Few-Boson systems. Few-Body Syst. **56**, 149 (2015)
11. Semay, C.: Improvement of the envelope theory with the dominantly orbital state method. Eur. Phys. J. Plus **130**, 156 (2015)

12. Semay, C., Sicorello, G.: Many-body forces with the envelope theory. Few-Body Syst. **59**, 119 (2018)
13. Semay, C.: General comparison theorem for eigenvalues of a certain class of Hamiltonians. Phys. Rev. A **83**, 024101 (2011)
14. Richard, J.-M., Fleck, S.: Limits on the domain of coupling constants for binding N-Body systems with no bound subsystems. Phys. Rev. Lett. **73**, 1464 (1994)

Chapter 128
Hyperspherical Harmonics Method with Particle Excitation Degrees of Freedom

Fabrizio Ferrari Ruffino, Winfried Leidemann and Giuseppina Orlandini

Abstract We introduce a few extensions to the Hyperspherical Harmonics (HH) approach in order to include intrinsic excitations degrees of freedom. To this end we adapt the HH expansion method, usually built up starting from a fixed mass-weighted set of Jacobi coordinates, to the more general case where an arbitrary weighted set of coordinates is adopted. We provide a few results for 3- and 4-body bound state calculations.

128.1 Introduction

The HH expansion method has been widely used for ab-initio calculations of nuclear systems. The method has been extended in different ways in order to treat a variety of interaction models like 3-body forces and non-local potentials. Here we introduce our recent extension in order to employ HH basis sets built over arbitrarily weighted sets of internal coordinates. This enables us to explicitly include particle transition degrees of freedom, as is the case for most of the models involving hyperons. In fact, with a few exceptions, almost all the available hyperon-nucleon (YN) interaction models are based on the explicit Λ-Σ coupling. Moreover in the $S=-2$ sector additional channels open up: $\Lambda\Lambda$, $\Lambda\Sigma$, $\Sigma\Sigma$ and $N\Xi$. This is similar to the NN case, where a few models include explicit Δ-resonance degrees of freedom, like, for example, the AV28 potential [1].

In line with our recent introduction of the HH method in the hypernuclear $S=-1$ sector [2] and in view of a future application in the $S=-2$ sector, we present here the extensions and the transformations needed in order to treat particle transition degrees of freedom. Separation energies of 3- and 4-body Λ-hypernuclei with a realistic Λ-Σ

F. Ferrari Ruffino · W. Leidemann (✉) · G. Orlandini
Department of Physics, University of Trento, 38123 Trento, Italy
e-mail: Winfried.leidemann@unitn.it

W. Leidemann · G. Orlandini
INFN-TIFPA Trento Institute of Fundamental Physics and Applications,
Via Sommarive, 14, 38123 Trento, Italy

N. A. Orr et al. (eds.), *Recent Progress in Few-Body Physics*,
Springer Proceedings in Physics 238,
https://doi.org/10.1007/978-3-030-32357-8_128

coupling YN interaction are calculated and already benchmarked with results from the literature in [2], here we give a further improvement of some of those results. A preliminary calculation of the triton binding energy and wave function is performed by considering the explicit $\Delta(1232)$ isobar degrees of freedom and including all the NN, $N\Delta$ and $\Delta\Delta$ channels. In this last case we adopt the AV28 potential.

128.2 Formalism

The explicit inclusion of particle excitation degrees of freedom implies that the particle masses are no longer fixed parameters, but they may depend on the state of the system. For example, the hamiltonian of Λ-hypernucleus is given by

$$H = T + \Delta M + \sum_{i<j} V_{N_i N_j} + \sum_{i} V_{N_i Y} \, . \tag{128.1}$$

Since the Λ hyperon has zero isospin, the strong $\Lambda\pi\Lambda$ vertex is forbidden due to isospin conservation. Thus the Λ particle can exchange one pion only via the $\Lambda\pi\Sigma$ vertex. This means that a YN force with pion exchange (OPE) must explicitly take into account the Λ-Σ coupling. In (128.1) the ΔM matrix is related to the mass difference in energy units between the Λ and the Σ particles. The kinetic energy T and a generic term of the hypernuclear part of the potential, V_{NY}, have the following block structure:

$$T = \begin{pmatrix} T_\Lambda & \\ & T_\Sigma \end{pmatrix} ; \qquad V_{NY} = \begin{pmatrix} V_{N\Lambda-N\Lambda} & V_{N\Sigma-N\Lambda} \\ V_{N\Lambda-N\Sigma} & V_{N\Sigma-N\Sigma} \end{pmatrix} , \tag{128.2}$$

where the upper diagonal block is defined over the subset of the basis where the hyperon is a Λ-particle, while, in the lower subset, a Σ-hyperon is considered. The potential operator couples the two subsets through the mixing terms $V_{N\Lambda-N\Sigma}$ and $V_{N\Sigma-N\Lambda}$.

Since the standard HH basis is built over the mass-weighted set of Jacobi coordinates, two different sets have to be included (one for the Λ and one for the Σ particle). Therefore the multidimensional integration of the mixing matrix elements cannot be reduced with standard HH techniques, due to the nontrivial relations between both HH coordinate sets. Obviously, the more particle excitation degrees of freedom are present, the more complex such a problem becomes.

In the following we briefly introduce the ways allowed to overcome the problem.

The transformation from cartesian to Jacobi coordinates is usually represented by means of the orthogonal matrix S:

$$\mathbf{v}_r = \frac{1}{\sqrt{m}} \mathcal{M} \cdot S^T \cdot \mathbf{v}_\eta \ ; \qquad \mathcal{M}_{ij} = \sqrt{m_i}\, \delta_{ij} , \tag{128.3}$$

where $\mathbf{v}_r$ and $\mathbf{v}_\eta$ are identical vectors defined, respectively, on the cartesian and on the Jacobi basis.

Each set of weights that defines a Jacobi system of coordinates is built over a set ζ of A real positive parameters:

$$\zeta = \{\mu_i > 0 \ : \ i = 1, \ldots, A\} . \tag{128.4}$$

In the standard case each parameter μ_i coincides with the physical mass m_i. The advantage of such a choice is the simpler form of the kinetic energy operator.

The transformation from one set of ζ-weighted coordinates to another ζ'-weighted is defined as:

$$W_{\zeta'\zeta} = S_{\zeta'} \, \mathcal{M}_{\zeta'}^{-1} \, \mathcal{M}_\zeta \, S_\zeta^T , \tag{128.5}$$

where S_ζ is the matrix defined in (128.3). The above matrix is upper triangular and it can be easily proved that each i-th subspace $\langle \boldsymbol{\eta}_1, \ldots, \boldsymbol{\eta}_i \rangle$ defined by the first i internal Jacobi coordinates is invariant under any transformation $W_{\zeta'\zeta}$ when ζ and ζ' differ only by the last i parameters. When $\mu_i' \neq \mu_i$ for all i, the subspace given by all the N internal coordinates is still invariant, therefore the separation of the $\boldsymbol{\eta}_0^\zeta$ coordinate from the internal set can be always preserved by keeping $\boldsymbol{\eta}_0^{\zeta'} = \boldsymbol{\eta}_0^\zeta$.

By means of a W transformation a generic A-body Hamiltonian can be expressed in terms of a ζ-weighted set of Jacobi coordinates:

$$H[\boldsymbol{\eta}(\boldsymbol{\eta}^\zeta)] = H(W_\zeta \cdot \boldsymbol{\eta}^\zeta) = H_\zeta(\boldsymbol{\eta}^\zeta) + \Delta T_\zeta , \tag{128.6}$$

where W_ζ transforms a ζ-weighted Jacobi set into a mass-weighted one. The operator H_ζ is the analogue of the Hamiltonian H but with the physical masses replaced by the weight parameters in ζ and ΔT_ζ is the difference between the kinetic energy operator T and the analogue operator T_ζ:

$$\Delta T_\zeta = T(\boldsymbol{\eta}^\zeta) - T_\zeta(\boldsymbol{\eta}^\zeta) . \tag{128.7}$$

By defining the ΔT_ζ operator one is free to work with the desired set of weight parameters without additional complications with respect to the mass-weighted case. However convergence is necessarily affected in some measure.

In this way, in order to calculate the mixing blocks of (128.2), one can simply adopt a fixed HH set of coordinates for both the Λ- and Σ-subsets and shift the mass-dependence into the kinetic energy operator:

$$T = \begin{pmatrix} T_\Lambda & \\ & T_\Lambda + \Delta T_{\Lambda-\Sigma} \end{pmatrix} . \tag{128.8}$$

Another way is to explicitly define the representation in the HH basis of the W transformations. It can be shown that the latter are compositions of kinematic rotations belonging to the orthogonal group $O(A-1)$ and real 1-d scaling transformations. Therefore, by defining the real scaling transformations in the HH basis, one is able to define any W transformation. We define the $\mathcal{W}$ coefficients representing a 1-d scaling transformation (over the last Jacobi coordinate) in the following way:

$$\left|\mathcal{L}_{n'}(\rho', \Omega_N)\right\rangle |\mathcal{Y}_{[K'_N]}(\Omega'_N)\rangle = \sum_{[K_N],n} \mathcal{W}_{[K_N],n}^{[K'_N],n'} |\mathcal{L}_n(\rho)\rangle \left|\mathcal{Y}_{[K_N]}(\Omega_N)\right\rangle , \qquad (128.9)$$

where we have stressed the fact that, in general, the dependence on ρ and Ω_N is no longer separate in the HH basis after a weight transformation.

128.3 Numerical Results

As a first step we have performed a number of numerical tests in order to verifiy the accuracy and numerical convergence of the approach shown above. We have obtained good results and convergence both for soft- and hard-core potentials. Then we have adopted the first approach described above for the calculation of binding and separation energies of 3- and 4-body Λ hypernuclei with explicit Σ degrees of freedom. We have also set up a calculation for the ^{3}H binding energy and wave function by including channels with one and two Δs, however, results are still preliminary in this case. The energies of the Λ-hypernuclei have been reported in (Table 128.1).

The accuracy in the 4-body case is lower due to much larger computational effort, however the agreement is good in bot 3- and 4-body cases.

In Table 128.2 we show the preliminar results of our 3-body calculation with the AV28 interaction model.

While the agreement between the wave function components related to the Δ components seems good, we still have a quite large discrepancy with the binding energy. A deeper study on the convergence of this last calculation is in progress.

Table 128.1 Binding energies and Λ separation energies (B_Λ) in MeV for various nuclear systems and interaction models

$\mathbf{V_{NN} + V_{YN}}$	System	NSHH	FY	GEM
AV8′	^{2}H	[−2.226(1)]	−2.226(1)	–
AV8′+sNSC97f	$^3_\Lambda$H	−2.413(3)	−2.415(1)	−2.42(1)
	B_Λ	0.187(3)	0.189(1)	0.19(1)
AV8′	^{3}H	−7.76(0)	−7.76(0)	−7.77(1)
AV8′+sNSC97f	$^4_\Lambda$H	−10.08(2)	–	−10.10(1)
	B_Λ	2.32(2)	–	2.33(1)

Results for Faddeev-Yakubovsky (FY) and Gaussian expansion method (GEM) as given in [2]

Table 128.2 Δ and Δ-Δ probabilities for ^{3}He with AV28 potential

Method	$\%(\Delta)$	$\%(\Delta\Delta)$	E_0
NSHH	1.6	1.1	[[−6.7]]
C.C	1.6	1.1	−7.30(1)[4]

128.4 Conclusions

The present aim is to obtain a definite result for the AV28 calculation of the ^{3}H nucleus by reaching full convergence. To this end a number of refinements of the above formalism are under development, in particular the inclusion of a suitable effective interaction procedure and the combination of both the $\mathcal{W}$ coefficients formalism and the ΔT one.

A deeper and general discussion on W-trnasformations and the $\mathcal{W}$ coefficients formalism will be presented soon.

References

1. Wiringa, R.B., Smith, R.A., Ainsworth, T.L.: Phys. Rev. C **29**, 1207 (1984)
2. Ferrari Ruffino, F., Lonardoni, D., Barnea, N., Deflorian, S., Leidemann, W., Orlandini, G., Pederiva, F.: Few-Body Syst. **58**, 113 (2017)
3. Hiyama, E., Ohnishi, S., Gibson, B.F., Rijken, T.A.: Phys. Rev. C **89**, 061302 (2014)
4. Picklesimer, A., Rice, R.A., Brandeburg, R.: Phys. Rev. C **45**, 2045

Chapter 129
Trions in Three-, Two- and One-Dimensional Materials

Roman Ya. Kezerashvili

Abstract The effect of reduction of dimensionality on the Coulomb potential and the binding energy of positively X^+ and negatively X^- charged trion is presented. In bulk semiconductors X^+ is unbound, while in 2D layered and 1D semiconductors both trions are bound and binding energies of X^+ are bigger than X^-.

129.1 Introduction

The study of a few-body electron-hole system in three-, two- and one-dimensional (3D, 2D, 1D) configuration spaces is of great fundamental significance in physics. Atomic or molecular thin 2D layers and 1D system such as quantum wires and nanotubes allow to address the role of Coulomb interactions in confined geometries. It was realized that for 2D semiconductors, the dielectric environment plays a crucial role and influences the effective strength of the Coulomb potential inside a semiconductor layer [1], which allows the formation of tightly bound two-, three- and four-body electron-hole complexes such as excitons, trions and biexcitons. Excitons and charged trions have been discovered in 2D transition metal dichalcogenides (TMDCs) monolayers. The strong exciton and trion binding in TMDC materials arises from the reduced screening in the 2D geometry as well as effective masses of both the electron and the hole. In fact, screening effects play a fundamental role in determining the electron dynamics and the binding energy of excitonic complexes. Moreover, the screening dictates the optical and transport properties of 2D devices [2].

Although the exciton complexes like trions in solid state physics are very similar to the three-body bound systems in atomic and nuclear physics, there are major differences: i. Screening effects, resulting from the host lattice, make the Coulomb

R. Ya. Kezerashvili (✉)
New York City College of Technology, City University of New York, New York City, NY 10212, USA
e-mail: rkezerashvili@citytech.cuny.edu

N. A. Orr et al. (eds.), *Recent Progress in Few-Body Physics*,
Springer Proceedings in Physics 238,
https://doi.org/10.1007/978-3-030-32357-8_129

force much weaker than in atomic systems; ii. Band effects, which make the effective masses of the electrons and holes smaller than the bare electron mass.

In this short note we address how the reduction of dimensionality affects the binding energy of trions in bulk and low-dimensional semiconductors.

129.2 Effects due to the reduction of dimensionality

Wannier-Mott trions in bulk and low-dimensional semiconductors represent a three-body system with two identical particle: AAB. Negative X^- and positive X^+ trions are formed in semiconductors when an exciton is correlated with an additional electron in a conduction band or hole in a valence band, respectively, and produces complexes with two identical particles: *eeh* or *ehh*. The Schrödinger picture within the framework of the effective-mass potential model approach is well suited to describe Wannier-Mott trions and quantum-confined structures. The corresponding Schrödinger equation for the trion in 3D, 2D and 1D configuration space reads as

$$\left({}^{jD}T + \sum_{i<k}^{jD} V(r_{ik})\right)\Psi_{jD}(r_{12}, r_{23}, r_{13}) = E_{jD}\Psi_{jD}(r_{12}, r_{23}, r_{13}), \qquad (129.1)$$

where index $j = 3, 2, 1$ presents the dimensionality of the space and the position vectors $\boldsymbol{r}_i$ are defined in the jD space. In (129.1) ${}^{jD}T = -\frac{\hbar^2}{2m_A}{}^{JD}_{A1}\Delta - \frac{\hbar^2}{2m_A}{}^{JD}_{A2}\Delta - \frac{\hbar^2}{2m_B}{}^{JD}_{B}\Delta$ is a three-body kinetic energy operator, where ${}^{JD}\Delta$ is the Laplace operator for each particle and ${}^{jD}V(r_{ik})$ is a pairwise potential in the jD configuration space. One can obtain the expectation value for the ground state energy as $E_{jD} = \langle {}^{jD}T\rangle + \left\langle \sum_{i<k}^{jD} V(r_{ik})\right\rangle$. The later expression could be viewed as the sum of the average value of the operators of kinetic and potential energies in 3D, 2D and 1D configuration space, respectively, obtained by using the corresponding eigenfunction of the trion $\Psi_{jD}(r_{12}, r_{23,}r_{13})$ in 3D, 2D and 1D configuration spaces. The decrease of dimensionality produces a variation in the energy spectrum of the system, as well as affects the Coulomb interaction between an electron and hole. For example, let us consider the effect of the reduction of the dimensionality on the kinetic energy. In the two-body problem when the interaction is described by the Coulomb potential due to the reduction of dimensionality only for the kinetic energy, one can observe that the spectrum of energy is changes from $E_{3D} \sim n^{-2}$ (Rydberg series) in the 3D case, to $E_{2D} \sim (n - 1/2)^{-2}$ in the 2D case. Therefore, for example, the ground state energy increases by a factor of 4. Thus, the reduction of dimensionality decreases the kinetic energy of the 2D exciton due to the decrease of the degrees of freedom from three to two. However, the reduction of dimensionality affects the potential energy of the charged carriers' interaction, while this interaction is still electromagnetic. In

particular, the Coulomb potential due to the decrease of dimensionality is modified to the potential [3] in a 2D system and cusp-type potential in a 1D system:

$$^{3D}V(r) = \frac{ke^2}{\varepsilon r} \Rightarrow {}^{2D}V(r) = \frac{\pi ke^2}{(\varepsilon_1 + \varepsilon_2)\rho_0}\left[H_0\left(\frac{r}{\rho_0}\right) + Y_0\left(\frac{r}{\rho_0}\right)\right] \Rightarrow {}^{1D}V(z) = \frac{\pi ke^2}{\varepsilon(a)}\frac{A}{z - z_0}. \tag{129.2}$$

In (129.2) $k = 9 \times 10^9\ \mathrm{Nm^2/C^2}$, ρ_0 is the screening length, ε, ε_1 and ε_2 are the dielectric constants of a bulk and two materials that the 2D layer is surrounded by, respectively, and $H_0(r/\rho_0)$ and $Y_0(r/\rho_0)$ are the Struve function and Bessel function of the second kind, respectively. For $r \gg \rho_0$ the potential has the 3D bare Coulomb tail and becomes $^{3D}V(r)$, while when $r \ll \rho_0$ it becomes a logarithmic potential: $\left(ke^2/\varepsilon\rho_0\right)\left[\ln(\mathrm{r}/2\rho_0) + \gamma\right]$, where γ is the Euler constant. Thus at $r \ll \rho_0$ the effect of the induced polarization becomes dominant—the $1/r$ singularity is replaced by a weaker logarithmic dependence. For a 1D case in (129.2) a dielectric constant depends on a radius a of a nanowire.

To better understand the difference between the screening in 3D and 2D materials let us follow [2] and consider the effect of the macroscopic polarization induced by a point charge surrounded by a 3D and 2D dielectric medium, respectively. The electric field at a distance r from the charge is the sum of the external field produced by the charge, e/r^2, and the induced field due to the polarization of the medium. This charge distribution produces a field of the same functional form $ke/\varepsilon r^2$ and the screening is given by a simple multiplicative renormalization by the dielectric constant ε. In contrast to the 3D case, in the 2D case the system is polarizable only on the plane and induced field is equivalent to the electric field produced by a uniform charge distribution on a circle of radius r. As a consequence it will be a function of r, but with a functional form substantially different from the electric field $ke/\varepsilon r^2$ and with a nonlocal macroscopic screening showing a logarithmic divergence for small distances [3]. Contrary to the 3D system, where the macroscopic screening is mapped in a dielectric constant, in 2D system the screening is nonlocal so that in the 2D Fourier space it is described by a q-dependent macroscopic dielectric function [2, 3], which should be contrasted to the multiplicative renormalization of the charge in 3D case, as seen in (129.2).

There are different models of the effective interaction potential in 1D quantum wire: (i) the singularity of the Coulomb potential is cut off at $r = a$, where a is the radius of a wire, and the effective potential is $^{1D}V(z) \sim (z^2 + a^2)^{-1/2}$; (ii) The effective 1D interaction is modeled as cusp-type Coulomb potential $^{1D}V(z) \sim A(z + r_0)^{-1}$, where the parameters A and r_0 are determined self-consistently by employing the eigenfunctions of the lateral confinement of electrons and holes. Figure 129.1 depicts the electric field lines between the interacting electron and hole. For 2D materials field lines are screened in plane and mainly lie unscreened in the vacuum. In 1D materials field lines lie mainly in the vacuum, hence screening is heavily suppressed.

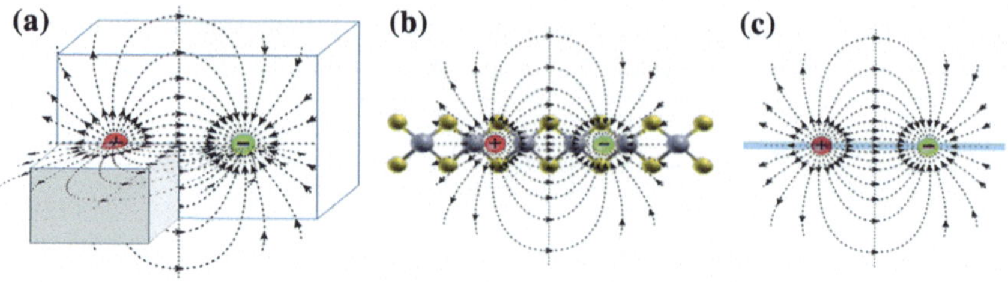

Fig. 129.1 The scheme of electric fields for two interacting particles in a uniform dielectric environment in 3D, 2D and 1D materials

129.3 Binding energy of trions

3D trions. Calculations of the binding energies of trions in different 3D materials using the Faddeev equations in configuration space [4] give the following results for X^-: 3.6 meV (InN), 0.5 mev (GaAs), 2.1 meV (ZnSe) and 0.6 meV (CdTe) and X^+ is unbound. In calculations the ratio of mass m_e/m_h for these materials is varied from 0.07 to 0.27 and for these ratios the exciton binding energies are in good agreement with experimental data and theoretical calculations. The binding energy for X^- and X^+ are equal if $m_e = m_h$ or if one ignores the interaction between two identical particles.

In [4] we considered the hypothetical model, which controls the strength of interaction between identical particles for both trions and leads to a weaker Coulomb interaction between the identical particles and hence an increased trion binding energy. Our calculations for the distributions of electrons and holes in $X^\pm$ based on the Faddeev equations shows that the X^- has a more extended distribution for two electrons, while the two holes in X^+ are considerably closer to each other. Therefore, comparing the interactions of three particles in $X^\pm$, one concludes that while the *eh* attraction is the same, the *hh* repulsion is stronger in X^+ than the *ee* one in X^- due to more close localization of the two holes. This repulsion reduces the energy of trions so that X^+ becomes unbound for all considered dielectric constants.

2D trions. Follow [5, 6] for the nonrelativistic trion (129.1) in a 2D configuration space can be written using the formalism of hyperspherical harmonics. Within this approach the binding energies of trions in the TMDCs monolayers are calculated and presented in Table 129.1. Analyses of the results shows that in contrast to 3D

Table 129.1 The ratio of binding energies B_{X^+} and B_{X^-} of charged trions for different TMDCs that have different ratio of effective electron m_e and hole m_h masses

	MoS_2	$MoSe_2$	$MoTe_2$	WS_2	WSe_2	WTe_2
m_e/m_h	0.819	0.863	0.821	0.739	0.756	0.602
ρ_0, Å	38.62	51.71	73.61	37.89	34.72	49.56
B_{X^+}/B_{X^-}	1.012	1.044	1.023	1.005	1.006	1.022

case X^+ is bound, moreover the binding energy of X^+ always exceeds the binding energy of X^-.

1D trions. The study of $X^\pm$ trions in 1D semiconductor wires within a variational approach [7] and of X^+ in the Born-Oppenheimer approximation [8] shows that trion binding energies are defined by two parameters: the mass ratio, m_e/m_h, and the radius of the quantum wire a. Both X^+ and X^- are bound. With the increase of mass ratio m_e/m_h and a the binding energies of $X^\pm$ trions are decreasing, but the $B_{X^+} > B_{X^-}$ and the binding energy B_{X^+} varies more dramatically than B_{X^-}.

To conclude, a decrease of dimensionality leads to a decrease of the degrees of freedom in the system, which, obviously, decreases the kinetic energy of the system. At the same time, a decrease of dimensionality reduces the screening that leads to weaker interaction. The combinations of these two effects finally leads to 2D and 1D bound trions because with the reduction of dimensionality the kinetic energy decreases faster than the potential energy, while in 3D X^- is bound and X^+ is unbound. Therefore, lowering the dimensionality of the system makes the effects of interaction between particles much more pronounced, resulting in an increase in the binding energy of trions.

Acknowledgements The work is supported by US Department of Defense under Grant No. W911NF1810433 and PSC CUNY under Grants No 61188-00-49 and 662261-00-50.

References

1. Wang, G., Chernikov, A., Glazov, M.M., et al.: Rev. Mod. Phys. **90**, 21001 (2018)
2. Cudazzo, P., Tokatly, I.V., Rubio, A.: Phys. Rev. B **84**, 085406 (2011)
3. Keldysh, L.V.: JETP Lett. **29**, 658 (1979)
4. Filikhin, I., Kezerashvili, R. Ya., Vlahovic, B.: Phys. Lett. A **382**, 787 (2018)
5. Kezerashvili, R. Ya., Tsiklauri, Sh. M.: Few-Body Syst. **58**, 18 (2017)
6. Filikhin, I., Kezerashvili, R. Ya., Tsiklauri, Sh. M., Vlahovic, B.: Nanotechnology **29**, 124002 (2018)
7. Semina, M.A., Sergeev, R.A., Suris, R.A.: Semiconductors **42**, 1427 (2008)
8. Kezerashvili, R. Ya., Machavariani, Z. S., Beradze, B., Tchelidze, T.: Physica E **109**, 228 (2019)

Chapter 130
A Statistical Analysis of the Nuclear Structure Uncertainties in μD

Oscar J. Hernandez, Sonia Bacca, Nir Barnea, Nir Nevo-Dinur, Andreas Ekström and Chen Ji

Abstract The charge radius of the deuteron (D), was recently determined to three times the precision compared with previous measurements using the measured Lamb shift in muonic deuterium (μD). However, the μD value is 5.6 σ smaller than the world averaged CODATA-2014 value (Pohl R et al. (2016) Science 353:669 [1]). To shed light on this discrepancy we analyze the uncertainties of the nuclear structure calculations of the Lamb shift in μD and conclude that nuclear theory uncertainty is not likely to be the source of the discrepancy.

O. J. Hernandez (✉) · S. Bacca
Institut für Kernphysik and PRISMA Cluster of Excellence,
Johannes-Gutenberg-Universität Mainz, 55128 Mainz, Germany
e-mail: javierh@phas.ubc.ca

O. J. Hernandez
Department of Physics and Astronomy, University of British Columbia,
Vancouver, BC V6T 1Z4, Canada

O. J. Hernandez · S. Bacca · N. Nevo-Dinur
TRIUMF, 4004 Wesbrook Mall, Vancouver, BC V6T 2A3, Canada

S. Bacca
Department of Physics and Astronomy, University of Manitoba,
Winnipeg, MB R3T 2N2, Canada

N. Barnea
Racah Institute of Physics, The Hebrew University, 9190401 Jerusalem, Israel

A. Ekström
Department of Physics, Chalmers University of Technology,
412 96 Gothenburg, Sweden

C. Ji
Institute of Particle Physics, Central China Normal University,
Wuhan 430079, China

N. A. Orr et al. (eds.), *Recent Progress in Few-Body Physics*,
Springer Proceedings in Physics 238,
https://doi.org/10.1007/978-3-030-32357-8_130

130.1 Introduction

The two-photon exchange (TPE) contribution is a crucial ingredient in the precision determination of the charge radius from Lamb shift (LS) measurements in muonic atoms. The charge radius is extracted from the measurements of the $2S$-$2P$ energy splitting ΔE_{LS} through

$$\Delta E_{LS} = \delta_{QED} + \delta_{TPE} + \delta_{FS}(r_d^2), \tag{130.1}$$

valid up to fifth order in $(Z\alpha)$, where Z is the charge number of the nucleus and α is the fine structure constant. The term δ_{QED} denote the quantum electrodynamic (QED) corrections, δ_{TPE} are the nuclear structure corrections dominated by the two-photon exchange, and $\delta_{FS}(r_d^2)$ is the finite size correction proportional to the deuteron charge radius r_d. The bottle-neck in the precise determination of r_d are the nuclear structure corrections. In this work, we overview the process of the uncertainty quantification of δ_{TPE} in μD using nucleon-nucleon (NN) potentials at various orders (from LO to N^4LO) in chiral effective field theory (EFT).

130.2 Analysis of Uncertainties

To quantify the total theoretical uncertainties of δ_{TPE}, all relevant uncertainty sources must be identified and estimated [2, 3]. These various sources are:

- σ_{stat}: uncertainties arising from the spread of the low-energy constants (LECs) $\tilde{\alpha}$ in the nuclear potential;
- $\sigma_{T_{Lab}^{Max}}$: uncertainties from the maximum lab energy T_{Lab}^{Max} used in the fits of the NN potential;
- σ_{Δ}: uncertainty due to the truncation of the chiral order;
- σ_{Λ}: uncertainty from the variations of the the cut-off Λ in the NN potentials;
- σ_{η}: uncertainty due to the expansion (on a parameter known as η) which we use in relating δ_{TPE} to the nuclear response functions;
- σ_{J}: uncertainties from systematic approximations in the electromagnetic operators $J^{\mu}(x)$;
- σ_{N}: uncertainties due to single nucleon physics;
- $\sigma_{Z\alpha}$: uncertainties arising from higher $(Z\alpha)$ corrections.

For an observable A, the statistical uncertainties $\sigma_{stat}(A)$ induced by variations in the LECs $\tilde{\alpha}$ of the NN potential are calculated around their optimal values $\tilde{\alpha}_0$ by assuming that the LECs follow a multivariate Gaussian probability distribution. Under these conditions the leading approximation to $\sigma_{stat}(A)$ will be given by

$$\sigma_{stat}^2(A) = \langle A^2\rangle - \langle A\rangle = \mathbf{J}_A \mathrm{Cov}(\tilde{\alpha}_0)\mathbf{J}_A^T, \tag{130.2}$$

where $\mathrm{Cov}(\tilde{\alpha}_0)$ represents the covariance matrix of the LECs at the optimum, and $\mathbf{J}_A$ is the Jacobian vector of A with respect to the LECs,

$$J_{A,i} = \left(\frac{\partial A}{\partial \tilde{\alpha}_i}\right)_{\tilde{\alpha}=\tilde{\alpha}_0}. \tag{130.3}$$

The systematic uncertainties $T_{\mathrm{Lab}}^{\mathrm{Max}}$ arise from the energy span in the NN scattering data used to fit the LECs. This uncertainty was estimated from the N^{k}LO$_{\mathrm{sim}}$ potentials ($k = 0, 1, 2$) [4] by varying the maximum lab energies of the fit from 125 to 290 MeV and their uncertainties $\sigma_{T_{\mathrm{Lab}}^{\mathrm{Max}}}$ where found to dominate over the statistical uncertainties σ_{stat}.

The chiral truncation uncertainties σ_Δ originate from the calculation of an observable $A(p)$ at a finite order ν, with associated momentum scale p. This observable is assumed to obey the same expansion as the underlying NN-force given by

$$A(p) = A_0 \sum_{\mu=0}^{\nu} c_\mu(p) Q^\mu, \tag{130.4}$$

where A_0 is the result at leading order, Q is the expansion parameter, and $c_\mu(p)$ are observable and interaction specific coefficients assumed to be independent and of natural size. Assuming that the next higher-order term $\Delta_\nu^{(1)} \equiv c_{\nu+1} Q^{\nu+1}$ dominates the truncation uncertainty in the calculation of $A(p)$, then the Bayesian posterior $P(\Delta_\nu^{(1)})$ is given by Furnstahl et al. [5]

$$P(\Delta_\nu^{(1)}) = \frac{\int d\bar{c}\ P(c_{\nu+1}|\bar{c}) P(c_0|\bar{c}) P(c_2|\bar{c}) \ldots P(c_\nu|\bar{c}) P(\bar{c})}{Q^{\nu+1} \int d\bar{c}\ P(c_0|\bar{c}) P(c_2|\bar{c}) \ldots P(c_\nu|\bar{c}) P(\bar{c})}, \tag{130.5}$$

where $P(c_\mu|\bar{c})$ is the distribution of c_μ conditioned on the scale parameter $\bar{c}$ and $P(\bar{c})$ is the prior. In this contribution we update the results in [2] by evaluating the 68% confidence intervals of the posteriors given in (130.5) that represent the chiral truncation uncertainty σ_Δ. The posterior distributions $A_0\Delta_\nu^{(1)}$ from N^{2}LO to N^{4}LO for δ_{TPE} using the chiral potentials from [6] are given in Fig. 130.1 for the priors A, B, C from Table I in [5].

Along with chiral truncation uncertainties, the chiral NN-potentials carry a parameter Λ that regulates the interactions. The systematic uncertainties σ_Λ arising from the regulators was probed using multiple cut-off values in the calculations of δ_{TPE}. These variations were found to be more significant than the uncertainties due to the chiral truncation.

The η-expansion arises from the calculation of δ_{TPE} as a power series of the dimensionless operator $\eta \ll 1$. In the work of [2, 3], this expansion was carried out to sub-sub-leading order in η and the truncation uncertainty σ_η from higher order terms was determined to be 0.3%.

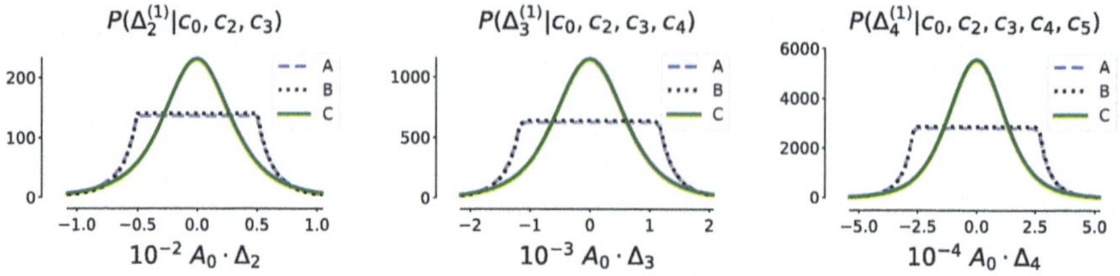

Fig. 130.1 Posterior distributions of the truncation uncertainties at different orders in the chiral EFT expansion ($A_0\Delta_2^{(1)}$, $A_0\Delta_3^{(1)}$, $A_0\Delta_4^{(1)}$) in meV units for δ_{TPE} in the leading-omitted term approximation using the potential in [6] with $(R_0, \Lambda) = (0.8, 600)$ [fm, MeV]. The expansion parameter is $Q = 0.23$, $\sigma = 1$ for prior B, $\bar{c}_< = 0.1$ and $\bar{c}_> = 10$

Uncertainties from approximations in the electromagentic operators $J^{\mu}(x)$, were estimated from the dipole response functions of Arenhövel [7] that included MEC and relativistic corrections. Both of these effects were of the order 0.05%.

The uncertainties σ_{N} from single nucleon contributions to the TPE are an input in our analysis and taken from [8, 9] and [10]. Lastly, there was an estimated 1% uncertainty from higher order $(Z\alpha)^6$ corrections, that include the three photon exchange.

130.3 Results and Conclusions

The results of the analysis outlined in the previous section are summarized in Table 130.1. The systematic nuclear physics uncertainty σ_{syst} is a combination of the σ_{Δ}, σ_{J} and $\sigma_{T_{\text{Lab}}^{\text{Max}}}$ uncertainties, while σ_{Total} is a quadrature sum of all items in Table 130.1. The calculation of σ_{Δ} through the explicit calculation of the 68% confidence interval of the Bayesian posteriors instead of the prescription in [6] increases the lower bound slightly in σ_{Total} from -0.024 meV in [2] to -0.023 meV when using the σ_{N} values of [8] since the values of σ_{Δ} at N^{4}LO for prior A are smaller

Table 130.1 The uncertainty breakdown of the δ_{TPE} at N^{4}LO

Source	% Uncertainty	Uncertainty in meV
σ_{syst}	+0.5	+0.008
	−0.6	−0.011
σ_{stat}	0.06	±0.001
σ_{η}	0.3	±0.005
σ_{N}	0.6/1.2	±0.0102 [8]/±0.0198 [10]
$\sigma_{Z\alpha}$	1.0	±0.0172
σ_{Total}	1.3/1.6-1.7	+0.022/+0.028
		−0.023/−0.029

when computed this way. The final value for the TPE correction was taken to be the average value of the calculations at N^4LO yielding $\delta_{TPE} = -1.715$ meV with the final uncertainty σ_{Total}. This value differs from the experimentally determined value from [1] of $\delta_{TPE} = -1.7638(68)$ meV by less than 2 σ, which is not significant. From Table 130.1 we find that the uncertainties arising from the nuclear model dependence, σ_{syst} and σ_{stat}, are small in comparison to the higher order $\sigma_{Z\alpha}$ or σ_N [10] uncertainties which dominate the total uncertainty. It is therefore unlikely that any differences between the experimental and theoretical determinations of δ_{TPE} stem from models of the NN-forces.

References

1. Pohl, R., et al.: Science **353**, 669 (2016)
2. Hernandez, O.J., et al.: Phys. Lett. B **778**, 377 (2018)
3. Ji, C., et al.: J. Phys. G: Nucl. Part. Phys. **45**, 093002 (2018)
4. Carlsson, B.D., et al.: Phys. Rev. X **6**, 011019 (2016)
5. Furnstahl, R.J., et al.: Phys. Rev. C **92**, 024005 (2015)
6. Epelbaum, E., et al.: Phys. Rev. Lett. **115**, 122301 (2015)
7. Arenhövel, H.: Priv. Commun.
8. Krauth, J.J., et al.: Ann. Phys. **366**, 168 (2016)
9. Carlson, C.E., et al.: Phys. Rev. A **89**, 022504 (2014)
10. Hill, R.J., Paz, G.: Phys. Rev. D **95**, 094017 (2017)

Chapter 131
Loss of Conformality in Efimov Physics

Abhishek Mohapatra and Eric Braaten

Abstract The loss of conformal invariance in Efimov physics is due to the merger and disappearance of an infrared and an ultraviolet fixed point of a three-body renormalization group flow as the spatial dimension d is varied. In the case of identical bosons at unitarity, it is known that there are two critical dimensions $d_1 = 2.30$ and $d_2 = 3.76$ at which there is loss of conformality. For $d < d_1$ and $d > d_2$, the beta function of the three-body coupling has real roots which correspond to infrared and ultraviolet fixed points. The fixed points merge and disappear into the complex plane at the critical dimensions d_1 and d_2. For $d_1 < d < d_2$, the beta function has complex roots and the renormalization group flow for the three-body coupling constant is a limit cycle.

131.1 Introduction

In relativistic systems, the conformal symmetry group includes Poincaré symmetry and continuous scaling symmetry as subgroups. From the renormalization group (RG) perspective, scale invariance in any system arises at the fixed points of a RG flow. A general mechanism for the loss of conformal invariance in the system is the merging of an infrared (IR) and an ultraviolet (UV) fixed point and their disappearance into the complex plane as an external parameter is varied [1]. In nonrelativistic systems with short-range interactions, Efimov discovered that the spectrum of a three-body system can have a sequence of infinitely many shallow three-body bound states with an accumulation point at the three-body threshold [2]. This phenomena is called the *Efimov effect*. In the case of identical bosons in three spatial dimensions, the

A. Mohapatra (✉)
Department of Physics, Duke University, Durham, NC 27705, USA
e-mail: am749@duke.edu

E. Braaten
Department of Physics, The Ohio State University, Columbus, OH 43201, USA
e-mail: braaten.1@osu.edu

N. A. Orr et al. (eds.), *Recent Progress in Few-Body Physics*,
Springer Proceedings in Physics 238,
https://doi.org/10.1007/978-3-030-32357-8_131

spectrum of Efimov states reflects a discrete scaling symmetry with discrete scaling factor $e^{\pi/s_0} \approx 22.7$, where $s_0 = 1.00624$.

In this article, we use the RG perspective to explicitly show how the loss of conformal invariance happens in Efimov physics in the case of identical bosons as the spatial dimension d is varied. We discuss briefly the fixed points in the two-body sector in Sect. 131.2. In Sect. 131.3, we discuss the fixed points and the loss of conformal invariance in the three-body sector in more detail. We summarize our results in Sect. 131.4.

131.2 Two-Body Sector

In three spatial dimensions, the low-energy scattering of two nonrelativistic particles with short-range interactions is characterized by the s-wave scattering length a only. The *zero-range* limit is defined by taking the range of interaction to zero with the scattering length a held fixed. In the *unitary* limit defined by $a \to \pm\infty$, the two-body system is scale invariant. From the RG perspective, scale invariance arises at zeros of the beta function, which correspond to RG fixed points.

We consider the behavior of identical bosons in the zero-range limit in d dimensions. We use the effective field theory (EFT) of Bedaque, Hammer and van Kolck (BHvK) that was used to calculate the behavior of three identical bosons in three spatial dimensions [3]. This EFT has a dynamical field ψ and an auxiliary diatom field Δ. The BHvK Lagrangian density is

$$\mathcal{L}_{\text{BHvK}} = \psi^\dagger \left(i\frac{\partial}{\partial t} + \frac{1}{2}\nabla^2\right)\psi + \frac{g_2}{4}\Delta^\dagger\Delta - \frac{g_2}{4}\left(\Delta^\dagger\psi^2 + \psi^{\dagger 2}\Delta\right) - \frac{g_3}{36}\left(\Delta\psi\right)^\dagger\left(\Delta\psi\right), \tag{131.1}$$

where g_2 and g_3 are the bare two-body and three-body coupling constants. The perturbative expansion of the off-shell two-body scattering amplitude in powers of g_2 is UV divergent. It can be regularized by imposing a cutoff Λ on the loop momenta. The bare coupling g_2 must depend on the cutoff Λ to exactly compensate for the Λ dependence in the two-body amplitude. The explicit dependence of the two-body coupling g_2 on the cutoff Λ is given in (5) of [4]. We define a dimensionless two-body coupling $\hat{g}_2$ by

$$\hat{g}_2(\Lambda) = \frac{1}{(d-2)(4\pi)^{d/2}\Gamma\left(\frac{d}{2}\right)}\Lambda^{d-2}g_2(\Lambda). \tag{131.2}$$

In terms of the dimensionless coupling $\hat{g}_2$, the RG equation is

$$\Lambda\frac{d}{d\Lambda}\hat{g}_2 = (d-2)\,\hat{g}_2\left(\hat{g}_2 + 1\right). \tag{131.3}$$

The β function defined by the right hand side of the above equation has zeros at $\hat{g}_2 = -1$ and $\hat{g}_2 = 0$, which are an UV and a IR fixed point of the RG flow. The IR fixed point corresponds to the trivial non-interacting limit, whereas the UV fixed point corresponds to the non-trivial unitary limit. There are scale-invariant interactions in the two-body sector at the UV fixed point.

131.3 Three-Body Sector

Discrete scale invariance in Efimov physics can be associated with an RG flow governed by a limit cycle [3]. The RG equation for the three-body coupling g_3 was first derived by Braaten and Hammer in $d = 3$ in [5]. The RG equation for g_3 in d spatial dimensions was derived by us in [4]. The detailed derivation of the results in this section are given in [4].

The dependence of the three-body coupling g_3 on the UV cutoff Λ can be determined from the off-shell atom-diatom amplitude. In the center-of-mass frame, the atom-diatom amplitude is a function of the incoming and outgoing relative momenta $\boldsymbol{p}$ and $\boldsymbol{k}$ and the total energy E. It satisfies the Skorniakov-Ter-Martirosian (STM) integral equation [6], which is given explicitly in (131.4) of [4]. The integral equation involves a dimensionless three-body coupling $\hat{G}(\Lambda)$ defined by

$$\hat{G}(\Lambda) = \frac{\Lambda^2 g_3(\Lambda)}{9 g_2(\Lambda)^2}. \tag{131.4}$$

In the limit $\Lambda \to \infty$, the solutions to the STM equation depend log-periodically on the cutoff Λ. The dependence of the three-body coupling g_3, or equivalently the dimensionless coupling $\hat{G}$, on Λ can be determined by demanding that the solutions to the integral equation are well behaved in the limit $\Lambda \to \infty$. Since $\hat{G}$ is the only coupling in the integral equation, the β function for $\hat{G}$ can only depend on $\hat{G}$. The RG equation for the dimensionless coupling $\hat{G}$ is

$$\Lambda \frac{d}{d\Lambda} \hat{G} = \frac{1 - s^2}{2} + \left(1 + s^2\right) \hat{G} + \frac{1 - s^2}{2} \hat{G}^2, \tag{131.5}$$

where s^2 is a function of dimension d that satisfies a transcendental equation:

$$2 \sin\left(\frac{d}{2}\pi\right) {}_2F_1\left(\begin{matrix} \frac{d-1+s}{2}, \frac{d-1-s}{2} \\ \frac{d}{2} \end{matrix} \middle| \frac{1}{4}\right) + \cos\left(\frac{s}{2}\pi\right) = 0. \tag{131.6}$$

The transcendental equation has infinitely many branches of solutions for s^2 as a function of d, but the physically relevant branch is shown in Fig. 131.1. At $d = 3$, the value of s^2 is $-s_0^2$, where $s_0 = 1.00624$ determines the discrete scale factor e^{π/s_0} of Efimov physics. There are two critical dimensions for which (131.6) is satisfied for $s^2 = 0$:

$$d_1 = 2.30, \quad d_2 = 3.76. \tag{131.7}$$

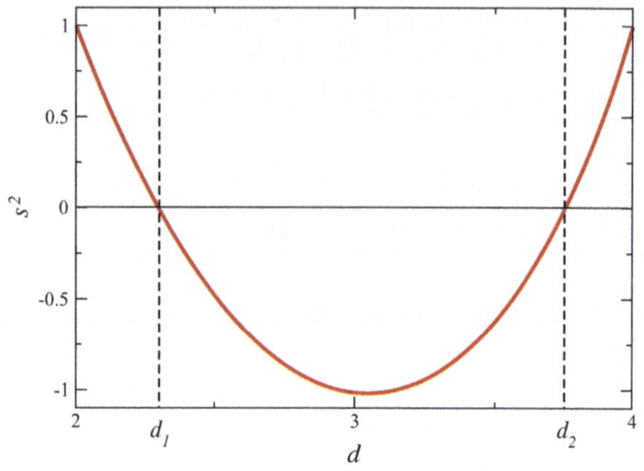

Fig. 131.1 The relevant branch of solutions of (131.6) for s^2 as a function of the dimension d. The value of s^2 decreases from 1 at $d = 2$ and $d = 4$ to a minimum value of -1.016 at $d = 3.04$. Adapted from [4]

These lower and upper critical dimensions for the Efimov effect were first derived in [7].

In the regions $d < d_1$ and $d > d_2$, the value of s^2 is positive. The β function in (131.5) for the dimensionless coupling $\hat{G}$ has zeros at $\hat{G}_+$ and $\hat{G}_-$, which correspond to a UV and an IR fixed point of the RG flow:

$$\hat{G}_\pm = -\frac{1 \pm s}{1 \mp s}. \tag{131.8}$$

As d approaches d_1 from below and d_2 from above, the value of s approaches 0, so the two fixed points approach -1.

In the region $d_1 < d < d_2$, the value of s^2 is negative. The β function for $\hat{G}$ has complex zeros, which implies that there are no fixed points. The RG flow for $\hat{G}$ is instead governed by a limit cycle:

$$\hat{G}(\Lambda) = -\frac{\cos\left[s_0 \log(\Lambda/\Lambda_*) + \arctan s_0\right]}{\cos\left[s_0 \log(\Lambda/\Lambda_*) - \arctan s_0\right]}, \tag{131.9}$$

where Λ_* is a constant momentum scale and $s_0 = \sqrt{-s^2}$. The expression for $\hat{G}$ in (131.9) is a log-periodic function of cutoff Λ with period e^{π/s_0}. At unitarity, the two-body sector is scale invariant ($\hat{g}_2 = -1$), but the three-body sector has discrete scale invariance with discrete scaling factor e^{π/s_0}.

131.4 Summary

In this work, we explicitly showed how conformality is lost in Efimov physics in the case of identical bosons as the spatial dimension d is varied. There are critical dimensions $d_1 = 2.30$ and $d_2 = 3.76$ at which conformality is lost [7]. For $d < d_1$ and for $d > d_2$, the RG flow has IR and UV fixed points that merge together at $d = d_1$ and at $d = d_2$, respectively. In the region $d_1 < d < d_2$, there are no fixed points and the RG flow is governed by a limit cycle with discrete scaling factor e^{π/s_0}.

Acknowledgements This research was supported by the National Science Foundation under grant PHY-1607190 and U.S. Department of Energy, Office of Science, Office of Nuclear Physics, under Award Number DE- FG02-05ER41368.

References

1. Kaplan, D., Lee, J.W., Son, D.T., Stephanov, M.A.: Phys. Rev. D. **80**, 125005 (2009). [arXiv:0905.4752]
2. Efimov, V.: Phys. Lett. B **33**, 563 (1970)
3. Bedaque, P.F., Hammer, H.-W., van Kolck, U.: Phys. Rev. Lett. **82**, 463 (1999) [nucl-th/9809025]; Nucl. Phys. A **646**, 444 (1999) [nucl-th/9811046]
4. Mohapatra, A., Braaten, E.: Phys. Rev. A **98**, 013633 (2018). [arXiv:1710.08447]
5. Braaten, E., Hammer, H.-W.: Phys. Rep. **428**, 259 (2006) [cond-mat/0410417]
6. Skorniakov, G.V., Ter-Martirosian, K.A.: Sov. Phys. JETP **4**, 648 (1957)
7. Nielsen, E., Fedorov, D.V., Jensen, A.S., Garrido, E.: Phys. Rep. **347**, 373 (2001)

Chapter 132
Neutron Matter in the Unitary Limit with Implicit Renormalization of Short Range Interactions

E. Ruiz Arriola, S. Szpigel and V. S. Timóteo

Abstract We study a strongly interacting many-fermion system in the unitary limit using an implicit renormalization framework and compute the Bertsch parameter ξ considering only contact interactions. The main ingredient of the calculation is the scale separation between low and high momentum degrees of freedom, which we take as the Fermi momentum k_F, and the assumption that the physics below this momentum scale can be re-parametrized into the low energy coefficients of the contact interactions. Once the unitary renormalization conditions are imposed on the two-body scattering amplitude obtained from the contact interactions, we evaluate ξ at leading order (LO), next-to-leading order (NLO) and next-to-next-to-leading order (NNLO).

132.1 Introduction

The S-wave nucleon-nucleon (NN) scattering length is negative and has a magnitude more than twenty times the nucleon size and almost ten times the typical nuclear force effective range. That means neutron matter at low densities is off the unitary limit only by the finite effective range. The first Quantum Monte Carlo calculation of a many-fermion system with short-range two-body attractive interaction with infinite scattering length was made in 2003 and estimated the energy per particle of such gases, in units of the energy of the non-interacting gas, to be $\epsilon/\epsilon_0 = \xi =$

E. Ruiz Arriola
Departamento de Física Atómica, Molecular Y Nuclear, Facultad de Ciencias, Universidad de Granada - UGR, Granada, Andalucía, Spain

S. Szpigel
Centro de Rádio-Astronomia e Astrofísica Mackenzie - CRAAM, Universidade Presbiteriana Mackenzie, São Paulo, SP, Brazil

V. S. Timóteo (✉)
Grupo de Óptica e Modelagem Numérica - GOMNI, Faculdade de Tecnologia - FT, Universidade Estadual de Campinas - UNICAMP, Limeira, SP, Brazil
e-mail: varese@unicamp.br; varese@ft.unicamp.br

N. A. Orr et al. (eds.), *Recent Progress in Few-Body Physics*, Springer Proceedings in Physics 238,
https://doi.org/10.1007/978-3-030-32357-8_132

0.44 ± 0.01 [1]. Later, in 2012, it was shown that neutron matter at very low densities behaves as a gas of ultra cold atoms and the energy is essentially given by the contribution from the S-wave [2].

Here we want to fix a set of contact interactions so that the unitary limit conditions for two nucleons are satisfied by using an implicit renormalization procedure [3, 4]. Then, using the Fermi momentum as our separation scale between high and low momentum degrees of freedom, we can restrict the calculation to the mean field level.

132.2 Unitary Limit and Contact Interactions

The two-body scattering amplitude is given by

$$T_2(k, k) \propto \frac{1}{k \cot \delta - i\, k}\,, \tag{132.1}$$

where k is the on-shell center-of-mass (CM) momentum and δ stands for the phase-shifts. By means of the Effective Range Expansion (ERE), the term $kcot\delta$ can be written as

$$k \cot \delta = -\frac{1}{a} + \frac{1}{2}\, r\, k^2 - \frac{1}{4}\, v\, k^4 + \cdots\,, \tag{132.2}$$

where a, r and v are respectively the scattering length, effective range and shape parameter. At low energies, the k^4 term can be neglected and the unitary limit corresponds to $-a \to \infty$ and $r \to 0$. In this case, the phase-shift is energy-independent: $\delta = \pi/2$ and the energy per particle of a strongly interacting many-fermion system is given by $\varepsilon = \xi \times \varepsilon_0$, where ξ is the Bertsch parameter and ε_0 is the energy per particle of the non-interacting Fermi Gas: $\varepsilon_0 = \frac{3}{5}\frac{\hbar^2 k_F^2}{2m}$.

Table 132.1 shows three different results from calculations of the Bertsch parameter based on ab initio Monte Carlo simulations.

The ab initio calculations are fundamental and precise but they require extreme computational resources. Here we want to provide a simple calculation using a set of contact interactions renormalized implicitly by the unitary limit conditions (instead of the physical S-wave scattering length and effective range).

Table 132.1 Bertsch parameter from ab initio Monte Carlo simulations

References	[5]	[6]	[7]
ξ	0.372(5)	0.3897(4)	0.388(1)

In momentum space, the set of contact interactions up to NNLO is given by

$$V_{\times}(p,p';X) = \underbrace{\underbrace{\underbrace{A_0}_{\text{LO}} + B_2\,(p^2+p'^2)}_{\text{NLO}} + C_4(p^4+p'^4) + D_4(p^4\times p'^4)}_{\text{NNLO}}\,, \tag{132.3}$$

where the strengths $X = (A_0, B_2, C_4, D_4)$ are determined by the the implicit renormalization method [3, 4], which consists of calculating the two-body scattering T-matrix with V by solving the Lippman-Schwinger equation and then matching it to the effective range expansion:

$$T_{\times}(p,p';k,X) = V_{\times}(p,p';X) + \frac{2}{\pi}\int_0^{k_{\mathrm{F}}} dq\, q^2\, V_{\times}(p,q;X)\frac{T_{\times}(q,p';k,X)}{k^2-q^2+i\epsilon}\,, \tag{132.4}$$

$$\tan\delta_{\times}(k;X) = -k\,T_{\times}(k,k;k,X)\,, \tag{132.5}$$

$$k\cot\delta_{\times}(k;X) = -\frac{1}{T_{\times}(k,k;k,X)} \simeq -\frac{1}{a} + \frac{1}{2}\,r\,k^2 + \cdots\,. \tag{132.6}$$

Note that the final renormalized interaction depends on the inputs for the scattering length a and the effective range r. So, if the unitary conditions are imposed, the phase-shifts will be energy independent and equal to $\pi/2$. Deviations from the unitary limit can be studied by departing from $-1/a = 0$ and $r = 0$ when matching the contact T-matrix to the effective range expansion. Here we present only results at the unitary limit $a = -\infty$ and $r = 0$.

132.3 Neutron Matter at LO, NLO and NNLO

An upper variational estimation for the energy per particle of neutron matter can be obtained through a mean field level calculation. At this level, the Hartree-Fock approximation gives

$$\varepsilon_{\times} = \varepsilon_0 + \frac{4}{\pi\,m}\int_0^{k_{\mathrm{F}}} dk\left(k^2 - \frac{3k^3}{2k_{\mathrm{F}}} + \frac{k^5}{2k_{\mathrm{F}}^3}\right)V_{\times}(k,k;X)\,, \tag{132.7}$$

where m is the fermion mass. At each order, once the strengths of the contact terms have been determined by the implicit renormalization with the unitary limit conditions on the scattering length and effective range $(-1/a = r = 0)$, the Bertsch parameter is simply given by:

Table 132.2 Bertsch parameter with only contact interactions at different orders

$V_\times$	LO	NLO	NNLO
$\xi_\times$	0.444	0.467	0.420

$$\xi_\times = 1 + \frac{4}{\pi \varepsilon_0 m} \int_0^{k_F} dk \left(k^2 - \frac{3k^3}{2k_F} + \frac{k^5}{2k_F^3} \right) V_\times(k, k; X) \,. \qquad (132.8)$$

The results for LO, NLO and NNLO are shown in Table 132.2. The LO result is close to the value obtained in [1]. Moving to NLO increases the value of $\xi_\times$ and adding the NNLO terms makes $\xi_\times$ go down to a value even smaller then the LO contribution but still doesn't reach the Monte Carlo values, which are closer to what is obtained in experiments with Fermi gases [8, 9]. It seems that higher order terms are still required for better convergence. However, considering the simplicity of our calculation at the mean field level with only two-body contact interactions and implicit renormalization, the results are quite reasonable.

132.4 Conclusions

We use an effective field theory approach to the unitary Fermi gas by applying an implicit renormalizing scheme on a set of contact interactions with the unitary limit conditions for the scattering length and the effective range. We then calculate the energy per particle of the many-fermion system at the mean field level for the first three orders of the momentum expansion: LO, NLO and NNLO. The values we obtain for the Bertsch parameter ξ are larger than those obtained through Quantum Monte Carlo simulations (see Tables 132.1 and 132.2). Nevertheless, higher order contact interactions (e.g. NNNLO) seem to be required to reach the experimental value $\xi = 0.370(5)(8)$ [9]. From the chiral effective field theory point of view, the relative success of our implicit renormalization approach to the unitary Fermi gas can be attributed to the dominance of the contact interactions over pion exchanges in the S-waves. Apart from including higher order terms, in our approach it is also possible to evaluate the effective range dependence of the Bertsch parameter by changing the renormalization conditions to support finite ranges. Work in these directions is in progress.

Acknowledgements Work supported by Spanish MINECO and European FEDER funds (grants FIS2014-59386-P and FIS2017-85053-C2-1-P), Junta de Andalucía (grant FQM-225), FAPESP (grant 2016/07061-3) and CNPq (grant 306195/2015-1).

References

1. Carlson, J., Chang, S.-Y., Pandharipande, V.R., Schmidt, K.E.: Superfluid Fermi gases with large scattering length. Phys. Rev. Lett. **91**, 050401 (2003)
2. Carlson, J., Gandolfi, S., Gezerlis, A.: Quantum Monte Carlo approaches to nuclear and atomic physics. Prog. Theor. Exp. Phys., **01A209** (2012)
3. Ruiz Arriola, E., Szpigel, S., Timóteo, V. S.: Implicit versus explicit renormalization and effective interactions. Phys. Lett. B **728**, 596–601 (2014)
4. Ruiz Arriola, E., Szpigel, S., Timóteo, V. S.: Implicit and explicit renormalization: two complementary views of effective interactions. Ann. Phys. **353**, 129–149 (2015)
5. Carlson, J., Gandolfi, S., Schmidt, E., Zhang, S.: Auxiliary-field quantum Monte Carlo method for strongly paired fermions. Phys. Rev. A **84**, 061602(R) (2011)
6. Forbes, M.M., Gandolfi, S., Gezerlis, A.: Effective-range dependence of resonantly interacting fermions. Phys. Rev. A **86**, 053603 (2012)
7. Schonenberg, L.M., Conduit, G.J.: Effective-range dependence of resonant Fermi gases. Phys. Rev. A **95**, 013633 (2017)
8. Ku, M.J.H., Sommer, A.T., Cheuk, L.W., Zwierlein, M.W.: Revealing the superfluid lambda transition in the universal thermodynamics of a unitary Fermi gas. Science **335**, 563 (2012)
9. Zürn, G., Lompe, T., Wenz, A.N., Jochin, S., Julienne, P., Hutson, J.M.: Precise characterization of ^6Li Feshbach resonances using trap-sideband-resolved RF spectroscopy of weakly bound molecules. Phys. Rev. Lett. **110**, 135301 (2013)

Chapter 133
Bosonic Drops with Two- and Three-Body Interactions Close to the Unitary Limit

A. Kievsky, A. Polls, B. Juliá-Díaz, N. Timofeyuk and M. Gattobigio

Abstract When the binding energy of a two-body system goes to zero the two-body system shows a continuous scaling invariance governed by the large value of the scattering length. In the case of three identical bosons, the three-body system in the same limit shows the Efimov effect and the scale invariance is broken to a discrete scale invariance. As the number of bosons increases correlations appear between the binding energy of the few- and many-body systems. We discuss some of them as the relation between the saturation properties of the infinite system and the low-energy properties of the few-boson system.

133.1 Introduction

The ground state properties of ^{4}He and ^{3}He droplets with N atoms have been studied in a series of papers [1–4]. The energy per particle, E_N/N can be described, as $N \to \infty$, using a liquid-drop formula in terms of $x = N^{-1/3}$

$$E_N/N = E_v + E_s x + E_c x^2 \tag{133.1}$$

where E_v, E_s and E_c, are the volume, surface and curvature terms respectively. Results for the infinite liquid can be obtained from calculations at finite values of N.

A. Kievsky (✉)
Istituto Nazionale di Fisica Nucleare, Largo Pontecorvo 3, 56100 Pisa, Italy
e-mail: kievsky@pi.infn.it

A. Polls · B. Juliá-Díaz
Departament de Física Quàntica i Astrofísica, Facultat de Física,
Universitat de Barcelona, 08028 Barcelona, Spain

N. Timofeyuk
Department of Physics, University of Surrey, Guildford, Surrey GU2 7XH, UK

M. Gattobigio
Université Côte d'Azur, CNRS, Institut de Physique de Nice,
1361 route des Lucioles, 06560 Valbonne, France

N. A. Orr et al. (eds.), *Recent Progress in Few-Body Physics*,
Springer Proceedings in Physics 238,
https://doi.org/10.1007/978-3-030-32357-8_133

Since the value at saturation can be obtained independently, these studies probe the validity of the extrapolation formulas used to predict the properties of the infinite system typically computed in droplets having a few hundred atoms.

More recently helium drops have been studied using modern helium-helium interactions [5, 6]. In [7] a diffusion Monte Carlo (DMC) method has been used to study clusters up to 10 atoms interacting through the Tang, Toennies, and Yiu (TTY) potential [8]. Helium trimers and tetramers have been studied around the unitary limit varying the potential strength [9–12]. It has been shown that with a very small reduction of the strength (about 3%) the binding energy of the helium dimer disappears. In fact the helium dimer is very close to the unitary limit having a two-body binding energy of about 1.3 mK and a large two-body scattering length of about 189 a_0 (a_0 is the Bohr radius).

We can observe two, very different, descriptions of light helium clusters. On one hand, several models of the helium-helium interaction are available. On the other hand, the large value of the helium-helium scattering length locates the small clusters of helium close to the unitary limit in which universal behavior can be observed. Accordingly, the particular form of the potential is not important, many properties are determined from a few parameters as the two-body scattering length a and the trimer ground state energy E_3^0 (or the first excited state E_3^1). Specific (soft) potential models can be constructed in order to reproduce those data and used to calculate binding energies of droplets and the saturation properties of the infinite system [13]. In this way, a direct link between the low energy scale (or long-range correlations) and the high energy energy scale (or short-range correlations) can be established.

133.2 Helium Dimer and Trimer with Soft Potential Models

In the following we study the ground state energy of the $N = 2, 3$ boson systems using a soft gaussian potential constructed to reproduce the low-energy behavior of the system. We define the two-body interaction as

$$V(r_{ij}) = V_0 e^{-r_{ij}^2/d_0^2} \tag{133.2}$$

with the two gaussian parameters, V_0 and d_0, determined from the dimer energy, E_2, and the two-body scattering length a. Realistic helium-helium potentials can be used to calculate E_2 and a, subsequently used to fix V_0 and d_0. In this way, the gaussian interaction results in a low-energy representation of the original potential. Using the LM2M2 interaction [14], widely used in the description of helium clusters, as the reference interaction, the values $V_0 = -1.2343566$ K and $d_0 = 10.0\, a_0$ can be used. To study correlations between observables we can start analyzing the Efimov radial law

$$E_3^n/(\hbar^2/ma^2) = \tan^2\xi \tag{133.3}$$

$$\kappa_* a = e^{(n-n^*)\pi/s_0} \frac{e^{-\Delta(\xi)/2s_0}}{\cos\xi}\,, \tag{133.4}$$

that gives, in the zero-range limit, the three-boson spectrum E_3^n in terms of the universal function $\Delta(\xi)$ and the three-body parameter κ_*, defined by the energy at the unitary limit of the reference level n^*, $E_3^{n^*} = \hbar^2\kappa_*^2/m$. Equation (133.4) indicates that the product $\kappa_* a$ is a function of the angle ξ. Assuming that for real systems the product $\kappa_* a$ is still a function of ξ we can propose:

$$\kappa_* a = [\kappa_* a]_G \tag{133.5}$$

where $[\kappa_* a]_G$ is the value of the product calculated with the gaussian potential at the angle ξ. To verify this hypothesis we consider the ground state binding energies of the dimer $E_2 = 1.303$ mK and trimer $E_3 = 126.4$ mK as given by the LM2M2 potential defining the angle ξ as $E_3/E_2 = \tan^2\xi = 97.0$. Modifying the strength of the gaussian potential to $V_0 = -1.24294$ K and calculating the dimer and trimer energies, the same angle is obtained. Moreover, a gaussian potential has the property that its three-body parameter verifies $\kappa_* = 0.488/d_0$ [15, 16]. The two-body scattering length using the modified strength is $a = 170.50$ a_0. Accordingly, we can estimate the three-body parameter κ_* of the LM2M2 interaction, knowing that the scattering length is 189.41 a_0, as

$$[\kappa_*]_{LM2M2} = \frac{170.5}{189.41} 0.488/d_0\,. \tag{133.6}$$

The obtained value is $\kappa_* = 0.044\ a_0{}^{-1}$ in complete agreement with the LM2M2 value given in the literature. We have shown that the three-body parameter can be determined by three quantities, the dimer and trimer energies and the two-body scattering length.

133.3 Saturation Properties of the N-Boson System

In the following we analyze correlations between the saturation properties of the infinite system and the low-energy behavior of the few-boson systems. To this end we use as the reference interaction the Aziz HFDHE2 potential used in [3] to compute binding energies of helium droplets. A low energy representation of the HFDHE2 potential is obtained by defining the parameters of the gaussian potential $V_0 = -1.208018$ K and $d_0 = 10.0485$ a_0, giving a trimer ground state binding energy of $E_3^0 = 139.8$ mK. This value is substantially greater than the value obtained using the HFDHE2 potential: $E_3^0 = 117.3$ mK. It is well known that to tune the trimer binding energy

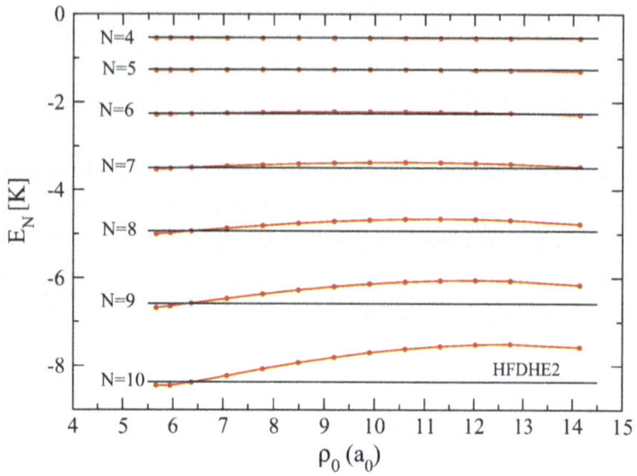

Fig. 133.1 Binding energy per particle up to $N = 10$ atoms. The results of the SGP for different values of the three-body force range ρ_0 are shown as red circles and compared to the HFDHE2 results

to the expected value a slightly repulsive three-body force has to be introduced. As proposed in Refs. [17–21] we define the following three-body force

$$W(\rho_{ijk}) = W_0 e^{-2\rho_{ijk}^2/\rho_0^2} , \tag{133.7}$$

where ρ_{ijk} is the hyperradius of particles i, j, k defined as $\rho_{ijk}^2 = (2/3)(r_{ij}^2 + r_{jk}^2 + r_{ki}^2)$. For selected values of the range ρ_0, the strength W_0 is fixed to reproduce the HFDHE2 trimer ground state binding energy E_3^0. The binding energy of the droplets E_N can be computed using this soft gaussian potential (SGP) and can be studied as a function of the range ρ_0. In Fig. 133.1 the binding energies of helium drops up to $N = 10$ are shown as a function of ρ_0 and compared to the HFDHE2 values from [3] using the Green Function Monte Carlo (GFMC) method. Though a small dependence on ρ_0 can be seen, an overall good description is obtained.

The ρ_0 dependence is analyzed in Fig. 133.2 in the case of the tetramer binding energy. It can be seen that there is a value of ρ_0, around 8.5a$_0$, that gives the best description of this quantity. The next step is to compute the droplets binding energies up to $N \approx 100$ and extract the saturation energy from (133.1). This is shown in Fig. 133.3 where the results for different values of ρ_0 form the dark band. The results using the optimum value of $\rho_0 = 8.5\text{a}_0$ are shown as (blue) points. They follow, with acceptable accuracy, the GFMC results using the HFDHE2 potential shown as the (red) solid line. Using the optimum value of ρ_0 it is possible to determine E_v, E_s and E_c defined in (133.1). From the results of the SGP in the range $20 \leq N \leq 112$ the following values are obtained (in K)

$$E_N/N = 6.98 - 18.6x + 10.3x^2 . \tag{133.8}$$

They can be compared to the values (in K) obtained with the GFMC method $E_v = 7.02$, $E_s = -18.8$ and $E_c = 11.2$ using the HFDHE2 interaction.

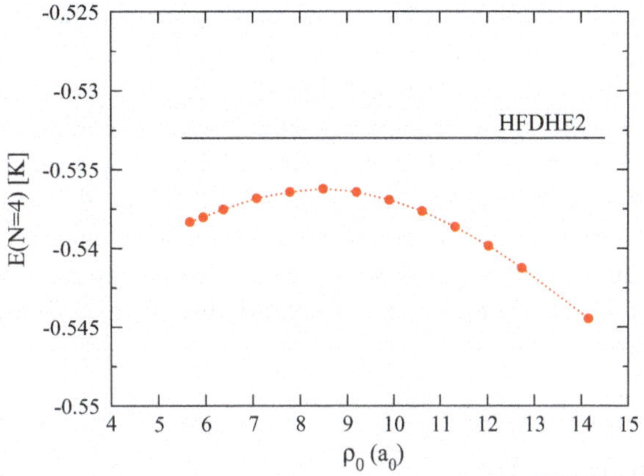

Fig. 133.2 The tetramer binding energy as a function of the three-body force range ρ_0. The value of the HFDHE2 potential is shown

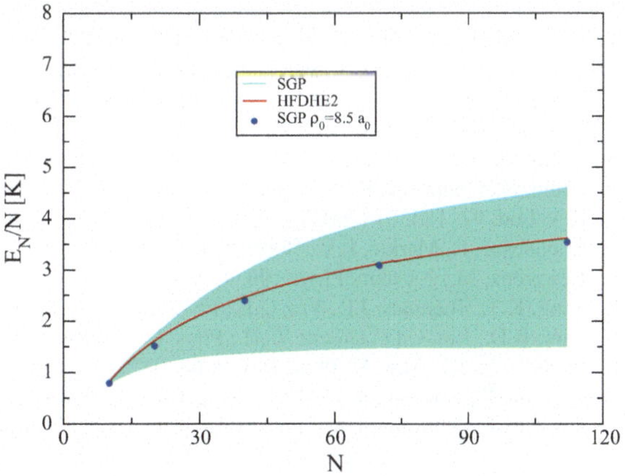

Fig. 133.3 The binding energy of the droplets as a function of the number of particles N

We conclude that after tuning the range of the three-body force to reproduce as better as possible the tetramer binding energy, the soft gaussian potential, consisting of a two- and a three-body term, with the four parameters determined by the dimer, trimer and tetramer binding energies and the two-body scattering length is able to estimate with good accuracy the energy per particle of the infinite system.

133.4 Conclusions

In the present work we have analyzed correlations between different observables imposed by the proximity of the system to the unitary limit. Due to the large value of the two-body scattering length, helium drops are well suited to study these phenomena. Correlations of this type can also be studied in nuclear systems, since the $n-n$ and $n-p$ scattering lengths are large [22, 23]. Here we have shown results for helium drops using a gaussian soft interaction to determine the three-body parameter κ_*. Noticeably, the result was in extremely good agreement with the values given in the literature calculated directly using the LM2M2 potential. Secondly, using the HFDHE2 as the reference potential, we have calculated binding energies for helium drops up to $N = 112$ and, using a liquid-drop formula, we have extracted the saturation energy. We have observed that using the optimum value for the range of the three-body interaction a good estimate of the experimental saturation energy is obtained. In this way we have clarified the existing correlations between different observables imposed by the unitary limit in many-body systems close to the unitary limit.

References

1. Kalos, M.H., Lee, M.A., Whitlock, P.A., Chester, G.V.: Phys. Rev. B **24**, 115 (1981)
2. Usmani, Q.N., Fantoni, S., Pandharipande, V.R.: Phys. Rev. B **26**, 6123 (1982)
3. Pandharipande, V.R., Zabolitzky, J.G., Pieper, S.C., Wiringa, R.B., Helmbrecht, U.: Phys. Rev. Lett. **50**, 1676 (1983)
4. Pandharipande, V.R., Pieper, S.C., Wiringa, R.B.: Phys. Rev. B **34**, 4571 (1986)
5. Brühl, R., Guardiola, R., Kalinin, A., Kornilov, O., Navarro, J., Savas, T., Toennies, J.P.: Phys. Rev. Lett. **92**, 185301 (2004)
6. Stipanović, P., Markić, L.V., Boronat, J.: J. Phys. B: At. Mol. Opt. Phys. **49**, 185101 (2016)
7. Lewerenz, M.: J. Chem. Phys. **106**, 4596 (1997)
8. Tang, K.T., Toennies, J.P., Yiu, C.L.: Phys. Rev Lett. **74**, 1546 (1995)
9. Esry, B.D., Lin, C.D., Greene, C.H.: Phys. Rev. A **54**, 394 (1996)
10. Barletta, P., Kievsky, A.: Phys. Rev. A **64**, 042514 (2001)
11. Hiyama, E., Kamimura, M.: Phys. Rev. A **85**, 062505 (2012)
12. Hiyama, E., Kamimura, M.: Phys. Rev. A **90**, 052514 (2014)
13. Kievsky, A., Polls, A., Juliá-Díaz, B., Timofeyuk, N.K.: Phys. Rev. A **96**, 040501(R) (2017)
14. Aziz, R.A., Slaman, M.J.: J. Chem. Phys. **94**, 8047 (1991)
15. Kievsky, A., Gattobigio, M.: Phys. Rev. A **92**, 062715 (2015)
16. Álvarez-Rodríguez, R., Deltuva, A., Gattobigio, M., Kievsky, A.: Phys. Rev. A **93**, 062701 (2016)
17. Kievsky, A., Garrido, E., Romero-Redondo, C., Barletta, P.: Few-Body Syst. **51**, 259 (2011)
18. Gattobigio, M., Kievsky, A., Viviani, M.: Phys. Rev. A **84**, 052503 (2011)
19. Gattobigio, M., Kievsky, A., Viviani, M.: Phys. Rev. A **86**, 042513 (2012)
20. Timofeyuk, N.K.: Phys. Rev. A **86**, 032507 (2012)
21. Timofeyuk, N.K.: Phys. Rev. A **91**, 042513 (2015)
22. Kievsky, A., Viviani, M., Gattobigio, M., Girlanda, L.: Phys. Rev. C **95**, 024001 (2017)
23. Kievsky, A., Viviani, M., Logoteta, D., Bombaci, I., Girlanda, L.: Phys. Rev. Lett. **121**, 072701 (2018)

Chapter 134
Evaluation of Correlations in Nuclear Matter by Using Spectral Expansion for the In-Medium Propagator

O. A. Rubtsova

Abstract A very brief description of the new technique which allows to calculate correlations in medium (e.g. in nuclear matter) by using a spectral expansion for the in-medium T-matrix is given. The case of the Galitskii–Feynman in-medium propagator is discussed and its relation to a paring instability.

134.1 Introduction

Correlations in nuclear matter (or some other medium) can be evaluated by using the integral T-matrix formalism [1] which has similar features with solving of scattering problems in vacuum. Recently, the diagonalisation technique [2] has been proposed for both type of calculations. In the approach, an effective in-medium Hamiltonian is introduced which allows finding the in-medium T-matrix at many energies and relative momenta by using discretized spectral expansion for the in-medium propagator.

In this paper, the case of the Galitskii–Feynman (GF) particle-particle and hole-hole (pphh) propagator within the mean field approximation is discussed. Here the effective Hamiltonian is non-Hermitian and may have complex eigenvalues signaling about a pairing instability and opening of a superfluid phase. Below the main results of the approach are reported briefly, in particular, reconstruction of momentum dependencies of pairing gaps in different partial NN channels without solving gap equations. The detailed explanation can be found in [2–4].

O. A. Rubtsova (✉)
Skobeltsyn Institute of Nuclear Physics, Moscow State University,
119991 Moscow, Russia
e-mail: rubtsova@nucl-th.sinp.msu.ru

N. A. Orr et al. (eds.), *Recent Progress in Few-Body Physics*,
Springer Proceedings in Physics 238,
https://doi.org/10.1007/978-3-030-32357-8_134

134.2 Diagonalisation Technique for the In-Medium T-matrix

In-medium T-matrix satisfies the integral equation:

$$T(E) = V + VG_0^{\rm II}(E)T(E), \tag{134.1}$$

where V is the bare NN interaction and $G_0^{\rm II}(E)$ is the propagator. Below, a case of the GF propagator is discussed which has a form:

$$G_0^{\rm II}(E) = \frac{Q}{E + i0 - H_{0Q}} - \frac{P}{E - i0 - H_{0P}}. \tag{134.2}$$

Here Q and P are Pauli-projectors to pp- and hh- subspaces correspondingly and H_{0Q} and H_{0P} are the mean-field free Hamiltonians [3].

It has been suggested [2, 3] to evaluate the in-medium T-matrix from the relation $T(E) = V + VG^{\rm II}(E)V$ in which the total in-medium propagator $G^{\rm II}(E)$ is introduced. This operator can be represented [3] in a form:

$$G^{\rm II}(E) = [E + i0 \cdot J - H_0 - JV]^{-1}, \tag{134.3}$$

where $H_0 = H_{0P} + H_{0Q}$ is the total pphh free Hamiltonian and J is a 'sign' operator in $(P + Q)$ space [3].

Thus, the operator $G^{\rm II}(E)$ can be considered as a 'generalized resolvent' of the effective Hamiltonian $H = H_0 + JV$ which is non-Hermitian due to the operator J. To find the spectral expansion for H and subsequently for $G^{\rm II}(E)$, the discrete wave-packet representation over the relative momentum k is employed. This discretization results in a finite-dimensional (matrix) form of all operators [3].

Finally, by using the constructed discretized spectral expansion for the total propagator, one gets an explicit expression for the fully off-shell in-medium T-matrix at any required energy [3, 4]. Apart from other advantages, this representation allows to separate forward- and backward-propagating parts of the T-matrix which are required in the complete self-consistent scheme [1].

Also, one can study a scattering problem for the in-medium total Hamiltonian H. The on-shell elements of the constructed T-matrix define amplitudes for the two-channel scattering problem with the hole-hole and particle-particle channels which are coupled by the interaction JV. The corresponding partial phase shifts satisfy the in-medium version of the Levinson theorem [5].

The total Hamiltonian may have complex eigenvalues $E_{\pm 1} = E_0 \mp i\Gamma_0$ which indicates a pairing instability. However these states can be considered in a strict manner [3]. In particular, the imaginary part of the corresponding eigenvalue coincides with the pairing gap at the Fermi-momentum $\Delta(k_F) = \Gamma_0$ [6, 7]. Recently, it has been shown [3, 4] that momentum dependence of the gap $\Delta(k)$ can be found form the corresponding eigenfunction $|z_1\rangle$ of H:

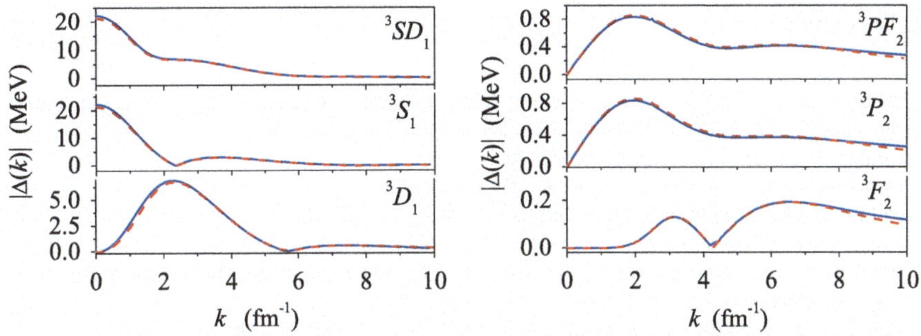

Fig. 134.1 Absolute values of total and partial pairing gaps in the coupled NN channels $^3SD_1$ at $k_F = 1.2$ fm^{-1} (left panel) and $^3PF_2$ at $k_F = 2.5$ fm^{-1} (right panel) found from the (134.4) (dashed curves) and from the solution of the gap equation [3] (solid curves). The CD Bonn NN potential [8] is used

$$\langle k|V|z_1\rangle/k = Be^{i\phi_k}\Delta(k), \tag{134.4}$$

where $|k\rangle$ is the plane-wave state, B is a real constant and ϕ_k is a phase factor.

Thus, the diagonalisation procedure for the effective Hamiltonian H allows finding pairing gaps in different spin-angular NN channels without solving the conventional gap equations. In Fig. 134.1 the partial as well as total pairing gaps for $^3SD_1$ and $^3PF_2$ channels are represented at zero center of mass momentum $K = 0$. This procedure can be generalized to a case of non-zero K as well [4].

Opening and closing of the superfluid regime correspond to the phase transitions in dependencies of the effective Hamiltonian eigenvalues on density [3]. These phase transitions can be observed in all three NN channels where the pairing takes place, i.e. $^1S_0$, $^3SD_1$ and $^3PF_2$. The most interesting picture is seen in $^3SD_1$ coupled NN channel, where real bound states of H are transformed into states with complex conjugated eigenvalues (see the details in [3]). This behavior might be related to the known problem of the BCS–BEC crossover.

134.3 Conclusions

The effective Hamiltonian treatment with the wave-packet continuum discretization occurs to be an efficient tool for the in-medium T-matrix calculations [2, 3]. The approach allows to study pairing beyond the conventional BCS treatment. The formalism will be generalized within the Self-Consistent Greens Function framework [1] for calculations of few-body correlations in nuclear matter.

Acknowledgements The author appreciate a partial financial support from the RFBR grants 19-02-00014 and 19-02-00011.

References

1. Dickhoff, W.H., van Neck, D.: Many-Body Theory Exposed! Propagator Description of Quantum Mechanics in Many-Body Systems. World Scientific, Singapore (2005)
2. Müther, M., et al.: Phys. Rev. C **94**, 024328 (2016)
3. Rubtsova, O.A., et al.: Phys. Rev. C **96**, 034327 (2017)
4. Rubtsova, O.A., Kukulin, V.I., Pomerantsev, V.N.: to be published
5. Bishop, R.F., Ghassib, H.B., Strayer, M.R.: Phys. Rev. A **13**, 1570 (1976)
6. Abrikosov, A.A., Gorkov, L.P., Dzyaloshinski, I.E.: Quantum Field Theoretical Methods in Statistical Physics, 2nd edn. Pergamon, New York (1965)
7. Ramanan, S., Bogner, S.K., Furnstahl, R.J.: Nucl. Phys. A **797**, 81 (2007)
8. Machleidt, R., et al.: Phys. Rev. C **53**, R1483 (1996)

Chapter 135
Universality and the Coulomb Interaction

Christiane H. Schmickler

Abstract We study the relationship between universal effects and the Coulomb interaction. Here, we present our approach and a first illustrative example. Understanding the relationship of universality and the Coulomb interaction is important for weakly bound nuclear few-body systems. The example nucleus we study is the excited state of ^{17}F which we model as a proton and an ^{16}O core. We use a Gaussian potential to represent short-range forces together with a Coulomb potential. Our calculation uses the Gaussian Expansion Method (GEM), which is a variational method well-suited to our problem. We find that we need to choose the range of the Gaussian potential in addition to the potential strength that reproduces the right scattering length. We propose to fit the range of the Gaussian potential such that the effective range is reproduced. We show that this approach leads to consistent results for ^{17}F.

135.1 Introduction

The Efimov effect has been in the focus of few-body research ever since its observation in cold atom experiments [1]. Even before that, there was discussion in the literature about whether and where the Efimov effect plays a role in nuclear physics [2].

However, the long-range Coulomb interaction complicates this picture in the nuclear sector and many studies to date focussed on the question of whether there would be an Efimov state without the Coulomb force.

In our studies, of which this paper is a first excerpt, we try to answer the question of which effect the Coulomb force has on would-be Efimov states and under which circumstances the Efimov picture is relevant in spite of the Coulomb interaction.

Christiane H. Schmickler (✉)
Institut Für Kernphysik, Technische Universität Darmstadt, 64289 Darmstadt, Germany

RIKEN Nishina Center, RIKEN, Saitama 351-0198, Japan
e-mail: christiane.schmickler@riken.jp

N. A. Orr et al. (eds.), *Recent Progress in Few-Body Physics*,
Springer Proceedings in Physics 238,
https://doi.org/10.1007/978-3-030-32357-8_135

To this end, we use well-tested methods of calculating universal properties of few-body systems, namely Gaussian potentials and variational methods, and add a Coulomb interaction to the system.

Here, we investigate a specific example system, $^{17}\mathrm{F}(\frac{1}{2}^{+})$, which we describe as an $^{16}\mathrm{O}$ nucleus plus a single proton. This state is a proton halo and a shallow bound state. As a dimer it is well-suited as a first test of our methods. The trimer and tetramer calculations will be presented in an upcoming more in-depth article.

135.2 Interaction and Methods

As we are interested in universal behaviour, we use a simple Gaussian potential plus the Coulomb potential.

$$H_2 = -\frac{\hbar^2}{2\mu}\frac{\partial^2}{\partial r^2} + V_0 e^{-\frac{r^2}{2r_0^2}} + \hbar c Z_1 Z_2 \frac{\alpha}{r} \tag{135.1}$$

We solved this system with the Gaussian Expansion Method (GEM) that was developed by Kamimura et al. [3]. It is a variational method with geometrically spaced Gaussian basis functions which is well-suited to this problem because the matrix elements can be calculated analytically and it gives accurate results without much fine tuning.

135.3 Coulomb-modified Effective Range Expansion

The usual effective range expansion has to be modified to account for the fact that the Coulomb interaction is long-range. This is achieved by simply replacing the free outside wave functions used to calculate the scattering length and effective range by Coulomb functions.

This leads to the Coulomb-modified effective range expansion:

$$C_{\eta,0}^2 p \cot \tilde{\delta}_0(p) + \gamma h(\eta) = -\frac{1}{a_C} + \frac{1}{2} r_{\mathrm{eff}}^C p^2 + \ldots, \tag{135.2}$$

where p is the wave number, $\eta = \mu c^2 \alpha Z_1 Z_2/(\hbar c p)$ and $\gamma = 2\frac{\mu c^2}{\hbar c}\alpha Z_1 Z_2$ with the reduced mass μ and charge numbers Z_1 and Z_2, $C_{\eta,0}^2 = \frac{2\pi\eta}{e^{2\pi\eta}-1}$, $h(\eta) = \Re \frac{\Gamma'(i\eta)}{\Gamma(i\eta)} - \log(\eta)$ and $\tilde{\delta}_0(p)$ is the phase shift between the ingoing and outgoing Coulomb wave functions. This determines the Coulomb-modified scattering length a_C and the Coulomb-modified effective range r_{eff}^C. For more detail, the reader is referred to [4–6].

135.4 Application to ^{17}F

We will concentrate on showing one application, namely ^{17}F as a dimer of ^{16}O and a proton, in this article and refer to another publication for more general results and more applications [7].

The first excited state of ^{17}F is a well-known proton halo. This means that in first approximation it can be described as a dimer of the core (^{16}O) and a proton [8]. From the values for the binding momentum $\gamma = 13.6\,\text{MeV}$ and the ANC $A \approx 80\,\text{fm}^{-\frac{1}{2}}$ presented in [8], we calculated the ^{16}O-p Coulomb-modified scattering length and the Coulomb-modified effective range to be $a_C = 4475.96\,\text{fm}$ and $r_{\text{eff}}^C = 1.18737\,\text{fm}$ according to the formulae derived in [8].

However, plugging in the parameters for this system, i.e. a_C, $Z_1 = 1$, $Z_2 = 8$ and the reduced mass $\mu \approx 873\,\text{MeV}$ will lead to an ambiguity because we obtain different curves for different r_0 as shown in the left panel of Fig. 135.1. This ambiguity has to be resolved in order to make any kind of prediction. From the point of view of the effective range expansion, (135.2), the most obvious parameter to fit after the scattering length is the effective range.

To this end, we calculated the effective range for different ranges r_0 of the Gaussian short-range potential. We selected results reproducing the scattering length of $a_C = 4475.96\,\text{fm} \pm 80\,\text{fm}$. These are shown in the right panel of Fig. 135.1. The value of r_{eff}^C rises monotonically with r_0 in the range we explored. Thus, we can uniquely identify a value for r_0 that reproduces the correct Coulomb-modified effective range for this system. For ^{16}O+p, $r_0 = 2.75\,\text{fm}$.

Going back to the left panel of Fig. 135.1, we can see that the determined value for r_0 reproduces the binding energy of the ^{17}F proton halo state well. The physical value is at the point where the horizontal and vertical lines cross, and the curve for $r_0 = 2.75\,\text{fm}$ crosses the vertical line very close to that point.

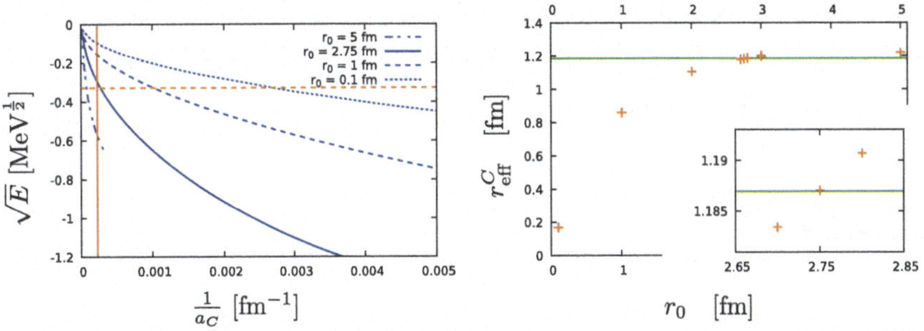

Fig. 135.1 *Left panel:* Different dimer predictions for different values of the range r_0 of the Gaussian short-range potential. The vertical orange line is the value of the scattering length in the ^{16}O+p system, the horizontal orange line is the binding energy of the ^{17}F($\frac{1}{2}^+$) state. *Right panel:* Determination of the effective range at the point of the ^{16}O-p scattering length for different r_0. The horizontal line shows the value of the Coulomb-modified effective range in the ^{16}O+p system

Note that this is only a consistency check, because in the formula for the scattering length from [8] the effective range and the binding energy is an input. The effective range in turn contains as additional input the asymptotic normalization coefficient, which can be measured. Therefore we use the binding energy as an input to determine the scattering length, which we in turn use as an input for obtaining the binding energy. Since the methods that were used to connect scattering length and binding energy are different, however, this is a good consistency check between the methods.

In addition, we also calculated the root-mean-square (rms) radius for $^{17}\mathrm{F}(\frac{1}{2}^+)$ and found $r_{\mathrm{rms}} = 5.3$ fm. This is in very good agreement with [9].

135.5 Summary and Conclusion

We showed in this paper a first demonstration of our method, i.e. using a Gaussian potential together with the Coulomb interaction to test universal properties of weakly bound states. We applied our method to a proton halo state, $^{17}\mathrm{F}(\frac{1}{2}^+)$, which we regarded as a dimer of a proton and a $^{16}\mathrm{O}$ core.

To make the result unique we have to fix r_0, which is the range of our Gaussian potential. This is unnecessary in systems without the Coulomb interaction, because the result would be the same for all r_0 after converting it into the right units. Having to fix r_0 is due to the Coulomb interaction introducing an additional scale to the system.

We chose to fix r_0 such that the Coulomb-modified effective range would be reproduced, which gave us the correct binding energy and rms radius for $^{17}\mathrm{F}(\frac{1}{2}^+)$. This is a promising first test of our method which we can build upon in future work.

We will expand this approach to trimers and tetramers of identical charged bosons in another publication [7].

Acknowledgements The author thanks H.-W. Hammer for a careful reading of the manuscript and helpful comments and E. Hiyama for teaching her the Gaussian Expansion Method.
Funded by the Deutsche Forschungsgemeinschaft (DFG, German Research Foundation)—Projektnummer 279384907—SFB 1245.

References

1. Kraemer, T., et al.: Nature **440**, 315 (2006). https://doi.org/10.1038/nature04626
2. Efimov, V.: Phys. Rev. C **44**, 2303 (1991). https://doi.org/10.1103/PhysRevC.44.2303
3. Hiyama, E., Kino, Y., Kamimura, M.: Prog. Part. Nucl. Phys. **51**, 223 (2003). https://doi.org/10.1016/S0146-6410(03)90015-9
4. König, S.: Effective quantum theories with short- and long-range forces. Ph.D. Thesis, Rheinische Friedrich-Wilhelms-Universität Bonn (2013)
5. Bethe, H.A.: Phys. Rev. **76**, 38 (1949). https://doi.org/10.1103/PhysRev.76.38
6. van Haeringen, H., Kok, L.P.: Phys. Rev. A **26**, 1218 (1982). https://doi.org/10.1103/PhysRevA.26.1218

7. Schmickler, C.H., Hammer, H.-W., Hiyama, E.: Eur. Phys. J. A **55**, 85 (2019). https://doi.org/10.1140/epja/i2019-12756-8
8. Ryberg, E., et al.: Ann. Phys. **367**, 13 (2016). https://doi.org/10.1016/j.aop.2016.01.008
9. Morlock, R., et al.: Phys. Rev. Lett. **79**, 3837 (1997). https://doi.org/10.1103/PhysRevLett.79.3837

Chapter 136
Effective Field Theory for the Heteronuclear Efimov Effect

Lucas Platter

Abstract I discuss recent results describing heteronuclear systems using effective field theory. I will focus on three-body systems with two identical particles that interact resonantly with a third particle. Specifically, I will consider recent calculations of universal scaling function that can be used to compute the temperature dependent recombination rate for positive scattering length and the inclusion of a finite effective range at zero temperature.

136.1 Introduction

Three identical bosons with a scattering length a much larger than the range R of the interaction will display the Efimov effect whose signature is an approximate discrete scaling invariance in three-body observables [1, 2]. For example, in the limit of infinite scattering length, successive states in three-body spectrum are characterized by binding energies whose ratio B^n/B^{n+1} approaches λ^2 with $\lambda = 22.7$. The parameter λ is the scaling factor and it can be different in other systems that display the Efimov effect such as in heteronuclear systems where it depends on the mass ratio.

Various approaches have been used to describe and analyze the Efimov effect. The zero-range model describes it in terms of zero-range interactions whose range is taken to zero while two-body properties such as the two-body scattering length. The effective field theory (EFT) for short-range interactions is a related approach. In general, EFTs turn a separation of scales into a small expansion parameter. The separation of scales is frequently associated with degrees of freedom integrated out of the theory. The EFT for short-range interactions employs the separation of scales between the scattering length and the range of the interaction as its expansion parameter. When the EFT for short-range interactions is applied to atoms, deep bound states associated with the van der Waals interaction with binding energies larger than $(ml^2)^{-1}$ are the

L. Platter (✉)
Department of Physics, University of Tennessee, Knoxville, TN 37996, USA
e-mail: lplatter@utk.edu

Physics Division, Oak Ridge National Laboratory, Oak Ridge, TN 37831, USA

N. A. Orr et al. (eds.), *Recent Progress in Few-Body Physics*,
Springer Proceedings in Physics 238,
https://doi.org/10.1007/978-3-030-32357-8_136

degrees of freedom that are integrated out. When it is applied to nucleons, it is the short-distance part of the pion exchange interactions that is integrated out.

Here we will consider the heteronuclear three-body system in which two identical particles interaction resonantly with a third one. This system was considered first by Helfrich et al. [3] who analyzed the properties of this system at leading order in the EFT. Below, I will discuss results that extend this work to recombination processes in ultracold gases at finite temperature and results that incorporate the effects of a finite effective range and that of a small scattering length between the identical atoms.

136.2 Effective Field Theory for Heteronuclear Atomic Systems

The universal features of the Efimov effect can easily be calculated with an effective field theory constructed for short-range interactions that lead to a large scattering length [2]. At the heart of an EFT is its Lagrange density that is in our case

$$\mathcal{L} = \sum_{i=1,2} \psi_i^\dagger \left(i\partial_t + \frac{\nabla^2}{2m_i} \right) \psi_i - d_{12}^\dagger \left(i\partial_t + \frac{\nabla^2}{2(m_1+m_2)} - \Delta_{12} \right) d_{12}$$
$$- g_{12} \left(d_{12}^\dagger \psi_1 \psi_2 + \psi_1^\dagger \psi_2^\dagger d_{12} \right) . \qquad (136.1)$$

The masses of the particles of type 1 and type 2 are denoted with m_1 and m_2, respectively. Implicit to this Lagrangian is the assumption that the scattering length a_{22} between particles with mass 2 is small and the scattering length a between particles 1 and 2 is large. The two-body coupling constant g_{12} is tuned to reproduce the scattering length a.

The resulting leading order of the EFT low-energy expansion is equivalent to the zero-range limit and leads therefore to a Skorniakov–Ter-Martorisian (STM) equation, that is a Faddeev integral equation for zero-range interactions, when the momentum space amplitude for atom-dimer scattering is constructed. The integral in this integral equation is over momenta up to a certain cutoff Λ and observables will remain strongly cutoff-dependent unless an additional three-body force is introduced to absorb this cutoff dependence [4, 5]. Here, the term

$$\mathcal{L}_3 = -h d_{12}^\dagger \psi_2^\dagger d_{12} \psi_2 , \qquad (136.2)$$

is added to the Lagrangian above. The coefficient h of this three-body force is then a function of the cutoff Λ and has to be adjusted such that the STM equation that includes this three-body force reproduces a desired observable.

136.3 Three-Body Recombination at Positive Temperature

Experiments searching for signatures of Efimov physics are frequently carried out at ultracold but non-zero temperatures. The finite temperature implies that the three-body recombination amplitude has to be averaged over different energies using a Boltzmann distribution to obtain the recombination rate α_T at finite temperature. The quantity that needs to be averaged is the so-called recombination rate K_3^J where the superscript J denotes the total angular momentum of the three-body channel under consideration. The recombination rate can also be expressed through the S-matrix element S_{A_2D,A_2D} which describes elastic atom-dimer scattering

$$K_3^{(J)}(E) = \frac{128\pi^2\mu^{3/2}}{\mu_{A_2D}^{3/2}} \frac{(2J+1)}{x^4} \left(1 - |S_{A_2D,A_2D}|^2\right) \frac{a^4}{2\mu}, \tag{136.3}$$

where μ denotes the reduced mass of the $(1, 2)$ two-body system, and μ_{A_2D} is the *three-body* reduced mass. Efimov's radial law provides a convenient way to factorize the S-matrix into functions that depend only on the scaled variable $x = E/E_2$ (E is scattering energy and $E_2 = (ma^2)^{-1}$ denotes the two-body binding energy) and a three-body parameter θ_*

$$S^{(J=0)}_{A_2D,A_2D} = s_{22}(x) + \frac{s_{21}(x)^2 e^{2i\theta_* - 2\eta_*}}{1 - s_{11}e^{2i\theta_* - 2\eta_*}}. \tag{136.4}$$

We can therefore easily calculate recombination amplitudes for any system (with arbitrary three-body parameter θ_* once the universal scaling functions s_{ij} are known. For three identical bosons that was done on the positive scattering length side in [6] and for negative scattering in [7].

For the heteronuclear system under consideration here, negative scattering lengths were considered in [8]. We considered positive scattering lengths in [9]. Figure 136.1 shows the recombination rate as a function of the scattering length a for the ^{40}K–^{87}Rb system as measured by Bloom et al. [10].

136.4 A Finite Effective Range in the Heteronuclear System

In the previous section, we noted already that experimental measurements are carried out at values of the scattering length at which finite range effects become sizeable. EFT provides a convenient way to include these corrections without specification of the underlying microscopic model. Following the previous work in [11, 12] we carried out a perturbative analysis of the heteronuclear system in [13]. We derived relations that can be used to predict the positions of recombination features in the presence of finite effective range and a *small* scattering length a_{22} between the two identical particles

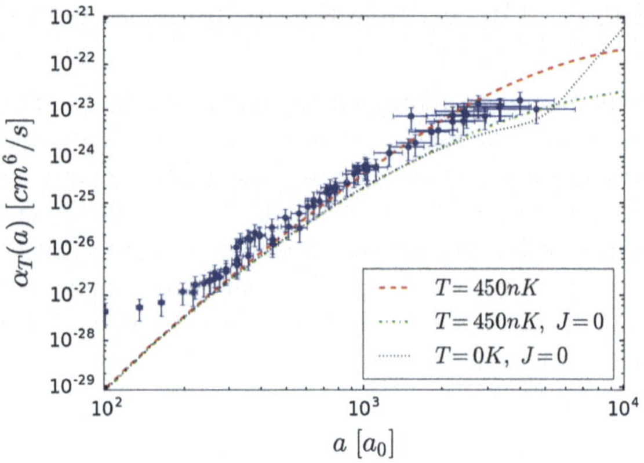

Fig. 136.1 The recombination rate constant α_T as a function of the scattering length a for $^{40}K-^{87}Rb$ with $\eta_* = 0.05(02)$ [3]. The three-body parameter is adjusted to reproduce a recombination minimum at $a_{*0} \approx 5000a_0$. The dotted red line at 450 nK corresponds to the average temperature at which the data of Bloom et al. was taken [10]

$$a_i(n) = \lambda^n \theta_i \kappa_*^{-1} + (J_i + n\sigma)r_0 + (Y_i + n\eta)a_{22}\ , \tag{136.5}$$

where $i = +, -, 0$ denotes a particular recombination feature (e.g. a_+ denotes a recombination minimum on the positive scattering length side). The parameters Y_i and J_i depend on the system under consideration, however, differences of two of these parameters such as $Y_i - Y_j$ are universal numbers.

136.5 Summary

I have discussed recent results for the heteronuclear three-body system with two identical particles that interact through a large scattering length with a third, different particle. We showed that the recombination rate at finite temperature dependence can be calculated with universal scaling functions and Efimov's radial law. I showed a comparison of our results with an experimental measurement by Bloom et al. [10].

I also discussed how the effects of a finite effective range can be included to describe its impact on observable features at zero temperatures using EFT. An analysis of the EFT results leads to universal relations that can be used to predict the positions of recombination features.

References

1. Efimov, V.: Phys. Lett. **33B**, 563 (1970)
2. Braaten, E., Hammer, H.-W.: Phys. Rept. **428**, 259 (2006)
3. Helfrich, K., Hammer, H.-W., Petrov, D.S.: Phys. Rev. A **81**, 042715 (2010)
4. Bedaque, P.F., Hammer, H.W., van Kolck, U.: Phys. Rev. Lett. **82**, 463 (1999)
5. Bedaque, P.F., Hammer, H.W., van Kolck, U.: Nucl. Phys. A **646**, 444 (1999)
6. Braaten, E., Hammer, H.-W., Kang, D., Platter, L.: Phys. Rev. A **78**, 043605 (2008)

7. Rem, B.S., et al.: Phys. Rev. Lett. **110**, 163202 (2013)
8. Petrov, D.S., Werner, F.: Phys. Rev. A **92**, 022704 (2015)
9. Emmons, S.B., Kang, D., Acharya, B., Platter, L.: Phys. Rev. A **96**(3), 032706 (2017)
10. Bloom, R.S., et al.: Phys. Rev. Lett. **111**, 105301 (2013)
11. Ji, C., Phillips, D.R., Platter, L.: Annals Phys. **327**, 1803 (2012)
12. Ji, C., Braaten, E., Phillips, D.R., Platter, L.: Phys. Rev. A **92**(3), 030702 (2015)
13. Acharya, B., Ji, C., Platter, L.: Phys. Rev. A **94**(3), 032702 (2016)

Chapter 137
Dipole-Dipole Interactions Between Neutrons

Renato Higa, James F. Babb and Mahir S. Hussein (in memoriam)

Abstract In this work we present results of the dipole-dipole interactions between two neutrons, a neutron and a conducting wall, and a neutron between two walls. As input, we use dynamical electromagnetic dipole polarizabilities fitted to chiral EFT results up to the pion production threshold and at the onset of the Delta resonance. Our work can be relevant to the physics of confined ultracold neutrons inside bottles.

Casimir-Polder forces between neutrons. The Casimir effect is a quite popular example of a non-trivial phenomenon arising from quantum fluctuations on the vacuum energy. The Casimir-Polder force was originaly devised to address the mismatch of the van der Waals $1/r^6$ tail of interatomic interactions and observations of Overbeek on quartz powder in colloid suspension. The correct asymptotic $1/r^7$ behavior is obtained by taking into account retardation effects due to the finiteness of the speed of light. Shortly after, Casimir related this force to quantum fluctuations of the vacuum between two neutral objects. Such force should appear due to the change in the zero-point electromagnetic energy between two neutral, conducting plates, an experimentally confirmed fact since then (see [1] and references therein).

R. Higa (✉) · M. S. Hussein (in memoriam)
(✉)
e-mail: higa@if.usp.br

J. F. Babb
M. S. Hussein
Instituto de FíSica, Universidade de São Paulo, R. Do Matão 1371,
São Paulo, SP 05508-090, Brazil

ITAMP, Center for Astrophysics | Harvard & Smithsonian,
MS 14, 60 Garden St., Cambridge, MA 02138, USA

M. S. Hussein
Instituto de Estudos Avançados, Universidade de São Paulo,
C. P. 72012, São Paulo, SP 05508-970, Brazil

Departamento de Física, Instituto Tecnológico de Aeronáutica CTA,
São José dos Campos, SP, Brazil

N. A. Orr et al. (eds.), *Recent Progress in Few-Body Physics*,
Springer Proceedings in Physics 238,
https://doi.org/10.1007/978-3-030-32357-8_137

Table 137.1 Parameters of (137.1), (137.2) fitted to the theoretical curves of [8]. $\alpha_n(0)$ and $\beta_n(0)$ units are 10^{-4}fm^3, the remaining ones in MeV

	$\alpha_n(0)$	a_1	a_2	$\beta_n(0)$	b_1	b_2	ω_Δ	Γ_Δ
Set 1	13.9968	12.2648	1621.63	4.2612	8.33572	22.85	241.484	66.9265
Set 2	11.6	2.2707	2721.47	3.7	8.67962	24.2003	241.593	68.3009
Set 3	12.5	5.91153	2118.79	2.7	9.27719	26.328	241.821	70.8674

A huge body of work has been devoted to this subject in atomic physics [2]. Feinberg and Sucher [3] derived the Casimir-Polder (CP) interaction between two neutral spinless particles via the two-photon exchange mechanism and general analytical properties of the related Compton scattering sub-amplitudes. The result is given in terms of atomic dipole polarizabilities reflecting the linear response to an external electromagnetic field. Arnold [4] was the first to calculate the CP interaction between two neutrons, however, at that time only the static, electric dipole polarizability data were available with nowadays outdated values. We extended Arnold's idea [5, 6] to include dynamic electric and magnetic dipole polarizabilities with updated information from low-energy chiral effective field theory analysis. We also performed calculations of the CP-interaction between a neutron and a wall, and one neutron between two walls. In the following we summarize our main results and present an outlook for future studies.

Compton scattering and neutron polarizabilities. Compton scattering on both proton and neutron became a wide subject in hadron physics comprising many theoretical and experimental efforts around the world. See recent review [7] for the current status of this line of research, where one also finds the intricate details on how to extract information on the polarizabilities from the Compton scattering amplitudes. Chiral effective field theory (χEFT), the effective theory of the underlying strong interactions (QCD), has being used to make rigorous and systematic predictions to Compton scattering observables at photon energies around and below the Δ-resonance excitation energy. The most recent χEFT calculations of Lensky et al. [8] for the electric $\alpha_n(\omega)$ and magnetic $\beta_n(\omega)$ dynamical dipole polarizabilities of the neutron is shown in Fig. 137.1 together with our low-energy parametrizations

$$\alpha_n(\omega) = \frac{\alpha_n(0)\sqrt{(m_\pi+a_1)(2M_n+a_2)}(0.2a_2)^2}{\sqrt{(\sqrt{m_\pi^2-\omega^2}+a_1)(\sqrt{4M_n^2-\omega^2}+a_2)\left[|\omega|^2+(0.2a_2)^2\right]}}, \quad (137.1)$$

$$\beta_n(\omega) = \frac{\beta_n(0)-b_1^2\omega^2+b_2^3\,\text{Re}(\omega)}{(\omega^2-\omega_\Delta^2)^2+|\omega^2\Gamma_\Delta^2|}, \quad (137.2)$$

with the set of parameters from Table 137.1 (more details in [5]).

Casimir-Polder forces. Between two neutrons, it is given by [2, 3, 5, 9]

$$\begin{aligned} V_{CP,nn}(r) &= -\frac{\alpha_0}{\pi r^6} I_{nn}(r), \\ I_{nn}(r) &= \int_0^\infty d\omega\, e^{-2\alpha_0\omega r}\Big\{\Big[\alpha_n(i\omega)^2 + \beta_n(i\omega)^2\Big] P_E(\alpha_0\omega r) \\ &\quad + \Big[\alpha_n(i\omega)\beta_n(i\omega) + \beta_n(i\omega)\alpha_n(i\omega)\Big] P_M(\alpha_0\omega r)\Big\}, \\ P_E(x) &= x^4 + 2x^3 + 5x^2 + 6x + 3, \quad P_M(x) = -(x^4 + 2x^3 + x^2), \end{aligned} \tag{137.3}$$

where $\alpha_0 \approx 1/137$ is the electromagnetic fine structure constant.

For the neutron-Wall (nW) CP potential one has [5, 10, 11]

$$\begin{aligned} V_{CP,nW}(r) &= -\frac{\alpha_0}{4\pi r^3} J_{nW}(r), \quad J_{nW}(r) = \int_0^\infty d\omega\, e^{-2\alpha_0\omega r}\alpha_n(i\omega) Q(\alpha_0\omega r), \\ Q(x) &= 2x^2 + 2x + 1. \end{aligned} \tag{137.4}$$

Finally, for two Walls separated by a distance L and one neutron in between, at a distance z from the midpoint [5, 10, 11],

$$\begin{aligned} & V_{CP,WnW}(z, L) = \\ & -\frac{1}{\alpha_0 \pi L^4}\int_0^\infty u^3 du\, \alpha\left(i\frac{u}{\alpha_0 L}\right)\int_1^\infty \frac{dv}{\sinh(uv)}\left[v^2\cosh\left(\frac{2z}{L}uv\right) - e^{-uv}\right]. \end{aligned} \tag{137.5}$$

Results. Our results are shown in Fig. 137.2. First row, $V_{CP,nn}$ as function of the separation distance r. In the left panel, the (orange) curves with smaller magnitudes stand for dynamical, ω-dependent polarizabilities while the (gray) ones with higher

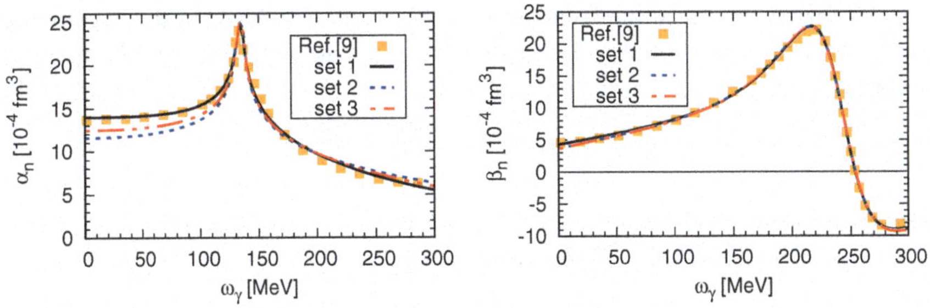

Fig. 137.1 Dynamic electric (left) and magnetic (right) polarizabilities, as functions of the photon energy ω_γ. The yellow circles are χEFT results of Lensky et al. [8] while sets 1, 2, and 3 correspond to our parametrizations using the numbers specified in Table 137.1. Adapted from [5]

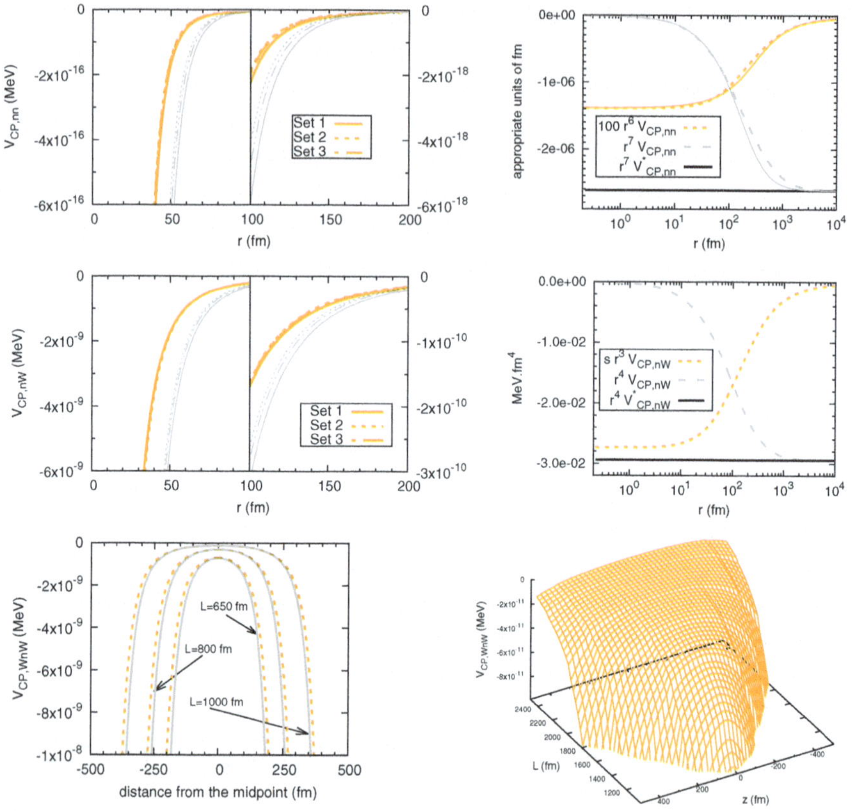

Fig. 137.2 Results for the various CP-interactions. See text for details. Adapted from [5]

magnitudes stand for the static limit. On the right panel, the (orange/gray) short-dashed/long-dashed curves are $V_{CP,nn}$ multiplied by appropriate factors $(100r^6/r^7)$, the (orange/gray) solid lines are the arctan parametrization [12] that phenomenologically makes the transition from short-distance van der Waals to the asymptotic $1/r^7$ Casimir-Polder behavior [4]. Second row, $V_{CP,nW}$ and shows similar qualitative behavior as $V_{CP,nn}$. Third row, $V_{CP,WnW}$ as function of L and z. For ultracold neutrons, these attractive CP forces may compete with repulsive Fermi pseudopotential, e.g., $V_F \approx 252\,\text{neV}$ for Ni [5].

Acknowledgements Work supported by INCT-FNA Proc. No. 464898/2014-5 (RH), US NSF through a grant for ITAMP at Harvard University and Smithsonian Astrophysical Observatory (JFB), and brazilian agencies FAPESP, CNPq and CAPES/ITA-PVS program (MSH).

References

1. Milonni, P.W., Shih, M.-L.: Contemp. Phys. **33**, 313 (1992)
2. Babb, J. F.: Casimir effects in atomic, molecular, and optical physics. In: Arimondo, E., Berman, P. R., Lin, C. C. (eds.) Advances in Atomic, Molecular, and Optical Physics, vol. 59, pp. 1. Academic Press, San Diego (2010)
3. Feinberg, G., Sucher, J.: Phys. Rev. A **2**, 2395 (1970)
4. Arnold, L.G.: Phys. Lett. B **44**, 401 (1973)
5. Babb, J.F., Higa, R., Hussein, M.S.: Eur. Phys. J. A **53**, 126 (2017)
6. Hussein, M.S., Babb, J., Higa, R.: Acta Phys. Polon. B **48**, 1837 (2017)
7. Hagelstein, F., et al.: Prog. Part. Nucl. Phys. **88**, 29 (2016)
8. Lensky, V., McGovern, J., Pascalutsa, V.: Eur. Phys. J. C **75**, 604 (2015)
9. Spruch, L., Kelsey, E.J.: Phys. Rev. A **18**, 845 (1978)
10. Zhou, F., Spruch, L.: Phys. Rev. A **52**, 297 (1995)
11. Yan, Z.-C., Dalgarno, A., Babb, J.F.: Phys. Rev. A **55**, 2882 (1997)
12. O'Carroll, M., Sucher, J., Phys. Rev. **187**, 85 (1969)

Chapter 138
Universal Phillips Lines for Identical Bosons and Particles of Different Masses

Vladimir Roudnev

Abstract The universal modification of the Phillips line is considered. We study the shape of the line for identical bosons and non-identical particles. The results suggest that the universal Phillips line is a manifistation of universality as ubiquitous as the Efimov effect.

138.1 Introduction

There are two observations that were made in the very end of 1960s and have been atrracting the few-body community's attention for a long time: the Phillips line [1] and the Efimov effect [2]. The Phillips line—the linear correlation between the neutron-deuteron scattering length and the triton bound state energy—constitutes a phenomenological observation for a very particular three-body system, whereas the Efimov effect is a universal theoretical prediction: it emerges for any three-body system which holds at least two subsystems with the two-body scattering lengths larger than all the other length scales. The latter case is often called "the universal interaction regime" in which only the leading orders of the two-body effective range expansions govern the three-body dynamics. It is exactly the regime which can be efficiently modeled on the base of Skorniakov-Ter-Martirosian (STM) equations. Application of the STM equations to studying the Phillips line by Efimov and Tkachenko [3] has demonstrated that the realistic models for the neutron-deuteron system produce results rather close to those based on the STM equations.

The question, however, whether the Phillips line is a manifestation of the universal interaction regime as ubiquitous as the Efimov effect remained unattended.

V. Roudnev (✉)
St. Petersburg State University, Ulyanovskaya, 3, St.petersburg 198504, Russian Federation
e-mail: v.rudnev@spbu.ru

N. A. Orr et al. (eds.), *Recent Progress in Few-Body Physics*,
Springer Proceedings in Physics 238,
https://doi.org/10.1007/978-3-030-32357-8_138

138.2 Identical Bosons

The study of the universal line for three identical bosons has been presented in detail earlier [4]. Let us briefly recall the major ideas and results of this work.

We shall assume that the two-body interaction supports a single bound state with the energy E_2, the three-body system may have one or more bound states, and the energy of the three-body bound state closest to the two-body threshold is E_3, $\mu_{12} = m_1 m_2/(m_1 + m_2)$ and $\mu_{1(23)} = m_1(m_2 + m_3)/(m_1 + m_2)$ are the reduced masses of the pair (12) and the particle 1 with respect to the pair (23). The two-body scattering length is referred to as a_2, and the particle-dimer scattering length a_3 stands for the particle-dimer scattering length. In the case of large positive two-body scattering length a_2 the following scaling holds $-E_2 \approx 1/(2\mu_{12}a_2^2)$. Similarly, at the two-body threshold of the three-body system, the very same scaling law takes the form of $E_2 - E_3 \approx \frac{1}{2\mu_{1(23)}a_3^2}$. This relationship suggests to introduce the following dimensionless parameters α and ω

$$\alpha \equiv a_3\sqrt{-2\mu_{1(23)}E_2} \propto \frac{1}{\sqrt{E_3/E_2 - 1}} \equiv \omega\,.$$

α can be thought of as a dimensionless particle-dimer scattering length, and the dimensionless parameter ω characterizes the proximity of the near-threshold three-body bound state to the two-body threshold.

The typical shape of the universal line is shown in Fig. 138.1. As the ratio of the two-body scattering length to the effective range increases (the values of the

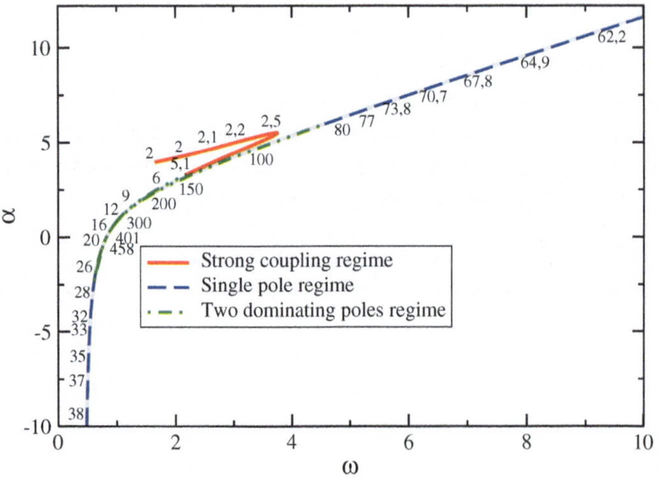

Fig. 138.1 Typical behavior of the dimensionless parameters α and ω for three-body bosonic systems (adapted from [4])

ratio are indicated along the line), the dimensionless parameters α and ω first follow an interaction-dependent half-loop, and for sufficiently large scattering length—approaching the universal interation regime—it starts following a universal line

$$\alpha \approx \frac{\alpha_1}{1/\omega - 1/\omega_0} + \omega + \alpha_0 \tag{138.1}$$

with $\alpha_0 = 5.5$, $\alpha_1 = 4$ and $\omega_0 = 0.419$. The critical value of $\omega = \omega_0$ corresponds to a new three-body bound state being formed near the two-body threshold, and the corresponding virtual state is responsible for the negative part of the universal line. As this state turns into a bound state, the system jumps into the far linear positive end of the diagram and continues to descend along the same universal line.

The universal regime manifest itself not only in the Efimov effect, but also in the proximity of the three-body system threshold parameters α and ω to the universal line. Two special points on this line corresponding to zero and infinite particle-dimer scattering length can be marked at $\frac{E_3}{E_2} \approx 2.54$ and $\frac{E_3}{E_2} \approx 6.7$ correspondingly.

138.3 Non-symmetric Systems

Unlike the fully symmetric systems of bosons, the non-symmetric system case might be complicated not only by the presence of multiple two-body thresholds, but also by their relative positions. Here we only consider a few scenarios where only one two-body threshold is present. In these scenarios we assume that two of the three particles are identical, but they do not form a bound state.

We start from the case of three equal mass particles which is the closest to the case of three identical bosons. In contrast with the latter case we assume that the scattering length in one of the pairs is negative. As we can see from Fig. 138.2a, the general shape of the line is similar to the bosonic case, but the position of the pole which corresponds to the appearance of the subsequent excited states of the three-body system is different. It is shifted to a lower value, which means that the ratio between the energies of the three-body Efimov state that just appeared and the one that already exists is bigger. A similar observation can be made for the particles of different masses (Fig. 138.2c, d). For the systems with one heavy and two light particles the subsequent Efimov states have larger energy ratio, and for the opposite case of two heavy and one light particle the subsequent Efimov states lie closer to each other. This property is evident in the behavior of the universal lines' poles: they consistently shift towards bigger or smaller values of ω.

One more example of non-identical particles is given by the three-nucleon system. This example is especially interesting: the triton has only one bound state. The proximity of the results obtained with realistic potential models [5] to the universal line indicates (Fig. 138.3) that the interaction in the system is dominated by the universal Efimov interaction, and the triton can be considered an example of a single bound state Efimov system.

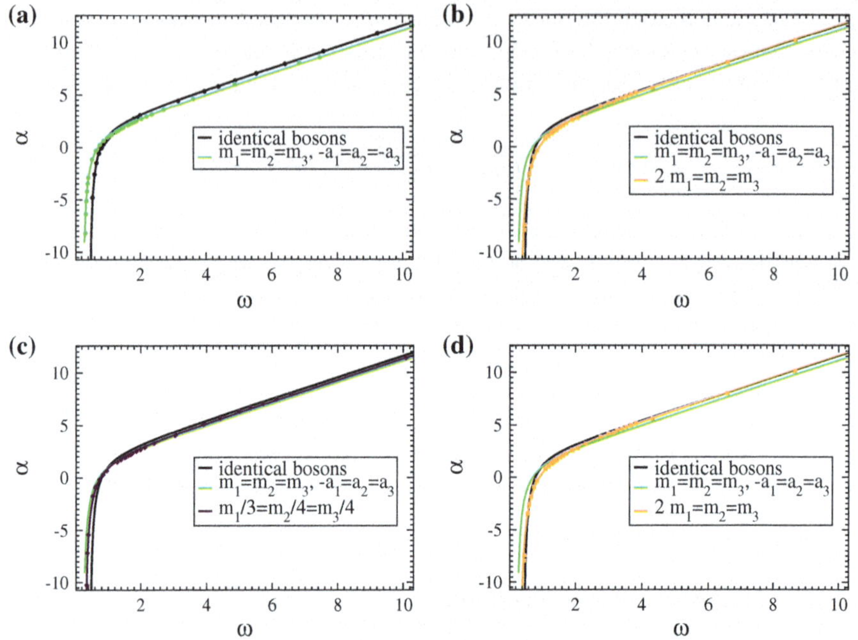

Fig. 138.2 Universal lines for non-identical particles: **a** equal masses, one unbound pair; **b** $m_1 = 1$, $m_2 = m_3 = 1/5$, light particles unbound; **c** $m_1 = 3$, $m_2 = m_3 = 4$, heavy particles unbound; **d** $m_1 = 1$, $m_2 = m_3 = 2$, heavy particles unbound

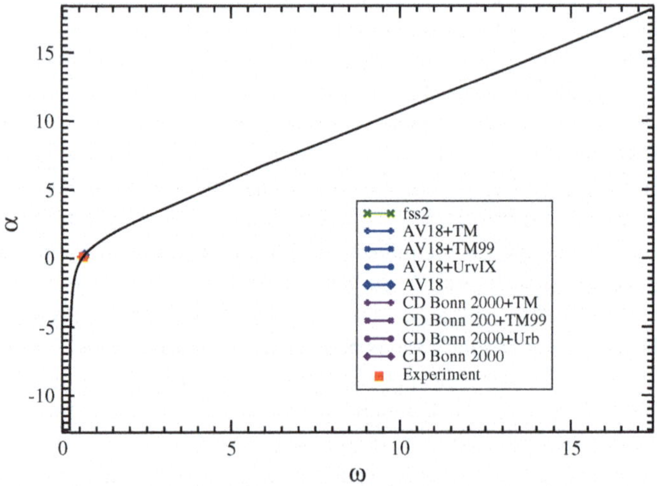

Fig. 138.3 Universal line for the nnp system. The proximity of the results for realistic models [5] to the universal line indicate that the three-nucleon system is close to the Efimov universality

References

1. Phillips, A.C.: Nucl. Phys. A **107**, 209 (1968)
2. Efimov, V.: Phys. Lett. B **33**(8), 563–564 (1970)
3. Efimov, V., Tkachenko, E.G.: Phys. Lett. B **157**, 108 (1985); Efimov, V., Tkachenko, E.G.: Few-Body Syst. **4**, 71 (1988)
4. Roudnev, V., Cavagnero, M.: Phys. Rev. Lett. **108**(11), 110402 (2012)
5. Witala, H. et. al.: Phys. Rev. C **68**, 034002 (2003)

Chapter 139
Efimov Physics Beyond Three Particles

Betzalel Bazak

Abstract Efimov physics originally refers to a system of three particles. Here we review recent theoretical progress seeking for manifestations of Efimov physics in systems composed of more than three particles. Clusters of more than three bosons are tied to each Efimov trimer, but no independent Efimov physics exists there beyond three bosons. The case of a few heavy fermions interacting with a lighter atom is also considered, where the mass ratio of the constituent particles plays a significant role. Following Efimov's study of the $(2 + 1)$ system, the $(3 + 1)$ system was shown to have its own critical mass ratio to become Efimovian. We show that the $(4 + 1)$ system becomes Efimovian at a mass ratio which is smaller than its sub-systems thresholds, giving a pure five-body Efimov effect. The $(5 + 1)$ and $(6 + 1)$ systems are also discussed, and we show the absence of 6— and 7—body Efimov physics there.

139.1 Introduction

Universal aspects of few-body systems with large scattering length have attracted attention in recent years from both theory and experiment perspectives [1]. Universality occurs when there is a large separation between the scale of the underlying physics and the scale of the phenomena observed. For example, if the inter-particle interaction range is much shorter than the spatial extent of the wave function governed by the scattering length a, most of the time the particles will be out of the potential range and therefore not sensitive to its fine details.

A few examples of relevant systems come to mind. In low-energy nuclear physics the scattering length of the singlet and triplet channels are $a_s \approx -23.4$ fm and $a_t \approx 5.42$ fm, while the long-range part of the nucleonic interaction, determined by the pion mass, is $R \approx \hbar/m_\pi c \approx 1.4$ fm.

B. Bazak (✉)
The Racah Institute of Physics, The Hebrew University, 9190401 Jerusalem, Israel
e-mail: betzalel.bazak@mail.huji.ac.il

N. A. Orr et al. (eds.), *Recent Progress in Few-Body Physics*,
Springer Proceedings in Physics 238,
https://doi.org/10.1007/978-3-030-32357-8_139

Larger scale separation can be found in ^{4}He atoms. Here the He-He scattering length is $a \approx 90$ Å, while the Van Der Waals interaction range is $r_{vdW} \approx 5.4$ Å.

Another interesting case is ultracold atoms near Feshbach resonance. Here the scattering length can be tuned to arbitrary value using, for example, external magnetic field,

$$a(B) = a_{bg}\left(1 + \frac{\Delta}{B - B_0}\right).$$

A fascinating effect was predicted by Efimov [2] for three identical bosons with resonating interaction: the existence of infinite tower of bound trimers. For a recent review see [3].

In this paper, we study Efimov physics beyond three particles. We start with a short review of universal features and Efimov physics in three identical bosons and in the $(2+1)$ system, which is a mixture of two identical fermions and distinguishable particle. Then we go beyond three particles and discuss the $N > 3$ identical bosons system, as well as the $(N+1)$ systems with $N \leq 6$.

139.2 Methods and Results

In order to study universality, one would like to neglect the system-specific details and concentrate on the universal features.

To do so one could use the zero-range limit, i.e. eliminate the spatial extent of the potential while applying the Bethe-Peierls boundary condition when two particles touch each other,

$$\frac{\partial \log(r_{ij}\psi)}{\partial r_{ij}} \xrightarrow[r_{ij}\to 0]{} -\frac{1}{a} \tag{139.1}$$

where $r_{ij} = |\boldsymbol{r}_j - \boldsymbol{r}_i|$ is the distance between any pair of interacting particles.

139.2.1 The Universal Dimer

A trivial example for universality is the existence of universal dimer composed of two identical bosons of mass m for $a > 0$. Working in the center-of-mass frame and taking the zero-range limit, one has to solve the free Schrödinger equation for the relative coordinate r and to apply the Bethe-Peierls boundary condition (139.1) at zero, giving for the bound state $\psi(r) \propto \exp(-r/a)/r$ corresponds to an energy of $-1/ma^2$, where here and thereafter $\hbar$ is set to 1.

This prediction is indeed valid for the three examples mentioned above. The deuteron binding energy, 2.22 MeV, is fairly close to the universal prediction $1/ma_t^2 \approx 1.4$ MeV. The ^{4}He atoms dimer binding energy was measured recently to be about 1.76(15) mK [4], where the universal prediction is $1/ma^2 \approx 1.48$ mK.

Since the next correction is of order of r_0/a, where r_0 is the effective range, one would expect the universal prediction to be even better with ultracold atoms, and indeed this is the case [5].

139.2.2 Efimov Physics in Three Identical Bosons

Adding another identical boson, the situation is changed dramatically, as Efimov has shown [2].

To see that one can start from the Faddeev equation for zero-range potential, and then transform to hyperspherical coordinates. In the unitary limit $a \longrightarrow \infty$, the energy is then determined by one-dimensional equation for the hyper-radius, $\rho^2 \propto r_{12}^2 + r_{23}^2 + r_{13}^2$,

$$\left(-\frac{d^2}{d\rho^2} + \frac{s^2 - 1/4}{\rho^2} \right) R(\rho) = E R(\rho), \tag{139.2}$$

where s is the eigenvalue of the corresponding hyper-angular equation. Equation (139.2) has two interesting features. First, the effective three-body potential has long range $\propto \rho^{-2}$, in contrast to the zero-range two-body interaction we start with. Second, it exhibits scale invariance, therefore if $R(\rho)$ is a solution with the corresponding energy E, $R(\lambda\rho)$ is also solution with the energy $\lambda^2 E$ for arbitrary λ.

At small ρ, E can be neglected, and the solution for (139.2) is $R_{\pm}(\rho) \propto \rho^{1/2 \pm s}$. The solution behavior is therefore determined by s. For $s^2 > 0$ the solution can be set to $R_{+}(\rho)$, while for $s^2 < 0$ the solution is a combination of two oscillating functions, whose relative phase is still needed to be fixed.

In the latter case the effective potential is attractive and one faces fall of a particle to the center of ρ^{-2} potential, i.e. the energy here is not bound from below [6].

Introducing three-body potential barrier at some finite ρ_0 saves us from this collapse by setting the system ground state. The scale invariance is now broken into discrete scale invariance, with $\lambda_n = e^{-\pi n/|s|}$, and therefore the energies are quantized, giving infinite series of bound states with geometric-series spectrum $E_n = E_0 e^{-2\pi n/|s|}$. Here ρ_0 is a three-body parameter, which sets the ground state energy and also fixes the relative phase of $R_{\pm}$. s is the scale factor which governs the scaling characters of the energies and the wave functions. For three identical boson it has the value $s = 1.00624\, i$.

This prediction had to wait about four decades before its verification in ultracold gases experiments, where particle loss from the trap is a three-body process, showing a significant signal when new Efimov state is formed. Studies of loss features of ultracold ^{133}Cs [7], ^{39}K [8], and ^{7}Li [9, 10] gases gave the first experimental verification for Efimov physics.

The existence of Efimov trimers in ^{4}He atoms was predicted long ago [11], where due to the finite scattering length only two trimers should exist. Only recently the excited trimer was seen experimentally [12], giving another verification for Efimov's prediction.

139.2.3 Efimov Physics Beyond Three Identical Bosons

Shortly after Efimov original paper, Amado and Greenwood have claimed that there is no Efimov effect for four or more particles [13].

However, three-body Efimov physics has a footprint in the four body system, were two tetramers are tied to each Efimov trimer [14], a prediction which was verified in ultracold atoms experiments [15]. The tetramer and trimer energies are correlated, similar to the correlation between triton and alpha binding energies known as the Tjon line [16, 17].

Larger clusters of identical bosons also exist, and their energies are correlated to the trimer energy, therefore not showing independent N-boson Efimov physics [18]. See, however [19].

139.2.4 Mass Imbalanced Fermionic Mixtures

Another system relevant to Efimov physics is a mixture of identical fermions and distinguishable particle with different mass.

Consider two heavy atoms with mass M interacting with a light atom with mass m. First, we apply the Born-Oppenheimer approximation, valid in the limit of $M \gg m$ [20, 21]. Here the motion of the light particle is first solved assuming the heavy particles position is fixed at $\pm R/2$, giving

$$\psi_{\boldsymbol{R}}^{\pm}(\boldsymbol{r}) \propto \frac{e^{-\kappa(R)|\boldsymbol{r}-\boldsymbol{R}/2|}}{|\boldsymbol{r}-\boldsymbol{R}/2|} \pm \frac{e^{-\kappa(R)|\boldsymbol{r}+\boldsymbol{R}/2|}}{|\boldsymbol{r}+\boldsymbol{R}/2|} \tag{139.3}$$

where $\boldsymbol{r}$ is the light particle position. Applying the boundary condition (139.1) gives

$$\kappa_{\pm}(R) \mp \frac{e^{-\kappa_{\pm}(R)R}}{R} = 1/a. \tag{139.4}$$

The energy of the light atom $\epsilon_{\pm}(R) = -\kappa_{\pm}^2(R)/2m$ is then considered as an effective potential between the heavy atoms. The minus state corresponds to repulsive effective potential, while the plus state induces attractive potential,

$$\epsilon_{+}(R) \approx \begin{cases} -\frac{0.16}{mR^2} & R/a \ll 1 \\ -\frac{1}{2m}\left(\frac{1}{a^2} + \frac{\exp(-R/a)}{aR}\right) & R/a \gg 1 \end{cases} \tag{139.5}$$

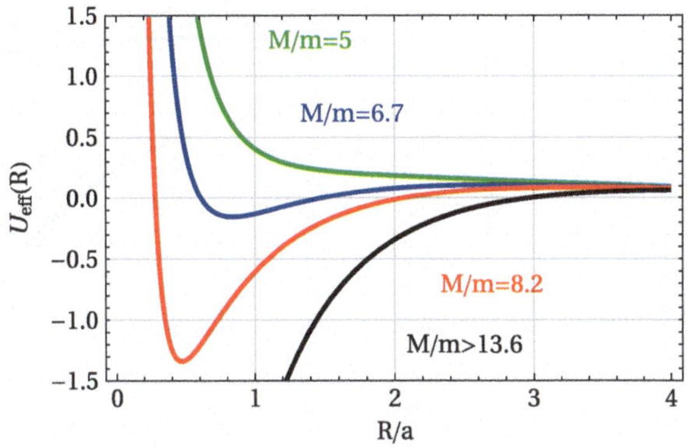

Fig. 139.1 The effective potential between heavy particles induced by the light particle as a function of their distance, for different mass ratios in the Born-Oppenheimer approximation. For a small mass ratio, the effective interaction is repulsive (green). As the mass ratio increases p-wave resonance occurs (blue), then the potential well becomes deep enough to support bound trimer (red), and finally, the attraction wins and Efimov states emerge (black)

The heavy-particles equation for $R \ll a$ is therefore identical to (139.2), replacing ρ by R. Here $s^2 = l(l+1) - 0.16M/m + 1/4$ for angular momentum l.

In the bosonic case, the ground state has $l = 0$, giving purely attractive $-1/R^2$ effective potential, and therefore Efimov physics.

In the fermionic case, the permutation symmetry dictates odd angular momentum and the ground state has $l = 1$. The centrifugal barrier $l(l+1)/MR^2$ therefore competes with the $-1/mR^2$ attraction, where the competition is governed by the mass ratio. Figure 139.1 shows the effective potential for various mass ratios M/m.

This simple picture indeed catches the physics here. For a small mass ratio, the effective potential is repulsive, and no bound trimer exists. As the mass ratio increases, the potential becomes more attractive, and a p-wave resonance occurs. Indeed this resonance was found in ultracold ^{40}K–^{6}Li mixture [22]. For larger mass ratio the potential well is deep enough to support a universal 1^- bound state [23]. For even larger mass ratio the system becomes Efimovian [24].

To proceed beyond this approximation, it is convenient to follow Skorniakov and Ter-Martirosian formalism [25, 26]. Here instead of solving the Schrödinger equation, one utilizes the zero-range potential to get an integral equation.

For the $(N+1)$ case, the STM equation in momentum space is [27],

$$\frac{1}{4\pi}\left(\frac{1}{a}-\kappa\right)F(\mathbf{q}_1,\ldots,\mathbf{q}_{N-1}) = \int \frac{d^3q_N}{(2\pi)^3}\frac{\sum_{i=1}^{N-1} F(\mathbf{q}_1,\ldots,\mathbf{q}_{i-1},\mathbf{q}_N,\mathbf{q}_{i+1},\ldots,\mathbf{q}_{N-1})}{-2\mu E + \frac{\mu}{M}\sum_{i=1}^{N} q_i^2 + \frac{\mu}{m}\left(\sum_{i=1}^{N}\mathbf{q}_i\right)^2}, \tag{139.6}$$

where $\mu = Mm/(M+m)$ is the reduced mass. The function $F(\mathbf{q}_1,\ldots,\mathbf{q}_{N-1})$ can be considered as the relative wave function of $N-1$ heavy atoms with momenta $\mathbf{q}_1,\ldots,\mathbf{q}_{N-1}$ and a heavy-light pair, the momentum of which equals $-\sum_{i=1}^{N-1}\mathbf{q}_i$.

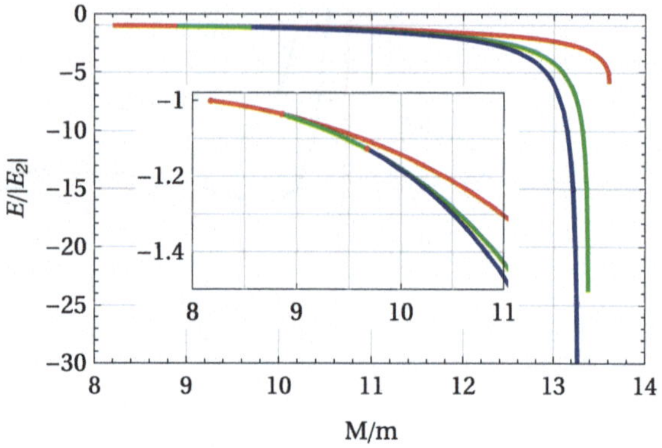

Fig. 139.2 The energies of the universal $(N+1)$ states in units of the dimer binding energy, as a function of the mass ratio. Shown are results for the 1^- $(2+1)$ state (red), the 1^+ $(3+1)$ state (green), and the 0^- $(4+1)$ state (blue). The inset shows zoom-in on the thresholds region. Adapted from [27]

Here

$$\kappa = \sqrt{-2\mu E + \frac{\mu}{M}\sum_{i=1}^{N-1} q_i^2 + \frac{\mu}{M+m}\left(\sum_{i=1}^{N-1} \mathbf{q}_i\right)^2}. \tag{139.7}$$

For a specific value of total angular momentum and parity L^π, the equation can be further simplified. For the $(2+1)$ case the relevant symmetry for both universal trimer and Efimov state is 1^-. The function F therefore takes the form $F(\mathbf{q}) = f(q)\hat{\mathbf{z}} \cdot \hat{\mathbf{q}}$, leaving one-dimensional equation which can be easily solved for E, as can be seen in Fig. 139.2. The universal trimer binding threshold could thus be obtained, $M/m = 8.173$, in agreement with the results obtained in [23] in the hyperspherical formalism.

To find the mass ratio where the system becomes Efimovian one would like to calculate the scale factor s which approaches zero at that point. For that, one can calculate the large-q asymptote of f, which has the form

$$f(q) \propto q^{-2-s}. \tag{139.8}$$

Solving (139.6) for f and fitting the results to extract s, the Efimov threshold can be found at $M/m = 13.607$, in agreement with the result of [24].

An interesting alternative, which may be more suitable for larger systems, is to utilize the mapping between the free-space system with finite a and the trapped system at unitarity [28, 29], whose energy is

$$E = \hbar\omega(s + 2n + 1) \tag{139.9}$$

where ω is the trap frequency, s is the same scale factor and n counts hyper-radial excitations. Hence, one can extract the scale factor s from the trapped energies.

Now that we have built our toolbox, we can face an interesting question: how many heavy fermions can be bound by a single light atom?

For the $(3+1)$ case the relevant symmetry is 1^+, and F takes the form $F(\mathbf{q}_1, \mathbf{q}_2) = f(q_1, q_2, \hat{\mathbf{q}}_1 \cdot \hat{\mathbf{q}}_2)\hat{\mathbf{z}} \cdot \hat{\mathbf{q}}_1 \times \hat{\mathbf{q}}_2$, leaving a three-dimensional integral equation which can be solved in deterministic method. It was shown that a universal 1^+ tetramer exists for a mass ratio $M/m \gtrsim 9.5$ [30]. Moreover, an 1^+ Efimov states exist above $M/m > 13.384$ [31].

Adding another fermion, the relevant symmetry is 0^-, therefore $F(\mathbf{q}_1, \mathbf{q}_2, \mathbf{q}_3) = f(q_1, q_2, q_3, \hat{\mathbf{q}}_1 \cdot \hat{\mathbf{q}}_2, \hat{\mathbf{q}}_1 \cdot \hat{\mathbf{q}}_3, \hat{\mathbf{q}}_2 \cdot \hat{\mathbf{q}}_3)\hat{\mathbf{q}}_1 \cdot \hat{\mathbf{q}}_2 \times \hat{\mathbf{q}}_3$, but the resulting six-dimensional integral equation is too hard to be solved with conventional method. Hence a novel method, which we call the STM-DMC method, is introduced [27, 32], where f is treated as density probability function for so-called walkers, whose stochastic dynamics is governed in such a way that their detailed-balance condition is (139.6). Given E, a is than changed in each iteration to keep the walkers' number constant.

Using this method, (139.6) can be solved for the $(2+1)$, $(3+1)$, and $(4+1)$ cases, getting both energies and scale factors. Figure 139.2 shows the energies of the universal states in these systems, its inset focuses on their thresholds. Known results are reproduced, i.e. the thresholds for the $(2+1)$ universal trimer and Efimov states. Moreover, we can locate better the threshold of the $(3+1)$ universal tetramer to be 8.862 [27], and confirm the threshold for four-body Efimov states [31].

We can now explore the *terra incognita* $(4+1)$ system. Here we find a 0^- universal pentamer as well as 0^- Efimov states. In Fig. 139.2 we also plot the energies for the universal 0^- pentamer, showing it is bound for mass ratio above 9.672. In Fig. 139.3 we show the scale factor for this system, showing Efimov 0^- states emerge here above $M/m = 13.279$ [27].

The different threshold for the $(N+1)$ states are summarized in Table 139.1.

We see that the $(N+1)$ systems with $N = 2, 3$ and 4 exhibit similar behavior, showing pure $N+1$-body Efimov physics. Does this pattern continue for $N \geq 5$?

The relevant symmetry for the $(5+1)$ ground state is 0^-, signaling that the additional fermion populates an excited s-shell, which has a radial node. This causes the stochastic method to suffer from a sign problem. Hence we choose here another approach, which is to extract the scale factor from the energies in a harmonic trap.

These energies were calculated using a Gaussian potential with a finite range R_0,

$$V(r) = -V_0\, e^{-\frac{r^2}{2R_0^2}} \tag{139.10}$$

Table 139.1 The thresholds for universal and Efimov states in the $(N+1)$ systems

System	L^π	Universal state	Efimov state
2+1	1^-	8.173	13.607
3+1	1^+	8.862	13.384
4+1	0^-	9.672	13.279
5+1	0^-	?	–
6+1	2^-	?	–

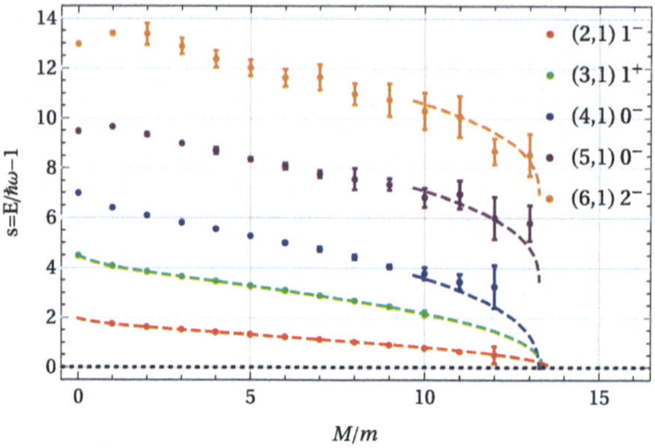

Fig. 139.3 The scale factor as extracted from the energies of the $(N + 1)$ state in a harmonic trap at unitarity, as a function of the mass ratio. The Efimov limit corresponds here to $s = 0$. Circles stand for the results extrapolated from finite-range gaussian potential to the zero-range limit. Dashed curves are for results acquired directly in the zero-range limit, by solving the STM equation on a grid (for $(2 + 1)$ and $(3 + 1)$ systems) or with stochastic method (for the $(4 + 1)$ case). The Dashed curves for the $(5 + 1)$ and $(6 + 1)$ cases taken from the $(4 + 1)$ case with appropriate shift, showing no Efimov effect exists in these cases

where the results are extrapolated to the zero-range limit $R_0 \longrightarrow 0$. The $(N + 1)$-body Schrödinger equation is solved using a basis of correlated gaussians, chosen by the stochastic variational method [33]. Using (139.9), the scale factor is then extracted [34].

In Fig. 139.3 we show the $(N + 1)$ system scale factor for $N \leq 6$. Efimov threshold here is signaled by $s = 0$. Results obtained with other methods are also shown. Indeed the scale factors for $N = 2, 3$ and 4 hit zero at the Efimovian threshold. However, no sign for Efimov physics is found in the $N = 5$ and 6 systems.

139.3 Conclusion

Efimov physics beyond three particles is studied here. For identical bosons, no independent Efimov effect exists beyond three particles, although bosonic clusters are tied to each Efimov trimer. For the $(N + 1)$ case of N identical fermions interact with distinguishable particle, Efimov states occur for mass ratio exceeds the relevant threshold for the $(2 + 1)$, $(3 + 1)$, and $(4 + 1)$ systems. However, no Efimov state exists for the $(5 + 1)$ and $(6 + 1)$ systems.

Acknowledgements I would like to thank Dmitry Petrov for useful discussions and communications.

References

1. Braaten, E., Hammer, H.-W.: Universality in few-body systems with large scattering length. Phys. Rep. **428**, 258 (2006)
2. Efimov, V.: Energy levels arising from resonant two-body forces in a three-body system. Phys. Lett. B **33**, 563 (1970)
3. Naidon, P., Endo, S.: Efimov physics: a review. Rep. Prog. Phys. **80**, 056001 (2017)
4. Zeller, S., et al.: Imaging the He_2 quantum halo state using a free electron laser. Proc. Natl. Acad. Sci. USA **113**, 14651 (2016)
5. Thompson, S.T., Hodby, E., Wieman, C.E.: Ultracold molecule production via a resonant oscillating magnetic field. Phys. Rev. Lett. **95**, 190404 (2005)
6. Thomas, H.: The interaction between a neutron and a proton and the structure of h^3. Phys. Rev. **47**, 903 (1935)
7. Kraemer, T., Mark, M., Waldburger, P., Danzl, J.G., Chin, C., Engeser, B., Lange, A.D., Pilch, K., Jaakkola, A., Nägerl, H.-C., Grimm, R.: Evidence for Efimov quantum states in an ultracold gas of caesium atoms. Nature **440**, 315 (2006)
8. Zaccanti, M., Deissler, B., D'Errico, C., Fattori, M., Jona-Lasinio, M., Müller, S., Roati, G., Inguscio, M., Modugno, G.: Observation of an Efimov spectrum in an atomic system. Nat. Phys. **5**, 586 (2009)
9. Gross, N., Shotan, Z., Kokkelmans, S., Khaykovich, L.: Observation of universality in ultracold ^{7}Li three-body recombination. Phys. Rev. Lett. **103**, 163202 (2009)
10. Pollack, S.E., Dries, D., Hulet, R.G.: Universality in three- and four-body bound states of ultracold atoms. Science **326**, 1683 (2009)
11. Lim, T.K., Duffy, S.K., Damert, W.C.: Efimov state in the ^{4}He trimer. Phys. Rev. Lett. **38**, 341 (1977)
12. Kunitski, M., et al.: Observation of the Efimov state of the helium trimer. Science **348**, 551 (2015)
13. Amado, R.D., Greenwood, F.C.: There is no Efimov effect for four or more particles. Phys. Rev. D **7**, 2517 (1973)
14. Hammer, H.-W., Platter, L.: Universal properties of the four-body system with large scattering length. Eur. Phys. J. A **32**, 113 (2007)
15. von Stecher, J., D'Incao, J.P., Greene, C.H.: Signatures of universal four-body phenomena and their relation to the Efimov effect. Nat. Phys. **5**, 417 (2009)
16. Tjon, J.A.: Bound states of ^{4}He with local interactions. Phys. Lett. B **56**, 217 (1975)
17. Platter, L., Hammer, H.-W., Meissner, U.-G.: On the correlation between the binding energies of the triton and the alpha-particle. Phys. Lett. B **607**, 254 (2005)
18. Bazak, B., Eliyahu, M., van Kolck, U.: Effective field theory for few-boson systems. Phys. Rev. A **94**, 052502 (2016)
19. Bazak B., Kirscher J., König S., Pavón Valderrama M., Barnea N., van Kolck U.: The four-body scale in universal few-boson systems, arXiv:1812.00387
20. Fonseca, A.C., Redish, E.F., Shanley, P.E.: Efimov effect in an analytically solvable model. Nucl. Phys. A **320**, 273 (1979)
21. Petrov D. S.: The few-atom problem. In: Many-Body Physics with Ultra-cold Gases: Lecture Notes of the 2010 Les Houches Summer School, vol. 94 (2012)
22. Jag, M., Zaccanti, M., Cetina, M., Lous, R.S., Schreck, F., Grimm, R., Petrov, D.S., Levinsen, J.: Observation of a strong atom-dimer attraction in a mass-imbalanced Fermi-Fermi mixture. Phys. Rev. Lett. **112**, 075302 (2014)
23. Kartavtsev, O.I., Malykh, A.V.: Low-energy three-body dynamics in binary quantum gases. J. Phys. B: At. Mol. Opt. Phys. **40**, 1429 (2007)
24. Efimov, V.: Energy levels of three resonantly interacting particles. Nucl. Phys. A **210**, 157 (1973)
25. Skorniakov, G.V., Ter-Martirosian, K. A.: Three body problem for short range forces. I. Scattering of low energy neutrons by deutrons, Sov. Phys. JETP **4**, 648 (1957)

26. Pricoupenko, L.: Isotropic contact forces in arbitrary representation: heterogeneous few-body problems and low dimensions. Phys. Rev. A **83**, 062711 (2011)
27. Bazak, B., Petrov, D.S.: Five-body Efimov effect and universal pentamer in fermionic mixtures. Phys. Rev. Lett. **118**, 083002 (2017)
28. Tan S.: Short range scaling laws of quantum gases with contact interactions, arXiv:cond-mat/0412764 (2004)
29. Werner, F., Castin, Y.: Unitary gas in an isotropic harmonic trap: symmetry properties and applications. Phys. Rev. A **74**, 053604 (2006)
30. Blume, D.: Universal four-body states in heavy-light mixtures with a positive scattering length. Phys. Rev. Lett. **109**, 230404 (2012)
31. Castin, Y., Mora, C., Pricoupenko, L.: Four-body Efimov effect for three fermions and a lighter particle. Phys. Rev. Lett. **105**, 223201 (2010)
32. Bazak, B., Petrov, D.S.: Energy of N two-dimensional bosons with zero-range interactions. New J. Phys. **20**, 023045 (2018)
33. Suzuki, Y., Varga, K.: Stochastic Variational Approach to Quantum-mechanical Few-body Problems. Springer (1998)
34. Bazak, B.: Mass-imbalanced fermionic mixture in a harmonic trap. Phys. Rev. A **96**, 022708 (2017)

Chapter 140
Ab Initio Calculation of Nuclear Structure Effects in Light Muonic Atoms

Chen Ji

Abstract Nuclear structure effects to the Lamb shift in light muonic atoms are driven by the two-photon exchange contribution, whose evaluation relies significantly on nuclear theory. Using ab initio calculations, we have studied nuclear structure effects in several light muonic atoms, including $\mu^{2,3}$H and $\mu^{3,4}$He$^+$ [1–5]. The two-photon exchange is associated with nuclear electromagnetic sum rules. The calculation was done with the implementation of state-of-the-art nuclear potentials, derived either phenomenologically or from chiral effective field theory. The statistical and systematic uncertainties from nuclear theory were analyzed by potential-model comparison, and by using power-counting and parameter-optimization techniques, that have been developed in chiral effective field theory. High precision of the ab initio results is crucial for accurately extracting nuclear charge radii in muonic atom spectroscopy.

140.1 Introduction

The proton charge radius, extracted by the CREMA collaboration (PSI) in the high-precision spectroscopy of muonic hydrogen [6, 7], revealed a 6σ deviation from the CODATA value, i.e., an average of experimental results in ordinary hydrogen spectroscopy and electron-proton scattering [8]. This mysterious discrepancy may originate from unknown systematic uncertainties in different experiments, and is also speculated due to missing quantum electrodynamics (QED) contributions or the breaking of lepton universality.

A tremendous theoretical and experimental efforts have been made to probe the proton radius puzzle. One direction towards resolving the discrepancy is to re-examine the proton charge radius measured in ordinary hydrogen spectroscopy. However, two new experiments on the hydrogen $2S$–$4P$ and $1S$–$3S$ transitions [9, 10] found contradictory results on the proton radius: one is consistent with muonic

C. Ji (✉)
Key Laboratory of Quark and Lepton Physics (MOE) and Institute of Particle Physics, Central China Normal University, Wuhan 430079, China
e-mail: jichen@mail.ccnu.edu.cn

N. A. Orr et al. (eds.), *Recent Progress in Few-Body Physics*,
Springer Proceedings in Physics 238,
https://doi.org/10.1007/978-3-030-32357-8_140

hydrogen and the other agrees with the CODATA value. The CREMA collaboration tackled the problem via a different path by extending their experiments to other light muonic atoms [11]. They performed a series of Lamb shift measurements in muonic deuterium (μ^2H) and helium ions ($\mu^{3,4}$He$^+$). By comparing the nuclear charge radii r_E extracted from muonic atom experiments and the ones from ordinary atom spectroscopy or electron–nucleus scattering, one can test how the radius discrepancy changes in lepton-nucleus systems with different light nuclei. The unveiled pattern can provide valuable hints for explaining the radius puzzle.

In muonic atom spectroscopy, the measurement of the Lamb shift has been reached at extremely high precision. However, to extract the value of r_E from the measured Lamb shift relies strongly on the theoretical inputs, which need to be calculated with very high accuracy. Those inputs include the QED contribution and the nuclear structure correction to the atomic spectrum. Since the muon is 200 times heavier than the electron, the muon orbits the nucleus about 200 times closer in an atom. Therefore, the nuclear structure correction in a muonic atom is considerably larger than that in an ordinary atom [12]. The Lamb shift δ_{LS}, measured in a muonic atom with high precision, is related to the nuclear charge radius r_E by

$$\delta_{\mathrm{LS}} = \delta_{\mathrm{QED}} + \mathcal{A}_{\mathrm{OPE}}\, r_{\mathrm{E}}^2 + \delta_{\mathrm{TPE}}. \tag{140.1}$$

The first term, δ_{QED}, is the QED contribution, which has been calculated with high accuracy. The rest two terms are nuclear structure corrections, which are depicted in Fig. 140.1 as the effects driven by the muon-nucleus one-photon exchange (OPE) and two-photon exchange (TPE) processes. The OPE term is proportional to r_E^2, where the coefficient $\mathcal{A}_{OPE}$, of order $m_\mu^3(Z\alpha)^4$, is precisely determined in theory. The TPE contribution, δ_{TPE}, is of order $(Z\alpha)^5$. Its evaluation relies on our accurate knowledge of the nuclear dynamics and structure.

The current bottleneck for accurately extracting r_E from (140.1) relies in the theoretical calculation of δ_{TPE}, with precision of a few percent to distinguish between the experimental results in muonic and electronic atoms. We present here an update to our recent progress of calculating δ_{TPE} in different light mounic atoms. The calculation utilized the nuclear ab initio methods, which have been greatly developed in the past few decades.

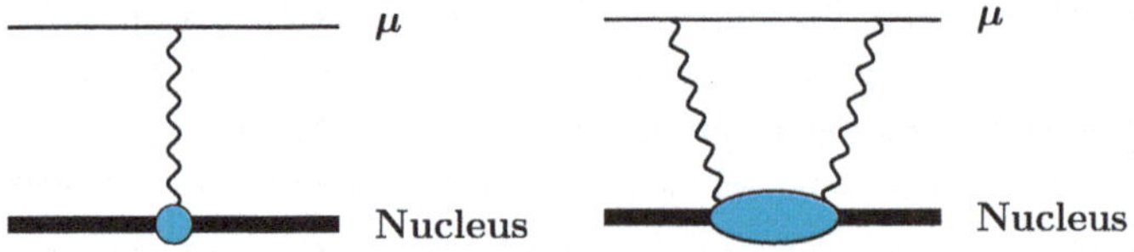

Fig. 140.1 The muon-nucleus one-photon and two-photon exchange effects

140.2 Ab Initio Theory for Two-Photon Exchange

The $\delta_{\rm TPE}$ contribution can be separated into elastic and inelastic parts, $\delta_{\rm TPE} = \delta_{\rm Zem} + \delta_{\rm pol}$. The elastic part $\delta_{\rm Zem}$, named Zemach/Friar term, is proportional to the electric Zemach moment [13, 14]. The inelastic part $\delta_{\rm pol}$, called the nuclear polarizability contribution, reflects the virtual excitation and de-excitation of the nucleus within the TPE process. The typical energy scales associated with the excitation of the nucleus and the nucleon are respectively around 30 MeV and 300 MeV. The scale separation indicates that one can study the nuclear and nucleonic parts of $\delta_{\rm TPE}$ separately using different effective theories. Therefore, for the elastic and inelastic parts of TPE, we have $\delta_{\rm Zem} = \delta^A_{\rm Zem} + \delta^N_{\rm Zem}$ and $\delta_{\rm pol} = \delta^A_{\rm pol} + \delta^N_{\rm pol}$. The superscripts A and N denote respectively the contributions from the atomic nucleus and the nucelon. The ab initio calculation is to solve the nuclear part of TPE, i.e., $\delta^A_{\rm Zem}$ and $\delta^A_{\rm pol}$. In this approach, the structure and reaction mechanism of the atomic nucleus is described in terms of constituent nucleons interacting with two- and three-nucleon forces.

The key ingredient for obtaining the TPE contributions is to calculate a series of generalized sum rules (GSR)

$$\Sigma_{\rm GSR} = \int d\omega\, g(\omega) S_O(\omega), \tag{140.2}$$

with different energy-weight functions $g(\omega)$ and nuclear structure functions $S_O(\omega)$. The dominant contribution involves the electric dipole structure function with an energy weight proportional to $\sqrt{m_\mu/\omega}$, where ω denotes the excitation energy of an intermediate nuclear state in the TPE process. Using the Laczos sum rule (LSR) approach [15], we can integrate over the full nuclear excitation spectrum in GSR. The method is embedded into an expansion over a discrete sets of localized functions. By doing so, one can numerically diagonalize the few-nucleon Hamiltonian. We used the harmonic oscillator (HO) basis expansion in the two-nucleon system and the hyperspherical harmonics (HH) basis expansion in the three-nucleon and four-nucleon systems. In the latter case, we utilized the effective interaction hyperspherical harmonics (EIHH) [16, 17] method to numerically achieve faster convergence.

The evaluation of the nucleonic part of the TPE contribution is beyond the scope of nuclear ab initio calculations, which do not describe the internal structure of the nucleon. Using dispersion relation analysis on electron-proton scattering data, $\delta^N_{\rm TPE}$ in μH was evaluated in [18, 19]. The corresponding nucleonic contributions in other light muonic atoms can be directly related to that in μH by simply multiplying some scaling coefficients.

A detailed calculation of each individual contribution to $\delta_{\rm TPE}$ can be found in our recent review paper [5].

140.3 Results and Error Estimates

We presented in 2014 [1] the calculation of $\delta_{\rm TPE}$ in μ^2H using χEFT nucleon-nucleon interactions with a series of different parameterizations. Based on the χEFT power counting, the order-by-order low-energy expansion of the nuclear interactions allows us to analyze the systematic uncertainty of χEFT predictions calculated at a given EFT order. In μ^2H, the nuclear-force uncertainty in $\delta_{\rm TPE}$ was estimated to be 0.6%. The result is highly consistent with calculations using the AV18 phenomenological potential [20, 21] and the zero-range potential [22], and also agrees with results from dispersion relation analysis [23]. However, the later has larger uncertainty due to the lack of electron-deuteron quasi-elastic data at low energies and forward angles.

The experimental results of the Lamb shift in muonic deuterium was published by the CREMA collaboration in 2016 [24]. A deuteron charge radius of $r_d = 2.12562(78)$ fm was extracted. Its uncertainty is dominated by the theoretical uncertainty of $\delta_{\rm TPE}$. The mean value is by 6.0σ smaller than the CODATA value [8]. It also deviates by 3.5σ from the result in ordinary e^2H spectroscopy [25]. Reference [24] also indicates a 2.6σ discrepancy in the measured isotope shift, $r_d^2 - r_p^2$, between ordinary atoms [26] and muonic atoms.

Recently, we re-analyzed uncertainty of the TPE contribution in μ^2H originating from the χEFT Hamiltonian [2]. By calculating TPE contributions within a set of simultaneously optimized chiral forces [27], we found that the statistical uncertainty due to variances of low-energy constants in χEFT are negligible compared to the systematic uncertainty. The later was estimated from the cutoff variation and order-by-order convergence in χEFT using potentials up to N^{4}LO. The systematic uncertainty was found well under control. The result is consistent with our previous calculation, but still deviates from $\delta_{\rm TPE}$ extracted from the μ^2H Lamb shift experiment.

In the past five years, we performed ab initio calculations of $\delta_{\rm TPE}$ in μ^3H, μ^3He$^+$, and μ^4He$^+$. By probing the three-nucleon and four-nucleon dynamics, one can thoroughly investigate the systematic nuclear structure uncertainty of $\delta_{\rm TPE}$ in these muonic atoms. For these calculations, we adopted two nuclear potential models. One is the phenomenological AV18+UIX two- and three-nucleon forces [28, 29]. The other one is from one parameterization of χEFT [30, 31].

The QED and nuclear structure contributions to the Lamb shift in $\mu^{3,4}$He$^+$ have been summarized in [32, 33] to prepare analyzing the experimental data and determining the ^{3,4}He charge radii in the future. These experiments can also provide an accurate isotope shift of charge radii in helium isotopes, which has raised great interest to the atomic physics community [34].

We list in Table 140.1, the elastic and inelastic, nuclear and nucleonic contributions to $\delta_{\rm TPE}$ calculated in several light muonic atoms, together with the estimated uncertainties. The result is an update to our previous studies (see [5] for details). One can observe that the nucleonic contribution to $\delta_{\rm TPE}$ is generally smaller than the corresponding nuclear part. Therefore, the reliable determination of the nuclear TPE and its uncertainty is key to this problem. From μ^2H to $\mu^{3,4}$He$^+$, we found that

Table 140.1 Contributions to δ_{TPE} in light muonic atoms (in meV), with separated contributions from individual nucleons and from the few-nucleon dynamics. The uncertainties of individual contributions $\delta_{\mathrm{pol}}^{N,A}$, $\delta_{\mathrm{Zem}}^{N,A}$ and δ_{TPE} are in the brackets. Table adapted from [5]

	$\delta_{\mathrm{Zem}}^{N}$	$\delta_{\mathrm{pol}}^{N}$	$\delta_{\mathrm{Zem}}^{A}$	$\delta_{\mathrm{pol}}^{A}$	δ_{TPE}
μ^2H	−0.030(02)	−0.020(10)	−0.423(04)	−1.245(13)	−1.718(17)
μ^3H	−0.033(02)	−0.031(17)	−0.227(06)	−0.480(11)	−0.771(22)
μ^3He$^+$	−0.52(03)	−0.25(13)	−10.49(23)	−4.23(18)	−15.49(33)
μ^4He$^+$	−0.54(03)	−0.34(20)	−6.14(31)	−2.35(13)	−9.37(44)

Table 140.2 Estimation of relative uncertainty (%) of $\delta_{\mathrm{Zem}}^{A}$, $\delta_{\mathrm{pol}}^{A}$, and $\delta_{\mathrm{TPE}}^{A}$, based on a compilation of various causes of uncertainty. The total uncertainty sums up each individual one in quadrature. Table adapted from [5]

	μ^2H			μ^3H			μ^3He$^+$			μ^4He$^+$		
	$\delta_{\mathrm{Zem}}^{A}$	$\delta_{\mathrm{pol}}^{A}$	$\delta_{\mathrm{TPE}}^{A}$	$\delta_{\mathrm{Zem}}^{A}$	$\delta_{\mathrm{pol}}^{A}$	$\delta_{\mathrm{TPE}}^{A}$	$\delta_{\mathrm{Zem}}^{A}$	$\delta_{\mathrm{pol}}^{A}$	$\delta_{\mathrm{TPE}}^{A}$	$\delta_{\mathrm{Zem}}^{A}$	$\delta_{\mathrm{pol}}^{A}$	$\delta_{\mathrm{TPE}}^{A}$
NA	0.0	0.0	0.0	0.0	0.1	0.1	0.1	0.4	0.1	0.3	0.4	0.4
NM	0.5	0.3	0.4	2.4	1.3	1.7	1.8	0.7	1.5	4.6	3.9	4.4
ISB	0.2	0.2	0.2	0.2	0.7	0.5	0.2	1.8	0.5	0.5	2.2	0.5
NS	0.8	0.3	0.0	0.9	0.6	0.2	1.3	1.2	0.9	2.0	2.7	1.2
R	–	0.0	0.0	–	0.1	0.1	–	0.4	0.1	–	0.1	0.0
C	–	0.4	0.3	–	0.5	0.3	–	3.0	0.9	–	0.4	0.1
ηE	–	0.4	0.3	–	1.3	0.9	–	1.1	0.3	–	0.8	0.2
$(Z\alpha)^6$	–	0.7	0.5	–	0.7	0.5	–	1.5	0.4	–	1.5	0.4
Total	0.9	1.0	0.8	2.2	2.3	2.0	2.2	4.2	2.1	5.1	5.5	4.6

the relative uncertainty due to nuclear structure models increases in systems with a heavier nucleus; while the relative uncertainty from atomic physics is of similar size in each muonic atom.

Because δ_{TPE} is the limiting factor in analyzing the Lamb shift experiments, it is crucial to quantify the uncertainty in the TPE calculations. We explore all possible causes of uncertainty in the predicted results. For the nuclear part, $\delta_{\mathrm{TPE}}^{A}$, the uncertainty comes from numerical accuracy (NA), nuclear-model uncertainty (NM), isospin symmetry breaking corrections (ISB), nucleon-size corrections (NS), relativistic corrections (R), Coulomb corrections (C), higher-multipole expansion (ηE), and $(Z\alpha)^6$ corrections. Detailed uncertainty evaluation for each muonic atom is given in [5]. The summary of individual uncertainty sources for $\delta_{\mathrm{TPE}}^{A}$ is listed in Table 140.2. For the nucleonic part, $\delta_{\mathrm{TPE}}^{N}$, the uncertainty in each muonic atom is obtained by scaling propagation of the corresponding uncertainty in μH from dispersion relation analysis. The total uncertainty of δ_{TPE} is obtained by summing the uncertainties of $\delta_{\mathrm{TPE}}^{A}$ and $\delta_{\mathrm{TPE}}^{N}$ in quadrature.

140.4 Conclusions

The TPE contribution is a crucial input from theory that determines the accessible accuracy for extracting the nuclear charge radii from the Lamb shift measurements in muonic atoms. Nuclear ab initio calculations have made reliable predictions on the δ_{TPE} contributions in light muonic atoms, which have substantially reduced the uncertainty of TPE compared to other methods and approaches. A precision of the order of a few percent has been reached in our studies of the TPE contributions in $\mu^{2,3}$H and $\mu^{3,4}$He$^+$. The high precision prediction will help the CREMA collaboration to achieve accurate determination of the nuclear charge radii and the isotope shift in helium isotopes. These upcoming experimental results may help to solve the proton radius puzzle.

References

1. Hernandez, J.O., Ji, C., Bacca, S., Nevo Dinur, N., Barnea, N.: Phys. Lett. B **736**, 344 (2014)
2. Hernandez, O., Ekström, A., Nevo Dinur, N., Ji, C., Bacca, S., Barnea, N.: Phys. Lett. B **778**, 377 (2018)
3. Nevo Dinur, N., Ji, C., Bacca, S., Barnea, N.: Phys. Lett. B **755**, 380 (2016)
4. Ji, C., Nevo Dinur, N., Bacca, S., Barnea, N.: Phys. Rev. Lett. **111**, 143402 (2013)
5. Ji, C., Bacca, S., Barnea, N., Hernandez, O.J., Nevo Dinur, N.: J. Phys. G **45**, 093002 (2018)
6. Pohl, R., et al.: Nature **466**, 213 (2010)
7. Antognini, A., et al.: Science **339**, 417 (2013)
8. Mohr, P.J., Newell, D.B., Taylor, B.N.: Rev. Mod. Phys. **88**, 035009 (2016)
9. Beyer, A., et al.: Science **358**, 79 (2017)
10. Fleurbaey, H., et al.: Phys. Rev. Lett. **120**, 183001 (2018)
11. Antognini, A., et al.: Can. J. Phys. **89**, 47 (2011)
12. Borie, E.: Ann. Phys. **327**, 733 (2012)
13. Friar, J.: Ann. Phys. **122**, 151 (1979)
14. Zemach, A.C.: Phys. Rev. **104**, 1771 (1956)
15. Nevo Dinur, N., Barnea, N., Ji, C., Bacca, S.: Phys. Rev. C **89**, 064317 (2014)
16. Barnea, N., Leidemann, W., Orlandini, G.: Phys. Rev. C **61**, 054001 (2000)
17. Barnea, N., Leidemann, W., Orlandini, G.: Nucl. Phys. A **693**, 565 (2001)
18. Carlson, C.E., Vanderhaeghen, M.: Phys. Rev. A **84**, 020102 (2011)
19. Birse, M., McGovern, J.: Eur. Phys. J. A **48**, 120 (2012)
20. Pachucki, K.: Phys. Rev. Lett. **106**, 193007 (2011)
21. Pachucki, K., Wienczek, A.: Phys. Rev. A **91**, 040503 (2015)
22. Friar, J.L.: Phys. Rev. C **88**, 034003 (2013)
23. Carlson, C.E., Gorchtein, M., Vanderhaeghen, M.: Phys. Rev. A **89**, 022504 (2014)
24. Pohl, R., et al.: Science **353**, 669 (2016)
25. Pohl, R., et al.: Metrologia **54**, L1 (2017)
26. Parthey, C.G., et al.: Phys. Rev. Lett. **104**, 233001 (2010)
27. Carlsson, B.D., et al.: Phys. Rev. X **6**, 011019 (2016)
28. Wiringa, R.B., Stoks, V.G.J., Schiavilla, R.: Phys. Rev. C **51**, 38 (1995)
29. Pudliner, B.S., Pandharipande, V.R., Carlson, J., Wiringa, R.B.: Phys. Rev. Lett. **74**, 4396 (1995)
30. Entem, D.R., Machleidt, R.: Phys. Rev. C **68**, 041001 (2003)
31. Navrátil, P.: Few-Body Syst. **41**, 117 (2007)

32. Franke, B., Krauth, J.J., Antognini, A., Diepold, M., Kottmann, F., Pohl, R.: Eur. Phys. J. D **71**, 341 (2017)
33. Diepold, M., Franke, B., Krauth, J.J., Antognini, A., Kottmann, F., Pohl, R.: Ann. Phys. **396**, 220 (2018)
34. Zheng, X., Sun, Y.R., Chen, J.J., Jiang, W., Pachucki, K., Hu, S.M.: Phys. Rev. Lett. **119**, 263002 (2017)

Chapter 141
Investigating Shakeoff Process in Precise Correlation Measurements in Nuclear β Decay

E. Liénard, P. Delahaye, X. Fléchard, B. Pons and G. Quéméner

Abstract Precise measurements of correlations in nuclear β decays are currently employed to probe the Standard Model. When the β particle and recoiling daughter ion are detected in coincidence, these measurements further allow the observation of the ion charge-state distribution which results from the atomic shakeoff process induced by the nuclear decay. This is of great interest for fundamental atomic physics. This paper presents the results obtained at GANIL by means of LPCTrap experiments with $^{6}He^{1+}$, $^{35}Ar^{1+}$ and $^{19}Ne^{1+}$ ions. Comparison with theoretical values is also presented and suggests further investigations.

141.1 Introduction

141.1.1 Correlations in Nuclear β Decay

Precision measurements in nuclear β decay constitute sensitive probes of the Standard Model of elementary particles (SM), complementarily to high energy physics experiments [1]. Correlations between particle momenta and/or spins enable to test the violation of fundamental symmetries or the V-A structure of the SM, according to the behavior of vectors involved in the correlation term under symmetry operations (P and T). In particular, the β-ν angular correlation term, $a_{\beta\nu}\frac{\vec{p}_e\cdot\vec{p}_\nu}{E_e E_\nu}$, is invariant under P and T operations and enables either to search for exotic scalar or tensor currents beyond the V-A theory in pure Fermi (F) or Gamow-Teller (GT) transitions, or to determine precisely the mixing ratio between the GT and F parts in mirror transitions.

E. Liénard (✉) · X. Fléchard · G. Quéméner
Normandie Univ, ENSICAEN, UNICAEN, CNRS/IN2P3, LPC Caen, 14000 Caen, France
e-mail: lienard@lpccaen.in2p3.fr

P. Delahaye
GANIL, CEA/DSM-CNRS/IN2P3, Bd Henri Becquerel, 14000 Caen, France

B. Pons
Université de Bordeaux - CNRS - CEA, CELIA, UMR5107, 33405 Talence, France

N. A. Orr et al. (eds.), *Recent Progress in Few-Body Physics*,
Springer Proceedings in Physics 238,
https://doi.org/10.1007/978-3-030-32357-8_141

Since neutrinos are difficult to detect, sensitive observables for β-ν angular correlation measurements are kinematic parameters affected by the recoiling daughter nucleus motion. Two methods are commonly used: (i) a *direct* method in which either the recoil energy distribution [2] or the recoil ion time of flight (ToF) distribution thanks to its detection in coincidence with the β particle [3], is measured or (ii) an *indirect* method in which the Doppler shift of a delayed particle (γ or p) emitted during the recoil motion [4–6] is measured. The direct method requires the use of very clean radioactive sources, ideally placed in vacuum, due to the very low recoil kinetic energy (~1–2 keV at most). The development of intense radioactive beams in new generation facilities coupled to efficient beam handling techniques has allowed the use of atom or ion traps to define excellent radioactive sources in very clean environment [3, 7–9]. However, the high precision level targeted (~0.1%) requires a deep knowledge of the experimental setup to properly manage any effect potentially affecting the shape of the recoil energy or ToF distribution. The analysis of data is performed with the help of realistic simulations considering a set of systematic effects as complete as possible [10, 11]. For instance, the effect due to the shakeoff process has to be considered since it modifies the charge state of the recoil ion. This process is introduced in the next section.

141.1.2 The Shakeoff Process

Electron shakeoff (SO) is a fundamental atomic process in which a bound electron is excited into the continuum because of a sudden change of the central potential. Nuclear β decay, which modifies the nuclear charge, may induce such an ionization process. If one neglects the SO process, the β decay of singly charged ions basically leads to:

$$\beta^-:\ {}^A_ZX^{1+} \rightarrow {}^{A}_{Z+1}Y^{2+} + \beta^- + \bar{\nu}_e \tag{141.1}$$

$$\beta^+:\ {}^A_ZX^{1+} \rightarrow {}^{A}_{Z-1}W^{0} + \beta^+ + \nu_e \tag{141.2}$$

As low energy neutrals are very difficult to detect, it is worth noting that in the second case, the recoil atom cannot be properly detected, while SO, yielding singly- and multi-charged ions W^{q+}, greatly facilitates the measurement. It is even worse with atom traps in which the radioactive source is neutral: detection of recoils in β^+ decay then requires a double SO process [12]. The probability of multiple SO processes depends obviously on the number of electrons available but also on the probability of other atomic processes such as electron shakeup (excitation) or Auger emission. This finally leads to charge state distributions that should emphasize the dominant atomic processes.

The β decay is a very fast process inducing a nuclear potential change with a typical time of 10^{-18} s, which justifies the use of the sudden approximation (SA) in the theoretical models describing SO. However this hypothesis had not been yet experimentally verified before the first experiment performed at GANIL with the LPCTrap setup described in the next section.

141.2 LPCTrap@GANIL

The LPCTrap setup is installed at the low-energy beam line LIRAT of the SPIRAL1-GANIL facility. Technical details on the setup are given in [13]. The central element of the device is a transparent Paul trap designed to confine singly charged radioactive ions, almost at rest, in a small volume. The trap consists of 6 concentric rings with an open geometry, allowing easy injection and extraction as well as an efficient detection in coincidence of the β particles and recoil ions.

The low energy radioactive beam is provided by the ECR source of the SPIRAL facility at 10 keV kinetic energy (typical energy dispersion: ~20 eV). In the LIRAT beam line, the installation is equipped with a Radio Frequency Cooler-Buncher (RFQCB) for the preparation of the beam by reducing the emittance and the production of ion bunches. In the transparent Paul trap, the ions, also cooled down by elastic collisions with an inert gas injected at very low pressure (10^{-6}–10^{-5} mbar), have energies of about 0.1 eV at equilibrium and the diameter of the ion cloud located at the centre of the trap is of the order of 2.5 mm.

In the detection setup, a telescope and a micro-channel plate position sensitive detector (MCPPSD) detect the β particles and the recoil ions respectively. The β telescope, located 10 cm away from the center of the trap, is made of a 60×60 mm^2 300 μm thick double-sided silicon strip detector for position readout, followed by a 7 cm thick plastic scintillator coupled to a photomultiplier to measure the β particle kinetic energy and to deliver the start signal for the recoil ion ToF measurement. The stop signal is provided by the MCPPSD, installed at the end of the ion spectrometer. This spectrometer is located in front of the telescope at the opposite side from the trap center. The ions emitted towards the spectrometer are accelerated by a −2 kV potential after they cross a collimator located in front of a 50 cm long free flight tube. An electrostatic lens allows the collection of all the ions on the MCPPSD. This spectrometer makes LPCTrap a unique and unprecedented setup for the measurement of charge-state distributions of ions associated to the β decay of singly charged radioactive ions. The performances of the whole setup for experiments performed with ^{6}He, ^{35}Ar and ^{19}Ne are given in [14].

141.3 The $^{6}He^{1+}$ Case

According to Eq. (141.1), one single electron remains bound to the $^{6}Li^{2+}$ recoil ion after the β^- decay of $^{6}He^{1+}$. Therefore the final ion charge-state distribution only depends on pure SO, and the basic assumptions underlying the usual descriptions of SO can be checked precisely by means of direct comparison with experiment. Full quantum calculations, employing hydrogen-like wavefunctions, were thus performed in the framework of the SA with the aim to probe the reliability of this latter fundamental approximation. The calculations led to the SO probability (see [15] for details):

$$P_{SO} = (2.33810 + 0.00412 E_{rec}(\text{keV})) \times 10^{-2}$$

where E_{rec} is the recoil kinetic energy.

Figure 141.1 shows the recoil ion ToF distribution measured at GANIL using LPCTrap. A total of 1.2×10^6 final coincidences has been collected and analyzed with realistic simulations including all the identified systematic effects [15]. The induced corrections and systematic uncertainties are summarized in Table 141.1. In the simulations, the SM value was used for $a_{\beta\nu}$, considering a variation of the parameter in a range compatible with the current experimental precision [16]. The difference in MCP efficiency due to different thresholds for the two charge states

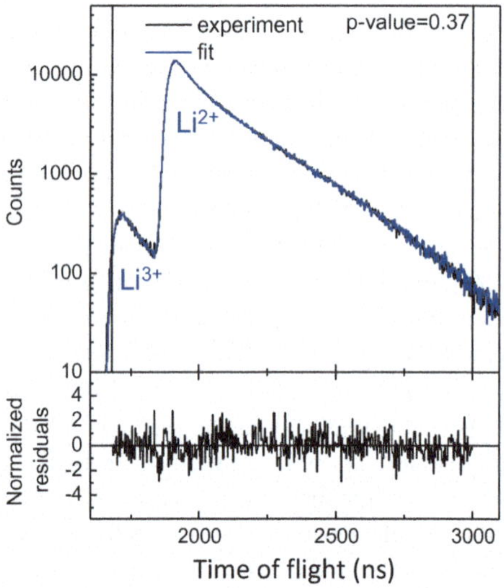

Fig. 141.1 Experimental ^{6}Li ions ToF distribution following $^6He^{1+}$ decay, with the fit superimposed (adapted from [15])

Table 141.1 Systematic effects considered in the analysis. (See text for details)

Source	Corr. (10^{-5})	Error (10^{-5})	Method
$a_{\beta\nu}$		4.0	[16]
β scattering	39	4.0	GEANT4
Background		3.5	Present data
E_β calibration		1.7	Present data
MCP efficiency	−9	1.2	Present data
Total	30	7.0	

was also taken into account. The agreement between the simulated and experimental distributions is excellent and it has enabled to deduce the following experimental SO probability:

$$P_{SO} = 0.02339(35)_{stat}(07)_{syst}.$$

The mean recoil energy corresponding to the events selected in the experimental data analysis has been substituted in the theoretical expression to compute the mean SO probability. A correction to the SA has been further estimated to account for the finite duration of the potential change, including also the direct collision mechanism between the bound electron and β particle, yielding $\delta P_{SO} = -20 \times 10^{-5}$. Finally, the mean theoretical SO probability is $\langle P_{SO} \rangle = 0.02322$, in perfect agreement with the experimental value.

141.4 The $^{35}Ar^{1+}$ Case

According to Eq. (141.2), the recoiling ^{35}Cl species are neutral after the β^+ decay of $^{35}Ar^{1+}$ ions but the SO process allows easier and direct detection of the recoil ions. Multielectron systems require a more sophisticated model to understand the charge-state distribution measured in the experiment. The calculation is performed in the framework of the SA, using an independent particle model (IPM) with Hartree-Fock wavefunctions to compute the ionization SO probabilities [17]. Subsequent Auger decays are also explicitly included in the calculation, and we discriminate between SO and Auger contributions to the production of charged daughter ions.

Figure 141.2 (*Left*) shows the recoil ion ToF distribution measured at GANIL using LPCTrap [17]. A total of 1.5×10^6 final coincidences has been collected. The experimental relative SO probability, indicated on the spectrum, was obtained by

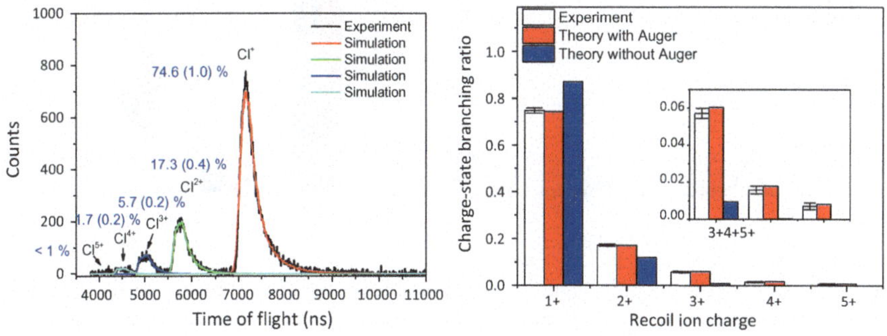

Fig. 141.2 *Left*: Experimental ^{35}Cl ions ToF distribution following $^{35}Ar^{1+}$ decay, with the simulated distributions superimposed (adapted from [17]). *Right*: Comparison between experimental and theoretical values with and without Auger emission (adapted from [18]). (See text for details)

peak integration including two corrections: the overlap of peak tails deduced from basic simulations (assuming SM $a_{\beta\nu}$) and the charge dependence of the recoiling ion detection efficiency. The comparison with the model is again excellent when including Auger emission (see Fig. 141.2 *Right*), and this confirms the paramount importance of Auger contribution to the production of high charge states. Moreover, the number of ^{35}Cl atoms produced during the experiment has been estimated using the number of detected β particles reduced by the number of coincidences corrected by the overall absolute detection efficiency for ions. This estimation leads to 72(10)% of neutral ^{35}Cl, which is also in good agreement with the 73.9% ratio obtained from the calculation.

141.5 The ^{19}Ne^{1+} Case

The last case studied with LPCTrap concerns the charge-state distribution of the recoiling ^{19}F ions after the β^+ decay of ^{19}Ne^{1+} ions [18]. A total number of 1.3×10^5 final coincidences has been collected in the spectrum shown in Fig. 141.3 (*Left*). The data were analyzed using the same method as for ^{35}Cl and the theoretical relative SO probabilities were obtained using the same approach as before. Here, the agreement seems better when Auger emission is neglected (see Fig. 141.3 *Right*), which is not satisfactory and clearly coincidental. The production of neutrals, experimentally estimated to be 69.5(4.2)%, is also badly reproduced by the model which provides a value of 76.5%. We traced back the root of the experimental/theoretical discrepancies to the inaccuracy of the multielectron SO probabilities computed by means of the IPM whose reliability worsens for systems with low nuclear charge [18].

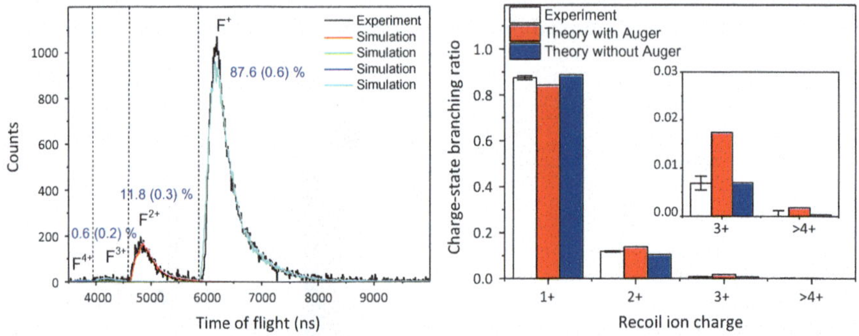

Fig. 141.3 *Left*: Experimental ^{19}F ions ToF distribution following ^{19}Ne^{1+} decay, with the simulated distributions superimposed. *Right*: Comparison between experimental and theoretical values with and without Auger emission. (See text for details) The figures are adapted from [18]

141.6 Conclusion

The SO process has been investigated in three nuclear β decays, starting from an ideal one-electron system which has enabled to confirm for the first time the reliability of the sudden approximation. In systems involving many electrons, primary SO ionization is followed by secondary processes which modify the recoil charge-state distribution. In $^{35}Ar^{1+}$ decay, Auger emission is dominant and monitors the production of the highest charge states, which is not observed in $^{19}Ne^{1+}$ decay. In this last case, Auger emission does not play such an important role so that the theoretical/experimental discrepancies which have been observed reveal the importance of electron correlations in the primary SO mechanism, beyond the independent electron assumption. Introducing such correlations in the calculation of SO probabilities constitutes a theoretical challenge and their importance as function of the nuclear charge Z could be directly gauged in future LPCTrap experiments performed at GANIL using beams (^{21}Na, ^{23}Mg, ^{33}Cl, ^{37}K) provided by the new SPIRAL1 sources [19].

References

1. González-Alonso, M., Naviliat-Cuncic, O., Severijns, N.: Prog. Part. Nucl. Phys. **109**, 165–223 (2019)
2. Johnson, C.H., Pleasonton, F., Carlson, T.A.: Phys. Rev. **132**, 1149–1165 (1963)
3. Gorelov, A., et al.: Phys. Rev. Lett. **94**, 142501 (2005)
4. Egorov, V., et al.: Nucl. Phys. A **621**, 745–753 (1997)
5. Adelberger, E.G., et al.: Phys. Rev. Lett. **83**, 1299–1302 (1999)
6. Severijns, N., Blank, B.: J. Phys. G: Nucl. Part. Phys. **44**, 074002 (2017)
7. Hong, R., et al.: Phys. Rev. A **96**, 053411 (2017)
8. Sternberg, M.G., et al.: Phys. Rev. Lett. **115**, 182501 (2015)
9. Fléchard, X., et al.: Phys. Rev. Lett. **101**, 212504 (2008)
10. Fléchard, X., et al.: J. Phys. G: Nucl. Part. Phys. **38**, 055101 (2011)
11. Fabian, X., et al.: Hyperfine Interact. **235**, 87–95 (2015)
12. Scielzo, N.D., Freedman, S.J., Fujikawa, B.K., Vetter, P.A.: Phys. Rev. A **68**, 022717 (2003)
13. Ban, G. et al.: Ann. Phys. (Berlin) **525**, 576–587 (2013) and references therein
14. Liénard, E., et al.: Hyperfine Interact. **236**, 1–7 (2015)
15. Couratin, C., et al.: Phys. Rev. Lett. **108**, 243201 (2012)
16. Glück, F.: Nucl. Phys. A **628**, 493–502 (1998)
17. Couratin, C., et al.: Phys. Rev. A **88**, 041403(R) (2013)
18. Fabian, X., et al.: Phys. Rev. A **97**, 023402 (2018)
19. Chauveau, P., et al.: Nucl. Instrum. Methods Phys. Res. B **376**, 35–38 (2016)

Part VI
Strange and Exotic Matter

Chapter 142
Hypernuclear Spectroscopy with Heavy-Ion Beams: Present Status and Perspectives

Christophe Rappold and Takehiko R. Saito

Abstract Research on the strange quark involves the study of the baryon-baryon interactions within $SU(3)_f$. The determination of the interaction strength between the hyperons (baryon containing a s-quark) and the nucleons is crucial for the description of the equation of state of neutron stars. The HypHI (**Hyp**ernuclei with **H**eavy **I**on) collaboration demonstrated the feasibility of the spectroscopy of hypernuclei produced in ion-induced reactions. A first experiment was performed with a ^{6}Li beam bombarding a carbon target at 2 AGeV. The experimental approach was developed to measure and study hypernuclei produced in the projectile rapidity region of the nuclear collisions. Such Λ-hypernuclei are produced by the coalescence between a Λ hyperon produced in the mid-rapidity region of the participant zone and a spectator fragment. The current status of the HypHI project are reviewed: the results on $^3_\Lambda$H and $^4_\Lambda$H are presented, together with the possible signal of nnΛ. The forthcoming experiment at the fragment separator FRS at GSI is described, aiming to improve the hypernuclear spectroscopy. The main goal is to verify the existence or not of a nnΛ bound state. The spectroscopy of $^3_\Lambda$H and study of its lifetime will be also pursued. The study of $^3_\Lambda$H and possibly of $^4_\Lambda$H will be used to demonstrate the feasibility of the new experimental approach of using a fragment separator for the measure of the decay fragments. Finally, further prospects within the HIAF facility on hypernuclear spectroscopy by heavy-ion reaction will be mentioned.

C. Rappold (✉) · T. R. Saito
GSI Helmholtz Center for Heavy Ion Research, Darmstadt, Germany
e-mail: c.rappold@gsi.de

T. R. Saito
Helmholtz Institute Mainz, Mainz, Germany

High Energy Nuclear Physics Laboratory RIKEN, Saitama, Japan

Institute of Modern Physics, Lanzhou, China

N. A. Orr et al. (eds.), *Recent Progress in Few-Body Physics*,
Springer Proceedings in Physics 238,
https://doi.org/10.1007/978-3-030-32357-8_142

142.1 Introduction

The hypernucleus, a bound state of hyperons and nucleons, has been one of the focus to comprehend the baryon-baryon interaction. One of main topic is the determination of the equation of state of baryonic matter up to the strangeness sector. Nuclear spectroscopy with heavy-ion beams and fixed nuclear targets has recently become a powerful tool to study hypernucleus.

142.2 HypHI Project: Results and Present Status

The first HypHI (**Hyp**ernuclei with **H**eavy **I**on) experiment at GSI was performed with a ^{6}Li beam bombarding a carbon target at 2 AGeV. The experimental approach was developed to measure and study hypernuclei produced in the projectile rapidity region of the nuclear collisions. The Λ-hypernuclei are yielded by the coalescence between a Λ hyperon from the mid-rapidity region of the participant zone and the spectator fragments. The produced hypernuclei in the projectile rapidity region have a velocity similar to the beam projectile, therefore are Lorentz boosted. Their decay length in the laboratory frame is thus long enough to allow the weak decay to happen well behind the target. The produced hypernuclei can then be observed and measured in-flight in a large decay volume located behind the target. By inferring the 4-vector of the decay particles induced from the hypernuclear decay, the invariant mass spectroscopy of hypernuclei can be achieved.

Using heavy-ion beam on a solid target material allows producing fragments with a large isospin distribution. Thus, the yield of exotic hypernuclei can be large enough to allow the precise spectroscopy of the hypernuclei of interest. One of the main objectives of the HypHI project is to study the high isospin effect in the hyperon-nucleon interaction by observing first off the mesonic weak decay of proton- and neutron-rich hypernuclei. The experiment was performed with ^{6}Li beam at 2 A GeV and 3×10^{6} ions per second on a graphite target of 8.84 g/cm^{2}. As discussed in [1], the experimental setup, exhibited in Fig. 142.1, consisted of three tracking stations, *TR0, TR1, TR2*, of scintillating fiber detector arrays, and two drift chambers, *BDC, SDC*, for the displaced vertex measurement. Additionally, three scintillating hodoscope walls, *TOF+, TFW, ALADiN TOF*, were used for tracking, energy loss and time-of-flight measurements of charged particles across the ALADiN dipole magnet. For the vertex reconstruction, the tracking system was placed in front of the dipole magnet around the decay volume of hypernuclei. Two separated detector sets, consisting of *TOF+* and *TFW & ALADiN TOF* hodoscope walls, were positioned behind the magnet to measure respectively the positively and negatively charged particles.

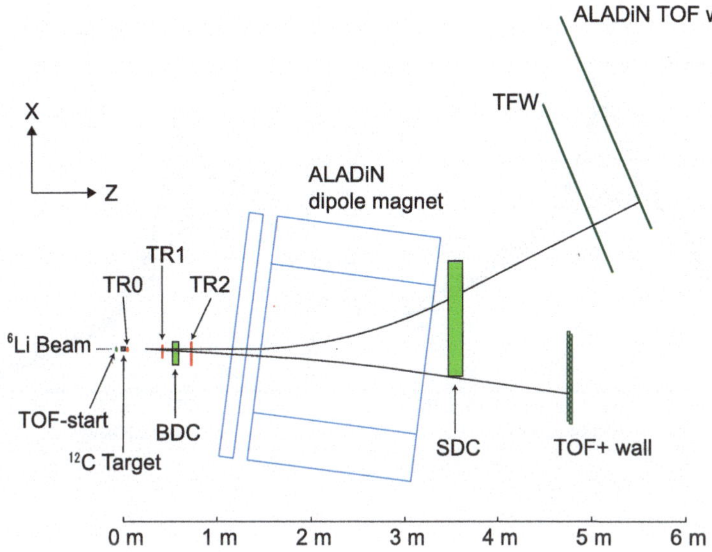

Fig. 142.1 Layout of the experimental setup

142.2.1 Spectroscopy of $^3_\Lambda$H and $^4_\Lambda$H light hypernuclei

The hypernuclear reconstruction focused on the two-body mesonic weak decay of $^3_\Lambda\mathrm{H} \rightarrow \pi^- + {}^3\mathrm{He}$, $^4_\Lambda\mathrm{H} \rightarrow \pi^- + {}^4\mathrm{He}$ and $\Lambda \rightarrow \pi^- + \mathrm{p}$. Then the measurement of the displaced vertex of the hypernuclear weak decay in the free space of the laboratory gives a clear lifetime measurement.

Those experimental results on the hypernuclear production are published [1, 2]. The $^3_\Lambda$H and $^4_\Lambda$H signals were extracted from the invariant mass distributions (Fig. 142.2). The $^3_\Lambda$H and $^4_\Lambda$H signal significance is respectively 4.7σ and 4.9σ [1]. With the $^3_\Lambda$H and $^4_\Lambda$H signal observation, the lifetimes were estimated via the unbinned maximum likelihood method. The profiled likelihood ratio of the lifetime estimator was used to infer the best estimate and the confidence interval at one-standard deviation. The experimental $^3_\Lambda$H and $^4_\Lambda$H lifetime was obtained as $183^{+42}_{-32} \pm 37$ ps and $140^{+48}_{-33} \pm 35$ ps respectively [1].

These inferred $^3_\Lambda$H and $^4_\Lambda$H lifetime was compared with the previous published results and a good agreement with the previous measurements was reported [1]. After a world data comparison of $^3_\Lambda$H and $^4_\Lambda$H lifetimes, the combination of existing measurements was performed in order to extract a more precise estimation of the experimental lifetime values [3]. The goal was to provide a more decisive perspective from experiments to be compared with theoretical estimations. The combined average and one-standard deviation of the lifetime for $^3_\Lambda$H and $^4_\Lambda$H were respectively 216^{+19}_{-16} ps and 192^{+20}_{-18} ps in 2014 with the available experimental data. The exclusion bands at 95% confidence level, which are suitable for the comparison with the theoretical

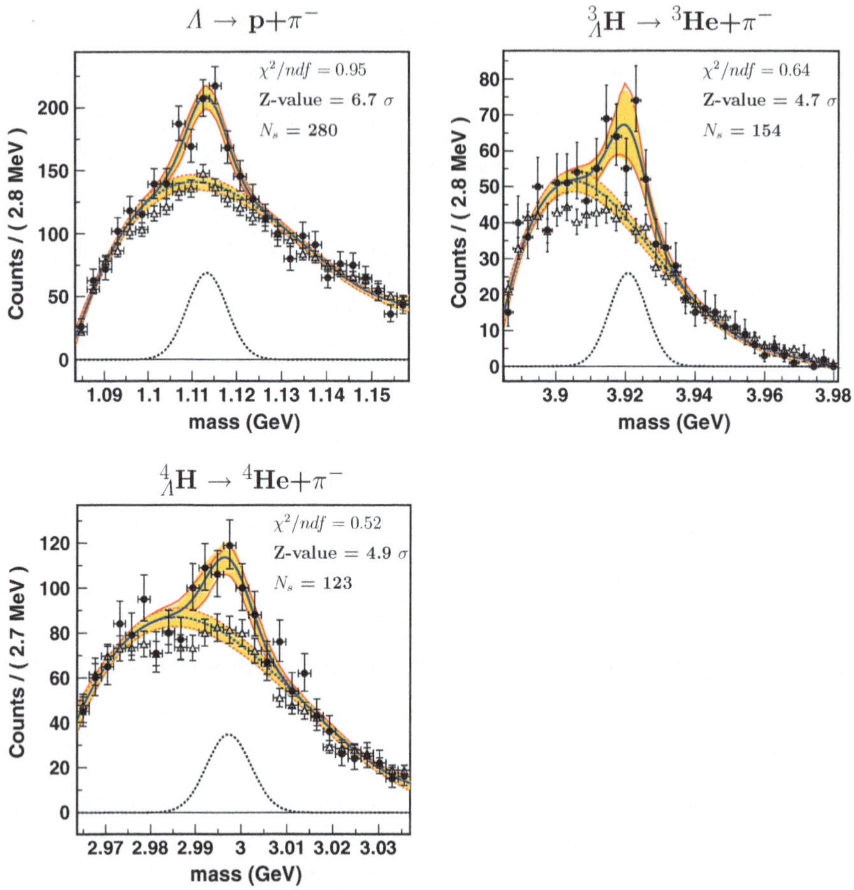

Fig. 142.2 Invariant mass distribution for candidates of Λ, $^3_\Lambda$H and $^4_\Lambda$H, respectively. The experimental data is shown by the filled circles. The opened triangles represent invariant mass distributions of the mixed event analysis to estimate the background contributions. Figure adapted from [1]

calculations, were then deduced to be [186 ps, 254 ps] for $^3_\Lambda$H or [158 ps, 233 ps] for $^4_\Lambda$H in 2014.

The obtained results reported a shorter lifetime of the $^3_\Lambda$H and $^4_\Lambda$H compared to their theoretical structure models [3], which had strong implications on the theoretical models of the hypernuclear structure.

142.2.2 Possible Signal of $^{3}_{\Lambda}n$ in the $D + \pi^-$ and $T + \pi^-$ Reconstruction

Separate results showed the indications of a possible bound state of two neutrons and one Λ hyperon, $^{3}_{\Lambda}$n or nnΛ [4]. Other decay channels were investigated, and in particular the decay channels d+π^- and t+π^- were reconstructed. Figure 142.3 shows the invariant mass spectra of d+π^- and t+π^- decay reconstruction when the decay vertex position is behind the target material. A clear signal peak can be seen, and the significance of the signal contribution was inferred to be 3.7σ and 5.2σ for the distributions in Fig. 142.3 [4].

For d+π^- and t+π^- decay channel, the lifetime of the mother candidates were extracted to be $181^{+30}_{-24} \pm 25$ps and $190^{+47}_{-35} \pm 36$ps, respectively. Those lifetime values are indicative of weak decay processes. It implied that a strangeness-changing weak interaction occurred.

The simplest bound system involving strangeness that could decay into a t+π^- channel was considered to be a bound system of two neutron and one Λ hyperon: nnΛ or $^{3}_{\Lambda}$n. Similarly, d+π^- could have been considered to be from the decay of nΛ, however previous experimental results on pΛ and nΛ have strongly disproved those states [4].

Meanwhile, recent theoretical calculations have shown that such a state is unlikely bound [5–7]. This possible bound state is thus open issue to be investigated in the forthcoming experiment.

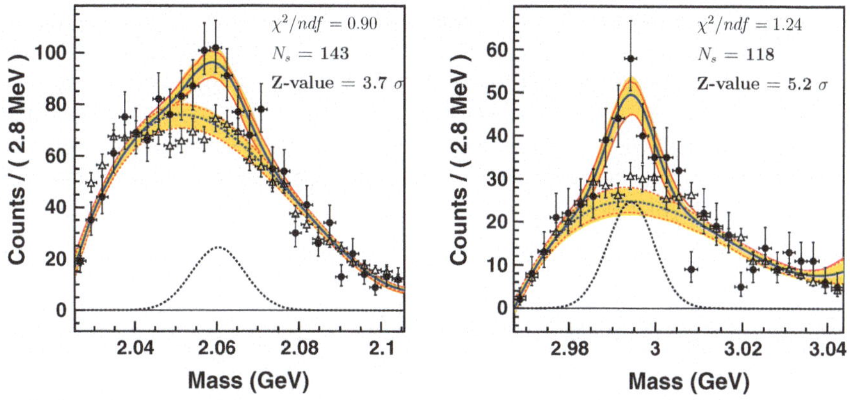

Fig. 142.3 Invariant mass distributions of d+π^- final state candidate in left panel, and of t+π^- in right panel. Those distributions were obtained when the target region was excluded. The experimental data is shown by the filled-in circles, while the open triangle distribution corresponds to invariant mass distribution of the mixed event analysis, estimating the background-only contribution. Figure adapted from [4]

142.3 Future Perspectives

With this current experimental approach of the HypHI project for the hypernuclear spectroscopy, the accuracy and statistics need to be improved. The experimental method also has to be further developed with different detection techniques and beams at higher energies.

142.3.1 Prospects in Hypernuclear Physics for FRS at GSI-FAIR

A new project to study hypernuclei has been proposed at GSI: it will employ the WASA central detector, currently at COSY in Jülich, Germany, and under transportation to GSI, for pion measurement combined with the high resolution fragment separator, FRS, for measuring decay residues. The novel experiment at the FRS fragment separator for the FAIR-Phase 0 beam-time period was proposed and approved in order to assess the existence of the nnΛ possible bound state. A new experimental concept has been under development [8]. Also to determine the feasibility of the experimental method, the experiment will be performed with similar conditions of the previously successful experiment. Light hypernuclei ${}^{3}_{\Lambda}$H, ${}^{4}_{\Lambda}$H and nnΛ will be the species of interest that are to be reconstructed and identified by invariant mass method.

After the two-body decay, a narrow magnetic rigidity acceptance window will be set for the FRS in order to measure precisely the momentum of outgoing decay fragment. The layout of the FRS is shown in top panel of Fig. 142.4. The detection apparatus at the S2 experimental area will be responsible for the measurement of a large portion of the emitted π^- as shown in bottom panel of Fig. 142.4. The acceptance and efficiencies have been evaluated and the invariant mass resolution is expected to be at least twice better with at worst an order of magnitude more of statistics than during the first HypHI experiment.

142.3.2 Further Perspective at HIAF

Another novel development at higher energies is in progress for the future heavy-ion accelerator facility in China, **H**igh **I**ntensity heavy ion **A**ccelerator **F**acility, HIAF [9]. The HIAF facility will consist of several heavy-ion storage rings as shown in Fig. 142.5: one ring of 34 Tm, *BRing*, for accumulation and acceleration of injected heavy-ion beam from the linear accelerator, *iLinac*. A second ring, *SRing*, is part of the HIAF facility for spectroscopy experiment based on storage ring, placed behind a High Energy Fragment Separator of 25 Tm that will provide rare-isotope beam to the *SRing*.

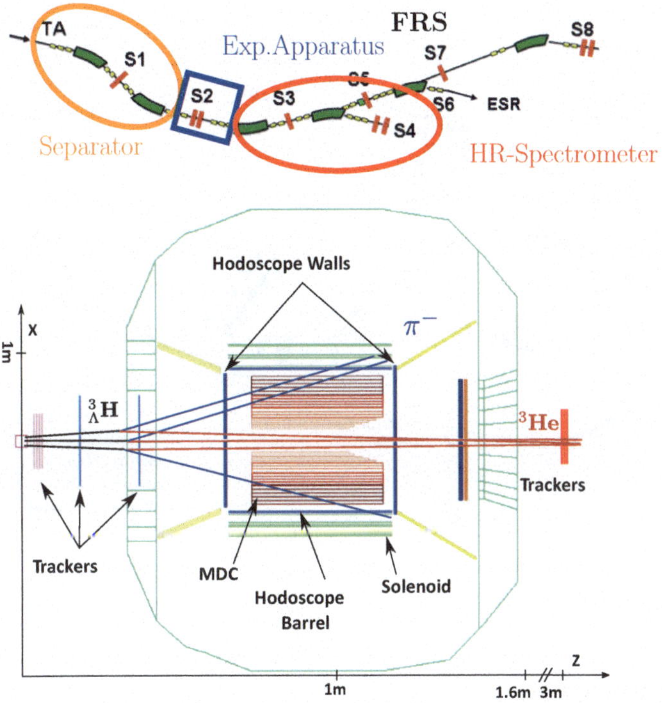

Fig. 142.4 Experimental layout of the forthcoming FAIR-Phase 0 experiment. In the top panel, the FRS layout: the beam is delivered to the experimental area S2 where the production of hypernuclei will occur and where the detection apparatus for the light particles will be placed. The FRS segment from S2 to S4 will be employed as a high-resolution forward spectrometer for the decay fragment of interest. In the bottom panel, the WASA cylindrical detector apparatus that will be placed in the S2 area. The light particles like pions will be measured within the solenoid magnet of the cylindrical detector

With the synchrotron of 34 Tm, *BRing*, the available beam energy is up to 4.25 *A*GeV for A/Z = 2. It will open opportunities to study in-flight double-strangeness hypernuclei with heavy-ion induced reaction on fixed target similarly to the HypHI project. The design of the future hypernuclear experimental setup is on-going. It would combine a solenoid magnet and a dipole magnet. The solenoid with its dedicated integrated detection system would detect the light hadrons produced in the mid- and projectile rapidity. While the dipole magnet placed well behind the solenoid with its planar detection system would measure precisely the out-going heavier decay fragments. It will be a new scope of work to us to perform an in-depth research program on $S = -1$ and $S = -2$ hypernuclei by means of precise invariant mass spectroscopy.

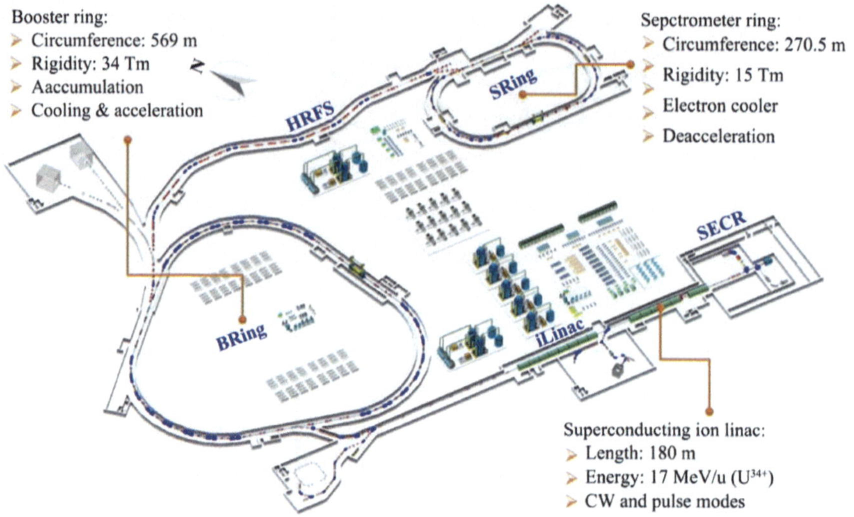

Fig. 142.5 Layout of the HIAF facility. The main characteristics of each sub-systems are shown. Figure adapted from [9]

142.4 Conclusion

After the completion of the first experimental campaign of the HypHI project, the feasibility of the hypernuclear spectroscopy via heavy-ion induced reaction was demonstrated. The experimental lifetime measurement of light hypernuclei ${}^{3}_{\Lambda}\mathrm{H}$ and ${}^{4}_{\Lambda}\mathrm{H}$ has become a hot topic in the recent years with the additional experimental results from the STAR and ALICE collaborations. Further theoretical calculations are needed to explain and comprehend the lifetime of light hypernuclei. Besides, new development and evidences are to be required to seal the fate of the experimental observation of the $nn\Lambda$. Improvement of the precision of the in-flight spectroscopy of light hypernuclei are on-going with the planned experiment WASA@FRS at GSI facility. As well, at the future accelerator HIAF novel experimental program on hypernuclear physics is currently under development.

References

1. Rappold, C., et al.: Hypernuclear spectroscopy of products from ${}^{6}\mathrm{Li}$ projectiles on a carbon target at 2 AGeV. Nucl. Phys. A **914**, 519 (2013)
2. Rappold, C., et al.: Hypernuclear production cross section in the reaction of ${}^{6}\mathrm{Li}+{}^{12}\mathrm{C}$ at 2 AGeV. Phys. Lett. B **747**, 129 (2015)
3. Rappold, C., et al.: On the measured lifetime of light hypernuclei ${}^{3}_{\Lambda}\mathrm{H}$ and ${}^{4}_{\Lambda}\mathrm{H}$. Phys. Lett. B **720**, 543 (2014)

4. Rappold, C., et al.: Search for evidence of $^{3}_{\Lambda}$n by observing d+π^- and t+π^- final states in the reaction of ^{6}Li+^{12}C at 2 *A* GeV. Phys. Rev. C **88**, 041001(R) (2013)
5. Hiyama, E., Ohnishi, S., Gibson, B.F., Rijken, ThA: Three-body structure of the $nn\Lambda$ system with $\Lambda N - \Sigma N$ coupling. Phys. Rev. C **89**, 061302(R) (2014)
6. Gal, A., Garcilazo, H.: Is there a bound $^{3}_{\Lambda}$n? Phys. Lett. B **736**, 93 (2014)
7. Garcilazo, H., Valcarce, A.: Nonexistence of a Λnn bound state. Phys. Rev. C **89**, 057001 (2014)
8. Rappold, C., López-Fidalgo, J.: Examination of experimental conditions for the production of proton-rich and neutron-rich hypernuclei. Phys. Rev. C **94**, 044616 (2016)
9. Zhou, X.: Physics opportunities at the new facility HIAF. Nucl. Phys. Rev. **34**, 339 (2018)

Chapter 143
Baryon-Baryon Interaction in Chiral Effective Field Theory

Johann Haidenbauer

Abstract The Jülich-Bonn-Munich Group has performed extensive studies of baryon-baryon interactions involving strange baryons (Λ, Σ, Ξ) within chiral effective field theory (EFT). A selection of the achieved results is presented. In particular, a calculation of the in-medium properties of a hyperon-nucleon interaction derived within chiral EFT and fitted to ΛN and ΣN scattering data is reviewed. Implications on the properties of neutron stars and, specifically, on the so-called hyperon-puzzle are addressed. In addition, results for baryon-baryon scattering in the strangeness $S = -2$ sector, obtained at next-to-leading order in the chiral expansion, are presented.

143.1 Introduction

The interaction between hyperons (Λ, Σ, Ξ) and nucleons is an important ingredient for microscopic calculations of few- and many body systems involving strangeness. This concerns light and heavy hypernuclei [1, 2] but also more exotic systems like neutron stars where the possible appearance of hyperons plays a crucial role for their properties and, specifically, their mass and size [3–5]. Motivated by these aspects, recently we examined the in-medium properties of a hyperon-nucleon (YN) interaction that has been derived within the modern approach of chiral effective field theory (EFT) [6–8]. In particular, we evaluated the single-particle potentials for the Λ, Σ and Ξ hyperons in nuclear matter in a conventional G-matrix calculation [9–11]. One issue of special interest is the Σ-nucleus potential. There is strong phenomenological evidence that it is repulsive [2]. However, conventional models of the YN interaction, fitted to ΛN and ΣN scattering data, often fail to produce a repulsive Σ-nuclear potential. Specifically, for models based on meson-exchange dynamics it

J. Haidenbauer (✉)
Institute for Advanced Simulation, Institut für Kernphysik,
and Jülich Center for Hadron Physics, Forschungszentrum Jülich, D-52425
Jülich, Germany
e-mail: j.haidenbauer@fz-juelich.de

N. A. Orr et al. (eds.), *Recent Progress in Few-Body Physics*,
Springer Proceedings in Physics 238,
https://doi.org/10.1007/978-3-030-32357-8_143

is rather difficult to obtain such a repulsion [12–14]. Another interesting question is the behavior of these effective interactions at high baryon density as realized in the core of neutron stars.

The *YN* interaction employed is the one derived up to next-to-leading order (NLO) in the chiral expansion by the Jülich-Bonn-Munich group [7, 8]. With that interaction an excellent description of available ΛN and ΣN scattering data could be achieved at NLO [7]. Also the $S = -2$ sector is well accounted for [8, 11]. Note, however, that in this case there are only a few data points and upper bounds for the ΞN elastic and inelastic cross sections that put constraints on the corresponding interactions.

The paper is structured in the following way: First we provide a brief review of the construction of the *YN* interaction within chiral EFT. We show also some selective results for the strangeness $S = -1$ $(\Lambda N, \Sigma N)$ and $S = -2$ (ΞN) sectors. In Sect. 143.3 we discuss in-medium results for those interactions. Specifically, we present predictions for the single-particle potential for the Λ and Σ at nuclear matter saturation density and we examine the behavior of the interactions for higher density in respect of the properties of neutron stars. The paper ends with a short summary.

143.2 The *YN* Interaction in Chiral EFT

The derivation of the chiral baryon-baryon potentials for the strangeness sector using the Weinberg power counting is outlined in [7, 8, 15]. The LO potential consists of four-baryon contact terms without derivatives and of one-pseudoscalar-meson exchanges while at NLO contact terms with two derivatives arise, together with contributions from (irreducible) two-pseudoscalar-meson exchanges. The contributions from pseudoscalar-meson exchanges (π, η, K) are completely fixed by the assumed SU(3) flavor symmetry. On the other hand, the strength parameters associated with the contact terms, the low-energy constants (LECs), need to be determined in a fit to data. How this is done is described in detail in [7] for the ΛN and ΣN interactions and in [8] for the $S = -2$ sector. Note that, in general, SU(3) symmetry is also imposed for those contact terms which reduces the number of independent LECs that can contribute.

The reaction amplitudes are obtained from the solution of a coupled-channels Lippmann-Schwinger (LS) equation for the derived interaction potentials:

$$T^{\ell''\ell',J}_{\nu''\nu'}(p'',p';\sqrt{s}) = V^{\ell''\ell',J}_{\nu''\nu'}(p'',p') + \sum_{\ell,\nu}\int_0^\infty \frac{dpp^2}{(2\pi)^3}\, V^{\ell''\ell,J}_{\nu''\nu}(p'',p)\frac{2\mu_\nu}{q_\nu^2 - p^2 + i\eta} T^{\ell\ell',J}_{\nu\nu',J}(p,p';\sqrt{s})\,.$$

The label ν indicates the particle channels and the label ℓ the partial wave. μ_ν is the pertinent reduced mass. The on-shell momentum in the intermediate state, q_ν, is defined by $\sqrt{s} = \sqrt{m^2_{B_{1,\nu}} + q^2_\nu} + \sqrt{m^2_{B_{2,\nu}} + q^2_\nu}$. Relativistic kinematics is used for relating the laboratory energy T_{lab} of the hyperons to the c.m. momentum.

We solve the LS equation in the particle basis, in order to incorporate the correct physical thresholds. The Coulomb interaction is taken into account appropriately via the Vincent-Phatak method. The potentials in the LS equation are cut off with a regulator function, $f_R(\Lambda) = \exp\left[-\left(p'^4 + p^4\right)/\Lambda^4\right]$, in order to remove high-energy components [16]. We consider cutoff values in the range $\Lambda = 550 - 700$ MeV (LO) and $\Lambda = 500 - 650$ MeV (NLO), similar to what was used for chiral NN potentials [16].

143.3 Results

143.3.1 ΛN, ΣN, and ΞN Scattering

Our results for ΛN and ΣN scattering are presented in Figs. 143.1 and 143.2. The bands (red for NLO, green for LO) represent the variation of the cross sections based on chiral EFT within the considered cutoff region stated above. For comparison also results for the Jülich '04 [14] meson-exchange model are shown (dashed line). Obviously, the available ΛN and ΣN scattering data are very well described by our NLO EFT interaction. In particular, and as expected, the energy dependence exhibited by the data is visibly better reproduced at NLO than at LO. This concerns especially the $\Sigma^+ p$ channel. But also for Λp the NLO results are well in line with the data

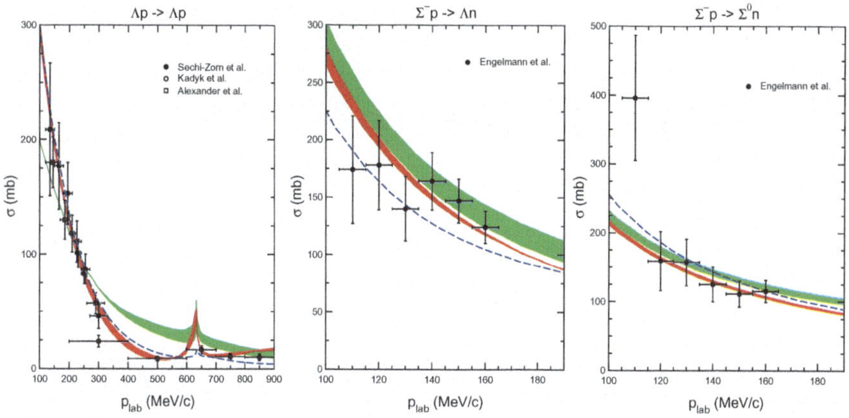

Fig. 143.1 Total cross sections for ΛN and ΣN scattering as a function of the laboratory momentum p_{lab}. The green (grey) band shows the chiral EFT results to LO for variations of the cut-off in the range $\Lambda = 550$–700 MeV, while the red (black) band are results to NLO for $\Lambda = 500$–650 MeV [7]. The dashed curve is the result of the Jülich '04 [14] meson-exchange potential

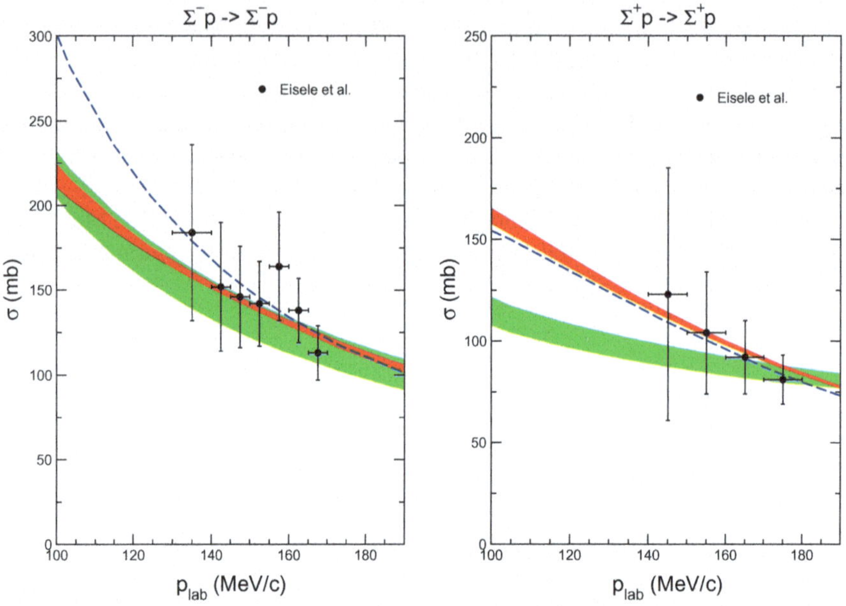

Fig. 143.2 Total cross sections for ΣN scattering as a function of the laboratory momentum p_{lab}. Same description of curves as in Fig. 143.1

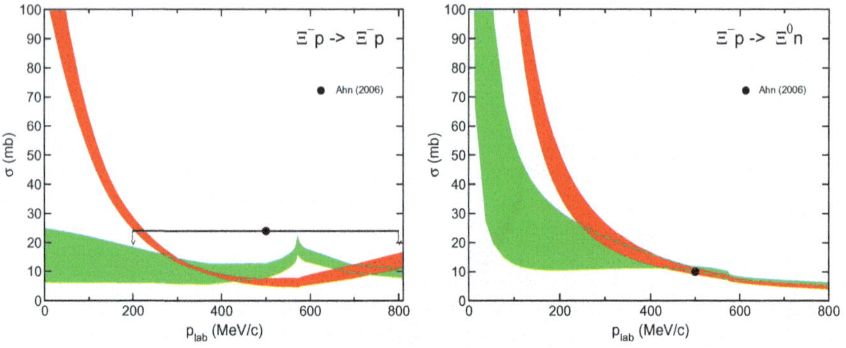

Fig. 143.3 Total cross sections for $\Xi^- p$ elastic and charge-exchange scattering as a function of the laboratory momentum p_{lab}. Same description of curves as in Fig. 143.1

even up to the ΣN threshold. Furthermore, one can see that the dependence on the cutoff mass is strongly reduced in the NLO case. Additional results, for differential cross sections and for phase shifts, can be found in [7]. Exemplary results for the ΞN interaction are show in Fig. 143.3, see also [8].

Besides an excellent description of the *YN* data the chiral EFT interaction yields a satisfactory value for the hypertriton binding energy, see [7]. Calculations for the four-body hypernuclei ${}^4_\Lambda$H and ${}^4_\Lambda$He based on the EFT interactions can be found in [17].

143.3.2 Λ and Σ in Nuclear Matter

In order to investigate the properties of our *YN* interactions in nuclear matter, we performed conventional first-order Brueckner calculations based on the standard (gap) choice of the single-particle (s.p.) potentials [9, 11] as well as with the continuous choice [10]. In the present work we focus on the results of [9], where one can also find details about how the Bethe-Goldstone equation is solved and how the s.p. potentials U_Λ, U_Σ, are determined self-consistently for a specific nuclear matter density ρ or Fermi momentum k_F. The latter are related by $\rho = (2/3\pi^2)\, k_F^3$ for symmetric nuclear matter.

Table 143.1 summarizes the predictions for the Λ and Σ potential depths, $U_\Lambda(p_\Lambda = 0)$ and $U_\Sigma(p_\Sigma = 0)$, evaluated at the saturation point of nuclear matter, i.e. for $\rho = \rho_0 = 0.17$ fm^{-3} ($k_F = 1.35$ fm^{-1}). Corresponding results obtained for the Jülich meson-exchange potentials from 2004 [14] and 1994 [12] are also included. In case of the EFT interactions we show the variation with the cutoff. These are comparable for U_Λ at LO and NLO, but noticeably reduced for U_Σ at NLO.

The predictions for $U_\Lambda(0)$ at NLO, for nuclear matter saturation density, are well in line with the 'empirical' value for the Λ binding energy in nuclear matter of about −27 to −30 MeV, deduced from the binding energies of finite Λ hypernuclei [18]. The predicted Σ s.p. potential is repulsive at NLO and also at LO. This property is in agreement with evidence from the analysis of level shifts and widths of Σ^- atoms and from measurements of (π^-, K^+) inclusive spectra related to Σ^--formation in heavy nuclei [2]. We could achieve a repulsive Σ s. p. potential because the interaction in the $^3S_1$ partial wave of the $\Sigma^+ p$ channel (which provides the dominant contribution, cf. Table 4 in [9]) is repulsive, for the LO potential but also for the NLO interaction. Note that a repulsive $^3S_1$ interaction is also in accordance with results from a recent lattice QCD calculation [19]. As exemplified by the predictions of the Jülich meson-exchange models, typically phenomenological potentials fail to produce a repulsive Σ-nuclear potential. Results for $U_\Xi(0)$ can be found in [11].

The dependence of the Λ potential depths on the Fermi momentum is displayed in Fig. 143.4, for symmetric nuclear matter (left) and for neutron matter (right). One can see that the density dependence predicted by the chiral EFT interaction is rather different from those of phenomenological models [13, 14]. Specifically, for the former the resulting Λ s.p. potential becomes already strongly repulsive for densities $\rho \gtrsim 2\rho_0$, whereas the one of the two models remains attractive or even becomes

Table 143.1 Results for the s.p. potentials $U_\Lambda(0)$ and $U_\Sigma(0)$ (in MeV) based on our EFT potentials and the Jülich meson-exchange interactions

	EFT LO	EFT NLO	Jülich '04 [14]	Jülich '94 [12]
Λ [MeV]	$550 \cdots 700$	$500 \cdots 650$		
$U_\Lambda(0)$	$-38.0 \cdots -34.4$	$-28.2 \cdots -22.4$	−51.2	−29.8
$U_\Sigma(0)$	$28.0 \cdots 11.1$	$17.3 \cdots 11.9$	−22.2	−71.4

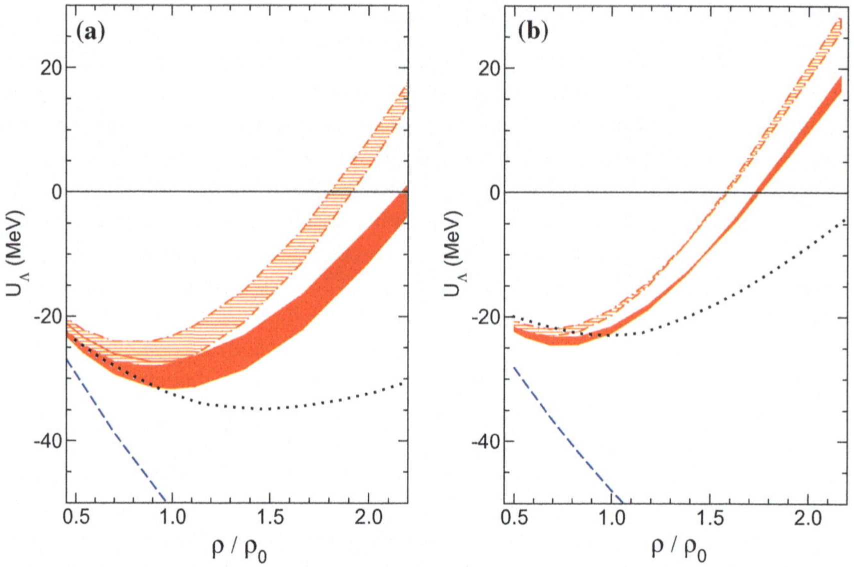

Fig. 143.4 The Λ single-particle potential $U_\Lambda(p_\Lambda = 0)$ as a function of ρ/ρ_0 in symmetric nuclear matter (a) and in neutron matter (b). The filled band is the chiral EFT results at NLO for the cutoffs $\Lambda = 450$ MeV (lower limiting line) and 500 MeV (upper limiting line), respectively. The dash-dotted lines include an effective density-dependent ΛN-interaction derived from the ΛNN three-body force [20]. The dashed curve is the result of the Jülich '04 meson-exchange model [14], the dotted curve that of the Nijmegen NSC97f potential [13, 22]

more attractive. Adding a density-dependent effective ΛN interaction constructed from consistently derived chiral ΛNN three-body forces [20] increases the repulsion further, cf. the hatched band in Fig. 143.4.

This feature should have consequences for neutron stars, as discussed in [21]: For hyperon-nuclear interactions with properties such as those deduced from the SU(3) EFT potentials, the onset for hyperon formation in the core of neutron stars could be shifted to much higher density which, in turn, could pave the way for resolving the so-called hyperon puzzle [4]. In standard calculations, based on *YN* interactions with properties similar to those of the Jülich '04 [14] or Nijmegen [13] models, hyperons appear in the core of neutron stars typically at densities around $(2-3)\rho_0$ [3–5]. This leads to a strong softening of the equation-of-state and, consequently, to a maximal mass of a neutron star that is far below the experimentally observerved values of $2M_\odot$ [4].

143.4 Summary

We have presented results for the in-medium properties of a hyperon-nucleon interaction derived within chiral effective field theory and fitted to ΛN and ΣN scattering data. The single-particle potentials for the Λ and Σ hyperons in nuclear matter were evaluated in a conventional G-matrix calculation.

The predictions for the Λ single-particle potential are found to be in good qualitative agreement with the empirical values inferred from hypernuclear data. A depth of about -25 MeV is predicted by the NLO interaction and of about -36 MeV by the LO potential. The Σ-nuclear potential turns out to be repulsive, in agreement with phenomenological information, with values around 15–20 MeV.

The density dependence of the s.p. potential U_Λ predicted by the NLO interaction [7] turned out to be remarkably different from the one of phenomenological potentials. For such an interaction the onset for hyperon formation in the core of neutron stars can be expected to be shifted to higher baryon density, which could potentially help to resolve the so-called hyperon puzzle.

Acknowledgements I would like to thank N. Kaiser, U.-G. Meißner, A. Nogga, S. Petschauer, and W. Weise for collaborating on the topics covered by my talk. This work is supported in part by the DFG and the NSFC through funds provided to the Sino-German CRC 110 "Symmetries and the Emergence of Structure in QCD" (DFG grant. no. TRR 110).

References

1. Botta, E., Bressani, T., Garbarino, G.: Eur. Phys. J. A **48**, 41 (2012)
2. Gal, A., Hungerford, E.V., Millener, D.J.: Rev. Mod. Phys. **88**, 035004 (2016)
3. Weissenborn, S., Chatterjee, D., Schaffner-Bielich, J.: Phys. Rev. C **85**, 065802 (2012)
4. Chatterjee, D., Vidaña, I.: Eur. Phys. J. A **52**, 29 (2016)
5. Tolos, L., Centelles, M., Ramos, A.: Astrophys. J. **834**, 3 (2017)
6. Epelbaum, E., Hammer, H.W., Meißner, U.-G.: Rev. Mod. Phys. **81**, 1773 (2009)
7. Haidenbauer, J., Petschauer, S., Kaiser, N., Meißner, U.-G., Nogga, A., Weise, W.: Nucl. Phys. A **915**, 24 (2013)
8. Haidenbauer, J., Meißner, U.-G., Petschauer, S.: Nucl. Phys. A **954**, 273 (2016)
9. Haidenbauer, J., Meißner, U.-G.: Nucl. Phys. A **936**, 29 (2015)
10. Petschauer, S., Haidenbauer, J., Kaiser, N., Meißner, U.-G., Weise, W.: Eur. Phys. J. A **52**, 15 (2016)
11. Haidenbauer, J., Meißner, U.-G.: Eur. Phys. J. A **55**, 23 (2019)
12. Reuber, A., Holinde, K., Speth, J.: Nucl. Phys. A **570**, 543 (1994)
13. Rijken, T.A., Stoks, V.G.J., Yamamoto, Y.: Phys. Rev. C **59**, 21 (1999)
14. Haidenbauer, J., Meißner, U.-G.: Phys. Rev. C **72**, 044005 (2005)
15. Petschauer, S., Kaiser, N.: Nucl. Phys. A **916**, 1 (2013)
16. Epelbaum, E., Glöckle, W., Meißner, U.-G.: Nucl. Phys. A **747**, 362 (2005)
17. Nogga, A.: Few Body Syst. **55**, 757 (2014)
18. Millener, D.J., Dover, C.B., Gal, A.: Phys. Rev. C **38**, 2700 (1988)
19. Nemura, H., et al.: EPJ Web of Conferences, vol. 175, p 05030 (2018)
20. Petschauer, S., Haidenbauer, J., Kaiser, N., Meißner, U.-G., Weise, W.: Nucl. Phys. A **957**, 347 (2017)
21. Haidenbauer, J., Meißner, U.-G., Kaiser, N., Weise, W.: Eur. Phys. J. A **53**, 121 (2017)
22. Yamamoto, Y., Nishizaki, S., Takatsuka, T.: Prog. Theor. Phys. **103**, 981 (2000)

Chapter 144
Hyperon and Hypernuclear Physics with PANDA at FAIR

Karin Schönning

Abstract Hyperons and hypernuclei are fruitful tools to shed light on the strong interaction. The PANDA experiment at FAIR in Germany will be a "strangeness factory" offering unique opportunities to study the largely unexplored multi-strange sector. In these proceedings, we will give an overview of the broad hyperon- and hypernuclear physics programme at PANDA, with focus on the first few phase of operation.

144.1 Introduction

Strangeness offers a new perspective of some of the most challenging problems in contemporary physics:

(1) Non-perturbative phenomena in QCD, manifested in the nucleon puzzles (mass [1], spin [2], radius [3] and structure [4]).
(2) The matter-antimatter asymmetry of the Universe. Baryogenesis is suggested as explanation but requires for instance CP violating particle decays [5].
(3) Quark-Gluon Plasma. Enhancement of strangeness in relativistic heavy-ion collisions was one of the first expected signals [6].
(4) Equation-of-state of neutron stars and the role of hypernuclei [7].

The Proton antiproton Annihilations at Darmstadt (PANDA) detector, under design for installation at the Facility for Antiproton and Ion Research (FAIR) in Germany, will provide a unique opportunity to study the strong interaction. During the first years of operation, focus will be on strangeness physics. The PANDA detector, a 4π spectrometer providing particle tracking and identification, vertex detection and calorimetry, will be an integrated part of the High Energy Storage Ring (HESR) antiproton ring. It is designed for a broad physics program, with focus on the strong

K. Schönning (✉)
Department. of Physics and Astronomy, Uppsala University, Box 516, 75120 Uppsala, Sweden
e-mail: karin.schonning@physics.uu.se

N. A. Orr et al. (eds.), *Recent Progress in Few-Body Physics*,
Springer Proceedings in Physics 238,
https://doi.org/10.1007/978-3-030-32357-8_144

interaction [8]. The first few years, the HESR will provide a luminosity of 10^{31} $cm^{-2}s^{-1}$ until the design luminosity of 1.5×10^{32} $cm^{-2}s^{-1}$ is achieved.

144.2 Hyperon Production and Decay

The strong interaction can be probed at several energy scales by studying systems with different quark content. The scale of strangeness production is governed by the strange quark mass, $m_s \approx 100$ MeV/c^2 which is rather close to the hadronization scale $\Lambda_{QCD} \approx 200$ MeV/c^2. As a consequence, the relevant degrees of freedom are unclear: quarks and gluons, or hadrons? Hence, strange hadrons are important probes of the confinement domain.

Past experiments focused mainly on single-strange hyperon production at low energies (see Fig. 144.1 and [9]). The cross sections of the single-strange production had been studied extensively by the PS185 collaboration at LEAR decomposing the complete spin structure of the $p\bar{p} \to \Lambda\bar{\Lambda}$ reaction. This triggered theoretical activity, mainly following three approaches to describe the $\bar{p}p \to \bar{Y}Y$ reaction: (i) meson exchange models [10], (ii) quark-gluon models [11] and (iii) models combining the two approaches [12]. The insights on the $\Lambda\bar{\Lambda}$ interaction [13] help understanding e.g. hyperon structure, measured in $e^+e^- \to \Lambda\bar{\Lambda}$ [14]. The next step towards a more coherent description of the strangeness onset is to map the multi-strange sector. However, no data beyond single-strangeness exist so far.

Simulation studies [15] show that large, clean samples of strange hyperons can be collected with PANDA already in the start-up phase of operation.

The expected copious production rates at the design luminosity will enable precision studies of weak and electromagnetic hyperon decays. The weak decays of hyperons are self-analyzing, i.e. the decay amplitude has a parity violating part that causes the daughter particles to be emitted according to the spin direction of the mother. In case of CP symmetry, the decay patterns of hyperons and anti-hyperons

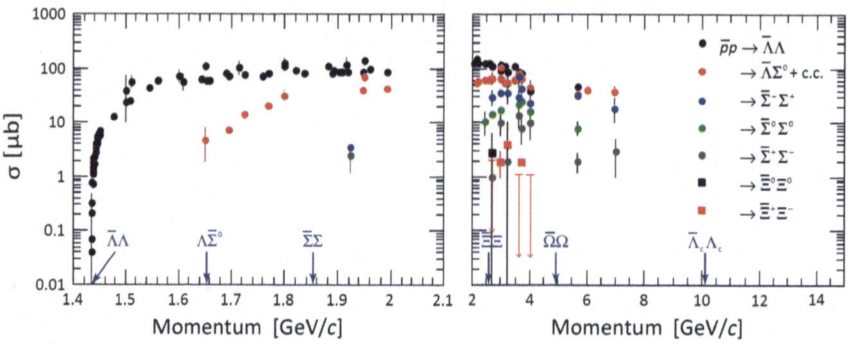

Fig. 144.1 The present data bank on hyperon production in $\bar{p}p$ annihilations. adapted from [9]. Thresholds of various reactions are marked

are identical. Precision measurements of decay asymmetries hence provide tests of CP violation, one of the requirements for baryogenesis [5]. The most precise measurement of CP violation in the Λ decay was provided by the BESIII collaboration and found to be consistent with zero [16]. The PANDA data samples will exceed those collected by BESIII by several orders of magnitude.

The electromagnetic Dalitz decay, i.e. $Y_1^* \to Y_2 e^+ e^-$, provide information of the hyperon structure in the low-energy region, parameterized by transition form factors [17]. The high production rates of e.g. Σ^0 hyperons in PANDA [15] imply excellent prospects for such measurements. Before the start-up of the full PANDA experiment, hyperon Dalitz decays can be studied with the predecessor PANDA@HADES [18].

144.3 Hyperon Spectroscopy

Instead of as in the previous section study the production and decay mechanisms of already established states, hyperon spectroscopy aims to search for and determine the properties of the less known or completely unknown states. We consider questions like (i) what is the role of SU(3) flavour symmetry in excited baryon spectra? (ii) what are the relevant degrees of freedom? (iii) what is the interaction dynamics inside baryons? (iv) are there exotic baryon states, e.g. penta-quarks or di-baryons?

So far, worldwide efforts in baryon spectroscopy have focused on N^* and Δ resonances, mainly discovered in πN scattering experiments and in recent years also in photon induced reactions [19]. These efforts have resulted in an impressive data bank for excited nucleon and Δ resonances. Nevertheless, several questions remain unsolved:

(1) Missing resonances: Predicted but yet unobserved baryonic states. Are they missing because they do not exist or because their coupling to γ and π is too weak? Different probes, e.g. anti-protons, could shed light on this puzzle.
(2) Parity doublets: The second lightest state after the positive parity nucleon is predicted to have negative parity. However, experiments found the contrary: the positive parity Roper resonance $N^*(1440)$, has a much smaller mass than the lightest negative parity state, the $N^*(1535)$.

In the single-strange sector, the puzzle of missing resonances remains. On the other hand, the situation for parity doublets is different: the parity partner of the Λ hyperon, the $\Lambda(1405)$, is indeed the next-to-lightest iso-singlet hyperon [20]. However, it is very light, even lighter than the $N^*(1440)$. The multi-strange data bank is very scarce: only one excited double-strange Ξ^* state and no Ω^* are well established [21]. A more complete picture of the multi-strange hyperon spectra would be very helpful in understanding to which extent SU(3) flavour symmetry is still valid, what the relevant degrees of freedom are and the possible role of di-quarks [22].

In PANDA, multi-strange hyperons Y^* can be produced in the two-body reaction $\bar{p}p \to \bar{Y}^* Y + c.c.$ In such processes it is easier to pin down the possible partial waves compared to in the case of e.g. photon- or kaon induced reaction where hyperons must be produced with two or three associated kaons. Furthermore, in $\bar{p}p$ annihilations, the production cross sections are expected to be large. Recent simulation studies of excited double-strange Ξ^* production in PANDA show that already during the start-up phase, large and clean exclusive data samples can be achieved [23].

144.4 Hyperons and Nuclei

The evolution of neutron stars is governed by the Equation-of-State (EoS) which results from an interplay of all four forces of Nature. However, the major roles are played by the strong interaction and gravity [24]. At the high densities of neutron stars, bound nuclear systems with strangeness—hypernuclei—should be formed. Their effect on the stiffness of EoS is however not yet understood [7]. A better understanding of the hyperon-hyperon and hyperon-nucleon interaction is required to solve this puzzle. The foreseen high production rates of hyperons and anti-hyperons in PANDA will provide an excellent environment for the formation of hyperatoms and hypernuclei. For this purpose, the PANDA hypernuclear setup comprises a high-precision germanium detector for γ-spectroscopy which is necessary to study the de-excitations of hyperatoms and hypernuclei [25]. Hyperatoms, as well as antihyperons in nuclei, will be studied already in the start-up phase of PANDA.

144.5 Summary

The future PANDA experiment at FAIR offers unique opportunities for various aspects of strangeness physics. During the start-up phase, focus will be on hyperon production and spectroscopy. Later phases includes hyperon decays and hypernuclear physics.

The author is grateful to the Knut and Alice Wallenberg Foundation for the support.

References

1. Scherer, S., Schindler, M.R.: Lect. Notes Phys. **830**, 1 (2012)
2. Aidala, C.A., et al.: Rev. Mod. Phys. **85**, 635 (2013)
3. Pohl, R., et al.: Nature **466**, 213; Carlson, C.E.: Prog. Part. Nucl. Phys. **82**, 59 (2015)
4. Miller, G.A.: Phys. Rev. Lett. **99**, 112001 (2007)
5. Sakharov, A.: Sov. Phys. Usp. **34**, 375 (1991)
6. Rafelski, J., Müller, B.: Phys. Rev. Lett. **48**, 1066 (1982)
7. Schaffner-Bielich, J.: Nucl. Phys. A **804**, 309–321 (2008)
8. The PANDA Collaboration: Physics Performance Report for PANDA: Strong Interaction Studies with Antiprotons, arXiv:0903.3905 [hep-ex]

9. Johansson, T.: Antihyperon-hyperon production in antiproton-proton collisions. In: AIP Proceedings of the 8th International Conference on Low Energy Antiproton Physics, p. 95 (2003)
10. Tabakin, F., Eisenstein, R.A.: Phys. Rev. C **31**, 1857 (1985)
11. Rubinstein, H.R., Snellman, H.: Phys. Lett. B **165**, 187 (1985)
12. Ortega, P.G., Entem, D.R., Fernandez, F.: Phys. Lett. B **696**, 352 (2011)
13. Haidenbauer, J., Meissner, U.-G.: Phys. Lett. B **761**, 456–461 (2016)
14. Li, C.: Spin polarisation of Λ hyperons in $e^+e^- \rightarrow \Lambda\bar{\Lambda}$ at BESIII. In: Proceedings of XIII:th International Conference on Beauty, Charm and Hyperon Hadrons (BEACH 2018)
15. Grape, S.: PhD thesis, Uppsala University (2009); Thomé, E.: PhD thesis, UU (2012)
16. Ablikim, M., et al.: Polarization and Entanglement in Baryon-Antibaryon Pair Production in Electron-Positron Annihilation. arXiv:1808.08917 [hep-ex]
17. Granados, C., Leupold, S., Perotti, E.: Eur. Phys. J. A **53**, 17 (2017)
18. Adamczewski-Musch, J., et al.: Proposal for experiments at SIS18 during FAIR Phase 0 (2017)
19. Crede, V., Roberts, W.: Rep. Prog. Phys. **76**, 076301 (2013)
20. Moriya, K., et al.: Phys. Rev. Lett. **112**, 082004 (2014)
21. Tanabashi, M., et al.: PDG. Phys. Rev. D **98**, 030001 (2018)
22. Anselmino, M., et al.: Rev. Mod. Phys. **65**, 1199 (1993)
23. Pütz, J.: Study of excited Ξ Baryons in $\bar{\text{p}}$p-collisions with the $\bar{\text{P}}$ANDA detector. J. Phys. Conf. Ser. **742**, 012028 (2016)
24. Psaltis, D.: Living Rev. Relativity **11**, 9 (2008)
25. Singh, B., et al.: Nucl. Phys. A **954**, 323–340 (2016)

Chapter 145
Low-Energy K$^-$ Nucleon/Multi-nucleon Interaction Studies by AMADEUS

Magdalena Skurzok, Massimiliano Bazzi, Gabriele Belotti, Alexandru Bragadireanu, Damir Bosnar, Arslan Butt, Michael Cargnelli, Catalina Curceanu, Luca de Paolis, Raffaele Del Grande, Laura Fabbietti, Carlo Fiorini, Francesco Ghio, Carlo Guaraldo, Ryugo Hayano, Mihai Iliescu, Masahiko Iwasaki, Paolo Levi Sandri, Johann Marton, Marco Miliucci, Pawel Moskal, Shinji Okada, Dorel Pietreanu, Kristian Piscicchia, Àngels Ramos, Alessandro Scordo, Hexi Shi, Michał Silarski, Diana Laura Sirghi, Florin Sirghi, Antonio Spallone, Oton Vazquez Doce, Eberhard Widmann, Sławomir Wycech and Johann Zmeskal

Abstract The AMADEUS collaboration aims to provide new experimental constraints to the K^-N strong interaction in the regime of non-perturbative QCD, exploiting low-energy K^- hadronic interactions with light nuclei (e.g. H, ^{4}He, ^{9}Be and ^{12}C). The low-momentum kaons ($p_K \sim 127$ MeV/c) produced at the DAΦNE collider are ideal to explore both stopped and in-flight K^- nuclear captures. The KLOE detector

M. Skurzok (✉) · P. Moskal · M. Silarski
Institute of Physics, Jagiellonian University, Kracow, Poland
e-mail: magdalena.skurzok@lnf.infn.it

M. Skurzok · M. Bazzi · C. Curceanu · L. de Paolis · R. D. Grande · C. Guaraldo · M. Iliescu · P. L. Sandri · M. Miliucci · K. Piscicchia · A. Scordo · H. Shi · D. L. Sirghi · F. Sirghi · A. Spallone
INFN, Laboratori Nazionali di Frascati, Rome, Frascati, Italy

G. Belotti · A. Butt · C. Fiorini
Politecnico di Milano, Dip. di Elettronica, Informazione e Bioingegneria, Milano, Italy

A. Bragadireanu · D. Pietreanu · D. L. Sirghi · F. Sirghi
Horia Hulubei National Institute of Physics and Nuclear Engineering, Magurele, Romania

D. Bosnar
University of Zagreb, Zagreb, Croatia

A. Butt · C. Fiorini
INFN Sezione di Milano, Milano, Italy

M. Cargnelli · J. Marton · E. Widmann · J. Zmeskal
Stefan-Meyer-Institut für subatomare Physik, Vienna, Austria

L. de Paolis · R. D. Grande · M. Miliucci
Universit degli Studi di Roma Tor Vergata, Rome, Italy

N. A. Orr et al. (eds.), *Recent Progress in Few-Body Physics*,
Springer Proceedings in Physics 238,
https://doi.org/10.1007/978-3-030-32357-8_145

is used as an active target, allowing to achieve excellent acceptance and resolutions for the data. In this work the results obtained from the study of $\Lambda\pi^-$ and Λp correlated production in the final state are presented.

145.1 Introduction

The theoretical investigation of the low-energy K^-N interaction predicts, in the energy region below the K^-N threshold, a sufficiently attractive interaction to form a bound state in the isospin I = 0 channel [1, 2].

In [3–7] the I = 0 $\Lambda(1405)$ is interpreted as a pure $\bar{K}N$ bound state, this leads to the prediction of deeply bound kaonic nuclear states. According to Chiral models [8–12] the $\Lambda(1405)$ emerges as a superposition of two states, as a consequence the K^-N interaction is much less attractive, which implies the prediction of only slightly bound kaonic nuclear states.

The experimental investigation of the K^-pp bound state properties in K^- induced reactions is strongly biased by the competing K^--multi-nucleon absorption processes, leading to the same final states (see e. g. [13, 14]). In [15, 16] a complete characterisation of the K^- two-, three- and four-nucleon absorptions (2NA, 3NA and 4NA) was performed for the first time in the Λp and $\Sigma^0 p$ final states exploiting low-energy K^- captures on a solid ^{12}C target. In particular, in [15] the corresponding low-energy cross sections are measured, these represent a crucial ingredient for the determination of the in-medium K^- optical potential [17, 18]. In Sect. 145.2 a brief summary of the analysis [15] is given.

L. Fabbietti · O. V. Doce
Excellence Cluster "Origin and Structure of the Universe", Garching, Germany

Physik Department E12, Technische Universität München, Garching, Germany

F. Ghio
INFN Sezione di Roma I, Rome, Italy

Istituto Superiore di Sanit, Rome, Italy

R. Hayano
The University of Tokyo, Tokyo, Japan

M. Iwasaki · S. Okada
RIKEN, The Institute of Physics and Chemical Research, Saitama, Japan

K. Piscicchia
Museo Storico della Fisica e Centro Studi e Ricerche Enrico Fermi, Rome, Italy

À. Ramos
Departament de Fisica Quantica i Astrofisica and Institut de Ciencies del Cosmos, Universitat de Barcelona, Marti i Franques 1, Barcelona, Spain

S. Wycech
National Centre for Nuclear Research, Warsaw, Poland

The experimental investigation of the $\Lambda(1405)$ properties is also challenging. The resonance line-shape is found to depend on both the production mechanism and the observed decay channel. Moreover in K^- induced reactions the non-resonant contribution to the final state $\Sigma\pi$ production has to be also taken into account. In Sect. 145.3 a brief summary of the results obtained in [19] is given, which could give important informations on the underlying $\bar{K}N$ interaction models.

The described analyses refer to a sample of 1.74 fb^{-1} integrated luminosity, collected by the KLOE collaboration [20] during the 2004/2005 data campaign. Low-energy K^-s are produced at the DAΦNE collider [21], from the phi-meson decay nearly at-rest, with a momentum of about 127 MeV/c. The K^- captures, at-rest and in-flight, on the materials of the KLOE detector, used as an active target, are investigated.

145.2 K^- Multi-nucleon Absorption Cross Sections and Branching Ratios in Λp and $\Sigma^0 p$ Final States

The possible existence of the K^-pp bound state can be investigated in low-energy K^- induced reactions by reconstructing the decays to $\Lambda(\Sigma^0)$p.

Recently, $\Lambda(\Sigma^0)$p decay modes were investigated by the AMADEUS collaboration in K^{-12}C absorption [15]. These studies allowed to perform the first comprehensive measurements of two, three and four nucleon absorption branching ratios (BRs) and cross sections for low-momentum kaons in Λp and Σ^0p channels. The BR of the Σ^0p direct production in K^- 2NA quasi free interaction is found to be greater than the corresponding Λp production, contrary to what is expected by comparing the pure phase spaces. This gives important indications on the underlying three-body interaction. The Λp spectra are entirely interpreted in terms of K^- multi-nucleon absorption processes, an eventual contribution due to the intermediate formation of a K^-pp bound state completely overlaps with the K^- 2NA in this channel, hence the corresponding yield is not extracted.

145.3 Resonant and Non-resonant $Y\pi$ Transition Amplitudes Below the $\bar{K}N$ Threshold

In the investigation of the $\Lambda(1405)$ properties, produced through the K^-p mechanism in light nuclear targets, two biases have to be taken into account. The first bias is the energy threshold imposed by the absorbing nucleon binding energy (for K^- capture at rest on ^{4}He the $\Sigma\pi$ invariant mass threshold is about 1412 MeV, while for ^{12}C it is about 1416 MeV). In order to access to the $\bar{K}N$ sub-threshold region corresponding to the $\Lambda(1405)$ high-mass predicted pole (about 1420 MeV), K^-N absorption in-flight has to be exploited. For a mean kaon momentum of 100 MeV/c, the $\Sigma\pi$ invariant mass threshold is shifted upwards by about 10 MeV.

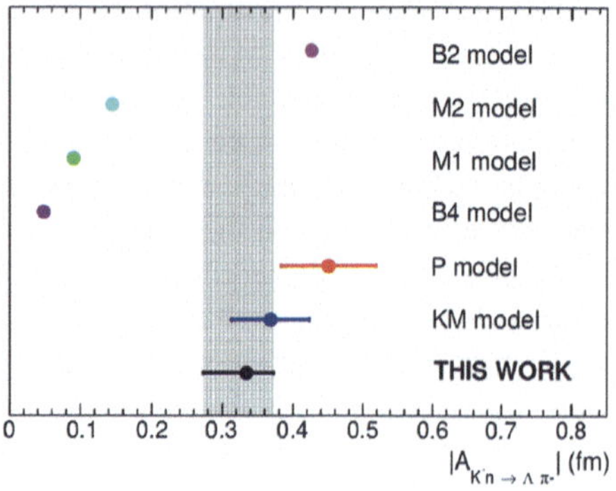

Fig. 145.1 Modulus of the measured non resonant $K^-n \to \Lambda\pi^-$ transition amplitude compared with theoretical calculations, see details in the text. Figure is adapted from [19]

The second bias is related to the non-resonant $K^-N \to Y\pi$ reaction, which was experimentally investigated for the first time in the $K^-n \to \Lambda\pi^-$ process, considering K^-n single nucleon absorptions in ^{4}He [19]. In this work the experimentally extracted $\Lambda\pi^-$ invariant mass, momentum and angular distributions were simultaneously fitted in by means of dedicated MC simulations. All the contributing reactions were taken into account: non-resonant processes, resonant processes and the primary production of a Σ followed by the $\Sigma N \to \Lambda N'$ conversion process. The simulations for non-resonant/resonant processes were based on the results of [22]. The analysis allowed the extraction of the non-resonant transition amplitude modulus $|A_{K^-n \to \Lambda\pi^-}|$ (33 ± 6) MeV below the $\bar{K}N$ threshold which is found to be 0.334 ± 0.018 *stat* $^{+0.034}_{-0.058}$ *syst* fm. The result of this analysis (with combined statistical and systematic errors) is shown in Fig. 145.1 and compared with the theoretical predictions (see [23] (P), [24] (KM), [25] (M1,M2), [26] (B2,B4)) rescaled for the $K^-n \to \Sigma\pi$ transition probabilities. This measurement can be used to test and constrain the S-wave $K^-n \to \Lambda\pi^-$ transition amplitude calculations.

145.4 Conclusion

In this work the low energy interaction between K^- and nucleons/nuclei in light nuclear targets are investigated with the aim to better understand the non-perturbative quantum chromodynamics QCD in the strangeness sector. Studies of low-energy K^- captures on a solid carbon target result in a complete characterisation of the two-, three- and four-nucleon absorptions in the Λp and $\Sigma^0 p$ final states (BRs and cross sections). The characterization of the non-resonant $K^-N \to Y\pi$ production was investigated for the first time for K^-n single nucleon absorption in ^{4}He. The result is crucial for the investigation of the $\Lambda(1405)$ characteristics.

Acknowledgements We acknowledge the KLOE/KLOE-2 Collaboration for their support and for having provided us the data and the tools to perform the analyses presented in this paper. We acknowledge the Centro Fermi - Museo Storico della Fisica e Centro Studi e Ricerche Enrico Fermi, for the project PAMQ. Part of this work was supported by the Austrian Science Fund (FWF): [P24756-N20]; the Austrian Federal Ministry of Science and Research BMBWK 650962/0001 VI/2/2009; the Ministero degli Affari Esteri e della Cooperazione Internazionale, Direzione Generale per la Promozione del Sistema Paese (MAECI), Strange Matter project PRG00892; the Polish National Science Center through the grant No. UMO-2016/21/D/ST2/01155

References

1. Wycech, S.: Nucl. Phys. A **450**, 399c (1986)
2. Akaishi, Y., Yamazaki, T.: Phys. Lett. B **535**, 70 (2002)
3. Akaishi, Y., Yamazaki, T.: Phys. Rev. C **65**, 044005 (2002)
4. Ikeda, Y., Sato, T.: Phys. Rev. C **76**, 035203 (2007)
5. Wycech, S., Green, A.M.: Phys. Rev. C **79**, 014001 (2009)
6. Revai, J., Shevchenko, N.V.: Phys. Rev. C **90**, 034004 (2014)
7. Maeda, S., Akaishi, Y., Yamazaki, T.: Proc. Jpn. Acad. B **89**, 418 (2013)
8. Dote, A., Hyodo, T., Weise, W.: Phys. Rev. C **79**, 014003 (2009)
9. Barnea, N., Gal, A., Liverts, E.Z.: Phys. Lett. B **712**, 132 (2012)
10. Ikeda, Y., Kamano, H., Sato, T.: Prog. Theor. Phys. **124**, 533 (2010)
11. Bicudo, P.: Phys. Rev. D **76**, 031502 (2007)
12. Bayar, M., Oset, E.: Nucl. Phys. A **914**, 349 (2013)
13. Suzuki, T., et al.: Mod. Phys. Lett. A **23**, 2520 (2008)
14. Sada, Y., et al.: PTEP **2016**(5), 051D01 (2016)
15. Del Grande, R., et al.: Eur. Phys. J. C **79**, 190 (2019)
16. Vazques Doce, O., et al.: Phys. Lett. B **758**, 134 (2016)
17. Friedman, E., Gal, A.: Nucl. Phys. A **959**, 66 (2017)
18. Hrtankova, J., Mares, J.: Phys. Rev. C **96**, 015205 (2017)
19. Piscicchia, K., et al.: Phys. Lett. B **782**, 339 (2018)
20. Bossi, F., et al.: Riv. Nuovo Cim. **31**, 531 (2008)
21. Gallo, A., et al.: Conf. Proc. **C060626**, 604 (2006)
22. Piscicchia, K., Wycech, S., Curceanu, C.: Nucl. Phys. A **954**, 75 (2016)
23. Cieply, A., Smejkal, J.: Nucl. Phys. A **881**, 115 (2012)
24. Ikeda, Y., Hyodo, T., Weise, W.: Nucl. Phys. A **881**, 98 (2012)
25. Guo, Z.H., Oller, J.A.: Phys. Rev. C **87**, 035202 (2013)
26. Mai, M., Meiner, U.G.: Eur. Phys. J. A **51**, 30 (2015)

Chapter 146
Search for the η-Mesic Helium in Proton–Deuteron and Deuteron–Deuteron Reactions

Magdalena Skurzok

Abstract This report presents the status of the search for η-mesic Helium nuclei with the WASA-at-COSY facility.

146.1 Introduction

The mesic nuclei are currently open issue in nuclear and hadronic physics, both from experimental [1–15] and theoretical points of view [16–35]. The existence of the mesic bound states has been postulated over thirty years ago [36, 37], however till now remains unconfirmed experimentally. A wide range of ηN scattering length values ($a_{\eta N}$) determined based on investigations of hadron- and photon-induced production of the η meson does not exclude creation of η-mesic bound states even with light nuclei like Helium [17–20, 38–40].

Some of the promising experiments related to η-mesic Helium nuclei have been performed with the COSY facility [41]. COSY-11 group carried out measurements to search for ^{3}He-η bound states in $dp \rightarrow ppp\pi^-$ and $dp \rightarrow {}^3\mathrm{He}\pi^0$ reactions. Excitation functions determined in the vicinity of the production threshold allowed to establish the upper limits of the total cross section to about 270 nb and 70 nb, respectively [15, 42]. The search for ^{4}He-η and ^{3}He-η mesic nuclei has been recently performed by WASA-at-COSY Collaboration in dd and pd collisions, respectively. This paper focuses on the results obtained for the search for η-mesic ^{4}He in $dd \rightarrow {}^3\mathrm{He}n\pi^0$ and $dd \rightarrow {}^3\mathrm{He}p\pi^-$ processes [1–4]. The analysis related to ^{3}He-η bound state is in progress [43].

The interested reader can find recent reviews on the η mesic bound state searches in [6, 7, 24, 26, 44–49].

M. Skurzok (✉)
M. Smoluchowski Institute of Physics, Jagiellonian University, Cracow, Poland
e-mail: magdalena.skurzok@uj.edu.pl

N. A. Orr et al. (eds.), *Recent Progress in Few-Body Physics*,
Springer Proceedings in Physics 238,
https://doi.org/10.1007/978-3-030-32357-8_146

146.2 Search for the η-Mesic ^{4}He

The WASA-at-COSY Collaboration performed two measurements dedicated to search for η-mesic ^{4}He nuclei, in 2008 and 2010. The measurements were carried out using the unique ramped beam technique which allows for the beam momentum to be changed slowly and continuously around the η production threshold in each of the acceleration cycles [2, 3, 47, 49]. The used technique allows to reduce systematic uncertainties with respect to separate measurements at fixed beam energies [3, 11].

The excitation functions for $dd \rightarrow {}^3\text{He}p\pi^-$ [1–4] and $dd \rightarrow {}^3\text{He}n\pi^0$ [1, 2, 4] reactions were investigated near the $^4\text{He}\eta$ production threshold to search for ^{4}He-η bound states. A detailed description of the data analysis one can find in [2, 3]. Since any narrow structure below the η production threshold, which could be a signature of the bound state, has not been observed, the upper limit of the total cross section for the η-mesic ^{4}He formation and its decay in proper channel at the 90% confidence level has been determined. For the $dd \rightarrow ({}^4\text{He-}\eta)_{bound} \rightarrow {}^3\text{He}p\pi^-$ reaction a sensitivity of the cross section of about 6 nb [2] was achieved which is about four times better in comparison with the result obtained in the previous experiment [3]. In case of the $dd \rightarrow ({}^4\text{He-}\eta)_{bound} \rightarrow {}^3\text{He}n\pi^0$ process, the upper limit was for the first time obtained experimentally and varies in the range from 2.5 to 3.5 nb. The data analyses were carried out assuming that the signal from the bound state is described by a Breit-Wigner shape with fixed binding energy and width while background is a second order polynomial. Also isospin relation between $n\pi^0$ and $p\pi^-$ pairs was taken into account.

Recently, a theoretical description of the cross sections in the excess energy range relevant to the η-mesic nuclear search proposed by [16] was compared with WASA-at-COSY excitation functions for the $dd \rightarrow {}^3\text{He}N\pi$ reactions. The authors of [16] used a phenomenological approach with an optical potential for the η-^{4}He interaction and determine the total cross sections for a broad range of real (V_0) and imaginary (W_0) parameters. Fitting the theoretical spectra convoluted with the experimental resolution of the excess energy (left panel of Fig. 146.1) to experimental data [2] allows to determine the upper limit of the total cross section (CL = 90%) for creation of η-mesic nuclei via the $dd \rightarrow {}^3\text{He}N\pi$ process varying from about 5.2 nb to about 7.5 nb. A comparison of the determined upper limits with the cross sections obtained in [16] allows to put a strong constraint on the η-^{4}He optical potential parameters (V_0, W_0). As it is presented in right panel of Fig. 146.1, the model allows only for extremely narrow and loosely bound states [16]. A detailed description of performed analysis interested reader can find in [1].

146.3 Summary and Perspectives

Experiments dedicated to the search for η-mesic ^{4}He in $dd \rightarrow {}^3\text{He}n\pi^0$ and $dd \rightarrow {}^3\text{He}p\pi^-$ reactions carried out with the WASA-at-COSY detection setup did not brought the observation of the peak structure related to the bound state. However, the upper limits of the total cross sections for η-mesic nuclei formation and decays were

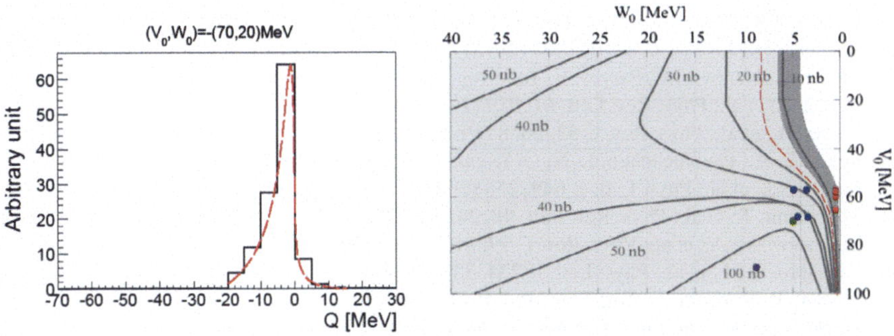

Fig. 146.1 (Left) Cross section of the $dd \rightarrow (^4\text{He-}\eta)_{bound} \rightarrow {}^3\text{He}N\pi$ reaction for the formation of the ^{4}He-η bound system calculated for η-^{4}He optical potential parameters $(V_0, W_0) = -(70,20)$ MeV, plotted as a function of the excess energy Q. The red dashed line shows the theoretical spectrum while the black solid line shows the convoluted spectrum. (Right) Contour plot of the theoretically determined conversion cross section in $V_0 - W_0$ plane [16]. Light shaded area shows the region excluded by our analysis, while the dark shaded area denotes systematic uncertainty of the $\sigma_{upp}^{CL=90\%}$. The red line extends the allowed region based on a new estimate of errors (see text for details). Dots correspond to the optical potential parameters corresponding to the predicted η-mesic ^{4}He states. Figures are adapted from [1]

determined to be in the order of a few nb [2, 3]. Moreover, a comparison between the theoretical model proposed in [16] and experimental data allowed, for the first time, to constrain the range of the η-^{4}He optical potential parameters [1].

Recently, a promising experiment dedicated to the search for η-mesic ^{3}He was performed by the WASA-at-COSY Collaboration. The bound state were searched for in three different mechanisms: (i) absorption of the η meson by one of the nucleons, which subsequently decays into N^*-π pair e.g.: $pd \rightarrow (^3\text{He-}\eta)_{bound} \rightarrow pp p\pi^-$, (ii) η -meson decay while it is still "orbiting" around a nucleus e.g.: $pd \rightarrow (^3\text{He-}\eta)_{bound} \rightarrow {}^3\text{He}2\gamma$ reactions and (iii) η meson absorption by few nucleons e.g.: $pd \rightarrow (^3\text{He-}\eta)_{bound} \rightarrow ppn$. The measurement allowed to collect the largest data sample in the world available up to now, for ^{3}He-η [43, 50, 51] (average luminosity $3{\cdot}10^{30}$ cm^{-2} s^{-1}). The data analysis is still in progress and the estimated upper limit value for $pd \rightarrow {}^3He2\gamma$ and $pd \rightarrow {}^3He6\gamma$ channels is on the level of a few nanobarns.

Acknowledgements We acknowledge the support from the Polish National Science Center through grants no. 2013/11/N/ST2/04152 and 2016/23/B/ST2/00784.

References

1. Skurzok, M., et al.: Phys. Lett. B **772**, 663 (2018)
2. Adlarson, P., et al.: Nucl. Phys. A **959**, 102 (2017)
3. Adlarson, P., et al.: Phys. Rev. C **87**, 035204 (2013)
4. Skurzok, M., Moskal, P., Krzemien, W.: Prog. Part. Nucl. Phys. **67**, 445 (2012)

5. Tanaka, Y.K., et al.: Phys. Rev. Lett. **117**, 202501 (2016)
6. Machner, H.: J. Phys. G **42**, 043001 (2015)
7. Metag, V., Nanova, M., Paryev, E. Ya.: Prog. Part. Nucl. Phys. **97**, 199 (2017)
8. Berger, J., et al.: Phys. Rev. Lett. **61**, 919 (1988)
9. Mayer, B., et al.: Phys. Rev. C **53**, 2068 (1996)
10. Sokol, G. A., Pavlyuchenko, L. N., *arXiv:nucl-ex/0111020* (2001)
11. Smyrski, J., et al.: Phys. Lett. B **649**, 258 (2007)
12. Mersmann, T., et al.: Phys. Rev. Lett. **98**, 242301 (2007)
13. Budzanowski, A., et al.: Phys. Rev. C **79**, 012201(R) (2009)
14. Papenbrock, M., et al.: Phys. Lett. B **734**, 333 (2014)
15. Moskal, P., Smyrski, J.: Acta Phys. Pol. B **41**, 2281 (2010)
16. Ikeno, N., et al.: Eur. Phys. J. A **53**(10), 194 (2017)
17. Xie, J.J., et al.: Phys. Rev. C **95**, 015202 (2017)
18. Fix, A., et al.: Phys. Lett. B **772**, 663 (2017)
19. Barnea, N., Bazak, B., Friedman, E., Gal, A.: Phys. Lett. B **771**, 297 (2017)
20. Barnea, N., Friedman, E., Gal, A.: Nucl. Phys. A **968**, 35 (2017)
21. Barnea, N., Friedman, E., Gal, A.: Phys. Lett. B **747**, 345 (2015)
22. Friedman, E., Gal, A., Mares, J.: Phys. Lett. B **725**, 334 (2013)
23. Kelkar, N.G.: Eur. Phys. J. A **52**, 309 (2016)
24. Kelkar, N.G., et al.: Rept. Progr. Phys. **76**, 066301 (2013)
25. Kelkar, N. G., Acta Phys. Pol. B **46**, 113 1 (2015)
26. Wilkin, C.: Acta Phys. Pol. B **47**, 249 (2016)
27. Wilkin, C.: Phys. Lett. B **654**, 92 (2007)
28. Bass, S.D., Thomas, A.W.: Phys. Lett. B **634**, 368 (2006)
29. Bass, S.D., Thomas, A.W.: Acta Phys. Pol. B **41**, 2239 (2010)
30. Hirenzaki, S., Nagahiro, H.: Acta Phys. Pol. B **45**, 619 (2014)
31. Nagahiro, H., Jido, D., Hirenzaki, S.: Phys. Rev. C **80**, 025205 (2009)
32. Nagahiro, H., et al.: Phys. Rev. C **87**, 045201 (2013)
33. Hirenzaki, S., et al.: Acta Phys. Pol. B **41**, 2211 (2010)
34. Wycech, S., Krzemien, W.: Acta Phys. Pol. B **45**, 745 (2014)
35. Niskanen, J.: Phys. Rev. C **92**, 055205 (2015)
36. Haider, Q., Liu, L.C.: Phys. Lett. B **172**, 257 (1986)
37. Bhalerao, R.S., Liu, L.C.: Phys. Rev. Lett. **54**, 865 (1985)
38. Green, A.M., Niskanen, J.A., Wycech, S.: Phys. Rev. C **54**, 1970 (1996)
39. Wilkin, C.: Phys. Rev. C **47**, 938 (1993)
40. Wycech, S., Green, A.M., Niskanen, J.A.: Phys. Rev. C **52**, 544 (1995)
41. Wilkin, C.: Eur. Phys. J. A **53**(6), 114 (2017)
42. Krzemien, W., et al.: Int. J. Mod. Phys. A **24**, 576 (2009)
43. Skurzok, M., et al.: EPJ Web of Conferences **181**, 01014 (2018)
44. Haider, Q., Liu, L.-C.: J. Phys. G **37**, 125104 (2010)
45. Krusche, B., Wilkin, C.: Prog. Part. Nucl. Phys. **80**, 43 (2014)
46. Moskal, P.: Few Body Syst. **50**, 667 (2014)
47. Skurzok, M., et al.: Acta Phys. Pol. B **47**, 503 (2016)
48. Moskal, P.: Acta Phys. Pol. B **47**, 97 (2016)
49. Moskal, P., Skurzok, M., Krzemien, W.: AIP Conf. Proc. **1753**, 030012 (2016)
50. Rundel, O., et al.: Acta Phys. Pol. B **48**, 1807 (2017)
51. Rundel, O., et al.: EPJ Web Conf. **130**, 02008 (2016)

Chapter 147
Energy Dependence of the $\bar{K}N$ Interaction and the Two-Pole Structure of the $\Lambda(1405)$—Are They Real?

János Révai

Abstract It is shown, that the energy-dependence of the chiral based $\bar{K}N$ potentials, responsible for the occurence of two poles in the $I = 0$ sector is the consequence of applying the on-shell factorization approximation [1]. When the dynamical equation is solved without this approximation, the T-matrix has only one pole in the region of the $\Lambda(1405)$ resonance.

147.1 Introduction

The $\Lambda(1405)$ is one of the basic objects of strangeness nuclear physics (SNP). Experimentally it is a well pronounced bump in the $\pi\Sigma$ missing mass spectrum in various reactions somewhat below the K^-p threshold with PDG resonance parameters $E - i\Gamma/2 = (1405 - 25i)$ MeV. Theoretically it is an $I = 0$ quasi-bound state in the $\bar{K}N - \pi\Sigma$ system, which decays into the $\pi\Sigma$ channel.

Constructing any multichannel $\bar{K}N$ interaction, the starting point of any SNP study, one of the first questions is: "What kind of $\Lambda(1405)$ it produces?" At present, it is believed, that theoretically substantiated $\bar{K}N$ interactions can be derived from the chiral perturbation expansion of the $SU(3)$ meson-baryon Lagrangian. For these interactions the widely accepted answer to the above question is, that the observed $\Lambda(1405)$ is the result of the interplay of two T-matrix poles. Our aim is to challenge this opinion.

J. Révai (✉)
MTA Wigner RCP, Budapest, Hungary
e-mail: revai.janos@wigner.mta.hu

N. A. Orr et al. (eds.), *Recent Progress in Few-Body Physics*,
Springer Proceedings in Physics 238,
https://doi.org/10.1007/978-3-030-32357-8_147

147.2 The Full and On-Shell Factorized WT Potentials $\hat{\text{V}}$ and $\hat{\text{U}}$

Our starting point is the lowest order Weinberg-Tomozawa (WT) term of the chiral Lagrangian (Eq. (7) from the basic paper [1]):

$$\langle q_i|v_{ij}|q_j\rangle \sim -\frac{c_{ij}}{4f_\pi^2}(q_i^0 + q_j^0), \tag{147.1}$$

where i and j denote the different meson-baryon channels $(i, j = 1, 2, 3, 4, 5 = [\bar{K}N]^{I=0}, [\bar{K}N]^{I=1}, [\pi\Sigma]^{I=0}, [\pi\Sigma]^{I=1}, [\pi\Lambda]^{I=1})$, q_i and $q_i^0 = \sqrt{m_i^2 + q_i^2}$ denote the meson c.m. momentum and energy, c_{ij} are $SU(3)$ Clebsch-Gordan coefficients and f_π is the pion decay constant, $m_i(M_i)$ are the meson (baryon) masses. Physical quantities can be derived from this expression via a certain dynamical framework, relativistic (BS equation, relativistic kinematics) or non-relativistic (LS equation, non-relativistic kinematics). We shall use the second option, having in mind applications for $n > 2$ systems. According to our choice and the usual practice, (147.1) has to be supplemented: adding appropriate normalization factors, applying a relativistic correction to meson energies and introducing two meson decay constants instead of f_π:

$$\langle q_i|v_{ij}|q_j\rangle = -\frac{c_{ij}}{64\pi^3 F_i F_j \sqrt{m_i m_j}}(q_i^{0'} + q_j^{0'}), \tag{147.2}$$

where

$$q_i^{0'} = q_i^0 + \frac{q_i^{0 2} - m_i^2}{2M_i} = q_i^0 + \frac{q_i^2}{2M_i} \underset{\text{nonrel}}{\approx} m_i + \frac{q_i^2}{2\mu_i} \tag{147.3}$$

with the reduced mass $\mu_i = m_i M_i/(m_i + M_i)$ and F_i, $i = \bar{K}, \pi$ are the new meson decay constants. In order to use the potential (147.2) in LS equation a regularization procedure has to be applied to ensure convergence of the occuring integrals. We use the separable potential representation of the interaction, which amounts to multiplying the potential (147.2) by suitable cut-off factors $u_i(q_i)$ and $u_j(q_j)$. Finally, the potential V_{ij} entering the LS equation for total energy W

$$\langle q_i|T_{ij}(W)|q_j\rangle = \langle q_i|V_{ij}|q_j\rangle + \sum_s \int \langle q_i|V_{is}|q_s\rangle G_s(q_s; W)\langle q_s|T_{sj}(W)|q_j\rangle d\mathbf{q}_s \tag{147.4}$$

has the form

$$\langle q_i|V_{ij}|q_j\rangle = u_i(q_i)\langle q_i|v_{ij}|q_j\rangle u_j(q_j) = \lambda_{ij}(g_{iA}(q_i)g_{jB}(q_j) + g_{iB}(q_i)g_{jA}(q_j)) \tag{147.5}$$

which is a two-term multichannel separable potential with form factors

$$g_{iA}(q_i) = u_i(q_i); \quad g_{iB}(q_i) = g_{iA}(q_i)\gamma_i(q_i); \quad \gamma_i(q_i) = (m_i + \frac{q_i^2}{2\mu_i}),$$

and coupling matrix

$$\lambda_{ij} = -\frac{c_{ij}}{64\pi^3 F_i F_j \sqrt{m_i m_j}}.$$

The non-relativistic propagator $G_s(q_s; W)$ has the form

$$G_s(q_s; W) = (W - m_s - M_s - \frac{q_s^2}{2\mu_s} + i\epsilon)^{-1} = \frac{2\mu_s}{k_s^2 - q_s^2 + i\epsilon}, \tag{147.6}$$

where $k_s = \sqrt{2\mu_s(W - m_s - M_s)}$ is the on-shell c.m. momentum in channel s.

A commonly used procedure before solving the integral equation (147.4) is to remove the inherent q-dependence of the potential by replacing q_i in $\gamma_i(q_i)$ by its on-shell value k_i: $\gamma_i(q_i) \to \gamma_i(k_i) = W - M_i$. This is the so-called on-shell factorization approximation, introduced in [1] and never checked afterwords. The separable potential representation of the interaction allows an exact solution of the LS equation (147.4) for both versions of the potential: the "full" WT potential

$$\hat{V} = \sum_{i,j} |g_{iA}\rangle\lambda_{ij}\langle g_{jB}| + |g_{iB}\rangle\lambda_{ij}\langle g_{jA}| \tag{147.7}$$

and its on-shell factorized energy-dependent counterpart

$$\hat{U}(W) = \sum_{i,j} |g_{iA}\rangle\lambda_{ij}(2W - M_i - M_j)\langle g_{jA}| \tag{147.8}$$

providing thus a check of the effects of this approximations.

147.3 Numerical results

Practical solution of (147.4) starts with an appropriate choice of the form- or cut-off factors $u_i(q)$, which ensures the convergence of all occuring integrals. In our case it was the dipole Yamaguchi form with adjustable cut-off (or range) parameters β_i:

$$u_i(q) = \left(\frac{\beta_i^2}{q^2 + \beta_i^2}\right)^2$$

The details of the formalism for the actual calculations can be found in [2]. Both potentials $\hat{V}$ and $\hat{U}$ depend on the same set of 7 parameters $F_{\bar{K}}, F_\pi, \beta_1, \beta_2, \beta_3, \beta_4$ and β_5 which have to be fitted to the available experimental data, which are the 6 low-energy elastic and inelastic K^-N cross sections, the 3 threshold branching ratios

Table 147.1 Calculated and experimental values of the discrete data for potentials $\hat{U}$ and $\hat{V}$

	$\hat{V}$	$\hat{U}$	Exp
γ	2.32	2.35	2.36 ± 0.04
R_c	0.671	0.664	0.664 ± 0.011
R_n	0.202	0.194	0.189 ± 0.015
$\Delta E\ (eV)$	$350 - 279\,i$	$302 - 294\,i$	$(283 \pm 36) - (271 \pm 46)\,i$

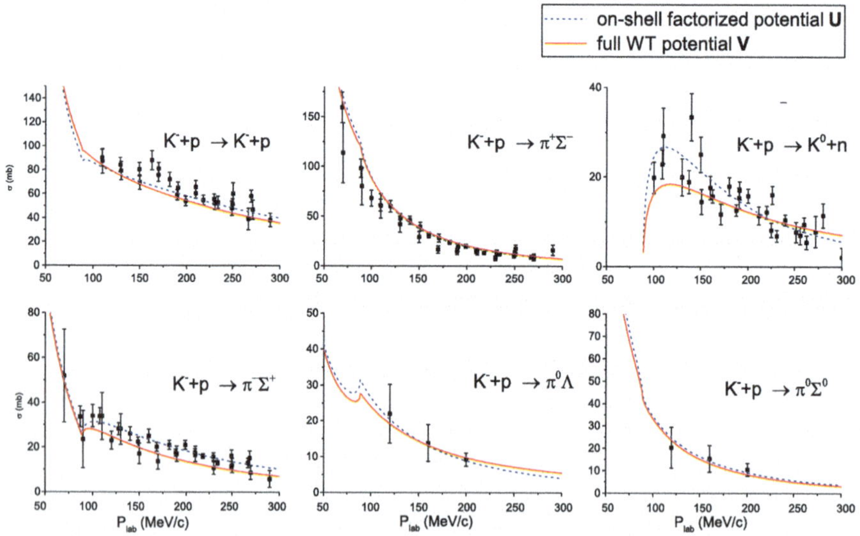

Fig. 147.1 Elastic and inelastic $K^{-}p$ cross sections for the potentials ***U*** and ***V***

Table 147.2 Parameters of the potentials $\hat{V}$ and $\hat{U}$ (all values are given in MeV

	F_π	F_K	β_1	β_2	β_3	β_4	β_5
$\hat{V}$	80.8	132	1094	960	516	537	629
$\hat{U}$	107	109	1247	1622	919	959	443

γ, R_n, R_c[1] and the 1 s level shift ΔE in kaonic hydrogen. The results of the fit for the two potentials are shown in Table 147.1 and Fig. 147.1. Table 147.2 shows the obtained parameter values.

More or less equal quality fits can be obtained for both potentials but for very different parameter values. This means, that $\hat{U}$ can not be considered as an approximation to $\hat{V}$—they are basically different interactions. Their most significant difference is, that, while the full WT potential $\hat{V}$ produces a single pole in the region of $\Lambda(1405)$,

[1]For their definition see [2].

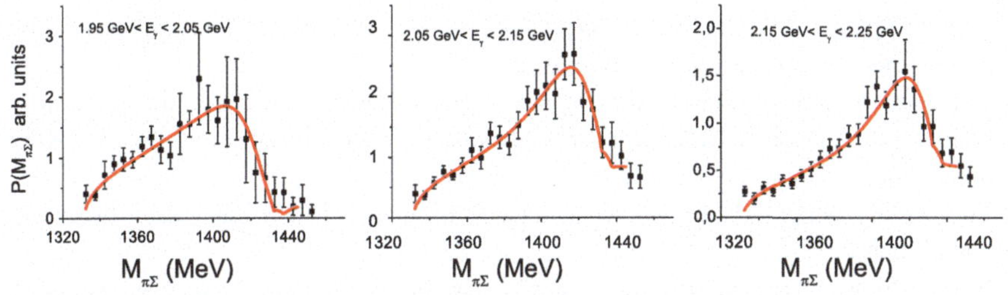

Fig. 147.2 Fit to the first 3 γ-energy bins of CLAS $M(\pi^0\Sigma^0)$ missing mass spectra in $\gamma + p \to K^+ + \pi^0 + \Sigma^0$ photoproduction

the on-shell factorized potential $\hat{U}$ for any reasonable combination of parameters produces the familiar two poles. The pole positions for the two potentials are

$$z1 = (1425 - 21i)\,\text{MeV} \quad \text{for} \quad \hat{V} \quad \text{and}$$

$$z1 = (1428 - 35i)\,\text{MeV} \quad \text{and} \quad z2 = (1384 - 62i)\,\text{MeV} \quad \text{for} \quad \hat{U}.$$

The position of the single $\hat{V}$ pole does not confirm the strong K^-p binding, which is the main feature of the phenomenological potentials adjusted to the PDG pole position.

Most recent and accurate information on the $\Lambda(1405)$ resonance comes from the CLAS photoproduction experiment $\gamma + p \to K^+ + \pi^0 + \Sigma^0$ [3] in which the $M(\pi^0\Sigma^0)$ missing mass spectra give the $\Lambda(1405)$ line shape. For the analysis of these spectra at present we have the semi-phenomenological final state interaction formula of Roca and Oset [4], which contains some adjustable parameters c_i and the T-matrix elements of the $\bar{K}N - \pi\Sigma - \pi\Lambda$ potential. Using energy-dependent (two-pole) $\bar{K}N$ potentials, with simultaneous variation of c_i and the potential parameters, acceptable fits to the CLAS data were obtained. This fact then was considered as the ultimate proof of the two-pole structure of the $\Lambda(1405)$.

Using the same formula (corrected for the repeated use of on-shell factorization), we have made a preliminary fit to the few lowest γ-energy bin CLAS data varying only the c_i-s with our unchanged single-pole potential $\hat{V}$. The results are shown in Fig. 147.2.

It does not seem, that another pole is necessarily needed to improve these fits. A complete analysis of the CLAS data, including the charged channels is the subject of a forthcoming work.

147.4 Conclusions

- It was shown, that the energy-dependence of the WT term of the $\bar{K}N$ interaction, derived from the chiral $SU(3)$ Lagrangian and responsible for the appearance of a second pole in the $\Lambda(1405)$ region, follows from the unjustified application of the on-shell factorization approximation.

- Without this approximation a new, chiral based, energy-independent $\bar{K}N$ interaction was derived, which supports only one pole in the region of the $\Lambda(1405)$ resonance.
- The widely accepted "two-pole structure" of the $\Lambda(1405)$ state thus becomes questionable.
- In coordinate space calculations for $n > 2$ systems the use of the new potential avoids the not easily (and not uniquely) surmountable difficulties arising from the energy dependence of the two-body interaction.

Acknowledgements The work was supported by the Hungarian OTKA grant 109462.

References

1. Oset, E., Ramos, A.: Nucl. Phys. A **635**, 99 (1998)
2. Revai, J.: Few-Body Syst. **59**, 49 (2018)
3. Moriya, K., et al.: (CLAS collaboration): Phys. Rev. C **87**, 035206 (2013)
4. Roca, L., Oset, E.: Phys. Rev. C **87**, 055201 (2013)

Chapter 148
Four-Body Faddeev-Type Equations for $\bar{K}NNN$ Quasi-bound State Calculations

Nina Shevchenko

Abstract The paper is devoted to the $\bar{K}NNN$ system, which is an exotic system consisting of an antikaon and three nucleons. Four-body Faddeev-type AGS equations, which are being used for evaluation of the possible quasi-bound state in the system are described.

148.1 Introduction

The attractive nature of $\bar{K}N$ interaction has stimulated theoretical and experimental searches for bound states of K^- with different number of nucleons. The interest in few-body antikaonic systems was stimulated by calculations, which predicted deep and relatively narrow quasi-bound K^- -nuclear states. Many theoretical calculations devoted to the lightest possible system $\bar{K}NN$ have been performed since then using different methods, see e.g. a review [1]. All of them agree that a quasi-bound state in the K^-pp system exists, but they yield quite diverse binding energies and widths.

Some theoretical works were devoted to the question of the quasi-bound state in the four-body $\bar{K}NNN$ system, but more accurate calculations within Faddeev-type equations are needed. Indeed, only these dynamically exact equations in momentum representation can treat energy dependent $\bar{K}N$ potentials, necessary for the this system, exactly.

The paper contains description of the four-body Faddeev-type AGS equations [2], written down for the $\bar{K}NNN$ system. We will solve the equations using our programs written for the three-body AGS calculations of the $\bar{K}NN$ system, described in [1], and our two-body potentials constructed for them.

N. Shevchenko (✉)
Nuclear Physics Institute, 250 68 Řež, Czech Republic
e-mail: shevchenko@ujf.cas.cz

N. A. Orr et al. (eds.), *Recent Progress in Few-Body Physics*,
Springer Proceedings in Physics 238,
https://doi.org/10.1007/978-3-030-32357-8_148

148.2 Four-Body Faddeev-Type AGS Equations

The four-body AGS equations contain three-body transition operatos, obtained from the three-body AGS equations. The three-body Faddeev-type equations in AGS form [3] written for separable potentials $V_\alpha = \lambda_\alpha |g_\alpha\rangle\langle g_\alpha|$ have a form

$$X_{\alpha\beta}(z) = Z_{\alpha\beta}(z) + \sum_{\gamma=1}^{3} Z_{\alpha\gamma}(z)\tau_\gamma(z)X_{\gamma\beta}(z) \tag{148.1}$$

with transition $X_{\alpha\beta}$ and kernel $Z_{\alpha\beta}$ operators

$$X_{\alpha\beta}(z) = \langle g_\alpha|G_0(z)U_{\alpha\beta}(z)G_0(z)|g_\beta\rangle, \tag{148.2}$$

$$Z_{\alpha\beta}(z) = (1-\delta_{\alpha\beta})\langle g_\alpha|G_0(z)|g_\beta\rangle. \tag{148.3}$$

Here $U_{\alpha\beta}(z)$ is the three-body transition operator, which describes process $\beta + (\alpha\gamma) \to \alpha + (\beta\gamma)$, while $G_0(z)$ is the three-body Green function. Faddeev partition indices $\alpha, \beta = 1, 2, 3$ simultaneously define a particle (α) and the remained pair ($\beta\gamma$). The operator $\tau_\alpha(z)$ in (148.1) is an energy-dependent part of a separable two-body T-matrix $T_\alpha(z) = |g_\alpha\rangle\tau_\alpha(z)\langle g_\alpha|$, describing interaction in the ($\beta\gamma$) pair; $|g_\alpha\rangle$ is a form-factor.

The four-body Faddeev-type AGS equations [2], written for separable potentials, have a form

$$\bar{U}^{\sigma\rho}_{\alpha\beta}(z) = (1-\delta_{\sigma\rho})(\bar{G_0}^{-1})_{\alpha\beta}(z) + \sum_{\tau,\gamma,\delta}(1-\delta_{\sigma\tau})\bar{T}^{\tau}_{\alpha\gamma}(z)(\bar{G_0})_{\gamma\delta}(z)\bar{U}^{\tau\rho}_{\delta\beta}(z), \tag{148.4}$$

$$\bar{U}^{\sigma\rho}_{\alpha\beta}(z) = \langle g_\alpha|G_0(z)U^{\sigma\rho}_{\alpha\beta}(z)G_0(z)|g_\beta\rangle, \tag{148.5}$$

$$\bar{T}^{\tau}_{\alpha\beta}(z) = \langle g_\alpha|G_0(z)U^{\tau}_{\alpha\beta}(z)G_0(z)|g_\beta\rangle, \;\; (\bar{G_0})_{\alpha\beta}(z) = \delta_{\alpha\beta}\tau_\alpha(z). \tag{148.6}$$

The operators $\bar{U}^{\sigma\rho}_{\alpha\beta}$ and $\bar{T}^{\tau}_{\alpha\beta}$ contain four-body $U^{\sigma\rho}_{\alpha\beta}(z)$ and three-body $U^{\tau}_{\alpha\beta}(z)$ transition operators, correspondingly. The last one ($U^{\tau}_{\alpha\beta}(z)$) differs from the three-body operator in (148.2) by additional upper index τ (σ, ρ), which defines a three-body subsystem of the four-body system. The free Green function $G_0(z)$ now acts in four-body space. If, in addition, the "effective three-body potentials" $\bar{T}^{\tau}_{\alpha\beta}(z)$ in (148.4) are presented in a separable form: $\bar{T}^{\tau}_{\alpha\beta}(z) = |\bar{g}^{\tau}_{\alpha}\rangle\bar{\tau}^{\tau}_{\alpha\beta}(z)\langle\bar{g}^{\tau}_{\alpha}|$, the four-body equations can be written as [4].

$$\bar{X}^{\sigma\rho}_{\alpha\beta}(z) = \bar{Z}^{\sigma\rho}_{\alpha\beta}(z) + \sum_{\tau,\gamma,\delta}\bar{Z}^{\sigma\tau}_{\alpha\gamma}(z)\bar{\tau}^{\tau}_{\gamma\delta}(z)\bar{X}^{\tau\rho}_{\delta\beta}(z) \tag{148.7}$$

with new transition $\bar{X}^{\sigma\rho}$ and kernel $\bar{Z}^{\sigma\rho}$ operators defined by

$$\bar{X}^{\sigma\rho}_{\alpha\beta}(z) = \langle \bar{g}^{\sigma}_{\alpha} | \bar{G}_0(z)_{\alpha\alpha} \bar{U}^{\sigma\rho}_{\alpha\beta}(z) \bar{G}_0(z)_{\beta\beta} | \bar{g}^{\rho}_{\beta} \rangle, \tag{148.8}$$

$$\bar{Z}^{\sigma\rho}_{\alpha\beta}(z) = (1 - \delta_{\sigma\rho}) \langle \bar{g}^{\sigma}_{\alpha} | \bar{G}_0(z)_{\alpha\beta} | \bar{g}^{\rho}_{\beta} \rangle. \tag{148.9}$$

The separabelization of the "effective three-body potentials" $\bar{T}^{\tau}_{\alpha\beta}(z)$ can be performed using e.g. the Hilbert-Schmidt expansion of the three-body AGS equations with separable potentials (148.1).

148.3 Four-Body Equations for the $\bar{K}NNN$ System

There are two types of partitions for a four-body system: 3 + 1 and 2 + 2. For the $\bar{K}NNN$ system they are: $|\bar{K} + (NNN)\rangle$, $|N + (\bar{K}NN)\rangle$ and $|(\bar{K}N) + (NN)\rangle$. At the begin we considered all three nucleons as different particles, so we started by writing down the four-body system of equations (148.7) for the following 18 channels σ_α (with $\alpha - NN$ or $\bar{K}N$):

$$\begin{aligned}
1_{NN} &: |\bar{K} + (N_1 + N_2N_3)\rangle, |\bar{K} + (N_2 + N_3N_1)\rangle, |\bar{K} + (N_3 + N_1N_2)\rangle, \\
2_{NN} &: |N_1 + (\bar{K} + N_2N_3)\rangle, |N_2 + (\bar{K} + N_3N_1)\rangle, |N_3 + (\bar{K} + N_1N_2)\rangle, \\
2_{\bar{K}N} &: |N_1 + (N_2 + \bar{K}N_3)\rangle, |N_2 + (N_3 + \bar{K}N_1)\rangle, |N_3 + (N_1 + \bar{K}N_2)\rangle, \\
&\quad |N_1 + (N_3 + \bar{K}N_2)\rangle, |N_2 + (N_1 + \bar{K}N_3)\rangle, |N_3 + (N_2 + \bar{K}N_1)\rangle, \\
3_{NN} &: |(N_2N_3) + (\bar{K} + N_1)\rangle, |(N_3N_1) + (\bar{K} + N_2)\rangle, |(N_1N_2) + (\bar{K} + N_3)\rangle, \\
3_{\bar{K}N} &: |(\bar{K}N_1) + (N_2 + N_3)\rangle, |(\bar{K}N_2) + (N_3 + N_1)\rangle, |(\bar{K}N_3) + (N_1 + N_2)\rangle
\end{aligned} \tag{148.10}$$

After this the operators and equations were antisymmetrized, and the system of operator equations was written in a form:

$$\hat{X} = \hat{Z}\hat{\tau}\hat{X}, \tag{148.11}$$

were $\hat{Z}$ and $\hat{\tau}$ are the 5 × 5 matrices containing the kernel operators $\bar{Z}^{\sigma\rho}_{\alpha}$ and $\bar{\tau}^{\rho}_{\alpha\beta}$, correspondingly. Since the initial state is assumed to be fixed, only one column of the 5 × 5 matrix $\hat{X}$, containing transition operators $\bar{X}^{\sigma\rho}_{\alpha\beta}$, is necessary:

$$\bar{X}^{\rho}_{\alpha} = \begin{pmatrix} \bar{X}^{1}_{NN} \\ \bar{X}^{2}_{NN} \\ \bar{X}^{2}_{\bar{K}N} \\ \bar{X}^{3}_{NN} \\ \bar{X}^{3}_{\bar{K}N} \end{pmatrix}, \qquad \bar{Z}^{\sigma\rho}_{\alpha} = \begin{pmatrix} 0 & \bar{Z}^{12}_{NN} & 0 & \bar{Z}^{13}_{NN} & 0 \\ \bar{Z}^{21}_{NN} & 0 & 0 & \bar{Z}^{23}_{NN} & 0 \\ 0 & 0 & \bar{Z}^{22}_{\bar{K}N} & 0 & \bar{Z}^{23}_{\bar{K}N} \\ \bar{Z}^{31}_{NN} & \bar{Z}^{32}_{NN} & 0 & 0 & 0 \\ 0 & 0 & \bar{Z}^{32}_{\bar{K}N} & 0 & 0 \end{pmatrix}, \tag{148.12}$$

$$\bar{T}^{\rho}_{\alpha\beta} = \begin{pmatrix} \bar{T}^{1}_{NN,NN} & 0 & 0 & 0 & 0 \\ 0 & \bar{T}^{2}_{NN,NN} & \bar{T}^{2}_{NN,\bar{K}N} & 0 & 0 \\ 0 & \bar{T}^{2}_{\bar{K}N,NN} & \bar{T}^{2}_{\bar{K}N,\bar{K}N} & 0 & 0 \\ 0 & 0 & 0 & \bar{T}^{3}_{NN,NN} & \bar{T}^{3}_{NN,\bar{K}N} \\ 0 & 0 & 0 & \bar{T}^{3}_{\bar{K}N,NN} & \bar{T}^{3}_{\bar{K}N,\bar{K}N} \end{pmatrix}. \tag{148.13}$$

148.4 Three-Body Subsystems and Two-Body Input

We are studying the $\bar{K}NNN$ system with the lowest value of the four-body isospin $I^{(4)} = 0$, which can be denoted as K^-ppn. Its total spin $S^{(4)}$ is equal to one half, while the orbital momentum is zero, since all two-body interactions are chosen to be zero. For the $\bar{K}NNN$ system with these quantum numbers the following three-body subsystems contribute:

- $\bar{K}NN$ with $I^{(3)} = 1/2$, $S^{(3)} = 0$ (K^-pp) or $S^{(3)} = 1$ (K^-d).
- NNN with $I^{(3)} = 1/2$, $S^{(3)} = 1/2$ (^{3}H or ^{3}He),

were $I^{(3)}$ and $S^{(3)}$ are the three-body isospin and spin.

The three-body antikaon-nucleon system $\bar{K}NN$ with different quantum numbers was studied in our previous works, see [1]. In particular, quasi-bound state pole positions in the K^-pp system ($\bar{K}NN$ with $I^{(3)} = 1/2$, $S^{(3)} = 0$) and near-threshold K^-d scattering amplitudes ($\bar{K}NN$ with $I^{(3)} = 1/2$, $S^{(3)} = 1$) were calculated (no quasi-bound states were found in the K^-d system). It was done using the three-body AGS equations with separable potentials (148.1) with three models of the $\bar{K}N$ interaction: two phenomenological potentials having one- or two-pole structure of the $\Lambda(1405)$ resonance and a chirally motivated model. All three potentials describe low-energy K^-p scattering and $1s$ level shift of kaonic hydrogen with equally high accuracy. We also used a two-term separable NN potential, which reproduces Argonne v18 NN phase shifts and scattering lengths. The same potentials are used in our four-body calculations.

The programs of numerical solution of the three-body AGS equations for the $\bar{K}NN$ systems can be used for separabelization of the "effective three-body potentials" (after some modifications). However, it is far not enough since the three-body system (148.1) was solved with the initial channel fixed by $\beta = 1$, which corresponds to the $\bar{K} + NN$ partition. It allows to calculate the three-body transition amplitudes, which in the four-body formalism are denoted as $\bar{T}^{2_i}_{NN,NN}$ and $\bar{T}^{2_i}_{\bar{K}N,NN}$. The other two three-body amplitudes, necessary for the four-body calculations: $\bar{T}^{2_i}_{NN,\bar{K}N}$ and $\bar{T}^{2_i}_{\bar{K}N,\bar{K}N}$—were calculated additionally by solving the AGS equations (148.1) with other two initial channels $\beta = 2, 3$, corresponding to the $N + \bar{K}N$ initial state (which were properly antisymmetrized).

The three-nucleon system NNN was treated by solving the system of AGS equations (148.1) with our separable two-term NN potential. The calculated binding energy was found to be 9.95 MeV for both ^{3}H and ^{3}He nuclei since Coulomb interaction was not taken into account.

148.5 Summary

The four-body Faddeev-type calculations of the $\bar{K}NNN$ system is very complicated and time-consuming task. Up to now the four-body AGS equations were written down for the $\bar{K}NNN$ system and antisymmetrized. All necessary calculations of the three-body (sub)systems were performed, their separable forms were numerically evaluated. The spin-isospin re-coupling parts of the four-body kernel functions $\bar{Z}_{\alpha}^{\sigma\rho}$ were also calculated. The work is in progress.

References

1. Shevchenko, N.V.: Three-body antikaon-nucleon systems. Few Body Syst. **58**, 6 (2017). https://doi.org/10.1007/s00601-016-1170-5
2. Grassberger, P., Sandhas, W.: Systematical treatment of the non-relativistic n-particle scattering problem. Nucl. Phys. B **2**, 181–206 (1967)
3. Alt, E.O., Grassberger, P., Sandhas, W.: Reduction of the three-particle collision problem to multi-channel two-particle Lippmann-Schwinger equations. Nucl. Phys. B **2**, 167–180 (1967)
4. Casel, A., Haberzettl, H., Sandhas, W.: Unitary pole approximations and expansions in few-body systems. Phys. Rev. C **25**, 1738 (1982). https://doi.org/10.1103/PhysRevC.25.1738

Chapter 149
Production of Hypernuclei and Strange Particles in Spallation Reactions at a Few GeV Using an Intranuclear Cascade Approach

Jean-Christophe David, Jason Hirtz, Jose Luis Rodríguez-Sánchez, Alain Boudard, Joseph Cugnon, Sylvie Leray, Ingo Leya, Davide Mancusi and Georg Schnabel

Abstract Motivated by a renewed interest in studies of hypernuclei, the strangeness degree of freedom was implemented in the intranuclear cascade model INCL. This model takes care of the first stage of reactions between a nucleon (or a light cluster) and a nucleus at energies from a few tens of MeV up to a few GeV. After emission of fast particles, a hot remnant nucleus is produced and another model, combined to INCL, handles the de-excitation (Abla in our case). The main ingredients are discussed and we compare the results to experimental data.

J.-C. David (✉) · J. Hirtz · J. L. Rodríguez-Sánchez · A. Boudard · S. Leray · G. Schnabel
IRFU, CEA, Université Paris-Saclay, 91191 Gif-sur-Yvette, France
e-mail: jcdavidb@free.fr; jean-christophe.david@cea.fr

J. Hirtz · I. Leya
Space Research and Planetary Sciences, Physics Institute, University of Bern, Sidlerstrasse 5, 3012 Bern, Switzerland

J. L. Rodríguez-Sánchez
Universidad de Santiago de Compostela, 15782 Santiago de Compostela, Spain

GSI-Helmholtzzentrum für Schwerionenforschung GmbH, 64291 Darmstadt, Germany

J. Cugnon
AGO department, University of Liège, allée du 6 août 19, bâtiment B5, 4000 Liège, Belgium

D. Mancusi
Den-Service d'étude des réacteurs et de mathématiques appliquées (SERMA), CEA, Université Paris-Saclay, 91191 Gif-sur-Yvette, France

N. A. Orr et al. (eds.), *Recent Progress in Few-Body Physics*, Springer Proceedings in Physics 238,
https://doi.org/10.1007/978-3-030-32357-8_149

149.1 Introduction

The hypernucleus is a suitable laboratory to study YN and YY interactions and new experiments, either in progress or planned, in several facilities (JPARC, MAMI, JLab, GSI, FAIR, ...), makes the topic more and more attractive. While light hypernuclei have been studied extensively, the heaviest ones are also interesting objects, for example to study the behavior of hyperons in nuclear matter (role of the Λ in neutron stars). Our nuclear reaction code, called INCL (Liège IntraNuclear Cascade), treats reactions between light particles and nuclei with incident energies from ~100 MeV up to ~15 GeV [1]. The renewed interest in hypernucleus studies and the sound bases of the model implemented in INCL were the motivations to add K's, Λ and Σ's as participant particles in INCL. The de-excitation code Abla was also upgraded. Below we give first the ingredients required to perform the calculations and, second, we compare the obtained results to experimental data and to predictions from other model codes.

149.2 Ingredients

An intranuclear cascade is a series of collisions between hadrons. The main ingredients to simulate such collisions are the elementary cross sections, the particle momenta and charges in the output channels and the nuclear potential felt by all particles. It must be stressed that in INCL the resonances are not considered as participant particles (except the $\Delta(1232)$); Only their decay products play a role.

Adding the four Kaons (K^+, K^-, K^0, $\bar{\mathrm{K}}^0$), the Λ and the three Σ (Σ^-, Σ^0, Σ^+), implies more than 400 new channels, when isospin is considered. All details are described in a recent published paper [2]. Here we only draw the attention to the difficulties to obtain all necessary information. As an example, if one excepts the Δ-induced and inclusive reactions, 382 isospin channels are involved and therefore 382 cross section parametrizations. Only 17% of them can be obtained by using experimental data. Considering isospin symmetry at the initial and final states of some binary collisions, an extra 18% is obtained from relations between known and unknown cross sections. Still relying on isospin symmetry, but, this time, at each vertex of tree Feynman diagrams used within a hadron exchange model, 37% of the cross sections is determined by ratios between known and unknown cross sections. The remaining cross sections are based on models or hypotheses.

Once a reaction between two hadrons in the nucleus is chosen, charges and momenta of the particles in the final state must be defined. Concerning the charges, Clebsch Gordan coefficients are used as far as possible if two or more particles are involved. When the number of particles increases, additional models are sometimes needed to remove ambiguities. Otherwise, as in the case of the inclusive reactions, results of other codes are used to determine the number and type of particles emitted. Momenta are taken from measurements or by assuming isotropy or considering a phase-space distribution. Here again, the reliability of our approach may differ from channel to channel.

For the potential, the values used in this study are: $V_{\Lambda} = -28$ MeV, $V_{\Sigma} = 16$ MeV, $V_{K^+} = 25$ MeV, $V_{K^-} = -60$ MeV, $V_{K^0} = 25$ MeV, $V_{\bar{K}^0} = -50$ MeV. The latest value implemented in INCL for V_{Λ} is actually mass dependent and will be discussed. However, the results presented here were performed with the value of −28 MeV, which is very close to the new $V_{\Lambda}(A)$ for the light and medium mass nuclei.

149.3 Results

Comparisons to experimental data and other models are the only way to test the reliability of a model and to try to understand remaining deficiencies. Most of the measured data in this domain are related to the K^+ production. However, some other data exist and were used to benchmark this first version of INCL considering strangeness.

149.3.1 Particle Production

The production of K^+ is well described by INCL. This is illustrated with Fig. 149.1, where we can see also the important role played by the Δ-induced reactions, especially at low energies. While at highest energies the parametrization used for those

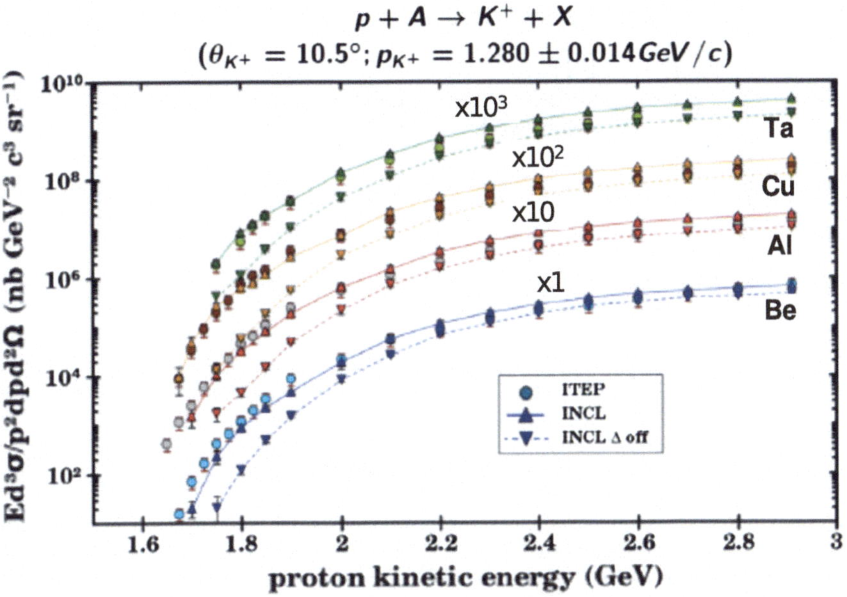

Fig. 149.1 Invariant production cross sections of K^+ emitted with a momentum of 1.28 ± 0.014 GeV/c at an angle of 10.5° for four targets as a function of proton projectile energy. Experimental data are from [4] (circles) and are compared to INCL with (up-oriented triangles) and without (down-oriented triangles) Δ-induced Kaon production

reactions [3] could be improved, benchmarks (not shown here) above 10 GeV still exhibit very good behaviors of INCL. In the forward direction more experimental data are needed to really know the situation. The conclusions for the K^- are different. The spectra are well reproduced except for low momenta. This seems to indicate that in INCL some production channels are missing, especially $YN \rightarrow \bar{K}NN$. Regarding the Λ, we compared the INCL results to the HADES rapidity spectrum for the reaction p(3.5 GeV) + Nb. INCL matches rather well the HADES data [5], except for rapidi-

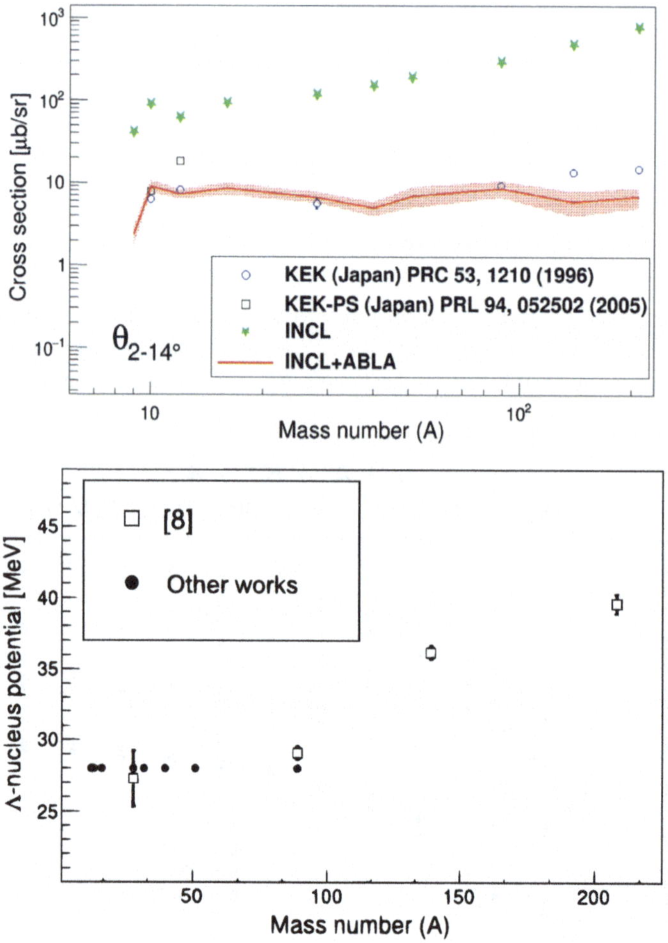

Fig. 149.2 Hypernucleus production cross section in function of the mass target for ${}^{A}X(\pi^+,K^+)^{A}_{\Lambda}X$ reactions, with incident energies of 1.06 and 1.048 GeV/c (upper part). Experimental data are from [6] (circles) and [7] (squares). Are plotted INCL hyperremnant production (green stars) and INCL-Abla result (solid red line surrounded by a red band for the uncertainties). Is plotted also Λ-Nucleus potential fitted with the above experimental data in function of the mass target (lower part)

ties larger than 0.8. This discrepancy, possibly due to the low transverse momenta, must be now carefully studied to understand the differences between experimental data and model predictions.

149.3.2 Hypernucleus Production

Fig. 149.2 shows the production of hypernuclei as a function of the mass target for cases in which one, and only one, K^+ is emitted by π^+-induced reactions (^{A}X(π^+,K^+)$^A_\Lambda$X). INCL fits very well the KEK experimental data, except an underestimate of the heaviest nuclei. This deviation has been used to redefine the potential, assuming a dependence on the mass of the nucleus. The thus updated INCL model produces a perfect fit of the measurements (not shown here). This increase of the potential with the mass, i.e., with the nucleon-isospin asymmetry, could indicate a role of the Λnn interaction. All details are in [8].

149.4 Conclusion

The strange particles K^+, K^-, K^0, $\bar{\mathrm{K}}^0$, Λ, Σ^-, Σ^0 and Σ^+ are implemented into the intranuclear cascade code INCL. Since the de-excitation code Abla is usually combined to INCL for a full simulation of spallation reactions, Abla has been also upgraded by adding Λ evaporation and hyperfission. The results obtained by the codes, and especially INCL, are very encouraging. The main mechanisms are incorporated and now it is time for improvements. The Δ-induced strangeness production is probably overestimated when energy goes up and some strangeness exchange reactions must be added, like $\Lambda N \rightarrow \bar{\mathrm{K}}NN$, to better reproduce K^- production. Other aspects must also be studied, like the momentum distribution of the emitted particles. However, the lack of experimental data to get better elementary ingredients and to benchmark carefully our model makes the task difficult.

References

1. Mancusi, D., et al.: Eur. Phys. J. A **53**, 80 (2017)
2. Hirtz, J., et al.: Eur. Phys. J. Plus **133**, 436 (2018)
3. Tsushima, K., et al.: Phys. Rev. C **59**, 369 (1999)
4. Akindinov, A.V., et al.: J. Exp. Theo. Phys. Lett. **72**, 100 (2000)
5. Agakishiev, G.: HADES collaboration. Eur. Phys. J. A **50**, 81 (2014)
6. Hasegawa, T., et al.: Phys. Rev. C **53**, 1210 (1996)
7. Saha, P.K., et al.: Phys. Rev. Lett. **94**, 052502 (2005)
8. Rodríguez-Sánchez, J.L., et al.: Phys. Rev. C **98**, 021602(R) (2018)

Chapter 150
Kaonic Deuterium Precision Measurement at DAΦNE: The SIDDHARTA-2 Experiment

M. Miliucci, A. Amirkhani, A. Baniahmad, M. Bazzi, G. Bellotti, C. Berucci, D. Bosnar, M. Bragadireanu, M. Cargnelli, C. Curceanu, A. Dawood Butt, L. De Paolis, R. Del Grande, L. Fabbietti, C. Fiorini, F. Ghio, C. Guaraldo, M. Iliescu, M. Iwasaki, P. Levi Sandri, J. Marton, P. Moskal, S. Niedźwiecki, S. Okada, D. Pietreanu, K. Piscicchia, A. Scordo, H. Shi, M. Silarski, D. Sirghi, F. Sirghi, M. Skurzok, A. Spallone, H. Tatsuno, O. Vazquez Doce, E. Widmann and J. Zmeskal

Abstract Light kaonic atoms spectroscopy offers the unique opportunity to perform experiments equivalent to scattering at vanishing relative energies. This allows the determination of the antikaon-nucleus interaction at threshold, without the need of extrapolation to zero energy, as in the case of scattering experiments. In this framework, the SIDDHARTA-2 collaboration aims to perform the first measurement of kaonic deuterium transition to the fundamental level, which is mandatory to extract the isospin dependent antikaon—nucleon scattering lengths. The experiment will be carried out at the DAΦNE collider of LNF-INFN in 2019–2020.

M. Miliucci (✉) · M. Bazzi · C. Berucci · C. Curceanu · L. De Paolis · R. Del Grande · C. Guaraldo · M. Iliescu · P. Levi Sandri · J. Marton · D. Pietreanu · K. Piscicchia · A. Scordo · H. Shi · D. Sirghi · F. Sirghi · A. Spallone · O. Vazquez Doce · J. Zmeskal
INFN, Laboratori Nazionali di Frascati, Frascati (Roma), Italy
e-mail: Marco.Miliucci@lnf.infn.it

M. Miliucci · L. De Paolis
Department of Physics, Faculty of Science MM.FF.NN., University of Rome 2 (Tor Vergata), Rome, Italy

A. Amirkhani · A. Baniahmad · G. Bellotti · A. Dawood Butt
Politecnico di Milano, Dipartimento di Elettronica, Informazione e Bioingegneria and INFN Sezione di Milano, Milan, Italy

C. Berucci · M. Cargnelli · C. Fiorini · J. Marton · D. Pietreanu · H. Shi · E. Widmann · J. Zmeskal
Stefan-Meyer-Institut für Subatomare Physik, Vienna, Austria

D. Bosnar
Department of Physics, Faculty of Science, University of Zagreb, Zagreb, Croatia

M. Bragadireanu · D. Sirghi · F. Sirghi
Horia Hulubei National Institute of Physics and Nuclear Engineering (IFIN-HH), Magurele, Romania

L. Fabbietti · O. Vazquez Doce
Excellence Cluster Universe, Technische Universiät München, Garching, Germany

N. A. Orr et al. (eds.), *Recent Progress in Few-Body Physics*,
Springer Proceedings in Physics 238,
https://doi.org/10.1007/978-3-030-32357-8_150

150.1 The Scientific Case

When a negatively charged kaon enters a target, it is slowed down and loses its kinetic energy through the interaction with the medium.

The kaon is then captured by an atom replacing an electron and forms a kaonic atom in a highly excited state. The kaonic atom cascades down to the low n-states where the strong interaction between the antikaon and the nucleus adds up to the electromagnetic one. The observables of interest are the shift (ϵ) and the width (Γ) of the atomic levels caused by the strong interaction. The SIDDHARTA collaboration measured, in 2009, the shift and the width of the kaonic hydrogen 1 s level [1], while SIDDHARTA-2, a major upgrade of SIDDHARTA, aims to perform the first measurement of the kaonic deuterium transition to the 1 s level. From the X-rays emitted by kaonic hydrogen and kaonic deuterium, is possible to extract the K^-p (and K^-d) scattering lengths using the Deser–Treumann type formulae with isospin-breaking corrections [2, 3]:

$$\epsilon_{1s} + \frac{i}{2}\Gamma_{1s} = 2\alpha^3\mu^2 a_{K^-p}[1 - 2\alpha\mu(ln\alpha - 1)a_{K^-p} + \ldots] \tag{150.1}$$

where:
μ: the reduced mass of the K^-p (K^-d) system;
α: the fine-structure constant.
These two quantities allow to determine the antikaon-nucleon isoscalar a_0 and isovector a_1 scattering lengths, trough the equations:

$$a_{K^-p} = \frac{1}{2}[a_0 + a_1] \quad ; \quad a_{K^-n} = a_1 \tag{150.2}$$

$$a_{K^-d} = \frac{4[m_N + m_K]}{[2m_N + m_K]}Q + C \tag{150.3}$$

$$Q = \frac{1}{2}[a_{K^-n} + a_{K^-p}] = \frac{1}{4}[a_0 + 3a_1] \tag{150.4}$$

F. Ghio
INFN Sez. di Roma I and Inst. Superiore di Sanità, Rome, Italy

M. Iwasaki · S. Okada
RIKEN, Tokyo, Japan

P. Moskal · S. Niedźwiecki · M. Silarski · M. Skurzok
Jagiellonian University, Kraków, Poland

K. Piscicchia
CENTRO FERMI – Museo Storico della Fisica e Centro Studi e Ricerche "Enrico Fermi", Rome, Italy

H. Tatsuno
Lund Univeristy, Lund, Sweden

Q: K^- scattering from each (free) nucleon of deuterium;

C: includes the K^-d three-body interaction, which can be studied by solving Faddeev-type equations.

The antikaon-nucleon scattering lengths are fundamental quantities to understand the QCD in the non-perturbative regime in the strangeness sector, with implications from particle and nuclear physics to astrophysics.

150.2 The SIDDHARTA-2 Experimental Setup

DAΦNE [4, 5] (Double Annular Φ Factory for Nice Experiments), a world-class electron-positron collider at LNF-INFN, Italy, is a unique low-energy kaon source via the decay of ϕ-mesons, produced with a momentum of 127 MeVc^{-1} and a spread δp/p below 0.1%. Due to the extremely low yield ($\simeq$0.1%) of the K^-d transition to 1 s level, an improved setup with respect to SIDDHARTA [6], has been developed, in order to perform the kaonic deuterium measurement with a precision similar to the kaonic hydrogen one. The new apparatus (Fig. 150.1a) increases drastically the signal-to-background ratio, by gaining in solid angle, taking advantage of the new SDDs geometry, and reducing the background thanks to the faster SDDs response and to the additional veto systems [7].

A cryogenic target cell, made by Kapton walls and reinforced with aluminium supports, operates below 30 K at a pressure of 0.4 MPa (3% LHD), optimizing the kaon stopping efficiency. A dedicated cooling system consisting in 4 CryoTigers reduces the temperature of the detectors down to 120 K, improving both the energy resolution (140 eV at 6 KeV) and the timing response (below 400 ns). Each element of the setup has been optimized using GEANT4 simulations and already tested during the SIDDHARTA run. Figure 150.1b shows the K^-d simulated spectrum, for an expected $\epsilon_{1s} = -800$ eV and $\Gamma_{1s} = 750$ eV and assuming an yield of 0.1% for the K_α transition, for an acquired luminosity of 800 pb^{-1}. The fit indicates that both ϵ_{1s} and Γ_{1s} are evaluated with a precision comparable with the kaonic hydrogen one measured by SIDDHARTA.

150.3 Conclusions

Light kaonic atoms spectroscopy allows to obtain fundamental informations for understanding the non-perturbative QCD with strangeness. The SIDDHARTA-2 experiment will perform the first measurement of the kaonic deuterium transitions, which allow to extract the isospin dependent antikaon-nucleon scattering lengths. Presently the SIDDHARTA-2 setup is under tests and debug, with the aim to be installed at DAΦNE and take data in 2019–2020.

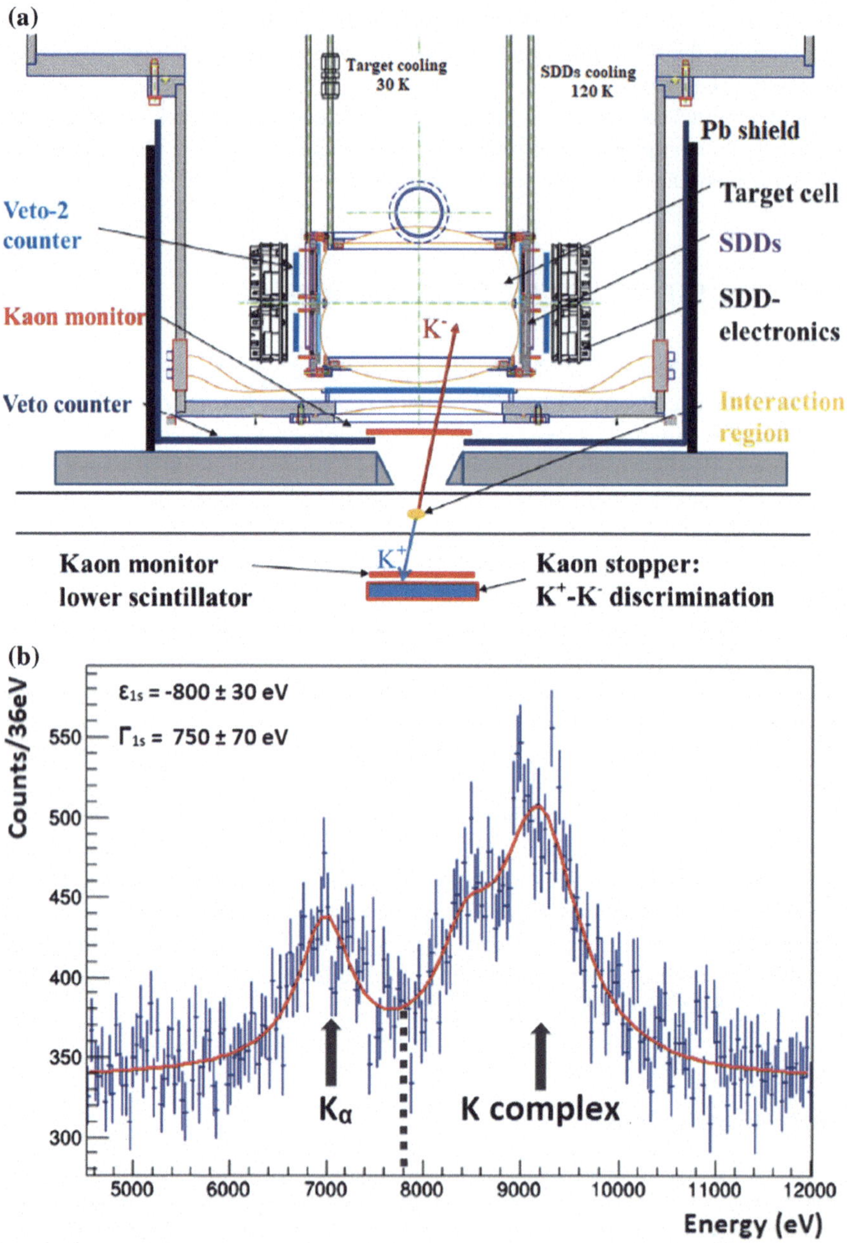

Fig. 150.1 **a** Cross section layout of SIDDHARTA-2 setup (adapted from [8]); **b** Simulated K^-d Monte Carlo spectrum corresponding to an integrated luminosity of 800 pb^{-1}, assuming $\epsilon_{1s} = -800$ eV and $\Gamma_{1s} = 750$ eV, and an yield of 0.1%. Dot line at 7834 eV corresponds to the pure QED K_α value (adapted from [8])

Acknowledgements We thank C. Capoccia and G. Corradi, from LNF-INFN; and H. Schneider, L. Stohwasser, and D. Pristauz-Telsnigg from Stefan-Meyer-Institut, for their fundamental contribution in designing and building the SIDDHARTA setup. We thank as well the DAΦNE staff for the excellent working conditions and permanent support. Part of this work was supported by the Austrian Science Fund (FWF): [P24756-N20]; Austrian Federal Ministry of Science and Research BMBWK 650962/0001 VI/2/2009; the Croatian Science Foundation, project IP-2018-01-8570; Minstero degli Affari Esteri e della Cooperazione Internazionale, Direzione Generale per la Promozione del Sistema Paese (MAECI), Strange Matter project; Polish National Science Center through grant no. UMO-2016/21/D/ ST2/01155; Ministry of Science and Higher Education of Poland grant no 7150/E-338/M/2018; CENTRO FERMI - Museo Storico della Fisica e Centro Studi e Ricerche 'Enrico Fermi' (*Open Problems in Quantum Mechanics project*).

References

1. Bazzi, M., et al.: Nucl. Phys. A **881**, 88 (2012)
2. Meißner, U.G., Raha, U., Rusetsky, A.: Eur. Phys. J. C **35**, 349 (2004)
3. Meißner, U.G., Raha, U., Rusetsky, A.: Eur. Phys. J. C **47**, 473 (2006)
4. Milardi, C.: Int. J. Mod. Phys. A **24**, 360 (2009)
5. Zobov, M., et al.: Phys. Rev. Lett. **104**, 174801 (2010)
6. Bazzi, M., et al.: Phys. Lett. B **704**, 113 (2011)
7. Bazzi, M., et al.: JINST **8**, T11003 (2013)
8. Scordo, A., et al.: EPJ Web Conf. **181**, 01004 (2017)

Chapter 151
In-Medium Properties of SU(3) Baryons

Ki-Hoon Hong, Ulugbek Yakhshiev and Hyun-Chul Kim

Abstract Changes of baryon properties in nuclear matter are investigated within the framework of an in-medium modified SU(3) Skyrme model. Introducing the medium functionals in the SU(2) sector and considering the alteration of kaon properties in nuclear medium, we are able to examine the medium modification of the nucleon and hyperons. The functionals introduced in the SU(2) sector are related to ordinary nuclear matter properties near the saturation point. The results indicate that the changes of the baryon properties in the strange sector are strongly correlated with the in-medium properties of kaons.

151.1 Introduction

Understanding how the hyperons undergo changes in nuclear matter is a very important issue in contemporary nuclear physics. In particular, it is of great interest to see how the hyperons are related to in-medium kaon properties at low densities and how they can be changed in higher densities that can be found in the interior of neutron stars [1, 2]. In the present contribution, we will discuss a recent work on the hyperon properties in nuclear matter, which was carried out in a simple but plausible framework of a chiral soliton approach to nonzero density phenomena in the SU(3)

K.-H. Hong (✉) · U. Yakhshiev · H.-C. Kim
Department of Physics, Inha University, Incheon 22212, Republic of Korea
e-mail: kihoon@inha.edu

U. Yakhshiev
e-mail: yakhshiev@inha.ac.kr

H.-C. Kim
e-mail: hchkim@inha.ac.kr

H.-C. Kim
Advanced Science Research Center, Japan Atomic Energy Agency, Shirakata, Ibaraki, Tokai 319-1195, Japan

School of Physics, Korea Institute for Advanced Study (KIAS), Seoul 02455, Republic of Korea

N. A. Orr et al. (eds.), *Recent Progress in Few-Body Physics*,
Springer Proceedings in Physics 238,
https://doi.org/10.1007/978-3-030-32357-8_151

sector [3]. Previously, a similar approach was developed in the non-strangeness sector to study various phenomena in medium (for example, see [4] and references therein) and the results were in qualitative agreement with those from other different approaches. In [3], we extended the work of [4] to the SU(3) including the hyperons. We discuss the main results and significance of the work.

151.2 The Model

The Lagrangian of the present model written by the following form [3]

$$\mathcal{L} = -\tfrac{F_\pi^2}{16}\alpha_2^t(\rho)\mathrm{Tr}L_0L_0 + \tfrac{F_\pi^2}{16}\alpha_2^s(\rho)\mathrm{Tr}L_iL_i - \tfrac{\alpha_4^t(\rho)}{16e^2}\mathrm{Tr}[L_0, L_i]^2 + \tfrac{\alpha_4^s(\rho)}{32e^2}\mathrm{Tr}[L_i, L_j]^2 + \tfrac{F_\pi^2}{16}\alpha_{\chi SB}(\rho)\mathrm{Tr}\mathcal{M}(U + U^\dagger - 2) + \mathcal{L}_{WZ}, \qquad (151.1)$$

where $L_\mu = U^\dagger\partial_\mu U$ and $U(\boldsymbol{x}, t)$ is a chiral field in SU(3). The Wess-Zumino term [5] $\mathcal{L}_{\rm WZ}$ in the Lagrangian constrains the soliton to be identified as a baryon and is expressed by a five-dimensional integral over a disk D

$$S_{\rm WZ} = -\frac{iN_c}{240\pi^2}\int_D d^5\mathbf{x}\,\epsilon^{\mu\nu\alpha\beta\gamma}\mathrm{Tr}(L_\mu L_\nu L_\alpha L_\beta L_\gamma). \qquad (151.2)$$

Here $\epsilon^{\mu\nu\alpha\beta\gamma}$ is the totally antisymmetric tensor defined as $\epsilon^{01234} = 1$ and $N_c = 3$ is the number of colors. The values of input parameters are defined in free space: $F_\pi = 108.783\,\mathrm{MeV}$ denotes the pion decay constant, $e = 4.854$ represents the Skyrme parameter, the masses of the π and K mesons are given respectively as $m_\pi = 134.976\,\mathrm{MeV}$ and $m_K = 495\,\mathrm{MeV}$, and the mass matrix of the pseudo-Nambu-Goldstone bosons $\mathcal{M}$ has the diagonal form $\mathcal{M} = (m_\pi^2, m_\pi^2, 2m_K^2 - m_\pi^2)$. The density-dependent functions $\alpha_2^t(\rho)$, $\alpha_2^s(\rho)$, $\alpha_4^t(\rho)$, $\alpha_4^s(\rho)$ and $\alpha_{\chi SB}(\rho)$ reflect the changes of the meson properties in nuclear medium. In an approximation of homogeneous infinite nuclear matter they are expressed in terms of the three linear density-dependent functions $f_i(\rho) = 1 + C_i\rho$, $(i = 1, 2, 3)$. The numerical values of C_i are fixed to be $C_1 = -0.279$, $C_2 = 0.737$ and $C_3 = 1.782$, respectively. They reproduce very well the equations of state (EoS) for symmetric nuclear matter near the normal nuclear matter density ρ_0 and at higher densities that may exist in the interior of a neutron star. The medium modification of the kaon properties is achieved by considering the following scheme

$$F_\pi m_K \rightarrow F_K^* m_K^* = F_\pi m_K(1 - C\rho/\rho_0) \qquad (151.3)$$

and can be explained in terms of the alteration of the kaon decay constant and/or of the kaon mass in nuclear environment.

The quantization of the model is performed by considering the time-dependent rigid rotation of a static soliton

$$U(\boldsymbol{r}, t) = \mathcal{A}(t) U_0(\boldsymbol{r}) \mathcal{A}(t)^\dagger, \tag{151.4}$$

where $U_0(\boldsymbol{r})$ denotes the static SU(3) chiral soliton with trivial embedding. The time-dependent rotational matrix $\mathcal{A}(t)$ is decomposed

$$\mathcal{A}(t) = \begin{pmatrix} A(t) & 0 \\ 0^\dagger & 1 \end{pmatrix} S(t), \tag{151.5}$$

in terms of the SU(2) isospin rotation $A(t) = k_0(t)\mathbf{1} + i \sum_{a=1}^{3} \tau_a k_a(t)$ and fluctuations into the strangeness sector given by the matrix $S(t) = \exp\left\{i \sum_{p=4}^{7} k_p \lambda_p\right\}$. Here $\tau_{1,2,3}$ denote the Pauli matrices, whereas λ_p stand for the strange part of the SU(3) Gell-Mann matrices. The time-dependent functions $k_a(t)$ $(a = 0, 1, 2, \ldots, 7)$ represent arbitrary collective coordinates. The more details of the approach can be found in [3].

151.3 Results and Discussions

All model parameters in free space and in nuclear matter, except for the parameter C in (151.3), are fixed in the SU(2) sector. The only remaining parameter C could be fixed by data on kaon-nucleus scattering and kaonic atoms. However, in the present work we carry out a qualitative analysis of the effects in the baryonic sector due to the modification of the kaon properties in nuclear medium. Consequently, we discuss the density dependence of the mass splittings among the various baryon multiplet members. In our calculation, the parameter value $C = 0$ corresponds to the case when the properties of kaon will not change in nuclear matter whereas a nonzero value of the parameter $C \neq 0$ indicates that the mass and/or kaon dynamics is alters in a dense nuclear environment.

The results show that in general the masses of the baryon octet tend to decrease in nuclear matter. Only Σ showed a different tendency if the parameter value is set to be $C = 0$. In the case of $C = 0.2$, m_Σ also tends to decrease as the density of nuclear matter increases [3]. In comparison, the results from SU(3) chiral effective field theory [6] show that m^*_Λ is decreased by about 17% at normal nuclear matter density ρ_0. The Ξ hyperon is behaved in a similar manner. At ρ_0 the change in the mass of Ξ was about 6 and 16% for the corresponding parameter values $C = 0$ and $C = 0.2$, respectively. The masses of the baryon decuplet increase in general as ρ increases. Changes are dramatic for $C = 0$ while for $C = 0.2$ they are less changeable.

We present the density dependence of the mass splittings among the multiplet members in Figs. 151.1 and 151.2. Figure 151.1 shows the density dependence of the mass splittings among the baryon octet members while Fig. 151.2 depicts the results corresponding to the mass splittings among the decuplet members. All the mass splittings in nuclear matter are normalized to the vaues of the corresponding

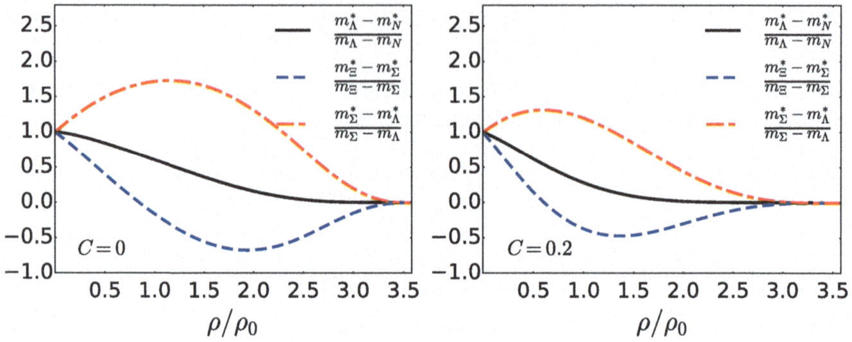

Fig. 151.1 Density dependence of the mass splittings among the baryon octet members. The mass splittings in nuclear matter are normalized to the corresponding free space mass splittings. The left and right panels in the figure corresponds to the results with $C = 0$ and $C = 0.2$, respectively

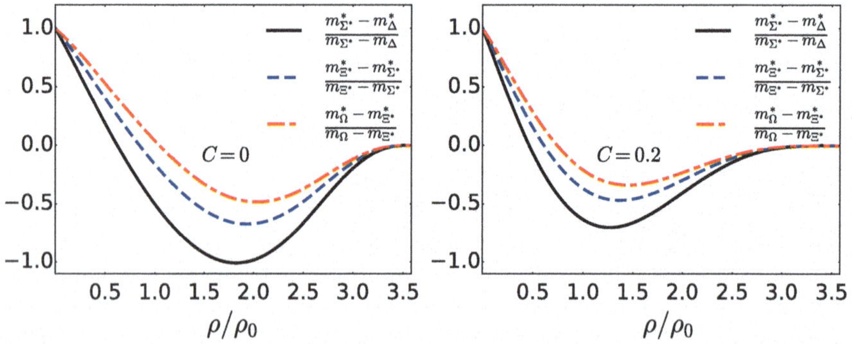

Fig. 151.2 Density dependence of the mass splittings among the baryon decuplet members. Notations are the same as in Fig. 151.1

ones in free space. The left and right panels in the figures illustrate the results with two different values of parameter C, respectively.

It is interesting to see that except $m^*_\Sigma - m^*_\Lambda$ all the mass splittings tend to decrease up to $(1.5 - 2)\rho_0$. This behavior can be explained in terms of the density-dependent functionals ω^*_- and c^* entering into the mass formula (see (36) in [3]). The first functional describes the fluctuations in the strangeness direction and comes into play for the mass splitting formula between the same strangeness members while all other mass splittings presented in the figures depend linearly on ω^*_-. This indicates that at large densities the fluctuations in strangeness direction gets weaker. From the figures one concludes also that at large densities SU(3) flavor symmetry tends to be restored.

The work is supported by Basic Science Research Program through the National Research Foundation (NRF) of Korea funded by the Korean government (Ministry of Education, Science and Technology, MEST), Grant No. 2016R1D1A1B03935053 (UY) and Grant No. NRF-2018R1A2B2001752 (HChK).

References

1. Gal, A., Hungerford, E.V., D. J. Millener, D.J.: Strangeness in nuclear physics. Rev. Mod. Phys. 88, 035004 (2016). https://doi.org/10.1103/RevModPhys.88.035004
2. Lattimer, J.M., Prakash, M.: The equation of state of hot, dense matter and neutron stars. Phys. Rept. **621**, 127 (2016). https://doi.org/10.1016/j.physrep.2015.12.005
3. Hong, K.H., Yakhshiev, U.T, Kim, H.C.: Modification of hyperon masses in nuclear matter. arXiv:1806.06504 [nucl-th]
4. Yakhshiev, U.T.: In-medium nucleons and nucleonic systems: infinite nuclear matter. Phys. Rev. C **88**, 034318 (2013). https://doi.org/10.1103/PhysRevC.88.034318
5. Wess, J., Zumino, B.: Consequences of anomalous Ward identities. Phys. Lett. B **37**, 95 (1971). https://doi.org/10.1016/0370-2693(71)90582-X
6. Petschauer, S., Haidenbauer, J., Kaiser, N., Meißner, U.G., Weise, W.: Hyperons in nuclear matter from SU(3) chiral effective field theory. Eur. Phys. J. A **52**, 15 (2016). https://doi.org/10.1140/epja/i2016-16015-4

Chapter 152
On the Bosonic Atoms—Structure and Processes

Miron Ya Amusia and Larissa V. Chernysheva

Abstract We study properties of atoms, in which substitute fermions–electrons by bosons, namely π^--mesons. We perform calculations in the frame of modified Hartree-Fock (HF) equation. The modification takes into account symmetry of the pair of identical bosons wave function, doubling instead of eliminating self-action as in the Fermion case. As examples of simple process with pion atoms, we consider their photoionization and collisions with other pions.

152.1 Introduction

Heavy ion colliders create very big numbers of $\pi^{\mp}$-mesons, or pions in each ion collision. In nuclear scale, the lifetime of $\pi^{\mp}(2.6 \times 10^{-8}\text{s})$ is rather big. So, an ion can capture negative pions during the nuclear interaction process and form so-called pion atoms. Each ion can capture several pions so the interaction between them can be essential and has to be considered. The structure of such an object is interesting.

Currently, the starting point in investigating ground state properties and processes in electronic atoms are Hartree-Fock (HF) equations. So, it is natural to apply to pion atoms the properly adjusted HF equations. However, for bosonic atoms they acquire unexpected features that we will consider below.

M. Y. Amusia (✉)
Racah Institute of Physics, Hebrew University, 91904 Jerusalem, Israel
e-mail: amusia@012.net.il

M. Y. Amusia · L. V. Chernysheva
A. F. Ioffe Physical-Technical Institute, St. Petersburg 194021, Russian Federation

N. A. Orr et al. (eds.), *Recent Progress in Few-Body Physics*,
Springer Proceedings in Physics 238,
https://doi.org/10.1007/978-3-030-32357-8_152

152.2 Main Formulas

The HF equation for multi-pion atoms looks like (see e.g. [1][1])

$$-\frac{\Delta}{2}\phi_j(x) - \frac{Z}{r}\phi_j(x) + \sum_{k=1}^{N_\pi}\int \phi_k^*(x')\frac{dx'}{|\mathbf{r}-\mathbf{r}'|}\left[\phi_k(x')\phi_j(x) + \phi_j(x')\phi_k(x)\right] = E_j\phi_j(x) \qquad (152.1)$$

Here Z is the nuclear charge, $\phi_j(x)$ is the one-pion wave function, $x \equiv \vec{r}, \vec{\sigma}$ are the combination of pion coordinate and spin variables, E_j—is the one-pion so-called HF energy; the summation is performed over all occupied pion states N_π. Sign plus in the square bracket takes into account that identical bosons wave function is symmetric against permutation of the coordinate/spin variables. As a result, this equation for bosons does not eliminate self-action as in Fermion atom case, but instead doubles it.

Indeed. if the interparticle interaction is not Coulombic, but short-range $u_0\delta(|\mathbf{r}-\mathbf{r}'|)$, and the interaction that holds the system together is some $U(r)$ that is more general than nuclear Coulomb potential, (152.1) reduces to the following

$$-\frac{\Delta}{2}\phi_j(\mathbf{r}) + U(r)\phi_j(\mathbf{r}) + 2u_0\sum_{k=1}^{N_\pi}|\phi_k(\mathbf{r})|^2\phi_j(\mathbf{r}) = E_j\phi_j(\mathbf{r}). \qquad (152.2)$$

We see that in HF for bosonic atoms the account of wave-function symmetrization leads to non-physical enhancement of the self-consistent field by a factor of 2.

Since all atomic pions can be in one lowest energy state, we obtain from (152.1), after eliminating the unphysical self-action "by hand", the following equation

$$-\frac{\Delta}{2}\phi_k(\mathbf{r}) - \frac{Z}{r}\phi_k(\mathbf{r}) + 2(N_\pi^k - 1)\int |\phi_k(\mathbf{r})|^2\frac{dr'}{|\mathbf{r}-\mathbf{r}'|}\phi_k(\mathbf{r}) = E_k\phi_k(\mathbf{r}). \quad (152.3)$$

152.3 The Results Obtained

After solving (152.3) numerically, we obtained HF energies of the 1s level, total energies and mean square radiuses for pionic atoms and negative pionic ions with Z from 2 to 80. It appeared that the total binding energy of a pion atom is much bigger than that for an ordinary atom, and the size of the pion atom is several times bigger than the size of a normal atom. We assume these quantities being in pionic and normal atomic units, respectively.

[1]We employ the system of units that defines $m = e = \hbar = 1$. Here m is the mass, in electron case $m = m_e$, in pion case $m = m_\pi$; $m_\pi/m_e \approx 273.13$.

A pion atom is able to form negative ions. For *He*, *Ne* and *Ar* formation of single negative ions He_{π}^{-}, Ne_{π}^{-}, Ar_{π}^{-} is possible, while for Zn formation of Zn_{π}^{-} and Zn_{π}^{-2} is allowed. *Kr* forms up to Kr_{π}^{-3}, while Xe binds up to 4 pions that leads to Xe_{π}^{-4}. We have extended calculations up to Z = 137. Note that at high Z values the Coulomb non-relativistic binding energy is close to the rest mass of a pion, and relativistic approach is required, along with accounting for the finite size of the nucleus.

For pion atoms notion "noble gases" has no sense, since pion shells do not exist.

Figure 152.1 presents the Z dependence of the number of extra bound pions.

As an example of pion atom process, consider photoionization. Figure 152.2 presents the results of calculations using for $\sigma_k(\omega)$ the computing codes [1]. For ordinary atoms, the cross-section monotonically decreases with photon energy growth from the maximum value at threshold. For pion atom the cross-section reaches its maximum value above ionization threshold, forming pronounced maxima, the height of which decreases while the width increases with Z growth [2].

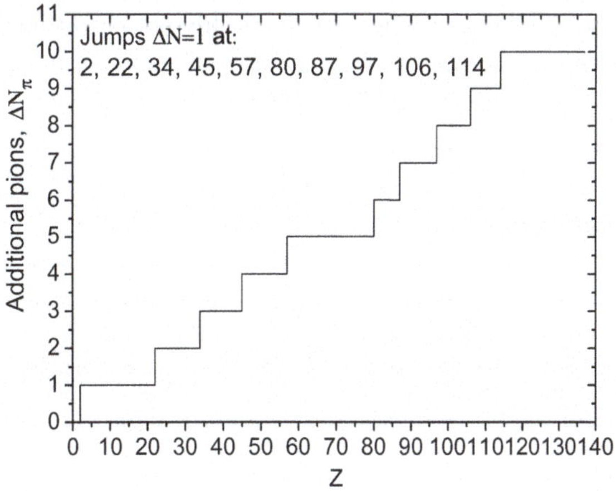

Fig. 152.1 Number of additional pions $\Delta N_{\pi} = N_{\pi}^{-n\max} - N_{\pi}$

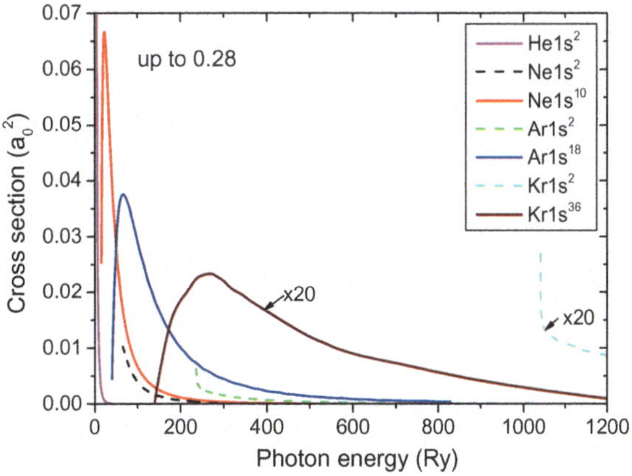

Fig. 152.2 Photoionization cross-sections for pion and ordinary atoms in respective atomic units as function of photon energy in Ry and Ry_{π}. Note that factor 20 multiply the values for $Kr1s^{36}$

We were unable to find discrete excitations in pion negative ions, contrary to the case of neutrals.

The elastic π^- scattering cross-section monotonically decreases with energy growth. Up to 100 Ry all but s scattering phases are negligibly small. In π^+ scattering upon pion atom the interaction of π^+ and π^- with formation of a bound pionium could be very important. Electron scattering cross-section upon pion atom is by a factor $\beta^2 \equiv (m_e/m_\pi)^2 \simeq 1.3 \times 10^{-5}$ smaller than the pion one (see [3]).

152.4 Conclusion

The results obtained for pion atoms are essentially different from that for ordinary atoms. In the consideration presented above, we neglected the nuclear forces between pions and nucleons. They can lead to sticking of extra pions to a nucleus but become decisively important only at $Z \approx 100$, when the pion's Bohr radius is about the range of nuclear forces action 10^{-13} cm. It would be of interest to look for data in multi-pion atoms formed in heavy atom collisions. In studies of a boson atom it would be meaningful also to go beyond the frame of self-consistent field approximation.

References

1. Amusia, M., Chernysheva, L.: Computation of Atomic Processes. Institute of Physics Publishing, Bristol and Philadelphia (1997)
2. Amusia, M., Chernysheva, L., Yarzhemsky, V.: Handbook of Theoretical Atomic Physics. Springer, Berlin (2012)
3. Amusia, M.: Quasi-three body systems—properties and scattering. Few Body Syst. **58**(88), 1–4 (2017)

Chapter 153
Local Meson-Baryon Coupled-Channels Potential for the $\Lambda(1405)$

Kenta Miyahara, Tetsuo Hyodo and Wolfram Weise

Abstract A local coupled-channels $\bar{K}N$-$\pi\Sigma$-$\pi\Lambda$ potential is constructed, equivalently reproducing the scattering amplitude in chiral SU(3) dynamics. By analyzing the wave function of the $\Lambda(1405)$, we show the $\bar{K}N$ dominance of the upper pole within the two-pole structure of the $\Lambda(1405)$. The resulting potential will be useful for investigating $\bar{K}$ few-nucleon systems including their coupled channels.

153.1 Introduction

Systems of an antikaon and few nucleons are of considerable interest in few-body physics because the low-energy $\bar{K}N$ interaction is considered to be sufficiently attractive to generate a quasibound state, the $\Lambda(1405)$, below $\bar{K}N$ threshold [1–3]. One of the recent theoretical achievements is the determination of the precise meson-baryon scattering amplitude, including the pole position of the $\Lambda(1405)$ [4]. Currently, a whole set of experimental data near the $\bar{K}N$ threshold is successfully described in the framework of chiral SU(3) coupled-channels dynamics [5, 6].

In order to study the $\bar{K}$ few-nucleon systems with established techniques for rigorous few-body calculations, it is desirable to construct a local meson-baryon potential. A detailed strategy for constructing such a local potential equivalent to chiral SU(3) dynamics has been developed in [7]. It is also shown that one can renormalize the effects of πY ($\equiv \pi\Sigma, \pi\Lambda$) channels to obtain a single-channel $\bar{K}N$ potential. By applying this strategy to the scattering amplitude in [5, 6], an elaborate

K. Miyahara
Department of Physics, Graduate School of Science, Kyoto University, Kyoto 606-8502, Japan

T. Hyodo (✉)
Yukawa Institute for Theoretical Physics, Kyoto University, Kyoto 606-8502, Japan
e-mail: hyodo@yukawa.kyoto-u.ac.jp

W. Weise
Physik-Department, Technische Universität München, 85748 Garching, Germany

N. A. Orr et al. (eds.), *Recent Progress in Few-Body Physics*, Springer Proceedings in Physics 238,
https://doi.org/10.1007/978-3-030-32357-8_153

single-channel $\bar{K}N$ potential has been derived in [8]. This potential is applied to the investigations of few-body kaonic nuclei [9] and kaonic deuterium [10].

While the single-channel $\bar{K}N$ potential exactly reproduces the two-body scattering amplitude, in the application to the few-body systems, dynamical effects of the coupled-channels (such as πY plus nucleons) are not properly included. For states near the $\bar{K}N$ threshold, the neglect of such components does not affect observables very much. However, the results in [9] suggest that the quasibound state of $\bar{K}$ and six nucleons appears near the threshold of $\pi\Sigma$ plus five nucleons. In such cases, the explicit treatment of coupled-channels dynamics is expected to give sizable contributions.

Here we construct the coupled-channels $\bar{K}N$-$\pi\Sigma$-$\pi\Lambda$ potential [11] equivalent to the scattering amplitude in [5, 6]. As an application within the two-body sector, we study the meson-baryon fractions in the $\Lambda(1405)$ by analyzing the wave function of the quasibound state.

153.2 Coupled-Channels Meson-Baryon Potential

We start from the coupled-channels Schrödinger equation for a nonrelativistic energy E,

$$\left[-\frac{\nabla^2}{2\mu_i}\delta_{ij} + \Delta M_i\,\delta_{ij} + V_{ij}^{\text{equiv}}(\boldsymbol{r}, E)\right]\psi_j(\boldsymbol{r}) = E\psi_i(\boldsymbol{r})\,, \tag{153.1}$$

where μ_i, ΔM_i, and ψ_i are the reduced mass, the energy difference from the reference threshold, and the wave function in channel i, respectively. The coupled-channels potential V_{ij}^{equiv} has a matrix form in channel basis with $i = \bar{K}N, \pi\Sigma$, and $\pi\Lambda$. In contrast to the single-channel $\bar{K}N$ potential which has a complex strength due to the absorption into the πY channels, the strengths of V_{ij}^{equiv} are real valued. In addition, V_{ij}^{equiv} has an energy dependence, which originates from the energy dependence of the chiral interaction and from the renormalization of higher energy channels such as $\eta\Lambda$ and $K\Xi$.

Our task is to determine V_{ij}^{equiv} such that the scattering amplitude from (153.1) reproduces the one in [5, 6]. The potential V_{ij}^{equiv} should be related to the interaction kernel in chiral SU(3) dynamics, but there is no direct way of converting the interaction kernel, given that the framework is different. We therefore introduce several matching conditions of V_{ij}^{equiv} and the interaction kernel, not only on the real energy axis but also in the complex energy plane, in order to systematically determine V_{ij}^{equiv}. For the details of the potential construction procedure, see [11].

For practical applications, it is useful to represent V_{ij}^{equiv} in a conveniently parametrized form. We use the following parametrization:

$$V_{ij}^{\text{equiv}}(\boldsymbol{r}, E) = e^{-r^2(1/2b_i^2+1/2b_j^2)} \sum_{\alpha=0}^{\alpha_{max}} K_{\alpha,ij} \left(\frac{E}{100\ \text{MeV}}\right)^{\alpha} . \tag{153.2}$$

The spatial distribution is assumed to be a Gaussian form with the range parameter b_i. The energy dependence is parametrized by a polynomial of order α_{max} with real coefficients $K_{\alpha,ij}$. These parameters are determined by the matching conditions mentioned above, and the results are tabulated in [11]. One remarkable fact is that the energy dependence of V_{ij}^{equiv} can be accurately parametrized by a second-order polynomial ($\alpha_{max} = 2$). This is in sharp contrast to the single-channel version in [11] where one needs α_{max} as large as 10 for sufficient accuracy. In other words, the explicit inclusion of the πY channels permits a parametrization of the potential strength with a more natural, much weaker energy dependence.

153.3 Compositeness of the $\Lambda(1405)$

In general, a resonance eigenstate is expressed by a pole of the scattering amplitude in the complex energy plane. In the case of the $\Lambda(1405)$, the resonance is expressed by two poles [4, 12]. From the coupled-channels potential, we find poles at $\sqrt{s} = 1424 - 27i$ MeV and $\sqrt{s} = 1380 - 81i$ with $\sqrt{s} = E + M_N + m_K$, consistently with the original scattering amplitude [5, 6]. The wave function of the $\Lambda(1405)$ can be obtained by evaluating (153.1) at these pole energies. Starting from the wave function it is possible to extract the properties of the resonance. It should be kept in mind that due to the unstable nature of resonances, the expectation value of operators (such as norm of the state) becomes complex.

Here we discuss the compositeness X_i and the "elementarity" Z [13] which are related to the norm of the wave function in channel i:

$$X_i = \langle \psi_i^\dagger \mid \psi_i \rangle, \quad Z = 1 - \sum_i X_i , \tag{153.3}$$

where $\langle \psi_i^\dagger \mid$ is a left eigenstate $\langle \psi_i^\dagger \mid H = \langle \psi_i^\dagger \mid E$. When we deal with resonances, $\langle \psi_i^\dagger \mid$ is different from $\langle \psi_i \mid$. The compositeness X_i represents the fraction of the channel i component in the total wave function. When the potential is energy dependent, the sum of X_is does not become unity, and the elementarity Z is introduced. Using the Feshbach projection method, Z can be expressed by the expectation value of the energy derivative of the potential [11]. Originally, Z has been introduced as a field renormalization constant of the bare state representing the elementary component of the composite system [14]. In the present context, it is understood as the contributions not explicitly included in the model space.

We show the results of $X_{\pi\Sigma}$, $X_{\bar{K}N}$, and Z for high-mass pole at $1424 - 27i$ MeV in Table 153.1. The results are consistent with those obtained by the single-channel potential [8], as well as with the evaluation by the residue of the pole [15]. In all cases, $X_{\bar{K}N}$ is close to unity and the others are almost zero. Hence, we conclude the $\bar{K}N$ dominance of the high-mass pole of the $\Lambda(1405)$.

Table 153.1 Compositeness X_i and elementarity Z for high-mass pole of the $\Lambda(1405)$ coupled-channels system.

Method	$X_{\pi\Sigma}$	$X_{\bar{K}N}$	Z
Coupled-channels potential	$-0.02-0.25i$	$1.01-0.13i$	$0.01+0.37i$
Single-channel potential [8]		$1.01-0.07i$	
Residue of the pole [15]	$-0.19-0.22i$	$1.14+0.01i$	$0.05+0.21i$

153.4 Summary

We have constructed a coupled-channels meson-baryon ($\bar{K}N \leftrightarrow \pi Y$) potential based on chiral SU(3) dynamics. The evaluation of the compositeness tells us that the high-mass pole of the $\Lambda(1405)$ is dominated by the $\bar{K}N$ component. This potential will be useful in order to shed new light on meson-baryon few-body systems with strangeness.

References

1. Dalitz, R.H., Tuan, S.F.: A possible resonant state in pion-hyperon scattering. Phys. Rev. Lett. **2**, 425 (1959). https://doi.org/10.1103/PhysRevLett.2.425
2. Dalitz, R.H., Tuan, S.F.: The phenomenological representation of $\bar{K}$-nucleon scattering and reaction amplitudes. Ann. Phys. **10**, 307 (1960). https://doi.org/10.1016/0003-4916(60)90001-4
3. Dalitz, R.H., Wong, T.C., Rajasekaran, G.: Model calculation for $Y_0^*(1405)$ resonance state. Phys. Rev. **153**, 1617 (1967). https://doi.org/10.1103/PhysRev.153.1617
4. Tanabashi, M., et al.: Review of particle physics. Phys. Rev. D **98**, 030001 (2018). https://doi.org/10.1103/PhysRevD.98.030001
5. Ikeda, Y., Hyodo, T., Weise, W.: Improved constraints on chiral SU(3) dynamics from kaonic hydrogen. Phys. Lett. B **706**, 63 (2011). https://doi.org/10.1016/j.physletb.2011.10.068
6. Ikeda, Y., Hyodo, T., Weise, W.: Chiral SU(3) theory of antikaon–nucleon interactions with improved threshold constraints. Nucl. Phys. A **881**, 98 (2012). https://doi.org/10.1016/j.nuclphysa.2012.01.029
7. Hyodo, T., Weise, W.: Effective $\bar{K}N$ interaction based on chiral SU(3) dynamics. Phys. Rev. C 77, 035204 (2008). https://doi.org/10.1103/PhysRevC.77.035204
8. Miyahara, K., Hyodo, T.: Structure of $\Lambda(1405)$ and construction of $\bar{K}N$ local potential based on chiral SU(3) dynamics. Phys. Rev. C **93**, 015201 (2016). https://doi.org/10.1103/PhysRevC.93.015201
9. Ohnishi, S., Horiuchi, W., Hoshino, T., Miyahara, K., Hyodo, T.: Few-body approach to structure of $\bar{K}$-nuclear quasi-bound states. Phys. Rev. C **95**, 065202 (2017). https://doi.org/10.1103/PhysRevC.95.065202
10. Hoshino, T., Ohnishi, S., Horiuchi, W., Hyodo, T., Weise, W.: Constraining the $\bar{K}N$ interaction from the $1S$ level shift of kaonic deuterium. Phys. Rev. C **96**, 045204 (2017). https://doi.org/10.1103/PhysRevC.96.045204
11. Miyahara, K., Hyodo, T., Weise, W.: Construction of a local $\bar{K}N-\pi\Sigma-\pi\Lambda$ potential and composition of the $\Lambda(1405)$. Phys. Rev. C **98**, 025201 (2018). https://doi.org/10.1103/PhysRevC.98.025201
12. Jido, D., Oller, J.A., Oset, E., Ramos, A., Meissner, U.G.: Chiral dynamics of the two $\Lambda(1405)$ states. Nucl. Phys. A **725**, 98, 181 (2003). https://doi.org/10.1016/S0375-9474(03)01598-7

13. Hyodo, T.: Structure and compositeness of hadron resonances. Int. J. Mod. Phys. A **28**, 1330045 (2013). https://doi.org/10.1142/S0217751X13300457
14. Weinberg, S.: Evidence That the Deuteron Is Not an Elementary Particle. Phys. Rev. **137**, B672 (1965). https://doi.org/10.1103/PhysRev.137.B672
15. Sekihara, T., Hyodo, T., Jido, D.: Comprehensive analysis of the wave function of a hadronic resonance and its compositeness. Prog. Theor. Exp. Phys. **2015**, 063D04 (2015). https://doi.org/10.1093/ptep/ptv081

Chapter 154
Studies of Hyperon Production in HADES—Cascade Production

Joanna Kuboś

Abstract Production of the doubly strange Ξ^- hyperon, known as a cascade particle, has been investigated in nuclear reactions in a wide energy range. However, only the results for the energies above the threshold for cascade production in elementary collisions were known for a long time. The first measurement of the subthreshold Ξ^- production was performed in heavy ion and nucleon-nucleus collisions by the HADES Collaboration. Because of the discrepancy between data and various model predictions deeper studies of the strangeness production, especially in elementary collisions near the threshold, are needed. In this paper simulations of the cascade production in proton-proton reaction at the beam kinetic energy of 4.5 GeV are presented. The results are relevant for the experiment planned by HADES with the upgraded setup during the FAIR-Phase 0 project.

154.1 Introduction

Processes of the strangeness production have been the subject of interest of nuclear physicists for a long time. Such studies are important for the investigations of the nuclear equation-of-state (EoS), neutron stars compositions or scattering processes in hot and dense matter. Especially the multistrange particles production yield, enhanced in nucleon-nucleus reactions in respect to the nucleon-nucleon, is a sign of the phase transition of the hadronic matter to the quark-gluon plasma state (QGP).

Production of cascade—a doublystrange Ξ^- hyperon—was measured by different experiments: in heavy ion collisions at LHC [2, 3], RHIC [4, 5], SPS [6, 7] and AGS [8], in nucleon-nucleon collisions at LHC [9], in nucleon-nucleus reactions at DESY [10] and SPS [11] and also at lowest energy in HADES: Ar+KCl@1.76 AGeV [12] and p+Nb@3.5 GeV [1] (experimental data shown in the Fig. 154.1). The latter

Joanna Kuboś: for the HADES Collaboration.

J. Kuboś (✉)
Institute of Nuclear Physics PAS, Kraków, Poland
e-mail: joanna.kubos@ifj.edu.pl

N. A. Orr et al. (eds.), *Recent Progress in Few-Body Physics*,
Springer Proceedings in Physics 238,
https://doi.org/10.1007/978-3-030-32357-8_154

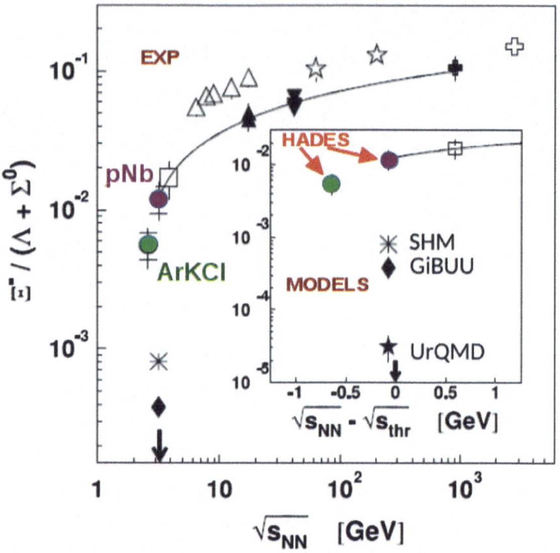

Fig. 154.1 The yield ratio $\Xi^-/(\Lambda + \Sigma^0)$ as a function of $\sqrt{s_{NN}}$ or $\sqrt{s_{NN}} - \sqrt{s_{thr}}$ (inset). Full dots—data measured by HADES. (figure from [1])

measurements have shown the enhanced Ξ^- production with respect to the theoretical predictions, obtained using various transport models [13–17]. The cascade production in the HADES energy range was initially modeled as processes of strangeness-exchange: $\bar{K}Y \rightarrow \pi\Xi$, where $Y = \Lambda, \Sigma$ [18]. However, the observed production of cascade in p+Nb reactions questions this hypothesis.

The unexpected high production rate of cascade in the Ar+KCl experiment [12] seems to be partially explained by adding an additional process to the UrQMD model—a hyperon-hyperon scattering $YY \rightarrow N\Xi$ [19, 20], but the theoretical Ξ^- abundance is still significantly lower than the experimental data. In the case of the p-Nb reaction differences between model predictions with and without YY scattering is negligible [1]. Therefore, a new mechanism of the cascade production in the nucleon-nucleus interaction was proposed—a subthreshold cascade production via high-mass baryon resonances [21].

To resolve this so called "HADES puzzle", i.e. the cascade overproduction near the threshold, and improve our knowledge about elementary processes responsible for the strangeness creation, it is necessary to measure the Ξ^- production in a nucleon-nucleon collision at the near-threshold energies.

In this proceedings result of simulations of the Ξ^- production in elementary collisions $pp \rightarrow \Xi K^+K^+p$ at $E = 4.5$ GeV will be presented.

154.2 HADES Detector

HADES (High Acceptance DiElectron Spectrometer) is a versatile magnetic spectrometer [22]. Its main purpose is to detect dielectrons and hadrons in pion, proton and heavy-ion induced reactions. Geometrically the detector is divided into six sectors placed symmetrically around the beam axis. HADES consists of a Ring Imaging gas Cherenkov detector (RICH) for electron-hadron discrimination, a set of six superconducting coils producing a toroidal field, four sets of multiwire drift chambers (MDC)—two before and two after the magnetic field serving as a tracking system, and a multiplicity and electron trigger array consisting of: a time-of-flight wall (TOF) and a RPC wall and the new Electromagnetic Calorimeter (ECAL). Identification of the charged particles—pions, kaons and protons is achieved by combining time-of-flight and energy loss measurements over a large momentum range. The detector system covers 85% of the polar angle in the range from 18° to 85° with the momentum resolution $\frac{\Delta p}{p} \approx 2 - 4\%$.

The physics program of HADES is mostly focused on studies of the QCD matter phase diagramme in the region of the high baryonic potential, the investigation of hadron properties in nuclei and in the hot and dense hadronic matter.

154.2.1 Forward Detector

Currently the HADES detector is upgraded by the new and faster readout electronics (DAQ) and the Forward Detector for tracking in very forward angles, which extends the angular coverage of the detection setup to the forward angles $\theta \in [0.5°; 6.5°]$. A schematic picture of the upgraded HADES is shown in Fig. 154.2.

The cascade in HADES is reconstructed from a two step decay process: (1) $\Xi^- \rightarrow \Lambda\pi^-$, (2) $\Lambda \rightarrow p\pi^-$. From the kinematic at $E = 4.5$ GeV, protons produced in the Λ decay are emitted in very forward angles (see Fig. 154.3). Therefore with the use of the currently operating HADES setup one is able to reconstruct only about 9% of all possible protons and, in consequence, only very small fraction of all cascade decays. To improve the count rate of the Ξ^- it is necessary to detect also the protons with scattering angles below $\theta < 18°$.

The Forward Detector consists of two stations of Straw Tube Trackers (STS1 and STS2) for a precise particles tracking [23] and a set of Resistive Plate Chambers (RPC) for the time-of-flight measurement. Due to the fact that the Forward Detector is placed in the region with no magnetic field full PID is not possible. Owing to this in the simulation scheme all charged particles detected in the Forward Detector are treated as protons (the mass of such particles is fixed at a proton mass).

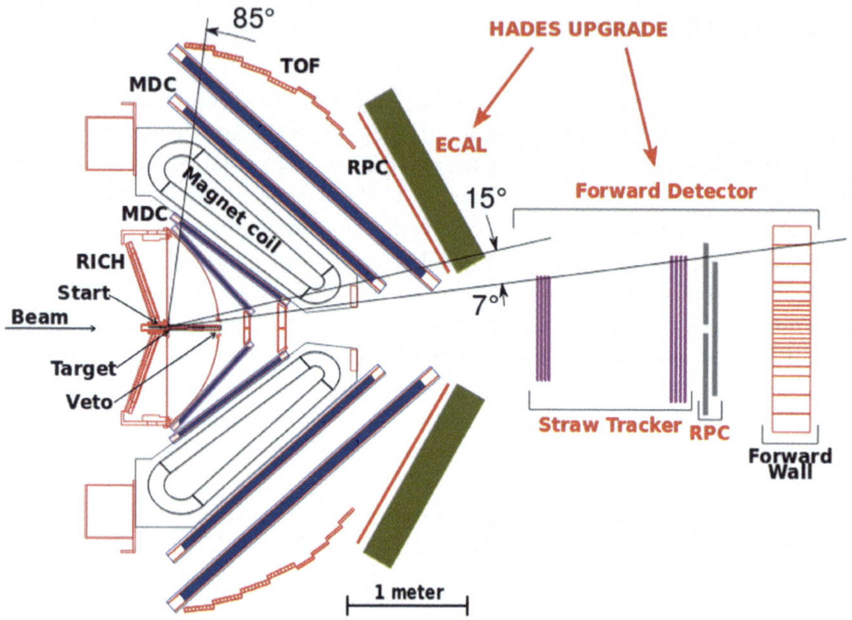

Fig. 154.2 Scheme of the upgraded HADES setup together with the Forward Detector

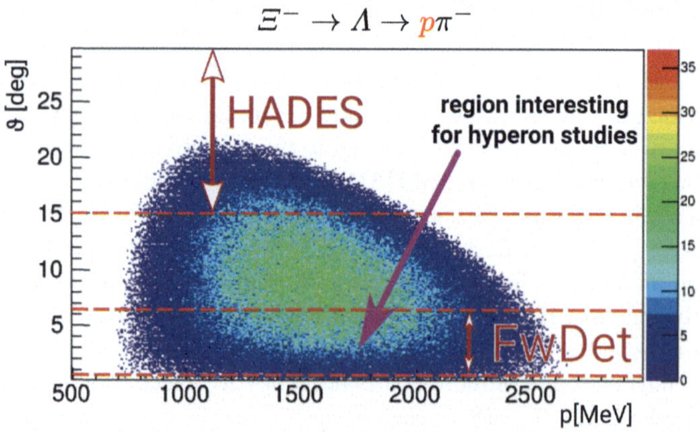

Fig. 154.3 Angular distribution of protons produced in the Λ decay, which is the second step of the cascade decay. The angular acceptance of the HADES detector before upgrade and of the Forward Detector is marked

Table 154.1 List of the simulated reaction channels together with the cross-sections.

No.	Reaction	Cross-section [μb]
0.	$K^+K^+p\Xi^-$	4.8
1.	$pp2\pi^-2\pi^+$	600
2.	$p\Lambda K_s^0\pi^+$	100
3.	$p\Lambda K^+\pi^+\pi^-$	30
4.	$n\Lambda K_s^0 2\pi^+$	30
5.	$p\Sigma^0 K_s^0\pi^+$	20
6.	$pp2K_s^0$	20

154.3 Simulations and Signal Reconstruction

The cascade production channel: $pp \rightarrow \Xi^- K^+ K^+ p$ as well as the most contributing background channels have been simulated with the use of the PLUTO Monte-Carlo event generator [24]. The simulated channels are listed in Table 154.1. The geometry and detector response have been simulated with the use of a GEANT3-based software considering the upgraded HADES setup.

154.3.1 Steps of the Analysis

The cascade is produced in a *pp* reaction accompanied by two positive kaons and one proton. Ξ^- decays to the negative pion and the Λ ground state, which subsequently decays into proton and π^-. Detection of two negative pions in HADES and one charged particle in the Forward Detector (treated as a proton) is considered as a candidate for the Ξ^- decay. In the first step the four-momentum and the decay vertex of the Λ hyperon ($\Lambda \rightarrow p\pi^-$) are reconstructed by the combination of the negative pion detected in HADES ($\theta > 18°$) with the proton detected in the Forward Detector ($\theta < 6.5°$). Next, the same observables are calculated for Ξ^- ($\Xi^- \rightarrow \Lambda\pi^-$) obtained from the momentum vectors of Λ reconstructed in the first step and another π^- detected in HADES (not used in the Λ reconstruction).

154.3.2 Topological Cuts

During the analysis following topological cuts have been applied to reduce background: (1) Minimum Tracks Distance for Λ decay vertex ($MTD\_L < 25$ mm)—a value of the shortest distance between p and π^- tracks; (2) the z coordinate of the Λ decay vertex (VERTz_L ∈ [-20;300] mm)—Λ decay vertex is the middle point of the shortest section drawn between p and π^- tracks; (3) Minimum Tracks Distance

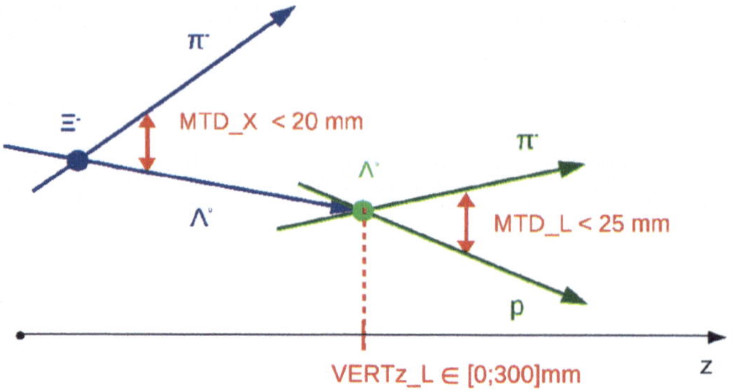

Fig. 154.4 The topological cuts applied to the cascade reconstruction

for Ξ^- decay vertex (MTD_X < 20 mm)—a value of the shortest distance between reconstructed Λ and π^- tracks. The applied cuts are presented in Fig. 154.4 together with the optimized values.

154.3.3 Results

The results of the analysis are presented in Fig. 154.5 with the reconstruction of the invariant mass of $p\pi^-$ in the top panel and $\Lambda\pi^-$ (after a cut on Λ peak in $p\pi^-$ invariant mass) in the bottom panel. The reconstructed candidates in the range of $\pm 3\sigma$ from the peak position have been accepted as a $\Lambda(1115)$ and Ξ^- signal candidates, respectively. Red dotted lines represent a combinatorial background—an appropriately normalized sum of the all miss-identified protons in the Forward Detector and pions in HADES.

The count rate for the reaction of the cascade production in *pp* scattering at E = 4.5 GeV considering the upgraded HADES setup (beam rate = 10^8 part/s, luminosity = $1.4 \cdot 10^{32}$ cm^{-2}s^{-1}) was estimated to $11.9 \cdot 10^4$ part/day.

154.4 Summary and Outlook

The Ξ^- production reaction has been simulated together with the most probable background sources. To reduce the background the topological cuts have been defined and optimized. The calculated count rate expected for the upcoming experiment shows that one can expect sufficient statistics.

Hardware and software for the HADES upgrade are currently being prepared. The Forward Detector is under construction and the first in-beam test with full-size prototypes are scheduled for the beginning of 2019. The experiment with a proton

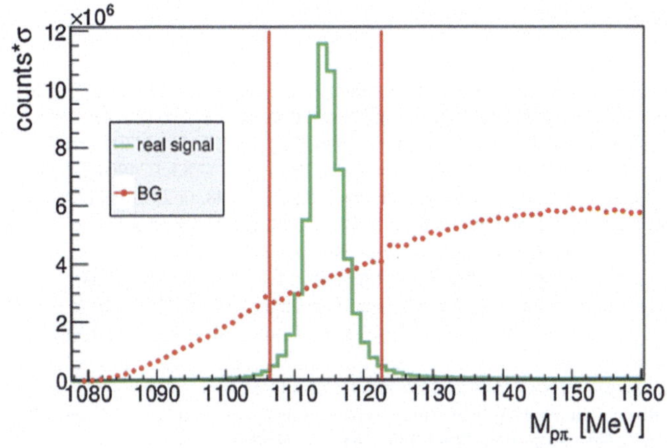

(a) invariant mass of the reconstructed Λ

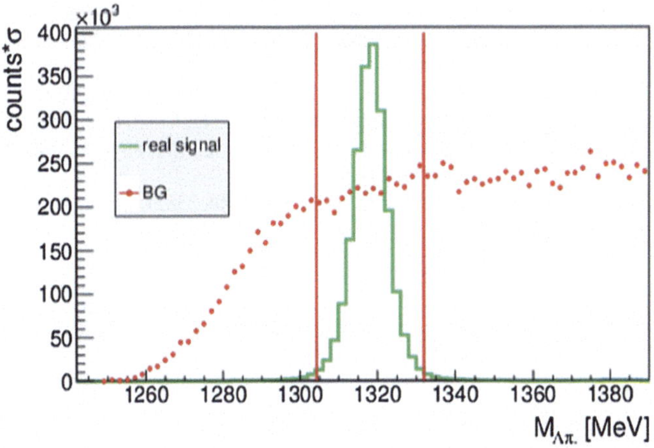

(b) invariant mass of the reconstructed Ξ^-

Fig. 154.5 Reconstructed signal of Λ (top panel) and cascade (bottom panel)—green solid line and the combinatorial background—red dotted line. Vertical lines—$\pm 3\sigma$ of the invariant mass peak

beam is planned for 2020. Measurement of the proton-proton reaction, investigated in this work, should lead us to the pioneering observation of the multistrangeness production in the elementary collisions.

Acknowledgements This project has received funding from the European Union's Horizon 2020 research and innovation programme under the Marie Skłodowska-Curie grant agreement No 665778 National Science Centre, Poland 2016/23/P/ST2/04066 POLONEZ.

References

1. Agakishiev, G., et al. (HADES Collaboration): Phys. Rev. Lett **114**, 212301 (2015)
2. Abelev, B., et al. (ALICE Collaboration): Phys. Lett. B **728**, 216 (2014)
3. Abelev, B., et al. (ALICE Collaboration): Phys. Rev. Lett. **111**, 222301 (2013)
4. Aggarwal, M.M., et al. (STAR Collaboration): Phys. Rev. C **83**, 024901 (2011)
5. Adams, J., et al. (STAR Collaboration): Phys. Rev. Lett. **98**, 062301 (2007)
6. Antinori, F., et al. (NA57 Collaboration): Phys. Lett. B **595**, 68 (2004)
7. Alt, C., et al. (NA49 Collaboration): Phys. Rev. C **78**, 034918 (2008)
8. Chung, P., et al. (E895 Collaboration): Phys. Rev. Lett. **91**, 202301 (2003)
9. Aamodt, K., et al. (ALICE Collaboration): Eur. Phys. J. C **71**, 1594 (2011)
10. Agari, M. (HERA-B), Ph.D. thesis, Universität Dortmund, [CERN-THESIS-2006-046, 2006 (unpublished)]
11. Antinori, F., et al. (NA57 Collaboration): J. Phys. G **32**, 427 (2006)
12. Agakishiev, G., et al. (HADES Collaboration): Phys. Rev. Lett. **103**, 132301 (2009)
13. Bass, S.A., et al.: Prog. Part. Nucl. Phys. **41**, 255 (1998)
14. Bleicher, M., et al.: J. Phys. G **25**, 1859 (1999)
15. Buss, O., et al.: Phys. Rep. **512**, 1 (2012)
16. Weil, J., van Hees, H., Mosel, U.: Eur. Phys. J. A **48**, 111 (2012)
17. Wheaton, S., Cleymans, J., Hauer, M.: Comput. Phys. Commun. **180**, 84 (2009)
18. Chen, Lie-Wen, Ko, Che Ming, Tzeng, Yiharn: Phys. Lett. B **584**, 269–275 (2004)
19. Li, F., et al.: Phys. Rev. C **85**, 064902 (2012)
20. Graef, G., Steinheimer, J., Li, F., Bleicher, M.: Phys. Rev. C **90**, 064909 (2014)
21. Steinheimer, J., Bleicher, M.: J. Phys. G: Nucl. Part. Phys. **43**, 015104 (2016)
22. The HADES Collaboration, Agakichiev, G., Agodi, C., et al.: Eur. Phys. J. A **41**, 243 (2009)
23. Smyrski, J., et al.: JINST **13**, P06009 (2018)
24. Frohlich, I., et al.: 2007 PoS ACAT2007 076 (Preprint 0708.2382)

Chapter 155
Universality of Two Neutrons and One Flavored Meson in Low-Energy Effective Theory

Udit Raha

Abstract We examine the universal physics associated with the s-wave system consisting of two neutrons and a flavored meson (K^- or D^0 meson) using the framework of low-energy effective theory. Through an extrapolation to an unphysical mass domain between the strange and the charm limits, we present an idealized scenario in which the system may be tuned to meson-neutron unitarity, and thereby Efimov effect can be manifest in such three-body systems.

155.1 Introduction

Few-body systems with a flavored meson (antikaon or the D^0 meson) and nucleons are of great interest in the study of nuclear physics. In the strangeness sector, the existence of the $\Lambda(1405)$ resonance below the $\bar{K}N$ threshold implies that the $\bar{K}N$ ($I = 0$) interaction is strongly attractive with which the antikaon could be bound in nuclei. Among others, the $\bar{K}NN$ system has been the most intensively studied three-body kaonic bound states. Analogous arguments hold for the charm sector where the charmed resonance $\Lambda_c(2595)$ suggests a fairly strong DN ($I = 0$) interaction. Here we focus on the K^-nn and D^0nn systems with $J = 0$ and $I = 3/2$. It is interesting to apply few-body techniques to study the low-energy universal aspects of kaonic and D meson systems, and in particular investigate whether *Efimov effect* [1, 2] can be responsible leading to possible three-body bound state formation.

Generally, in low-energy hardron physics where momentum scales are much smaller than m_π, all physics involving explicit pion degrees of freedom remain unresolved leading to the idea of a *pionless* effective theory [3–5]. Such a EFT framework is most suitable for the study of universal features of the three-body system under the assumption that the two-body physics is determined only by the scattering lengths that is *finely tuned*. Incorporated with a renormalization group analysis of short-distance contact interactions, an EFT framework allows a natural mechanism of fine-tuning

---

U. Raha (✉)
Department of Physics, Indian Institute of Technology Guwahati, Assam, India
e-mail: udit.raha@iitg.ac.in

N. A. Orr et al. (eds.), *Recent Progress in Few-Body Physics*,
Springer Proceedings in Physics 238,
https://doi.org/10.1007/978-3-030-32357-8_155

two-body parameters to probe the *unitary limit* of low-energy two-body interactions. In this way, complex three-body universal dynamics are easily investigated, and in particular, many recent EFT based analyses have used similar frameworks to predict exotic bound states in a variety of three-body nuclear and hypernuclear cluster systems. Our primary objective of this contribution is to present a brief report of our EFT analysis in an idealized limit to illuminate certain aspects of remnant two- and three-body universal physics that may be associated with bound *dimer* and *trimer* state formation in the K^-nn and D^0nn systems. Details of the methodology can be found in our earlier works [6, 7], and references therein.

155.2 Two- and Three-Body Dynamics

In the two-body meson-neutron sectors, we have employed a contact interaction EFT based model with flavor symmetry. In the strangeness sector, the *Weinberg-Tomozawa* contact interaction model is known to successfully describes the chiral SU(3) dynamics of the $\bar{K}N$ system where the $\Lambda(1405)$ resonance is generated dynamically in the $I = 0$ channel. Analogously, in the charm sector, the SU(4) generalization of this approach is used to dynamically generate the $\Lambda_c(2595)$ resonance ($I = 0$). In this work we describe both $\bar{K}N$ and DN systems in the $I = 1$ channel within a common unified framework of a *dynamical coupled-channel model*. Here we introduce a parameter $0 \leq x \leq 1$ which controls the extrapolation from strangeness ($x = 0$) to charm ($x = 1$) limits. The details of the model construction are given in [7].

Our model analysis reveals that, while on the one hand the K^-n system has a weakly attractive scattering length, not strong enough to form a bound state, on the other hand the D^0n system can support the quasi-bound state $\Sigma_c(2800)$ below the threshold. We, however, note that the model extracted meson-neutron scattering lengths are not large enough to exhibit resonant phenomena. If we perform an extrapolation of the K^-n interaction to the D^0n interaction by changing the flavored meson mass (or the quark mass) from the strange to the charm limits, we can expect the existence of an unphysical mass region where a very shallow bound state is formed when the magnitude of the scattering length becomes infinitely large. This yields an universal window around the unitary limit of the meson-neutron interactions.

To elucidate a possible scenario to access the unitary limit, we consider a *zero coupling limit* (ZCL) of our model in which the channel couplings are switched off artificially. With this idealization, the coupled-channel problem reduces to a single-channel scattering of the K^-n (D^0n) system. In this scenario, the K^-n system is again found to have no bound state, while the previous full-model quasi-bound D^0n state now becomes bound. Correspondingly, the full-model complex scattering lengths, namely, $a_{0,K^-n} = -0.135 - i0.410$ fm and $a_{0,D^0n} = 0.764 - i0.615$ fm, now become real and negative (positive) in the K^-n (D^0n) channel, namely, $a^{\text{ZCL}}_{0,K^-n} = -0.394$ fm and $a^{\text{ZCL}}_{0,D^0n} = 4.141$ fm. These values of the ZCL scattering lengths strongly suggest that the meson-neutron interactions become resonant in the intermediate unphysical region $0 < x < 1$.

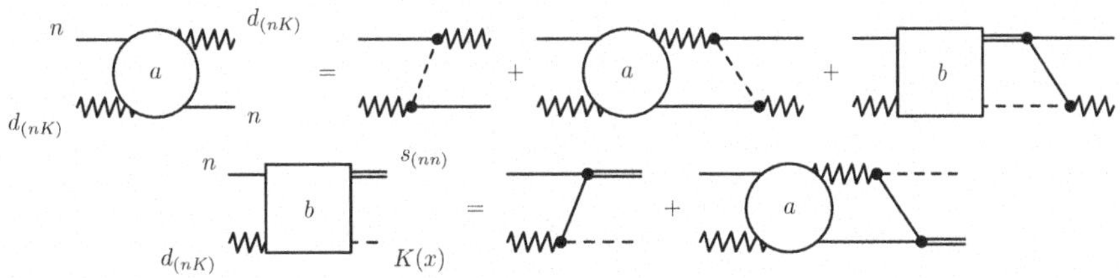

Fig. 155.1 Integral equations (strangeness sector) with the three-body contact terms omitted for brevity. A solid (dashed) line represents the $n(K^-)$ field, and the double (zigzag) line stand for the dressed field $s_{(nn)}$ ($d_{(nK)}$). Figure adapted from [7]

In the three-body sector, we report our results for the meson-neutron-neutron dynamics in the idealized ZCL limit, neglecting possible influence of decay channels. Here we use a non-relativistic *leading order* (LO) pionless EFT [3–5] with the physical $^1S_0$ nn scattering length, $a_{s(nn)} = -18.63$ fm, and the aforementioned $^2S_{1/2}$ ZCL meson-neutron scattering lengths as the input parameters. In particular, it is convenient to introduce the *dihadron* fields [2] to unitarize the two-body sector, especially in the vicinity of a non-trivial RG fixed point of the two-body couplings whenever the system is close to the unitary limit. For example, in our case there is a spin-singlet *nn-dibaryon* $s_{(nn)}$, and a spin-doublet nK^-*-dihadron* $d_{(nK)}$ in the strangeness sector. The resulting set of integral equations is diagrammatically represented in Fig. 155.1 and very similar to the so-called *Skornyakov-Ter-Martirosian* (STM) equations [8].

In addition, a LO three-body interaction counter-term must be introduced for renormalization purpose. This is because the asymptotic analysis of the integral equations leads to a RG limit cycle associated with the breakdown of scale invariance into a discrete scaling symmetry, parametrized in terms of a transcendental number, $s_0^\infty = 1.03069\ldots(1.02387\ldots)$, for the K^-nn (D^0nn) system. In fact by an extrapolation of the integral equations to the unphysical domain of the meson mass, $m(0 \leq x \leq 1)$ between the strange and the charm limits, the asymptotic limit cycle parameter values $s_\infty^0(x)$ can be shown to be *continuously connected*. We thereby conclude that *formally* Efimov effect must be manifest in the three-body system, not only at the physical limits ($x = 0, 1$), but at *all* unphysical points in the intermediate quark mass domain. The explicit expressions of the non-relativistic effective Lagrangian, integral equations and the details of the three-body asymptotic analysis can be found in [7].

Finally, we present our non-asymptotic results obtained by numerical solving the homogeneous part of the coupled integral equation by introducing a sharp momentum cut-off Λ to regularize the otherwise ill-defined integral equations in the UV limit ($\Lambda \to \infty$). In this regime Λ is chosen low enough where scales of the scattering lengths and three-body binding energies become significant. Furthermore, the LO three-body counter-term Lagrangian has a scale dependent a priori unknown coupling g_3 which can not be currently fixed due to the absence of a three-body datum. Thus, Λ and g_3 constitute two extra free parameters in the theory, without a knowl-

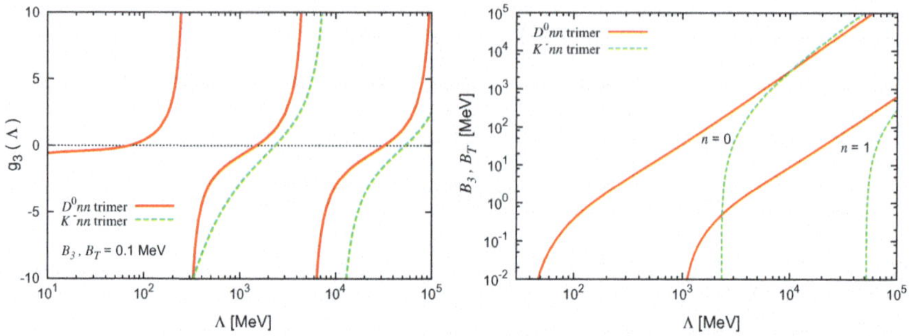

Fig. 155.2 Left: RG Limit cycle for $g_3(\Lambda)$ for fixed three-body binding energy (B.E) $B_3 = B_T = 0.1$ MeV in ZCL. Right: The B.E. of the ground ($n = 0$) and first ($n = 1$) excited level states in ZCL as a function of Λ, excluding the three-body counter-term. For the K^-nn system B_3 is measured with respect to the three-particle break-up threshold, and for the D^0nn system $B_T = B_3 - E_D$ MeV is measured with respect to the particle-dimer break-up threshold energy $E_D = 1.82$ MeV. Figure adapted from [7]

edge of which no definitive prediction of three-body bound state formation is possible. Instead, here we have pursued an RG analysis by studying the scale variation of $g_3(\Lambda)$, as shown in Fig. 155.2 (left). Thus, we confirm an approximate RG limit cycle behavior of $g_3(\Lambda)$ with a quasi-log-periodicity of the form, $g_3(\Lambda^{(1)}) = g_3(\Lambda^{(n+1)})$ with $\Lambda^{(n+1)} \approx \Lambda^{(1)} \exp(n\pi/s_0)$, where s_0 is a real three-body parameter reminiscent of the corresponding asymptotic limit cycle value of s_0^∞. In the same figure (right) we also display our results for the cut-off scale dependence of the three-body binding energies in the physical strange and charm limits, with the unknown coupling g_3 excluded for simplicity. We find that the spectrum has a typical Efimov-like character with increasing binding energies with increasing Λ. The various level states emerge in order from the zero energy threshold starting from a deepest (ground) state that appears at a certain critical cut-off value, i.e $\Lambda_c^{(0)} \simeq 38\,\text{MeV}(2.3\,\text{GeV})$ for the D^0nn (K^-nn).

155.3 Conclusion

It clearly emerges from our results that with the much smaller critical cut-offs for the D^0nn level states than K^-nn, under the idealized assumptions the D^0nn Efimov-like trimer states are manifested much more easily. While for the K^-nn system, no physically realizable mechanism in the context of a low-energy EFT can generate sufficient interaction strength to form Efimov bound states. The much steeper Efimov spectrum for the K^-nn system is consistent with the fact that such Borromean trimers are extremely difficult to form.

References

1. Efimov, V.: Phys. Lett. B **33**, 563 (1970)
2. Braaten, E., Hammer, H.-W.: Phys. Rep. **428**, 259 (2006)
3. Kaplan, D.B., Savage, M.J., Wise, M.B.: Phys. Lett. B **424**, 390 (1998)
4. Kaplan, D.B., Savage, M.J., Wise, M.B.: Nucl. Phys. B **534**, 329 (1998)
5. van Kolck, U.: Nucl. Phys. A **645**, 273–302 (1999)
6. Ando, S.-I., Raha, U., Oh, Y.: Phys. Rev. C **92**, 024325 (2015)
7. Raha, U., Kamya, Y., Ando, S.-I., Hyodo, T.: Phys. Rev. C **98**, 034002 (2018)
8. Skornyakov, G.V., Ter-Martirosyan, K.A.: Sov. Phys. JETP **4**, 648 (1957)

Chapter 156
Dark Matter Bound States from Three-Body Recombination

Eric Braaten, Daekyoung Kang and Ranjan Laha

Abstract The small-scale structure problems of the universe can be solved by self-interacting dark matter that becomes strongly interacting at low energies. A particularly predictive model is resonant short-range self-interactions, with a dark-matter mass of about 19 GeV and a large S-wave scattering length of about 17 fm. Such a model makes definite predictions for the few-body physics of weakly bound clusters of the dark-matter particles. We calculate the production of two-body bound clusters by three-body recombination in the early universe under the assumption that the dark matter particles are identical bosons, which is the most favorable case for forming larger clusters. The fraction of dark matter in the form of two-body bound clusters can increase by as much as 4 orders of magnitude when the dark-matter temperature falls below the binding energy, but its present value remains less than 10^{-6}.

156.1 Introduction

The model with collisionless cold dark matter and a cosmological constant provides an excellent description of the large-scale structure of the universe, but it has encountered problems at smaller scales associated with galaxies and clusters of galaxies. The problems involve the dark-matter distribution in the cores of galaxies and the

E. Braaten (✉)
Department of Physics, The Ohio State University, Columbus, OH 43201, USA
e-mail: braaten.1@osu.edu

D. Kang
Key Laboratory of Nuclear Physics and Ion-beam Application (MOE)
and Institute of Modern Physics, Fudan University, Shanghai 200433, China
e-mail: dkang@fudan.edu.cn

R. Laha
PRISMA Cluster of Excellence and Mainz Institute for Theoretical Physics,
Johannes Gutenberg-Universität Mainz, 55099 Mainz, Germany
e-mail: ranjalah@uni-mainz.de

N. A. Orr et al. (eds.), *Recent Progress in Few-Body Physics*,
Springer Proceedings in Physics 238,
https://doi.org/10.1007/978-3-030-32357-8_156

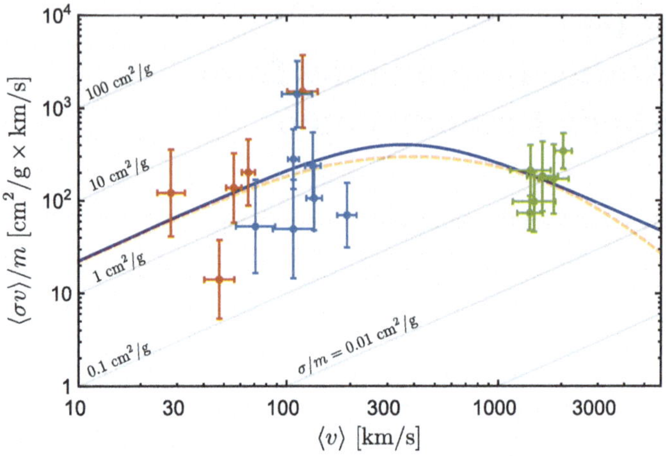

Fig. 156.1 Self-interaction reaction rate $\langle v\, \sigma_{\text{elastic}} \rangle$ for dark matter particles as a function of their mean velocity $\langle v \rangle$ (adapted from [3]). The data points from [2] are for dwarf galaxies (red), low-surface-brightness galaxies (blue), and galaxy clusters (green) [2]. The curves are the best fits to the model of [2] (dashed) and to (156.1) (solid). The diagonal lines are energy-independent cross sections

properties of satellite galaxies. They can all be solved by self-interacting dark matter that is strongly interacting at low energies [1].

In [2], Kaplinghat, Tulin, and Yu deduced self-interaction reaction rates $\langle v\sigma_{\text{elastic}} \rangle$ and mean velocities $\langle v \rangle$ of dark matter particles for a number of dwarf galaxies, low-surface-brightness galaxies, and clusters of galaxies. Their results are shown in Fig. 156.1. They fit their results with a simple self-interacting dark matter model with 3 parameters: the dark matter mass m_χ, a dark mediator mass μ, and a Yukawa coupling α'. Their best fit for the masses with fixed Yukawa coupling $\alpha' = 1/137$ was $m_\chi = 15$ GeV and $\mu = 17$ MeV [2].

In [3], we showed that the results in Fig. 156.1 can be fit equally well by a simpler self-interacting dark matter model with 2 parameters. The model has resonant short-range interactions with an S-wave resonance close to the scattering threshold [4]. The parameters are the dark matter mass m_χ and the scattering length a. This model has been applied previously to the direct detection of dark matter [5, 6]. The self-interaction reaction rate as a function of the velocity v is

$$v\, \sigma_{\text{elastic}}(v) = \frac{8\pi a^2 v}{1 + (a m_\chi/2)^2 v^2}. \tag{156.1}$$

The best fit to the results in Fig. 156.1 is $m_\chi = 19$ GeV and $a = \pm 17$ fm [3].

The dark matter could all be in the form of individual dark matter particles d, but some (or all) of it could be bound into few-body clusters d_N, which we call *dark nuclei*. There are two basic formation mechanisms for larger dark nuclei. If there is a light mediator γ_d for dark matter self-interactions, larger dark nuclei can be formed by radiative reactions: $d + d_{N-1} \to d_N + \gamma_d$. If there is no light mediator, as in our resonant short-range interaction model, larger dark nuclei must be formed instead by rearrangement reactions, such as 3-body recombination: $d + d + d_{N-1} \to d_N + d$. The formation of dark deuterons d_2 is a bottleneck for formation of larger dark nuclei d_N.

156.2 Few-Body Physics

The low-energy two-body physics of particles with resonant short-range interactions is very simple. It is completely determined by the large scattering length a. The cross section for the elastic scattering reaction $d + d \to d + d$ is given in (156.1). If a is negative, there are no 2-body bound states. If a is positive, there is a single 2-body bound state d_2 that we call the dark deuteron. Its binding energy is $E_2 = 1/m_\chi a^2$.

If the particles are identical bosons, the 3-body physics is much more intricate [3]. It is determined not only by the large scattering length a, but also by a 3-body parameter. There is a sequence of 3-body bound states called Efimov states. In the limit $a \to \pm\infty$, there are infinitely many Efimov states with an accumulation point at the 3-boson threshold and with the binding energy of each successive Efimov state smaller by a factor of $22.7^2 = 515$. Three-boson reaction rates also have remarkable behavior. They depend log-periodically on a 3-body parameter a_+ with discrete scaling factor 22.7. If $a > 0$, a simple example is the rate for the 3-body recombination reaction $d + d + d \to d_2 + d$ at 0 collision energy:

$$R(E = 0) = \frac{399.8 \sin^2[s_0 \log(a/a_+)]}{1 - 0.00717 \sin^2[s_0 \log(a/a_+)]} a^4/m_\chi, \tag{156.2}$$

where $s_0 = 1.00624$. The 3-body recombination rate at nonzero collision energy E has been calculated in [7] and in [8].

We consider a gas consisting of dark matter particles d with number density n_1 and dark deuterons d_2 with number density n_2 in thermal equilibrium at temperature T. The rate of change in the number density of dark deuterons is

$$\frac{d}{dt} n_2 = +K_3(T)\, n_1^3 - K_2(T)\, n_1 n_2, \tag{156.3}$$

where $K_3(T)$ and $K_2(T)$ are the rate constants for 3-body recombination and for the dark deuteron breakup reaction $d + d_2 \to d + d + d$. These rate constants were calculated in [8]. The results for $K_3(T)$ are shown in Fig. 156.2.

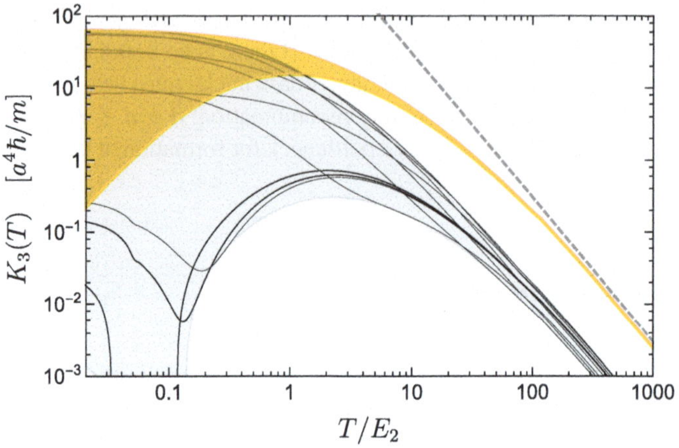

Fig. 156.2 Rate coefficient $K_3(T)$ for three-body recombination as a function of the temperature T (adapted from [3]). The upper band is the envelope of $K_3(T)$ for all possible values of the three-body parameter a_+. The dashed line is the extrapolation from the scaling behavior at high temperature. The lower band is the envelope of the $J = 0$ contribution to $K_3(T)$ for all possible values of a_+. The curves inside the lower band are for 8 values of a_+

156.3 Dark Matter in the Early Universe

We calculate the formation of dark deuterons in the Hubble expansion of the early universe, taking into account the 3-body recombination and dark deuteron breakup reactions. We calculate the number densities n_1 and n_2 as functions of the redshift z, which is a convenient time variable. The initial condition is that n_2 is negligible when dark matter decouples at a redshift of about $z_{dc} \approx m_\chi/20kT_{cmb}$, where T_{cmb} is the present temperature of the cosmic microwave background [9]. For $m_\chi = 19$ GeV, this redshift is $z_{dc} \approx 10^{13}$. The dark-matter temperature as a function of z is

$$T(z) = T_{cmb}\,\frac{(1+z)^2}{1+z_{dc}}\,. \tag{156.4}$$

The total number density of dark matter is determined by the present mass density ρ_{cdm} of cold dark matter:

$$n_1(z) + 2n_2(z) = \frac{\rho_{cdm}}{m_\chi}\,(1+z)^3\,. \tag{156.5}$$

It is convenient to express our results in terms of the dark deuteron mass fraction:

$$f_2(z) = 2\,n_2(z)/[n_1(z) + 2n_2(z)]\,. \tag{156.6}$$

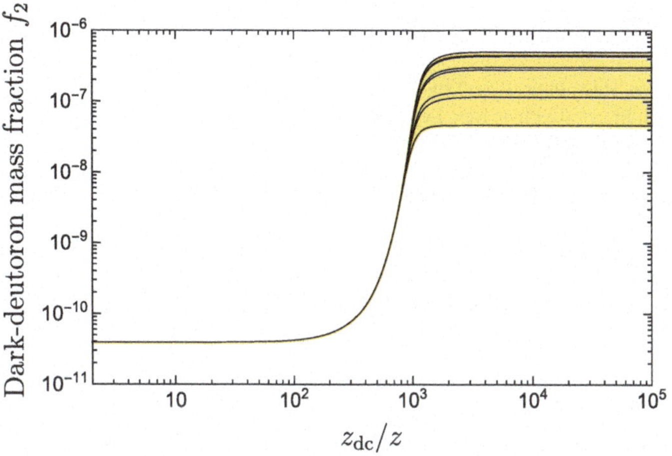

Fig. 156.3 Dark-deuteron mass fraction $f_2(z)$ as a function of z_{dc}/z for $m_\chi = 19$ GeV and $a =$ 17 fm (adapted from [3]). The band is the envelope of all possible values of a_+. The curves are for 8 values of a_+

We assume the dark matter particles are identical bosons with mass $m_\chi = 19$ GeV, large scattering length $a = 17$ fm, and unknown 3-body parameter a_+. The dark deuteron mass fraction $f_2(z)$ is shown as a function of the redshift variable z_{dc}/z in Fig. 156.3. The fraction increases by 3 or 4 orders of magnitude around the redshift $10^{-3} z_{dc}$ when kT is equal to the binding energy $E_2 = 7$ keV of the dark deuteron. At smaller $1/z$, there is a plateau in f_2 at about 4×10^{-11} from equilibrium between recombination and breakup. The final fraction $f_2(0)$ depends log-periodically on a_+, ranging from 5×10^{-8} to 5×10^{-7}. It can be increased by ignoring the data from clusters of galaxies in Fig. 156.1, which means keeping $a = 17$ fm but allowing m_χ to decrease. The final fraction $f_2(0)$ can be increased to about 10^{-2} for $m_\chi = 0.4$ GeV. If m_χ is smaller, $f_2(0)$ is sensitive to the range of self-interactions.

In summary, the small-scale structure problems of the universe can be solved by a self-interacting dark matter model with resonant S-wave interactions, with parameters $m_\chi \approx 19$ GeV and $a \approx \pm 17$ fm. Dark nuclei d_N must be produced by rearrangement reactions, such as 3-body recombination: $d + d + d_{N-1} \to d_N + d$ The most favorable case for producing dark nuclei larger than the dark deuteron is for the dark matter particles to be identical bosons. We found that a significant fraction of dark deuterons cannot be formed in the early universe by 3-body recombination. Since the formation of the dark deuteron d_2 is a bottleneck for the formation of larger dark nuclei d_N, they cannot be formed either.

Acknowledgements This research was supported in part by U.S. National Science Foundation, U.S. Department of Energy, Natural Science Foundation of China, German Research Foundation, and European Research Council.

References

1. Tulin, S., Yu, H.B.: Phys. Rep. **730**, 1 (2018)
2. Kaplinghat, M., et al.: Phys. Rev. Lett. **116**, 041302 (2016)
3. Braaten, E., et al.: JHEP **1811**, 084 (2018)
4. Braaten, E., Hammer, H.-W.: Phys. Rev. D **88**, 063511 (2013)
5. Laha, L., Braaten, E.: Phys. Rev. D **89**, 103510 (2014)
6. Laha, R.: Phys. Rev. D **92**, 083509 (2015)
7. Suno, H., et al.: Phys. Rev. A **65**, 042725 (2002)
8. Braaten, E., et al.: Phys. Rev. A **78**, 043605 (2008)
9. Steigman, G., et al.: Phys. Rev. D **86**, 023506 (2012)

The manufacturer's authorised representative in the EU is Springer Nature Customer Service Centre GmbH, Europaplatz 3, 69115 Heidelberg, Germany. If you have any concerns regarding our products, please contact ProductSafety@springernature.com

Printed and bound by CPI Group (UK) Ltd, Croydon, CR0 4YY
30/09/2026
02225464-0001